赖绍聪论著集

LAISHAOCONG LUNZHUJI

赖绍聪 等著

国家教学名师学术文库

西北大学出版社
·西安·

图书在版编目（CIP）数据

赖绍聪论著集 / 赖绍聪等著. -- 西安：西北大学出版社，2024. 7. -- ISBN 978-7-5604-5446-7

Ⅰ. P5-53

中国国家版本馆 CIP 数据核字第 20247FB229 号

赖绍聪论著集

作　　者	赖绍聪　等著
出版发行	西北大学出版社
地　　址	西安市太白北路 229 号
邮　　编	710069
电　　话	029－88302590
经　　销	全国新华书店
印　　装	西安华新彩印有限责任公司
开　　本	787mm × 1 092mm　1/16
插　　页	9
印　　张	170
字　　数	3700 千
版　　次	2024 年 7 月第 1 版　2024 年 7 月第 1 次印刷
书　　号	ISBN 978－7－5604－5446－7
审 图 号	GS 陕（2024）122 号
定　　价	950.00 元

本版图书如有印装质量问题，请拨打电话 029-88302966 予以调换。

赖绍聪简介

赖绍聪，男，1963 年生。西北大学二级教授、博士生导师。历任西北大学地质学系系主任，西北大学研究生院院长，西北大学党委常委、副校长。国家高层次人才特殊支持计划领军人才，国家级教学名师，国务院政府特殊津贴获得者，全国优秀博士学位论文指导教师，陕西省首批“三秦学者”特聘教授，陕西省“三五”人才。国务院学位委员会学科评议组成员，教育部高等学校地质学类专业教学指导委员会副主任，教育部高等学校地球科学教学指导委员会秘书长，教育部高等学校地球科学教学指导委员会地球物理学与地质学专业教学指导委员会秘书长，陕西省学位委员会委员，陕西省决策咨询委员会委员，陕西省高等学校教学指导委员会地矿类工作委员会主任委员、专业设置与教学指导委员会委员，陕西省学科（专业）建设和研究生教育教学指导委员会常务委员，陕西省理农医类学科评议组成员，重庆市普通本科高等学校地矿与环境安全类专业教学指导委员会副主任委员。首批“全国高校黄大年式教师团队”负责人，国家级教学创新团队负责人，国家级精品课程/国家级精品资源共享课程/国家级一流本科课程负责人。曾获得国家级教学成果奖二等奖（3 项），陕西省教学成果奖特等奖（3 项）、一等奖（1 项）、二等奖（2 项），陕西省优秀教材一等奖（2 项），陕西省自然科学奖一等奖（1 项），陕西省科学技术奖一等奖（2 项）、二等奖（1 项），并获得教育部高校青年教师奖、青藏高原青年科技奖、陕西青年科技奖，以及青年地质科技工作者最高荣誉奖“黄汲清青年地质科学技术奖”、地矿行业最高终身荣誉奖“李四光地质科学奖”等多项奖励。发表学术论文 200 余篇，110 篇论文被 SCI 收录，30 篇论文被 EI 收录，出版著作 5 部、教材 7 部。

出版说明

西北大学是一所具有丰厚文化底蕴和卓越学术声望的综合性大学。在百年发展历程中,学校致力于传承中华灿烂文明,融汇中外优秀文化,追踪世界科学前沿,在人才培养、科学研究、文化传承创新等方面成绩卓著,涌现出了一批蜚声中外的学术巨匠。

自2003年教育部启动高等学校教学名师奖评选工作以来,西北大学相继有史启祯、唐宗薰、耿国华、赖绍聪、陈峰、李浩等教授获此殊荣。国家级教学名师不仅要主动承担本学科基础课教学任务,在教学实践中努力探索教育教学规律,运用现代教育教学思想改革传统教育教学过程,还要在引领教学内容和方法改革、创新课程教材和教学模式、创建合理教学梯队等方面取得突出成绩,亦要在学术研究中成就卓著。

在加快建设教育强国、贯彻落实"强教必先强师"精神的时代背景下,西北大学出版社启动了"国家教学名师学术文库"编纂计划。《赖绍聪论著集》作为该学术文库中的一部,主要收录了赖绍聪教授教学及科学研究成果中对现代教学改革有促进参考作用、对目前地质学科建设仍有启示意义的教学研究论文和中、英文专业论文149篇。本论著集的出版,旨在对地质学科发展提供借鉴,对教师队伍建设提供指导参考。

《赖绍聪论著集》编辑组

2022年12月

目　录

上　册

教学研究论文

英文专业论文

中 册

下　册

中文专业论文

附　录

教学研究论文

J I A O X U E Y A N J I U L U N W E N

如何做好课程教学设计[①]

赖绍聪

摘要:教学的宗旨在于培养学生的独立人格、探索精神以及学习能力和实践能力。教的内容需要科学选择和高度凝练,教的过程需要合理安排和精心设计。结合作者自身33年从教经历,从教学的内涵、教学设计的概念、教学设计的指导思想、教学设计的核心内容4个方面进行了归纳总结,探讨了教学设计各个环节的核心内涵,初步构建了课程教学设计的构架。

教学要讲究方法,要引起学生的兴趣,要掌控课堂氛围。教学设计的目标是通过教学内容的高度提炼、教学方式方法的合理有效运用、教学过程的精心安排,深入浅出,旁征博引,把复杂的问题讲简单,使其通俗易懂,便于理解。教学设计通过基础信息的分析、教学环节及活动的设计、教学反思与评价的促进,使以教师的"教"为主的传统课堂,转变为以学生的"学"为主的课堂,让每堂课都能深入人心,让学生站在教师的肩膀上看世界。

那么,应该如何设计生动有趣的高质量的课程教学呢?要实现这一目标,我们必须真正地站在学习者的角度来思考问题,强化知识的无形渗透,让学习者充分体验知识的形成过程,加强学习方法的指导,激发学习者的创造性思维和求知欲望,这样才能使学习者"知其然,知其所以然",从而达到提高教学质量、培养出有创造力的优秀人才的目的。

一、教学的内涵

在我国历史上,关于"教学"的最早文字记载可以追溯到公元前1271—前1213年的殷商时代。在商朝出现的甲骨文中已经出现了"教"字,同时,在甲骨文中也找到了"学"字。从甲骨文中的字形特征分析来看,"教"是从"学"派生出来的。按照《说文解字》中的解释,"教,上所施,下所效也"。显然,在我国传统文化中,"教"主要强调了教授、教诲、教化、告诫、令使等含义,而"学"则强调了模仿和仿效。这种理念至今仍然在一定程度上左右着我们的教育教学思想。然而,随着时代的发展,教学的内涵已经发生了重要的转变,我们对教学内涵的认识也应该在继承的基础上有所发展。把"教"从以知识传授为主逐步转向以创新为主,把"学"从以模仿和仿效为主转向以探究、实践和合作式学习为主,应该更加强调学生的自主学习,逐步实现"要我学"向"我要学""我想学"和"我会

① 原载于《中国大学教学》,2016(10)。

学”的转变,以培养学生的创新精神和创新能力。

目前,教育教学理论中一般将教学的特征与属性归纳为以下4个方面:①教学是有目的、有计划、有组织地进行人类已有经验和知识传授的活动形式;②教学将传授的知识,按照逻辑关系和认知规律形成系统完整的教学体系;③教学是在教师的引导和精心安排过程中进行的;④教学所要实现的不仅仅是知识的传授,更有人格的完善、智力的发展和能力的培养[1-4]。

值得指出的是,实现教学目标需要借助一定的教学形式。到目前为止,我们在运用教学这一途径时可供选择的教学组织形式主要有课堂教学(教学的基本形式)、现场教学(教学的辅助形式)、复式教学(教学的特殊形式)和信息化教学(教学的电化形式)。

显然,对于教育教学工作者而言,了解和明确教学的内涵是十分必要的。

二、教学设计的概念

教学是师生之间、学生之间交往互动与共同发展的过程。对于课堂细节的研究,与自然科学和人文科学研究是一样的,深入研究教学过程、精心设计课程(课堂)教学是教师教学能力提升和专业成长的必由之路。

那么,教学设计到底是什么?教学设计对于课程(课堂)教学到底意味着什么?课程(课堂)教学设计到底有多重要?这些问题可以从教学设计概念中得到初步的诠释:教学设计就是在正确的教育理念指导下,关于“教什么”和“怎么教”的一种方案设计,它是教师根据学科教学的原理和教学目标要求,运用系统的方法,对参与教学活动的诸多要素所进行的一种行之有效的分析和策划[1-4]。

三、教学设计的指导思想

1.正确的教育理念——什么是教育

教育不是简单的理性知识的堆积,教给学生专业知识虽然重要,但更需要的是培养学生正确的价值观和探索、思考、学习以及践行的能力。鲁迅先生曾明确指出:“教育是要立人。”教育家斯金纳用简洁且极富哲理的语言深刻地揭示了教育的内涵,他认为:“教育就是将学过的东西忘得一干二净时,最后剩下来的东西。”那么,所谓“剩下来的东西”究竟是什么?作者认为,“剩下来的东西”就是教育的本质和精髓,就是“独立的人格、探索的精神、学习的能力和践行的能力”[5]。

作为教育工作者,弄清楚什么是教育这个问题,对于认清教育的本质、明确自己的职能和职责、找准前进的方向是非常必要的,这是实现高质量教学设计最为核心的指导思想。

2.精心凝练的教学内容——教什么

教师要精心凝练自己的教学内容,应该将学科知识融会贯通,“基于教材,高于教材”,讲授自己数年乃至数十年科学研究所获得的感悟、自己对学科的认识过程和体验、自己在领悟学科内涵过程中走过的“弯路”、自己对学科体系认识和感悟中提炼的“精华”。要让学生站在“你”的肩上认识世界,坚决杜绝“照本宣科”。

3.合理有效的教学方式与方法——怎么教

教学过程是“教”与“学”双边互动的过程,教学是“教”与“学”的统一,“教”为“学”而存在,“学”又要靠“教”来引导,二者相互依存、相互作用。课堂教学不能拘泥于某一固定的形式,而是要根据具体的教学内容选择适当的、合理的、有效的方式与方法。因此,教学方式与方法设计的核心思想是“内容决定形式”。

总之,教学设计的精髓是要以学习者为出发点,强化思想方法的渗透,体验知识的形成过程,加强学法指导,激发创新思维。而科学家自身对学科的感悟是教学设计的灵魂。

四、教学设计的核心内容

1.基础信息分析

(1)明确“教学质量国家标准”的要求。本科专业类教学质量国家标准乃是该专业类人才培养、专业建设等应达到的基本要求,是人才培养和专业建设的指导性规范,是质量评价的基本依据[6]。作为教师,我们必须对授课对象——学生的专业属性、特点以及该专业教学质量国家标准对人才培养的基本要求有充分的了解,并以此为基本依据,引导教学工作,按照基于“标准”、高于“标准”、突出“特色”的指导思想,凝练形成先进的、高质量的教学设计。

(2)专业知识结构体系分析。不同的本科专业,由于其学科长期发展的积累,必定形成其特定的知识结构体系[7]。因此,明确授课对象(学生所属专业)应该具备的知识结构体系、所授课程在学生知识结构体系中的地位和作用、所授课程上游课程需要为学生奠定的知识基础和技能要求,以及所授课程需要为学生下游课程学习奠定的知识基础和技能要求,乃是做好课程教学设计必不可少的重要环节。

(3)教材及阅读材料分析。教材、辅助教材以及主要阅读材料的选择绝不是盲目的,它必须基于所属专业学生知识结构体系的需要,有目的、有针对性地选择教材和相关辅助材料。要达到这一目的,就必须分析把握所属专业学科发展总体趋势,分析“课程教学大纲”对本课程的基本要求,分析选用教材知识结构体系的优势与不足,分析每堂课教材的内容在整个课程教学中和每个模块中的地位与作用。在此基础上,确定辅助教材及阅读材料选取的必要性和选取标准。

(4)学情分析。不可忽视的是,随着科学技术和时代的发展,不同年代大学生的学习需要、学习行为、学习风格已经发生了重大变化。现今大学生的学习风格带有显著的信息化时代风格。因此,客观地分析学生的学习行为和风格、学生存在的学习问题、学生的认知特点、学生已有的认知水平和能力状况,将对课程教学设计产生重要影响。我们应该在上述分析的基础上,客观地、科学地凝练教学内容,选择最为合理和有效的教学方式与方法。

(5)教学目标。从教学的内涵中我们可以清楚地看到,教学所要实现的绝不仅仅是知识传授,它还应该包括知识目标、能力目标、情感目标三大核心目标。因此,分析和明确课程教学以及每堂课教学的知识目标,确定能力培养目标,确定引导学生情感、态度、价值观目标,以及实现这些目标的教学选点设计及教学实施策略,乃是课程(课堂)教学设计的必备内容。

(6)教学重点与难点。明确分析并确定课程以及每堂课的教学重点、教学难点是十分必要的,即在分析的基础上科学、客观地设计出处理教学重点与难点的措施和方法,从而达到有效解决教学过程中疑难问题、实质性提高教学质量的目的。

2.教学内容的凝练

事实上,我国目前高等院校仍然在不同程度上存在着先进的科学理论(技术)与陈旧的课程体系设置及教学内容(体系)之间的矛盾。教材是以经典和成熟的知识体系为核心内容的,如果课程教学内容仅仅是基于对教材的"照本宣科",显然将与学科发展脱节。因此,按照"基于教材,高于教材,突出特色"的指导思想,以先进的人才培养方案为指导,凝练符合创新人才培养需要的课程教学内容就成为提升人才培养质量、实现教育教学目标的核心要素之一。

我们的建议是,教师应该密切结合学科发展国际趋势,梳理课程教学核心知识,使教学内容尽可能跟上时代发展的步伐。从人才培养目标定位出发,通过教学与科研的结合,将教师的研究成果融入教学内容中,充分体现教学与科研的深度融合,实质性地形成特色鲜明的课程教学体系,保证课堂教学内容的先进性。在这个环节中,教师的科研能力和学术水平是教学内容凝练提升的基石。

3.教学环节和活动设计

为了保证课程(课堂)教学的有效性、先进性和开放性,我们应该在明确每堂课的知识目标、能力目标和情感目标的前提下,精心设计出每堂课教学内容的结构体系,如简要回顾—问题引入—展开内容—适度归纳—引起思考。这样的课堂教学内容结构设计,要尽可能充分地融入教师自身对学科知识体系的感悟。而实现每堂课高质量教学内容结构设计的前提,是要对本堂课核心教学内容的知识点进行分析,建立核心知识点之间的内在联系,发现核心知识点之间的逻辑关系,形成以核心知识点为节点的知识导航图。而每堂课的知识导航图共同构成本门课程的知识概念图。

在教学内容结构设计的基础上,进一步完成与各部分教学内容相匹配的多媒体结构设计、板书结构设计,建立多媒体与板书之间的协同关系。在此过程中,必须明确教学过程设计的设计目的和设计意图,不能为了设计而设计。

在课程(课堂)教学设计过程中,应该根据课程性质的不同和教学内容的不同,实事求是,按照客观实际,考虑是否需要设计合理的教学活动。教学活动设计的指导思想是突出学生的主体地位,从学生的问题出发营造教学情境,设计教学问题并引导学生探究、解决问题,设计师生互动方式,增加学生的自主探究。

为了增强学生的自主学习能力,还应该尽可能设计出针对教学内容的系统巩固方案,设计出引导学生进行课后阅读和探究的方案,提出扩展、升华学生思维的问题,为学生留下思考。事实上,关于引导学生课后自主学习的有关方案设计,在我国目前高等学校教学过程中仍是有待进一步加强的环节。

4.教学方式与方法

在教学设计中,关于教学方式与方法的设计是必不可少且十分重要的内容,它是“怎么教”的具体实施方案。然而,长期以来,源于我国传统观念的“教,上所施,下所效也”这一教学理念,仍然禁锢着我们的思想。这种以知识传授为主要目的的教学方式是以教师为中心,教师是主动的施教者,学生则是被动的知识接受者。这种传统的教学方式曾对教学活动产生过强烈的进步性冲击,且至今价值犹存。然而,随着社会的发展,课程教学的核心任务已经不再局限于知识传授,而是更加强调教会学生从学习中“学会思想”“生成智慧”“生成正确的人生观和价值观”。因此,“思想”应当是贯穿于各类、各门课程教学过程的灵魂。显然,传统教学方式和学习理论已经越来越不适应当前教育教学发展的趋势。

高等教育教学方式的改革创新已经迫在眉睫,探索更加合理、有利于促进学生自主学习和能力提升的教学方式是当前高等教育改革的重要任务之一。然而,迄今为止,我国高等教育领域关于教学方式的改革仍然处在逐步探索之中,尚未形成完整的系统理论。诸如讲授和问题研讨相结合的研究性教学方式、案例教学和问题驱动相结合的教学方式、概念图教学方式、专业教育与前沿讲座有机结合的合作式教学方式等[8],值得我们进一步探索与实践。

关于课堂教学方法,通常会因教师而异,因对象而异,因课程性质而异,因课程教学内容而异,并无统一模式。但是,以下 8 个方面值得教师在教法设计中予以高度重视:①教师的语言艺术。苏霍姆林斯基说过,教师的教学语言修养在极大程度上决定着学生在课堂上脑力劳动的效率。教师的语言是一种技术更是一种艺术,教师的语言是一种知识更是一种思想。因此,教师的语言要精练、规范、激励、动情、启智。语言艺术不仅要体现在教法设计中,更多的则是需要长期锤炼。②2 个近 1/3 位。0. 618 被公认为是最具有审美意义的比例数字,即站在讲台 1/3 位、目视前 1/3 位学生,以视阈控制全教室,充分掌控课堂氛围。理论研究表明,当眼睛看到黄金分割位(前 1/3 学生)时,所有学生都会觉得你在看着他,这样就照顾到了班上所有学生的趣味和目光。③要提倡和体现质疑与挑战精神,对问题多分析、多论证,不让学生接受单一结论,不让学生对书本和老师产生迷信。④教学是一项“留白”的艺术。课堂教学需要给学生思考的空间,不能变成教师一个人的讲演。⑤要启发学生学会提问,会提问是最具创造力的品质。鲜明而强烈的现象是激发学生提问的重要驱动因素。⑥合理把握适时适度的赏识教育,好孩子是夸出来的。卡耐基说过,使人发挥最大能力的方法就是赞美和鼓励。赏识教育是世界著名的 6 种教育方法之一。⑦充分发挥教师的人格魅力,人格力量是教师的基石。俄国教育家乌申斯基认为,只有人格才能影响人格的形成和发展,只有人格才能形成性格。理想的人格具有崇高的价值,具有巨大的感召力、凝聚力和渗透力,是吸引学生的主要源泉。⑧丰富的课堂过程。包括充沛的精力投入,熟练的多媒体与板书,启智的语言,活跃的课堂氛围,依据内容不同而采用多种课堂形式。

在教法设计中，关键是要把自己摆进去，讲自己认识的过程，引导学生思考。教师在讲台上展示的是教师的声音、姿态、神情和心境。成功的讲课是若干年后，当听过的人忘了你讲的具体内容时，仍然赞扬你的课讲得好。因为，你给学习者留下的不仅是理性知识，同时也是一种思想。

5.教学反思与评价

当我们每完成一门课程的教学任务时，都应该深刻反思，评价每堂课教学设计的实施结果，对每堂课的教学设计进行及时修改、补充与完善，写出自己的教学感想、心得与体会。而这个过程恰恰就是教师教学能力提升与专业成长的必然途径。

6.教学改革与研究

教师在教学过程中常常会遇到各种各样的疑难、矛盾与困境，并且没有现成成功的解决方法可供借鉴，教学过程中的问题是教学研究课题的主要来源。我们应该把那些重要的、迫切需要解决的教学问题凝练出来，转化为研究课题。对这些教学实践过程中凸显出来的现实问题进行研究，最能充分发挥教师自身的优势并能直接提高教学质量。

教学设计是一门严谨的学问，它既要有知识传授的科学性，又包含了知识的艺术再创造性。教学设计作为连接教学理论与教学实践的桥梁，是针对学生学习需要，从教学思想、教学内容、教学方式方法以及教学过程的科学性和整体性出发，有目的、有计划地制定最优教学方案的系统决策过程。课程教学设计的质量总体体现了教师对教学思想、教学内容、教学方法的理解与把握，是教师对教学内容达到懂、透、化的程度，是教师主导作用的最充分体现。大学不是工厂，教育教学本质上是人与人之间的交流，是精神、观念、人格、方法的交流。因此，作为高等教育战线的每一位教师，都应该精心做好课程（课堂）教学设计，让每堂课都能深入人心。

参考文献

[1] 乌美娜.教学设计[M].北京：高等教育出版社，1994.

[2] 钟志贤.面向知识时代的教学设计框架：促进学习者发展[M].北京：中国社会科学出版社，2006.

[3] 戴风明.教学设计：有效教学的关键[J].教育理论与实践，2012，32（5）：6-8.

[4] 杨开城，李文光，胡学农.现代教学设计的理论体系初探[J].中国电化教育，2002（2）：12-18.

[5] 赖绍聪.创新教育教学理念 提升人才培养质量[J].中国大学教学，2016（3）：22-26.

[6] 赖绍聪.建立教学质量国家标准 提升本科人才培养质量[J].中国大学教学，2014（10）：56-61.

[7] 赖绍聪.地球科学实验教学改革与创新[M].西安：西北大学出版社，2010.

[8] 赖绍聪，华洪.课程教学方式的创新性改革与探索[J].中国大学教学，2013（1）：30-31.

高等学校教师教学团队建设的策略与路径[①]

赖绍聪

摘要：教师教学团队是最基层的教学组织，在教育教学研究、教学内容凝练、教学方法创新、推动人才培养质量实质性提升等方面发挥着十分重要的作用。进一步强化教师教学团队建设，探索团队建设的有效路径与策略，是当前我国高等学校面临的重要任务。本文结合作者40年高等教育从教经历和教师教学团队建设经验，从教师教学团队的建设意义、建设的指导思想和关键因素等方面进行了深入分析，探讨了教师教学团队建设的策略，初步构建了教师教学团队建设的有效路径。

党的二十大报告指出，要加强师德师风建设，培养高素质教师队伍，弘扬尊师重教社会风尚。教育部《关于深化本科教育教学改革全面提高人才培养质量的意见》中明确要求，高等学校必须将坚持以立德树人作为根本任务，把立德树人的成效作为检验学校一切工作的根本标准，要努力构建德智体美劳全面培养的教育体系，要形成更高水平的人才培养体系[1]。教育部教师工作司2022年工作要点中提出，要推进全国高校黄大年式教师团队的示范创建，引领各地加强教师团队建设，推进协同育人，集智攻关。

事实上，我国高等学校在基层教学组织建设过程中，仍然存在定位不明确、形式单一、教师参与度不高、激励机制不健全等问题，这些问题已经成为进一步提高人才培养质量的制约因素。教师教学团队作为最基层的教学组织，在教育教学、人才培养过程中发挥着越来越重要的作用。因此，强化教师教学团队建设将有助于激发基层教学组织的活力，改善教育教学环境，有效促进我国高等教育教学和人才培养质量的提升。

那么，教师教学团队应该是一个什么样的组织？如何建设教师教学团队？这显然已经成为我们必须深入探讨和研究的课题。

一、教师教学团队的定义

教师教学团队就是由有互补技能、愿意为了共同的目的，建立业绩目标和工作方法，相互承担责任的教师组成的群体；或者是以教书育人为共同目标，为了完成具体教学目标而明确分工协作，相互承担责任的教师群体。也可以认定为以提高教学质量和效果为目的，以推进教学改革为主要任务，为了实现共同的教学改革目标而相互承担责任的教

---

① 原载于《中国大学教学》，2023(5)。

师群体[2-5]。

当前,我国高等学校教师队伍表现出一些独有的特征:①高学历教师的比例越来越高。统计表明,我国高水平研究型大学中具有博士学位教师的比例已经由 1997 年的 17. 5%发展到现在的高于 75%,甚至有的高校已经高于 95%。②教师承担教学和科研双重任务。大学教师既要承担教书育人的责任,又要努力进行科学研究、发展学术、产出创新性的成果。③中青年教师已经成为本科教学的主力军,甚至部分进入高校时间不长的青年教师开始成为教师团队的重要组成人员。④国际学术交流与广泛的合作已经成为常态,教师参加国际学术交流与合作的机会明显增多,学术水平和国际化视野得到明显的提升[2-5]。

上述分析表明,高校教师面临教学、科研双重压力,教师团队中中青年教师已经占有较大的比例。在这种情况下,如何整体提升教师的教育教学水平和能力?显然,聚焦教师教学团队建设,构建老、中、青相结合,能够有效实现传帮带的、结构合理的教师教学团队,已经成为当前我国高等教育实现创新性发展的重要抓手和切入点。

二、教师教学团队建设的意义

通过组建教师教学团队,提升团队成员各方面的素质,从而达到确保和提高教学质量的目的。教师教学团队的优势在于,它是以共同的教学理念和目标为牵引,形成的结构合理、相互帮助、相互学习、共同提高的一个群体。这样的群体能够有效提高团队教师的整体素质,起到推动教学研究、提高教学质量、产生良好育人效果的目的[2-5]。

教师教学团队成员之间会产生密切的联系,在学科建设和学术科研方面,可以共同利用优质资源,实现资源共享;学科和学术方向能够相互补充、相互帮助。团队教师有更多的机会开展各种研讨活动,交流教学科研经验,探讨学术思想。通过这样一个团队的组织,团队内的教师能够得到知识与技能的互补,相互促进,共同提高教学水平。同时,优秀教师作为团队的领头人或团队骨干,他们的优良学风及优良品格,能够使团队全体成员的师德师风、工作风格、教学理念、教学方法、学术思想得到提升,团队成员之间的优良品格实现相互渗透和影响。另外,团队有激励约束,有利于形成优良的师德师风,这些都是团队建设的重要意义和价值[2-5]。

三、教师教学团队建设的指导思想

1.师德师风

牢记心有大我、志诚报国的理念。全面贯彻党的教育方针,坚持教书和育人相统一,言传和身教相统一,潜心问道和关注社会相统一,学术自由和学术规范相统一。以德立身,以德立学,以德立教。模范践行社会主义核心价值观[6-7]。

2.教育教学

将立德树人、教书育人放在首位。把思想政治工作贯穿到教育教学的全过程,实现全程育人、全方位育人。教育理念先进,教学科研深度融合,努力创新教学内容和方法[6-7]。

3.科学研究

倡导敢为人先、开拓创新的精神。聚焦国家重大战略和地方经济社会的发展,在学术研究方面培育持续创新的能力,在构建中国特色学术体系方面做出重要探索和创新。

4.服务社会

坚持知行统一、甘于奉献的原则。注重科研成果的转化,突出社会效益,积极开展社会实践,主动弘扬中华优秀传统文化和革命红色文化,发展社会主义先进文化。

5.体制机制

聚焦团结协作,实现团队可持续发展。提出团队建设模式、团队结构素质、发展目标的明确要求。团队带头人要注重提升学术造诣、创新学术思想、强化组织协调能力,在团队群体中发挥凝聚作用。

总之,教师教学团队的专业结构和年龄结构应该合理,发展目标应该明确,发展规划应该清晰。要注重学习共同体的建设,在团队内部要实现老、中、青"传帮带"的健全机制,从而为团队教师专业发展搭建通畅的平台,整体提升教师的教学和科研能力。

四、教师教学团队建设的关键因素

教师教学团队建设是一项系统工程,团队建设的目标应该指向师德优良、理念先进、目标远大、结构合理、水平高、能力强。因此,师德师风建设是先导和保障,结构是基础,水平和能力建设是目标。团队建设的关键要素包括5个结构、5种能力和5项保障。

1.5个结构

教师教学团队要保持可持续的发展力,就要关注结构问题。①知识结构。团队教师的组成应该在知识结构上不仅有共同的优势和特点,而且更为重要的是应该有一定的互补性。②学缘结构。教师团队应该具有交叉型的学缘结构,不能够"近亲繁殖",因为交叉型的学缘结构更有利于学术思想的创新。③学历结构。形成梯度型的学历结构将更有利于团队的可持续发展。④职称结构。就职称结构而言,正高级—副高级—中级—初级呈正态分布的结构较为合理,这将更有利于团队的协调与可持续发展。⑤年龄结构。老、中、青合理配置的教师团队,不仅能够充分发挥"传帮带"的作用,而且还是团队保持可持续发展的基本要求。

2.5种能力

提升教师教学团队全体成员的能力和水平乃是团队建设的核心目标。①知识创新能力。国家在推进"双一流"建设总体方案中将知识创新能力放在了首位。作为教师,我们必须有较强的学术能力和较高的学术造诣,必须能够把握学科发展的前沿,因为这是培养一流人才最基本的要素。我们只有拥有较强的学术能力和较高的学术造诣,学生才能站在我们的肩膀上认识世界,他们才能够成为优秀的学生。我们有多"高",学生才能有多"高"。因此,知识创新能力是教师教学团队建设的关键要素。②知识传授能力。在团队建设中,提升团队成员的素质、能力和水平,教学与科研不可偏废。强大的科研实力能够为我们的教学发展奠定坚实的基础。同时,研究教育与社会发展关系之规律、教育与人的发展关系之规律,探索认知心理学、发展心理学,熟悉学生由认知—探索—创新的能力发展路径,以及学生由认知—认理—认同的情感意识形成路径,充分掌握"教"与"学"之间充分互动的规律,将极大增强我们对知识传授客观规律的把握,提升我们的知识传授能力。显然,知识创新能力和知识传授能力是团队能力建设中最为重要的两项基本能力。③人才培养能力。卢梭说:"教育,我们从 3 个来源接受,即自然天性、人和事物。我们的器官和才能自然而然的发展是自然天性的教育;别人教我们如何利用这个发展,是人的教育;我们从影响我们的周围事物中获得个人经验,是事物的教育。"[8]教育并非是一件"告知"和"被告知"的事情,而是一个主动的建构过程。随着社会的发展,教育的功能要从以继承为主转向以创新为主,以传授知识为主转向以培养创造力为主,以训练标准化的个性为主转向以培养多样化的个性为主。作为教师,必须能够较好地把握大学的培养目标(培养什么人)、课程目标(怎样培养人)、教育目标(为谁培养人)、专业目标(培养能干什么的人),并努力构建符合当代教育理念、聚焦学生身心发展的教学范式,熟悉教学论与课程论的基本原理,有能力开发建设具有高阶性、创新性、挑战度的高质量一流本科课程,具备培养高素质、创新型人才的基本能力。④服务社会能力。服务社会是大学 5 项基本功能之一。中国的大学要服务于国家建设,我们要把论文写在祖国的大地上,大学要为国家发展、民族复兴做出应有的贡献,我们要为国家培养社会主义建设事业的合格接班人。因此,教师教学团队在建设过程中,团队成员能力建设必然包含服务社会能力的提升。应该倡导开门办学,不将教育教学囿于教室,要更加主动、深入地了解社会需求,提升我们服务社会的能力。⑤教学管理能力。教学管理能力的建设和提升在传统的教师教学团队建设中没有得到足够重视。事实上,"教育"就是"教"与"育"的有机统一,教书与育人不可分割。作为一名教师,应该具备一定的教学管理能力。当代高等教育更加注重由提高"教"的质量向提高"学"的质量发展,由关注"灌输学生的教学质量"向关注"激发学生的教学质量"发展,由教师主要作为讲解者向教师主要作为学习方法和学习环境的设计者发展。对于学生的学业评价,更加注重学生在整个学习过程中长期的学业表现,以课堂表现和学习档案为典型做法;注重教学过程中的持续性评估,使学业评价始终与教育教学目标保持一致;学业评价应将学生置于核心地位,通过学业评价发现问题、帮助学生、促进学习、提高学业成绩;通过评价、评估为教师提供及时准确的信息,以帮助教师改进教学方法[9]。因此,班级管理、课程管理与教学质量密切相关,熟悉教学效果评价体系,创新学生学业成绩评价策略,乃是当代优秀教师必备的教学管理能力。

3.5 项保障

要保障教师教学团队良好的可持续发展,需要有切实可行的团队运行机制。①团队负责人的学术能力和人格魅力。团队负责人能不能够服众、能不能够得到大家的尊重,建设一支高质量的教师教学团队,对团队负责人有很高的要求。作为团队负责人,不仅要有较高的学术造诣和较强的学术领导能力,而且还要有一定的人格魅力和高尚的品格。团队负责人的学术能力和人格魅力是团队凝聚力的核心指标之一。②团队全体成员的事业心和向心力。团队全体成员对事业的追求和热爱,团队协作、互帮互助、团结奉献,将团队和集体利益放在首位,心往一处想、劲往一处使,倡导奉献精神,凝聚向心力,推进协同育人、集智攻关,乃是保障教师教学团队可持续健康发展的重要方面。③有效的团队运行机制。作为一个团队、一个集体,应该有纪律约束、有制度约束,有全体成员必须共同遵守的严格的规章制度体系。因此,教师教学团队建设必须构建一套有效的团队运行机制。④合理的评价体系。形成客观、合理、公正的评价体系乃是保障教师教学团队良性发展的一个重要方面。中共中央、国务院关于《深化新时代教育评价改革总体方案》(中发〔2020〕19号文件)为高校教育评价改革指明了方向[10]。教师教学团队建设应该将师德师风建设作为首要任务,通过制定政策、构建师德考核评价体系、建立师德师风预警监督机制等措施,完善师德建设评价体系。努力探索、构建教师分类评价体系,结合团队定位及发展目标,不断完善团队教师分类评价办法,突出教书育人实绩,坚持教学质量评价创新,深化教师科研评价改革,从而逐步构建并完善更加注重实际贡献、突出“破五维”评价理念的评价体系。⑤资源和资金保障。教师教学团队的建设与发展,需要一定的资源和资金保障,学校、学院以及相关职能部门应该给予团队有力的资金支持和必要的资源保障。

五、教师教学团队建设的路径

统一团队目标—提高团队成员的认同度—培育优良的师德品格—提升学术水平和整体素质—提升教学能力和水平—实现团队合作与协同创新,这就是团队建设的基本逻辑和路径。具体而言,教师教学团队建设必然要落实到师德师风建设、科研能力建设、教学能力建设、课程体系建设和教学资源建设5个方面[2-5]。

1.师德师风建设

杜威提出,教育即生活,教育即生长,教育即经验的改造[11]。教育不仅要使人学会做事,更重要的是要使人学会做人。教师的行为示范、学习环境及学习氛围,对于学生健康成长极为重要。环境对于一个人的语言习惯、礼貌、美感和对美的欣赏会产生潜移默化的影响,这些用不着思考的习以为常,恰恰是我们通过与他人日常交往的接受关系形成的。因此,环境会通过它的各种活动来塑造个人的行为、知识、能力倾向和情感倾向。

理想和信念教育对于学生正确的世界观、人生观、价值观的养成十分重要。当然,需要学生形成的理想和信念,不能够硬性灌输,需要学生形成的态度也无法复制粘贴。

我们在和学生共同生活、学习的时候,比较容易忽视我们的言行举止对他们的社会倾向可能产生的重要影响。或者说,我们不会认为这种隐性影响和其他某种显性影响同样重要。品格是通过习惯进行刺激以形成习惯性的行为来获得的,而不是通过知识的传授来获得的。虽然我们一直不停地在对学生进行着意识的修正和教导,但其周围的气氛和风气是最终形成品格的主要力量。对于他人传授得来的审美知识,只是用来提醒人们对别人的审美有何种看法,它不可能变成自发的审美品质,也不可能对人产生根深蒂固的影响。审美是一种习惯,比较深刻的审美价值判断标准是一个人在日常活动情境中逐渐形成的。因此,凡是不经过我们的研究或思考而被视为理所当然的东西,恰恰就是决定我们明确思想和结论的东西[11]。

境界和价值观决定着学生最终的成才高度。教师与学生共同学习、研究、交流和研讨,对学生的世界观、人生观和价值观的养成,对学生知识结构的形成,对学生能力的塑造和智慧的获得具有重大影响。因此,作为教师,我们的专业知识和技术、思想境界、思维方式、言行举止、待人接物,都会深远地影响着学生和他们的未来。教师在教育教学过程中、在与学生相处过程中,不仅要重视言传更要重视身教,要为学、为事、为人相统一,要留下故事可以被流传。

中国高等教育要培养品德高尚、学术精良、人格完善的人,要求其树立远大的理想,具有严谨的学风,爱党、爱国、爱人民、爱社会主义、爱集体,追求真善美。要达到这样的目标,作为教师,我们就要深刻理解教育的内涵,做到为学、为事、为人相统一。

《中华人民共和国教师法》明确指出,教师是履行教育教学职责的专业人员,承担教书育人,培养社会主义事业建设者和接班人、提高民族素质的使命。从历史溯源来看,在古代,教师是指年老资深的学者,后来把教学生的人称为老师。中国有悠久的尊师重道传统,古代就有"人有三尊,君、父、师"的说法。教师这个称谓经历了一个历史的演变过程,并有多种说法,先生——历史最悠久的尊称,夫子——最古老的尊称,园丁——最质朴无华的褒称,蜡烛——最温馨动人的称谓,春蚕——最纯挚的称谓,春雨——最生动形象的默称,人梯——最高评价的专称,孺子牛——最具中国特色的喻称,而最贴切的称谓应该是"人类灵魂的工程师",这是最富有哲理的一种称谓。

教师要清楚地知道自己的职责。教师是仅次于学生的父母、对学生影响最大的人。教师对待学生如同父母对待子女,是不图回报、不辞辛劳的。教师的言行举止、思维方式等,都将潜移默化、深远地影响着学生和他们的未来。因此,教师理当为人师表[12]。

教师共同的格言:凡是要求学生做到的事,教师都必须首先做到;凡是不让学生做的事,教师必须带头遵守。教师应有的教学思维是要为学生的终身做准备,教师考虑的不是我想教什么,而是学生需要学习什么。我们应该做受学生欣赏、追慕、景仰的师长。

大学不是工厂,教育、教学本质上是人与人之间的交流,是精神、观念、人格、方法的交流,而不仅仅限于业务的交流。教师务必与学生真诚对话,教师要留下故事可以被流传,要有音容笑貌可以被追忆,应该给学生留下一个可以仰望的身影[13-14]。

2.科研能力建设

"你在大学工作,不教书干什么?教书育人是教师的根本任务。你在大学教书,不搞科研你教什么?"这就是教学和科研之间的关系,它们应该是相互支撑、互为唇齿的依存关系。教学与科研并重,赋能创新人才培养。如果没有雄厚的学术实力,不可能成为一名优秀教师。优秀教师的课程教学绝不是背诵教材、照本宣科,必须能够把握学科前沿,有扎实的学科基础和很高的学术造诣。要培养一流的优秀人才,就要让学生站在我们的肩膀上去认识世界,我们有多高学生才能有多高,甚至学生将来比我们更高。因此,作为教师,长期不懈的科学研究奠定了我们的学术水平,保障我们能够站在学科的前沿,而学生站在我们的肩膀上才能够看到世界的学科前沿,才能够保证具有培养学生的高起点。

(1)科研能力的内涵。什么是科研能力?简单地说,就是发现问题、分析问题和解决问题的能力;就是确定科研课题、设计实验方案、组织实验和研究、收集实验资料,最后进行科学分析、得出正确结论的能力;就是科学思维加上适当的方法,对未知领域进行科学探索的能力,包括创新能力和观察能力、思维能力、实操能力几个重要的方面。

一般认为,科研能力包含 4 个层次:第一个层次是基本功,就是个人的基本学术素养,对学科充分了解的程度,学科基本知识、基本技能掌握、积累的程度和完备性;第二个层次是科研敏锐性,是指发现、感知学科未知领域存在的关键科学问题的敏锐程度;第三个层次是学术交流能力,是指与他人进行充分学术沟通、交流,以及能够有效参与团队合作、协同攻关的能力;第四个层次是价值观,也就是我们的学术价值观和科研道德观,这是科研能力的最高层次,它通常决定了一个学者最终成才的高度。

(2)科研能力提升的意义。人才培养、科学研究、社会服务、文化传承和合作交流是大学的 5 项基本功能。其中,人才培养是核心,科学研究是高质量人才培养的重要支撑,而社会服务、文化传承和合作交流是人才培养和科学研究两项基本功能的进一步延伸。因此,教师教学团队科研能力建设具有以下几方面的重要意义:一是培养创新型人才的必然要求。培养高素质创新型人才是高等学校的历史使命,是实现创新型国家建设战略目标的重要保障,是民族复兴的重要基础。教师肩负着教书育人和培养学生创新能力的重要职责。因此,教师教学团队科研能力建设对于培养高质量的创新型人才意义重大。二是提升教学质量的关键。教学与科研是教师职业能力的两翼。深入的科学研究能够更新知识结构,丰富教学内容。只有教学与科研深度融合,才能够提升教学内容的品质,保障教学内容体系的前沿性、先进性;才能够开拓学生的视野,培养创新能力,促进教学质量提升。教学与科研协同发展是优秀教师成长的必由之路。三是高等学校学科内涵发展的重要支撑。创新性、引领性科研成果的不断产出,乃是国家“双一流”建设的核心要求之一。因此,教师教学团队科研能力建设必然成为推动高等学校学科内涵发展的根本要素。

(3)科研能力提升的途径。正是由于科研能力的提升具有重要意义,导致如何提升科研能力成为广受关注的重要议题。具体而言,我们可以从以下几个方面深入开展实践。

找准学术生长点。学术发展之路,明确发展方向、找准学术生长点是前提。学术研究要专一,不可能什么方向都做。对有限目标要有所为有所不为,特色发展不可替代是原则。因此,需要明确国内外学科发展趋势,立足学科发展前沿,契合国家、地方重大需求,结合个人志趣和知识结构,脚踏实地、实事求是,找准自身的学术生长点。同时,个人学术生长点和发展方向的选择,还可以充分依托所在学科的长期发展优势、学科领军人才的科研特色、学科平台的优势,以及行业或者地域优势。

优化知识结构。每个人在学术发展的不同阶段,都需要不断优化自身的知识结构,不断学习新的学科知识,提升、改造原有知识结构体系,尽力融入所在团队的主流发展方向。随着科学技术的不断发展,每个人原有的知识储备都需要进一步优化、丰富。这是每一位教师在提升自己科研能力过程中都必然经历的一个过程,有时候甚至是一个痛苦的过程。

强化团队意识。团队比个体更有力量,团队是实现重大科学突破、促进交流学习、提升竞争力和创造力、发挥科研资源最大效益的一个重要的载体。团队是学术共同体,共同承担科学研究项目,联合发表论文,共同拥有学术成果,团队对于每位成员学术影响力的共同提升均有重要价值和推动力。因此,强化团队意识是提升科研能力的一个重要方面。

提倡质疑与挑战精神。不能迷信书本和权威,要敢于挑战,倡导多因素非线性思维,而不是非此即彼的线性思维。对问题要多分析、多论证,不轻易接受单一结论。这是提升我们科研能力的另一个重要方面。

文献阅读。阅读文献是提高我们自身学术水平的最佳途径。只有努力阅读本学科主要研究论文、学科经典文献、国际主流期刊最新文献,通过充分地文献阅读,才有可能瞄准学科国际研究前沿。只有充分了解这一领域国际一流科学家在想什么、在干什么,我们才有可能把握学科前沿,在这样的基础上从事学术研究才能有所创新[13-14]。

在反思中学习。重要的学术思想一般都是通过探索、反思,再探索、再反思形成的。已有的研究思路是否可行,是否需要修正?在学术研究过程中要不断地回头看、不断反思。

学会提问。善于找出问题,发现并聚焦关键科学问题是关键。科研能力就是发现问题、分析问题和解决问题的能力。如果不能提出问题,那我们分析什么、解决什么?所以,提问是创造性思维的起点和核心,是学习隐性知识最好的途径,会提问是最具创造力的品质。我们要学会提问、敢于质疑,逐渐培养自己的学术敏锐性。

增强行动力。发现了问题就要深入分析问题的本质,为厘定解决问题的方案奠定基础。一个人的思维能力是在不断分析的过程中进步和发展的。通过不断地深入分析,才能逐渐找到解决问题的正确路径。

制定合理的解决方案,形成可行的研究计划。要尽最大努力让我们的技术路线和研究方案具有切实的可行性,能够破解关键科学问题。

扎实推进研究方案的实施。发现问题—分析问题—提出解决方案—实施方案—破解问题—形成新的创新性学术思想,这就是科学研究的基本底层逻辑。

在这个过程中,我们的基本功和知识结构得到进一步完善,学术敏锐性得到进一步增强。通过团队协作,我们的交流合作能力进一步得到提升。同时,在一个优秀和谐的科研群体中,在师德师风建设引领下,团队成员正确的科研价值观和学术道德观逐渐养成;团队成员的学术影响力不断得到增强,逐渐成长为一流的学术人才。

3.教学能力建设

(1)构建先进的教学范式。高等教育正面临重大改革,新的教育理念不断融入教育、教学中,传统认知主义的教育理念和行为主义的教育理念已经逐渐被建构主义的教育理念所替代。一个教学团队首先应探索符合当代教育学发展趋势的教学范式。一是倡导教学与科学研究密切结合,形成教学和科研相互融通的氛围。二是寓教于学、寓教于研,在教学过程中,教师和学生都是其中的学习者和探索者。三是尊重学生的主体地位,以学生为主体、教师为主导。四是教师与学生相互交流与合作,构建"教"与"学"充分互动的探究式、合作式教学范式。大学的课堂应该是教师与学生基于人格平等前提下的共同探讨和学习,这种教学范式更有利于促进学术和治理环境的健康发展[9]。

(2)精心凝练课程教学内容。课程教学内容应该包括三大部分:一是学科的基本知识体系。课程首先要为学生构建本课程最为核心的专业知识体系。这一部分的内容可以从教材中精心凝练和选择,这是课程最基本的内容。二是教师的学术思想及典型的研究案例。凝练能够反映教师学术认识、学术感悟、学术思想的教学内容和研究案例,将其融入课程基本知识体系中,成为构成课程教学内容的重要组成部分。三是学科发展趋势与前沿。可以从文献阅读以及我们自身所做的前沿科学研究中提炼。这样的教学内容体系不仅能够为学生构建最为核心的专业知识体系,还能够引导学生学会科学研究的方法和辩证唯物主义的思维模式,并让学生通过课程学习感知、触摸学科前沿。

(3)选择合理、有效的教学策略。"怎么教"是值得每位教师深入思考的重要课题。教学策略选择因为学习内容不同、教师个性不同、学生群体不同、学习环境不同、学习方式不同而有所不同,有效的策略就是最好的策略。教学策略选择的原则应该是突出学生的主体地位,活跃课堂气氛,引导学生探究解决问题,激发学生兴趣并能够留下思考。

(4)有效运用现代教育技术。教育就是为了促进学习者学习,"教"是为了"不教","学"是为了"会学"。充分利用互联网先进技术以及丰富的数字化资源,采取线上与线下相融合的混合式教学模式,激发学生学习的主动性,让学生学会如何学习、逐渐拥有获取知识的能力。

(5)不断深入开展教育教学实践。教师教学团队是一个优质平台,互助协作的氛围为每一位团队成员教学能力的提升提供了充分的机会,为做中学、学中做开辟了更大的空间。教学团队中的教学观摩、教学研讨、课程说课、新课试讲等一系列教学活动的常态化开展,将有效践行互帮互学、老中青"传帮带",为团队教师教学能力建设提供不断深入实践的平台。

(6)形成教学反思。在实践中不断总结反思,对课程教学思想、教学内容、教学策略和教学方法的合理性和先进性进行过程性评价和形成性评价,进一步改进教学方法。团队教师在这个过程中教学能力不断得到提升。

4.课程体系建设

作为一个教师教学团队,通常需要承担一个课程群或者是相关联的一个课程体系的教学任务。因此,团队建设还有一个很重要的任务,就是构建更加合理的课程体系。

每一所大学都应当有自己的办学理念、有自己的精神追求,每一所大学的课程价值都应当是独特的、富有个性特征的。大学课程由于自身的复杂性,对教师的要求和依赖度较高。大学主要以校本课程为主体。因此,对大学课程而言,教师的人格、学识、志趣和专业水平等,在一定程度上决定着课程内容的实质。正是由于这些原因,使得同一科目的大学课程,可能因为主持教师不同而有很大的差别。而教师教学团队建设则有利于把每位教师最优秀的、最好的教育教学思想结合起来,从而整体提升课程体系或课程群的建设质量[15]。

20世纪认知心理学进一步向建构主义发展,建构主义的课程理论已得到更多的认同。课程教学更加重视认识主体的作用和认识活动中的主观能动性,把探索与发现作为发展认知的重要手段。在这种教育观念指导下,课程建设常常不再追求知识体系的完整性,而是更多地追求对知识的理解和关联[15],这就为教师教学团队课程体系建设、课程群内课程之间相互关联性建设奠定了重要的理论基础。因此,教师教学团队应该努力对团队所承担的课程体系尽可能进行创新和改造。在课程体系建设过程中,应该更加突出思维方法训练,加强实践方法创新,打破课程教学分门别类、单课独进的课程教学和课程建设模式。

在过去一个时期内,由于基层教学组织建设被淡化,教研室建设和教研室常态化教学研讨活动逐渐被弱化,基层教师教学团队建设尚未广泛兴起,教师各上各的课、谁跟谁都没有关系,我讲什么、怎么讲你不知道,你在教什么、怎么教我也不了解,课程教学缺失左邻右舍,学生从各门课程中获取的知识缺失相互关联,失去关联的知识就不能形成知识结构,没有结构体系的知识就不能发挥功能!这样的课程教学怎么可能培养出优秀的学生?

教师教学团队建设为破解这一困境提供了有效路径。团队建设建立起教师之间的密切联系,互帮互学和常态化教学研讨活动将教师群体和他们承担的课程群相互密切关联起来,从而逐渐形成完整、科学的课程体系,为高质量人才培养奠定基础。

课程体系建设应该更加强调能力发展,把科学研究的思维引入教学中,更加关注思维方法的训练和实践方法的创新,更加全面地体现团队所承担的课程群、课程体系的设计性和综合性。

5.教学资源建设

优质教学资源的建设和积累是课程体系建设的重要支撑。因此,教学资源建设必然成为教师教学团队建设的另一项重要任务。我们应该努力凝练和提升课程中的优质资源,形成一批经典的、先进的教学资源积累。我们应该结合国际发展趋势,优化课程的核心知识,参考国际教材内容融合我们自身的科研优势,最终形成我们的教学团队所承担的课程体系、课程群水平高、质量优,凸显国际化视野和学科研究特色的优质教学资源体系。优质的教学资源建设与积累可以为学生的自主学习、教师的教学研究、对外资源服务,以及提高教学保障能力起到重要的支撑作用。

六、结语

教师教学团队是高等学校最基层的教学组织,是提高教育教学质量、实现高质量人才培养目标最基本的教学组织单元。教师教学团队建设理应受到各级教育行政主管部门和各高校各级组织的高度重视。21世纪初我国国家级教学创新团队建设计划,以及近年来教育部大力推动的全国高校"黄大年式教师教学团队"建设和国家级"虚拟教研室"建设,均充分表明了国家对高等学校基础教学组织建设的高度重视。我们相信,在各级教育行政主管部门和各高等学校的共同努力下,我国高等学校基层教学组织建设一定能够取得重要新突破。让我们共同努力,为我国高等教育事业的发展进步、为早日实现高等教育强国梦而努力奋斗!

参考文献

[1] 中华人民共和国教育部.关于深化本科教育教学改革 全面提高人才培养质量的意见[Z].教高〔2019〕6号.

[2] 赖绍聪."双一流"背景下高等学校学科建设策略分析[J].中国地质教育,2021(1):18-22.

[3] 鲁卫平,王润孝.研究型大学教学团队建设的研究与思考[J].中国大学教学,2010(3):71-73.

[4] 刘广虎,关成尧,王承洋,等.资源勘查工程专业教学团队建设研究[J].科技创新导报,2018(3):234-236.

[5] 邱海军,李书恒,曹明明,等.高校自然地理学教学团队建设思路与途径[J].教育教学论坛,2016(30):228-231.

[6] 中华人民共和国教育部.教育部关于公布首批全国高校黄大年式教师团队的通知[Z].教师函〔2018〕1号.

[7] 中华人民共和国教育部.教育部关于开展第二批"全国高校黄大年式教师团队"创建活动的通知[Z].教师函〔2021〕2号.

[8] 让-雅克·卢梭.爱弥儿[M].叶红婷,译.北京:台海出版社,2016.

[9] 赖绍聪.聚焦学习者身心发展构建教育教学新范式[J].中国大学教学,2020(6):7-10.

[10] 中共中央办公厅.深化新时代教育评价改革总体方案[Z].中发〔2020〕19号.

[11] 杜威.民主主义与教育[M].北京:中国轻工业出版社,2019.

[12] 赖绍聪.寓教于学 寓教于研[J].中国大学教学,2017(12):37-42.

[13] 赖绍聪.创新教育教学理念提升人才培养质量[J].中国大学教学,2016(3):22-26.

[14] 赖绍聪.如何做好课程教学设计[J].中国大学教学,2016(10):14-18.

[15] 孙莱祥,张晓鹏.研究型大学的课程改革与教育创新[M].北京:高等教育出版社,2005.

有效构建以问题为导向的课堂教学范式[①]

赖绍聪

摘要: 高等教育阶段是学生基础认知能力、思考能力和创新能力奠基的重要时期。本文聚焦当前高等教育课堂教学改革热点,紧紧围绕能力提升目标,从课堂教学目标、教学内容凝练以及策略方法选择3个维度,探讨如何有效构建以问题为导向的课堂教学范式。

高等教育阶段是人生成长的重要阶段,这一阶段不仅为学生今后从事专业工作奠定知识基础,同时也是学生人格、品德、修养等方面完善的重要时期,更为重要的是,这一阶段是学生基础认知能力、思考能力和创新能力奠基的重要时期。当前,我国高等教育面临的核心任务之一,就是要着力提高教育教学质量,培养德智体美劳全面发展的社会主义建设者和接班人。提高教育教学质量是一项系统工程,涉及高等教育过程中的诸多关键环节。一线教师上好每一堂课,让每堂课都能够深入人心,乃是提升教育教学质量,培养高素质、有能力优秀人才最为基础的工作。也就是说,我们仍然要注重向第一课堂要质量。

我国高校长期以来高度重视人才培养质量的提升,在人才培养目标中均有全面提升学生能力的核心要求。然而,我国高等教育在学生能力提升方面取得一定成效的同时,也面临着严峻的挑战。我国传统的高等教育课堂在教学观念、教学内容、教学方法等方面均存在一定局限性:①专业课程教学主要将关注点聚焦在知识传授方面,对能力提升和正确的价值观养成重视程度不足,对"教"与"育"的辩证关系思想认识模糊。②"照本宣科"式的教学内容很难让学生接触学科前沿、了解学科发展趋势,难以拓展学生的学术视野。③教师居于主导的学术氛围未能充分尊重学生学习的主体地位,在课堂教学过程中并非所有学生都能充分表达自己的见解。④"知识传授型课堂"教学方式以及教师在课堂教学中的权威不利于改善学生盲目崇拜的价值观念,难以提升学生的思辨能力。

显然,传统的课堂教学范式已经明显不适应社会发展对提升学生思辨能力、创新能力所带来的新要求。在高校课堂教学过程中,如何坚持以学生为中心,深化教学改革,构建理念先进、体系完善、教法合理、行之有效的课堂教学新范式,已经成为我们必须从理论以及方法上进行深入探讨和研究的重要课题。

1969年,Barrows教授首创以问题为导向的教学范式(Problem-Based Learning,简称

---

① 原载于《中国大学教学》,2021(9)。

PBL),又称为问题教学法。这是基于建构主义学习理论,强调个体的主动性,本质上是一种以学生为中心的教育方式[1]。

本文基于Barrows的Problem-Based Learning教学理论,结合作者38年的教学经历,探讨如何有效构建以问题为导向的课堂教学范式;如何充分运用问题教学法,通过课堂教学过程实质性地提升学生的能力。

一、课堂教学目标

课堂教学的目标绝不仅仅是知识的传授。知识传授是手段和过程,通过知识的传授和学习,要逐渐内化为学生的能力,并在这个过程中养成学生正确的世界观、人生观和价值观[2],从而达成由教知识向教能力、教思维、教品德的转变。也就是说,课堂教学应该聚焦知识、能力、情感三大目标。明确并深刻理解课堂教学目标,深入分析教学目标的实质内涵乃是改革和创新课堂教学范式的重要前提。

按照当代建构主义的教育理论,教育应该是形成有意义的人的实践,是对人的价值的发现、挖掘、形成、提升和规定。教育的目的就是要促进学习者的学习,教育教学过程是提升学生的能力和促进学生的身心发展[2]。因此,提升学生的能力是课堂教学的核心任务。

人创造了知识,知识又规范和引导着人,人都有求知的本能。然而,知识并不等同于能力,能力是依靠本能又超越本能的人在认识和实践中形成的主体性能动力量[3]。人类不仅需要"学习",从学习中认知并积累知识;更为重要的是,人类需要学会"如何学习"。而"如何学习"不仅仅是知识,更是一种能力。

从科学研究和学术探索的角度来看,学术探索能力就是指发现问题、分析问题和解决问题的能力;是指个人在其所从事的专业中,以科学的思维和适当的方法,对未知领域进行科学探索的能力[4]。可以看到,"问题"是推动思维发展从不确知向确知、从朦胧向清晰、从局部向整体、从未知向已知的桥梁和原动力。发现问题是创造性思维的起点和核心,是学习隐性知识最好的途径,善于发现问题是最具创造力的品质。

显然,以问题为导向的课堂教学范式将为课堂教学创新开辟广阔的空间,将成为通过课堂教学实质性提升学生能力的重要抓手,是值得我们深入探讨的重要课堂教学范式研究课题。

二、以问题为导向的课堂教学内容设计

1.凝练教学内容

无论采用哪种课堂教学范式,课堂教学内容的精心凝练均是首要任务。高等教育的课堂教学内容必须做到让学生感知前沿、触摸前沿。保持课堂授课内容的先进性、前沿性,是开展课堂教学范式创新的前提和成功保障。换句话说,课堂教学范式是授课策略和方式的优选,而课堂授课内容才是课堂授课的核心和实质。"照本宣科"式的教学内容很难让学生接触学科前沿、了解学科发展趋势,难以拓展学生的学术视野。

教学内容的选择与凝练将直接关系到学生学习的深入程度。合理选择与精心凝练的教学内容能够让学习者通过内容的学习完全清楚教学内容所要表达的论点,并且可以与用来支持论点的论据有机联系起来[5]。这样的教学内容才能够真正为以问题为导向的课堂教学范式创新奠定基础。因此,在课堂教学内容凝练中,要强调授课内容的层次性和逻辑关联性,依据“基于教材,高于教材,突出特色”的原则[5],结合相关学科研究方向以及教师自身长期学术研究的成果与感悟,融入学科发展的最新成果和最近进展,打破照本宣科、按部就班的传统教学方式,形成高质量课堂教学内容体系。这样的教学内容体系才能够引导学生开展探究性学习,有效开展以问题为导向的课堂教学范式改革创新。

同时,要尽可能凝练本学科门类在长期发展中形成的价值取向、伦理规范和科学精神等思想政治教育元素,为课程教学过程中开展“课程思政”探索做好准备。

2.提出关键问题

提出问题是开展以问题为导向的课堂教学范式改革创新的关键。教师对课堂教学内容加以充分解析,将本堂课教学内容分解为若干个核心知识点,通过追索核心知识点的思想来源、所反映的特定现象或所延伸的前沿领域,凝练出该核心知识点的溯源问题、现象问题或前沿问题,从而将课堂教学内容由“知识形态”转化为“问题形态”,进而以问题为“教”与“学”的切入点和抓手,贯穿教学全过程。需要指出的是,问题设计要尽可能密切结合学科发展的前沿与趋势,最大限度实施教学科研深度融合。问题设计要因事而化、因时而进、因势而新。同时,问题设计要充分考虑学生的已有知识基础,能够引起学生的兴趣,与现实生活及经济社会发展相契合。可以将课堂教学的专业知识与一些社会发展、需要解决的热点问题结合起来,提升课堂教学的前沿性,学以致用,在服务我国社会主义建设事业中提高认识和体现价值。

3.分析问题关联

对问题之间关联性的深入分析是以问题为导向的课堂教学范式能否有效实施并取得实质性效果的核心。对已经构建的课堂教学问题系列进行逻辑梳理,区分宏观问题、中观问题和微观问题,或“章”问题、“节”问题和“点”问题,进一步厘清问题之间的并列关系、从属关系或交叉关系等逻辑关系,从而形成本堂课思路明确、逻辑清晰的问题链。这对于教学过程中帮助和引导学生逐步形成符合逻辑规律的正确思维形式将会有重要的帮助。

4.形成教学逻辑

形成每堂课完整的教学逻辑,是以问题为导向的课堂教学范式实施过程中能否拓展学生思维、增强学生学术敏锐度的重要抓手。正确的认识是通过正确的思维获得的,根据经过实践验证过的真实知识,运用正确的逻辑推理,也可以获得间接的知识,获得原来不知道的新知识[6]。教师在课堂教学过程中要善于润物无声地引导学生学习和运用形式逻辑与辩证逻辑的思维形式,逐步引导学生在学习自然科学、哲学与社会科学知识过程中自觉地进行思维的逻辑训练,提高学生的逻辑思维能力,增强逻辑论证的力量,这对于整体提高学生的理论思维水平具有重要意义。

教学内容体系的逻辑性是逐步引导学生形成严密逻辑思维模式的最有效催化剂,它能够润物无声地对学生逐步养成严密的辩证思维逻辑起到有效的促进作用,并最终实现提升学生“发现问题—分析问题—解决问题—形成思想”的能力。

因此,在有效运用以问题为导向的课堂教学范式过程中,教师应紧紧围绕每堂课的若干关键问题,分析问题的内涵,发现问题间的逻辑关联,并最终构建起本堂课以问题为导向的主体教学内容体系的知识导航图(知识导图),这是一件有意义、有价值的工作,它能够增强以问题为导向的课堂教学范式的实际效果,切实体现课堂教学由教知识逐步向教能力、教思维、教品德的转变。构建课堂教学知识导图也是教学法中“概念图教学策略”的精髓,将其有机融入也是拓展以问题为导向的课堂教学范式宽度、增强其有效性的一个发展方向。

三、以问题为导向的课堂教学策略与方法

1.情景创设,激发兴趣

教学策略与方法的选择与运用必须突出学生的主体地位,目的是要引导学生思考问题、探究问题、分析问题和解决问题,并能够在这个过程中引导学生“观察—归纳—形成概念—对新观察进行推理判断—形成新的认识”,反复训练、润物无声,逐渐养成学生“概念+判断+推理”的正确逻辑思维形式,从而在知识学习的过程中实质性地提高学生的能力,达成课堂教学应该从教知识向教能力、教思维、教品德的转变。

教学过程是教与学的统一,是教与学的双边互动过程,教为学而存在,学又要靠教来引导,二者相互依存、相互作用,任何单边思维都很难达成教学目标。教师对教学策略与方法的精心设计必须得到学生的心灵响应,而能够达成互动双边有效的关键点就是学生对学习内容的兴趣问题。兴趣是最好的老师,是驱动学习者主动学习和深入思考的第一原动力。因此,激发学生的学习兴趣是教法设计中的首要任务。激发学生的学习兴趣有很多种方法。作者认为,依据课堂教学内容中凝练的关键问题,以问题为导向,通过教师教学与科研深度融合,充分结合学科发展前沿,紧密联系社会经济发展中的热点,精心创设能够让学生感同身受,在学习和生活中有所了解,但又未知其详的“问题情景”,通常可以成为一堂高质量课堂教学的良好开端。有效的情景创设能够起到先声夺人进而引人入胜的效果,这是课堂教学设计中有效的开课方法之一。情景创设教学策略已有较为广泛的应用,创设方式灵活多样,自然科学、哲学与社会科学课堂教学均可根据自身需要,结合课程类型及本堂课的教学内容实际需要灵活使用。情景创设易调动感情、点燃激情,激发学生对学习内容的强烈兴趣和迫切的求知欲,进而培养学生的想象力与创造性思维能力,不失为课堂教学方法应用中的一种有较强实用性的策略选择,它可以为以问题为导向的课堂教学范式的进一步有效展开奠定良好基础。

2.促进协作,初步认知

分组学习是以问题为导向的课堂教学范式常用的教学方法。学生在教师指导下,按

照一定的原则(如自由组合原则等)划分成若干学习小组,各学习小组的学生在对课堂教学关键问题有着浓厚的兴趣和强烈的求知欲驱动下开展小组学习活动,达到如下效果:①互助合作。这样的学习方式有助于培养学生的团队精神,增强学生的协作意识。而科研团队协同攻关是未来科学技术发展的必然趋势。②共享资源。小组学习有利于学生之间共享学习资源,小组成员课前预习、课中阅读以及自身已有知识、经验、感悟均可以得到比较充分的共享,从而能拓展小组成员的整体视野。③分享观点。在小组学习中,学生之间可以开展资料的查阅、搜集、整理等工作,分工协作,并充分表达自己的观点和认识。通过小组集体学习与交流,学生的观察、归纳能力将得到进一步提升,在这个过程中逐步对"问题"的来源、属性、内涵形成初步认知。

3.沟通交流,逐步认理

在小组学习的基础上,进一步开展各小组之间观点、认识的充分交流、沟通甚至激烈的辩论,将起到良好的相互启发、反思不足、拓展思维的作用。教师在这个过程中的作用主要是引导和启发,激励学生大胆发表个人观点,但不要急于总结、归纳或做出结论。对于学生在讨论、辩论过程中可能出现的不足和认识上的偏差要有足够的容忍度,要让学生的主体地位得到充分尊重,使学生在心理上实实在在地感受尊重感、信任感和获得感,增强学生的自信心。这种主题引导与开放式讨论相结合的教学方式,让师生共同享有话语权,较大限度地实现了由教师"独白式教学"到师生之间、学生之间"对话式教学"的转变。在这个过程中,学生对"问题"的来源、属性、内涵的理解得到进一步深化,达成逐步认理的目标。

4.深刻辨析,形成认同

在小组学习、交流、讨论、辩论的基础上,教师依据问题属性进行归纳总结,在这个过程中尤其要关注以下几点:①对学生在学习交流过程中产生的思维亮点,教师应给予肯定性评价,并依托教师自身更加深厚的知识积淀和更强的认知能力,从更高和更广的角度加以拓展,这将对学生学习积极性的提高起到强有力的推动作用,能极大地增强学生的自信心和学习获得感与成就感。②对学生在学习交流过程中产生的思维偏差,教师要理性分析产生的原因,找出症结所在,并列举正确的理解和认识与之进行对比分析。这有助于引导学生逐步学会"概念+判断+推理"的正确逻辑思维形式。③依据课堂教学内容的系统性,综合学习小组学习、小组间研讨所形成的初步认识架构。教师以关键问题为切入点,进一步全面、系统、深刻地辨析"问题"的来源、属性与内涵,从而让学生能够比较深刻地理解本堂课教学内容体系所要表达的思想,达成本堂课教学内容体系认同的目标。

5.意义建构,提升能力

当课堂教学把学生的注意力引向教学内容所要表达的论点和思想时,学生就能够把新的信息与他们已有的信息和实践经历联系起来,从而引发学生自身的对比、分析、思考与反思,甚至能引发学生对新信息的质疑与挑战[7]。以问题为导向的课堂教学范式正是以此为课堂教学目标的。

通过上述以问题为导向,由“认知—认理—认同”的学习过程,针对教学内容的深入学习与研讨,使学生逐渐构建起他们自己头脑中关于知识的意义,而这样的“知识”才能够成为学生自己的、能够被学生灵活运用的、能够为学生提出问题和质疑论点时提供论据的有效知识[2]。在这个过程中,学生获取知识的能力以及运用知识解决实际问题的能力将得到提高,学生逐步养成思考的习惯,能够运用“观察—归纳—形成概念—对新观察进行推理判断—形成新的认识”的逻辑思维形式。最终,学生的思考能力和创新能力能得到实质性提升,并且实现了把“教”从知识传授为主逐步转向以能力提升、价值观引领为主,把“学”从模仿、仿效为主转向以探究、实践与合作式学习为主,把被动学习转变为学生的主动学习。

四、结语

以问题为导向的课堂教学范式属于开放式的教学范式,其指导思想、基本架构和核心要素清晰,使用方式灵活,形式多样,有较为广泛的适应性。无论是自然科学还是哲学与社会科学的不同学科门类、不同课程类型,均可依据课程自身特色,灵活地运用以问题为导向的课堂教学范式。

当前,“双一流”建设对中国高等教育的发展提出了更高的要求。根据《统筹推进世界一流大学和一流学科建设总体方案》的相关要求,至 21 世纪中叶,我国一流大学和一流学科的数量和质量将进入世界前列,将实现由高等教育大国向高等教育强国的巨大跨越。根据新时代的新要求,作为高等学校教师,我们必须不断探索,改革创新,积极推动课堂革命,进一步提升课堂教学质量,努力让我们的每堂课都能够深入人心,为国家培养更多素质高、能力强、拥护中国共产党领导和我国社会主义制度、立志为中国特色社会主义事业奋斗终生的优秀人才。

参考文献

[1] 佟晓丽,刘红,李纪周.问题导向教学模式的探索与实践[J].科教导刊,2019(8):41-42.

[2] 赖绍聪.聚焦学习者身心发展构建教育教学新范式[J].中国大学教学,2020(6):7-10.

[3] 郝文武.教育哲学研究[M].北京:教育科学出版社,2009.

[4] 何志杰,李儒新,冯明丽.专业化教练员的素质要求及其实现对策[J].体育科技文献通报,2011(2):124-126.

[5] 赖绍聪.论课堂教学内容的合理选择与有效凝练[J].中国大学教学,2019(3):54-58.

[6] 华东师范大学哲学系逻辑学教研室.形式逻辑[M].5 版.上海:华东师范大学出版社,2016.

[7] 玛丽埃伦·韦默.以学习者为中心的教学:给教学实践带来的五项关键变化[M].洪岗,译.杭州:浙江大学出版社,2006.

论课堂教学内容的合理选择与有效凝练①

赖绍聪

摘要：结合作者自身从事高校专业课程教学36年的经历，从课堂教学内容的功能、课堂教学内容构建的指导思想、课堂教学内容的选择、课堂教学内容的凝练4个方面进行了归纳总结，探讨了当前我国高校课堂教学过程中存在的问题及其改革方向。

作为大学教师，在课堂教学过程中究竟应该给学习者教什么，这是值得每一位教师深入思考的重要问题。教育不是简单的理性知识的堆积，知识不等于能力，知识传授是课堂教学的起点，更为重要的是要通过知识的传授内化为学生的能力，并在这个过程中逐渐形成学生正确的人生观、世界观和价值观。

教学内容是指教师以学科知识体系为依据，以教材、教学资料、社会文化为基础，密切结合学科发展趋势与前沿，充分融入教师自身长期学术研究积累之精华，以服务于教学过程中知识、能力、情感三大目标为目的，以促进"教"与"学"的互动并充分结合学习者的学习经验为导向而精心选择、凝练而生成的课程教学基本教学资源，是教师课堂教学的施教蓝本。

高质量的课堂教学要讲究方法，要引起学习者的兴趣，引发学习者的思考。要达到这一目标，我们必须真正站在学习者的角度来思考问题，通过教学内容的合理选择与高度凝练、教学方式方法的有效运用、教学过程的精心安排，深入浅出，旁征博引。这样才能打破以教师的"教"为主的传统课堂，转变为以学习者的"学"为主的新型课堂，让每堂课都能深入人心。而选择合理的课堂教学内容、构建合理的教学内容体系乃是课堂教学过程中的核心环节，这将直接关系到教学质量和人才培养质量。

一、课堂教学内容的功能

课堂教学内容在学习者学习过程中的作用是认知和教育心理学研究中的重要课题[1]。构建教学内容体系的核心，是要让学习者通过内容的学习，理解"内容"所要传达的"思想"和"智慧"，而不是仅仅去关注学习者能够"记住多少"。合理选择与精心凝练的教学内容能够让学习者通过内容的学习完全清楚教学内容所要表达的论点，并且可以与用来支持论点的论据有机联系起来。这样的教学内容才能够真正实现课堂教学的知

---

① 原载于《中国大学教学》，2019(3)。

识、能力、情感三大目标。

教学内容的选择与凝练将直接关系到学习者学习的深入程度。当教学内容把学习者的注意力引向记忆事实和具体的知识性细节上时,学习者就很难区别论据与信息,缺失了反思,甚至把问题看成是外在的设置[1]。那么,这样的学习只能是"浅表学习"。当教学内容把学习者的注意力引向教学内容所要表达的论点和思想时,学习者就可能把新的信息与他们已有的信息和实践经历联系起来,从而引发学习者自身的对比、分析、思考与反思,甚至引发学习者对新信息的质疑与挑战。这样的学习才属于我们应该倡导的"深层学习"[1]。

建构主义理论认为[1],教学内容与学习者之间的核心关系是通过教学内容的学习和研讨,逐渐构建起学习者自己的知识体系、能力体系和情感体系,而不是仅仅被动地接受教师和课本传递给他们的信息。也就是说,知识不能简单地给予学习者,而是要通过对知识的学习让学习者构建起在他们的头脑中关于知识的意义。这样的"知识"才能够成为学习者自己的、能够被学习者灵活运用的、能够为学习者提出问题和质疑论点时提供论据的有效知识。

事实上,在教学过程中,我们并非一定要等到学习者完全掌握了基本技能后再去接触教学内容,而是应该鼓励学习者去主动探索、逐步掌握教学内容,并尽可能地将教学内容与学习者的亲身经历相联系。无论学习者已经具备了什么水平的专门知识,教师在教学过程中都应该积极鼓励他们对教学内容提出质疑与挑战。

总之,教学内容的功能绝不仅仅是传递知识,更为重要的是要引发思考。只有在我们对教学内容的功能有了深刻理解后,才能够思考教学内容合理选择及有效凝练的指导思想与技术方法。

二、课堂教学内容构建的指导思想

大学教育理应让学习者感知和触摸学科前沿。然而,我国高等院校长期以来仍然在不同程度上存在先进的科学理论(技术)与陈旧的课程体系设置及教学内容(体系)之间的矛盾。究其原因,是由于长期以来存在落后的教育理念以及"照本宣科"式的教学方式[2-5]。

作者认为[6-7],当前大学专业课程教学过程中关于教学内容的构建大体存在以下3种不同状态:①理想状态。教师拥有体现自己学术感悟、学科特色的教学内容体系和自编的教材。②通常状态。教师依据学科发展趋势、专业知识结构需求,选取合适教材,融入自身学术感悟,形成教学内容体系。③较差状态。教师选取一部教材,照本宣科。

显然,第一种状态是我们所应该倡导的,而第三种照本宣科式的状态是我们应该坚决反对的。那么,大学课堂教学内容构建应该遵循什么原则呢?我们认为,"基于教材,高于教材,突出特色"是我们可以参考的一个基本原则。基于教材是指教师可以从一部优秀的教材中提取基础的学科知识体系;高于教材是指教师在构建教学内容过程中必须融入学科发展前沿与趋势,能够让学习者触摸前沿、感知前沿,而这一点常常是经典教材很难做到的,这主要是由于教材出版周期的限制,以及"无争议的内容才能进教材"等教材自身属性的限制;突出特色是指教师在教学内容构建过程中必须融入自身的学科感悟及学术认识。

在教学内容构建过程中还应该关注以下 3 点基本要求:①教学内容应该能够被学习者用作发展知识的基础;②教学内容应该能够被用来发展学习技巧,能够帮助学习者掌握学习策略、方式与技能,使学习者在类似科目的学习中可以有效运用;③教学内容应该能够被用来提升学习者自我学习的意识,培养学习者处理学习事务的能力和自信心。

三、课堂教学内容的选择

1.教学科研深度融合,形成高质量的教学内容

大学课堂教学要求教师要合理选择和构建每堂课的教学内容。教师应该将学科知识融会贯通,讲授自己多年在科学研究中所获得的感悟,自己对学科的认识过程和体验,自己在领悟学科内涵过程中走过的"弯路",自己对学科体系认识和感悟的"精华",要让学生站在你的肩膀上去认识世界,坚决杜绝照本宣科[8]。

在实践中,教师应该密切结合学科国际发展趋势,梳理课程教学核心知识,使教学内容尽可能跟上时代发展的步伐,让学生获得最新的学科专业发展动态;从人才培养目标定位出发,通过教学与科研的结合,将教师的研究成果充分融入教学内容中,实质性地形成特色鲜明的课程教学体系,保证课堂教学内容的先进性。而教师的科研能力和学术水平则是提升教学内容质量的基石。

作者依据自身 36 年的从教经历,在此提出实现教学科研深度融合、形成高质量课堂教学内容的"五步"法则:①分析本堂课的核心知识点;②发现核心知识点之间的逻辑关系;③建立核心知识点之间的内在联系;④形成以核心知识点为节点的知识导航图;⑤通过教学科研深度融合,发掘教师在多年科学研究中积累的感悟和典型案例,融入相应的核心知识节点,形成知识节点科学论据的强大支撑。这样的教学内容才能够充分反映学科前沿,保证教学内容的先进性,体现大学专业教育的学科特色。

2.教学内容的多与少

对于课程教学内容,我们一直认为是越多越好,但是现在已经开始质疑这种观点了。

在大学传统的教育理念中,通常会认为,一门课程的教学内容既丰富又复杂,而且逻辑严密,就意味着讲授这门课程的教师的教学水平要高且教学态度要端正。这样的理念使得我们的教师常常会追求多而且全的教学内容体系。例如,在大学里常常会出现这样的情况,任课教师主动向院系或教务部门申诉:我所承担的课程很重要,有很多内容是必须教给学生的,而这门课程的学时数又太少,远远不能够完成这些教学内容,所以要给这门课程增加学时数! 这样的情况在大学并非个例,作者在大学担任二级学院负责人期间就曾无数次处理过类似事例。有调查资料表明,在我国大学的一门计算机专业类主干课程设定为 80 课堂教学学时,而同样的这门课程在美国 MIT 却只有 20 个课堂教学学时。其结果是:①中国学生非常忙,一直忙于听课;②美国学生同样非常忙,但却主要忙于思考问题、查阅资料、阅读文献、小组讨论、撰写课程小论文等。

显然,对教学内容功能的理解不同在极大程度上影响了我们对教学策略的选择,其结果是形成学习者机械式的被动学习和自主式的主动学习两种不同的方式。不言而喻,自主式的主动学习将更加有利于学习者发现问题、分析问题、解决问题以及创新等多方面能力的提升。我们之所以会囿于多而全的教学内容的约束,关键是我们的传统教学理念过分注重对知识性教学内容的全面覆盖,过多地强调了注重记忆、机械重复和遗忘过程的学习策略。这样的学习策略必然会对学习者产生消极的影响,学习者常常不能够对课程教学内容进行持续、有效和深入的理解。那么,这样的教学内容(知识)就很难在学习者的头脑中形成关于知识的意义。这样的"知识"就不能够成为学习者自己的、能够被学习者灵活运用的、能够为学习者提出问题和质疑论点时提供论据的有效知识。

在作者36年的从教经历中曾很多次遇到这样的情况:当我们向高年级本科生进行专业课程教学时,对他们提出问题,而这些问题需要学习者充分利用以前课程中已经学习过的知识才能够正确回答时,相当一部分学生常常会一脸茫然、困惑或不知所措。这样的结果使得我们必须对教学内容的功能做深入理解,并合理选择教学内容。

总之,在思考教学内容的"多与少",进而对教学内容的选择做出正确抉择时,应该关注以下3点:①当今时代是一个知识爆炸的时代,各类知识均呈现巨量增长趋势,各类学科的知识量是如此之多,以至我们的教师不需要、也不可能将本学科所有知识体系都在课程教学过程中完整地传递给学习者。因此,教师必须深刻理解教学内容的功能,选择并构建有利于学习者能力提升和正确价值观养成,高度凝练、体现教师自身学术特色,能够让学习者触摸前沿和感知前沿的教学内容体系。②教学内容体系必须有利于引导学习者的主动学习精神。因为在当前社会发展形势下,对于学习者的将来,学习必将伴随他们终身。③教学内容体系的构建必须有利于学习者通过学习能够掌握复杂而精湛的学习技能,因为这些技能可能在学习者的一生中都会被用到。

3.给学习者一定的选择权

对于教学内容的选择权,不能完全掌握在教师手中。依据教学过程中以教师为主导、以学习者为主体的原则,充分尊重学习者的主体地位,在教学内容构建过程中应该给予学习者一定程度的选择权,这将符合建构主义的理论。建构主义并不反对教师专门知识的有效性,而是反对其专有权。根据建构主义理论,学习者并非要等到掌握了专门知识以后再去接触内容,应该鼓励学习者去探索、掌握内容,并将内容与他们的经历相联系,无论拥有什么水平的专门知识,都应该鼓励他们对内容提出挑战。很明显,鼓励学习者参与教学内容的构建是明智之举。实践中,我们可以在教学过程中更加灵活地采取随机应变的课堂范式,也就是合理使用K型、S型等不同课堂范式。在S型课堂中,给予学习者在阅读文献、演讲主题、课程小论文选题等方面充分的自主权,从而达到让学习者主动参与教学内容构建的目的。

四、课堂教学内容的凝练

合理选择的教学内容体系将成为教师施教的蓝本[8]。然而,教学内容体系的构建一般会尊重学科的知识体系。也就是说,内容的构建关注了学科知识构成的逻辑顺序。需要指出的是,这样看似十分合理的“学科知识逻辑顺序”(章—节—点)与学习者“获取知识的规律”并不一定完全吻合。这就给我们提出了一个新的问题:在每堂课的具体教授过程中,是否还应该按照学习者获取知识的认知规律,对依据“学科知识逻辑顺序”(章—节—点)构建的教学内容进行进一步深加工?答案是肯定的。为了保证每堂课教学的有效性、先进性和开放性,我们应该在明确每堂课的知识目标、能力目标和情感目标的前提下,精心凝练每堂课教学内容的结构体系。应该依据学习者获取知识的认知规律,对每堂课的教学内容进行新的结构安排,以形成更有利于学习者接受知识、提升能力、引发思考的教学体系。

作者依据自身36年的从教经历,提出以下“五层次”课堂教学内容结构设计建议:简要回顾—论点引入—展开论据—适度归纳—引起思考。这样的课堂教学内容结构设计,要尽可能充分地融入教师自身对学科知识体系的感悟。而要实现每堂课高质量教学内容结构设计的前提,是应对本堂课核心教学内容的关键论点进行分析,建立核心论点之间的内在联系,发现核心论点之间的逻辑关系,形成以核心论点为节点的逻辑导航图,每堂课的逻辑导航图共同构成本门课程的概念图。

1.简要回顾

根据现在大学的通常管理模式,一般将2小时作为一堂课的标准时长。因此,一门数十课时的课程将被安排在半学期、一学期,甚至一学年内完成。然而,与相对独立的学术讲座不同,课堂授课具有连续的知识体系。也就是说,一门课程的每堂课之间都有其必然的关联性,这就给予每堂课开篇时需要“简要回顾”的必要性。从认知心理学的角度来看,回顾的功能就是让学习者在很短的时间内(回顾需要高度凝练,一般不能耗时太长)迅速建立起知识体系的连续性。从宏观层面来看,回顾是对课程已经学习过的内容的扼要小结;从微观层面来看,回顾是对上堂课已经学习过的内容的扼要小结。无论宏观还是微观,“回顾”都是教师高度概括能力以及对课程内容感悟深度的考量。

2.论点引入

回顾之后需要进行的下一项工作就是“引入”,也就是本堂课核心论点的引入。教师并非一定要在论据展示充分以后再推出论点。首先引入论点可以起到先声夺人的效果,对于吸引学习者的注意力、引发关注、引起兴趣常常会起到良好效果。当然,有效提炼每堂课的核心论点仍然是对教师学术水平和能力的考量,它同样要求教师自身对课堂教学内容有深刻的理解和感悟。引入的方法很多,可以开门见山提出问题,可以是一个典型的现象展示,也可以是教师在学术研究过程中亲身经历的一个典型案例,等等。

3.展开论据

成功的“引入”可能已经引起了学习者的兴趣、关注、疑惑甚至质疑,这是引起学习者深入思考、主动参与学习过程的重要前提。教师在此基础上即可展开论据,从不同的角度加以论证。论据的展开决不能局限于教材,而是应该充分体现教学与科研的深度融合,充分融入教师自身对学科知识体系的感悟。这一环节也是教师能否形成自身课程教学特色的关键环节,是“基于教材、高于教材、突出特色”的践行过程。教师可以根据学习者的理解状态,确定展开程度,从而做到课堂授课收放自如。

需要再次指出的是,展开论据的过程一定要尽可能地让学习者能够把新的信息与他们已有的信息和实践经历联系起来,从而引发学习者自身对教师提出的论据以及他们自身已经掌握和了解的论据进行对比分析、思考与反思,这将使学习成为真正的“深层学习”,有利于学习者创新与探索意识的培养,并能促进教学内容与学科前沿接轨。

4.适度归纳

一般认为,归纳法是在认识事物过程中所普遍使用的思维方法之一。它是指人们以一系列经验事物或知识素材为依据,寻找其中的基本规律或共同特点,并假设同类事物中的其他事物也服从这些规律,从而将这些规律作为预测同类事物中的其他事物的基本原理的一种认知方法。事实上,科学研究通常要经过观察、分析、质疑、论证、归纳,进而获得规律,再将规律带回实践中去检验。在课堂教学中也应该充分利用这一人类的自然认知规律。在论据得以充分展开之后,即可引导学习者与教师一起共同进行归纳,从而拟合起始时提出的论点。需要指出的是,课堂教学中的归纳应该适度。因为,科学研究和人类对真理的追求永无止境,科学论点一般都是科学发展到现在最为合理的认识,而不是终极认识或绝对真理。教师在这个环节中尤其要关注的是,依据自身长期学术研究的积累,尽可能地发掘和梳理本堂课核心论点存在的不足或不完善之处,并结合国际学科发展趋势,指出将来可能的发展方向,这样才能够实现课堂教学应该达到的一个重要教学目标——引起思考。

5.引起思考

“引起思考”,不能简单地理解为课堂教学结束时给学习者留思考题,而是要将“引起思考”融入课堂教学的全过程。就算课堂教学结束时没有留思考题,本堂课的教学也能够引起学习者思考以及对下一步学习内容的期盼。毫无疑问,我们对每堂课教学内容的结构设计,其目的都是为了突出学习者的主体地位,引导学习者探究、解决问题,提出发散、扩展、升华思维的问题。因此,无论是教学内容的合理选择与有效凝练,还是教学策略的选择,其目标都是营造互助合作的学习气氛,激发学习者的学习兴趣,培养其自主学习的能力,提高独立思考、逻辑思维、发现问题、分析问题、解决问题的能力。

在大学课堂教学中,构建合理、有效的教学内容是达成课堂教学目标的核心和关键。选择了教学内容,也就是选择了课堂发展的基点,就等于为学生选择了发展的基础。无论多么先进的教学策略均要以合理、有效的教学内容为基础,失去合理、有效教学内容支撑的教学策略只能是无本之源。因此,作为大学教师,我们必须高度关注教学内容的合理选择与有效凝练,以达成高质量课堂教学的目的,为培养一流创新人才,为我国高等教育实现内涵式发展,为建设高等教育强国做出我们应有的贡献。

参考文献

[1] 玛丽埃伦 · 韦默.以学习者为中心的教学:给教学实践带来的五项关键变化[M].洪岗,译.杭州:浙江大学出版社,2006.

[2] 曾毅.语文教学内容的选择与有效生成[J].教育理论与实践,2014,34(8):45-47.

[3] 许锴杰.思想品德课堂教学内容的选择策略[J].思想政治课研究,2014(12):104-106.

[4] 黄建军.教育 · 教学 · 教师[J].教育实践与研究,2015(3):8-9.

[5] 余昱.语文课堂教学内容的选择与设计[J].广西师范大学学学报(哲学社会科学版),2007,43(6):82-85.

[6] 赖绍聪.创新教育教学理念提升人才培养质量[J].中国大学教学,2016(3):22-26.

[7] 赖绍聪.建立教学质量国家标准提升本科人才培养质量[J].中国大学教学,2014(10):56-61.

[8] 赖绍聪.寓教于学 寓教于研[J].中国大学教学,2017(12):37-42.

聚焦学习者身心发展 构建教育教学新范式①

赖绍聪

摘要:从教育教学目标、知识体系构建、"教"与"学"的策略、教学氛围与学业评价等方面,分析了当前我国高校教育教学过程中存在的问题,探讨了以学生为中心、能够更加有效促进学生身心发展的教育教学范式的核心构成。

迄今教育理论大体经历了3个主要发展阶段[1]:①基于认知主义的传统教育理论。重视人的自然性,认为天理是什么、自然是什么,人就应该是什么。这种教育理论强调教育的"培养"作用和教师的主体地位。古代人性论教育学说就是这种教育理论的典型代表。②基于行为主义的近现代教育理论。这一理论强调环境决定论,重视教育和教师的"造就"功能,重视教师的主体地位,轻视学生的主动性、自主性和创造性,强调教育的程序化和标准化。在教育过程中,学校是"工厂",教师是"工程师",学生是"产品"。应试教育就是这种教育理论的典型代表。③基于建构主义的当代教育理论。这一理论认为人是交往和建构的人,而教育则是主体间的指导学习。强调"教"与"学"的双边互动、平等交往、主动对话、相互理解,强调教育对学生价值的发现、挖掘、形成、提升和限定。

因此,当代教育理论给予教育更为客观科学的定义,教育应该是形成有意义的人的实践,是对人的价值的发现、挖掘、形成、提升和规定[2]。换言之,教育是要立人的,"立人教育"应该成为当代高等教育所拥有的教育思想,而教育教学规律研究应该把学生的身心发展放在首位。

显然,在传统教育理论指导下构建的以知识灌输和传授为主导的大学教育教学范式已经明显不适应社会发展对当代创新性人才培养的需要,传统的高等学校人才培养过程片面强调学生智力的发展和"教"一方的单边作用,在一定程度上忽略了"教"与"学"的双边互动、学生人格的完善和个性化发展。在人才培养过程中,如何深化大学教育教学改革,构建理念先进、体系完善、能够有效促进学生身心发展、符合基于建构主义教育理论的大学教学新范式,全面实现"立人教育"的教育教学目标,已经成为我们必须从理念、理论、方法上进行深入探讨和研究的重要课题。

一般情况下,教育教学范式(Model)可以简单定义为是在一定教育教学理论或教育教学思想指导下建立起来的具有一定逻辑关系的教育教学活动结构框架和活动程序[3]。那么,如何构建一个聚焦学生身心发展,更加符合"立人教育"教育思想的大学教育教学

① 原载于《中国大学教学》,2020(6)。

新范式？这样的一个新范式应该涉及哪些核心要素？本文依据建构主义的基本理论并结合作者37年从事高等教育的经历，从教育教学目标、知识体系构建、教与学的策略、教学氛围以及学业评价等几个方面进行了初步探索与讨论。

一、教育教学目标

按照当代建构主义的教育思想，教育的目的就是要促进学习者的学习，教育教学过程要提升学生的学习能力和促进学生的身心发展。因此，教育教学的目标绝不仅仅是知识的传授。知识传授是手段和过程，通过知识的传授和学习，内化为学生的能力，并在这个过程中逐渐养成学生正确的世界观、人生观和价值观。也就是说，教育教学应该聚焦知识、能力和情感三大目标[4]。①知识目标。深入了解课程的知识体系，对课程体系中的基本知识、基本技能和理论认识达到一定的深度和广度。通过知识的学习，构建起在学生头脑中关于知识的意义，并成为学生自己的、能够被学生灵活运用的、能够为学生提出问题和质疑论点时提供论据的有效知识。②能力目标。培养学生获取知识的能力，能够理论联系实际，融会贯通，举一反三，通过“发现问题—分析问题—解决问题—形成思想”，培养探索精神，提高创新能力。③情感目标。重视树立学生正确的价值观，严谨求实的工作态度，润物无声，促进学生确立对真善美的价值追求。

显然，聚焦学习者身心发展的大学教育教学范式，应该更加注重把“教”从以知识传授为主逐步转向以能力提升、价值观引领为主，把“学”从以模仿、仿效为主转向以探究、实践与合作式学习为主，把学生的被动学习转变为主动学习[2]。

二、知识体系构建

长期以来，我们的教育似乎已经习惯于由教师向学生传输知识的范式，学院、专业、教师牢牢掌握课程知识体系构建的权力。学院制定宏观规划，确定专业培养目标；专业负责人或专业教师团队主导专业知识结构、专业课程体系的制定；专业课程教师掌握课程知识体系以及课程教学大纲的制定。在这样的机制下，遵循的是行为主义的教育思想，课程知识体系的构建高度强调客观化、统一化、程序化和标准化，按照一种规定的格式构建课程知识体系，学校成为工厂，学生则被视为工厂生产的标准化“产品”，学生的主体地位没有得到充分尊重，学生的个性化发展被严重忽略。

事实上，每个学生都是不同的个体。当代高等教育体系中，应该充分考虑学生的个性化发展，倡导由师生共同构建知识体系，允许或容忍个性化培养方案的存在。专业及专业知识结构、课程体系及课程教学内容可以依据学生的不同采用个性化方案。“专业”可以是依据学生个性化发展需要而定制的、独一无二的，这样的“专业”实际上在国内外部分高校中已经存在。

南京大学“三三制”人才培养体系已经明确提出并切实践行了“以学生发展为中心”的人才培养理念，给予学生充分的选择权，为学生自主构建课程模块和知识体系搭建平台，拓宽了学生成长的途径，从而充分激发了学生的学习能动性，促使其变被动学习为主动学习。西北大学近年来实施的完全学分制改革,同样践行了尊重学生主体地位、服务于学生个性化发展的教育理念。

总之,大学教育知识体系构建不能完全掌握在教师手中,在教学过程中应该给予学生一定程度的选择权,鼓励学生参与知识体系、教学内容的构建,这样才能符合建构主义的理论。

三、教师教学策略

随着教育理论的不断深化提高与教育思想的不断进步,教师教学策略大体经历了3个典型的发展阶段。

(1)传统的教学策略是典型的以教师为中心的范式。教师传授、灌输知识,居于主体地位,掌握教学进程和绝对话语权。教师利用讲解、板书和各种媒体作为教学手段,教师遵循教科书、教材组织教学内容,并以知识构成的逻辑顺序向学生灌输教学内容。学生只能处于被动从属地位,机械地接受知识的灌输。教师的讲解常常与教科书、教材中的文字表述和逻辑顺序几乎完全一致,学生的学习过程演化为对教科书、教材及教师讲解的简单校验,学生主要依靠机械的重复记忆来达到获取知识或信息的目的。这种教学策略最典型的结构模型就是“讲—听—读—记—练”,或其升级版的“感知—记忆—理解—判断”结构模型[5]。

(2)基于行为主义的近现代教育理论顺应了19世纪科学技术的发展和科学实验的快速兴盛,在心理学研究的基础上,提出学生在学习过程中,只有在新的知识与已有知识有机联系起来时,学生才能更好地学习和掌握知识。由此产生了“明了—联合—系统—方法”或“预备—提示—联合—总结—应用”的教师教学策略结构模型[5]。

需要指出的是,以上这些教学策略都有一个不可回避的共同特性,就是在很大程度上忽视了学生在学习过程中的主体地位,其结果是压抑和阻碍了学生的个性化发展。

(3)随着20世纪至21世纪科学技术的发展,教育教学理论也进入了一个新的发展阶段。强调个性化发展思想的普遍深入与流行,建构主义的教育理论得到社会的推崇,同时教学策略也向前推进了一步。

建构主义的教育理论打破了以往教学策略和教学范式单一化的倾向,强调学生的主体地位,强调活动教学,培养学生发现探索的技能,获得探究问题和解决问题的能力,开辟了现代大学教师教学策略的新路径。

当前,随着科学技术的发展,教育面临着新科技革命的挑战,促使人们利用新的理论和技术去研究学校教育教学问题。现代心理学和思维科学对人脑活动机制的揭示,发生认识论对个体认识过程的概括、认知心理学对人脑接受和选择信息活动的研究,特别是系统论、控制论、信息加工理论等的产生,对教学实践产生了深刻的影响,也给教学模式提出了许多新的课题[5]。

因此,当代大学教育教学新范式应该注重培养学生的能力和才干,注重交流合作和解决问题。应该由单一教学策略向多样化教学策略发展,由归纳型向演绎型发展,由提高“教”的质量向提高“学”的质量发展,由关注“灌输学生的教学质量”向关注“激发学生的教学质量”发展[6],由教师主要作为讲解者向教师主要作为学习方法和学习环境的设计者发展。

也就是说,教师正确的角色应该是向导。向导是给人指路甚至随行,但绝不会代替旅行者长途跋涉;向导指明地点,但不会代替旅行者感受领略美景时的激动与喜悦;向导提供建议、指出陷阱,但不会去代替旅行者探险。教师紧随学生的每一步,但是真正产生意义的还是学生自己和他们所实施的行为[7-9]。这样的教育思想才应该是教师教学策略选择的方向与遵循。

新教育教学范式必然要求创建有别于传统的、新的教育教学氛围。聚焦学习者身心发展的教育教学范式,应该强调教育教学过程中的民主氛围,倡导教育教学过程是基于教师和学生人格平等前提下的共同探讨和学习的理念。学生与教师之间是学习与促进学习的关系,学生与学生之间是合作学习的关系。课堂文化则应该倡导多元和谐的文化氛围,并强调对教师和每一个学生个体的文化尊重。教学过程中,教师和学生共同享有探究、创造、交流、合作以及质疑的权利。

四、学生学习策略

长期以来,在基于认知主义和行为主义的教育理念环境下,灌输式教育教学范式和应试教育教学范式广泛应用,教师在教育教学过程中拥有绝对的主导权,掌控着学习的进程,学生的主体地位被严重忽略,学生已经将教师的权威视为理所当然。多数学生在课堂教学过程中关注的是“听—读—记—练”,通常不会对教师列举的知识体系进行分析、对比、反思甚至质疑,学习成为典型的对知识或信息的记忆。学生在学习过程中,只要不被教师点名,就不会在课堂上发言,已经习惯了等待教师明确告诉他们应该“做什么”以及“怎么做”,教育对于他们来说就是天经地义的强制性过程[7]。

很显然,高等教育发展到今天,我们必须改变传统的教育教学范式。基于建构主义的教育理论强调学生在教育教学过程中的主体地位,学生的学习观念和学习策略必须转变。当课堂教学以学生为主体时,学生应该从被动的、依赖性的学习者转变为自主的、积极的具有自我管理能力的学习者。学生应该改变仅仅是“接受教师知识灌输的被动容器”,积极参与知识体系的共同构建,成为教育教学过程中积极的知识构建者、发现者和改造者。改变记忆式的学习方式,积极践行关联式学习策略,将自身的成长目标定义为能力和才干的提升,以及正确价值观的养成。更为重要的是,学生的学习目标应该超越传统的“竭力完成教学要求,获得学业文凭”,更加关注持续的终身学习。这是一个学生成长的过程,这一过程不会自动完成,也不会立刻完成,而是需要一个自我觉醒的逐渐成长过程[6]。

作者认为,就我国传统文化、高等教育发展历史以及当前高等教育现状而言,学生的自我觉醒和学习策略的转变还需要一个较为艰难的漫长过程。

五、学业评价策略

传统的教育教学范式中对学生学业评价体系过分看重考试成绩,忽略了对学生潜在创新能力的开发与评价。这样的学生学业评价策略不能很好反映学生的实际能力和才干的提升程度,不能帮助学生更好地学习,对学生学业成绩的提升推动力不强。

聚焦学习者身心发展的教育教学范式,应该采取更加全面综合的学生学业评价策略,更加注重学生在整个学习过程中长期的学业表现,以课堂表现和学习档案为典型做法,并注重教学过程中的持续性评估,使学业评价始终与教育教学目标保持一致[8-11]。学业评价应将学生置于核心地位,通过学业评价发现问题,帮助学生提高学业成绩。同时,通过评价和评估为教师提供及时准确的信息,以帮助教师改进教学。

2019 年 10 月 8 日,教育部发布了《关于深化本科教育教学改革全面提高人才培养质量的意见》,明确指出高等学校必须坚持以立德树人作为根本任务,把立德树人的成效作为检验学校一切工作的根本标准,努力构建德智体美劳全面培养的教育体系,形成更高水平的人才培养体系;并提出了让"学生忙起来、教师强起来、管理严起来、效果实起来"的改革方向。因此,加快我国高等教育教学改革,构建理念更加先进、有利于促进学生主动学习、有利于人才培养质量实质性提升的教育教学范式,是我们面临的不可回避的严峻挑战。我们有责任加快探索与改革的步伐,积极推动新的教育理念不断深入人心,为培养一流创新人才,为我国高等教育的健康发展,实现我国由高等教育大国向高等教育强国的转型做出我们应有的贡献。

参考文献

[1] 郝文武.教育哲学研究[M].北京:教育科学出版社,2009.

[2] 刘晓伟.情感教育:理性的回归与顺应[J].杭州师范学院学报(医学版),2005(6):212-215.

[3] 罗靖.英语专业教学中语音教学模式的研究[D].山东师范大学,2009.

[4] 赖绍聪.如何做好课程教学设计[J].中国大学教学,2016(10):14-18.

[5] 许世坚,朱前星.论思想政治教育研究型教学法的教学模式[J].职业时空,2007(11):41-42.

[6] 周仕德.美国大学教学范式的转型研究及启示[J].重庆高教研究,2014,2(5):102-107.

[7] 赖绍聪.以学习者为主体的课堂教学[J].西北工业大学学报(社会科学版),2018,137(3):39-42.

[8] 迪·芬克.创造有意义的学习经历:综合性大学课程设计原则[M].胡美馨,刘颖,译.杭州:浙江大学出版社,2006.

[9] 刘桂辉.论以学习者为中心的教学策略转换研究[J].教育与职业,2013(24):103-105.

[10] 赖绍聪.寓教于学 寓教于研[J].中国大学教学,2017(12):37-42.

[11] 赖绍聪,华洪.课程教学方式的创新性改革与探索[J].中国大学教学,2013(1):30-31.

以学习者为主体的课堂教学①②

赖绍聪

摘要：大学课堂教学应该以学习者为主体，只有在充分尊重学习者主体地位的前提下，通过教师对课堂教学的精心组织与策划，课堂教学才能够真正实现知识、能力、情感三大教学目标。本文结合作者35年从事高校专业课程教学的经历，从教育理念、教师角色、学习者角色以及课堂教学内容凝练和教学策略选择等方面阐述了以学习者为主体的课堂教学模式，探讨了当前我国高校课堂教学过程中存在的问题及其改革方向。

教学是有目的、有计划、有组织地进行人类经验传授的活动形式。更为重要的是，教学过程中，教师必须将需要传授的内容，经过科学选择，依据知识构成的逻辑顺序和学习者获取知识的认知规律形成教学体系。长期以来，在我国高等教育课堂教学体系中，教师通常高度关注了"知识构成的逻辑顺序"，教学内容体系的构建一般会尊重学科的知识体系。也就是说，内容的构建关注了学科知识构成的逻辑顺序。然而，这样看似十分合理的"学科知识逻辑顺序"（章—节—点）与学习者"获取知识的规律"并不一定是完全吻合的。正是由于这样的教育理念，使得长期以来学习者在学习过程中所应该拥有的主体地位未能得到充分体现。在课堂教学过程中，"知识"仍然主宰着一切，学科知识逻辑顺序引领课堂进程，教师代表知识的权威，掌控课堂的主导权，居于课堂的主体地位，而学习者始终处于被动接受知识的从属地位。

高质量的课堂教学要讲究方法，要能引起学习者的兴趣，引发学习者思考。要达到这一目标，我们必须真正站在学习者的角度来思考问题，打破以教师的"教"为主的传统课堂，转变为以学习者的"学"为主的课堂，让每堂课都能深入人心。

一、以学习者为主体的教学理念

以学习者为主体的课堂教学，其核心理念是将课堂教学的功能定位为"促进学习、学会思想、生成智慧"，而绝不能只局限于传递知识。以学习者为主体的课堂教学充分尊重学习者的主体地位，将学习者作为课堂教学活动的聚焦点，在整个教学过程中，高度关注学习者的感受，关注他们在学习什么、怎样学习，在什么样的环境下学习，学习者目前的

① 原载于《西北工业大学学报》社会科学版，2018(3)。
② 国家自然科学基金项目"国家理科地质学人才培养基地建设"(J1310029,J1210021)。

学习内容是否能够为他们将来的终生学习做好准备[1-5]。

以学习者为主体的课堂教学并不反对教师专门知识的有效性，也并不忽视教师在教学过程中的组织策划作用，而是更加强调教学过程中“教”与“学”的双边互动。因为，教学是“教”与“学”的统一，“教”为“学”而存在，“学”又要靠“教”来引导，二者相互作用、相互依存。作为教师，我们的核心任务不仅仅是传递知识，而是要通过课堂教学的精心组织与策划，培养学习者终生学习的技能，以及使用这些技能的信心，并最终实现由“教知识”向“教技能”“教智慧”的转变。

以学习者为主体的课堂教学重视培养学习者的学习兴趣，倡导学习者在兴趣的驱使下主动获取知识。注重提高学习者的学习效率，提升学习者的综合素质。而教师的主要职责是激发学习者的学习兴趣，为学习者引领学习方向，培养学习者观察、发现、分析、研究以及解决问题的能力。

以学习者为主体的课堂教学打破了以教师为中心的被动接受式、以记忆为主的机械学习和单纯的知识学习，而是以学习者为中心的主动参与式学习，以深刻理解为主的意义学习，包含知识、方法、能力、情感态度与价值观的多方面发展学习者的学习。

以学习者为主体的课堂教学在教学目标中体现以学习者为中心，在活动的设计上考虑学习者相关能力的培养以及情感态度与价值观的形成；在时间安排上让学习者从课堂上的“看客”成为“参与者”；在组织形式上给予学习者自主学习、自由讨论的课程形式；在教学过程中鼓励学习者参与到活动中来，鼓励学习者发表观点[1-5]。

二、教师角色的转变

当课堂教学以学习者为主体时，教师的角色将发生显著转变。教师不再仅仅是内容的专家或者课堂上掌控进程的权威，不再仅仅是通过提高复杂的传授技能去改善教学。教师主要起指导和促进学习的作用，教师将成为学习经验的引导者、激励者和设计者，教学行为的标志性特征是“促进学习”。

在课堂教学过程中，教师不再是唯一的主导者和表演者，而是将聚焦点转向教育教学理论研究、优质教育教学资源建设以及精心细致的课程教学设计，包括对教育理念、教学思想、教学内容、教学方法与教学策略的精心设计，以及对学习者的经验分享、互动活动、阅读理解、作业与思考等方面的精心设计。教学设计的目的是要让学习者通过这些活动、思考、分享、讨论等来学习内容。教师还要在学习内容时提供引领、指导、解释、批评及鼓励。教师在这个过程中的指导、评价、综述或建设性的批评将能够促进学习者下一次的学习活动。

三、学习者角色的转变

在传统的、以教师为主体的课堂教学中，学习者已经习惯于教师主导和控制学习进程，并将教师的权威视为理所当然。学生在学习过程中，只要不被教师点名，就不会在课堂上发言，都已经习惯等待教师明确告诉他们该做什么以及怎么做，教育对于他们来说就是天经地义的强制性过程[6-12]。

当课堂教学以学习者为主体时,学习者从被动的依赖性的学习者转变为自主的积极的具有自我管理能力的学习者。这是一个成长的过程,不会自动完成,也不会立刻完成,而是需要一个自我觉醒的逐渐成长过程。作者认为,就我国传统文化、高等教育的发展历史以及当前高等教育的现状而言,学习者的这种自我觉醒和主动学习、自我管理意识、理念的转变将是一个艰难而又漫长的过程,这个过程可能比教师角色的转变更为漫长和困难。

四、以学习者为主体的教学内容凝练

建构主义理论认为[1],教学内容与学习者之间的核心关系是通过教学内容的学习和研讨,逐渐构建学习者自己的知识体系、能力体系和情感体系,而不是仅仅被动地接受教师和课本传递给他们的信息。也就是说,知识不能简单地给予学习者,而是要通过对知识的学习让学习者构建起在他们自己的头脑中关于知识的意义,这样的“知识”才能够成为学习者自己的、能够被学习者灵活运用的、能够为学习者提出问题和质疑论点时提供论据的有效知识。

在以学习者为主体的课堂教学中,我们并非一定要等到学习者完全掌握了基本技能以后再去接触教学内容,而是应该鼓励学习者去主动探索、逐步掌握内容,并尽可能地将教学内容与学习者的亲身经历相联系。无论学习者已经具备了什么水平的专门知识,教师在教学过程中都应该积极鼓励他们对教学内容提出质疑与挑战。

以学习者为主体的课堂教学内容凝练应该重点关注以下 3 个方面[1]:①教学内容应该能够被学习者用作发展知识的基础。②教学内容应该能够被用来发展学习技巧,能够帮助学习者掌握学习策略、方式和技能,使学习者在类似科目的学习中可以有效运用。③教学内容应该能够被用来提升学习者自我学习的意识,培养学习者处理学习任务的能力和自信心。

五、以学习者为主体的教学策略选择

传统的以教师为主体的课堂教学策略具有五大显著特点[1-5]:①教师利用讲解、板书和各种媒体作为教学手段传授知识,学生被动接受知识。②教师是主动的施教者(知识的传授者、灌输者)。③学生是外界刺激的被动接受者和知识灌输的对象。④教材是教师向学生灌输的内容。⑤教学媒体则是教师向学生灌输的方法与手段。显然,这样的传统学习理论和教学策略已经不能够适应以学习者为主体的教学活动。

当课堂教学以学习者为主体时,教学策略的选择将紧紧围绕突出学习者的主体地位,引导学习者思考、分析与探究,提出发散、扩展与升华学习者思维的问题。因此,教学策略选择的原则是激发兴趣、促进学习,养成学习者主动学习的意识,逐步形成学习者自我管理的能力。

具体来说,以学习者为主体的教学策略多种多样,而且仍然在不断发展和完善之中。

然而,一般情况下常见的教学策略大体有以下几种[1-12]:情境创设策略、探究性学习策略、合作学习策略、自主学习策略、STSE 策略(Science Technology Society Environment)、科学史教学策略、概念图教学策略和网络探究性学习教学策略。

以合作学习策略为例,它更加强调教师与学习者之间、学习者与学习者之间的面对面互动,以及学习者的分组学习。探究性学习策略则强调在任务驱动下的思考、分析、讨论和探索。而所有这些教学策略的设计均体现了一个明确的指向,即以学习者为主体。

六、结语

以学习者为主体的教学改革已经不是一个新颖的话题,这种理念已经提出多年,并在教育界已有较长时间的研究与探索历史。然而,在我国高等教育领域,这样的改革还远远不够深入,传统的以教师为主体的教学模式仍然在大学课堂教学中占据着主导地位。尤为突出的是,学习者主动学习、自我管理意识的觉醒任重而道远。作为高等教育领域的一线教师,我们有责任、有义务不断加快探索与改革的步伐,通过我们不懈的努力,积极推动新的教育理念不断深入人心,为培养一流创新型人才,为我国高等教育的健康发展,为实现我国由高等教育大国向高等教育强国的转型做出我们应有的贡献。

参考文献

[1] 玛丽埃伦·韦默.以学习者为中心的教学:给教学实践带来的五项关键变化[M].洪岗,译.杭州:浙江大学出版社,2006.
[2] 戴扬,鄢浩."以学生为中心"的教学理念在化学工艺学教学上的探索与实践[J].广东化工,2018(8):246-251.
[3] 王明利,韩振丽,范力茹."以学生为中心"的《大学物理》课程教育教学改革与实践[J].教育教学论坛,2018(18):103-104.
[4] 卫建国.大学课堂教学改革的理念与策略[J].高等教育研究,2018(4):66-70.
[5] 窦祥胜.以学生为中心的经贸类专业教学模式改革与创新[J].黑龙江教育学院学报,2018(5):42-44.
[6] 赖绍聪.创新教育教学理念提升人才培养质量[J].中国大学教学,2016(3):22-26.
[7] 赖绍聪.建立教学质量国家标准 提升本科人才培养质量[J].中国大学教学,2014(10):56-61.
[8] 赖绍聪.地球科学实验教学改革与创新[M].西安:西北大学出版社,2010.
[9] 赖绍聪,华洪.课程教学方式的创新性改革与探索[J].中国大学教学,2013(1):30-31.
[10] 赖绍聪.如何做好课程教学设计[J].中国大学教学,2016(10):14-18.
[11] 赖绍聪.岩浆岩岩石学课程教学设计[M].北京:高等教育出版社,2017.
[12] 赖绍聪.寓教于学 寓教于研[J].中国大学教学,2017(12):37-42.

寓教于学 寓教于研[①]

赖绍聪

摘要：教育不是简单的理性知识的堆积，更需要的是培养学生正确的价值观。结合作者长期从事高校专业课程教学的经历，从价值观的内涵、价值观的形成过程、价值观的教育现状，以及怎样通过深化高校专业课程教学改革，将价值观教育实质性地融入专业课程教学过程，实现价值引领等4个方面进行了归纳总结，探讨了当前我国高校专业课程教学体系中价值观教育存在的问题及其改革方向。

长期以来，在我国高等教育教学体系中，尤其是专业课程教学体系中，仍然是以系统的专业知识传授为主导，教师在教学过程中高度关注知识目标和能力目标，教师依旧将学科知识的传递灌输以及专业实践能力提升作为其主要职责，但却在一定程度上忽视了价值观教育，部分专业课程教师甚至存在“专业课程教学与价值观教育关系不大”的模糊认识。

事实上，教学有3个主要目标，包括知识目标、能力目标和情感目标[1]。在我国现行高等教育教学体系中，亟须进一步加强对学生的精神关怀，将价值观教育实质性纳入课堂教学，尤其是专业课程课堂教学情感目标体系，培养合格的社会主义建设者和接班人。

习近平总书记在全国高校思想政治工作会议上强调，高校思想政治工作关系高校培养什么样的人、如何培养人以及为谁培养人这个根本问题。要坚持把立德树人作为中心环节，把思想政治工作贯穿教育教学全过程，从而实现全程育人、全方位育人，努力开创我国高等教育事业发展新局面[2]。2017年7月26日，习近平总书记再次强调，中国特色社会主义是改革开放以来党的全部理论和实践的主题，全党必须高举中国特色社会主义伟大旗帜，牢固树立中国特色社会主义道路自信、理论自信、制度自信、文化自信，确保党和国家事业始终沿着正确方向胜利前进。习近平总书记的重要阐述，为我国高校坚持社会主义办学方向，培养德才兼备、德智体美全面发展的社会主义事业建设者和接班人指明了方向。2017年9月23日，在“中国共产党创办新型高等教育八十年”论坛上，教育部负责人针对如何办好中国特色高等教育提出了6个问题，这些问题的提出都是为了解决高校培养什么样的人、如何培养人以及为谁培养人这一根本问题，为我国高等教育事业发展指出了具体路径。

① 原载于《中国大学教学》，2017(12)。

课堂教学中如何把“立德树人”作为中心环节,这不仅仅是高校思政教育的重要课题,同时也是专业课程教学中面临的重要挑战。要成为合格的社会主义事业接班人,就必须具有正确的价值观。近年来,世界各国高等教育中出现的一个引人注目的重要新动向,就是教育取向从能力导向逐渐地朝着价值观导向转变。而所谓的价值观导向,归根结底就是教育学生如何对待自己、对待他人,以及对待国家、社会和世界。

一般来说,遵照教育规律,依据知识目标、能力目标、情感目标三大教学目标要求,教师大多会通过课程设置、大纲厘定、教材编写、教学内容凝练、典型案例引入等多种方式,在课堂教学中渗透一定的价值观内容。然而,不尽如人意的是,由于传统的教学方式仍然是以知识传授为核心导向,这种“教知识”的教学方式不仅淡化了知识学习过程中的能力提升,而且严重制约了教学过程中情感目标的实现,使得教学过程中价值观教育的渗透,以及对学生正确价值观形成的引领效果甚微。因此,不断全面深化教学改革,探索价值观的内涵,高度重视价值性课堂文化对学生价值观形成的引领与促进作用,积极创建良好的课堂文化与环境[3],将价值观教育实质性地融入专业课程课堂教学中去,实现价值引领,培养社会主义合格接班人,乃是当前我国高等教育必须面对的重要课题。

一、价值观及其形成过程

现有的理论体系认为,价值观是基于人思维感官之上而做出的认知、理解、判断或抉择,也就是人认定事物、辨别是非的一种思维或取向[4]。通常情况下,一个人的价值观是从出生开始,在家庭、环境和社会的多种复杂因素共同影响下逐步形成的[5]。个人所处时代的政治生态、文化生态、社会生产方式,以及个人在社会中所处的政治地位、社会地位和经济地位,均对一个人价值观的最终形成和稳定起到决定性作用。所以,在不同时代以及不同社会生活环境中形成的价值观是有所不同的[6]。因此,可以说价值观一旦形成,它就常常具有相对的稳定性和持久性[3,7-9]。

一个人的价值观通常包含2个重要方面[3,7-9]:一是如何看待社会,如何看待世界,如何与人和谐相处,如何处理人生历程中个人与社会、现实与理想、付出与收获之间的关系。也就是选择什么样的人生,走什么样的道路,如何从价值角度考虑人生问题,即所谓的“人生价值观”。它决定了人的自我认识,人的理想、信念、生活目标以及追求方向,这对自身行为的定向与调节起到重要的作用。二是一个人对职业的认识与态度,对职业目标的追求与向往[10],即所谓的“职业价值观”。它决定了一个人是干一行爱一行还是爱一行干一行的职业取向。

显然,价值观是人们对客观世界及行为结果的评价和看法,是人际关系的基石和核心。价值观信息上的沟通是健康人际关系形成的关键,价值观实践上的一致是人际关系的保证[11]。形成正确的价值观是比知识积累和能力提升更为重要的、决定人生态度和事业成败的核心关键要素。因此,培养合格的社会主义事业接班人必须将价值观教育置于首要位置。

价值观的形成、发展和变化始终与社会生活的发展、变化息息相关,它是个体在自身不断社会化的进程中逐渐形成的[12]。社会的教化与个体的内化之间的双向互动共同成就了每一个社会成员鲜活的价值观念体系。已有的研究资料表明,家庭、学校、群体、榜样、传媒和社会环境等是个体价值观形成过程中的重要影响因素,它们共同作用于个体价值观的形成过程中[3,7-9]。

从理论上来讲,大学期间是价值观基本形成的核心阶段,高校教育中的价值引领是大学生能否形成正确价值观的决定性因素。

按照我国现行教育体制,学生一般在18岁左右完成基础教育(高中毕业)。在此期间,他们通常生活在父母身边,父母、学校、社会共同影响其价值取向,由于年龄因素,其价值观常常处于朦胧和不确定阶段。大学生的年龄通常为18~22岁,这个年龄段是个体价值观基本形成的最为核心的阶段。在此期间,父母通常不在身边,学校教育和社会影响,尤其是学校教育成为影响个体价值观形成的决定性因素,这个阶段也是个体价值观取向由朦胧、不确定逐步走向清晰、明确的关键时期。不言而喻,高校教育中正确的价值观引领,在极大程度上决定了我国高等教育能否培养出一批又一批社会主义合格接班人。

因此,高校教育必须践行社会主义核心价值观。无论是高校管理者、思政工作者,还是专业教师,都务必明确教育观念,将引领正确价值观有机融入高校教育的每一堂课、每一个教学环节中。

二、价值观的教育现状

1.思想政治课程教育

高度重视思想政治教育是我国高等教育办学的鲜明特色。党和国家历来对大学生的思想政治教育都很重视,同时,全国高校也在积极努力探索思想政治教育的有效方式方法,从而加强对大学生正确价值观形成的有效引导。然而,就近年来高校实际情况来看,思想政治教育和价值观引领在取得一定成效的同时,也面临着严峻的挑战,导致课堂教学效果不佳。思想政治教育所面临的困难主要反映在以下几个方面:①大学生的生理年龄(一般在18~22岁)决定了他们正处在独立意识和叛逆心理并存的特殊成长期,常常会对正面的价值观教育持有疑惑与怀疑,有时甚至是排斥的态度。②由于社会和就业竞争压力增大,使得大学生在一定程度上存在功利主义思想,更加关注专业知识学习和专业能力提升,并将学习简单视作未来求生的手段。而对价值观教育重视不够,热情不高,部分学生学习思政课只是为了获取学分或考研的需要。③现行价值观教育的内容与部分社会不良现象之间的反差,在一定程度上影响了当代大学生对高校价值观教育的认同度。④在互联网时代,由于商业文化、消费文化和网络文化的兴起,以及异域文化和媒体文化的渗透[13-14],使得当代大学生对价值取向是与非的判别难度明显增加。

2.专业课程教育

我国传统的专业课程教育教学观念与教学内容的凝练、教学方式方法的选择均存在一定局限性,在一定程度上制约了专业课程教学过程中价值观教育的融入。①专业课程教学主要将关注点聚焦在知识传授方面,专业课程教师对价值观教育重视程度不足,思想认识模糊。②"照本宣科"式的教学内容很难让学生接触学科前沿、了解学科发展趋势,难以提升学生的思辨能力。③"灌输式"的教学方式不利于改善学生盲目崇拜的价值观念。④"一言堂"式的课堂氛围未能充分尊重学生学习的主体地位,学生不能充分表达自己的见解。⑤教师对师生关系、课堂环境、课堂秩序、课堂文化以及校园文化等因素与正确价值观形成的密切关系认识不足。

显然,高校教育中占据主体学时的专业课程教学体系在一定程度上忽略了价值观教育的融入,传统的专业课程教学方式已经明显不适应社会发展对价值观引领所带来的新挑战。在专业课程教学过程中,教师如何深化教学改革,充分发挥自身的价值引领作用,已经成为我国高等教育领域必须从理念、理论及方法上进行深入探讨和研究的重要课题。

三、深化教学改革,实现价值引领

师者,所以传道授业解惑也。教师一词有双重含义,既指一种社会角色,又指这一角色的承担者[15-18]。根据《中华人民共和国教师法》,教师被定义为:"履行教育教学的专业人员。承担教书育人,培养社会主义事业建设者和接班人,提高民族素质的使命。"

在我国历史上,教师曾经有许多不同的称谓,比如夫子、先生、园丁、蜡烛、春蚕、春雨、孺子牛、人梯等。新中国给予了教师最为神圣的称谓——人类灵魂工程师。

事实上,教师是仅除学生父母以外,对学生影响最大的人。教师的言行、举止和思维方式等,都将潜移默化、深远地影响着学生本身和他们的未来[19]。对每一个教师而言,无论是从事思政教育还是专业教育,都对学生正确价值观形成负有义不容辞的责任,教师理应成为正确价值观的引领者。

然而,教师尤其是专业课程教师,作为价值观的引领者需要具备一定的能力,掌握一定的教育教学方法,并且能在教学过程中挖掘出那些对学生具有思想启迪、行为导向和心灵震撼的有益价值因素,从而能够发现学生在生活中的价值冲突与危机,同时结合知识的教学来引导学生进行正确的辩证分析、比较、判断和选择[3,7-9]。这就要求教师对学科知识和生活经验中的有效价值因素有积极的体验和敏锐的感悟,并以自己的智慧进行精心的课程教学设计[20-25]。

1.转变教育观念

价值观教育应该贯穿高等教育教学全过程,高校思政教育体系和专业教育体系必须相互融通、互为支撑。专业课程教师在学生价值观教育过程中担负着义不容辞的重要责任,每一位专业课程教师都必须牢固树立教书育人、育人为先的教育教学观念,通过教学内容的精心凝练、教学方法的合理选择以及良好教学氛围的营造,将价值观教育实质性地融入专业课程教学的每一堂课、每一个教学环节中。

2.转换教师角色

传统课程教学模式"以教师为中心",教师定位为知识的传递者,"知识"变成了主宰一切的绝对力量,教师因依附于大大小小的知识点,从而自然而然地成为课程教学的唯一强势权威。在这种学习过程中,学生没有发言权,而且缺少积极主动的学习体验,导致学生所拥有的主体地位未能得到充分体现,教师和学生都只不过是知识的简单传递工具与机械接受容器而已。如此一般的教育教学观念严重忽视了教学参与体验过程中情感与价值观的引领[20-23]。因此,正视价值冲突,抓住教学契机,实现角色转换,教师角色由简单的知识传递者向有效的价值引领者转变,乃是当前全面深化教学改革的核心任务和目标[20-21]。

3.凝练教学内容

①专业课程教学内容需要精心凝练,按照“基于教材,高于教材,突出特色”的原则,依据教师从事专业研究的亲身经历,努力发掘出那些有利于学生养成严谨的治学态度、不懈追求真理的精神的典型案例,融入教学内容之中。以学科发展历史中一代又一代学人孜孜不倦、求真务实、献身科学、献身事业的感人事迹,引导学生正确对待学术探索与研究过程中的成功、失败、付出、收获、成绩、荣誉等,引导学生逐步形成正确的价值观。②教师需要结合学科前沿与自己的科研成果,实现教学与科研的深度融合,保证教学内容的先进性和前沿性,让学生获得最新的专业科学发展动态。这种教育氛围不仅能提升学生的辩证思考能力,而且还能逐步改善学生盲目崇拜的价值观念。教师讲授用理论联系实际的方式,不仅能使学生真正感受到知识的魅力与理论的魅力,而且也达到了调动学生思维的目的。③在专业课程课堂教学过程中,应该在明确每堂课的知识目标、能力目标和情感目标的前提下,对本堂课核心教学内容的知识点进行分析,建立核心知识点之间的内在联系,发现核心知识点之间的逻辑关系,尽可能融入教师自身对学科知识体系的感悟,从学习者的角度出发,精心设计每堂课的内容结构[24-25]。④在专业知识讲授时,要强化“思想方法”的渗透。在这个过程中,实际上渗透了认识自然、理解社会的正确方法。⑤让学生能够从课堂教学的教学内容中深刻地体验“知识的形成过程”。这个过程实际上渗透了对自然、社会发展规律的正确理解。⑥不仅要关注知识的传授,更应该加强“学法指导”。这个过程实际上渗透了人类获取知识、追求真理的正确方法。⑦以先进的课堂教学内容为载体,实现激发学生创新思维的教学目的。这个过程将有利于引导学生正确理解人类发展、社会进步的必然性。通过这一系列教学内容的精心设计和教学改革,才能有效推动专业知识教育与价值观教育的有机融合,真正实现知识、能力、情感三大教学目标。

4.改革教学方法

在专业课程教学过程中,教师必须进一步明确和牢固树立“学生主体”的观念,着重培养学生的主体意识,积极倡导相互讨论研究的教学氛围,必须最终从根本上改变教师的“一言堂”和“满堂灌”的教学方式方法。①教师要勇于突破传统观念,将传统教学方法的精华与现代教育技术相结合,跟踪当代教育学方法与理论研究的新进展,不断更新教学模式,根据每一堂课教学内容的不同,采用最为合理的教学模式,努力构建和实践符合教育科学发展规律的多模态课堂教学模式,凸显教学方式与方法“多元化”的改革特点。比如,根据专业课程教学过程中不同章节教学内容的差异,合理有效地引入 PBL (Problem Based Learning)、TBL(Team Based Learning)、CBE(Competency Based Education)等基于建构主义学习理论的方法,将更加有效地调动学生主动学习的积极性,提升专业教学质量,更好地融入价值观教育,实现专业课程教学过程中的价值引领。②教师在专业课程课堂教学中,应该积极鼓励和引导学生大胆质疑并且勇于表达自己的见解,不断培养学生发现问题、分析问题和解决问题的自主能力[22-23]。这种教学方式体现了对学生价值的肯定和尊重,体现了"教"与"学"之间积极的思维共鸣和教师的主导作用与学生的主体地位的和谐统一,是融合专业教育与价值观教育、引领正确价值观的重要改革方向。③充分利用互联网先进技术和数字化资源,采取线上学习线下辅导、线上线下"混合式"教学等多种新方式,改变传统的"教"与"学"方法,积极倡导和推动学生学习的主动性,学生在获得学习能力、践行能力的过程中逐渐形成探索精神和独立人格。④改革实验教学模式,坚持实验教学向研究性、探索性、设计性和综合性教学方向转变和发展,鼓励学生自己或以小组为单位根据实验原理自行设计探索性、研究性实验方案并予以实施,充分体会自我实践的价值。⑤拓展实践教学途径,坚持鼓励和支持学生积极参加社会实践和课外学术科技创新活动,通过多途径提升自身的实践能力。进一步让学生了解社会、了解国情,以此激发自身的社会情感、求知欲、创造欲和成功欲,这样学生就会在实践中不断运用所学知识积极回报社会,从而明确自己的社会责任,形成正确的价值观,实现自己的人生价值。⑥引导学生自学和从反思中学习。在课程教学过程中,要时刻强调学生通过课堂中的知识学习逐渐更好地反思自我和外部世界,使得学生从被动接受转变为主动审思和探索过程,学生灵活地运用知识学习中学到的道理、观念来回顾和审思自身的生活经历,不断地深入发现存在的意义和价值,形成正确的价值观。

5.营造良好氛围

建立和营造良好的教学环境与教学氛围,是实现高质量专业教育和价值观教育的重要条件保障。①建立和谐的师生关系。师生关系是教师与学生在教学过程中以"教"和"学"为中介而形成的一种特殊的人际关系,是基于相互尊重、共同探索的平等关系。只有双方相互尊重、互相认可,有高度的信任和默契,才能最终顺利完成培养过程,达到较高的预期目标。一种温暖、有人情味、平等融洽的师生关系,可以使学生积极地参与课堂学习活动,也更容易接受来自教师的价值引导。②充分发挥教师的人格魅力。人格是尊严、价值和品格的总和,人格是个体内在的以及行为上的倾向性,人格也是对知识、事业、日常生活的态度,人格力量是教师的基石。教师端正的行为、高尚的品德,作为一种榜样,本身就是一种人格力量的宝贵价值资源,也是价值存在和实现的典范。因此,在课程教学过程中,教师必须高度重视"身教",即以自身的良好品格、文化和科学素养、心理素质和责任心,潜移默化,引领正确的价值观。教师应竭尽全力,努力成为受学生欣赏、追慕、景仰的师长。③制定合理的课堂纪律。纪律不仅是一个为了教室里表面的平静而设计的规范,更是作为一个小社会(教室)里应具有的美德,其主要反映了师生共同的价值诉求。因此,制定合理的课堂纪律,教师在率先垂范并要求和引导学生共同遵守,这本身就是引发价值观教学的一个机会。课堂纪律作为课堂学习的行为准则与规范,对学生的思想观念和行为活动将起到重要的价值导向作用,使他们明确哪些是可以做的、哪些是不应该做的。④营造积极向上的校园及课堂文化氛围。事实上,校园及课堂(教室)文化氛围能够通过种种价值观的传达使之成为影响价值观教学过程的有效因素,它将对学生的价值取向产生潜移默化的影响。着力建设优良健康的大学文化,积极营造传递正确价值观的课堂(教室)文化氛围,已经成为不可忽视的、引领正确价值观的重要方式之一,并已经引起各高校的高度重视,在价值观教育中正在逐步发挥其推动作用。

高校专业课程教学所要实现的目标不仅仅是知识的传授,更应该是能力的培养与人格的完善。教师是除父母之外对学生价值观形成影响最大的人,引领正确的价值观是我们每一个专业课程教师义不容辞的责任。我国高等教育的基本任务就是为国家培养合格的社会主义建设者和接班人,知识传授、能力提升必须与价值观引领有机融合。我们必须改变高校教育中专业课程教育与价值观引领脱节的模糊认识,更新教育教学观念,将价值引领融入专业课程教育的每一堂课、每一个教学环节中,用我们的实际行动践行社会主义核心价值观。

参考文献

[1] 韩凡.《高效课堂教学模式》在生物教学中的实践[C]//中国教师发展基金会“全国教师队伍建设研究”科研成果集,2013.
[2] 朱炳文.加快建设有特色高水平大学[J].实践:党的教育版,2014(3):6-7.
[3] 姚林群.课堂中的价值观教学[D].武汉:华中师范大学,2011.
[4] 李金宵.高校民主党派思想建设问题及对策研究[D].锦州:辽宁工业大学,2015.
[5] 包利萍,廉惠丽.把思想政治教育落到实处[J].黑龙江科技信息,2009(22):87-87.
[6] 蒿萍.体育院校价值观教育研究[D].长春:吉林大学,2015.
[7] 陈章龙.冲突与构建:社会转型时期的价值观研究[M].南京:南京师范大学出版社,1997.
[8] 侯华莉.当代大学生价值观形成机制研究[J].安徽职业技术学院学报,2013,12(4):61-64.
[9] 田益玲,林杨,张会彦,等.教学方式对大学生价值观形成的影响[J].中国校外教育,2010(9):56.
[10] 程建军,李欢欢.试论当代高校教师核心价值观的构建[J].江苏第二师范学院学报,2012(2):1-4.
[11] 马鑫一.社会主义核心价值体系引领我国当代大学生价值观教育研究[D].西安:西安工程大学,2013.
[12] 侯华莉.当代大学生价值观形成机制研究[J].安徽职业技术学院学报,2013,12(4):61-64.
[13] 龙宝新.价值商谈与学校道德生活的建构[J].华东师范大学学报(教育科学版),2005,23(3):17-23.
[14] 姚林群.论课堂教学中价值观目标的达成[J].中国教育学刊,2014(4):64-68.
[15] 黎旭,李早华.审视教师幸福感[J].新学术,2007(6):8.
[16] 杜耿杰.中美教师专业发展比较分析[J].河南教育:高教版,2006(10):21-22.
[17] 田原.论教师的修养[J].辽宁教育行政学院学报,2004.
[18] 黄建军.教育·教学·教师[J].教育实践与研究,2015(3):8-9.
[19] 贾莉.素质教育中的师爱艺术[J].美与时代月刊 2004(4):89-90.

[20] 赖绍聪.创新教育教学理念提升人才培养质量[J].中国大学教学,2016(3):22-26.

[21] 赖绍聪.建立教学质量国家标准提升本科人才培养质量[J].中国大学教学,2014(10):56-61.

[22] 赖绍聪.地球科学实验教学改革与创新[M].西安:西北大学出版社,2010.

[23] 赖绍聪,华洪.课程教学方式的创新性改革与探索[J].中国大学教学,2013(1):30-31.

[24] 赖绍聪.如何做好课程教学设计[J].中国大学教学,2016(10):14-18.

[25] 赖绍聪.岩浆岩岩石学课程教学设计[M].北京:高等教育出版社,2017.

高等学校课程建设的理念与路径分析①②

赖绍聪

摘要:课程建设是高等教育体系中重要的核心环节之一,契合学科发展前沿、符合人才成长规律的高质量课程教学体系是创新型人才培养的最有力保障。本文基于对课程建设内涵的分析,结合作者39年从事高等教育和课程建设的经验,从理念、思路、内容、方法、评价、机制等多个方面,针对高质量课程建设关键环节及建设路径进行了深入分析。

近年来,随着高等学校"双一流"建设规划的有序推进,我国高等教育事业进入了全新的快速发展阶段,实现由高等教育大国向高等教育强国的跨越式发展已经成为我国高等教育事业发展与改革的最强音,而高质量的课程建设是高等教育体系中至关重要的核心环节之一,是推动高等教育教学改革、提高人才培养质量最为重要的核心抓手。教育部《关于一流本科课程建设的实施意见》以及《关于深化本科教育教学改革 全面提高人才培养质量的意见》中明确提出[1-2],"在全国高校范围内全面开展一流本科课程建设,一段时期内,建成万门左右的国家级和万门左右的省级一流本科课程。通过课程建设,提高教学质量,最终实现管理严起来、课程优起来、教师强起来、学生忙起来、效果实起来的效果"。显然,深入思考课程建设的基本理念与思路,探索课程建设的有效方法与路径,有效构建契合学科发展前沿、符合人才成长规律、适应社会经济发展的高质量课程教学体系,既是我国高等学校每一位一线教师都必须认真思考的重要课题,也是我国高等学校当前形势下共同面临的严峻挑战。

一、课程建设的指导思想

吴岩在2018年曾经指出,"课程虽然是教育的微观问题,却是关乎宏观的战略大问题,是高校人才培养的核心要素,是立德树人成效根本标准的具体化、操作化、目标化,是当前中国大学带有普遍意义的短板、瓶颈和关键"。同时他还指出,一流课程的建设要紧紧围绕高阶性、创新性和挑战度设置。高阶性是指课程应该具备知识、能力、情感(素质)三大教育教学目标有机融合的特点,课程要能够成为培养学生解决复杂问题的综合能力

---

① 原载于《法学教育研究》,2022(39)。

② 国家"双一流"世界一流学科建设(地质学)项目;"国家级一流本科课程(岩浆岩岩石学)""高等学校专业综合改革试点""基础学科拔尖学生培养计划2.0基地"建设项目。

和高级思维的重要推动力。创新性是指课程教学内容选择应该契合学科学术发展前沿趋势,反映当代科学技术进步以及人文社会科学的时代性和先进性,课程组织和教学方法能够充分体现“教”与“学”的双边互动,激发学生的学习兴趣,促进学生主动学习,而课程的学习结果应该有利于促进学生的个性化发展,激发学生探究和解决实际问题。挑战度是指课程学习内容与学习方法应该具有一定的难度,任课教师备课和学生课下学习应该有较高的要求。

泰勒在《课程与教学的基本原理》中指出[3],课程建设和教学计划开发需要面对4个最为基本的问题:①课程试图实现或达到什么样的教育目标;②课程需要提供什么样的教育经验(教学内容)才最有可能达成这些目标;③课程采用什么样的方式对这些教育经验(教学内容)进行有效组织;④课程采用什么样的评学评教策略才能够准确判断预定教育教学目标正在或已经得以有效实现。也就是说,课程建设和教学计划开发实质上需要着重考虑为什么教、教什么、怎么教以及教得怎么样4个最为核心的关键问题。

值得注意的是,高质量的课程体系还必须有良好的条件保障。一是重视团队建设。高素质、高水平的教师团队是课程建设最为核心的条件保障。二是强调资源整合。要整合各类资源,为课程提供充分的线上线下资源保障。三是注重技术创新。课程应该是传统教育教学方式与现代教育教学技术充分融合的结果。四是倡导文化引领。高等教育的课程体系,无论是自然科学类还是人文社会科学类,均应该拥有丰富的文化底蕴,课程建设应该注重先进的文化意识和内涵的融入,形成先进的高等教育课程文化,引领时代发展。

因此,我们可以认为,教学目标、教学内容、教学方法、教学评价、教学资源以及师德师风和执教能力是高质量课程建设的核心要素。其中,明确目标是前提,选择内容是核心,创新方法是策略,评教评学是结果,师德师风、执教能力和资源建设是保障。

二、一流课程建设路径分析

(一)明确课程教学目标

一般情况下,课程教学或者说教育教学主要涉及学生在知识、能力、情感(素质)3个方面的进步与发展。因此,课程教学目标应该包括理论教学、实践技能教学、情感及价值观教育三大领域[3]。

1.理论教学目标(认知领域)

理论教学目标包括对知识的记忆→领会→运用→分析→综合→评价6个层次。①知识的“记忆”是最低水平的认知学习,学生通过课程学习,能够对知识点形成长效的记忆是课程知识学习最基本、最低的目标;②对知识的“领会”,超越了记忆,学生对知识内涵逐步形成了自身的初步感悟,这是最低水平的理解;③有效地“运用”知识是较高水平的理解,学生能够将知识体系较为合理地运用于生产、生活实践;④在知识学习、领会、运用的基础上,能够对事物进行“分析”,则代表知识学习中比运用更高的智能水平,学生已经逐步形成发现问题、分析问题的能力;⑤“综合”需要产生新的模式结构,它强调的是创造能力,是认知领域的较高境界,代表学生已经开始具备解决问题和创新创造的能力;⑥“评价”蕴含价值的判断,是最高水平的认知学习目标,代表学生已经逐渐形成了辩证唯物主义的思维逻辑。

显然,课程教学绝不仅仅是让学生知道、了解或记住某些知识,而应该是在记忆知识的基础上能够领会知识的内涵,并能够有效地运用知识进行分析和综合,最终能够对相关问题做出正确的价值判断。

2.实践技能教学目标(技能领域)

对于学生实践动手能力的提升,同样需要经历一个循序渐进的过程。这个过程包括对动作技能的知觉→定向→有指导的反应→机械动作→复杂的外显反应→适应6个层次。①“知觉”是指学生在实践技能教学过程中,通过看、听等感官获得实践操作动作技能的基本信息,并为下一步的实践操作做好准备;②“定向”是指学生对一套程式化的实践动手操作程序有了较为系统的了解,并为实践操作做好了充分准备;③“有指导的反应”是指学生在教师的指导下能够模仿教师的操作初步完成一套程式化的实践操作,但操作可能不够熟练,甚至在操作过程中会出现错误;④“机械动作”是指学生通过反复训练,对实践操作的反应已经成为一种习惯,已经能够熟练、准确、自信地完成实践技能操作;⑤“复杂的外显反应”是指学生已经能够对程式化,甚至非程式化的复杂实践动手技能进行熟地操作,这是动作技能领域的较高层面;⑥“适应”是指学生不仅能够熟练完成已知的实践技能操作,而且还能够根据仪器、设备型号的不同,或外界环境的变化创新性地自主修正自己的实践动作模式,从而达到对特殊装置或变化的外界环境的自主适应,这是动作技能领域教学目标的最高层面。

3.情感及价值观教育目标(情感领域)

情感及价值观教育目标包括接受→反应→价值化→组织→价值与价值体系的性格化5个层次。①“接受”是指低级的价值化水平,它仅仅代表学生对课程所传递的价值标准的认知;②“反应”是指学生能够主动参与,代表学生对课程所传递的价值标准的逐步认识与理解;③“价值化”则是代表学生对课程所传递的价值标准的认同和欣赏;④“组织”是指学生已经能够将多种价值标准进行比较、关联和系统化;⑤“价值与价值体系的性格化”则是指学生对课程所传递的价值标准达到深度认同。

总之,准确设定并深刻理解理论教学、实践技能教学、情感及价值观教育的目标是高质量课程建设的必要前提。有了明确的目标,才能够发现课程建设中存在的问题,从而通过目标牵引和问题导向,为教学内容选择、教学方法创新、教学评价改革奠定良好基础。

(二)合理选择教育经验

选择教育经验的实质就是课程教学内容的选择与凝练。知识是教不完的!什么知识最有价值?这是每门课程都必须回答的重要问题。

传统的课程教学是典型的教材导向的教学,照本宣科是经典的教学内容组织模式。在传统的课程教学中,教师首先考虑的是用什么教材。教材怎么选？长学时选一部长学时的教材,短学时选一部短学时的教材,没有教材就编一部教材。然后,按照教材的章—节—点体例形成教学内容体系。也就是说,课程讲什么,每部分讲多少学时,决定于选用什么教材。因此,传统的课程教学内容体系构建必然存在一定的问题。比如,认识明显滞后,专业知识体系陈旧,知识体系构建过于倚重教材,按部就班,照本宣科,注重知识的灌输,忽略科学精神、科学思维和科学方法的融入,教学内容的选择与社会经济发展及产业发展不匹配,不能很好体现现代科学技术发展前沿[4-5]。

我们认为,高质量的课程建设可以参考目标导向(Outcome Based Education)的教育理念,紧扣反向设计、学生中心、持续改进 3 个关键点。依据学科国际国内发展前沿、社会经济及产业发展现状、学生能力发展需求,充分尊重学生的主体地位,严谨梳理与论证课程教学目标。在明确课程教学目标的前提下,以目标为牵引,问题为导向,反向设计,精心选择凝练教学内容,构建教学内容体系,从而决定课程讲什么、怎么讲、讲多少学时,并在课程教学实施过程中进行不断优化和改进。这样的内容构成体现了最有价值的知识构成,符合课程高阶性、创新性、挑战度的要求以及认知、技能、情感三大领域的课程目标定位。

总之,在课程建设过程中,选择教育经验是核心,高质量的课程建设应该形成先进的课程教学内容体系。一是要强化基础,选择知识体系中最为重要、不可或缺的核心基础知识。二是要突出前沿知识理论和方法,体现课程的高阶性。三是要凸显特色,突出教师的学术感悟,增强课程教学内容体系的创新性和挑战度。四是要知行合一,加强理论教学体系与实践教学体系的更好结合。另外,课程的教学内容体系应该具有国际化的视野,应该对国内外同类课程进行充分的研究、对比与分析。在此基础上,通过教学与科研的深度融合,形成课程自身教学内容体系的比较优势。应该更加突出思维方法的训练,加强实践方法的创新,加强学生综合分析问题能力的培养,建立起特色鲜明的自然认知—方法实践—综合分析—探索创新的循序渐进的课程教学内容体系。

(三)有效组织教学过程

课程组织是指在一定的教育价值观指导下,将所选出的各种课程要素妥善组织成课程结构,使各种课程要素在动态运行的课程结构系统中形成合力,以有效实现课程目标。课程组织就好比是一台智能“编织机”,它将零散的课程要素编织成课程智慧的彩缎,以更好地促进人的发展。显然,课程要素是课程组织的基本线索和脉络,课程要素好比大厦的钢筋结构,尽管它看不见,但它对大厦的强固却极为重要。

一般情况下,课程要素包括宏观要素和微观要素两部分。其中,宏观要素主要指学生、教师、教材、环境、范围 5 个方面,微观要素主要指概念、原理、技能、方法、价值观 5 个方面。

课程组织的原则应该是充分调动“教”与“学”的双边互动,充分尊重学生的主体地位,通过教学过程的有效组织,也就是教学过程中教学策略、教学方法的合理运用,使我们的课程教学能够引导学生探究并解决问题,能够让学生发散、扩展、升华他们的思维,激发他们的学习兴趣,并给他们留下思考。因此,在课程组织过程中,要着重培养学生的主体意识,彻底改变"一言堂"和"满堂灌"的课程组织方式;要更多地运用PBL、TBL、CBE等多种不同的教学策略,有效地调动学生主动学习的意识,提升课程的教学质量;要引导学生大胆质疑;要改革实验教学模式,鼓励学生根据实验原理自行设计实验方案,使实验教学向研究性、探究性和设计性方向转变;要鼓励学生积极参与社会实践和课外学术活动,多途径提升学生的自身素质和能力;要鼓励和引导学生自学和进行研究;要引导学生从反思中学习,将学生的被动接受逐渐转变为主动探索的过程。

课程组织方式通常采用垂直组织和水平组织2种方式[3]:①垂直组织是指按照符合学习者身心发展阶段的序列以及符合学科知识逻辑演进的序列,将各种课程要素按纵向发展序列组织起来;②水平组织是指将各种重要的课程要素按照横向(水平)关系有效地组织起来。

有效的课程组织应该具有以下特点和属性:①课程组织应该能够激发学生的学习兴趣,充分给予学生主动学习的空间,允许学生在教学活动过程中做出自己的选择,并对选择所带来的结果做出自己的反思;②应该让学生在学习情境中充当主动角色,而不是被动角色;③应该要求学生探究各种观念,探究智力过程的应用,探究当前个人和社会关注的一些问题;④应该能够促使学生在课程学习过程中尽可能地涉及实物教具,与课程学习密切关联的真实物体、材料或人工制品;⑤应该使学生的学习过程能够由处于不同能力水平的学生成功地共同完成;⑥应该让学生在一个新的背景下审查观念、智力活动,或探讨在以前的研究过程中存在的一些问题;⑦应该要求学生审查一些题目或问题,这些题目或问题是社会生活中被忽略的;⑧应该尽可能促使学生与教师共同参与冒险,这里所说的"冒险"不是冒生命或肢体之险,而是冒成功与失败之险;⑨应该尽可能要求学生改写、重温或完善他们已经开始的一些学习尝试;⑩应该尽可能促使学生应用与掌握有意义的规则或标准;⑪应该给学生提供一个与别人分享制订计划、执行计划及活动结果的机会;⑫应该让学习过程与学生所表达的目的密切关联起来。

(四)改革课程学习评价

传统的根据试卷定分数,由分数定学生学习效果的评价策略,忽略了对学生潜在的创新能力开发和价值观升华的评价,这是长期以来传统课程评价的一个严重弊端。因此,传统的课程评价策略不能够很好地反映学生的实际才干提升和核心价值观的养成。

关注学生在整个课程学习过程中所表现出来的创造力、创新力和思想意识形态的进步,而不是仅仅通过期中考试、期末考试的试卷成绩来判定学生的学业和思想表现,这是课程教学评价改革的一个重要方向。

课程学习评价应该采取更加全面综合的学生评价策略,注重理论教学、实践教学、社会实践等不同教学环节中的持续性评价,注重学生在整个学习过程中长期的学习表现、行为表现、思想意识表现,这样才能使课程学习评价与课程教学目标保持一致。课程学习评价应该将学生置于核心地位,通过评价发现问题,帮助学生提高学业成绩,促进学生的能力发展、素质提高和社会主义核心价值观的逐步养成。同时,通过课程评估和评价,为教学内容选择、教学方法改进提供及时准确的信息,帮助教师改进教学[6]。

(五)强化课程思政元素

课程思政元素的合理融入是高质量课程建设的必然要求。事实上,课程思政建设的核心就是将“教”与“育”有机统一,以教学改革为抓手,强化教学内容中社会主义核心价值观导向,在专业知识教学过程中善于挖掘那些对学生具有思想启迪、行为导向和心灵震撼的价值因素,引导学生正确对待学术探索与研究过程中的成功、失败、付出、收获、成绩、荣誉等,使我们的专业课程教学在润物无声中有效践行育人功能,达到知识、能力、情感目标的有机融合,养成学生正确的世界观、人生观和价值观。

同时,对于不同的课程类型,课程思政的侧重点是不同的。根据教育部教高〔2020〕3号文件《高等学校课程思政建设指导纲要》的要求[7]:公共基础课应该重点关注理想信念、爱国主义情怀、品德修养、奋斗精神等,提升学生的综合素质;专业课程要依据课程类型的不同,研究不同专业的育人目标,依托不同学科专业的特色优势,挖掘专业知识体系中蕴含的思政价值和精神内涵,拓展专业课程的广度、深度和温度;实践类课程则要更加注重学思结合、知行统一;创新创业类课程应该注重增强创新精神、创业意识和创业能力;社会实践类课程应该注重弘扬劳动精神;理学类课程要注重科学思维方法训练和科学伦理教育;工学类课程应该注重强化学生工程伦理教育,培养学生大国工匠精神,激发科技报国的使命担当。

因此,在专业课程教学过程中教师应该如何合理选择、精心凝练课程思政内容,并有效融入专业教学体系,强化专业课程“育人” 功能,需要我们每位教师依据不同的课程类型深入研究和思考。

(六)重视课程条件保障

1.重视团队建设

学术水平高、教学能力强、结构合理的课程教学团队是一流课程建设的最基本保障。一流课程建设应该通过组建老中青有机结合的教学团队,确保课程建设的可持续发展。课程教学团队的优势在于以共同的教育理念和目标为牵引,能够有效提升课程教师团队的整体素质,推动课程教学研究,提高课程教学质量,保证课程教学质量[8]。

2.强调资源整合

教学资源建设意义重大,具有不可估量的重要价值,是一流课程建设和开发的重要条件保障。优质的教学资源建设,能够为学生的自主学习、教师的教学研究、对外的资源服务提供高效能的保障。一流课程应该有充分充足的优质资源保障。在一流课程建设过程中,应该尽力整合和提升课程中的优质资源,形成一批经典鲜活的教学资源和实验资源,形成线上、线下多种形式的优质资源体系。同时,应该建设一流的教学实验室,通过教学、科研一体化的建设思路,让学科建设与实验室建设同步发展,从而带动实验室建设达到较高的层次。一流课程的实验资源应该尽力面向学生全面开放,实践教学管理模式应该逐步国际化。

3.注重技术创新

在一流课程建设过程中,必须合理运用教育技术,将现代教育技术与传统的教育教学优势有机结合,充分利用互联网先进技术和数字化资源,采取线上学习、线下辅导,线上线下混合式教学等多种新的教学方式,提高学生学习的主动性。

4.倡导文化引领

大学有大学文化,院系有院系文化,一流课程也应该有先进的课程文化。优良的课程文化主要包括以下几个方面:①崇尚学术的课程文化。课程虽然是教育的微观问题,却是关乎宏观的战略大问题。因此,学术性理应是大学课程建设的一个核心要素。②批判与创新的课程文化。创新是大学文化的显著特征,失去了创新,大学就成了无源之水,无本之水,失去了动力与活力。同样,课程教学是实现大学人才培养基本功能的核心载体,因此课程理应承袭大学文化之精髓,倡导批判与创新精神。③追求真理的课程文化。传播和追求真理是大学课程教学的灵魂。大学课程教学的目的是为了完善学生的心智,因此,追求真理应该是大学课程教学理应崇尚的一种课程文化。④自由包容的课程文化。大学课程教学应该拥有民主的课堂氛围,倡导多元和谐的课堂文化,强调对教师和每个学生个体的文化尊重。在课程教学过程中,教师和学生共同拥有探究、创造、交流、合作甚至质疑的权利。⑤开放多元的课程文化。大学的课程文化应该是开放的文化体系,课程建设要面向世界、不断学习和吸收世界的先进经验,培养造就富民强国的高质量人才。

三、结语

综合上述分析,我们可以看到,当前形势下高等学校课程建设应该从理念、内容、方法、机制等多个方面进行深入思考。聚焦立德树人,坚持德智体美劳全面发展,以促进学生发展为中心,是高等学校高质量课程建设最为核心的理念。注重反映学科前沿动态、社会经济发展现状以及行业产业需求,精心选择凝练教学内容是课程教学内容体系构建的关键;有效组织教学内容,重视教学策略创新,实施更为有效的学生学业评价改革是课程方法创新的精髓;重视团队建设,强调资源整合,注重技术创新,构建优良的课程文化氛围是课程建设的重要机制保障。

参考文献

[1] 中华人民共和国教育部.关于一流本科课程建设的实施意见[Z]. 教高〔2019〕8 号.

[2] 中华人民共和国教育部.关于深化本科教育教学改革 全面提高人才培养质量的意见[Z]. 教高〔2019〕6 号.

[3] 张华.课程与教学论[M].上海:上海教育出版社,2000.
[4] 赖绍聪.聚焦学习者身心发展 构建教育教学新范式[J].中国大学教学,2020(6).
[5] 赖绍聪.论课堂教学内容的合理选择与有效凝练[J].中国大学教学,2019(3).
[6] 玛丽埃伦·韦默.以学习者为中心的教学:给教学实践带来的五项关键变化[M].洪岗,译.杭州:浙江大学出版社,2006.
[7] 中华人民共和国教育部. 高等学校课程思政建设指导纲要[Z].教高[2020]3号.
[8] 鲁卫平,王润孝.研究型大学教学团队建设的研究与思考[J].中国大学教学,2010(3).

课程教学方式的创新性改革与探索①

赖绍聪 华 洪

摘要：课程教学方式的改革对于培养创新性人才具有重要作用。为此，我们在以学生为主体、以能力培养为目标的理念下，针对传统课程教学方式普遍存在的内容单调、过程呆板、缺乏师生交流等不利于调动学生积极性的弊端，基于地质学学科课程教学，进行了初步的课程教学方式改革与探索，目的是让学生变被动为主动，使课程教学方式多元化，实现大学专业课程教学的设计性及创新性。

现代教育理论认为，教学过程是“教”与“学”双边互动的过程。教学是“教”与“学”的统一，“教”为“学”而存在，“学”又要靠“教”来引导，二者相互依存，相互作用。从静态来看，“教”与“学”是整个教学活动的2个方面，教学方式的改革必然要求和引起学习方式的改革，如果教学方式改革关注的只是“教”的改革，而忽视了学生“学”的改革，教学改革就很难深入。

对于高素质创新人才，科研意识、科学素质、科研能力的培养十分重要。一个人的科研意识、科学素质和科研能力，是很难通过某一门专门的课程来培养的，必须在教学过程中有意识地进行渗透与感染，在教学中改革教学方式，以加强学生的科研意识和科研能力，提高科学素质，培养学生的创新意识，充分调动学生自我学习的能力。不把教学活动囿于教室与课堂上，而是要将其扩展到整个校园甚至校园之外。要努力调动学生自我学习的能力，注重教育的个性和适应性。因此，课程教学方式的改革探索是目前我国高等教育面临的重要任务之一。

一、传统课程教学方式的局限性

所谓传统课程教学方式是“以教师为中心”的教学模式，教师利用讲解、板书和各种媒体作为教学的手段和方法向学生传授知识，学生被动地接受教师传授的知识。在这种模式中，教师是主动的施教者、知识的灌输者；学生是被动的接受者、知识灌输的对象；教材是教师向学生灌输的内容；教学媒体则是教师向学生灌输的方法与手段。这种传统的教学模式曾对教学活动产生过积极的进步性作用，且至今价值犹存。但是随着社会的发展、时代的进步、知识经济时代的到来，传统学习理论的问题越发明显起来，越来越不适

① 原载于《中国大学教学》，2013(1)。

合当前高科技迅速发展所提出的新要求,其存在的主要问题是:①学生在学习过程中所拥有的主体地位未能得到充分体现。在课程里,所谓"知识"变成了主宰一切的绝对力量,教师因依附于大大小小的知识点而自然而然地成了课程教学的唯一强势权威。在这样的课程教学过程中,教师和学生都只不过是知识的简单传递工具与机械接受容器而已。②课程教学的一个重要任务,就是教学生从学习中"学会思想",生成智慧,形成正确的人生观和价值观。这就决定了"思想"应当是贯穿于各类、各门课程的灵魂。而传统课程教学基本上是采用灌输式和强制式的方式,这种教学方式所造就的只不过是机械的"短时记忆"而已,并非"有意义"的学习,从而忽视了学生的主观能动性。

二、地质学学科的特殊性及课程教学方式改革的必要性

地球科学是以地球系统(包括大气圈、水圈、岩石圈、生物圈和日地空间)的过程与变化及其相互作用为研究对象的基础学科,其研究的尺度大到全球甚至整个宇宙,研究的时间长达数十亿年,其中有高等生命出现的时间也长达5亿~6亿年,无法用实验的方法重现。与其他经典自然科学相比,地质学对大部分问题只能用定性的方式来表述,采用将今论古的推理方法。研究方法包括岩石、矿物、化石的鉴定,逻辑推理,以及将今论古的类比法等。这表明地质学与数、理、化等经典自然科学之间存在着差别,因此在研究手段和思维方式上,地学及地学教育与其他学科间也有明显的不同。

(1)地质学是"推理和历史的科学"。地质事实往往是对地质作用结果的描述而不是作用过程的描述,地质作用历史的漫长性和地质客体的巨大性及其演化的不可逆性决定了主体不能亲身历验地质作用过程及作用范围,地质实验也只能反映一种极为理想的边界条件下的情形。因此,它的教学和研究手段必须建立在倒推性思维(对过去事件的预测和推断)、宏观尺度思维和综合大量不完备数据资料的基础上。由于地质学的推论常常存在多解性和不确定性,造成了地质学理论的普遍假说化倾向,这为基于问题的讨论和探究性学习提供了很好的素材。

(2)地质学在整体地球系统科学思维的发展中起着决定性的作用。地球自身就是一个由相互作用的地核、地幔、岩石圈、水圈、大气圈、生物圈和人类社会等构成的统一巨系统。因此,在地质教学中必须注重整体性思维和推理能力的培养。

(3)地质学的很多理论模型必须建立在大尺度、长时间、大空间的框架下。因此,地质学需要较高的空间想象能力和对长尺度时间的理解能力。

(4)地质学具有非常强的地域性。地质学所研究的客体——地球在空间上的巨大性、非均衡性、复杂性,使得地质学事实普遍存在不完整性。

地质学的这些特点,决定了一般地质研究必须通过一定比重的野外实际调查,配合相应的室内研究。野外调查和室内研究,构成一次观察、记录(包括制图)、采样、初步综合、试验分析、总结提高以至复查验证的完整地质研究过程。因此,必须具备从简单观察记录到高水平分析综合的能力。

地质学学科的特殊性要求我们在课程设置和教学过程中要改变传统的以知识传授

为目的的课程教学方式。在以培养学生的实际技能,培养学生解决问题的能力,培养学生的综合性、创新性思维和团队协作能力为主要目标的基础上,对地质学课程教学方式进行系统的改革。

三、课程教学方式的改革与探索

1.讲授和问题研讨相结合的研究性课程教学方式

为了提高教学质量,让学生最大限度地参与教学过程,我们在古生物学与地层学、地史学、岩石物理化学等相关课程的教学中尝试推行课堂讲授与分组讨论相结合的教学方式。在课堂讲授的基础上,根据学生人数,将全班分成6~10组,每个组确定1个组长。组长负责将教师布置的课程作业进行分解,安排小组内每个同学的任务,并限定时间将每个学生完成的结果集中,组织全组同学进行讨论,然后完成总结报告。

教师在整门课程的教学过程中,共布置2次课程作业,每次作业有2种类型。第一种类型是课程总结作业。要求学生根据教材、课件及教师提供的教学园地中的网络参考资料,写一篇1 000字左右的总结报告或课程设计报告。第二种类型是课程汇报。教师根据课程进度,设计与课程教学相关的几个综合性选题,组长抽签,决定选题,全组学生共同收集整理与主题相关的资料,汇总讨论,然后以PPT的形式上台汇报。汇报过程中,平均每人2分钟左右的汇报时间。这些尝试激发了学生的学习兴趣,培养了学生的自主学习能力,提高了学生解决问题的能力,同时也营造了互助合作的学习气氛,取得了良好的教学效果。

2.案例教学和问题驱动相结合的课程教学方式

案例教学的模式是教师选择一些古生物、岩浆岩、岩石物理化学典型真实的研究成果案例,以书面形式发给学生,或以视频形式展示给学生。教师首先讲解案例的有关事实,对案例进行分析,然后组织学生进行讨论。

而问题驱动教学是以问题为中心,在教师的引导下,学生通过独立思考、讨论、交流等形式,对教学问题进行思考、探索、求解、延伸和发展。问题驱动和案例教学相结合的方式对培养学生主动学习、分析和解决问题的能力以及独立思考能力和逻辑思维能力非常有效,取得了很好的教学效果。

3.概念图教学方式的尝试

概念图(conceptmaps)作为一种教学、学习策略和评估工具,可使知识模块形象化、结构化、可视化和清晰化,能够协助学习者更好地整合知识和内化知识,促进有意义的“教”与“学”。概念图作为一种教学和学习策略,在国外特别是欧美国家尤为盛行。

近年来,结合地质学专业课程的教学,我们首次把概念图教学法引入岩石物理化学的教学中,在课程总结、期末串讲环节中得到了很好的应用。使用概念图进行学习,学生可以积极地对关键字进行加工、分析和整理,并与教师积极对话,同时概念图还非常有利于开发学生的空间智能,在地学人才培养中显得尤为重要。

4.专业教育与学科前沿讲座有机结合的合作教学方式

我们每年有计划地邀请国内外学术专家来校为本科生开设学科前沿讲座。近年来，我们已先后邀请了美国南爱达荷州立学院的 Timothy Gunderson 教授、堪萨斯州威奇塔州立大学地质学系的 Toni Jackman 教授、麦卡莱斯特学院的 John P. Craddock 教授、德国哥廷根大学的 Joachim Reitner 教授等，给基地班学生开设国际化的岩石学、矿物学、地球化学前沿进展讲座，极大地开拓了学生的眼界，提高了学生利用英语语言和英语文化掌握学科专业知识，在专业和科技方面运用英语进行听说和交流的能力，提高了学生阅读外文专业文献的能力，使得学生在本科阶段就能够较顺利地查阅有关本专业的外文期刊与资料，跟踪地质学前沿知识。该项措施加强了创新人才的培养，加快了与国际接轨的步伐，取得了不错的效果。同时，我们还先后邀请了香港大学赵国春教授、中国科学院地质研究所翟明国院士、中国科学院广州地球化学研究所孙卫东研究员、北京大学张立飞教授、南京大学王汝成教授等，给学生开设了岩石学、板块构造与火山活动、大陆地壳的形成及其演化等学科前沿讲座。“专业教育与学科前沿讲座”有机结合、相互融通的合作教学方式已经成为增强学生专业适应性的特色教学措施，丰硕的科研成果通过学术讲座带进了课堂，使学生在多样化的教学方式中学习与进步。

课程教学方式的改革对于培养创新性人才具有重要作用。为此，我们在以学生为主体、以能力培养为目标的理念下，针对传统教学方式普遍存在的教学内容单调、教学过程呆板、缺乏师生交流等不利于调动学生积极性和创新思维的弊端，进行了初步的专业课程教学方式改革与探索。我们的改革思路是让学生在学习过程中变被动为主动，减少课堂讲授学时，增加讨论课和实践课，使课程教学方式多元化。使用现代化教学手段，增加课堂信息量。改变以往以验证为目的的课程教学内容，加强新思维、新技术和新方法在课程教学中的应用，全面体现专业课程教学方式的设计性及创新性，取得了良好的教学效果。

创新教育教学理念　提升人才培养质量[①]

赖绍聪

摘要：随着教育改革和发展规划纲要的全面实施，我国高等教育进入了一个全新的发展阶段。如何创新教育教学理念，实质性地提高人才培养质量，已经是我国高等教育面临的不可回避的严峻挑战。优秀人才培养是一项系统工程，本文归纳了人才培养的10个重要环节，并探讨了各个环节的核心内涵，初步构建了优秀人才培养体系的总体构架。

《国家中长期教育改革和发展规划纲要(2010—2020年)》指出，教育要发展，根本靠改革。要以体制机制改革为重点，鼓励地方和学校大胆探索和试验，加快重要领域和关键环节改革的步伐。"十二五"期间，我国高等教育教学改革不断深化，人才培养质量稳步提高，科学研究水平全面提升，社会服务能力显著增强，国际合作交流日益广泛，国际地位明显提高，各项改革取得突破性进展，高等教育迎来了生机勃勃的崭新局面。但是，目前我国高校人才培养仍然存在不少问题，还有许多薄弱环节，深化改革的任务相当艰巨。

一、我国高等教育人才培养中存在的问题

我国高等教育由于受到传统观念的影响，许多高校的教育理念并未跟上时代的发展步伐，教育目标长久处在使受教育者获得一套对特定行业和职位有用的知识和技术的层面上，较多地把注意力放在对于知识要点的掌握上，缺乏培养学生独立性和批判性思维能力的有效措施，学生创新能力不足。

由于教育观念落后，必然带来课程设置的不合理。课程结构单一，内容陈旧，不能及时反映时代、学科发展的前沿水平，课程开设的先后顺序不尽合理，各课程之间的联系不密切。

教学渠道单一，大多仍然沿袭班级授课的单一传统模式，难以照顾学生的个别差异，严重限制了学生个性的发展。

教学方式方法缺乏创新。"讲授＋考试"构成了教学活动主体，学生被动地接受和学习。这种模式忽视了学生素质的提高和创新能力的培养。

教学评价体系偏颇。对学生的评价过分看重考试成绩，考试测验仍然是教师评价学生的几乎唯一的法宝，忽略了对学生潜在创新能力的开发和评价。这种单一形式既不利

---

① 原载于《中国大学教学》，2016(3)。

于考查学生的综合素质,又不利于学生独立性、创造性的培养。对教师的考核评价方法单一,内容不尽合理,也难以真实地评判教师的水平和成绩。

随着社会的发展,教育观念和教育功能正在发生显著的变化。这种变化主要表现在以下几个重要方面:①从以继承为主转向以创新为主。传统的以传授知识为主的教育教学模式已不能适应当前社会的发展形势,各国教学模式的改变几乎都朝着通过探究式学习、实践式学习和合作式学习的方向发展,以培养学生的创新精神和创新能力。②从以能力为导向到以价值观为导向的转变。而价值观导向归根结底就是教育学生如何对待自己,对待他人以及对待社会、国家和世界。③从以课程为中心转向以学生为中心,以训练标准化的个性为主转向以培养多样化的个性为主。以学生为中心正在成为很多国家提升教育质量的核心导向。而以学生为中心主要体现在 2 个方面:一是全员化发展,即每个学生都是重要的;二是个性化发展,即每个学生都是不同的。④从信息工具使用到教学模式的改变。近 20 年来,信息技术在教育领域得到了广泛应用,但仍然处于信息工具与技术使用的初级阶段,信息技术与教育教学的深度融合仍然欠缺。⑤评估评价观念的重大转变。新的理念提出,评价评估要与教育目标保持一致;评价评估的重点应放在改进课堂实践上,确保所有利益相关者尽早参与以及将学生置于核心地位。通过评价评估发现问题,为教师提供及时准确的信息,帮助他们改进教学,提高学生的学业成绩。

因此,面对当前国际教育教学改革新的发展,抓住机遇,应对挑战,紧紧围绕高等教育质量这一生命线,乃是当前我国高等教育面临的最根本任务。我们需要正确认识我国高等教育发展的历史成就,客观深入分析我国高等教育发展面临的困难与问题。我们必须把高等教育工作的重心放在更加注重提高质量上来,深入思考、探索并实践提高教学质量的有效方法和途径。

二、教育教学理念与教学改革思路

不断提高教学质量,培养创新性人才,是高等学校永恒的主题。教学工作始终是高校的中心工作,教学质量始终是高校的生命线。

面对当前我国高等教育教学质量存在的问题,我们应该深入讨论与思考。在理论上,要对国家急需的各类人才培养的思路、方案、模式等进行更深入的探讨,形成先进而完整的体系,形成办学特色;在实践中,要形成兼顾服务于学生全面发展和个性发展的思路,形成科学的适用于不同层次人才培养的课程体系、实践教学体系,并及时实施,最终达到实质性提高人才培养质量的目标。

基于这一基本思想,本文提出在人才培养和具体教学过程中,我们需要密切关注的 10 个重要方面。

1.明确人才培养的指导思想——立人教育

国际 21 世纪教育委员会向联合国教科文组织提交的教育研究报告称,教育是“保证人人享有他们为充分发挥自己的才能和尽可能牢牢掌握自己的命运而需要的思想、判断、感情和想象方面的自由”。马克思和恩格斯指出:教育是促进“个人的独创的自由发展”。孔子说:“大学之道,在明德,在亲民,在止于至善。”而鲁迅先生则明确指出:“教育是要立人。”著名教育家蔡元培先生认为:“教育是帮助被教育的人给他能发展自己的能力,完成他的人格,于人类文化上能尽一分子的责任,不是把被教育的人造成一种特别器具。”在国际教育界有着同样类似的对教育的理解和认识。著名教育家雅斯贝尔斯认为:“教育是人的灵魂的教育,而非理性知识的堆积。”教育家斯金纳则用更加简洁且极富哲理的语言深刻地揭示了教育的内涵,他认为:“教育就是将学过的东西忘得一干二净时,最后剩下来的东西。”所谓“剩下来的东西”究竟是什么?我们认为,“剩下来的东西”就是教育的本质和精髓,就是“独立的人格、探索的精神、学习的能力、践行的能力”。

因此,大学不是“职业培训班”,专业知识虽然重要,但更需要的是培养学生正确的价值观和探索、思考、学习以及践行的能力。

作为教育工作者,弄清楚什么是教育这个问题,对于认清教育的本质、明确自己的职能和职责、找准前进的方向是非常必要的。因为没有理性的自觉,是不可能在实践中做个自觉而清醒的教育者的。

2.以“本科专业类教学质量国家标准”引导教学工作,形成先进的人才培养方案

人才培养方案是在一定的教育思想和理念指导下,为实现特定的教育目标,通过专业培养计划、课程体系、评价体系、管理制度等实施人才培养的方案与计划。简单地说,人才培养方案就是学校为学生构建知识、能力、素质结构的实施计划。

很显然,构建合理的、先进的、适应当代科学技术发展和社会需求的人才培养方案是决定人才培养质量的关键因素。要提高人才培养质量,就必须将人才培养方案的精心研制和凝练作为重点。

根据教育部高教司理工处 2013 年 7 月关于《理学本科专业类教学质量国家标准框架说明》的通知精神,本科专业类教学质量国家标准乃是该专业类人才培养、专业建设等应达到的基本要求。它主要适用于 3 个方面:一是作为设置专业的参考;二是作为人才培养和专业建设的指导性规范;三是作为质量评价的参考。

因此,“标准”是设置专业的基本准则,是人才培养和专业建设的指导性规范,是质量评价的基本依据。我们应该以“本科专业类教学质量国家标准”为基本依据引导教学工作,按照基于“标准”、高于“标准”、突出“特色”的指导思想,依据各校各专业办学条件的实际,凝练形成先进的人才培养方案。

3.合理的课程体系与实践教学体系

目前,我国高等院校仍然在不同程度上存在先进的科学理论、技术与老旧的课程体系设置及教学内容、体系之间的矛盾。因此,要密切结合国际科学技术发展趋势,统筹本科教育不同阶段、不同课程教学内容,梳理核心知识,优化课程体系,使教学内容尽可能跟上时代发展的步伐。从人才培养目标定位出发,建立与之相适应的课程体系,通过教学与科研的结合,将教师的研究成果充实到教学内容中,实质性地形成特色鲜明、与国际接轨的理论教学体系,以保证教学内容的先进性。这些都是亟待进一步加强与深化的工作。

良好的实践动手能力对创新性人才培养尤为重要。传统教学在进行知识传授的同时,也注重学生的实践,但它强调实践教学的验证性过程,注重对经典学说的认知和接受,创新性实践环节较少,制约了学生的创造性。我们需要在已有实践性教学改革的基础上,按照实践教学"认知—综合—探索"的循序渐进规律,统筹本科4年的不同阶段、不同课程的教学内容和计划,构建课堂教学与实践教学内容协调配置、时间穿插的循序渐进的实践性教学新体系,完善本科实践教学环节。

总之,在构建合理的课程体系与实践教学体系过程中,我们应该突出思维方法训练,加强实践方法创新,打破课堂教学分门别类、自成体系的教学过程,使不同课程内容自然交融、互相关联。加强学生综合分析问题的能力培养,最终建立特色鲜明的"自然认知—方法实践—综合分析—探索创新"循序渐进的课程新体系。

4.教学方法与方式的创新与探索

教学过程是"教"与"学"双边互动的过程。教学是"教"与"学"的统一,"教"为"学"而存在,"学"又要靠"教"来引导,两者相互依存,相互作用。教学方式的改革必然要求和引起学习方式方法的改革。教学方法改革的基本思路是让学生变被动为主动,减少课堂讲授学时,增加讨论课和实践课,使课堂教学方式多元化。使用现代化教学手段,增加课堂信息量。积极鼓励学生直接参加部分教学活动等。我们应该以学生为主体,以能力培养为目标,针对传统教学过程中普遍存在的教学内容单一、考核方式简单、教学内容单调、教学过程呆板、缺乏师生交流等不利于调动学生积极性和创新思维等弊端进行系统改革,形成创新的教学方法与方式。

对于课堂教学,应该根据教学内容的不同,灵活采用孔子型课堂(知识驱动、教师主动、学生能动),苏格拉底型课堂(问题驱动、学生主动、师生互动),翻转型课堂(基于MOOC的在线学习、课堂质疑),自主型课堂(程)(兴趣目标驱动、自主自觉行动、课内课外联动)等多种教学形式。

通过教学方法与方式的创新,使教学过程实现5个转变,即从验证性到设计性的转变,从灌输式到启发、讨论式教学的转变,从传统技术到新技术与传统技术结合的转变,从单科性到综合性的转变,从认识性、继承性到创新性的转变。

5.建设高素质教师和管理队伍

推进教育创新和培育人才,既离不开教师的辛勤工作,也取决于教师队伍的素质。优秀的师资团队和教学管理团队是践行正确的教育教学理念、先进的教学体系以及创新教学方法的关键要素。

应该以深化教师聘任制度改革为重点,加强政策调控,科学设置机构和岗位,实行教师资格制度,实行"资格准入、竞争上岗、全员聘任",完善教师岗前培训、在岗学习、国内外进修体系,从而建设一支相对稳定的师资队伍。

师资队伍应具有合理的年龄结构。重视师资队伍学缘结构优化,采取有效措施促进学缘交叉,改善知识结构和教育背景。对于理、工、农、医类专业,还要不断加强实验教学队伍建设,形成能够满足实验教学需求、队伍稳定、专/兼结合的实验教学队伍。

专职教辅/教学管理人员应当队伍稳定、结构合理、团结协作、素质高、服务意识和质量意识强。

高度重视师德师风建设,完善制度,落实措施,不断提高教师的思想道德素质和业务素质,不断促进教师教育教学水平的提高。教师还应积极开展教学研究,积极参与教学改革与教学建设。

6.建立一流的教学实验室

在实验室建设过程中,始终坚持培养高素质创新型人才、不断提高教学质量是专业发展的生命线,牢固树立实验室与学科建设同步发展的思路,带动实验室建设上台阶。逐步实现学生在教师指导下进入实验室进行测试,为学生参加科研训练、提高科研素质提供平台和条件,这对促进本科教学质量和培养创新意识将起到积极的推动作用。

在专业实验室建设过程中,应该大力倡导"教学科研一体化"的主导思想,使专业实验室不仅能够全方位地服务于本科及研究生教学工作,而且能够在较高层次上进行科学研究工作。在科学研究工作中不断积累经验,逐步改进实验方法、提高实验技术,这就极大地提高了专业教学实验室的水平和档次,以最直接和最有效的方式将科研成果实时转化为教育资源。高水平科研资源向教学资源的转化,奠定了教学的高起点、高标准。激发创新意识,训练创新能力,达到培养学生观察事物、思考问题、自我设计、研究解决问题的素质,形成实践教学从理念到形式全面改革。同时,实践教学条件、实践教学方法、实践管理模式应该逐步国际化,这对促进本科教学质量和培养创新意识将起到重要作用。

7.推行教育信息化,以信息资源服务教学工作

教育信息化是教育实现现代化和适度超前发展的重要途径。教高〔2015〕3 号文件指出,大规模在线开放课程等新型信息化教学和学习平台在世界范围内的迅速兴起,拓展了教学时空,增强了教学吸引力,激发了学习者的学习积极性和自主性,扩大了优质教育资源的受益面,正在促进教学内容、方法、模式和教学管理体制机制发生变革,给高等教育教学改革发展带来了新的机遇和挑战。

信息技术在教育领域的应用可分为 3 个阶段:工具与技术的改变,教学模式的改变,最终可能产生学校形态的改变。而我们目前仍然处在"工具与技术改变"的初级阶段。近年来,随着 MOOC 的冲击,有可能真正迎来"教学模式的改变"。

很显然,信息化已经成为当前高等教育面临的重要挑战与机遇。我们必须积极应对当前面临的教育国际化和信息化趋势,以传统教学模式与当代信息化技术的深度融合为抓手,在信息化的基础设施、信息化的资源配置、信息化的应用水平等 3 个方面加强建设,为学生自主学习、教师教学研究、师生资源服务提供高水平、高效能的保障。

8.特色鲜明的第二课堂

对于高素质优秀创新人才,探索精神、科学素质、科研能力的培养十分重要。但是,这些很难通过某门专门课程进行培养,必须在教学过程中进行有意识地熏陶培养,在教学中改革内容、改进方法,以加强学生的科研能力,提高其科学素质,培养学生的创新意识,充分调动学生自我学习的能力。

第二课堂是指在教学计划所规定的教学活动之外,引导和组织学生开展各种有意义的学习、研究和探索活动。它的形式多样,内容丰富多彩,具有协作性、多样性、实践性和创新性等显著特点。第二课堂对于开阔学生的学术视野,提高学生的学习兴趣,加深学生对事物的理解和认识,帮助学生发掘自身的潜能,发现适合于自己的学习方法,激发他们的创新意识和探索精神作用显著。同时,对于加强学生的团队意识和协作精神,提高学生的综合素质也有重要的促进作用。

不要把教学活动囿于教室、课堂,而要把它们扩展到整个校园以至校园之外。因此,建立特色鲜明、形式多样的第二课堂就显得尤为重要。第二课堂对于调动学生自我学习能力,注重教育的个性和适应性意义重大,同时也是培养优秀人才、提升本科教学质量的重要手段和途径。

9.高效务实的教学质量监督保障体系

高效务实的教学质量监督保障体系的关键是实施教学管理制度创新,建立富有特色的和有效的教学质量标准以及教学质量监督机制。教学质量考核标准应该以兼顾规范性和个性化为原则,从而形成完善的自我约束机制。

(1)教学质量标准。教学质量标准应该兼顾规范性和个性化 2 个方面。在具体做法上,应该根据学科的发展,集体建立可供检查的教学大纲,加强教学大纲的评审机制,实行教学大纲听证制度,以确保教学大纲(教学质量评估的重要标准)的先进性和引领性。

(2)教学质量监控体系。为了规范教学管理,应该对课程教学实行三段式教学管理模式,即教前管理(大纲听证、大纲公示、教学纪律公示),教中管理(教案公开、教学督导、听课检查)和教后管理(教考分离、教学质量评估),加强过程管理等。通过这些措施的实施,进一步加强教学质量的管理和监控,逐步形成“社会—管理层—教师—学生”多层面教学质量监控体系和自我约束、自我发展的教学管理机制。

10.毕业生跟踪反馈机制及专业的持续改进机制

采用多种途径定期或不定期跟踪调查毕业生的工作情况及升学后的学习情况,广泛收集信息,并以此作为改革人才培养模式、修订人才培养方案、调整招生计划等的参考依据。及时跟踪社会和科学技术进展,根据学科的内涵和功能变化,及时调整和提高人才培养目标,推动专业建设的持续稳步发展。

教育的目的应该是让一个人变得健全和完善,形成正确的价值观,具备探索的精神,获得学习的能力以及实践的能力。大学教育不能局限于理性知识的堆砌,更不能将大学教育视作“职业培训班”。

优秀人才培养是一项系统工程。作者认为,正确的教育理念、先进的人才培养方案、合理的课程体系、创新的教学方法、一流的师资团队、教学科研一体化的实验实践教学平台、先进的信息化教学、特色鲜明的第二课堂、高效务实的质量监控体系以及逐步完善的人才培养质量跟踪反馈与专业持续改进机制 10 个方面,初步构成了逻辑关系明确、相辅相成、互为支撑的优秀人才培养的完整体系。只有牢牢地抓住每个关键环节,切实深化教育教学改革,才能实质性地提升人才培养质量,推进我国高等教育健康发展。

坚持教学质量国家标准　突出本科专业办学特色

——“地质学专业教学质量国家标准”解读[①]

赖绍聪

摘要：介绍了制定“地质学专业教学质量国家标准”的历史背景，详细分析了地质学专业教学质量国家标准的特点和基本要求，阐述了当前形势下坚持教学质量国家标准的必要性，同时提出了在正确理解和使用教学质量国家标准的前提下，如何充分体现办学单位自身优势、突出地质学本科专业办学特色的思想和理念。

建立本科专业教学质量基本标准，规范本科专业办学基本要求，严格专业基本办学条件，确保本科人才培养质量，是当前我国高等教育面临的不可回避的严峻挑战。教育部不久前颁布的《普通高等学校本科专业类教学质量国家标准》，对于规范我国本科专业教育、推进高等教育教学改革、实质性提升本科人才培养质量必将产生重大的引领、推动和促进作用。

一、地质学专业教学质量国家标准的制定背景

20 世纪 80 年代，随着国家改革开放政策的不断深入推进，地质矿产行业再次获得新的发展机遇，地质学专业点不断增加。随着科学技术的不断发展，21 世纪地质学的内涵已经发生了重大变化，服务领域也越来越宽。

在地球科学的研究历史中，最初把对地球的研究称为地质科学或地质学；20 世纪末，科学的发展表明，要从整体上对地球进行各种时间和空间演化尺度的研究，整体研究地球内部各圈层及圈层间运动变化过程、形成机制以及可能发生的变化趋势。由此，地质学逐渐发展成为以地球整体为研究对象的一门学科。不久之后，地球科学研究就进入了以地心到地球外层空间，包括固态、液态、气态圈层和生物圈、人类圈在内的地球系统为研究对象的地球系统科学阶段，逐步形成了以“上天、入地、下海、登高、探极”为特征的空间格局。地质科学进入了一个以建立地球系统知识体系为标志的新的转折时期。

在新的历史背景下，地质学本科专业的专业内涵、知识体系、理论教学及实践教学体系、基本办学条件等亟待明确。地质学专业教学质量国家标准正是在这样的历史背景下应运而生的。

---

① 原载于《中国大学教学》，2018(3)。

二、地质学专业教学质量国家标准的特点

1.充分反映地质学学科特点

地质学是推理和历史的科学,地质学的很多理论模型必须建立在大尺度、长时间、大空间的框架下,地质学具有非常强的地域性。地质学的这些特点,对专业人才的实际技能、解决问题的能力、综合性与创新性思维、团队协作能力均提出了明确要求。地质学专业教学质量国家标准充分考虑了地质学科的上述特点,在培养目标、课程体系、实践体系、条件支撑等方面提出了必须满足的基本要求[1-3]。

2.强调地质学专业知识结构体系

不同的本科专业,由于其学科长期发展的积累,必定形成其特定的知识结构体系[4-12]。地质学专业教学质量国家标准进一步明确了地质学专业本科学生在当前科学技术发展新形势下应该具备的知识结构体系。同时,将坚实的基础理论、基础知识以及宽广的知识面,人文社会科学知识和文化素质,专业基础知识和专业理论知识,专业技能知识和地质实践能力 4 个方面作为地质学专业本科学生的基本知识结构构成。

与此同时,将地质学专业理论教学知识体系划分为 5 个子体系:①地球系统科学体系。包括地球科学概论、地理信息系统、环境科学概论、海洋科学概论和大气科学导论等。②地质学体系。包括普通地质学、结晶学与矿物学、晶体光学、岩石学、矿床学、古生物学、地史学和构造地质学等。③地球物理学体系(地球物理学基础)。④地球化学体系(地球化学)。⑤其他。包括矿相学、GIS 和遥感技术、大地构造与中国区域地质学、层序地层学、同位素地球化学等。

将地质学专业实践教学体系划分为 4 个方面:①野外教学实习。包括一年级地质认识实习、二年级地质填图实习、三年级综合毕业实习。②课程实验和实习。包括课程教学实验课、课程课间教学实习和课程小论文。③科研训练。包括学生论文报告会、科学讲座、导师指导下的科研活动和毕业论文。④社会实践。包括参加社会活动及校园活动和假期集中社会实践。

上述知识结构体系符合当前地质学国际发展趋势,乃是 21 世纪地质学本科专业学生应该具备的基本知识体系。

3.严格要求专业基本办学条件

地质学专业教学质量国家标准对本专业应该具备的基本办学条件提出了最低要求。包括:①师资队伍;②实验教学队伍;③专职教辅/教学管理人员队伍;④教师背景和教师发展环境;⑤教学条件(各类功能教室、教材选用和评价制度、公共图书馆、基础课程实验室、实习基地以及校园网等,具体要求参见《地质学专业教学质量国家标准》)。

这些基本办学条件乃是办学必备的基本要素,也是人才培养质量的基本保障。

4.高度重视教学质量监督保障

规范专业办学条件,不断提升本科教育教学质量乃是我国高等教育当前形势下面临的不可回避的严峻挑战。

地质学专业教学质量国家标准把教学质量监督保障体系建设作为专业办学的必备条件之一。要求办学学校把提高教学质量作为生存与发展的生命线,不断建立和健全教学质量保障体系,制定和完善各教学环节的质量标准,实施校、院、系(专业教研室)三级教学督导制,教学管理队伍稳定、结构合理,教学质量监控体系科学、合理,运行有效,并将以下 6 个方面作为教学质量监督保障体系的基本构成:①管理决策体系;②教学质量标准体系;③教学质量监控体系;④信息反馈体系;⑤教学质量改进体系;⑥教学质量激励体系。

通过制定相应的规章制度,激发教师“教”的热情和学生“学”的热情,促进良性循环,使教学过程更加规范,确保教学质量。

5.推动建立专业持续改进长效机制

地质学专业教学质量国家标准特别关注专业的长期持续发展,并对此提出了明确要求。

专业建设与专业发展必须密切结合学科国际发展趋势,适应社会经济建设与发展的实际需求。当今世界正处在一个知识爆炸和人类社会快速进步的新时代。因此,本专业所在院/校应遵循“教育要面向现代化、面向世界、面向未来”的当代地学教育办学基本方针。在严格执行本专业国家标准的前提下,根据自身办学条件,提出高于国家标准的本专业建设发展和持续改进计划与方案,及时跟踪社会和科学技术发展,根据地质学的内涵和功能变化,及时调整和提高人才培养目标,推动专业建设的持续稳步发展,最终能够形成专业建设与专业发展的持续改进长效机制。

6.倡导深化教育教学改革

在地质学专业教学质量国家标准中虽然没有提出教育教学改革的具体要求,但是明确强调了各高校应密切关注和顺应高等教育发展趋势,加强和深化教育教学改革的理念和思想,给各办学高校留下了极大的突出特色和深化改革的空间。

在课程教学内容凝练方面,密切结合学科国际发展趋势,通过教学与科研的结合,使教学内容尽可能跟上时代发展的步伐。倡导教师在教学过程中,树立“学生主体”观念,改变“一言堂”和“满堂灌”的教学方式方法,构建符合教育规律的多模态课堂教学模式,提升专业教学质量。倡导改革实验教学模式,实验教学向研究性、探索性、设计性方向转变,鼓励学生根据实验原理自行设计实验方案,充分体会自我实践的价值[9]。

这些先进的教育教学思想和理念将有助于推动我国各高校地质学专业办学点的专业建设和发展。

三、基于质量标准,突出办学特色

本科专业类教学质量国家标准是专业类人才培养、专业建设等应达到的最基本要求,国标的出台是为了规范基本的本科专业教学行为[10-12]。当前,我国高等教育已步入提高质量、内涵发展的新阶段,根据国标进一步规范我国本科教育的各个环节,促进世界一流大学和一流学科的建设与发展。

地质学专业教学质量国家标准,已经充分考虑了研究型大学、教学科研型大学、教学型本科院校和高等专科学校或高等职业学校等不同类型高校的差异性与灵活性。同时,明确强调了“基于标准,高于标准,突出特色”的专业建设思想,为我国各高校地质学专业办学点留下了充分的办学空间。不同高校可以在符合地质学专业教学质量国家标准基本要求的前提下,根据自己的人才培养目标、特色与定位,制定自己进一步细化、量化的质量标准。鼓励各高等学校决策者根据自身的类型定位,选择差异化发展策略,办出特色,有效规避“千校一面、千人一面”的局面。

根据教育部要求,下一阶段各高校地质学专业办学点的主要任务是,根据地质学专业教学质量国家标准修订人才培养方案,尽快提升教学水平,培养多样化、高质量人才。

参考文献

[1] 赖绍聪.建立教学质量国家标准提升本科人才培养质量[J].中国大学教学,2014(10):56-61.

[2] 赖绍聪.改革实践教学体系创新人才培养模式[J].中国大学教学,2014(8):40-44.

[3] 赖绍聪.深化课程体系改革构建人才培养新模式[J].西北高教论评,2015(1):15-20.

[4] 赖绍聪.创新教育教学理念提升人才培养质量[J].中国大学教学,2016(3):22-26.

[5] 赖绍聪.地球科学实验教学改革与创新[M].西安:西北大学出版社,2010.

[6] 赖绍聪,华洪.课程教学方式的创新性改革与探索[J].中国大学教学,2013(1):30-31.

[7] 赖绍聪.如何做好课程教学设计[J].中国大学教学,2016(10):14-18.

[8] 赖绍聪.岩浆岩岩石学课程教学设计[M].北京:高等教育出版社,2017.

[9] 赖绍聪.寓教于学 寓教于研[J].中国大学教学,2017(12):37-42.

[10] 教育部高等学校化学类专业教学指导委员会.“化学类专业本科教学质量国家标准”的研制与解读[J].中国大学教学,2015(2):31-33.

[11] 蒋宗礼.关于研制计算机类专业教学质量国家标准的思考[J].中国大学教学,2014(10):52-55.

[12] 樊丽明,刘小兵,姚玲珍.研制财政学类专业教学质量国家标准的实践及思考[J].中国大学教学,2014(7):21-25.

改革实践教学体系　创新人才培养模式

——以西北大学地质学国家级实验教学示范中心为例[①]

赖绍聪

摘要：随着教育规划纲要的全面实施，实践教学环节的系统改革已成为当前我国高等学校深化教学改革、提高人才培养质量的关键。实践教学体系的改革，不仅要更新人才培养观念，更要加强与之相应的课程体系改革。本文简要介绍了近年来西北大学地质学国家级实验教学示范中心在实践教学体系改革方面的一系列举措和经验。

长期以来，西北大学地质学系高度重视实践教学环节的系统改革，并将科学研究成果实质性地融入实践教学体系，将现代高新技术成果适时地引入实践教学过程中。尤其是地质学系"地质学国家级实验教学示范中心"获准建设以来，我们认真遵循教育部关于示范中心建设的有关要求，结合当代高等教育人才培养规律、教学改革精神和实践教学特点，不断深化对实践教学重要性的认识，探讨实践教学理念，以先进的教学理念为指导，建设科学的实践教学内容体系，培养学生的实践能力和创新能力。示范中心建设和实验教学改革均取得了明显进展。

一、实践教学改革的基本思路与方案

示范中心实践教学改革秉承"加强基础、拓宽知识、培养能力、激励个性、提高素质"的人才培养理念，坚持把知识传授、能力培养和素质提高贯穿于实验教学全过程，以培养学生实践能力和创新能力为核心，逐步建立起分层次、相互衔接、科学系统的创新实验教学体系，大力推进人才培养的国际化步伐，为培养面向21世纪需要的地质学本科人才奠定了良好的基础。

一是实验内容统筹安排，构建地球物质、地球动力学、地球演化及数字地球等分层次的实验内容体系。各门实验课程内容之间合理衔接，使实验教学内容体系具有基础性、系统性和先进性。

二是加强实验技术和方法的综合，设立基础型、综合设计型、研究创新型实验，引入科研成果和与地方经济相结合的实验项目，使实验内容与科研、工程、社会应用项目密切联系，充分体现基础与前沿、经典与现代的有机结合。

三是加强与精品课程配套的精品教材建设，引入国内外方法成熟、适合本科生实验

---

① 原载于《中国大学教学》，2014(8)。

教学的先进实验项目,使教材建设有利于学生自主训练与创新能力的培养。

四是实验教学采用开放模式,基础地质实验室每周开放6天,每天开放15小时,其他实验室随时预约开放;专设一门现代分析实验技术课程,为本科生创新基金的实施和科学训练创造条件,同时大陆动力学国家重点实验室尖端的实验仪器全面为本科生开放。

五是探索符合学生认知规律的实验教学方法,加强信息化、网络化实验教学平台建设,激发学生的实验兴趣,调动学生的积极性和主动性。建立实验习惯、实验过程、实验结果和实验考试综合评定学生实验成绩的考核方法,引导学生知识、能力和素质协调发展。

六是采取在校学生、毕业生、助教研究生和教师综合问卷、调查表、网上反馈、座谈和教学检查等方式进行教学效果评价,保证实验教学质量不断提高。

二、实践教学体系的基本构成

地质学专业以实践性强为突出特点,野外实践教学与课堂实习实验教学构成了传统的本科实践教育的两大体系。同时,在新的形势下,本科生科学研究基础训练和社会实践已成为本科教育必不可少的重要环节。在此基础上,我们全面统筹、协调贯穿本科教学全过程的理论教学和实践教学,构建了以秦岭造山带及其相邻稳定地块的大陆地质为天然实验园地,协调不同课程的课间实践教学和不同阶段的野外实践教学,穿插实施科研训练和必要的社会实践活动,科学、合理、循序渐进并符合“地质学专业教学规范”要求的实践教学体系。

1.课程实习与实验

地质学课程体系中专业类课程一般均有相应的实习实验。以理科地质学专业为例,地球科学概论、结晶学与矿物学、晶体光学与岩石学、古生物学、地史学、构造地质学等课程大多具有很强的实践性与经验性,且知识更新快。随着科学技术的迅速发展,课程内容变化较大。因此,我们近年来在高度重视传统基础性实习安排的前提下,对于专业课程的课程实习与实验课程适度强调了其先进性与前缘性,努力与当代地球科学发展、国内外学科发展同步,将新的科学理论、观念与新的科研成果融入实习实验教学中,将教师及学科群体的科研成果融入教学中。

2.课间实习

为了强化野外现场教学在地质学专业课程教学中的重要作用,我们在新的实验教学体系中扩大了课间实习的力度,在10余门课程中实施课程间穿插野外实习的方法,将科研小课题、课程大作业等明显具有研究性、设计性的实习教学内容融入课程中,取得了良好的效果。比如,矿床学课程课间野外教学实习,就很好地纳入了科学研究的教学思想。我们充分利用地域优势,优选世界著名超大型金堆城斑岩钼矿床为教学实习基地,改变描述型、填鸭灌输式和传统的被动教学模式,采用现场考察研讨式、设计式教学的模式,模拟工作访问式科研活动进行教学,训练学生的综合专业素质和专业科研能力。通过这样的野外实践,使学生掌握野外考察、研究的程序以及研究内容与工作方法,提高综合分析的能力。

3.野外综合实习

在地质学本科专业教学计划中一般包括三大野外实习:一年级地质认识实习;二年级地质填图实习;三年级生产实习(毕业实习)。我们在充分考虑当前地球科学发展新趋势以及学生认知能力逐步提升的科学规律前提下,系统改革了地质学专业本科4年的野外教学实习计划,将新技术、新方法和科学研究引入教学,培养学生的创新能力,同时教学方式方法也发生了重要变化。

(1)由认知向实证的转变。我们选择自然、人文、科研条件均佳的秦皇岛地区为地质认识实习基地,将现代技术纳入一年级野外地质认识实习教学。为每个实习小组配备全球卫星定位系统GPS,将GPS在现代地质、地形测量中的应用技术纳入教学内容,提升了野外技术方法,丰富了地质认识实习的内涵。学生需完成4张地质剖面图,并对4个地质剖面的地层、岩石、岩石组合、岩性变化规律、沉积特征、沉积环境分析等方面进行详细论述,以学生自己的野外记录和实际观察内容为主。总结报告要求学生在充分发挥主观能动性、创造性的前提下,充分消化野外实际观察内容,用自己的理解方式将实习区地层体系和岩石类型进行总结,全面归纳分析实习区的地层序列和特征。正是由于这一系列具体有效的改革,使地质学系一年级实习已不再是通常概念上的地质认识实习,其难度、内涵均较传统的地质认识实习产生了飞跃性的变化,初步实现了由认知向实证的转变。

(2)传统地质填图与现代数字化地质填图的有机结合。在地质学本科专业教学计划中,二年级地质填图实习以1:50 000区域地质调查总则为基本要求,对学生进行野外踏勘、地层剖面选择与实测、地质填图和地质报告编写、地质图件编制等全面训练,要求学生将地质基础理论知识与野外实践很好地结合,掌握地质学野外工作的基本内容和工作方法;培养学生的野外工作和综合分析能力,为以后的毕业实习及未来的地质科研、生产实践和教学打下坚实基础。

近年来,随着GPS、GIS等高新技术在地学的广泛应用,区域地质调查(填图)的效率大大提高,地质填图质量明显改进,并使区域地质调查实现了信息化。因此,我们必须改革传统的教学计划,实现传统地质填图与现代数字化地质填图的有机结合。为此,我们自主开发了与科研教学密切结合的地质数字填图软件系统,以安徽巢湖地区为实践教学基地,在扎实开展传统地质填图训练基础上,实施数字化地质填图,使学生受到严格的技术方法训练,并使已学到的基础知识应用到实际中。

(3)将生产(毕业)实习提升为科学研究实习,实现由综合向研究性实习的转变。在新的实践教学体系中,我们将三年级的生产(毕业)实习明确提升为科学研究实习。三年级的科学研究实习是学生在本科期间进行独立科研训练,以培养创新能力为目的的重要教学环节。该项教学工作是在学生经过了一、二年级的不同阶段,分别以认识、方法为重点的基础训练,在初步认识、观察、分析和研究客观地质事物的基本能力基础上,对学生进行综合运用基本知识和技能,独立研究和解决实际问题,实现从选题到实践、实验(测试)再到总结、提高、理论化的完整科研过程训练,以培养学生从事科学研究的基本能力和方法。三年级实践教学依照学生的兴趣、爱好(张扬个性),通过教师与学生的双向选择,学生或参与不同专业方向教师的科研项目,或申请本科生创新基金项目,完成研究性学习。

(4)专门增设地质创新综合实习。在学生完成了华北区秦皇岛一年级实习、扬子区巢湖地区二年级实习和三年级不同学科方向的研究性实习基础上,我们专门增设了四年级地质创新综合实习。四年级以鄂尔多斯盆地-秦岭造山带地质走廊为教学基地,在对不同构造单元、不同时代、不同类型和不同层次地层系统和构造变形为主的实际观察、研究、分析基础上,在点、线、面结合的同时进行异岩、异相、异构的对比分析,构建区域时、空地层-构造格架,研究地质事件、分析区域地质结构构造特征,探讨区域地质演化过程。以此使学生树立新的地球观,训练学生的基本素质和技能,培养学生多学科交叉观察、分析、解决问题的能力,特别是综合分析问题的能力,激发其创新思维。

4.实习实验教学改革的创新举措

为了配合实习实验教学体系的全面改革,我们还采取了一系列相应的辅助措施。

(1)实施学生科研能力培训计划,形成"教师—研究生—本科生"学术团队。通过本科学生科学研究能力培训计划的有效实施,采取师生双向选择最终确定的方式,自三年级起,将本科学生逐步融入教师的科研团队中。由不同研究方向的教师根据自己的科研特色和研究实际,提出科研小课题(有限时间、有限经费、有限目标),学生根据自己的兴趣和特长选择课题,将导师制与创新基金有机地结合在一起,初步形成了教师—研究生—本科生研究群体以及教师科研项目—研究生论文选题—本科生创新基金多层面的课题组,从而将导师制、创新基金研究计划、实验室开放及本科毕业论文有机融为一体。这一措施使部分高年级本科学生实质性地独立承担小课题、加入教师的科研群体中。

(2)面向本科生和研究生全面开放基础实验室、大陆动力学国家重点实验室,培养学生的实践能力。长期以来,我们示范中心基础教学实验室坚持面向地质学系学生和教职员工全面开放,承担着地质学系本科生、研究生的实习和教学任务,每周开放6个工作日,为学生和教职员工免费使用。对于精密显微镜及高档显微结构摄(照)相系统,我们采用分段管理模式。教师直接负责学生显微镜操作培训和资格论证,以及毕业论文设计、学生创新基金研究工作、学生参加导师科学研究项目所涉及的实验工作等,并由教师直接负责实验设备的使用安全。对于学生自主选定的实验工作(如课程实验报告、实习附加作业、选修课程、开放性课程作业等),直接由实验室专职管理人员负责论证学生的上机操作资格,经登记注册后,进入实验室自主完成实验工作。中心各专业实验室配备了专职实验管理人员和具有高级职称(教授或副教授)的教师,负责实验室的技术开发和长远规划设计,同样实行全天及全方位对本系教师和学生开放。

(3)营造优良的学术氛围。为了活跃学术气氛,提供更多的互相交流及学习机会,搭建信息资源共享平台,我们有计划、有目的地建立了示范中心教师—研究生—本科生联合学术沙龙,一般每月举办一次。地学沙龙已逐渐成为西北大学地质学国家级实验教学示范中心青年教师交流学术心得,研究生扩大学术视野,高年级本科生培育学术思想,师生之间、学科之间广泛交流和相互启发的重要渠道。

为配合学生科研能力培训计划的实施，使学生能够积极参与科研项目的研究和学术交流活动，为本科生搭建一个较好的学术交流平台，示范中心积极鼓励学生创办了自己的学术刊物。2004 年，由学生自行撰稿、自主编辑的学生学术刊物《地学新苑》创刊，学生从此有了一个采撷思想火花、交流学习心得、进行学术交流的园地。

（4）实施多校联合实习，实现优质资源共享，推进人才培养的国际化。近年来，西北大学地质学系，以及北京大学、南京大学、中国地质大学（武汉）、吉林大学、南京师范大学、成都理工大学、东华理工大学等高校以国家级实验教学示范中心建设为契机，结合自身的地域特点和长期科研、教学积累，以具有传统地学实践教学优势的河北秦皇岛实习区、北京周口店实习区、安徽巢湖实习区、秦岭造山带实习区和鄂尔多斯盆地实习区为基础，充分发挥各校的教学资源优势，加大投入力度，建设完成了在地域上相互关联，在教学内容上循序渐进，涵盖中国北方典型华北型地层系、中国南方典型扬子型地层系、秦岭典型碰撞型造山带、鄂尔多斯典型沉积盆地等具有丰富大陆地质内涵，具有科学性、综合性和前瞻性的不同年级野外实践教学基地群。在此基础上，全面推行不同高校本科学生联合交叉野外实习，并初步尝试实施了中国—美国、中国—俄罗斯、中国—日本等学生联合野外实习，初步实现了各地各校特色优质资源的互补与共享，推动了不同院校、不同国别师生之间的交叉融合，为地球科学创新型人才培养探索了重要新途径。

此外，西北大学还积极实施中美学生联合野外实习计划。美国堪萨斯州威奇塔州立大学地质学系多次与西北大学地质学系本科生一起共同进行鄂尔多斯盆地-秦岭造山带地质走廊野外教学实习，并达成了“美国威奇塔州立大学—美国威斯康星大学—中国西北大学鄂尔多斯-秦岭野外地质教学联合实习”协议。通过这样的合作实习，西北大学地质学系将“鄂尔多斯盆地-秦岭造山带野外地质教学”作为一个对外开放的窗口，为西北大学和地质学系请进来、走出去，加快国际化进程创造条件，同时为地质学基地班学生提供了与国外学生学习交流的机会，使他们逐步掌握国外的先进教学理念，推进了人才培养的国际化。

三、实验教学体系的特色

1.先进的人才培养机制和管理体制

在学校的支持下，示范中心对相关的教学和科研资源进行了整合优化，实现了 4 个打通。

一是打通了教学实验室与科研实验室。示范中心与大陆动力学国家重点实验室共建了大型仪器公共实验室，实现了空间共享、资源共享。

二是打通了基础教学实验室与专业教学实验室，建立了教学科研一体化实验室，避免了重复建设，提高了建设水平。

三是打通了教学仪器与科研仪器的使用。由“211 工程”、基地基金投资购买的大型科研仪器全部面向实验教学开放，为开设综合性、创新性实验提供了条件，做到了科研反哺教学。示范中心特种显微镜室就是在原有教师科研用显微镜基础上建立和扩展的。

四是打通了本科生与研究生培养通道。通过大学生科学研究训练计划的实施,使本科生尽早进入科研实验室,融入教师的科研团队中,促进了创新型人才的培养和选拔。

通过这些措施,理顺了示范中心管理体制,建立了教学和科研设备共享机制,探索建立了教学与科研上下贯通、相互支撑的创新型人才培养机制。先进的管理体制,优化了资源配置,拓展了实验教学资源。

2.符合现代地质科学认知规律的实验教学体系

示范中心通过建立"一体化、多层面、开放式、重特色"的现代地质科学实验体系,为开设探索性、综合性实验创造了条件。例如,根据不同的专业定位和高素质人才培养需求,设置了基础、专业、创新3个层面的实验;将实验教学和科学研究有机融通,通过实施本科生导师计划、本科生创新基金计划等项目,使本科实验教学与科学研究紧密衔接。

3.富有现代地质学特色的实验内容

示范中心以模拟科学研究的方式设置实验,减少了验证性实验,增加了综合性实验,并从科研课题中精选部分实验项目,将科研成果和社会应用项目引入实验教学,营造出有利于自主学习、合作学习、研究性学习的环境。

4.强大的学科群与硬件条件支撑

示范中心所依托的地质学系,是国家"211工程"大学重点建设单位,设有4个长江学者特聘教授岗位。具有地质学、地质资源与地质工程2个一级学科、12个二级学科博士点和12个硕士点、2个工程硕士点,设有地质学、地质资源与地质工程2个博士后科研流动站。拥有国家地质学基础科学研究与教学人才培养基地,拥有地质学国家一级重点学科、矿产普查与勘探国家二级重点学科、10个省级重点学科、1个国家重点实验室、2个国家级特色专业。强大的学科支撑、优秀的教学和科研队伍为学生的成才提供了智力保障。

学高为师　身正为范

——研究生导师的责任与义务①

赖绍聪

摘要：培养与指导研究生是一门艺术，对导师的学术水平和教学能力都是一种考验。本文结合作者长期担任高校研究生导师的经历，从研究生导师的行为规范、学术创新以及在研究生培养过程中应该如何创新机制、科学育人等方面进行了分析，总结了高校研究生培养过程中的一些有益经验，探讨了我国高校研究生培养的改革方向。

我国科技发展的方向就是创新、创新、再创新。实施创新驱动发展战略，最根本的是要增强自主创新能力。创新的事业呼唤创新的人才，人才就是未来。我国要在科技创新方面走在世界前列，必须在创新实践中发现人才、在创新活动中培育人才、在创新事业中凝聚人才。研究生已日益成为我国拔尖创新人才的重要来源和科学研究的生力军。导师作为研究生培养的第一责任人，承担着培养学生创新技能的重要职责，肩负着重大的历史使命。因此，如何成为一名合格的研究生导师，如何针对目前研究生的具体情况培养创新型人才，已经成为当前我国研究生教育教学领域必须认真思考的重要问题。

一、严于律己，身正为范

研究生阶段是人生成长的重要阶段，它不但奠定了学生今后从事学术研究的基础，也是一个人的人格、品德、修养等方面完善的重要时期。我们要造就的是学术精良、品德高尚、人格完善的人。因此，树立远大的理想、形成严谨的学术风范是研究生学习的核心内容。

随着近年来研究生教育的迅猛发展，在校学生规模也在不断扩大。处于改革开放背景下的研究生群体，由于受到多元思想文化的影响，在心态、目标、理想等方面都表现出一些新的特点。对学术和学术研究的态度成为一个比较突出的问题。学术研究，有时已经被当作谋生的简单工具和晋升的敲门砖。毋庸讳言，对美好生活的向往是人类与生俱来的本性，是正当合理的。但是，如果把学术降低为简单的谋生手段，“学术”就只剩下了形而下的“术”，而没有了形而上的“学”的神圣意义。在纷繁复杂的社会环境下，要能够抵御外来的诱惑，克服内心的浮躁，并保有一颗坚守信念的心。这样的态度，对研究生的培养至关重要，甚至从某种程度上决定了每位研究生求学的最终高度。

---

① 原载于《法学教育研究》，2018，20。

在研究生培养过程中,导师是第一责任人,数年来面对面共同学习、研究与交流,使导师常常成为可能仅次于学生父母、对学生影响最大的人。导师对待学生就如同父母对待子女一样,是不图回报、不辞辛劳的。作为研究生导师,除传播专业知识和技术外,导师的待人接物、言行举止、思维方式等,都将潜移默化地影响着其所指导的研究生和他们的未来,甚至还会被研究生所模仿。因此,让我们常怀敬畏之心,以一种"战战兢兢""如履薄冰"、小心谨慎的态度去对待"师德"和"学风"问题。严于律己,为人师表,为研究生形成严谨的学风起到示范作用。因此,在研究生培养过程中,我们应该努力提高自身修养,高度关注、高度重视、充分发挥导师的人格魅力。人格力量是导师的基石。人格是尊严、价值与品格的总和,是个体内在的、行为上的倾向性,是对知识、事业、日常生活的态度。大学不是工厂,教育、教学本质上是人与人之间的交流,是精神、观念、人格、方法的交流,而不只是限于业务。导师务必与学生真诚对话,必须竭尽全力,努力成为受学生欣赏、追慕与景仰的师长。作为一名优秀的研究生导师,要留下故事可以被流传,要有音容笑貌可以被追忆。

严谨的学风也是导师人格的重要组成部分。中国古语云:"蓬生麻中,不扶而直;白沙在涅,与之俱黑。"导师群体要从自身做起,坚守和营造严谨的学风,为研究生培养创设一个风清气正的学术氛围。崇尚科学精神、增强自律意识,是养成良好学术品德、建设优良学风的基本保障。崇尚科学就是求真务实,不研究假问题、伪问题,选择科学适切的研究方法,对假设进行充分而谨慎的论证。崇尚科学也包含着尊重学术自由,涵养学术良知,具有学术风骨,不媚权势,不盲目跟风。增强自律就是坚持学术定力,要甘坐冷板凳,耐得住寂寞,下得了苦功,不为名利所扰。导师严谨的学风是研究生教育中重要的部分,对研究生的影响是无声而深刻的。

二、为人师表,学高为师

高等教育的任务就是要为国家培养优秀的创新型人才。随着创新型国家概念的提出及我国高等教育事业的不断发展,创新型人才的培养已成为高等教育面临的必须解决的重大课题。培养研究生创新精神与创新能力是研究生教育的核心任务,创新是研究生教育的本质属性。优秀的研究生是各类高层次科研与技术人才的预备军,是国家创新体系的后备力量。

创新型人才培养应具有一种教育教学与科学研究密切结合的氛围,导师和研究生都是其中的学习者和研究者,在研究生培养过程中理应构建一种基于研究探索的学习模式,导师和研究生的相互交流与合作,促进学术和智力环境的健康发展。研究生在学科与知识创新体系中具有重要的作用。在教学互动、教研互动中,研究生教育遇到的问题常常成为导师学科与知识创新的源头,促使导师的研究更为深入或者另辟蹊径。这是教学相长的自然结果。在导师所涉及的各种应用研究中,研究生作为创新研究的骨干力量,常常直接参与技术创新的生成。显然,这一重任对研究生导师提出了极高的要求。首先,导师必须不断提高自身的科学素质与学术水平,才能胜任研究生导师的历史使命。

作为一名研究生导师，不从事科学研究、不追踪学科发展最前沿，就很难提高自身的学术水平，也就无法胜任培养高质量研究生的重任。因此，研究生导师必须在学科研究方面不断探索创新，牢牢把握学科前沿进展，关注和解决社会生产生活实践中的学科问题，磨砺和提升学科探究与创新能力，持续形成有影响力的学术成果，为研究生树立科学求真的榜样与典范。其次，研究生导师必须不断提高自身的教学能力。对于研究生而言，这种教学能力突出体现在对研究生科研能力的培养上。研究生导师必须通过自身高超的教学艺术与教学智慧，不断打磨和锤炼研究生的创新技能，提升研究生的创新能力。即使具备较高的学术水平，但如果不用心，不研究探索人才培养的规律，就很难培养出一流的研究生。此外，研究生导师还必须在教育教学方面不断更新知识储备，强化实践性反思能力，强化自身的教学学术素养，从教育学、心理学等相关学科视角，不断提升自身言传身教的水平与影响力，从而为创新人才培养奠定良好的教育教学基础。只有这样，才能有效培养学生的深层学习动力、较高的自学能力和科学研究的方法与能力，使学生的学术之路走得坚定、长远和富有成效。

三、创新机制，科学育人

在研究生培养过程中，如何创新机制、科学育人，需要导师花费更多的心血，需要一系列切实具体可行的措施。对此，每一位研究生导师都有许多“独家秘籍”。

1.深刻理解教育的本质与内涵，为科学育人奠定理论基础

什么是教育？“教育就是当你将学过的东西忘得一干二净时，最后剩下来的东西”。所谓“剩下来的东西”就是学习的能力、“自修”或“无师自通”的能力，这应该就是教育的本质与内涵。大学不是“职业培训班”，而是一个能学会适应社会、做各种不同工作的平台，专业知识虽然重要，但更需要的是思考的能力、学习新东西的能力以及举一反三的能力。“能力培养”是贯穿研究生培养过程始终的核心元素，是研究生教育的本质与内涵，作为研究生导师，我们必须对此有深刻的理解，明确我们在研究生培养过程中的基本任务是什么[1-2]。

2.明确导师应有的教学思维，为学生的终生做准备

在培养研究生过程中，导师应当考虑的不仅仅是想给学生教什么，而是学生需要学习什么。除了专业知识与素养培养外，应该把培育研究生良好的情感智力和阳光心理（个人成功的决定因素）放在至关重要的位置。对于研究生而言，良好的情感智力和阳光心理不仅有助于学业的成功，更有利于健康人格与社会责任感的形成。从而为社会培养出真正有方向、有动力、有水平的创新型人才。

3.相互尊重，形成师生默契的良好合作关系

研究生的培养在一定程度上带有“口耳相传”、手把手教授的性质。学生和导师之间的联系，是双方自愿选择的结果。充分尊重和相互包容，才能真正潜心学术研究，让研究生在求学过程中真正学有所得。借用佛教术语来讲，研究生导师和学生之间应该有一种“缘分”，这种缘分包含了学生对导师的学识与人品的认同，也包含了导师对学生的潜能与性格的肯定。只有双方互相认可,有高度信任与默契,才能最终顺利完成培养过程,达到较高的预期目标。因此,在课程教学过程中,教师首先必须高度重视"身教",即以自身的良好品格、文化和科学素养、心理素质和责任心,潜移默化,引领正确的价值观。教师应竭尽全力,努力成为受学生欣赏、追慕和景仰的师长[1-2]。

另外,导师应当尊重和包容学生。研究生的学习需要某种悟性,由于存在各种原因,不是每个学生都能完全认同导师的指导方法、完全按照导师的指导进行学习的,这时就需要充分尊重和包容学生。毕竟不是每一个人都能在学术上有所成就,只要导师在研究生的求学过程中真正有所得即可。

4.提倡质疑,培养研究生的挑战精神

会提问是最具创造力的品质,应倡导由多因素决定的非线性思维,而不是非此即彼的线性思维。鲜明而强烈的现象是激发学生提问的驱动因素。问题是推动思维发展,从不确知向确知、从朦胧向清晰、从局部向整体、从未知向已知的桥梁和原动力。也就是说,思想总是以系统的问题形式表现出来,成为推动探索的动力[3-4]。因此,要鼓励研究生多质疑,教授他们提出假设的方法,引导他们运用科学的方法去验证或证伪所提出的研究假设。

5.赏识是导师手里奇妙的工具

要注重对研究生进行适时、适度的赏识教学。敬畏他们的天真,一句表扬胜过千句批评,赏识是教师手里奇妙的雕刻人生的最好工具。"好孩子是夸出来的"。以赏识激发学生内在的信心与勇气,鼓励他们产生更多的创新性思维与行为。

6.因材施教,制定个性化培养方案

作为研究生导师,我们每个人都应该认真思考和回答一个问题,那就是研究生培养究竟应该"以教师为中心""以科研项目为中心",还是"以学生为中心"。

我们认为,在研究生培养中,要特别注重因材施教,根据每个学生的基础和悟性制定一对一的培养方案。作为研究生导师,在充分遵照学生意愿的前提下,通常要耗费很多时间和精力去研究每个学生的基础特点、能力特点和兴趣特点,为每位学生制定个性化的培养方案,寻找最适合他们的学术生长点,选择最适合他们的论文选题。事实证明,这样的培养方式是行之有效的,真正实现了从"以教师为中心""以科研项目为中心"的管理模式向"以学生自主个性发展"管理模式的转变[5]。

四、结语

培养指导研究生是一门艺术,对导师的学术水平和教学能力都是一种考验。创新之路永无止境,只有在科学面前永不言倦的人,才能成为真正结出创新成果的缔造者。

清代学者章学诚讲,史家有四长:史才、史学、史识和史德。研究生的指导培养也是这样,需要导师在才、学、识和德方面都能达到一定的高度。"学高为师,身正为范",愿和每位研究生导师以此共勉,共同为研究生教育贡献自己的绵薄之力。

参考文献

[1] 赖绍聪.寓教于学 寓教于研[J].中国大学教学,2017(12).
[2] 赖绍聪.改革实践教学体系 创新人才培养模式[J].中国大学教学,2014(8).
[3] 赖绍聪,华洪.课程教学方式的创新性改革与探索[J].中国大学教学,2013(1).
[4] 赖绍聪.如何做好课程教学设计[J].中国大学教学,2016(10).
[5] 赖绍聪.创新教育教学理念提升人才培养质量[J].中国大学教学,2016(3).

“双一流”背景下高等学校学科建设策略分析①②

赖绍聪

摘要:学科建设是高等学校事业发展的重要动力。依据国家“双一流”建设总体方针,结合西北大学地质学科长期以来的建设经验,基于对学科定义及学科建设内涵的分析,提出高等学校学科建设包含“奠定学科建设基础—实施学科建设过程—形成学科建设效果”3个重要阶段,进而针对当前我国高等学校“双一流”建设背景下学科建设关键环节的建设策略与举措进行了初步分析。

国家“双一流”建设《统筹推进世界一流大学和一流学科建设总体方案》(以下简称《总体方案》)的出台,拉开了中国高等学校建设发展的新序幕。根据《总体方案》的相关要求,我国“双一流”建设工作将实施三步走的战略目标。2020年,我国将有若干所大学和一批优秀学科进入世界一流;2023年,更多大学和学科进入世界一流,一批优秀学科进入世界一流学科前列;至21世纪中叶,我国一流大学和一流学科的数量和质量将进入世界前列,实现由高等教育大国向高等教育强国的巨大跨越[1]。显然,中国高等教育当前所面临的最重大任务就是“双一流”建设。面对当前科学技术的快速发展以及复杂多变的国际形势,我国高等学校“双一流”建设,尤其是一流学科建设应该如何更为有效地实施,我们应该有什么样的指导思想,应该采取何种有效的建设策略,这已经成为我国高等学校每位管理者和每位教师都应该认真思考和探索的重大课题。本文依据国家“双一流”建设总体方针,结合西北大学地质学科长期的建设经验,主要针对一流学科建设过程中关键环节的建设策略与举措进行了初步的分析与总结。

一、学科建设的指导思想

《总体方案》明确了“双一流”建设的5个核心内涵:①以一流的人才汇聚为核心;②以一流的科学研究为基础;③以一流的教育治理为支撑;④以一流的社会服务为动力;⑤以一流的经费投入为保障[1]。

一流大学建设和一流学科建设是“双一流”建设过程中2个重要的方面。就一流大学建设而言,《总体方案》明确了7个方面的建设指导思想:①明晰的办学理念与目标定

---

① 原载于《中国地质教育》,2021(1)。

② 国家“双一流”世界一流学科建设(地质学)项目;“地质学国家级特色专业”“高等学校专业综合改革试点”“基础学科拔尖学生培养计划2.0基地”建设项目。

位;②先进的培养模式与质量保障;③优质的学科建设与科学研究;④开放的国际视野与交流合作;⑤精良的人力资源与物质资源;⑥科学的治理体系与运行机制;⑦优良的校园文化与育人环境[1]。

就一流学科建设而言,《总体方案》也提出了 6 个方面的重要战略指引:①优化学科规划。重点包括突出学科交叉融合、调整学科结构、进一步凝练学科方向。在学科规划过程中,尽可能实现重点突破。②汇聚学科队伍。队伍建设始终是学科建设的重中之重,高质量学科人才队伍的培育和引进,以及团队合作精神、奉献精神等优良团队文化构建是学科队伍汇聚的核心。③搭建学科基地。有了明确的学科规划,汇聚了一支优秀的学科队伍,紧接着就是要围绕学科队伍的学术专长和实际需求,有目的、有计划、有针对性地搭建适合他们的工作平台,包括高水平专业实验室建设、先进仪器设备购置与建设以及建设资金的有力保障等多个方面。④创新学科管理。要激发科技人员的工作积极性和学术创造创新力,就必须实施治理结构体系的系统改革,倡导协同创新,突出绩效评价,着力构建良好的激励奖励机制。⑤培养拔尖人才。培养高质量一流拔尖人才是高等学校学科建设最为核心的目标导向,面向未来的人才培养,更加强调创新能力、综合素质的提升。同时,在人才培养过程中,更加关注科教融合、产教融合。我们培养的人才,必须能够服务于国家社会主义建设事业。⑥产出原创性成果。对于原创性成果的认定,将遵循“国际标准,中国学派”的基本理念,突出成果对推动人类文明与进步、促进国家重大战略与经济社会发展的贡献度[1]。

总之,我国新一轮高等学校“双一流”建设的战略目标大体可以用 8 个字来概括,那就是“中国特色,世界一流”。

需要特别强调的是,在“双一流”建设过程中,学科建设一定要凸显自身的特色,绝不能照搬西方模式,要坚持“将论文写在祖国的大地上”。同时,高质量的学科建设也离不开国际标准。由此,在一流学科建设过程中,仅有特色是不够的,“特色”并不等于“高水平”,特色学科不能够简单地等同于一流学科。在特色的基础上,一定要形成自身学科在国际同类学科中的比较优势,形成核心竞争力,这样的学科才能真正成为符合“国际标准,中国学派”检验标准的世界一流学科。

二、学科建设的策略分析

(一)现代大学的基本功能

现代大学主要包括 5 项最基本的功能:人才培养是大学的核心工作;科学研究是大学的重要职能,是人才培养的重要载体;而服务社会、传承文化、国际合作与交流是人才培养和科学研究功能的重要延伸。但是,归根结底,高水平的学科建设是大学基本功能的综合载体。可以说,没有一流学科的大学不是一流大学,一流学科群是一流大学的重要标志。因此,建设一流学科是“双一流”建设的核心和基础。

(二)学科及学科建设的内涵

什么是学科？一般认为,学科就是相对独立的知识体系,也就是自然科学、哲学与社会科学范畴内学术领域或学术方向相对独立的知识体系。然而,在我国高等学校学科建设体系中,学科不仅仅是指知识体系,它大体包含了2个重要的含义:一是指科学领域或一门科学的分支;二是指高等学校教学科研等的功能单位。因此,本文所讨论的学科建设体系中的“学科”应该是知识形态与组织形态的共同体。

那么,什么是学科建设？我们是否可以这样来理解,学科是知识形态和组织形态的共同体,其知识形态就是自然科学、哲学与社会科学范畴内学术领域或学术方向、科目或分支相对独立的知识体系,而组织形态就是高等学校开展教学、科研等活动的功能单位和组织,是实现办学职能的基础和平台。因此,高等学校的“学科建设”,实际上就是一种基于知识形态的组织行为。

(三)学科建设策略与举措分析

按照上述对高等学校学科建设内涵的理解,结合西北大学地质学科80多年的建设历史,尤其是近年来地质学世界一流学科建设经验,我们认为学科建设主要应该包括“奠定学科建设基础—实施学科建设过程—形成学科建设效果”3个重要的阶段(图1)。

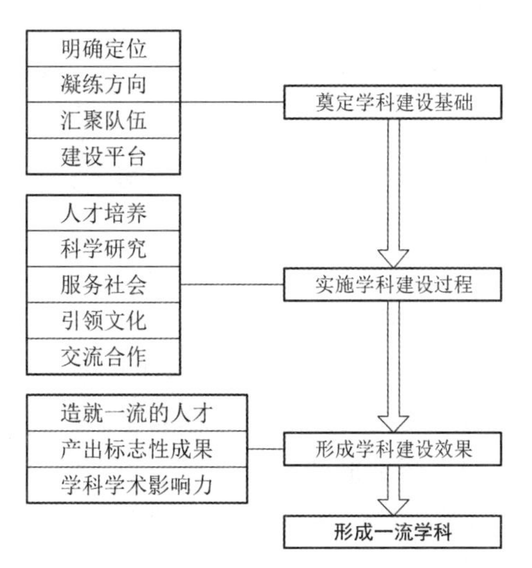

图1　高等学校一流学科建设的3个重要阶段

奠定学科建设基础是前提,是后续实施建设过程的条件保障,没有坚实的建设基础就很难展开有效的建设过程。而奠定建设基础的核心包含了“明确定位、凝练方向、汇聚队伍、建设平台”4个环环相扣、密切关联的环节。“明确定位”将为我们的学科建设勾勒宏观蓝图、谋划长远愿景、明确战略目标,它是确保学科长期稳定、健康发展的总指南;“凝练方向”必然要契合学科建设的总体定位,依据宏观蓝图准确选择和凝练学科发展的重点学术方向,而重点突破的学术方向将为我们指明学科建设的战术目标;围绕重点学术方向汇聚和建设具有知识技能互补、层次结构合理、水平高、能力强、能够承担重大科技攻关任务的学科学术团队,就成为我们的必然选择;立足学科学术团队的研究特色和实际需求,有计划、有目的、有针对性地购置仪器设备,建设实习、实验科技平台,从而完成学科建设基础奠定的完整逻辑闭环。

实施学科建设过程是关键。强有力地推进高质量的学科建设是大学实现其办学功能的最重要途径,是高等学校事业发展的总引擎。学科建设过程的实施,必然体现为一流的人才培养、符合国际前沿发展趋势的科学研究、契合国家重大战略需求和地方经济社会发展急需的社会服务,以及先进文化引领和广泛的国际交流与合作。

高质量的学科建设过程必然形成良好的学科建设效果,形成学科重大学术影响力,并最终实现建设一流学科的目标。

1.奠定学科建设基础

(1)明确定位。在学科建设过程中,明确学科建设的定位是首要任务。我们究竟要把我们的学科建设成为一个什么样的学科,也就是学科建设发展的长远规划与宏观定位问题。在思考论证学科建设定位过程中,应主要考虑以下 3 个方面的原则:第一个重要原则是要依据特色发展、不可替代的原则。也就是有限目标,有所为、有所不为。第二个重要原则是要立足前沿,引领方向。要依据学科的国际国内发展趋势,明确我们的学科建设定位,学科建设愿景一定要契合学科的国际发展前沿,学科建设所达成的目标要能够在学科学术领域起到国际引领作用。第三个重要原则是要脚踏实地,实事求是。学科建设定位要依据所在高校的办学总体定位,学科建设定位与学校总体办学定位与目标应该有很高的契合度。

(2)凝练方向。在明确了学科建设总体定位的前提下,凝练学科建设的方向就显得极为重要。

学科方向凝练,主要应该考虑 3 个核心要素:一是本学科学术发展的国际前沿领域;二是国家急需与地方经济社会发展的需要;三是本学科长期建设发展过程中形成和积累的特色优势。具体而言,可以分解为以下 5 个方面:①依托学科长期发展的优势。高等学校的学科体系常常都有长期的历史积累,在长期的发展过程中,形成了自身独有的研究特色和优势,这是我们选择学科方向的一个重要依托点。也就是要基于自身优势,发扬自身长处。②依托学科领军人才的研究特色。在我们的学科队伍中,常常拥有一批高水平的领军型人才,因此,他们的研究方向也可以作为我们学科建设过程中选择方向的一个重要依据。③依托学科平台的特色与优势。在我们的学科体系中,如果已经拥有一些重要的学科学术科研平台(比如国家重点实验室、国家工程技术中心、国家级创新引智基地等),那么这些平台将为学科建设提供重要的基础支撑,同时它们也成为我们凝练和选择学科方向时需要考虑的重要依据之一。④依托行业优势。在我们的高等教育体系中,不同高校常常具有自身独有的特色,高校也存在不同的办学类型,通常可以划分为综合性大学、行业性大学和专科性大学。对于一些行业性大学和专科性大学,其在本行业往往具有较强的研究特色和优势。因此,依托自身在本行业中的优势,凝练和选择学科方向是可取的。⑤依托地域优势。我们的学科建设要服务于国家战略和地方经济社会发展。因此,我们也可以将高校所在地域所具有的特色和优势作为学科方向凝练和选择时的重要依据。

总之,学科方向的凝练十分关键。正确合理的学科方向应该是在符合该学科领域国际发展趋势前提下,自身学科特色优势最大化的体现,这才是我们应该选择的学科方向。需要指出的是,学科方向不能是一个领域的名称(如地球科学),也不应该是一个一级学科或二级学科的名称。比如,我们不能把"地质学"或"矿物学岩石学矿床学"这样的一级学科名称或二级学科名称简单地作为我们学科建设的学术方向,这样的方向实际上并不代表学科学术研究方向。因为,它既没有反映我们所在学科长期积累的优势,也没有反映该领域的国际前沿发展趋势。以西北大学地质学科为例,学科总方向是"中国中西部典型造山带盆山耦合动力学及其资源能源和环境效应"。这样一个学科方向,既体现了当前地学领域的国际前沿,同时也反映了西北大学地质学科 80 多年发展历程积累的优势与特色。

(3)汇聚队伍。学科团队建设一定要紧紧围绕学科方向,建设结构合理的高质量学科团队,是当前"双一流"建设过程中的核心任务,也是实现"双一流"建设目标的最根本保障。通常,学科学术团队有 3 种不同的定义:①由部分具有互补技能,愿意为了共同的目标而努力,且相互承担责任的人们组成的群体。②以科学研究、人才培养为共同目标,为完成共同目标而分工协作、相互承担责任的知识技能互补的个体所组成的团队。③以提高学术研究水平、产出高质量学术成果、提高人才培养质量为目的,以推进学术发展和教学改革为主要任务而建立起来的相互承担责任的教师群体[2-4]。

学科学术团队应该具有师德优良、理念先进、目标远大、结构合理、水平高、能力强 6 项基本特质。学科学术团队建设是一项系统工程,师德师风建设是先导和保障,结构是基础,水平和能力建设是目标。团队建设的关键要素包括 5 个结构、5 种能力和 5 项保障。5 个结构主要是指知识结构、学缘结构、学历结构、职称结构和年龄结构。5 种能力主要是指知识创新能力、知识传授能力、人才培养能力、服务社会能力和教学管理能力。而 5 项保障主要是指团队负责人的学术能力和人格魅力,团队成员的事业心和向心力,有效的团队运行机制,客观、合理的评价体系以及有力的资源、资金保障[2-4]。

在汇聚学科队伍、建设学科团队过程中,统一团队目标、提高团队成员认同度、培养优良师德师风、提升学术水平和整体素质、提升教学能力和水平是团队建设的一个基本路径。通过激发全体成员的能力和潜力,形成凝聚力,营造人人皆是人才、人人皆可成才、青年人才脱颖而出的优良环境,强化队伍梯队的培育力度,促进团队的整体进步,实现科学研究与人才培养质量的提升[2-4]。

(4)建设平台。学科平台建设同样要紧密围绕我们的学术团队,应该坚持有所为与有所不为的建设思路,充分发挥有限资金、有限资源的最大效益,依据学术队伍的技能特点和研究特色构建我们的学科研究平台。

以西北大学地质学科为例,我们主要依托大陆动力学国家重点实验室为主要建设平台,根据地质学科的 3 个学科方向和 3 支学术团队展开建设。我们根据实验室的技术特长,集中有限的财力投入,打造了设备、技术和理念先进的国际一流激光微区原位分析测试和研究中心,并实现与国内其他重点实验室大型仪器设备的远程联网资源共享,积极推进公用社会技术平台建设。大陆动力学国家重点实验室 15 次参加国际地质分析家协会组织的、有全球 63~87 个实验室参加的地球化学分析水平测试,测试结果有 12 次名列或并列全球第一,从而为学科建设奠定了一个高质量的学科平台。这样的平台才能够有效支撑学科核心研究方向。

2.实施学科建设过程

明确学科建设定位,精心凝练学科方向,围绕学科方向汇聚和建设学科团队,围绕团队搭建平台。这些工作为全面开展高质量的学科建设奠定了坚实基础,而学科建设的核心任务是实施高质量的人才培养、科学研究、服务社会以及文化引领和交流合作。

我们要充分认识人才培养在高等教育中的首要地位。进一步完善培养方案,优化课程设置,打造品牌金课,推进国际化课程建设,创造成才条件,培养一流人才。科学研究要契合国际学科发展前沿,团队协同攻关,不断进行理论创新与技术突破,产出具有世界影响的标志性成果。要着力服务国家重大需求和地方经济社会发展,建立学校与行业企业全面互动的良性机制。同时,还要加强国际交流与合作,扩大学科的学术影响与学术自信[5-6]。

3.形成学科建设效果

(1)造就一流人才。拥有明晰的学科建设理念,具备良好的学科建设基础,实施强有力的学科建设举措,学科建设必然能取得良好的建设成效。以西北大学地质学科为例,高质量的学科建设助推了高质量的人才培养和高质量的专业建设,地质学专业成为国家一流本科专业、国家级特色专业、基础学科拔尖学生培养计划 2.0 基地;地质学人才培养基地在全国仅有的 3 次评估中均被评为全国优秀基地,6 篇博士学位论文入选全国优秀博士学位论文,多人次获得地矿行业学生最高荣誉奖“李四光优秀博士研究生奖”“李四光优秀硕士研究生奖”和“李四光优秀本科生奖”,地质学系荣获“全国教育系统先进集体”称号。近 5 年,本科学生研究生考取率保持在 85%以上,本科生承担大学生创新项目国家级 25 项、省级 18 项,理科基地科研小课题 256 项,在读本科生公开发表论文 98 篇,获得省部级以上学生竞赛奖励 183 人次,毕业生中涌现出一批优秀人才。

目前,西北大学地质学科共有教师 110 余人。其中,中国科学院院士 4 人,长江学者 12 人,国家杰出青年科学基金获得者 6 人,国家优秀青年科学基金获得者 3 人,“万人计划”入选者 6 人,国家级教学名师 1 人;国家自然科学基金委员会创新群体 2 个,教育部创新团队 3 个,国家级教学创新团队 3 个,国家级人才在教师中占比高达 30%以上。

(2)产出标志性成果。通过我们长期不懈的努力,学科建设产出一系列标志性成果。多年来,西北大学地质学科获得了国家自然科学一等奖、国家自然科学二等奖、国家科技进步二等奖、陕西省科学技术最高成就奖、长江学者成就奖一等奖。研究成果两次入选全国十大科技进展。以第一作者身份在 *Nature* 和 *Science* 期刊上发表学术论文 15 篇。

2016—2020 年首轮国家一流学科建设周期内,西北大学地质学科获得国家级教学成果奖二等奖 1 项,2 个专业获批国家一流本科专业,2 个专业通过工程教育认证。新增国家级高层次人才 18 人次。其中,中国科学院院士 2 人,长江学者 8 人,“杰青”项目获得者 1 人。获批国家自然科学基金创新群体 1 个,入选“全国高校黄大年式教师团队”1 个、科技部重点领域创新研究团队 1 个。获得国家自然科学二等奖 1 项,获批国家重大、重点项目 19 项,获批陕西地方高校首个国家“111”引智基地,新增国家地方联合工程研究中心1 个。2 篇学术论文分别在 *Nature* 和 *Science* 期刊上发表,CNS 发文总数达 15 篇,高水平论文篇均引文数位列中国高校前列。积极参与国际大科学计划与工程,发现并命名距今 5. 18 亿年的寒武纪特异埋藏软躯体化石库“清江生物群”,被誉为进化古生物学里程碑事件。CCUS 研究团队多次作为中国政府代表参与应对气候变化重要国际会议,交流研发进展、分享治理经验。西北大学地质学科排名最高位列世界第 107 位。

(3)形成学科学术影响力。面向科学前沿与国家重大战略需求,西北大学地质学科已经在大陆构造、早期生命演化及盆山系统及其资源环境效应三大主干方向取得了系列重大成果,形成了鲜明的学科特色优势和国际影响力。

通过上述分析,我们可以看到,学科建设应该包括“奠定学科建设基础—实施学科建设过程—形成学科建设效果”3 个方面。学科建设的基础是明确定位、凝练方向、汇聚队伍、建设平台;人才培养、科学研究、服务社会以及文化引领、交流合作是实施学科建设的过程;经过不懈努力,学科建设一定能够产出标志性的成果,造就一流的人才,形成重要学术影响力,并最终实现建成一流学科的目标。

三、结语

在高等学校建设和发展过程中,学科建设是龙头,学科建设是学校事业发展的总引擎。学科建设需要明确定位,持之以恒,久久为功。在学科建设过程中,要有甘于奉献的精神风貌,要形成良好的学科文化;要突出自身特色,坚持有所为有所不为的发展思路;要群策群力,谋定而后动;要通过策划重大科研项目,不断提高科研水平,以项目凝练团队;要坚持可持续发展的队伍建设思路,重心前移,狠抓青年师资队伍建设;要树立教学质量是生命线的思想意识和管理基调,狠抓教学质量,构建创新型人才培养模式;要本着以服务求支持、以贡献求发展的思路,主动适应国家发展目标,不断为国家的经济社会发展做出重要贡献。

我们相信,在国家宏观规划的正确指导下,在各高校师生的不懈努力下,中国高等学校“双一流”建设一定能够取得重要的进展,建成满足国家重大需求、特色更加鲜明、能够有效凝聚和稳定一流人才、在国际上占有一席之地的一流学科群。

参考文献

[1] 中华人民共和国国务院.统筹推进世界一流大学和一流学科建设总体方案[EB/OL].(2015-10-24)[2020-10-10].http://www.moe.gov.cn/jyb_xxgk/moe_1777/moe_1778/201511/t20151105_217823.html.

[2] 鲁卫平,王润孝.研究型大学教学团队建设的研究与思考[J].中国大学教学,2010(3):71-73.
[3] 刘广虎,关成尧,王承洋,等.资源勘查工程专业教学团队建设研究[J].科技创新导报,2018(3):234-236.
[4] 邱海军,李书恒,曹明明,等.高校自然地理学教学团队建设思路与途径[J].教育教学论坛,2016(30):228-231.
[5] 赖绍聪.建立教学质量国家标准提升本科人才培养质量[J].中国大学教学,2014(10):56-61.
[6] 赖绍聪.创新教育教学理念提升人才培养质量[J].中国大学教学,2016(3):22-26.

“岩石学”系列课程建设的改革与探索[①②]

赖绍聪

摘要：岩石学是地质学、固体地球科学教学结构体系中的一门专业基础课，根据加强基础、拓宽专业、增强适应性的国家理科基地人才培养方针，10年来我们围绕着面向21世纪人才培养模式，从社会需求出发，在岩石学教学内容以及学生综合素质培养等方面进行了一系列有益的探索和实践，取得了一定成效，并已逐步形成了一个特色鲜明的课程体系。

一、课程体系改革

“岩石学”是地质学、固体地球科学和地质类院校（系）教学结构体系中的一门专业基础课，无论对地质学的发展还是培养学生创新能力方面都具有不可替代的功能。根据加强基础、拓宽专业、增强适应性的人才培养方针及国家理科基地培养出学生有较强创新能力的总要求，近年来，西北大学地质学系在原有“岩石学”教学内容的基础上，对理科基地班“岩石学”课程进行了较大幅度的调整。具体做法如下：

（1）将“岩石学”与“晶体光学及光性矿物学”作为一个统一的整体对待，强调两门课程的统一性与协调性，将“光性矿物学”作为岩石学的前期基础，由岩石学任课教师提出具体要求，并在共同讨论的基础上，集体修编了晶体光学及光性矿物学教学大纲，使得其教学内容更加适合当代岩石学新发展的需要，不仅要重视晶体光学基本理论的学习，而且更加强调其学生显微镜下识别鉴定各类主要造岩矿物的实际能力，从而为下一步的岩石学教学奠定坚实的基础。

（2）保留“岩石学”中岩浆岩、沉积岩和变质岩三大体系，但更加强调三大岩类的内在联系及协调一致，大幅度压缩重复内容，在体现各岩类自身独特性的同时，切实注意三大岩类的内在联系及相互衔接。以物质组成、成岩条件的系统变化为纲，达到三大岩类教学内容的协调一致。在此基础上于1996年、1999年、2000年、2003年4次大幅度修订了教学大纲，取得了较好效果。

（3）由于当代地球科学的迅猛发展，使得改革开放初期出版的统编教材，在一定程度

---

① 原载于《高等理科教育》，2004（3）。

② 本文得到教育部高等学校优秀青年教师教学科研奖励计划及教育部创建名牌课程（岩石学）计划资助。

上出现了知识老化、课程体系陈旧的严重不足。针对这一问题,地质学系配合学校的教学改革统一步骤,并借“211工程”建设的东风,申报并获准了岩石学教材及实验室建设、岩浆岩岩石学教材及多媒体课件开发、变质岩岩石学教材及多媒体课件开发、沉积岩岩石学教材及多媒体课件开发、晶体光学课程改革及多媒体课件开发5项校级教改项目,并分别于1998年和2000年启动,总投入达到12万元,从而初步开始了自编教材的进程。其中,由李文厚教授完成的《沉积岩教程》和《沉积岩实习》已在教学中全面使用,《岩浆岩岩石学》和《变质岩岩石学》电子教材的雏形也已形成,推动了教学质量的大幅度提高。

二、全面开展多媒体教学方式

多媒体辅助教学是计算机多媒体技术在教育领域中的应用,它将计算机多媒体技术与教学内容相结合,生动、形象、直观,信息量大,极大地提高了教学效率和教学质量,推动了教学改革的进一步发展和深化。地质学系自1997年以来,十分重视多媒体教学的研究和实践,在系内全面实施多媒体辅助教学计划,要求国家理科基地教学课程全面逐步实现多媒体辅助教学,并投入大量资金为承担理科基地教学的教师每人配备东芝(或戴尔)笔记本电脑1台。系内教学室、实验室现已全部装备高精度激光投影器和台式电脑控制设备,并为此全面改建了地质学系网络教学体系,局域网站服务器、CAI教学、CAD教学研究和控制室,使地质学系的教学手段跨入了一个全新的时代。经过岩石学教研组的共同努力,目前岩浆岩岩石学、变质岩岩石学和沉积岩岩石学已全部实现多媒体辅助教学,并已将各门课程的多媒体教案、讲稿、教学大纲、课程进度表等全部上传至西北大学校园网,全校师生可以随时阅读、观看和下载,从而大大地提高了地质学系课程的教学质量,深受学生的欢迎。

值得强调的是,岩石学的任课教师始终坚持追踪学科前沿,不断改进教学手段和更新教学内容。比如,岩浆岩岩石学、沉积岩岩石学等多媒体课件自2000年上传至校园网后,至今已进行了3次改版和更新,使得教学内容不断完善。

三、加强实验室建设

岩石学是一门实践性很强的学科,有近40%~50%的学时是在实验室完成的。因此,实验室建设对于保证课程的教学质量至关重要。自1995年起,地质学系利用国家理科基地建设经费以及“211工程”建设经费,对岩石学实验室进行了全面更新建设,显微镜全部更新为日产双筒尼康镜,配合各实验室装备了莱卡数码显微照(摄)相投影系统,使得地质学系的实验室硬件条件已居国内领先地位,并与国外大学教学设备处于类似水平,为确保教学质量提供了物质基础。在岩石学实验课中,教师利用10~15分钟时间把本节课的核心内容、要达到的目的用多媒体、激光投影等形式向学生展示,并把典型薄片向学生进行简明扼要和精辟地讲解和示范。有了这种直观的演示,学生很快就能了解本节实验课的内容和目的,这比学生自己在显微镜下无目的地摸索效果要好得多。比如,由我们研制的岩浆岩岩石学实验纲要电子教案,投入使用后效果很好。

岩石学实验室建设的另一个极为重要的方面,就是典型岩石标本库和薄片库的建立。数十年来,地质学系全体教师投入了巨大的精力,在标本库及薄片库建设方面做出了重大贡献,岩石学标本库中至今还陈列有 20 世纪初由张伯声等老一辈地质学家采集的典型标本。自 1993 年国家理科基地建立后,地质学系在这方面更是加大了力度,并取得了重大进展。自 1994 年起,全教研室曾 6 次召集全体教师会议,布置、落实标本、薄片库建设的具体方案和措施,先后投入大量资金,专程派出教师赴典型岩区采集岩石、矿物标本,以及向国内标本厂订购成套的标本和薄片,特别强调教师在承担科研项目和出国考察进修期间注意收集典型岩石、矿物标本,并于 1995 年年初在地质楼二层和三层开辟了专门场所,将原有的岩矿陈列室改建为岩石陈列室和矿物陈列室,收集、积累了一大批典型岩石、矿物标本,从而为全面推进实验室建设和提高教学质量提供了物质基础和保障。

四、改革考试方式,全面推行考教分离

由于岩石学属专业基础课程,因而目前仍然沿袭了闭卷考试的基本方式。但强调了实际操作和岩石鉴别能力的重要性,从而将学生总成绩构成区分为两部分,即理论考试占 70%,实验课成绩占 30%。重要的是,我们对于考题类型和出题方式进行了系统改革,参照国外一些著名大学的试卷类型(如澳大利亚 Macquarie 大学等),增加了相对较为灵活的思考性题型,减少和压缩了较为死板的考题类型,并集全体任课教师的集体智慧,于 1996 年年初建设至今已初步形成了岩石学试题库,使考试体制更加科学化和系统化。另外,地质学系于 1997 年起,首先在岩石学学科实施考教分离试点,任课教师管教不管考,考试由除任课教师之外的教研室其他教师共同组成命题小组,以教学大纲为基本准则,独立命题并完成课程的考试,由此检验教师的授课质量,避免了考试过程中的倾向性、暗示性和可能出现的不规范操作,确保了教学质量。目前,考教分离制度已在地质学系全面正式实施。

五、师资队伍建设

由于科学技术的迅速发展,如何适应现代化教学手段、丰富教学内容、追踪学科发展趋势、拓宽知识面,这对教师提出了更新、更高的要求。加强师资队伍建设,必须走在职培训、提高和发挥现有教师潜力之路,并大力提倡教学、科研并重,教师在承担国家级科研项目过程中,大力度地提高了自身的学术水平和科学素质,从而为保证教学工作的创新性、前沿性和科学性提供了前提条件。地质学系岩石学主讲教师共有 7 人,其中教授 5 人(博士研究生导师 2 人),副教授 2 人。7 人中有 4 人具有博士学位,1 人具有博士后研究经历。3 位教授曾赴国外进修和合作研究。7 位主讲教师近 5 年来分别承担完成国家自然科学基金项目、305 项目、省部级重点项目等科研项目多项,在 SCI、EI 收录期刊及核心期刊发表相关学术论文 40 余篇,充分反映了岩石学课程教研组的学术实力,确保了本课程教学质量的大幅度提高。

六、强调学生综合素质、科研素质培养

作为培养高级基础地质学人才的国家基地，几年来，我们围绕着面向21世纪人才培养模式，从社会需求出发，在专业设置和教学内容以及学生综合素质的培养等方面进行了一些有益的探讨和实践，取得了一定成效。主要可以概括为以下几个方面：

(1)实施导师制。为了加强本科学生的专业素质教育，培养学生的创新精神和科研能力，我们针对三至四年级学生为每1~3名有针对性地聘请一位具有博士学位或副教授以上职称的教师做导师，主要负责学生专业思想教育和思维、动手能力的培养，言传身教，以教师的工作热情及严谨求实的工作态度带动学生，培养其良好的科学素质。让学生参与导师的科研项目，从书本走向实践，提高其独立思考问题、解决问题的能力，激发他们发表论文的意识。这一措施取得了可喜的成效，在岩石学教研组指导的学生中，已有8名学生在本科阶段作为第一作者在公开刊物上发表了学术论文。同时，导师还像家长一样对学生在生活和思想上给予关心和帮助，以自己良好的风范和行为教学生，不但指导他们学会"认知"，同时还要学会"做人"，以适应将来复杂的社会环境和工作。

(2)加强实践和动手能力的训练，培养学生艰苦奋斗为科学事业勇于奉献的精神。在教学过程中注重素质培养与业务教育相结合，既要求学生基本具备能独立进行实际工作的能力，也把培养学生树立良好的思想作风、实事求是的工作态度、吃苦耐劳和为科学事业献身的精神作为重要内容。

(3)举办"名人学者谈治学"系列讲座。为了培养学生的科学素质和专业素质，几年来，我们一直坚持举办名人学者谈治学系列讲座。被邀请做专题讲座的都是在各学科学术界卓有成绩、颇具知名度的专家学者，他们循循善诱的报告以及他们本身在学术界所取得的成绩和为祖国科学事业顽强不息的奋斗精神，深深地打动了学生，也使学生了解和学习了其他学科的知识，在专业教育中起到了积极推动作用。

研究性教学改革与创新型人才培养①

赖绍聪　华　洪　王震亮　张成立　张云翔

摘要：研究型、创新型本科生—研究生教育与一般大众型、适应型人才培养有明显不同，更注重学生的科研意识、科研能力和创新意识的培养，属于“精英型”教育模式。所以，要给本科生—研究生提供更多的科研训练机会，以期学生尽早进入科研领域，接触学科的前沿，了解学科的发展动态，培养学生的科研创新能力。

一、改革思路与教育理念

将科学研究实质性地纳入教学过程，实现优势科研成果、科研资源向教学的转化，达到高等学校教学与科研的完美有机结合，实现创新型人才培养，乃是当前高等理科教育面临的重大课题[1]。为了进一步加强基地学生实践能力、科学研究能力和创造性思维能力的培养，我们必须对研究性教学改革的规律进行努力探索。因此，如何正确处理科研与教学的关系，将科学研究实质性地纳入教学过程，使科研成果有效地转化为教育资源，本科生—研究生教育应创建何种教学与研究相结合的氛围，如何在教学过程中通过优化课程结构建立一种基于研究探索的学习模式，如何有效地建立“教师—研究生—本科生”学术群体，是亟待解决的重要问题。

我们的思路是，充分利用地域优势和学科优势，构建贯穿本科生—研究生教育全过程，形成在教学上循序渐进，在教学内容上密切协调，在实践教学的地域上相互关联、特色鲜明、科学合理的研究性教学体系。具体做法如下：

(1)根据专业基础课、专业课和专业选修课的不同性质，采用不同的教学新模式，将研究探索的思维意识融入教学。

(2)根据认知—技能—探索—综合的不同阶段，系统改革本科4个年级的野外实践教学模式。

(3)实施本科生科学研究培训计划，实现本科—硕士的贯通培养。

(4)从本科抓苗子—硕士阶段稳定研究方向—博士阶段特别注重发挥学科群的指导作用，在此过程中培育研究型精英人才。

---

① 原载于《中国大学教学》，2007(8)。

二、课程教学过程中实施研究性教学

在课程研究性教学改革过程中,我们注重改变以往以验证为目的的课程教学内容,培养学生全新的地学观和综合分析问题及创新性的能力,加强新思维、新技术和新方法在课程教学中的应用,建立特色鲜明、科学合理、循序渐进的课程教学新体系,全面体现研究性教学课程的设计性、综合性及创新性。由于不同课程内容的差异,我们以课程的性质为依据,建立了具有特色的研究性课程教学模式。

1.专业基础课程——综合教学模式

以往的地质专业基础课程教学多是以观察为主,以达到理解课堂中理论的阐述和验证课程中对地质现象形态的描述的目的。我们在新的地质专业基础课程课堂教学中,一方面突出重点,精简原来过于烦琐的记忆性内容,另一方面提出一些学科发展中具有代表性的问题及相应的参考文献,通过学生自己阅读,写出该方面学科发展综述及自己对这些问题的认识。同时,在专业基础课程的实习教学中只提供基本的地质素材,让学生综合运用所学知识,采用多种方法对所提供的地质现象、地质剖面进行分析,最终写出地质分析报告。教学中还提出一些学科前沿的热点问题进行讨论,学生最终以课程论文的形式完成教学的考核。实践证明,这些是加强学生动手能力、综合能力训练行之有效的方法,对于提高学生思维能力有着良好的作用。

2.专业课程——现场实践教学模式

部分专业课程,如构造地质学、矿床学、第四纪地质学等的课间野外现场教学,是自20世纪70年代以来我们长期坚持的教学环节,并在长期教学过程中,不断继承、调整、改革而建立起来了新的教学模式。通过课间穿插的现场教学,使学生认识和掌握野外考察研究的程序、研究内容与工作方法,提高综合分析的能力,为今后的科研工作进行方法、思路、综合能力和科研报告编写的训练,提高学生的科研能力。

3.专业选修课程——研究性教学模式

对于本科高年级学生的专业选项课程,在充分掌握课程基本知识点的前提下,以教师命题和学生根据学科发展自己提出问题相结合的方式设立专题,给学生更多的空间在课外利用各种渠道收集与专题相关的资料,并严格按照科技论文的形式撰写各种形式的读书报告(学科发展综述、存在的科学问题以及自己的看法与认识等)。这种做法不但培养了学生独立收集、阅读文献和综合分析资料的能力,同时还锻炼了他们科学研究的方法和技能。

通过上述系列课程教学方式改革,在一定程度上改变了传统的教学体系和教学方式方法,完善了本科阶段的研究性教学环节,学生的素质、综合能力得到全面增强。

三、野外实践教学过程中研究性教学改革的实施

实践是一切真知的源泉。对以实践性强为突出特点的地球科学而言,地质演化历史之长,各种地质现象及形成过程之复杂,绝大部分是无法模拟或不可再现的。因此,地质

学的特点要求必须要有与之相适应的实践教学体系[2]。实践过程就是认知思辨、训练能力、提高素质、启发思维、激发创新的过程。对地质现象的不同观察和不同认识、不同学术观点的讨论是培养正确的地球科学思维观、独立思考和创造性思维的过程。地球科学的发展和社会进步对人才的需求促使国内外地学工作者和教育者努力探索实践性教学环节和课程设置。随着思维科学和新技术、新方法的不断发展,将科学研究引入教学、培养学生创新能力,教学方式和方法正在发生着重要变化,并取得了显著成效。

实践性教学过程在完成从单科性向综合性,从认识性、继承性向研究性,从验证向创新,从灌输式向启发式、讨论式,从传统向高科技的思想观念和方法转变的基础上,突出以学生为主体,全面改造实践性教学环节的教学方法和方式,实现知、辩、行的全面训练,培养基本素质和综合思维,激发创新精神。具体内容包括了以下几点:

(1)在已有实践性教学改革的基础上,统筹本科4年不同阶段、不同课程的教学内容和计划,构建课堂教学与实践教学内容协调配置、时间穿插及循序渐进的实践性教学新体系,完善本科实践教学环节。

(2)建设部分理论课程室内实践和课间野外实践教学的实践教学模式。

(3)建设秦岭造山带与相邻地区在地域上相互关联,具有丰富大陆地质内涵,在教学内容上循序渐进,具有科学性、综合性和前瞻性的不同年级野外实践教学基地和不同理论课程的课间野外实践基地。

(4)发挥学科优势,实现科研资源向教育资源的转化,完成不同年级、不同实践教学基地的教学内容建设和教材建设。

(5)探索实践教学认识—技术方法—多学科综合思维与创新的实践—研究性实践—创新性实践的循序渐进的实践教学方式和方法。

(6)建立数字地球实验室,在实现硬件和软件建设的基础上,开发新技术、新方法与传统的、行之有效的地质方法密切结合,便于操作的研究方法。

(7)建立与实践性教学体系配合的科学、自律的管理体制。

(8)在实践教学过程中,加强人文素质培养,增强相互协作的团队精神,形成既有严谨的学习风气,又有集体性与个性共存、生动活泼的人文环境。

四、学生科学研究训练计划的实施

为了加强学生科研能力的培训,西北大学地质学系设立了地质学本科学生创新课题研究基金。我们设立创新基金的宗旨:强化学生的开拓精神和创新意识,培养他们的创新思维、创业精神和实践能力,使其能尽早地参与科学研究、技术开发和社会实践等课外学术科技活动,并得到基本的科研训练。创新基金的资助原则为"理实结合、突出重点、鼓励创新、注重实效"。资助办法为:自主申请、公平立项、择优资助、规范管理。为此,我们成立了学生创新基金管理领导小组,对基金重要事项和基金项目资助经费进行管理。

2003年度,地质学系批准创新基金立项21项,总资助金额达到10.25万元,研究领域涉及岩石学、矿物学、矿床学、地球化学、古生物学、古生态学、石油地质学、环境地质与灾害地质学、工程地质学等10余个学科领域。2004年度，批准创新基金立项17项，投入总经费9.8万元。2005年度，批准创新基金立项18项，总资助金额达到10.50万元。2006年度，批准创新基金立项16项，总资助金额达到11.50万元。

目前，2003年度创新基金项目已全部结题，提交学术论文42篇，其中28篇分别发表在《岩石学报》(SCI源期刊)、《地质通报》《地球科学与环境学报》《西北大学学报》等学术期刊上。另有3项成果分获西北大学挑战杯特等奖和二等奖。这充分表明，创新基金计划的实施成效十分显著，对提高高年级本科学生的创新能力，以及实现本科生—硕士研究生的贯通培养有极大的推动作用。

2004年度，创新基金共提交学术论文23篇，部分论文分别发表在《科学通报》(SCI源期刊)、《地质通报》和《西北大学学报》等学术期刊上。

2005年度、2006年度的创新基金也在有条不紊地执行中。

五、强化学生科学研究学术氛围

为配合创新基金的实施，使学生能够积极参与科研项目的研究和学术交流活动，为本科生搭建一个较好的学术交流平台，地质学系积极鼓励学生创办了自己的学术刊物，进一步提高了学生的理论水平、写作能力和综合素质。2004年，由地质学系学生自行撰稿、自主编辑的学生学术刊物《地学新苑》创刊，地质学系400多名学生从此有了一个采撷思想火花、交流学习心得、进行学术交流的园地。至今该刊物已出刊5期，发表学术小论文50余篇。

六、从本科抓起，为“百篇优秀博士学位论文”培养后备人才

西北大学地质学系建立了完整而有特色的本科生和研究生培养体系。高层次的人才培养从本科生抓起，实施导师制和创新基金研究计划，筛选出有培养潜力的苗子；硕士研究生阶段稳定研究方向，注重科研能力培养；博士研究生阶段发挥整个学科的指导作用，重点放在创新能力的培养上。只要学生根据自己的特长稳定在某一研究方向上长期不懈地努力，就能凝练出具有重要意义的科学问题和取得创新性研究成果。同时，抓好教学各个环节，实施教学改革和加强课程建设，取得了丰硕成果。

在本科筛选培养阶段，采取师生双向选择、最终确定的方式进行。即由不同研究方向的教师根据自己的科研特色和研究实际，提出科研小课题(有限时间、有限经费、有限目标)，学生根据自己的兴趣和特长选择课题，将导师制与创新基金有机地结合在一起，初步形成了教师—研究生—本科生研究群体与教师的科研项目—研究生的论文选题—本科生的创新基金等多层面的课题组，从而将导师制、创新基金研究计划、实验室开放及本科毕业论文有机地融为一体。这一措施，使部分高年级本科学生实质性地独立承担起小课题或加入教师的科研群体中。本科生—研究生—教师共同开展野外工作、同场参与学术报告和学术讨论，形成了颇具西北大学特色的科研群体模式，真正实现了将科学研究实质性地纳入教学过程，实践教学由综合向研究性的转变，从而使本科科研训练实践

教学产生了质的飞跃,学生以第一作者身份公开发表的论文数量明显增加。

在博士、硕士研究生培养阶段,采取以导师为主、导师负责与指导小组集体培养相结合的方式,充分发挥整个学科的指导作用。以参加科研课题为主线,着重培养研究生独立进行野外和室内创新性科学研究的素质和能力。培养过程中采取理论学习与科学研究实践相结合,知识传授与素质教育相结合,基本训练与能力培养相结合的原则,特别注重对于创新能力、科学道德、严谨学风和敬业精神的培养。此外,出台了硕、博连读政策,以稳定部分学生的研究方向;举办研究生论坛和科学沙龙,为研究生搭建学术平台。实践证明这套培养方案行之有效,先后有4篇博士论文被评为全国优秀博士学位论文。

七、结语

(1)将科学研究实质性地纳入教学过程,实现了优势科研成果、科研资源向教学转化,达到高等学校教学与科研的完美有机结合。

(2)通过高年级本科学生科学研究培训计划的大力度实施,初步实现了本科生—硕士研究生的贯通培养,极大地提高了学生的创新能力和独立研究能力。

(3)实质性地形成了颇具特色的本科生—研究生—教师科研群体,为新一代“精英型”人才培养奠定了坚实基础,并为西北大学地质学系整体创新能力的大幅度提高提供了人才保障。

(4)有利于为教师、学生营造一种紧张有序、严谨求实的良好学习氛围,为本科生—研究生教育注入新的活力。

参考文献

[1] 杨承运,张大良.地学教育总体改革研究报告[M].北京:高等教育出版社,2003:101.

[2] 杜远生,刘世勇,杨坤光,等.国家地质学理科基地创新人才培养模式[M].武汉:中国地质大学出版社,2004:94.

完善教学质量监督保障体系 培养高质量创新型人才①②

赖绍聪 张云翔 曹 珍 汪海燕

摘要：文章详细介绍了西北大学地质学系国家基础科学人才基地教学质量监督保障体系的构成、实施方法、实施效果及其对创新型人才培养的有效促进和保障作用。

质量是教育的生命线，面对新的形势，高等教育必须积极探索提高质量的新思路、新途径，树立科学的质量观，推动培养模式、课程体系、教学内容和教学方法的改革与创新，并把教育的信息化作为提高教育质量的新手段，在教学和教学质量管理中广泛地应用起来[1-3]。教育部历来重视高等教育质量，2001 年 8 月专门下发了《关于加强高等学校本科教学工作，提高教学质量的若干意见》，提出了 12 条加强本科教学工作、提高教学质量的措施和意见，得到了全国高等教育战线的广泛拥护和认真落实。西北大学地质学系人才培养基地教学质量监督保障体系的建立和完善正是遵循了国家的需求。

一、教学质量监督保障体系的构成

教学过程不能完全放任自流，要根据国家的大政方针和专业人才培养规律来规范教学过程，约束教学行为[4-5]。根据高等教育教学过程和特点，我们以教学质量标准和教学质量监控体系的建立为重点，实施了实践证明行之有效的科学教学管理模式。

地质学系人才培养基地教学质量监督保障体系的最大特点就是使影响教学质量的全部因素在教学全过程中实现可控，认真研究、详细厘定并建立了教学质量反馈机制，在 10 年时间内持续改进，从机制和制度上保证教学工作在基地建设中的主体地位。为了达到教学过程实现可控的目标，我们将影响教学质量的主要方面分为教学质量标准，管理职责，教学资源管理，教学过程管理，教学质量监控、分析和改进，基本上做到了每个重要环节都有责任人、执行人或执行部门及具体执行内容。我们在制定教学质量标准的同时，十分注重建立教学质量反馈机制及分析改进措施。通过完善制度和组织机构，来促进有效的质量管理。在做好监控和分析的基础上，改进教学质量，制定纠正措施和预防

① 原载于《高等理科教育》，2007(6)。

② “教育部高等学校优秀青年教师教学科研奖励计划”资助；国家基础科学人才培养基金“创新型地学人才培养机制中教学方法改革的研究与实践”项目(J0105)资助。

措施,进行持续改进。教学质量的持续改进始终以学生及社会用人单位的要求和满意度作为首要标准。在教学管理规范的建立过程中,我们充分考虑了制度的规范性和个性化,实践中又要尊重教师对教学内容的把握,充分发挥特色。

近10年来,我们按照"巩固、深化、提高、发展"的思路,巩固成果,深化改革,提高质量,持续发展,坚持有所为有所不为,扎扎实实地推进地质学系人才培养基地的教学质量。我们在全系范围内倡导科学的、全面的质量观,从严治教,规范管理,把质量工程作为一个系统工程来对待,精心组织,认真实施。在系统完善传统质量监控手段的基础上,顺应现代科学技术发展趋势,深刻理解"教育信息化是我国高等教育适度超前发展的重要途径"的内涵,将传统质量监控手段与现代信息技术有机结合,全方位实施"教育资源上网工程",购进千兆交换机,改造局域网,搭建地质学系教育资源共享数字化平台,率先在全国实施全部教育资源对外开放。在这一过程中,接受社会的反馈意见和监督,提高地质学系基地人才培养质量,并在教学质量监控方面初步形成了具有特色的、较为系统的"社会—管理层—教师—学生"理论体系和具体实施措施[6]。

(1)社会层面。实施教学档案上网工程,构建网络教学与管理平台,将地质学系70门课程的教学大纲、电子教案、多媒体课件、教学进度表和教材等教学文档全部上传发布在网络平台,从而形成全社会的教学质量监督保障体系。

(2)管理层面。地质学系党政一把手对教学质量负直接责任,系务会、系务扩大会专题研究本科教学工作,充分发挥教学督导组、教学委员会、本科教学关键岗在教学质量监督中的作用,从而构成管理层面的监督保障体系。

(3)教师层面。对教师实施三段式教学管理模式。即教前必须举行教学大纲听证会、大纲上网、教学纪律上网;教中实行教师课程教案网上公开、教学督导、听课检查;教后则全面实施教考分离、教学质量评估制。从而构成教师层面的自我约束机制,并充分调动教师对基地建设的责任感和使命感,构成强大的自律机制。事实上,地质学系在2002—2003年,由教师起草、教研室讨论、系内讲评、系教学委员会最终审定,完成了70门课程全新的教学大纲,奠定了创新型人才培养方案的基础。为了确保教学大纲的质量,我们对主干课程大纲进行了外审,分送至南京大学及中国地质大学同行专家审评。为了确保教考分离的真实有效,我们建立了全部课程的试题库,并对主干课程的教考分离试卷进行了外审。为了鼓励教师在教学工作中投入更多的精力,取得更大的成果,我们还在校系岗位聘任、奖金发放、出国交流学习及教学设备配给中向教学大力度倾斜。我们将教学作为衡量教师工作的主要标准,在职务、岗位聘任及考核中实行教学考核一票否决制。对教学内容陈旧、教学效果差、学生反映强烈、教学投入精力不足者,下调考核等级及奖金。发生教学事故,考核为不合格者停发津贴。发生严重教学事故,当年不得晋升职务,降低岗位聘任级别。同时,以凝聚力、使命感进一步加强教师的自律精神,形成更为有效的保障机制。

(4)学生层面。每门课程进行学生评教。据西北大学教务处2005年对地质学系12门课程进行学生问卷调查,学生评分大多在85分以上,其中4门课程得分90分以上。为了全面了解教师的教学态度及水平等,每学期开学后地质学系还会及时召开各类学生座谈会,特别是对上一学期成绩出现不及格的学生,针对学习成绩,分析原因和教学中可能存在的问题。我们还在地质学系局域网上建立了主任信箱,广泛听取全系、全校乃至全社会的批评、建议及各种反馈意见,从而有效地保证了教学质量的不断提高。

(5)规范的教学档案管理。我们一贯重视教学档案管理。1993 年基地建设以来,对档案管理更加规范,建有完备的系教学档案室。保存有 1955 年以来的学生成绩册,1993 年基地建立以来的各种重要教学档案。规范的教学档案管理为质量监督提供了有力保障,为教学研究与改革提供了基本依据,为学科发展奠定了坚实基础。

二、教学质量监督保障体系的实施效果

由于地质学系人才培养基地实施和建立了完善的教学质量监督保障体系,全系基地班学生形成了良好的学风,学生综合素质与技能培养得到了整体提高。良好的学生培养质量使地质学系毕业生供不应求,成为我国地质科学研究人才培养的重要基地之一。地质学系基地班学生 1998—2005 年以第一作者身份发表学术论文 60 余篇。基地班学生一次就业率连续 7 年达到 100%。研究生升学率达到 80%,出现了中国科学院地质与地球物理研究所和中国科学院地球化学研究所争相在地质学系设立奖学金,争抢地质学系毕业生的喜人局面。

近年来,教学质量的提高使地质学基地点涌现出了一批特优学生,他们表现出对地质的特殊喜好和钻研,受到教师的普遍好评。通过系统教学训练和严格的教学质量监控,基地班学生基本具备自己检索文献、自己完成论文的能力,毕业论文达到了一定水平。近年来,我们承担了大量的教学改革研究项目。发表教学改革研究论文 95 篇,研究报告及交流论文 6 篇。这些成果的取得是地质学系严格教学质量监控的重要体现。

截至目前,我们共获得国家级教学成果奖 3 项,省部级教学成果奖 8 项,校级教学成果奖 25 项。其中,“高等地质教育的创新、改革与实践”获得 2001 年度国家普通高校高等教育教学成果奖一等奖。梁山实习基地经过 30 余年的建设,已成为地质学系地质技能训练的良好场所,形成了一整套完整的地质资料,1997 年获得国家普通高校教学成果奖二等奖。地质学系“地质学实践教学新体系”获得 2005 年度国家级教学成果奖二等奖。

三、辐射与示范作用

自地质学系基地建设以来,我们始终以教学工作为中心,把教学质量作为院系生存的生命线。将提高学生的创新能力贯彻到整个教学过程中,全面深化教学改革,不断加强教学建设,取得了一系列显著的教学成果,教学质量得到明显提高,地质学系基地开拓性努力和创新性实践使得我们正在成为学校教学建设和改革实践的“排头兵”和“示范田”。长期以来,地质学系基地主动适应社会、科技、经济发展对人才培养提出的新要求,及时更新教育理念,确立了“加强基础、突出创新、提高素质、体现特色”的人才培养指导

思想,深入开展教学内容与课程体系改革,不断提高教学质量。在改革过程中,及时融入最新的学科发展动态及教学改革成果,保证了教学内容的先进性;为了培养学生的创新意识,提高学生的创新能力,我们积极推行了实践教学新体系。

地质学系人才培养基地"社会—管理层—教师—学生"多层面立体化教学质量监督保障体系具有一定的系统性和先进性。该管理系统率先在全国高校实行所有教学档案上网工程,构建网络教学与管理平台,形成自我约束、自我发展机制。将教学大纲、电子教案、多媒体课件、教学进度表、教材等全部上网发布,配合主任信箱等,不仅可以及时了解教学动态,而且还能主动接受全校师生和社会的监督,为建立健全自我约束、持续发展机制发挥了重要作用。教学档案上网工程,极大地方便了教师之间的相互交流与相互督促,进一步增强了教师的自律性,为学生课后自学、深刻理解课程内容、真正参与教学互动提供了良好条件,为社会层面的教学质量监督保障体系构建了平台[7]。

参考文献

[1] 徐建平.关于提高人才培养质量问题的思考[J].中国地质教育,2002(1):30-32.
[2] 李正.若干高等教育质量观述评[J].高等理科教育,2004(1):35-39.
[3] 徐向艺.高等学校教学质量管理的范畴、原则与体系[J].高等理科教育,2004(1):40-45.
[4] 孔锐,黄启,王兰兰.高等教育过程中质量管理的途径分析[J].高等理科教育,2004(3):43-47.
[5] 逄增苗,郑红.高等院校教学质量评价研究与实践[J].高等理科教育,2004(5):43-47.
[6] 张云翔,赖绍聪.国家"理科人才培养基地"的创新教育[J].高等理科教育,2004(3):9-11.
[7] 赖绍聪."岩石学"系列课程建设的改革与探索[J].高等理科教育,2004(3):58-60.

寻找契机 融合优势
逐步形成特色显著的教学团队①

赖绍聪

摘要：西北大学地质学系晶体光学与岩石学国家级教学团队在多年的教学实践、教学研究、教学改革与教学建设过程中，结合自身特色，强化师资队伍建设，打造老、中、青相结合以中青年骨干教师为主体的课程群教学团队。教学科研相互促进，实施课程教学体系的创新性改革，将科学研究实质性地纳入教学过程，实现优势科研成果、科研资源向教学转化，奠定课程教学的高起点。逐渐摸索出一条适合自身发展规律的晶体光学与岩石学教学团队建设方案，并逐步形成了特色显著的教学团队。

西北大学地质学系诞生至今已有70余载，历经曲折坎坷，在艰苦创业的历程中，逐渐形成了强烈的团队意识和甘于奉献的精神。在地域、体制、条件都不占优的情况下，抢抓机遇，突出自身特色，坚持“有所为，有所不为”的发展思路，变劣势为优势，走好了“关键”几步，开创了地质学系今天人才培养体系和教学团队建设的良好局面。

一、教学团队建设思路

长期以来，我们总是面临两难问题：一方面，虽然学校对地质学系给予了重点扶持，但受条件所限，与国内地学兄弟院系经费投入和政策支持相去甚远；另一方面，随着学科发展，学科平台不断拓展，学科层次不断提升，势必要求地质学系在人才培养及师资团队方面与国内一流甚至是国际一流地质院系相互竞争。要破解这样的难题，只有一条出路——突出特色，以有限的人力、物力、财力做重点突破。

由此，在多年的教学改革与教学建设中，我们结合自身特色，逐渐摸索出一条适合自身发展规律的教学团队建设方案，并逐步形成了特色显著的教学团队。

1.依据课程性质，融合特色团队

晶体光学与岩石学课程群是地质学的主要支柱课程之一，是地质学本科各专业必修的专业基础课，是地质学创新人才培养中不可缺少的重要基础主干课程。它包括晶体光学、光性矿物学、岩浆岩岩石学、沉积岩岩石学、变质岩岩石学以及岩石物理化学。它要求教师必须具有扎实的结晶学与矿物学基础，同时在教学过程中又要充分考虑矿床学、

---

① 原载于《中国大学教学》，2010(10)。

地球化学等后续课程的实际需要。由此,晶体光学与岩石学教学团队长期以来在地质学系岩矿(矿物学、岩石学、矿床学、地球化学)教研室中逐渐孕育成长。团队中1/3的教师承担过或仍在承担结晶学与矿物学课程教学工作,1/3的教师承担过或仍在承担矿床学、地球化学课程教学工作;新加入团队的青年教师,首先要跟听结晶学与矿物学本科课程,并参加矿床学、地球化学课程的辅助教学工作,协助矿床学、地球化学课程教师完成实习实验、批改作业等教辅工作,从而使本团队教师对上、下游课程的课程体系、教学内容、教学方法有较为深入的了解。这对于本团队教师全面正确地把握晶体光学与岩石学课程群的课程体系与教学内容具有十分重要的意义,对于本课程群与上、下游课程的相互配合与合理衔接有直接帮助。

2.构建合理的教师知识结构

长期以来,我们一直坚持本团队教师对晶体光学与岩石学课程群总体教学内容与课程体系要有较好的了解与把握。团队中教师均必须首先跟听晶体光学与光性矿物学课程或承担相应教学工作,形成对前端课程的充分了解以及教师个人良好的知识储备;并在此基础上跟听非地质专业(资源勘查工程、地质工程专业)矿物岩石学、晶体光学与岩石学课程或承担相应教学工作,逐步提高自身的业务水平与教学能力。经过这一过程培训,团队教师才能根据自身特点承担国家理科人才培养基地晶体光学与光性矿物学、岩浆岩岩石学、沉积岩岩石学、变质岩岩石学以及岩石物理化学课程教学工作。

3.依据教师科研特长,分担专业课程教学任务

地质学系的教师长期以大别-秦岭-祁连-昆仑-天山造山带及其两侧盆地为主要研究对象和“天然实验室”,并在长期的实践中形成了优势科研团队。在本教学团队的建设过程中,我们充分考虑教研室教师的科研主攻方向,从岩矿教研室长期从事矿物学研究、中国中西部高亚超高亚变质带研究、秦岭及青藏高原岩浆作用研究、鄂尔多斯盆地沉积体系研究以及中国中西部金属矿产研究的教师中自然组合,形成晶体光学与岩石学教学团队。同时,通过吸纳青年教师参加科研工作,共同进行科学研究,帮助他们申请科研项目等方法,在实践中选拔对岩石矿物学研究方向有浓厚兴趣的青年教师逐渐融入我们的教学、科研团队,为本团队培养后备人才。正是由于这样的运行机制,使得地质学系晶体光学与岩石学教学团队在长期的教学科研工作中自然形成并和谐发展。团队中的教师有共同的研究方向和研究内容相近的科研项目,在教学、科研活动中有一致的目标和共同的探索内容。

4.实施优质科研资源向教学资源转化,奠定课程教学的高起点

长期以来,我们十分重视优质科研资源向教学资源的转化,主要表现在以下几个方面:

(1)将科研成果适时转化为教学内容。结合教师自己的科研特色,加入一些课本上没有的新鲜知识及近年来学科进展等内容,实现科研向教学转化。事实证明,将科研成果和科研中获得的新认识用于教学,对于提高教学水平起到了极大的推动作用。

(2)科研标本、图件、照片、仪器设备等,在一定条件下,直接用于教学,服务学生,为人才培养发挥重要作用。

(3)通过科研,提高了教师素质、稳定了教师队伍。通过出高水平的科研成果,不仅可以逐步解决或缓解教师的职称及收入等问题,同时也营造出一种蓬勃向上的学术气氛,增强了凝聚力、稳定了教师队伍。

(4)团队的科学研究为提高学生的科研素质提供了条件。学生有机会直接参加教师的科研项目,在科研中不仅能加深对理论知识的理解,而且在动手能力、科学思维等诸多方面也得到了锻炼和提高,为进一步学习和研究奠定了良好的基础。

在保证教学的前提下,我们通过加强科学研究,提高了教师的业务素质,稳定了教师队伍,改善了教学和科研条件,营造了良好的学术氛围。不仅为学生提供了较好的学习条件,同时还给学校的科学研究和教学工作注入了新的活力。

5.发挥学科优势,形成教育资源特色

本科教育与学科建设同步发展,学科建设带动本科教育上层次。地质学系拥有地质学一级学科国家重点学科和矿物岩石学长江学者特聘教授岗位,有一批著名的学者,他们工作在科研第一线,活跃在本学科的国际舞台上。发挥学科优势,设计学科研究教学板块,开设紧密结合国际学科发展前沿内容的课程,使本团队课程教学具有前沿性和国际性。我们以国家重点学科为依托,以地质学系 70 年科研积累为基础,逐步形成了融入中国西部地质特色的教育资源。

6.突出专业特点,形成实践特色

实践教学是地质科学人才培养环节中不可替代的重要环节。我们在实践教学环节中特别注重加强新技术、新方法对原有体系的改造,突出实践教学的综合性与创新性。以提高学生的动手能力、分析能力、提取信息能力和创新能力为目标,课程实习实验教学中所使用的标本、薄片均充分融入了中国西部地质元素,使实践教学从理念到形式均有其自身的特色。

二、实施课程教学体系的创新性改革

1.课程体系改革

根据加强基础、拓宽专业、增强适应性的人才培养方针及国家理科基地培养学生有较强创新能力的总要求,近年来我们在原有基础上,对课程体系和教学内容进行了较大幅度的调整。具体做法:①将岩石学与晶体光学及光性矿物学作为一个统一的整体对待,强调课程设置和教学内容的统一性与协调性,将光性矿物学作为岩石学的前期基础,由岩石学任课教师提出具体要求,并在共同讨论的基础上,集体修编了晶体光学及光性矿物学教学大纲,使得其教学内容更加适合当代岩石学新发展的需要。②更加强调三大岩类的内在联系及协调一致,大幅度压缩重复内容,在体现各岩类自身独特性的同时,切实注意三大岩类的内在联系及相互衔接。

2.教学方法和考核方式的全面革新

教学方法的改革已经刻不容缓,其基本思路是让学生变被动为主动,减少课堂讲授学时,增加讨论课和实践课,使课堂教学方式多元化。使用现代化教学手段,增加课堂信息量。积极鼓励学生直接参加部分教学活动等。针对传统教学过程中普遍存在的教学内容单一、考核方式简单、教学内容单调、教学过程呆板、缺乏师生交流等不利于调动学生积极性和创新思维等弊端,我们提出如下教学方法和考核方式的系统改革新方案:①分块授课和阶段考核的教学方式。在教学过程中,根据课程特点,按学科最新进展与成果归类划分出相对独立的次级"小学科"。将教材内容与学科新成果相结合,着重对基本理论、研究方法、科学思维、学科进展和当前存在问题等方面的讲授。同时推荐最新参考文献供学生课后阅读,并以思考题的方式启发和鼓励学生课后查找和阅读文献,获取新知识。②将专题讨论作为阶段考核的重要形式,激发学生学习的积极性。根据教学内容自然分出的"小学科",组织学生进行专题讨论。既做到专题与课程的紧密结合,又给学生一定空间允许他们独立思考并设定与课程内容相差不大的专题,将专题讨论作为平时考核的一种重要形式纳入整个教学过程,使学生从被动接受转向主动学习,达到从"要我学"向"我要学""我愿学"和"我会学"转变。③开展实用技能锻炼,提高学生通过多种形式获取知识的能力。课外查阅科技文献,不仅是对课堂教学内容的补充,也是对学生实用技能的有效训练,而这也是以往教学中只重视读书不重视能力培养的一个软肋。④综合训练,提高学生分析问题和解决问题的能力。培养学生综合分析能力是课程改革的最终目的。通过平时专题报告编写训练,对学生提出更高的标准,课程结束后的期终考核以一定形式的综合性论文为主,集中体现学生综合分析和解决问题的能力,使学生得到一次较全面的锻炼。总之,教学方法和考核方式的系统改革,将学生从一个知识的被动接受者转变成主动的探索者。

3.加强实验室建设

晶体光学与岩石学是实践性很强的学科,其40%~50%的学时是在实验室完成的。因此,实验室建设对于保证课程的教学质量至关重要。自2005年起,我们利用国家理科基地建设经费以及"211工程"建设经费,对晶体光学与岩石学实验室进行了全面更新建设,使得实验室硬件条件已居国内领先地位,为确保教学质量提供了物质基础。实验室建设的另一个极为重要的方面就是典型岩石标本库和薄片库的建立。数十年来,教师投入了巨大的精力,在标本库及薄片库建设方面做出了重大贡献,岩石学标本库中至今还陈列有20世纪初由张伯声等老一辈地质学家采集的典型标本。自1993年国家理科基地建立后,我们在这方面更加大了力度,并取得了重大进展。近年来,团队教师曾数次召开全体会议,布置落实标本、薄片库建设的具体方案和措施,先后投入大量资金,专程派出教师赴典型岩区采集岩石、矿物标本,以及向国内标本厂订购成套的标本和薄片,特别强调教师在承担科研项目和出国考察进修期间注意收集典型岩石、矿物标本,并建设了岩矿陈列室,从而为全面推进实验室建设和提高教学质量提供了物质基础和保障。

4.强调学生专业素质培养

为了加强本科生的专业素质教育,培养学生的创新精神和科研能力,我们有意识地在教学过程中加入了一些研究性的、讨论性的内容,引导学生从书本走向实践,提高独立思考问题、解决问题的能力，激发学生的创新意识。在教学过程中注重素质培养与业务教育相结合，利用优秀教师的亲身经历循循善诱，以教师本身在学术界所取得的成绩和为祖国科学事业努力拼搏的奋斗精神深深地打动学生，也使学生了解和学习了其他学科的知识，在专业教育中起到了积极推动作用。

三、强化梯队建设，打造一流师资队伍

以学科和科研为纽带的团队建设，是我们师资队伍建设的一大特色。提倡合作共事，凝聚学科方向，整合实力，组成科研配套、优势互补、梯队结构合理、富有朝气与活力的教学团队。团队中青年学者初显才华，中年骨干勇挑重担，老一辈科学家运筹帷幄，老中青学者共同营造出人际关系宽松和谐、学术创新氛围浓厚的研究与教学环境，使得团结、合作、严谨、求实的精神与传统得以延续和传承。

总之，10 多年来，我们在创新型教学团队以及晶体光学与岩石学课程群的教学内容、教学手段、实验室建设、师资队伍建设等方面进行了一系列探索和改革，晶体光学与岩石学教学团队于 2009 年被评为国家级教学团队。放眼未来，任重而道远，如何进一步加强教学团队建设，深入推进课程体系改革，真正适应当代科学技术的发展趋势，仍是摆在我们面前的艰巨任务。

国家理科地质学人才基地研究性教学改革的探索与实践[①②]

赖绍聪 张云翔 周鼎武 张成立 张复新

摘要:文章系统介绍了西北大学地质学系国家基础科学人才基地研究性教学改革的思想、理念、具体措施、实施方案及初步实施效果,以及研究性教学过程对创新型人才培养的促进作用。

随着我国高等教育事业的不断发展,国家对研究型人才的培养越来越重视,特别是“国家理科基础科学研究和教学人才培养基地”的建设,教育部和地方政府斥巨资重点共建各理科人才培养基地,力争使其成为我国研究型、创新型人才培养的典范[1]。理科人才培养基地要给本科生提供更多的科研训练机会,以期学生尽早进入科研领域,接触学科前沿,了解学科发展动态,培养学生的科研创新能力[2]。为了进一步加强基地学生实践能力、科学研究能力和创造性思维能力的培养,我们必须对研究性教学改革的规律进行努力探索[3]。我们的思路是,通过研究性教学改革的探索与实践,强化基地班学生的开拓精神和创新意识,培养他们的创新思维、创业精神和实践能力,使其尽早地参与科学研究、技术开发和社会实践等课外学术科技活动,并得到基本的科研训练,在此过程中探索21世纪创新型、研究型精英人才培养模式与规律。

一、基地学生科研课题的设立和实施措施

为了加强基地学生科研能力的培训,地质学系从“国家理科基础科学研究和教学人才培养基金”中单列6%的经费,设立地质学基地学生创新课题研究(该措施完全符合国家自然科学基金委关于基地人才培养基金使用的规定)。我们设立创新基金的宗旨:强化基地班学生的开拓精神和创新意识,培养他们的创新思维、创业精神和实践能力,使其尽早地参与科学研究、技术开发和社会实践等课外学术科技活动,并得到基本的科研训练。创新基金的资助原则为“理实结合、突出重点、鼓励创新、注重实效”,资助办法为“自主申请、公平立项、择优资助、规范管理”。为此,我们成立了基地班学生创新基金管理领

① 原载于《高等理科教育》,2008(1)。

② “教育部高等学校优秀青年教师教学科研奖励计划”资助;“第三轮陕西高等教育教学改革研究项目”资助;国家基础科学人才培养基金“创新型地学人才培养机制中教学方法改革的研究与实践”项目(J0105)资助。

导小组，对基金的重要事项和基金项目资助经费进行管理。领导小组由相关系领导、专家学者和教学秘书组成。主管基地建设工作的系主任兼任领导小组组长。领导小组的具体职责：审定创新基金评审专家组人员；批准各类项目的资助金额；研究决定基金实施中的重要问题。教学秘书负责基金项目的组织、实施和日常管理工作。创新基金项目资助经费是指基金直接用于资助基地班学生科学研究项目的经费。创新基金重点资助学术思想新颖、目标明确，具有创新性和探索性、研究方案可行的项目，资助范围包括：①专业性研究及创新项目；②实践教学中的综合性、设计性和创新性实验项目；③其他有研究与实践价值的项目。

符合条件的在校二、三年级基地班学生均可向基金管理领导小组提出项目资助申请，并按规定如实填写《地质学基地学生创新基金申请书》，项目资助申请经地质学系基金专家组初评并签署意见后报基金管理领导小组。基金管理领导小组对专家组初评并建议立项的申请项目进行评审，确定并下达项目资助额度。每项获准资助项目的资助额度一般为 0.5 万~0.7 万元。

项目执行过半，项目主持人向基金管理领导小组提交项目研究进展报告，接受中期检查。中期检查合格者，拨付剩余经费；中期检查不合格者，终止项目资助。

在基地学生本科毕业前 2 个月（每年 5 月中旬），实施创新基金项目结题验收工作，各项目提交项目结题报告，由 5~7 位教授组成的专家组听取各项目 15 分钟的多媒体汇报，并对项目完成情况给出评价意见。2003 年度，地质学系批准创新基金立项 21 项，总资助金额达到 10.25 万元，研究领域涉及岩石学、矿物学、矿床学、地球化学、古生物学、古生态学、石油地质学、环境地质与灾害地质学、工程地质学等 10 余个学科领域，并以基地为龙头，逐步向非基地专业辐射。2004 年度，基地班立项 13 项，资源勘查工程专业获准 2 项，勘查技术与工程专业获准 2 项，投入总经费 8.8 万元。2005 年度，批准创新基金立项 18 项，总资助金额达到 10.50 万元。

目前，2003 年度创新基金项目已全部结题，提交学术论文 42 篇，其中 28 篇已分别发表于《岩石学报》《地质通报》《地球科学与环境学报》《西北大学学报》等学术期刊。另有 3 项成果分别获得西北大学“挑战杯”特等奖和二等奖。可见，创新基金计划的实施成效十分显著，对提高基地学生的创新能力以及实现本科生—硕士研究生的贯通培养有极大的推动作用。

二、基地学生野外实践教学过程中研究性教学改革的实施措施

1.三年级科学研究实习实施措施

三年级科学研究实习是激发本科期间学生创新性的重要教学环节。该项教学工作是在学生经过一、二、三年级不同阶段分别以认识、方法和综合为重点的基础训练，并具备了初步认识、观察、分析和研究客观地质事物基本能力的基础上，对学生进行综合运用基本知识和技能，独立研究和解决实际问题，实现从选题到实践、实验（测试）再到总结、提高、理论化的完整科研训练过程，以培养和掌握从事科学技术研究的基本能力和方

法[4]。三年级研究性实践教学依照学生的兴趣和爱好,以教师与学生双向选择的方式参与各专业方向教师的科研项目,或申请本科生创新基金项目,完成研究性学习。

实践的方式是以学生承担的科研小课题(创新基金项目)或参与指导教师的科研工作,亦可师生共同研究,设定研究专题。研究过程是实施研究实习教学的重要环节之一,该环节既体现在野外实践的过程中,又体现在野外实践基础上的室内研究中,包括不同目的、要求下的样品处理、观察、测试和实验;数据和参数的整理、处理和分析;图件的制作和分析等。

在室内研究工作中,系里全部教学、科研实验室均为学生提供全面的服务,并由学生实际操作完成,从而使学生在研究工作中训练了独立的动手能力和思考能力。

2.四年级鄂尔多斯–秦岭地质大剖面实习研究性教学改革实施措施

四年级以鄂尔多斯盆地–秦岭造山带地质走廊为教学基地,在对不同构造单元、不同时代、不同类型和不同层次地层系统和构造变形为主的实际观察、研究和分析基础上,进行点、线、面的同时异岩、异相、异构的对比分析,构建区域时空地层–构造格,研究地质事件、分析区域地质结构构造特征,探讨区域地质演化过程。以此训练学生的基本素质和技能,培养多学科交叉观察、分析、解决问题,特别是综合分析问题的能力,激发他们的创新热情。

秦岭及邻区是国内外关注的大陆动力学研究、解剖的重要地区,也是西北大学几代学人长期致力的教学科研基地,具有丰厚的研究积累和跟踪国际地球科学前缘的理论与方法的实践探索,并具有与时俱进的持续研究和探索价值。野外教学基地建设遵循由易到难、由简单到复杂的教学规律,进行了点、线、面相结合的基地建设。特别是鄂尔多斯盆地–秦岭造山带地质走廊教学基地建设,本课程建立了以主干剖面地质观察研究为主,辅助剖面和典型区段地质解析为辅的点—线—面相结合的区域调研的野外地质综合教学方案,并针对不同剖面的地质特点实施系统的重新地质调研。对主干、辅助剖面和典型解剖区段的选择,兼顾了区域的贯通性,地质现象的代表性、典型性和可对比性,以及地壳结构的层次性。

三、重要方向课程和特色选修课程教学过程中,研究性教学改革的实施措施

在课程研究性教学改革过程中,我们注重改变以往以验证为目的的课程教学内容,培养学生全新的地学观及综合分析问题与创新能力,加强新思维、新技术和新方法在课程教学中的应用,建立特色鲜明、科学合理、循序渐进的课程教学新体系,全面体现研究性教学课程的设计性、综合性及创新性。由于不同课程内容存在的差异,我们以课程的性质为依据,建立了具有特色的研究性课程教学模式。

1.古生物学课程研究性教学改革的实施措施

在古生物学课程研究性教学改革过程中,我们普遍结合课程内容进行综合教学。通过资料阅读,由教师引导,学生自行提出问题、自行设计解决问题的技术途径,达到提高

学生分析问题与解决问题能力的目的,完成自我设计、自我解决的综合性、设计性、研究性教学过程[5]。在课程教学过程中安排科研性质的教学实践,为学生提供基本地质素材(包括未定名化石标本及化石产地地质剖面图),学生综合运用所学知识,采用多种方法对所提供的未定名化石标本进行鉴定。在标本鉴定的基础上,对所提供的地质剖面进行分析,最终提供化石鉴定报告和地质分析报告。提出古生物学新近研究的热点问题进行讨论,要求内容基本正确,有新意,参考资料丰富,最终以课程论文的形式完成研究性教学的考核。

2.环境地球化学课程研究性教学改革的实施措施

根据环境地球化学课程的特点,以及考虑到它是一门完成了地质学主要基础课程之后为高年级学生开设的指选课,我们将教学方向重点放在研讨、综合分析和实际动手的互动交流能力的培养上。如鉴于近年来出现的大量新成果以及信息技术的广泛应用,课程制定激发学生主观能动地利用各种资源的阶段式考核,将教师命题和学生根据学科发展自己提出问题相结合的方法设立专题,给学生更多的空间在课外利用各种渠道收集与专题相关的资料,并严格按照科技论文的形式撰写各种形式的读书报告。这样做,不仅可以培养学生独立收集阅读文献和综合分析资料的能力,而且还能锻炼学生科学研究的方法和技能。另外,改革传统教学考核方式,改进教学方法,采取课堂讲解与讨论的互动教学方法,给学生创造上讲台的机会,以小组为单位对大家关心的问题和当前本学科的研究热点进行讲解答辩,锻炼学生的表达和辩解能力。此外,本课程还注重开发学生实际科学研究的动手和应用能力,并在 2 个方面做了一些有益的尝试:一是进行模拟式科研训练,主要是通过收集前人对特定地区研究所得的实际资料,去除其论证和观点,给出地质背景和分析数据等,让学生根据自己的知识撰写科研论文;二是野外观察和采样以及实验室实际操作的研究过程培养,具体做法是结合本课程设计的课间野外现场教学,让学生取得第一手实际资料,并在室内研究的基础上进入学校大陆动力学国家重点实验室开展各种样品处理和测试工作。这一工作的开展基本是以小组为单位并与四年级学生毕业论文工作相结合,成果突出者,鼓励在正式刊物上发表。

通过多种技能训练的研究性授课方式,使学生在有限授课学时内最大限度地掌握环境地球化学的学科体系、新知识、新内容和发展趋势,同时培养学生独立思考和综合分析、解决问题的能力,学会解决实际问题的科研方法和基本技能,满足创新性人才培养的目的。

四、研究性教学改革的实践意义

研究性教学改革的核心主要体现在教育教学过程的设计性、综合性及创新性上,使教学内容实现了一系列转变。这对于培养学生创新意识、提高学生科研素质将起到实质性的推动作用。这种开拓性的努力和创新性实践必将成为深化教学改革、不断加强教学建设、促进教学建设和改革实践的“排头兵”和“示范田”。研究性教学改革的实施,一方面使学生的科研素质和能力得到了锻炼和提高,学生的学习由被动变为主动,提高了多方面的能力,开阔了视野,完善了自己的知识结构,意志力得到了磨炼。同时,使大学生有更多的时间和机会与导师、博士研究生、硕士研究生接触,在这样充满学术氛围的研究集体中,大学生的个性品质得到了锻炼,对科学研究的态度、对工作的认真踏实作风、对事业的敬业精神、对他人和社会的责任感等方面都得到了培养和提高。另一方面是教师的教育观念得到了转变。通过研究性教学改革项目的实施,学校教师的教育理念和对人才的培养目标有了新的认识,对能力培养和推进素质教育的重要性和必要性认识得到了进一步提高,"以学生为主体、教师为主导"的观念进一步增强,对完善人才培养模式和加强教学与科研相结合、推进学生科研训练、培养学生创新精神和实践动手能力更为关注。

参考文献

[1] 杨承运,张大良.地学教育总体改革研究报告[M].北京:高等教育出版社,2003:101.

[2] 杜远生,刘世勇,杨坤光,等.国家地质学理科基地创新人才培养模式[M].武汉:中国地质大学出版社,2004:94.

[3] 李昌年,杜远生,欧阳建平,等.国家理科基地地质学专业研究型人才培养模式探索及课程体系和教学内容改革[J].中国地质教育,2002(2):25-27.

[4] 张云翔,赖绍聪.国家"理科人才培养基地"的创新教育[J].高等理科教育,2004(3):9-11.

[5] 赖绍聪.谈硕士研究生学位论文的准备和设计[J].高等理科教育,2004(4):11-116.

地球科学高等教育改革与发展的若干建议①②

赖绍聪　何　翔　华　洪

摘要：随着当代地球科学的迅猛发展，21 世纪地球科学高等教育面临新的形势与挑战，地学界更需要素质高、基础扎实、知识面宽、综合能力强的创新型人才。地球科学高等教育如何适应新形势，是当前地学高等教育面临的重大课题。本文阐述了对当前地球科学高等教育改革与发展战略的思考与建议。

地球科学是探索地球起源、演化和发展趋势的科学，为人类认识自然提供基础科学知识、认知途径和研究工具，与人类社会生活和经济发展密切相关。在跨入 21 世纪的今天，人类生存和社会发展对地球科学提出了比以往任何时期都更加广阔也更加急迫的需求，地学界更需要素质高、基础扎实、知识面宽、综合能力强的创新型人才。为了适应这种新要求，我们必须对传统的地球科学高等教育格局进行全面的改革，重新构建新时期中国地球科学高等教育新体系。

一、地球科学高等教育改革的主导思想

教学发展依赖于不断的改革与建设，教学建设永远处于学校发展的优先地位。地球科学的根本任务在于认识地球，并利用这种认识保证人类生存和发展所需要的自然资源，保护和改善人类的居住环境。因此，需要加强新的地学观教育，加速对人类赖以生存的地球的发生、发展和演化趋势的认识。近代地学研究表明，全球变化、深部地质与地球动力学、地球表层各圈层系统的相互作用与联系将是未来地球科学研究的主要问题。我国是一个发展中国家，在经济快速发展进程中，面临着环境、资源、灾害、人口等问题的严峻挑战，地质学家对此负有义不容辞的历史责任。我国地球科学高等教育的核心任务是要为我国基础科学培养出一批献身科学事业，具有扎实的基础理论，宽广的知识面，合理的知识结构，较强的获取与综合运用知识能力，富有创新精神的高素质优秀人才，为我国基础科学研究人才储备力量。因此，地学类人才的基本培养目标应该是经过本科阶段的系统培养，使学生牢固树立热爱祖国、献身地质事业的坚强信念，具有扎实的基础理论知识、宽广的知识面、合理的知识结构，较强的获取知识、综合运用知识能力以及较高的综合素质，富有创新精神并具特色的优秀人才，成为我国地球科学基础科学研究顶尖人才

① 原载于《中国地质教育》，2009(4)。

② 教育部高等学校地球科学教学指导委员会“地球科学学科发展战略研究”项目资助。

的后备力量。

1.强化素质教育

全面推进素质教育,以培养学生的创新精神和实践能力为重点,加强和改进德育教育,深化课程改革和考试评价制度改革,切实提高学生的身体健康水平、心理素质和艺术素养,努力培养德智体美等全面发展的一代新人。

2.建立教学质量监督保障体系,以政策法规保障教学工作

巩固和提高教学质量,关系到高等教育发展的全局,必须长期坚持狠抓教学改革和教学质量评估保障机制等环节,建立较为完善的教学质量监督保障体系。在教学管理方面进一步实施教学管理制度创新。建立富有特色的教学质量标准和有效的教学质量监督机制,利用网络技术,达到社会—管理层—教师—学生多层面的管理模式,形成完善的自我约束机制。

3.推进教育信息化建设,以信息资源服务教学工作

教育信息化是我国教育实现现代化和适度超前发展的重要途径,我们在教育信息化的基础设施、资源建设、人才培养以及教育系统应用水平等方面需做出更大的努力。

4.建设高素质教师和管理队伍

推进教育创新和培育人才离不开教师的辛勤工作,也取决于教师队伍的素质。以深化教师聘任制度改革为重点,加强政策调控,科学设置机构和岗位,实行教师资格制度,实行“资格准入、竞争上岗、全员聘任”,完善教师岗前培训、在岗学习及国内外进修体系。

5.以教学、科研一体化为主导思想,建立一流的教学实验室

在实验室建设过程中,以教学、科研一体化为主导思想,坚持培养高质量创新型人才,不断提高教学质量,是今后地球科学高等教育发展的生命线。应牢固树立实验室、学科建设同步发展的思路,带动实验室建设上档次。逐步实现学生在教师指导下进入实验室进行测试,为学生直接参加科研训练、提高科研素质提供条件,这对促进本科教学质量和培养创新意识将起到重要作用。

6.以专业规范引导教学工作,以教改项目支持教学工作

随着教学改革的深入,深层次的问题需要进一步解决。理论上,应对地球科学创新性基础研究人才培养的思路、方案和模式等进行较为深入的探讨,形成较为先进而完整的体系。在实践中,逐步形成服务于学生全面发展和个性发展的思路,形成科学的培养创新人才的课程体系、实践教学体系并及时实施。配套建设一批先进的专业实验室,使高校能够更好地培养出高水平的基础科学研究人才。

7.以基地建设促进教学工作

理科地质学人才培养基地建设对于我国重点高校地质系来说,是十分难得的发展机遇,一开始就应紧抓不放,将各建设项目落实到人,一抓到底。

总之,应该继续在改善办学条件,优化师资结构,加大教改力度,树立良好教风、学风,强化素质教育,突出办学特色,培养优秀地学人才上下功夫。经过若干年的建设,在我国地球科学基础理论研究人才培养的思路、方案及模式等方面,形成先进而完整的体系,使我国地质院校(系)成为具有先进的教学设备、完善的教学管理体制和合理的课程体系,拥有一支高水平的师资队伍,能够培养适应21世纪国家发展需求的优秀人才,具有中国特色的、名副其实的中国地球科学人才培养园地。

二、地球科学高等教育的课程体系改革

近年来,地学的发展进入了一个大变革时期,研究的重点发生了重大转折。据对美国Georef收录的全世界文献统计,地球科学单学科的纵向深入总体上已走向衰落,地球科学已由单学科的纵向深入向横向交叉和渗透的方向发展,代表当代地球科学重大前沿的一批交叉学科、横断学科和综合学科将主宰未来地球科学的发展,并使地球科学朝着整体化方向迈进。近10年来,地球科学的前沿研究由立足于学科本身的科学问题转移到多学科之间共同关注的科学问题。在社会可持续性发展中,环境、资源、灾害、人口等问题日益严峻,许多重大问题亟待地球科学家去解决。地学界更需要的是素质高、基础扎实、知识面宽、综合能力强的基础研究人才。过去的课程体系和教学内容基本上是按照学科和专业的人才培养模式制定的,已不能满足新世纪人才培养的需要,必须改革。在教学改革中必须处理好继承与发扬、"老"知识与"新"知识、素质教育与业务教育、共性与个性培养的关系。应突出办学特色,特色应该与我国社会发展相适应,与科研优势、师资优势和地域优势相结合。

课程体系是实现人才培养的重要方面。坚持加强基础,体现专业特色,放宽专业选修课,突出地球、资源和环境的整体协调性,重视实践课,以素质教育为主,培养德、智、体、能全面发展的优秀人才的总原则。经过对教学计划的修订,应该在总课时大幅度压缩的同时,加大选修课的学时,尽量发挥综合性大学的优势,给学生发展提供较大的自主空间,以利于学生个性的发展。为了培养学生的科研素质、扩大知识面,应加大力度开设科学系列讲座,并在管理方面与此相配合。

改革陈旧的教学内容是教学改革的中心环节。教学内容改革应按照现代和未来地球科学的发展对教学内容进行重新组合、分配和取舍。在课程设置中应加强野外地质实习,并适当加强双语课程的建设。

教学方法的改革已经刻不容缓,其基本思路是让学生变被动为主动,注重学生能力的培养。结合学科优势,增加课堂讨论和实验课时,使课堂教学方式多元化。鼓励学生直接参加部分教学活动,参加多媒体教学课件的制作,从而加深对课程的理解。野外教学可考虑采用综合性的集体实习。实习过程中坚持以基本技能训练入手,在地质现象典型的观察点,让学生自己观察、描述地质现象;在构造复杂的地点学生之间或与教师一起讨论研究;在一些重要区段教师只负责介绍,由学生寻找证据,从而使学生变被动为主动。培养学生灵活运用所学知识进行综合分析的能力。通过这种方式的实习,使学生学会野外素材的收集、整理、分析和研究的方法。

三、地球科学高等教育改革的实验室建设与实践教学

在专业实验室建设过程中,应该倡导"教学、科研一体化"的主导思想,使专业实验室

不仅能够全方位服务于本科及研究生教学工作,而且还能够在较高层次进行科学研究工作,并在科学研究工作中不断积累资料和经验,逐步改进实验方法、提高实验技术,从而大大提高专业实验室的实验水平和档次,以最直接和最有效的方式将科研成果实时转化为教育资源,这对于迅速提升学生的创新能力、科研能力、分析问题和解决问题的能力将起到重要的促进作用。高水平科研资源向教学资源的转化,奠定了教学的高起点和高标准。激发创新意识,训练创新能力,达到培养学生观察事物、思考问题、自我设计、研究与解决问题的素质,形成实践教学从理念到形式的全面改革。

实践教学是地质科学人才培养环节中不可替代的重要环节,也是实现创新人才培养目标的有效途径。原有实践教学体系突出认识与方法的锻炼,具有明显的单科性和验证性,启发学生自主思维的作用不足。针对地质学实践教学中存在的问题,目前的教学计划在实践教学环节中应加强新技术、新方法对原有体系的改造,将培养学生提出问题、分析问题及解决问题的能力,把培养创新思维放在首要位置,突出实践教学的综合性与创新性,打破课堂教学分门别类、自成体系、单课独进的教学过程,使不同课程内容互相交融。

在实践教学环节的改革过程中,应大力加强实践教学方法和实践教学手段的改革,增大综合性和设计性实验教学内容。把科学研究实质性地纳入教学环节。以教学、科研一体化为主导思想,改造现有实验室,促进高水平科研成果向教育资源的实时转化,提高实验室的实验技术、实验水平和实验档次,使实验室在提高教学质量、培育创新型基础科学研究型人才中起到关键性的推动作用。

地质学实践教学体系主要包括室内课程实习、实验和野外综合性实习、实验两部分。近年来由于新技术、新手段的广泛使用,地质科学逐步从传统的地球表层地质和定性研究发展到深部地质和定量研究。建立相应新实验体系以适应对学科创新性的需要是当务之急。因此,实验课教学体系应向跟踪学科发展、具有设计性和综合性的方向发展,使学生的学习由被动变为主动。这就要求学生不但具有扎实的学科基础,还要有多学科知识运用的综合能力。如古生物学和地史学实习,以前以认识化石的形态为主,记忆其时代与分类位置;地史学实习则以各时代地层标本与古生物组合为主。现在的实习,一方面应突出重点,精简原来过于烦琐的记忆性内容;另一方面提出了一些学科发展中具有代表性的问题及相应的参考文献,通过学生的阅读,写出学科发展综述及自己对这些问题的认识。在标本的实习中,给出一些未定名化石和其产出地的地层剖面,学生自己鉴定,在分类的基础上,进而进行地层的划分和对比,完成地史发展的分析。如此一来,极大地锻炼学生的分析能力和综合能力,收到了事半功倍的效果。

对野外实习教学给予了足够的重视,提出了更高的要求。野外教学过程应坚持自始至终贯穿以学生为主体,以培养学生能力、激发主动性和独立思考、创造性思维为目的的启发式教学。野外实习之后的室内总结是培养学生对野外工作全面整理、归纳、综合分析能力的过程,也是对区域地质时空演变的动态思考分析和理性升华能力的培养过程,室内总结除全面整理、归纳野外观察、研究资料外,还应该要求学生收集区域地球物理、地球化学等资料,以弥补野外工作的不足,最后以完成地质报告为终结。这一过程对学生综合能力的提高有着极大的促进作用。

在实习过程中,还应十分注重素质培养与业务教育的结合,把树立良好的思想作风、实事求是的工作态度、吃苦耐劳和为地质科学事业献身的精神作为重要内容。

四、地球科学高等教育改革的教学质量保障体系

高等教育的核心是创新教育,教学环节是创新教育的重要过程。教师教学思想的发散性和前沿性决定了对其管理不能管得过死,要鼓励教师将科研资源向教育资源转化,为教师发挥自身特色和优势留下充分的空间。但教学过程也不能完全放任自流,要根据国家的大政方针和专业人才培养规律来规范教学过程,约束教学行为。

为了规范教学管理,应对课程教学实行三段式教学管理模式,即教前管理(大纲听证、大纲上网、教学纪律上网)、教中管理(教案公开、教学督导、听课检查)和教后管理(教考分离、教学质量评估),加强过程管理等,通过这些措施的实施,进一步加强教学质量的管理和监控,逐步形成社会—管理层—教师—学生多层面的教学质量监控体系和自我约束、自我发展的教学管理机制。

1.教学大纲听证与评审

编写一个好的教学大纲,不仅是教师讲好课的重要基础,也是教学内容改革的重要标志。为做好教学内容和课程体系改革,应在对教学大纲进行修订的基础上,对课程的教学大纲进行公开讲评。这对深化教学改革、提高教学质量将起到明显的促进作用。

2.公开制

地质院校(系)应利用网络技术,将主干课程的教学大纲、多媒体课件、电子讲稿、教案纳入网络,上网公布,接受全社会的监督。广泛听取兄弟院校(系)乃至全社会的批评、建议及各种反馈意见,从而有效地保证教学质量的不断提高。

3.教学督导制

督导的重点应放在“导”字上,请督导员做青年教师的知心朋友,通过督导使青年教师尽快成长起来。

4.教考分离制

教师管教不管考,由教研室组织相关教师按照教学大纲的要求命题,在考试过程中由教学管理人员随机抽取一份试卷实施考试。这一制度对于提高教师的责任心和学生的学习积极性将起到重要作用。

5.听课制

全面推行系务会人员、教学督导员、教研室主任、课程负责人听课与评估制度。

6.教学质量评估制

所有课程均需接受校、系两级的随机抽查和评估。在具体操作过程中,主要由教研室或教学督导组组织听课、学生座谈、收集学生评教意见表等,进而对课程教学质量给予评价。对于评估结果为一般者,予以通报批评并限期整改;评估结果不合格者,取消其任

课资格。

7.试卷抽查制

每学期考试结束后，由主管教学工作的领导从各门课程的试卷中随机抽取若干份，分别交给相关领域的非本学期本门课程任课教师进行重新评阅，从而在一定程度上杜绝试卷评阅过程中的人情分、印象分等种种弊端，保证试卷评阅的严肃性和公正性。

8.教学质量一票否决

建立科学、有效的教学管理，制定教学秩序等教学管理细则，实施教学质量一票否决制。教学工作的好坏是衡量教师工作的主要标准，在职务聘任、岗位聘任及年终考核中实行教学考核一票否决制，这将对教学质量起到有效的保障作用。

五、地球科学高等教育改革应突出办学特色

1.突出理想教育，形成就业特色

对学生进行献身地质事业、树立崇高理想的教育，是我们在地学人才培养过程中应该坚持不渝的工作。地质行业是一个艰苦的行业，在市场经济的冲击下，世俗的偏见和行业本身的问题使得学生专业思想普遍不够稳定。如果地质专业的毕业生大部分不从事专业工作，不能不认为是地质教育的一大损失。因此，学生进校的第一节课，就应该是理想信念教育，把教育学生树立远大理想，热爱社会主义祖国，热爱地质专业，献身地质事业，报效祖国作为素质教育的重中之重一抓到底。把教书育人作为每个教师不可推卸的神圣职责。在教学工作中要坚持以学风建设为中心的工作思路，营造出一种蓬勃向上、积极进取的氛围，这些做法对学生将产生潜移默化的教育作用。

2.结合区域特点，形成具有特色的课程体系

发挥区域优势、突出特色教育是我国地质学专业教育行之有效的方法。中国地域广袤，资源丰富，是21世纪世界经济快速增长的热点地区，是社会可持续发展最重要的资源储备和生存发展空间。中国地质，尤其是青藏高原构造、环境以及改造型盆地、陆相生油盆地等领域在全球具有鲜明的独特性和典型性，在解决全球大陆动力学、中新生代环境演变、寒武纪生命大爆发和复杂油气勘探技术与方法等方面具有重要的科学意义和显著的学术地位。特有的区域条件形成了我国造山带与盆地、古生物、陆相石油与天然气等学科的优势与特色，凝聚着几代中国地质人的心血。这些成果以不同形式融入了新的教学计划，将形成鲜明的教学特色。特别是广袤的西部大地是我国今后开发的重中之重，保持国民经济可持续发展为我们的研究提出了新的课题，开发西部需要大量的合格地学人才，研究西北、教授西北，为开发西北培养所需人才是我们义不容辞的责任，各地质院校（系）在条件允许的前提下，在地质学专业教育和人才培养过程中应充分体现这种优势与特色。

3.发挥学科优势，形成教育资源特色

本科教育与学科建设应该同步发展，各高校应大力提倡科研资源向教育资源转化，带动本科教育建设上档次。全国各大地质院校（系）有系列国家级重点学科和省级重点

学科,在这些学科中有一批著名的学者,他们工作在科研第一线,活跃在本学科的国际舞台。发挥学科优势,设计学科研究教学板块,开设紧密结合国际学科发展前沿内容的课程,使基础科学教育具有前沿性和国际性。这些优质的教育资源对人才培养质量无疑将发挥重要作用。

4.突出专业特点,形成实践特色

实践教学是地质科学人才培养环节中不可替代的重要环节,各院校(系)在地质学实践教学环节中应加强新技术、新方法对原有体系的改造,突出实践教学的综合性与创新性。在新地质学实践教学体系实践中,以提高学生的动手能力、分析能力、提取信息能力和创新能力为目标,将地质学科研和生产实践中的新理论、新技术、新方法引入实践教学。使实践教学实现从理念到形式的转变,即从验证性到设计性的转变;从单科性到综合性的转变;从认识性、继承性到研究性、创新性的转变;从传统地质教学到利用高科技改造传统专业的转变;从灌输式到启发式、讨论式实践教学的转变。改变灌输—认知—验证的教学模式,贯穿以学生为主体,激发学生主动性和独立思考、创造性思维为目的启发式教学,实施提问—观察与思考—讨论—总结的教学方法。特别是通过跨单元区域地质、多学科交叉综合分析研究的综合教学训练,形成实践教学中的鲜明特色。

参考文献

[1] 杨承运,张大良.地学教育总体改革研究报告[M].北京:高等教育出版社,2003.

[2] 杜远生,刘世勇,杨坤光,等.国家地质学理科基地创新人才培养模式[M].武汉:中国地质大学出版社,2004.

[3] 中华人民共和国教育部.高等学校中长期科学和技术发展规划(2006—2020)[M].北京:清华大学出版社,2005.

[4] 国家自然科学基金委地球科学部.21 世纪初地球科学战略重点[M].北京:中国科学技术出版社,2002.

[5] 毕孔彰,胡轩魁.关于地学教育的思考和建议[J].中国地质教育,2002(2):1-3.

[6] 刘瑞珣.回顾地质事业的发展,思考理科地质教育改革[J].中国地质教育,2002(2):4-5.

[7] 李昌年,杜远生,欧阳建平,等.国家理科基地地质学专业研究型人才培养模式探索及课程体系和教学内容改革[J].中国地质教育,2002(2):25-27.

[8] 王德滋,赵连泽.关于地球科学人才培养的实践与思考[J].中国地质教育,2002(1):9-13.

[9] 徐建平.关于提高人才培养质量问题的思考[J].中国地质教育,2002(1):30-32.

[10] 曾广策.俄罗斯地质类某些专业的课程体系及其对我国同类课程教学的启示[J].中国地质教育,2002(2):74-79.

[11] 毕孔彰,胡轩魁.高等教育与国土资源可持续发展[J].中国地质教育,2002(4):9-15.

[12] 于在平,张云翔,刘永昌,等.以学科建设和教学改革促进地质学基地建设[J].高等理科教育,2004(1):18-20.

[13] 李正.若干高等教育质量观述评[J].高等理科教育,2004(1):35-39.

[14] 徐向艺.高等学校教学质量管理的范畴、原则与体系[J].高等理科教育,2004(1):40-45.

[15] 房力明,梁方君.我国办学体制改革发展历程综述[J].高等理科教育,2004(2):4-10.

[16] 孔锐,黄启,王兰兰.高等教育过程中质量管理的途径分析[J].高等理科教育,2004(3):41-43.

[17] 张云翔,赖绍聪.国家“理科人才培养基地”的创新教育[J].高等理科教育,2004(3):9-11.

[18] 赖绍聪.“岩石学”系列课程建设的改革与探索[J].高等理科教育,2004(3):58-60.

[19] 赖绍聪.谈硕士研究生学位论文的准备和设计[J].高等理科教育,2004(4):113-116.

[20] 赖绍聪.重视国家基础科学人才基地高年级学生文献阅读能力的培养[J].高等理科教育,2004(5):33-34.

[21] 王子贤,王恒礼.简明地质学史[M].郑州:河南科学技术出版社,1985.

本科生科研能力训练的问题与思考①②

赖绍聪 何 翔 华 洪

摘要：将科学研究实质性地纳入教学过程，实现优势科研成果和科研资源向教学的转化，达到高等学校教学与科研的完美有机结合，实现高素质创新型人才培养。本文初步讨论了本科生科研能力培养的现状、理念以及本科生科研训练计划对高素质创新型人才培养的促进作用。

随着我国高等教育事业的不断发展，国家对创新型人才的培养越来越重视。创新型本科生教育注重学生的探索意识、科研能力和创新意识的培养。要给本科生提供更多的训练培训机会，以期尽早进入科学领域，接触学科前沿，了解学科发展动态，培养学生的探索创新能力[1-2]。高等教育能否培养出一批真正具有探索意识、科研能力和创新精神的一流人才，将直接关系到我国人才素质教育全面推进，关系到科学技术能否在更高的层次上取得进展，乃至于祖国现代化建设的兴衰成败[3-4]。

一、本科生科研能力训练的现状

20世纪90年代初，教育部与国家基金委共同建立了"国家理科基础科学研究和教学人才培养基地"。基地建立以后，极大地推动了研究型、创新型高等教育教学的改革与发展。一批大学，尤其是一些知名大学，教学基本条件和实验设备条件得到明显改善。同时，由于科研项目和经费扩增，实验室条件大为改善，已经有一批本科生得以与研究生一道参与课题研究。开放实验室经验普遍推行，学生课外科研活动非常活跃，各校在学生科研能力训练方面已积累了一些初步经验，探索出一些初步规律[1-6]。

但是，在如何把本科人才培养成高素质的创新型人才，以及如何更为有效地实施本科生科研能力培训、提高科学素养方面，我们目前仍然没有一个成熟的、统一可行的实施模式。这是一个需要不断研讨的任务，需要我们进一步探索。这里既有思想观念和精神意识层次的熏陶，也有科技知识和实际能力层面的训练。比如，确立唯物主义世界观，反对愚昧迷信，提倡科学态度，发扬科学精神；增加当代科技知识，培养理性思维和定量分析的能力。这些对提高学生科研素质有重要意义。为此，通过什么样的教学形式与方式

---

① 原载于《中国地质教育》，2011(4)。

② 教育部地球科学教学指导委员会学科发展战略研究项目资助；第三轮陕西高等教育教学改革研究项目资助。

来达到目标,是值得思考和探讨的。是通过开设研究性课程,还是利用第二课堂——讲座、课外习作、社会实践社团活动、科研小课题的实际实施等?若要开设研究性课程,开设什么样的课,怎样开课,怎样进行研究性教学,都是需要研究的问题[7-9]。

总之,新形势给我们提出的新要求、新任务可以归结为4个字:深化、拓宽。深化,就是要进一步深入探索培养专业拔尖创新人才的规律和经验;拓宽,就是要扩展高等教育的观念,把高等学校自然科学基础教学的任务揽过来。为此,我们必须进行新的思考和探索。现在是高等教育改革和发展的大好时机,我们应当把握机遇,把高等教育促上去,为我国科学技术的跨越式发展、为本科生创新能力的提高做出应有的贡献。

二、本科学生科研能力训练的发展趋势

统计结果表明,美国3 500多所高等院校中有125所研究型大学,大约占大学总数的3%,这些研究型大学科学研究优先,研究实力强大,研究生和博士后的数量远超出其他类型的院校。研究型大学的独特之处决定了研究型大学本科生教育的特点:研究型大学本科生教育应具有一种教学与研究相结合的氛围,教师和学生都是其中的学习者和研究者,在教学过程中通过优化课程结构建立一种基于研究探索的学习模式,教师和学生的相互交流和合作促进学术和智力环境的健康发展[1-5]。

国内部分高校在培养大学生科研能力上已取得了一些初步的成功经验。近年来,清华大学、浙江大学、西北大学、南京大学等重点高校,在本科生教育中开展的“大学生科研训练计划”“学生科研能力培训课题研究计划”,经过多年的实践证明是一种非常有效的培养和提高学生科学素养与科研能力的方法及载体。出发点在于充分利用学校综合教学资源和雄厚的科研实力,给学生提供更多的科研训练空间及平台,把学生从“以教师为中心”的管理模式转化为“学生自主个性发展”的管理模式。给学生提供科研训练机会,使学生深入科研与创新过程,去了解科研的基本理论和实验技能与方法等,激发他们对科学研究的兴趣[6-9]。

目前,随着教育部“大学生创新计划”的推行与实施,国内开展学生科研训练的高校越来越多,其实施方法也大体相同。如清华大学、浙江大学最早分别于1996年、1998年就开始实施,每年一期,在教师指导下,以学生为主体开展科研训练。实施方法:学校建立组织机构,制定工作职责,成立专门的校、院系两级指导工作组,负责项目的整体规划,并制定了项目审查、验收、评价等相关标准和规范;教务处对项目实施制定有关政策、条例、管理目标、工作程序等,确保研究性教学改革的组织与实施[6-9]。

学生科研能力培训计划的实施,一方面使学生的科研素质和能力得到锻炼和提高,学生的学习由被动变为主动,提高了多方面的能力,如查阅文献资料能力、实践操作能力、学习能力、研究能力、创新能力等,开阔了视野,完善了自己的知识结构,意志力也得到了磨炼,大学生有更多的时间和机会与导师、博士研究生、硕士研究生接触。在这样充满学术氛围的研究集体中,大学生的个性品质得到了锻炼,对科学研究的态度、对工作的认真踏实作风、对事业的敬业精神、对他人和社会的责任感等方面都得到了培养和提高。

另一方面是教师的教育观念得到了转变。通过本科生科研能力培训计划的实施，学校教师的教育理念和对人才培养的目标有了新的认识，对能力培养和推进科学素质教育的重要性和必要性的认识得到进一步提高，“以学生为主体、教师为主导”的观念进一步增强，对完善人才培养模式和加强教学与科研相结合、推进学生科研训练、培养学生创新精神和实践动手能力更为关注[6-9]。

三、本科学生科研能力训练计划实施的目标定位

我国高等教育的核心是要为我国社会主义建设培养出一批献身事业、富有创新精神的高素质优秀人才，经过本科阶段的培养，使学生牢固树立热爱祖国、献身事业的坚强信念，具有扎实的基础理论知识、宽广的知识面和合理的知识结构，较强的获取知识和运用知识的能力，成为我国社会主义建设事业的中坚力量。在教学实践中如何实现这一目标，是值得深入探讨的。因此，我们应该始终把教育学生树立远大理想，热爱祖国，热爱专业，献身事业，报效祖国，不断提高创新能力作为素质教育的重中之重一抓到底。为了进一步加强本科学生实践能力、科学研究能力和创造性思维能力的培养，应将科学研究实质性地纳入教学过程，必须对其规律进行努力探索。我们认为，当前本科学生科研能力培训计划实施的目标应该是强化学生的开拓精神和创新意识，培养他们的创新思维、创业精神和实践能力，使其尽早参与科学研究、技术开发和社会实践等课外学术科技活动，得到基本的科研训练，在此过程中探索出 21 世纪创新型一流人才培养模式与规律。

四、本科学生科研能力训练计划应解决的主要问题

(1)如何正确处理科研与教学的关系，将科学研究实质性地纳入教学过程，使科研成果有效地转化为教育资源。

(2)高等教育应具有一种教学与探索相结合的氛围，教师和学生都是其中的学习者和探索者，如何在教学过程中通过优化课程结构、设立科研小课题建立一种基于研究探索的学习模式乃是亟待解决的重要问题。

(3)教师和学生的相互交流和合作将大大促进学术和智力环境的健康发展，如何有效地建立“教师—研究生—本科生”学术群体，是另一个值得探索的重要课题。

总之，我国高等教育应该对本科生能力训练给予足够的重视。目前，国内部分高校在本科生科研能力培训方面已经先行一步，并取得了一些初步的经验。但是，如何把这些规律和经验用实、用足，还有哪些新的东西需要摸索，有待我们进一步努力探索。这方面的任务将是国家赋予高等教育的中坚，我们务必切实做好，以提高中国的科学技术地位，为世界文明做出新贡献。

参考文献

[1] 杨承运，张大良.地学教育总体改革研究报告[R].北京：高等教育出版社，2003：101.
[2] 杜远生，刘世勇，杨坤光，等.国家地质学理科基地创新人才培养模式[M].武汉：中国地质大学出版

社,2004:94.

[3] 中华人民共和国教育部.高等学校中长期科学和技术发展规划(2006—2020)[M].北京:清华大学出版社,2005:156.

[4] 国家自然科学基金委地球科学部.21世纪初地球科学战略重点[M].北京:中国科学技术出版社,2002:4.

[5] 李昌年,杜远生,欧阳建平,等.国家理科基地地质学专业研究型人才培养模式探索及课程体系和教学内容改革[J].中国地质教育,2002(2):25-27.

[6] 赖绍聪,华洪,王震亮,等.研究性教学改革与创新型人才培养[J].中国大学教学,2007(8):12-14.

[7] 赖绍聪,何翔,华洪.地球科学高等教育改革与发展的若干建议[J].中国地质教育,2009(4):35-39.

[8] 赖绍聪,华洪,常江.探索创新性实践教学新模式 稳步提高教学质量:以"鄂尔多斯盆地-秦岭造山带野外地质教学基地"建设为例[C]//大学地球科学课程报告论坛组委会.大学地球科学课程报告论坛论文集.北京:高等教育出版社,2009:307-310.

[9] 赖绍聪,常江,华洪,等.理科地质学高等教育课程体系的重构与创新型人才培养[C]//大学地球科学课程报告论坛组委会.大学地球科学课程报告论坛论文集.北京:高等教育出版社,2010:141-146.

谈硕士研究生学位论文的准备和设计[①②]

赖绍聪

摘要：硕士研究生学位论文的质量乃是研究生培养质量的最直接反映。本文详细介绍了西北大学地质学系近年来在硕士研究生培养过程中，从选题、资料收集、开题报告论证、野外工作、室内分析测试、分段小结，直至论文设计过程中的一些做法和思考。

西北大学地质学系成立于1939年，60多年来为国家培养了6 000多名各类地学人才，其中包括一批高质量的硕士研究生和博士研究生。作为研究生导师，回顾几年来指导硕士研究生的经历，深感研究生教育与本科教育存在很大区别。硕士研究生学位论文也因为研究生培养目标的不同而与本科毕业论文有所不同，有其自身的特殊性。在此，作者就硕士研究生学位论文的准备和设计谈一些体会，希望能对提高地质学系研究生教育质量有所裨益。

一、论文的选题

众所周知，硕士研究生的任务不单纯是接受知识，而且要进行科研工作，逐步培养其独立工作、独立科研和分析问题、解决问题的能力。攻读硕士学位的研究生均要求在3年内相对独立地完成一篇学位论文，也就是要把自己的科研成果记录下来，并升华到一定的高度。因此，硕士研究生必须在完成课程学习的基础上选定自己的研究课题。选题工作必须在导师的直接指导下进行，选题的好坏、恰当与否将直接影响后来科研工作的顺利进行。我们认为，论文的范围不宜太大，主要是因为时间不够，更重要的是研究经费有限。一年半写一篇很大的论文，很难写好，就是勉强写出来了，也不会写得好。范围大了就很难深入。选题过程中必须考虑的另一个重要问题就是经费的来源。坦率地说，仅仅依靠有限的一点硕士研究生经费很难做好一篇高质量学位论文。因此，硕士研究生的论文题目必须尽可能地与科研项目挂钩，最好能与导师以及地质学系的有关科研项目紧密配合，作为科研项目下的一个子题目或一个部分，这样就能在研究经费上获得更大的余地，且使论文成果具有实际价值。

根据西北大学地质学系现行教学计划，硕士研究生一般是在第二学年上学期进行第

① 原载于《高等理科教育》，2004(4)。

② 教育部高等学校优秀青年教师教学科研奖励计划及教育部创建名牌课程(岩石学)计划资助。

一次野外实习。因此,选题工作应该尽量抢在野外实习之前。选好题目后,还应制定一个初步的论文纲要,对所要研究的基本内容有一个大体的安排。这样就能最大限度地避免盲目性,增强野外工作的目的性,对于广泛收集资料、野外踏勘及采集样品有重要指导意义。

二、广泛收集资料

《中华人民共和国学位条例》明确规定:硕士学位论文要反映出研究生在本门学科上已掌握坚实的理论和系统的专门知识;对所研究的课题应当有新见解;要求在导师指导下独立完成,表明作者具有从事科学研究工作或独立担负专门技术工作的能力。显然,要做到这一点并不容易。

首先,我们必须占有充分的资料,应该尽可能多地了解前人对于这一问题已发表过的意见,这些意见往往可以在一定程度上给我们一些启发,是进一步做好该领域研究工作的重要基础。他们已经取得的成果,正确的可以吸取和继承;他们走过的弯路我们可以避免和防止。只有在吃透了前人已有研究资料的基础上才有可能发现问题,找出不足,从而使我们的工作更上一层、更进一步、有所创新。正所谓"吾尝歧而望矣,不如登高之博见也"。站在巨人的肩上,才能登高远望。

其次,在阅读资料的过程中,还有一个技巧问题。现代社会正处在一个知识爆炸的时代,各种资料浩如烟海,仅仅是每年出版的期刊就有数万种之多,与地质专业有关的期刊也不下百种。如此多的材料,如果不加选择地阅读,就是花费平生精力也读不完的。因此,在收集、阅读资料的过程中,必须时时记住自己的目的和任务,时时联系自己研究的问题,有目的地阅读。重要的先读,次要的后读,无关的暂不读,这样方能取得事半功倍的效果。

三、扎实的野外工作是写好论文的基础

一方面,地质科学是一门实践性很强的学科,任何一项研究工作都脱离不了野外实践,最终的研究结果也必须回到实践中去,接受实践的检验。正如马克思主义认识论所指出的,实践—认识—再实践—再认识乃是人类认识发展的根本途径。这一规律在我们的研究工作中得到了充分体现。任何脱离野外、脱离实际、凭空想象的理论都是不切实际的,不可信的。正是由于地质科学的这一特点,决定了攻读地质专业硕士学位的研究生必须具备一定的野外工作能力,要有坚实的野外工作作为硕士学位论文的基础。

另一方面,由于研究生学制的限制,不可能有很多的野外工作时间,这就构成了一对相互制约的矛盾。解决这一矛盾最好的方法,只能是在最短的时间内做最有效的野外工作。因而,野外工作的目的性、有效性对于研究生来说就显得尤为重要。首先,出野外之前必须对研究区的区域地质资料有初步了解,以便确定野外工作的重点和突破口。其次,要在导师的指导下结合自己的研究课题制订一个周密的野外工作计划。计划越全面、越细致,野外工作就会越顺利。相反,盲目的、无目的的野外工作只能造成人力、物

力、财力和时间的浪费。当然,室内所制订的计划毕竟与野外实际情况存在一定的差异,有一定的出入。因此,野外工作中还需要有一个不断调整、不断完善的过程,必须经过边干、边学、边记、边整理、边调整的过程,才能最终顺利完成研究课题的野外工作,取得完备的第一手资料,采集到系统的、有代表性的室内分析测试样品。

四、样品测试和分析结果的获得

野外工作结束后,紧接着就要进行系统的室内分析研究。主要是对野外获得的分析测试样品进行处理,包括自己动手测定的样品和送外单位测试的样品。

提高科研能力的一个重要手段就是增强自己动手的能力。根据地质学系目前的实际情况,许多分析项目是可以自己完成的。因此,研究生必须消除依赖思想,尽可能多地自己动手,通过自身艰苦的劳动去获得数据和资料。这样既可以节省经费又能够使自己得到锻炼,增强了分析测试工作的目的性和测试结果的实用性,同时还可以避免不必要的浪费。当然,并非所有项目都能自己测试,还有一部分测试工作需要送外单位进行分析。对于硕士研究生来说,经费是十分有限的。因而,一切外送样品都必须遵循少花钱、多办事的原则,花最少的钱获得最有用的测试结果。这就要求样品必须具有较好的代表性,测试项目必须具有明确的目的性。任何盲目送样的做法,都无疑是对国家资金的重大浪费。这个问题必须引起硕士研究生的高度重视。

五、测试结果的整理

测试数据获得以后,接下来就要对这些结果进行全面的分析、归纳与总结,找出规律、发现问题,最后得出研究结果。在这个过程中必须实事求是,尽量避免人为因素。同时,必须遵循马克思主义认识论,运用科学的思维方法方能得出正确的认识。自然科学在自己的发展过程中,逐渐形成了一种传统,它要求科学思维有严密的逻辑性。爱因斯坦曾经指出:“科学家的目的是要得到关于自然界的一个逻辑上前后一贯的摹写。”事实上,在我们的研究工作中,时刻都在自觉和不自觉地运用一些科学思维的方法。因此,掌握和运用科学的思维方法乃是认识正确与否、研究工作成功与否的关键所在。

在岩矿、地质专题研究中,经常用到的自然科学逻辑思维方法有以下几种:

(1)比较和分类。比较是确定对象之间差异点和共同点的逻辑方法,而分类则是根据对象的共同点和差异点将对象区分为不同种类的逻辑方法。比较是分类的前提,分类是比较的结果,它们是地质学研究中常用的方法。

(2)类比。类比是根据2个(或两类)对象之间在某些方面的相似或相同而推出它们在其他方面也可能相似或相同的一种逻辑方法。

(3)归纳和演绎。归纳是从个别事实中概括出一般原理,演绎则是从一般到个别的推理。进行归纳往往不是一帆风顺的,常常会遇到一些例外,这在地质学研究中表现得尤为明显。因此,在我们的研究中更多地采用了不完全归纳法。我们往往从岩石或矿床的各方面特征中得出岩石的成岩模式、矿床的成矿模式,反过来又用这些模式去指导未知地区的岩石学研究和找矿工作。所以,归纳和演绎方法运用得恰当与否直接影响结论的正确性。

(4)分析和综合。分析是把整体分解为部分,把复杂事物分解为简单要素分别加以研究的一种思维方法。而综合则是把对象的各个部分、各个方面和各种因素联系起来考虑的一种思维方法。

(5)证明和反驳。在阐述一种思想时,往往举出一些事实或科学原理作为根据来论证这一思想的正确性,这就是证明。而反驳则是用已知为真的判断揭露另一个判断的虚假性。证明和反驳为我们检验认识正确与否提供了科学的方法论。研究所得出的结论必须有充分的证据,同时还需要从不同的角度推敲和考证,找出不足和不全面之处。如果不明白这一点,所得结论将是片面的甚至是错误的。

在这些方法的基础上进行科学抽象、得出科学的概念,从而完成我们对问题的认识过程。一些著名的科学家之所以能取得较大的成就,往往在于他们能巧妙地把所需要的各种方法结合起来加以运用,通过一系列的科学思维方法,从现象深入到本质、从感性上升到理性,最后获得对自然界的规律性认识,形成科学理论。只有用科学的、逻辑的思维方法,在大量测试、观察结果的基础上总结出的认识才能是有意义、有价值的。

六、论文的撰写

学位论文能否顺利完成,主要取决于2个因素:其一是基础工作是否扎实,材料是否完备,数据结果是否可靠;其二是论文写作过程中是否运用了科学的思维方法。

基础工作不完备就很难写出一篇好的论文。这里主要是指对于所研究的题目、所要说明的问题,是否已经有了充分的证据和足以说明这一问题的数据和资料。只有在大量观察、实验、测试结果的基础上才能得出充分的论据,其论点才能让人信服。任何没有根据的推断和不切实际的假想都是不可取的。当然,并不是说论文不能推断和设想,关键是要看推断和设想有没有科学根据,能不能让人信服。德国气象学家魏格纳于1910年提出了大陆漂移的假说。当时,很多人怀疑这一说法的正确性。但让人们感到惊奇的是,他列举了大量的调查材料,对各大陆之间的一系列相似性、可对比性、可拼复性进行了科学的论证,终于写成了《海陆起源》这一名著,开创了大陆漂移说的新领域。诚然,对于硕士研究生来说,不可能完成这样大的研究项目,但道理却是相同的,那就是要在大量实验材料的基础上去认识事物。搞科研最忌讳的就是先有结论然后再去找例证,这样做害处很多,因为它与马克思主义认识论相违背,不是从实践到认识,而是采取了从认识到实践的错误方法。

在撰写论文的过程中还必须注意层次分明,逻辑合理,详略得当。应该按照研究过程来引导读者的思路,怎么研究的就怎么写,从头讲起,引导读者逐渐深入到自己的结论上来。同时,文章中的摘抄成分不宜太多,引用材料太多就会削弱自己所做的工作。对于地质专业的硕士研究生,不管是对哪个地区进行研究,都免不了要对其研究区域地质做介绍,这一部分主要是引用前人的资料,自己所做的工作相对较少,因而必须简明扼要,以较少的语言来表达较多的信息。相反,对于自己所做的工作则应介绍得细致一些。文章最后的结论和认识必须恰如其分、清楚明白,工作做到哪种程度结论就写到哪种程度,不能言过其实,也不能言未尽意。语言必须质朴无华、真实严谨。

开始撰写论文的时间会因人而异。但总的来说,动笔太早往往没有内容可写,太晚了又往往觉得时间不够,写出来的论文比较粗糙。作者认为,可以采取在阶段性总结的基础上进一步归纳,最后整理成文的办法。实践证明,这一方法有一定的可行性。也就是说,在研究过程中应该及时小结,做完一部分就总结一部分。这样既可以发现这一阶段的工作是否已经完成,避免重复性工作,同时又可以为下一步的工作提供线索,找到进一步研究的关键和突破口。而且,阶段性小结也是撰写论文最必要的素材,是能否成文的基础。

最后一个问题就是论文的修改。出口成章很难做到,写论文往往都需要一个不断修改、不断完善的过程。首先自己要对论文初稿进行反复修改,然后广泛征求、虚心听取他人意见,特别要对不同的意见甚至相反的看法进行反复琢磨,以发现问题,找出不足,使论文的论点明确、论据充分、论证合理,具有逻辑性、科学性和创新性。

需要特别指出的是,无论在基础研究工作中还是在论文撰写过程中,都必须与导师保持密切联系。事实上,研究生的研究工作无时无刻不是在导师的精心指导下完成的,导师给予了具体的方法论的指导。对于一些重大问题和关键性认识,导师起着“把关”的作用。因此,研究生学位论文乃是导师和学生共同劳动的结晶。

重视国家基础科学人才基地高年级学生文献阅读能力的培养[①②]

赖绍聪

摘要:本文简要介绍了国家基础科学人才基地高年级学生在课程学习和毕业论文设计过程中,如何培养和提高中、外文文献阅读能力的方法。

国家基础科学人才基地人才培养目标是经过本科阶段的培养,使学生牢固树立热爱祖国、献身地质事业的坚强信念,具有扎实的基础理论知识,宽广的知识面和合理的知识结构,较强的获取知识、运用知识能力及较高的综合素质、富有创新精神并具特色的优秀人才,成为我国地质学基础研究顶尖人才的后备力量。因此,基地班学生应是具有一定专业基础理论水平、一定独立工作能力和科研能力的高层次专门技术人才。接受新知识,及时了解和掌握本专业各种最新科技动态,进行学术交流,是基地班学生能力训练的重要项目,这就要求基地班学生具有一定的外语水平和阅读本专业中、外文资料的能力。为了加强本科学生的专业素质教育,培养学生的创新精神和科研能力,西北大学地质学系为三、四年级学生有针对性地聘请一位具有博士学位或副教授以上职称的教师做导师,主要负责学生专业思想教育和思维、动手能力的培养。言传身教,以教师的工作热情及严谨求实的工作态度带动学生培养良好的科学素质。让学生参与导师的科研项目,从书本走向实践,提高独立思考问题、解决问题的能力,激发学生发表论文的意识,这一措施在几年的实施过程中已取得了可喜的成效。本文就近年来指导基地班学生过程中,关于基础课学习和毕业论文准备期间阅读能力培养的问题谈一点体会和建议。

一、基础课教学应与培养学生文献阅读能力紧密结合

基础课教学是培养基地班学生的重要环节,开设的课程一般都是与本专业关系较为密切的课程,包括基础课(公共课)、专业基础课和专业课,课程的要求一般略高于非基地班本科专业,教学方法、教学要求及师资配备与普通本科专业也有一定的区别。基地班课程除了学习基础理论外,还应该反映本学科的最新成果、国内外的新动向。因此,介绍国内外有关方面的最新研究成果就成为基地班教学的重要环节之一。就公共外语而言,作者认为,除进一步精通语法和适当加强写作、听力、口语外,还应该着重加强阅读能力

① 原载于《高等理科教育》,2004(5)。
② “教育部高等学校优秀青年教师教学科研奖励计划”资助。

的培养,任课教师应该在考虑学生课程负担的前提下,适当指定或安排一定分量的相关文献,让学生阅读并写出读书笔记,在一定情况下可作为必须完成的作业来完成。同时,考虑基地班学生学习上的主动性,还可在更广泛的范围内介绍一些中、外文读物,让学生根据自身的实际情况自由选择阅读,这样就可做到"指令性任务"与"自由调节"相结合,既使学生感到阅读能力培养的重要性,又不至学生负担过重。而且还应该定期或不定期组织学生进行交流,对于阅读和理解中所遇到的问题和语法难点进行答疑和自由讨论,充分发挥基地班学生在学习上的主动性和独立性。

就专业课和专业基础课教学而言,应该在掌握基本理论的基础上,力求反映本课程有关方面的国内外最新动向,不能局限于国内已有资料,要着眼于一个"新"字。事实上,由于种种因素的限制,使得国际最新科技动态往往不能及时得以翻译和出版。这样,就需要基地班学生自己直接去阅读最新的本专业外文期刊,从中汲取养分,使他们能更快地成长。因此,任课教师的重要作用之一,就是向学生介绍本课程研究领域的国内外最新动向,国际上目前正在做什么,已做到什么程度,然后让学生自己去挑选和阅读有关文献资料。实践中,在我们所开设的专业基础课程中大都已关注到了这一点,比如岩浆岩和计算机应用技术。无疑,这将对基地班学生接受最新知识以及阅读能力的提高起到很好的促进作用。

二、论文准备及编写过程中应特别强调阅读有关最新中、外文文献

基地班学生的任务不单纯是接受知识,而且还要进行科研工作。基地班四年级学生有一项很重要的任务就是要完成毕业论文。一篇论文,不管是教授、专家写的或是学生写的,均应尽可能地体现其创新性,也就是说在其研究范围内,前人或者没有接触过,或者虽有接触而语焉未尽,能在他们的基础上进一步加以研究,提出新的看法,论据确凿,言之成理。那么,怎样才能使论文具有创新性呢?因素固然很多,但其中很重要的一点就是要广泛地搜集和阅读与本课题有关的中、外文文献,充分了解和掌握前人已做工作的程度。因此,一名合格的基地班学生必须具有相当地阅读有关中、外文文献资料的能力。

事实上,西北大学地质学系对新入学的基地班学生的阅读能力都有较严格的要求,专业文献阅读是基地班学生必备的基本能力之一。地质学系近年来开设的双语教学,以及直接从美国、加拿大聘请外籍教授承担系基地班 Geosystem 等课程教学的目的就是为了提高基地班学生的外语交流、阅读和写作能力。另外,地质学系许多导师也有意识地安排基地班学生参加一些文献的整理工作,以提高其接受新知识和阅读文献的能力。有的基地班学生甚至在进入硕士研究生阶段之前就已经阅读了相当数量的相关文献,并有一篇综述性的小论文作为成果得以公开发表。作者以为,这些好的经验值得各院校在培养基地班学生时借鉴。

阅读文献还可使我们在准备毕业论文的过程中少走弯路。有时,一个冥思苦想而尚

未得到解决的问题,或一个较为复杂的实验,有时可能是前人已经部分解决或已有人做过并得出初步实验结果的。这样,通过阅读中、外文文献,就可使我们从中取得捷径、受到启发、少做无用功。同时,只有在精研前人文献资料的前提下,才能使我们的毕业论文更有价值,才能真正做到站在巨人的肩上,更上一层、更进一步。因此,导师应该指导基地班学生阅读与毕业论文题目有关的最新中、外文文献,这一点尤为重要和必不可少。而且,西北大学图书馆和系资料室目前已具有很好的条件,订购了较多、较全面的中、外文期刊,并对基地班学生开放。加之网络技术和互联网的普及,均为开展这项工作提供了良好的环境,这些都是值得充分利用的有利条件。

最后,希望基地班学生导师和任课教师在公共外语、基础课教学和毕业论文指导过程中,更加重视基地班学生外语水平的提高和阅读能力的训练,使基地班学生具备较高的接受新知识、了解国内外科技最新动向的能力,为今后承担更高层次的科研任务打下坚实的理论基础。

注重基础 强化实践
以国际化视野构建矿物岩石学“434”教学新体系①②

赖绍聪 刘养杰 刘林玉 陈丹玲 康 磊

摘要：以西北大学地质学系80年科研成果和优质资源积累为基础，密切结合国际地学发展趋势，统筹矿物岩石学不同阶段教学内容，形成了基础—理论—前沿—探索“四层次”理论教学新体系，保障了教学内容的先进性；构建基础训练—能力提升—探索创新“三维度”实践教学新体系，提升了实践教学国际化水平；以传统教学模式与当代信息化技术深度融合为抓手，自主开发，建成了晶体三维结构—矿物形态3D可视化教学平台、全球典型矿物岩石信息库平台、虚拟偏光显微镜教学平台以及显微数码互动实验教学平台等含资源、互动、交流为一体的教学“四平台”，为实质性提高人才培养质量奠定了重要基础。

一、引言

地质学就是研究“石头”的科学。“石头”就是岩石，岩石主要由矿物组成。因此，矿物岩石学在地质学专业知识结构体系中占据核心基础地位。然而，长期以来，我国高等学校地质学专业矿物岩石学教学体系存在明显不足，主要表现在以下3个方面：①矿物岩石学传统理论教学体系知识陈旧，科教融合不足，强调知识灌输，相对忽视能力及思维训练。具体表现：21世纪国际地学正处于发展板块构造、构建大陆构造新理论的关键时期，知识体系快速更新。然而，局限于“教材”的单一课堂教学严重缺乏教学与科研的深度融合，探索研究的思维意识难以融入。结果是学生难以“触摸”学科前沿，缺乏国际视野，分析、解决问题的能力差，创新能力不强。②矿物岩石学传统实验形式单一，内容局限，严重缺乏国际视野，学生主动性差。具体表现：传统实验体系局限于教师指导下的验证性课堂实验，学生局限于被动接受和验证。受地质学地域特性制约，实验标本、岩石薄片大多来自我国境内。结果是实习实验能动性和积极性差，学生学习了“中国地质”，不了解“世界地质”。③矿物岩石学传统实验体系技术落后，难以满足当代地质学创新人才培养需求。具体表现：基于“静态图片—模具—手标本—岩石薄片—单体显微镜”的传统

① 原载于《高校人才培养的理论与实践探索》，刘建林主编，西北大学出版社，2019。

② 国家自然科学基金“国家理科地质学人才培养基地建设”项目（J1310029，J1210021）资助。教育部“地质学国家级特色专业”“高等学校专业综合改革试点”建设项目资助。

实验，无法解决学生对地质学大尺度（宏观—微观—超微观）—长时间（数亿年）—大空间（整体地球）格架下非均衡性—复杂性的深刻理解。有限学时/有限实验室空间等难以满足强化实践训练的需求。结果是学生实践训练不足，空间思维、整体思维能力差。

西北大学晶体光学与岩石学教学团队自2003年岩石学获准为教育部立项的国家理科人才培养基地“名牌课程”建设项目以来，在7个教育部质量工程项目、一个陕西省质量工程项目和6个校级教改项目的支持下，不断探索实践，构建了特色显著的矿物岩石学“四层次—三维度—四平台”教学新体系，即矿物岩石学“434”教学新体系。

二、基础—理论—前沿—探索“四层次”理论教学新体系

根据注重基础、强化实践、突出国际化视野的理念，我们对矿物岩石学教学体系和教学内容进行了大幅度调整。将结晶学、矿物学、晶体光学、光性矿物学和岩石学作为一个统一的整体对待，将结晶学作为矿物学的前导课程，晶体光学与光性矿物学作为岩石学的前期基础，强调课程设置和教学内容的统一性与协调性。密切结合地球科学国际发展趋势，参考澳大利亚 Macquarie 大学、加拿大 Toronto 大学矿物岩石学教学体系，重新梳理优化教学核心知识，引进了 Shelley 编写的 *Optical Mineralogy*，Heirich 编写的 *Microscopic Petrography*，Raymond 编写的 *The Study of Igneous, Sedimentary and Metamorphic Rocks*，Blatt 编写的 *Sedimentary Petrology*，Winter 编写的 *An Introduction to Igneous and Metamorphic Rocks*，RobinGill 编写的 *Igneous Rocks and Processes* 等欧美发达国家教材中的部分内容。同时，融合地质学系80年来矿物岩石学领域的科研成果和优质资源，尤其是团队在青藏高原、华北、中央造山系等典型地质区域的研究成果，构建了全新的矿物岩石学理论教学体系，实质性地形成了特色鲜明、与国际接轨的，以7门描述矿物岩石学课程、3门理论矿物岩石学课程、11个凸显国际化视野和学科研究特色的专题、5类探索创新活动为主线的基础—理论—前沿—探索“四层次”矿物岩石学理论教学新体系（图1），保证了教学内容的先进性。

三、基础训练—能力提升—探索创新“三维度”实践教学新体系

矿物岩石学是实践性很强的学科，建立合理的实践教学体系是高质量完成矿物岩石学教学任务的重要环节。我们以当代地质学知识体系为基础，根据不同尺度矿物岩石对象具有不同的教学重点和教学目标，充分利用最先进的信息化技术，构建符合当代地质学发展趋势、含实践教学全部核心知识、导航整个矿物岩石学实践教学知识地图的基础训练—能力提升—探索创新“三维度”实践教学新体系。该体系包括宏观—微观—超微观3个尺度的基础训练实践，实验课—自选实验课—研讨实验课3类能力提升实践，以及课堂实验—网络实验—数码互动实验三型深度融合信息化技术手段的探索性实践。其理念先进、体系完备，涵盖实践教学“认知—综合—探索”3个主要阶段（图2）。该体系提升了矿物岩石学实践教学的国际化水平，为实质性地提高人才培养质量奠定了重要基础。

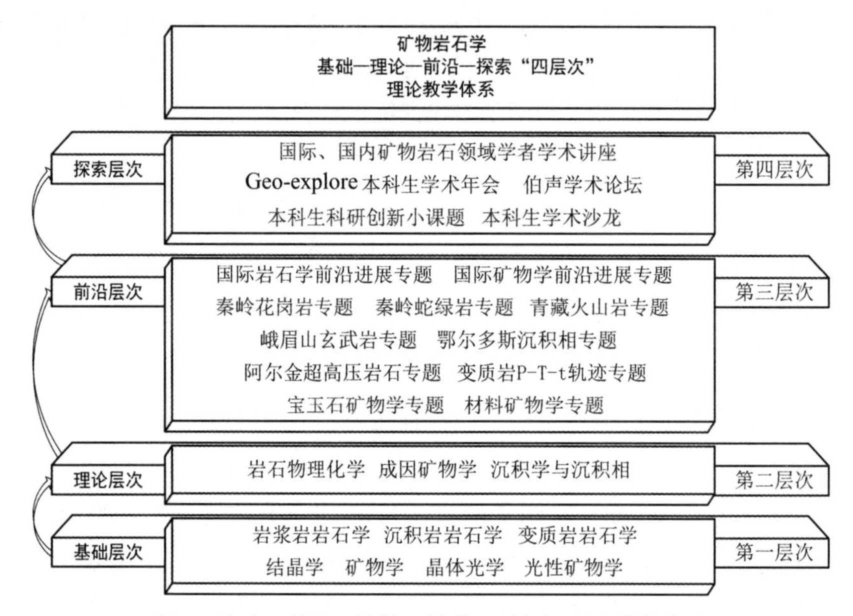

图 1 基础—理论—前沿—探索“四层次”理论教学体系

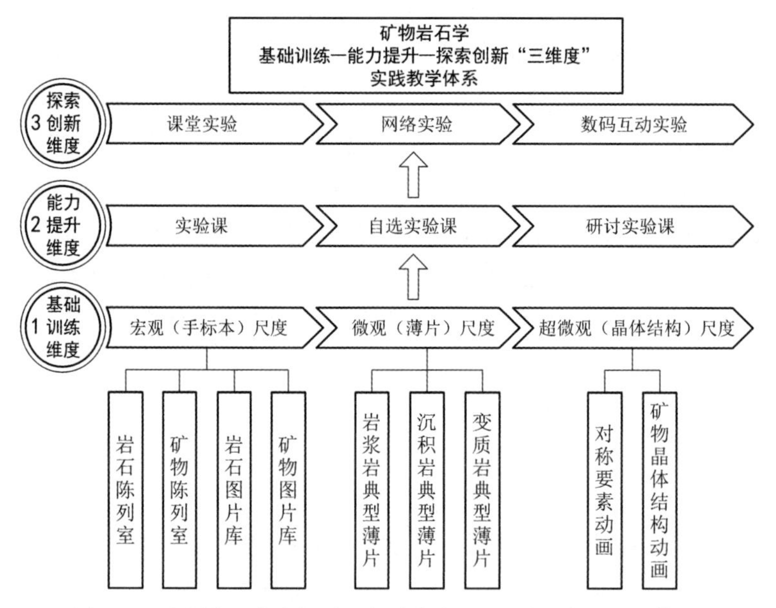

图 2 基础训练—能力提升—探索创新“三维度”实践教学体系

四、自主开发四大数字化教学平台

1.晶体三维结构 3D 可视化教学平台

矿物晶格、对称型(肉眼)不可见,内容抽象,学生理解困难,严重影响了学生对结晶学与矿物学核心知识的掌握,成为教学中亟待解决的难点。经过 10 余年的不断探索,我们利用计算机信息化技术重点解剖矿物的晶体结构，成功地开发了晶体三维结构3D可视化教学平台，建立了矿物晶体结构的三维模型以及晶体结构中质点运动的三维动画。学生通过自主操作，能够直观地观察质点在三维空间的位置、位移及动态的对称操作等。建立了矿物中典型结构的三维动画分析。目前，已经可以进行123个典型矿物结构、14种布拉维格子、32种对称型、47种单形、300余种典型矿物结构—形态等的三维虚拟实时表达。

这一平台的建成使原本十分抽象、晦涩难懂的结晶学教学内容大部分实现了三维交互式可视化，创造性地解决了矿物晶格、对称型(肉眼)不可见这一教学过程中长期存在的难题，使学生在学习结晶学与矿物学时兴趣倍增。更为重要的是，该平台在教学中的使用极大地提升了学生的空间(地质)思维能力。

2.典型矿物岩石信息库平台的开发建设

由于地质学具有显著的地域特性，各院校矿物岩石教学标本资源主要来自我国，世界其他地域的标本缺乏，稀缺、珍贵或新发现矿物岩石标本很难获得，从而使学生的知识面和国际化、全球化视野受到限制。针对这一问题，我们以信息化技术为支持，经过10年努力，建成了矿物岩石信息库。该信息库以西北大学岩矿陈列室现有标本资源的数字化收集为基础，同时以国际化、全球化视野从国内外院校、国际国内文献和网络资源中广泛收集世界各地各类矿物岩石信息以及稀缺标本的信息资源，建成了具有国际化属性的矿物岩石信息库。目前，信息库已有涵盖全球不同区域的各种矿物岩石信息6 000余种，图片20 000余幅，基本涵盖了教学过程中涉及的各类矿物岩石。

该信息库对现有标本资源形成有效扩充，缓解了教学标本资源紧缺的问题，使学生能够及时了解全球范围内矿物岩石领域的新发现和新进展，突破了地质学地域性约束，提升了学生国际化、全球化视野。

3.建设虚拟偏光显微镜教学平台

显微镜是地质学教学和科研过程中最重要的基本观察工具之一，被视为地质学专业人员的必备“武器”。19世纪，显微镜引入地质学领域，对地质学的发展产生了革命性的推动。因此，强化训练学生利用显微镜观察分析矿物岩石，提取矿物岩石成岩成矿关键信息的能力，已经成为当前地质学专业学生基础实践能力提升中最为核心的目标之一。然而，由于受课程学时数，实体显微镜实验室空间大小、开放时间、设备台套数的限制，本科生显微镜观察能力严重不足，这已经成为地质学高等教育界的广泛共识，严重制约了学生实践能力和创新能力的提高。

为破解这一难题，我们利用计算机网络技术，原创虚拟显微镜(发明专利申请公布及进入实质审查阶段：发文序号2016122101543109)。将显微镜下实际现象虚拟为可在计算机上自主操作，通过浏览器访问的虚拟显微镜平台，学生即可对岩石薄片进行镜下学习。该虚拟显微镜的最大特色是具备真实显微镜的实时交互式操作：可以360°旋转薄片，可以切换单偏光和正交偏光，完成不同光路系统下薄片中矿物特征的观察，其操作使用与真实显微镜十分接近。

该平台突破了传统显微镜实习受学时数、实体实验室空间大小以及实体显微镜台套数量的限制,实现了显微镜实习随时随地常态化,弥补了教学资源的不足,扩展了学生的学习时间和空间,对于大幅度提升学生的基础实践能力具有重要的实用价值。

4.显微数码互动实验教学平台

传统显微镜设计是单人单镜观察,教学过程中,每位学生一台,教师只能一对一指导,师生单独互动,交流效率低,实验教学效果较差。针对这一教学难题,我们建设了信息化数码偏光显微互动实验室,从而使显微镜实验教学模式发生了根本性变化。

数码显微互动实验教学平台(图3)的主要功能在于其直观性和互动性,教师可将偏光显微镜下的现象实时传送到任何一位学生的显示屏上;同时可以实时动态监控每位学生偏光显微镜下的观察内容,并可将任意学生的观察内容演示给所有学生;通过语音系统对学生进行集体授课或单独答疑,学生也可以分组讨论;还能对图像进行处理及共享等。

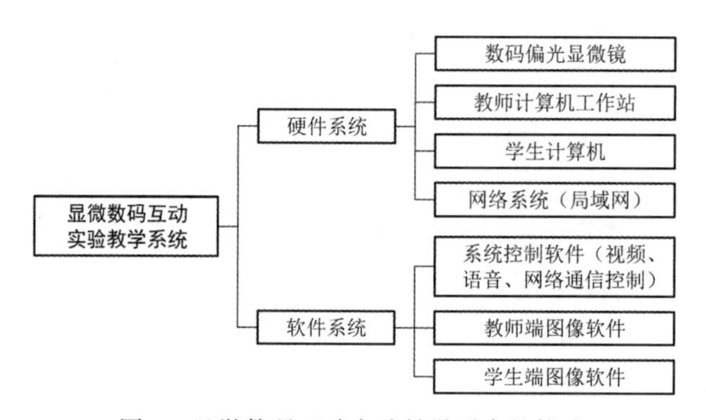

图3　显微数码互动实验教学平台的构成

显微数码互动实验教学平台将偏光显微镜、CCD、电脑、工作站、IDB、录播、手拉手会议音频和网络有机结合,使显微镜实验教学形式、教学观念和教学方法得到根本转变。突破了传统矿物岩石学单人单镜显微镜实习实验师生互动效率低、教学效果较差的瓶颈,全面实现了显微镜实习实验师生之间、学生之间的全方位多点互动、信息交流与交换,极大地提高了显微镜实习实验的教学质量。

五、结语

矿物岩石学是地质学专业核心基础课程体系,面对当前国际地学发展趋势,以科教深度融合为抓手,将研究探索的思维融入教学,实施教学体系的深度改革创新,必将有力推动我国地质学类本科生教育教学质量的显著提升。矿物岩石学434教学新体系的创新性探索与实践,将有利于打破传统教学体系囿于教室课堂的观念,推动实验教学由“要我做”向“我要做”“我想做”“我会做”转变,逐步实现矿物岩石学基础实践技能训练随时随地常态化,以及矿物岩石学实习实验过程中师生之间、学生之间的全方位多点互动。这些改革,符合当前高等教育改革的方向和地学人才培养的实际需求,对于培养高素质、具有创新精神的新型地学人才具有重要的现实意义。

参考文献

[1] 赖绍聪."岩石学"系列课程建设的改革与探索[J].高等理科教育,2004(3):58-60.

[2] 刘养杰,李立宏,马维峰."矿物学"教学的好助手:"矿物学"CAI课件[J].高等理科教育,2004(3):81-83.

[3] 刘林玉.沉积岩石学实验教学改革与多媒体课件制作[J].高等理科教育,2004(3):104-106.

[4] 刘养杰,刘良,陈丹玲,王焰.坚持不断改革,提高"矿物学"教学质量[J].高等理科教育,2004(3):87-90.

[5] 陈丹玲.强化基础,突出技能,因材施教:"矿物学"教学过程的几点体会[J].高等理科教育,2004(3):84-86.

[6] 刘养杰.研究型实验课程的设计与教学实践:以"结晶学与矿物学"实验课程教学为例[J].高等理科教育,2008(6):106-108.

[7] 赖绍聪.创新教育教学理念提升人才培养质量[J].中国大学教学,2016(3):22-26.

[8] 赖绍聪.建立教学质量国家标准提升本科人才培养质量[J].中国大学教学,2014(10):56-61.

[9] 赖绍聪.地球科学实验教学改革与创新[M].西安:西北大学出版社,2010.

[10] 赖绍聪,华洪.课程教学方式的创新性改革与探索[J].中国大学教学,2013(1):30-31.

[11] 赖绍聪.如何做好课程教学设计[J].中国大学教学,2016(10):14-18.

[12] 赖绍聪.寓教于学 寓教于研[J].中国大学教学,2017(12):37-42.

岩石学实验教学创新
——虚拟偏光与数码互动显微镜教学系统的开发与应用①②

赖绍聪　康　磊

摘要：显微镜实验教学对于培养地质学类专业本科生实践创新能力具有重要意义。传统显微镜实验教学难以实现优质教学资源的共享、师生互动，学生与学生之间互动差，教学效率低。虚拟偏光显微镜和数码显微互动教学系统的开发利用，使岩石学显微镜实验彻底打破了传统岩石学实验中学时、实体实验室空间及实体显微镜台套数量的限制，实现了显微镜实习实验的随时随地常态化，有效地促进了教学互动，使教学手段产生了深刻的变革，对于大幅度提升学生的实践能力具有重要的实用价值。该教学方法可广泛推广应用于地质学类专业涉及显微镜教学的各门课程，同时对生物学、医学等相关专业的显微镜课程教学也具有重要的借鉴价值。

显微镜是地质学教学和科研过程中最重要的基本观察工具之一，被视为地质学专业人员的必备“武器”。19 世纪，显微镜引入地质学领域，对地质学的发展产生了革命性的推动。因此，反复强化训练学生利用显微镜观察分析矿物岩石微观组构、提取矿物岩石成岩和成矿关键信息的能力，已经成为当前地质学专业学生基础实践能力提升计划中最为核心的目标之一，受到地质学高等教育界的广泛重视。

然而，由于受课程学时数、实体显微镜实验室空间、开放时间、设备台套数的限制，地质学专业本科生显微镜观察能力严重不足，这已经成为地质学高等教育界的广泛共识，严重制约了学生实践能力和创新能力的提高。

为了破解这一教学难题，进一步提高教学质量及充分发挥优质教学资源的利用率，近年来我们尝试利用最新的计算机和网络技术，开展了虚拟显微镜平台和数码多通道显微镜教学系统的探索和建设工作，并在实验教学实际应用中取得了显著效果。

一、虚拟偏光显微镜教学系统的开发应用

显微镜对岩石学的研究产生了有力的推动作用，是岩石学发展过程中一个突破性转折点。它可以观察及研究宏观地质体（标本）无法看到的细节和显微结构，弥补了宏观观

---

① 原载于《高校地球科学课程教学序列报告会论文集（2016）》，高等教育出版社，2017。

② 国家级地质学特色专业建设项目（337010001）和国家级地质学专业综合改革试点项目（331040089）资助。

察的不足。因此,了解显微镜的基本原理和构造,掌握使用显微镜观察地质标本的方法,是地质教学工作的一个重要组成部分。

地质学专业涉及使用显微镜的课程有晶体光学、光性矿物学、三大岩类(岩浆岩、沉积岩和变质岩)、矿相学和微体古生物学等。其中,晶体光学与岩石学是地学类专业本科学生最早学习的专业基础课程,是地学类专业的必修课程。课程不但理论课时多,而且还有大量的实验课时配套,基本要求是每位学生都会使用偏光显微镜,对矿物岩石标本进行微观形貌分析;通过显微镜观察描述各种矿物晶体及岩石薄片的结构组成与形态特征,加强理论与实际的结合,为后续专业课的学习打下基础。

传统的显微镜理论教学课,教师在备课时需先观察岩石薄片,找出典型矿物和现象后拍照,保存成图像文件后制成 PowerPoint 幻灯片,在理论讲解时再使用该幻灯片。这种教学方式的优点是照片素材丰富多彩,但也有缺点。首先,因每一块岩石标本薄片普遍具有差异性和唯一性,教师备课费时费力,需要对不同的现象拍摄大量照片;其次,在讲述矿物多色性、消光角、贝克线等内容时,由于所展示的照片是静态的,不能直观地阐述这类需要转动物台从不同角度观察的学习内容,而常常使学生感到抽象和难以理解。另外,虽然我们在长期教学过程中积累了大量典型的教学标本,但由于标本在长期使用过程中难免会发生破损,以及学生人数逐年增加,原有的教学资源已较为紧缺。

(一)虚拟偏光显微镜的开发

为了解决上述传统教学过程中存在的严重不足,进一步提高教学质量,以及充分发挥优质教学资源的利用率,近年来我们尝试利用最新的计算机和网络技术,开展了虚拟显微镜系统的探索和建设工作。即将显微镜下实际观察到的真实现象虚拟为可在计算机网络上自主操作观察,通过浏览器访问的虚拟显微镜网站,学生可随时对典型岩石薄片进行自主镜下学习,观察各类岩石在显微镜下的矿物组成、光性特征、鉴定标志及显微结构构造。该虚拟显微镜的最大特色是模拟真实显微镜的实时交互式操作,可以 360°旋转薄片,并且可以切换单偏光和正交偏光,完成不同光路系统下薄片中矿物特征的观察。由于操作使用体验与真实显微镜十分接近,并且所展示的薄片是任课教师精心挑选的典型标本,因此,可以弥补优质教学资源的不足,同时扩展了学生的学习时间和空间,大大提高了优质教学资源的利用率和教学效果。

(二)虚拟偏光显微镜教学系统的构成

虚拟偏光显微镜教学系统的内容是基于我们多年来教学过程中积累的典型代表性岩石薄片而构成的。系统主要由岩浆岩、沉积岩、变质岩三大部分构成,每一部分又主要由各类代表性岩石的矿物组成、光性特征、鉴定标志和显微结构构造等方面组成(图 1)。目前,我们已经完成了教学必需的典型岩石标本显微镜观察的数字虚拟化工作并发布到网站上。此外,三大岩(沉积岩、变质岩、岩浆岩)更多岩石薄片的虚拟显微镜系统库正在进一步扩充中。

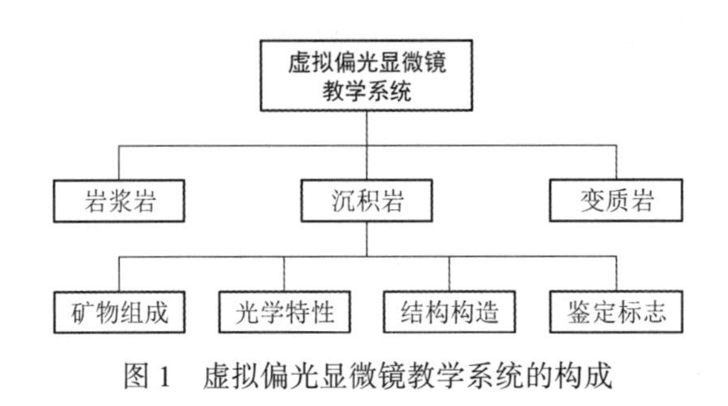

图 1　虚拟偏光显微镜教学系统的构成

(三) 虚拟偏光显微镜教学系统的教学效果

与传统显微镜教学模式相比,虚拟偏光显微镜教学系统使得教学效果显著提高。

(1) 由于能够模拟真实显微镜 360°旋转薄片的实时交互式操作,因此突破了传统课堂理论教学中静态展示的矿物多色性、消光角、贝克线等需要转动物台从不同角度观察的学习内容,从而使学生在学习相关内容时,感受更加生动、直观、真实。

(2) 虚拟显微镜系统中的每一组现象都是教师精心选择的代表性岩石中的典型现象,即使教学资源有限,每一位学生也可以登录系统对同一内容进行观察学习,因此该系统可以节约现有教学资源,并充分发挥最优质资源的利用率。

(3) 该系统基于网络运行,因此只要在有网络的条件下,学生均可在任何时间、任何地点访问网站进行学习或课后复习,突破了传统教学的时间和空间限制,使学生能够远程自主学习。

二、显微数码互动实验教学系统

地质类专业的矿物岩石学实验最常用的仪器设备是偏光显微镜。但由于传统显微镜设计是单人单镜单筒观察,在教学过程中,每位学生一台,教师也只能一对一进行指导,师生单独互动交流效率低,实验教学效果较差。针对这一教学难题,我们建设了信息化数码偏光显微互动实验室,从而使显微镜实验教学发生了根本变化。

(一) 显微数码互动实验教学系统的开发

数码显微互动实验系统的主要功能在于其可真实、直观、互动地展示。教师将偏光显微镜下的所见实时地传送到任何一位学生的显示屏上,便于观察和讲解;同时,可以实时动态监控每位学生偏光显微镜下的画面,并可以将任意学生的观察内容演示给所有学生;通过语音系统对学生进行集体授课或单独答疑辅导,学生也可以分组讨论;还能对图像进行处理及共享等。所有功能都在教师的监控与授权情况下实现,提高了网络的安全性。通过这一网络平台,可以将教师与学生联系在一起,使教师和学生之间、学生和学生之间实现了真正意义上的双向互动、多向互动。显微数码互动实验系统将计算机、偏光显微镜、数码摄像装置和网络有机地结合起来,使得显微镜实验教学形式、教学观念和教学方法等得到了根本性转变。

(二)显微数码互动实验教学系统的构成

显微数码互动实验教学系统由30台三通偏光显微镜、1台示教三通偏光显微镜、CCD数控中心、30台电脑、1台工作站、IDB互动系统、录播系统、手拉手会议音频系统等硬件系统和软件系统共同构成(图2),系统分为教师端和学生端。教师端配备有高端Nikon-LV100POL透、反射偏光显微镜1台,Dell工作站1台;学生端配备有Nikon-50iPOL型高级透、反射偏光显微镜30台,Dell计算机30台,配套的成像系统CCD 30套。通过网络交换机组成数字网络显微互动教学系统,可以实现教师和学生之间、学生与学生之间实时互动,进行图像、语音及文字等传输,实现了实验教学的多媒体化。

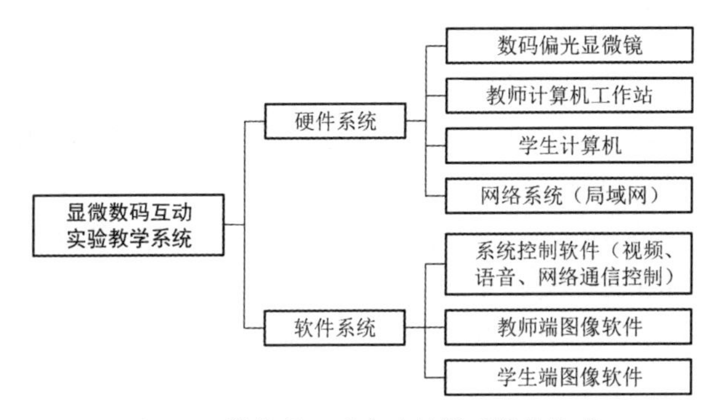

图2 显微数码互动实验教学系统的构成

(1)硬件配置:主要包括偏光显微镜、微型计算机(包括教师端和学生端用计算机)、显微摄像装置、网络系统等。

(2)软件配置:分为教师端控制软件和学生端控制软件。教师端与每个学生端都是独立的图像处理单元,各自具备强大的图像分析、处理功能,还可通过系统自带的局域网,实现实时动态图像的共享和语音的交流。教师通过教师端软件可以观察所有学生的偏光显微镜图像与电脑屏幕图像,随时掌握学生的学习情况,并可通过屏幕控制对学生进行独立辅导。学生端软件在教师端软件的操作下,可以显示教师端或其他任意一个学生端的图像,实现师生之间(学生之间)真正的互动。该系统还具备完善的语音和文字交流功能,提供广播、对话、问答等多种交流方式,全面方便地满足互动式教学的需求。

(三)显微数码互动实验教学系统的教学效果

1.提高课堂效率,激发学习兴趣

由于教师可以随时观察到学生的实验进展,因而可以指引学生快速、准确地找到要观察的结构或内容。而传统的实验课,岩石薄片的唯一性和差异性导致部分学生需要教师的帮助,因此教师需要花费大量时间来帮助学生在显微镜下找图像。但由于课堂时间有限,教师不能顾及每个学生,造成有些学生因实验中的疑问得不到教师的及时解答,自己又确定不了所观察的结构,其观察、判断能力的提高都会受到影响,久而久之会导致学生失去对实验的兴趣。使用数码显微互动系统,所有问题都可通过大屏幕显示。对于共性的问题,教师一次讲解,全班学生就都能得到解答,再不必花费时间重复,这样教师会留有更多时间来解答其他问题。对于个别问题,学生则可以单独和教师通过系统交流却不会影响他人。因此,数码显微互动系统能提高课堂效率,激发学生的学习兴趣,提高教学质量。

2.实时互动交流

通过监控屏幕教师可以随时掌握学生的学习情况,师生可以进行互动交流。例如,教师发现薄片中有典型的矿物,或者学生在偏光显微镜下观察到特殊的现象,教师可以对部分或全班学生示教并进行讨论,这样就避免了传统显微镜实验教学中可能出现的找不准或看错等现象。如果学生在偏光显微镜中发现自己不认识或不能确定的矿物,可对图像进行标记并提问,也可以请求发言,师生之间可以通过耳机用单独对讲的方式进行解答。显微数码互动实验系统的应用,不仅建立了师生间互动交流的平台,更重要的是在提高实验教学效果的同时减轻了授课教师的工作强度及重复性。另外,在每一次实验课程中可观察学习的薄片标本都不再局限于手边的几块,同一实验组的学生之间可以相互观察所见到的矿物特征,为实验小组进行内部讨论提供了良好的平台,促使学生之间相互交流,既提高了学生的学习兴趣,又培养了他们发现问题、解决问题的能力。

3.图像分析及资源共享

学生在偏光显微镜下观察的矿物光学特征能够被及时显示到计算机显示器上,可以对观察到的矿物形态的局部进行预览、放大,使观察效果更好。图像处理系统还可对图像进行各种测量(长度、面积、灰度值等)、录像、编辑、剪切、放大、对比等。这些功能使教学内容得到了极大的拓展。在实验过程中,学生还可把多个易混淆的矿物图像集中放在同一个屏幕上进行对比分析,如钾长石和斜长石、普通辉石和紫苏辉石、黑云母和普通角闪石等,在同一画面上比较它们的异同,找出它们的区别点,从而加深对知识的理解,提高自主分析、理解能力。

此外,由于在磨制岩石薄片过程中切片的类型、部位、角度等一系列原因,导致每个学生所观察到的矿物形态不尽相同。例如,观察薄片中垂直光轴的矿物时,由于矿物切片方位不一样,一些矿物很可能是斜交光轴的形态。利用数码显微互动系统,教师可随时通过教师显微镜给学生示教,也可随时把某一学生显微镜画面切换到大屏幕上展示给全班学生看,保证每个同学都能看到任何一台显微镜下的画面,实现了全班图像资源共享,打破了传统显微镜只能一对一观察的局限性,从而实现更高效的学习与交流。尤其对不容易见到的特殊现象或典型结构还可拍照保留,以备随时调出观看。另外,经过长时间积累的图件可以丰富教学资源,可用到后续的实验课和理论课的教学、学生课外自学、复习以及科研中,从而提高教学质量。整个实验全程,师生之间、学生之间互为借鉴,取长补短,不但节约了课堂时间,还使得教学更为直观。通过计算机还可存储大量的教学资料,如图片、影片、动画、文献等,都可随时给学生演示。课堂上可进行相关内容之间的对照比较,使教学内容得到极大的丰富和充实,教学资源得以充分利用,在有限的时间内可获取更多的知识,拓宽了学生的思路,增加了学生的信息量。

三、结语

虚拟偏光显微镜教学系统可以使地质类专业学生十分便捷、直观地学习显微镜岩石学,且不受时间和地点的限制,彻底打破了传统晶体光学、岩石学实习实验中学时、实体实验室空间及实体显微镜台套数量的限制,实现了显微镜实习实验随时随地常态化,这对于大幅度提升学生的实践能力具有重要的实用价值。同时,该系统对其他使用显微镜的相关学科(如古生物学、材料学、医学等)的教学改革也有重要的参考与借鉴价值。

数码显微互动系统为现代化显微镜实验教学提供了优质的技术支持,使教学手段产生了深刻的变革。该教学方法可广泛推广应用于地质学类专业涉及显微镜教学的各门课程,同时对生物学、医学等相关专业的显微镜课程教学也具有重要参考价值。

参考文献

[1] 赖绍聪."岩石学"系列课程建设的改革与探索[J].高等理科教育,2004(3):58-60.

[2] 赖绍聪.改革实践教学体系创新人才培养模式[J].中国大学教学,2014(8):40-44.

[3] 赖绍聪,华洪.课程教学方式的创新性改革与探索[J].中国大学教学,2013(1):30-31.

[4] 赖绍聪,华洪,王震亮,等.研究性教学改革与创新型人才培养[J].中国大学教学,2007(8):12-14.

[5] 赖绍聪,张国伟,张云翔.理科地质学高等教育改革的实验室建设与实践[J].高教发展研究,2006,88(3):32-34.

[6] 刘一飞,张景华,俞海洋.实施大学生创新性实验计划,构建基于能力培养的创新人才培养体系实践与探索[J].实验技术与管理,2012,29(7):21-23.

[7] 韩芝侠,魏辽博,韩宏博,等.仿真虚拟实验教学的研究与实践[J].实验技术与管理,2006,23(2):63-65.

[8] 韩响玲,金一粟,穆克朗,等.以实践创新能力培养为核心 全面推进实验室面向本科生开放[J].实验技术与管理,2013,30(2):10-13.

[9] 李琰,吴建强,齐凤艳.开放与自主学习模式下的实验教学体系[J].实验室研究与探索,2012,31(1):134-137.

深化课程体系改革　构建人才培养新模式[①]

赖绍聪

摘要：随着教育规划纲要的全面实施，人才培养模式的改革与创新成为当前我国高等学校深化教学改革、提高人才培养质量的关键。人才培养模式的改革，不仅要更新人才培养观念，更要加强与之相应的课程体系改革。本文讨论了创新型人才培养模式改革的必要性与重要性，并探讨了与之相关的通识课程、专业核心课程、专业课程以及实践教学的改革方案与建议。

人才是强国之本。高等学校作为人才培养的重要基地，在培养创新型人才、建设创新型国家的进程中担负着重要的使命。近年来，我国的人才培养和高校的软、硬件设施建设都取得了显著的成效，我国高等教育人才培养模式也发生了一些可喜的转变，国内部分高校进行了多方面的探索。比如，同济大学的“知识/能力/人格三位一体模式”、北京大学的“元培计划”、清华大学的“基础科学计划”、南京大学的“3 个融为一体”模式、西北大学地质学系的“研究型人才培养新方案”等，均在创建新型培养模式、提高人才培养质量方面实施了一系列有效的改革措施，并取得了重要的具有一定示范效应的新进展。但是，与国家经济和社会发展要求尤其是与发达国家相比，还存在许多不适应的地方。

中国缺乏创造性人才已是不争的事实，“钱学森之问”的提出使这个问题成为全社会瞩目的问题。究其根本，主要是人才培养模式仍然相对较为落后，由此所造成的直接结果就是人才创新能力不够强，缺乏勤于思考、敢于怀疑、勇于创新的思想和动力。因此，中国高等教育教学改革仍然面临重大的挑战。

一、创新型人才与创新型人才培养模式

什么是创新型人才？这本身就是一个值得深思且存在不同看法和认识的问题。一般认为，创新型人才是指具有创新意识、创新精神、创新能力并能够取得创新成果，能开创新局面，对社会发展做出创造性贡献的人才。多数学者认为，创新型人才主要应该具有如下几个特征：①有很强的好奇心和求知欲望；②有很强的自我学习与探索能力；③基础知识和专业知识扎实，并对历史、文学、地理等人文学科有一定的涉猎，能够快速实现知识的更新和迁移；④有良好的道德修养和团队合作精神；⑤有健康的体魄和良好的心理素质，能承担艰苦的工作；⑥对各种问题有高度的敏感性和较强的洞察力。

---

① 原载于《西北高教论评》，2015(1)。

人才培养模式是在一定的教育思想和理念指导下,为实现特定的教育目标,通过专业培养计划、课程体系、评价体系、管理制度等实施人才培养的方式。简单地说,人才培养模式就是学校为学生构建的知识、能力、素质培养结构,以及实现这种结构的方式。

很显然,构建合理的、先进的、适应当代科学技术发展和社会需求的人才培养模式,是决定人才培养质量的关键因素。要提高人才培养质量,就必须将人才培养模式的改革与创新作为重点。当前我国高等教育人才培养模式尚存在一系列的不足和问题,主要表现在以下几个方面:①高等教育理念滞后。当前,许多高校的教育理念并未跟上时代的步伐,仍然在培养普通的继承性人才,较多地把注意力放在了对于知识要点的掌握上,从而缺乏开发学生独立性和批判性思维能力的有效措施。②课程设置严重不合理。课程结构安排欠妥,开设的课程种类过繁、过多,课程开设的先后顺序不尽合理,各课程之间的联系不够密切。课程内容陈旧,许多知识都是被社会淘汰的知识,不能及时反映时代、学科发展的前沿水平。③教学组织形式单一。绝大多数高校仍然沿袭班级授课的单一传统教学模式。教学组织形式明显缺乏灵活性和多元性,不利于因材施教,难以照顾学生的个别差异,在一定程度上限制了学生的独立性与创新性的发展。④教学方式不当。注入式、灌输式教学仍然在现今中国高等教育中占据主导,即教师课堂讲授、学生被动接受。这种简单单一的方式明显不利于激发学生的主动性、积极性和创造性。⑤教学评价体系偏颇。考试测验仍然是教师评价学生几乎唯一的法宝。这种单一的形式不利于对学生综合素质的考核,不利于学生独立性、创造性的发挥。这些问题的存在制约了我国高等教育的发展以及高质量创新型人才的培养,需要我们共同努力,采取有效措施逐步去解决。

二、课程体系的全面革新

符合创新人才培养需要的课程体系优化和教学内容改革是改革的核心,也是人才培养模式改革的重点与难点。目前,我国高等院校仍然在不同程度上存在先进的科学理论、技术与老的课程体系设置与旧的教学内容、体系的矛盾。因此,优化课程体系,使教学内容尽可能跟上时代发展的步伐是当前教学改革面临的重要任务。我们必须从人才培养目标定位出发,建立相适应的课程体系,组织力量编写有特色、高质量的教学参考书及选修课教材。创造条件,引进有特色的原版教材,推进双语教学,并让学生尽可能多地接触学科发展新成果的最新动态信息。同时,通过教学与科研的结合,将教师研究成果充实到教学内容中等,都是亟待进一步加强与深化的。

1.关于通识课程

创新型人才的培养最终要靠科学合理的课程设置来实现。课程设置是教学计划的核心内容,是实现人才培养的关键所在。通识课程是基于通识教育在构建人才创造性思维能力中的重要性而设置的课程体系。设置通识课程和实施通识教育将有助于构建创新人才的知识结构层次。人类科学发展的历史过程表明,现代科学研究前沿的重大突破、重大原创性科研成果的产生,通常都是在不同学科领域的相互交叉、相互渗透中形成和发展起来的。因此,通识教育的内容主要应该涵盖人文科学、自然科学和社会科学三大领域,且强调学科间的交叉、集成和融合,培养学生具有较完整的、合理的知识结构和多维思维能力。设置通识课程和实施通识教育将有助于创新型人才良好品格的形成。

在通识课程设置方面,要实现文、理科相互渗透,扩大选修课范围,为学生提供更大的选择、想象和创造空间;要以前沿知识为先导,以相邻学科知识为辅助,使培养的人才成为各个领域的多面手;要注重基础教育和课程内容的综合化,加强基础知识和专业知识的结合,以求克服大学教育过分专业化的倾向,使普通教育与专业教育并重;要以学科大类或学科群为背景,通过改革和修订教学计划,对课程进行合理的归并、缩减和革新。实施人文素养和科学精神相结合的通识课程教育,完成学科交叉和综合背景下的宽口径专业教育和个性化培养。

2.关于专业核心课程

一个学科的核心课程应该是为实现该专业培养目标,对学生掌握专业核心知识和培养核心能力、提高该专业核心竞争力起决定性作用且不宜过多的课程。核心课程的教学质量具有非常重要的实际意义,同时也是培养创新型人才的基本保障。核心课程构建的指导思想是将过去的专才培养模式引向一定领域的通才教育,使专业在课程设置上更适应快速的经济发展和科技进步。因此,加强学科基础、拓宽专业面、夯实核心课程符合当代学科前缘的教学内容,乃是专业核心课程体系构建的基点。

以理科地质学专业为例,该专业通常将地球科学概论、地球化学、地球物理学导论、结晶学与矿物学、晶体光学与岩石学、古生物学、地史学、构造地质学等课程设置为专业核心课程。这些课程是整个专业课程体系的核心,在总课程学分中占据了较大的份额。这些课程教学内容应该充分体现经典与前缘的结合、理论与实践的结合,更加注重对学生能力的培养。这样的课程体系结构,有利于学生掌握地质学核心理论与实践技能,也为以后的再学习打下扎实基础。

在地质学专业核心课程体系构建和优化过程中,我们还应该特别注重改变以往以验证为目的的课程教学内容,培养学生全新的地学观及综合分析问题与创造性能力,加强新思维、新技术和新方法在课程教学中的应用,建立特色鲜明、科学合理、循序渐进的地质学专业核心课程教学体系,全面体现专业核心课程教学的知识性、技能性、思维性、综合性与创新性。

3.关于专业课程

理工科类专业的专业课程常常涉及面较宽。以理科地质学专业为例,开设有地球科学前沿、矿床学与矿相学、大地构造与中国区域地质学、层序地层学、环境地球化学、现代测试技术与方法等课程。这些课程大多具有很强的实践性和经验性,且知识更新快,随着科学技术的迅速发展,课程内容变化较大。因此,面对当前的新形势,对于专业课程的设置和教学内容,应该适度强调其先进性和前缘性,必须建设符合当代地球科学发展趋势,与国内外学科发展同步的优质专业课程群。要打破陈旧的课程教学体系,将新的科学理论、观念与新的科研成果融入教学,将教师及学科群体的科研成果融入教学,这些优质的教学与科研相结合的教育资源对提高人才培养质量将发挥着重要作用。在专业课程教学过程中,必须因课制宜,努力寻求有效的方法让学生在课程学习中获得良好的感性认识,提高志趣与学习的主动性,有利于消化理论知识,并体现课程教学的设计性和综合性,启发、引导学生学习的创造性思维以及实践动手能力。要结合课程内容进行综合教学,通过资料阅读,由教师引导,学生自行提出问题、自行设计解决问题的教学途径,达到提高学生分析问题、解决问题能力的目的,完成自我设计、自我解决的综合性、探索思维的教学过程。

4.关于实践教学

良好的实践动手能力对创新型人才培养尤为重要。传统教学在进行知识传授的同时,也注重学生的实践,但它强调实践教学的验证性过程,注重对经典学说的认知和接受,创新性实践环节较少,制约了学生的创造性。现在需要在已有实践性教学改革的基础上,努力统筹本科 4 年的不同阶段、不同课程的教学内容和计划,构建课堂教学与实践教学内容协调配置、时间穿插的循序渐进的实践性教学新体系,完善本科实践教学环节。同时,还应该在实现硬件和软件建设相结合的基础上,开发新技术、使用新方法,使之与传统的行之有效的科学研究方法密切结合,建立与当前自然科学发展相适应的实习、实践教学实验室和现场实践教学基地。在实践教学过程中,还必须加强人文素质培养,增强相互协作的团队精神,形成既有严谨的学习风气,又有集体性与个性共存、生动活泼的人文环境,这就是大学所需要的一种造就人才的环境氛围。

三、结语

新世纪创新型人才培养和高等教育改革,是我国科教发展从科教大国走向科教强国的关键问题之一,任重而道远,我们都肩负着重任。可喜的是,近年来全国各高校、院、系在教育教学改革方面都取得了很好的成果,有很多经验与体会。我们相信,通过广泛地相互交流、相互学习、相互借鉴,促进全国高校间的互动,一定能够更加有效地推动我国教育教学改革的深入发展,开创我国高等教育的全新局面!

探索创新性实践教学新模式 稳步提高教学质量

——以“鄂尔多斯盆地-秦岭造山带野外地质教学基地”建设为例①

赖绍聪 华 洪 常 江

摘要:“鄂尔多斯盆地-秦岭造山带野外地质教学基地”具有丰富的大陆地质内涵,极富研究性、启迪性和创新性,是教学成果和科研成果互动、理论和实践相结合的天然实验室,是学科融会贯通、多学科交叉的区域地质综合实践教学的有益尝试和探索,开拓了创新型人才培养的新途径。本文系统地介绍了西北大学地质学系10余年来,在创新性野外综合实践教学改革方面的探索与成果。

西北大学地质学科创建于1939年,杨钟健、张伯声等著名学者曾在此任教,为地球科学和中国地质事业的发展培养了大批人才。地质学科通过长期建设,已经形成了特色鲜明、结构合理、富有创新能力的高层次学科体系。地质学科长期把构造地质学、古生物学、环境与资源研究作为重点发展领域,以中央造山带及两侧相关盆地作为重点野外研究基地,主动适应国家重大战略需求,通过广泛的国际合作研究,瞄准学科前沿,持续攻关、重点突破,在中央造山带与盆地、早期生命与寒武纪大爆发、新生代环境变迁、中西部大陆深俯冲作用及其地球动力学等方面取得了重要研究成果。

本学科始终把人才培养质量作为生命线,把培养学生的创新意识和能力、提高学生的综合素质贯穿到整个教学过程中,全面深化教学改革,不断加强教学建设,建立了完整而有特色的本科生培养体系,取得了一系列开拓性和创新性的教学改革与研究成果,已有4篇博士学位论文入选“全国百篇优秀博士学位论文”,并先后获得国家级教学成果奖一等奖1项,国家级教学成果奖二等奖4项,陕西省教学成果奖特等奖3项。实力雄厚的师资力量,成效显著、硕果累累的学科建设,长期的科学研究积累,系列标志性成果,奠定了西北大学地质学系人才培养坚实的基础。

一、实习基地建设的指导思想

不断提高教学质量、培养创新型人才是高等学校永恒的主题。教学工作始终是地质学系的中心工作,教学质量是地质学系的生命线。地质学系国家理科人才培育基地的主导办学思想是立足西部,面向全国,放眼世界,培养少而精的、德智体能全面发展的、优秀创新型、研究型地学人才,并将创新能力培养放到突出地位。

创新是一个民族的灵魂,是一个国家兴旺发达不竭动力,创新教育是我国当代高等

① 原载于《大学地球科学课程报告论坛论文集(2008)》,高等教育出版社,2009。

教育发展的主题。如何将科学研究实质性地纳入教学过程、将优势科研成果、科研资源向教学转化,实现创新型人才培养,是当前高等理科教育面临的重大课题。作为中国高等理科教育改革创新的先行者,探索21世纪创新型研究型人才培养的新模式,乃是国家理科人才培养基地所面临的义不容辞的重大任务。

众所周知,地球科学是一门实践性学科,实践教学环节在人才培养过程中具有极为重要的作用。因此,长期以来实践教学环节的系统改革就成为地质学系教学改革的重中之重。根据地球科学发展和人才培养现状,我们把西北大学地质学系实践教学改革的目标确定为建设国内地学一流的,具系统性、科学性和示范性的创新人才培养实践教学新体系,为中国地球科学人才培养做出应有贡献。为此,我们必须从根本上改变以往的灌输式、验证性、单科性、继承性教学方式方法,实施从灌输式到开放式,从验证性到设计性,从单科性到综合性,从继承性到创新性的教学模式。在具体措施上,我们注重突出实践教学体系的科学性和前瞻性,实施多学科实践教学的交叉融合,体现人才培养的系统性;将地球科学的新理论、新方法融入实践教学,更新教学内容,实施研究性、开放性教学,全面提高学生的综合分析能力和创新能力。基于上述基本思想,自地质学系国家理科人才培养基地建立以来,我们致力于"鄂尔多斯盆地-秦岭造山带野外地质教学基地"的建设和实践教学的创新性改革。

西北大学坐落在渭河地堑,北邻华北地块,南接秦岭-大别造山带和扬子地块,具有得天独厚的地质与地理优势。我们最大限度地利用了西北大学的地理优势和地学研究长期积累的学科优势,将优势科研资源、地质资源转化成为教育资源。我们选择鄂尔多斯盆地-秦岭造山带为实习基地,实施跨大地构造单元、多学科交叉的区域地质综合教学。完善本科实践教学环节,完成课程内容和基本知识的融会贯通。实现地球科学复合型、创新型基础人才的培养。因此,"鄂尔多斯盆地-秦岭造山带野外地质教学基地"的基本特色可以概括为纵横不同地质单元,精细解读地质现象,培养综合分析能力,激发学生创新意识。

正是由于10多年来坚定不移地实施了鄂尔多斯盆地-秦岭造山带野外地质实践教学环节,我们改变了传统的大学本科"地质认识—地质填图—生产实习"这样的传统实践教学体系,逐步形成了"由认知—技能培训—到科研素质培训—综合分析能力、创新能力提高"的实践教学新体系。新的实践教学体系更加突出了地质综合与创新能力培养的实践教学过程,实现了课程内容和基本知识的融会贯通,学生的素质尤其是综合创新能力明显增强。因此,"鄂尔多斯盆地-秦岭造山带野外地质教学基地"建设,无疑是学科融会贯通、多学科交叉的区域地质综合实践教学的有益尝试和探索,开拓了创新型人才培养的新途径。

二、实习基地的区域特色

"鄂尔多斯盆地-秦岭造山带野外地质教学基地"跨越华北、秦岭两大一级构造单元,浓缩西北大学地质学系数十年研究成果之精华,通过对不同构造单元、不同类型地质现象的时空对比,完成多学科交叉的综合实践教学。

秦岭是在中国中心地带崛起的庞大山系,是世界著名的造山带。它是中国地质、地理、气候和环境的南北天然分界线。以秦岭山脉和淮河为界,其南北两侧在气候、自然地理环境等方面均存在显著差别。秦岭造山带及其相邻的鄂尔多斯盆地是中国大陆地质和古气候环境变迁研究的关键地区,陕北黄土高原的黄土堆积记录了第四纪气候变迁的历史,它与全球海洋沉积和极地冰川一起构成全球古气候变化的三大支柱。实习基地内包含3个国家级地质公园它们分别是洛川黄土国家地质公园、翠华山山崩国家地质公园,以及黄河壶口瀑布国家地质公园。同时,还包含两条省级自然保护地质剖面。小秦岭中-新元古代地层剖面自然保护区是研究元古代华北区与秦岭造山带区域地质的纽带;陕西柞水石瓮子中、上泥盆世地层和沉积相自然保护区,记录了由浅海台地相向深水相转变,发育滑塌、重力流沉积等一系列特征现象。

实习区内包含有跨构造单元的不同级序、不同动力学、不同深度层次的构造变形及其构造组合。鄂尔多斯盆地表层构造层次为弱变形域,华北陆块南缘构造带属于浅-表构造层次,北秦岭属于中深层次的韧性构造变形,而南秦岭构造带则属于浅-中层次的冲断褶皱组合。同时,实习区内还有多期次岩浆侵入。以上事实充分表明,本实习基地有显著的区域特色,为盆-山结合、多学科综合教学实习提供了理想的自然条件,是地学野外教学难得的"天然实验室"。

三、实习基地的教学体系

"鄂尔多斯盆地-秦岭造山带野外地质教学基地"于1993年开始建设,1998年入选国家教育委员会"创建名牌课程"项目,2002—2003年度入选陕西省精品课程,2004年度入选国家级精品课程。经历10多年来的反复认识、实践、再认识,已初步建成集高年级区域地质综合实习与创新教育于一体、具有显著特色的野外教学基地。

1.实习基地实践教学体系的教学目标

充分利用西北大学所处的地域优势,通过跨不同大地构造单元、理论和实践密切结合、多学科交叉的区域地质野外教学,实质性地改变课堂单科独进、自我封闭的教学体系,实现不同课程的融会贯通,完善本科阶段的知识结构,训练野外工作素质,培养综合分析能力,激发创新意识,实现"基础扎实、知识面宽、能力强、素质高、具创新意识"的新型人才的培养目标。

2.教学内容

采用室内与野外、理论与实践、宏观与微观的密切结合,以及不同学科交叉融合的教学过程,实际考察、研究、对比分析不同构造单元的沉积建造、岩浆建造、变质建造及其相应的不同层次构造变形和不同类型成矿特征,并结合对不同地质体的地球化学、地球物理等的综合分析,探索区域地质的演化过程。

3.教学方式

以区域地质实践教学为主,实施分阶段、协调配合的教学过程。第一阶段是野外实习前的室内教学,采用教师讲授、播放录像等方式,讲授区域地质综合研究的思路和方法,介绍野外实习地区不同大地构造单元的基本地质特征和相互关系,使学生了解教学内容和教学方法,了解野外地质概况。第二阶段是野外实践教学,它是教学的主体阶段。教学过程以学生为主体,实施自我管理、主动参与为主的管理方式。采用启发式、讨论式、研究式的教学方法,充分调动学生的积极性、主动性和趣味性,自觉观察、分析野外地质现象,收集地质素材,思考相关问题,以便训练工作能力、培养综合素质、激发创新精神。第三阶段是野外实践结束后的室内教学,采用教师和学生共同回顾总结、讨论和使用多媒体教材的不同方式,全面进行区域地质的比较综合分析,完成野外资料和地质素材的系统整理,进行室内研究和分析,编写地质报告,撰写专题论文。地质报告不拘形式,给学生提供张扬个性、体现自我的空间。

4.教学条件

在实践教学环境方面,为实施区域地质点、线、面相结合的实践教学,我们已经完成了北起延安,南至镇安全长 1 200 km,1 条主干观察路线、9 条辅助观察路线和 16 个典型区段及 80 余个观察点的野外地质研究和区域对比综合分析。主干剖面保证了对不同大地构造单元地质特征及其时空展布的理解;辅助剖面补充因露头限制的缺憾,弥补由地质作用的特殊性造成的地质现象的差异;而典型区段则兼顾区域贯通性、地质现象代表性和地壳结构层次性;在教材建设方面,实习基地已经建设了一套理论与实践相结合,具有科学性、实用性和指导性的"立体化"教材系列,包括录像教材、多媒体教材、中文野外实习指导书、英文野外实习指导书以及精品课程网站。

5.教学方法与手段

通过不断的实践、探索和改进,我们已建立了一套以学生为主体,以培养素质和能力、激发创新思维为目标的教学方法和管理体系,形成了教学特色。按照认识—实践—再认识—再实践的认知深化过程,我们实施了野外准备阶段的室内教学—野外实践教学—室内总结教学的 3 阶段教学模式及其相应的管理体系。我们营造了良好的教学环境,既建设不同构造单元自然配置、不同学科交叉融合的天然实验基地,又营造师生情感交融、宽松、自由、民主、开放的精神氛围,提供实践教学的良好环境,为学生张扬个性、激发创新思维提供优越的条件。

我们强化开展研究性教学,培养研究和创新能力。针对野外讨论、争论的地质现象,由教师启发、学生主动思考,提出既有兴趣又有科学意义的地质问题,鼓励并指导学生进行专题性研究,充分利用西北大学大陆动力学国家重点实验室的先进技术进行研究性实践,撰写研究论文,在此过程中培养学生的研究创新能力。我们还特别强调加强质量管理,改革考核方式,"以考促学、以考促教"。

6.教学体系的主要特色

本课程以现代教育理念构建了跨不同大地构造单元、多学科交叉综合的新的野外教学课程体系,实质性地完成了不同课程的融会贯通,实现了学科的交叉综合和知识结构的一体化。实习基地具有丰富的大陆地质内涵,极富研究性、启迪性和创新性,是教学成果和科研成果互动、理论和实践相结合的天然实验室,具有与时俱进的开发前景。

尽管我们在“鄂尔多斯盆地-秦岭造山带野外教学基地”建设方面已经具备了坚实的基础条件,但展望未来仍任重而道远。在未来的实习基地建设工作中,我们将进一步主动适应当代地球科学创新型人才培育的需求,立足国内,放眼世界,以服务求支持,以贡献求发展,为国家理科人才培育基地高质量优秀人才培育做出我们应有的贡献。

英文专业论文

YINGWEN ZHUANYE LUNWEN

Geochemistry and petrogenesis of Cenozoic andesite-dacite associations from the Hoh Xil region, Tibetan Plateau①

Lai Shaocong Liu Chiyang Yi Haisheng

Abstract: Volcanic units with ages in the range 4.27–44.60 Ma in the Zhentouya area south to Ulan Ul Lake, Hoh Xil region, are mainly lava, predominantly dacite in composition, together with lesser amounts of andesite. The high-potassium calc-alkaline dacite series is characterized by enriched LIL elements such as Rb, Ba, Th, U, and K, and depleted HFSE such as Ti, Nb, Ta, and Sr, whereas the andesite series containing high-potassium calc-alkaline and shoshonitic components is enriched in the LIL elements Cs, Rb, Ba, Th, U, and Sr, and depleted in HFSE Ti, Nb, and Ta. La/Sm-La and Zr/Sm-Zr discrimination diagrams indicate that both the dacite and the andesite series, with their contrasting evolutionary trends, were generated from partial melting of the continental crust rather than by fractional crystallization of a primary basaltic magma derived from the mantle.

Nd, Sr, and Pb isotopic compositions of the dacite (radiogenic $^{87}Sr/^{86}Sr$ = 0.709 227–0.709 578, $^{208}Pb/^{204}Pb$ = 39.135 6–39.262 2, $^{207}Pb/^{204}Pb$ = 15.667 5–15.706 3, $^{206}Pb/^{204}Pb$ = 18.763 5–18.797 0 and nonradiogenic $^{143}Nd/^{144}Nd$ = 0.512 287–0.512 591, ε_{Nd} = −6.85 to −0.92) provide evidence that crustal anatexis in the Tibetan Plateau played an important role in the petrogenesis of the dacite series. Moreover, the presence of granulite xenoliths in the dacite series places *P-T* constraints on the magma source at a temperature range of 780–820 ℃ and pressure of 0.86 GPa (equal to a depth of 28 km). Combined with *P-T* conditions inferred from the granulite-facies xenoliths, geochemistry of the dacite and isotopic signature of crustal anatexis suggest that partial melting of the middle part of thickened Tibetan crust could have produced the primary dacite magma series.

The andesite, to some degree, shows an adakitic compositional signature involving high Sr > 1 000 ppm, Sr/Y ratios > 50 and low Yb contents < 2 ppm. Evidently plagioclase broke down in the andesitic magma source region under the *P-T* conditions of partial melting. Compared with the dacite series, it appears that partial melting of the lower part of the thickened, eclogite-facies Tibetan crust was responsible for generation of the andesitic magma.

① Published in *International Geology Review*, 2003, 45(11).

Introduction

Over the past 50 million years, collision between India and Eurasia has produced the highest, most extensive plateau on Earth today, and the greatest present-day topographic relief. India has moved northwards with respect to Asia at a substantial rate, apparently accommodated by internal deformation of India and Asia; seismic data indicate that the crust is twice normal thickness (75 km) (Molnar, 1988). The Tibetan Plateau has a great significance for the study of the mechanics of mountain belts (England and Houseman, 1988, 1989). The present elevation and extensional deformation of the plateau are generally regarded as a response to the convective removal of the lower portion of thickened Asian lithosphere (Platt and England, 1994). This removal is also thought to be responsible for E-W extension that has taken place during the India-Asia collision (Coleman and Hodges, 1995). Despite numerous studies of the region, processes responsible for formation of the plateau and the surrounding mountain ranges are still controversial. Three hypotheses (as summarized by Miller et al., 1999) have been suggested to explain the crust thickening and the elevation of Tibet. In the first, the entire Tibetan Plateau is underthrust by the Indian lithosphere (Argand, 1924; Powell and Conaghan, 1973; Ni and Baranzagi, 1984). In the second, thickening results from inflow of material from India (Zhao and Morgan, 1987). In the third, the crust of Tibet is interpreted as having been thickened by shortening (Dewey and Burke, 1973).

Cenozoic volcanic rock associations that are widely distributed in the Tibetan Plateau can contribute to our understanding of the uplift mechanism, formation processes, and development of the plateau. They provide information on past tectonic environments as an aid to the paleo-tectonic reconstruction of the plateau. Their geochemical and isotopic signatures provide a foundation to understand the magma source region characteristics, mantle/crust interaction, and continental collision processes. Such extrusive rocks have been studied for a long time in order to provide constraints on the uplift mechanism of the Tibetan Plateau (Ugo, 1990; Arnaud et al., 1992; Turner et al., 1996; Deng, 1998; Liu, 1998; Wang et al., 1999; Deng and Sun, 1998, 1999; Chi et al., 1999; Lai, 1999; Miller et al., 1999). Up to now, three volcanic series have been identified: ultrapotassic, shoshonitic, and high-potassium calc-alkaline. Petrologic, geochemical, isotopic, and genetic features of the ultrapotassic and shoshonitic volcanic rock series have been studied (Ugo, 1990; Arnaud et al., 1992; Turner et al., 1996; Deng, 1998; Liu, 1998; Wang et al., 1999; Deng and Sun, 1998, 1999; Chi et al., 1999; Lai, 1999; Miller et al., 1999).

The Tibetan Plateau has a crust as thick as 70−75 km (Wortel et al., 1992). What happened in the middle-lower crust of the plateau during the Cenozoic period? Do geological conditions in the middle-lower crust of the plateau favor generation of intermediate and silicic volcanic rocks? If so, what kinds of magma were generated? Available data suggest that processes responsible for generation of the intermediate-silicic volcanic rock association in the Hoh Xil and north Qiangtang areas are still unclear (Lai and Liu, 2001). This paper documents the presence of Cenozoic andesite-dacite associations in the Zhentouya area, Hoh Xil region, and presents detailed geochemical compositional and Sr-Nd-Pb isotopic data for these rocks. In addition, granulite xenoliths contained mainly in the dacite are described in detail. The main objective is to better understand the middle-lower crustal magmatic processes attending Cenozoic thickening of the Tibetan Plateau. We then discussed their source and origin, and proposed a preliminary model for the generation of the intermediate-silicic crust-derived andesite-dacite associations in the Hoh Xil region.

General geological setting

Distributed from the Shishui River, via the Yudai Mountain, to the Yanshiping area in northern Tibet, Cenozoic volcanic rocks crop out as 50–200 m-thick lava sheets and long flows. The rocks overlying the Neogene Suona Lake group (N_1s) and the Jurassic Yanshiping group (J_2ys) belong to the Shipingding group (N_1sh) defined by the north Qiangtang stratigraphical classification (Zhang and Zheng, 1994). The Cenozoic volcanic series is characterized by overflow volcanics with minor pyroclastic rocks, representing a subaerial eruption of the central type. Some subvolcanic rocks generated in the volcanic conduit or nearby the crater also support this type of eruption (Fig. 1).

This study focuses on the volcanic rocks exposed in the Zhentouya area, south to the Ulan Ul Lake area, Hoh Xil region (Fig. 1). The main rock type is andesite and dacite, belonging to a typical intermediate-silicic magma series. Basalts have not been reported in this area. According to field observation, the volcanic rocks were generated in two eruption cycles, between which there is a separation of a 20–30 m-wide layer of volcanic breccia (Fig. 2). The first eruption cycle is dominated by andesite, and the second by dacite. Granulite xenoliths have been identified from both the andesitic series and dacitic series, but are mainly contained in the later (Fig. 2). The volcanic rocks discordantly overlie the Tertiary Tuotuo River group (Et) (Fig. 2), so the age of the volcanic rocks in the area lies in between the Tertiary and Quaternary. Isotope ages determined by the K-Ar method range from 4.27 Ma to 44.66 Ma (Zhang and Zheng, 1994; Deng, 1998), which fall into the age range from the above stratigraphical constraint. Therefore, the major part of the volcanic rocks in the Zhentouya area was probably formed during the period from the Late Paleogene to Neogene.

Sample and analytical techniques

All samples were examined first by microscopice study. Weathered and altered parts of the

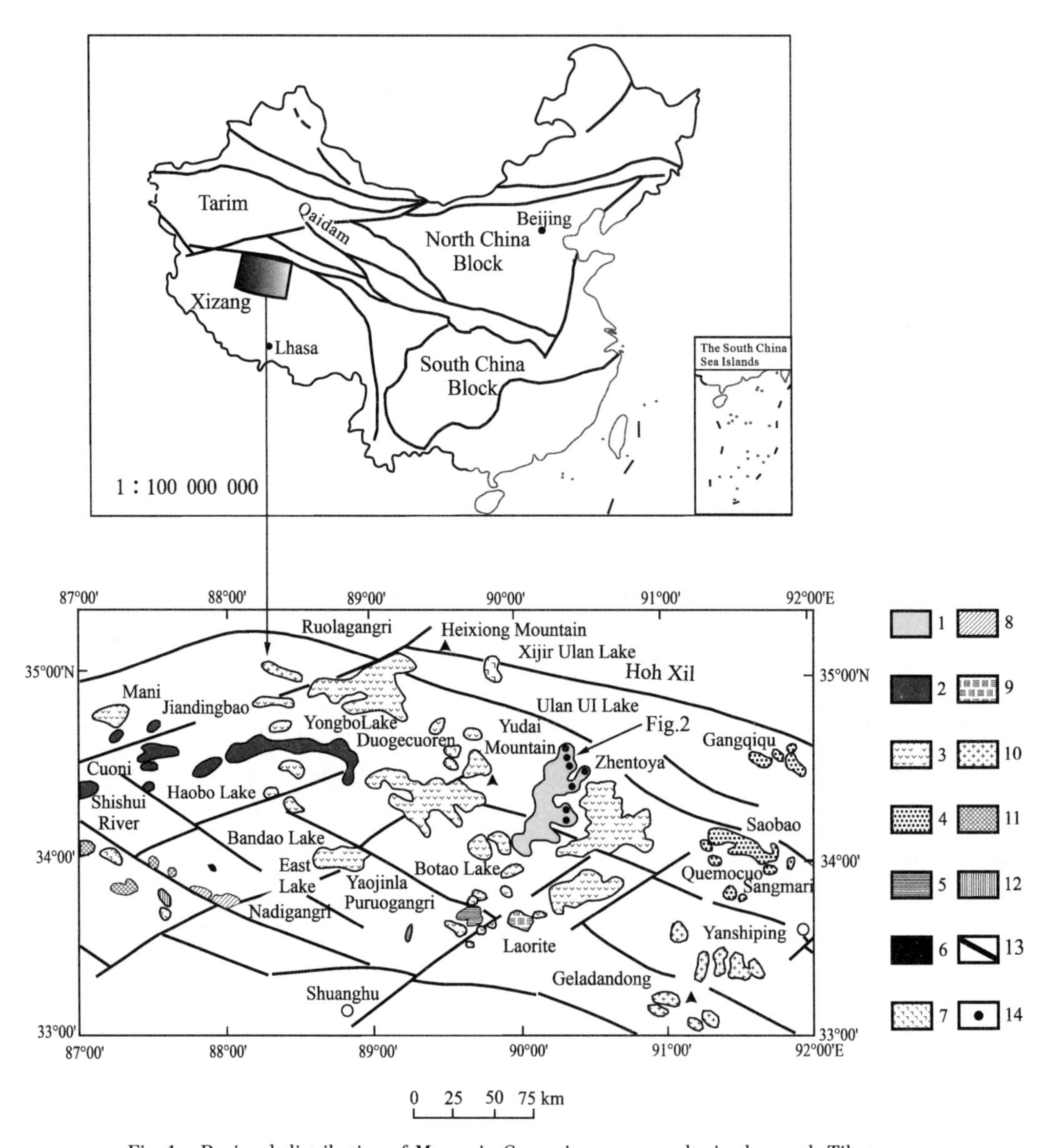

Fig. 1 Regional distribution of Mesozoic-Cenozoic magma rocks in the north Tibet area.

1.Late Paleogene to Neogene andesite and dacite; 2. Neogene adakitic rocks (Lai et al., 2003); 3.Neogene ultrapotassic and shoshonitic volcanic rocks; 4. Paleogene quartz-syenite porphyry; 5. Paleogene quartz-monzodiorite porphyry; 6. Paleogene basaltic-trachyandesite porphyry; 7. Cretaceous volcanic rocks; 8. Late Cretaceous rhyolite porphyry; 9.Late Creta-ceous adamellite porphyry; 10.Late Cretaceous adamellite; 11.Late Cretaceous granodiorite; 12.hypabyssal intrusive; 13.fault; 14.sampling locality.

samples were carefully removed and the unaltered specimens were cut into small chips with a diamond saw. The surface of the small chips was carefully cleaned in deionized water, then dried and further hand crushed. Powder of each sample was made from 50 g to 100 g splits with a tungsten carbide shatterbox at the Key Laboratory of Continental Dynamics, Northwest University, Xi'an.

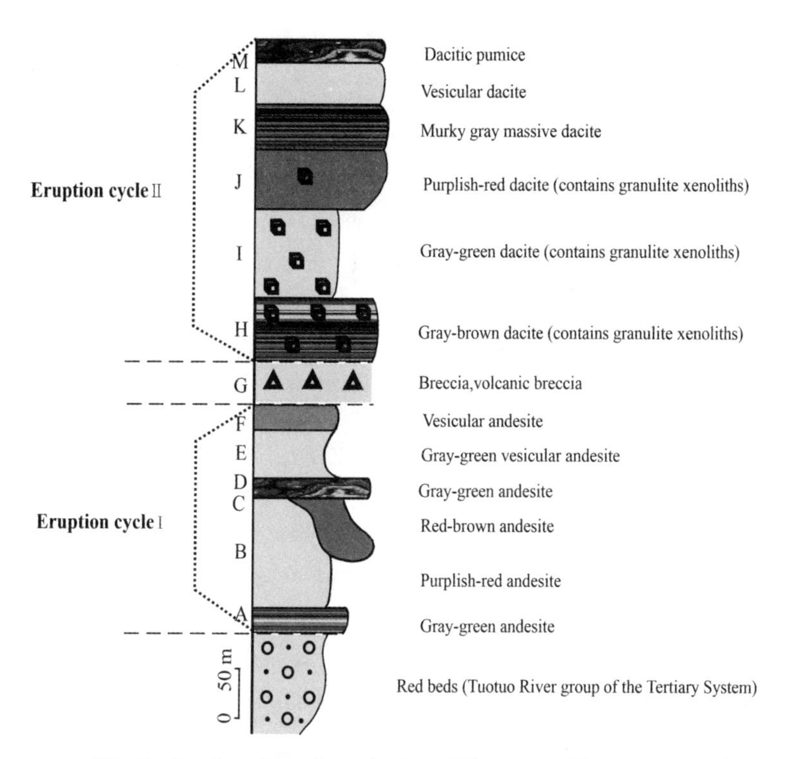

Fig. 2 Stratigraphic column for Late Paleogene to Neogene age andesite and dacite associations distributed in the Zhentouya area, Hoh Xil region.

Major elements were analyzed by means of wet chemistry at the Institute of Geochemistry, Chinese Academy of Sciences. Analysis of trace elements, including rare earth element (REE), was performed by a Finnigan MAT ELEMENT magnetic sector inductively coupled plasma mass spectrometry at the Institute of Geochemistry, Chinese Academy of Sciences. PTFE screw-top bombs were used to serve sample dissolution. Analytical reagent-grade HF and HNO_3 were purified prior to use by sub-boiling distillation. The PTFE bombs were cleaned by 20% HNO_3(v/v) and heated to 100 ℃ for 1 hour in an oven. The procedure for the ICP-MS analysis is similar to that described by Qi et al (2000), and Qi and Gregoire (2000). Trace element concentrations were drift corrected to those of standard BCR-1; analytical uncertainties are generally better than 5% for most elements.

Sr-Nd isotope ratios and contents were determined through isotope dilution techniques on a Finnigan MAT-261 multicollector mass spectrometer at the Chinese Academy of Geological Sciences. Nd isotopic ratios were measured in metal form, mass fractionation corrected to $^{146}Nd/^{144}Nd = 0.721\ 9$, and are reported relative to $^{143}Nd/^{144}Nd = 0.511\ 125$ for the J.M.Nd_2O_3 standard and $^{143}Nd/^{144}Nd = 0.512\ 725$ for the GBW04419 standard. Analytical uncertainty for $^{143}Nd/^{144}Nd$ measurements is $\pm 2\sigma$. Sr isotopic ratios were also measured in metal form, mass fractionation corrected to $^{88}Sr/^{86}Sr = 8.375\ 21$ and are reported in reference to $^{87}Sr/^{86}Sr =$

0.710 25 for the NBS987 $SrCO_3$ standard. The level of background contamination for Sm and Nd is $<5\times10^{-11}$ g and for Rb and Sr $<10^{-9}-10^{-10}$g.

For Pb analysis, 200 mg of rock powder was dissolved with bi-distilled hydrofluoric and perchloric acids in a clean Teflon bomb. Pb was separated from the silicate matrix employing HCl ion-exchange procedures. The Pb isotopes were measured on a Finnigan MAT-261 mass spectrometer in a static mode with Re single filaments and the silica-gel-phosphoric acid technique. Pb isotopic fractionation was corrected off-line with correction factors derived from NBS SRM987 standard measurements. The level of background contamination for Pb is $<2\times10^{-10}$ g.

Petrology and chemical composition

The studied andesite is characterized by phenocrysts of plagioclase ± amphibole ± biotite. Pyroxene phenocrysts occur in some samples. The groundmass shows a fine-grained texture, with a little volcanic glass. Small amounts of apatite and magnetites are restricted to the groundmass. The andesite is slightly altered, with feldspars kaolinized, mafic minerals chloritized, and carbonates invading vacuoles and fractures. However, alteration is rather limited. Judged by the presence of vesicular and amygdaloidal structures, and aphyric and porphyro-aphanitic textures, the dacite is relatively fresh. The porphyro-aphanitic dacite in many samples contains phenocrysts of quartz, amphibole, biotite and plagioclase. The groundmass with interlocking and hyalopilitic textures is fine-grained with abundant volcanic glass.

Major and trace element data of representative samples from the Zhentouya area, Hoh Xil region are given in Table 1. Two types of intermediate-silicic rocks were identified based solely on SiO_2 contents, namely andesite and dacite, with the latter dominant in volume over the former. Both andesite and dacite belong to the sub-alkaline series in the $K_2O + Na_2O$ vs. SiO_2 diagram (Fig. 3a) and show a calc-alkaline trend when plotted in the Alk-FeO-MgO diagram (Fig. 3b). In addition, the dacitic rocks plot as the high-potassium calc-alkaline series on the SiO_2 vs. K_2O diagram (Fig. 4a), whereas the andesitic rocks are variable, from high-potassium calc-alkaline to shoshonite. In general, the TAS classification schemes (Fig. 4b) of Le Bas (1986) can be used to verify rock names. Thus the analytical samples from the studied area were mainly classified into two groups, dacite and andesite (with a small amount of trachyandesite).

SiO_2 is the most abundant constituent in the andesite (56.59% – 60.74%, average 58.21%), followed by Al_2O_3 (13.78% – 17.66%, average 15.37%); MgO content is less than 6.10%, varying from 2.01% to 6.10% (average 3.97%); $Na_2O > 2.50\%$ (2.50% – 3.90%, average 3.03%). The Na_2O/K_2O ratio ranges from 0.58 to 1.39 (average 0.96). In the $Mg^{\#}$-SiO_2 diagram (Fig. 5), andesite samples from the Zhentouya area were plotted very close to the adakites from the Adak and Cook islands, and the N-AVZ (Defant et al., 2002). Therefore, the chemical composition of these rocks is principally comparable to that of the typical adakite; they are most probably related to an adakitic melt-generation process from eclogitic lower crust (Defant et al., 2002).

Table 1 Major(wt%) and trace(ppm) element analyses of volcanic rocks from the Zhentouya area, Hoh Xil region, Tibetan Plateau.

Sample	P22H5	P10H5	P14H6	P11H6	P11H7	P7H4	P16H6	P10H6	P13H3	P12H5	P3H6
Rock	Andesite	Andesite	Andesite	Andesite	Andesite	Andesite	Andesite	Andesite	Andesite	Andesite	Andesite
SiO_2	56.59	56.65	57.43	57.53	57.68	57.94	58.56	58.59	58.98	59.57	60.74
TiO_2	0.65	0.75	0.50	0.85	0.87	0.82	0.97	0.77	0.52	0.55	0.50
Al_2O_3	15.31	14.21	14.21	14.75	14.21	13.78	15.09	15.31	17.50	17.06	17.66
Fe_2O_3	2.60	1.80	6.93	1.80	1.62	1.00	2.50	1.50	1.15	1.10	1.40
FeO	2.90	3.50	1.97	3.60	2.68	3.60	3.00	3.10	3.05	3.00	3.50
MnO	0.13	0.03	0.16	0.12	0.06	0.01	0.10	0.04	0.14	0.01	0.34
MgO	6.10	5.31	4.57	3.66	3.33	4.62	3.87	2.68	4.31	3.26	2.01
CaO	5.45	5.03	4.56	4.13	4.73	4.25	3.68	4.63	4.43	4.88	2.80
Na_2O	3.30	2.60	3.30	2.60	2.60	2.80	2.50	2.80	3.70	3.90	3.20
K_2O	2.50	3.90	2.70	4.20	4.20	3.90	4.30	3.70	2.80	2.80	2.30
P_2O_5	0.40	0.80	0.43	0.90	0.87	0.77	0.87	0.77	0.50	0.43	0.27
LOI	3.86	4.85	2.87	4.34	6.82	5.83	3.90	5.80	2.30	3.10	4.67
Total	99.79	99.43	99.63	99.48	99.67	99.32	99.34	99.69	99.38	99.66	99.39
$Mg^{\#1}$	0.685	0.668	0.493	0.577	0.609	0.672	0.589	0.539	0.674	0.616	0.449
Li	34.9	42.6	21.2	16.3	15.7	18.1	16.0	15.3	14.0	16.0	10.5
Be	1.19	3.32	2.12	3.60	3.44	4.19	3.75	3.24	1.97	2.12	1.90
Sc	17.8	15.3	12.8	16.2	16.1	15.0	16.5	14.3	13.1	13.0	14.4
V	115	97.2	86.3	113	110	107	113	104	92.7	87.9	95.8
Cr	416	298	215	363	366	348	368	303	223	221	155
Co	39.8	29.8	34.9	35.8	33.5	36.0	37.6	35.6	32.9	38.7	51.7
Ni	232	196	115	206	230	220	211	163	112	117	85.8
Cu	25.9	38.9	14.1	52.6	47.0	41.1	60.2	36.3	15.9	17.0	17.1
Zn	54.8	79.4	56.8	74.2	94.2	66.8	83.4	55.7	57.9	54.0	66.5
Ga	12.4	13.3	14.0	14.2	13.3	13.4	13.7	14.4	15.2	14.5	14.9
Ge	1.21	1.54	1.18	1.45	1.25	1.48	1.36	1.13	1.30	1.26	1.07
As	17.1	13.8	18.3	15.9	13.6	15.3	27.9	13.6	15.5	7.33	36.6
Rb	82.9	184	98.7	219	221	197	223	197	82.8	103	120
Sr	1 165	1 379	1 653	1 548	1 546	1 611	1 555	1 476	1 629	1 554	1 003
Y	17.6	18.6	13.9	21.9	22.5	21.2	23.1	20.5	14.5	14.3	19.6
Zr	181	303	221	344	362	344	363	340	227	228	147
Nb	8.90	13.4	10.3	14.2	14.5	14.4	15.2	14.3	10.3	9.95	7.71
Mo	1.67	0.664	1.59	0.654	0.916	0.821	0.807	1.20	1.38	1.21	1.61
Cd	0.097	0.199	0.120	0.144	0.197	0.178	0.158	0.153	0.121	0.14	0.200
In	0.031	0.048	0.029	0.044	0.056	0.044	0.049	0.043	0.040	0.033	0.023
Sn	2.19	4.12	2.84	3.76	4.75	3.95	3.79	4.00	2.71	3.00	2.19

Continued

Sample	P22H5	P10H5	P14H6	P11H6	P11H7	P7H4	P16H6	P10H6	P13H3	P12H5	P3H6
Rock	Andesite	Andesite	Andesite	Andesite	Andesite	Andesite	Andesite	Andesite	Andesite	Andesite	Andesite
Sb	1.15	1.11	0.811	1.61	1.43	1.71	1.34	1.09	0.747	1.72	0.556
Cs	3.08	2.62	3.99	3.39	3.29	3.94	3.30	4.09	3.64	3.26	8.67
Ba	1 481	1 641	1 437	1 806	1 807	1 627	1 789	1 753	1 552	1 441	1 530
Hf	5.02	8.97	6.48	10.3	10.6	9.88	10.6	9.55	6.53	6.49	4.49
Ta	0.547	0.770	0.573	0.877	1.08	0.786	0.852	0.803	0.614	0.567	0.587
W	62.3	28.4	107	62.5	64.8	78.9	73.3	112	102	131	87.0
Tl	0.307	0.822	0.566	0.795	0.771	0.860	0.837	0.618	0.432	0.441	0.706
Pb	30.1	27.5	32.0	27.7	29.3	28.2	28.1	28.5	30.8	25.3	32.0
Bi	0.102	0.067	0.402	0.109	0.171	0.049	0.075	0.103	0.194	0.193	0.042
Th	17.6	18.2	21.1	19.7	19.6	18.5	19.3	19.3	21.7	20.8	13.0
U	3.57	3.54	4.16	3.79	3.78	3.45	3.54	3.57	4.24	4.42	3.94
La	57.5	70.7	70.1	74.4	74.1	69.1	74.6	74.7	76.6	71.8	43.3
Ce	108	138	133	152	152	139	148	147	143	136	84.7
Pr	11.0	14.8	13.2	16.4	16.8	15.5	16.1	16.0	14.1	13.5	9.07
Nd	41.0	56.6	48.5	62.6	64.2	59.1	61.2	59.3	51.0	48.8	33.3
Sm	7.16	9.15	7.30	10.4	10.5	9.38	10.1	9.49	7.74	7.24	5.89
Eu	1.61	2.10	1.66	2.21	2.37	2.27	2.30	2.10	1.73	1.76	1.42
Gd	5.19	6.78	4.95	7.66	7.58	6.74	7.52	6.91	5.49	5.14	4.60
Tb	0.660	0.784	0.596	0.924	0.993	0.837	0.951	0.885	0.653	0.591	0.586
Dy	3.26	3.96	3.01	4.58	4.60	4.13	4.66	4.02	3.05	3.02	3.31
Ho	0.633	0.679	0.492	0.762	0.795	0.727	0.805	0.694	0.500	0.500	0.634
Er	1.70	1.85	1.31	2.12	2.18	1.88	2.25	1.92	1.47	1.35	1.82
Tm	0.221	0.231	0.167	0.270	0.275	0.263	0.300	0.244	0.172	0.162	0.230
Yb	1.49	1.59	1.08	1.82	1.81	1.68	1.95	1.70	1.22	1.16	1.52
Lu	0.229	0.263	0.167	0.267	0.258	0.243	0.302	0.241	0.180	0.157	0.230

Sample	P4H	P25H11	P25H6	P1H5	P3H1	P2H4	P8H	P1H2	P2H9	P1H1	P24H3	P25H8
Rock	Dacite	Dacite	Dacite	Dacite	Dacite	Dacite	Dacite	Dacite	Dacite	Dacite	Dacite	Dacite
SiO_2	61.91	62.24	62.31	62.43	62.76	62.83	62.86	62.94	62.96	63.03	63.20	63.46
TiO_2	1.10	1.37	1.30	1.30	1.15	1.12	1.20	1.48	1.26	1.22	1.30	1.38
Al_2O_3	16.40	15.53	16.62	17.06	17.28	16.18	15.75	16.18	16.48	16.40	15.75	15.96
Fe_2O_3	2.21	2.20	1.90	1.45	1.40	1.85	1.65	1.40	1.50	1.45	2.73	1.80
FeO	3.19	2.50	3.50	3.25	3.20	3.51	3.55	3.20	3.50	3.49	2.27	2.90
MnO	0.06	0.14	0.13	0.12	0.08	0.10	0.05	0.12	0.07	0.13	0.08	0.12
MgO	1.55	1.26	1.62	1.52	1.61	1.66	1.51	1.52	1.82	1.54	1.61	1.42
CaO	4.09	4.88	3.41	2.61	2.97	3.08	3.85	3.08	2.76	3.14	3.13	3.28
Na_2O	3.50	3.40	3.30	4.00	3.20	3.50	3.30	3.40	3.31	3.40	3.90	3.50
K_2O	3.80	3.70	3.60	3.40	3.70	3.60	3.60	3.71	3.50	3.70	3.70	3.80
P_2O_5	0.73	0.93	0.83	0.90	0.83	0.87	0.73	0.90	0.76	1.13	0.83	0.84
LOI	0.90	1.42	1.02	1.50	1.20	1.05	1.30	1.30	1.30	1.19	1.33	1.03

Continued

Sample	P4H	P25H11	P25H6	P1H5	P3H1	P2H4	P8H	P1H2	P2H9	P1H1	P24H3	P25H8
Rock	Dacite	Dacite	Dacite	Dacite	Dacite	Dacite	Dacite	Dacite	Dacite	Dacite	Dacite	Dacite
Total	99.44	99.57	99.54	99.54	99.38	99.35	99.35	99.23	99.22	99.82	99.83	99.49
$Mg^{\#}$	0.363	0.347	0.374	0.394	0.413	0.383	0.369	0.400	0.422	0.386	0.389	0.377
Li	22.9	21.0	23.4	19.0	22.6	16.1	19.4	20.5	18.6	22.6	23.6	21.4
Be	3.75	2.77	7.48	2.03	2.77	2.28	2.68	2.55	2.04	3.16	2.53	2.38
Sc	9.70	7.92	15.1	8.57	8.74	9.15	8.77	8.85	9.08	9.66	8.41	8.40
V	68.9	69.7	65.8	54.7	63.3	57.1	60.7	60.8	70.4	63.2	66.8	71.7
Cr	16.2	13.6	56.0	17.5	12.9	12.5	10.6	9.61	10.7	10.4	13.2	15.7
Co	32.6	18.2	75.6	29.1	38.7	18.7	24.0	25.7	30.7	50.8	38.0	61.9
Ni	9.42	6.81	9.13	10.2	6.85	6.99	5.37	6.20	7.14	8.02	6.05	7.71
Cu	8.09	7.12	28.8	9.38	18.9	12.7	11.3	13.9	15.6	12.4	10.7	9.49
Zn	117	111	118	106	115	92.0	112	121	107	116	130	121
Ga	20.6	18.7	23.3	20.5	20.0	21.1	18.6	23.7	20.5	21.5	19.6	19.6
Ge	1.54	1.45	3.38	1.48	1.67	1.57	1.51	1.50	1.47	1.45	1.65	1.66
As	7.66	856	204	74.8	5.51	4.22	4.65	54.9	0.074	68.5	26.1	206
Rb	179	163	176	162	179	182	171	183	178	168	168	172
Sr	611	617	630	505	558	526	550	545	554	539	647	634
Y	24.0	22.8	24.0	22.0	23.5	23.0	23.2	24.6	24.8	23.2	24.7	23.8
Zr	513	538	595	537	534	546	527	599	523	557	615	521
Nb	40.8	41.6	40.0	36.0	36.3	38.1	37.2	41.5	39.4	37.8	43.1	42.8
Mo	3.16	4.44	3.33	3.67	3.22	4.90	3.10	4.06	3.45	3.89	3.24	3.21
Cd	0.164	0.192	0.698	0.275	0.163	0.230	0.168	0.283	0.218	0.248	0.245	0.195
In	0.053	0.040	0.054	0.056	0.052	0.059	0.048	0.062	0.048	0.060	0.045	0.045
Sn	4.43	4.51	7.48	4.23	4.25	5.03	4.33	5.25	4.84	4.49	4.97	5.01
Sb	0.580	0.393	0.566	0.437	0.427	0.192	0.424	0.739	0.293	0.571	0.447	0.357
Cs	2.41	3.04	4.42	3.46	2.93	2.74	2.23	3.65	2.52	3.42	2.99	3.03
Ba	1 633	1 686	1 715	1 762	1 646	1 701	1 548	1 823	1 630	1 675	1 750	1 775
Hf	13.9	14.0	16.0	16.2	14.4	14.6	12.9	16.0	13.5	14.7	15.6	15.2
Ta	1.92	1.92	1.81	1.92	1.84	1.96	1.72	2.17	2.00	2.01	1.92	1.97
W	119	54.5	443	162	185	80.8	91.0	126	137	237	199	275
Tl	0.698	0.745	0.661	0.754	0.694	0.639	0.703	0.989	0.567	0.731	0.869	0.762
Pb	38.7	37.0	43.0	37.0	36.1	33.8	40.5	41.5	32.9	38.6	38.1	38.2
Bi	0.124	0.081	0.118	0.217	0.091	0.155	0.142	0.102	0.073	0.101	0.077	0.099
Th	33.4	33.1	29.9	31.4	30.6	35.7	31.7	34.0	31.8	30.1	34.7	33.4
U	4.19	4.12	3.59	3.83	3.99	4.33	4.00	3.92	4.17	3.85	4.13	3.97
La	129	132	146	142	135	140	129	156	132	142	145	146
Ce	223	246	271	176	222	258	235	272	224	256	247	256
Pr	25.4	26.3	27.1	26.1	26.5	26.3	24.6	29.6	25.5	27.9	28.4	28.2
Nd	91.4	93.7	93.3	94.0	92.7	88.3	87.5	103	88.9	96.6	99.9	102
Sm	13.8	13.7	12.4	14.1	13.9	13.0	13.8	15.1	14.1	14.3	15.0	15.2

Continued

Sample	P4H	P25H11	P25H6	P1H5	P3H1	P2H4	P8H	P1H2	P2H9	P1H1	P24H3	P25H8
Rock	Dacite	Dacite	Dacite	Dacite	Dacite	Dacite	Dacite	Dacite	Dacite	Dacite	Dacite	Dacite
Eu	2.57	2.76	3.00	3.05	2.55	2.77	2.39	2.98	2.66	2.76	2.83	2.83
Gd	9.66	10.0	9.87	11.3	10.1	10.5	9.68	11.2	10.5	10.3	10.8	10.6
Tb	1.09	1.13	1.10	1.29	1.18	1.21	1.13	1.28	1.23	1.21	1.28	1.20
Dy	5.33	5.02	5.33	5.44	5.19	5.71	4.95	5.56	5.43	5.45	5.29	5.55
Ho	0.835	0.798	0.904	0.812	0.811	0.885	0.846	0.842	0.832	0.832	0.860	0.882
Er	2.19	1.99	2.19	2.13	2.06	2.14	2.03	2.22	2.20	2.04	2.12	2.16
Tm	0.268	0.229	0.193	0.217	0.251	0.234	0.238	0.241	0.240	0.248	0.244	0.246
Yb	1.69	1.46	1.84	1.49	1.57	1.49	1.41	1.52	1.58	1.47	1.61	1.44
Lu	0.232	0.190	0.137	0.176	0.220	0.211	0.200	0.203	0.204	0.197	0.212	0.198

Sample	P25H	P7H	P24H4	P1H3	P25H5	P2H8	P3H10	P3H16	P6H	P2H12	P3H3	P2H11
Rock	Dacite	Dacite	Dacite	Dacite	Dacite	Dacite	Dacite	Dacite	Dacite	Dacite	Dacite	Dacite
SiO_2	63.47	63.48	63.50	63.51	63.65	63.78	63.92	63.95	64.21	64.40	64.59	65.70
TiO_2	1.32	1.25	1.40	1.15	1.35	1.25	1.27	2.55	1.40	1.22	1.35	1.23
Al_2O_3	17.06	15.75	15.31	15.81	15.53	16.4	15.31	14.87	15.09	15.87	16.18	15.09
Fe_2O_3	1.70	1.60	2.10	1.50	2.20	1.50	1.45	1.50	1.60	1.30	1.25	1.30
FeO	2.80	3.60	3.10	3.40	3.00	3.40	3.35	3.30	3.40	3.40	3.05	3.00
MnO	0.16	0.04	0.09	0.08	0.11	0.15	0.05	0.05	0.02	0.10	0.03	0.05
MgO	1.44	1.58	1.52	1.61	1.68	1.67	1.55	1.78	1.50	1.53	1.49	1.62
CaO	3.06	3.59	3.22	2.93	3.25	2.63	2.81	3.96	3.66	2.42	2.33	2.77
Na_2O	3.60	3.40	3.60	3.60	3.20	3.30	3.40	2.40	3.40	3.31	3.30	3.30
K_2O	3.70	3.50	3.90	3.50	3.71	3.30	3.70	2.50	3.50	3.40	3.60	3.40
P_2O_5	0.57	0.80	0.77	0.83	0.87	0.77	0.70	0.80	0.77	0.80	0.77	0.83
LOI	1.00	0.85	1.23	1.40	1.09	1.15	1.98	1.79	0.95	1.77	1.54	1.02
Total	99.88	99.44	99.74	99.32	99.64	99.30	99.49	99.45	99.50	99.52	99.48	99.31
$Mg^{\#}$	0.390	0.379	0.369	0.398	0.391	0.406	0.395	0.422	0.376	0.396	0.412	0.432
Li	22.3	21.9	22.3	21.1	22.2	19.7	23.0	21.0	21.1	24.8	21.1	21.0
Be	3.28	3.20	3.19	2.50	2.69	2.48	2.62	3.43	3.26	2.71	2.67	2.22
Sc	9.45	9.19	7.99	9.56	9.80	9.34	8.71	11.0	8.95	9.25	8.37	8.65
V	66.3	71.9	64.4	59.4	71.2	60.8	66.7	72.4	68.7	64.0	60.6	60.3
Cr	18.0	10.1	40.1	15.3	17.5	14.8	10.9	28.5	10.8	19.0	11.3	10.8
Co	101	23.4	45.7	39.2	45.0	35.2	40.6	36.5	22.2	86.2	40.1	32.0
Ni	7.90	5.71	21.2	6.26	9.66	9.45	5.14	15.4	5.10	6.47	5.39	6.31
Cu	11.4	4.54	5.99	12.5	13.3	17.1	8.89	10.7	4.22	18.2	6.81	13.8
Zn	141	116	127	101	138	97.2	121	121	111	119	108	108
Ga	20.5	19.1	18.6	21.6	19.3	19.7	18.6	20.0	19.4	20.7	19.2	20.6
Ge	1.54	1.66	1.59	1.59	1.66	1.32	1.56	1.66	1.40	1.77	1.40	1.55
As	25.3	3.15	47.0	3.76	39.7	2.32	5.00	4.23	3.53	4.55	1.70	3.68
Rb	174	171	164	181	171	187	181	167	170	171	177	180
Sr	658	573	576	527	567	550	582	567	558	530	556	499

Continued

Sample	P25H	P7H	P24H4	P1H3	P25H5	P2H8	P3H10	P3H16	P6H	P2H12	P3H3	P2H11
Rock	Dacite	Dacite	Dacite	Dacite	Dacite	Dacite	Dacite	Dacite	Dacite	Dacite	Dacite	Dacite
Y	21.7	23.9	23.8	23.2	24.7	24.1	24.2	38.0	23.4	23.4	23.2	23.1
Zr	531	520	490	544	606	553	561	534	522	534	524	519
Nb	42.6	39.7	41.0	38.3	43.8	38.3	37.1	41.5	38.3	38.6	37.2	35.8
Mo	3.06	3.05	4.85	3.78	3.58	3.70	3.58	3.23	2.97	3.36	3.60	3.17
Cd	0.207	0.191	0.198	0.180	0.249	0.256	0.254	0.194	0.174	0.221	0.189	0.166
In	0.051	0.051	0.050	0.046	0.051	0.054	0.054	0.042	0.045	0.049	0.046	0.046
Sn	5.03	4.03	4.91	4.06	4.68	5.71	5.18	3.50	4.31	5.37	4.16	4.18
Sb	0.631	0.216	0.498	0.354	0.588	0.536	0.572	0.590	0.255	0.294	0.289	0.471
Cs	3.31	1.91	2.98	3.13	6.44	3.07	3.06	2.28	2.16	3.05	3.00	3.13
Ba	1 843	1 513	1 632	1 696	1 781	1 609	1 653	1 571	1 525	1 609	1 631	1 584
Hf	16.8	12.8	14.5	15.6	15.7	15.2	14.4	13.2	12.9	14.7	12.9	14.0
Ta	2.16	1.86	1.90	2.05	2.13	2.05	1.89	2.03	1.76	1.93	1.78	1.82
W	370	68.3	258	241	208	225	202	138	77.0	362	193	181
Tl	0.726	0.721	0.774	0.751	1.28	0.762	0.714	0.534	0.703	0.742	0.77	0.683
Pb	40.1	35.9	35.2	37.3	36.9	38.4	38.1	36.4	36.8	37.8	37.0	38.1
Bi	0.174	0.085	0.075	0.095	0.109	0.103	0.098	0.039	0.073	0.080	0.091	0.094
Th	33.7	34.7	34.8	31.9	33.2	31.6	31.4	37.1	34.7	32.4	30.7	30.8
U	4.04	4.18	3.83	4.03	3.97	4.03	4.02	4.60	4.08	4.17	3.76	4.03
La	139	128	142	142	150	142	140	133	127	135	130	135
Ce	253	233	252	141	261	238	239	248	228	231	198	214
Pr	26.5	25.3	27.7	26.0	29.2	25.5	26.6	26.5	24.9	26.3	24.8	26.4
Nd	96.3	92.2	96.3	89.4	102	88.9	94.6	97.0	90.4	93.0	88.6	91.8
Sm	13.2	13.8	14.4	13.2	14.7	13.2	13.4	13.9	13.6	13.9	13.5	13.5
Eu	2.69	2.53	2.70	2.76	2.73	2.90	2.58	2.52	2.42	2.57	2.54	2.66
Gd	9.94	9.56	10.3	10.8	10.5	10.6	10.2	10.8	9.75	10.1	10.0	9.94
Tb	1.14	1.13	1.18	1.20	1.20	1.12	1.22	1.46	1.14	1.17	1.14	1.16
Dy	4.93	5.36	5.19	5.01	5.66	5.01	5.57	7.58	5.05	5.14	5.20	4.99
Ho	0.778	0.802	0.833	0.859	0.878	0.79	0.849	1.45	0.804	0.822	0.797	0.815
Er	1.87	2.21	2.06	2.06	2.16	2.11	2.23	3.94	2.12	2.14	2.12	2.11
Tm	0.222	0.262	0.240	0.237	0.252	0.233	0.361	0.549	0.225	0.251	0.254	0.229
Yb	1.43	1.67	1.38	1.64	1.63	1.47	1.57	3.50	1.63	1.49	1.43	1.41
Lu	0.202	0.235	0.192	0.210	0.217	0.202	0.234	0.531	0.199	0.214	0.226	0.205

[1] $Mg^{\#}$ is the molecular proportion of $MgO/(MgO+FeO)$, assuming 90% of total iron as ferrous iron.

The dacite exhibits $SiO_2 > 61\%$ (61.91%–65.70%, average 63.38%), $Al_2O_3 > 14\%$ (14.87%–17.28%, average 15.99%), $MgO < 2\%$ (1.26%–1.82%, average 1.57%), $Na_2O > 2.40\%$ (2.40%–4.00%, average 3.40%), $Na_2O/K_2O = 0.86-1.18$ (average 0.95). In the $Mg^{\#}$-SiO_2 diagram (Fig. 5), dacite samples plot between the adakites from N-AVZ and the amphibolitic + eclogitic melt (Defant et al., 2002).

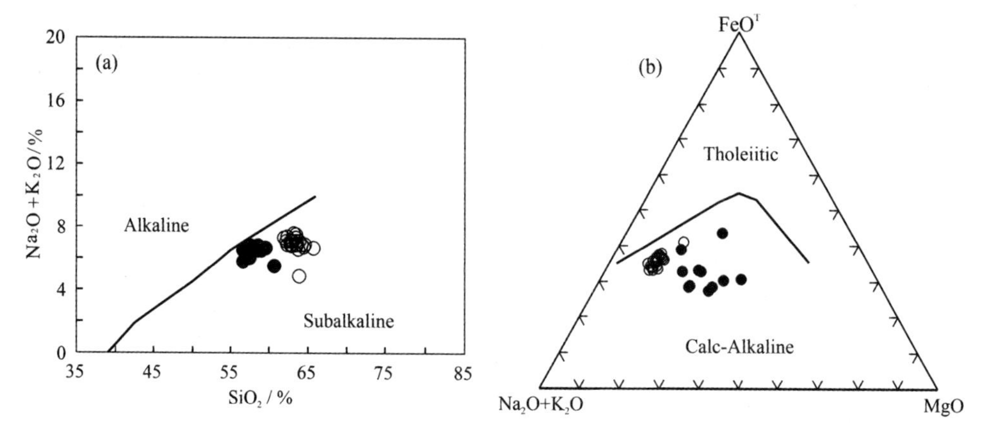

Fig. 3 SiO_2-(K_2O+Na_2O) (a) and Alk-FeO-MgO (b) diagrams for the andesite (solid circles) and dacite (open circles) from Zhentouya area, Hoh Xil region.

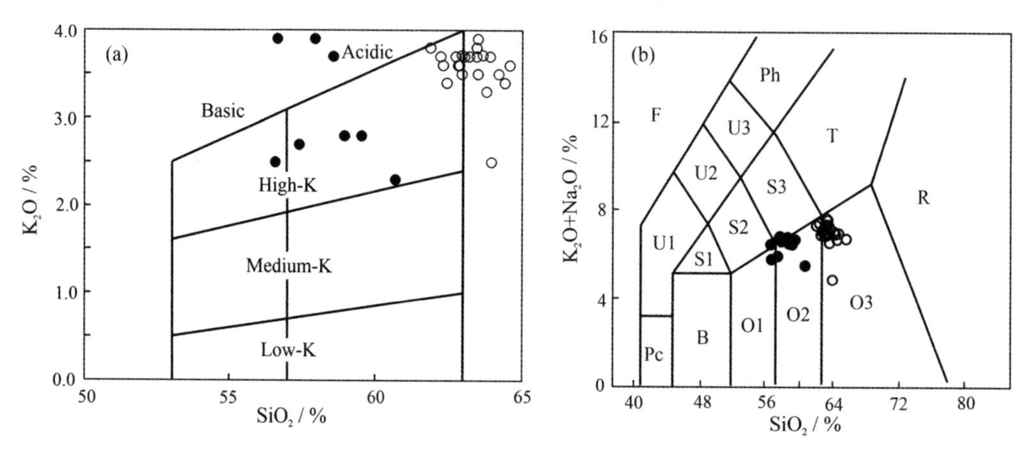

Fig. 4 SiO_2 vs. K_2O diagram (a) and TAS classification schemes (b) (after Le Bas, 1986) for the andesite (solid circles) and dacite (open circles) form the Zhentouya area, Hoh Xil region.

Low-K: low-potassium calc-alkaline series; Medium-K: medium-potassium calc-alkaline series; High-K: high-potassium calc-alkaline series; SHO: shoshonitic series; F: foidite; U1: tephrite (Ol < 10%), basanite (Ol > 10%); U2: phonotephrite; U3: tephriphonolite; Ph: phonolite; S1: trachybasalt; S2: basalitic trachyandesite; S3: trachyandesite; T: trachyte (q<20%), trachydacite (q>20%); Pc: picrobasalt; B: basalt; O1: basaltic andesite; O2: andesite; O3: dacite; R: rhyolite.

$Mg^{\#}$ is lower in dacites (0.347–0.432) than in the andesites (0.449–0.685). The oxide variation diagrams for SiO_2, TiO_2, Al_2O_3, Fe_2O_3, FeO, MgO, CaO, Na_2O, K_2O and P_2O_5 vs. $Mg^{\#}$ are shown in Fig. 6. The majority of the dacite samples show a trend of approximately constant SiO_2, TiO_2, Al_2O_3, MgO, and P_2O_5, and slightly decreasing CaO, Fe_2O_3, K_2O, over a $Mg^{\#}$ of 0.3–0.5. The andesite is relatively higher in $Mg^{\#}$, and a positive correlation between $Mg^{\#}$, MgO, and CaO can be seen in Fig. 6. Andesite defines a trend of slightly increasing K_2O and Fe_2O_3 over a $Mg^{\#}$ of 0.5–0.7, exhibiting a compositional trend that is different from that of the dacite assemblage. That is to say, the dacite was not exactly plot on the continuation from andesite on the oxide variation diagrams, especially on the $Mg^{\#}$ vs. K_2O, Fe_2O_3, and CaO plots. Accordingly, dacite and andesite from the studied area probably have different genetic processes and are less likely the results of fractional crystallization from a comagmatic system.

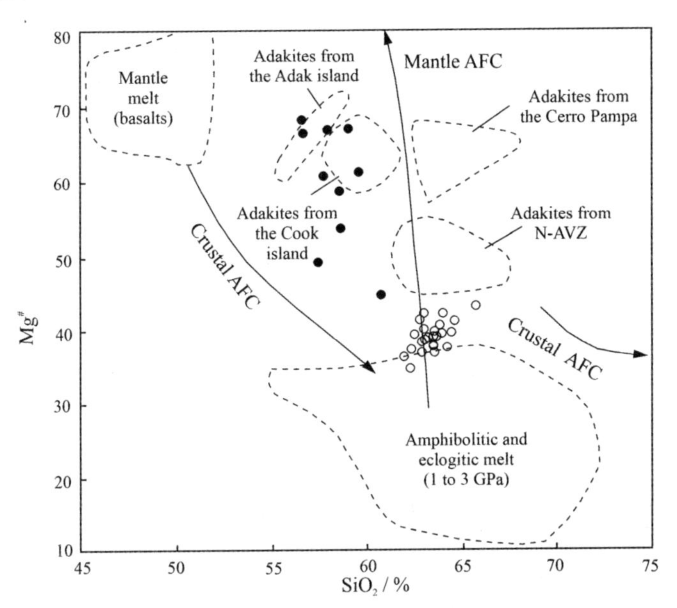

Fig. 5 $Mg^{\#}$-SiO_2 diagram for andesite (solid circles) and dacite (open circles) from the Zhentouya area, Hoh Xil region.

The original diagram is after Stern and Kilian (1996).

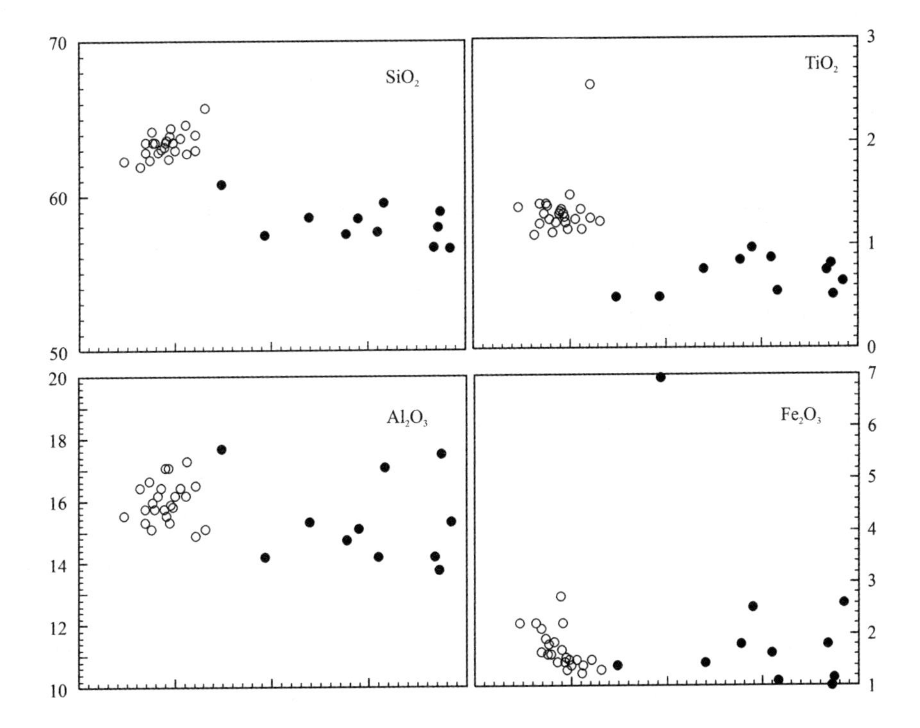

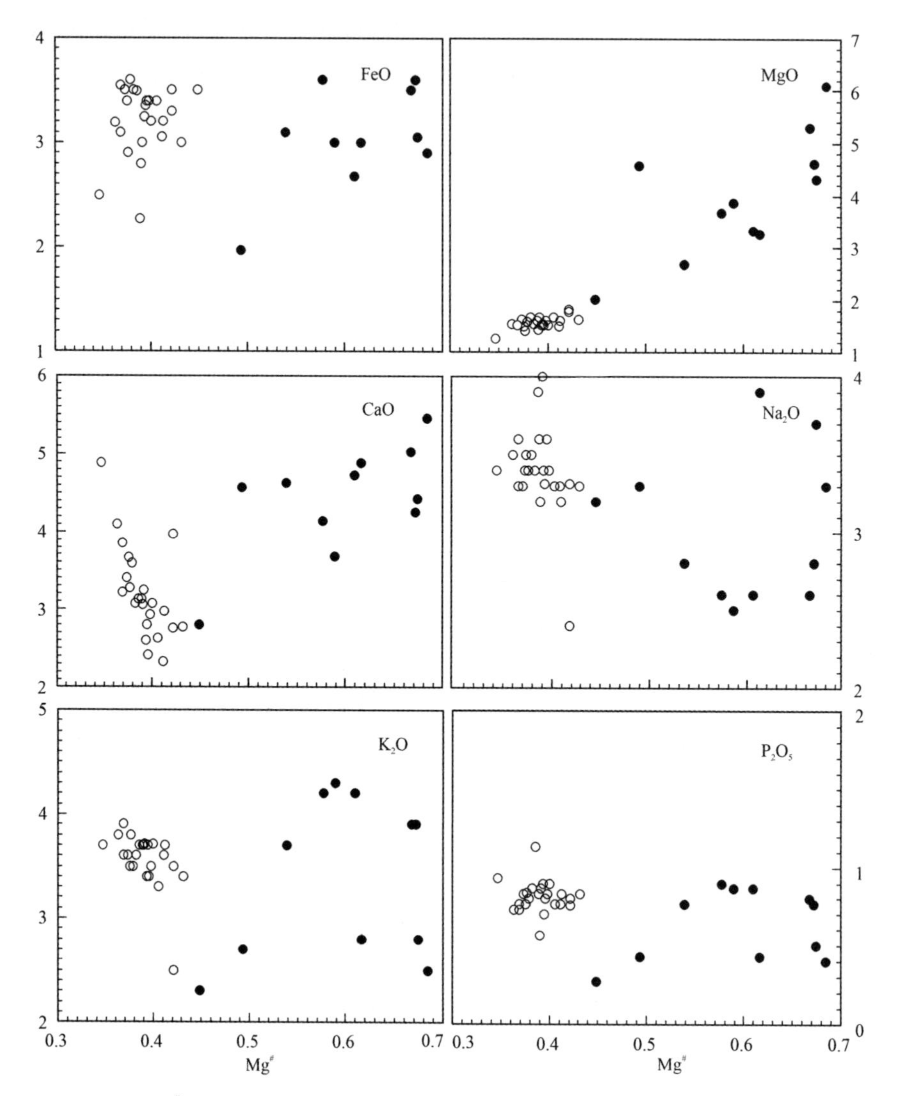

Fig. 6 Mg# vs. major-element variation diagrams for andesite (solid circles) and dacite (open circles) from the Zhentouya area, Hoh Xil region.

Trace- and rare-earth-element geochemistry

Fig. 7 shows the variation of trace elements as a function of Mg# values, which serves as a rough index of magmatic differentiation. In the andesite from the studied area, the positive correlation between Mg# and Ni, Cr, and Sc is evident. The andesite shows geochemical characteristics of high Sr and Sr/Y and slightly low Y and heavy rare earth elements (HREE), exhibiting Sr > 1 000 ppm (1 165–1 653 ppm, average 1 465 ppm), Y < 24 ppm (13.9–23.1 ppm, average 18.9 ppm), Yb < 2.00 ppm (1.08–1.95 ppm, average 1.55 ppm), Sr/Y > 50 (51–119, average 80.5), La/Yb > 28 (28–65, average 46). In the Sr/Yb vs. Yb diagram (Fig. 8), andesite samples plot within the typical adakite field. The chemical compositions of the rocks is very close to that of typical adakites from N-AVZ (Yogodzinski et al., 1995; Stern and Kilian, 1996).

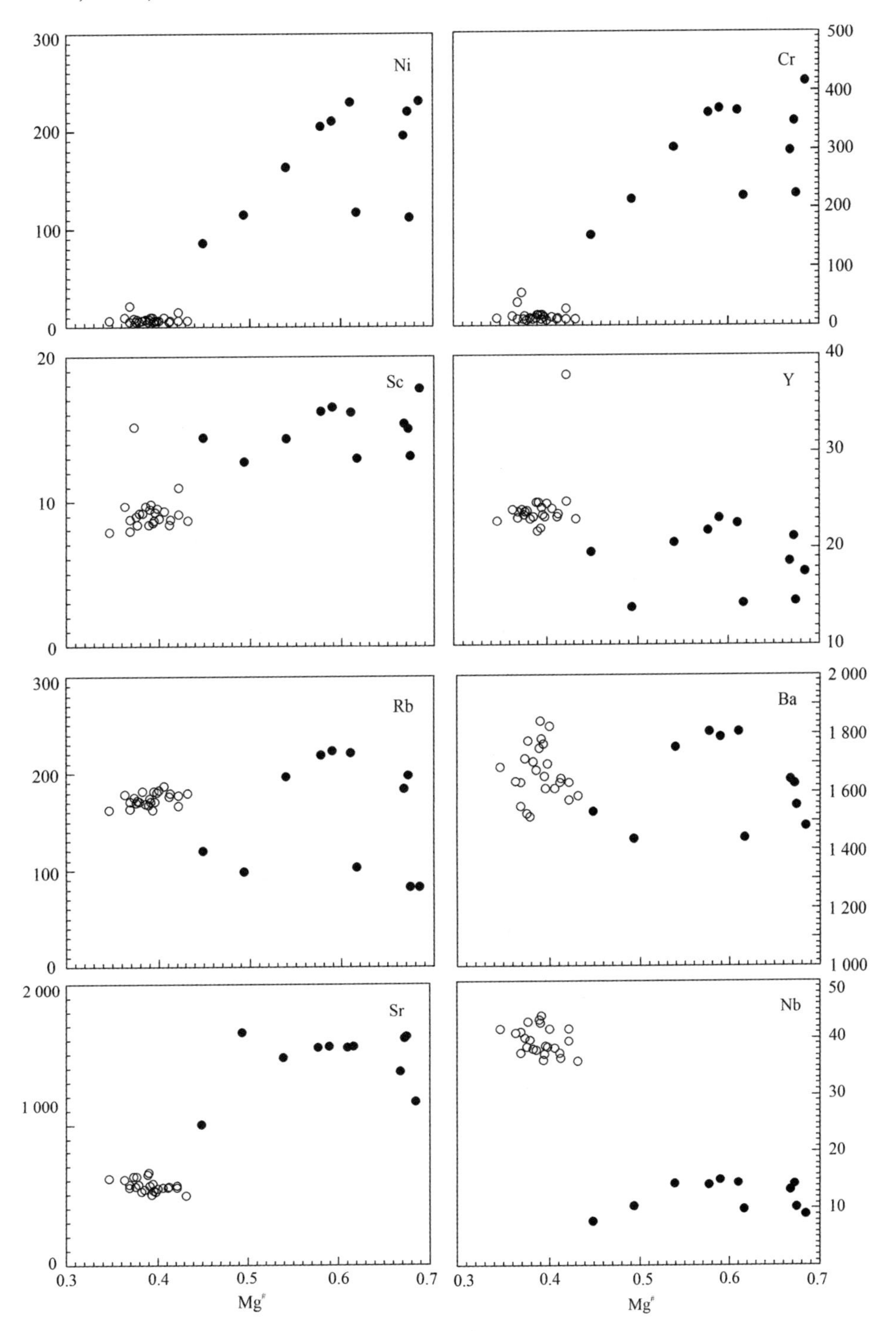

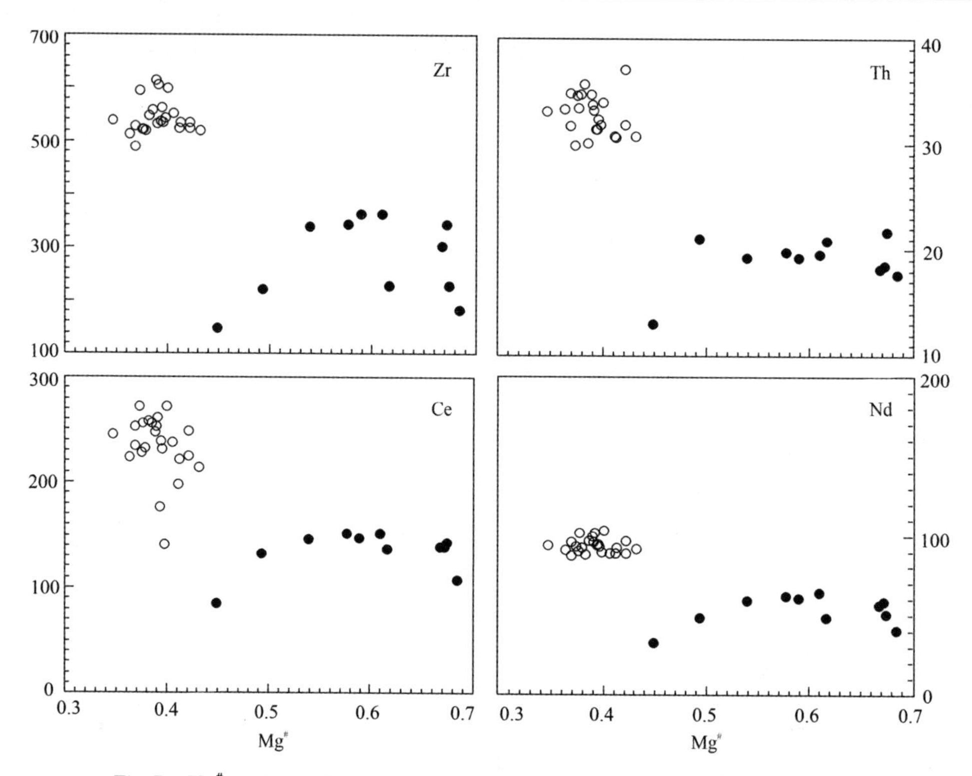

Fig. 7 $Mg^{\#}$ vs. trace-element variation diagrams for andesite (solid circles) and dacite (open circles) from the Zhentouya area, Hoh Xil region.

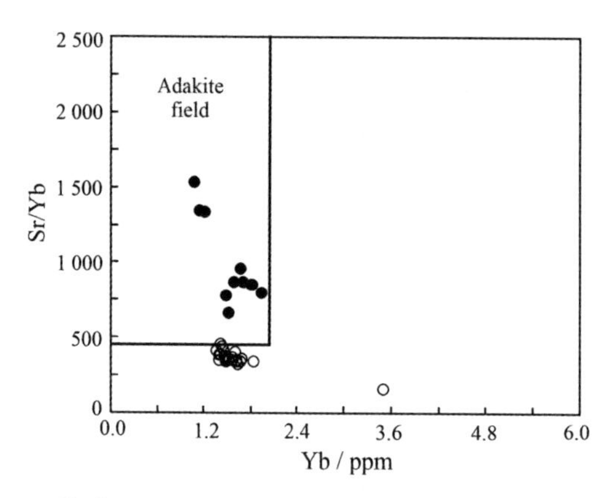

Fig. 8 Sr/Yb vs. Yb diagram for andesite (solid circles) and dacite (open circles) from the Zhentouya area, Hoh Xil region.
The original diagram is after Kay and Kay (2002).

It is clear that the Ni and Cr contents are much lower in the dacite. On the $Mg^{\#}$ vs. trace-element variation diagrams (Fig. 7), andesite and dacite are obviously different from each other, showing different trends on the $Mg^{\#}$ vs. Nb, Zr, Th, Ce and Nd, and different compositional trends as well. Compared with andesite, the dacite shows a relatively lower Sr (<660 ppm, 499–658 ppm, average 569 ppm), Y (>20 ppm, 21.7–38.0 ppm , average 24 ppm), Yb (<3.5 ppm, 1.38–3.50 ppm, average 1.62 ppm), Sr/Y (<30, 15–30, average 24), La/Yb>38 and variable (38–103, average 88).

In primary mantle-normalized trace-element spider diagrams, andesite plots (Fig. 9a) show a right sloping pattern. All samples display a strong depletion in some high-field-strength elements (Nb, Ta, and Ti) compared to the primary mantle. On the other hand, some strongly incompatible elements such as Ba, Th, U, K, and Sr are strongly enriched in the andesite. The plots of the dacite from the studied area (Fig. 9b, c) also show a right-sloping pattern. The depletion of Nb, Ta, and Ti is clear even though the Nb, Ta depleting degree is slightly lower than that of the andesite. A striking difference between andesite and dacite is that Sr is relatively

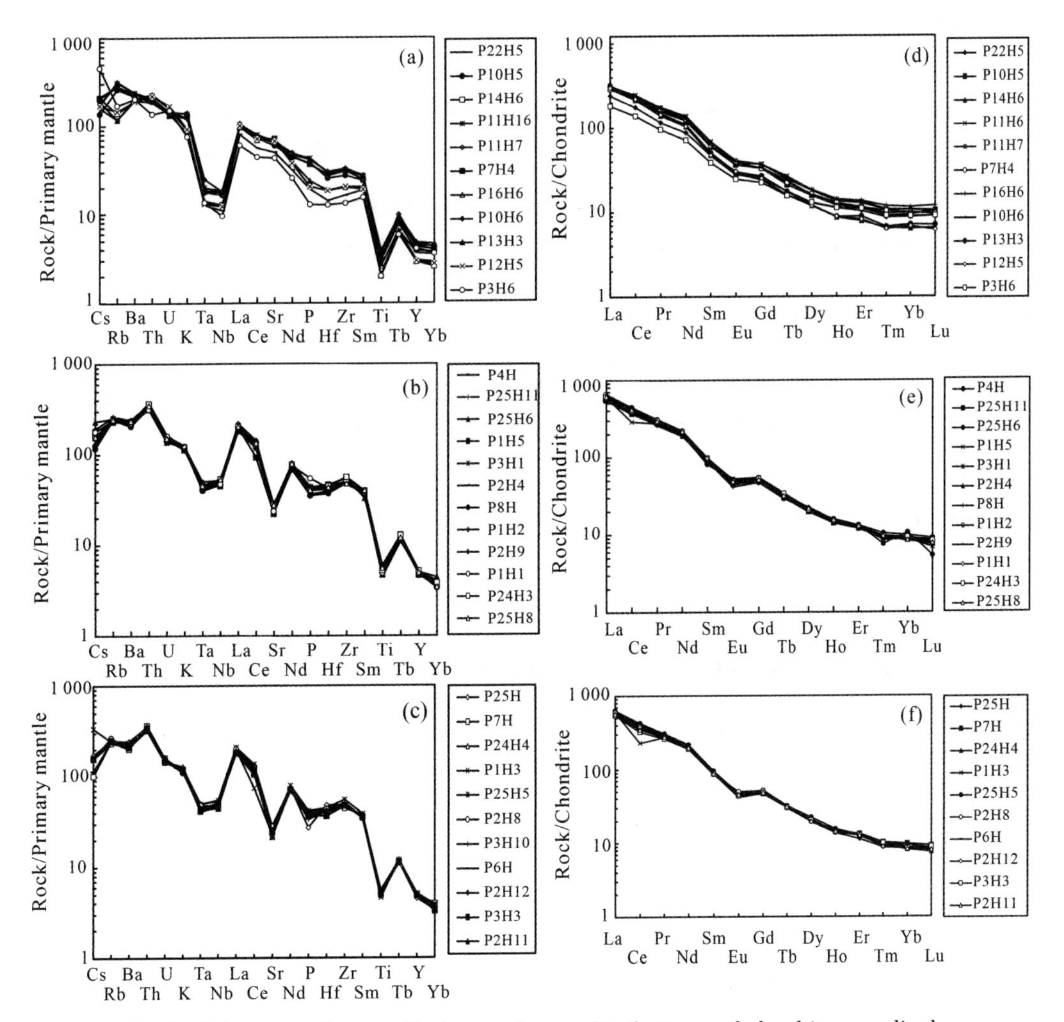

Fig. 9 Primary mantle-normalized trace element distributions and chondrite-normalized rare-earth element distributions.

(a) Andesite. (b) Dacite. (c) Dacite. (d) Andesite. (e) Dacite. (f) Dacite. Primary mantle values are from Wood et al. (1979); chondrite values are from Sun and McDonough (1989).

rich in the former but depleted in the latter. It is an important geochemical signature because it indicates that the plagioclase must have broken down in the andesitic magma sources at the temperature and pressure of melting. On the contrary, depletion of Sr in the dacite implies that the plagioclase is still stable in the dacitic magma source region at the pressure and temperature of partial melting. Therefore, dacite and andesite from the studied area should have different genetic processes and have been generated from different depths of the Tibetan Plateau crust.

The andesite from the studied area shows $(La/Yb)_N$ = 20.46–46.36 (average 33.01), $(Ce/Yb)_N$ = 15.49–33.98 (average 24.86), and δEu = 0.72–0.84 (average 0.78). The dacite shows $(La/Yb)_N$ = 27.32 – 74.12 (average 63.47), $(Ce/Yb)_N$ = 19.68 – 50.83 (average 41.56), δEu = 0.60–0.79 (average 0.66).

According to the chondrite-normalized REE distributions (Fig. 9d,e,f), all the patterns for andesite and dacite are similar to one another. It indicates that both andesite and the dacite are strongly rich in light rare-earth elements (LREE). Moreover, the andesite exhibits a slight depletion in Eu ($\delta Eu > 0.70$), similar to typical adakites (Drummond and Defant, 1990), whereas the dacite exhibits a low to intermediate degree Eu depletion.

Sr-Nd-Pb isotopic systematics

Four samples of the dacite from the studied area were analyzed for their Sr, Nd, and Pb isotopic ratios (Table 2). The isotopic compositions exhibit higher radiogenic Sr, $^{87}Sr/^{86}Sr$ is 0.709 227–0.709 578 (average 0.709 445 7), and ε_{Sr} is +146.45 to +151.48 (average 149.58). However, their Nd isotopic ratio is slightly low. $^{143}Nd/^{144}Nd$ is 0.512 287 – 0.512 591 (average 0.512 390); ε_{Nd} is negative (between –6.85 to –0.92, average –4.84).

Table 2 Sr-Nd-Pb isotopic data of volcanic rocks from the Zhentouya area, Hoh Xil region, Tibetan Plateau[1].

Sample Rock type	P2H3 dacite	P8H dacite	P25H8 dacite	P2H12 dacite
U	4.325	3.997	3.972	4.174
Th	35.744	31.727	33.397	32.387
Pb	33.76	40.465	38.185	37.75
$^{206}Pb/^{204}Pb$	18.772 4±10	18.797 0±15	18.779 5±8	18.763 5±13
$^{207}Pb/^{204}Pb$	15.681 8±7	15.706 3±10	15.688 4±7	15.667 5±10
$^{208}Pb/^{204}Pb$	39.177 3±19	39.262 2±26	39.214 6±18	39.135 6±27
Rb	158.6	168.9	164.7	159.0
Sr	757.0	557.3	696.0	526.4
$^{87}Rb/^{86}Sr$	0.606 7	0.877 3	0.685 1	0.874 4
$^{87}Sr/^{86}Sr$	0.709 227±14	0.709 578±14	0.709 412±14	0.709 566±13

Continued

Sample Rock type	P2H3 dacite	P8H dacite	P25H8 dacite	P2H12 dacite
ε_{Sr}	+146.45	+151.48	+149.10	+151.30
Sm	13.237	13.142	13.891	13.426
Nd	90.191	89.899	96.813	92.144
$^{147}Sm/^{144}Nd$	0.088 78	0.088 43	0.086 79	0.088 14
$^{143}Nd/^{144}Nd$	0.512 287±8	0.512 348±5	0.512 591±14	0.512 334±8
ε_{Nd}	−6.85	−5.66	−0.92	−5.93

[1]U, Th, and Pb concentrations were analyzed by ICP-MS; Sm, Nd, Rb, Sr, and isotopic ratios were measured through isotopic dilution. ε_{Nd} was calculated using $\varepsilon_{Nd}=[(^{143}Nd/^{144}Nd)_m/(^{143}Nd/^{144}Nd)_{CHUR}-1]\times 10^4$, $(^{143}Nd/^{144}Nd)_{CHUR}=0.512\,638$. ε_{Sr} was calculated using $\varepsilon_{Sr}=[(^{87}Sr/^{86}Sr)_m/(^{87}Sr/^{86}Sr)_{UR}-1]\times 10^4$, $(^{87}Sr/^{86}Sr)_{UR}=0.698\,990$. ε_{Nd} and ε_{Sr} and isotopic ratios of these rocks were not corrected for age; because this age is small, the difference between corrected and uncorrected values falls in the range of the analytical error.

According to the $^{143}Nd/^{144}Nd$ vs. $^{87}Sr/^{86}Sr$ diagram (Fig. 10), the Sr-Nd isotopic composition of the dacite, to some degree, is similar to that of the Cordillera Blanca Batholith which derived from newly underplated basaltic crust (Petford and Atherton, 1996), but clearly different from that of subduction-related adakites (e.g. Aleutian, Cook, and Cerro Pampa adakites; Kay, 1978; Stern and Kilian, 1996; Kay et al., 1993).

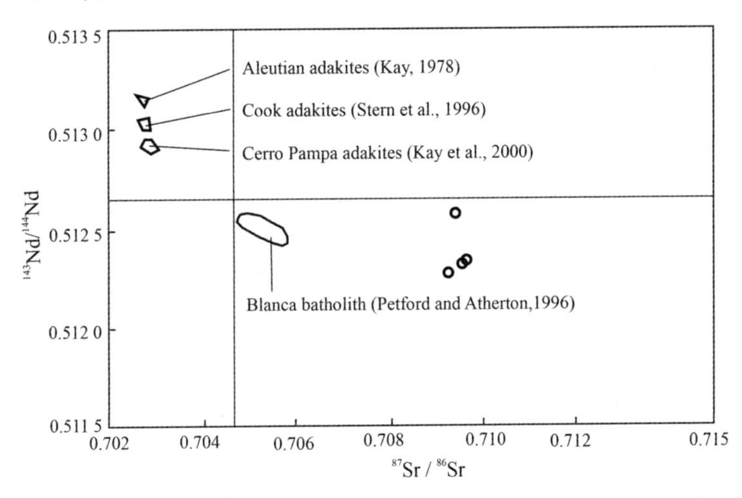

Fig. 10 $^{143}Nd/^{144}Nd$ vs. $^{87}Sr/^{86}Sr$ diagram for dacite from the Zhentouya area, Hoh Xil region.

The dacite shows $^{206}Pb/^{204}Pb$ = 18.763 5−18.797 0 (average 18.778 1), $^{207}Pb/^{204}Pb$ = 15.667 5−15.706 3 (average 15.686 0), $^{208}Pb/^{204}Pb$ = 39.135 6−39.262 2 (average 39.197 4), which indicates that the Pb isotopic ratios are rather consistent in the dacite. The Pb isotope data presented in Fig. 11 were superimposed upon Zartman and Doe (1981) Pb composition diagram. In both the $^{207}Pb/^{204}Pb$ vs. $^{206}Pb/^{204}Pb$ and $^{208}Pb/^{204}Pb$ vs. $^{206}Pb/^{204}Pb$

diagrams, samples were plotted between or adjacent to the lower crust and upper crust. This isotopic geochemical feature suggests that the magma was derived from a source typified by crust compositions.

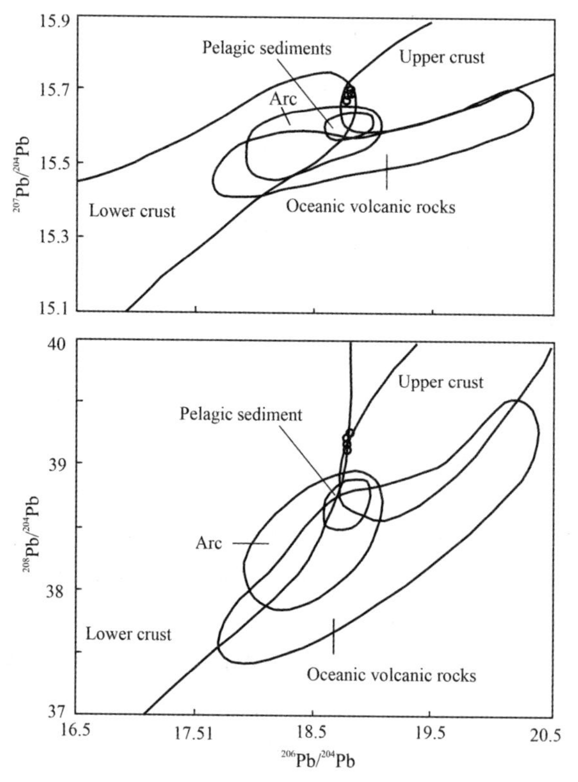

Fig. 11 Pb isotope data for dacite from Zhentouya area, Hoh Xil region.
The diagram is adapted from Zartman and Doe(1981).

In addition, their Sr-Pb and Nd-Pb isotopic systematics (Fig. 12) are quite different from DMM, PREMA, MORB, HIMU, and EM Ⅰ, though samples were plotted close to the region of EM Ⅱ. These isotopic characteristics imply that this series of rocks could have been generated by the partial melting of continental crust, but the magma source region might be contaminated by enriched mantle material that probably provided melts that intruded the lower Tibetan Plateau crust.

Granulite xenoliths included in the volcanic rocks

Granulite xenoliths were identified mainly from the second (dacite series) eruption cycle. The granulites appear fresh in a drab red and/or greyish black color. The size generally ranges from 2 cm to 6 cm in diameter, the biggest reaching 12 cm. Massive structures (some of the xenoliths show slight gneissic structures) and medium-fine grained granoblastic textures are present. On the basis of the mineral assemblage features, the granulite xenoliths can be divided

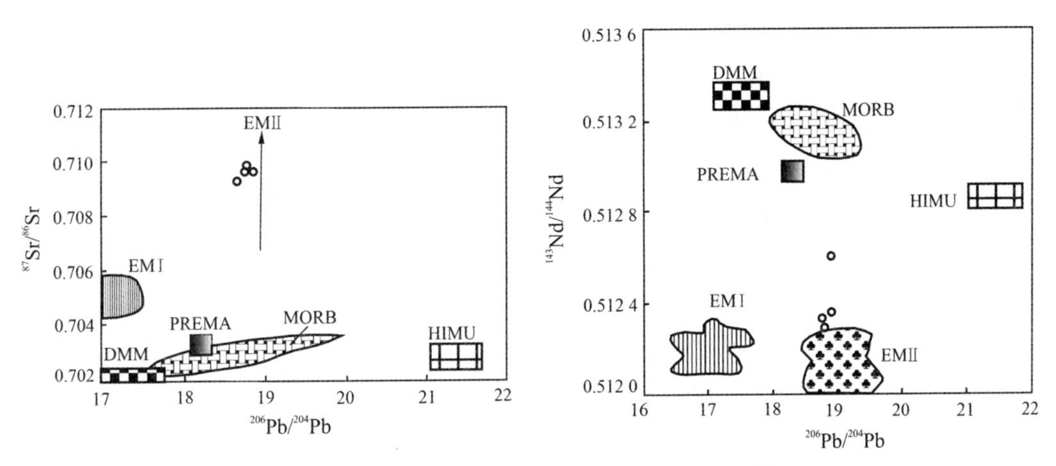

Fig. 12 $^{87}Sr/^{86}Sr$ vs. $^{206}Pb/^{204}Pb$ and $^{143}Nd/^{144}Nd$ vs. $^{206}Pb/^{204}Pb$ diagrams for dacite from the Zhentouya area, Hoh Xil region.

DMM: depleted mantle; PREMA: primary mantle; EM Ⅰ: enriched mantle (type Ⅰ); EM Ⅱ: enriched mantle (Type Ⅱ); HIMU: anomalously high $^{238}U/^{204}Pb$ mantle. The diagram is adapted from Zindler and Hart (1986).

into two categories, two-pyroxene granulite and clinopyroxene granulite. The former is composed of orthopyroxene, clinopyroxene, plagioclase, alkali feldspar, and biotite, and the latter consists of clinopyroxene, plagioclase, and quartz.

Two granulite xenoliths included in the dacite were studied by an electron microprobe. Results are shown in Table 3. According to the analysis data, the xenolith in Sample P26H10 is a typical two-pyroxene granulite, composed of Cpx + Opx + Pl + Af + Q. The orthopyroxene exhibits a light red color and obvious pleochroism under the microscope. Its component is Wo (wollastonite) 1.27 – 1.34, En (enstatite) 64.92 – 66.64, and Fs (ferrosilite) 32.09 – 33.73, a typical hypersthene. The clinopyroxene is light green, with Wo 44.76, En 27.84, and Fs 27.40, and therefore assigned to salite. The biotite shows a brown-red color, reflecting Fe, Ti enrichment, $Mg/(Mg+Fe^{T})$ (MF) >0.5 (0.73–0.75), Ti>0.3 (0.770 0–0.773 4), and is a titanium-ferrous biotite. The plagioclase is allotriomorphic-granular, An = 51.03 – 61.49, belonging to labradorite. The alkali feldspar exhibits Or = 63.62, Ab = 30.77, soda-sanidine. Based on the electron microprobe data (Table 3), the granulite was formed at a temperature between 783 ℃ and 818 ℃, as determined by the two-pyroxene thermometry (Wood and Banno, 1973).

The xenolith in Sample P6H belongs to the typical clinopyroxene granulite, composed of Cpx + Pl + Q. The clinopyroxene is salite, with Wo = 44.80–44.99, En = 33.19–33.34, Fs = 21.82–21.86. The plagioclase An = 50.88–65.90 is labradorite.

According to the clinopyroxene-plagioclase-quartz mineral assemblage geobarometer suggested by Ellis (1980), the pressure of formation of the clinopyroxene granulite xenolith has been calculated, as follows:

Table 3 Electron microprobe mineral analysis (wt%) for granulites from Cenozoic volcanic rocks, Hoh Xil region[1].

Sample	P26H10	P26H10	P26H10	P26H10	P26H10	P26H10	P26H10	P26H10	P6H	P6H	P6H	P6H	P6H
Point	1	2	3	4	5	6	7	8	9	10	11	12	13
Mineral	Cpx	Cpx	Cpx	Pl	Pl	Af	Bi	Bi	Cpx	Cpx	PI	PI	Q
SiO_2	51.20	49.26	51.30	54.26	54.91	64.82	36.51	35.33	47.63	48.15	57.87	57.04	99.53
TiO_2	0.15	0.11	0.11	0.00	0.04	0.19	7.11	7.09	0.42	0.38	0.00	0.08	0.00
Al_2O_3	1.01	2.30	2.81	30.61	29.45	19.24	15.56	16.50	4.12	4.59	26.01	26.09	0.03
Cr_2O_3	0.03	0.08	0.00	0.00	0.00	0.05	0.00	0.00	0.03	0.00	0.00	0.00	0.18
FeO	16.40	21.64	21.55	0.27	0.13	0.88	10.50	11.47	13.31	13.23	0.51	0.55	0.00
MnO	0.70	0.80	0.94	0.00	0.01	0.00	0.06	0.16	0.51	0.56	0.00	0.05	0.06
MgO	9.35	25.22	23.27	0.14	0.09	0.14	17.32	17.40	11.36	11.32	0.13	0.08	0.19
CaO	20.91	0.67	0.67	11.61	10.89	1.13	0.06	0.00	21.42	21.16	11.93	10.79	0.00
Na_2O	0.45	0.07	0.04	2.89	3.19	3.42	0.46	0.44	1.17	1.00	3.09	5.34	0.00
K_2O	0.00	0.00	0.00	0.76	0.88	10.75	8.97	8.98	0.00	0.00	0.49	0.66	0.02
NiO	0.00	0.02	0.00	0.07	0.18	0.00	0.00	0.03	0.06	0.00	0.01	0.00	0.00
Total	100.20	100.17	100.68	100.61	99.79	100.63	96.55	97.40	100.04	100.38	100.04	100.68	100.01
[O]=	6	6	6	8	8	8	22	22	6	6	8	8	2
Si	1.975 2	1.847 1	1.900 1	2.427 0	2.472 6	2.940 9	5.280 8	5.102 3	1.833 8	1.839 7	2.594 5	2.562 1	0.996 9
Al	0.045 9	0.101 6	0.122 7	1.613 7	1.563 0	1.028 8	2.652 6	2.808 5	0.187 0	0.206 7	1.374 4	1.381 2	0.000 4
Ti	0.004 4	0.003 1	0.003 1	0.000 0	0.001 4	0.006 5	0.773 4	0.770 0	0.012 2	0.010 9	0.000 0	0.002 7	0.000 0
Cr	0.000 9	0.002 4	0.000 0	0.000 0	0.000 0	0.001 8	0.000 0	0.000 0	0.000 9	0.000 0	0.000 0	0.000 0	0.001 4
Fe	0.529 1	0.678 7	0.667 5	0.010 1	0.004 9	0.033 4	1.270 1	1.385 3	0.428 6	0.422 7	0.019 1	0.020 7	0.000 0
Mn	0.022 9	0.025 4	0.029 5	0.000 0	0.000 4	0.000 0	0.007 4	0.019 6	0.016 6	0.018 1	0.000 0	0.001 9	0.000 5
Mg	0.537 6	1.409 5	1.284 7	0.009 3	0.006 0	0.009 5	3.734 0	3.745 5	0.651 9	0.644 7	0.008 7	0.005 4	0.002 8
Ca	0.864 3	0.026 9	0.026 6	0.556 4	0.525 4	0.054 9	0.009 3	0.000 0	0.883 6	0.866 2	0.573 1	0.519 3	0.000 0
Na	0.033 7	0.005 1	0.002 9	0.250 6	0.278 5	0.300 9	0.129 0	0.123 2	0.087 3	0.074 1	0.268 6	0.465 1	0.000 0
K	0.000 0	0.000 0	0.000 0	0.043 4	0.050 6	0.622 2	1.655 1	1.654 4	0.000 0	0.000 0	0.028 0	0.037 8	0.000 3
Ni	0.000 0	0.000 6	0.000 0	0.002 5	0.006 5	0.000 0	0.000 0	0.003 5	0.001 9	0.000 0	0.000 4	0.000 0	0.000 0
Sum	4.013 9	4.100 4	4.037 0	4.913 1	4.909 2	4.998 8	15.512	15.612	4.103 8	4.083 1	4.866 7	4.996 1	1.002 3
Wo	44.76	1.27	1.34						44.99	44.80			
En	27.84	66.64	64.92						33.19	33.34			
Fs	27.40	32.09	33.73						21.82	21.86			
An				51.03	61.49	5.61					65.90	50.80	
Ab				29.47	32.59	30.77					30.88	45.50	
Or				5.10	5.92	63.62					3.20	3.70	
Ca							1.86	0.00					
Mg							74.48	73.00					
Fe							25.33	27.00					

[1] Analyzed by Electron Microprobe Section, Xi'an Institute of Geology and Mineral Resources, 2001.

(1) Plagioclase (average composition of the analysis points 11 and 12 listed in Table 3): Si^{4+}(0.956 1), Ti^{4+}(0.000 5), Al^{3+}(0.511 0), Fe^{3+}(0.007 4), Mg^{2+}(0.002 6), Mn^{2+}

(0. 000 4), Ca^{2+}(0. 202 6), Na^{+}(0. 136 0), K^{+}(0. 012 2). Molecular formula: $(Na_{0.367}K_{0.033})_{0.4}Ca_{0.546}Fe^{3+}_{0.02}Al_{1.378}Si_{2.578}O_8$, end member components: Ab = 38. 80, An = 57. 70, Or = 3. 5.

(2) Clinopyroxene (average composition of the analysis points 9 and 10 listed in Table 3): Si^{4+}(1. 837), Ti^{4+}(0. 012), $Al^{3+(IV)}$(0. 163), $Al^{3+(VI)}$(0. 034), Fe^{3+}(0. 080), Fe^{2+}(0. 346), Mg^{2+}(0. 647), Mn^{2+}(0. 017), Ca^{2+}(0. 782), Na^{+}(0. 080), K^{+}(0. 000). Molecular formula: $T_{Sch16.3}(Ac+Jd)_8Hd_{36.6}Di_{39.2}$.

(3) Based on the temperature obtained from two-pyroxene granulite in Sample P26H10 determined by geothermometry, we take the equilibrium temperature of the clinopyroxene granulite as T = 1 073 K = 800 ℃. Then the lnKd = −2. 727 and the pressure of formation should be 0. 842 GPa. According to the pressure correction method suggested by Ellis (1980), we can arrive at the corrected pressure range as follows:

P_{max}(corrected) = 0. 858 GPa (equal to a depth of 28. 31 km)

P_{min}(corrected) = 0. 845 GPa (equal to a depth of 27. 90 km)

Discussion

Petrogenesis of the volcanic rocks

One of the critical issues in igneous petrology is regarding the generation of intermediate-silicic magmas. In general, two basic petrogenesis models has been suggested for the origin of such volcanic rocks: ①they are fractional crystallization products of mantle-derived basaltic magma; and ② they originated by partial melting of the lower continental crust.

Based on research done by Allerger and Minster (1978), La/Sm-La discrimination diagram can be effectively used to identify the genesis for those volcanic rocks. As shown in Fig. 13a, andesite and dacite associations from the studied area possess the lithogenesis characteristics of the source rock that underwent partial melting. The Zr/Sm-Zr diagram in Fig. 13b has illustrated these characteristics.

It is well known that intermediate-silicic magmas can be generated by partial melting of the continental crust. The degree of compositional diversity among these magmas is determined by the pressure of melting, the availability of aqueous fluids, and the original composition of the protolith. Because crustal rocks are subjected to extremely high pressure and temperature during continental collision, a contractional orogen potentially is the place where such magmas can be produced. The most dramatic effect of continental collision is that it routinely transports large volumes of supracrustal rocks to a depth of at least 60 km. Transport is accomplished by a combination of thrusting, homogeneous shortening, and subduction processes. In addition, the lithosphere mantle under a continental collision zone can be either thickened (by continued

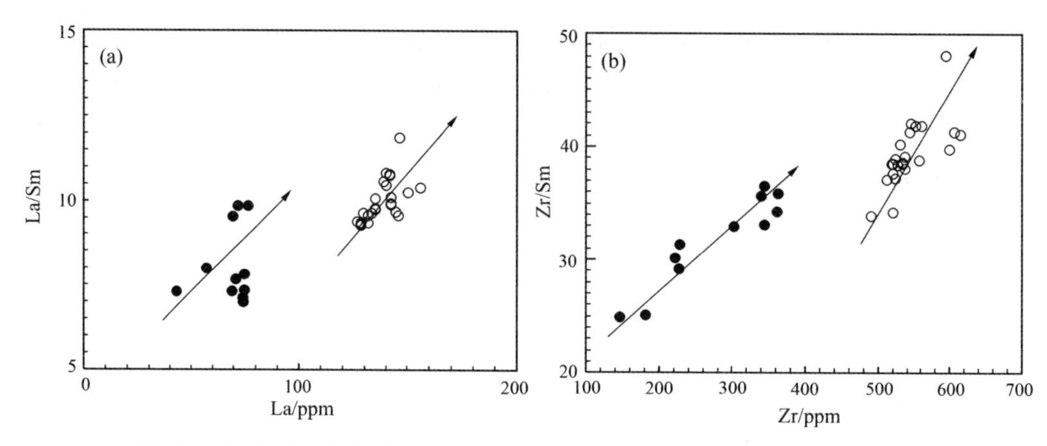

Fig. 13 La-La/Sm (a) and Zr-Zr/Sm (b) diagrams for andesite (solid circles) and dacite (open circles) from the Zhentouya area, Hoh Xil region.

subduction) or thinned (by gravitational instability or thermal erosion), which strongly affects the heat flux into the base of the orogen. If the lithospheric mantle underlying the collisional orogen detaches and sinks, it will raise the temperature of the base of the crust to the values characteristic of asthenospheric mantle (Butler et al., 1997; Patino and McCarthy, 1998), and then partial melting of the lower crust would be initiated.

A recent study conducted by Wei et al. (2001) indicated that the middle and lower crust is anomalously conductive across most of the Tibetan Plateau from south to north. The integrated conductivity of the Tibetan crust ranges from 3 000 to greater than 20 000 siemens. Such pervasively high conductance suggests that partial melts and/or aqueous fluids are widespread within the Tibetan crust. In northern Tibet, the conductive layer, formed due to partial melting, is situated at a depth of 30–40 km (Wei et al., 2001).

Atherton and Petford (1993) proposed that the melting of the lower crust occurs when basaltic melts underplate the lower crust; another model suggests that in areas where the continental crust is thick, the lower crust can become garnet-granulitic to eclogitic, separate from the upper crust, and sink into the mantle (Kay et al., 1993, 1994). This delamination process would bring deeply buried sections of the lower crust or the upper part of the delaminated lower crust in contact with a relatively hot mantle, which could initiate melting, and in turn bring about the formation of intermediate-silicic crust-derived volcanic rocks. Thus andesite and dacite associations from the studied area could have originated from the lower part of the thickened crust of the Tibetan Plateau.

Source regions of the andesite and dacite magmas

As shown in Fig. 13, it is clear that the andesite and dacite samples from the studied area have different La/Sm and Zr/Sm ratios, and plot along two different equilibrium partial melting compositional evolutionary trend lines. In addition, on the Mg# vs. Sr, Nb, Zr, Th, Ce, and Nd diagrams (Fig. 7), the andesite and dacite samples define contrasing trends, representing two different compositional evolution models. It suggests that the dacite and andesite probably were subjected to different genetic processes and are unlikely to be the result of fractional crystallization from a comagmatic system.

It is noteworthy that the dacite exhibits a relatively lower Sr (Sr<660 ppm) compared with andesite (Fig. 9b,c). This may imply that plagioclase was stable in the dacitic magma source region at the pressure and temperature of partial melting (Rapp et al., 1999, 2002; Rapp, 2001). As mentioned above, the granulite xenoliths included in the dacite indicate a formation pressure of about 0.845–0.858 GPa, equal to a depth of about 27–30 km. At this depth (<30 km), plagioclase is still stable. Therefore, the dacitic magma probably was generated by partial melting of the thickened Tibetan Plateau crust at the depth of about 30 km. Its source region was attended by granulite-facies *P-T* conditions.

In contrast, the andesite from the studied area exhibits high Sr (Sr > 1 000 ppm) (Fig. 9a). Additionally, the andesite exhibits Y<24 ppm, Yb<2.00 ppm, Sr/Y>50, located in the adakite region on the Sr/Yb-Yb diagram (Fig. 8). These facts imply that the plagioclase must have been broken down in the andesitic magma source region at the pressure and temperature of partial melting. The andesitic magma source region was attended by eclogitic-facies *P-T* conditions. Thus, generation depth of the andesitic magma was deeper than that of the dacitic magma.

Experimental petrology and phase equilibrium research (Huang and Wyllie, 1986; Deng et al., 1996; Patino and McCarthy, 1998) indicate that when the pressure is ≤1 GPa, the liquidus line mineral of the andesitic magma system is plagioclase ± pyroxene; when >1 GPa, it is pyroxene ± garnet ± amphibole. In cases where the buried depth is more than 45 km, continental crust will be transferred into eclogitie (Yardley and Valley, 1997). The andesitic magma system that was generated from the lower part of the thickened crust of the Tibetan Plateau would not be in equilibrium with plagioclase during the partial melting process and differentiation, and was evidently generated from an eclogitic source at a depth of at least 45 km.

Volcanic rock generation model

Mantle-derived alkaline volcanic rocks have erupted on the Tibetan Plateau since at least 45 million years ago. The shoshonitic to ultrapotassic volcanic rock series is the most widespread and common type in the northern Qiangtang and the Hoh Xil areas (Liu, 1989; Ugo, 1990; Arnaud et al., 1992; Turner et al., 1996; Deng, 1998; Wang, 1999; Deng and Sun, 1999; Yang et al., 1999; Chi et al., 1999; Lai, 1999; Miller et al., 1999). The generation of andesitic and dacitic magmas in the thickened crust of the Tibetan Plateau requires addition of heat that was probably provided by mantle-derived melts intruded into the crust. In fact, the isotopic characteristics of the dacite from the studied area imply that this series was generated by crust partial melting. But the magma source region might have been contaminated by mantle-derived melts intruded into the thickened lower crust. Therefore, mantle-derived alkaline magma may have provided extra heat for the thickened Tibetan crust.

As we know, when basaltic magmas are emplaced into the continental crust, the melting and generation of intermediate-silicic magma can be expected (Herbert and Sparks, 1988). Convincing evidence exists that most of the large granite complexes in the continental crust are the result of crustal anatexis (Pitcher, 1987). There is also widespread evidence that basaltic magma from the mantle in many cases is intimately associated with the generation of intermediate-silicic magmas (Hildreth, 1981). Petrological and geochemical features of many intermediate-silicic igneous rocks are also convincingly explained by admixture of a mantle-derived (mafic) component with a crustal melt. According to experimental studies (Herbert and Sparks, 1988), basalt intruded into the crust causes melting and high heat flow. Indeed, the basalt underplating of the crust is currently a popular idea to explain both large-scale crustal melting and the strongly layered character of the lower crust. The Tibetan thickened crust is strongly layered in terms of its composition, density, and mechanical behaviour (Molnar, 1988). The upper crust is relatively cold and brittle, whereas the middle-lower crust is hotter (Wei et al., 2001), has a higher density, deforms in a ductile manner, and is characterized by prominent horizontal layering. Basalt magma is emplaced into the continental crust as dikes and sills. In some cases, where the rate of magma input is high, these intrusions can coalesce to form large magma chambers (Fig. 14). In general, sills provide a more promising situation in which extensive crustal melting can occur. Horizontal intrusions concentrate their heat at a particular level in the crust and do not dissipate it over a large depth range (Herbert and Sparks, 1988). A large or region of partially molten crust provides an effective density barrier. Herbert and Sparks (1988) suggested that basalt magma reaching such a level spreads out as horizontal intrusions (Fig. 14).

In as much as the Qiangtang and Hoh Xil areas constitute the core of the Tibetan Plateau, the crust obviously was thickened during the Late Paleogene to Neogene period (Deng, 1998). It is a typical continental collision orogenic belt. In addition, the intrusion of mantle-derived magma (shoshonitic and ultrapotassic series) and the underplating in the lower crust led to partial melting of the crustal rock sequence. Therefore, the lower part of the thickened Tibetan crust, to some degree, had been favorable to generate crust-derived magmas during the Cenozoic Period. Especially, when the mantle-derived alkaline magma intensively intruded the Tibetan lower crust and/or underplated the Moho, the initial melting of intermediate-silicic

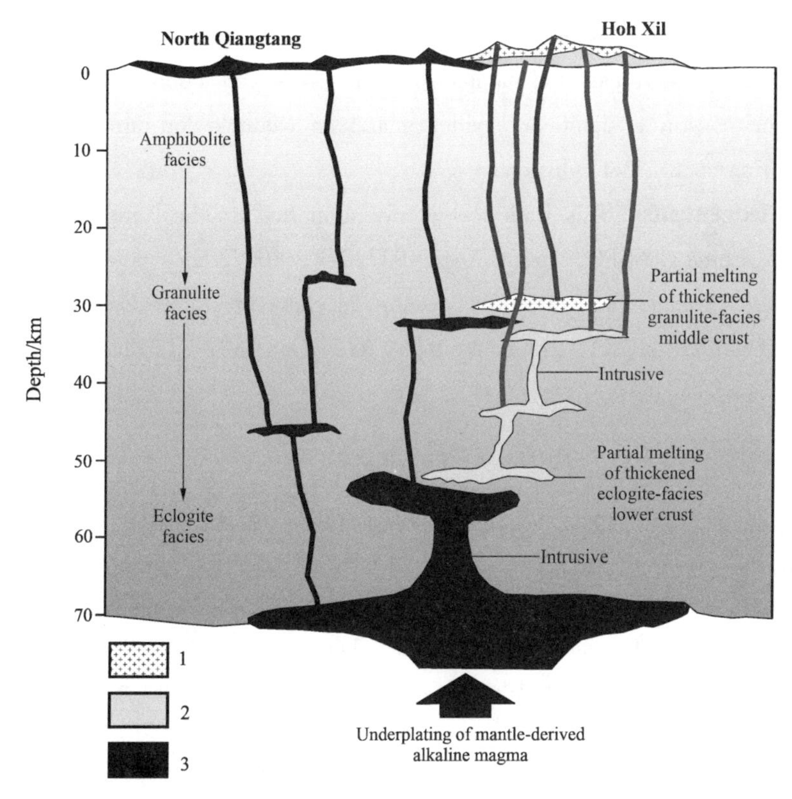

Fig. 14 Sketch model for the andesitic and dacitic magmas formation by emplacement of mantle-derived alkaline magma into thickened Tibetan crust.

1. crust-derived dacite in the studied area; 2. crust-derived andesite in the studied area; 3. mantle-derived alkaline (ultrapotassic and shoshonitic series) magma.

crust-derived magmas would be expected (Fig. 14).

At the depth of >45 km in eclogitic facies, partial melting of the thickened Tibetan crust would result in high-Sr, low-Y andesitic magma showing a somewhat adakite-like signature. At the depth of about 30 km under granulitic-facies conditions, where plagioclase is stable, partial melting of the crust would lead to a dacitic magma that shows a strong depletion in Sr and contains many granulite xenoliths. The formation of the andesite and dacite associations developed in the studied area should be ascribed to this kind of typical tectonic setting. A sketch of the petrogenetic model is summarized in Fig. 14.

Magma generation during continental collisions requires the convergence of appropriate source rocks and requisite thermal conditions. Rocks of sedimentary and igneous origin are the most abundant in the continental crust, and can partially melt to yield intermediate-silicic magmas. Belts of continental collision thus are fertile environments for magma generation on a large scale. Whether this potential is realized or not depends on the thermal structure of the orogen. Processes such as thermal blanketing resulting from tectonic thickening, delamination of the subcrustal lithosphere, underplating of mantle-derived magma, and extensional collapse of thickened crust favor the attainment of temperature high enough to cause widespread crustal melting. Therefore, identification of the andesite and dacite associations in the Zhentouya area, Hoh Xil region constitute significant evidence and an example for further study of magma generation during continental collisions.

Acknowledgements This work was jointly supported by the National Natural Science Foundation of China (NSFC, Grant Nos. 40272042, 40072029) and the Teaching and Research Award Program for Outstanding Young Teachers in Higher Education Institutions, Ministry of Education (MOE), China. We thank A. L. Guo and X. L. Zhang for assitance with English exposition.

References

Allegre, C.J., Minster, J.F., 1978. Quantitative method of trace element behavior in magmatic processes. Earth & Planetary Science Letters, 38, 1-25.

Argand, E., 1924. La tectonique de L'Asie. Proc. 13th Int. Geol. Congress, 7, 171-172.

Arnaud, N.O., Vidal, P., Tapponnier, P., Matte, P., Deng, W.M., 1992. The high K_2O volcanism of northwestern Tibet: Geochemistry and tectonic implications. Earth & Planetary Science Letters, 111, 351-367.

Atherton, M.P., Petford, N., 1993. Generation of sodium-rich magmas from newly underplated basaltic crust. Nature, 362, 144-146.

Butler, R.W.H., Harris, N.B.W., Whittington, A.G., 1997. Interactions between deformation, magmatism and hydrathermal activity during active crustal thickening: A field example from Nanga Parbat, Pakistan Himalayas. Mineralogical Magazine, 61, 37-68.

Chi, X.G., Li, C., Jin, W., Liu, S., Yang, R.H., 1999. The Cenozoic volcanism evolutionary and uplifting mechanism of the Qinghai-Tibet plateau. Geological Review, 45, 978-986 (in Chinese).

Coleman, M., Hodges, K., 1995. Evidence for Tibetan plateau uplift before 14 Myr ago from a new minimum age for east-west extension. Nature, 374, 49-52.

Defant, M.J., Xu, J.F., Kepezhinskas, P., Wang, Q., Zhang, Q., Xiao, L., 2002. Adakites: Some variations on a theme. Acta Petrologica Sinica, 18, 129-142.

Deng, J.F., Zhao, H.L., Mo, X.X., Wu, Z.X., Luo, Z.H., 1996. Continental roots-plume tectonics of China: Key to the continental dynamics. Beijing: Geological Publishing House, 112 (in Chinese).

Deng, W.M., 1998. Cenozoic intraplate volcanic rocks in the Northern Qinghai-Xizang plateau. Beijing: Geological Publishing House, 168 (in Chinese).

Deng, W.M., Sun, H.J., 1998. Features of isotopic geochemistry and source region for the intraplate volcanic rocks from northern Qinghai-Tibet plateau. Earth Science Frontiers, 5, 307-317 (in Chinese).

Deng, W.M., Sun, H.J., 1999. Cenozoic volcanism and the uplifting mechanism of the Qinghai-Tibet plateau. Geological Review, 45, 952-958 (in Chinese).

Dewey, J.F., Burke, K., 1973. Tibetan variscan and precambrian basement reactivation: Products of continental collision. Journal of Geology, 81, 683-692.

Drummond, M.S., Defant, M.J., 1990. A model for trondhjemite-tonalite-dacite genesis and crustal grouth via slab melting. Journal of Geophysical Research, 95, 21503-21521.

Ellis, D.J., 1980. Osumilite-sapphirine-quartz granulites from Enderby Land, Antarctica: *P-T* conditions of metamorphis , implications for garnet-cordierite equilibria and the evolution of the deep crust. Contrib. Mineral. Petrol., 74, 201-210.

England, P.C., Houseman, G.A., 1989. Extension during continental convergence, with application to the Tibetan plateau. J. Geophys. Res., 94, 17561-17579.

England, P.C., Houseman, G.A., 1988. The mechanics of the Tibetan plateau. Philosophical Transactions of the Royal Society of London, Series A, 326, 301-319.

Herbert, E.H., Sparks, R.S.J., 1988. The generation of granitic magmas by intrusion of basalt into continental crust. J. Petrol., 29, 599-624.

Hildreth, W., 1981. Gradients in silicic magma chambers: Implication for lithospheric magmatism. J. Geophys. Res., 86, 92-153.

Huang, W. L., Wyllie, P. J., 1986. Phase relationships of gabbro-tonalite-granite H_2O at 15 kbar with applications to differentiation and anatexis. American Mineralogist, 71, 301-316.

Kay, R. W., 1978. Aleutian magnesian andesites: Melts from subducted Pacific Ocean crust. J. Volcanol Geotherm Res., 4, 117-132.

Kay, S.M., Coira, B., Viramonte, J., 1994. Young mafic back-arc volcanic rocks as guides to lithospheric delamination beneath the Argentine Puna Plateau, Central Andes. J. Geophys. Res., 99, 14323-14339.

Kay, S.M., Ramos, V.A., Marquez, Y.M., 1993. Evidence in Cerro Pampa volcanic rocks for slab-melting prior to ridge-trench collision in southern South America. J. Geol., 101, 703-714.

Kay, R.W., Kay, S.M., 2002. Andean adakites: Three ways to make them. Acta Petrologica Sinica, 18, 303-311.

Lai, S.C., Liu, C.Y., 2001. Enriched upper mantle and eclogitic lower crust in north Qiangtang, Qinghai-Tibet plateau: Petrological and geochemical evidence from the Cenozoic volcanic rocks. Acta Petrologica Sinica, 17, 459-468 (in Chinese).

Lai, S.C., 1999. Petrogenesis of the Cenozoic volcanic rocks from the northern part of the Qinghai-Tibet plateau. Acta Petrologica Sinica, 15, 98-104 (in Chinese).

Lai, S.C., O'Reilly, S.Y., Zhang, M., 2003. Geochemistry of Cenozoic Adakitic Rocks from Qiangtang of the Tibetan Plateau: Evidence of Partial Melting of the Thickened Tibetan Crust. Earth Planet. Sci. Lett. (submitted).

Le Bas, M.J., 1986. A chemical classification of volcanic rocks based on the total alkaline-silica diagram. J. Petrol., 27, 745-750.

Liu, J.Q., 1998. Volcanoes in China. Beijing: Science Press, 110 (in Chinese).

Miller, C., Schuster, R., Klotzli, U., Frank, W., Purtscheller, F., 1999. Post-collisional potassic and ultrapotassic magmatism in SW Tibet: Geochemical and Sr, Nd, Pb, O isotopic constraints for mantle source characteristics and petrogenesis. J. Petrol., 40, 1399-1424.

Molnar, P., 1988. A review of geophysical constraints on the deep structure of the Tibetan plateau, the Himalaya and the Karakoram, and their tectonic implications. Philosophical Transactions of the Royal

Society of London, Series A, 326, 33-88.

Ni, J., Barazangi, M., 1984. Seismotectonics of the Himalayan collision zone: Geometry of the underthrusting Indian plate beneath the Himalaya. J. Geophys. Res., 89, 1147-1163.

Patiño, D.A.E., McCarthy, T.C., 1998. Melting of crustal rocks during continental collision and subduction. Netherlands: Kluwer Academic Publishers, 55.

Petford, N., Atherton, M.P., 1996. Na-rich partial melting from newly underplated basaltic crust: The Cordillera Blanca Batholith, Peru. J. Petrol., 37, 1491-1521.

Pitcher, W.S., 1987. Granites and yet more granites forty years. Geol. Rundschau, 76, 51-79.

Platt, J.P., England, P.C., 1994. Convective removal of lithosphere beneath mountain belts: Thermal and mechanical consequences. American Journal of Science, 294, 307-336.

Powell, C.M., Conaghan, P.G., 1973. Plate tectonics and the Himalayas. Earth Planet. Sci. Lett., 20, 1-12.

Qi, L., Gregoire, D.C., 2000. Determination of trace elements in twenty six Chinese geochemistry reference materials by inductively coupled plasma-mass spectrometry. The Journal of Geostandards and Geoanalysis, 24, 51-63.

Qi, L., Hu, J., Gregoire, D.C., 2000. Determination of trace elements in granites by inductively coupled plasma mass spectrometry. Talanta, 51, 507-513.

Rapp, R.P., Shimizu, N., Norman, M.D., Applegate, G.S., 1999. Reaction between slab-derived melts and peridotite in the mantle wedge: Experimental constraints at 3. 8 GPa. Chem. Geol., 160, 335-356.

Rapp, R.P., 2001. A review of experimental constraints on adakite petrogenesis. In: Symposium on adakite-like rocks and their geodynamic significance, Beijing, China, 10-13.

Rapp, R.P., Xiao, L., Shimizu, N., 2002. Experimental constraints on the origin of potassium-rich adakites in eastern China. Acta Petrologica Sinica, 18, 293-302.

Stern, C.R., Kilian, R., 1996. Role of the subducted slab, mantle wedge and continental crust in the generation of adakites from the Andean Austral volcanic zone. Contrib. Mineral. Petrol., 123, 263-281.

Sun, S.S., McDonough, W.F., 1989. Chemical and isotopic systematics of oceanic basalts: Implications for mantle composition and processes. In: Saunders, A.D., Norry, M. J. Magmatism in the Ocean Basin. Geol. Soc. Special Publ., 42, 313-345.

Turner, S., Arnaud, N., Liu, J., Rogers, N., Hawkesworth, C.J., Harris, N., Kelley, S., van Calsteren, P., Deng, W., 1996. Post-collision, shoshonitic volcanism on the Tibetan plateau: Implications for convective thinning of the lithosphere and the source of ocean island basalts. J. Petrol., 37, 45-71.

Ugo, P., 1990. Shoshonitic and ultrapotassic post-collisional dykes from northern Karakorum (Sinkiang, China). Lithos, 26, 305-316.

Wang, B.X., Ye, H.F., Peng, Y.M., 1999. Isotopic geochemistry features and its significance of the Mesozoic-Cenozoic volcanic rocks from Qiangtang basin, Qinghai-Tibet plateau. Geological Review, 45, 946-951 (in Chinese).

Wei, W.B., Unsworth, M., Jones, A., Booker, J., Tan, H.D., Nelson, D., Chen, N.S., Li, S.H., Solon, K., Bedrosian, P., Jin, S., Deng, M., Ledo, J., Kay, D., Roberts, B., 2001. Detection of widespread fluids in the Tibetan crust by magnetotelluric studies. Science, 292, 716-718.

Wood, B.J., Banno, S., 1973. Garnet-orthopyroxene relationships in simple and complex systems. Contrib.

Mineral. Petrol., 42, 109-124.

Wood, D.A., Joron, J.L., Treuil, M., Norry, M., Tarney, J., 1979. Elemental and Sr isotope variations in basic lavas from Iceland and the surrounding sea floor. Contrib. Mineral. Petrol., 70, 39-319.

Wortel, M.J.R., Hansen, U., Sabadini, R., 1992. Convective removal of thermal boundary lager of thickened continental lithosphere: A brief summary of causes and consequences with special reference to the Cenozoic tectonics of the Tibetan plateau and surrounding regions. Tectonophysics, 223, 67-73.

Yang, D.M., Li, C., He, Z.H., 1999. Petrochemistry and tectonic settings of the volcanic rocks of Songwori in Nima county, Tibet. Geological Review, 45, 972-977 (in Chinese).

Yardley, B.W.D., Valley, J.W., 1997. The petrologic case for a dry lower crust. J. Geophys. Res., 102, 12173-12185.

Yogodzinski, G.M., Kay, R.W., Volynets, O.N., Koloskov, A.V., Kay, S.M., 1995. Magnesian andesite in the western Aleutian Komandorsky region: Implications for slab melting and processes in the mantle wedge. Geol. Soc. Am. Bull., 107, 505-519.

Zartman, R.E., Doe, B.R., 1981. Plumbotectonics: The model. Tectonophysics, 75, 135-162.

Zhang, Y.F., Zheng, J.K., 1994. An introduction to the geological evolution of Hoh Xil and its adjacent region. Beijing: Seismological Press, 1-86 (in Chinese).

Zhao, W., Morgan, W.J., 1987. Injection of Indian crust into Tibetan lower crust: A two dimentional finite element model study. Tectonics, 6, 489-504.

Zindle, A., Hart, S.R., 1986. Chemical geodynamics. Annu. Rev. Earth Planet. Sci., 14, 493-573.

Partial melting of thickened Tibetan crust: Geochemical evidence from Cenozoic adakitic volcanic rocks①

Lai Shaocong　Qin Jiangfeng　Li Yongfei

Abstract: Major and trace elements and Sr-Nd-Pb isotopic data are presented for the Late Paleogene to Neogene age (9.4–26.9 Ma) adakitic volcanic rocks from the northern Qiangtang region, Tibetan Plateau. Lacking Eu depletion, this group of rocks exhibits high SiO_2 (>58%), Sr (>350 ppm) and La/Yb (>22) ratios, and low Y (<18 ppm) and Yb (<1.5 ppm). Samples plot within the adakite field on the Sr/Y vs. Y diagram. Isotopic compositions of this group rocks have radiogenic Sr and Pb ($^{87}Sr/^{86}Sr$ = 0.706 365–0.708 156; $^{208}Pb/^{204}Pb$ = 38.955–39.052; $^{207}Pb/^{204}Pb$ = 15.651–15.672; $^{206}Pb/^{204}Pb$ = 18.679–18.839), and non-radiogenic Nd ($^{143}Nd/^{144}Nd$ = 0.512 411–0.512 535; ε_{Nd} = −4.43 to −2.01). These geochemical features suggest magma generation by partial melting of an eclogitic mafic lower crust beneath the Tibetan Plateau. On the basis of $Mg^{\#}$, the Qiangtang "adakites" can be divided into two groups—a high-Mg group ($Mg^{\#} \geqslant 0.45$) and a subordinate low-Mg group ($Mg^{\#} \leqslant 0.45$). These groups are attributed to convective thinning of the enriched lithospheric mantle beneath Qiangtang region and southward intracontinental subduction of Kunlun-Qaidam block along the Jinshajiang suture (JRS) zone, respectively. Our study indicated that both convective thinning of the mantle lithosphere and intracontinental subduction played important roles in the Cenozoic magmatism of northern Tibet.

Introduction

Convergence and subsequent collision of the Indian and Eurasian continents created the Tibetan Plateau, and the Himalaya and the Karakoram ranges during the past 40–50 Ma (Klootwijk and Radhakrishnamurty, 1981; England and Houseman, 1988, 1989; Molnar, 1988). The present elevation and extensional deformation of the plateau are generally regarded as a response to the convective removal of the lower portion of the thickened Asian lithosphere (Platt and England, 1994). Despite numerous studies of the region, processes responsible for formation of the plateau and the surrounding mountain ranges are still controversial. In fact, study of Cenozoic volcanic rocks widely distributed in the Tibetan Plateau can contribute to our

① Published in *International Geology Review*, 2007, 49(4).

understanding of the processes of its formation and development. Their geochemical and isotopic signatures provide basic information about magma source characteristics, mantle/crust interaction, and continental collision processes.

Cenozoic volcanism in the Tibetan Plateau includes three volcanic series: ultrapotassic, potassic, and high-potassium calc-alkaline (Deng, 1989, 1991, 1992, 1993, 1998; Liu, 1998). Petrology, geochemistry, and petrogenesis of the ultrapotassic and potassic volcanic rock series have been studied extensively (Ugo, 1990; Arnaud et al., 1992; Turner et al., 1996; Deng, 1998; Deng and Sun, 1998; Chi et al., 1999; Lai, 1999; Miller et al., 1999; Wang et al., 1999; Nomade et al., 2004). However, little information exists on the adakitic volcanic rocks in the northern Qiangtang region (Lai and Liu, 2001). This study therefore deals with postcollisional northern Qiangtang volcanic rocks found in an active continental collision zone. Our new geochemical and Sr-Nd-Pb isotopic data suggest the magmas to be, in some degree, comparable to adakites from modern subduction zones. Our main objective is to improve understanding of the lower crust magmatic processes of the Tibetan Plateau in response to Cenozoic crustal thickening. We then discuss the source and origin of these rocks.

Geological background

From north to south, the interior of the Tibetan Plateau comprises the roughly E-W-trending Songpan-Ganzi, Qiangtang and Lhasa terranes. Volcanic rocks have been mapped by many workers throughout the Tibetan Plateau. These rocks, ranging in age from 65 Ma to <1 Ma (Fig. 1), are mainly ultrapotassic, potassic, and minor calc-alkaline lavas with a range of compositions (mafic, intermediate and silicic) (Arnaud et al., 1992; Turner et al., 1996; Chung et al., 1998; Nomade et al., 2004).

The oldest Cenozoic volcanic rocks that have been studied in detail occur in the southern Lhasa terrane. They are represented by the widespread, voluminous Linzizong Formation, which ranges in age from 65 Ma to 37 Ma (Maluski et al., 1982; Xu et al., 1985; Coulon et al., 1986; Zhou et al., 2001; Ding et al., 2003). It consists of up to 2 500 m of calc-alkaline andesitic flows, tuff and breccias, and dacitic to rhyolitic ignimbrites (Wang, 1980; Coulon et al., 1986; Pearce and Deng, 1988; Miller et al., 2000).

Cenozoic volcanic rocks in the Songpan-Ganzi terrane are volumetrically minor but widely distributed (Fig. 1), are mainly ultrapotassic to mafic-potassic in composition, and range in age from 17 Ma to Recent (Deng, 1989; Turner et al., 1993, 1996; Zheng et al., 1996; Cooper et al., 2002). An exception is present in the Ulugh Muztagh area (Fig. 1), where ~4 Ma rhyolites have been documented (Burchfiel et al., 1989; McKenna and Walker, 1990).

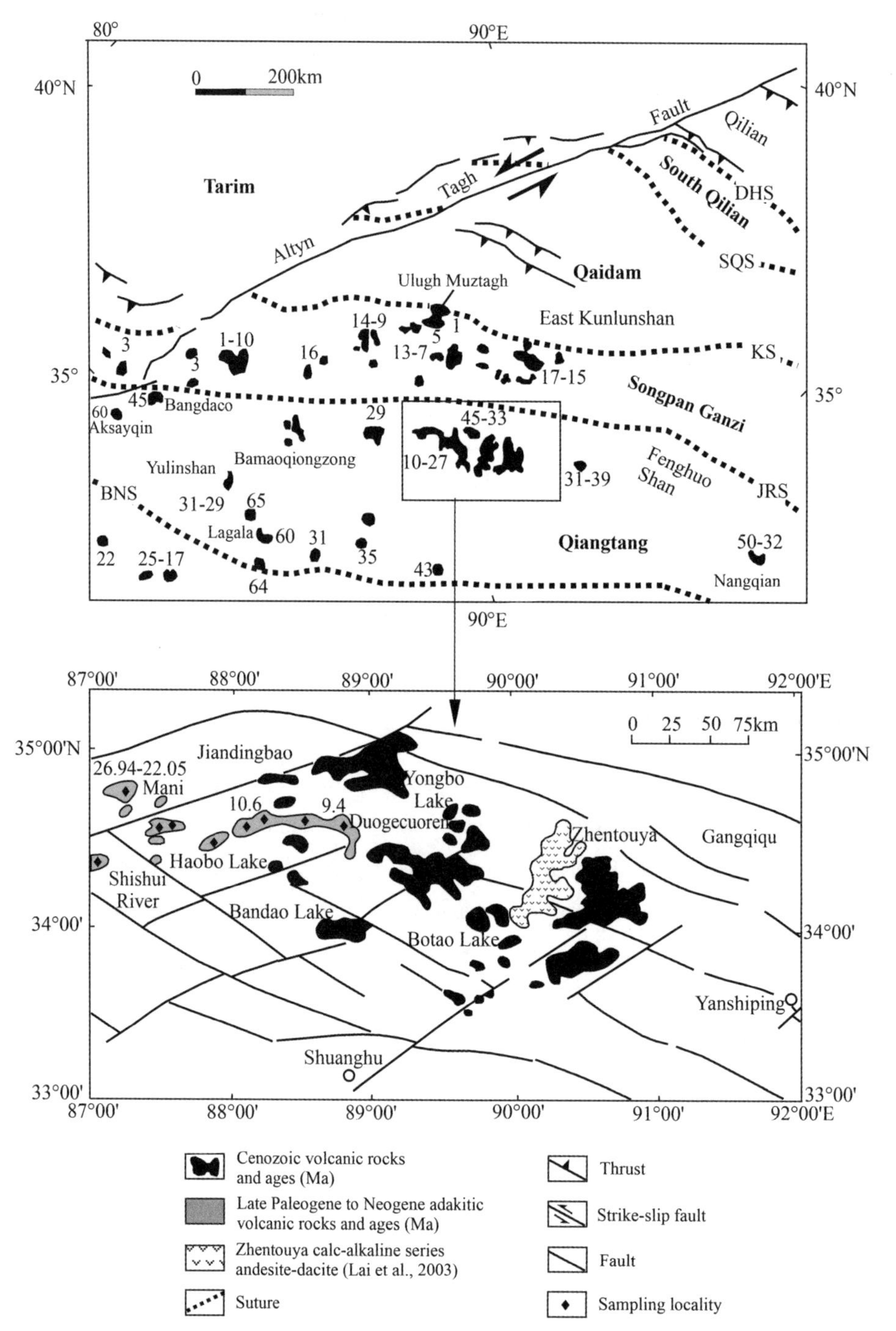

Fig. 1 Regional distribution of Cenozoic volcanic rocks in the northern Tibet.

Ages compiled from Coulon et al. (1986), Burchfie et al.(1989), Deng (1998), Turner et al. (1996), Zheng et al. (1996), Chung et al. (1998), Miller et al.(1999,2000), Deng et al.(2000), Hacker et al.(2000), Tan et al. (2000), Horton et al.(2002) and Kapp et al.(2002). From north to south, the main suture zones between the terranes are: DHS: Danghe Nan Shan; SQS: Southern Qilian; KS: Kunlun; JRS: Jinshajiang; BNS: Banggong-Nujiang.

In the central Qiangtang terrane, Lower Tertiary (60–45 Ma) alkaline basalts occur in the northwesternmost part of the Aksayqin area. Eocene-Early Oligocene (50–29 Ma) shoshonitic to ultrapotassic volcanic and subordinate intrusive rocks are widely distributed throughout the Qiangtang terrane (Fig. 1). At least five localities of Cenozoic volcanic rocks have been studied in detail —Duogecuoren, Bamaoqiongzong, Yulinshan, Lagala, and Zhengtouya (Deng, 1993,1998; Hacker et al., 2000; Tan et al., 2000; Ding et al.,2003; Lai et al., 2003). Cenozoic volcanism has been demonstrated to be associated with extensional features (Ding et al., 2003) but seems to be older (45–30 Ma) in the southern than in the northern edge of the Qiangtang region (Turner et al., 1996; Fig. 1).

In the Tibetan Plateau, the most recent volcanism, continuing into the Quaternary (along the northern margin of the plateau), is concentrated in the northwestern part along the Kunlun Mountains and along the northern branch of the Kunlun and Altyn-Tagh fault, where volcanism is typically associated with small pull-apart structure (Tapponnier et al.,2001; Turner et al., 1996; Nomade et al.,2004).

This study focuses on geochemistry and petrogenesis of Cenozoic adakitic rocks in the northern Qiangtang region (Fig. 1). The rocks extend discontinuously over 150 km along an E-W-trending fault in the Shishui River and Duogecuoren areas, and are exposed as 50 m to 200 m thick lava sheets and long flows in low-relief hills. In the field, the rocks discordantly overlied the Neogene Suona Lake group and are locally interstratified with Neogene sandstones and Jurassic flysches and limestones. Two distinct volcanic rock sequences, separated by a weathered horizon, have been recognized. The older sequence consists of basaltic trachyandesite and trachyandesite and belongs to a shoshonitic volcanic series that originated from an enriched continental mantle source (Lai and Liu, 2001). The younger sequence includes two eruption cycles, between which there is a volcanic breccia layer of 15 m to 40 m in thickness. The first eruption cycle is dominated by pyroclastic rocks (crystal tuff) and andesite, whereas the second consists of dacite-rhyolite with minor pyroclastic rocks. Samples studied in this paper were collected from the younger sequence and were identified as andesitic, dacitic, rhyolitic lavas through microscopic examination.

K-Ar dating of the above rocks has yielded a number of ages from 9.4 Ma to 26.94 Ma. For example, K-Ar whole-rock apparent ages of 22.05 Ma and 26.94 Ma were obtained for the andesite and pyroxene-andesite in the Mani area, respectively, 10.6 Ma for the dacite in the Jiandingbao area, and 10.6 Ma and 9.4 Ma for dacite and andesite in the Duogecuoren and Yongbo Lake areas (Deng, 1998).

Samples and analytical techniques

Only unweathered samples were selected for chemical analysis. Whole-rock powder was obtained from 50 g to 100 g splits using an agate mortar in order to minimize potential contamination.

Major-element compositions were determined using a fully automated Rigaku X-ray fluorescence spectrometer (XRF), with a method described by O'Reilly and Griffin (1988). Analysis for trace elements, including the rare-earth elements (REE), was performed by inductively coupled plasma mass spectrometry (ICP-MS) on a Perkin-Elmer Sciex ELAN-5100 spectrometer, employing the method described by Eggins et al. (1997). Trace-element concentrations were corrected in reference to the BCR-2 standard; analytical uncertainties are generally less than 5% for most elements. Sample preparation, XRF, and ICP-MS analyses were all carried out at the National Key Center of GEMOC, Macquarie University, Australia in 1999.

Sr and Nd isotopes and Sm, Nd, Sr, and Rb concentrations were determined using isotope dilution techniques on a Finnigan MAT-261 multicollector mass spectrometer at the Chinese Academy of Geological Sciences. Nd isotope ratios were measured in metal form, and correction for mass fractionation was carried out relative to $^{146}Nd/^{144}Nd = 0.721\ 9$. The ratios are reported relative to $^{143}Nd/^{144}Nd = 0.511\ 125$ for the J. M. Nd_2O_3 standard. Analytical uncertainty for $^{143}Nd/^{144}Nd$ measurements was ±2 sigma. Sr isotope ratios were also measured in metal form and mass fractionation was corrected to $^{88}Sr/^{86}Sr = 8.375\ 21$. The total procedural blanks were $<5\times10^{-11}$ g for Sm and Nd, and $<10^{-9}-10^{-10}$ g for Rb and Sr.

For Pb analysis, 200 mg of rock powder was dissolved using double-distilled hydrofluoric and perchloric acids in Teflon bombs. Pb was separated from silicate matrix using HCl-HBr ion exchange procedures. Pb isotopes were analyzed on a Finnigan MAT-261 mass spectrometer in a static mode with Re single filaments using the silica-gel-phosphoric acid technique. Mass fractionation was corrected off-line using correction factors derived from NBS SRM987 standard measurements. Total procedural Pb blacks are $<2\times10^{-10}$ g.

Petrology and chemical composition

The rocks in study area basically are unweathered. Some rocks are slightly altered, showing kaolinization of feldspar and chloritization of mafic minerals. The samples include both aphyric and porphyritic lavas. The latter contain phenocrysts of amphibole, biotite, and plagioclase, whereas pyroxene phenocrysts are present in the andesites. The fine-grained groundmass for the rocks is characterized by pilotaxitic or hyalopilitic textures.

Major-and trace-element data of representative samples from the northern Qiangtang region are given in Table 1. As shown in Fig. 2a, b, the oxide variation diagrams of (K_2O+Na_2O) vs. SiO_2 and K_2O vs. SiO_2, most samples define a trend characterized by approximately constant (K_2O+Na_2O) and slightly increasing K_2O within a SiO_2 range from

58.18 wt% to 71.69 wt%. In the (K_2O+Na_2O) vs. SiO_2 diagram (Fig. 2a), all the samples belong to the sub-alkaline series and most belong to a high-potassium calc-alkaline association. Also, the rocks are classified as andesite, dacite, and rhyolite, respectively, based on the (K_2O+Na_2O) vs. SiO_2 classification (Fig. 2a) of Le Bas (1986) and the former two are dominant in the region.

Table 1 Major- (wt%) and trace-element (ppm) analyses of adakitic volcanic rocks from the Northern Qiangtang region, Tibetan Plateau.

Sample	Lai02	Lai07	Lai10	Lai11	Lai14	Lai39	Lai41	Lai35	Lai16	Lai20	Lai22	Lai24
Rock	Andesite	Andesite	Andesite	Andesite	Andesite	Andesite	Andesite	Dacite	Dacite	Dacite	Dacite	Dacite
SiO_2	61.66	58.18	60.59	60.89	61.75	60.53	58.39	63.11	66.06	66.18	63.43	66.25
TiO_2	0.82	0.75	0.80	0.77	0.82	0.77	0.83	0.64	0.47	0.59	0.57	0.57
Al_2O_3	15.49	14.37	15.42	15.59	15.61	14.93	16.91	16.26	15.43	15.69	14.94	16.07
Fe_2O_3	1.36	1.44	2.38	1.54	1.27	2.92	5.13	3.35	1.98	2.92	3.07	3.13
FeO	3.00	3.80	1.16	3.12	3.10	1.54	0.32	0.50	0.83	0.60	0.27	0.33
MnO	0.08	0.09	0.05	0.08	0.07	0.07	0.06	0.04	0.04	0.02	0.05	0.01
MgO	4.07	5.82	2.21	4.51	4.03	4.11	2.79	1.38	1.35	1.03	2.32	1.09
CaO	5.18	5.94	5.65	5.69	5.14	5.42	6.97	5.13	3.46	3.55	5.15	3.50
Na_2O	4.69	3.86	4.68	4.03	4.70	2.83	4.12	3.73	3.87	4.53	4.40	4.38
K_2O	2.92	3.06	2.82	2.48	2.93	4.36	2.40	2.76	3.13	3.15	2.84	3.13
P_2O_5	0.28	0.28	0.33	0.27	0.28	0.44	0.34	0.21	0.13	0.18	0.19	0.18
H_2O^+	0.32	0.75	1.65	0.98	0.46	1.16	0.65	1.57	2.12	1.46	1.38	1.45
H_2O^-	0.04	0.17	0.47	0.15	0.02	0.17	0.25	0.80	0.28	0.56	0.12	0.60
CO_2	0.08	1.10	1.36	0.29	0.17	0.24	0.99	0.86	0.25	0.21	1.48	0.28
Total	99.99	99.61	99.57	100.39	100.35	99.50	100.16	100.34	99.40	100.67	100.21	100.97
$Mg^{\#1}$	0.63	0.67	0.55	0.64	0.63	0.64	0.50	0.41	0.48	0.36	0.58	0.38
Li	11.6	19.1	19.6	11.8	11.1	14.9	10.0	13.1	5.85	10.1	7.62	11.6
Be	2.69	2.77	2.47	2.79	2.83	4.60	3.04	2.90	3.14	2.93	2.87	2.96
Sc	10.3	13.0	8.32	12.7	9.64	10.6	12.2	9.46	5.87	7.18	7.23	7.45
Ti	5 540	4 679	5 251	4 957	5 417	4 743	4 709	4 093	3 275	3 967	3 694	3 879
V	82.0	95.5	74.7	105	81.2	74.9	72.3	84.2	61.7	56.9	54.9	50.5
Cr	152	268	150	150	150	111	68.1	58.9	41.1	76.5	50.8	55.3
Co	16.2	20.9	14.5	16.9	15.9	15.8	13.2	11.6	6.31	5.88	10.3	4.92
Ni	96.4	148	75.5	87.1	93.4	65.6	57.3	42.5	25.2	41.4	41.1	34.7
Cu	26.4	27.9	19.4	19.7	24.3	21.8	20.1	20.0	13.3	16.5	21.6	20.7
Zn	63.5	54.2	40.6	59.4	62.1	60.1	46.0	55.3	52.4	55.9	57.2	54.4
Ga	18.5	15.8	18.7	18.7	19.0	17.9	17.6	20.4	20.5	19.0	19.0	20.6
Rb	80.1	79.2	59.1	60.3	81.4	140	56.2	71.5	99.2	81.3	78.0	85.6
Sr	1 249	1 130	1 289	1 177	1 282	1 415	1 243	1 234	1 117	946	955	991
Y	12.8	15.8	11.5	17.5	12.9	15.4	17.1	12.5	9.23	10.3	10.3	9.35
Zr	194	182	175	172	193	258	156	181	171	234	214	237

Continued

Sample	Lai02	Lai07	Lai10	Lai11	Lai14	Lai39	Lai41	Lai35	Lai16	Lai20	Lai22	Lai24
Rock	Andesite	Andesite	Andesite	Andesite	Andesite	Andesite	Andesite	Dacite	Dacite	Dacite	Dacite	Dacite
Nb	13.4	8.89	13.6	9.30	13.5	13.7	12.5	6.71	5.03	7.79	7.45	7.46
Mo	1.49	1.23	0.92	1.36	1.49	1.90	0.99	0.68	1.31	1.08	1.11	0.89
Cd	0.05	0.04	0.04	0.05	0.04	0.05	0.04	0.02	0.03	0.02	0.05	0.02
Sn	1.73	1.75	1.62	1.94	1.70	2.42	2.88	1.59	1.80	1.95	1.85	1.82
Sb	0.24	0.61	0.22	0.29	0.25	0.27	0.38	0.19	0.30	0.20	0.13	0.15
Cs	2.44	2.42	2.09	7.18	2.48	5.87	1.34	1.29	3.75	1.85	2.05	1.44
Ba	1 217	1 517	1 269	1 095	1 197	1 900	1 135	1 247	1 222	1 700	1 224	1 263
La	48.9	60.5	49.3	44.6	48.6	87.3	55.7	43.1	38.0	54.2	55.1	56.5
Ce	91.5	112	89.3	83.3	89.9	157	104	79.9	69.4	93.1	97.2	94.6
Pr	10.5	13.2	10.3	9.58	10.4	18.4	11.9	9.38	7.89	10.5	10.9	10.8
Nd	37.9	46.8	35.8	34.1	36.9	62.7	43.4	33.1	28.3	34.8	37.4	36.3
Sm	6.02	7.34	5.87	6.05	5.82	9.37	7.38	5.09	4.74	5.32	5.45	5.08
Eu	1.75	2.08	1.75	1.72	1.73	2.46	1.97	1.57	1.37	1.64	1.41	1.41
Gd	4.56	5.35	4.28	4.86	4.51	6.72	5.76	3.74	3.53	3.96	3.85	3.66
Tb	0.59	0.71	0.55	0.67	0.57	0.82	0.78	0.48	0.44	0.49	0.49	0.45
Dy	2.39	2.97	2.26	3.17	2.43	3.20	3.53	2.06	1.73	1.85	2.02	1.76
Ho	0.44	0.56	0.40	0.61	0.43	0.55	0.63	0.40	0.31	0.33	0.38	0.33
Er	1.12	1.49	1.03	1.69	1.13	1.35	1.67	1.06	0.82	0.86	0.96	0.87
Yb	0.99	1.29	0.86	1.45	0.94	1.09	1.47	0.93	0.67	0.73	0.82	0.70
Lu	0.14	0.19	0.12	0.22	0.14	0.16	0.21	0.14	0.10	0.10	0.12	0.10
Hf	4.47	4.30	4.15	4.26	4.42	6.43	4.32	4.63	4.25	5.36	4.91	5.39
Ta	0.73	0.50	0.74	0.50	0.72	0.77	0.64	0.34	0.29	0.38	0.34	0.35
W	1.56	1.87	1.58	2.53	1.63	1.83	1.09	0.92	1.93	1.44	0.77	0.68
Pb	31.7	39.9	25.9	39.3	31.4	30.5	34.6	39.3	43.4	28.6	27.7	31.0
Th	13.7	23.3	13.2	12.9	13.8	28.7	13.2	15.0	16.5	19.2	18.2	19.5
U	3.07	4.99	2.87	3.18	3.09	6.26	1.77	3.47	4.63	3.39	3.35	3.04
Eu/Eu^{*2}	1.02	1.01	1.07	0.97	1.03	0.95	0.92	1.10	1.02	1.09	0.94	1.00
ΣREE	206.8	254.5	201.8	192.0	203.5	351.1	238.4	181.0	157.3	207.9	216.1	212.6

Sample	Lai28	Lai33	Lai34	Lai37	Lai38	Lai29	Lai32	Lai18	Lai25	Lai27	Lai40
Rock	Dacite	Dacite	Dacite	Dacite	Dacite	Dacite	Rhyolite	Rhyolite	Rhyolite	Rhyolite	Rhyolite
SiO_2	69.05	64.58	67.37	66.34	68.07	64.31	69.58	71.69	71.39	70.61	71.26
TiO_2	0.35	0.57	0.50	0.46	0.36	0.56	0.29	0.28	0.27	0.27	0.29
Al_2O_3	14.53	15.30	15.73	15.58	15.38	15.95	15.11	14.55	15.31	15.54	15.60
Fe_2O_3	2.31	2.43	2.81	3.07	2.14	2.66	1.66	0.78	1.37	1.22	0.38
FeO	0.29	0.87	0.06	0.06	0.09	0.98	0.21	0.70	0.34	0.18	0.22
MnO	0.01	0.05	0.03	0.05	0.04	0.04	0.02	0.04	0.01	0.00	0.00
MgO	0.50	2.42	0.72	1.71	0.85	2.35	0.67	0.47	0.62	0.73	0.28
CaO	2.54	3.91	3.60	3.75	3.70	3.95	2.86	2.12	2.38	1.81	2.92
Na_2O	3.94	3.44	4.23	3.82	4.20	4.06	3.73	3.93	4.02	3.28	3.86

Continued

Sample Rock	Lai28 Dacite	Lai33 Dacite	Lai34 Dacite	Lai37 Dacite	Lai38 Dacite	Lai29 Dacite	Lai32 Rhyolite	Lai18 Rhyolite	Lai25 Rhyolite	Lai27 Rhyolite	Lai40 Rhyolite
K_2O	2.83	3.27	3.40	3.06	3.37	3.80	3.55	3.71	3.78	4.13	3.58
P_2O_5	0.11	0.19	0.16	0.14	0.10	0.33	0.07	0.07	0.05	0.05	0.02
H_2O^+	1.39	1.96	0.54	1.23	0.94	0.74	1.25	1.04	0.76	1.76	0.67
H_2O^-	0.99	0.20	0.36	0.17	0.12	0.28	0.71	0.38	0.28	0.84	0.16
CO_2	0.19	0.23	0.33	0.21	1.00	0.24	0.61	0.20	0.20	0.02	0.74
Total	99.04	99.41	99.84	99.65	100.35	100.24	100.32	99.96	100.78	100.47	99.99
$Mg^{\#}$	0.28	0.59	0.33	0.52	0.43	0.56	0.41	0.38	0.42	0.51	0.47
Li	8.65	11.7	10.4	7.63	14.2	9.28	17.8	14.3	12.4	20.0	13.9
Be	3.03	2.94	3.45	2.88	3.17	3.34	3.13	3.05	3.16	3.77	3.02
Sc	4.16	6.77	5.67	6.77	4.40	9.27	4.39	2.48	3.80	2.50	2.33
Ti	2 071	3 622	3 144	2 845	2 288	3 531	1 677	1 449	1 794	1 743	1 531
V	38.0	58.3	33.4	51.1	30.6	76.2	31.8	20.9	26.0	21.9	21.4
Cr	11.5	49.8	19.0	19.1	7.34	19.2	5.68	4.08	5.49	2.87	4.62
Co	3.20	9.05	5.57	7.22	4.63	9.77	3.50	2.85	3.09	2.45	1.12
Ni	5.96	40.8	15.4	10.9	6.91	19.6	2.59	2.51	2.45	1.23	1.03
Cu	9.59	16.1	13.5	10.4	12.7	17.2	14.2	7.68	10.6	14.1	6.59
Zn	23.2	59.8	46.3	44. 9	39.7	45.9	35.6	29.2	38.7	46.2	13.4
Ga	15.3	19.8	19.4	18.3	18.8	19.3	17.9	11.4	18.8	20.0	15.2
Rb	93.1	83.1	111	116	141	113	157	119	176	186	136
Sr	777	982	825	850	545	1 634	420	340	379	357	380
Y	8.63	11.5	8.59	12.0	7.37	15.5	5.46	5.66	4.16	3.80	3.65
Zr	83.5	226	180	80.3	119	141	86.0	72.1	87.0	78.1	84.8
Nb	4.46	7.60	5.82	5.77	4.59	6.98	4.17	3.70	4.24	4.53	3.81
Mo	0.50	1.44	0.70	0.76	1.02	0.59	0.55	1.30	0.90	0.37	1.04
Cd	0.02	0.03	0.01	0.01	0.02	0.02	0.01	0.04	0.01	0.01	0.02
Sn	1.07	1.86	1.22	2.02	1.50	1.74	1.17	2.16	2.06	1.78	1.49
Sb	7.52	0.14	0.25	0.54	0.50	0.50	0.31	0.52	0.44	0.14	0.30
Cs	52.8	2.11	2.71	6.44	7.16	4.84	5.71	14.4	5.83	6.48	5.52
Ba	6 116	1 181	1 309	1 131	1 116	1 704	1 156	1 175	1 164	1 056	1 157
La	23. 9	57.1	41.4	31.4	26.5	71.8	19.4	25.9	24.5	31.4	25.8
Ce	44.3	96.9	74.8	59.6	47.8	137	35.4	47.2	41.4	53.9	43.8
Pr	4.97	11.7	8.56	6.77	5.38	16.2	3.94	5.03	4.58	5.73	4.80
Nd	17.5	40.9	30.1	24.0	18.7	59.2	13.8	17.0	15.1	18.7	15.5
Sm	2.88	5.97	4.70	4.28	3.41	8.79	2.32	2.94	2.33	2.81	2.70
Eu	1.53	1.57	1.35	1.25	0.98	2.32	0.75	0.87	0.72	0.77	0.77
Gd	2.12	4.30	3.46	3.44	2.57	5.91	1.84	2.30	1.68	1.79	1.89
Tb	0.31	0.53	0.40	0.46	0.34	0.77	0.24	0.30	0.22	0.24	0.24
Dy	1.45	2.14	1.67	2.11	1.45	3.01	1.02	1.23	0.89	0.84	0.88
Ho	0.30	0.39	0.30	0.41	0.26	0.54	0.19	0.21	0.16	0.13	0.14

Continued

Sample	Lai28	Lai33	Lai34	Lai37	Lai38	Lai29	Lai32	Lai18	Lai25	Lai27	Lai40
Rock	Dacite	Dacite	Dacite	Dacite	Dacite	Dacite	Rhyolite	Rhyolite	Rhyolite	Rhyolite	Rhyolite
Er	0. 89	1. 00	0. 73	1. 14	0. 69	1. 40	0. 49	0. 55	0. 37	0. 34	0. 35
Yb	0. 87	0. 83	0. 63	0. 99	0. 54	1. 21	0. 43	0. 44	0. 29	0. 27	0. 29
Lu	0. 13	0. 12	0. 09	0. 15	0. 08	0. 18	0. 06	0. 06	0. 04	0. 03	0. 04
Hf	2. 28	5. 24	4. 64	2. 39	3. 36	3. 84	2. 70	2. 51	2. 71	2. 60	2. 89
Ta	0. 31	0. 35	0. 32	0. 40	0. 35	0. 46	0. 37	0. 35	0. 39	0. 39	0. 34
W	0. 95	0. 88	0. 83	1. 08	1. 51	1. 09	1. 89	2. 99	1. 51	1. 26	1. 49
Pb	26. 1	25. 6	33. 3	28. 4	32. 2	44. 5	34. 2	30. 6	33. 8	34. 9	30. 7
Th	11. 2	18. 3	18. 6	13. 6	17. 3	25. 5	14. 6	18. 1	19. 0	27. 6	19. 9
U	3. 43	3. 61	3. 82	4. 35	5. 02	4. 30	2. 84	7. 95	3. 47	4. 59	5. 75
Eu/Eu*	1. 89	0. 95	1. 02	1. 00	1. 01	0. 98	1. 01	1. 02	1. 11	1. 05	1. 04
ΣREE	77. 3	223. 5	168. 2	136	108. 7	308. 3	79. 9	104. 0	92. 3	117. 0	97. 2

[1] $Mg^{\#}$ calculation method: $Mg^{\#}=Mg^{2+}/(Mg^{2+}+TFe^{2+})$, $FeO^{T}=FeO+0.8998Fe_2O_3$.

[2] $Eu/Eu^{*}=Eu_N/(Sm_N+Gd_N)^{1/2}$.

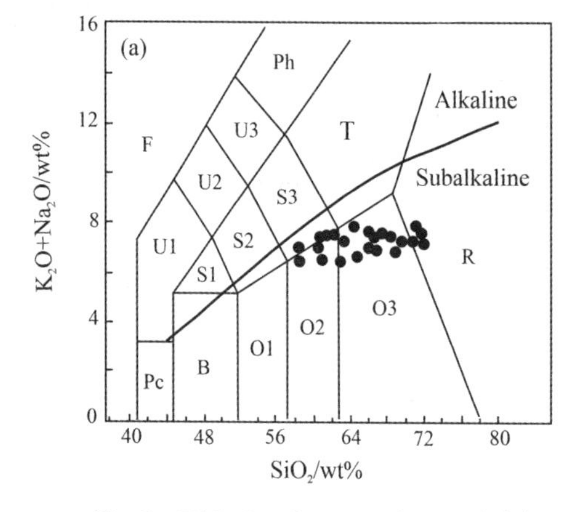

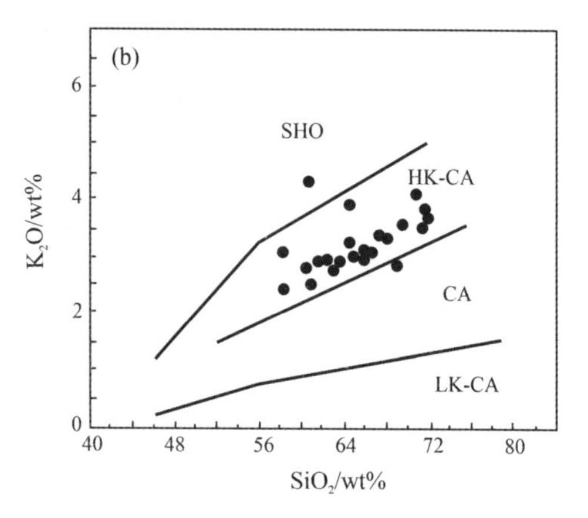

Fig. 2 TAS classification schemes (a) (after Le Bas, 1986) and SiO_2 vs. K_2O diagram (b) (after Le Maitre et al., 1989) for the Qiangtang adakitic volcanic rocks.

F: foidite; U1: tephrite (Ol < 10%), basanite (Ol > 10%); U2: phonotephrite; U3: tephriphonolite; Ph: phonolite; S1: trachybasalt; S2: basaltic trachyandesite; S3: trachyandesite; T: trachyte ($q<20\%$), trachydacite ($q>20\%$); Pc: picrobasalt; B: basalt; O1: basaltic andesite; O2: andesite; O3: dacite; R: rhyolite; LK-CA: low-potassium calc-alkaline series; CA: calc-alkaline series; HK-CA: high-potassium calc-alkaline series; SHO: shoshonitic series.

The samples exhibit $SiO_2>58$ wt%, $Al_2O_3>14\%$, MgO from 0. 28% to 5. 82%, and Na_2O from 2. 83% to 4. 70%, with Na/K ratios of 0. 99–2. 61 (most ratios > 1). In the Na-K-Ca ternary diagram (Fig. 3), the samples straddle the boundary between the adakite, trondhjemite-tonalite-granodiorite (TTG), and the island-arc fields.

When comparing the rocks from the study area with the experimental melts from metabasalts (Rapp et al., 1999), the Archean trondhjemites-tonalites-dacites (TTD) and the glass inclusions in sub-arc mantle xenoliths in the MgO versus SiO_2 diagram (Fig. 4; Kay et al., 1993), samples from the northern Qiangtang region all plot within the adakite field with slightly higher MgO than the normal calc-alkaline volcanic rock series. The Qiangtang "adakites" can be divided into two groups on the basis of $Mg^{\#}$–i.e., a high-Mg group ($Mg^{\#} \geqslant 0.45$) and a subordinate low-Mg group ($Mg^{\#} \leqslant 0.45$).

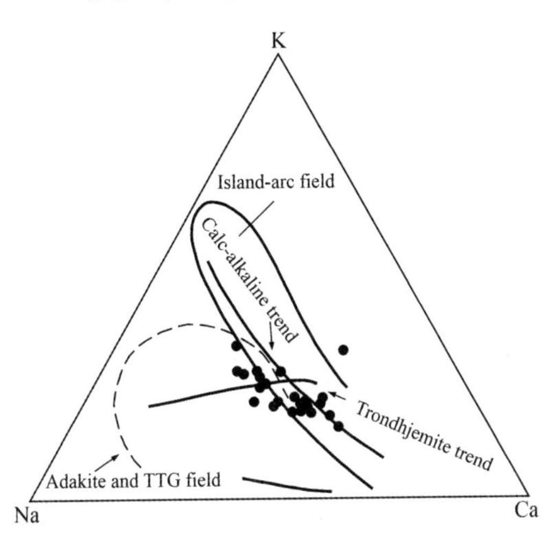

Fig. 3 Na-K-Ca ternary diagram (after Kay et al., 1993) with calc-alkaline (Tatsumi and Eggins, 1995) and trondhjemitic (Francalanci et al., 1993) trends superimposed. The island arc field and the adakite + trondhjemite-tonalite-granodiorite (TTG) field are adapted from Defant and Drummond (1993).

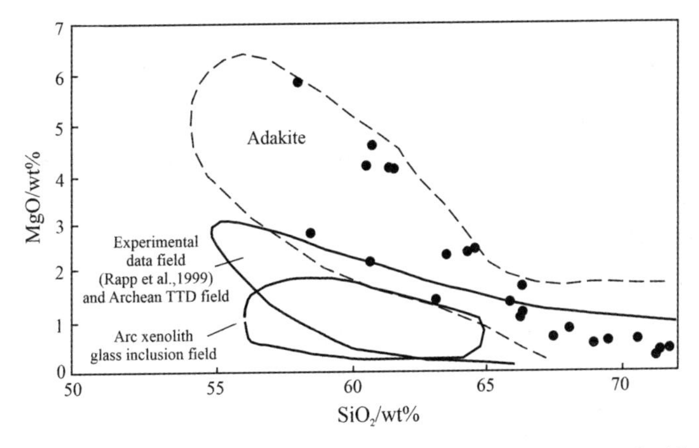

Fig. 4 MgO vs. SiO_2 diagram (after Kay et al., 1993) in adakites compared with the fields for experimental data (after Rapp et al., 1999) and Archean trondhjemites-tonalites-dacites (TTD) and glass inclusions in sub-arc mantle xenoliths (after Nockholds and Allen, 1953).

Trace and rare-earth-element geochemistry

Table 1 demonstrates that both Cr and Ni contents for all the samples are negatively correlated with SiO_2. The andesites have the highest Cr and Ni contents (up to 148 ppm and

268 ppm, respectively), whereas the rhyolites have the lowest (< 10 ppm). Sc contents are low for analyzed rocks (< 15 ppm).

The samples also show distinctive geochemical characteristics of high Sr, low heavy rare earth elements (HREE) and Y, and high Sr/Y. Sr ranges from 340 ppm to 1 634 ppm, Y from 3.65 ppm to 17.5 ppm, and Yb from 0.27 ppm to 1.47 ppm. Sr/Y ratios are > 60 (67–121), La/Yb > 27 (27–118). On the Sr/Y vs. Y diagram (Fig. 5), all samples fall in the adakite field.

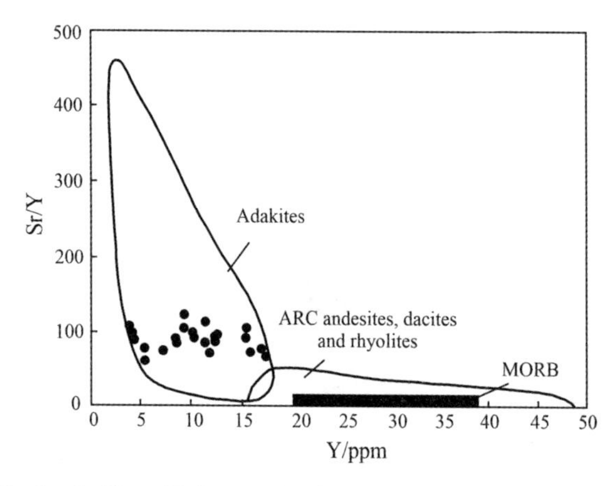

Fig. 5 Sr/Y vs. Y diagram for the Qiangtang adakitic volcanic rocks. The original diagram is after Defant and Drummond (1990).

The Qiangtang rocks exhibit strong enrichment in incompatible elements, such as K, Rb, Ba and Th, and slight enrichment in Ce in the N-MORB-normalized (Sun and McDonough, 1989) incompatible element patterns (Fig. 6). Otherwise, the plots show a strong depletion in Nb, Ta and Ti, similar to the rock series in typical island-arc tectonic settings (e.g. Francalanci et al., 1993).

The samples have high $(La/Yb)_N$ ratios ranging from 20 to 85 (average 43) and $Eu/Eu^* = 0.89–1.81$ (average 1.01). This indicates that the rocks are strongly enriched in light rare earth elements (LREE) but without remarkable Eu anomalies (Fig. 6), similar to the typical adakites studied by Defant and Drummond (1990). Significantly, the total content of rare earth element ($\sum$REE) in andesite and dacite is higher than that in rhyolite.

Sr-Nd-Pb isotopic systematics

Four samples from the study area were analyzed for Sr, Nd, and Pb isotopes (Table 2). These samples exhibit high radiogenic Sr, with $^{87}Sr/^{86}Sr$ ratios ranging from 0.706 365 to 0.708 156 and ε_{Sr} from +105.5 to +128.7. Nd isotope ratios are low, with a relatively limited variation from 0.512 411 to 0.512 535, corresponding to negative ε_{Nd} values between −4.43 and −2.01.

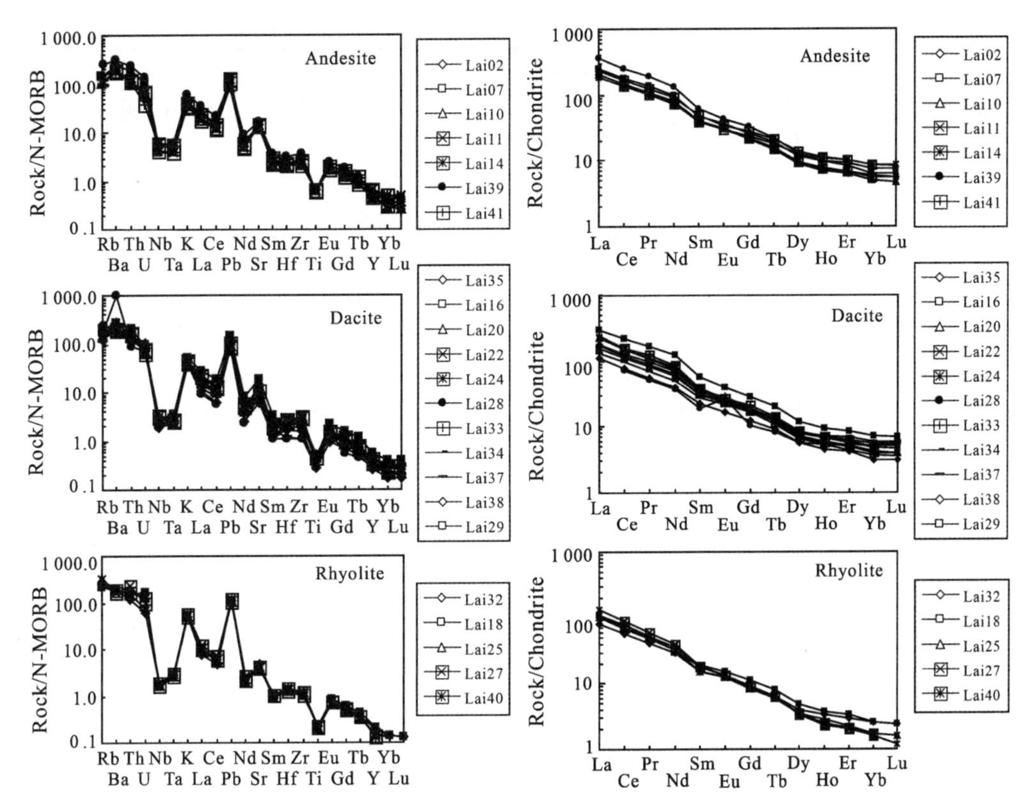

Fig. 6 N-MORB normalized trace-element distributions and chondrite-normalized rare-earth-element distributions for the Qiangtang adakitic volcanic rocks.
N-MORB values and Chondrite values are from Sun and McDonough (1989).

Table 2 Sr-Nd-Pb isotopic Data of Adakitic Volcanic Rocks from the Northern Qiangtang region, Tibetan Plateau[1].

Sample	Lai07	Lai11	Lai29	Lai40
Rock type	Andesite	Andesite	Dacite	Rhyolite
U	4.99	3.18	4.30	5.75
Th	23.26	12.98	25.45	19.89
Pb	39.86	39.31	44.49	30.73
$^{206}Pb/^{204}Pb$	18.779±1	18.679±13	18.758±10	18.839 7±10
$^{207}Pb/^{204}Pb$	15.651±15	15.672±11	15.660 1±10	15.657 2±9
$^{208}Pb/^{204}Pb$	39.012 2±36	38.955 5±27	38.965 0±24	39.052 5±25
Rb	85.69	84.37	109.2	176.8
Sr	1 316	1 246	1 625	391.5
$^{87}Rb/^{86}Sr$	0.188 6	0.196 1	0.067 21	1.308
$^{87}Sr/^{86}Sr$	0.706 950±12	0.708 156±11	0.706 365±17	0.707 991±14
ε_{Sr}	+113.88	+131.13	+105.51	+128.77
Sm	6.979	5.615	8.263	2.536
Nd	46.560	33.477	55.277	15.391

Continued

Sample	Lai07	Lai11	Lai29	Lai40
Rock type	Andesite	Andesite	Dacite	Rhyolite
$^{147}Sm/^{144}Nd$	0. 090 67	0. 101 4	0. 090 42	0. 099 67
$^{143}Nd/^{144}Nd$	0. 512 411±8	0. 512 450±8	0. 512 500±7	0. 512 535±9
ε_{Nd}	−4. 43	−3. 67	−2. 69	−2. 01

[1]U, Th, and Pb concentrations were analyzed by ICP-MS; Sm, Nd, Rb, Sr, and isotopic ratios were measured through isotopic dilution. ε_{Nd} was calculated using $\varepsilon_{Nd}=[(^{143}Nd/^{144}Nd)_m/(^{143}Nd/^{144}Nd)_{CHUR}-1]\times10^4$, $(^{143}Nd/^{144}Nd)_{CHUR}=0.512638$. ε_{Sr} was calculated using $\varepsilon_{Sr}=[(^{87}Sr/^{86}Sr)_m/(^{87}Sr/^{86}Sr)_{UR}-1]\times10^4$, $(^{87}Sr/^{86}Sr)_{UR}=0.698990$. ε_{Nd} and ε_{Sr} and isotopic ratios of these rocks were not corrected for age; because this age is small, the difference between corrected and uncorrected values falls in the range of the analytical error.

In the $^{143}Nd/^{144}Nd$ vs. $^{87}Sr/^{86}Sr$ diagram (Fig. 7), the studied rocks show similar Sr-Nd isotopic compositions to the Cordillera Blanca Batholith which derived from newly underplated basaltic crust (Petford and Atherton, 1996), but obvious differences from those of subduction-related adakites (Kay, 1978; Defant and Drummond, 1990; Stern and Kilian, 1996).

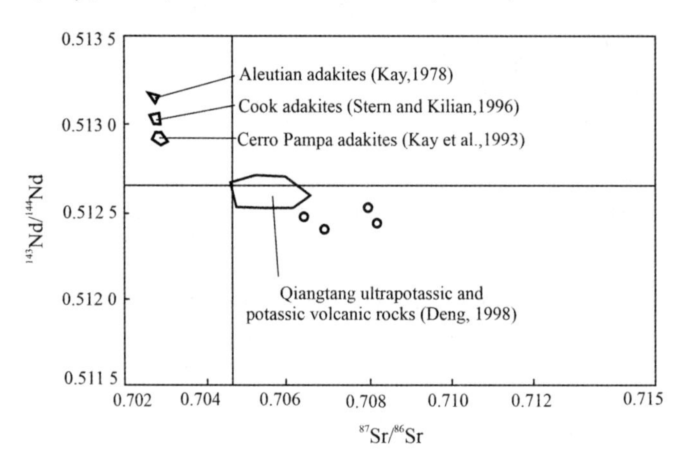

Fig. 7 $^{143}Nd/^{144}Nd$ vs. $^{87}Sr/^{86}Sr$ diagram for the Qiangtang adakitic volcanic rocks.

All the samples are also more radiogenic in Pb, their $^{206}Pb/^{204}Pb$ = 18. 679 − 18. 839, $^{207}Pb/^{204}Pb$ = 15. 651 − 15. 672, and $^{208}Pb/^{204}Pb$ = 38. 955 − 39. 052. The Pb isotope data presented in Fig. 8 were superimposed upon the Hugh (1993) Pb composition diagram. In both the $^{207}Pb/^{204}Pb$ vs. $^{206}Pb/^{204}Pb$ and $^{208}Pb/^{204}Pb$ vs. $^{206}Pb/^{204}Pb$ diagrams, samples plot above the Northern Hemisphere reference line (NHRL), with Th/U = 4. 0, and fall in the lower continental crust and/or EM Ⅱ fields.

Sr, Nd, and Pb isotopic data on Fig. 9 demonstrate that the samples from this study are quite different from DMM, PREMA, HIMU, and EM Ⅰ, but similar to EM Ⅱ in nature (Zindle and Hart, 1986).

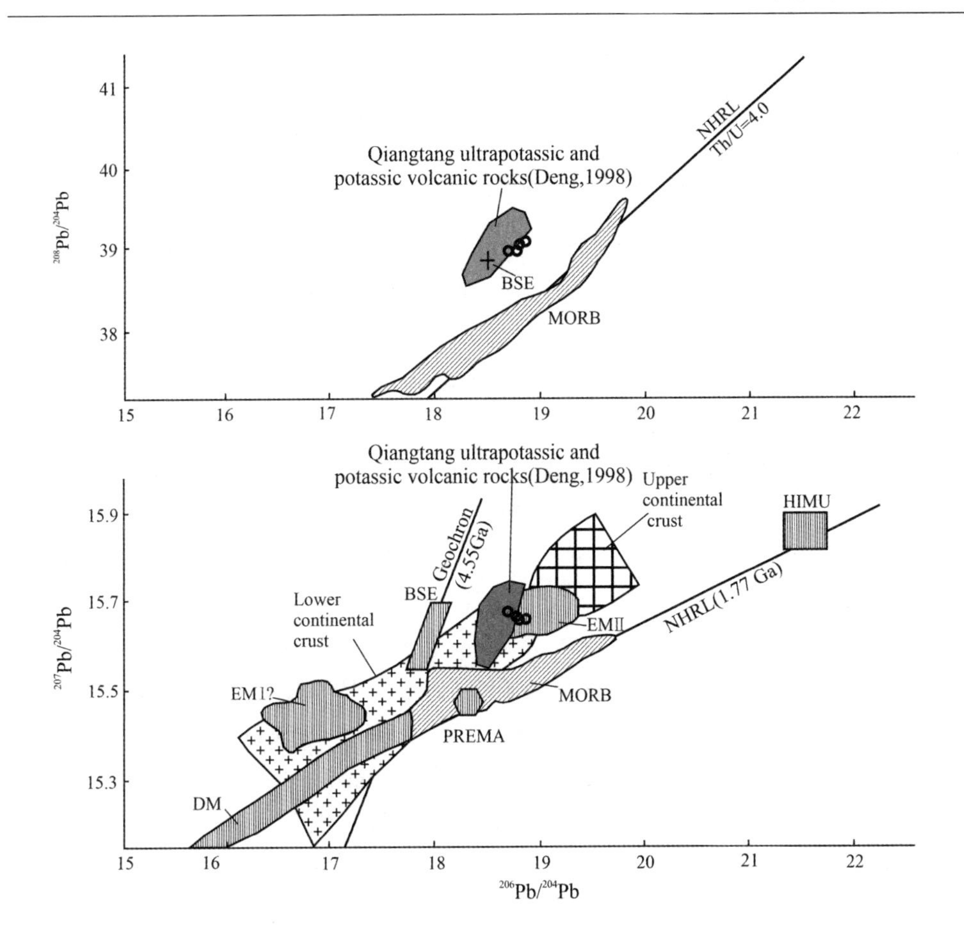

Fig. 8 Pb isotope data for the Qiangtang adakitic volcanic rocks.

The original diagram is adapted from Hugh (1993).

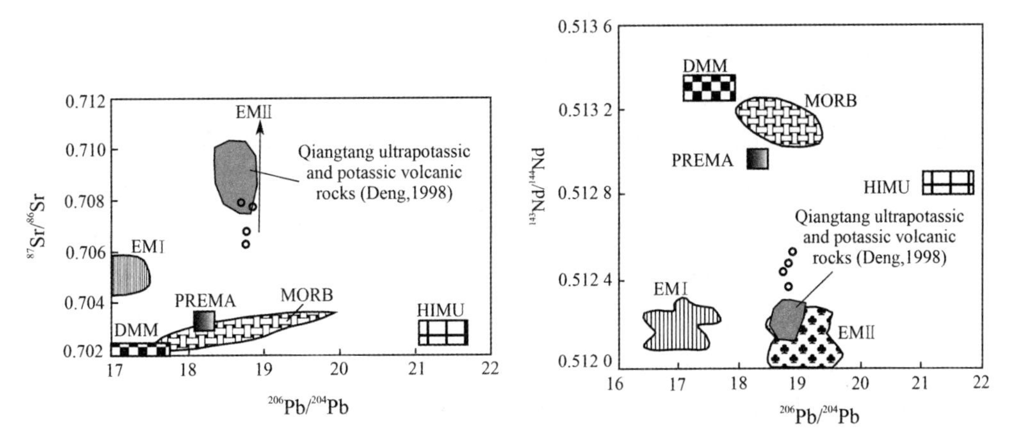

Fig. 9 $^{87}Sr/^{86}Sr$ vs. $^{206}Pb/^{204}Pb$ and $^{143}Nd/^{144}Nd$ vs. $^{206}Pb/^{204}Pb$ diagrams for the Qiangtang adakitic volcanic rocks.

DMM: depleted mantle; PREMA: primary mantle; EM Ⅰ: enriched mantle (type Ⅰ); EM Ⅱ: enriched mantle (type Ⅱ); HIMU: anomalously high $^{238}U/^{204}Pb$ mantle. The diagram is adapted from Zindler and Hart (1986).

Discussion

Genesis of the Northern Qiangtang volcanics

Our study of major- and trace-element geochemistry has led to a new finding that the Qiangtang intermediate-felsic volcanic rocks are geochemically similar to those of modern adakites from subduction zones (Defant and Drummond, 1990). Exploring the origin of these adakitic rocks will be greatly significant for tracing the tectonic evolution of the Tibetan Plateau.

To date, four origins have been proposed for adakites, high-Mg andesites and TTGs: ① partial melting of young subducted oceanic crust, followed by interaction with an overlying mantle wedge (Kay, 1978; Defant and Drummond, 1990; Martin, 1998); ② an assimilation-fractional crystallization (AFC) process involving basaltic magma (Castillo et al., 1999); ③ partial melting of thickened mafic lower continental crust (Atherton and Petford, 1993); and ④ partial melting of foundered lower crust (Kay et al., 1993; Xu et al., 2002; Gao et al., 2004). All of these models except for ② are characterized by depletion in heavy rare earth elements (HREEs) and Y, which requires melting of a mafic source within the stability field of garnet, most probably under eclogite-facies conditions. We will employ field geology and geochemistry to discuss which model may be appropriate to explain the origin of the Qiangtang adakitic rocks.

Partial melting of a subducted oceanic slab cannot be the source of these rocks because the Qiangtang block has been in an intracontinental setting since Mesozoic time. Thus, the subducted oceanic slab could not exist beneath the region during 26.9–9.4 Ma, when these adakitic rocks were formed. Also, geophysical evidence (Owens and Zandt, 1997; Tilmann and Ni, 2003) indicates that the northward-subducted Indian plate and the new Tethyan slab have not reached the southern boundary of the Qiangtang block (Bangong-Nujiang suture zone), in opposition to the idea that "flat subduction" in Cenozoic time has been responsible for the genesis of the Qiangtang magmas. Furthermore, the high K_2O (Fig. 2b), high $^{87}Sr/^{86}Sr$ ratios, and low ε_{Nd} (Fig. 7) of the Qiangtang adakitic rocks do not provide any indication of the participation of a MORB component, which means the source of the Qiangtang adakitic rocks lacked an involvement of the subducted slab or sediments.

Castillo et al. (1999) proposed that the assimilation-fractional crystallization (AFC) process involving basaltic magma produced rocks of adakitic characteristics in Camiguin Island, Philippines. Our evidence, however, argues that the similar AFC process is unlikely to have generated the Qiangtang adaktic magma since the Miocene potassic or ultrapotassic mafic and intermediate lavas in the northern Qiangtang region have much lower ε_{Nd} (−8.08 to −5.21), and higher $^{87}Sr/^{86}Sr$ (0.707 9–0.709 3) and K_2O content (3.85%–7.68%) (Turner et al., 1996).

A final possibility is that the Qiangtang adakitic magma could have been generated by partial melting of the thickened Tibetan crust. Generally, it is believed that adakitic magmas produced by crustal melting were derived directly from a lower crust source heated by a underplating basaltic melt at the bottom of continental crust (e.g., Antherton and Petford, 1993). If so, this kind of adakitic magma should have a relatively low MgO content (or $Mg^{\#}$), similar to the experimental melts of Rapp and Watson (1995). However, the Qiangtang "adakites" includes a high-Mg group ($Mg^{\#} \geq 0.45$) and a subordinate low-Mg group ($Mg^{\#} \leq 0.45$). The high-Mg group may be explained by pristine adakitic melts that interacted to some degree with mantle materials (e.g., Rapp et al, 1999). The best scenario that can explain the elevated $Mg^{\#}$ seems to be delamination of the lower crust composed of amphibole-bearing eclogitic materials, coinciding with dehydration melting in an enriched subcrustal mantle; as the pristine adakitic melts rise, they pass through the mantle, elevating their $Mg^{\#}$ via reaction (e.g., Xu et al., 2002; Gao et al., 2004).

Generally, coexisting andesite-dacite-rhyolite suite may be considered to result from fractional crystallization (FC) of a parental magma, and the rhyolite should have a higher $\sum$REE content and lower Eu^{*}/Eu value, whereas it is notable that $\sum$REE in the studied andesite and dacite is higher than that in the rhyolite, and yet the Eu^{*}/Eu value is stable from andesite to rhyolite in both high-Mg and low-Mg groups.

We propose that, from the andesite to rhyolite in the study area, the compositional correlation between SiO_2 and MgO, TiO_2, P_2O_5, Sr, and Eu cannot be attributed to fractional crystallization (FC). The reasons as follows: ① the stable Eu^{*}/Eu and positive Sr anomaly from andesite to rhyolite indicate that the plagioclase must have broken down in the magma source at the temperature and pressure of partial melting (Lai et al., 2003); ② from andesite to rhyolite, the negative correlation between SiO_2 and $\sum$REE suggest that the magma was contaminated in homogenously by mantle-derived rocks represented by the Miocene shoshonitic volcanics in the northern Qiangtang region (Turner et al., 1996); and ③ the higher $\sum$REE in dacite of low-$Mg^{\#}$ group reveals dacitic magma was contaminated by fluids enriched in REE and large ion lithos elements (LILE). Furthermore, the low $Mg^{\#}$ (< 0.45) easily rules out the probability that the fluids were derived from mantle. Coupled with the previous work (e.g., Wang et al., 2005), the fluids may be attributed to the southward intracontinental subduction of the Qaidam block along the Jinshajiang suture (JRS) zone.

Geodynamic process and tectonic implication

A critical question raised here is what mechanism caused the partial melting of the thickened Tibetan lower crust. Radiogenic heating could be responsible for the partial melting of mafic materials. However, due to the low thermal conductivity of thickened crust, heating caused by thickening will only be developed on the time scale of the thermal time constant of the lithosphere (~240 Ma, Turner et al., 1996). The thickening of lithosphere and low degree of extension also rule out the possibility of decompression melting, because the decrease in pressure only result in phase transformation of eclogite to granulite in the lower crust, as observed in the eastern Himalayan syntaxis (Ding et al., 2001; Hou et al., 2004).

Another possibility for melting of thickened Tibetan crust is upwelling of asthenosphere materials caused by convective thinning of lithospheric mantle (i.e., Turner et al., 1996) or the participation of fluids generated by southward intracontinental subduction of the Qaidam block (i.e., Wang et al., 2005).

Upwelling asthenosphere could cause partial melting of the enriched lithospheric mantle to produce ultrapotassic magma (i.e., Turner et al., 1996), which then could pool at the bottom of the lower crust, in turn triggering melting of the thickened Tibetan lower crust (Hou et al., 2004). The new major, trace-element, and isotopic Sr-Nd-Pb data suggest that the input of Cenozoic ultrapotassic lava components made an important contribution to the generation of high-Mg adakitic lavas in the northern Qiangtang region.

Our geochemical data indicates that contamination by upper-crustal components was essential to the generation of low-Mg adakitic magma. Available geologic and geophysical evidence indicates that the north Kunlun-Qaidam continental lithosphere was partially subducted southward beneath the Songpan-Ganzi (SG) and Qiangtang blocks since 20 Ma (Tapponnier et al., 2001; Arnaud et al., 1992; Deng, 1998; Meyer et al., 1998). Wang et al. (2005) argued that "burial of continental crust was accompanied by the release of fluids that ascended into the overlying lower crust of the SG block, triggering partial melting of the lower crust and resulting in the formation of the Hoh Xil potassium-rich adakitic magmas. The buried continental crust may have included a small sedimentary components, fluids from which could produce the K, Rb, and Th enrichment in the Hoh Xil potassium-rich adakitic rocks". We argue that this geodynamic process is also applicable for the generation of the low-Mg adakitic rocks in the northern Qiangtang (Fig. 10).

Based on the about disscussion, we propose that the generation of the Qiangtang adakitic rocks was due to to convective thinning of the mantle lithosphere (Turner et al., 1996) and intracontinental subduction (Deng, 1998; Hacker et al., 2000; Tapponnier et al., 2001) simultaneously.

Conclusions

The Upper Paleogene to Neogene volcanic rocks in the northern Qiangtang region are

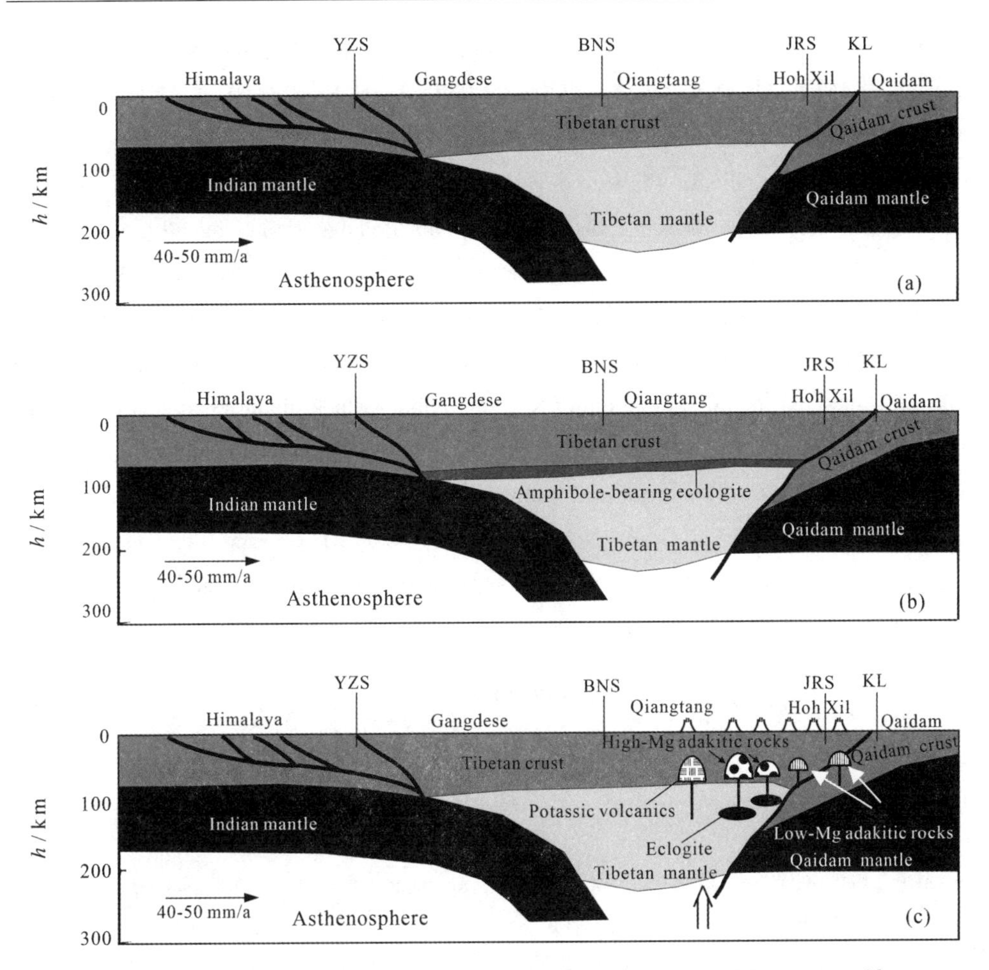

Fig. 10 Proposed model to produce Qiangtang adakitic rocks via delamination of lower crust and southward intracontinental subduction of Kunlun-Qaidam Block.

(a) Relatively thick crust and lithospheric mantle in Early Paleogene (?) time. (b) Because of the continuing northwand subduction of Indian plate and southward intrasubduction of the Kunlun-Qaidam Block, the Tibetan crust and lithospheric mantle become thicker, and the lower crust is composed of amphibole-bearing eclogite that is not yet fully dehydrated. (c) Thickened crust was thinned after Paleogene time. As amphibole-bearing eclogite bodies sank into the underlying mantle, adakitic melt was produced by dehydration melting of amphibole-bearing eclogite materials. The adakitic melt passes through mantle, elevating MgO contents (and $Mg^{\#}$) via reaction. At the same time, participation of fluids generated by the southward-subducted continental lithosphere of Kunlun-Qaidam Block triggered partial melting of the un-delaminated lower crust, and then produced low-Mg adakites. Heat flux from upwelling asthenosphere induced partial melting of the Tibetan enriched lithospheric mantle and produced the potassic-ultrapotassic lavas (Tuener et al, 1996). YZS: Yarlung-Zanbo sutre; BNS: Bangong-Nujiang suture; JRS: Jinshajiang suture; KL: Kunlun fault.

andesite, dacite, and rhyolite. These rocks exhibit distinctive geochemical features similar to adakites in modern subduction zones, including high SiO_2 ($SiO_2 > 58\%$), high Sr and Sr/Y (Sr > 340 ppm, Sr/Y > 60), low Y and HREE (Y < 18 ppm, Yb < 1.5 ppm), radiogenic Sr and Pb ($^{87}Sr/^{86}Sr = 0.706\ 365 - 0.708\ 156$; $^{208}Pb/^{204}Pb = 38.955 - 39.052$; $^{207}Pb/^{204}Pb =$

15. 651 – 15. 672; $^{206}Pb/^{204}Pb$ = 18. 679 – 18. 839), and non-radiogenic Nd isotopic ratios ($^{143}Nd/^{144}Nd$ = 0. 512 411 – 0. 512 535), K and LREE enrichment and lack of Eu anomaly, suggesting their source should be a thickened mafic lower crust. The data shows that the Qiangtang region of the Tibetan Plateau developed an eclogitic and mafic lower crust in a tectonic setting involving continental collision during Late Paleogene to Neogene time.

On the basis of $Mg^{\#}$, the Qiangtang "adakites" can be divided into two groups – i. e., a high-Mg group ($Mg^{\#} \geqslant 0.45$) and a subordinate low-Mg group ($Mg^{\#} \leqslant 0.45$), which were attributed to convective thinning of the enriched lithospheric mantle beneath Qiangtang region, and southward intracontinental subduction of Kunlun-Qaidam Block along the Jinshajiang suture (JRS) zone, respectively.

Acknowledgements This work was jointly supported by the National Natural Science Foundation of China (NSFC, Grant Nos. 40572050, 40272042) and the Teaching and Research Award Program for Outstanding Young Teachers in Higher Education Institutions, Ministry of Education (MOE), China.

References

Arnaud, N.O., Vidal, P., Tapponnier, P., Matte, P., Deng, W.M., 1992. The high K_2O volcanism of northwestern Tibet: Geochemistry and tectonic implications. Earth & Planetary Science Letters, 111, 351-367.

Atherton, M.P., Petford, N., 1993. Generation of sodium-rich magmas from newly underplated basaltic crust. Nature, 362, 144-146.

Burchfiel, B.C., Molnar, P., Zhao, Z., Liang, K.U., Wang, S., Huang, M., Sutter, J., 1989. Geology of the Ulugh Muztagh area, northern Tibet. Earth & Planetary Science Letters, 94, 57-70.

Castillo, P.R., Janney, P.E., Solidum, R.U., 1999. Petrology and geochemistry of the Camiguin island, southern Philippines: Insight to the source of adakites and other lavas in a complex arc setting. Contributions to Mineralogy and Petrology, 134, 33-51.

Chi, X.G., Li, C., Jin, W., Liu, S., Yang, R.H., 1999. The Cenozoic volcanism evolutionary and uplifting mechanism of the Qinghai-Tibet plateau. Geological Review, 45, 978-986 (in Chinese).

Chung, S.L., Lo, Z.H., Lee, T.Y., Zhang, Y.Q., Xie, Y.W., Li, X.H., Wang, K.L., Wang, P.L., 1998. Diachronous uplift of the Tibetan plateau starting 40 Myr ago. Nature, 394, 769-773.

Cooper, K.M., Reid, M.R., Dunbar, N.W., McIntosh, W.C., 2002. Origin of mafic magmas beneath northwestern Tibet: Constraints from ^{230}Th-^{238}U diseqilibria. Geochemistry, Geophysics, Geosystems, 3, 10.

Coulon, C., Maluski, H., Bollinger, C., Wang, S., 1986. Mesozoic and Cenozoic volcanic rocks from central and southern Tibet: $^{39}Ar/^{40}Ar$ dating, petrological characteristics and geodynamical significance. Earth & Planetary Science Letters, 79, 281-302.

Defant, M.J., Drummond, M.S., 1990. Derivation of some modern arc magmas by melting of young subducted lithosphere. Nature, 347, 662-665.

Defant, M.J., Drummond, M.S., 1993. Mount St Helens: Potential example of the partial melting of the subducted lithosphere in a volcanic arc. Geology, 21, 547-550.

Deng, W.M., 1989. The Cenozoic volcanic rocks in north Ali area, Xizang. Acta Petrologica Sinica, 3, 1-11 (in Chinese).

Deng, W.M., 1991. The geology, geochemistry and forming age of the shoshonitic volcanic rock in middle Kunlun orogenic belt. Scientia Geologica Sinica, 3, 201-213 (in Chinese).

Deng, W.M., 1992. The intracontinental subduction zone and its magmatism in the Qinghai-Tibet plateau. Beijing: Science Press, 262 (in Chinese).

Deng, W.M., 1993. Trace element and Sr, Nd isotopic features of the Cenozoic potassium-volcanic rocks from northern Qinghai-Tibet plateau. Acta Petrologica Sinica, 9, 379-387 (in Chinese).

Deng, W.M., 1998. Cenozoic intraplate volcanic rocks in the Northern Qinghai-Xizang plateau. Beijing: Geological Publishing House, 168 (in Chinese).

Deng, W.M., Sun, H.J., 1998. Features of isotopic geochemistry and source region for the intraplate volcanic rocks from northern Qinghai-Tibet plateau. Earth Science Frontiers, 5, 307-317 (in Chinese).

Deng, W.M., Sun, H.J., Zhang, Y.Q., 2000. K-Ar age of the Cenozoic volcanic rocks in the Nangqen Basin, Qinghai Province and its geological significance. Chinese Science Bulletin, 45, 1015-1019.

Ding, L., Kapp, P., Zhong, D.L., Deng, W.M., 2003. Cenozoic volcanism in Tibet: Evidence for a transition from oceanic to continental subduction. Journal of Petrology, 44, 1833-1865.

Ding, L., Zhong, D.L., Yin, A., Harrison, T.M., 2001. Cenozoic structural and metamorphic evolution of the eastern Himalayan syntaxis(Namche Barwa). Earth & Planetary Science Letters, 192, 423-438.

Eggins, S.M., Woodhead, J.D., Kinsley, L.P.J., Sylvester, P., McCulloch, M.T., Hergt, J.M., Handler, M.R., 1997. A simple method for the precise determination of 40 or more trace elements in geology sample by ICP-MS using enriched isotope internal standardization. Chemical Geology, 132, 311-326.

England, P.C., Houseman, G.A., 1988. The mechanics of the Tibetan plateau. Philosophical Transactions of the Royal Society of London: Series A, 326, 301-319.

England, P.C., Houseman, G.A., 1989. Extension during continental convergence, with application to the Tibetan plateau. Journal of Geophysical Research, 94, 17561-17579.

Francalanci, L., Taylor, S.R., McCulloch, M.T., 1993. Geochemical and isotopic variations in the calc-alkaline rocks of Aeolian arc, southern Tyrrhenian Sea, Italy: Constraints on magma genesis. Contributions to Mineralogy and Petrology, 113, 300-313.

Gao, S., Roberta, L.R., Yuan, H.L., Liu, X.M., Liu, Y.S., Xu, W.L., Ling, W.L., John, A., Wang, X. C., Wang, Q.H., 2004. Recycling of lower continental crust in the North China Craton. Nature, 424, 392-397.

Hacker, B.R., Edwin, G., Ratschbacher, L., Grove, M., McWilliams, M., Sobolev, S., Jiang, W., Wu, Z.H., 2000. Hot and dry deep crustal xenoliths from Tibet. Science, 287, 2463-2466.

Horton, B.K., Yin, A., Spurlin, M.S., Zhou, J., Wang, J., 2002. Paleocene-Eocene syncontractional sedimentation in narrow lacustrine-dominated basins of east-central Tibet. Geological Society of America Bulletin, 114, 771-786.

Hugh, R.R., 1993. Using geochemical data. Singapore: Longman Singapore Publishers, 240.

Hou, Z.Q., Gao, Y.F., Qu, X.M., Rui, Z.Y., Mo, X.X., 2004. Origin of adakitic intrusives generated during mid-Miocene east-west extension in southern Tibet. Earth & Planetary Science Letters, 220, 139-155.

Kapp, P., Yin, A., Harrison, T.M., Ding, L., 2002. Cretaceous-Tertiary deformation history of central Tibet. Geological Society of America, Abstract with Programs, 34, 487.

Kay, R. W., 1978. Aleutian magnesian andesites: melts from subducted Pacific Ocean crust. Journal of Volcanology Geothermal Research, 4, 117-132.

Kay, S.M., Ramos, V.A., Marquez, Y.M., 1993. Evidence in Cerro Pampa volcanic rocks for slab-melting prior to ridge-trench collision in southern South America. Journal of Geology, 101, 703-714.

Klootwijk, C. T., Radhakrishnamurty, C., 1981. Phanerozoic paleo-magnetism of the Indian plate and the India-Asia collision. In: McElhinny, M. W., Valencio, D. A. Paleoreconstruction of the Continent. American Geophysical Union, Washington DC, 93-105.

Lai, S.C., Liu, C.Y., 2001. Enriched upper mantle and eclogitic lower crust in north Qiangtang, Qinghai-Tibet plateau: Petrological and geochemical evidence from the Cenozoic volcanic rocks. Acta Petrologica Sinica, 17, 459-468 (in Chinese).

Lai, S. C., 1999. Petrogenesis of the Cenozoic volcanic rocks from the northern part of the Qinghai-Tibet plateau. Acta Petrologica Sinica, 15, 98-104 (in Chinese).

Lai, S.C., Liu, C.Y., Yi, H.S., 2003. Geochemistry and petrogenesis of Cenozoic andesite-dacite association from the Hoh Xil region, Tibetan plateau. International Geology Review, 45, 998-1019.

Le Bas, M.J., 1986. A chemical classification of volcanic rocks based on the total alkaline-silica diagram. Journal of Petrology, 27, 745-750.

Le Maitre, R.W., Bateman, P., Dudek, A., Keller, J., Le Bas, M.J., Sabine, P.A., Schmid, R., Sorensen, H., Streckeisen, A., Woolley, A.R., Zanettin, B., 1989. A classification of igneous rocks and glossary of terms. Oxford: Blackwell Science Press, 120

Liu, J.Q., 1998. Volcanoes in China. Beijing: Science Press, 110 (in Chinese).

Maluski, H., Proust, F., Xiao, X.C., 1982. $^{39}Ar/^{40}Ar$ dating of the trans-Himalayan calc-alkaline magmatism of southern Tibet. Nature, 298, 152-154.

Martin, H., 1998. Adakitic magmas: Modern analogues of Archaean granitoids. Lithos, 46, 411-429.

McKenna, L.W., Walker, J.D., 1990. Geochemistry of crustally derived leucocratic igneous rocks from the Ulugh Muztagh area, Northern Tibet and their implications for the formation of the Tibetan plateau. Journal of Geophysical Research, 95, 21483-21502.

Meyer, B., Tapponnier, P., Bourjot, L., Metivier, F., Gaudemer, Y., Peltzer, G., Shuinmin, G., Zhitai, C., 1998. Crustal thickening in Gansu-Qinghai, lithosphic mantle subduction, and oblique, strike slip controlled growth of the Tibetan plateau. Geophys. J. Int., 135, 1-47.

Miller, C., Schuster, R., Klotali, U., Frank, W., Grasemann, B., 2000. Late Cretaceous-Tertiary magmatic and tectonic events in the Transshimalaya batholith (Kailas area, SW Tibet). Schweizerische Mineralogische und Petrographische Mitteilungen, 80, 1-20.

Miller, C., Schuster, R., Klotzli, U., Frank, W., Purtscheller, F., 1999. Post-collisional potassic and ultrapotassic magmatism in SW Tibet: Geochemical and Sr, Nd, Pb, O isotopic constraints for mantle source characteristics and petrogenesis. Journal of Petrology, 40, 1399-1424.

Molnar, P., 1988. A review of geophysical constraints on the deep structure of the Tibetan plateau, the Himalaya and the Karakoram, and their tectonic implications. Philosophical Transactions of the Royal Society of London: Series A, 326, 33-88.

Nockholds, S.R., Allen, R., 1953. The geochemistry of some igneous rock series. Geochim. Cosmochim. Acta, 4, 105-142.

Nomade, S., Paul, R.R., Mo, X.X., Zhao, Z.D., Zhou, S., 2004. Miocene volcanism in the Lhasa block, Tibet spatial trends and geodynamic implications. Earth & Planetary Science Letters, 221, 227-243.

O'Reilly, S.Y., Griffin, W.L., 1988. Mantle metasomatism beneath Victoria, Australia I: Meta somatic processes in Cr-diopside lherzolites. Geochim. Cosmochim. Acta, 52, 433-447.

Owens, T.J., Zandt, G., 1997. Implication of the crustal property variations for models of Tibetan plateau evolution. Nature, 387, 37-43.

Petford, N., Atherton, M.P., 1996. Na-rich partial melting from newly underplated basaltic crust: The Cordillera Blanca Batholith. Peru. J. Petrol., 37, 1491-1521.

Pearce, J.A., Deng, W.M., 1988. The ophiolites of the Tibetan geotraverses, Lhasa to Golmud (1985) and Lhasa to Kathmandu (1986). Philosophical Transactions of the Royal Society of London, 327, 215-238.

Platt, J.P., England, P.C., 1994. Convective removal of lithosphere beneath mountain belts: Thermal and mechanical consequences. Am. J. Sci., 294, 307-336.

Rapp, R.P., Shimizu, N., Norman, M.D., Applegate, G.S., 1999. Reaction between slab-derived melts and peridotite in the mantle wedge: Experimental constraints at 3. 8 GPa. Chemical Geology, 160, 335-356.

Rapp, R.P., Watson, E.B., 1995. Dehydration melting of metabasalt at 8-32kbar: Implication for continental growth and crust-mantle recycling. Journal of Petrology, 36, 891-931.

Stern, C.R., Kilian, R., 1996. Role of the subducted slab, mantle wedge and continental crust in the generation of adakites from the Andean Austral volcanic zone. Contributions to Mineralogy and Petrology, 123, 263-281.

Sun, S.S., McDonough, W.F., 1989. Chemical and isotopic systematics of oceanic basalts: Implications for mantle composition and processes. In: Saunders, A.D., Norry, M.J. Magmatism in the Ocean Basin. Geol. Soc. Spec. Publ., 42, 313-345.

Tan, F.W., Pan, G.T., Xu, Q., 2000. The uplift of Qiangtang-Xizang plateau and geochemical characteristics of Cenozoic volcanic rocks from the center of Qiangtang, Xizang. Acta Petrologica et Mineralogica, 19, 121-130 (in Chinese).

Tapponnier, P., Xu, Z.Q., Roger, F., Meyer, B., Arnaud, N., Wittlinger, G., Yang, J.S., 2001. Oblique stepwise rise and growth of the Tibet plateau. Science, 294, 1671-1677.

Tatsumi, Y., Eggins, S., 1995. Subduction zone magmatism. New York: Blackwell Science Press, 138.

Tilmann, F., Ni, J., 2003. Seismic imaging of the downwelling Indian lithosphere beneath central Tibet. Science, 300, 1424-1427.

Turner, S., Arnaud, N., Liu, J.Q., Rogers, N., Hawkesworth, C.J., Harris, N., Kelley, S., Van Calsteren, P., Deng, W.M., 1996. Post-collision, shoshonitic volcanism on the Tibetan plateau: Implications for convective thinning of the lithosphere and the source of ocean island basalts. Journal of

Petrology, 37, 45-71.

Turner, S., Hawkesworth, C. J., Liu, J., Rogers, N., Kelley, S., Van Calsteren, P., 1993. Timing of Tibetan uplift constrained by analysis of volcanic rocks. Nature, 364, 50-53.

Ugo, P., 1990. Shoshonitic and ultrapotassic post-collisional dykes from northern Karakorum (Sinkiang, China). Lithos, 26, 305-316.

Wang, B. X., Ye, H. F., Peng, Y. M., 1999. Isotopic geochemistry features and its significance of the Mesozoic-Cenozoic volcanic rocks from Qiangtang basin, Qinghai-Tibet plateau. Geological Review, 45, 946-951 (in Chinese).

Wang, Q., McDermott, F., Xu, J.F., Bellon, H., Zhu, Y.T., 2005. Cenozoic K-rich adakitic volcanics in the Honxil area, northern Tibet: Lower crustal melting in an intracontinental setting. Geology, 33, 465-468.

Wang, S., 1980. The features of the Linzizong volcanic series in the eastern section of the Gangdise volcanic arc in Xizang (Tibet). In: Mission Franco-Chinois au Tibet. Beijing: Geological Publishing House, 320 (in Chinese).

Xu, J.F., Shinjo, R., Defant, M.J., Wang, Q., Rapp, R.P., 2002. Origin of Mesozoic adakitic intrusive rocks in the Ningzhen area of east China: Partial melting of delaminated lower continental crust? Geology, 30, 1111-1114.

Xu, R.H., Scharer, U., Allegre, C.J., 1985. Magmatism and metamorphism in the Lhasa block (Tibet): A geochronological study. Journal of Geology, 93, 41-57.

Zheng, X.S., Bian, Q.T., Zheng, J.K., 1996. On the Cenozoic volcanic rocks in Hoh Xil district, Qinghai province. Acta Petrologica Sinica, 12, 530-545 (in Chinese).

Zhou, S., Fang, N., Dong, G., Zhao, Z., Liu, X., 2001. Argon dating on the volcanic rocks of the Linzizong group, Tibet. Bulletin of Mineralogy, Petrology and Geochemistry, 20, 317-319.

Zindler, A., Hart, S.R., 1986. Chemical geodynamics. Annu. Rev. Earth Planet. Sci., 14, 493-573.

Geochemistry of ophiolites from the Mian-Lue suture zone: Implications for the tectonic evolution of the Qinling orogen, central China①

Lai Shaocong Qin Jingfeng Chen Liang Rodney Grapes

Abstract: The Mian-Lue suture zone along the southern margin of the Qinling orogenic belt, contains abundant ophiolite fragments that record collision between the North and South China terranes. The ophiolites are well exposed in the areas of Lueyang, Pipasi, and Derni. In Lueyang, ultramafic rocks comprise serpentinised harzburgite and dunite exhibiting LREE depletion and marked positive Eu anomalies, gabbroic rocks enriched in Sr and Ba and positive Eu anomalies, diabase dikes enriched in LREE with slightly negative Eu anomalies, and tholeiite metabasalts with MORB-type characteristics of LREE depletion, low Nb (3.9–8.4 ppm), TiO_2 (0.92–1.86 wt%,), Ti/V (18.14–23.76), Th/Ta (0.04–1.86), Th/Y (0.01–0.02), and Ta/Yb (0.01–0.13). Ophiolites from Pipasi comprise tholeiitic basalt with average low $(La/Yb)_N$ (0.65–0.97), Ti/V (26.98), Th/Ta (1.06), Th/Y (0.01), and Ta/Yb (0.05). Ophiolites from Derni are represented by basalt, gabbro, and metaperidotites. The basalts are LREE depleted and lack prominent Eu anomalies, indicating a depleted asthenosphere source. Isotopic ages of 350–245 Ma imply the existence of a Paleozoic ocean basin from Devonian to Permian time in central China. The MORB-type ophiolites in the Mian-Lue suture zone are regarded as the fragments of the Paleo-Tethys oceanic lithosphere that separated the Qinling and South China terranes during middle Paleozoic to early Mesozoic.

Introduction

The Qinling orogenic belt extends for more than 1 500 km in central China and forms the middle portion of the major E-W-trending Central orogenic belt (Zhang et al., 1995, 1996). This orogenic belt connects with the UHP Dabie-Sulu metamorphic belt to the east and with the North Qilian and Kunlun orogenic belts to the west, separating the North China and South China terranes (Fig. 1). The North Qinling orogenic belt is separated into the North and South Qinling terranes by the Shangxian-Danfeng (Shang-dan) suture zone that is interpreted as the main boundary between the North China and South China terranes (Zhang et al., 1995, 1996; Li et al., 1996) (Fig. 1).

---

① Published in *International Geology Review*, 2008, 50(7).

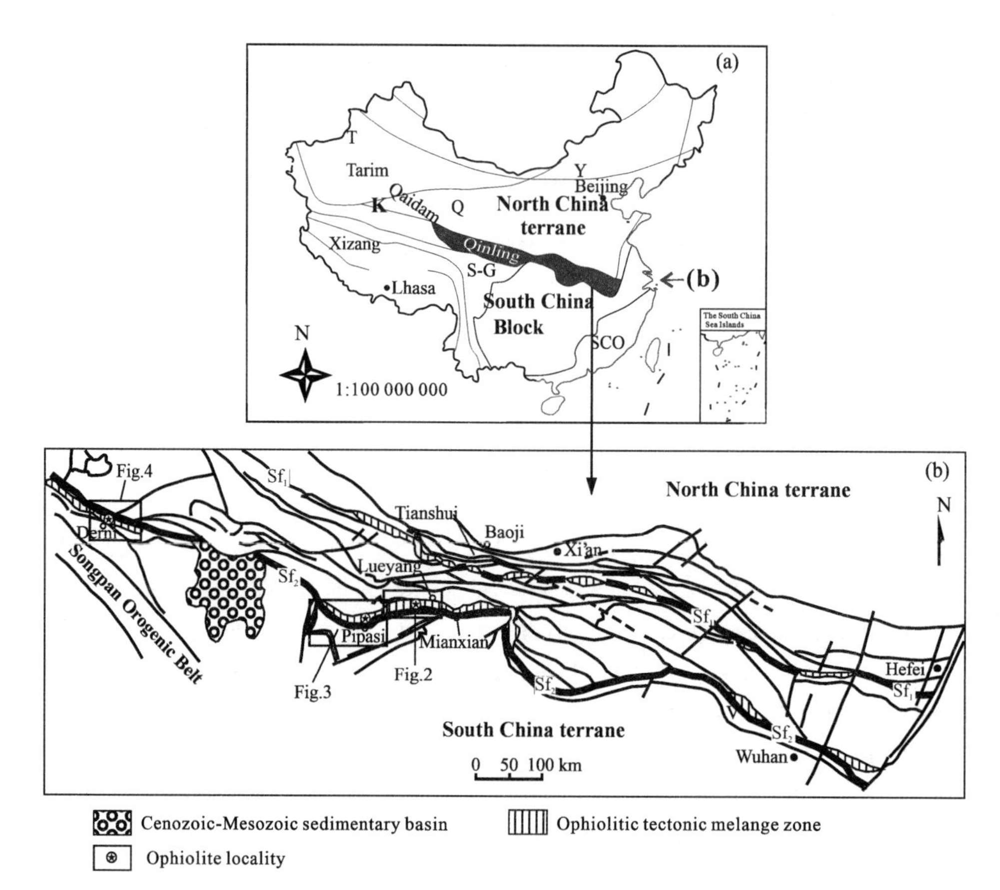

Fig. 1 (a) Index map of China. (b) Geological map showing the location of ophiolites in Mian-Lue suture zone, Qinling orogenic belt, central China.

K: Kunlun orogen; Q: Qilian orogen; SCO: South China orogen; T: Tianshan orogen; Y: Yanshanian orogen; S-G: Songpan-Ganzi terrane; Sf_1: Shangdan suture; Sf_2: Mian-Lue suture zone.

Recently, a series of ophiolite slices have been found in the Mianxian and Lueyang areas of the western portion of the South Qinling terrane. The ophiolites have different ages and geochemical characteristics from ophiolites in the Shang-Dan suture zone (Zhang et al., 1995, 1996; Meng and Zhang, 1999, 2000; Chen et al., 2000; Xu et al., 2000), and are considered to represent an additional suture zone in the south Qinling terrane, termed the Mianxian-Lueyang (Mian-Lue) suture zone (Zhang et al., 1995). This suture zone suggests that the Qinling orogenic belt may have been the result of collision of at least three major terranes, namely, the North China, Qinling-Dabie, and South China terranes, and challenges the idea that the Qinling orogenic belt formed by the collision of the North China and South China terranes (Wang et al., 1982).

Meng and Zhang (1999) have proposed a model involving a two-stage collision for the formation of the Qinling oregenic belt, with the Mian-Lue suture zone representing the final amalgamation (Yang and Hu, 1990; Zhang et al., 1995; Lai and Zhang, 1996; Xu and Han, 1996; Lai et al., 1997). According to this model, middle Paleozoic collision along the Shang-Dan suture zone, led to accretion of the South Qinling terrane to the North Qinling terrane, rather than a mutual amalgamation of the two terranes, and that the South China terrane was simultaneously rifted from the South Qinling terrane by formation of the Paleo-Tethyan Qinling ocean. Finally, in the Late Triassic, collision between the South China and the South Qinling terranes occurred along the Mian-Lue suture zone, leading to amalgamation of the two terranes to form the Qinling orogenic belt. However, the tectonic significance of the Mian-Lue suture zone in the evolution the Qinling orogenic belt remains a matter of debate.

New geochemical data from the Mian-Lue ophiolites presented in this paper help provide a better understanding of the nature and evolution of the Mian-Lue suture zone and its relationship to collision of the North and South China terranes (Yang and Hu, 1990; Zhang et al., 1995; Lai and Zhang, 1996; Lai et al., 1997; Xu and Han, 1996) that allows reconstruction of the geologic framework and tectonic evolution of the Qinling orogenic belt in the Phanerozoic.

Geologic Setting

The Qinling orogenic belt is a multistage orogenic belt composed of two mountain chains and two sutures. The north suture is the Shang-Dan tectonic belt that formed as a result of the Middle Paleozoic collision of the North China and South Qinling terranes (Zhang et al., 1995). The south suture zone is exposed along the Mian-Lue tectonic belt and is the result of Late Triassic collision of the South Qinling and South China terranes. The Mian-Lue suture zone is an E-W-trending mélange zone (Fig. 1) consisting of numerous tectonic slices within an intensely sheared matrix that have been imbricated by south-directed thrust faults (Zhang et al., 1995). The tectonic slices comprise: Sinian-Cambrian mudstones, pyroclastic rocks and limestones; Devonian turbidites, limestones, and mudstones; and Carboniferous limestones and abundant ophiolite fragments (Yang and Hu, 1990). Ophiolitic rocks in the Lueyang, Pipasi, and Derni areas are particularly well preserved (Fig. 1), and are described below.

Ophiolites in the Mian-Lue suture zone

Lueyang ophiolite

Ophiolite in the Lueyang area comprises basalt, gabbro, diabase dikes, and ultramafic rocks. These rocks have faulted contacts and constitute a typical tectonic mélange (Fig. 2). Ultramafic rocks are represented by highly serpentinized harzburgite and dunite consisting mainly of antigorite with minor chrysotile. Pseudomorphs after orthopyroxenes (bastite) and olivine (serpentine) are recognizable. Most of the gabbros have been intensively deformed and exhibit cataclastic and mylonitic textures, although an original cumulate texture is locally preserved as indicated by igneous layering ranging in thickness from several millimeters to several centimeters, reflecting variable proportions of pyroxene and plagioclase. Sub-parallel diabasic dikes with individual thicknesses of 0. 5 m to 3. 5 m cut the gabbroic rocks.

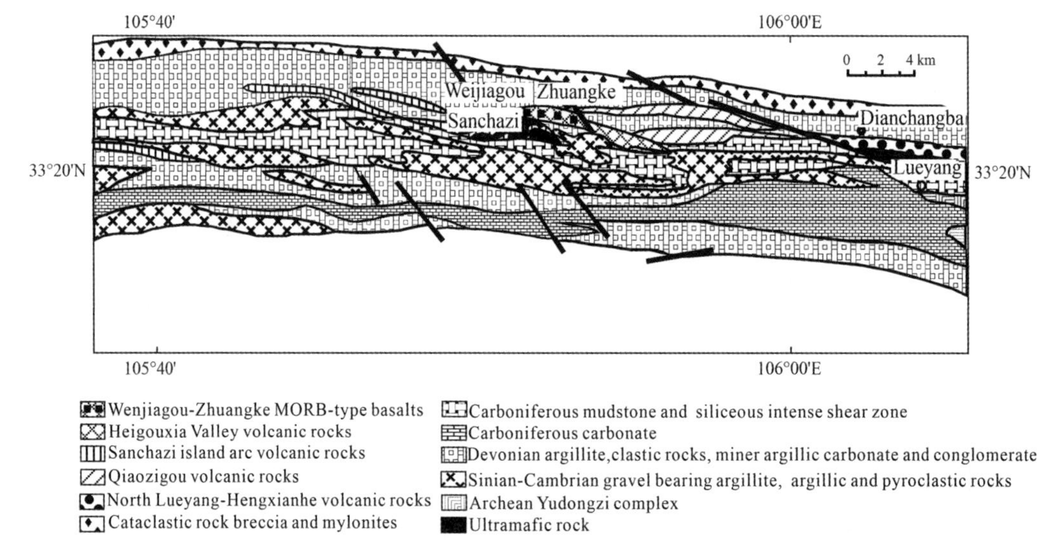

Fig. 2 Geologic map of the Lueyang area, Mian-Lue suture zone.

Basalt of the ophiolite suite occurs as a WNW-trending belt exceeding 5 km in length and 300 m to 700 m wide (Fig. 2). The basalts are faulted against carbonaceous mudstones to the south and Devonian clastic sedimentary rocks and limestones to the north. All these rocks have undergone a lower grade greenschist-facies metamorphism.

Pipasi ophiolite

The Pipasi ophiolite is located in southwestern Kangxian County (Fig. 3), and the eastern part is connected with the Lueyang ophiolite, with both ophiolites exhibiting similar lithologic and structural features. The Pipasi ophiolite is mainly composed of basalts with a few mafic dikes. Between the Nanping-Kangxian and Wenxian-Pipasi faults, the basalts occur as four slabs, each between about 2–15 km long and 0. 2–3 km wide (Fig. 3) and they are intensively sheared to form chlorite schist (Lai et al., 2004). The Pipasi ophiolite is associated with numerous imbricated thrust slices of Sinian, Devonian, and Carboniferous sedimentary rocks. The Sinian thrust sheets comprise mainly gravelly mudstones, fine-grained clastic rocks, and carbonates with some volcaniclastic units that are present as exotic tectonic terranes. The Devonian thrust sheets are characterized by turbidites, argillaceous limestones and mudstones, whereas the Carboniferous sheets are predominantly carbonates (Fig. 3).

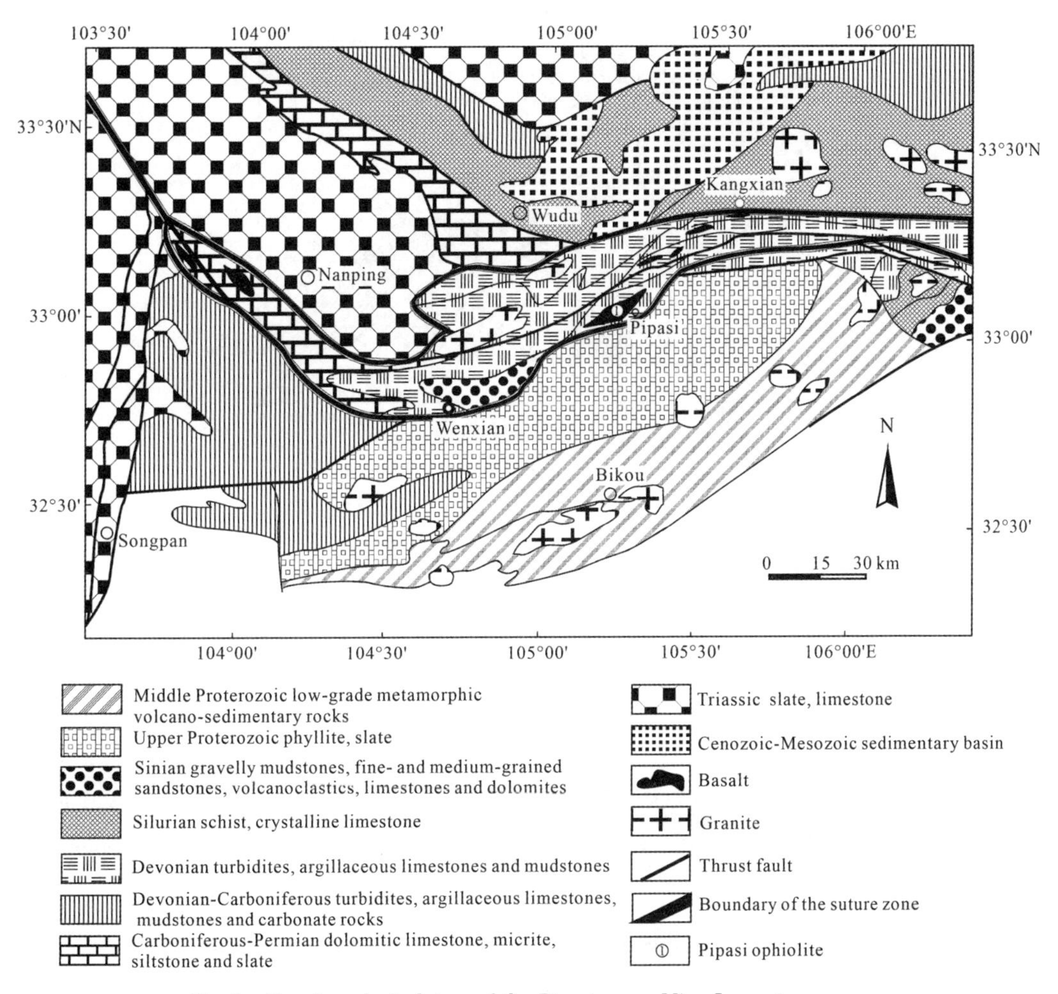

Fig. 3 Sketch geological map of the Pipasi area, Mian-Lue suture zone.

Derni ophiolite

The Derni ophiolite is located in the Derni and Gialige area, Maqin County, Qinhai Province (Fig. 4). The ophiolite forms a linear terrane about 2 km wide and consists of massive basalt, coarse-grained massive gabbro, and meta peridotite with lenses of pyroxenite. Radiolarian chert and mudstone are locally associated with the basalt and are weakly deformed and metamorphosed to low-grade greenschist-facies. Geochemical data from Chen et al. (2000) imply that the Derni ophiolite was formed in a mid-ocean ridge setting.

Analytical methods

Weathered surfaces and veins were carefully removed before the rock samples were powdered with a tungsten carbide mill. Major elements for the Lueyang samples (Table 1) were analyzed by wet chemical methods at the Chinese Academy of Geological Sciences, Beijing.

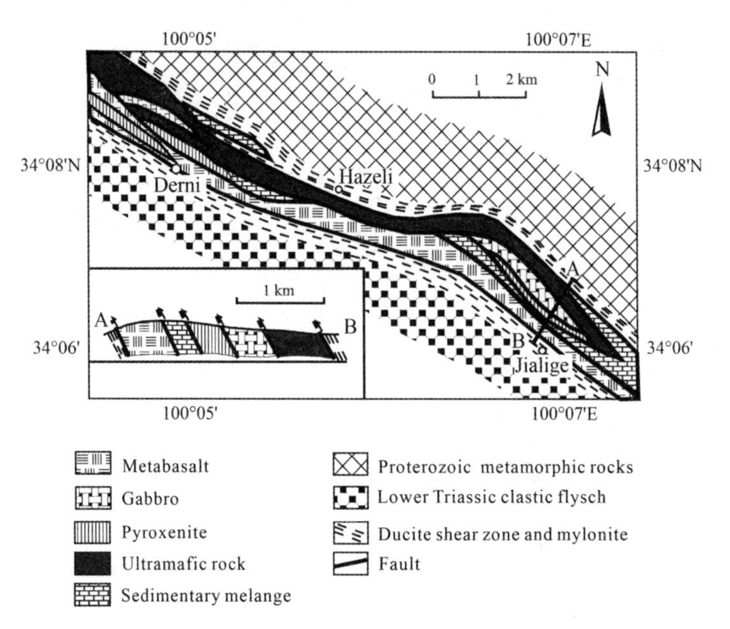

Fig. 4 Sketch geological map of the Derni area, Mian-Lue suture zone.

Trace elements (Ba, Co, Cr, Nb, Ni, Rb, Sr, V, Y, and Zr), were determined using a fully automated Rigaku X-ray fluorescence (XRF) spectrometer at the Beijing Metallurgical Institute of Nonferrous Metals. Sc, Cs, Hf, Ta, Th, U and REEs were determined by neutron activation analysis (NAA) at the Institute of High Energy Physics, Chinese Academy of Sciences, Beijing.

Table 1 Major(wt%) and trace(ppm) element compositions of Lueyang ophiolite.

Sample	LQ14	LQ33	LQ34	M15	M19	LQ36	LQ45	LQ46	LQ48	LQ49	LQ50	LQ51	M40
Rock	Serp.[1]	Serp.	Serp.	Gabbro	Gabbro	Diabase	Diabase	Diabase	Diabase	Basalt	Basalt	Basalt	Basalt
SiO_2	39.60	41.00	41.10	46.99	41.54	54.30	51.20	50.60	55.00	49.60	49.90	48.50	47.28
TiO_2	0.03	0.07	0.01	1.20	4.71	0.88	0.76	1.28	0.85	1.32	0.92	1.13	1.86
Al_2O_3	1.20	2.74	1.27	20.78	20.29	14.00	15.60	15.70	14.60	12.00	13.90	14.40	14.68
Fe_2O_3	8.56	3.57	4.07	3.65	9.17	2.87	2.22	1.27	2.17	3.50	4.81	6.15	3.89
FeO	3.81	3.81	3.45	6.04	3.52	5.03	6.11	9.63	5.46	10.10	7.19	6.90	11.25
MnO	0.19	0.11	0.11	0.14	0.13	0.14	0.15	0.16	0.13	0.15	0.15	0.18	0.21
MgO	35.10	36.20	37.30	5.86	4.75	6.73	8.29	4.84	7.20	7.00	6.91	8.36	4.69
CaO	0.12	0.11	0.07	9.79	12.60	6.94	7.89	5.35	6.65	3.63	5.55	7.53	8.84
Na_2O	0.02	0.37	0.02	4.16	2.15	3.78	1.67	4.72	3.52	0.83	3.19	3.22	2.72
K_2O	0.19	0.24	0.29	0.20	0.20	2.35	3.51	1.22	2.36	0.95	0.70	0.04	0.20
P_2O_5	0.09	0.09	0.09	0.38	0.15	0.40	0.22	0.21	0.41	0.15	0.14	0.15	0.12
LOI	10.9	11.60	11.60	1.47	0.84	1.58	2.18	3.16	1.42	5.58	5.02	3.44	3.49
CO_2	0.19	0.19	0.05	0.19	0.38	0.39	0.39	1.74	0.58	3.39	1.45	0.05	1.23
Total	100.00	100.10	99.43	100.85	100.43	99.39	100.19	99.88	100.35	98.20	99.83	100.05	100.46

Continued

Sample	LQ14	LQ33	LQ34	M15	M19	LQ36	LQ45	LQ46	LQ48	LQ49	LQ50	LQ51	M40
Rock	Serp.[1]	Serp.	Serp.	Gabbro	Gabbro	Diabase	Diabase	Diabase	Diabase	Basalt	Basalt	Basalt	Basalt
Sc	7.25	13.5	8.88	26.9	28.6	24.2	43.6	34.0	24.0	44.4	43.7	52.2	51.1
Cs	1.93	1.38	1.07	7.53	6.75	3.12	1.70	1.24	2.08	4.95	4.30	1.08	11.0
Hf	0.16	0.17	0.15	5.70	2.00	3.45	1.93	3.08	3.40	2.02	1.97	1.64	1.78
Ta	2.45	1.56	1.37	0.21	0.98	0.40	2.45	0.36	2.63	3.65	0.10	0.11	0.28
Th	0.25	0.21	0.21	0.22	0.50	3.65	1.60	2.68	3.48	0.15	0.15	0.15	0.52
U	0.25	0.41	0.39	0.21	0.20	1.00	0.39	0.53	1.18	0.24	0.25	0.26	0.25
Ba	11	10	530	214	701	1 110	936	408	907	318	210	17	65
Co	114	96	87.2	26	56	27	34	32	28	43	44	55	48
Cr	5 610	3 560	2 030	91	13	286	158	3.0	362	184	160	155	7.50
Nb	2.8	3.1	6.7	6.7	16	7.2	5.9	9.0	7.1	4.2	4.0	3.9	8.4
Ni	1 900	1 960	2 110	44	25	101	35	9.9	127	69	65	80	26
Rb	2.6	2	29	16	7.2	46	102	48	55	44	28	3.2	11
Sr	2.9	7.8	274	1 300	1 410	652	239	164	452	75	55	132	438
V	58	70	148	207	427	211	204	332	196	333	304	291	478
Y	3.3	3.5	28	28	13	21	25	32	21	29	25	28	29
Zr	11	13	140	203	64	133	72	115	131	75	54	64	72
La	0.43	0.36	0.16	17.3	15.3	17.3	11.3	18.2	15.7	1.89	1.23	1.59	5.20
Ce	1.13	0.98	0.86	39.5	29.8	33.8	21.2	28.0	29.8	4.63	4.46	5.42	13.1
Nd	1.07	0.89	0.84	28.5	17.7	19.7	11.4	16.2	17.0	3.98	5.42	6.87	10.2
Sm	0.35	0.29	0.33	7.07	4.40	4.96	2.99	4.39	4.73	2.27	2.23	2.63	3.51
Eu	0.48	0.48	0.37	2.14	2.04	1.45	1.07	1.19	1.61	1.32	0.78	0.96	1.47
Gd	0.57	0.56	0.53	6.73	3.50	4.86	3.90	5.32	4.55	5.32	3.65	3.89	4.73
Tb	0.11	0.10	0.10	1.03	0.48	0.79	0.75	0.93	0.76	0.99	0.72	0.88	0.78
Ho	0.14	0.15	0.15	1.29	0.55	0.89	1.04	1.31	0.86	1.59	1.06	1.45	1.10
Tm	0.04	0.08	0.05	0.45	0.16	0.30	0.43	0.51	0.34	0.62	0.43	0.63	0.49
Yb	0.22	0.50	0.26	2.51	0.84	1.72	2.37	2.94	1.98	3.41	2.59	3.48	3.13
Lu	0.03	0.08	0.03	0.33	0.12	0.26	0.33	0.42	0.24	0.46	0.35	0.42	0.45

[1] Serp. = Serpentinite.

Major elements of the Pipasi samples were analyzed by wet chemical techniques at the Institute of Geochemistry, Chinese Academy of Sciences. Trace elements on these samples, including REEs, were analyzed using a Finnigan MAT ELEMENT magnetic sector inductively coupled plasma mass spectrometer (ICP-MS) also at the Institute of Geochemistry, Chinese Academy of Sciences. The procedure for ICP-MS analysis is similar to that described by Qi et al. (2000) and Qi and Gregoire (2000). Analytical uncertainties are generally better than 5% for most elements. Major oxide and trace element compositions of the Pipasi basalts are listed in Table 2.

Table 2 Major(wt%) and trace(ppm) element compositions of Pipasi ophiolite.

Sample	PBS-01	PBS-06	PBS-13	PBS-30	PBS-35
Rock	Basalt	Basalt	Basalt	Basalt	Basalt
SiO_2	49.64	49.06	48.63	50.94	49.25
TiO_2	1.52	1.65	1.10	1.37	1.38
Al_2O_3	14.22	12.18	16.06	15.23	16.25
Fe_2O_3	7.33	6.00	4.60	4.35	4.62
FeO	6.37	8.20	5.20	7.05	7.18
MnO	0.20	0.21	0.24	0.18	0.20
MgO	5.60	5.40	4.90	6.20	6.90
CaO	7.30	10.00	9.50	8.40	8.80
Na_2O	4.56	3.83	3.84	1.07	2.76
K_2O	0.07	0.07	0.13	1.02	0.21
P_2O_5	0.23	0.20	0.07	0.10	0.11
LOI	2.55	2.66	5.50	3.68	2.21
Total	99.59	99.46	99.77	99.59	99.87
Sc	51.5	55.0	42.1	47.4	54.4
Cs	0.24	0.11	0.31	2.02	0.42
Hf	4.10	3.69	1.58	2.59	3.18
Ta	0.18	0.18	0.13	0.22	0.14
Pb	179	200	71.2	100	243
Th	0.20	0.16	0.13	0.22	0.17
U	0.14	0.08	0.08	0.09	0.08
Ba	176	57.3	55.2	451	81.3
Co	43.5	47.2	44.5	50.5	59.0
Cr	85.0	89.1	228	155	169
Nb	2.49	2.61	1.59	3.09	2.22
Ni	49.5	50.9	108	69.9	90.4
Rb	0.58	0.62	2.10	28.1	4.49
Sr	155	259	128	771	130
V	362	439	203	262	354
Y	43.9	47.6	23.2	31.4	34.9
Zr	112	133	50.3	78.9	96.6
La	4.24	4.35	3.34	3.89	3.87
Ce	13.7	14.1	8.08	11.0	12.6
Pr	2.36	2.47	1.27	1.82	2.30
Nd	12.7	14.11	6.97	10.3	13.3
Sm	4.90	5.29	2.35	4.00	4.37
Eu	1.47	1.71	1.04	1.30	1.56
Gd	6.50	6.40	3.15	4.65	5.66
Tb	1.16	1.23	0.58	0.85	1.00
Dy	8.19	8.28	4.00	5.68	6.63

Continued

Sample	PBS-01	PBS-06	PBS-13	PBS-30	PBS-35
Rock	Basalt	Basalt	Basalt	Basalt	Basalt
Ho	1. 65	1. 62	0. 80	1. 20	1. 27
Er	4. 76	5. 21	2. 60	3. 46	3. 81
Tm	0. 66	0. 65	0. 36	0. 42	0. 50
Yb	4. 67	4. 81	2. 47	3. 40	3. 62
Lu	0. 69	0. 65	0. 37	0. 48	0. 50

Geochemistry

Lueyang ophiolite

Most ultramafic rocks from Lueyang are low in CaO (< 0. 12 wt%) , total alkalis (K_2O + Na_2O < 0. 61 wt%) and Al_2O_3 (1. 20 – 2. 74 wt%) (Table 1). They are highly magnesium-depleted peridotites (MgO = 35. 10–37. 30 wt%). The ultramafic rocks are also slightly LREE depleted with ($(La/Yb)_N$ = 0. 4 – 1. 24 (average 0. 70), and slightly HREE enriched with $(Ce/Yb)_N$ = 0. 48 – 1. 23 (average 0. 84) (Fig. 5a). Chondrite-normalized REE patterns of most ultramafic rocks show slightly positive Eu anomalies (Eu = 2. 72–3. 61, average 3. 22) and compared with the primitive mantle (Fig. 6a), they are depleted in La, Ce, Nd, Hf, Ti, and Y (Fig. 6a).

The gabbros have SiO_2 = 41. 54–46. 99 wt% and $Mg^{\#}$ [100 MgO/(MgO + FeO^T)] of 60–70. Chondrite-normalized REE patterns show LREE-enrichment with $(La/Yb)_N$ = 4. 46–11. 73 (average 8. 10), slightly positive Eu anomalies (Eu = 0. 94–1. 55, average 1. 25) (Fig. 5a). Primitive mantle-normalized trace element patterns of the gabbros show positive anomalies in Cs, Ba, Nb and Sr (Fig. 6b).

The diabase dikes have higher SiO_2 (50 –55 wt%) and total alkalis (5. 18–6. 13 wt%), and lower Al_2O_3, CaO, and FeO^T than the gabbros, with MgO ranging between 4. 04–8. 29 wt%. The dikes show LREE enrichment with $(La/Yb)_N$ = 3. 09–6. 51 (average 4. 69), without Eu anomalies (δEu = 0. 76 – 1. 06) (Fig 5c). Primitive mantle-normalized trace element patterns indicate slight depletion in Th, U, Nb, La, and Ti (Fig. 6c).

The basalts have a tholeiitic affinity with moderate TiO_2 (0. 92 – 1. 86 wt%), high FeO^T (12. 00–15. 14 wt%) and MgO (4. 69–8. 36 wt%). They are moderately LREE depleted and HREE enriched, without significant Eu anomalies (δEu = 0. 84 – 1. 13) (Fig. 5d) and with $(La/Yb)_N$ = 0. 30 – 1. 07. Primitive mantle-normalized trace element patterns show Nb-enrichment and slight Ti-depletion, unlike typical island-arc basalts (Fig. 6d). Ratios of Ti/V (18. 14–23. 76), Th/Ta (0. 04–1. 86), Th/Y (0. 01–0. 02), and Ta/Yb (0. 01–0. 13),

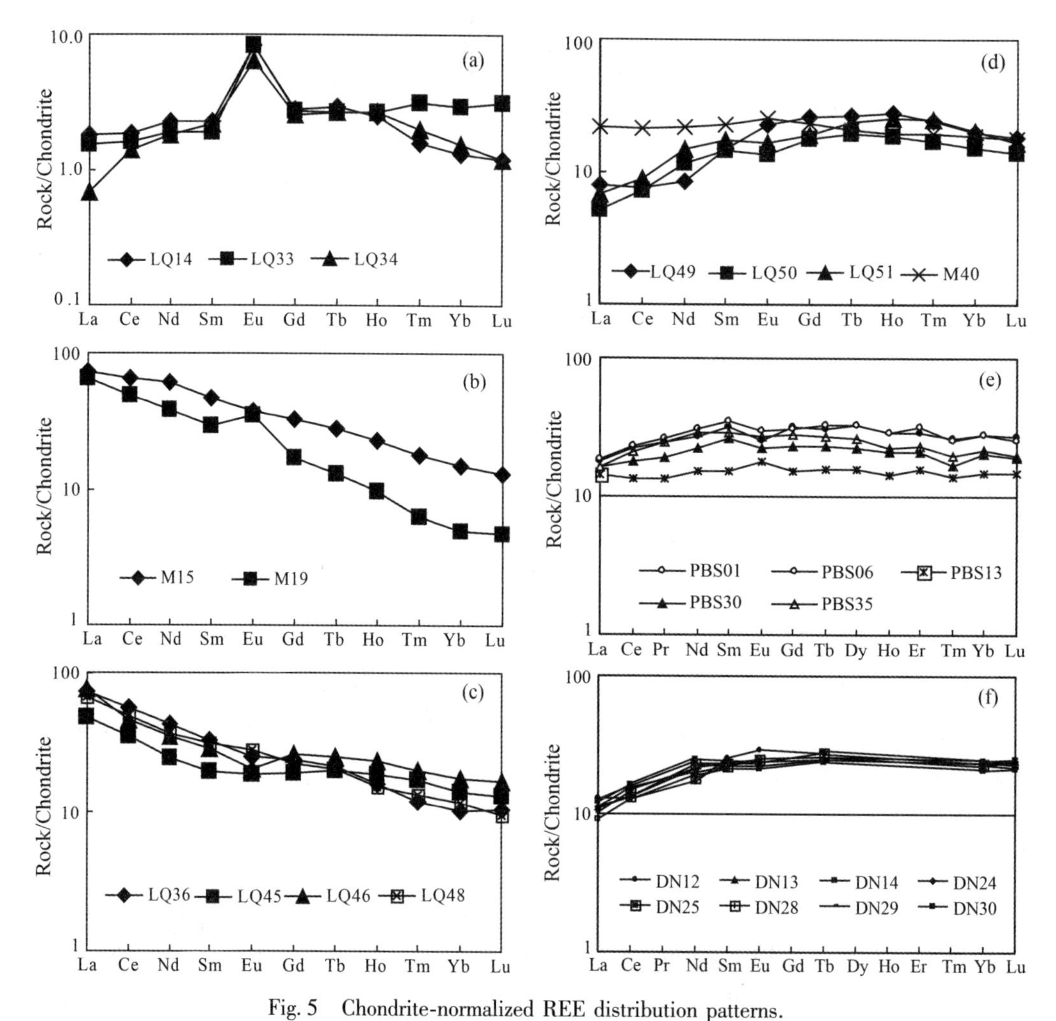

Fig. 5 Chondrite-normalized REE distribution patterns.

(a) ultramafic rock, Lueyang; (b) gabbro, Lueyang; (c) diabase, Lueyang; (d) basalt, Lueyang; (e) basalt, Pipasi; (f) basalt, Derni. Chondrite normalized values are from Sun and McDonough (1989). Sample numbers refer to those in Tables 1, 2 and 3.

are similar to those of typical MORB (Pearce, 1983).

Pipasi ophiolite

Basaltic rocks of the Pipasi ophiolite have SiO_2 = 48.6–50.9 wt%, high $Fe_2O_3^T$ (9.80–14.20 wt%), low MgO (4.90–6.90 wt%), and moderate TiO_2 (1.10–1.65 wt%). On SiO_2 vs. Nb/Y and SiO_2 vs. Zr/TiO_2 diagrams (Fig. 7a, b), they show a clear tholeiitic affinity. Chondrite-normalized REE patterns of the basalts are relatively flat with $(La/Yb)_N$ = 0.65–0.97 (average 0.77), and lack Eu anomalies (Fig. 5e). Primitive mantle-normalized trace element patterns show no obvious depletion in Nb (Fig. 6e) and Ti/V = 22.53 – 32.49 (average 26.98), Th/Ta = 0.89 – 1.21 (average 1.04), Th/Y = 0.003 – 0.007 (average

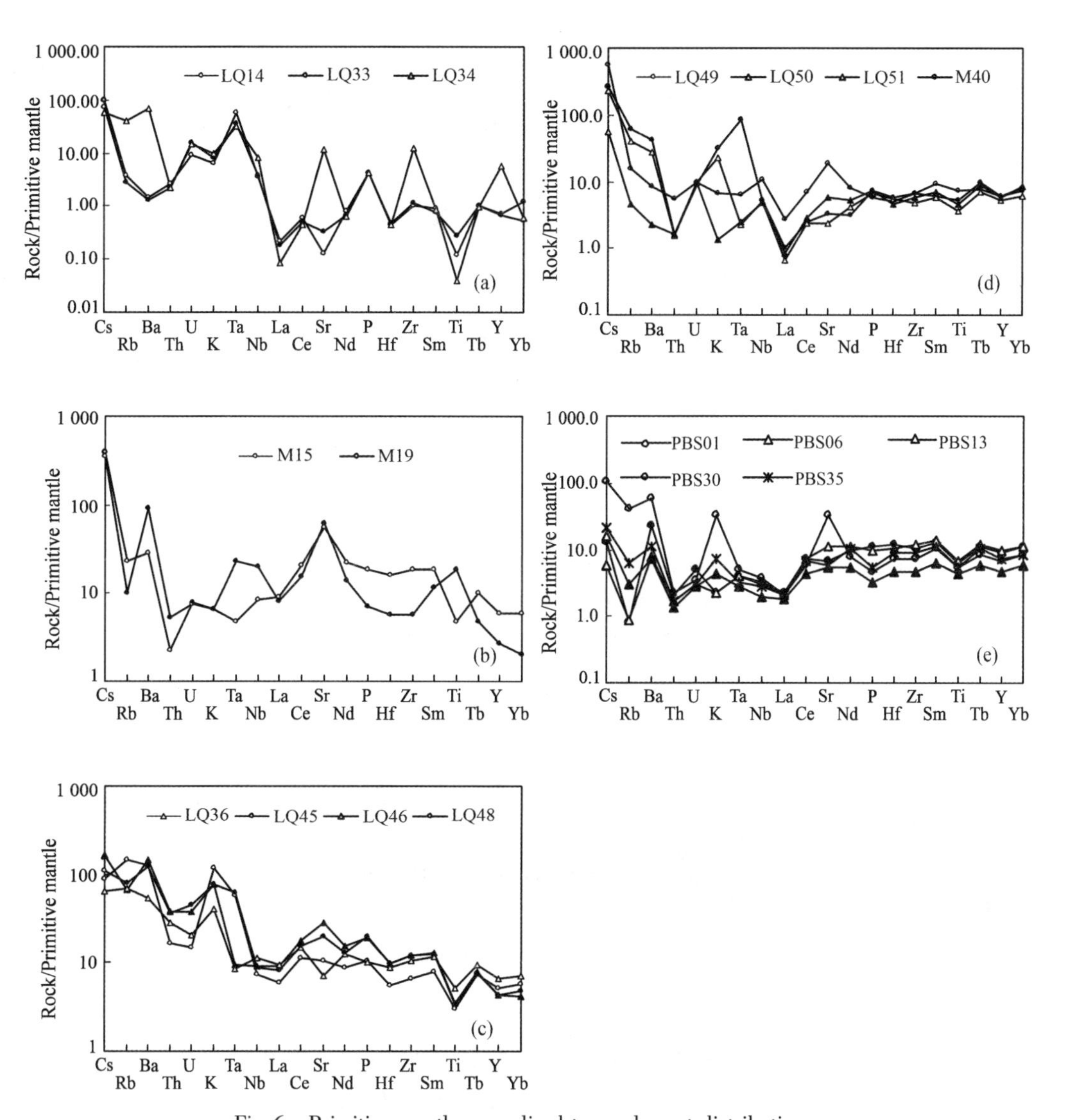

Fig. 6 Primitive-mantle normalized trace element distributions.

(a) Ultramafic rock, Lueyang. (b) Gabbro, Lueyang. (c) Diabase, Lueyang. (d) Basalt, Lueyang. (e) Basalt, Pipasi. Primitive mantle normalized values are from Taylor and McLennan (1985). Sample numbers refer to those in Tables 1 and 2.

0.005), and Ta/Yb = 0.04–0.06 (average 0.05), resembling those of typical MORB.

Derni ophiolite

Metabasalts of the Derni ophiolite have typical N-MORB compositions (Chen et al., 2000). Chondrite-normalized REE diagrams (Fig. 5f) show moderate depletion in LREE with $(La/Yb)_N = \sim 0.45$, and no significant Eu anomalies. The N-MORB-normalized trace element patterns shown in Fig. 8 resemble that of a typical N-MORB (Pearce, 1983).

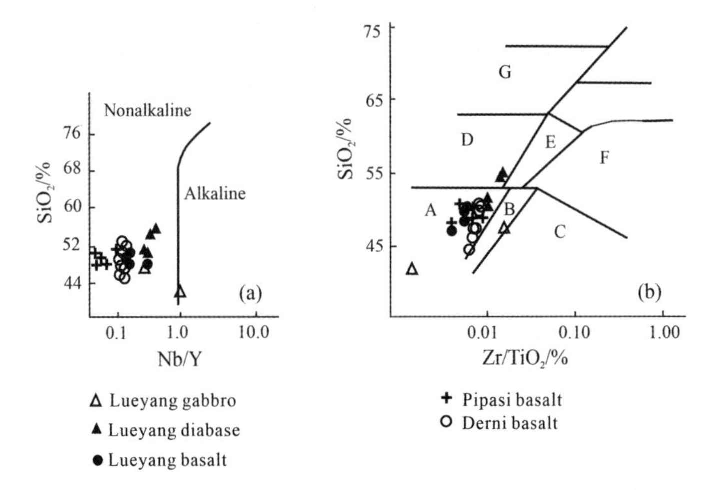

Fig. 7 Plots of SiO_2 vs. Nb/Y (a) and SiO_2 vs. Zr/TiO_2 (b) for meta-volcanic rocks from the Mian-Lue ophiolite. Reference fields are from Winchester et al. (1977).
A: subalkaline basalt; B: alkaline basalt; C: trachybasalt; D: andesite; E: trachyandesite; F: phonolite; G: dacite and dacitic rhyolite.

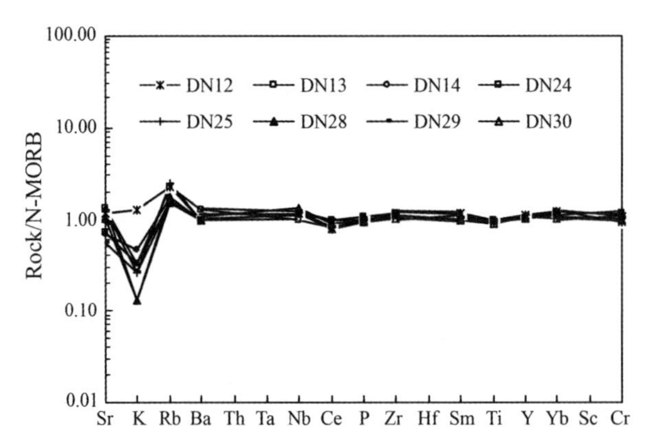

Fig. 8 N-MORB-normalized trace element distributions for basalt from Derni area.
N-MORB values from Pearce (1983). Sample numbers refer to those in Table 3.

The Sm-Nd-Pb isotopic composition of the Derni ophiolite show a narrow range, i.e., $^{143}Nd/^{144}Nd$ = 0.513 1–0.513 2, $^{147}Nd/^{144}Nd$ = 0.21–0.22, $\varepsilon_{Nd}(t)$ (t = 340 Ma) = +8.4 to +9.6, with $^{208}Pb/^{204}Pb$ = 37.70–38.14, $^{207}Pb/^{204}Pb$ = 15.14–15.52 and $^{206}Pb/^{204}Pb$ = 17.78–18.21 (Chen et al., 2001).

On 2Nb-Zr/4-Y and Ti/100-Zr-3Y diagrams (Fig. 9), tholeiitic basalts from the Lueyang, Pipasi, and Derni ophiolites plot in the MORB field, suggesting a typical mid-ocean ridge setting. The geochemical characteristics of the basalts from the Mian-Lue suture zone are summarized in Table 3.

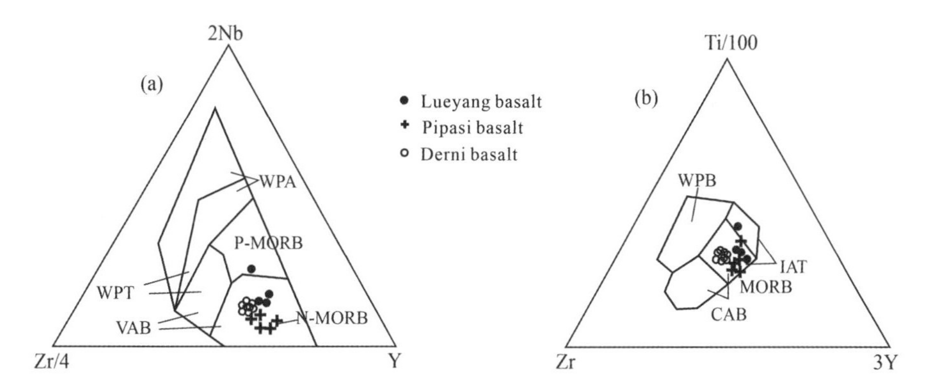

Fig. 9 Plots of 2Nb-Zr/4-Y (a) and Ti/100-Zr-3Y (b) for mafic rocks of the Mian-Lue ophiolite. Reference fields are from Meschede (1986) and Pearce and Cann (1973). WPA: within-plate alkaline basalt; WPT: within-plate tholeiite; WPB: within-plate basalt; P-MORB: P-type mid ocean ridge basalt; N-MORB: N-type mid ocean ridge basalt; MORB: mid ocean ridge basalt; VAB: volcanic arc basalt; CAB: calc-alkaline basalt; and IAT: island arc tholeiite.

Table 3 Characteristics of Mian-Lue suture zone ophiolites.

Region	Rock assemblage	Basalt geochemical features
Lueyang	Ultramafic rocks, gabbro, cumulate gabbro, diabase dike swarm, basalt, radiolarian chert	LREE-depleted N-MORB basalt
Pipasi	Basalt dominated	Typical LREE-depleted N-MORB basalt
Derni	Ultramafic rock, gabbro, basalt, radiolarian chert, and radiolaria-bearing argillite	Basalts of N-MORB affinityand LREE depletions

Discussion

Petrogenesis of the Mian-Lue suture zone ophiolites

Depletion of the Mian-Lue suture zone ultramafic rocks in CaO, TiO_2 and LREEs can be attributed to a high degree of partial melting of a depleted mantle source. Three samples with obvious positive Eu anomalies suggest possible late-stage crystallization of interstitial plagioclase.

Gabbros of the Lueyang ophiolite exhibit geochemical features of cumulate rocks in that they are depleted in Zr and Y and extremely enriched in Sr and Ba (Fig. 6b). Slight LREE enrichment and positive Eu anomalies indicate plagioclase accumulation. Ti depletion (Fig. 6c), slight enrichment in LREEs with a weak negative Eu anomaly (δEu = 0.76–1.06), indicate that the diabase dikes of the Lueyang ophiolite are the result of fractional crystallization.

MORB-type basalts of the Lueyang, Pipasi, and Derni ophiolites (Fig. 9) display steep left-sloping REE patterns with significant depletion in LREEs and other compatible trace elements. Zr/Nb, and Zr/Nd ratios as well as Nb, Nd, Zr contents of the basalts are very similar to those of typical MORB (Table 4), although their $(Ce/Yb)_N$ ratios are somewhat lower than expected, suggesting relatively high degrees of partial melting of a depleted mantle source. Nd-Pb isotopic features of Derni ophiolite MORB basalts indicates derivation from a mantle source intermediate between DMM and EMⅡ components, comparable to modern Indian MORB (Chen et al., 2004). Based on the above evidence, it is concluded that the ophiolitic rocks of the Mian-Lue suture zone are fragments of oceanic lithosphere formed in a mid-ocean ridge environment.

Table 4 Nb, Nd, and Zr contents[1].

Rock	Nb	Nd	Zr	Zr/Nb	Zr/Nd	Ce/Y
Basalt, Lueyang	5.13	6.62	66.3	12.9	10.0	0.25
Basalt, Pipasi	2.40	11.48	94.16	39.32	8.08	0.33
Basalt, Derni	4.10	9.77	99.87	24.48	10.37	0.28
MORB[2]	3.5	8.79	90	25.7	10.23	—
Peridotite from DM[2]	0.27[3]	0.544	5	18.5	9.2	—
Subduction component[2]	1.55[5]	8	21.6	13.9	2.7	—
Crust[4]	20	28	165	8.25	5.9	—

[1] ppm and ratios of Mian-Lue suture zone-ophiolites.

[2] After Ewart and Hawkesworth, 1987.

[3] After Wasson, 1985.

[4] After Henderson, 1984.

[5] After Hole et al., 1984.

Age ofthe Mian-Lue suture zone ophiolites

Radiolarian fauna of Early Carboniferous age are well preserved in cherts of the Lueyang ophiolite. Late Devonian to Carboniferous radiolarians has also been reported in the Pipasi area (Yin et al., 1996; Feng et al., 1996; Lai et al., 1997). MORB-type basalts of the Lueyang ophiolite have a whole-rock Rb-Sr isochron age of 286 Ma and a $^{40}Ar/^{39}Ar$ plateau age ranging from 326 Ma to 344 Ma (Lai et al., 2004). Basalts of the Derni ophiolite have a $^{40}Ar/^{39}Ar$ plateau age of 345 Ma and a whole-rock Sm-Nd isochron age of 320 Ma (Chen et al., 2001). High-precision U-Pb zircon dating of six I-type granite bodies in the South Qinling terrane by Sun et al (2000) yielded ages of 220 Ma to 205 Ma, which are interpreted as the younger age limit of collision between the North China and Yangtze terranes. These ages are in a good agreement with fossil Carboniferous-Permian ages in the studied regions (Bian et al., 1999). Accordingly, it is likely that the Mian-Lue ocean developed in the middle Paleozoic.

Implications for tectonic evolution of the Mian-Lue suture zone

Devonian deep-water turbidites in the Mian-Lue suture zone suggest that rifting in the South Qinling terrane began in the Devonian (Zhang et al., 1995), whereas radiolarian ages of cherts associated with the ophiolites imply the presence of oceanic crust in the Early

Carboniferous (Feng et al., 1996). Thus, the Paleo-Tethyan Qinling Ocean might have begun to form in the Early Carboniferous as a result of Devonian rifting along the southern margin of the Qinling orogenic belt (Meng and Zhang, 2000). Metavolcanic rocks from the Heigouxia Valley in the Mianxian-Lueyang area have yielded an Sm-Nd whole-rock isochron age of 242 ± 21 Ma and a whole rock Rb-Sr isochron age of 221 ± 13 Ma, indicating the timing of peak greenschist facies metamorphism (Li et al., 1996), which is in agreement with the inferred closure time of the Mian-Lue oceanic basin.

The ophiolite of the Mian-Lue suture zone could also imply that the Qinling orogenic belt developed as an isolated lithospheric microplate in the late Paleozoic. Due to northward subduction of the Paleo-Tethyan Qinling ocean crust, the southern margin of this microplate was accreated into an active margin in the Early Triassic (Yang and Hu, 1990; Zhang et al., 1995; Meng and Zhang, 1999). As a result of this process, continental margin-arc volcanic rocks could also be expected in the suture zone (Lai et al., 1998). It is considered that the Qinling microplate could have been separated from the South China terrane in the Early Carboniferous, and that it experienced a unique evolution history until its collision in the late Triassic (Xu and Han, 1996). Thus, a model for the tectonic evolution of the Mian-Lue paleo-ocean basin can be summarized as follows (Fig. 10):

(1) Initial extension and subsidence consequent on the initiation of continental rifting (D_1): Opening of the east paleo-Tethys oceanic basin from west to east along the Mian-Lue suture zone in the passive continental margin that formed part of the northern margin of the

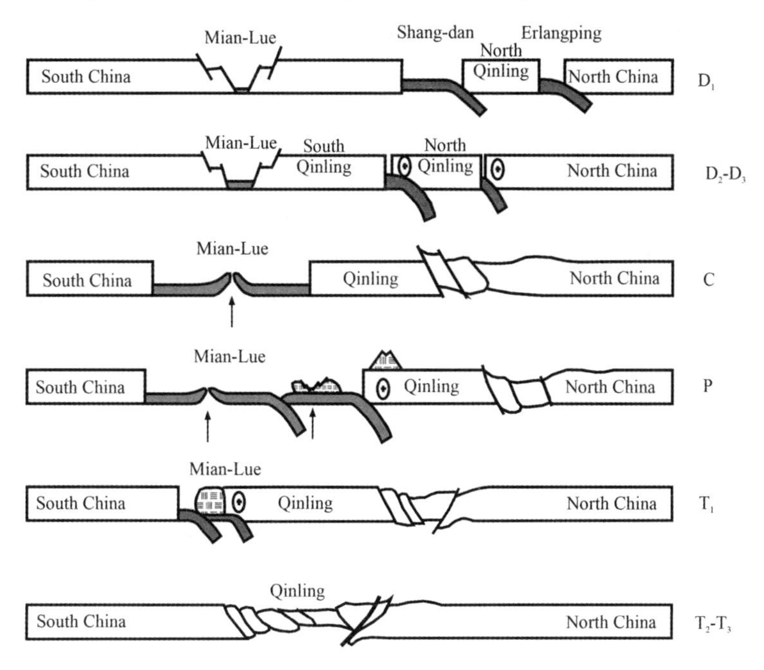

Fig. 10 Suggested tectonic model for evolution of the Mian-Lue suture.

South China terrane.

(2) Small-scale juvenile oceanic basin development (D_2-D_3): Rifting continued with formation of a series of west to east unconnected juvenile basins.

(3) Mian-Lue oceanic basin development (C): A unified limited Mian-Lue oceanic basin formed during the Carboniferous.

(4) Subduction of oceanic crust (P): Island-arc volcanics, subduction-related (island-arc) granites, subduction-associated deformation and metamorphism (Zhang et al., 1995; Lai et al., 1997, 1998; Sun et al., 2000) indicate that the Mian-Lue oceanic basin began to be subducted during the Permian.

(5) Remnant oceanic basin (T_1): By Early Triassic, the Mian-Lue oceanic basin existed as small remnant basins.

(6) Continent-continent collision (T_2-T_3): Continental collision associated with deformation, metamorphism, magmatism, formation of a foreland fold-thrust zone involving pre-Triassic successions, paired metamorphism, and exhumation of deep-seated granulite facies rocks (Zhang et al., 1995), heralded the disappearance of the Mian-Lue oceanic basin during the Middle-Late Triassic.

Conclusions

(1) The ophiolitesin the Mian-Lue suture zone are well exposed in the Lueyang, Pipasi, and Derni areas of central China. In Lueyang, ophiolite is composed of ultramafic rocks, diabase dikes, and basalt. The metabasalts are tholeiite with MORB-type characteristics such as depleted LREE, Nb, TiO_2 and low Ti/V, Th/Ta, Th/Y, and Ta/Yb ratios. The Pipasi and Derni ophiolites are mainly composed of basalt, also of tholeiitic affinity with LREE depletions comparable to MORB.

(2) The North and South China terranes were separated by the Paleo-Tethys Qinling (Mian-Lue) ocean. This branch of the Tethys formed in the Devonian and closed in the Triassic. The Mian-Lue suture zone can be regarded as the boundary between the North and South China terranes.

(3) The ophiolites represent fragments of the Qinling (Main-Lue) ocean lithosphere incorporated within the Mian-Lue suture zone that represents Triassic amalgamation of the North and South China terranes.

Acknowledgements This work was jointly supported by the National Natural Science Foundation of China (NSFC, Grant No. 40234041) and the Teaching and Research Award Program for Outstanding Young Teachers in Higher Education Institutions, Ministry of Education (MOE), China. We thank Y. Wang and M.F. Zhou for their comments on an early draft of this paper.

References

Bian, Q.T., Luo, X.Q., Li, H.S., Chen, H.H, Zhao, D.S., Li, D.H., 1999. Discovery of early Paleozoic and early Carboniferous-early Permian ophiolites in the A'nyemaqen, Qinling Province, China. Scientia Geologica Sinica, 34, 523-529 (in Chinese).

Chen, L., Sun, Y., Liu, X. M., Pei, X. Z., 2000. Geochemistry of Derni ophiolite and its tectonic significance. Acta Petrologica Sinica, 16, 106-110 (in Chinese).

Chen, L., Sun, Y., Pei, X.Z. Gao, M., Feng, T., Zhang, Z.Q., Chen, W., 2001. Northernmost paleo-Tethyan oceanic basin in Tibet: Geochronolgical evidence from $^{40}Ar/^{39}Ar$ age dating of Der'ngoi ophiolite. Chinese Science Bulletin, 46, 1203-1205.

Chen, L., Sun, Y., Pei, X.Z., Feng, T., Zhang, G.W., 2004. Comparison of eastern paleo-Tethyan ophiolites and their geodynamic significance. Science in China: Series D, 47, 378-384.

Ewart, A., Hawkesworth, C.J., 1987. The Pleistocene-Recent Tonga-Kermadec arc lavas: Interpretation of new isotopic and rare earth data in terms of a depleted mantle source model. Journal of Petrology, 28, 495-530.

Feng, Q.L., Du, Y.S., Yin, H.F., Sheng, J.H., Xu, J.F., 1996. Carboniferous radiolaria fauna firstly discovered in the Mian-Lue ophiolitic melange belt of south Qinling. Science in China: Series D, 39, 87-92.

Henderson, P., 1984. Rare earth element geochemistry. Amsterdam: Elsevier Science Publishers, 103-106.

Hole, M.J., Saunders, A.D., Marriner, G.F., 1984. Subduction of pelagic sediments: Implications for the origin Ce-anomalous basalts from the Mariana island. Journal of the Geological Society, London, 141, 453-472.

Lai, S.C., Zhang, G.W., 1996. Geochemical features of ophiolite in Mianxian-Lueyang suture zone, Qinling orogenic belt. Journal of China University of Geosciences, 7, 165-172.

Lai, S.C., Zhang, G.W., Dong, Y.P., Pei, X.Z., Chen, L., 2004. Geochemistry and regional distribution of ophiolites and associated volcanics in Mian-Lue suture zone, Qinling-Dabie. Science in China: Series D, 47, 289-299.

Lai, S.C., Zhang, G.W., Yang, Y.C., Chen, J.Y., 1997. Petrology and geochemistry features of the metamorphic volcanic rocks in Mianxian-Lueyang suture zone, south Qinling. Acta Petrologica Sinica, 13, 563-573 (in Chinese).

Lai, S.C., Zhang, G.W., Yang, Y.C, Chen, J.Y., 1998. Geochemistry of the ophiolite and island-arc volcanic rocks in the Mianxian-Lueyang suture zone, southern Qinling and their tectonic significance. Geochimica, 27, 283-293 (in Chinese).

Lai, X.L., Yang, F.Q., Du, Y.S., Yang, H.S., 1997. Triassic stratigraphic sequence and depositional environment in the Nanping-Zoige area, northwestern Sichuan. Regional Geology of China, 16, 193-202 (in Chinese).

Li, S.G., Sun, W.D., Zhang, G.W., Chen, J.Y., Yang, Y.C., 1996. Chronology and geochemistry of metavolcanic rocks from Heigouxia Valley in the Mian-Lue tectonic zone, South Qinling-evidence for a Paleozoic oceanic basin and its time of closure. Science in China: Series D, 39, 301-310.

Meng, Q.R., Zhang, G.W., 2000. Geologic framework and tectonic evolution of the Qinling orogen, central

China. Tectonophysics, 323, 183-196.

Meng, Q.R., Zhang, G.W., 1999. Timing of collision of the North and South China terranes: Controversy and reconciliation. Geology, 27, 123-126.

Meschede, M.A., 1986. A method of discriminating between different types of mid-ocean basalts and continental tholeiites with the Nb-Zr-Y diagram. Chemical Geology, 56, 207-218.

Pearce, J.A., 1983. The role of sub-continental lithosphere in magma genesis at destructive plate margins. In: Hawkesworth, et al. Continental Basalts and Mantle Xenoliths. Nantwich Shiva, 230-249.

Pearce, J.A., Cann, J.R., 1973. Tectonic settings of basic volcanic rocks determined using trace element analysis. Earth & Planetary Science Letters, 19, 290-300.

Qi, L., Gregoire, D.C., 2000. Determination of trace elements in twenty-six Chinese geochemistry reference materials by inductively coupled plasma-mass spectrometry. Geostandards Newsletters, 24, 51-63.

Qi, L., Hu. J., Gregoire, D.C., 2000. Determination of trace elements in granites by inductively coupled plasma mass spectrometry. Talanta, 51, 507-513.

Sun, S.S., McDonough, W.F., 1989. Chemical and isotopic systematics of oceanic basalts: Implications for mantle composition and processes. In: Saunders, A.D., Norry, M.J. Magmatism in the Ocean Basin. Geological Society, London, Special Publications, 42,313-345.

Sun, W.D., Li, S.G., Chen, Y.D., Li, Y.J., 2000, Zircon U-Pb dating of granitoids from south Qinling, Central China and their geological significance. Geochimica, 29, 209-216 (in Chinese).

Taylor, S.R., McLennan, S.M., 1985. The continental crust: Its composition and evolution. Oxford: Blackwell Scientific Press, 124.

Wang, H.Z., Xu, C.Y., Zhou, Z.G., 1982. Tectonic development of two sides of continental margins of East Qinling paleo-seas. Acta Geologica Sinica, 56, 270-279 (in Chinese).

Wasson, J.T., 1985. Meteorites: Their record of early solar system history. New York: Freeman Publishing Company, 47-67.

Winchester, J.A., Floyd, P.A., 1977. Geochemical discrimination of different magmas series and their differentiation products using immobile elements. Chemical Geology, 20, 325-343.

Xu, J.F., Han, Y.W., 1996. High radiogenic Pb-isotope composition of ancient MORB-type rocks from Qinling area-Evidence for the presence of Tethyan-type oceanic mantle. Science in China: Series D, 39, 34-41.

Xu, J.F., Yu, X.Y., Li, X.H., Zhang, B.R., Han, Y.W., 2000. Geochemistry of the Anzishan ophiolitic complex in the Mian-Lue belt of Qinling Orogen: Evidence and implication of the palaeoocean crust. Acta Geologica Sinica, 74, 39-50 (in Chinese).

Yang, Z.R., Hu, Y.X., 1990. Indication of the paleoplate suture in Mian-Lue, Shaanxi Province and its relation to the Southern Qinling plate. Geology of Northwestern China, 2, 13-20 (in Chinese).

Yin, H.F., Du, Y.S., Xu, J.F., Sheng, J.H, Feng, Q.L., 1996. Carboniferous radiolaria fauna firstly discovered in Mian-Lue ophiolitic melange belt of south Qinling. Earth Sciences, 21, 184 (in Chinese).

Zhang, G.W., Guo, A.L., Liu, F.T., Xiao, Q.H., and Meng, Q.R., 1996. Three-dimensional architecture and dynamic analysis of the Qinling orogenic belt. Science in China: Series D, 39, 1-9.

Zhang, G.W., Meng, Q.R., Lai, S.C., 1995. Structure and tectonics of the Qinling orogenic belt. Science in China: Series B, 25, 994-1003.

Petrochemistry of granulite xenoliths from the Cenozoic Qiangtang volcanic field, northern Tibetan Plateau: Implications for lower crust composition and genesis of the volcanism①

Lai Shaocong Qin Jiangfeng Rodney Grapes

Abstract: High-K calc-alkaline magmas from the Cenozoic Qiangtang volcanic field, northern Tibetan Plateau, contain lower crustal two-pyroxene and clinopyroxene granulite xenoliths. The petrology and geochemistry of six mafic and three felsic xenoliths from the Hol Xil area south of Ulan Ul Lake are discussed. Mafic granulites (Pl, Opx, Cpx, Ksp and Bt) contain 48.76%-58.61% SiO_2, 18.34%-24.50% Al_2O_3, 3.16%-5.41% Na_2O, 1.58%-3.01% K_2O, low $Mg^{\#}$(30-67), LREE and LILE enrichment, high Rb/Sr (0.09-0.21), $(La/Yb)_N$ (17.32-49.35), low Nb/Ta (9.76-14.92), and variable Eu anomalies (Eu = 0.19-0.89). They also have more evolved Sr-Nd-Pb isotopic compositions in comparison with the host dacites $^{87}Sr/^{86}Sr$ (0.710 812-0.713 241), ε_{Sr} (+169.13 to +203.88), $^{143}Nd/^{144}Nd$ (0.512 113-0.512 397); ε_{Nd}(-10.05 to -4.70), $^{206}Pb/^{204}Pb$ (18.700 0-18.956 5), $^{207}Pb/^{204}Pb$ (15.713 5-15.766 2), and $^{208}Pb/^{204}Pb$ (39.109 0-39.473 3). Felsic granulites (Qtz, Pl, Ksp, Bt, and Cpx) show enrichment of LREE and LILE and have evolved Sr-Nd-Pb isotopic compositions with $(La/Yb)_N$ (2.04-10.82), $^{87}Sr/^{86}Sr$ (0.712 041-0.729 088), ε_{Sr} (+180.71 to +430.59), $^{143}Nd/^{144}Nd$ (0.512 230-0.512 388); ε_{Nd}(-7.96 to -4.74), $^{206}Pb/^{204}Pb$ (18.925 0-19.171 7), $^{207}Pb/^{204}Pb$ (15.766 2-15.772 0), and $^{208}Pb/^{204}Pb$ (39.210 9-39.646 7). These geochemical data suggest that the protolith of the mafic granulites could have been a hybrid mafic magma (e.g. enriched mantle type II) or metasomatized restite derived from the partial melting of metamafic-intermediate rocks rather than basaltic cumulates, whereas the felsic granulite protolith was a quartzofeldspathic S-type granitic rock. We argue that the lower crust of the northern Tibetan Plateau is hot and heterogeneous rather than wholly gabbroic. Interaction between the mantle-derived magma and metasedimentary/granitic lower crust of the Tibetan Plateau may have played an important role in the generation of shoshonitic and high-K calc-alkaline andesite-dacite rocks.

1 Introduction

Xenolith-bearing volcanic rocks and kimberlites occur in many different tectonic settings

---

① Published in *International Geology Review*, 2011,53(8).

and provide extensive sampling of the continental crust. Compositionally diverse xenoliths typically record complex thermal and deformation histories (Mattie et al., 1997 and references therein). Compared with meta-igneous xenoliths, metasedimentary xenoliths are minor. In general, mafic xenoliths are considerably more abundant than felsic xenoliths, suggesting that the composition of the lower continental crust is dominantly mafic. Many mafic xenoliths appear to have originated as cumulates from basaltic magma intruded into or underplated beneath the continental crust, and most have recrystallized under granulite-facies conditions reflecting equilibration depths > 20 km, and in some cases more than 40 km. A small portion of the metasedimentary and gneissic xenoliths also record similar depths of recrystallization, suggesting interlayering of felsic and mafic rocks in the lower crust (Mattie et al., 1997).

Knowledge of the lower continental crust has been significantly improved through studies of granulites, emplaced at the Earth's surface either as high-grade terranes or as xenoliths in volcanic rocks. Granulite-facies terranes tend to be dominated by intermediate to felsic compositions, whereas granulite xenoliths are predominantly mafic (e.g. Griffin and O'Reilly, 1986; Rudnick, 1992; Rudnick and Fountain, 1995).

The India-Asia collision and resultant Himalaya orogen provide the best modern example of continental collision plate tectonics. The problematic Tibetan Plateau is associated with the development of this orogen and formed as a consequence of the continuing indentation of India into Asia. The plateau has had a central role in the development of recent models for the formation of mountain belts (e.g. Dewey et al., 1988; England and Houseman, 1988, 1989) as well as for Cenozoic climate change (e.g. Raymo and Ruddiman, 1992; Kutzbach et al., 1993). However, the formation processes of the Tibetan Plateau remain unclear. Models of wholescale underthrusting of India (Barazangi and Ni, 1982), lower crustal flow (Zhao and Morgan, 1987), intracontinental subduction (Meyer et al., 1998; Deng and Sun, 1999), or distributed shortening followed by convective thinning of thickened mantle lithosphere (Houseman et al., 1981) make predictions about the thickening, heating, and melting histories of the plateau, but we have little direct knowledge of the thickness, thermal regime, and composition of the Tibetan crust during much of the Cenozoic. The timing of crustal extension (Edward and Harrison, 1997), crustal shortening (Murphy et al., 1997), rapid upper crustal cooling, and potassic volcanism (Arnaud et al., 1992) have all been used to infer when the Tibetan Plateau attained its present thickness of 60–75 km, but few studies quantify crustal thickness or thermal gradient due to the remoteness of this region and the total absence of deep crustal exposures. Outcrops of metamorphic basement rocks are also limited in the northern Tibetan area. Moreover, the presence of potassic and ultra-potassic volcanism has been causally linked to convective thinning of overthickened mantle lithosphere, and this seems to be a plausible explanation after volcanism was dated at only 13 Ma and younger (Turner et al., 1996). However, recent dating (Chung et al., 1998) indicates that potassic volcanism has occurred in Tibet for about 45 million years. This long-sustained volcanic activicy calls into question the role of convective thinning, which is considered to be a short-term event of approximately 5 million years(Lenardic and Kaula, 1995).

Alkaline and high-K calc-alkaline volcanic rocks in the Tibetan Plateau are characterized by shoshonitic and ultra-potassic trachybasalt to trachyte compositions (Turner et al., 1996; Miller et al., 1999; Ding et al., 2003) with minor high-K calc-alkaline andesite to dacite (Lai et al., 2003, 2007). Granulite xenoliths and xenocrysts have been recently discovered in potassic volcanic rocks at eight localities (Deng et al., 1996; Miller et al., 1999). Hacker et al. (2000) reported granulite xenoliths in potassic volcanic rocks from the Taipinghu area of northern Qiangtang terrane. These granulite xenoliths suggest unusually high temperatures (>800 ℃) of the northern Tibetan lower crust, although its composition is a matter of debate. Hence, understanding of the composition and structure of the thickened northern Tibetan crust will continue to rely on further studies of granulite xenoliths in the Cenozoic volcanic rocks. In this paper, we present mineralogical, geochemistry, and isotopic (Sr-Nd-Pb) data of mafic and felsic granulite xenoliths in Late Paleogene to Neogene high-K calc-alkaline dacite from the south Ulan Ul Lake, Hol Xil area, northern Qiangtang terrane. The role played by a granulite lower crust in the generation of shoshonitic and high-K calc-alkaline andesite to dacite is also discussed.

2 Geological setting

The interior of the Tibetan Plateau is composed by EW-trending Songpan-Ganzi, Qiangtang, and Lhasa terranes from north to south (Fig. 1a). The Qiangtang terrane is bounded by the Jinsha suture to the north and Bangong-Nujiang suture to the south. Structurally, Qiangtang is an anticlinorium, with the central part occupied by an anticline of pre-Jurassic strata or metamorphic rocks and the northern and southern limbs by synclines of mainly Jurassic sedimentary rocks. Cenozoic volcanic rocks in the Qiangtang terrane are composed of ultra-potassic, potassic and minor calc-alkaline (CA) varieties (Lai et al., 2003, 2007; Ding et al., 2003, 2007; Guo et al., 2006). Geochemical data suggested that these post-collisional potassic and ultra-potassic rocks were probably derived from a low degree of partial melting of the metasomatized lithospheric mantle (phlogopite-bearing peridotite) (Turner et al., 1996). The calc-alkaline volcanics also include "adakites" which comprise intermediate and felsic rocks and are considered to be derived from partial melting of thickened lower crust.

Volcanic rocks hosting granulite xenoliths are mainly distributed in Zhentouya area, south to Ulan Ul Lake, Qiangtang terrane (Fig. 1). The volcanic rocks occur as lava sheets that range in thickness from 50 m to 200 m and can be subdivided into two eruption cycles separated

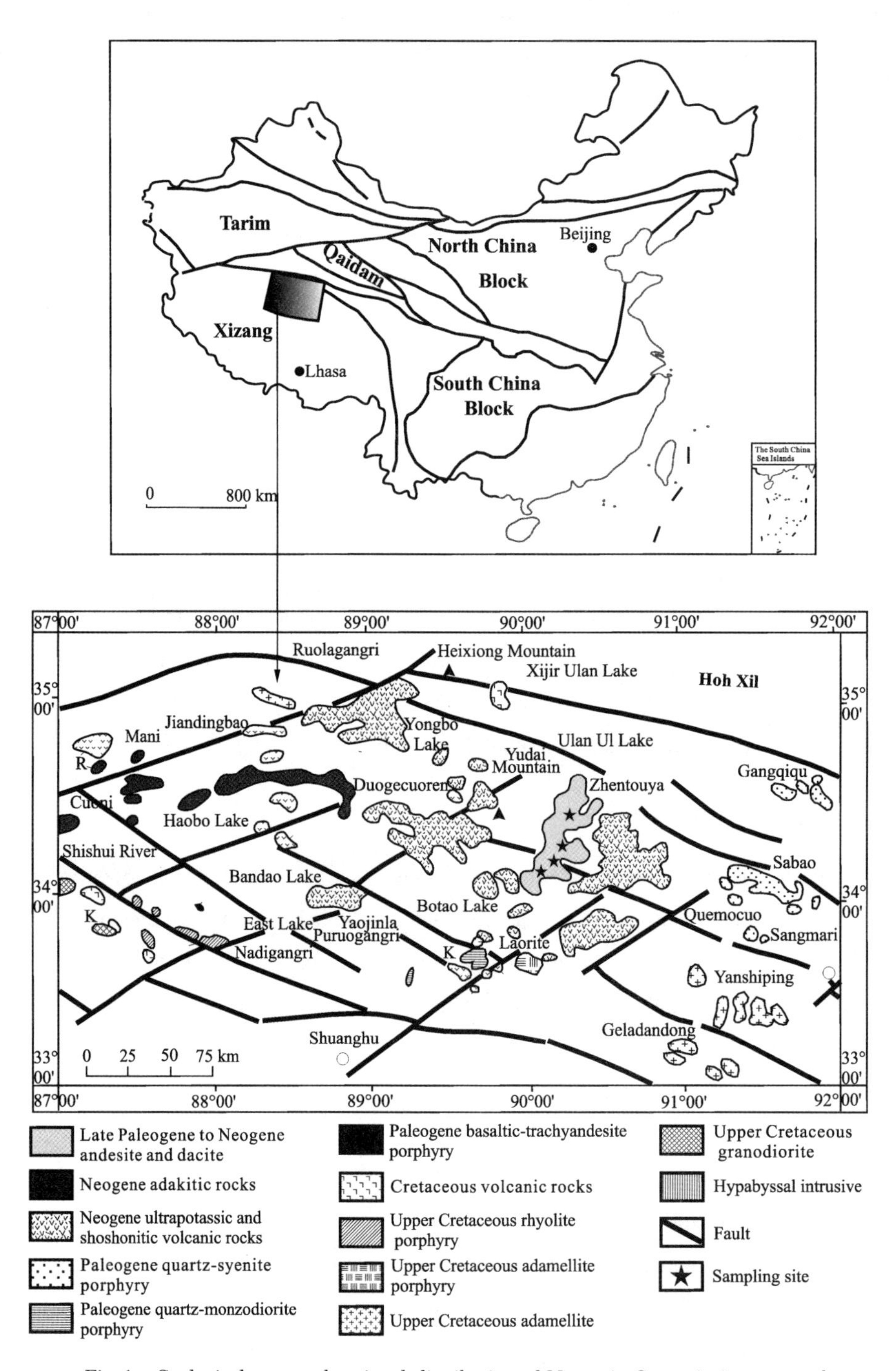

Fig. 1　Geological map and regional distribution of Mesozoic-Cenozoic igneous rocks in Zhentouya area, Qiangtang terrane.

by a 20–30 m-thick volcanic breccia. The first eruption cycle is dominantly andesite and the second eruption cycle is dacite (Fig. 2). Andesites are characterized by phenocrysts of plagioclase ± amphibole ± biotite and some clinopyroxene in a fine-grained, sometimes glassy matrix containing minor apatite and magnetite. Dacites are typically fresh, vesicular, and amygdaloidal with aphyric and porphyro-aphanitic textures. Porphyro-aphanitic dacites contain phenocrysts of quartz, amphibole, biotite, and plagioclase in a glass-bearing an ophitic to hyalopilitic groundmass. In addition, possible xenocrystal sanidine rimmed by K-oligoclase (0.5–5.0 cm) occurs in both andesite and dacite. K-Ar dating indicates a late Paleogene to Neogene age for the volcanic rocks (Ding et al., 2007).

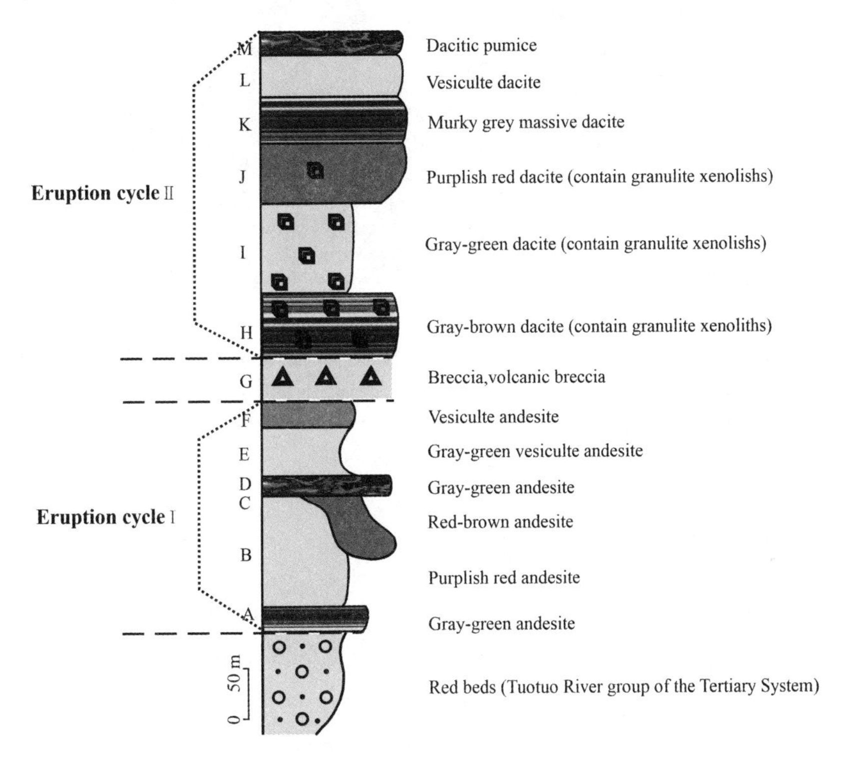

Fig. 2 The stratigraphic column for the late Paleogene to Neogene andesite and dacite association, Qiangtang region, northern Tibetan Plateau

Granulite xenoliths are mainly enclosed in dacite. The xenoliths are fresh and mafic varieties have a drab-red and/or greyish-black colour. Their size generally ranges from 2 cm to 6 cm, and the largest xenolith collected is 12 cm in diameter. The xenoliths are typically massive (some show a weak gneissic structure) and have a medium to fine-grained granoblastic texture (Fig. 3a, b). The xenolith consists of mafic and siliciclastic varieties. Mafic granulite xenoliths include plagioclase + orthopyroxene + clinopyroxene granulite and biotite-bearing plagioclase + K-feldspar + clinopyroxene granulite. Biotite is an interstitial phase (Fig. 3c) and also occurs along fractures in clinopyroxene and orthopyroxene. Glass inclusions are present in plagioclase and biotite (Fig. 3a). The felsic granulite xenoliths are composed of plagioclase + clinopyroxene + quartz + zircon, and clinopyroxene contains some glass inclusion (Fig. 3c, d).

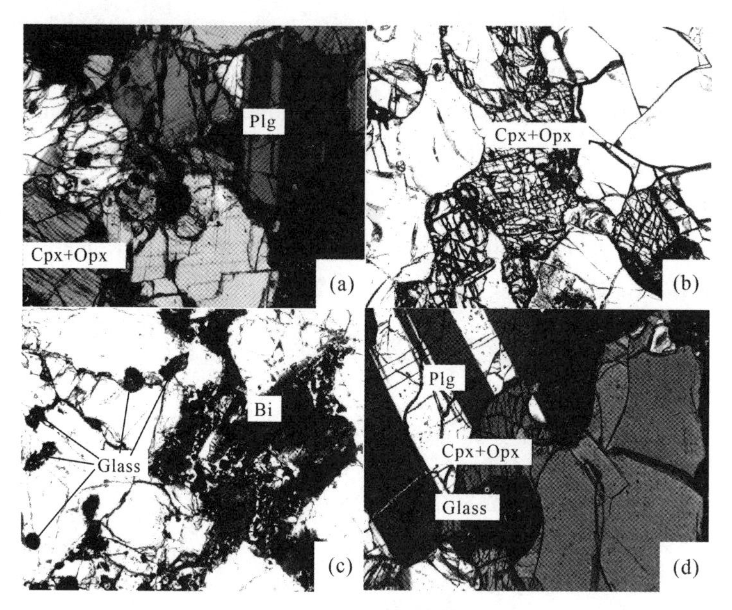

Fig. 3 Photomicrographs of granulite xenoliths in Cenozoic volcanics, Northern Tibet.
(a) Clinopyroxene (Cpx) + plagioclase (Plg) + potassium feldspar (Af) granulite (P2126H10); (b) orthopyroxene (Opx) + clinopyroxene + plagioclase granulite (P2106H); (c) glass inclusions in plagioclase (clear grains) and partial melting of biotite (Bi) to form Fe-oxide, orthopyroxene, and glass (P2126H10); (d) glass inclusions in clinopyroxene (P2106H).

3 Analytical methods

Fresh granulite xenoliths samples were selected for elemental analyses. Major and trace elements were analyzed by X-ray fluorescence (Rikagu RIX 2100) and inductively coupled plasma mass spectroscopy (ICP-MS) (Agilent 7500a), respectively. Analyses of USGS and Chinese national rock standards (BCR-2, GSR-1, and GSR-3) indicate that analytical precision and accuracy for major elements are generally better than 5%. For trace element analysis, sample powders were digested using an $HF + HNO_3$ mixture in high-pressure Teflon bombs at 190 ℃ for 48 hours. Analytical precision is better than 10% for most trace elements.

Whole-rock Sr-Nd-Pb isotopic data were obtained using a Nu Plasma HR mutli-collector mass spectrometer at the State Key Laboratory of Continental Dynamics, Northwest University. Sr and Nd isotopic fractionations were corrected to $^{87}Sr/^{86}Sr = 0.1194$ and $^{146}Nd/^{144}Nd = 0.7219$, respectively. During the period of analysis, the NIST SRM 987 standard yielded an average value of $^{87}Sr/^{86}Sr = 0.710250 \pm 12$ (2σ, $n = 15$) and the La Jolla standard gave an average of $^{143}Nd/^{144}Nd = 0.511859 \pm 6$ (2σ, $n = 20$). Whole-rock Pb was separated by anion exchange in HCl-Br columns, and Pb isotopic fractionation was corrected to $^{205}Tl/^{203}Tl = 2.3875$. Within the period of analysis, 30 measurements of NBS981 gave average values of $^{206}Pb/^{204}Pb =$

16.937±1 (2σ), $^{207}Pb/^{204}Pb$ = 15.491±1 (2σ), $^{208}Pb/^{204}Pb$ = 36.696±1 (2σ) and the BCR-2 standard gave $^{206}Pb/^{204}Pb$ = 18.742±1 (2σ), $^{207}Pb/^{204}Pb$ = 15.620±1 (2σ), and $^{208}Pb/^{204}Pb$ = 38.705±1(2σ). Total procedural Pb blanks were in the range of 0.1–0.3 ng.

4 Whole-rock chemistry

Major and trace element data of representative mafic and felsic granulite xenoliths are given in Tables 1 and 2, respectively.

Table 1 XRF analyses, CIPW and Niggli norms of mafic and felsic granulite xenoliths in Cenozoic volcanics, Hol Xil area, northern Tibetan Plateau.

Sample	LBT5	LBT18	LBT20	LBT41	LBT42	LBT46	LBT17	LBT36	LBT54
Rock	Mafic-granulite	Mafic-granulite	Mafic-granulite	Mafic-granulite	Mafic-granulite	Mafic-granulite	Felsic-granulite	Felsic-granulite	Felsic-granulite
SiO_2	53.58	54.51	52.67	48.76	58.61	52.15	71.23	71.71	75.93
TiO_2	0.87	0.83	1.03	0.95	0.28	0.78	0.39	0.40	0.02
Al_2O_3	19.44	18.68	21.74	24.50	22.51	18.34	13.71	13.91	12.75
FeO	6.05	5.16	5.08	5.99	3.40	4.34	2.38	2.53	1.98
Fe_2O_3	2.38	3.58	2.58	2.41	0.19	3.99	1.26	0.87	0.12
MnO	0.12	0.14	0.08	0.07	0.06	0.24	0.05	0.05	0.04
MgO	3.52	3.99	3.06	1.83	0.96	8.58	0.95	0.83	0.14
CaO	5.61	5.72	5.62	7.89	4.82	5.22	2.85	3.02	0.89
Na_2O	4.45	4.32	4.20	3.92	5.41	3.16	3.02	3.13	3.26
K_2O	2.39	2.06	2.51	1.58	3.01	1.85	2.80	2.53	4.84
P_2O_5	0.27	0.26	0.21	0.22	0.09	0.12	0.09	0.09	0.03
LOI	0.83	0.32	0.75	0.80	0.40	0.75	1.13	0.92	0.50
Total	99.51	99.57	99.51	98.92	99.74	99.52	99.86	99.99	100.50
$^{a}Mg^{\#}$	43.50	45.70	42.20	28.30	33.20	65.30	32.90	31.20	11.40
Q	0.00	1.83	0.05	0.00	1.52	1.32	34.85	35.17	34.78
C	0.00	0.00	2.43	2.57	1.82	1.96	0.75	0.76	0.60
Or	14.31	12.27	15.02	9.51	17.91	11.07	16.76	15.09	28.60
Ab	38.16	36.83	35.98	33.80	46.08	27.07	25.88	26.73	27.59
An	26.36	25.68	26.83	38.43	23.48	25.43	13.73	14.53	4.22
Di(FS)	0.02	0.31	0.00	0.00	0.00	0.00	0.00	0.00	0.00
Di(MS)	0.03	0.66	0.00	0.00	0.00	0.00	0.00	0.00	0.00
Hy(MS)	5.75	9.71	7.72	0.81	2.41	21.64	2.40	2.09	0.35
Hy(FS)	5.20	5.28	5.71	1.35	5.77	3.88	2.82	3.39	3.58
Ol(MS)	2.19	0.00	0.00	2.69	0.00	0.00	0.00	0.00	0.00
Ol(FS)	2.18	0.00	0.00	4.92	0.00	0.00	0.00	0.00	0.00
Mt	3.50	5.23	3.79	3.56	0.28	5.86	1.85	1.27	0.17
al	32.55	30.93	36.48	39.00	44.52	30.67	40.13	40.99	45.39
fm	33.78	36.40	30.20	25.17	14.09	41.42	21.33	19.59	11.10

Continued

Sample	LBT5	LBT18	LBT20	LBT41	LBT42	LBT46	LBT17	LBT36	LBT54
c	17.07	17.22	17.16	22.84	17.34	15.87	15.16	16.20	5.77
alk	16.59	15.45	16.15	12.98	24.05	12.04	23.39	23.23	37.74
Si	152.22	153.18	149.97	131.72	196.67	147.98	353.67	358.65	458.64
Ti	1.86	1.76	2.21	1.93	0.71	1.67	1.46	1.50	0.11
k	0.26	0.24	0.28	0.21	0.27	0.28	0.38	0.35	0.49
mg	0.44	0.46	0.43	0.29	0.34	0.53	0.33	0.32	0.11

[a] $100(MgO/40)/(MgO/40+FeO/72+Fe_2O_3/80)$.

Table 2 ICP-MS trace elements and REE (ppm) of mafic and felsic granulites in Cenozoic volcanics, Hoh Xil area, northern Tibetan Plateau.

Sample	LBT5	LBT18	LBT20	LBT41	LBT42	LBT46	LBT17	LBT36	LBT54
Rock	Mafic-granulite	Mafic-granulite	Mafic-granulite	Mafic-granulite	Mafic-granulite	Mafic-granulite	Felsic-granulite	Felsic-granulite	Felsic-granulite
Li	20.9	27.4	24.0	23.1	17.3	22.7	20.0	17.4	18.9
Be	3.06	3.73	2.62	6.11	2.88	1.35	1.27	1.10	0.78
Sc	16.8	23.2	21.0	17.6	5.93	30.6	9.01	8.83	5.10
V	174	166	231	163	39.3	249	45.4	44.3	4.66
Cr	27.8	39.1	44.0	61.8	11.7	222	27.3	6.12	1.69
Co	75.3	27.1	6.13	75.9	3.64	76.2	137	137	240
Ni	27.1	47.0	12.7	20.8	6.30	24.0	7.15	4.87	2.83
Cu	11.3	8.00	29.7	18.9	4.80	104	8.34	8.75	3.08
Zn	101	125	173	159	40.7	179	32.6	29.8	20.3
Ga	19.2	19.7	27.6	39.4	22.0	19.2	13.7	14.0	13.4
Ge	1.16	1.16	1.25	2.80	0.96	2.05	1.23	1.28	1.06
Rb	75.9	61.9	57.8	99.9	48.5	65.9	190	183	103
Sr	427	450	422	309	492	304	179	189	80.9
Y	23.6	26.8	25.4	41.5	17.9	29.3	25.4	30.1	39.6
Zr	241	158	243	343	136	79.0	251	274	64.1
Nb	13.0	12.3	12.0	25.2	7.91	4.22	8.44	9.56	0.22
Cs	6.57	4.57	1.75	12.4	1.47	5.45	5.41	4.23	1.35
Ba	1 294	1 099	1 039	149	935	1 134	797	878	347
Hf	5.21	4.29	5.49	8.08	3.56	2.15	6.19	6.84	2.28
Ta	1.03	1.05	0.80	1.82	0.66	0.43	1.00	1.13	0.16
Pb	34.3	34.6	52.2	31.1	33.2	50.4	36.2	35.8	33.7
Th	24.4	24.4	21.4	30.4	17.3	5.62	39.0	44.4	4.69
U	4.32	4.60	2.13	7.28	3.72	1.32	8.21	9.40	0.53
La	55.4	54.3	70.8	142	50.8	140	35.3	36.9	11.5
Ce	114	103	144	349	100	259	64.7	68.5	23.6
Pr	11.7	11.0	14.3	36.7	10.0	24.9	6.45	6.92	2.44
Nd	42.4	41.9	50.0	124	34.5	90.7	23.2	25.1	8.88

Continued

Sample	LBT5	LBT18	LBT20	LBT41	LBT42	LBT46	LBT17	LBT36	LBT54
Sm	7. 13	7. 34	8. 00	17. 7	5. 25	14. 1	4. 32	4. 78	2. 19
Eu	1. 49	1. 69	1. 36	1. 08	1. 50	1. 10	0. 64	0. 77	0. 30
Gd	6. 40	6. 70	7. 27	15. 6	4. 88	12. 8	4. 23	4. 75	2. 56
Tb	0. 79	0. 86	0. 87	1. 71	0. 58	1. 30	0. 62	0. 70	0. 59
Dy	4. 04	4. 53	4. 54	8. 20	2. 91	5. 74	3. 61	4. 17	4. 66
Ho	0. 78	0. 87	0. 87	1. 37	0. 55	0. 94	0. 80	0. 93	1. 23
Er	1. 98	2. 26	2. 19	3. 35	1. 40	2. 37	2. 17	2. 57	3. 51
Tm	0. 28	0. 34	0. 31	0. 41	0. 20	0. 30	0. 35	0. 41	0. 59
Yb	1. 83	2. 18	2. 02	2. 40	1. 29	2. 03	2. 34	2. 83	4. 05
Lu	0. 29	0. 34	0. 32	0. 35	0. 20	0. 32	0. 39	0. 47	0. 65
$(La/Yb)_N$	21. 72	17. 82	25. 10	28. 16	42. 53	49. 35	10. 82	9. 35	2. 04
$(Ce/Yb)_N$	17. 29	13. 06	19. 78	21. 56	40. 35	35. 33	7. 69	6. 73	1. 62
δEu	0. 66	0. 72	0. 53	0. 89	0. 19	0. 25	0. 45	0. 49	0. 39

4. 1 Major elements

The mafic granulites (e.g. LBT42) contain normative quartz and corundum and most are hypersthene-, olivine-, plagioclase-normative (Table 1). They have 48. 76%–58. 61% SiO_2, 0. 78%–1. 03% TiO_2, high Al_2O_3(18. 34%–24. 50%), 4. 82%–7. 89% CaO, and their high alkali contents (Na_2O = 3. 16% – 5. 41%, K_2O = 1. 58% – 3. 01%) with σ = 3. 54 – 5. 25, suggesting that they could be of alkali basaltic parentage. The mafic granulites have variable MgO contents of 0. 96–8. 58 wt%, with $Mg^{\#}$ ranging from 30 to 67; this is lower than those of primary mantle-derived melts, so that the mafic granulites do not simply represent primary mantle-derived melts. If exclusively of igneous origin, they must have experienced significant degrees of fractionation before or during emplacement in the crust.

Felsic granulite xenoliths (n = 3) are quartz-, corundum-, hypersthene-normative and have metasedimentary quartzofeldspathic or "granitic" compositions with SiO_2 = 71. 23% – 75. 93%, TiO_2=0. 02%–0. 40%, relatively low Al_2O_3 of 12. 75% – 13. 91% and an A/CNK [molar Al_2O_3/($CaO + K_2O + Na_2O$)] = 1. 04. indicating that they are weakly peraluminous. Alkali contents range from Na_2O=3. 02%–3. 26%, K_2O=2. 53%–4. 84%.

4. 2 Trace and rare earth elements

The granulite xenoliths show large variation in trace and REE element patterns (Table 2). Given the overall lack of consistent element correlations, it is clear that neither mafic nor felsic granulites can be regarded as representing a single rock suite. The mafic granulites are extremely enriched in LREE, with $(La/Yb)_N$ = 17. 32 – 49. 35 and $(Ce/Yb)_N$ = 13. 06 –

40.35. Their δEu = 0.19–0.89 (Fig. 4), and no correlation has been found between δEu and SiO_2 or $Mg^{\#}$, suggesting that plagioclase fractional crystallization cannot cause the variable δEu. Felsic granulite xenoliths are less LREE-enriched, with $(La/Yb)_N$ = 2.04 – 10.82 and $(Ce/Yb)_N$ = 1.62–7.69, δEu = 0.39–0.49.

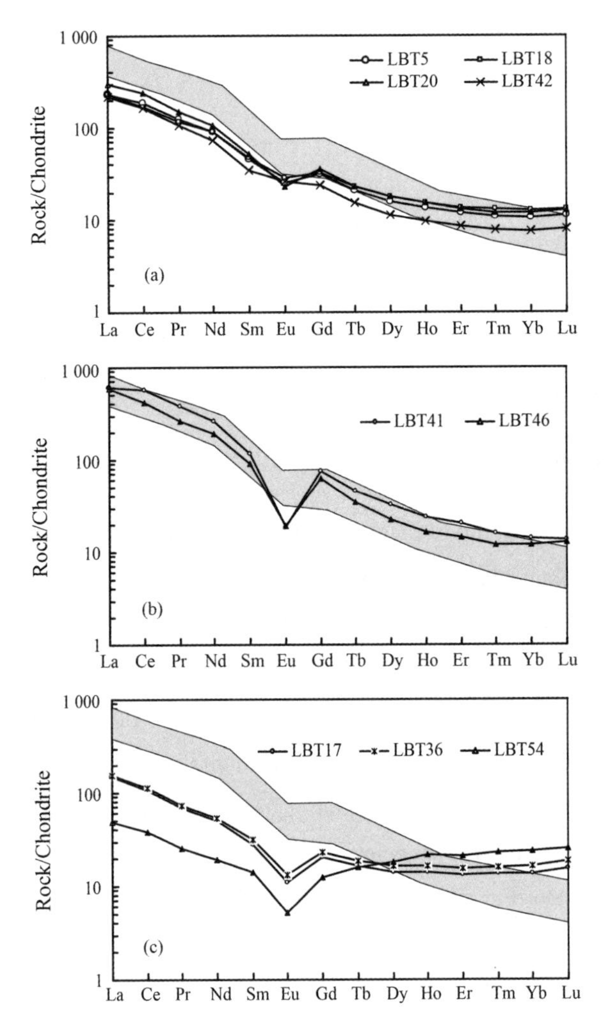

Fig. 4 Chondrite-normalized rare earth element distributions.
(a) Mafic granulite; (b) mafic granulite; (c) felsic granulite.
Chondrite values are from Sun and McDonough (1989).

The mafic granulite xenoliths exhibit a large range of LILE, with Cs ranging from 1.47–12.14 ppm, Rb 48.5–99.9 ppm, Ba 149–1 294 ppm, Sr 304–450 ppm, Th 5.62–30.4 ppm, U 1.32–7.28 ppm, and relatively high contents of V (39.3–249 ppm), Ni (6.3–47 ppm) and Cr (11.7–222 ppm). The mafic granulite xenoliths display strong depletion in high-field-strength elements (Nb, Ta and Ti) and are enriched in LILE (Fig. 5). However, unlike worldwide mafic granulites (Rudnik, 1992), they have higher Rb/Sr (0.09–0.21) and lower

Nb/Ta (9.76–14.92). The felsic granulites also display a large variation in LILE with Cs = 1.35–5.41 ppm, Rb = 103–190 ppm, Ba = 347–878 ppm, Sr = 80.9–179 ppm, Th = 4.69–44.4 ppm, and U = 0.53–9.40 ppm. They have lower V (4.69–45.4 ppm), Ni (2.83–7.15 ppm), and Cr (1.69–27.3 ppm) than those of the mafic granulites.

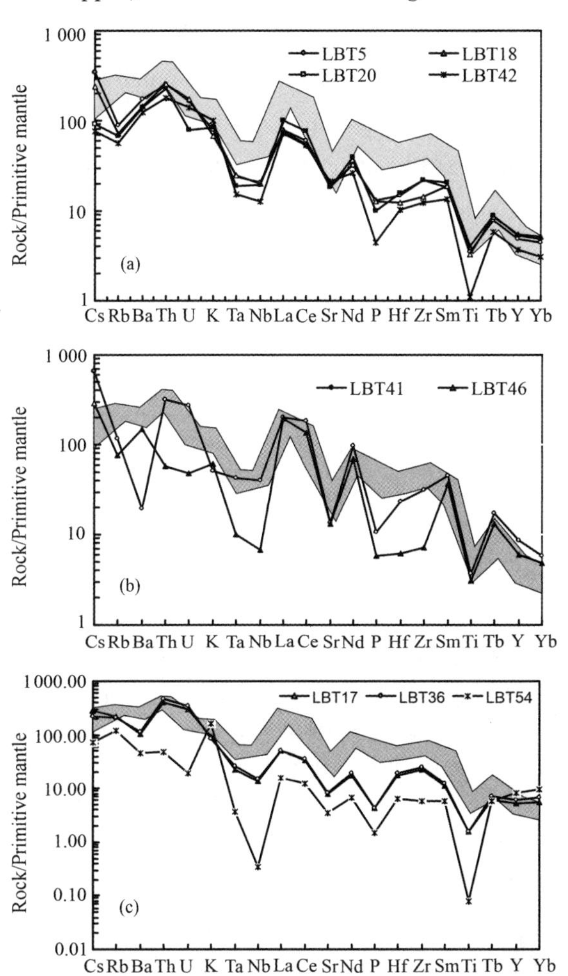

Fig. 5 Primitive mantle-normalized trace element distributions.

(a) Mafic granulite; (b) mafic granulite; (c) felsic granulite.

Primitive mantle values are from Sun and McDonough (1989).

4.3 Sr-Nd-Pb isotopic composition

Sr, Nd, and Pb isotopic data are given in Table 3 and plotted in Figs. 6 and 7 in relation to the isotopic composition fields of Cenozoic host dacite and shoshonitic rocks in northeastern Tibet (Turner et al., 1996; Lai et al., 2003). The mafic granulite xenoliths have evolved Sr and Nd isotopic compositions with $^{87}Sr/^{86}Sr$ of 0.710 812–0.713 241, ε_{Sr} = +169.13 to +203.88, and lower Nd isotopic ratios of $^{143}Nd/^{144}Nd$ = 0.512 113–0.512 397; ε_{Nd} is negative

(−10. 05 to −4. 70), and $^{206}Pb/^{204}Pb$ = 18. 700 0–18. 956 5, $^{207}Pb/^{204}Pb$ = 15. 713 5–15. 766 2, and $^{208}Pb/^{204}Pb$ = 39. 109 0–39. 473 3. The felsic granulite has evolved radiogenic Sr and Nd ratios of $^{87}Sr/^{86}Sr$ = 0. 712 041–0. 729 088, ε_{Sr} = + 180. 71 to + 430. 59, and $^{143}Nd/^{144}Nd$ = 0. 512 230–0. 512 388; ε_{Nd} = −7. 96 to −4. 74, and they also have higher Pb isotopic values with $^{206}Pb/^{204}Pb$ = 18. 925 0–19. 171 7, $^{207}Pb/^{204}Pb$ = 15. 766 2–15. 772 0, and $^{208}Pb/^{204}Pb$ = 39. 210 9–39. 646 7. In comparison with the host dacite and shoshonitic rocks, the granulite xenoliths have more evolved Sr-Nd isotopic ratios and similar Pb isotopic compositions (Fig. 6). In addition, their Sr-Pb and Nd-Pb isotopic systematics (Fig. 7) are distinctly different from depleted mantle (DMM), primary mantle (PREMA), mid-ocean ridge basalt

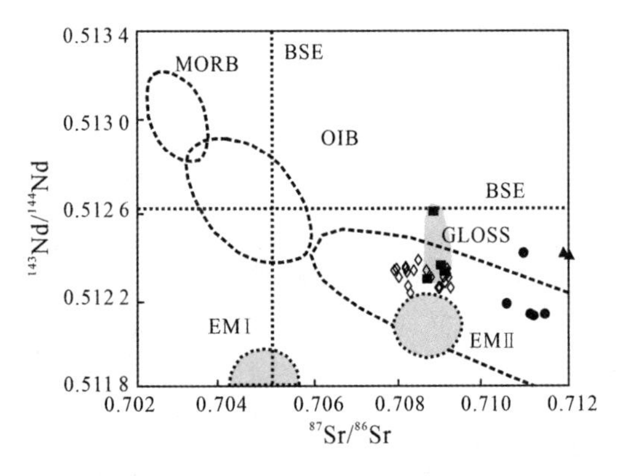

Fig. 6 Plot of $^{143}Nd/^{144}Nd$ vs. $^{87}Sr/^{86}Sr$ for mafic (solid circles), felsic (solid triangles) granulite xenoliths, dacite, Qiangtang region, and shoshonitic rocks from northern Tibetan Plateau.

Data for dacite from Lai et al. (2003), and data for shoshonotic rocks from Turner et al. (1996). MORB: midocean ridge basalt; OIB: oceanic island basalts; BSE: bulk silicate Earth; GLOSS: global subducting sediments; EM Ⅰ: enriched mantle (type Ⅰ); EM Ⅱ: enriched mantle (type Ⅱ).

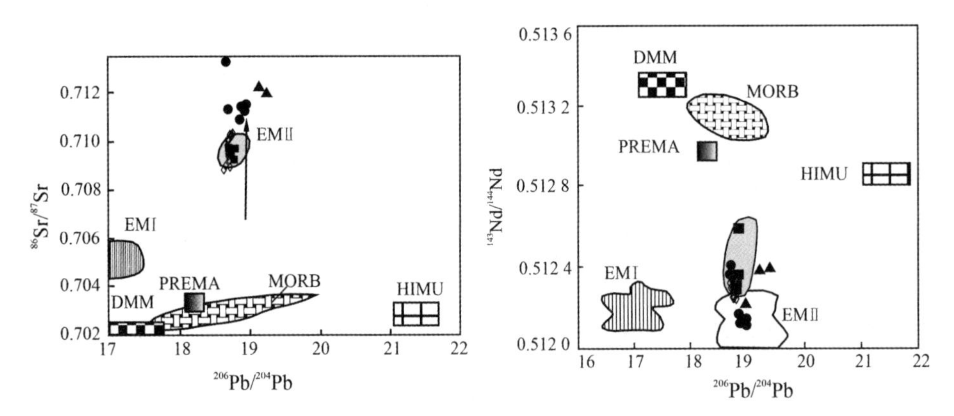

Fig. 7 Plots of $^{87}Sr/^{86}Sr$ vs. $^{206}Pb/^{204}Pb$ and $^{143}Nd/^{144}Nd$ vs. $^{206}Pb/^{204}Pb$ diagrams for mafic (solid circles), felsic (solid triangles) granulite xenoliths, compared with dacite, Qiangtang region, and shoshonitic rocks from northern Tibetan Plateau.

DMM: depleted mantle; PREMA: primary mantle; EM Ⅰ: enriched mantle (type Ⅰ); EM Ⅱ: enriched mantle (type Ⅱ); HIMU: anomalouly high $^{238}U/^{204}Pb$ mantle. Original diagram from Zindler and Hart (1986).

Table 3 Sr-Nd-Pb isotope data for mafic and felsic granulites in Cenozoic volcanics, Hoh Xil area, northern Tibetan Plateau.

Sample	LBT5	LBT18	LBT20	LBT41	LBT42	LBT46	LBT17	LBT36	LBT54
Rock	Mafic-granulite	Mafic-granulite	Mafic-granulite	Mafic-granulite	Mafic-granulite	Mafic-granulite	Felsic-granulite	Felsic-granulite	Felsic-granulite
U	4. 32	4. 60	2. 13	7. 28	3. 72	1. 32	8. 21	9. 40	0. 53
Th	24. 4	24. 4	21. 4	30. 4	17. 3	5. 62	39. 0	44. 4	4. 69
Pb	34. 3	34. 6	52. 2	31. 1	33. 2	50. 4	36. 2	35. 8	33. 7
$^{206}Pb/^{204}Pb$	18. 956 5±6	18. 918 9±8	18. 675 5±6	18. 897 9±9	18. 885 3±6	18. 700 0±4	19. 171 7±7	19. 191 5±6	18. 925 0±6
$^{207}Pb/^{204}Pb$	15. 766 2±6	15. 759 5±7	15. 714 0±5	15. 731 5±8	15. 713 6±4	15. 717 8±4	15. 766 2±6	15. 772 0±6	15. 767 0±6
$^{208}Pb/^{204}Pb$	39. 473 3±14	39. 441 1±17	39. 109 0±11	39. 326 1±25	39. 249 2±11	39. 147 6±10	39. 613 4±16	39. 646 7±16	39. 210 9±12
Rb	75. 9	61. 9	57. 8	99. 9	48. 5	65. 9	190	183	103
Sr	427	450	422	309	492	304	179	189	80. 9
$^{87}Rb/^{86}Sr$	0. 515 15	0. 397 59	0. 396 79	0. 935 98	0. 285 61	0. 627 89	3. 068 65	2. 793 07	3. 686 66
$^{87}Sr/^{86}Sr$	0. 711 736±7	0. 711 470±6	0. 713 241±7	0. 711 382±17	0. 710 812±8	0. 711 227±10	0. 712 087±9	0. 712 041±9	0. 729 088±9
ε_{Sr}	+182. 35	+178. 54	+203. 88	+177. 28	+169. 13	+175. 07	+187. 37	+186. 71	+430. 59
Sm	7. 13	7. 34	8. 00	17. 7	5. 25	14. 1	4. 32	4. 78	2. 19
Nd	42. 4	41. 9	50. 0	124	34. 5	90. 7	23. 2	25. 1	8. 88
$^{147}Sm/^{144}Nd$	0. 101 64	0. 105 95	0. 096 70	0. 086 38	0. 092 02	0. 094 27	0. 112 39	0. 115 28	0. 148 99
$^{143}Nd/^{144}Nd$	0. 512 123±5	0. 512 112±4	0. 512 372±4	0. 512 123±5	0. 512 165±4	0. 512 397±11	0. 512 388±7	0. 512 395±5	0. 512 230±9
ε_{Nd}	−10. 05	−10. 26	−5. 19	−10. 05	−9. 23	−4. 70	−4. 88	−4. 74	−7. 96
T_{DM}/Ma	1 395	1 466	1 013	1 228	1 178	961	1 146	1 168	2 159

U, Th, Pb Sm, Nd, Rb, and Sr concentrations were analysed by ICP-MS, and isotopic ratios were measured througy isotopic dilution. $\varepsilon_{Nd}(t)$ was calculated using $\varepsilon_{Nd}=[(^{143}Nd/^{144}Nd)_m/(^{143}Nd/^{144}Nd)_{CHUR}-1]\times10^4$, $(^{143}Nd/^{144}Nd)_{CHUR}=0.512\ 638$. ε_{Sr} was calculated using $\varepsilon_{Sr}=[(^{87}Sr/^{86}Sr)_m/(^{87}Sr/^{86}Sr)_{UR}-1]\times10^4$, $(^{87}Sr/^{86}Sr)_{UR}=0.698\ 990$. $T_{DM}=\frac{1}{\lambda}\ln[1+(^{143}Nd/^{144}Nd)_m/-0.513\ 15]/[(^{147}Sm/^{144}Nd)_m/-0.213\ 71]$, $\lambda_{Rb}=1.42\times10^{-11}a^{-1}$, $\lambda_{Sm}=6.54\times10^{-12}a^{-1}$; ε_{Nd}, ε_{Sr}, T_{DM}, and isotopic ratios of these rocks were not corrected for age.

(MORB), anomalously high $^{238}U/^{204}Pb$ mantle (HIMU), and enriched mantle (EM Ⅰ) compositions but plot close to the region of EM Ⅱ. These unique isotopic features can be attributed to presence of K-feldspar and biotite and also to dacitic melt (glass) introduced when the xenoliths were entrained in dacite magma.

5 Discussion

5.1 Origin of granulite xenoliths and nature of the lower crust in northern Qiangtang terrane

One of the most difficult problems encountered in the study of lower crustal xenoliths is to determine whether they represent cumulates, restites or solidified melts and yet the role of a mafic lower crust in crustal evolution models rests heavily on this interpretation. Although an unambiguous means of distinguishing between the above possibilities has not been found, some geochemical parameters can be used to restrict the choice (Kempton et al., 1997).

Mafic granulite xenolith has $Mg^{\#}$ of 30–67 in plot of $Mg^{\#}$ vs. SiO_2/Al_2O_3 (Fig. 8). Most of them plot close to the fields of mafic-intermediate granulites, except for LBT-46, which has the highest $Mg^{\#}$ and plots in the field of primitive basaltic magmas. The geochemical features of the mafic granulite suggest that they may be solidified crustally contaminated melts or restites of partial melting of metamorphic equivalents, which produce a clinopyroxene-orthopyroxene-plagioclase assemblage, for example, quartz amphibolite (Patiño-Douce and Beard, 1995) or metabasalt-andesite (Beard and Lofgren, 1991; Springer and Seck, 1997), rather than

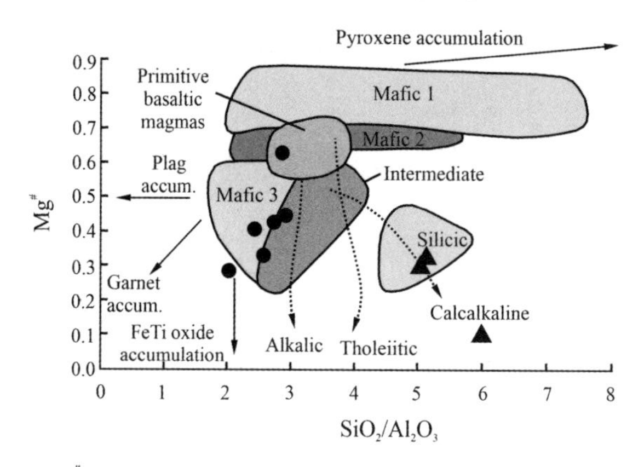

Fig. 8 $Mg^{\#}$ vs. SiO_2/Al_2O_3 variation diagram (after Kempton et al., 1997) for granulite xenoliths, Qiangtang region, northern Tibetan Plateau.

Solid circles: mafic granulites of the present work; Solid triangles: felsic granulites of this work; Composition fields from Kempton et al. (1997): Mafic 1, mafic granulites (SiO_2<54 wt%; $Mg^{\#}$>0.7); Mafic 2 (SiO_2<54 wt%; $Mg^{\#}$ 0.6–0.7); Mafic 3 (SiO_2<54 wt%; $Mg^{\#}$<0.6); intermediate (SiO_2>54 wt% and <66 wt%); felsic (SiO_2>66 wt%); Dashed arrows indicate average differentiation trends for alkalic, tholeiitic and calc-alkaline melts. Continuous arrows show generalized direction of compositional change with accumulation of indicated mineral phase.

cumulates of basaltic magma. Pressures < 1.0 GPa are consistent with the formation of a mafic granulite assemblage from partial melting of a quartz amphibolite composition according to the reaction: hornblende + quartz ± garnet = orthopyroxene + clinopyroxene + plagioclase + granitic melt at 850 ± 50 ℃ (e.g. Pattison et al., 2001). In this case, a metasomatic origin may be required for the presence of biotite and possibly K-feldspar in the xenoliths. The following geochemical and mineralogical features clearly preclude a cumulate or an unmodified resitite origin:

(1) Clinopyroxene, orthopyroxene, and biotite in the mafic granulite have lower $Mg^{\#}$ than those of corresponding minerals in Cenozoic shoshonite (Turner et al., 1996).

(2) The mafic granulite samples have extremely high Al_2O_3 (18.34% - 24.50%), significant negative Eu anomalies (0.19–0.89), and no correlation between Al_2O_3 and δEu, indicating that the high Al_2O_3 content cannot be attributed to plagioclase accumulation during fractional crystallization of basaltic magma, but rather to K-feldspar, biotite.

(3) In general, mantle-derived basaltic rocks have high Nb/Ta ratios (17.5 ± 2), whereas the mafic granulites have low Nb/Ta ratios (9.76–14.92), inconsistent with direct derivation from a mantle basalt protolith; only continental crustal rocks or material derived from them have such low Nb/Ta ratios and in this case the "dilution effect" of K-feldspar and biotite.

(4) The overall lack of consistent element correlations indicates that the mafic granulites cannot be regarded as belonging to a single igneous rock suite.

(5) Compared with the Cenozoic shoshonitic rocks in northern Tibet and their host dacite, the mafic granulites have extremely evolved Sr-Nd-Pb isotopic compositions. This again implies a "continental crustal component". High Rb (48.5–99.9 ppm) and Rb/Sr (0.09–0.21) is not consistent with the mafic granulite xenoliths being exclusively the restites of partial melting, but reflects the presence of K-feldspar and biotite, which could be taken to represent a continental crustal component of an EM Ⅱ-type magma, alkali metasomatism during granulite facies recrystallization or when the xenoliths became entrained in dacite magma.

The geochemistry of the felsic granulites also indicates a continental crustal origin, in this case a peraluminous siliceous (quartzofeldspathic/granitic) protolith, for example, as indicated by their low Nb/Ta ratios (~8.4), especially LBT54 with a Nb/Ta ratio of 1.4. Two of the three analyzed xenoliths plot in the silicic granulite field of the $Mg^{\#}$ vs. SiO_2/Al_2O_3 diagram (Fig. 8).

Hacker et al. (2000) calculated *P-T* conditions from siliciclastic metasedimentary xenoliths from the Lake Dogai Coring area and obtained equilibration temperatures of 800–1 100 ℃ and depths of 30–50 km, indicating a hot lower crust beneath the northern Tibetan Plateau. Jolivet et al. (2003) also describes granulite xenoliths from near the Kunlun Fault in northern Tibet and suggest similar high lower crustal temperatures.

Thermobarometric estimates of 783 – 818 ℃/0.85 GPa from the granulite xenoliths described in this paper indicate derivation from middle to lower crustal depths of ~30 km, supporting the idea of a hot lower crust beneath the northern Qiangtang terrane. Our geochemical data also suggest that the lower crust in northern Tibet includes metasedimentary rocks and is heterogeneous rather than being uniformly mafic.

5.2 Implication for the genesis of Cenozoic shoshonitic and high-K calc-alkaline rocks in northern Qiangtang terrane

The genesis of the Cenozoic potassic and calc-alkaline felsic rocks in the northern Qiangtang terrane is still a matter of debate (Turner et al., 1996; Lai et al., 2003, 2007; Guo et al., 2006; Ding et al., 2007). Turner et al. (1996) interpreted Cenozoic shoshonitic and ultra-potassic lavas from the northern Qiangtang terrane to have been generated by partial melting of ancient and enriched lithospheric mantle in part because of high K_2O, LREE concentration, radiogenic Sr and Pb isotopic ratios and relatively non-radiogenic Nd. However, based on the exhumed belt of blueschist-bearing mélange within the middle of the Qiangtang terrane, the northern Qiangtang mantle is inferred to have been completely removed by Mesozoic southward low-angle oceanic subduction along the Jinsha suture to the north, during which Songpan-Ganzi sedimentary rocks and metasedimentary-matrix mélange underthrust Qiangtang continental margin rocks (Kapp et al., 2000, 2003). This hypothesis predicts that the northern Qiangtang terrane is underlain by a metasedimentary-bearing lower crust. This prediction has been directly validated through documentation of metasedimentary lower crustal xenoliths in northern Qiangtang lavas (Hacker et al., 2000), they suggested that the lower crust of northern Tibet is partially metasedimentary, rather than wholly gabbroic, and is hot. Interaction with rising magma generated from a non-enriched mantle source could result in the formation of shoshonitic and ultra-potassic magma (high K_2O and LREE).

Our observations suggest that interaction between hot metasedimentary/granitic lower crust and mantle-derived magma has played an important role in the generation of shoshonites, ultra-potassic rocks and the andesite-dacite that host the granulite xenoliths. First, as discussed above, the lower crust in northern Tibet is hot and partially metasedimentary and/or granitic. Second, indications of *in situ* melting of biotite together with feldspar and quartz in the granulite xenoliths (Fig. 3c), glass inclusions in some of the xenoliths (Fig. 3a,c,d), and resorption of biotite, feldspar and quartz in the host dacite and andesite indicate that partial melting has occurred that could have contributed silicic (granitic) melt to mantle-derived magma. The widespread occurrence of fine-grained, undigested xenocrysts in the dacite and andesite suggest that the unusual geochemical composition of some Cenozoic Tibet lavas may be the result of mixing of silicic (granitic-granodioritic-trondhjemitic) melts derived from partial melting of lower crustal Qiangtang terrane metasedimentary-metamafic rocks and mantle-derived magma. Third, compared to shoshonitic rocks in northern Tibet documented by Turner et al. (1996), Cenozoic dacite in the study area has different major and trace element geochemical features and similar Sr-Nd-Pb isotopic composition. In general, Cenozoic calc-alkaline dacite and shoshonitic rocks from the northern Qiangtang terrane are considered to be derived from partial melting of thickened lower crust (Lai et al., 2003, 2007; Wang et al., 2005; Ding et al., 2007), and mantle lithosphere (Turner et al., 1996; Miller et al., 1999; Guo et al., 2006), respectively. Then, what mechanism can account for the similar Sr-Nd-Pb isotopic composition between the Cenozoic dacite and shoshonites in Northern Qiangtang terrane (Figs. 6 and 7). As there is no evidence of magma mixing, the most plausible explanation is that different batches of melt from lower crust and mantle lithosphere interacted with hot metasedimentary lower crust in northern Tibet during their ascent with the result that they have similar Sr-Nd-Pb isotopic compositions. Fourth, Patiño-Douce and McCarthy (1998) have shown that dehydration melting of biotite —the melting documented in the Qiangtang xenoliths—produces potassic melts and a garnet-orthopyroxene-cordierite-rich restite. The Qiangtang xenoliths are considered to have previously undergone H_2O-saturated melting to yield trondhjemitic liquids and biotite-rich residues. This inferred early melting event could have occurred either during deep burial of low-grade metasediments such as the Songpan-Ganze flysch or possibly much earlier if the metasedimentary protolith was already a granulite. Surface exposures have been used to infer that upper crustal lithologies are widespread beneath central and northern Tibet (Hacker et al., 2000).

Ding et al. (2007) reported post-collisional calc-alkaline lavas (28–43 Ma) and xenoliths in the southern Qiangtang terrane, which has coeval age with the lavas that host granulite xenoliths in the northern Qiangtang terrane. They suggest that both suites of Qiangtang lavas were derived from a primitive mantle source and that the enriched nature of the northern Qiangtang lavas reflects contamination by partial melts of metasedimentary lower crust. Major and trace element geochemistry, Sr-Nd-Pb isotopic composition, as well as mineralogical features of the xenoliths in our study clearly indicate that the lower-middle crust is metasedimentary and enriched in nature. Although removal of lithospheric mantle seems to be required to produce the high-temperature melts, Eocene-Oligocene volcanism was coeval with thrust reactivation along bounding suture zones, implying that mantle dynamics were linked to intracontinental subduction.

6 Conclusions

Mineralogy and geochemistry of the granulite xenoliths indicate a metasedimetary quartzofeldspathic or granitic component (K-feldspar and biotite in mafic granulites) in their protolith and this is supported by their more evolved Sr-Nd-Pb compositions than those of the host andesite-dacite. The granulite xenoliths indicate a heterogeneous mafic/quartzofeldspathic lower crust in northern Tibet. Interaction between mantle-derived basaltic magma and lower crustal quartzofeldspathic metasedimentary/granitic rocks is inferred to have played an important role in the generation of shoshonitic rocks and high-K calc-alkaline andesite-dacite. In light of these results, the currently popular model that the Tibetan ultra-potassic-shoshonitic rocks were derived by partial melting of enriched lithospheric mantle and the nature of the lithospheric mantle in northern Tibet should be reconsidered.

Acknowledgements This work is jointly supported by the National Natural Science Foundation of China (Grant Nos. 40572050, 40272042, and 40234041) and the Teaching and Research Award Programme for Outstanding Young Teachers in Higher Education Institutions of MOE, PR China. Northwest University Graduate Innovation and Creativity Funds (08YYB01), the MOST Special Fund from the State Key Laboratory of Continental Dynamics, Northwest University and Province Key Laboratory Construction Item(08JZ62).

References

Arnaud, N.O., Vidal, P., Tapponnier, P., Matte, P., Deng, W.M., 1992. The high K_2O volcanism of northwestern Tibet: Geochemistry and tectonic implications. Earth & Planetary Science Letters, 111, 351-367.

Barazangi, M., Ni, J., 1982. Velocities and propagation characteristics of Pn and Sn beneath the Himalayan Arc and Tibetan Plateau; possible evidence for underthrusting of Indian continental lithosphere beneath Tibet. Geology, 10(4),179-185.

Beard, J. S., Lofgren, G. E., 1991. Experimental melting of and water-saturated melting of basaltic and andesitic greenstones and amphibolites at 1, 3 and 6. 9 kbar. Journal of Petrology, 32,365-401.

Chung, S.L., Lo, C.H., Lee, T.Y., Zhang, Y.Q., Xie, Y.W., Li, X.H., Wang, K.L., Wang, P.L., 1998. Dia-chronous uplift of the Tibetan plateau starting 40 Myr ago. Nature, 394, 769-773.

Deng, J.F., Zhao, H.L., Mo, X.X., Wu, Z.X., Luo, Z.H., 1996. Continental roots-plume tectonics of China: Key to the continental dynamics. Beijing: Geological Publishing House, 112 (in Chinese).

Deng, W.M., Sun, H.J., 1999. The Cenozoic volcanism and the uplifting mechanism of the Qinghai-Tibet Plateau. Geology Review, 45,952-958 (in Chinese).

Dewey, J.F., Shackleton, R.M., Chang, C.F., Sun, Y.Y., 1988. The Tectonic Evolution of the Tibetan Plateau. Philosophical Transactions of the Royal Society of London, Series A, Mathematical and Physical Sciences, 327(1594). The Geological Evolution of Tibet: Report of the 1985 Royal Society — Academia Sinica Geotraverse of Qinghai-Xizang Plateau (12 December 1988), 379-413

Ding, L., Kapp, P., Yue, Y.H., Lai, Q.Z., 2007. Postcollisional calc-alkaline lavas and xenoliths from the southern Qiangtang terrane, central Tibet. Earth & Planetary Science Letters, 254, 28-38.

Ding, L., Kapp, P., Zhong, D.L., Deng, W.M., 2003. Cenozoic volcanism in Tibet: Evidence for a transition from oceanic to continental subduction. Journal of Petrology, 44, 1833-1865.

Edwards, M.A., Harrison, T.M., 1997. When did the roof collapse? Late Miocene north-south extension in the high Himalaya revealed by Th-Pb monazite dating of the Khula Kangri Granite. Geology, 25, 543-546.

England, P.C., Houseman, G.A., 1989. Extension during continental convergence, with application to the Tibetan plateau. Journal of Geophysical Research, 94, 17561-17579.

England, P.C., Houseman, G.A., 1988. The mechanics of the Tibetan plateau. Philosophical Transactions of the Royal Society of London: Series A, 326, 301-319.

Griffin, W.L., O'Reilly, S.Y., 1986. The lower crust in Eastern Australia: xenoliths evidence. In: Dawson, J. B., Carswell, D. A., Hall, J., Wedepohl, K. H. The nature of the lower continental crust. Oxford Blackwell, Geological Society Special Publications, 24, 363-374.

Guo, Z.F., Wilson, M., Liu, J.Q., Mao, Q., 2006. Post-collisional, potassic and ultrapotassic magmatism of the northern Tibetan Plateau: Constraints on characteristics of mantle source, geodymamic setting and uplift mechanisms. Journal of Petrology, 47,1177-1220.

Hacker, B.R., Edwin, G., Lothar, R., Marty, G., Michael, M.C., Sobolev, S.V., Wang, J., Wu, Z.H., 2000. Hot and dry deep crustal xenoliths from Tibet. Science, 287, 2463–2466.

Houseman, G.A., Mckenzie, D.P., Molnar, P.C., 1981. Convective instability of a thickened boundary layer and its relevance for the thermal evolution of continental convergent belts. Journal of Geophysical Research, 86, 6115-6132.

Jolivet, M. Brunel, M., Seward, D., Xu, Z., Yang, J., Malavielle, J., Roger, F., Leyreloup, A., Arnaud, N., Wu, C., 2003. Neogene extension and volcanism in the Kulun Fault Zone, northern Tibet: New constraints on the age of the Kulun Fault. Tectonics, 22,1052-1062.

Kapp, P., Yin, A., Manning, C.E., Harrison, T.M., Taylor, M.H., Ding, L., 2003. Tectonic evolution of the early Mesozoic blueschist-bearing Qiangtang metamorphic belt, central Tibet. Tectonics, 22,1043-1053.

Kapp, P., Yin, A., Manning, C.E., Murphy, M., Harrison, T.M., Spurlin, M., Ding, L., Deng, X.G., Wu, C.M., 2000. Blueschist-bearing metamorphic core complexes in the Qiangtang Block reveal deep crustal structure of northern Tibet. Geology, 28, 19-22.

Kempton, P.D., Downes, H., and Embey-Isztin, A., 1997. Mafic granulites xenoliths in Neogene alkali basalts from the western Pannonian Basin: Insight into the lower crust of a collapsed orogen. Journal of Petrology, 38,941-970.

Kutzbach, J.E., Prell, W.L., Ruddiman, W.F.,1993, Sensitivity of Eurasian climate to surface uplift of the Tibetan Plateau. Journal of Geology, 101, 177-190.

Lai, S.C., Liu, C.Y., Yi, H.S., 2003. Geochemistry and petrogenesis of Cenozoic andesite-dacite association from the Hoh Xil region, Tibetan plateau. International Geology Review, 45, 998-1019.

Lai, S.C., Qin, J.F., Li, Y.F., 2007. Partial melting of thickened Tibetan crust: Geochemical evidence from Cenozoic adakitic volcanic rocks. International Geology Review, 49, 357-373.

Lenardic, A., Kaula, W.M., 1995. More thoughts on convergent crustal plateau formation and mantle dynamics with regard to Tibet. Journal of Geophysical Research, 100, 15193-15204.

Mattie, P.D., Condie, K.C., Selverstone, J., Kyle, P.R., 1997. Origin of the continental crust in the Colorado Palteau: Geochemical evidence from mafic xenoliths from the Navajo Volcanic Field, southwestern USA. Geochimica et Cosmochimca Acta, 61, 2007-2021.

Miller, C., Schuster, R., Klotzli, U., Frank, W., Purtscheller, F., 1999. Post-collisional potassic and ultrapotassic magmatism in SW Tibet: Geochemical and Sr, Nd, Pb, O isotopic constraints for mantle source characteristics and petrogenesis. Journal of Petrology, 40, 1399-1424.

Meyer, B., Tapponnier, T., Bourjot, L., Metivier, F., Gaudemer, Y., Peltzer, G., Guo, S., Chen, Z., 1998. Crustal thickening in Gansu-Qinghai, lithospheric mantle subduction, and oblique, strike slip controlled growth of the Tibet Plateau. Journal of Geophysics, 135, 1-47.

Murphy, M.A., Yin, A., Harrison, T.M., Durr, S.B., Chen, Z., Ryerson, F.J., Kidd, W.S.F., Wang, X., Zhou, X., 1997. Did the Indo-Asian collision alone create the Tibetan plateau? Geology, 25, 719-722.

Patiño Douce, A.E., Beard, J.S., 1995. Dehydration melting of biotite gneiss and quartz amphibolite from 3 kbar to 15 kbar. Journal of Petrology, 36, 707-738.

Patiño Douce, A. E., McCarthy, T. C., 1998. Melting of crustal rocks during continental collision and subduction. In: Hacker, B.R., and Liou, J.G. When continents collide: Geodynamics and geochemistry of ultra-high pressure rocks. Dordrecht Kluwer, 27-55.

Pattison, D.R.M., Chacko, T, Farquhar, J., McFarlane, C.R.M., 2001. Temperatures of granulite-facies metamorphism: Constraints from experimental phases equilibria and thermometry corrected for retrograde exchange. Journal of Petrology, 44, 867-900.

Raymo, M.E., Ruddmian, W.F.,1992. Tectonic forcing the Late Cenozoic climate. Nature, 359, 117-122.

Rudnick, R.L., 1992. Xenoliths sample of lower continental crust. In: Fountain, R.J., Arculus, R.W., Kay, R.W. The continental lower crust. Amsterdam: Elsevier, 269-316.

Rudnick, R. L., Fountain, D. M., 1995. Nature and composition of the continental: A lower crustal perspective. Reviews of Geophysics, 33, 267-309.

Spinger, W., Seck, A., 1997. Partial fusion of basic granulite at 5 kbar to 15 kbar: Implications for the origin of TTG magmas. Contrib. Mineral. Petrol., 127, 30-45.

Sun, S.S., McDonough, W.F., 1989. Chemical and isotopic systematics of oceanic basalts: Implications for mantle composition and processes. In: Saunders, A.D., Norry, M. J. Magmatism in the Ocean Basin. Geological Society Special Publication, 42, 313-345.

Turner, S., Arnaud, N., Liu, J., Rogers, N., Hawkesworth, C., Harris, N., Kelley, S., Calsteren, P.V., Deng, W., 1996. Post-collision, shoshonitic volcanism on the Tibetan Plateau: Implication for convective thinning of the lithospheric and source of ocean island basalts. Journal of Petrology, 37, 45-71.

Wang, Q., McDermott, F., Xu, J., Belloon, H., Zhu, Y., 2005.Cenozoic K-rich adakitic volcanic rocks in the Hohxil area, northern Tibet: Lower-crustal melting in an intracontinental setting. Geology, 33,465-468.

Zhao, W., Morgan, W.J., 1987. Injection of Indian crust into Tibetan lower crust: A two dimentional finite element model study. Tectonics, 6, 489-504.

Zindle, A., Hart, S.R., 1986, Chemical geodynamics. Annual Review of Earth and Planetary Sciences, 14, 493-573.

Permian high Ti/Y basalts from the eastern part of the Emeishan Large Igneous Province, southwestern China: Petrogenesis and tectonic implications[①]

Lai Shaocong Qin Jiangfeng Li Yongfei Li Sanzhong M Santosh

Abstract: The source characteristics of the widespread Permian high Ti/Y basalts in the Emeishan Large Igneous Province (LIP) constitute important themes to evaluate the possible connection with a mantle plume beneath the Yangtze Block in southwestern China. Here we investigate the geochemical and isotopic signature of basalts from the Guangxi and Guizhou regions in the eastern margin of the Emeishan LIP and report the occurrence of high Ti/Y basalts in Guangxi. The zircons separated from Guangxi basalts yield a U-Pb concordia age of 257 Ma, which is consistent with the age of the plume-related eruption of the Emeishan LIP. Both the Guangxi and Guizhou basalts studied here display evolved Sr-Nd isotopic composition and Dupal Pb isotopic composition, with $(^{87}Sr/^{86}Sr)_i = 0.705\,231$ to $0.706\,147$, positive to slightly negative $\varepsilon_{Nd}(t)$ values of -0.13 to $+0.68$, and $\Delta 7 = 5-11$, $\Delta 8 = 70-84$. The Guangxi basalts possess low SiO_2 (44.82–49.71 wt%), and TiO_2 (2.29–3.54 wt%), but are enriched in LREE and LILE with $(La/Sm)_N$ values of 2.2–2.6, high Ce/Yb ratios (19.6–30.0) and slightly negative Nb and Ta anomalies. The Guizhou basalts display higher SiO_2 (47.49–50.27 wt%), TiO_2 (3.93–4.92 wt%), Zr, Nb and La contents. They also show higher Ce/Yb (30.2–37.8) and Sm/Yb (3.56–4.19) ratios as compared to those of the Guangxi basalts, and do not possess negative Nb and Ta anomalies. The distinct differences between the Guangxi and Guizhou basalts may be caused by different degrees of partial melting of the garnet peridotite in their source region. We propose that the Guizhou basalts were derived from partial melting of metasomatized veins in the Yangtze continental lithosphere which were heated by the upwelling Emeishan plume, and the Guangxi basalts represent higher degrees of partial melting of the sub-continental lithospheric mantle and undergone minor crustal contamination during their ascent.

1 Introduction

The Large Igneous Provinces (LIPs) constitute voluminous mafic rock formations together with felsic rocks, and are generally linked to processes associate with mantle plume or hotspot,

---

① Published in *Journal of Asian Earth Sciences*, 2012, 47.

and unrelated to "normal" sea-floor spreading and subduction (Coffin and Eldholm, 1994). Continental flood basalts (CFBs) are the dominant rock type in most of the LIPs, and are generally regarded as products of adiabatic decompression melting of anhydrous mantle peridotite and their formation is correlated to the head of mantle plumes (White and McKenzie, 1989; Garfunkel, 2008, and references therein). The "continental" signature in their chemistry might have resulted from the assimilation of crustal or other lithospheric material (McKenzie and Bickle, 1988; McKenzie, 1989; White and McKenzie, 1989). However, the involvement of lithospheric mantle and continental crust in CFB has not been adequately evaluated (Xiao et al., 2004 and references therein). Gallagher and Hawkesworth (1992) suggested that CFB forms by the melting of hydrated (metasomatized) lower continental lithosphere through heating by plumes, while the plumes themselves do not melt.

The Late-Permian Emeishan Large Igneous Province (LIP) at the western margin of the Yangtze block, southwestern China, has been considered to have formed above a mantle plume (Ali et al., 2005; Chung and Jahn, 1995; Chung et al., 1998; Xu et al., 2001). Various mafic and felsic rocks, including continental flood basalts (Xu et al., 2001), mafic intrusions and picritic basalts (Zhang et al., 2006; Zi et al., 2008), as well as trachyte and rhyolite (Xu et al., 2001) have been recognized to be associated with the Emeishan LIP. Xu et al. (2001) proposed that the CFB from the Emeishan LIP can be divided into high-Ti (HT) and low-Ti (LT) basalts, with the former having high TiO_2 (> 2.8 wt%) contents and Ti/Y (>500) ratios, whereas the latter shows low TiO_2 (< 2.8 wt%) contents and Ti/Y (< 500) ratios. The petrogenesis of these rocks remains controversial (Chung and Jahn, 1995; Xu et al., 2001, 2004; Xiao et al., 2003, 2004; Song et al., 2005, 2008). In general, the domal region of the Emeishan LIP comprises thick (2 000–5 000 m) sequences of low-Ti volcanic rocks and subordinate picrites (Chung and Jahn, 1995; Zhang et al., 2006). In contrast, thin sequences (< 500 m) of high-Ti volcanic rocks mainly occur on the periphery of the domal structure. The major debates concerning these rocks are focused on: ① the role of continental lithosphere and mantle plumes in the petrogenesis of the HT and LT basalts (Xiao et al., 2004); and ② the geodynamic process and their relationship with the evolutionary history of the Paleo-Tethys ocean in the Sanjiang area (Fig. 1).

In this study, we report new data relating to the major- and trace element geochemistry, Sr-Nd-Pb isotopic composition and zircon LA-ICP-MS U-Pb analysis from the Late-Permian basalts of Guizhou and Guangxi Provinces at the eastern margin of the Emeishan LIP. Our geochemical data indicate that the basalts have high TiO_2 contents and a Ti/Y ratios, classifying as HT basalts. These data provide important insights into understanding of petrogenesis and tectonic setting of the Emeishan LIP.

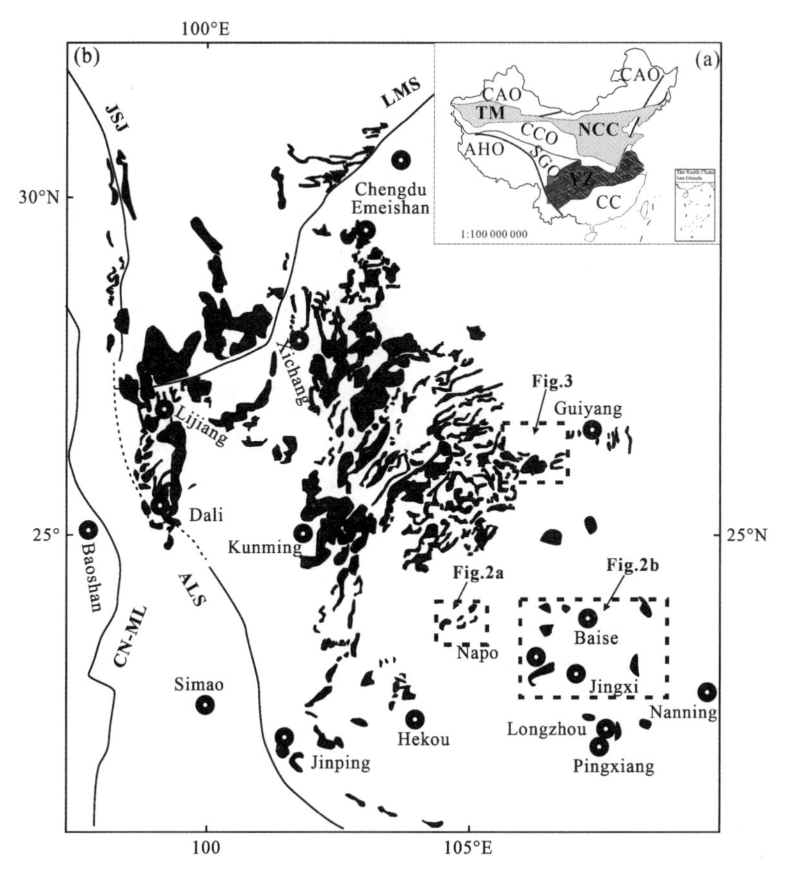

Fig. 1 Spatial distribution pattern of the Permian Emeishan flood basalts.

CAO: Central Asia Orogen; TM: Tarim Block; NCC: North China Craton; CCO: Central China Orogen; SGO: Songpan-Ganzi Orogen; YC: Yangtze Craton; AHO: Alpine-Himalaya Orogen; CC: Cathaysia Craton; JSJ: Jinshajiang; LMS: Longmenshan; CN-ML: Changning-Menglian; ALS: Ailaoshan.

2 Field geology and petrography

The Emeishan LIP is considered to represent a major late Permian basaltic magmatism in the western margin of the Yangtze Block in South China (Xu et al., 2001). The previous geochronological studies on mafic intrusion (Zhou et al., 2002a-c) and felsic ignimbrite (He et al., 2007) have assigned an eruption age of 259–263 Ma for the Emeishan LIP, which is coincident with the end-Guadalupian mass extinction (He et al., 2007; Isozaki, 2009, 2010). The Emeishan LIP is considered to have been disrupted by the Ailaoshan-Red River Fault zone (Zhou et al., 1988), with the principal remaining volcanic outcrop covering an area of about 250 000 km^2 (Fig. 1). However, recent studies have revealed that some basalts and mafic complexes exposed in the Simao Basin and Qiangtang Terrane are possible extensions of the Emeishan LIP. Xiao et al. (2003) argued that the Emeishan LIP may extend beyond the Yangtze Block, and that basalts and mafic-ultramafic rocks that outcrop at Jinping and in the

Zhongza micro-block are also part of the Emeishan LIP. If this is true, a revised estimate of the extent of the Emeishan LIP is in excess of 500 000 km^2.

The Emeishan LIP can be divided into a central part and an outer zone (Song et al., 2005, 2008). The extensive erosion and thinning of the Middle Permian limestone in the central part of the LIP indicates kilometer-scale regional uplift linked to the rising plume prior to eruption of the Emeishan CFB (He et al., 2003; Xu et al., 2004). The central part is characterized by low-Ti/Y basalts overlain by high-Ti/Y basalts, and the presence of large layered intrusions hosting giant V-Ti-magnetite deposits and numerous syenite and alkaline granite plutons (Xiao et al., 2003). On the other hand, the outer zone consists of high-Ti/Y basalts (e.g., Song et al., 2005, 2006). The Emeishan LIP volcanic sequence generally decreases in thickness from the central part to the margins (Xu et al., 2004; Song et al., 2004). The identification of the involvement of mantle plumes in the Emeishan LIP is therefore important for deciphering the dynamic trigger of the Emeishan volcanism.

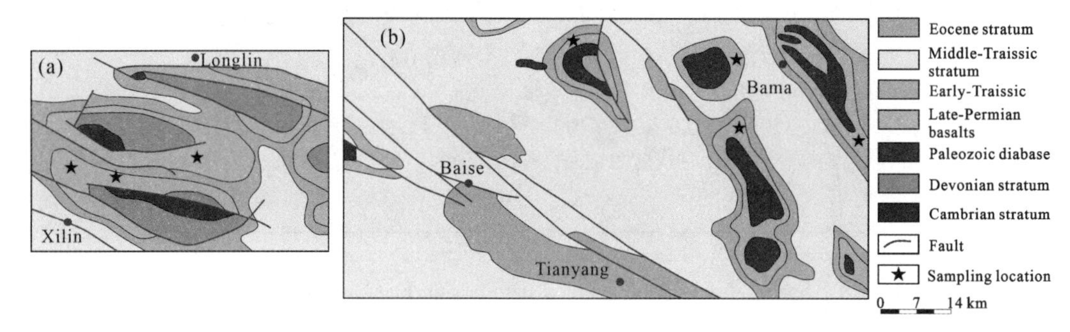

Fig. 2 Geologic sketch map of the northwestern part of Guangxi Province.
(a) Longlin and Xilin area; (b) Bama, Baise and Tianyang area.

In the northwestern part of Guangxi Province, the Late-Permian basalts are mainly occurred in the Longlin, Xilin, Bama, Tianyang and Baise area, these basalts have thickness of tens of meters, they are scattered in the Cambrian and Devonian system and have unconformable relation with the Mid-Permian Maokou Formation, they are overlain by early Late Permian Longtan Formation or Triassic sedimentary rocks (Fig. 2a,b). The Late-Permian basaltic lava piles (~80 m) from the Baise area interbedded with silicate rock and diatom perlite. Basalts from this area display vesicular and amygdaloidal structures (Fig. 4a), with a near constant thickness. Under polarizing microscope, these basalts display diabasic texture, divergent plagioclase laths with interstitial glass or intergrown with pyroxene granules (Fig. 4b,c).

Late-Permian basalts from the western part of Guizhou Province are mainly distributed in the Zhijin, Nayong and Puding area, they overlie the Carboniferous system and Middle Permian Maokou Formation, and are, in turn, overlain by the Late Permian Xuanwei (terrestrial clastic rocks), Longtan (marine clastic rocks) (He et al., 2007) and Triassic system (Fig. 3), they show pillow structures, however, there is no radial joints in the pillow lavas(Fig. 4d), suggesting they were erupted in continental setting rather than subaqueous. These basalts are black and fine grained, under polarizing microscope, they also display diabasic texture(Fig. 4e,f).

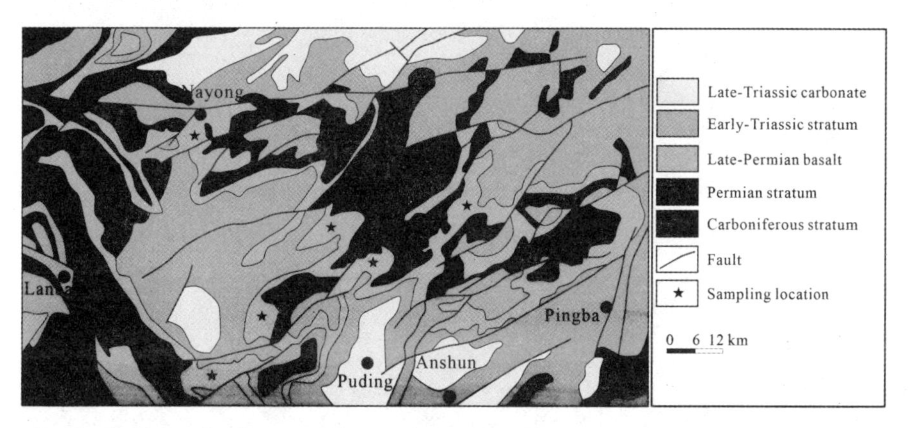

Fig. 3 Geologic sketch map of the sampling area in the western part of Guizhou Province.

3 Analytical techniques

Samples of the basalts from Guangxi and Guizhou provinces were analyzed at the State Key Laboratory of Continental Dynamics, Northwest University in Xi'an, China. Fresh rock chips were powdered to 200 mesh size using a tungsten carbide ball mill. Major and trace elements were analyzed using XRF (Rikagu RIX 2100) and ICP-MS (Agilent 7500a), respectively. Analyses of USGS and Chinese national rock standards (BCR-2, GSR-1 and GSR-3) indicate that both analytical precision and accuracy are generally better than 5% and 10%, for major and trace elements, respectively.

Whole-rock Sr-Nd-Pb isotopic data were obtained using a Nu Plasma HR multi-collector mass spectrometer. Sr and Nd isotopic fractionation was corrected to $^{87}Sr/^{86}Sr = 0.1194$ and $^{146}Nd/^{144}Nd = 0.7219$, respectively. Whole-rock Pb was separated by an anion exchange in HCl-Br columns, Pb isotopic fractionation was corrected to $^{205}Tl/^{203}Tl = 2.3875$.

For cathodoluminescent (CL) imaging, representative zircon grains were handpicked and mounted in epoxy resin discs, and then polished and coated with carbon. Internal morphology was examined using CL prior to U-Pb isotopic analyses. Laser ablation ICP-MS zircon U-Pb analyses were performed on an Agilent 7500a ICP-MS equipped with a 193 nm laser, which is housed at the State Key Laboratory of Continental Dynamics, Northwest University in Xi'an, China. The diameter of the analytical spot was 30 μm. Details of the analytical technique are given in Yuan et al. (2004). Common Pb contents were evaluated using the method described

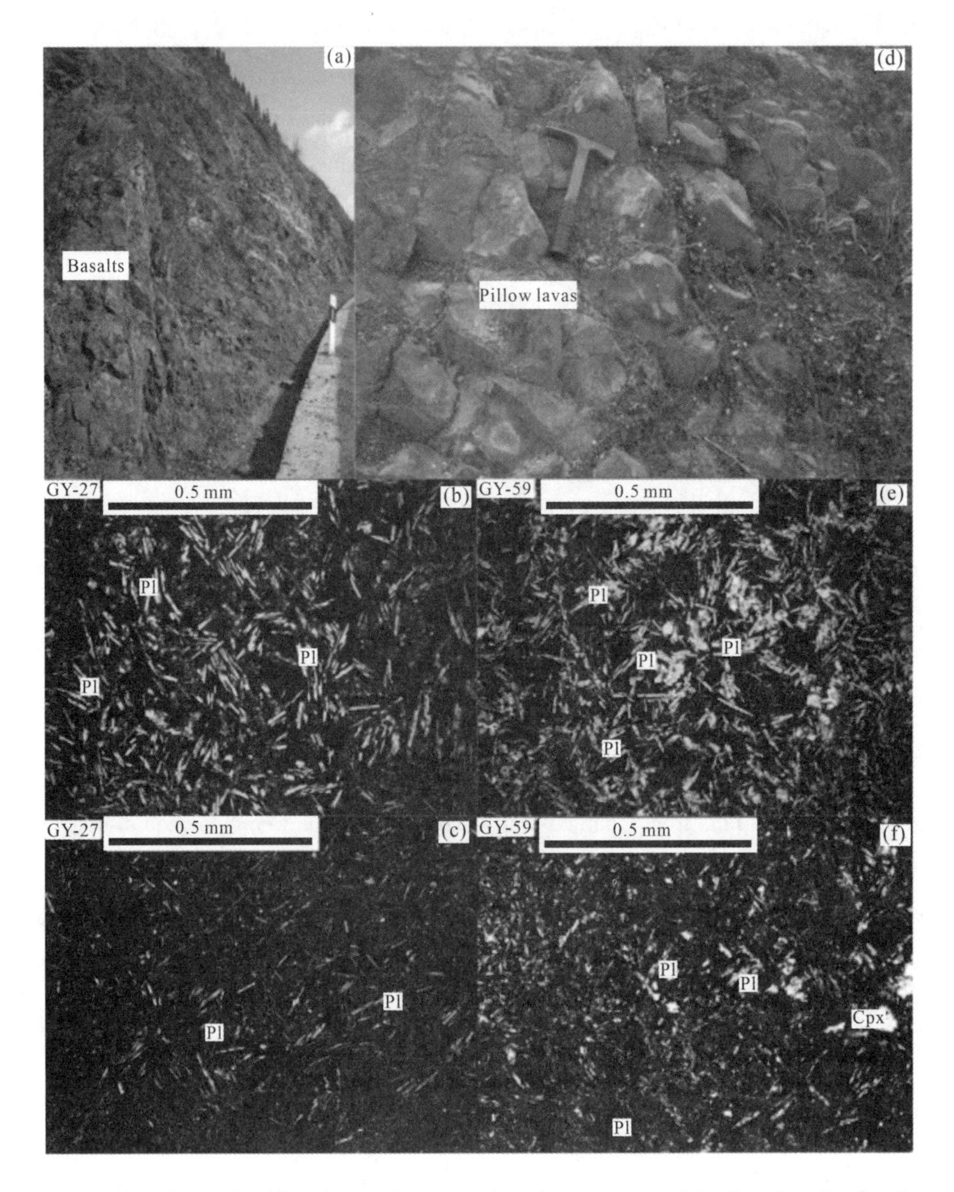

Fig. 4 The photos of field geology and petrographic characteristics of the Late-Permian basalts from Guangxi and Guizhou provinces.

(a) Basaltic laves in the Tianlin area; (b) and (c) intergranular and tholeiitic texture of the Guangxi basalts; (d) pillow laves from the Zhijin area; (e) and (f) intergranular and tholeiitic texture of the Guizhou basalts.

by Andersen (2002). The age computations and concordia diagrams were made using ISOPLOT (version 3.0; Ludwig, 2003). The errors quoted in tables and figures are at the 2σ levels.

4 Analysis results

4.1 Zircon LA-ICP MS U-Pb dating

Zircons from the basalts in Guangxi Province (GYD) are mainly prismatic, and do not show any marked oscillatory zoning. The length of the zircon grains is about 100 μm. Nine spots made on nine grains were selected for U-Pb isotopic analysis, and the results are listed in Table 1.

Among these, one spot (G10) displays a Paleoproterozoic $^{206}Pb/^{207}Pb$ age of 1 828 ± 40 Ma. The U = 192 ppm and Th = 49 ppm yield a moderate Th/U value (0.47), suggesting derivation from a magmatic protolith. The other eight spots record Paleozoic ages, with $^{206}Pb/^{238}U$ ages ranging from 248 Ma to 278 Ma. The U (447–1 312 ppm), Th (1 440–2 384 ppm), markedly high Th/U (1.40 – 3.22) values are typical of magmatic zircons (e.g., Rubatto, 2002). A $^{206}Pb/^{238}U$ weighted mean age of 257 ± 9 Ma (MSWD = 2.8, n = 8) derived from this data is regarded as the crystallization age of the basalts from Guangxi Province (Fig. 5).

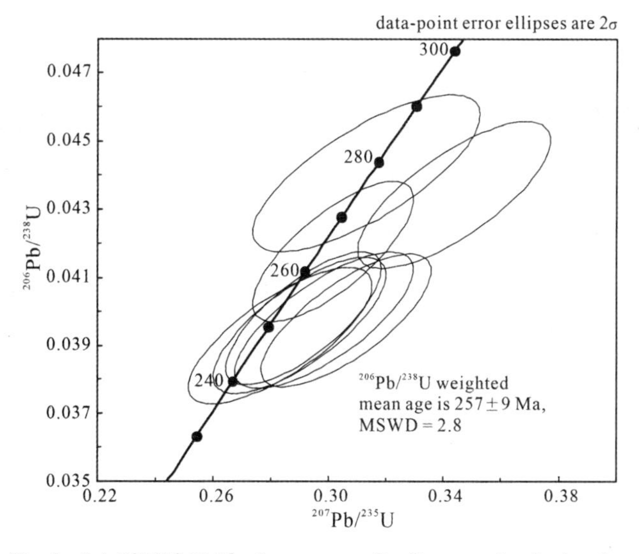

Fig. 5 LA-ICPMS U-Pb zircon concordia diagrams for the basalts from Guangxi Province. Ellipse dimensions are 2σ.

4.2 Major and trace element geochemistry

Both of the Guangxi and Guizhou basalts have varied K_2O, Na_2O and contents (Tables 2 and 3) and LOI. These variations suggest some degree of surface alteration, because of which the TAS diagrams may be unsuitable in the present case to classify the volcanic series. Nb and

Table 1 Results of zircon LA-ICP-MS U-Pb results for the Permian basalts from Guangxi Province.

Analysis	Content/ppm			Ratios							
	Th	U	Th/U	$^{207}Pb/^{206}Pb$	2σ	$^{207}Pb/^{235}U$	2σ	$^{206}Pb/^{238}U$	2σ	$^{208}Pb/^{232}Th$	2σ
G01*	2 192	863	2.54	0.053 160	0.002 340	0.290 050	0.012 280	0.039 580	0.000 820	0.012 960	0.000 800
G02*	2 213	1 007	2.2	0.054 490	0.002 400	0.298 500	0.012 680	0.039 730	0.000 820	0.015 230	0.000 960
G04*	1 888	1 348	1.4	0.053 130	0.002 100	0.291 450	0.011 040	0.039 790	0.000 800	0.017 350	0.001 160
G05*	2 003	805	2.49	0.052 380	0.002 080	0.301 600	0.011 420	0.041 760	0.000 840	0.015 890	0.001 080
G06*	1 976	991	1.99	0.057 410	0.002 380	0.343 590	0.013 640	0.043 410	0.000 880	0.017 270	0.001 260
G07*	2 385	1 607	1.48	0.055 890	0.002 320	0.306 160	0.012 100	0.039 730	0.000 800	0.017 650	0.001 340
G08*	1 958	1 313	1.49	0.052 330	0.002 520	0.283 410	0.013 040	0.039 280	0.000 820	0.012 330	0.000 980
G09*	1 440	447	3.22	0.051 580	0.002 760	0.313 210	0.016 140	0.044 050	0.000 940	0.014 320	0.001 160
G10	91	192	0.47	0.111 740	0.004 980	4.838 180	0.204 540	0.314 060	0.006 540	0.088 660	0.007 600

Analysis	Age/Ma							
	$^{207}Pb/^{206}Pb$	2σ	$^{207}Pb/^{235}U$	2σ	$^{206}Pb/^{238}U$	2σ	$^{208}Pb/^{232}Th$	2σ
G01*	336	59	259	10	250	5	260	16
G02*	391	59	265	10	251	5	306	19
G04*	334	50	260	9	252	5	348	23
G05*	302	50	268	9	264	5	319	21
G06*	507	52	300	10	274	5	346	25
G07*	448	53	271	9	251	5	354	27
G08*	1 300	67	253	10	248	5	248	20
G09*	267	79	277	12	278	6	287	23
G10	1 828	47	1 792	36	1 761	32	1 717	141

Table 2 Analytical results of major(%) and trace(ppm) element from the Permian basalts from Guangxi Province.

Sample	GY34	GY37	GY42	GY44	GY53	GY54	GY55	GY58	GY64	GY66	GY68	GY69	GY72
SiO_2	47.11	49.71	47.15	45.54	47.01	46.85	47.85	47.31	45.91	45.86	45.61	45.77	45.39
TiO_2	3.53	3.3	3.54	3.48	2.35	2.38	2.31	2.29	3.04	3.17	2.98	3.07	3.14
Al_2O_3	13.45	13.24	13.33	13.6	14.19	14.15	14.08	14.17	13.67	13.58	13.75	13.72	13.69
$Fe_2O_3^T$	14.67	12.59	15.6	15.56	13.11	13.54	13.04	13.11	15.18	15.32	15.29	15.68	15.66
MnO	0.23	0.17	0.21	0.23	0.18	0.19	0.17	0.18	0.21	0.21	0.21	0.22	0.22
MgO	4.5	5.2	5.01	5.04	6.28	6.42	5.95	6.12	5.59	5.82	5.78	5.8	5.81
CaO	8.86	8.32	7.76	8.54	10.08	9.34	10.43	9.76	9.96	9.58	9.76	9.87	9.52
Na_2O	2.17	4.4	2.43	1.94	2.71	2.61	2.23	2.52	2.54	2.69	2.64	2.98	3.04
K_2O	2.17	0.51	1.99	2.64	1.01	1.1	1.01	1.16	1.06	0.91	1.08	0.78	0.72
P_2O_5	0.64	0.58	0.65	0.63	0.3	0.3	0.3	0.29	0.56	0.55	0.55	0.55	0.55
LOI	2.21	2.3	2.44	2.33	2.61	2.65	2.32	2.6	1.93	2.03	1.85	1.13	2.05
Total	99.54	100.32	100.11	99.53	99.83	99.53	99.69	99.51	99.65	99.72	99.5	99.57	99.79
Li	25.4	17.8	27.4	26.8	14.2	16.5	13.8	15	17.6	18.6	19.3	19.7	19.2
Be	1.45	1.17	1.36	1.3	1.17	1.42	1.27	1.29	1.13	1.14	1.17	1.13	1.16
Sc	26.2	28.5	25.8	26.7	30.7	31.3	30.7	29.8	15.2	28.7	28.1	28.7	28.7
V	401	367	386	411	343	345	336	334	362	380	377	384	385
Cr	32.7	82.5	20.7	22.5	125	127	124	128	68.7	75.6	74.2	72.6	74.3
Co	64.1	60.8	59.8	65	63.7	62.8	65.8	63.2	66.4	61.1	64.8	63	59.1
Ni	49.1	59.4	44	46.9	106	110	106	114	72.9	73.3	73.8	75.8	74.7
Cu	77.7	87.2	95.3	78.2	160	152	157	155	115	109	116	89.2	61.9
Zn	128	101	138	123	102	108	97.6	102	108	108	116	119	110
Ga	21.5	15.7	20.5	21.9	20.6	19.8	20.4	20.4	20.2	20.8	20.6	20.7	20.4
Ge	1.8	1.66	1.65	1.69	1.61	1.5	1.68	1.57	1.59	1.61	1.64	1.64	1.58
Rb	38.5	9.95	42.1	49.8	31	35.7	29.9	38.5	15	26.5	25.8	22.7	21.6
Sr	611	441	623	794	406	335	460	387	558	543	563	527	521
Y	32.3	28.3	32.5	32.6	29.4	29.4	28.5	28.5	24.6	28.2	28.3	27.7	27.6
Zr	186	156	190	187	169	171	167	165	151	148	150	144	149
Nb	28.9	24.7	29.3	16.3	20.9	21.4	20.8	20.3	23.2	23.9	24.1	23.5	23.2
Cs	12.8	1.26	13.5	21.7	0.7	1.09	0.64	0.95	1.05	1.23	1.21	1.65	1.09
Ba	982	202	705	1182	311	332	321	351	736	524	742	480	417
Hf	4.44	3.78	4.55	4.5	4.18	4.21	4.2	4.03	3.62	3.62	3.63	3.53	3.57
Ta	1.83	1.57	1.86	0.78	1.34	1.36	1.34	1.3	1.48	1.5	1.51	1.47	1.44
Pb	1.11	1.03	2.66	1.14	2.68	2.4	2.8	2.46	3.36	2.32	3.87	2.9	2.63
Th	3.71	3.08	3.82	3.75	3.34	3.35	3.29	3.21	1.86	2.95	2.99	2.88	2.87
U	0.93	0.84	0.96	0.94	0.84	0.81	0.83	0.8	0.76	0.73	0.75	0.74	0.74
La	34.6	23.6	35.1	34.7	22.4	22.8	21.7	21.8	26.8	28.9	28.8	28.5	27.7
Ce	77.5	55.3	77.7	77.8	49.2	50.1	48.3	47.9	60.6	64.3	64.3	63.4	61.3
Pr	9.77	7.24	9.81	9.81	6.12	6.25	6.05	5.95	7.69	8.13	8.14	8.01	7.65
Nd	44.3	33.7	44.3	44.4	26.9	27.3	26.4	26.1	35	36.7	36.9	36.2	35.9

Continued

Sample	GY34	GY37	GY42	GY44	GY53	GY54	GY55	GY58	GY64	GY66	GY68	GY69	GY72
Sm	8.91	7.06	8.91	8.87	6.12	6.22	6.06	5.92	7.14	7.5	7.43	7.47	7.29
Eu	3.37	2.34	2.93	3.34	1.97	1.95	1.9	1.87	2.69	2.75	2.85	2.85	2.68
Gd	8.18	6.61	8.16	8.21	6.05	6.11	5.95	5.86	6.57	6.93	6.92	6.89	6.71
Tb	1.15	0.97	1.17	1.16	0.93	0.94	0.91	0.9	0.93	0.98	0.98	0.98	0.95
Dy	6.41	5.43	6.38	6.45	5.48	5.51	5.38	5.24	5.23	5.5	5.46	5.44	5.35
Ho	1.24	1.07	1.23	1.24	1.11	1.11	1.09	1.07	1.02	1.06	1.07	1.06	1.04
Er	3.11	2.69	3.13	3.12	2.85	2.89	2.8	2.75	2.55	2.68	2.67	2.67	2.60
Tm	0.43	0.38	0.43	0.43	0.41	0.42	0.41	0.4	0.36	0.38	0.37	0.37	0.37
Yb	2.56	2.25	2.59	2.59	2.5	2.5	2.46	2.39	2.12	2.23	2.2	2.18	2.19
Lu	0.36	0.32	0.37	0.37	0.35	0.36	0.35	0.34	0.31	0.32	0.32	0.31	0.31
La/Nb	1.20	0.96	1.20	2.13	1.07	1.07	1.04	1.07	1.16	1.21	1.20	1.21	1.19

Sample	GY74	GY75	GY78	GY80	GY81	GY85	GY86	GY87	GY88	GY89	GY90	GY91	GY92
SiO_2	45.91	45.31	45.26	45.17	45.28	44.82	45.34	45.81	46.19	45.68	45.57	45.89	45.76
TiO_2	3.02	2.77	2.88	2.75	2.73	2.7	2.62	2.97	3.21	3.12	3.14	2.82	3.01
Al_2O_3	14.17	14.1	14.27	14.33	14.28	13.74	14.23	14.06	14.04	14.07	13.67	14.12	14.22
$Fe_2O_3{}^T$	14.85	14.47	14.18	14.5	14.55	14.81	14.46	14.95	14.43	14.64	14.57	14.46	15.23
MnO	0.21	0.21	0.2	0.21	0.2	0.22	0.2	0.2	0.2	0.2	0.19	0.19	0.19
MgO	5.56	6.3	6.13	6.11	6.14	7.53	6.65	5.51	5.31	5.2	5.94	5.64	5.5
CaO	9.49	10.07	10.34	9.89	9.7	8.91	9.49	9.87	10.24	10.02	10.08	10.02	9.16
Na_2O	2.79	2.51	2.58	2.44	2.39	2.34	2.53	2.98	3.24	3.22	2.55	2.8	2.73
K_2O	1.08	1.3	1.11	1.44	1.47	1.52	1.48	1.24	0.77	0.76	1.63	1.2	1.81
P_2O_5	0.56	0.53	0.47	0.52	0.54	0.48	0.47	0.51	0.53	0.57	0.52	0.51	0.61
LOI	2	2.08	2.18	2.2	2.5	2.48	2.4	1.5	1.82	2.03	1.68	1.92	1.58
Total	99.64	99.65	99.6	99.56	99.78	99.55	99.87	99.6	99.98	99.51	99.54	99.57	99.8
Li	18.8	20.4	22.2	21.8	22.3	27.7	27.2	9.9	8.52	8.8	12.8	12.7	13.5
Be	1.16	1.05	1.04	1.04	1.05	1.03	1.02	1.06	1.08	1.11	1.03	1.07	1.18
Sc	27.4	26.5	28.1	24.7	24.8	24	26.6	29	31.1	28.8	33	29	23.6
V	371	350	366	335	332	332	332	421	424	417	442	359	323
Cr	67.2	75	84.5	69.5	73.6	85.1	86.7	69.5	94.8	81.5	118	98.4	38.8
Co	80.6	65.1	68.6	66.2	63.3	68	64.8	62.5	62.8	69.5	63.6	62.7	62.4
Ni	69.5	94.7	95.3	101	101	138	107	72.9	65	70.4	78.5	71.7	71.1
Cu	121	101	90.6	122	117	102	96.2	101	107	138	139	142	124
Zn	104	120	104	149	123	117	117	120	107	115	122	113	121
Ga	21.5	19.8	19.5	19.8	19.7	18.8	18.8	20.4	20.6	20.8	20	20.3	21
Ge	1.65	1.56	1.57	1.55	1.52	1.47	1.51	1.6	1.65	1.65	1.65	1.6	1.58
Rb	26.1	30.7	27	30.4	32.7	37.4	28.7	25.4	15.1	15.8	39.3	27.9	44.4
Sr	568	548	579	542	520	405	499	603	717	723	446	533	481
Y	28.6	26.1	24.5	25.6	26.2	23.8	23.9	27	27.7	28.9	26.8	27	29.5
Zr	155	147	131	137	142	135	131	144	148	161	137	146	159
Nb	23.8	21.6	21.4	21.2	21.9	21.5	20.8	21.6	24.2	24	22.8	23.4	26.1

Continued

Sample	GY74	GY75	GY78	GY80	GY81	GY85	GY86	GY87	GY88	GY89	GY90	GY91	GY92
Cs	1.13	9.2	9.41	10.1	11	7.81	3.89	4.39	4.42	4.81	4.88	8.25	7.76
Ba	664	614	542	732	640	611	638	528	379	383	646	535	686
Hf	3.76	3.49	3.21	3.3	3.41	3.22	3.18	3.46	3.61	3.84	3.35	3.51	3.79
Ta	1.51	1.38	1.35	1.34	1.36	1.35	1.3	1.38	1.49	1.53	1.41	1.46	1.62
Pb	2.72	3.68	2.47	3.09	4.56	4.88	8.65	3.75	2.97	2.99	1.78	2.78	2.23
Th	3.12	2.9	2.55	2.72	2.83	2.64	2.66	2.9	2.79	3.24	2.53	2.89	3.21
U	0.78	0.74	0.64	0.71	0.72	0.66	0.67	0.74	0.72	0.83	0.65	0.73	0.81
La	29.4	27.4	25	26.6	27.7	25.7	25.1	27.6	27.9	30.2	27.3	28.2	31.5
Ce	65.4	60.9	55.9	58.9	61.3	56.1	54.6	61.1	62.1	66.7	60.5	62.3	69.1
Pr	8.1	7.63	7.01	7.47	7.69	7.12	6.91	7.7	7.81	8.34	7.7	7.79	8.61
Nd	36.8	34.1	31.1	33.2	35	30.8	30.2	34.4	35.6	37.6	33.9	34.8	39.6
Sm	7.6	6.99	6.54	6.79	6.97	6.37	6.34	7.04	7.27	7.62	7.09	7.18	7.7
Eu	2.92	2.57	2.49	2.54	2.56	2.42	2.43	2.63	2.69	2.78	2.66	2.69	2.79
Gd	6.95	6.47	6.05	6.22	6.44	5.85	5.89	6.54	6.74	7.05	6.58	6.54	7.21
Tb	0.99	0.92	0.85	0.88	0.91	0.83	0.83	0.92	0.96	0.99	0.93	0.93	1.01
Dy	5.48	5.06	4.77	4.91	5.08	4.6	4.64	5.16	5.35	5.52	5.16	5.23	5.63
Ho	1.07	0.99	0.93	0.97	0.98	0.89	0.9	1.01	1.04	1.07	1	1.02	1.1
Er	2.7	2.46	2.34	2.4	2.48	2.26	2.3	2.51	2.61	2.75	2.51	2.56	2.78
Tm	0.38	0.35	0.32	0.34	0.34	0.31	0.32	0.36	0.36	0.38	0.34	0.36	0.39
Yb	2.24	2.08	1.96	2	2.04	1.89	1.9	2.08	2.15	2.27	2.07	2.14	2.3
Lu	0.33	0.29	0.28	0.29	0.29	0.26	0.27	0.3	0.3	0.33	0.29	0.31	0.33
La/Nb	1.24	1.27	1.17	1.25	1.26	1.20	1.21	1.28	1.15	1.26	1.20	1.21	1.21

Table 3 Analytical results of major(%) and trace(ppm) element from the Permian basalts from Guizhou Province.

Sample	GY03	GY09	GY12	GY20	GY21	GY29	GY30	GY31	GY32	GY33
SiO_2	48.84	49.4	49.28	47.49	50.27	49.63	49.95	49.2	49.18	48.76
TiO_2	4.19	3.95	3.93	4.16	4.15	4.2	4.31	4.22	4.21	4.32
Al_2O_3	13.09	12.85	12.95	13.11	13.01	13.04	13.15	13.29	12.93	13.2
$Fe_2O_3^T$	14.75	14.72	14.69	15.55	14.04	14.31	14.25	14.64	14.44	14.86
MnO	0.17	0.19	0.19	0.21	0.19	0.18	0.19	0.21	0.17	0.21
MgO	4.56	4.43	4.45	4.67	3.75	4.36	4.6	4.49	4.3	4.79
CaO	8.2	7.08	7.73	8.37	8.85	8.71	7.69	8.58	8.82	7.7
Na_2O	2.02	3.14	2.97	2.17	2.32	2.63	2.88	2.47	2.92	2.83
K_2O	1.78	1.71	1.61	0.92	1.1	0.67	0.43	0.51	0.42	0.37
P_2O_5	0.47	0.43	0.44	0.48	0.46	0.47	0.47	0.47	0.47	0.47
LOI	1.45	1.67	1.69	2.58	1.45	1.89	2.14	2.13	1.89	2.37
Total	99.52	99.57	99.93	99.71	99.59	100.09	100.06	100.21	99.75	99.88
Li	6.9	14.9	14.9	11.6	6.3	10.1	13.2	10.8	10.1	12.7
Be	1.92	1.96	1.99	1.77	2.23	1.85	1.9	1.88	2.07	1.76

Continued

Sample	GY03	GY09	GY12	GY20	GY21	GY29	GY30	GY31	GY32	GY33
Sc	25	26.6	26.3	25.1	24.5	25	25.4	24.9	24.8	25.3
V	400	400	392	394	376	392	405	396	392	401
Cr	23.2	41.9	42.8	22.1	23.1	22.9	23.1	22.6	22.3	25.3
Co	72.4	52.3	79.1	78.5	83.8	64.4	65.3	59.2	67.1	69.8
Ni	56.3	52.8	55.5	57.3	52.2	54	58.1	53.7	53.3	59.6
Cu	197	210	208	206	186	206	195	197	198	201
Zn	139	136	129	140	120	114	131	122	130	127
Ga	26.1	25.3	24.7	27.6	24.1	25	26.1	25.7	25.2	26.4
Ge	1.75	1.8	1.77	1.78	1.75	1.79	1.72	1.73	2.01	1.76
Rb	34.4	57.9	49.2	19.5	22	13.2	9.28	10.1	7.36	6.86
Sr	614	617	496	591	656	725	633	678	647	613
Y	38.2	40.6	39.5	38.3	37.3	38.1	38.3	37.2	37.3	38.2
Zr	350	342	332	357	341	350	354	346	346	352
Nb	42.8	39	37.9	43.5	41.4	42.7	43.3	42.2	42	42.8
Cs	1.13	0.62	0.99	0.39	0.28	0.31	0.18	0.22	0.24	0.19
Ba	716	452	352	409	581	431	353	376	317	303
Hf	8.56	8.32	8.17	8.67	8.23	8.48	8.62	8.34	8.32	8.48
Ta	2.85	2.55	2.5	2.89	2.77	2.81	2.84	2.76	2.76	2.82
Pb	6.76	19.4	4.57	5.55	6.76	4.75	5.93	6.8	7.02	6.32
Th	6.87	6.92	6.78	7.03	6.61	6.82	6.95	6.65	6.66	6.84
U	1.64	1.66	1.62	1.67	1.61	1.67	1.67	1.61	1.63	1.65
La	46.8	42.1	41.4	44.7	45.7	46.7	46.1	44.9	46.5	46.9
Ce	105	95.2	93.3	103	103	106	105	101	104	107
Pr	13	11.8	11.6	12.9	12.6	13.1	13	12.5	12.9	13.3
Nd	57.8	53.1	52.4	57.5	56.2	57.3	57.3	55.1	57	58.1
Sm	11.9	11.2	11.1	11.8	11.4	11.8	11.8	11.4	11.6	11.7
Eu	3.45	3.19	3.06	3.31	3.38	3.37	3.22	3.22	3.34	3.39
Gd	10.5	10.2	10	10.5	10.1	10.3	10.4	10	10.1	10.3
Tb	1.46	1.47	1.44	1.46	1.41	1.44	1.47	1.41	1.42	1.45
Dy	7.89	8.15	7.95	7.95	7.59	7.76	7.82	7.6	7.63	7.81
Ho	1.47	1.56	1.52	1.48	1.42	1.46	1.47	1.43	1.44	1.45
Er	3.6	3.86	3.77	3.63	3.47	3.56	3.62	3.49	3.51	3.56
Tm	0.5	0.54	0.53	0.5	0.48	0.5	0.5	0.48	0.48	0.49
Yb	2.84	3.14	3.08	2.89	2.77	2.84	2.88	2.77	2.79	2.83
Lu	0.4	0.43	0.42	0.4	0.38	0.39	0.4	0.38	0.38	0.39
La/Nb	1.09	1.08	1.09	1.03	1.10	1.09	1.06	1.06	1.11	1.10

Y are generally considered to be immobile during surface alternation. Here we discriminate the volcanic series by means of Nb/Y ratios. As shown in the Nb/Y vs. Zr/0.000 1 TiO_2 diagram (Fig. 6), both the Guangxi basalts and Guizhou basalts plot in the alkaline basalts field.

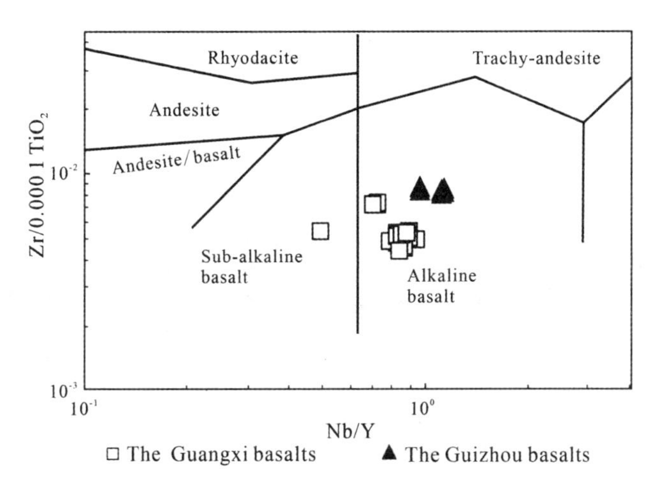

Fig. 6 Nb/Y vs. Zr/0. 000 1 TiO_2 diagram for the Permian basalts from Guangxi and Guizhou provinces.

The Guangxi basalts have SiO_2 = 44. 82–49. 71 wt%, TiO_2 = 2. 29–3. 54 wt%, high alkali contents ($K_2O + Na_2O$ = 3. 24–4. 91 wt%). They are characterized by low MgO (4. 50–5. 73 wt%, with $Mg^{\#}$ ranging from 41. 6 to 54. 2), P_2O_5 = 0. 29–0. 65 wt%, and low CaO/Al_2O_3 ratios of 0. 56–0. 74. The Guizhou basalts display relatively higher SiO_2(47. 49–50. 27 wt%) and TiO_2(3. 93–4. 92 wt%) contents, lower MgO (3. 75–4. 79 wt%) content, and $Mg^{\#}$ values of 38. 4–42. 9. Their alkali contents, P_2O_5(0. 43–0. 47 wt%), and low CaO/Al_2O_3 ratios (0. 55–0. 68) are similar to those of the Guangxi basalts.

As shown in the trace element primitive mantle normalized spider diagrams, the Guangxi basalts are enriched in LILE. They show Nb = 16. 3–29. 3 ppm, Ta = 0. 78–1. 86 ppm, and Zr = 131–190 ppm; when compared to typical oceanic island basalts (OIB), they display slight depletion in Nb and Ta (Fig. 7b). They also display Ti/Y ratios of 479–741, mostly greater than 500, and can be considered to be high-Ti basalts (Xu et al., 2001; Xiao et al., 2004), with Nb/La ratios of 0. 46–1. 04, and Sm/Yb ratios of 2. 4–3. 5. Most of the samples have Th/Ta ratios ranging from 1. 8–2. 5; these values are slightly lower or similar to those of the primitive mantle. They possess high $(La/Yb)_N$ ratios of 6. 3–9. 8, $(La/Sm)_N$ ratios of 2. 1–2. 6, Eu/Eu^* anomalies of 0. 95–1. 20. These features are distinctly different from those of N-MORB, but similar to those of OIB.

The Guizhou basalts are also enriched in LILE, and they have higher Nb (37. 9–42. 8 ppm), Ta (2. 50–2. 89 ppm) and Zr (332–357 ppm) than those of the Guangxi basalts. In the primitive mantle normalized trace element patterns (Fig. 8b), they are enriched in Nb and Ta compared to the typical OIB. This feature is clearly different from that of the Guangxi basalts. The Guizhou basalts also have high Ti/Y ratios of 653–680. Compared to the Guangxi basalts, the Guizhou basalts have higher Sm/Yb (3. 5–4. 2), Nb/La (0. 90–0. 94) and

Th/Ta (2.4–2.7) ratios. They also display high $(La/Yb)_N$ ratios of 9.6–11.8, and $(La/Sm)_N$ ratios of 2.4–2.6, suggesting their LREE enriched nature. In addition, they also have slight negative Eu anomalies of 0.86–0.92.

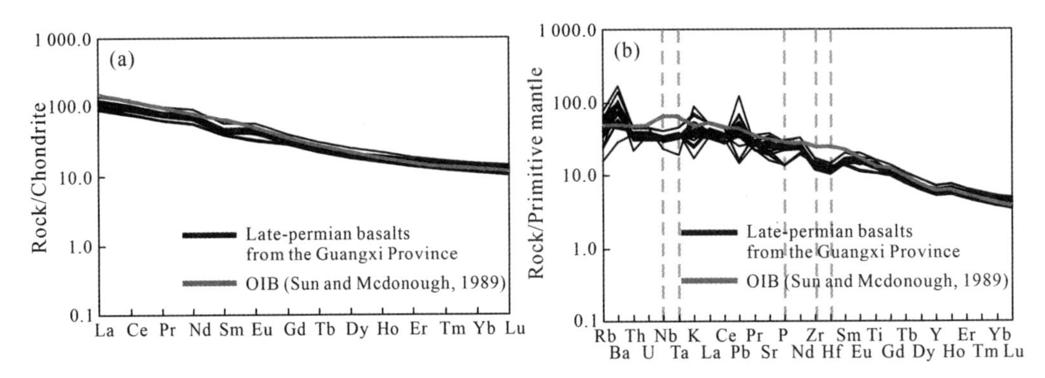

Fig. 7 Chondrite-normalized REE patterns (a) and primitive mantle (PM) normalized trace element spider diagram (b) and for the Permian Guangxi basalts.

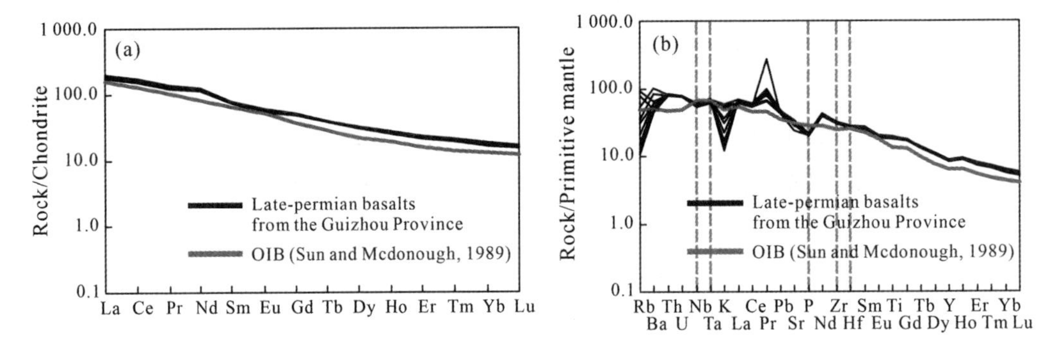

Fig. 8 Chondrite-normalized REE patterns (a) and primitive mantle (PM) normalized trace element spider diagram (b) for the Permian Guizhou basalts.

4.3 Sr-Nd-Pb isotopic composition

Whole-rock Sr-Nd-Pb isotopic results for the Permian basalts from the eastern part of the Emeishan LIP are given in Table 4 and plotted in Figs. 9 and 10 in relation to the isotopic composition fields of the Permian Emeishan basalts (Xu et al., 2001; Xiao et al., 2004).

The Guangxi basalts have high $(^{87}Sr/^{86}Sr)_i$ ratios of 0.705 231–0.706 147, and lower Nd isotopic ratios of $^{143}Nd/^{144}Nd$ = 0.512 513–0.512 550 with positive to slightly negative $\varepsilon_{Nd}(t)$ values of −0.13 to +0.68, they have $^{206}Pb/^{204}Pb$ = 18.374 2–19.415 1, $^{207}Pb/^{204}Pb$ = 15.596 9–15.679 7, $^{208}Pb/^{204}Pb$ = 38.543 3–39.888 5. The Guizhou basalts have $(^{87}Sr/^{86}Sr)_i$ ratios of 0.705 600–0.706 930, and higher Nd isotopic ratios of $^{143}Nd/^{144}Nd$ = 0.512 564–0.512 586 with positive $\varepsilon_{Nd}(t)$ values of +0.92 to +1.30, they have $^{206}Pb/^{204}Pb$ = 19.099 4–19.037 4, $^{207}Pb/^{204}Pb$ = 15.644 6–15.645 8, $^{208}Pb/^{204}Pb$ = 39.539 3–39.577 7. Both the

Guangxi and Guizhou basalts have DUPAL Pb isotopic compositions with Δ7/4 > 5, Δ8/4 > 70 (Hart, 1984).

Table 4 Sr-Nd-Pb isotopic analyses results of Permian basalts from Guangxi and Guizhou provinces.

Sample	GY03	GY21	GY32	GY42	GY74	GY86	GY88
Rock	Basalt	Basalt	Basalt	Basalt	Basalt	Basalt	Basalt
Pb	6.76	6.76	7.02	2.66	2.72	8.65	2.97
Th	6.87	6.61	6.66	3.82	3.12	2.66	2.79
U	1.64	1.61	1.63	0.96	0.78	0.67	0.72
$^{206}Pb/^{204}Pb$	19.137 4	19.121 8	19.099 4	19.415 1	19.201 3	18.374 2	19.135 2
2σ	0.000 7	0.001 5	0.000 9	0.001 3	0.001	0.000 8	0.001 7
$^{207}Pb/^{204}Pb$	15.644 6	15.645 8	15.645 8	15.642 9	15.679 7	15.596 9	15.626 3
2σ	0.000 6	0.001 3	0.000 8	0.001 1	0.000 9	0.000 7	0.001 7
$^{208}Pb/^{204}Pb$	39.595	39.577 7	39.539 3	39.888 5	39.618 5	38.543 3	39.497 2
2σ	0.001 7	0.003 2	0.002 4	0.003 7	0.002 5	0.002 4	0.004 2
$\Delta7^a$	8	8	8	5	11	11	6
$\Delta8^b$	83	83	82	79	78	70	74
Sr/ppm	614	656	647	623	568	499	717
Rb/ppm	34.4	22	7.4	42.1	26.1	28.7	15.1
$^{87}Rb/^{86}Sr$	0.162 316	0.096 87	0.032 918	0.195 641	0.132 985	0.166 409	0.061 022
$^{87}Sr/^{86}Sr$	0.706 205	0.706 122	0.706 048	0.706 243	0.706 62	0.705 823	0.705 619
2σ	0.000 079	0.000 027	0.000 025	0.000 118	0.000 017	0.000 026	0.000 011
$(^{87}Sr/^{86}Sr)_i$	0.705 6	0.705 777	0.705 93	0.705 548	0.706 147	0.705 231	0.705 402
ΔSr^c	62	61	60	62	66	58	56
Nd/ppm	57.8	56.2	57	44.3	36.8	30.2	35.6
Sm/ppm	11.9	11.4	11.6	8.91	7.6	6.34	7.27
$^{147}Sm/^{144}Nd$	0.124 353	0.122 694	0.122 569	0.121 674	0.124 658	0.126 848	0.123 494
$^{143}Nd/^{144}Nd$	0.512 586	0.512 564	0.512 57	0.512 55	0.512 513	0.512 544	0.512 53
2σ	0.000 009	0.000 007	0.000 007	0.000 008	0.000 008	0.000 008	0.000 011
T_{DM2}/Ga	0.82	0.84	0.83	0.86	0.91	0.88	0.89
$\varepsilon_{Nd}(t)$	1.3	0.92	1.04	0.68	-0.14	0.4	0.23

$^a\Delta7/4=[(^{207}Pb/^{204}Pb)_S-0.1084(^{206}Pb/^{204}Pb)_S-13.491]\times100$.

$^b\Delta^{208}Pb/^{204}Pb(\Delta8/4)=[(^{208}Pb/^{204}Pb)_S-1.209(^{206}Pb/^{204}Pb)_S-15.627]\times100$.

$^c\Delta Sr=[(^{87}Sr/^{86}Sr)_S-0.7]\times10\,000$.

When plotted in the $(^{87}Sr/^{86}Sr)_i$ vs. $\varepsilon_{Nd}(t)$ diagram (Fig. 9), both the Guizhou and Guangxi basalts display similar Sr-Nd isotopic composition with the Emeishan flood basalts (Zhou et al., 2008) and clearly differ from the depleted mantle (DM) and Emeishan picrites, suggesting enriched components in their source region. In the $^{206}Pb/^{204}Pb$ vs. $^{207}Pb/^{204}Pb$ and $^{206}Pb/^{204}Pb$ vs. $^{208}Pb/^{204}Pb$ (Fig. 10a,b) diagrams, all the samples display a similar Pb isotopic composition with EM Ⅱ, in distinct contrast from DMM and MORB.

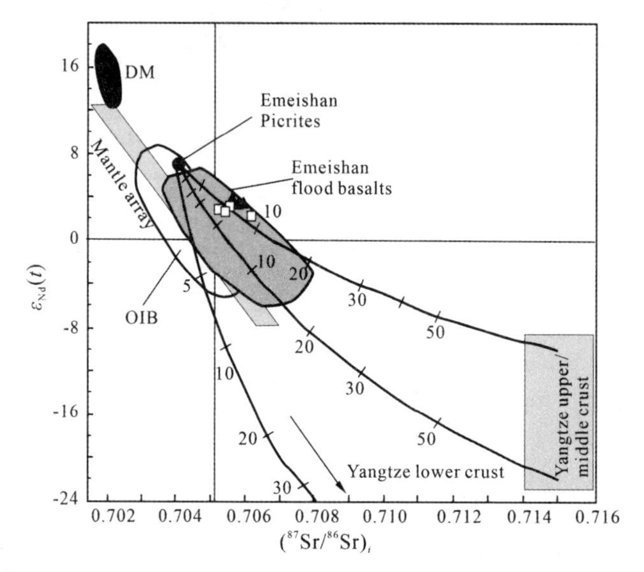

Fig. 9 $(^{87}Sr/^{86}Sr)_i$ vs. $\varepsilon_{Nd}(t)$ plot for the Permian basalts from Guangxi and Guizhou provinces. Symbols as in Fig. 6.

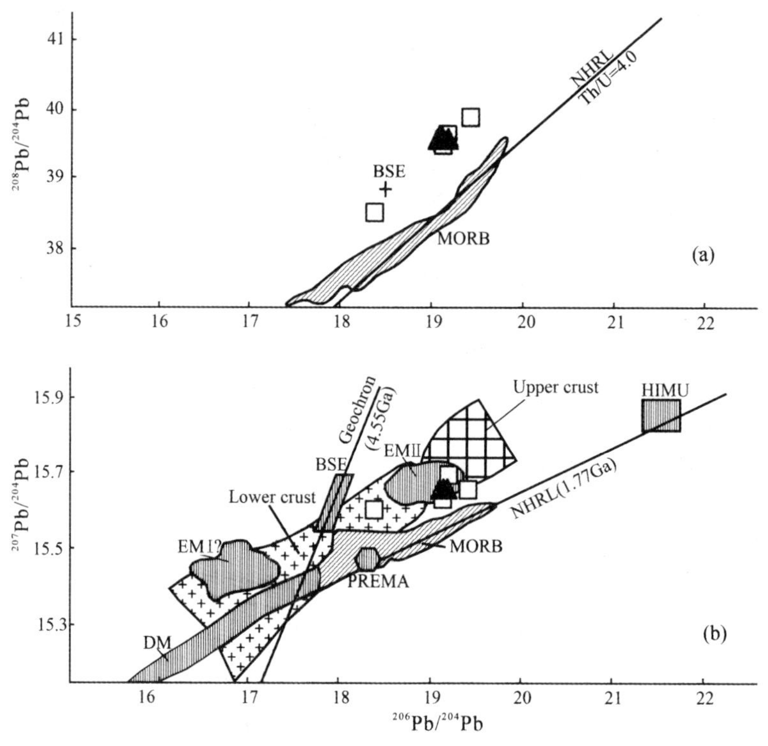

Fig. 10 $^{208}Pb/^{204}Pb$ vs. $^{206}Pb/^{204}Pb$ (a) $^{207}Pb/^{204}Pb$ vs. $^{206}Pb/^{204}Pb$ (b) diagrams for Permian basalts from Guangxi and Guizhou provinces (revised from Rollinson, 1993).

NHRL: northern hemisphere reference line (Th/U = 0.4); BSE: bulk silicate Earth value; MORB: mid-ocean ridge basalt; DM: depleted mantle; EM Ⅰ and EM Ⅱ: enriched mantle; HIMU: mantle with high U/Pb ratios; PREMA: frequently observed prevalent mantle composition (Zindler and Hart, 1986). Symbols as in Fig. 6.

5 Discussion

5.1 Crustal contamination fractional crystallization

Both Guangxi and Guizhou basalts have low MgO, Cr and Ni contents (Table 2), suggesting that they were not derived from a primary melt that was in equilibrium with mantle peridotites. As shown in the variation of major-and trace-element vs. SiO_2 wt%(Fig. 11), with increasing SiO_2, the MgO, CaO, Al_2O_3 and Ni contents are decreasing, while TiO_2 contents and $(La/Yb)_N$ ratios are increasing. Furthermore, the Guangxi basalts show lower SiO_2 and

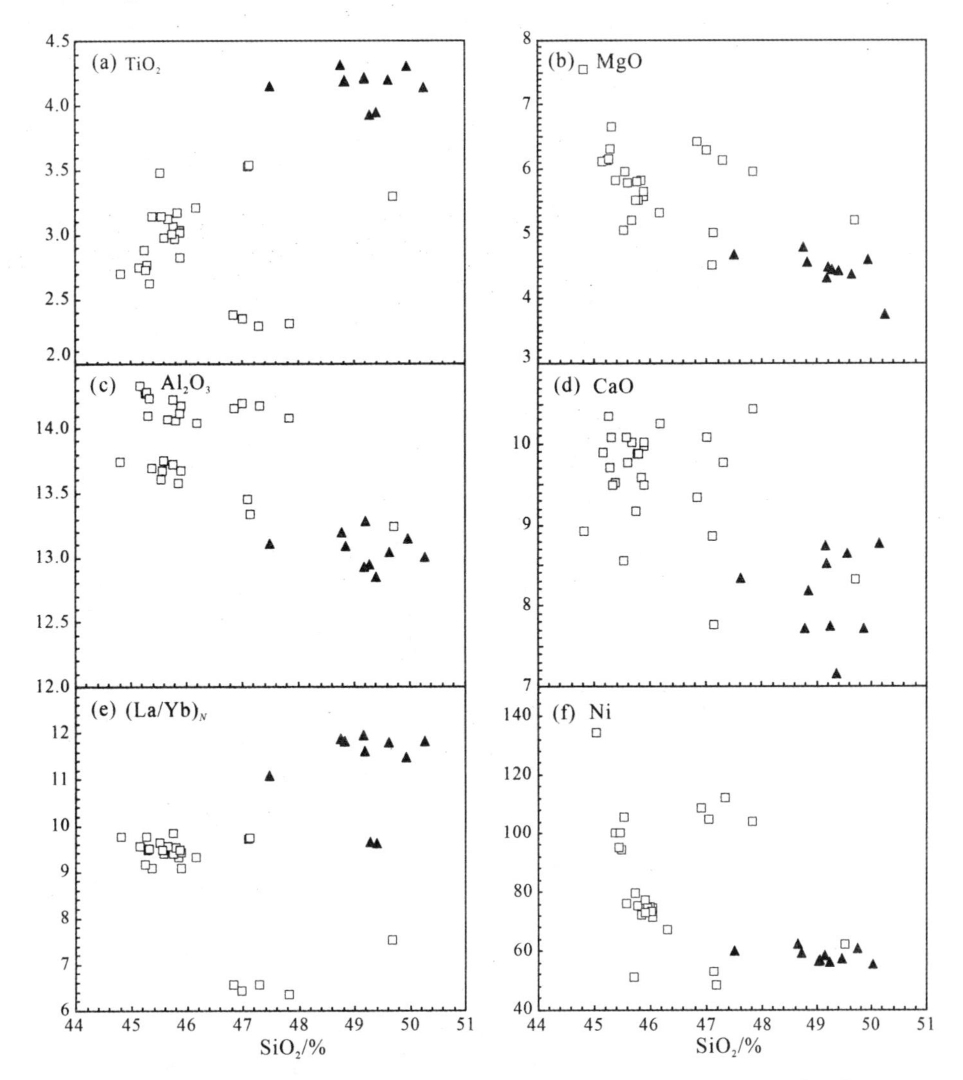

Fig. 11 Major- and trace-element variation vs. SiO_2 for the Late-Permian basalts from Guangxi and Guizhou provinces.

Symbols as in Fig. 6.

positive Eu anomalies of 0.95 to 1.20, while the Guizhou basalts have higher SiO_2 and negative Eu anomalies of 0.86–0.92. These features may be caused by fractionation of olivine and plagioclase or different degrees partial melting of their source region. In general, during fractional crystallization, the TiO_2 content increases where as the Ti/Y ratios remain consistent (Peate et al., 1992; Peate and Hawkesworth, 1996). As shown in the plot of SiO_2 vs. Nb/La diagram (Fig. 12a), both the Guangxi and Guizhou basalts have low and constant Nb/La ratios with increasing SiO_2 contents, both the Guangxi and Guizhou basalts have low and consistent Th/Ta ratios (< 2.7), suggesting minor degree of crustal contamination (Sobolev, 2005; Sobolev et al., 2007). However, the studied basalts have relatively higher Zr/Nb and Th/Nb ratios (6.0–11.5 and 0.08–0.23, respectively) than those of the OIB (4.2 and 0.06, respectively) (Hoffmann, 1998; Ionov et al., 1997) and display weakly negative anomalies of Nb and Ta in the spider diagrams (Figs. 7 and 8), these features suggest minor crustal contamination during their ascent (Sun, 1980), furthermore, their varied Pb anomalies also indicate crustal contamination. Then, it can be considered that the Guangxi and Guizhou basalts may have experienced fractional crystallization process of olivine and plagioclase or derived from different degrees partial melting of their source region, with minor crustal contamination during their ascent.

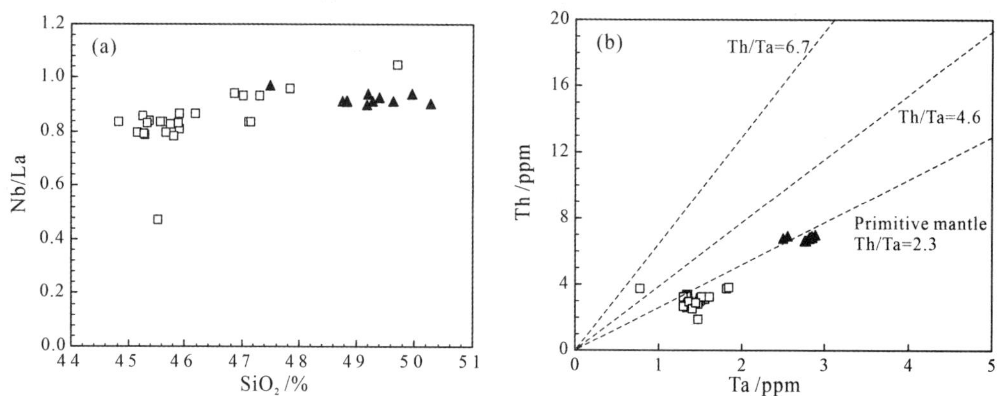

Fig. 12 SiO_2 vs. Nb/La (a) and Ta vs. Th (b) diagrams for the Permian basalts from Guangxi and Guizhou provinces.
Symbols as in Fig. 6.

5.2 Origin of the Permian high Ti/Y basalts from the eastern margin of the Emeishan LIP

Zircon LA-ICP-MS U-Pb dating performed in this study reveals that the Guangxi basalts have a crystallization age of 257 Ma, which is identical to the formation age of the Emeishan mantle plume (He et al., 2007). We therefore consider the basalts from Guangxi as a constituent of the Emeishan LIP. Both of Guangxi and Guizhou basalts are enriched in LILEs and LREE (Figs. 7 and 8), clearly differing from typical N-MORB, but similar to OIB or island arc basalts. They display no significant negative anomalies in HFSEs, similar to those of the OIB or continental flood basalt (CFB). In Th/Yb vs. Nb/Yb diagram (Fig. 13), both the Guangxi and Guizhou basalts show high Nb/Yb and Th/Yb ratios and are plotted in the OIB region. The Nb/Yb vs. TiO_2/Yb relationship also illustrates a similar character. These features suggest that the basalts from both regions may have originated from a primitive mantle source. The Ti/Y ratios are consistent during fractional crystallization process (Peate et al., 1992), and it is usually used as a discriminator of rock types (Xu et al., 2001; Xiao et al., 2004). Both of the Guangxi and Guizhou basalts have high TiO_2 contents (Table 2) with Ti/Y ratios greater than 500. In Ti/Y vs. $Mg^{\#}$ and Ti/Y vs. Sm/Yb diagrams (Fig. 14), all the samples plot in the high-Ti basalts field, then they can be classified into high-Ti basalts.

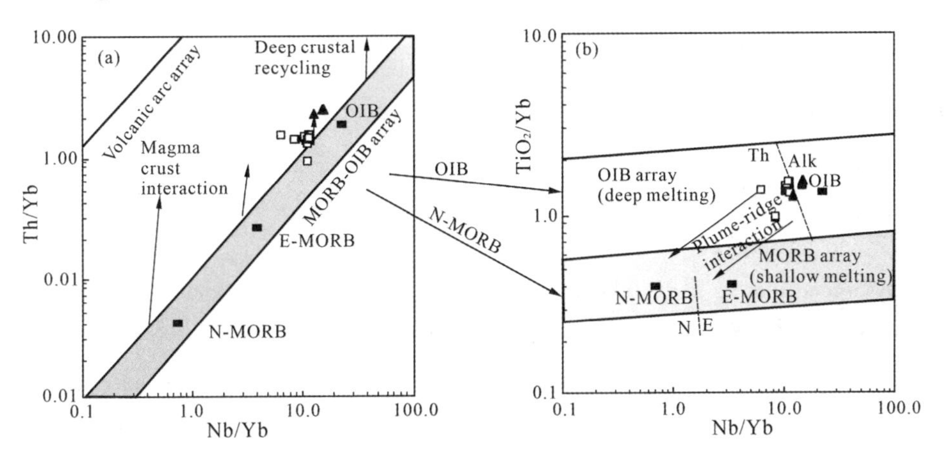

Fig. 13 Nb/Yb-Th/Yb and Nb/Yb-TiO_2/Yb diagrams for the Permian basalts from the Guangxi and Guizhou basalts (Pearce, 2008).

Symbols as in Fig. 6.

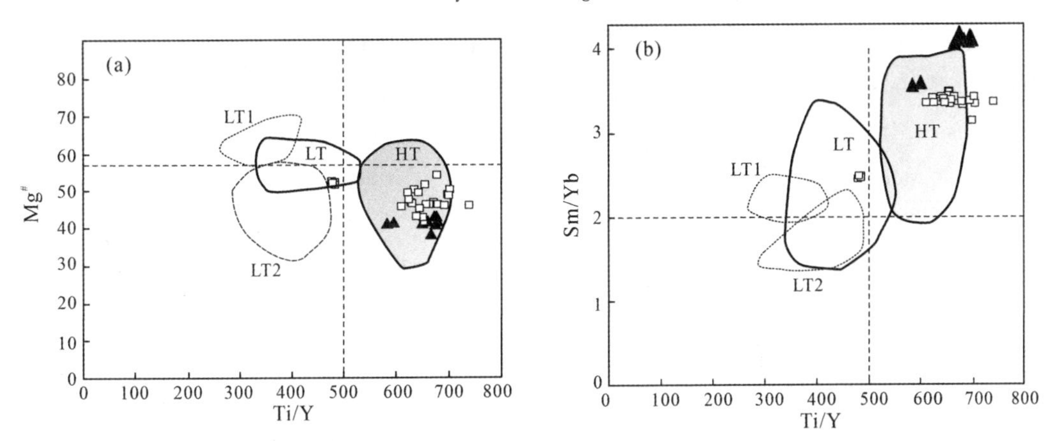

Fig. 14 Ti/Y vs. $Mg^{\#}$(a) and Ti/Y vs. Sm/Yb (b) diagrams for the Permian basalts from Guangxi and Guizhou provinces.

Symbols as in Fig. 6.

Origin of the high-Ti/Y basalts in the Emeishan LIP is still controversial. Xu et al. (2007) proposed that the low-Ti basalts have the highest Os concentration with γOs (*t*) values of +6.5, suggesting that they might have originated from a mantle plume reservoir. In contrast, the high-Ti basalts have relative lower Os contents with γOs (*t*) values of −1.4 to −0.8, suggesting that a sub-continental lithospheric mantle (SCLM) component most likely contributed to the generation of these magmas. However, other workers (e.g., Song et al., 2001, 2008; Xu et al., 2004; Qi et al., 2010) argued that the high-Ti basalts were generated by low degrees (< 8%) partial melting of the Emeishan mantle plume in the garnet stability field, while the low-Ti basalts were derived from either partial melting of sub-continental lithosphere mantle (SCLM) or pictiric magma that assimilated upper crust. Recently, Shellnutt and Jahn (2011) argued that the Late-Permian the high- and low-Ti basaltic rocks are likely derived from the same garnet-bearing source and represent different degrees of partial melting with or without crustal assimilation, based on trace element modeling and experimental works, they proposed that the high-Ti basalts were originated from low degree (1.2%–1.5%) partial melting of garnet peridotite source.

The Guangxi basalts display low SiO_2 (44.82–49.71 wt%), TiO_2 (2.29–3.54 wt%), low Ni and Cr and low $Mg^{\#}$ of 41.6–54.2. They possess high $(La/Sm)_N$ values of 2.2–2.6. It is commonly argued that La and Nb behave in a similar way during melt generation, and that their partition coefficients and abundances in parts per million in the mantle are similar. Metasomatic melts from the primitive mantle source have La/Nb ratios of 0.53, whereas those derived from a mid-ocean ridge basalt (MORB) source are 1.02 (McKenzie and O'Ninos, 1995). The Guangxi basalts have high La/Nb ratios of 1.0–2.1, suggesting that they might have originated from an enriched mantle source that was previously metasomatized. In the primitive mantle normalized trace element patterns, they show slight negative anomalies in Nb and Ta, indicating minor crustal contamination. They have higher Nb (16.3–29.3 ppm) and Zr (131–190 ppm) contents than those of the N-MORB (Nb = 2.33 ppm, Zr = 74 ppm), albeit lower than those of OIB (Nb = 48 ppm, Zr = 280 ppm), suggesting that they could be derived from an enriched mantle source (Sun and McDonough, 1989). Furthermore, they have high Ce/Yb ratios of 19.6 – 30.0. In general, Yb is compatible in garnet, whereas La and Sm are incompatible, La/Yb and Sm/Yb will be strongly fractionated when melting degree is low. In contrast, La/Yb is only slightly fractionated and Sm/Yb is nearly unfractionated during the melting in the spinel stability field (White and McKenzie, 1995; Yaxley, 2000; Xu et al., 2005), the La/Yb vs. Sm/Yb diagram are usually used to distinguished between melting of spinel and garnet peridotite. As shown in the plot of La/Yb vs. Sm/Yb (Fig. 15), the Guangxi basalts can be generated by 10% – 15% batch melting of a hypothetical light REE-enriched mantle source [$(La/Yb)_N > 1$] in the garnet stability field. The Guangxi basalts also display evolved and varied Sr-Nd-Pb isotopic composition, $(^{87}Sr/^{86}Sr)_i$ = 0.705 231 – 0.706 147, positive to slightly negative $\varepsilon_{Nd}(t)$ values of −0.13 to +0.68, $\Delta7$=5−11, $\Delta8$= 70−79, in the suggesting that these rocks were derived by low degree (10%−15%) partial melting of a thick lithosphere (garnet stability field) and were subjected to a minor degree of crustal contamination. The Guizhou basalts display higher SiO_2(47.49−50.27 wt%) and TiO_2 (3.93−4.92 wt%), lower $Mg^{\#}$ values of 38.4−42.9, and high La/Nb ratios (1.0−1.1), similar to those of the metasomatic melts that from N-MORB source (McKenzie and O'Ninos, 1995). Their Th/Ta (2.4−2.7) ratios are similar to those of the primitive mantle. They have higher Ce/Yb (30.2−37.8) and Sm/Yb (3.56−4.19) ratios. The plot of La/Yb vs. Sm/Yb (Fig. 15) ratios indicate that the Guizhou basalts may be generated by low degree (<10%) partial melting of garnet peridotite. Their high Nb (37.9−43.5 ppm) and Zr (332−357 ppm) contents are similar to those of OIB (Nb=48 ppm, Zr=280 ppm), indicating that these rocks originated from an enriched mantle source (Sun and McDonough, 1989). Compared to the Guangxi basalts, the Guizhou basalts display limited Sr-Nd-Pb isotopic composition, suggesting limited crustal contamination.

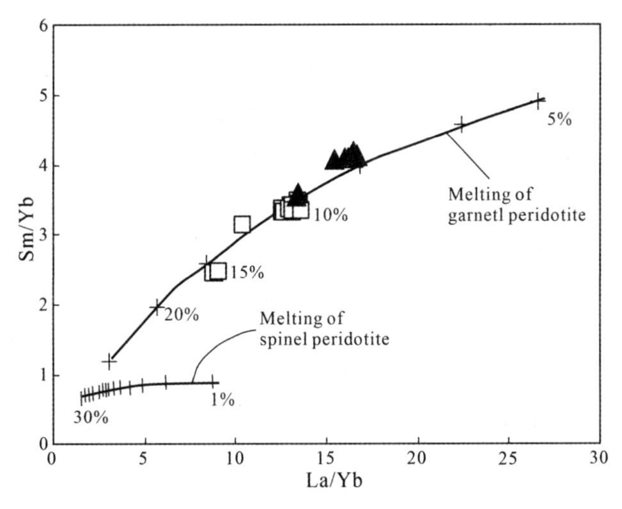

Fig. 15 La/Yb vs. Sm/Yb diagram for the Permian basalts from Guangxi and Guizhou provinces.
Symbols as in Fig. 6.

5.3 Petrogenetic implications

According to the definition of Xu et al. (2001), both the Guangxi and Guizhou basalts can be classified into high-Ti basalts in the Emeishan LIP. Xu et al. (2004) proposed that the high-Ti basalts are predominant in the outer zone of the Emeishan LIP and represent the waning stage of the volcanism; Xiao et al. (2004) argued that the high-Ti basalts was directly generated from the head of the Emeishan mantle plume, by large scale and low degrees of partial melting. However, based on the study of basalts and mafic dykes in the Panxi region, Shellnutt and Jahn (2011) argued that there is no correlation between the spatial distribution and the composition of the basalts, because they identified early stage high-Ti basaltic rocks in the inner zone of the Emeishan LIP.

The Guangxi and Guizhou basalts, like elsewhere in the Emeishan high-Ti basalt provinces (Song et al., 2001, 2008; Xiao et al., 2004), display geochemical features identical to those of OIB. Generally, the OIB source materials are considered to be recycled oceanic crust (Hofmann and White, 1982; Hofmann, 1997) or recycled continental crust (Chauvel et al., 1992; White and Duncan, 1996; Eisele et al., 2002). Takahashi et al. (1998) proposed a model that envisages the melting of a heterogeneous plume head to explain rapid production of large volumes of lavas in the Columbia River Basalts (CRB) of NW USA. However, Niu (2009) argued that neither the recycled oceanic crust nor recycled continental crust can account for the geochemical features of typical OIB, because the oceanic crust is far too depleted with $(La/Sm)_N \ll 1$, whereas the continental crust shows depletion in Nb and Ta. Such features are inconsistent with typical OIB.

Pilet et al. (2008) proposed the existence of metasomatized veins in mature oceanic lithosphere, which consist of amphiboles. In their melting experiments on natural amphibole-rich veins at 1.5 GPa, they observed that partial melts of metasomatic veins can reproduce key major- and trace-element features of oceanic and continental alkaline magmas. Their experiments with hornblendite plus lherzolite showed that reaction of melts of amphibole-rich veins with surrounding lherzolite can explain the observed compositional trends from nephelinites to alkali olivine basalts. Pilet et al. (2008) thus conclude that melting of metasomatized lithosphere is a viable alternative to models of alkaline basalt formation by melting of recycled oceanic crust with or without sediments. When hot plume rises up from the deep mantle, these metasomatized veins in the lithosphere, instead of the recycled oceanic crust, produces melts. The melts thus derived are enriched in H_2O and CO_2, as well as incompatible elements, with elemental compositions matching with those of OIB. The vein melts may be altered through the addition of surrounding material and mixing with plume melt. The degree of mixing determines which type of alkaline magma is formed and erupted (Niu, 2008).

In combination with the regional geological setting, we propose the following model Fig. 16 to explain the generation of the high Ti/Y basalts from the eastern margin of the Emeishan LIP. The study area is located in the eastern margin of the Emeishan LIP (Fig. 1). The sedimentary sequences (He et al., 2003, 2007) and geophysical data (Xu et al., 2004) reveal that this area is characterized by a thick lithosphere and experienced insignificant uplift during the Late-Permian Emeishan mantle plume event. The upwelling of a hot plume from deep mantle triggered the partial melting of metasomatic veins and pockets in the Yangtze continental

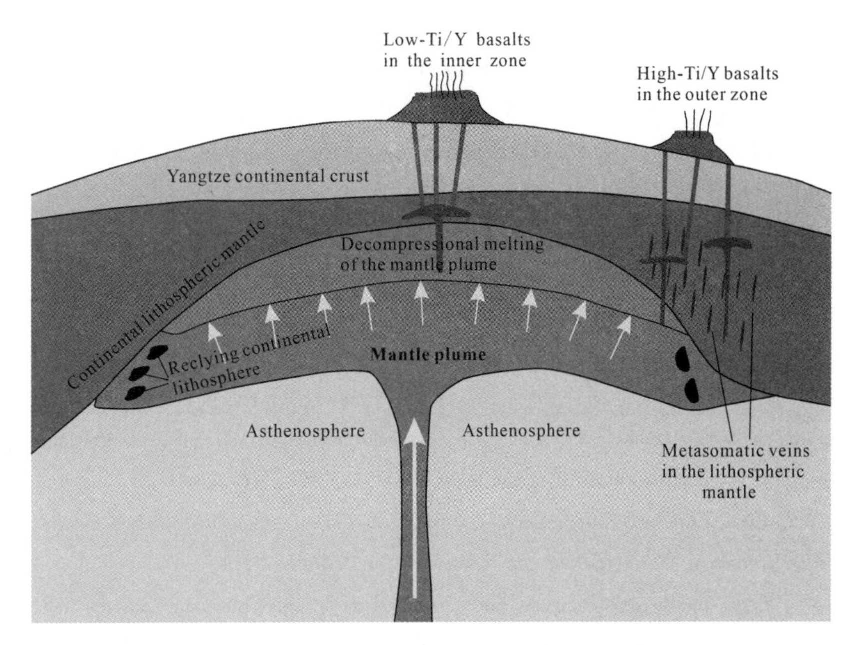

Fig. 16 Petrogenetic model for the origin and evolution of the Late-Permian Guangxi and Guizhou basalts and the related Emeishan mantle plume.

lithosphere, generating melts with high SiO_2 and TiO_2 contents which represent the Guizhou basalts, The Guangxi basalts formed by higher degrees partial melting of the sub-continental lithospheric mantle and undergone minor crustal contamination during their ascent.

References

Ali , J.R., Thompson, G.M., Zhou, M., Song, Y., 2005. Emeishan large igneous province, SW China. Lithos, 79, 475-489.

Andersen, T., 2002. Correction of common lead in U-Pb analyses that do not report ^{204}Pb. Chemical Geology, 192, 59-79.

Chauvel, C., Hofmann, A.W., Vidal, P., 1992. HIMU-EM: The French Polynesian connection. Earth & Planetary Science Letters, 110, 99-119.

Chung, S.L., Jahn, B.M., 1995.Plume-lithosphere interaction in generation of the Emeishan flood basalts at the Permian-Triassic boundary. Geology, 23, 889-892.

Chung, S.L., Jahn, B.M., Wu, G.Y., Lo, C.H., Cong, B.L., 1998. The Emeishan flood basalt in SW China: a mantle plume initiation model and its connection with continental break-up and mass extinction at the Permian-Triassic boundary. In: Flower, M.F.J., Chung, S.L., Lo, C.H., Lee, T.Y. Mantle Dynamics and PlateInteraction in East Asia AGU Geodyn. AGU, Washington DC, 27, 47-58.

Coffin, M.F., Eldholm, O., 1994. Large igneous provinces: Crustal structures, dimensions and external consequences. Reviews in Geophysics, 32,1-36

Eisele, J., Sharma, M., Galer, S.J.G., Blicher-Toft, J., Devey, C.W., Hofmann, A.W., 2002. The role of sediment recycling in EM Ⅰ inferred from Os, Pb, Hf, Nd, Sr isotope and trace element systematics of the

Pitcairn hotspot. Earth & Planetary Science Letters, 196(3-4), 197-212.

Gallagher, K., Hawkesworth, C.J., 1992. Dehydration melting and the generation of continental flood basalts. Nature, 358, 57-59.

Garfunkel, Z., 2008. Formation of continental flood volcanism: The perspective of setting of melting. Lithos, 100, 49-65

Hart, S.R., 1984. A large-scale isotope anomaly in the Southern Hemisphere mantle. Nature, 5971, 753-757.

He, B., Xu, Y., Xiao, L., Chung, S., Wang, Y., 2003. Sedimentary evidence for a rapid, kilometer-scale crustal doming prior to the eruption of the Emeishan flood basalts. Earth & Planetary Science Letters, 213, 391-405.

He, B., Xu, Y.G., Huang, X.L., Luo, Z.Y., Shi, Y.R., Yang, Q.J., Yu, S.Y., 2007. Age and duration of the Emeishan flood volcanism, SW China: Geochemistry and SHRIMP zircon U-Pb dating of silicic ignimbrites, post-volcanic. Earth & Planetary Science Letters, 255, 306-323.

Hofmann, A.W., 1988. Chemical differentiation of the Earth: The relationship between mantle, continental crust, and oceanic crust. Earth & Planetary Science Letters, 90, 297-314.

Hofmann, A.W., 1997. Mantle geochemistry: The message from oceanic volcanism. Nature, 385,219-229.

Hofmann, A.W., White, W.M., 1982. Mantle plumes from ancient oceanic crust. Earth & Planetary Science Letters 57, 421-436.

Ionov, D.A., Griffin, W.L., O'Reilly, S.Y., 1997. Volatile-bearing minerals and lithophile trace elements in the upper mantle. Chemical Geology, 141, 153-184.

Isozaki, Y., 2009. Illawara Reversal: The fingerprint of a superplume that triggered end-Guadallupian (Permian) mass extinction. Gondwana Research, 15, 431-432.

Isozaki, Y., 2010. Reply to the comment by J.R. Ali on "Illawara Reversal: The fingerprint of a superplume that triggered end-Guadallupian (Permian) mass extinction" by Yukio Isozaki. Gondwana Research, 17, 718-720.

Ludwig, K. R., 2003. ISOPLOT 3. 0: A geochronological toolkit for Microsoft Excel. Berkeley Geochronology Center, Special Publication, 4.

McKenzie, D., Bickle, M. J., 1988. The volume and composition of melt generated by extension of the lithosphere. Journal of Petrology, 29, 625-679.

McKenzie, D., 1989. Some remarks on the movement of small melt fractions in the mantle. Earth & Planetary Science Letters, 95, 53-72.

McKenzie, D., O'Ninos, K., 1995. The Source Regions of Ocean Island Basalts. Journal of Petrology, 36(1), 133-159

Niu, Y.L., 2008. The origin of alkaline lavas. Science, 320, 883-884.

Niu, Y.L., 2009. Some basic concepts and problems on the petrogenesis of intra-plate ocean island basalts. Chinese Science Bulletin, 54, 4148-4160.

Peate, D.W., Hawkesworth, C.J., 1996. Lithospheric to asthenospheric transition in Low-Ti flood basalts from southern Parana, Brazil. Chemical Geology, 127, 1-24.

Peate, D.W., Hawkesworth, C.J., Mantovani, M.S.M., 1992. Chemical stratigraphy of the Parani lavas (South America): Classification of magma types and their spatial distribution. Bulletin of Volcanology, 55,

119-139.

Pearce, J.A., 2008.Geochemical fingerprinting of oceanic basalts with applications to ophiolite classification and the search for Archean oceanic crust. Lithos, 100, 14-48

Pilet, S., Baker, M.B., Stolper, E.M., 2008. Metasomatized lithosphere and the origin of alkaline lavas. Science, 320, 916-919.

Qi, H., Xiao, L., Balta, B., Gao, R., Chen, J., 2010. Variety and complexity of the Late-Permian Emeishan basalts: Reappraisal of plume-lithosphere interaction processes. Lithos, 119, 91-107.

Rollinson, H.R., 1993. Using Geochemical Data: Evaluation, Presentation, Interpretation. Chichester: John Wiley, 352.

Rubatto, D. 2002. Zircon trace element geochemistry: Partitioning with garnet and the link between U-Pb ages and metamorphism. Chemical Geology, 184, 123-138.

Shellnutt, J.G., Jahn, B.M., 2011. Origin of Late Permian Emeishan basaltic rocks from the Panxi region (SW China): Implications for the Ti-classification and spatial-compositional distribution of the Emeishan flood basalts. Journal of Volcanology and Geothermal Research, 199, 85-95.

Sobolev, A.V., et al., 2007. The amount of recycled crust in sources of mantle-derived melts. Science, 316, 412-417.

Sobolev, A.V., Hofmann, A.W., Sobolev, S.V., Nikogosian, I.K., 2005, An olivine-free mantle source of Hawaiian shield basalts. Nature, 434, 590-597.

Song, X., Zhou, M., Hou, Z., Cao, Z., Wang, W., Li, Y., 2001. Geochemical constraints on the mantle source of the upper Permian Emeishan continental flood basalts, southern China. International Geological Review, 43, 213-225.

Song, X.Y., Zhong, H., Zhou, M.F., Tao, Y., 2005a. Magmatic sulfide deposits in the Permian Emeishan Large Igneous Province, SW China. In: Mao, J.W., Bierlein, F.P. Mineral Deposit Research: Meeting the Global Challenge. Berlin: Springer.

Song, X.Y., Zhou, M.F., Cao, Z.M., Robinson, P.T., 2004. Late Permian rifting of the South China Craton caused by the Emeishan mantle plume. Journal Geological Society London, 161, 773-781.

Song, X.Y., Qi, H.W., Robinson, P.T., Zhou, M.F., Cao, Z.M., Chen, L.M., 2008. Melting of the subcontinental lithospheric mantle by the Emeishan mantle plume: Evidence from the basal alkaline basalts in Dongchuan, Yunnan, Southwestern China. Lithos, 100, 93-111

Song, X.Y., Zhou, M.F., Keays, R.R., Cao, Z., Sun, M., Qi, L., 2006. Geochemistry of the Emeishan flood basalts at Yangliuping, Sichuan, SW China: Implications for sulfide segregation. Contributions to Mineralogy and Petrology, 152, 53-74.

Sun, S.S., 1980. Lead isotopic study of young volcanic rocks from mid-oceanic ridges, ocean islands and island arcs. Philosophical Transactions of the Royal Society, A297, 409-445.

Sun, S.S., McDonough, W.F., 1989. Chemical and isotopic systematics of oceanic basalts: implications for mantle composition and processes. In: Saunders, A.D., Norry, M.J. Magmatism in the Ocean Basins. Geological Society, London, Special Publications, 42, 313-345.

Takahashi, E., Nakajima, K., Wright, T.L., 1998. Origin of the Columbia River basalts: Melting model of a heterogeneous plume head. Earth & Planetary Science Letters, 162, 63-80.

White, W.M., Duncan, R.A., 1996. Geochemistry and geochronology of the Society Islands: New evidence for deep mantle recycling. American Geophysical Union Geophys Monogr, 95, 183-206.

White, R.S., McKenzie, D., 1989. Magmatism at rift zones: The generation of volcanic continental margins and flood basalts. Journal of Geophysical Research, 94, 7685-7729.

White, R.S., McKenzie, D., 1995. Mantle plumes and flood basalts. Journal of Geophysical Research, 100, 17543-17585.

Xiao, L., Xu, Y.G., Chung, S.L., He, B., Mei, H., 2003. Chemostratigraphic correlation of Upper Permian lava succession from Yunnan Province, China: Extent of the Emeishan igneous province. International Geology Review, 45, 753-766.

Xiao, L., Xu, Y.G., Mei, H.J., Zheng, Y.F., He, B., Pirajno, F., 2004. Distinct mantle sources of low-Ti and high-Ti basalts from the western Emeishan large igneous province, SW China: Implications for plume-lithosphere interaction. Earth & Planetary Science Letters, 228, 525-546.

Xu, Y., Chung, S., Jahn, B., Wu, G., 2001. Petrologic and geochemical constraints on the petrogenesis of Permian-Triassic Emeishan flood basalts in southern China. Lithos, 58, 145-168.

Xu, Y., He, B., Chung, S.L., Menzies, M.A., Frey, F.A., 2004. Geologic, geochemical, and geophysical consequences of plume involvement in the Emeishan flood-basalt province. Geology, 32, 917-920.

Xu, Y.G., Ma, J.L., Frey, F.A., Feigenson, M.D., Liu, J.F., 2005. Role of lithosphere-asthenosphere interaction in the genesis of Quaternary alkali and tholeiitic basalts from Datong, western North China Craton. Chemical Geology, 224 (4),247-271.

Xu, J.F., Suzuki, K., Xu, Y.G., Mei, H.J., Li, J., 2007. Os, Pb, and Nd isotope geochemistry of the Permian Emeishan continental flood basalts: Insights into the source of a large igneous province. Geochim. Cosmochim. Acta, 71, 2104-2119.

Yaxley, G.M., 2000. Experimental study of the phase and letting relations of homogeneous basalt + peridotite mixtures and implications for the petrogenesis of flood basalts. Contributions to Mineralogy and Petrology, 139, 326-338.

Yuan, H.L., Gao, S., Liu, X.M., Li, H.M., Gunther, D., Wu, F.Y., 2004. Accurate U-Pb age and trace element determinations of zircon by laser ablation—inductively coupled plasma mass spectrometry. Geostandards Newsletter, 28, 353-370.

Zhang, Z.C., Mahoney, J.J., Mao, J.W., Wang, F.S., 2006. Geochemistry of picritic and associated basalt flows of the western Emeishan flood basalt province, China. Journal of Petrology, 47, 1997-2019.

Zhou, M.F., Arndt, N.T., Malpas, J., Wang, C.Y., Kennedy, A.K., 2008. Two magma series and associated ore deposit types in the Permian Emeishan large igneous province, SW China. Lithos, 103, 352-368.

Zhou, L., Kyte, F.T., 1988. The Permian-Triassic boundary event: A geochemical study of three Chinese sections. Earth & Planetary Science Letters, 90, 411-421.

Zhou, M.F., Malpas, J., Song, X.Y., Robinson, P.T., Sun, M., Kennedy, A.K., Lesher, C.M., Keays, R. R., 2002a. A temporal link between the Emeishan large igneous province (SW China) and the end-Guadalupian mass extinction. Earth & Planetary Science Letters, 196, 113-122.

Zhou, M.F., Yan, D.P., Kennedy, A.K., Li, Q.I., Ding, J., 2002b. SHRIMP zircon geochronological and geochemical evidence for Neo-Proterozoic arc-related magmatism along the western of the Yangtze Block, South China. Earth & Planetary Science Letters, 196, 51-67.

Zhou, M.F., Yang, Z.X., Song, X.Y., Lesher, C.M., Keays, R.R., 2002c. Magmatic Ni-Cu-(PGE) sulfide deposits in China. In: Cabri, L.J. The Geology, Geochemistry, Mineralogy and Mineral Beneficiation of Platinum-Group Elements, Canadian Inst. Mining, Metal. Petrol. Spec., 54,619-636.

Zi, J., Fan, W., Wang, Y., Peng, T., Guo, F., 2008. Geochemistry and petrogenesis of the Permian dykes in Panxi region, SW China. Gondwana Research, 14, 368-382.

Zindler, A., Hart, S.R., 1986. Chemical dynamics. Annual Review of Earth and Planetary Sciences, 14, 493-571.

Adakitic rocks derived from the partial melting of subducted continental crust: Evidence from the Eocene volcanic rocks in the northern Qiangtang Block①

Lai Shaocong Qin Jiangfeng

Abstract: Eocene is a critical time for the elevation of Tibetan Plateau and global climate change, and previous studies suggested that the Eocene elevation was caused by intra-continental subduction of the Songpan-Garze Block beneath the Qiangtang Block. This paper reports zircon U-Pb age and geochemistry of the Eocene volcanic rocks from the Zuerkenwula mountain area in the northern part of Qiangtang Block, and proposes that both slab break-off of the Neo-Tethys oceanic slab along the Bangong-Nujiang suture and intra-continental subduction of the Songpan-Garze Block beneath the Qiangtang Block caused the extensive partial melting of lithospheric mantle and subducted Songpan-Garze continental crust, which resulted in the significant elevation of the Tibetan Plateau. The volcanic rocks have LA-ICP MS U-Pb zircon age of 40.25±0.15 Ma (MSWD=2.1, 2σ), which is contemporaneous with the Eocene eclogites in the Great Himalayan and K-rich lavas in the southeastern Tibet. They display some adakitic characteristics with SiO_2 = 57.44%-68.72%, TiO_2 = 0.38%-0.81%, Na_2O = 2.89%-4.35%, K_2O = 2.77%-4.48%, Al_2O_3 = 13.92%-18.22%, A/CNK = 0.69-1.03, MgO = 0.27%-5.86% with $Mg^{\#}$ ranging from 13.2 to 72.0, strongly depleted in heavy rare earth elements (HREEs) (Yb = 0.92-1.51 ppm and Y = 10.1-24.1 ppm), in combination with their positive Sr anomalies, high Sr/Y ratios and no significant Eu anomalies, which suggest a garnet-in and plagioclase-free source residue. These volcanic rocks can be divided into high-$Mg^{\#}$(>45) and low-$Mg^{\#}$(<45) groups. Both of the two groups share evolved Sr-Nd-Pb isotopic compositions with $^{87}Sr/^{86}Sr$ = 0.707 412-0.708 284; $\varepsilon_{Nd}(t)$ = -5.7 to -4.0; $^{206}Pb/^{204}Pb$ = 18.749 9-18.818 9, $^{207}Pb/^{204}Pb$ = 15.718 9-15.738 4; $^{208}Pb/^{204}Pb$ = 39.166-39.262. The geophysical data and regional geological setting suggest that the low-$Mg^{\#}$ adakitic rocks were derived from the decompression melting of a subducted lower continental crust, when low-$Mg^{\#}$ adakitic melts in the overlying peridotite mantle wedge captured some olivine crystals, resulting in their elevated $Mg^{\#}$ and MgO values.

1 Introduction

Adakitic rocks are sodium-rich and characterized by high Sr, Sr/Y and depletion in heavy

① Published in *Gondwana Research*, 2013, 23.

rare earth element (HREE). Previous studies revealed that these rocks were formed by the partial melting of a young subducted oceanic crust (Defant and Drummond, 1990) or a thickened lower continental crust (Rapp and Watson, 1995; Petford and Atherton, 1996). However, a recent progress in continental geology and experimental petrology (Rapp and Watson, 1995; Rapp et al., 1999) has revealed that Archean TTG and igneous rocks from continental collisional orogens also display adakitic signatures (Defant et al., 2002; Condie, 2005; Zhang et al., 2006; Wang et al., 2008). Thus, the petrogenesis of adakitic rocks is a key for our understanding the continental growth, tectonic evolution of continental collision orogen in continental dynamics (Martin, 1999; Condie, 2005; Martin et al., 2005).

The Tibetan Plateau is unique phenomena on the Earth, and its composition in deep lithosphere and uplifting mechanism is a key for our understanding of the tectonic evolution of continental collision orogen and global climate changing process (Wang et al., 2010; Xia et al., 2011; Zhang and Santosh, 2011). Cenozoic potassic and high-K calc-alkaline rocks from the northern Tibetan Plateau may provide important information about these issues (Turner et al., 1996; Guo et al., 2006; Lai et al., 2007; Lai and Qin, 2008). Recent studies indicate that Cenozoic adakitic rocks from the northern Tibetan Plateau can be divisible into high-$Mg^{\#}$ and low-$Mg^{\#}$ types (Wang et al., 2008). Resolution of the petrogenesis of high-$Mg^{\#}$ and low-$Mg^{\#}$ adakitic rocks can promote our understanding of the composition of thickened lower crust, partial melting mechanism and crust/mantle interaction process of the northern Tibetan Plateau (Lai and Liu, 2001; Lai et al., 2003; Guan et al., 2011).

In this paper, we present new data of zircon U-Pb age, petrography, major and trace element and Sr-Nd-Pb isotopic composition for a suite of Eocene high-K calc-alkaline volcanic rocks from the Zuerkenwula Mountain area, northern Qiangtang block. The results indicate that adakitic rocks in this area include high-$Mg^{\#}$ and low-$Mg^{\#}$ types, both of which are meta-aluminous. We argue that the low-$Mg^{\#}$ adakitic rocks were derived from decompression melting of subducted lower continental crust, and when these low-$Mg^{\#}$ adakitic melt throughout the overlying peridotite mantle wedge, they would capture some olivine crystals and elevated their $Mg^{\#}$ and MgO values.

2 Regional geological setting and petrology

The Qiangtang Block is bounded by the Jinsa suture to the north and Bangong-Nujiang suture to the south. Structurally, Qiangtang is an anticlinorium, with the central part occupied by an anticline of pre-Jurassic strata or metamorphic rocks and the northern and southern limbs by synclines of mainly Jurassic sedimentary rocks (Fig. 1a). The Jinsha suture marks the Triassic-early Jurassic closure of the Paleotethys ocean between the Songpan-Garze and Qiangtang blocks, while the Bangong-Nujiang zone marks the Jurassic-Early Cretaceous

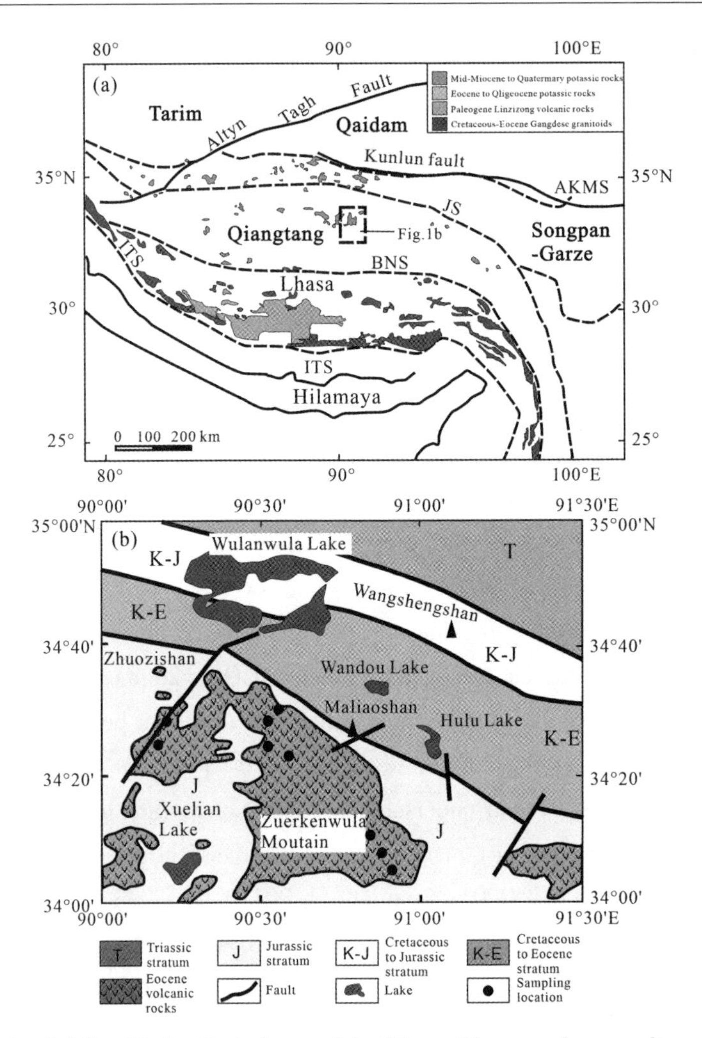

Fig. 1 (a) Simplified geological map of the Tibetan Plateau and surrounding areas.
(b) Distribution of Cenozoic volcanic rocks in the Qiangtang Block, northern Tibetan Plateau.
MCT: Main Central Thrust; STDS: South Tibet Detachment System; ITS: Indus-Tsangpo Suture;
BNS: Bangong-Nujiang Suture; JS: Jinsha Suture; AKMS: Ayimaqin-Kunlun-Muztagh Suture; KF: Kunlun Fault.
Fig. 1a modified from Chung et al. (2005).

collision between the Qiangtang and Lhasa terrane (Zhang and Santosh, 2011). The Eocene-Early Oligocene (50–29 Ma) volcanic and intrusive rocks in the Qiangtang Block include mafic, intermediate and felsic ultra-potassic, potassic and minor calc-alkaline (CA) varieties (Roger et al., 2000; Guo et al., 2006; Lai et al., 2003, 2007; Ding and Lai, 2003; Ding et al., 2007; Wang et al., 2008, 2010; Xia et al., 2011).

The Zuerkenwula Mountain is located in the northern part of the Qiangtang Block, headstream of Tuotuo River, where Cenozoic volcanic rocks cover an area of 2 500 km^2 (Lin et al., 2003, 2004). These volcanic rocks occur as lava sheets that range in thickness from 10 m to 425 m. They are unconformably overlying on the Jurassic and Cretaceous sedimentary rocks, which consisted mainly of trachyte and andesitic trachyte. Volcanic rocks in this area are mainly distributed in the Zuerkenwula Mountain, Wulanwula Mountain, Zhuozi Mountain and Xuelian Lake (Fig. 1b). Previous studies reveal that these rocks have $^{40}Ar/^{39}Ar$ ages of 39-45 Ma, belonging to Eocene Epoch (Lin et al., 2003), which is older than the volcanic rocks from the Hoh Xil and Kunlun Mountains (Fig. 1a), but younger than the Na-rich basalts from the western Qiangtang (Ding and Lai, 2003). The Tuotuohe formation, which overlies the volcanic rocks, has a paleo-magnetism ages of 46.2-51.7 Ma. Volcanic rocks in this study were sampled from the Wulanwula Mountain section.

Most of the volcanic rocks are fresh, displaying dark or gray in color, with a massive structure and porphyritic or cryptocrystalline textures. According to their petrographic features, the rocks can be divided into two types: ① Porphyritic textures of which the matrix is composed mainly of plagioclase microlite (0.05-0.2 mm), because of the later alteration, the plagioclase has changed into epidote. In addition, there are many olivine xenocrysts (0.2-0.4 mm) in these rocks (Fig. 2a,b), and other mafic minerals, i.e., amphibole (pyroxene?) underwent evaporation, and changed in to opaque tiny iron grains, though their crystal shapers still remains unchanged. And ②porphyritic textures, of the matrix is cryptocrystalline, and the phenocryst are composed mainly of andesine, which are euhedral and have a good zonary structure (Fig. 2c,d), but some andesine crystals have fractures, suggesting that they have crystallized at the earlier stage and were captured by hot magmas during the volcanic eruption.

Fig. 2 Petrography of the Eocene volcanic rocks from the Zuerkenwula Mountain area.

3 Analytical methods

Fresh samples of volcanic rocks were selected for elemental analyses. All of our zircon U-Pb dating, element and Sr-Nd-Pb isotopic analysis are performed in the State Key Laboratory of Continental Dynamics, Northwest University Major elements were determined by X-ray fluorecence spectrometry with analytical errors of less than 2%. Trace element and REE concentrations were determined by a Perkin-Elmer ELAN 6100 inductively coupled plasma source mass spectrometer (ICP-MS) of deputized solutions using a VG Plasma-Quad Excel ICP-MS, after a 2-day closed-beaker digestion using a mixture of HF and HNO_3 acids in high-pressure bombs. Pure element standard solutions were used for external calibration and BHVO-1(basalt) and SY-4 (syenite) were used as reference materials. Analytical error for most elements is less than 2% (Liu et al., 2007).

Sr-Nd-Pb isotopic data were obtained using a Nu Plasma HR mutli-collector mass spectrometer. Sr and Nd isotopic fractionations were corrected to $^{87}Sr/^{86}Sr = 0.1194$ and $^{146}Nd/^{144}Nd = 0.7219$, respectively. During the period of analysis, the NIST SRM 987 standard yielded an average value of $^{87}Sr/^{86}Sr = 0.710250 \pm 12$ (2σ, $n = 15$) and the La Jolla standard gave an average of $^{143}Nd/^{144}Nd = 0.511859 \pm 6$ (2σ, $n = 20$). Whole-rock Pb was separated by anion exchange in HCl-Br columns, and Pb isotopic fractionation was corrected to $^{205}Tl/^{203}Tl = 2.3875$. Within the period of analysis, 30 measurements of NBS981 gave average values of $^{206}Pb/^{204}Pb = 16.937 \pm 1$ (2σ), $^{207}Pb/^{204}Pb = 15.491 \pm 1$ (2σ), $^{208}Pb/^{204}Pb = 36.696 \pm 1$ (2σ), the BCR-2 standard gave $^{206}Pb/^{204}Pb = 18.742 \pm 1$ (2σ), $^{207}Pb/^{204}Pb = 15.620 \pm 1$ (2σ), $^{208}Pb/^{204}Pb = 38.705 \pm 1$ (2σ). Total procedural Pb blanks were in the range of 0.1–0.3 ng.

Laser ablation ICP-MS zircon U-Pb analyses were conducted on an Agilent 7500a ICP-MS equipped with a 193 nm laser, following the method of Yuan et al. (2004). The $^{207}Pb/^{206}Pb$ and $^{206}Pb/^{238}U$ ratios were calculated using the GLITTER program, which was corrected using the Harvard zircon 91500 as external calibrant. These correction factors were then applied to each sample to correct for both instrumental mass bias and depth-dependent elemental and isotopic fractionation. The detailed analytical technique is described in Yuan et al. (2004). Common Pb contents were therefore evaluated using the method described by Andersen (2002). The age calculations and plotting of concordia diagrams were made using ISOPLOT (version 3.0; Ludwig, 2003). The errors quoted in tables and figures are at the 2σ levels.

4 Results

4.1 Zircon LA-ICP-MS U-Pb chronology

The zircon was selected from a fine-grained sample (Z19H4), which is located in the volcanic neck and cooled slowly, and thus retained many large plagioclase crystals. Petrography observations reveal that most zircons are enclosed within the plagioclase crystals. Forty analyses were obtained for sample Z19H4, and according to the CL images (Fig. 3) and U-Pb isotopic composition, the zircons from a fine-grained diorite sample can be divided into three groups. Zircons in the first group are tiny crystals, with grain sizes ranging from 60 μm to 80 μm, and most of the crystals are gray and have good oscillatory zonings (Fig. 3). These zircons have various U (235–2 024 ppm) and Th (75–1 224 ppm) contents, with Th/U ratios ranging from 0.3 to 1.13, typical of magmatic origin (Hoskin and Schaltegger, 2003). The $^{206}Pb/^{238}U$ ages of these zircons range between 38.1±1 Ma and 41.5±0.8 Ma, yielding a weighted mean age of 40.25±0.15 Ma (MSWD=2.1, 2σ) (Fig. 4), which is interpreted as the crystallization age of the volcanic rocks from the Zuerkenwula Mountain. Zircons in the second group are bright in the CL images, have various U and Th contents, and display discordant ages. Zircons in the third group are large crystals, with lengths of 100–200 μm and good oscillatory zonings, which are clearly different from the zircons of the first group. As shown in Table 1, these zircons have Mesozoic to Paleozoic $^{206}Pb/^{238}U$ ages of 211±4 Ma to 438±7 Ma, indicating that they are xenocrysts from wall rocks or source regions.

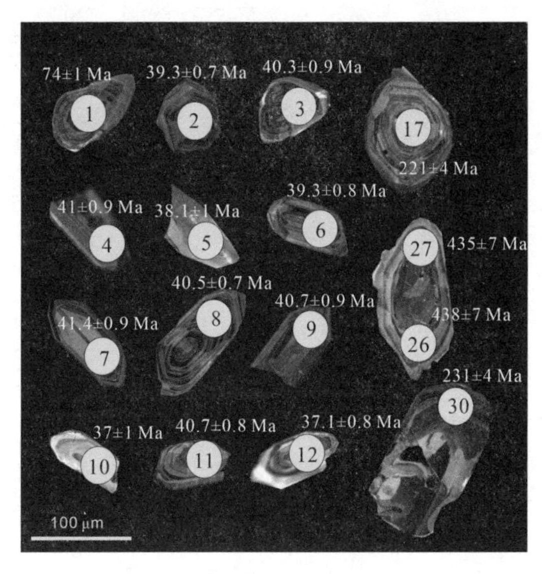

Fig. 3 Cathodoluminescence (CL) images of representative zircons from the Eocene volcanic rocks from the Zuerkenwula Mountain area.

Table 1 Zircon LA-ICP MS U-Pb isotopic analysis for the volcanic rocks in the Zuerkenwula Moutain.

	Isotopic ratios								Content			Age/Ma							
	$^{207}Pb/^{206}Pb$	1σ	$^{207}Pb/^{235}U$	1σ	$^{206}Pb/^{238}U$	1σ	$^{208}Pb/^{232}Th$	1σ	Th	U	Th/U	$^{207}Pb/^{206}Pb$	1σ	$^{207}Pb/^{235}U$	1σ	$^{206}Pb/^{238}U$	1σ	$^{208}Pb/^{232}Th$	1σ
LAI-01	0. 051 93	0. 003 90	0. 082 93	0. 006 01	0. 011 58	0. 000 23	0. 003 64	0. 000 06	327	491	0. 66	283	173	81	6	74	1	73	1
LAI-02	0. 046 05	0. 002 63	0. 038 82	0. 002 10	0. 006 11	0. 000 11	0. 00 1 97	0. 000 05	365	683	0. 53		124	39	2	39. 3	0. 7	40	1
LAI-03	0. 046 06	0. 005 06	0. 039 87	0. 004 28	0. 006 28	0. 000 14	0. 002 01	0. 000 06	1 224	1 085	1. 13	1	221	40	4	40. 3	0. 9	41	1
LAI-04	0. 046 05	0. 003 33	0. 040 50	0. 002 80	0. 006 38	0. 000 13	0. 002 04	0. 000 06	364	592	0. 62		160	40	3	41	0. 9	41	1
LAI-05	0. 046 05	0. 011 29	0. 037 67	0. 009 18	0. 005 93	0. 000 15	0. 003 56	0. 001 23	75	252	0. 30		400	38	9	38. 1	1	72	25
LAI-06	0. 050 52	0. 005 05	0. 042 58	0. 004 16	0. 006 11	0. 000 13	0. 001 93	0. 000 04	228	353	0. 64	219	228	42	4	39. 3	0. 8	38. 9	0. 7
LAI-07	0. 051 46	0. 004 41	0. 045 66	0. 003 80	0. 006 44	0. 000 14	0. 002 02	0. 000 03	304	377	0. 80	261	196	45	4	41. 4	0. 9	40. 9	0. 6
LAI-08	0. 046 05	0. 002 26	0. 040 05	0. 001 83	0. 006 31	0. 000 11	0. 002 05	0. 000 05	446	750	0. 60		105	40	2	40. 5	0. 7	41	1
LAI-09	0. 046 64	0. 003 76	0. 040 77	0. 003 07	0. 006 34	0. 000 14	0. 002 09	0. 000 08	392	614	0. 64	31	123	41	3	40. 7	0. 9	42	2
LAI-10	0. 066 01	0. 022 40	0. 052 22	0. 017 61	0. 005 74	0. 000 21	0. 001 75	0. 000 20	122	225	0. 54	807	693	52	17	37	1	35	4
LAI-11	0. 046 05	0. 003 24	0. 039 88	0. 002 68	0. 006 28	0. 000 13	0. 002 04	0. 000 06	483	678	0. 71		156	40	3	40. 4	0. 8	41	1
LAI-12	0. 046 05	0. 002 15	0. 036 68	0. 001 56	0. 005 78	0. 000 11	0. 002 04	0. 000 11	501	605	0. 83		100	37	2	37. 1	0. 7	41	2
LAI-13	0. 046 05	0. 012 12	0. 038 31	0. 010 03	0. 006 03	0. 000 17	0. 001 99	0. 000 30	140	235	0. 59		429	38	10	39	1	40	6
LAI-14	0. 050 79	0. 004 81	0. 042 91	0. 003 96	0. 006 13	0. 000 13	0. 001 93	0. 000 04	363	627	0. 58	231	217	43	4	39. 4	0. 8	39	0. 7
LAI-15	0. 055 09	0. 005 79	0. 046 52	0. 004 79	0. 006 12	0. 000 13	0. 001 91	0. 000 03	1 177	1 001	1. 18	416	240	46	5	39. 4	0. 8	38. 6	0. 6
LAI-16	0. 046 05	0. 002 47	0. 040 69	0. 002 05	0. 006 41	0. 000 12	0. 002 10	0. 000 07	484	744	0. 65		116	40	2	41. 2	0. 8	42	1
LAI-17	0. 054 5	0. 002 27	0. 262 12	0. 008 59	0. 034 88	0. 000 62	0. 011 36	0. 000 22	231	317	0. 73	392	42	236	7	221	4	228	4
LAI-18	0. 046 44	0. 002 27	0. 039 78	0. 001 81	0. 006 21	0. 000 11	0. 001 98	0. 000 04	1 061	2 024	0. 52	20	105	40	2	39. 9	0. 7	40	0. 8
LAI-19	0. 046 05	0. 002 20	0. 040 07	0. 001 77	0. 006 31	0. 000 11	0. 002 06	0. 000 05	468	742	0. 63		102	40	2	40. 6	0. 7	42	1
LAI-20	0. 050 74	0. 007 16	0. 039 49	0. 005 49	0. 005 64	0. 000 14	0. 001 78	0. 000 06	229	326	0. 70	229	296	39	5	36. 3	0. 9	36	1
LAI-21	0. 047 51	0. 002 39	0. 042 28	0. 001 82	0. 006 46	0. 000 12	0. 002 16	0. 000 05	836	1 072	0. 78	75	63	42	2	41. 5	0. 8	44	1
LAI-22	0. 049 51	0. 002 82	0. 044 03	0. 002 21	0. 006 45	0. 000 13	0. 002 16	0. 000 05	738	829	0. 89	172	79	44	2	41. 4	0. 8	44	1
LAI-23	0. 204 69	0. 011 52	0. 176 02	0. 008 87	0. 006 24	0. 000 16	0. 001 70	0. 000 04	286	503	0. 57	2 864	94	165	8	40	1	34. 3	0. 8

Continued

	Isotopic ratios								Content			Age/Ma							
	$^{207}Pb/^{206}Pb$	1σ	$^{207}Pb/^{235}U$	1σ	$^{206}Pb/^{238}U$	1σ	$^{208}Pb/^{232}Th$	1σ	Th	U	Th/U	$^{207}Pb/^{206}Pb$	1σ	$^{207}Pb/^{235}U$	1σ	$^{206}Pb/^{238}U$	1σ	$^{208}Pb/^{232}Th$	1σ
LAI-24	0. 199 02	0. 014 08	0. 173 93	0. 011 35	0. 006 34	0. 000 17	0. 001 73	0. 000 06	294	401	0. 73	2 818	119	163	10	41	1	35	1
LAI-25	0. 062 08	0. 012 92	0. 056 35	0. 011 63	0. 006 58	0. 000 18	0. 002 02	0. 000 09	203	306	0. 66	677	443	56	11	42	1	41	2
LAI-26	0. 060 6	0. 002 07	0. 582 68	0. 017 30	0. 069 73	0. 001 17	0. 021 51	0. 000 35	198	757	0. 26	625	75	466	11	435	7	430	7
LAI-27	0. 057 86	0. 002 76	0. 560 98	0. 024 80	0. 070 31	0. 001 24	0. 021 81	0. 000 35	184	441	0. 42	525	107	452	16	438	7	436	7
LAI-28	0. 052 37	0. 007 10	0. 044 52	0. 005 93	0. 006 17	0. 000 16	0. 001 94	0. 000 06	168	319	0. 53	302	300	44	6	40	1	39	1
LAI-29	0. 049 94	0. 002 41	0. 044 02	0. 001 79	0. 006 39	0. 000 12	0. 002 12	0. 000 06	296	754	0. 39	192	60	44	2	41. 1	0. 8	43	1
LAI-30	0. 053 01	0. 001 73	0. 266 09	0. 005 53	0. 036 42	0. 000 6	0. 011 73	0. 000 19	307	613	0. 50	329	22	240	4	231	4	236	4
LAI-31	0. 055 20	0. 003 34	0. 060 84	0. 003 27	0. 008 00	0. 000 16	0. 002 01	0. 000 07	652	1 143	0. 57	420	84	60	3	51	1	41	1
LAI-32	0. 047 44	0. 003 55	0. 040 31	0. 002 93	0. 006 16	0. 000 11	0. 001 96	0. 000 03	562	582	0. 97	71	167	40	3	39. 6	0. 7	39. 5	0. 7
LAI-33	0. 046 05	0. 002 73	0. 039 79	0. 002 22	0. 006 27	0. 000 12	0. 002 04	0. 000 06	438	690	0. 63		129	40	2	40. 3	0. 8	41	1
LAI-34	0. 056 32	0. 008 83	0. 043 99	0. 006 81	0. 005 66	0. 000 14	0. 001 76	0. 000 03	4876	2 711	1. 80	465	351	44	7	36. 4	0. 9	35. 6	0. 5
LAI-35	0. 051 38	0. 002 50	0. 045 82	0. 001 87	0. 006 47	0. 000 12	0. 002 12	0. 000 05	442	685	0. 64	258	60	45	2	41. 6	0. 8	43	1
LAI-36	0. 048 58	0. 002 34	0. 042 98	0. 001 74	0. 006 42	0. 000 12	0. 002 16	0. 000 05	306	608	0. 50	128	60	43	2	41. 3	0. 8	44	1
LAI-37	0. 046 79	0. 002 65	0. 040 01	0. 002 16	0. 006 20	0. 000 11	0. 001 97	0. 000 04	401	675	0. 59	38	124	40	2	39. 9	0. 7	39. 9	0. 7
LAI-38	0. 049 86	0. 004 15	0. 042 07	0. 003 40	0. 006 12	0. 000 12	0. 001 93	0. 000 03	244	440	0. 55	188	189	42	3	39. 3	0. 8	39	0. 7
LAI-39	0. 046 05	0. 002 00	0. 039 32	0. 001 57	0. 006 19	0. 000 11	0. 002 00	0. 000 04	340	605	0. 56		93	39	2	39. 8	0. 7	40. 4	0. 9
LAI-40	0. 052 59	0. 002 72	0. 240 98	0. 011 72	0. 033 24	0. 000 60	0. 010 43	0. 000 16	218	459	0. 48	311	121	219	10	211	4	210	3

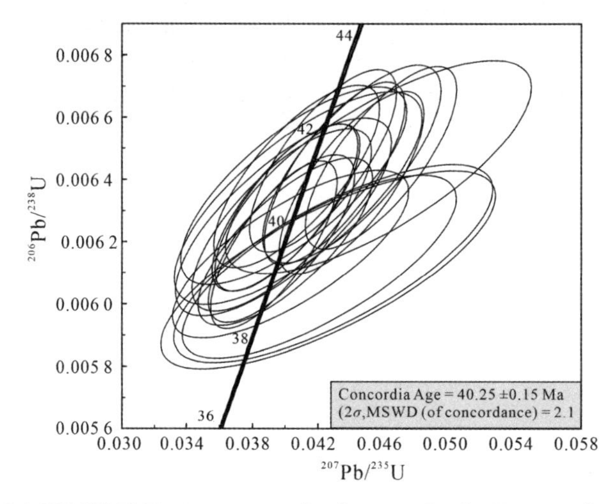

Fig. 4 LA-ICP-MS U-Pb zircon concordia diagrams for the Eocene volcanic rocks from the Zuerkenwula Mountain area.

4.2 Major element geochemistry

The result of major (wt%) and trace element (ppm) analysis of the studied volcanic rocks from the Zuerkenwula Mountain are listed in Table 2. The volcanic rocks show SiO_2 = 57.00%–68.72%, TiO_2 = 0.38%–0.81% (with exception of TiO_2 = 1.72% in one sample), Na_2O = 2.89%–4.35%, K_2O = 2.77%–4.48%, and $Na_2O + K_2O$ = 6.12%–8.16%. All these samples fall into the field of trachyte and dacite in TAS diagram (Fig. 5a), belonging to high-K calc-alkaline intermediate rocks (Fig. 5b). In addition, the volcanic rocks have high Al_2O_3 (13.92%–18.22%), with A/CNK values [$= Al_2O_3/(Na_2O + K_2O + CaO)$] of 0.69–1.03. They have MgO = 0.27%–5.86%, $Fe_2O_3{}^T$ = 2.14%–6.35%, and $Mg^{\#}$ = 13.2–72.0. Based on $Mg^{\#}$ values, the volcanic rocks can be divided into high-$Mg^{\#}$ ($Mg^{\#}$ > 45; including 23 samples) and low-$Mg^{\#}$ ($Mg^{\#}$ < 45; including 9 sample) groups. It is should be noticed that the high-$Mg^{\#}$ and low-$Mg^{\#}$ samples have similar $Fe_2O_3{}^T$ contents and different MgO contents (Table 2), indicating the existence of some Mg-rich minerals in the high-$Mg^{\#}$ samples.

Table 2 Major (wt%) and trace (ppm) element analysis result of the volcanic rocks from Zuerkenwula Mountain area.

Sample	Z02H1	Z07H	Z07H1	Z07H2	Z07H3	Z07H4	Z07H5	Z07H6	Z08H1	Z08H2	Z08H3
Lithology	Dacite	Dacite	Trachyte	Trachyte	Dacite	Dacite	Dacite	Dacite	Dacite	Dacite	Dacite
SiO_2	61.13	66.04	58.71	58.3	62.09	58.44	62.31	63.58	62.54	63.33	63.22
TiO_2	0.7	0.43	0.72	0.71	0.59	0.71	0.65	0.65	0.69	0.67	0.7
Al_2O_3	15.62	15.58	14.03	13.95	15.24	13.92	15.3	15.28	15.3	15.78	15.49
$Fe_2O_3{}^T$	4.6	3.29	5.28	5.31	5.09	5.14	4	4.85	4.18	4.61	4.44
MnO	0.06	0.08	0.08	0.08	0.08	0.08	0.06	0.05	0.06	0.03	0.05

Continued

Sample	Z02H1	Z07H	Z07H1	Z07H2	Z07H3	Z07H4	Z07H5	Z07H6	Z08H1	Z08H2	Z08H3
Lithology	Dacite	Dacite	Trachyte	Trachyte	Dacite	Dacite	Dacite	Dacite	Dacite	Dacite	Dacite
MgO	3.75	1.42	5.42	5.86	4.18	5.47	4.23	3.61	3.79	2.76	2.89
CaO	5.6	3.24	5.99	5.9	4.89	6.31	4.62	4.58	4.52	4.04	4.42
Na_2O	3.79	3.62	2.89	3.09	3.47	2.96	3.12	3.41	3.61	3.55	3.67
K_2O	3.27	3.18	3.48	3.5	3.29	3.4	3.33	3.19	3.3	3.33	3.4
P_2O_5	0.38	0.14	0.35	0.36	0.21	0.36	0.26	0.25	0.31	0.26	0.31
LOI	1.28	2.65	2.89	3.03	1.14	2.78	1.94	0.83	1.22	1.36	1.31
Total	100.18	99.67	99.84	100.09	100.27	99.57	99.82	100.28	99.52	99.72	99.72
$Mg^{\#}$	65.5	50.1	70.5	72.0	65.7	71.3	71.1	63.4	67.9	58.3	60.3
Sc	11.8	8.15	12.7	12.7	13.3	13	12.4	13.1	11.3	12.7	11.4
V	93	58.4	97.9	98	97.1	100	98.5	103	82.5	94.7	96.4
Cr	200	92.8	285	283	241	289	203	209	243	208	259
Co	76.9	74.5	69.8	66.2	69.7	66.7	57.8	75.9	58.7	48.2	47.8
Ni	115	43.8	192	200	172	185	108	135	148	121	128
Cu	15.7	8.54	28.7	28.2	30.4	28.7	24.6	16.7	30.6	24.3	25.1
Zn	82.3	61.7	77.4	70.5	73.6	133	66.5	64	61.1	67.3	70.4
Rb	100	147	126	125	108	122	140	116	115	122	117
Sr	1 397	632	1 238	1 230	1 068	1 257	998	978	1 151	1 010	1 192
Y	15.2	14.9	17.4	17.1	16.1	17.7	16.5	15.9	17.7	13.3	16.7
Zr	233	134	204	205	189	206	200	196	204	200	211
Nb	10.6	6.13	9.23	9.21	7.01	9.29	8.35	8.07	8.79	8.27	8.93
Cs	3.23	9.89	9.41	7.86	11.5	10.3	15.5	10.6	3.98	12.9	4.43
Ba	1 531	1 027	1 690	1 629	1 877	1 757	1 589	1 505	1 732	1 680	1 736
Hf	5.71	3.56	5.14	5.14	4.71	5.07	4.9	4.82	5.08	4.96	5.26
Ta	0.68	0.62	0.63	0.62	0.47	0.64	0.58	0.6	0.59	0.57	0.6
Pb	28.4	36.2	39.4	39.1	44.7	39.1	38.4	32.2	37.6	34.3	37.9
Th	22.6	11.4	22.7	22.7	21.4	22.5	22.9	22.3	24	23	24.7
U	4.67	4.53	4.38	4.36	4.43	4.36	4.55	4.25	4.58	3.82	4.68
La	75.1	24.7	68.8	68.7	50.4	69.4	59.2	58	68.2	58.4	71.1
Ce	137	46	127	127	90.8	128	107	104	123	103	124
Pr	14.3	5.05	13.6	13.6	9.55	13.6	11.1	11	13	10.7	13.4
Nd	52.9	18.6	51.6	51.6	35.1	51.4	41.3	40.8	48.9	39.3	49.9
Sm	7.57	3.36	7.94	7.83	5.51	7.9	6.27	6.29	7.36	5.93	7.54
Eu	1.84	0.89	1.92	1.9	1.42	1.9	1.55	1.54	1.82	1.56	1.88
Gd	5.76	2.82	6.15	6.09	4.5	6.15	4.94	4.98	5.79	4.59	5.84
Tb	0.63	0.42	0.71	0.7	0.55	0.7	0.6	0.6	0.68	0.55	0.68
Dy	3.04	2.25	3.42	3.35	2.83	3.38	3.01	3.02	3.37	2.66	3.35
Ho	0.53	0.5	0.62	0.61	0.56	0.61	0.56	0.55	0.6	0.48	0.58
Er	1.39	1.47	1.63	1.57	1.51	1.62	1.53	1.47	1.62	1.25	1.54
Tm	0.2	0.24	0.23	0.23	0.23	0.23	0.22	0.21	0.24	0.18	0.22

Continued

Sample	Z02H1	Z07H	Z07H1	Z07H2	Z07H3	Z07H4	Z07H5	Z07H6	Z08H1	Z08H2	Z08H3
Lithology	Dacite	Dacite	Trachyte	Trachyte	Dacite	Dacite	Dacite	Dacite	Dacite	Dacite	Dacite
Yb	1.22	1.48	1.37	1.35	1.43	1.39	1.38	1.29	1.43	1.08	1.34
Lu	0.18	0.23	0.21	0.2	0.21	0.21	0.2	0.19	0.21	0.15	0.19
$(La/Yb)_N$	44.2	12.0	36.0	36.5	25.3	35.8	30.8	32.3	34.2	38.8	38.1
Sr/Y	91.9	42.4	71.1	71.9	66.3	71.0	60.5	61.5	65.0	75.9	71.4
Y/Yb	12.5	10.1	12.7	12.7	11.3	12.7	12.0	12.3	12.4	12.3	12.5
Nb/Ta	15.6	9.9	14.7	14.9	14.9	14.5	14.4	13.5	14.9	14.5	14.9
Nb/U	2.3	1.4	2.1	2.1	1.6	2.1	1.8	1.9	1.9	2.2	1.9

Sample	Z08H4	Z08H6	Z12H3	Z15H1	Z15H2	Z15H3	Z15H5	Z15H6	Z19H1	Z19H4	Z19H5
Lithology	Dacite	Dacite	Dacite	Dacite	Dacite	Dacite	Dacite	Dacite	Dacite	Dacite	Dacite
SiO_2	63.39	62.66	59.1	61.15	61.63	61.46	65.74	65.36	63.7	65.18	65.27
TiO_2	0.72	0.68	0.72	0.53	0.53	0.55	0.45	0.44	0.48	0.4	0.4
Al_2O_3	15.96	15.38	15.71	15.22	15.27	15.36	15.34	15.18	14.91	15.1	15.13
$Fe_2O_3^T$	3.97	4.76	5.25	4.61	4.61	4.8	3.47	3.42	3.8	3.1	3.07
MnO	0.04	0.07	0.09	0.14	0.08	0.18	0.03	0.05	0.06	0.05	0.05
MgO	1.87	3.08	3	2.86	2.59	2.89	1.89	2.43	3.56	2.49	2.5
CaO	4.42	4.39	6.69	5.76	5.57	5.62	3.75	3.68	4.12	3.51	3.51
Na_2O	3.74	3.65	3.63	3.88	3.77	3.76	3.99	3.65	3.72	3.77	3.68
K_2O	3.45	3.36	3.03	3.14	3.17	3.16	3.44	3.52	3.55	3.69	3.75
P_2O_5	0.32	0.31	0.27	0.21	0.21	0.2	0.15	0.15	0.19	0.15	0.17
LOI	1.82	1.36	2.28	2.01	2.1	1.85	1.34	1.62	1.59	2.32	2.45
Total	99.7	99.7	99.77	99.51	99.53	99.83	99.59	99.5	99.68	99.76	99.98
$Mg^{\#}$	52.3	60.1	57.1	59.1	56.7	58.4	55.9	62.3	68.6	65.2	65.5
Sc	11.7	12	15.9	12	11.7	12.1	7.75	8.21	9.21	7.82	7.72
V	93.7	93.5	117	81	84.3	83.8	58.1	63.5	67.5	58.4	60.1
Cr	262	245	239	97.7	111	103	60.1	65.6	122	67.6	69.2
Co	50.9	53.4	42.2	42.5	38.1	43.9	45.7	60.5	56.4	63.6	45.7
Ni	105	137	110	76.8	75.6	80.3	41.2	41.1	75.3	41.6	43.1
Cu	24.2	23.1	11.9	23.4	23.2	24.2	13.4	12	12.9	12.1	15.2
Zn	131	153	60.2	59.5	56.9	60	45.3	58.7	90.4	50.5	54.2
Rb	121	113	110	121	117	118	149	152	141	150	153
Sr	1 218	1 173	914	1 317	1 269	1 308	717	650	891	724	728
Y	16.4	19.1	16	17.7	16.4	17.3	10.6	12	13.6	11	11
Zr	216	207	177	153	154	160	146	152	160	137	138
Nb	9.15	8.83	8.33	7.43	7.35	7.5	6.35	6.25	6.87	6.04	6.12
Cs	4.61	4.1	4.82	4.6	4.25	4.57	5.32	8.07	7.42	8.42	8.58
Ba	1 748	1 695	1 249	1 755	1 640	1 815	1 247	1 158	1 303	1 160	1 161
Hf	5.47	5.25	4.48	4.04	4.08	4.17	3.92	3.99	4.15	3.69	3.66
Ta	0.62	0.58	0.58	0.6	0.6	0.58	0.56	0.59	0.59	0.57	0.54
Pb	38.7	37.8	30.3	39	36.2	39.4	32	38.6	37.6	38.8	39

Continued

Sample	Z08H4	Z08H6	Z12H3	Z15H1	Z15H2	Z15H3	Z15H5	Z15H6	Z19H1	Z19H4	Z19H5
Lithology	Dacite	Dacite	Dacite	Dacite	Dacite	Dacite	Dacite	Dacite	Dacite	Dacite	Dacite
Th	25.5	24.5	16.8	19.2	18.7	19.1	15.2	15.2	15.9	15	14.9
U	4.37	4.63	4.4	5	5	5.07	2.96	5.11	5.32	5.44	5.4
La	73.7	70.3	45.3	56.9	55.6	56.1	34.6	33.3	41.4	30.4	30.7
Ce	128	123	83.6	106	101	104	60.5	58	75.9	53.6	54.2
Pr	13.8	13.1	9.03	11.5	11.1	11.3	6.46	6.25	8.26	5.85	5.9
Nd	51.6	49	34.5	44.8	42.4	43.6	23.2	22.8	30.9	21.5	21.4
Sm	7.74	7.35	5.71	7	6.56	6.69	3.76	3.9	4.94	3.61	3.57
Eu	1.94	1.8	1.51	1.71	1.64	1.68	1.03	0.98	1.3	0.97	0.94
Gd	6.02	5.7	4.63	5.22	5.02	5.23	2.87	3.12	3.88	2.93	2.84
Tb	0.7	0.68	0.6	0.63	0.6	0.62	0.37	0.42	0.49	0.37	0.38
Dy	3.35	3.25	3.06	3.19	2.98	3.14	1.78	2.08	2.47	1.88	1.84
Ho	0.58	0.58	0.56	0.6	0.56	0.58	0.35	0.41	0.46	0.37	0.37
Er	1.51	1.54	1.52	1.65	1.53	1.6	0.94	1.08	1.22	0.98	0.94
Tm	0.21	0.22	0.22	0.25	0.23	0.24	0.15	0.17	0.19	0.15	0.15
Yb	1.25	1.32	1.34	1.51	1.41	1.47	0.92	1.03	1.14	0.93	0.93
Lu	0.18	0.19	0.2	0.23	0.21	0.22	0.14	0.15	0.17	0.14	0.14
$(La/Yb)_N$	42.3	38.2	24.2	27.0	28.3	27.4	27.0	23.2	26.0	23.4	23.7
Sr/Y	74.3	61.4	57.1	74.4	77.4	75.6	67.6	54.2	65.5	65.8	66.2
Y/Yb	13.1	14.5	11.9	11.7	11.6	11.8	11.5	11.7	11.9	11.8	11.8
Nb/Ta	14.8	15.2	14.4	12.4	12.3	12.9	11.3	10.6	11.6	10.6	11.3
Nb/U	2.1	1.9	1.9	1.5	1.5	1.5	2.1	1.2	1.3	1.1	1.1

Sample	Z19H6	Z06H	Z08H5	Z10H	Z11H1	Z11H2	Z12H1	Z12H2	Z15H7	Z15H11
Lithology	Dacite	Trachyte	Dacite	Dacite	Trachyte	Dacite	Dacite	Dacite	Dacite	Dacite
SiO_2	64.99	57.44	64.69	60.78	59.92	60.12	60.24	62.94	68.72	61.09
TiO_2	0.39	0.76	0.72	0.8	0.81	0.78	0.73	0.78	0.41	1.72
Al_2O_3	14.95	18.22	16.23	17.79	17.54	16.88	15.82	17.09	16.22	14.89
$Fe_2O_3^T$	3.07	6.05	3.21	5.23	5.56	6.35	6.01	3.07	2.14	5.7
MnO	0.05	0.04	0.02	0.01	0.01	0.01	0.06	0.06	0.01	0.05
MgO	2.51	0.85	1.05	0.34	0.48	0.99	2	1.06	0.27	1.2
CaO	3.51	5.02	4.09	4.74	4.7	4.8	6.02	5.28	3.2	4.51
Na_2O	3.59	3.35	3.86	4.11	4.09	3.88	3.66	4.12	4.35	3.68
K_2O	3.76	2.77	3.55	3.41	3.39	3.14	2.98	3.24	3.3	4.48
P_2O_5	0.18	0.3	0.33	0.29	0.3	0.3	0.27	0.29	0.13	0.68
LOI	2.54	4.72	1.79	2.16	2.76	2.29	2.08	1.62	1.06	1.51
Total	99.54	99.52	99.54	99.66	99.56	99.54	99.87	99.55	99.81	99.51
$Mg^{\#}$	65.6	24.7	43.3	13.2	16.7	26.7	43.7	44.6	22.7	32.9
Sc	7.96	11.3	9.38	9.11	10.5	14.2	16	11	5.64	7.35
V	61	69.8	92.1	96.6	116	119	117	105	47	74
Cr	71.4	218	278	248	226	255	241	258	45.7	9.81

Continued

Sample	Z19H6	Z06H	Z08H5	Z10H	Z11H1	Z11H2	Z12H1	Z12H2	Z15H7	Z15H11
Lithology	Dacite	Trachyte	Dacite	Dacite	Trachyte	Dacite	Dacite	Dacite	Dacite	Dacite
Co	46. 5	23. 1	42. 4	32. 8	29	37. 2	38. 8	44	43. 3	30. 6
Ni	42. 9	88. 8	81	76. 5	115	148	125	78. 5	18. 5	5. 47
Cu	14	28. 7	22. 9	23. 2	29. 9	25. 1	17. 6	25. 6	6. 81	12. 9
Zn	75. 5	53. 2	72. 8	40	64	56	71. 3	105	37. 2	124
Rb	153	91. 7	118	126	124	116	110	117	149	175
Sr	716	1 081	1 228	985	963	971	927	972	602	681
Y	11. 1	17. 2	16. 8	13. 9	14. 4	16	17. 1	14. 6	10. 1	24. 1
Zr	144	213	217	190	191	189	179	184	129	629
Nb	6. 19	9. 52	9. 25	9. 2	9. 09	8. 9	8. 31	8. 81	5. 91	41. 7
Cs	8. 55	17. 4	3. 99	8. 6	5. 09	4. 85	4. 71	9. 09	6. 9	3. 19
Ba	1 172	1 623	1 665	1 418	1 455	1 533	1 280	1 358	1 083	1 784
Hf	3. 79	5. 47	5. 5	4. 85	4. 86	4. 85	4. 51	4. 73	3. 4	13. 6
Ta	0. 56	0. 64	0. 63	0. 66	0. 63	0. 61	0. 59	0. 63	0. 59	2. 26
Pb	38. 8	40	39. 8	34. 2	31. 9	33. 4	28. 4	29. 8	34. 2	36. 3
Th	14. 9	27. 3	25. 9	19. 1	18. 9	18	17	17. 7	11	33. 9
U	5. 41	3. 69	4. 36	4. 53	4. 2	3. 8	3. 83	3. 1	4. 65	4. 14
La	30. 6	63. 8	76. 5	49. 3	49	49. 2	46. 2	48. 5	24. 3	146
Ce	54. 3	118	132	90. 4	88. 1	89. 9	85	87. 9	43. 3	264
Pr	5. 9	12. 1	14. 3	9. 62	9. 47	9. 74	9. 26	9. 51	4. 68	28. 2
Nd	21. 7	45. 1	54	36. 2	35. 4	37. 1	35. 4	36. 1	17. 1	103
Sm	3. 68	6. 92	8. 09	5. 8	5. 63	6. 13	5. 84	5. 76	3. 1	14. 3
Eu	0. 96	1. 59	1. 97	1. 57	1. 57	1. 66	1. 55	1. 62	0. 89	2. 96
Gd	2. 94	5. 45	6. 32	4. 59	4. 49	4. 92	4. 83	4. 63	2. 5	11. 1
Tb	0. 38	0. 65	0. 73	0. 57	0. 57	0. 61	0. 61	0. 57	0. 35	1. 23
Dy	1. 92	3. 23	3. 5	2. 75	2. 81	3. 16	3. 25	2. 82	1. 76	5. 35
Ho	0. 37	0. 6	0. 61	0. 5	0. 5	0. 57	0. 61	0. 52	0. 35	0. 85
Er	0. 95	1. 64	1. 54	1. 28	1. 32	1. 52	1. 67	1. 32	0. 91	2. 04
Tm	0. 15	0. 24	0. 22	0. 18	0. 19	0. 21	0. 24	0. 2	0. 15	0. 27
Yb	0. 94	1. 41	1. 28	1. 05	1. 13	1. 23	1. 5	1. 1	0. 92	1. 47
Lu	0. 14	0. 21	0. 18	0. 15	0. 16	0. 18	0. 22	0. 16	0. 13	0. 2
$(La/Yb)_N$	23. 4	32. 5	42. 9	33. 7	31. 1	28. 7	22. 1	31. 6	18. 9	71. 2
Sr/Y	64. 5	62. 8	73. 1	70. 9	66. 9	60. 7	54. 2	66. 6	59. 6	28. 3
Y/Yb	11. 8	12. 2	13. 1	13. 2	12. 7	13. 0	11. 4	13. 3	11. 0	16. 4
Nb/Ta	11. 1	14. 9	14. 7	13. 9	14. 4	14. 6	14. 1	14. 0	10. 0	18. 5
Nb/U	1. 1	2. 6	2. 1	2. 0	2. 2	2. 3	2. 2	2. 8	1. 3	10. 1

In the Harker diagram, high-$Mg^{\#}$ and low-$Mg^{\#}$ volcanic rocks display a similar evolutionary trend, but the low-$Mg^{\#}$ volcanic rocks have higher TiO_2 content and plot in the field of adakites from a delaminated lower crust in the SiO_2-TiO_2 diagram (Fig. 6a). The high-$Mg^{\#}$ and low-$Mg^{\#}$ volcanic rocks have similar P_2O_5 (Fig. 6b), Al_2O_3 (Fig. 3), Yb (Fig. 6d) and Th/Ce ratios (Fig. 6f).

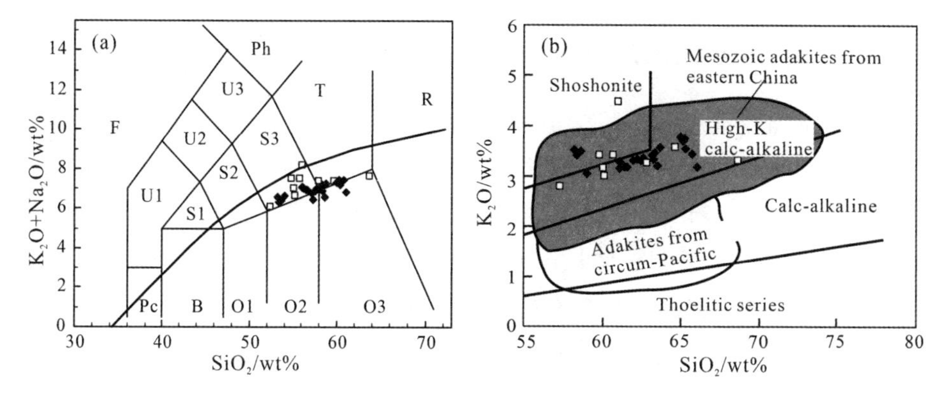

Fig. 5 (a) TAS (Peccerillo and Taylor, 1977) and (b) SiO_2-K_2O (Wu et al., 2002) diagrams for the volcanic rocks from the Zuerkenwula Mountain area.

F: foidite. U1: tephrite (Ol< 10%), basanite (Ol> 10%). U2: phonotephrite. U3: tephriphonolite. Ph: phonolite. S1: trachybasalt. S2: basaltic trachyandesite. S3: trachyandesite. T: trachyte ($q < 20\%$), trachydacite ($q > 20\%$). Pc: picrobasalt. B: basalt. O1: basaltic andesite. O2: andesite. O3: dacite. R: rhyolite. ◆high-Mg# volcanic rocks; □low-Mg# volcanic rocks. Fields in the total alkaline-silica diagram are from Le Bas and Streckeisen (1991).

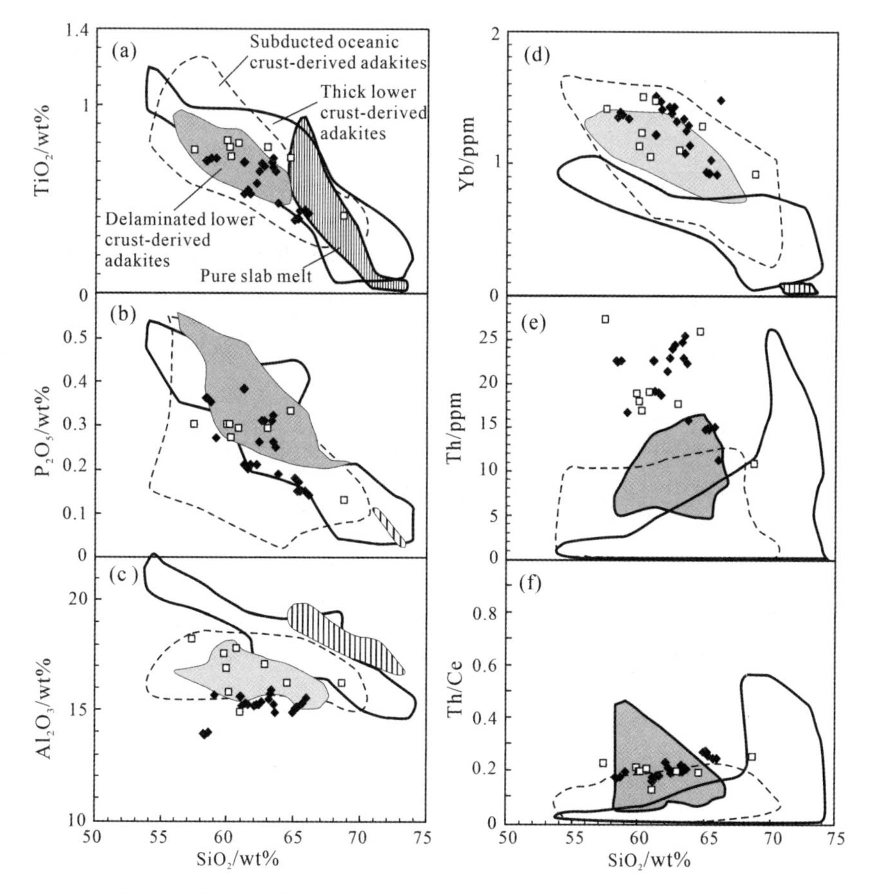

Fig. 6 Harker diagrams for the volcanic rocks from the Zuerkenwula Mountain area.

◆high-Mg# volcanic rocks; □low-Mg# volcanic rocks.

4.3 Trace and rare earth element geochemistry

The high-$Mg^{\#}$ and low-$Mg^{\#}$ rocks display similar trace element patterns in the primitive mantle normalized trace spider diagram (Fig. 7), and both of them are enriched in LILEs and depleted in Ti, Nb, Ta, Zr and Hf. The high-$Mg^{\#}$ samples have Sr = 632–1 397 ppm, Ba = 1 027–1 877 ppm, Y = 10.6–19.1 ppm, and Sr/Y ratios of 42–91 with a mean value of 68, whereas the low-$Mg^{\#}$ samples have Sr = 602–1 228 ppm, Ba = 1 083–1 784 ppm, Y = 10.1–24.1 ppm, and Sr/Y ratios of 28–73 with a mean value of 60.

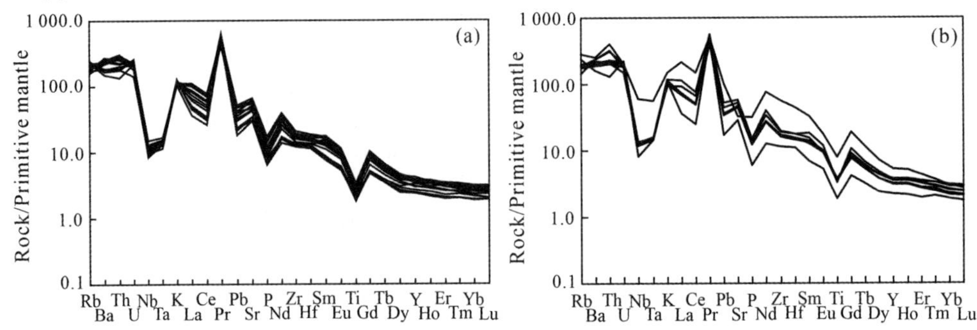

Fig. 7 Primitive mantle normalized trace element spider diagram for the high-$Mg^{\#}$(a) and low-$Mg^{\#}$ (b) adakitic volcanic rocks from the Zuerkenwula Mountain. Normalized value was cited from Sun and McDonough (1989).

The high-$Mg^{\#}$ and low-$Mg^{\#}$ samples display similar REE patterns in the chondrite-normalized REE diagrams (Fig. 8), and both of them are enriched in LREE, but depleted in HREE, without significant Eu anomalies. The high-$Mg^{\#}$ volcanic rocks have $(La/Yb)_N$ = 11–44 with a mean value of 30, Eu/Eu^* = 0.83–0.95, and Yb = 0.92–1.51 ppm (< 1.8 ppm), whereas the low-$Mg^{\#}$ volcanic rocks have $(La/Yb)_N$ = 18–71 with a mean value of 34, Eu/Eu^* = 0.72–0.98, and Yb = 0.92–1.50 ppm (< 1.8 ppm).

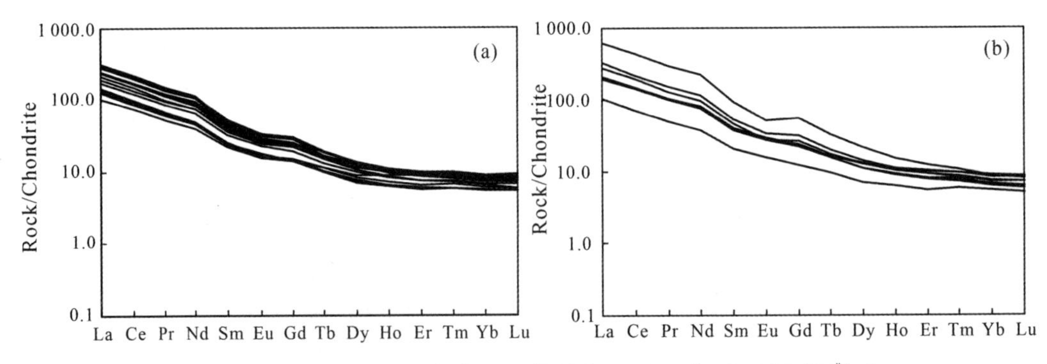

Fig. 8 Chrondrite rare earth element (REE) patterns for the high-$Mg^{\#}$(a) and low-$Mg^{\#}$(b) adakitic volcanic rocks from the Zuerkenwula Mountain. Normalized value was cited from Sun and McDonough (1989).

4.4 Sr-Nd-Pb isotopic composition

Based on the major- and trace-element geochemistry, we selected for three low-Mg# sample (Z08H5, Z12H2 and Z15H7) and five high-Mg# volcanic rock samples (Z07H2, Z07H3, Z07H8, Z08H1 and Z08H6) for Sr-Nd-Pb isotopic analysis (Table 2).

In the $^{87}Sr/^{86}Sr$-$^{143}Nd/^{144}Nd$ (Fig. 9) and $^{206}Pb/^{204}Pb$-$^{207}Pb/^{204}Pb$ (Fig. 10) diagram, both the high-Mg# and low-Mg# volcanic rocks from Zuerkenwula Mountain display similar Sr-Nd-Pb isotopic compositions, and they also share similar Sr-Nd-Pb isotopic compositions with the data of Lin et al. (2004). In the $^{87}Sr/^{86}Sr$-$^{143}Nd/^{144}Nd$ diagram, the volcanic rocks from Zuerkenwula Mountain fall into the field of the Linzizong and Qiangtang Cenozoic volcanic rocks. The volcanic rocks also display similar Pb isotopic compositions to those of the high-K calc-alkaline and potassic rocks from northern Tibetan Plateau and display the feature of EM Ⅱ-type mantle.

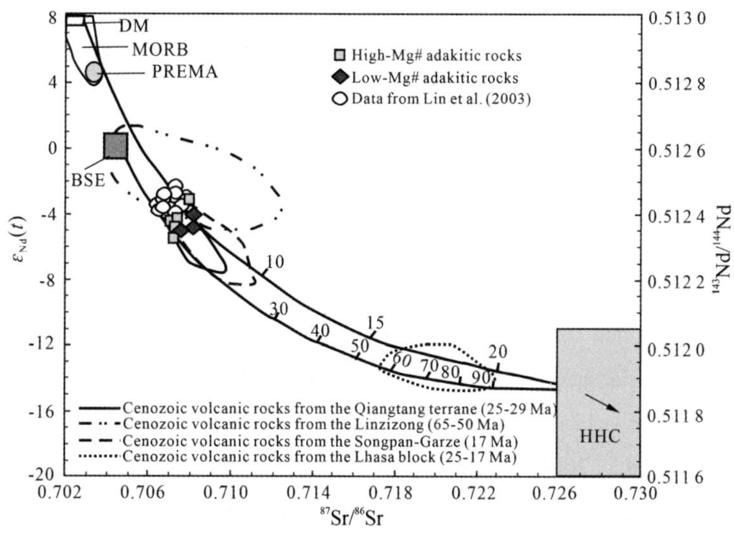

Fig. 9 A $^{143}Nd/^{144}Nd$-$^{87}Sr/^{86}Sr$ diagram for the volcanic rocks from Zuerkenwula Mountain area. The data of high Himalayan basement (HHC) cited from Harrison et al. (1995), Parrish and Hodges (1996), Robinson et al. (2001), Yang (2002), Ding and Lai (2003); data of MORB, DM, PREMA and BSE were cited from Zinder and Hart (1986); field for volcanic rocks from other area are based on Ding et al (2003) and references therein.

5 Discussion

5.1 Origin of the adakitic volcanic rocks from the Zuerkenwula Mountain

The volcanic rocks from the Zuerkenwula Mountain display typical features of continental crust-derived magma with enriched in LILEs and depletion in Nb, Ta and Ti (Wilson, 1989). They are strongly depleted in HREE (Yb = 0.92 - 1.51 ppm, < 1.8 ppm), Y (10.1 - 19.1

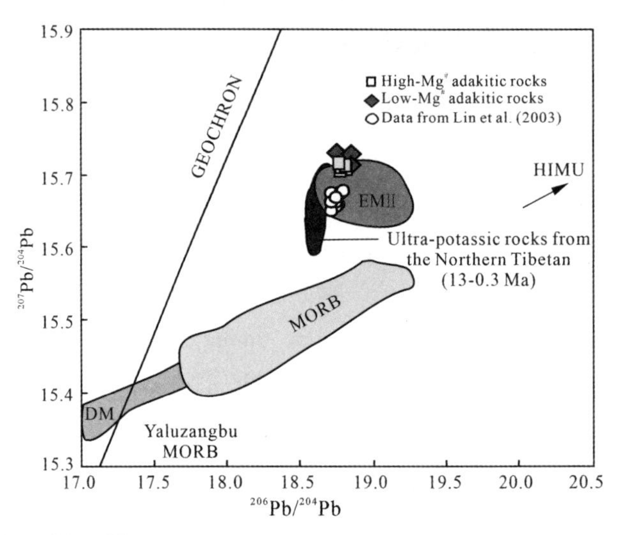

Fig. 10 A $^{207}Pb/^{204}Pb$-$^{206}Pb/^{204}Pb$ diagram for the volcanic rocks from the Zuerkenwula Mountain area. The mantle sources of MORB and the mantle end-members DM, PREMA and BSE are from Zindler and Hart (1986), fields for ultra-potassic rocks from northern Tibetan (13–0.3 Ma) are from Guo et al. (2006).

ppm) and enriched in Sr (>600 ppm), and in the primitive mantle normalized trace element spider diagram, they display weak positive Sr anomalies. In combination with their high Sr/Y and La/Yb ratios, as well as no significant Eu anomalies, these features are consistent with the definition of adakite (Defant and Drummond, 1990). In the Sr/Y-Y and $(La/Yb)_N$-Yb_N diagrams (Fig. 11a, b), both high-$Mg^{\#}$ and low-$Mg^{\#}$ samples are plotted in the field of typical adakite. Their high Ba and Sr contents and no significant Eu anomalies, as well as extremely depleted HREE and Y patterns, indicate garnet (Castillo, 2008; Yu et al., 2012) residue and the absence of plagioclase in their source region, which suggest that the crust was thickened up to 50 km (Xiong et al., 2005). Experimental petrology suggests that partial melts from a basaltic lower crust generally have low $Mg^{\#}(<45)$ (Rapp et al., 1999), and in the SiO_2-$Mg^{\#}$ (Fig. 12a) and SiO_2-MgO (Fig. 12b) diagrams, the high-$Mg^{\#}$ and low-$Mg^{\#}$ volcanic rocks from the studied area mainly plot in the field of adakites from a delaminated lower crust and the field of adakites from a thickened lower crust, respectively. Present petrogentic models for adakitic rocks include: ① partial melts from a young and hot subducted oceanic crust (Defant and Drummond, 1990); ② partial melting of a basaltic lower crust that was underplated by mantle materials (Petford and Atherton, 1996; Castillo et al., 1999; Prouteau and Scaillet, 2003); and ③ partial melting of a thickened (Petford and Atherton, 1996; Ernst, 2010) or delaminated lower crust (Xu et al., 2002; Gao et al., 2004).

The Zuerkewula Mountain is located in the northern part of Qiangtang Block, which is bounded by the Jinshajiang suture and the Bangong-Nujiang suture in the north and south, respectively. The rocks have a LA-ICP MS U-Pb zircon age of 40.25±0.15 Ma, belonging to

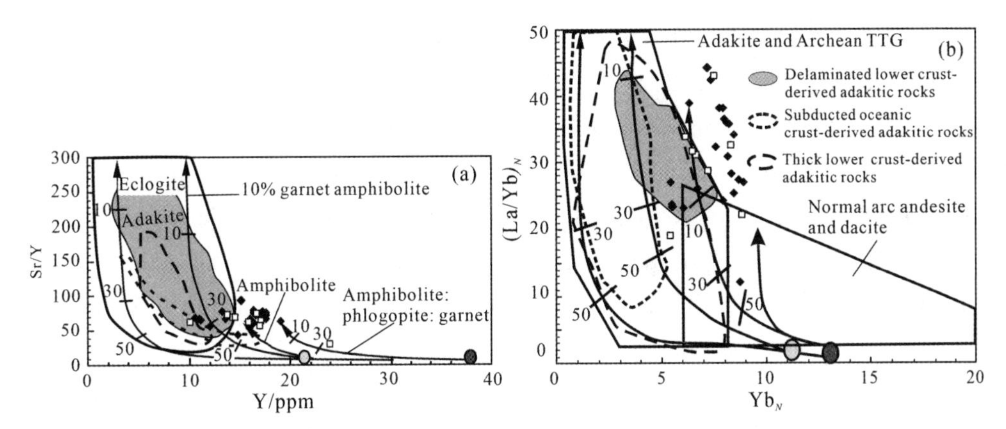

Fig. 11 Sr/Y-Y (a) and $(La/Yb)_N$-Yb_N (b) diagrams for the volcanic rocks from the Zuerkenwula Mountain area (Defant and Drummond, 1990; Castillo, 2008).

◆high-Mg# volcanic rocks; □low-Mg# volcanic rocks. The adakite fields were cited from Martin (1999), the three partial melt curves were cited from Castillo (2008).

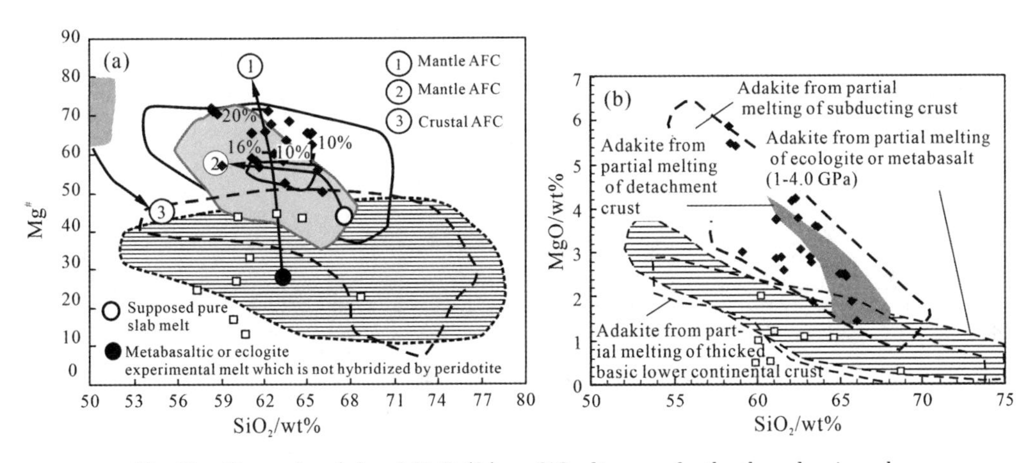

Fig. 12 Mg-number (a) and MgO (b) vs. SiO_2 diagrams for the the volcanic rocks from the Zuerkenwula Mountain area.

◆high-Mg# volcanic rocks; □low-Mg# volcanic rocks. Mantle AFC curves are after Stern and Killian (1996) (Curve 1) and Rapp et al. (1999) (Curve 2); the proportion of assimilated peridotite is also shown. The crustal AFC curve is after Stern and Killian (1996) (Curve 3). The starting point of Curve 1 represents the composition of a pure slab melt, which is supposed by Stern and Killian (1996). The starting point of Curve 2 represents the composition of a metabasaltic or eclogite experimental melt, which is not hybridized with peridotite (Rapp et al., 1999). Fields of delaminated lower crust-derived adakitic rocks, subducted oceanic crust-derived adakites and thick lower crust-derived adakitic rocks are constructed using the same data sources as those in Fig. 4. The field of metabasaltic and eclogite experimental melts (1-4.0 GPa) is from the following: Rapp et al. (1999); Rapp and Watson (1995); Prouteau et al. (2003); and references therein. The field of metabasaltic and eclogite experimental melts hybridized with peridotite is after Rapp et al. (1999).

Middle-Eocene, which is obviously younger than the Linzizong volcanic rocks (65–50 Ma) (Fig. 1a) that is considered to be related to the Neo-Tethys subduction (Deng, 1998). In addition, these volcanic rocks have more evolved Sr-Nd-Pb isotopic compositions (Figs. 9 and

10) and higher K_2O (Fig. 5b) contents than those of the subduction-related adakites. Geophysical data (Tilmann and Ni, 2003; Zhang and Santosh, 2011) indicate that the northern edge of the Neo-Tethys oceanic lithosphere only reaches to the southern margin of the Bangong-Nujiang suture. Thus, it can be concluded that the adakitic volcanic rocks from the Zuerkenwula Mountain cannot be resulted from the partial melting of a subducted oceanic crust.

Castillo et al. (1999) proposed that the assimilation-fractional crystallization (AFC) of basaltic magmas can produce adakitic rocks in the Camiguin Island, Philippines.The La/Sm-La discrimination diagram can be effectively used to identify the genesis for volcanic rocks (Allergre and Minster, 1978). As shown in the La/Sm-La diagram (Fig. 13a), andesite and dacite associations possess the partial melting trend, while the Zr/Sm-Zr diagram display a similar trend. In addition, potassic mafic rocks in the northern Tibetan Plateau have ages of 20 Ma (Turner et al., 1996; Miller et al., 1999; Guo et al., 2006), and fractional crystallization of these rocks cannot account for the Eocene adakitic rocks in the Zuerkenwula Mountain.

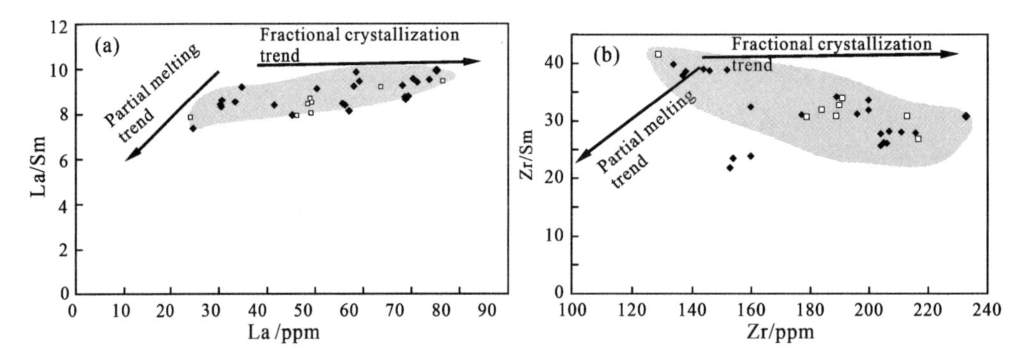

Fig. 13 La-La/Sm (a) and Zr-Zr/Sm (b) diagrams for the volcanic rocks from the Zuerkenwula Mountain area (Allegre and Minster, 1978).
◆high-Mg# volcanic rocks; □low-Mg# volcanic rocks.

The evolved Sr-Nd-Pb isotopic composition [$\varepsilon_{Nd}(t)=-5.1$ to -3.5, whole-rock Nd two-stage model ages of 0.93–1.11Ga] as well as high K_2O contents of the volcanic rocks indicate a lower continental crustal origin. Geophysical evidence indicates that there is no an enriched lithospheric mantle and a mafic lower crust beneath the northern Qiangtang block (Kapp et al., 2000). Further geophysical data suggest that the lower crust beneath the Qiangtang Block consists of meta-sedimentary rocks (Galve et al., 2006; Li et al., 2012). It should be noted that the partial melting of meta-sedimentary rocks could not form the high-Mg# andesitic to dacitic adakitic magma in the Zuerkenwula Mountain area.

5.2 Petrogenesis of the high-Mg# and low-Mg# adakitic rocks

Wang et al. (2008) reported Eocene high-Mg# and low-Mg# adakitic rocks in the Duogecuoren area, west to the Zuerkenwula Mountain, and these rocks have relative higher

SiO_2(66.82%-72.94%) and similar Al_2O_3(Al_2O_3 = 14.94% - 15.32%, A/CNK = 1.02-1.20), belonging to aluminous rhyolite-dacite. They proposed that these rocks were derived from the partial melting of a thickened sedimentary-bearing lower continental crust with eclogite residues. However, different from the low-$Mg^{\#}$ aluminous adaktic rocks in the Duogecuoren area, the low-Mg adakitic rocks from the Zuerkenwula Mountain area display lower SiO_2(mostly ranging from 57.44% to 64.69%), Al_2O_3(Al_2O_3=14.89%-18.22%, and A/CNK=0.78-1.03) and Sr contents, indicating that the partial melting of a basaltic lower crust may be more plausible for their origin. In addition, the low-$Mg^{\#}$ adakitic rocks have geochemical features that were consistent with the melts that were derived from the partial melting of basaltic rocks at high temperature (>1 050 ℃) and high pressure (>1.2 GPa) (Petford and Atherton, 1996). The mafic xenlithos from the Cenozoic volcanic rocks have more evolved Sr-Nd-Pb isotopic composition than those of the adakitic rocks in the Zuerkenwula Mountain, which may be caused by fluids metasomatism (Lai and Qin, 2008), and their similar Th content and Th/La ratios suggest that these mafic xenoliths can represent the lower crust beneath the Qiangtang Block. The low-$Mg^{\#}$ adakitic rocks may be derived from the partial melting of a thickened lower continental crust in the Qiangtang block.

Experimental work (Rapp et al., 1999) indicates that when the melts interacted with between mantle peridotite, their $Mg^{\#}$ can be elevated significantly. The high-$Mg^{\#}$ adakitic rocks from the Zuerkenwula Mountain also display low SiO_2(58.30%-65.74%) and Al_2O_3(Al_2O_3=13.92%-15.96%, A/CNK=0.75-1.05), which is consistent with the high-$Mg^{\#}$ andesite in an island arc setting (Martin, 1999). They also have similar Sr/Y, La/Yb and Dy/Yb ratios with the low-$Mg^{\#}$ adakitic rocks in the studied area. Thus, it is considered that the high-$Mg^{\#}$ adakitic rocks may be resulted from an interaction between low-$Mg^{\#}$ adakitic melts and mantle peridotite, and in this process, the $Mg^{\#}$ can be elevated significantly, while the Sr/Y, La/Yb ratios remain changed (Rapp et al., 1999). Various dynamic models have been proposed to explain the mechanism of interaction between crust-derived melts and mantle peridotite, including: ① adakitic melts from subducted oceanic crust contaminated by mantle peridotite during emplacement in an island arc setting (Rapp et al., 1999; Tatsumi et al., 2003); ② adakitic melts from a subducted continental crust interacted with mantle peridotite during emplacement in an intra-continental subduction setting (Wang et al., 2008); ③ delaminated lower crust foundered in to asthenosphere mantle, adakitic melts from the delaminated lower crust interacted with mantle peridotite (Xu et al., 2002; Gao et al., 2004); and ④ basaltic magmas underplated beneath the lower crust and magma mixing between adakitic melts from thickened lower crust and mantle-derived mafic magma cause the elevated $Mg^{\#}$ (Qin et al., 2010).

Geophysical data have suggested that the subduction of an oceanic crust cannot form the adakitic rocks in the Zuerkenwula Mountain area. However, the other three mechanisms can account for the genesis of the high-Mg# adakitic rocks in the Zuerkenwula Mountain area. Therefore, it still needs further constraints from the regional geology and geophysics on the dynamic models for the petrogenesis of the high-Mg# adakitic rocks from the Zuerkenwula Mountain area.

5.3 Dynamic implications of the adakitic rocks

The Eocene volcanic and intrusive rocks in the central Qiangtang Block are considered as a result of slab break-off of the Neo-Tethys oceanic slab (Chung et al., 2005; Zhang and Santosh, 2011) or the intra-continental subduction of the Songpan-Garze Block beneath the Qiangtang Block (Roger et al., 2000; Wang et al., 2010). New geophysical data indicate that lithosphere mantle beneath the Qiangtang Block is hot and its velocity is low, which is obviously different from those of the Lhasa terrane. Based on the detailed study of blueschists in the central Qiangtang, Kapp et al. (2000) proposed that the Late-Mesozoic southward subduction of the Jinshajiang oceanic lithosphere made the lower crust beneath the Qiangtang Block with hydrated basaltic components after Late-Mesozoic time. However, present geophysical data indicate the absence of cold, high-velocity crust beneath the Qiangtang Block, and it may be composed by hot felsic components (Galve et al., 2006), suggesting that the lower crust beneath the Qiangtang Block had undergone extensive partial melting. Cenozoic volcanism and tectonic process were closely related to this particular crust-mantle structure in northern Tibetan Plateau. According to the INDEPTH-Ⅲ data, Tailmann and Ni (2003) find a cold lithosphere beneath 100 km to 400 km depth, south to the Bangong-Nujiang suture, and thus they argued that this cold lithosphere represented the northern edge of the subducted Neo-Tethys oceanic lithosphere.

Both the low-Mg# and high-Mg# samples are metaluminous, and they have evolved Sr-Nd-Pb isotopic compositions (Table 3), low SiO_2 contents an and high K, Sr contents (Table 1), clearly indicating that they were derived from basaltic lower continental crust, and the small scale partial melting of sedimentary or basaltic continental crust during continental subduction (Wang et al., 2008) could not account for the formation of the large scale Eocene adakitic rocks in the Qiangtang Block. Moreover, the existence of some olivine xenocrysts in the high-Mg# samples (Fig. 2) clearly suggests that the primary adakitic melts have captured some olivine crystals during their ascent process. Wang et al. (2010) reported Eocene (47–38 Ma) high-Mg# adakitic andesitic porphyry and diabase dike in the central Qiangtang Block, and they argued that the occurrence of these mafic rocks indicates lithosheric mantle partial melting and east-west extension and uplift in the central Qiangtang Block, which also suggests that the lithospheric mantle beneath the Qiangtang Block underwent extension during Eocene time.

Table 3 Sr-Nd-Pb isotopic composition of the Eocene volcanic rocks from the Zuerkenwula mountain.

Sample	Z07H2	Z08H5	Z12H2	Z15H7	Z07H3	Z07H6	Z08H1	Z08H6
Lithology	Trachyte andesite	Trachyte	Trachyte	Trachyte	Trachyte	Trachyte	Trachyte	Trachyte
U	4. 36	4. 36	3. 1	4. 65	4. 43	4. 25	4. 58	4. 63
Th	22. 7	25. 9	17. 7	11	21. 4	22. 3	24	24. 5
Pb	39. 1	39. 8	29. 8	34. 2	44. 7	32. 2	37. 6	37. 8
$^{206}Pb/^{204}Pb$	18. 749 9 ± 10	18. 756 9 ± 6	18. 841 7 ± 6	18. 818 9 ± 7	18. 780 2 ± 7	18. 777 7 ± 7	18. 762 0 ± 6	18. 760 6 ± 10
$^{207}Pb/^{204}Pb$	15. 719 7 ± 10	15. 718 9 ± 6	15. 729 0 ± 5	15. 731 4 ± 6	15. 719 1 ± 7	15. 709 7 ± 6	15. 710 9 ± 5	15. 710 7 ± 10
$^{208}Pb/^{204}Pb$	39. 166 ± 3	39. 178 ± 2	39. 262 ± 1	39. 183 ± 2	39. 205 ± 2	39. 181 ± 2	39. 170 ± 1	39. 169 ± 2
Rb	125	118	117	149	108	116	115	113
Sr	1 230	1 228	972	602	1 068	978	1 151	1 173
$^{87}Rb/^{86}Sr$	0. 294 02	0. 278 01	0. 348 25	0. 716 14	0. 292 57	0. 34 316	0. 289 07	0. 278 71
$^{87}Sr/^{86}Sr$	0. 707 500 ± 10	0. 707 502 ± 14	0. 707 412 ± 10	0. 708 284 ± 17	0. 707 509 ± 10	0. 707 258 ± 9	0. 707 259 ± 10	0. 707 279 ± 11
ε_{Sr}	122	122	120	133	43	40	40	40
Sm	7. 83	8. 09	5. 76	3. 1	5. 51	6. 29	7. 36	7. 35
Nd	51. 6	54	36. 1	17. 1	35. 1	40. 8	48. 9	49
$^{147}Sm/^{144}Nd$	0. 091 73	0. 090 57	0. 096 46	0. 109 59	0. 094 90	0. 093 20	0. 090 99	0. 090 68
$^{143}Nd/^{144}Nd$	0. 512 347 ± 6	0. 512 350 ± 4	0. 512 433 ± 6	0. 512 393 ± 6	0. 512 375 ± 7	0. 512 337 ± 6	0. 512 369 ± 15	0. 512 321 ± 5
ε_{Nd}	−5. 7	−5. 6	−4	−4. 8	−4. 1	−4. 8	−4. 2	−5. 2

U, Th and Pb concentrations were analysed by ICP-MS; Sm, Nd, Nd, Sr and isotopic ratios were measured through isotopic dilution. ε_{Nd} was calculated using $\varepsilon_{Nd} = [(^{143}Nd/^{144}Nd)_m / (^{143}Nd/^{144}Nd)_{CHUR} - 1] \times 10^4$, $(^{143}Nd/^{144}Nd)_{CHUR} = 0.512\ 638$. ε_{Sr} was calculated using $\varepsilon_{Sr} = [(^{87}Sr/^{86}Sr)_m / (^{87}Sr/^{86}Sr)_{UR} - 1] \times 10^4$, $(^{87}Sr/^{86}Sr)_{UR} = 0.698\ 990$. ε_{Nd}, ε_{Sr} and isotopic ratios of these rocks were not age corrected, because the rocks are late Eocene.

Thus we argue that neither a model involving the simple slab break-off of the Neo-Tethys oceanic slab (Chung et al., 2005) nor a model invoking the intra-continental subduction of the Songpan-Garze block beneath the Qiangtang Block (Roger et al., 2000; Wang et al., 2005) can reasonably explain the large-scale formation of the Eocene volcanic and intrusive rocks in the central block. In combination with regional geology, a scenario for the origin of Cenozoic adakitic rocks in Zuerkenwula Mountain area is summarized as follows(Fig. 14):

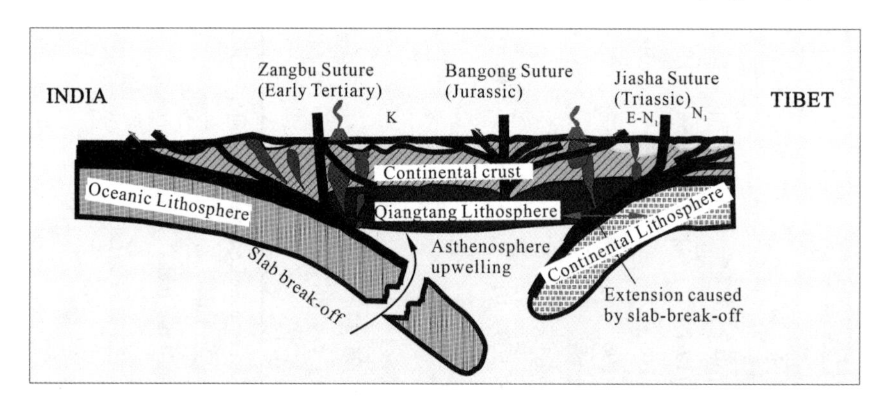

Fig. 14 Aproposed tectonic model for the generation of the Eocene high-Mg# and low-Mg# adakitic rocks in the Qiangtang Block.

Slab-breakoff of the India oceanic lithosphere caused asthenosphere upwelling and extension in the Qiangtang Block; the heat and melts from the lithospheric mantle caused extensive melting of the subducted continental crust and produced low-Mg# adakitic melt, interaction with overlying peridotite would elevated their Mg# values. Revised from Roger et al. (2000).

(1) The collision betweenthe India and Eurasian blocks occurred at ca.55 Ma (Yin and Harrison, 2000), and due to the draggle force of the dense oceanic lithosphere, the northern edge of the Neo-Tethys oceanic lithosphere reached to the southern margin of the Bangong-Nujiang suture during Late-Paleocene to Eocene time (Tailmann and Ni, 2003; Zhang et al., 2010).

(2) The southward intra-continental subduction of the Jinshajiang terrane beneath the Qiangtang Block (Ding et al., 2007) made the continental crust subducted to a certain depth. Given the Neo-Tethys oceanic slab detached from the Indian continental lithosphere at 45 Ma, this would cause large scale extension in the Qiangtang Block and led to the emplacement of diabase dikes in the central Qiangtang block (Wang et al., 2010).

(3) As the continental crust are relatively dry and cold, it is impossible that the extensive partial melting occurred during continental subduction (Zheng, 2008). However, under the circumstance of lithospheric extension, in combination with the heat and melts from the lithospheric mantle (Fig. 14), the subducted continental crust would undergo extensive melting, producing low-Mg# adakitic melt, and when these low-Mg# adakitic melts rose up to the overlying peridotite mantle wedge, they would capture some olivine crystals and elevated their Mg# and MgO values.

Acknowledgements This work is jointly supported by the National Natural Science Foundation of China (Grant Nos. 41190072,41072052 and 40572050) and the Teaching and Research Award Program for Outstanding Young Teachers in Higher Education Institutions of MOE, P.R. China. The MOST Special Funds were from the State Key Laboratory of Continental Dynamics, Northwest University and the Province Key Laboratory Construction Item (08JZ62).

References

Allegre, C.J., Minster, J.F., 1978. Quantitative method of trace element behavior in magmatic processes. Earth & Planetary Science Letters, 38,1-25.

Andersen, T., 2002. Correction of common lead in U-Pb analyses that do not report ^{204}Pb.Chemical Geology, 192,59-79.

Castillo, P.R., 2008. Origin of the adakite high-Nb basalt association and its implications for postsubduction magmatism in Baja California, Mexico. Geological Society of America Bulletin, 120,451-462.

Castillo, P.R., Janney, P.E., Solidum, R.U., 1999. Petrology and geochemistry of Camiguin island, southern Philippines: Insights to the source of adakites and other lavas in a complex arc setting. Contributions to Mineralogy and Petrology, 134,33-51.

Chung, S.L., Chu, M.F., Zhang, Y., Xie, Y., Lo, C.H., Lee, T.Y., Lan, C.Y., Li, X., Zhang, Q., Wang, Y., 2005. Tibetan tectonic evolution inferred from spatial and temporal variations in post-collisional magmatism. Earth-Science Reviews, 68(3-4),173-196.

Condie, K.C., 2005. TTGs and adakites: Are they both slab melts? Lithos, 80,33-44.

Defant, M.J., Drummond, M.S., 1990. Derivation of some modern arc magmas by partial melting of young subducted lithosphere. Nature, 347,662-665

Defant, M.J., Xu, J.F., Kepezhinskas, P., Wang, Q., Zhang, Q., Xiao, L., 2002. Adakites: Some variations on a theme. Acta Petrologica Sinica, 18(2),129-142

Deng, W.M., 1998. Cenozoic intraplate volcanic rocks in the Northern Qinghai-Xizang plateau: Beijing, China

Ding, L., Lai, Q., 2003. New geological evidences of crust thickening in the Gangdese block prior to the Indo-Asian collision. Chinese Science Bulletin, 48,36-842

Ding, L., Kapp, P., Yue, Y.H., Lai, Q.Z., 2007. Post-collisional calc-alkaline lavas and xenoliths from the southern Qiangtang block, central Tibet. Earth & Planetary Science Letters, 254,28-38

Ernst, W.G., 2010. Subduction-zone metamorphism, calc-alkaline magmatism, and convergent-margin crustal evolution. Gondwana Research, 18(1),8-16

Galve, A., Jiang, M., Hirn, A., Sapin, M., Laigle, M., De, Voogd. B., Gallart, J., Qian, H., 2006. Explosion seismic P and S velocity and attenuation constraints on the lower crust of the North-Central Tibetan Plateau, and comparison with the Tethyan Himalayas: implications on composition, mineralogy, temperature, and tectonic evolution. Tectonophysics, 412,141-157.

Gao, S., Rudnick, R.L., Yuan, H.L., et al., 2004. Recycling lower continental crust in the North China craton. Nature, 432,892-897.

Guan, Q., Zhu, D.C., Zhao, Z.D., Dong, G.C., Zhang, L.L., Li, X.W., Liu, M., Mo, X.X., Liu, Y.S., Yuan, H.L., 2011. Crustal thickening prior to 38 Ma in southern Tibet: Evidence from lower crust-derived adakitic magmatism in the Gangdese Batholith. Gondwana Research, 21 (1),88-99.

Guo, Z.F., Wilson, M., Liu, J.Q., Mao, Q., 2006. Post-collisional, potassic and ultrapotassic magmatism of the Northern Tibetan Plateau: Constraints on characteristics of the mantle source, geodynamic setting and uplift mechanisms. Journal of Petrology, 47(6),1177-1220.

Harrison, T.M., Copeland, P., Kidd, W.S.F., et al., 1995. Activation of the Nyainqentanghla Shear Zone: Implications for uplift of the southern Tibetan Plateau. Tectonics, 14,658-676.

Hoskin, P.W.O., Schaltegger, U., 2003. The composition of zircon and igneous and metamorphic petrogenesis. Reviews in Mineralogy and Geochemistry, 53(1),27-62.

Kapp, P., Yin, A., Manning, C.E., Murphy, M., Harrison, T.M., Spurlin, M., Ding, L., Deng, X.G., Wu, C.M., 2000. Blueschist-bearing metamorphic core complexes in the Qiangtang block reveal deep crustal structure of northern Tibet. Geology, 28,19-22.

Lai, S.C., Liu, C.Y., 2001. Enriched upper mantle and eclogitic lower crust in north Qiangtang, Qinghai-Tibet Plateau: Petrological and geochemical evidence from the Cenozoic volcanic rocks. Acta Petrologica Sinica, 17,459-468 (in Chinese).

Lai, S.C., Qin, J.F. 2008. Petrology and geochemistry of the granulite xenoliths from Cenozoic Qiangtang volcanic field: Implication for the nature of the lower crust in the northern Tibetan plateau and the genesis of Cenozoic volcanic rocks. Acta Petrologica Sinica, 24(2),25-36

Lai, S.C., Liu, C.Y., Yi, H.S., 2003. Geochemistry and petrogenesis of Cenozoic andesite-dacite associations from the Hoh Xil Region,Tibetan Plateau. International Geology Review, 45(11),998-1019

Lai, S.C., Qin, J.F., Li, Y.F., 2007. Partial melting of thickened Tibetan crust: Geochemical evidence from Cenozoic adakitic volcanic rocks. International Geology Review, 49(4),357-373

Le Bas, M.J., Streckeisen, A.L., 1991. The IUGS systematics of igneous rocks. Journal of the Geological Society, 148 (5),825-833

Li, Y., Wang, C., Zhao, X., Yin, A., Ma, C., 2012. Cenozoic thrust system, basin evolution, and uplift of the Tanggula Range in the Tuotuohe region, central Tibet. Gondwana Research, http://dx. doi. org/10. 1016/j.gr.2011. 11. 017.

Lin, J.H., Yin, H.S., Zhao, B., Li, B.H., Shi, Z.Q., Huang, J.J., 2003. ^{40}Ar-^{39}Ar isotopic dating and its implication of cenozoic volcanic rocks from Zuerkengwula mountain area, Northern Tibetan. Journal of Mineralogy and Petrology, 23(3),31-34.

Lin, J.H., Yin, H.S., Shi, Z.Q., Zhao, B., Li, B.H., Huang, J.J., 2004. Study on isotopic geochemistry of cenozoic high-K calc-alkaline volcanic rocks in the Zuerkengwula mountain area, northern Tibet. Journal of Mineralogy and Petrology, 24(4),59-64.

Liu, Y., Liu, X.M., Hu, Z.C., Diwu, C.R., Yuan, H.L., Gao, S., 2007. Evaluation of accuracy and long-term stability of determination of 37 trace elements in geological samples by ICP-MS. Acta Petrologica Sinica, 2007,23(5),1203-1210.

Ludwig, K.R., 2003. ISOPLOT 3.0: A geochronological toolkit for Microsoft Excel. Berkeley Geochronology Center, Special publication, 4.

Martin, H., 1999. Adakitic magmas: Modern analogues of Archaean granitoids. Lithos, 46(3),411-429.

Martin, H., Smithies, R.H., Rapp, R., et al., 2005. An overview of adakite, tonalite-trondhjemite-granodiorite (TTG), and sanukitoid: Relationships and some implications for crustal evolution. Lithos, 79 (5),1-24.

Miller, C., Schuster, R., Klotzli, U., et al., 1999. Postcollisional potassic and ultrapotassic magmatism in SW Tibet: Geochemical and Sr-Nd-Pb-O isotopic constraints for mantle source characteristics and petrogenesis. Journal of Petrology, 40,1399-1424.

Parrish, R.R., Hodges, K.V., 1996. Isotopic constraints on the age and provenance of Lesser and Greater Himalayan sequences, Nepalese Himalaya. Geological Society of America Bulletin, 108,904-911.

Peccerillo, R., Taylor, M., 1976. Sr Geochemistry of Eocene calc-alkaline volcanic rocks from the Kastamonu area, northern Turkey. Contributions to Mineralogy and Petrology, 58,63-81.

Petford, N., Atherton, M., 1996. Na-rich partial melt from newly underplated basaltic crust: The Cordmera Blanca Batholith, Peru. Journal of Petrology, 37,491-521.

Prouteau, G., Scaillet, B., 2003. Experimental constraints on the origin of the 1991 Pinatubo dacite. Journal of Petrology, 44,2203-2241.

Qin, J.F., Lai, S.C., Diwu, C.R., Ju, Y.J., Li, Y.F., 2010. Magma mixing origin for the post-collisional adakitic monzogranite of the Triassic Yangba pluton, Northwestern margin of the South China block: Geochemistry, Sr-Nd isotopic, zircon U-Pb dating and Hf isotopic evidences. Contributions to Mineralogy and Petrology, 159(3),389-409.

Rapp, P.R., Watson, E.B., 1995. Dehydration Melting of Metabasalt at 8 – 32 kbar: Implications for Continental growth and crust-mantle recycling. Journal of Petrology, 36(4),891-931.

Rapp, R.P., Shimizu, N., Norman, M.D., et al., 1999. Reaction between slab-derived melts and peridotite in the mantle wedge: Experimental constraints at 3. 8 GPa. Chemical Geology, 160(4),335-356.

Robinson, D.M., DeCelles, P.G., Patchett, P.J., et al., 2001. The kinematic evolution of the Nepalese Himalaya interpreted from Nd isotopes. Earth & Planetary Science Letters, 192,507-521.

Roger, F., Tapponnier, P., Arnaud, N., Schärer, U., Brunel, M., Zhiqin, X., Jingsui, Y., 2000. An Eocene magmatic belt across central Tibet: mantle subduction triggered by the Indian collision? Terra Nova, 12,102-108.

Stern, C.R., Kilian, R., 1996. Role of the subducted slab, mantle wedge and continental crust in the generation of adakites from the Andean Austral Volcanic Zone. Contributions to Mineralogy and Petrology, 123,263-281.

Sun, S.S., McDonough, W.F., 1989. Chemical and isotopic systemmatics of oceanic basalts: Implication for the mantle composition and process. In: Saunder, A.D., Norry, M.J. Magmatism in the ocean basins. London: Geological Society of London and Blackwell Scientific Publications, 313-345.

Tatsumi, Y., Shukuno, H., et al., 2003. The petrology and geochemistry of high-magnesium andesites at the Western Tip of the Setouchi Volcanic Belt, SW Japan. Journal of Petrology, 44(9),1561-1578.

Tilmann, F., Ni, J., 2003. Seismic imaging of the downwelling Indian lithosphere beneath central Tibet. Science, 300,1424-1427.

Turner, S., Arnaud, N., Liu, J., Rogers, N., Hawkesworth, C., Harris, N., Kelley, S., van Calsteren, P., Deng, W., 1996. Post-collision, Shoshonitic Volcanism on the Tibetan Plateau: Implications for convective thinning of the lithosphere and the source of ocean island basalts. Journal of Petrology, 37(1),45-71.

Wang, Q., McDermott, F., Xu, J.F., Bellon, H., Zhu, Y.T., 2005. Cenozoic K-rich adakitic volcanic rocks in the Hohxil area, northern Tibet: Lower-crustal melting in an intracontinental setting. Geology, 33 (6),465-468.

Wang, Q., Wyman, D.A., Xu, J., et al., 2008. Eocene melting of subducting continental crust and early uplifting of central Tibet: Evidence from central-western Qiangtang high-K calc-alkaline andesites, dacites and rhyolites. Earth & Planetary Science Letters, 272,158-171.

Wang, Q., Wyman, D.A., Li, Z.X., Sun, W., Chung, S.L., Vasconcelos, P.M., Zhang, Q., Dong, H., Yu, Y., Pearson, N., Qiu, H., Zhu, T., Feng, X., 2010. Eocene north-south trending dikes in central Tibet: New constraints on the timing of east-west extension with implications for early plateau uplift? Earth & Planetary Science Letters, 298(1-2),205-216.

Wilson, M., 1989. Igneous petrogenesis. London: Unwin Hyman Press, 295-323.

Wu, F.Y., Ge, W.C., Sun, D.Y., 2002. The definition, discrimination of adakites and their geological role. In: Xiao, Q.H., Deng, J.F., Ma, D.Q., et al. The Ways of Investigation on Granotoids. Beijing: Geological Publishing House, 2002, 172-191 (in Chinese with English abstract).

Xia, L., Li, X., Ma, Z., Xu, X., Xia, Z., 2011. Cenozoic volcanism and tectonic evolution of the Tibetan plateau. Gondwana Research, 19(4), 850-866.

Xiong, X.L., Adam, J., Green, T.H., 2005. Rutile stability and rutile/melt HFSE partitioning during partial melting of hydrous basalt: Implications for TTG genesis. Chemical Geology, 218,339-359.

Xu, J.F., Shinjio, R., Defant, M.J., et al., 2002. Origin of Mesozoic adakitic intrusive rocks in the Ningzhen area of east China: Partial melting of delaminated lower continental crust? Geology, 12,1111-1114.

Yang, X.S., 2002. Intracrustal Partial Melting and Its Significances, Exemplified by Tibetan Plateau. Ph.D. dissertation, Institute of Geology, China Seismological Bureau, 1-109.

Yin, A., Harrison, T.M., 2000. Geologic evolution of the Himalayan-Tibetan orogen. Annual Review of Earth and Planetary Sciences, 28 (1),211-280.

Yu, S., Zhang, J., Del Real, P.G., 2012. Geochemistry and zircon U-Pb ages of adakitic rocks from the Dulan area of the North Qaidam UHP terrane, north Tibet: Constraints on the timing and nature of regional tectonothermal events associated with collisional orogeny. Gondwana Research, 21 (1),167-179.

Yuan, H.L., Gao, S., Liu, X.M., Li, H.M., Gunther, D., Wu, F.Y., 2004. Accurate U-Pb age and trace element determinations of zircon by laser ablation-inductively coupled plasma mass spectrometry. Geostandard Newsletters, 28,353-370.

Zhang, Z., Santosh, M., 2011. Tectonic evolution of Tibet and surrounding regions. Gondwana Research, 21 (1),1-3.

Zhang, Q., Wang, Y., Li, C.D., Wang, Y.L., Jin, W.J., Jia, X.Q., 2006. Granite classification on the basis of Sr and Yb contents and its implications. Acta Petrologica Sinica, 22(9),2249-2269.

Zhang, Z., Zhao, G., Santosh, M., Wang, J., Dong, X., Shen, K., 2010. Late Cretaceous charnockite with adakitic affinities from the Gangdese batholith, southeastern Tibet: Evidence for Neo-Tethyan mid-ocean ridge subduction? Gondwana Research, 17(4),615-631.

Zheng, Y., 2008. A. perspective view on ultrahigh-pressure metamorphism and continental collision in the Dabie-Sulu orogenic belt. Chinese Science Bulletin, 53(20),3081-3104.

Zindler, A., Hart, S., 1986. Chemical geodynamics. Annual Review of Earth and Planetary Sciences, 14,93-571.

The carbonated source region of Cenozoic mafic and ultra-mafic lavas from western Qinling: Implications for eastern mantle extrusion in the northeastern margin of the Tibetan Plateau[①]

Lai Shaocong Qin Jiangfeng Jahanzeb Khan

Abstract: Cenozoic Indo-Asian collision caused significant crustal shortening in central Tibet. The strike-slip faults around the Tibetan Plateau (TP) are generally attributed to extrusion tectonics, resulting from lower crust flow. Therefore, the mantle extrusion site corresponding to the Cenozoic elevation of the TP needs to be identified. This paper reports the petrology and geochemistry of Cenozoic mafic and ultra-mafic volcanic rocks in the Xiahe and Lixian areas, at the northeastern margin of the TP. Detailed analysis indicates a regular change in partial melting conditions and source regions of the volcanic rocks from west to east, revealing a Cenozoic eastward mantle extrusion in the eastern margin of the TP. The Xiahe volcanic rocks display ocean island basalt affinity with negative K anomalies and positive Nb and Ta anomalies. They are alkaline with extremely high Na_2O/K_2O ratios and relatively enriched Sr-Nd-Pb isotopic compositions, indicating that these basalts were derived from partial melting of carbonated pyroxenite. The Lixian picro-basalts are closely associated with igneous carbonatites. They have relatively high TiO_2 (3.47%–4.66%) and MgO (11.24%–18.88%) contents and low SiO_2 (41.14%–44.82%) and Al_2O_3 (5.84%–9.18%) contents. Based on the depleted Sr-Nd-Pb isotopic compositions, we propose that the Lixian picro-basalts may have originated from the partial melting of carbonated lithosphere mantle peridotites at relatively high pressure (>3 GPa). Minor hornblendite in their source region can account for the high TiO_2 and Na_2O+K_2O contents. Thus, we argue that these volcanic rocks were formed by episodic decompression melting of the carbonated mantle lithosphere during the eastward extrusion of the Tibetan lithosphere, in contrast to the conventional view that they were formed in a continental rift setting. Their partial melting and eruption processes may be closely related to the Cenozoic strike-slip fault activities in the northeastern margin of the TP.

1 Introduction

The Cenozoic Indo-Asian collision caused significant N-S crustal shortening of about 2 000 km (Yin and Harrison, 2000; Xia et al., 2011; Zhang and Santosh, 2011) and

① Published in *Gondwana Research*, 2014, 25.

significant uplift of the Tibetan Plateau up to 5 000 m. It is the most spectacular example of mountain building, plateau development, and continental-scale strike-slip faulting on Earth (Schoenbohm et al., 2006; Xu et al., 2012). Several lines of evidence from geological, geochronological, and seismic studies indicate that strike-slip faults (Tapponnier et al., 1982, 2001; Bowman et al., 2003; Cao et al., 2011; Zhang et al., 2011) and lateral extrusion of blocks of Asian continent around the Tibetan Plateau (TP) (Clark and Royden, 2000) are significant controlling factors in the growth of the TP. Wang and Burchfiel (2004) proposed that there are three regions of different late Cenozoic and currently active deformations in the eastern margin of the TP: ① the northeastern region, where displacement along the sinistral east-northeast-striking Altyn Tagh fault is transferred to northeast-southwest shortening in the Qilian Shan; ② the central region, where the Longmen Shan rises to more than 6 km above the Sichuan basin, forming one of the steepest mountain fronts along any margin of the TP; and ③ the southwest region, where there is no prominent topographic break that defines the eastern margin of the plateau (Cao et al., 2011).

Cenozoic volcanic rocks in the TP display obvious and regular migration corresponding to the different stages of the Indo-Asian collision (Chung et al., 2005; Mo et al., 2006). The 65–70 Ma Linzizong arc volcanic rocks represent the start of the collision between India and Asia; 45–40 Ma post-collisional igneous rocks are subsequently found in the Qiangtang and northern part of the Sanjiang area, and the Miocene to Quaternary volcanic rocks in the Hoh Xil, Tengchong, and western Qinling are believed to result from the Indo-Asian collision (Mo et al., 2006; Hu et al., 2012). These migrations clearly suggest the lateral extrusion of the Tibetan lithosphere. However, whether the lateral extrusion only involved the lower crust, or the lithospheric mantle and asthenosphere were also involved is yet to be determined.

The TP is surrounded by the Pamirs Block to the west, the Tarim Block to the north, the Ordos Block to the northeast, and the Yangtze Block to the east. These blocks have thick lithosphere (Mo, 2011 and references therein; Li et al., 2012), which makes lateral extrusion against these blocks difficult. However, the weak channel along the Qinling and Sanjiang orogenic belt may have facilitated the lateral extrusion that resulted from Indo-Asian collision (Mo, 2011). Cenozoic potassic rocks (Huang et al., 2010) and igneous carbonatites (Hou et al., 2006) provide direct petrological evidence that mantle extrusion took place along the Sanjiang orogenic belt. Paleogene potassic and ultra-potassic rocks (Huang et al., 2010) in the Tengchong area, at the southeastern margin of the TP, were inferred to be derived from very low-degree decompression melting of metasomatized mantle lithosphere. Hou et al. (2006) reported Cenozoic igneous carbonatites at the eastern margin of the TP. They proposed that the carbonatites were formed by partial melting of the metasomatized mantle and that they represent the transition from transpression to transtension at the Eocene/Oligocene boundary at the eastern margin of the TP.

Miocene lavas in western Qinling are believed to be kamafugite (Yu, 1994; Yu and Zhang, 1998; X. Yu et al., 2001; X.H. Yu et al., 2001; Yu et al., 2003, 2004, 2009, 2011). They are ultra-potassic and were formed in a continental rift setting. Previous $^{39}Ar/^{40}Ar$ dating on these lavas indicates that they were formed in the Miocene (22–23 Ma) (Yu et al., 2006). Mineralogy of mantle xenoliths from the Cenozoic kamafugite revealed that the lithospheric mantle beneath western Qinling underwent partial melting, re-crystallization, deformation, and metasomatism due to asthenospheric upwelling and decompression resulting from the Cenozoic extensive tectonic setting (Su et al., 2009, 2011). Yu et al. (2004) argued that these lavas resulted from interaction between mantle plumes and the lithospheric mantle. The geodynamic implications of these Cenozoic lavas in western Qinling remain controversial.

This paper reports major- and trace-element geochemistry, and Sr-Nd-Pb isotopic compositions for the Cenozoic mafic and ultra-mafic rocks from the Xiahe and Lixian area, western Qinling. Volcanic rocks from the Xiahe and Lixian areas display different petrological and geochemical features. Detailed work indicates that the Xiahe and Lixian volcanic rocks were derived from the partial melting of carbonated pyroxenite and carbonated peridotite, respectively. This suggests that the source region of the northeastern margin of the TP regularly changes from west to east. This conclusion will help in understanding Cenozoic mantle extrusion processes in the northeastern margin of the TP.

2 Geological setting and petrography

The western Qinling-Songpan region, at the eastern margin of the TP, is a convergence tectonic transform region of some major blocks and orogenic belts. Not only is it a vertical converging zone of the E-W Qilian-Qinling-Tongbai-Dabie orogenic belt and N-S Helan-Chuandian tectonic belts, but it is also an extensional region of the eastern margin of the TP (Fig. 1a). The Cenozoic lavas in western Qinling are mainly located in the Lixian, Dangchang, and Xiahe areas of the Gansu Province (Fig. 1b).

The Xiahe volcanic rocks outcrop in Madang Town (N35°18′, E102°46′), about 4 km to the southwest of Xiahe county (Fig. 2). Stratigraphic characterization of the study area is relatively simple, since it is mainly composed of the Jurassic and Cretaceous rocks. The Jurassic upper layer mainly consists of yellow-gray phyllitic argillaceous slates, silty slates, conglomerates, feldspar-quartzose sandstones, phyllitic tuffaceous slates, and lenticular gravel-bearing limestone. The lower layer mainly consists of gray feldspathic-quartzose sandstones, quartz arkose, silty slates, conglomerates, and gravel-bearing limestones. The Cretaceous stratum is dominated by andesites, argillaceous siltstones, tuffaceous sandstones, mudstones and breccia. The Xiahe volcanic rocks outcrop as a small W-E extended ellipsoid hillock

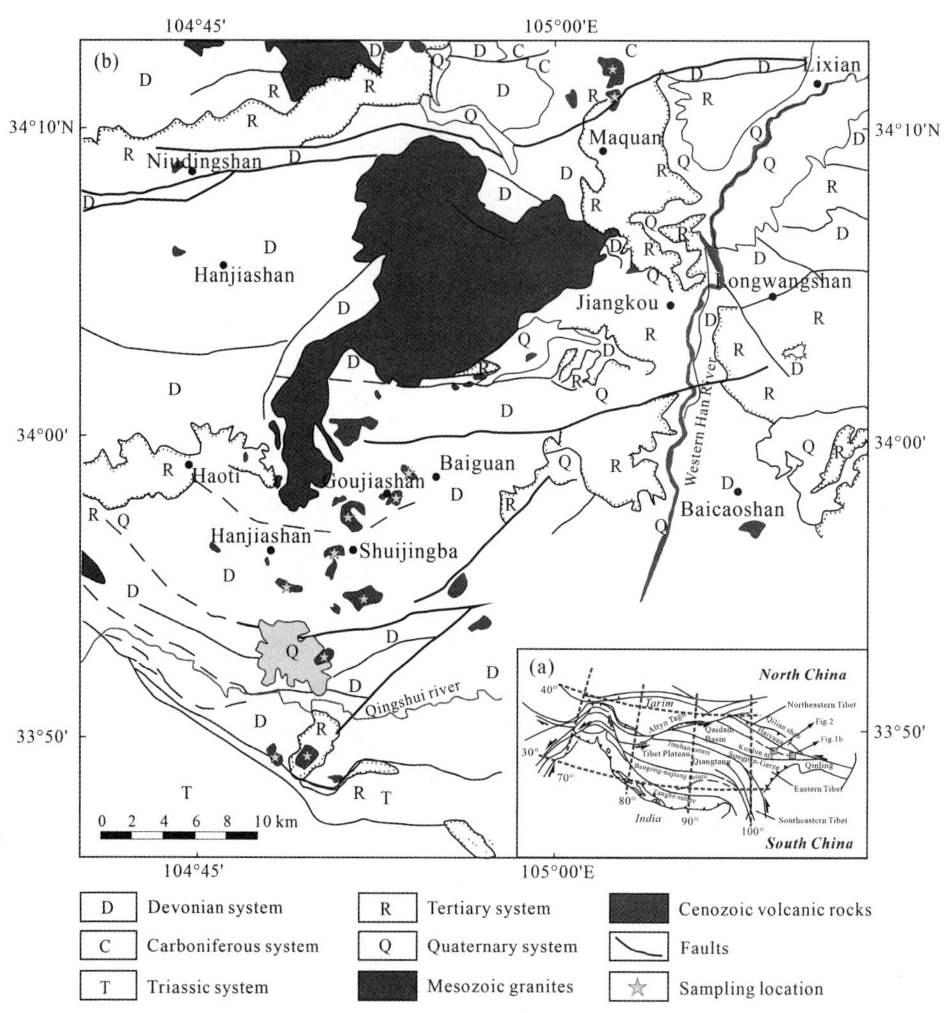

Fig. 1 (a) Simplified geological map of the Tibetan Plateau and surrounding areas; (b) Sketch geological map of the Lixian area.
Panel (a) is modified from Enkelmann et al. (2006).

covering an area of about 0.3 km^2. These volcanic rocks discordantly overlie the Cretaceous taupe, dark purple argillaceous siltstone, sandstone and breccia. In addition, a large amount of Cretaceous sandstone and brecciated xenoliths have been found to underlie these volcanic rocks. Therefore, the Xiahe volcanic rocks are inferred to be younger than the Cretaceous rocks (Fig. 2). The Xiahe volcanic rocks are gray-black and exhibit a porphyritic texture with massive structures, and are brecciated. Most of the samples are fresh and the phenocrysts include euhedral tabular plagioclase, clinopyroxene and rare olivine phenocrysts. The fine-grained groundmass is composed of tabular plagioclase microlite, fine-grained clinopyroxene and volcanic glass with pilotaxitic or hyalopilitic textures (Fig. 3a, b).

The Cenozoic lavas in Lixian area are mainly distributed within or at the margin of the

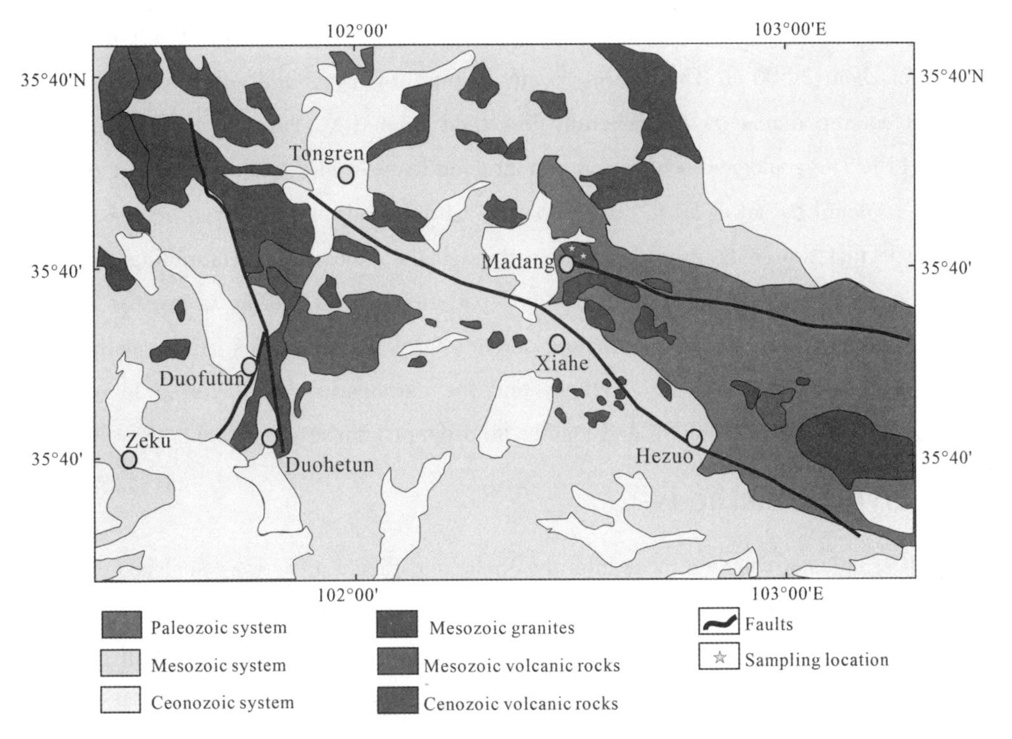

Fig. 2 Sketch geological map of the Xiahe area.

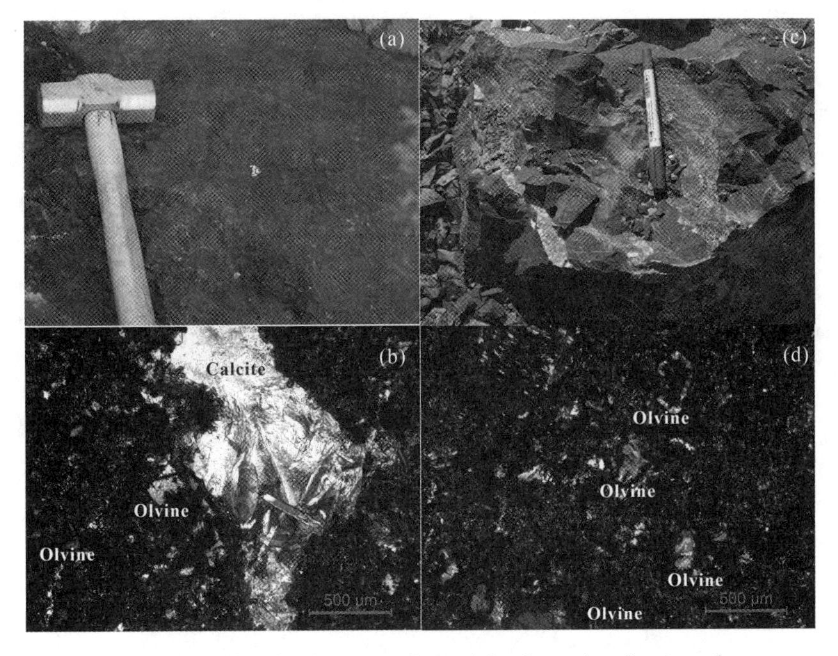

Fig. 3 Field and microscope photo of the Cenozoic volcanic rocks from Xiahe and Lixian area, northeastern margin of the Tibetan Plateau.

(a) sodic alkaline volcanic rocks from the Xiahe area; (b) microscope photo of the Xiahe sodic alkaline volcanic rocks, minor olivine and large calcite crystals crystal occurred in matrix; (c) picro-basalt from the Lixian area; (d) microscope photo of the Lixan picro-basalts, many olivine crystals occurred in matrix.

Cenozoic Tianshui-Lixian basin, where a cluster of 30 outcrops occur in a plateau at an elevation of about 2 500 m. They mainly occur as pipe tuff rings and sub-volcanic intrusions, and single outcrop diameters are generally less than 1 km (X. Yu et al., 2001; Yu et al., 2003, 2011). The geological setting and petrographic features of the Lixian volcanic rocks were described in detail by Yu et al. (2003), so only a brief description is given here. Geophysical data indicate that two NNE blind faults cross through the lithosphere, controlling not only the formation of the basin, but also the distribution patterns of the Cenozoic lavas. The Baihe and Haoti lavas contain large amounts of mantle-derived peridotite xenolithos. Additionally, most of the lavas are associated with carbonatites, and the carbonatites are believed to result from interaction between asthenosphere and mantle lithosphere (Yu et al., 2004).

3 Analytical methods

Fresh samples of volcanic rocks from the Xiahe and Lixian areas were analyzed at the State Key Laboratory of Continental Dynamics, Northwest University in Xi'an, China. Fresh chips of whole rock samples were powdered to 200 mesh-size using a tungsten carbide ball mill. Major and trace elements were analyzed using XRF (Rikagu RIX 2100) and ICP-MS (Agilent 7500a), respectively. Analyses of USGS and Chinese National Rock Standards (BCR-2, GSR-1 and GSR-3) indicate that both the analytical precision and the accuracy are generally better than 5% and 10%, for major and trace elements, respectively.

Whole-rock Sr-Nd-Pb isotopic data were obtained using a Nu Plasma HR multi-collector mass spectrometer. Sr and Nd isotopic fractionation was corrected to $^{87}Sr/^{86}Sr = 0.119\ 4$ and $^{146}Nd/^{144}Nd = 0.721\ 9$, respectively. During the analysis, the NIST SRM987 standard yielded an average value of $^{87}Sr/^{86}Sr = 0.710\ 250 \pm 12$ (2σ, $n = 15$) and the La Jolla standard gave an average of $^{143}Nd/^{144}Nd = 0.511\ 859 \pm 6$ (2σ, $n = 20$). Whole-rock Pb was separated by an anion exchange in HCl-Br columns and the Pb isotopic fractionation was corrected to $^{205}Tl/^{203}Tl = 2.387\ 5$. Within the analytical period, 30 measurements of NBS981 gave average values of $^{206}Pb/^{204}Pb = 16.937 \pm 1 (2\sigma)$, $^{207}Pb/^{204}Pb = 15.491 \pm 1 (2\sigma)$, and $^{208}Pb/^{204}Pb = 36.696 \pm 1 (2\sigma)$. The BCR-2 standard gave $^{206}Pb/^{204}Pb = 18.742 \pm 1 (2\sigma)$, $^{207}Pb/^{204}Pb = 15.620 \pm 1 (2\sigma)$ and $^{208}Pb/^{204}Pb = 38.705 \pm 1 (2\sigma)$. All procedural Pb blanks were in the range of 0.1–0.3 ng.

4 Major and trace element geochemistry

Major and trace element data of the Xiahe and Lixian volcanic rocks are given in Tables 1 and 2, respectively. The volcanic rocks from both the Xiahe and Lixian areas are strongly alkaline, with the Xiahe volcanic rocks lying mainly in the picrobasalt, basanite, and trachybasalt fields, and the Lixian volcanic rocks, with lower SiO_2 contents, lying mainly in

Table 1 Major(wt%) and trace(ppm) element analysis result of the volcanic rocks from Xiahe area.

	HQ-08	HQ-12	HQ-13	HQ-16	HQ-17	HQ-18	HQ-19	HQ-23	HQ-26	HQ-27	MD-01[a]
SiO_2	50.07	43.47	43.42	42.33	42.32	42.53	47.62	49.10	45.16	48.67	45.28
TiO_2	2.65	3.12	3.60	3.49	3.57	3.63	2.49	2.46	3.34	2.50	1.84
Al_2O_3	15.97	16.29	18.40	18.30	18.25	18.45	15.35	15.42	17.63	15.62	16.22
$Fe_2O_3^T$	10.08	12.09	12.18	12.87	12.48	12.59	10.97	10.84	11.98	10.92	9.37
MnO	0.04	0.05	0.05	0.04	0.04	0.04	0.07	0.07	0.03	0.06	0.12
MgO	6.33	8.00	7.30	7.99	7.17	7.58	7.02	6.62	7.54	6.40	7.64
CaO	3.24	5.76	4.03	4.06	3.83	3.92	3.99	3.39	3.63	3.90	8.09
Na_2O	3.03	2.47	2.88	2.77	2.64	2.70	3.47	3.43	2.68	3.29	4.32
K_2O	1.14	0.58	0.21	0.20	0.23	0.20	1.06	1.04	0.19	1.07	0.76
P_2O_5	0.51	0.53	0.45	0.52	0.37	0.37	0.56	0.54	0.41	0.47	0.76
LOI	6.80	7.48	7.36	7.36	9.47	8.04	7.39	7.12	7.54	7.06	5.89
Total	99.86	99.84	99.88	99.93	100.37	100.05	99.99	100.03	100.13	99.96	99.83
$Mg^{\#}$	59.4	60.7	58.3	59.1	57.2	58.4	59.9	58.7	59.5	57.7	65.5
CaO/Al_2O_3	0.2	0.4	0.2	0.2	0.2	0.2	0.3	0.2	0.2	0.2	0.5
Li	32.0	30.8	31.0	32.8	31.3	33.2	38.9	40.3	34.2	37.3	32.50
Be	2.91	2.49	2.40	2.52	2.36	2.58	3.41	3.25	2.43	3.05	[b]
Sc	18.0	23.8	24.5	24.6	23.6	23.7	18.7	18.0	22.5	17.4	17.70
V	169	230	243	252	257	257	168	172	249	169	152.00
Cr	220	309	334	327	339	329	210	223	314	241	222.00
Co	45.5	56.7	57.4	60.0	66.6	64.9	58.4	46.3	59.4	47.3	44.20
Ni	110	159	161	173	168	173	114	111	176	105	86.00
Cu	30.0	50.9	58.0	60.4	50.1	44.8	41.3	39.7	43.2	37.3	52.00
Zn	89.7	97.5	109	100	116	115	90.1	87.0	95.8	87.4	75.10
Ga	23.0	21.4	22.4	23.4	23.7	23.2	22.2	21.8	22.9	21.3	[b]
Ge	1.36	1.46	1.53	1.56	1.60	1.60	1.35	1.33	1.60	1.31	[b]
Rb	13.6	3.59	1.75	1.91	1.68	1.35	10.3	11.1	1.71	10.2	4.11
Sr	1 125	986	829	830	853	840	1 123	1 103	804	1 101	882.00
Y	20.1	22.6	23.5	25.7	23.2	23.7	20.5	20.2	22.4	19.6	21.00
Zr	393	297	341	335	351	348	382	382	323	390	320.00
Nb	69.1	48.8	56.3	54.8	57.7	57.2	67.6	67.5	52.8	68.1	54.00
Cs	2.76	1.80	0.89	0.78	0.89	0.91	3.95	3.89	0.74	4.35	8.32
Ba	285	206	155	150	159	150	270	270	144	272	217.00
La	27.1	23.9	22.4	25.5	21.0	20.9	29.8	28.8	20.3	26.3	38.60
Ce	50.8	46.5	43.7	50.8	42.2	42.5	55.2	53.8	40.2	49.0	75.50
Pr	5.92	5.81	5.53	6.28	5.35	5.41	6.43	6.37	4.98	5.83	8.93
Nd	23.8	23.8	22.7	25.8	21.9	22.1	26.0	25.3	20.3	23.2	36.60
Sm	5.13	5.18	4.89	5.65	4.76	4.90	5.58	5.41	4.49	5.12	7.24
Eu	1.91	1.88	1.86	2.05	1.84	1.85	1.99	1.97	1.71	1.93	2.29
Gd	4.88	5.05	4.90	5.55	4.75	4.85	5.24	5.12	4.53	4.87	5.97

Continued

	HQ-08	HQ-12	HQ-13	HQ-16	HQ-17	HQ-18	HQ-19	HQ-23	HQ-26	HQ-27	MD-01[a]
Tb	0.71	0.76	0.77	0.86	0.74	0.78	0.75	0.74	0.73	0.71	0.95
Dy	4.12	4.54	4.73	5.20	4.66	4.77	4.31	4.22	4.54	4.10	4.84
Ho	0.79	0.91	0.95	1.03	0.95	0.96	0.83	0.81	0.92	0.79	0.92
Er	1.98	2.30	2.42	2.59	2.41	2.44	2.02	1.97	2.32	1.92	2.38
Tm	0.27	0.31	0.34	0.35	0.34	0.34	0.28	0.28	0.33	0.27	0.29
Yb	1.69	1.91	2.09	2.19	2.09	2.14	1.72	1.69	2.01	1.68	1.94
Lu	0.26	0.29	0.32	0.33	0.33	0.33	0.26	0.26	0.30	0.25	0.28
Hf	7.45	6.04	7.08	6.94	7.29	7.23	7.48	7.49	6.69	7.61	7.55
Ta	4.49	3.22	3.72	3.65	3.83	3.79	4.42	4.43	3.52	4.44	4.18
Pb	4.57	2.56	3.55	3.52	10.6	3.82	4.37	3.88	3.26	3.85	3.53
Th	4.15	2.65	2.84	2.94	2.75	2.69	4.47	4.41	2.67	4.39	4.53
U	1.45	0.94	1.04	1.11	1.03	1.03	1.53	1.49	1.02	1.52	1.87
Nb/Ta	15.4	15.2	15.1	15.0	15.1	15.1	15.3	15.3	15.0	15.3	12.9
Nb/U	47.8	51.7	53.9	49.5	56.1	55.5	44.3	45.3	51.8	44.8	28.9
Ce/Pb	11.1	18.1	12.3	14.4	4.0	11.1	12.6	13.9	12.3	12.7	21.4
La/Yb	11.51	8.99	7.68	8.38	7.19	6.99	12.44	12.24	7.22	11.24	14.27
Eu[a]	1.15	1.11	1.15	1.10	1.17	1.15	1.11	1.12	1.15	1.16	1.03
Zr/Hf	52.8	49.2	48.2	48.3	48.2	48.1	51.1	51.0	48.2	51.2	42.4
Th/U	2.87	2.81	2.72	2.66	2.67	2.61	2.93	2.95	2.62	2.89	2.42

	MD-02[a]	MD-04[a]	MD-06[a]	MD-08[a]	MD-15[a]	MD-16[a]	MD-17[a]	MD-19[a]	MD-20[a]	MD-22[a]	MD-23[a]
SiO_2	42.44	43.94	45.44	50.76	45.71	44.53	50.70	45.55	45.83	50.62	49.45
TiO_2	1.46	2.40	1.82	1.81	1.81	1.46	2.00	1.86	1.87	2.00	1.82
Al_2O_3	13.01	15.66	15.68	15.54	15.65	13.22	15.47	16.43	17.60	15.68	16.74
$Fe_2O_3^T$	8.74	10.39	10.92	10.00	9.37	9.35	10.02	10.27	10.94	9.28	10.85
MnO	0.16	0.13	0.15	0.11	0.14	0.15	0.11	0.14	0.13	0.10	0.15
MgO	7.68	7.71	7.80	4.91	7.43	7.77	4.82	7.92	7.41	5.00	7.29
CaO	11.36	10.00	8.65	8.47	8.50	11.78	8.43	8.80	8.24	8.81	8.82
Na_2O	2.53	3.12	4.17	4.00	4.35	2.56	4.10	5.14	4.34	4.41	4.14
K_2O	0.60	0.61	0.96	0.28	1.08	0.38	0.27	0.73	0.72	0.53	0.71
P_2O_5	0.81	0.94	0.64	0.71	0.86	0.75	0.80	0.68	0.82	0.73	0.85
LOI	11.01	5.75	4.20	3.36	5.50	8.00	3.31	2.84	2.70	2.86	2.57
Total	99.67	100.12	99.89	99.75	99.87	99.77	99.83	99.89	100.10	99.82	99.90
$Mg^{\#}$	67.2	63.4	62.5	53.4	64.9	65.9	52.9	64.2	61.2	55.7	61.0
CaO/Al_2O_3	0.9	0.6	0.6	0.5	0.5	0.9	0.5	0.5	0.5	0.6	0.5
Li	29.30	41.20	42.50	44.40	39.70	31.70	31.40	42.00	40.50	40.80	44.60
Be	[b]	[b]	[b]	[b]	[b]	[b]	[b]	[b]	[b]	[b]	[b]
Sc	15.20	23.60	19.60	19.40	17.30	15.50	20.20	18.70	20.30	20.80	18.40
V	68.40	191.00	172.00	114.00	154.00	76.00	140.00	147.00	177.00	131.00	148.00
Cr	159.00	197.00	179.00	232.00	174.00	180.00	207.00	173.00	201.00	236.00	175.00
Co	50.80	57.70	52.50	47.00	42.20	53.80	60.20	44.00	42.60	55.10	44.90

Continued

	MD-02[a]	MD-04[a]	MD-06[a]	MD-08[a]	MD-15[a]	MD-16[a]	MD-17[a]	MD-19[a]	MD-20[a]	MD-22[a]	MD-23[a]
Ni	96.30	120.00	103.00	108.00	75.50	117.00	109.00	79.20	91.90	103.00	85.90
Cu	34.40	50.80	48.40	39.00	46.20	34.60	50.60	45.60	50.70	44.80	43.00
Zn	51.70	63.10	73.20	49.50	75.20	56.30	55.00	78.40	76.90	54.90	79.20
Ga	[b]	[b]	[b]	[b]	[b]	[b]	[b]	[b]	[b]	[b]	[b]
Ge	[b]	[b]	[b]	[b]	[b]	[b]	[b]	[b]	[b]	[b]	[b]
Rb	4.45	1.94	5.97	3.69	4.32	4.04	5.69	2.99	2.47	5.29	4.68
Sr	675.00	833.00	1067.00	387.00	962.00	645.00	648.00	969.00	955.00	573.00	1055.0
Y	18.00	24.70	23.90	20.60	22.10	19.90	22.50	23.50	23.20	21.80	22.30
Zr	198.00	271.00	354.00	310.00	324.00	207.00	305.00	334.00	354.00	315.00	318.00
Nb	29.00	43.90	59.20	43.90	56.00	30.40	41.50	55.10	60.60	43.60	51.10
Cs	0.49	1.74	4.55	0.11	3.93	0.40	0.97	6.67	3.44	0.22	4.86
Ba	113.00	224.00	245.00	72.80	253.00	123.00	143.00	219.00	218.00	264.00	243.00
La	23.10	34.30	44.30	30.40	41.60	25.10	35.70	42.30	42.20	36.00	40.30
Ce	47.10	66.40	86.40	64.50	80.80	50.60	66.80	79.40	85.80	70.00	75.70
Pr	5.20	7.60	9.40	7.53	9.30	5.65	7.27	8.82	9.99	7.72	8.57
Nd	21.00	32.00	37.60	30.50	37.80	23.10	28.30	36.30	38.50	30.80	36.50
Sm	4.70	7.00	7.33	6.24	7.61	5.17	6.36	7.85	7.75	6.45	7.82
Eu	1.66	2.50	2.61	2.02	2.51	1.81	2.27	2.66	2.61	2.15	2.51
Gd	4.52	6.47	6.96	5.77	6.49	4.89	5.94	6.55	7.19	5.99	6.34
Tb	0.69	0.92	1.02	0.87	0.97	0.77	0.86	0.95	1.09	0.89	0.92
Dy	3.68	5.05	5.19	4.50	5.11	3.97	4.40	4.97	5.56	4.67	4.82
Ho	0.70	0.99	0.94	0.83	0.94	0.76	0.87	0.97	0.98	0.91	0.95
Er	2.02	2.82	2.67	2.30	2.47	2.25	2.63	2.82	2.62	2.55	2.63
Tm	0.24	0.34	0.32	0.28	0.29	0.27	0.31	0.33	0.33	0.31	0.31
Yb	1.58	2.21	2.19	1.90	2.08	1.76	2.04	2.15	2.36	2.13	1.98
Lu	0.21	0.28	0.31	0.26	0.29	0.23	0.26	0.28	0.31	0.27	0.28
Hf	4.48	5.98	7.62	6.96	7.58	4.72	6.74	7.65	7.89	7.30	7.49
Ta	2.38	3.54	4.41	3.56	4.39	2.51	3.61	4.74	4.47	3.91	4.47
Pb	3.94	3.88	3.89	4.89	4.46	4.21	4.96	3.60	4.37	4.27	3.13
Th	2.91	3.22	5.11	5.96	4.86	3.30	5.27	5.02	5.30	5.88	4.85
U	0.87	1.31	2.11	1.68	1.98	0.94	2.03	1.93	2.44	2.06	2.00
Nb/Ta	12.2	12.4	13.4	12.3	12.8	12.1	11.5	11.6	13.6	11.2	11.4
Nb/U	33.3	33.5	28.1	26.1	28.3	32.3	20.4	28.5	24.8	21.2	25.6
Ce/Pb	12.0	17.1	22.2	13.2	18.1	12.0	13.5	22.1	19.6	16.4	24.2
La/Yb	10.49	11.13	14.51	11.48	14.35	10.23	12.55	14.11	12.83	12.12	14.60
Eu[a]	1.09	1.12	1.10	1.01	1.06	1.08	1.11	1.10	1.05	1.04	1.06
Zr/Hf	44.2	45.3	46.5	44.5	42.7	43.9	45.3	43.7	44.9	43.2	42.5
Th/U	3.34	2.46	2.42	3.55	2.45	3.51	2.60	2.60	2.17	2.85	2.43

[a] The data were cited from Lai et al. (2007b).

[b] No value.

Table 2 Major(wt%) and trace(ppm) element analysis result of the volcanic rocks from Lixian area.

	FSL-01	FSL-03	FSL-05	FSL-07	FSL-09	FSL-10	FSL-12	FSL-13	ZLG-01	ZLG-03	ZLG-05	ZLG-08	ZLG-10	ZLG-11	ZLG-12	ZLG-13	MQ-01
SiO_2	41.43	41.15	40.32	39.21	41.68	40.57	40.10	40.65	41.37	42.12	40.97	42.47	40.72	41.14	41.72	41.36	40.95
TiO_2	3.55	3.59	3.45	3.28	3.55	3.37	3.44	3.54	3.52	3.44	3.32	3.54	3.41	3.46	3.32	3.62	4.05
Al_2O_3	6.01	6.48	6.69	6.85	6.28	6.55	6.76	6.55	7.55	7.79	8.18	7.78	8.68	8.49	8.11	8.04	7.25
$Fe_2O_3{}^T$	11.18	11.16	11.59	11.07	11.12	11.01	11.30	11.47	12.20	11.28	11.56	11.72	11.44	11.66	11.49	12.03	12.18
MnO	0.14	0.16	0.15	0.15	0.16	0.15	0.15	0.15	0.13	0.13	0.15	0.16	0.15	0.16	0.15	0.17	0.17
MgO	16.73	15.91	16.29	17.10	15.88	17.90	16.47	16.65	11.33	11.67	12.11	11.81	12.40	11.09	12.33	11.88	12.18
CaO	12.50	12.90	12.79	12.80	12.92	12.15	12.83	12.66	14.81	14.59	14.35	13.91	13.72	14.56	13.95	14.44	14.81
Na_2O	1.50	1.43	1.91	1.82	1.29	1.26	1.83	1.64	1.93	1.82	2.02	1.54	1.82	1.64	1.65	1.96	1.03
K_2O	0.81	0.83	1.09	1.04	0.76	0.70	1.09	0.95	1.36	1.27	1.38	1.05	1.36	1.13	1.18	1.37	1.17
P_2O_5	1.11	1.08	1.10	1.28	1.04	1.13	1.28	1.14	0.75	0.84	0.81	0.77	0.90	0.93	0.81	0.88	1.05
LOI	4.56	4.77	4.04	4.77	4.73	4.66	4.18	3.99	4.62	4.61	4.66	4.84	4.91	5.27	4.86	3.82	4.59
Total	99.52	99.46	99.42	99.37	99.41	99.45	99.43	99.39	99.57	99.56	99.51	99.59	99.51	99.53	99.57	99.57	99.43
$Mg^{\#}$	77.7	76.9	76.6	78.3	76.9	79.1	77.3	77.2	68.4	70.7	70.9	70.1	71.6	68.9	71.4	69.7	70.0
CaO/Al_2O_3	2.1	2.0	1.9	1.9	2.1	1.9	1.9	1.9	2.0	1.9	1.8	1.8	1.6	1.7	1.7	1.8	2.0
Li	10.7	11.8	13.7	10.6	10.6	10.1	11.8	13.5	35.2	27.8	35.4	48.2	28.6	57.4	34.3	18.4	25.8
Be	2.07	2.13	2.22	2.22	2.13	2.09	2.17	2.26	1.92	1.78	1.91	1.85	2.11	1.89	1.93	2.15	2.82
Sc	20.1	20.3	19.3	19.0	19.8	19.2	19.3	20.9	24.4	24.9	23.1	23.7	21.0	21.7	23.4	21.5	22.8
V	162	172	146	151	171	154	164	171	199	203	205	206	200	207	199	214	199
Cr	715	704	632	670	670	668	700	666	318	347	307	291	247	276	296	227	252
Co	75.9	67.1	79.8	75.1	69.2	76.5	69.5	68.3	70.6	71.1	76.1	61.3	67.6	64.2	66.8	63.3	86.0
Ni	469	460	487	503	464	500	486	476	254	273	272	241	229	240	240	209	178
Cu	80.1	81.6	85.2	75.6	82.6	71.9	80.1	75.6	80.3	60.6	68.1	77.7	69.0	72.5	71.1	53.4	71.4
Zn	119	123	121	152	122	118	119	121	122	119	120	127	123	125	121	131	143
Ga	15.8	16.3	16.3	15.8	16.0	15.3	16.8	16.7	17.6	17.0	17.4	18.1	18.5	18.7	17.4	22.2	17.0
Ge	1.42	1.43	1.36	1.36	1.41	1.37	1.37	1.40	1.41	1.36	1.34	1.42	1.33	1.34	1.35	1.41	1.51

Continued

	FSL-01	FSL-03	FSL-05	FSL-07	FSL-09	FSL-10	FSL-12	FSL-13	ZLG-01	ZLG-03	ZLG-05	ZLG-08	ZLG-10	ZLG-11	ZLG-12	ZLG-13	MQ-01
Rb	28. 3	26. 2	34. 4	34. 1	25. 0	23. 8	34. 7	30. 3	30. 1	34. 3	38. 8	25. 5	36. 7	26. 8	32. 8	40. 8	46. 5
Sr	1 182	1 171	1 158	1 328	1 127	1 397	1 219	1 184	1 001	1 111	1 311	1 077	1 164	1 100	1 105	1 049	1 885
Y	31. 4	32. 0	31. 8	31. 9	31. 0	30. 6	31. 4	31. 4	26. 7	26. 1	25. 7	26. 3	28. 2	26. 8	26. 8	30. 0	37. 9
Zr	426	442	414	394	435	400	408	429	337	313	309	335	330	331	328	372	526
Nb	148	150	145	141	149	143	141	143	124	114	118	126	127	128	121	146	153
Cs	0. 44	0. 40	0. 43	0. 45	0. 37	0. 32	0. 40	0. 35	1. 08	0. 71	0. 54	0. 84	0. 55	0. 87	0. 43	1. 26	2. 34
Ba	955	1 834	1 124	1 443	2 160	979	1 636	1 080	995	1 825	1 590	492	716	1812	1 115	428	1 591
La	137	136	130	131	132	134	130	126	87. 7	87. 3	87. 1	92. 1	93. 8	90. 3	91. 1	106	138
Ce	254	250	240	241	246	251	238	228	165	162	163	170	170	168	168	195	259
Pr	29. 4	28. 9	27. 9	27. 8	28. 4	29. 2	27. 4	26. 4	18. 7	18. 8	18. 9	19. 5	19. 6	19. 5	19. 3	22. 2	29. 8
Nd	111	109	104	106	107	109	105	99. 9	72. 4	71. 5	71. 5	74. 4	73. 9	73. 4	73. 7	84. 1	114
Sm	20. 3	20. 2	19. 3	19. 5	19. 7	19. 9	19. 1	18. 5	13. 7	13. 6	13. 5	14. 0	13. 9	14. 0	13. 8	15. 6	21. 6
Eu	5. 56	5. 61	5. 29	5. 37	5. 54	5. 48	5. 38	5. 18	3. 95	3. 97	3. 96	3. 98	4. 05	4. 09	4. 01	4. 44	6. 01
Gd	16. 1	16. 1	15. 5	15. 5	15. 9	15. 9	15. 5	15. 1	11. 5	11. 5	11. 5	11. 6	11. 8	11. 9	11. 8	13. 0	17. 7
Tb	1. 81	1. 82	1. 76	1. 76	1. 80	1. 80	1. 75	1. 70	1. 33	1. 31	1. 31	1. 35	1. 38	1. 36	1. 36	1. 53	2. 01
Dy	8. 55	8. 70	8. 40	8. 39	8. 55	8. 55	8. 29	8. 19	6. 57	6. 50	6. 46	6. 65	6. 89	6. 69	6. 74	7. 54	9. 74
Ho	1. 32	1. 36	1. 34	1. 32	1. 32	1. 32	1. 30	1. 29	1. 08	1. 06	1. 05	1. 07	1. 14	1. 10	1. 10	1. 23	1. 55
Er	2. 82	2. 90	2. 88	2. 82	2. 81	2. 85	2. 80	2. 78	2. 35	2. 28	2. 28	2. 36	2. 49	2. 41	2. 40	2. 69	3. 37
Tm	0. 32	0. 32	0. 33	0. 33	0. 32	0. 32	0. 31	0. 31	0. 28	0. 27	0. 28	0. 28	0. 30	0. 28	0. 28	0. 32	0. 40
Yb	1. 66	1. 72	1. 73	1. 69	1. 65	1. 67	1. 63	1. 66	1. 60	1. 56	1. 55	1. 59	1. 70	1. 64	1. 64	1. 82	2. 23
Lu	0. 22	0. 23	0. 23	0. 22	0. 22	0. 22	0. 22	0. 22	0. 23	0. 21	0. 22	0. 22	0. 24	0. 23	0. 22	0. 25	0. 30
Hf	8. 78	9. 14	8. 68	8. 12	9. 08	8. 55	8. 27	8. 73	7. 20	6. 73	6. 60	7. 18	6. 79	6. 89	6. 93	7. 53	10. 3
Ta	6. 88	6. 79	6. 48	6. 39	6. 72	6. 80	6. 41	6. 29	5. 96	5. 44	5. 60	6. 11	6. 08	6. 11	5. 78	6. 90	6. 52
Pb	7. 05	6. 89	3. 44	10. 5	5. 30	5. 02	5. 79	5. 45	4. 52	3. 79	4. 74	5. 47	6. 23	5. 30	4. 63	5. 77	7. 63
Th	18. 3	17. 8	16. 7	16. 9	17. 3	18. 3	16. 7	16. 1	12. 6	12. 1	12. 3	13. 4	13. 3	13. 2	13. 0	15. 2	19. 1

Continued

	FSL-01	FSL-03	FSL-05	FSL-07	FSL-09	FSL-10	FSL-12	FSL-13	ZLG-01	ZLG-03	ZLG-05	ZLG-08	ZLG-10	ZLG-11	ZLG-12	ZLG-13	MQ-01
U	3. 51	3. 66	3. 60	3. 43	3. 60	3. 55	3. 37	3. 34	2. 17	2. 39	2. 45	1. 78	1. 90	1. 98	2. 40	3. 09	3. 64
Nb/Ta	21. 6	22. 1	22. 3	22. 1	22. 2	21. 1	21. 9	22. 7	20. 9	20. 9	21. 1	20. 7	20. 9	21. 0	20. 9	21. 2	23. 5
Nb/U	42. 3	41. 0	40. 2	41. 2	41. 4	40. 5	41. 7	42. 7	57. 4	47. 7	48. 4	71. 0	66. 9	64. 7	50. 3	47. 4	42. 0
Ce/Pb	36. 0	36. 3	69. 7	23. 0	46. 5	50. 0	41. 1	41. 9	36. 5	42. 7	34. 4	31. 1	27. 3	31. 7	36. 2	33. 7	33. 9
La/Yb	59. 21	56. 54	53. 92	55. 29	57. 30	57. 57	57. 03	54. 40	39. 32	40. 22	40. 23	41. 54	39. 51	39. 59	39. 81	41. 93	44. 26
Eu*	0. 91	0. 92	0. 91	0. 91	0. 93	0. 91	0. 93	0. 92	0. 93	0. 95	0. 95	0. 93	0. 94	0. 94	0. 94	0. 92	0. 91
Zr/Hf	48. 5	48. 4	47. 7	48. 5	47. 9	46. 8	49. 3	49. 2	46. 9	46. 5	46. 9	46. 6	48. 6	48. 1	47. 3	49. 5	50. 9
Th/U	5. 22	4. 86	4. 64	4. 93	4. 81	5. 17	4. 96	4. 82	5. 81	5. 06	5. 02	7. 56	6. 99	6. 64	5. 40	4. 94	5. 26
	MQ-02	MQ-05	MQ-06	MQ-07	MQ-09	MQ-11	MQ-12	LP-03	LP-04	LP-06	LP-07	LP-08	LP-10	LP-13	LP-15	LP-16	LP-17	LP-18
SiO_2	40. 86	41. 84	41. 39	40. 96	40. 40	41. 19	40. 88	42. 82	42. 33	42. 60	42. 54	41. 72	42. 60	42. 12	42. 03	42. 37	41. 98	41. 99
TiO_2	4. 05	4. 36	4. 18	4. 23	3. 73	4. 39	4. 56	3. 61	3. 57	3. 57	3. 58	3. 46	3. 71	3. 60	3. 57	3. 63	3. 62	3. 61
Al_2O_3	6. 93	6. 22	7. 32	6. 65	7. 65	7. 86	6. 16	6. 00	6. 42	6. 35	6. 51	6. 75	5. 61	5. 82	6. 13	5. 75	5. 94	5. 92
$Fe_2O_3^T$	12. 25	12. 65	12. 22	12. 46	12. 53	12. 86	13. 35	12. 17	12. 21	12. 14	12. 03	11. 84	12. 24	11. 75	11. 75	11. 67	11. 79	11. 70
MnO	0. 17	0. 18	0. 18	0. 19	0. 17	0. 17	0. 18	0. 15	0. 16	0. 16	0. 16	0. 16	0. 16	0. 15	0. 16	0. 16	0. 16	0. 15
MgO	12. 24	12. 13	11. 38	12. 67	13. 05	10. 81	12. 83	15. 47	15. 24	14. 93	15. 38	15. 70	15. 40	15. 95	16. 06	16. 24	16. 12	15. 75
CaO	14. 90	15. 71	14. 75	14. 85	13. 74	13. 65	14. 46	12. 77	12. 82	12. 85	12. 65	12. 66	13. 25	12. 95	12. 73	12. 78	12. 93	12. 87
Na_2O	1. 02	1. 02	0. 80	0. 57	2. 59	2. 10	1. 57	1. 37	1. 37	1. 34	1. 48	1. 63	1. 51	1. 25	1. 10	0. 95	1. 28	1. 25
K_2O	1. 07	0. 57	0. 37	0. 45	3. 24	1. 93	3. 04	0. 73	0. 71	0. 72	0. 80	0. 90	0. 80	0. 59	0. 51	0. 48	0. 61	0. 60
P_2O_5	1. 07	1. 07	1. 17	0. 90	1. 11	1. 20	0. 79	0. 68	0. 78	0. 77	0. 67	0. 81	0. 84	0. 89	0. 91	0. 86	0. 89	0. 88
LOI	4. 83	3. 67	5. 67	5. 52	1. 20	3. 30	1. 50	4. 10	4. 45	4. 29	4. 16	4. 09	3. 61	4. 46	4. 75	4. 80	4. 43	4. 81
Total	99. 39	99. 42	99. 43	99. 45	99. 41	99. 46	99. 32	99. 87	100. 06	99. 72	99. 96	99. 72	99. 73	99. 53	99. 70	99. 69	99. 75	99. 53
$Mg^{\#}$	70. 0	69. 1	68. 5	70. 3	70. 8	66. 2	69. 1	74. 8	74. 4	74. 1	74. 9	75. 6	74. 6	76. 0	76. 1	76. 4	76. 1	75. 8
CaO/Al_2O_3	2. 2	2. 5	2. 0	2. 2	1. 8	1. 7	2. 3	2. 1	2. 0	2. 0	1. 9	1. 9	2. 4	2. 2	2. 1	2. 2	2. 2	2. 2
Li	23. 9	27. 4	46. 3	60. 7	13. 6	14. 3	17. 2	17. 7	18. 0	18. 1	18. 4	18. 8	13. 1	16. 2	18. 3	14. 3	16. 4	16. 2
Be	2. 78	2. 44	2. 47	2. 58	2. 84	2. 70	2. 63	2. 44	2. 50	2. 50	2. 45	2. 53	2. 56	2. 36	2. 43	2. 15	2. 33	2. 37

Continued

	MQ-02	MQ-05	MQ-06	MQ-07	MQ-09	MQ-11	MQ-12	LP-03	LP-04	LP-06	LP-07	LP-08	LP-10	LP-13	LP-15	LP-16	LP-17	LP-18
Sc	23.2	24.9	21.6	22.6	21.8	21.4	24.3	20.7	20.5	21.0	20.2	20.0	21.6	20.8	20.6	20.1	20.4	20.8
V	209	218	202	204	203	215	244	136	131	130	130	131	151	171	168	174	170	172
Cr	252	271	273	280	307	236	292	656	643	649	627	618	641	689	688	670	687	665
Co	76.4	76.9	72.3	59.9	77.5	92.6	80.2	76.1	96.4	74.3	73.7	73.4	78.3	76.6	78.8	73.4	70.3	131
Ni	175	191	207	193	272	188	224	488	476	470	479	498	469	493	491	487	491	487
Cu	70.1	64.5	56.3	58.5	83.7	129	102	88.0	85.1	88.1	87.1	87.4	93.0	94.4	93.5	72.2	90.6	91.1
Zn	149	157	148	149	141	146	151	128	127	126	124	126	131	127	126	121	127	124
Ga	16.8	22.4	18.3	18.4	19.1	21.6	19.3	16.9	17.2	17.4	17.1	17.2	17.1	16.9	16.9	16.4	16.6	16.4
Ge	1.53	1.69	1.52	1.53	1.52	1.44	1.61	1.46	1.44	1.45	1.46	1.43	1.50	1.48	1.47	1.46	1.45	1.46
Rb	43.7	27.8	15.5	31.9	83.1	77.0	77.6	27.5	27.0	27.1	28.9	31.5	29.6	22.0	19.5	17.5	21.6	21.4
Sr	1 810	2 083	2 366	1 625	1 759	1 705	1 612	1 094	1 113	1 108	1 084	1 177	1 283	1 465	1 502	1 555	1 432	1 453
Y	37.4	38.4	37.7	37.6	41.4	38.2	39.6	31.5	31.7	31.5	31.6	31.9	31.7	31.6	31.0	31.1	31.2	30.7
Zr	530	565	533	538	511	527	590	453	448	457	443	437	474	449	438	445	442	438
Nb	153	163	152	154	152	156	184	152	148	150	148	146	155	150	147	147	145	144
Cs	2.30	3.90	11.4	4.45	0.88	0.81	1.38	0.52	0.53	0.51	0.52	0.54	0.63	0.45	0.44	0.35	0.43	0.45
Ba	1 455	1 020	1 555	1 312	1 916	1 429	1 570	399	418	420	400	446	853	937	1 054	495	552	795
La	137	144	140	144	150	124	159	127	129	128	128	126	129	130	126	130	126	124
Ce	260	277	262	267	267	231	289	239	240	241	238	238	246	243	236	241	234	232
Pr	30.1	31.9	30.0	31.0	30.3	26.9	32.1	27.3	27.6	27.4	27.2	26.9	27.8	27.7	26.9	27.7	26.7	26.3
Nd	114	122	116	119	116	103	121	104	104	105	104	103	105	106	103	105	101	101
Sm	21.7	22.9	21.9	22.5	21.7	19.8	21.9	19.3	19.3	19.4	19.3	18.9	19.3	19.5	18.8	19.3	18.7	18.5
Eu	6.05	6.29	6.09	6.27	6.11	5.59	6.04	5.31	5.31	5.32	5.34	5.25	5.34	5.40	5.28	5.38	5.15	5.14
Gd	17.8	18.6	17.8	18.4	18.1	16.5	18.1	15.5	15.4	15.5	15.4	15.3	15.7	15.7	15.2	15.5	15.1	14.9
Tb	2.01	2.10	2.02	2.09	2.10	1.94	2.07	1.77	1.76	1.77	1.76	1.75	1.76	1.77	1.72	1.78	1.70	1.70
Dy	9.71	10.2	9.81	9.97	10.3	9.53	10.1	8.43	8.48	8.54	8.44	8.39	8.41	8.41	8.22	8.50	8.15	8.11

Continued

	MQ-02	MQ-05	MQ-06	MQ-07	MQ-09	MQ-11	MQ-12	LP-03	LP-04	LP-06	LP-07	LP-08	LP-10	LP-13	LP-15	LP-16	LP-17	LP-18
Ho	1.56	1.60	1.56	1.58	1.69	1.57	1.64	1.34	1.34	1.35	1.35	1.34	1.34	1.34	1.31	1.33	1.29	1.27
Er	3.40	3.51	3.35	3.37	3.66	3.40	3.58	2.86	2.85	2.87	2.86	2.86	2.84	2.81	2.78	2.81	2.75	2.65
Tm	0.39	0.40	0.39	0.39	0.42	0.40	0.41	0.33	0.32	0.33	0.32	0.33	0.33	0.32	0.31	0.32	0.31	0.31
Yb	2.26	2.32	2.15	2.19	2.29	2.19	2.24	1.82	1.79	1.80	1.81	1.79	1.82	1.74	1.72	1.74	1.71	1.68
Lu	0.30	0.31	0.29	0.29	0.30	0.29	0.30	0.25	0.24	0.25	0.24	0.24	0.25	0.24	0.23	0.24	0.23	0.23
Hf	10.5	11.4	10.7	10.8	10.3	10.9	12.1	9.21	9.06	9.35	9.12	8.83	9.55	9.15	8.88	9.15	8.84	8.74
Ta	6.56	7.23	6.60	6.73	6.69	7.32	8.58	6.97	6.94	7.04	6.98	6.74	7.01	6.97	6.71	6.90	6.64	6.64
Pb	7.46	7.30	7.51	6.05	3.75	6.98	6.12	4.66	5.78	5.23	4.70	4.58	6.61	5.89	6.35	4.85	5.86	6.05
Th	19.1	20.9	18.4	19.0	19.4	16.8	21.2	17.0	16.9	17.1	16.9	16.2	16.8	17.0	16.1	16.9	16.3	16.2
U	3.58	3.73	3.67	3.61	3.90	3.32	4.20	3.65	3.59	3.67	3.54	3.45	3.51	3.55	3.50	3.43	3.44	3.40
Nb/Ta	23.3	22.5	23.0	23.0	22.7	21.3	21.5	21.7	21.3	21.3	21.3	21.6	22.1	21.5	21.8	21.2	21.8	21.8
Nb/U	42.7	43.6	41.5	42.8	38.9	47.0	43.8	41.5	41.2	41.0	42.0	42.2	44.1	42.2	41.9	42.8	42.1	42.5
Ce/Pb	34.8	38.0	34.9	44.0	71.2	33.1	47.2	51.4	41.5	46.1	50.7	51.9	37.2	41.3	37.2	49.6	39.9	38.3
La/Yb	43.60	44.60	46.61	47.31	46.91	40.84	51.03	50.09	51.62	50.99	50.87	50.55	50.84	53.69	52.83	53.41	52.62	53.04
Eu^*	0.91	0.90	0.91	0.91	0.91	0.92	0.90	0.91	0.91	0.91	0.92	0.91	0.91	0.91	0.92	0.92	0.91	0.92
Zr/Hf	50.5	49.6	49.8	50.0	49.9	48.5	48.8	49.2	49.5	48.9	48.6	49.5	49.6	49.1	49.3	48.7	50.0	50.1
Th/U	5.34	5.61	5.01	5.27	4.96	5.05	5.04	4.66	4.70	4.67	4.78	4.69	4.80	4.78	4.60	4.93	4.75	4.76

the picrobasalt, foidite, and basanite fields (Fig. 4). The Xiahe volcanic rocks have low SiO_2 contents (45.73%-53.80%), Al_2O_3 = 14.38%-20.08%, MgO = 4.98%-8.66%, $Mg^{\#}$ = 53-67; CaO contents varying from 3.48% to 12.81%, and CaO/Al_2O_3 ratios of 0.24-0.64. Na_2O+K_2O = 3.10%-6.02% and Na_2O/K_2O = 2.66-15.19. These geochemical features indicate that the Xiahe volcanic rocks are sodic alkaline volcanic rocks. The rocks have higher TiO_2 content (1.59%-3.95%) than that of island arc volcanic rocks (0.58%-0.85%) and MORB (1.5%), but similar to that of OIB alkaline basalts (2.20%). Compared to the Xiahe volcanic rocks, the Lixian volcanic rocks display lower SiO_2 (41.14%-44.82%) and Al_2O_3 (5.84%-9.18%) contents, and CaO/Al_2O_3 ratios of 1.43 to 2.37. They also have lower Na_2O/K_2O ratios (0.52-2.16) and higher TiO_2 (3.47%-4.66%) and MgO (11.24%-18.88%) contents, with $Mg^{\#}$ values of 66 to 79. In the major elements vs. SiO_2 diagrams (Fig. 5), the Lixian picro-basalts display lower SiO_2, Al_2O_3, and Na_2O, and higher MgO and CaO contents than those of the Xiahe sodic alkaline basalts.

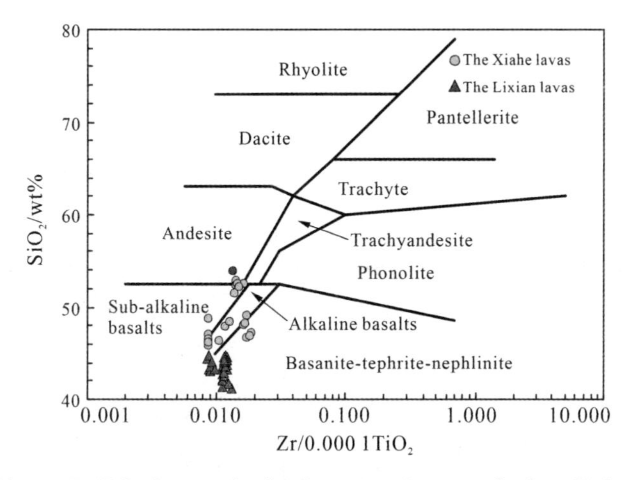

Fig. 4 SiO_2 vs. Zr/TiO_2 diagram for the Cenozoic volcanic rocks from Xiahe and Lixian area. After Winchester and Floyd (1977).

The Xiahe volcanic rocks have high total rare earth elements (∑REE) contents of 134-231 ppm, with remarkable REE fractions of $(La/Yb)_N$ = 10.2-14.6 and $(Ce/Yb)_N$ = 8.0-10.9, and are slightly enriched in Eu (Eu^*/Eu = 1.01-1.12) (Fig. 6a). They are enriched in Ba, Th, and U, and particularly in Nb and Ta relative to Nd, Hf, Sm, Y, and Yb (Fig. 6b), while the obvious negative anomalies in K and Rb are quite distinct from those of the Cenozoic potassic-ultrapotassic volcanic rocks in the northern TP. The rocks' geochemical ratios are Ti/V = 63.4-128, Th/Ta = 0.9-1.7, Th/Y = 0.1-0.3, and Ta/Yb = 1.4-2.3. The volcanic rocks are enriched in the LILEs, especially in Nb and Ta, resembling source region characteristics of the OIB-type. The Lixian picro-basalts display higher ∑REE contents of 382-668 pm, also yielding higher $(La/Yb)_N$ (39.3-59.2) and $(Ce/Yb)_N$ (27.8-42.4)

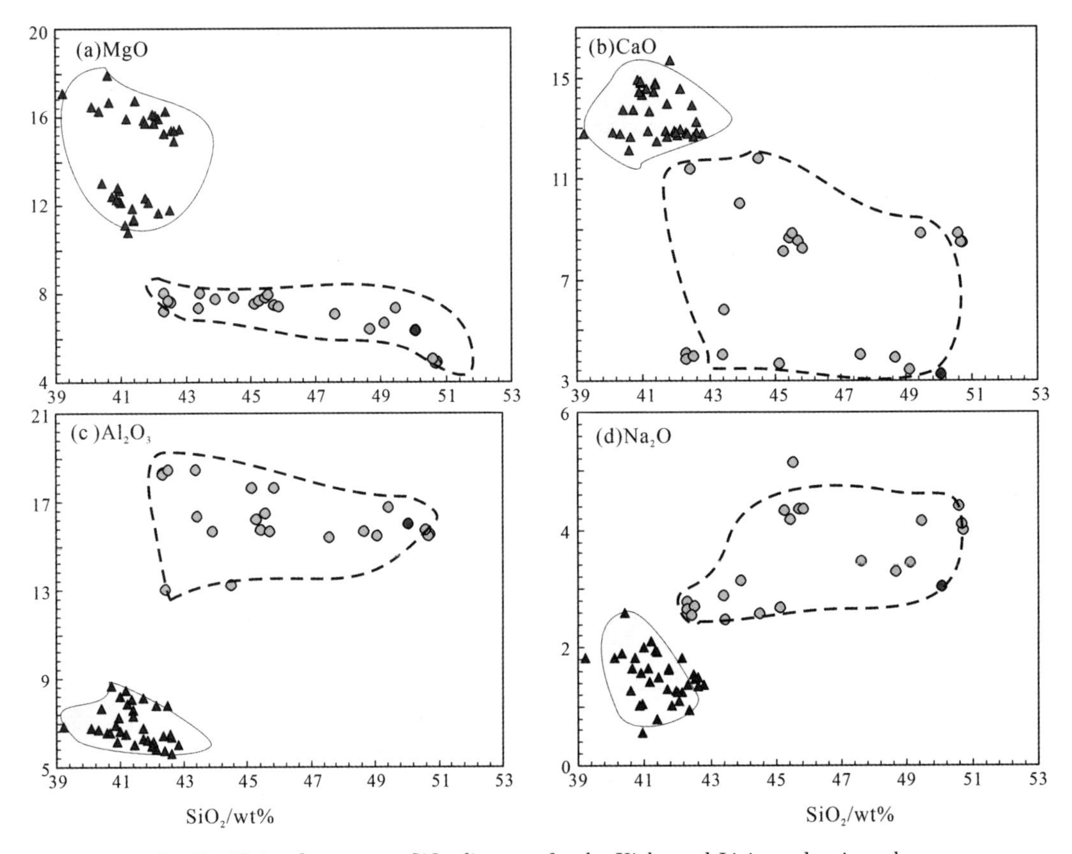

Fig. 5 Major elements vs. SiO_2 diagrams for the Xiahe and Lixian volcanic rocks.

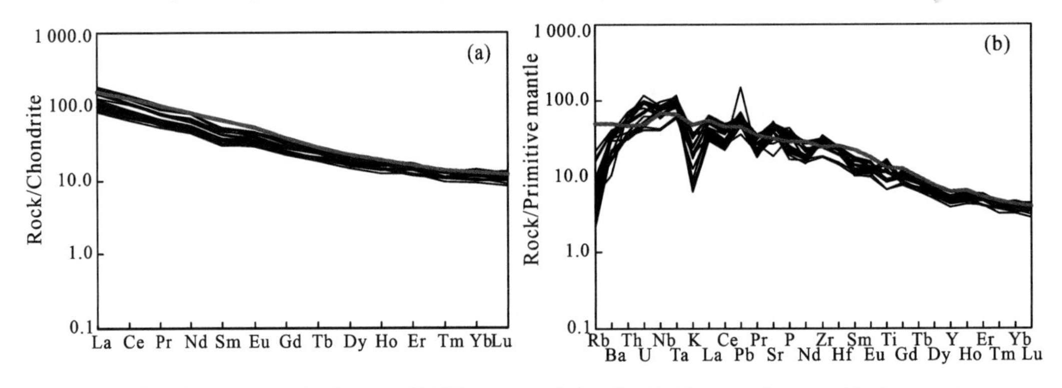

Fig. 6 Chrondrite rare earth element (REE) pattern (a) and primitive mantle normalized trace element spider diagram (b) for the Xiahe alkaline basalts.

Normalized value was cited from Sun and McDonough (1989).

ratios (Fig. 7a), and are enriched in LREE, with $Eu^*/Eu = 0.90-0.94$. These rocks display high Cr (237–715 ppm) and Ni (175–503 ppm) contents, which are significantly higher than those of the Xiahe volcanic rocks. The Lixian picro-basalts also display negative anomalies in K and Rb, no negative anomalies in Nb and Ta, Ti/Ti^* values of 0.3–0.6, Zr/Nb ratios of 2.5–3.5, and Zr/Hf ratios of 46.5–50.9 (Fig. 7b).

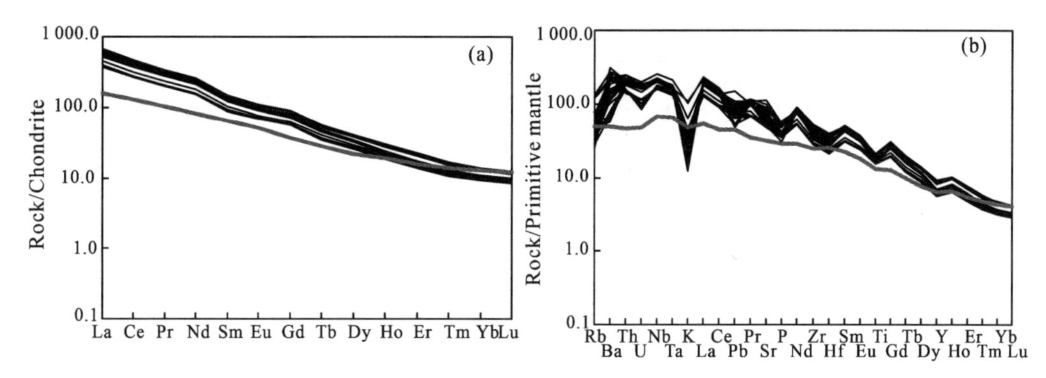

Fig. 7 Chrondrite rare earth element (REE) pattern (a) and primitive mantle normalized trace element spider diagram (b) for the Lixian picro-basalts.

Normalized value was cited from Sun and McDonough (1989).

5 Sr-Nd-Pb isotopic composition

The Sr-Nd and Pb isotopic compositions for the Xiahe and Lixian lavas are presented in Tables 3 and 4, respectively. Given that the lavas erupted after the Cenozoic, and that their isotopic composition had not evidently changed with time, the present isotopic ratios could be used as evidence to place a constraint on their source regions.

The Xiahe volcanic rocks have $^{87}Sr/^{86}Sr$ values of 0. 704 071–0. 704 693, low $^{143}Nd/^{144}Nd$ ratios of 0. 512 409–0. 512 513, and $\varepsilon_{Nd}(t)$ values of −2. 4 to −4. 5. In the $^{143}Nd/^{144}Nd$ vs. $^{87}Sr/^{86}Sr$ diagram (Fig. 9), the rocks plotted in the region between EM Ⅰ and BSE, implying that their source region has been metasomatized. The samples have $^{206}Pb/^{204}Pb$ = 18. 699 0–18. 849 8, $^{207}Pb/^{204}Pb$ = 15. 846 2–15. 973 6, and $^{208}Pb/^{204}Pb$ = 39. 305 3–39. 819 8. The Pb isotope data were superimposed over the Hugh Pb composition diagram (Fig. 10). In both the $^{207}Pb/^{204}Pb$ vs. $^{206}Pb/^{204}Pb$ and $^{208}Pb/^{204}Pb$ vs. $^{206}Pb/^{204}Pb$ diagrams, samples are plotted above the northern hemisphere reference line (NHRL) marked by Th/U = 4. 0, and shown to have isotopic compositions that accord with EM Ⅱ, BSE, or PREMA.

The Lixian picro-basalts display similar Sr isotopic compositions ($^{87}Sr/^{86}Sr$ = 0. 704 090–0. 704 668) and depleted Nd isotopic compositions to the Xiahe volcanic rocks. They have high $^{143}Nd/^{144}Nd$ ratios of 0. 512 768 to 0. 512 904, and $\varepsilon_{Nd}(t)$ values of +2. 6 to +5. 3. In the $^{143}Nd/^{144}Nd$ vs. $^{87}Sr/^{86}Sr$ diagram (Fig. 9), the rocks plotted in the region between PREMA and BSE, clearly differing from the Xiahe volcanic rocks. They have $^{206}Pb/^{204}Pb$ = 18. 353 8–18. 852 5, $^{207}Pb/^{204}Pb$ = 15. 494 8–15. 603 0, and $^{208}Pb/^{204}Pb$ = 38. 074 6–39. 074 7. These features are similar to those of kamafugite from the Lixian area and are distinct from the carbonatites (Yu et al., 2003). The Pb isotopic data presented in Fig. 10 were superimposed upon the Hugh Pb composition diagram (Zindler and Hart, 1986). In the $^{208}Pb/^{204}Pb$ vs.

Table 3 Sr-Nd isotopic composition of the Cenozoic Xiahe and Lixian volcanic rocks, western Qinling.

Rock type	Sample	Sr /ppm	Rb /ppm	$^{87}Sr/^{86}Sr$	2σ	$^{87}Rb/^{86}Sr$ [a]	$^{87}Sr/^{86}Sr$ (20 Ma)	Nd /ppm	Sm /ppm	$^{143}Nd/^{144}Nd$	2σ	$^{147}Sm/^{144}Nd$ [a]	T_{DM} /Ga [b]	$\varepsilon_{Nd}(t)$ [c]	$^{143}Nd/^{144}Nd$ (20 Ma)
Tephrite	FSL-01	1 182	28. 3	0. 704 128	0. 000 013	0. 069 247	0. 704 128	111	20. 3	0. 512 777	0. 000 004	0. 109 959	0. 51	2. 74	0. 512 776
Tephrite	FSL-05	1 158	34. 4	0. 704 133	0. 000 010	0. 085 841	0. 704 132	104	19. 3	0. 512 806	0. 000 011	0. 111 679	0. 47	3. 31	0. 512 804 9
Tephrite	FSL-12	1 219	34. 7	0. 704 136	0. 000 009	0. 082 376	0. 704 135	105	19. 1	0. 512 794	0. 000 029	0. 110 272	0. 48	3. 06	0. 512 792 1
Tephrite	FSL-13	1 184	30. 3	0. 704 193	0. 000 013	0. 074 054	0. 704 193	99. 9	18. 5	0. 512 791	0. 000 004	0. 111 911	0. 49	3. 00	0. 512 789 2
Tephrite	LP06	1 108	27. 1	0. 704 400	0. 000 012	0. 070 844	0. 704 400	105	19. 4	0. 512 876	0. 000 031	0. 111 532	0. 37	4. 67	0. 512 874 6
Tephrite	LP07	1 084	28. 9	0. 704 380	0. 000 011	0. 077 006	0. 704 380	104	19. 3	0. 512 881	0. 000 033	0. 111 534	0. 37	4. 76	0. 512 879 5
Tephrite	LP-17	1 432	21. 6	0. 704 668	0. 000 011	0. 043 708	0. 704 668	101	18. 7	0. 512 831	0. 000 012	0. 111 740	0. 43	3. 78	0. 512 829 3
Tephrite	MQ05	2 083	27. 8	0. 704 556	0. 000 013	0. 038 631	0. 704 555	122	22. 9	0. 512 869	0. 000 034	0. 113 318	0. 38	4. 52	0. 512 867 2
Tephrite	MQ11	1 705	77. 0	0. 704 266	0. 000 010	0. 130 579	0. 704 265	103	19. 8	0. 512 849	0. 000 057	0. 116 682	0. 41	4. 13	0. 512 847 2
Tephrite	MQ12	1 612	77. 6	0. 704 090	0. 000 013	0. 139 218	0. 704 090	121	21. 9	0. 512 770	0. 000 007	0. 109 774	0. 52	2. 60	0. 512 768 6
Tephrite	ZLG-01	1 001	30. 1	0. 704 189	0. 000 010	0. 087 094	0. 704 189	72. 4	13. 7	0. 512 832	0. 000 034	0. 114 633	0. 43	3. 80	0. 512 830 2
Tephrite	ZLG-03	1 111	34. 3	0. 704 229	0. 000 011	0. 089 265	0. 704 229	71. 5	13. 6	0. 512 906	0. 000 025	0. 114 807	0. 33	5. 26	0. 512 904 9
Tephrite	ZLG-13	1 049	40. 8	0. 704 418	0. 000 010	0. 112 566	0. 704 418	84. 1	15. 6	0. 512 803	0. 000 004	0. 112 526	0. 47	3. 23	0. 512 801 2
Trachy-basalt	MD-20	955	2. 48	0. 704 071	0. 000 008	0. 007 511	0. 704 071	38. 5	7. 75	0. 512 43	0. 000 007	0. 1216 91	0. 98	−4. 04	0. 512 428 4
Trachy-basalt	MD-22	573	5. 29	0. 704 693	0. 000 009	0. 026 703	0. 704 693	30. 7	6. 45	0. 512 409	0. 000 008	0. 1270 10	1. 00	−4. 45	0. 512 407 3
Trachy-basalt	MD-23	1 055	4. 68	0. 704 265	0. 000 01	0. 012 830	0. 704 265	36. 5	7. 82	0. 512 513	0. 000 005	0. 129 521	0. 86	−2. 42	0. 512 511 3

[a] $^{87}Rb/^{86}Sr$ and $^{147}Sm/^{144}Nd$ ratios were calculated using Rb, Sr, Sm and Nd contents, measured by ICP-MS.

[b] T_{DM} values were calculated using present-day $(^{147}Sm/^{144}Nd)_{DM}=0.2137$ and $(^{143}Nd/^{144}Nd)_{DM}=0.51315$.

[c] $\varepsilon_{Nd}(t)$ values were calculated using present-day $(^{147}Sm/^{144}Nd)_{CHUR}=0.1967$ and $(^{143}Nd/^{144}Nd)_{CHUR}=0.512638$.

Table 4 Pb isotopic composition of the Cenozoic Xiahe and Lixian volcanic rocks, western Qinling.

Rock type	Sample	U	Th	Pb	$^{206}Pb/^{204}Pb$	2σ	$^{207}Pb/^{204}Pb$	2σ	$^{208}Pb/^{204}Pb$	2σ	$^{238}U/^{204}Pb$ [a]	$^{232}Th/^{204}Pb$ [a]	$^{206}Pb/^{204}b$ (20 Ma) [b]	$^{207}Pb/^{204}Pb$ (20 Ma) [b]	$^{208}Pb/^{204}Pb$ (20 Ma) [b]	Δ7	Δ8
Tephrite	FSL-01	3. 51	18. 3	7. 05	18. 763 253	0. 001 106	15. 599 497	0. 000 954	39. 168 525	0. 002 520	32	170	18. 753 389	15. 599 042	39. 151 596	7. 52	85. 17
Tephrite	FSL-05	3. 60	16. 7	3. 44	18. 850 889	0. 001 464	15. 603 952	0. 001 226	39. 277 235	0. 003 040	67	319	18. 830 124	15. 602 995	39. 245 517	7. 08	85. 29
Tephrite	FSL-12	3. 37	16. 7	5. 79	18. 747 502	0. 001 096	15. 595 601	0. 001 002	39. 129 308	0. 002 520	37	189	18. 735 969	15. 595 069	39. 110 502	7. 31	83. 17
Tephrite	FSL-13	3. 34	16. 1	5. 45	18. 752 492	0. 000 786	15. 597 309	0. 000 712	39. 134 579	0. 001 854	39	194	18. 740 340	15. 596 749	39. 115 295	7. 43	83. 12
Tephrite	LP06	3. 67	17. 1	5. 23	18. 779 184	0. 000 828	15. 602 454	0. 000 792	39. 181 458	0. 002 100	45	215	18. 765 264	15. 601 812	39. 160 067	7. 67	84. 59
Tephrite	LP07	3. 54	16. 9	4. 70	18. 729 871	0. 000 540	15. 595 391	0. 000 522	39. 097 372	0. 001 296	48	235	18. 714 972	15. 594 704	39. 073 965	7. 50	82. 06
Tephrite	LP-17	3. 44	16. 3	5. 86	18. 760 983	0. 000 818	15. 591 395	0. 000 722	39. 207 882	0. 001 826	37	183	18. 749 348	15. 590 859	39. 189 699	6. 74	89. 47
Tephrite	MQ05	3. 73	20. 9	7. 30	18. 363 698	0. 000 716	15. 495 292	0. 000 612	38. 092 958	0. 001 610	32	184	18. 353 771	15. 494 835	38. 074 653	1. 43	25. 79
Tephrite	MQ11	3. 32	16. 8	6. 98	18. 758 433	0. 000 688	15. 595 121	0. 000 670	39. 202 591	0. 001 718	30	158	18. 748 996	15. 594 686	39. 186 910	7. 13	89. 24
Tephrite	MQ12	4. 20	21. 2	6. 12	18. 866 220	0. 000 714	15. 602 144	0. 000 652	39. 399 417	0. 001 690	44	228	18. 852 549	15. 601 514	39. 376 744	6. 69	95. 70
Tephrite	ZLG-01	2. 17	12. 6	4. 52	18. 643 446	0. 002 020	15. 591 105	0. 001 802	39. 042 692	0. 004 520	31	182	18. 63 3971	15. 590 669	39. 024 582	7. 97	86. 91
Tephrite	ZLG-03	2. 39	12. 1	3. 79	18. 656 486	0. 001 332	15. 592 384	0. 001 152	39. 051 474	0. 003 280	40	208	18. 644 034	15. 591 810	39. 030 762	7. 98	86. 31
Tephrite	ZLG-13	3. 09	15. 2	5. 77	18. 685 488	0. 000 760	15. 594 871	0. 000 908	39. 077 318	0. 002 200	34	173	18. 674 903	15. 594 384	39. 060 120	7. 90	85. 52
Trachy-basalt	MD-20	2. 44	5. 3	4. 37	18. 699 000	0. 000 700	15. 846 200	0. 000 600	39. 305 300	0. 001 600	36	80	18. 687 884	15. 845 688	39. 297 358	32. 89	107. 67
Trachy-basalt	MD-22	2. 06	5. 88	4. 27	18. 849 800	0. 000 400	15. 973 600	0. 000 400	39. 819 800	0. 001 200	31	92	18. 840 094	15. 973 153	39. 810 687	43. 99	140. 60
Trachy-basalt	MD-23	2	4. 85	3. 13	18. 747 900	0. 000 600	15. 850 700	0. 000 500	39. 355 200	0. 001 300	41	102	18. 735162	15. 850113	39. 345039	32. 82	106. 72

[a] Calculated by measured whole-rock U, Th and Pb contents and present-day whole-rock Pb isotopic ratios.

[b] Initial Pb isotopic ratio at $t = 20$ Ma, calculated using single-stage model.

$^{206}Pb/^{204}Pb$ diagram, these samples are plotted in the region close to MORB, PREMA and EMⅡ, suggesting that they were derived from a depleted mantle source and metasomatized by some enriched fluids (Fig. 10).

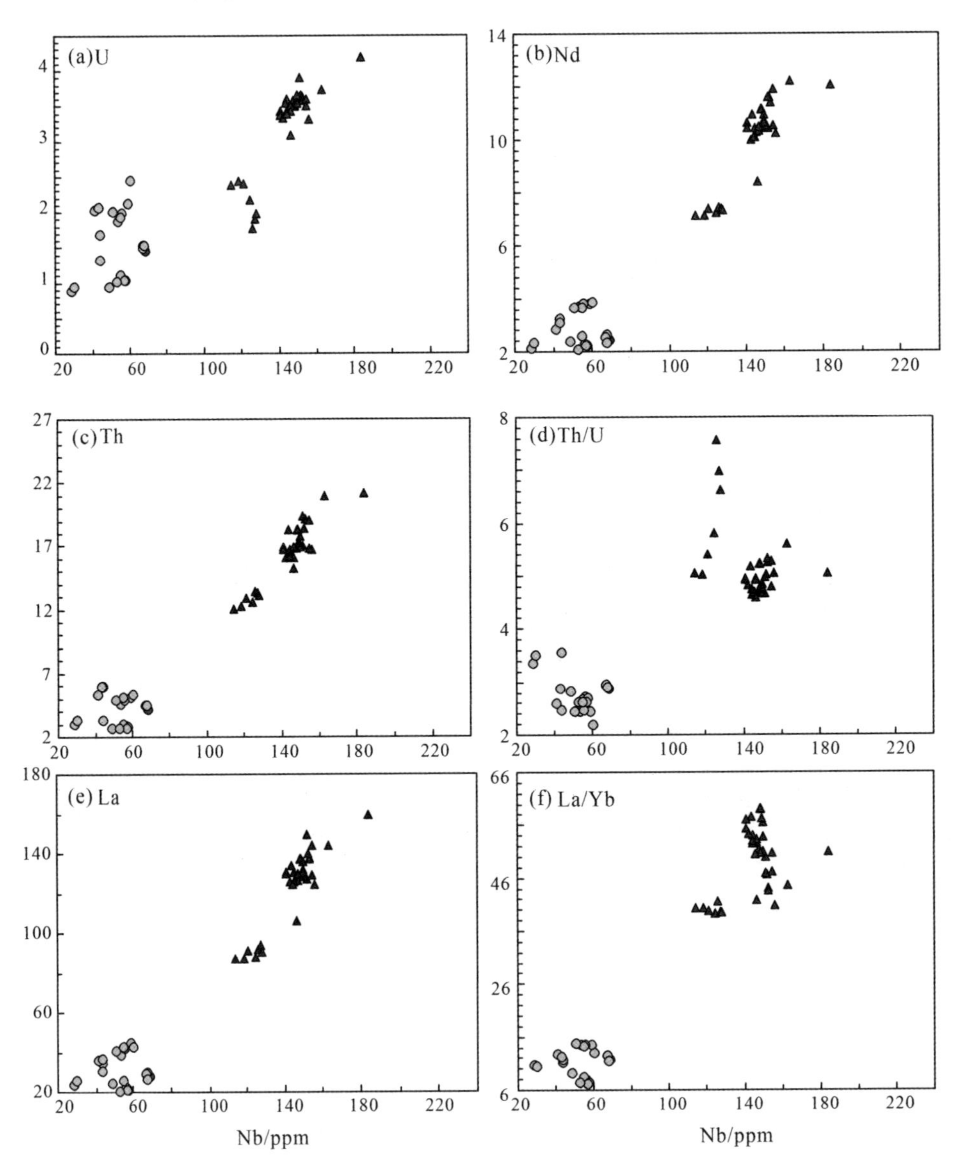

Fig. 8 Trace elements vs. SiO_2 diagrams for the Xiahe and Lixian volcanic rocks.

6 Petrogenesis of the Xiahe and Lixian lavas

6.1 Lateral variations in the Cenozoic lavas in western Qinling

The Xiahe and Lixian volcanic rocks display different geochemical features, as shown in

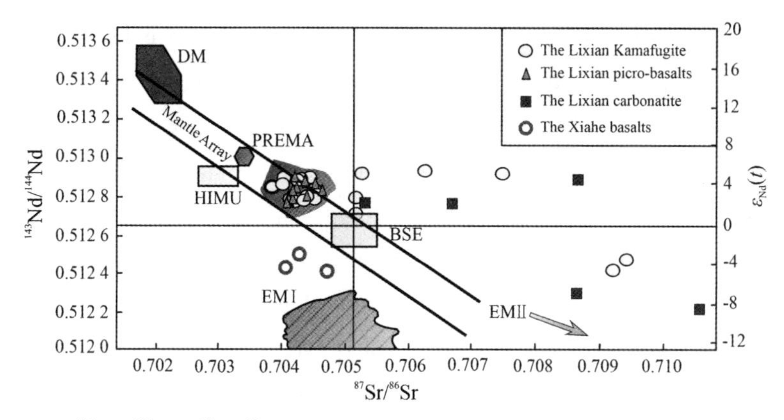

Fig. 9 $^{143}Nd/^{144}Nd$ -$^{87}Sr/^{86}Sr$ diagram for the Xiahe and Lixian volcanic rocks.

The data of Lixian kamafugite and carbonatites were cited from X. H. Yu et al. (2001), X. Yu et al. (2001), and Yu et al. (2004), and data of MORB, DM, PREMA and BSE were cited from Zinder and Hart (1986).

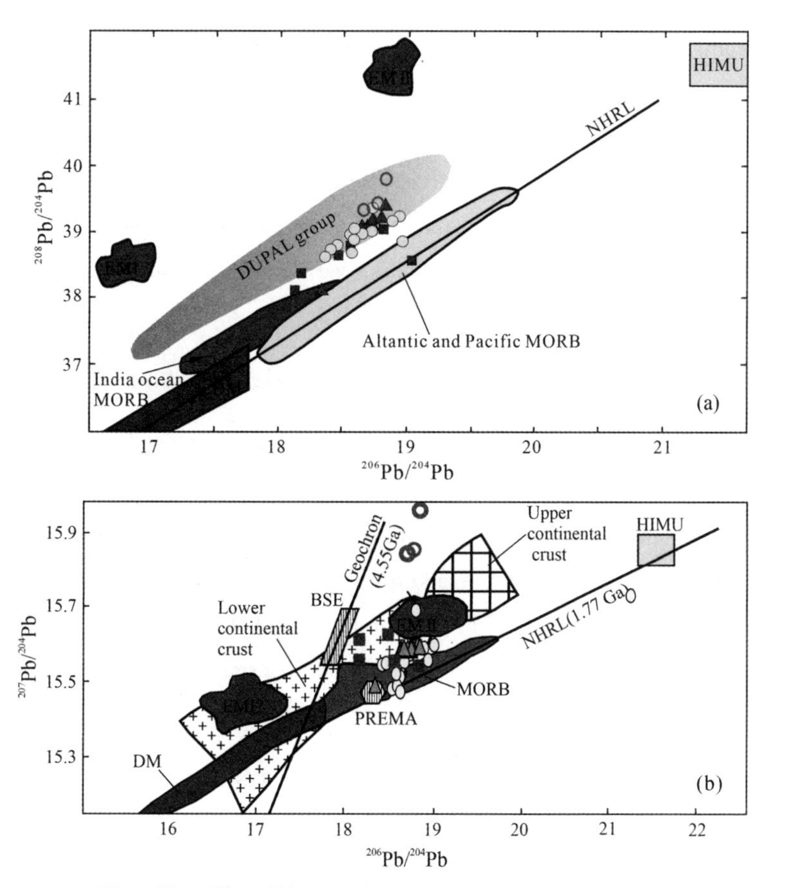

Fig. 10 $^{207}Pb/^{204}Pb$-$^{206}Pb/^{204}Pb$ diagram for the Xiahe and Lixian volcanic rocks.

The data of Lixian kamafugite and carbonatites were cited from X. H. Yu et al. (2001), X. Yu et al. (2001), and Yu et al. (2004). The mantle sources of MORB and the mantle end-members DM, PREMA and BSE are from Zindler and Hart (1986).

the oxideHarker diagrams (Fig. 5). The Lixian picro-basalts have lower SiO_2, Al_2O_3, and Na_2O contents, and higher MgO, $Mg^{\#}$, CaO, and TiO_2 contents. This obvious difference is also demonstrated by the trace elements (ratios) (Fig. 8), with the Lixian picro-basalts displaying higher U, Th, La, Nd, Cr, Ni, Nb, and Ba contents, higher Ce/Pb, Th/U and La/Yb ratios, and lower Zr/Nb ratios. Furthermore, the Xiahe and Lixian lavas have different isotopic compositions. The former have $^{87}Sr/^{86}Sr$ ratios of 0.703 948 – 0.705 059, low $^{143}Nd/^{144}Nd$ ratios of 0.512 409–0.512 513, ε_{Nd} values of −0.53 to −0.66 and T_{DM} ages of 0.74–0.75 Ga, while although the Lixian picro-basalts display similar $^{87}Sr/^{86}Sr$ ratios of 0.704 071–0.704 693, they have higher $^{143}Nd/^{144}Nd$ ratios of 0.512 770–0.512 847, higher ε_{Nd} values of +2.9 to +4.8 and lower T_{DM} ages of 0.38–0.55 Ga.

These variations can neither be explained by crustal contamination, nor by fraction crystallization of co-genetic melt from a homogeneous source region, because crustal contamination would cause significant negative Nb and Ta anomalies, and this is not observed in all the samples. In addition, fractional crystallization cannot produce the increasing incompatible elements and increasing MgO with decreasing SiO_2 content. Furthermore, Zr/Nb ratios can remain constant during fractional crystallization. In combination with the different isotopic compositions, it can be inferred that the Xiahe and Lixian lavas were derived from different source regions. In Sections 6.2 to 6.4, we discuss the source regions and partial melting process of the Xiahe and Lixian lavas in detail.

6.2 Possible source region of the Xiahe and Lixian lavas

Both the Xiahe and Lixian lavas are enriched in LREE and LILEs, display negative K anomalies and positive Nb and Ta anomalies, and have depleted Sr-Nd isotopic compositions. These features are identical to OIB rocks (Wilson, 1989). The following rocks have been proposed as possible source rocks of the alkaline mafic magma: ① silica-deficient eclogite-garnet pyroxenite (Kogiso et al., 2003); ② hornblendite (Pilet et al., 2008 and references therein; Niu et al., 2011); ③ carbonated peridotite (Dasgupta et al., 2007; Brey et al., 2008); and ④ carbonated MORB-like pyroxenite (Gerbode and Dasgupta, 2010).

6.3 Carbonated MORB-like pyroxenite source region for the Xiahe volcanic rocks

As mentioned in Sections 4 and 6.2, the Xiahe volcanic rocks display an affinity with OIB, with negative K anomalies and positive Nb and Ta anomalies, alkaline pH (Fig. 6), and extremely high Na_2O/K_2O ratios. It is believed that these lavas derived from partial melting of asthenosphere mantle, and that the interaction with lithospheric mantle is insignificant because such interaction would cause significant negative Nb and Ta anomalies.

The Xiahe volcanic rocks display higher SiO_2(45.73%–53.80%) and Al_2O_3(14.38%–20.08%) values, and lower MgO (4.98%–8.66%) values than those of the Lixian lavas. They have $Mg^{\#}$ values of 53 to 67 and extremely high Na_2O/K_2O ratios of 2.66–15.19. The following geochemical features suggest that carbonated MORB-like pyroxenite (Gerbode and Dasgupta, 2010) may be the most plausible source region: ① melts that derived from partial melting of carbonated peridotite at 3 GPa (Dasgupta et al., 2007) or 6–10 GPa (Brey et al., 2008) have extremely low SiO_2(<45%), TiO_2(<1%) and Al_2O_3(<10%) contents, but these features are not consistent with the Xiahe volcanic rocks, which have high SiO_2(45.73% to 53.80%), TiO_2(1.59%–3.95%) and Al_2O_3(14.38%–20.08%) contents; ② the Xiahe volcanic rocks have TiO_2 contents of 1.59%–3.95%, and varied $(K_2O+Na_2O)/TiO_2$ ratios, with most of the samples having $(K_2O+Na_2O)/TiO_2>1$, which is similar to that of the melts from carbonated MORB-like pyroxenite (Fig. 11), while hornblendite melts have $(K_2O+Na_2O)/TiO_2<1$ (Pilet et al., 2008); ③ partial melting of silica-deficient eclogite-garnet pyroxenite at 5.0 GPa (Kogiso et al., 2003) can produce melt with low Al_2O_3(<12%), high CaO (>12%) and varying TiO_2 contents, but these features also do not match the geochemical features of the Xiahe sodic alkaline basalts; and ④ melts derived from the carbonated MORB-

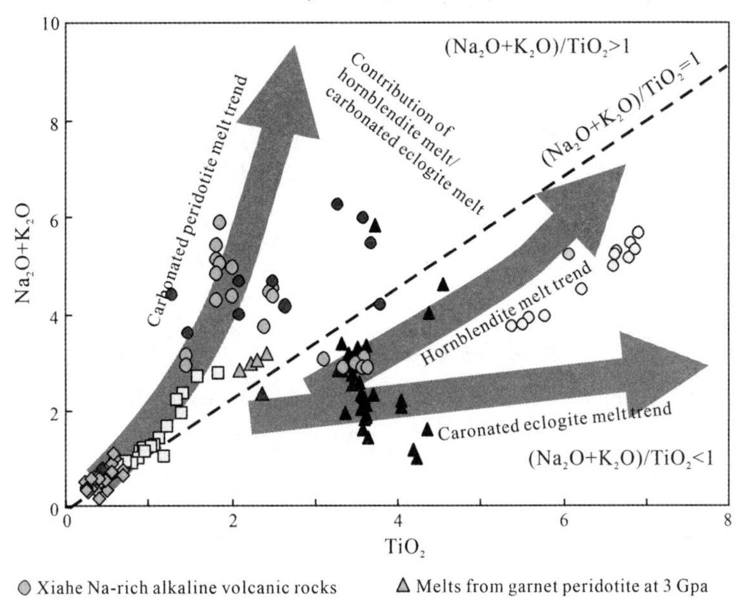

Fig. 11 Variations in Na_2O+K_2O vs. TiO_2 for Cenozoic volcanic rocks from the Xiahe and Lixian area (after Zeng et al., 2010).

Melts from carbonated peridotite at 3.0 GPa (Dasgupta et al., 2007), melts from carbonated peridotite at 6–10 GPa (Brey et al., 2008); melts from garnet peridotite (Kogiso et al., 2003); melts from carbonated MORB-like pyroxenite at 2.9 GPa (Gerbode and Dasgupta, 2010); and melts from hornblendite (Pilet et al., 2008).

like pyroxenite (Gerbode and Dasgupta, 2010) at 2.9 GPa have high SiO_2, TiO_2, and Al_2O_3 contents, but have low MgO contents, which match the geochemical features of the Xiahe volcanic rocks well. For these reasons, we argued that the Xiahe volcanic rocks were derived from partial melting of carbonated MORB-like pyroxenite at 3 GPa.

6.4 Carbonated peridotite source region for the Lixian picro-basalts

Cenozoic volcanic rocks from the Lixian area are traditionally considered to be kamafugite (Yu, 1994; X. H. Yu et al., 2001; Yu et al., 2003, 2004, 2011), which is always associated with igneous carbonatites (Yu et al., 2003, 2004). Based on C-O and Sr-Nd isotopic compositions, Yu et al. (2004) proposed that the volcanic carbonatites were closely related mantle plume activity in the western Qinling area. The occurrence of igneous carbonatites is also a strong indication of the involvement of CO_2 in their source region. Experimental work has revealed that the participation of CO_2 is important for the genesis of alkaline lavas (Dasgupta et al., 2010). The kamafugite-carbonatite combination was inferred to be related to formation from a mantle plume (Bell and Simnetti, 1996). Mantle xenoliths of garnet lherzolite and garnet websterite in the Lixian picro-basalts have equilibrium temperatures and pressures of 1 172–1 226 ℃ with 2.9–3.6 GPa, and 1 169–1 248 ℃ with 2.8–3.2 GPa, respectively (X. H. Yu et al., 2001). These results indicate that the temperature of mantle lithosphere in western Qinling is high enough to produce kamafugitic magma. They also reveal that the kamafugite in the western Qinling was derived from upwelling of asthenosphere at a depth of 90 km.

Though some of the Lixian picro-basalts are potassic ($K_2O/Na_2O > 1$), the majority are not. They also have relatively high TiO_2 (3.47% – 4.66%) and MgO (11.24% – 18.88%) contents, and low SiO_2 (41.14% – 44.82%) and Al_2O_3 (5.84% – 9.18%) contents. In addition, the Lixian picro-basalts are enriched in LREE and LILEs, with high $(La/Yb)_N$ ratios of 39.3–59.2 and $(Ce/Yb)_N$ ratios of 27.8–42.4. In combination with their depleted Sr-Nd-Pb isotopic compositions (Tables 3 and 4), it can be inferred that the Lixian picro-basalts were derived from the metasomatized mantle peridotite. Experimental melts of carbonated peridotite at 6 – 10 GPa (Brey et al., 2008) have extremely low SiO_2 (1.7% – 25.7%) and TiO_2 (0.24% – 0.7%) contents. Furthermore, these melts also have low K_2O (0.02%–0.3%) and Na_2O (0.14%–0.9%) contents, with K_2O/Na_2O ratios ranging from 0.05 to 0.43. These features are clearly distinct from the Lixian picro-basalts. Equilibrium temperatures and pressures of mantle xenoliths from the Lixian picro-basalts also indicate that they were formed at a depth of 90 km (X. Yu et al., 2001). Partial melting of mantle peridotite at shallow depth may thus be the most plausible source for the genesis of the Lixian picro-basalts.

In the upper mantle, the Al_2O_3 contents in melts were affected by the composition of source rocks and pressures. With increasing pressure, formation of garnet will decrease Al_2O_3 contents in the melts (Dasgupta et al., 2007). The Lixian picro-basalts display low and consistent Al_2O_3(5.61%–8.68%) contents, which is markedly lower than those of the melts from silica-deficient eclogite-garnet pyroxenite (Kogiso et al., 2003), hornblendite (Pilet et al., 2008) and carbonated MORB-like pyroxenite (Gerbode and Dasgupta, 2010), but close to melts from carbonated peridotite at a moderate degree of partial melting at 3.0 GPa (Dasgupta et al., 2007; Brey et al., 2008). However, melts from carbonated peridotite have much lower TiO_2 contents (0.75%–1.61%) (Dasgupta et al., 2007) than those of the Lixian picro-basalts (3.47%–4.66%). While the melts from hornblendite (Pilet et al., 2008) have higher SiO_2, TiO_2 and Al_2O_3 contents. Prytulak and Elliott (2007) argued that addition of minor (1%–10%) recycled mafic crust in peridotite source can account for the high TiO_2 contents in OIB rocks.

Melts derived from carbonated peridotite at 3 GPa (Dasgupta et al., 2007) have varied compositions at different temperature conditions; as temperature increases, the melts display increments in SiO_2, Al_2O_3, MgO contents and a decline in CaO content. At temperatures ranging from 1 475 ℃ to 1 525 ℃, the melt has moderate SiO_2(40.2%–42.6%) and Al_2O_3 (9.4%–10.8%) contents, similar to those of the Lixian picro-basalts (Table 2), while their TiO_2, Na_2O, and K_2O contents are significantly lower than those of the Lixian picro-basalts. Thus, we argued that CO_2 is not the unique metasomatizing agent in the source region of the Lixian picro-basalts, but that the fluids or melts from mafic lower crust also play an important role in their genesis. Isotopic results reveal that the Lixian picro-basalts have depleted Sr-Nd isotopic compositions, with Nd model ages of 0.33–0.51 Ga, which is consistent with the formation age of the Proto-Tethys oceanic crust in the western Qinling area (Dong et al., 2011). However, the oceanic crust in this region has lower $\varepsilon_{Nd}(t)$ values at given times, thus the oceanic lithosphere mantle of the Paleo-Tethys may be the most plausible source rocks for the Lixian picro-basalts.

Accordingto their geochemical features, we proposed that the Lixian picro-basalts may be derived from partial melting of the Paleozoic oceanic lithosphere mantle peridotite at relatively high pressure (>3 GPa), and that the metasomatism of CO_2 and fluids or melts from mafic lower crust can account for the following geochemical features of the Lixian picro-basalts: ① high TiO_2, Na_2O, and K_2O contents; ② extremely enriched in LREE and LILEs; and ③ high Nb/Ta and Nb/U ratios.

7 Cenozoic mantle extrusion process in the northeastern margin of the Tibetan Plateau

Cenozoic lavas in the western Qinling area are widely believed to be sourced from a mantle plume (X. H. Yu et al., 2001; Yu et al., 2003, 2004). Yu et al (2011) proposed that the Cenozoic bimodal volcanism in the western Qinling area was formed in a continent rift setting, which is related to the eastward migration of upwelling asthenosphere during the India and Asia collision. However, the Cenozoic lavas are small scale and discontinuous, and are controlled by several strike-slip faults (Fig. 1a). These features are not consistent with the mantle plume model.

Extrusion tectonics models have been applied to the east-west extension in Tibet. These models include: ① rigid body extrusion (Armijo et al., 1986, 1989); ② oroclinal bending (Ratschbacher et al., 1994); ③ a Late Miocene-Early Pliocene large scale mantle flow beneath Asia (Yin and Harrison, 2000); ④ conjugate strike-slip faulting (Taylor et al., 2003); and ⑤ eastward flow of lower crust (Royden et al., 1997, 2008; Zhao et al., 2008). Wang et al (2010) argued that the Middle-Late Eocene (47 – 35 Ma) partial melting of eclogitic lower crust (Wang et al., 2008) and lithospheric mantle in central Tibet could have triggered crustal weakening and eastward flow of deep crust. The lower crust flow model has been used to explain the elevation process around the Tibet Plateau (Royden et al., 1997; Clark and Royden, 2000). In the eastern margin of the TP, the strong South China Craton beneath the Sichuan Basin caused the flow to "pile up" and create a narrow, steep margin; diversion of the flow to the north into northeastern Tibet and the Qinling and south into southeastern Tibet created broad, gentle margins (Enkelmann et al., 2006). Harrison et al. (1992) proposed that the Late-Miocene uplift of the TP caused E-W extension of central Tibet, which is bounded by the 105° N-S trending Helan and Chuandian orogenic belt. Geophysical data also indicate that western Qinling has thin lithosphere with a depth of just 80–120 km (X. Yu et al., 2001 and references therein).

As mentioned in Section 6, the Xiahe and Lixian volcanic rocks were derived from different source regions. This lateral variation in source region and partial melting conditions reflect eastward mantle extrusion in this area. In comparison with the Eocene volcanic rocks in the northern Qiangtang block (Lai et al., 2007a, 2011; Wang et al., 2008, 2010), we propose the following model (Fig. 12) to explain the Cenozoic volcanic rocks in the northeastern margin of the TP: ① Collision between India and Asia caused significant shortening of central Tibet, resulting in extrusion tectonics along the southwestern and northeastern margin of TP; while the occurrence of shoshonitic and adakitic lavas across the Qiangtang (c. 50–29 Ma) (Turner et al., 1996; Guo et al., 2006; Lai et al., 2007a; Wang et al., 2008, 2010) and Lhasa blocks (c. 30–10 Ma) indicate a hot mantle, thick crust and eclogitic root during this period; ② the extrusion of hot materials would have followed the rheologically weak crustal corridor along the Paleozoic-Mesozoic Qinling suture (Enkelmann et al., 2006), but the limited obstruction along the Qinling suture would have caused a "pile

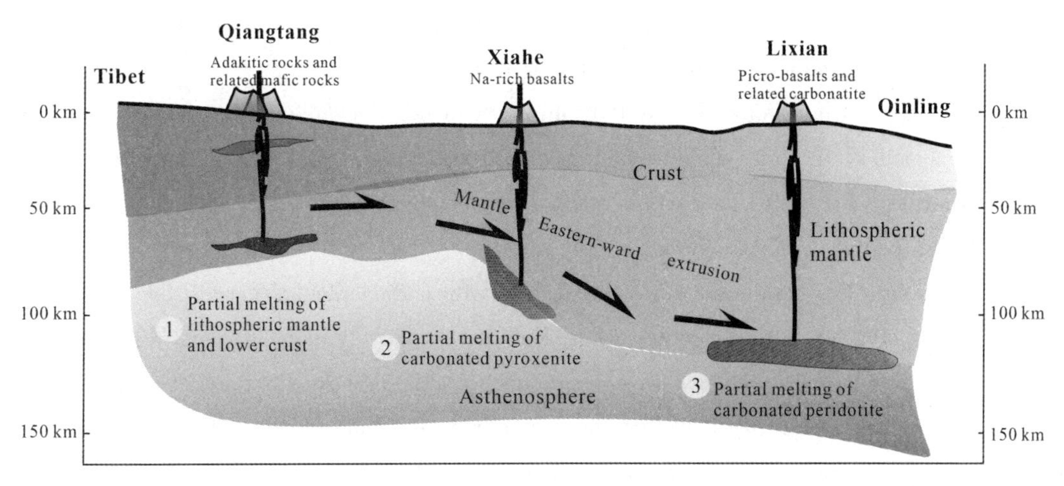

Fig. 12 A proposed tectonic model for the generation of the Cenozoic volcanic rocks from the northeastern margin of the Tibetan Plateau.

Collision between India and Asia caused large-scale mantle extrusion along the southeastern and northeastern margin of the Tibetan Plateau, the lithosphere in northeastern margin of the Tibetan Plateau will have a gradually decreasing geothermal gradient and increasing thickness from west to east; extension along the Qinling suture would cause low degree partial melting of mantle and produce some fluids that are enriched in LREE, LILEs, CO_2 and H_2O; these fluids would metasomatized their overlaying lithosphere, partial melting of these carbonated source region will form Na-rich alkaline basalts in the Xiahe area and picro-basalts in the Lixian area.

up" of solid lithosphere at the front edge; and ③ considering thermal diffusion, the geothermal gradient of the lithosphere at the northeastern margin of the TP will gradually decrease and its thickness will increase from west to east; the resultant extension along the Qinling suture would cause low degree partial melting of the asthenosphere and produce some fluids that are enriched in LREE, LILEs, CO_2, and H_2O; these fluids would metasomatize their overlying lithosphere mantle. Partial melting of the carbonated pyroxenite would generate sodic alkaline basalts in the Xiahe area, and partial melting of carbonated peridotites at greater depth could generate the picro-basaltic melts found in the Lixian area.

The occurrence and geochemistry of the Cenozoic volcanic rocks in the western Qinling area provide an excellent example to discuss the Cenozoic eastward mantle extrusion process in the northeastern margin of the TP. Thus, we argued that these volcanic rocks were formed by episodic decompression melting of the mantle lithosphere during the eastward extrusion process, instead of the established theory that they were formed in a continental rift setting. The partial melting and eruption process may be closely related to the strike-slip fault activities in the northeastern margin of the TP.

8 Conclusion

Geochemistry and Sr-Nd-Pb isotopic compositions of the Xiahe and Lixian lavas suggest differences in their source regions and partial melting processes. These features suggest a mantle extrusion process in the northeastern margin of the TP.

The Xiahe volcanic rocks display OIB affinity with negative K anomalies and positive Nb and Ta anomalies. They are alkaline and have extremely high Na_2O/K_2O ratios, which, in combination with the Sr-Nd-Pb isotopic compositions, indicate that these volcanic rocks were derived from partial melting of carbonated pyroxenite.

The Lixian picro-basalts are closely associated with carbonatites, giving a strong indication of CO_2 being involved in their source region. They have relatively high TiO_2 and MgO contents and low SiO_2 and Al_2O_3 contents. In combination with their Sr-Nd-Pb isotopic composition, these geochemical results indicate that the Lixian picro-basalts may be derived from partial melting of carbonated oceanic lithosphere mantle peridotite at a relatively high pressure (>3 GPa). Their higher TiO_2 and Na_2O+K_2O contents may be caused by minor hornblendite in the source region.

Acknowledgements This work is jointly supported by the National Natural Science Foundation of China (Grant Nos. 41072052, 41190072 and 40572050) and the Teaching and Research Award Program for Outstanding Young Teachers in Higher Education Institutions of MOE, P.R. China. The MOST Special Fund from the State Key Laboratory of Continental Dynamics, Northwest University and Province Key Laboratory Construction Item (08JZ62). We thank two anonymous reviewers for their constructive comments. We are also grateful to Prof. Wenjiao Xiao for his kind help.

References

Armijo, R., Tapponnier, P., Mercier, J.L., Tong-Lin, H., 1986. Quaternary extension in southern Tibet: Field observations and tectonic implications. Journal of Geophysical Research, 91, 13803-13872.

Armijo, R., Tapponnier, P., Tonglin, H., 1989. Late Cenozoic right-lateral strike-slip faulting in Southern Tibet. Journal of Geophysical Research, 94, 2787-2838.

Bell, K., Simnetti, A., 1996. Carbonatite magmatism and plume activity: Implications from the Nd, Pb and Sr isotope systematics of Oldoinyo Lengai. Journal of Petrology, 37, 1321-1339.

Bowman, D., King, G., Tapponnier, P., 2003. Slip partitioning by elastoplastic propagation of oblique slip at depth. Science, 300, 1121-1123.

Brey, G.P., Bulatov, V.K., Girnis, A.V., Lahaye, Y., 2008. Experimental melting of carbonated peridotite at 6–10 GPa. Journal of Petrology, 49, 797-821.

Cao, S., Liu, J., Leiss, B., Neubauer, F., Genser, J., Zhao, C., 2011. Oligo-Miocene shearing along the Ailao Shan-Red River shear zone: Constraints from structural analysis and zircon U/Pb geochronology of magmatic rocks in the Diancang Shan massif, SE Tibet, China. Gondwana Research, 19, 975-993.

Chung, S.L., Chu, M.F., Zhang, Y., Xie, Y., Lo, C.H., Lee, T.Y., Lan, C.Y., Li, X., Zhang, Q., Wang, Y., 2005. Tibetan tectonic evolution inferred from spatial and temporal variations in post-collisional magmatism. Earth-Science Reviews, 68, 173-196.

Clark, M.K., Royden, L.H., 2000. Topographic ooze: Building the eastern margin of Tibet by lower crustal flow. Geology, 28, 703-706.

Dasgupta, R., Hirschmann, M.M., Smith, N.D., 2007. Partial melting experiments of peridotite + CO_2 at 3 GPa and genesis of alkaline ocean island basalts. Journal of Petrology, 48, 2093-2124.

Dasgupta, R., Jackson, M.G., Lee, C.T.A., 2010. Major element chemistry of ocean island basalts: Conditions of mantle melting and heterogeneity of mantle source. Earth & Planetary Science Letters, 289, 377-392.

Dong, Y., Genser, J., Neubauer, F., Zhang, G., Liu, X., Yang, Z., Heberer, B., 2011. U-Pb and $^{40}Ar/^{39}Ar$ geochronological constraints on the exhumation history of the North Qinling terrane, China. Gondwana Research, 19, 881-893.

Enkelmann, E., Ratschbacher, L., Jonckheere, R., Nestler, R., Fleischer, M., Gloaguen, R., Hacker, B. R., Zhang, Y. Q., Ma, Y.S., 2006. Cenozoic exhumation and deformation of northeastern Tibet and the Qinling: Is Tibetan lower crustal flow diverging around the Sichuan Basin? Geological Society of America Bulletin, 118, 651-671.

Gerbode, C., Dasgupta, R., 2010. Carbonate-fluxed melting of MORB-like pyroxenite at 2.9 GPa and genesis of HIMU ocean island basalts. Journal of Petrology, 51, 2067-2088.

Guo, Z., Wilson, M., Liu, J., Mao, Q., 2006. Post-collisional, potassic and ultra-potassic magmatism of the Northern Tibetan Plateau: Constraints on characteristics of the mantle source, geodynamic setting and uplift mechanisms. Journal of Petrology, 47, 1177-1220.

Harrison, T.M., Copeland, P., Kidd, W.S.F., Yin, A., 1992. Raising Tibet. Science, 255, 1663-1670.

Hou, Z., Tian, S., Yuan, Z., Xie, Y., Yin, S., Yi, L., Fei, H., Yang, Z., 2006. The Himalayan collision zone carbonatites in western Sichuan, SW China: Petrogenesis, mantle source and tectonic implication. Earth & Planetary Science Letters, 244, 234-250.

Hu, J., Yang, H., Xu, X., Wen, L., Li, G., 2012. Lithospheric structure and crust-mantle decoupling in the southeast edge of the Tibetan Plateau. Gondwana Research, 22, 1060-1067.

Huang, X.L., Niu, Y., Xu, Y.G., Chen, L.L., Yang, Q.J., 2010. Mineralogical and geochemical constraints on the petrogenesis of post-collisional potassic and ultra-potassic Rocks from Western Yunnan, SW China. Journal of Petrology, 51, 1617-1654.

Kogiso, T., Hirschmann, M.M., Frost, D.J., 2003. High-pressure partial melting of garnet pyroxenite: possible mafic lithologies in the source of ocean island basalts. Earth & Planetary Science Letters, 216, 603-617.

Lai, S.C., Qin, J.F., Li, Y.F., 2007a. Partial melting of thickened Tibetan crust: Geochemical evidence from Cenozoic adakitic volcanic rocks. International Geology Review, 49, 357-373.

Lai, S.C., Zhang, G.W., Li, Y.F., Qin, J.F., 2007b. Genesis of the Madang Cenozoic sodic alkaline basalt in the eastern margin of the Tibetan Plateau and its continental dynamic implications. Science in China: Series D, Earth Sciences, 50, 314-321.

Lai, S., Qin, J., Grapes, R., 2011. Petrochemistry of granulite xenoliths from the Cenozoic Qiangtang volcanic field, northern Tibetan Plateau: Implications for lower crust composition and genesis of the volcanism. International Geology Review, 53, 926-945.

Li, Y., Wang, C., Zhao, X., Yin, A., Ma, C., 2012. Cenozoic thrust system, basin evolution, and uplift of the Tanggula Range in the Tuotuohe region, central Tibet. Gondwana Research, 22, 482-492.

Mo, X. X., 2011. Magmatism and evolution of the Tibetan Plateau. Geological Journal of China Universities, 17, 351-367 (in Chinese).

Mo, X., Zhao, Z., Deng, J., Flower, M., Yu, X., Luo, Z., Li, Y., Zhou, S., Dong, G., Zhu, D., Wang, L., 2006. Petrology and geochemistry of post-collisional volcanic rocks from the Tibetan Plateau: Implications for lithosphere heterogeneity and collision-induced asthenospheric mantle flow. Geological Society of America Special Papers, 409, 507-530.

Niu, Y., Wilson, M., Humphreys, E. R., O'Hara, M.J., 2011. The origin of intra-plate ocean island basalts (OIB): the lid effect and its geodynamic implications. Journal of Petrology, 52, 1443-1468.

Pilet, S., Baker, M.B., Stolper, E.M., 2008. Metasomatized lithosphere and the origin of alkaline lavas. Science, 320, 916-919.

Prytulak, J., Elliott, T., 2007. TiO_2 enrichment in ocean island basalts. Earth & Planetary Science Letters, 263, 388-403.

Ratschbacher, L., Frisch, W., Liu, G., Chen, C., 1994. Distributed deformation in southern and western Tibet during and after the India-Asia collision. Journal of Geophysical Research, 99, 19917-19945.

Royden, L.H., Burchfiel, B.C., King, R.W., Wang, E., Chen, Z., Shen, F., Liu, Y., 1997. Surface deformation and lower crustal flow in Eastern Tibet. Science, 276, 788-790.

Royden, L.H., Burchfiel, B.C., van der Hilst, R.D., 2008. The geological evolution of the Tibetan Plateau. Science, 321, 1054-1058.

Schoenbohm, L.M., Burchfiel, B.C., Liang, Z.C., 2006. Propagation of surface uplift, lower crustal flow, and Cenozoic tectonics of the southeast margin of the Tibetan Plateau. Geology, 34, 813-816.

Su, B.X., Zhang, H.F., Ying, J.F., Xiao, Y., Zhao, X.M., 2009. Nature and processes of the lithospheric mantle beneath the western Qinling: Evidence from deformed peridotitic xenoliths in Cenozoic kamafugite from Haoti, Gansu Province, China. Journal of Asian Earth Sciences, 34, 258-274.

Su, B.X., Zhang, H.F., Sakyi, P. A., Ying, J.F., Tang, Y.J., Yang, Y.H., Qin, K.Z., Xiao, Y., Zhao, X.M., 2010. Compositionally stratified lithosphere and carbonatite metasomatism recorded in mantle xenoliths from the Western Qinling (Central China). Lithos, 116, 111-128.

Sun, S.S., McDonough, W. F., 1989. Chemical and isotopic systematics of oceanic basalts: Implications for mantle composition and processes. Geological Society, London, Special Publications, 42, 313-345.

Tapponnier, P., Peltzer, G., Le Dain, A. Y., Armijo, R., Cobbold, P., 1982. Propagating extrusion tectonics in Asia: New insights from simple experiments with plasticine. Geology, 10, 611-616.

Tapponnier, P.,Xu, Z.Q., Roger, F., Meyer, B., Arnaud, N., Wittlinger, G., Yang, J.S., 2001. Oblique stepwise rise and growth of the Tibet Plateau. Science, 294, 1671-1677.

Taylor, M., Yin, A., Ryerson, F.J., Kapp, P., Ding, L., 2003. Conjugate strike-slip faulting along the Bangong-Nujiang suture zone accommodates coeval east-west extension and north-south shortening in the interior of the Tibetan Plateau. Tectonics, 22, 1044.

Turner, S., Arnaud, N., Liu, J., Rogers, N., Hawkesworth, C., Harris, N., Kelley, S., Van Calsteren,

P., Deng, W., 1996. Post-collision, shoshonitic volcanism on the Tibetan Plateau: Implications for convective thinning of the lithosphere and the source of ocean island basalts. Journal of Petrology, 37, 45-71.

Wang, E., Burchfiel, B.C., 2004. Late Cenozoic right-lateral movement along the Wenquan Fault and associated deformation: Implications for the kinematic history of the Qaidam Basin, Northeastern Tibetan Plateau. International Geology Review, 46, 861-879.

Wang, Q., Wyman, D.A., Xu, J., Dong, Y., Vasconcelos, P.M., Pearson, N., Wan, Y., Dong, H., Li, C., Yu, Y., Zhu, T., Feng, X., Zhang, Q., Zi, F., Chu, Z., 2008. Eocene melting of subducting continental crust and early uplifting of central Tibet: Evidence from central-western Qiangtang high-K calc-alkaline andesites, dacites and rhyolites. Earth & Planetary Science Letters, 272, 158-171.

Wang, Q., Wyman, D.A., Li, Z.X., Sun, W., Chung, S.L., Vasconcelos, P.M., Zhang, Q., Dong, H., Yu, Y., Pearson, N., Qiu, H., Zhu, T., Feng, X., 2010. Eocene north-south trending dikes in central Tibet: New constraints on the timing of east-west extension with implications for early plateau uplift? Earth & Planetary Science Letters, 298, 205-216.

Wilson, M., 1989. Igneous petrogenesis: A global tectonic approach. Netherland: Springer, 466.

Winchester, J. A., Floyd, P. A., 1977. Geochemical discrimination of different magma series and their differentiation products using immobile elements. Chemical Geology, 20, 325-343.

Xia, L., Li, X., Ma, Z., Xu, X., Xia, Z., 2011. Cenozoic volcanism and tectonic evolution of the Tibetan plateau. Gondwana Research, 19, 850-866.

Xu, Z., Ji, S., Cai, Z., Zeng, L., Geng, Q., Cao, H., 2012. Kinematics and dynamics of the Namche Barwa Syntaxis, eastern Himalaya: Constraints from deformation, fabrics and geochronology. Gondwana Research, 21, 19-36.

Yin, A., Harrison, T.M., 2000. Geologic evolution of the Himalayan-Tibetan Orogen. Annual Review of Earth and Planetary Sciences, 28, 211-280.

Yu, X.H., 1994. Cenozoic potassic alkaline ultrabasic volcanic rocks and its genesis, in Lixian-Dangchang area, Gansu Province. Tethyan Geology, 18, 114-129.

Yu, X.H., Zhang, C.F., 1998. Sr-Nd isotopic and trace elements geochemical features of the Cenozoic volcanic rocks from west Qinling, Gansu Province. Earth Science Frontiers, 5 (4), 319-328.

Yu, X., Mo, X., Liao, Z., Zhao, X., Su, Q., 2001. Temperature and pressure condition of garnet lherzolite and websterite from west Qinling, China. Science in China: Series D, Earth Sciences, 44, 155-161.

Yu, X. H., Mo, X.X., Martin, F., 2001. Cenozoic kamafugite volcanism and tectonic meaning in west Qinling area, Gansu Province.Acta Petrology Sinica, 17, 366-377 (in Chinese).

Yu, X.H., Mo, X.X., Su, S.G., Dong, F.L., Zhao, X., Wang, C., 2003. Discovery and significance of Cenozoic volcanic carbonatite in Lixian, Gansu Province. Acta Petrologica Sinica, 19, 105-112 (in Chinese).

Yu, X.H., Zhao, Z.D., Mo, X.X., Wang, Y.L., Xiao, Z., Zhu, D.Q., 2004. Trace elements, REE and Sr, Nd, Pb isotopic geochemistry of Cenozoic kamafugite and carbonatite from west Qinling, Gansu Province: Implication of plume-lithosphere interaction. Acta Petrologica Sinica, 20, 483-494 (in Chinese).

Yu, X., Zhao, Z., Zhou, S., Mo, X., Zhu, D., Wang, Y., 2006. $^{40}Ar/^{39}Ar$ dating for Cenozoic kamafugite from the western Qinling in Gansu Province. Chinese Science Bulletin, 51, 1621-1627.

Yu, X.H., Mo, X.X., Zhao, Z.D., Huang, X.K.,Li, Y., Wei, Y.F., 2009.Two types of Cenozoic potassic volcanic rocks in Western Qinling, Gansu Province: Their petrology, geochemistry and petrogenesis. Earth Science Frontiers, 16, 79-89 (in Chinese).

Yu, X.H., Mo, X.X., Zhao, Z.D., He, W.Y., Li, Y., 2011.Cenozoic bimodal volcanic rocks of the West Qinling: Implication for the genesis and nature of the rifting of north-south tectonic belt. Acta Petrologica Sinica, 27, 2195-2202 (in Chinese).

Zeng, G., Chen, L.H., Xu, X.S., Jiang, S.Y., Hofmann, A.W., 2010. Carbonated mantle sources for Cenozoic intra-plate alkaline basalts in Shandong, North China. Chemical Geology, 273, 35-45.

Zhang, Z., Santosh, M., 2011. Tectonic evolution of Tibet and surrounding regions. Gondwana Research, 21, 1-3.

Zhang, Z., Klemperer, S., Bai, Z., Chen, Y., Teng, J., 2011. Crustal structure of the Paleozoic Kunlun orogeny from an active-source seismic profile between Moba and Guide in East Tibet, China. Gondwana Research, 19, 994-1007.

Zhao, G., Chen, X., Wang, L., Wang, J., Tang, J., Wan, Z., Zhang, J., Zhan, Y., Xiao, Q., 2008. Evidence of crustal "channel flow" in the eastern margin of Tibetan Plateau from MT measurements. Chinese Science Bulletin, 53, 1887-1893.

Zindler, A., Hart, S., 1986. Chemical geodynamics. Annual Review of Earth and Planetary Sciences, 14, 493-571.

Neoproterozoic quartz monzodiorite-granodiorite association from the Luding-Kangding area: Implications for the interpretation of an active continental margin along the Yangtze Block (South China Block)①

Lai Shaocong　Qin Jiangfeng　Zhu Renzhi　Zhao Shaowei

Abstract: Neoproterozoic magmatism along the western margin of the Yangtze Block can provide important information about the evolution of the Supercontinent Rodinia. For this reason, this study is focused on Neoproterozoic high-$Mg^{\#}$ quartz monzodiorite and granodiorite from the Luding-Kangding area. Detailed zircon LA-ICP MS U-Pb dating indicates that the high-$Mg^{\#}$ quartz monzodiorite and the granodiorite have identical ages of 754±10 Ma (MSWD=0.51, 2σ) and 748±11 Ma (MSWD=0.33, 2σ), respectively. The high-$Mg^{\#}$ quartz monzodiorite are characterized by low SiO_2 contents (60.76%–63.78%) and high TiO_2 contents (0.41%–0.56%), and they are sodic and meta-aluminous. They are enriched in LREE, with $(La/Yb)_N$ values of 4.14–8.51, and without significant negative Eu anomalies (Eu^*/Eu = 0.79 – 0.92). The high-$Mg^{\#}$ quartz monzodiorite have relative low initial $(^{87}Sr/^{86}Sr)_i$ ratios of 0.703 513–0.704 519, and positive $\varepsilon_{Nd}(t)$ values of +2.4 to +4.8, and are depleted in HFSEs (Nb and Ta). Thus, it can be inferred that these high-$Mg^{\#}$ quartz monzodiorite were formed by high degree (~40%) partial melting of newly formed mafic lower crust, and that their high $Mg^{\#}$ values were the result of assimilation of some residual mafic minerals in their source region. The granodiorite are characterized by higher SiO_2 contents (65.32% – 67.59%), and they are also sodic and meta-aluminous. They have more evolved Sr-Nd isotopic compositions and higher Sr (425–537 ppm) and Ba (705–1 074 ppm) contents than the quartz monzodiorite, suggesting that they were derived from a plagioclase-rich source region. Based on their field occurrence and geochemical features, we propose that both the Neoproterozoic high-$Mg^{\#}$ quartz monzodiorite and the granodiorite were formed in an active continental arc setting, but they derived from two distinct source regions.

1　Introduction

Precambrian crystalline basement in the western margin of the Yangtze Block is mainly composed of Neoproterozoic (750–830 Ma) granitic rocks, deformed granitic gneisses, and

① Published in *Precambrian Research*, 2015, 267.

massive granodiorites and monzogranites, which were formed during the Neoproterozoic crustal evolution, the reworking of continental crust, and the evolution of the Supercontinent Rodinia (Zhao and Zhou, 2008; Liu and Zhao, 2012). However, their petrogenesis and geodynamic context remain poorly understood. Mantle plume model (Li et al., 2002, 2003), continental rifting model and active continental margin model (Zhou et al., 2006; Zhao et al., 2010a) have been proposed to explain their genesis. Detailed geochemical and Sr-Nd isotopic composition analyses of the intermediate and felsic igneous rocks indicate that they were formed by partial melting of lower continental crust (Li et al., 2003; Zhou et al., 2006). Although recent works have mainly focused on the geodynamic model of these igneous rocks (Zhou et al., 2002, 2006; Li et al., 2002, 2003; Zhao and Zhou, 2008; Zhao et al., 2010a), the detailed subduction geodynamic process, partial melting mechanism of the lower arc crust, and the role of mantle components in the genesis of the Neoproterozoic granitoids are still poorly understood. In this paper, we present new geochemical data, zircon U-Pb ages, and Sr-Nd-Pb isotopic compositions of the Neoproterozoic (750 Ma) high-$Mg^{\#}$ quartz monzodiorite and granodiorite of the Luding-Kangding area, situated in the western margin of the Yangtze Block, for information supporting the active continental margin model and information regarding the partial melting process of lower arc crust during the subduction process.

2 Geological background and field geology

South China Block (SCB) consists of the Yangtze Block in the northwest and Cathaysian Block in the southeast (Fig. 1), bounded by the Neoproterozoic Jiangnan orogenic belt (Wang et al., 2014). The Yangtze Block consists of a Mesoproterozoic basement composed of low-and medium-grade metamorphosed sedimentary rocks, overlain by a Neoproterozoic to Cenozoic cover (Zhou et al., 2002). Neoproterozoic granitioids, granitic gneisses, and associated calc-alkaline volcanic rocks are found in the Yangtze block.

The studied Kangding Group, which is located in the Luding-Kangding area, in the western margin of the Yangtze Block, is a 100-km-long and 30-km-wide block, tectonically bound and cut by several faults (Fig. 2), and is predominantly composed of granitic gneisses and migmatitic granites. This igneous complex was distributed almost continuously along the Kangdian Rift, which includes granites, granodiorites, tonalite, diorite, gabbros, mafic dykes, and minor ultramafic bodies. Due to the ambiguous contact relationships between the different rocks, the Kanding Complex was traditionally mapped as a part of the Archean-Paleoproterozoic basement of the Yangtze Block. However, new chronology works (Zhou et al., 2002; Li et al., 2002; Lin et al., 2007; Liu et al., 2009) reveal that the metamorphic rocks, granitoids and related mafic dykes that compose this complex have ages between 750 Ma and 830 Ma.

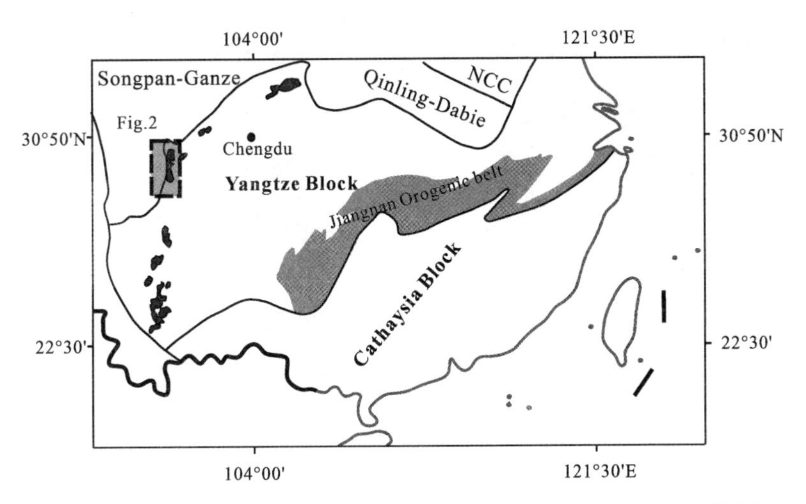

Fig. 1 Sketch geological map of the Yangtze Block (after Wang et al., 2014).

The quartz monzodiorite samples from the Luding area are fine-grained, and consist mainly of plagioclase (40–45 vol%), amphibole (20–25 vol%), alkali feldspar (5–10 vol%), biotite (10–15 vol%), and quartz (5–10 vol%), accessory minerals including zircon, apatite, and magnetite. The plagioclase crystals are 1.0–3.0 mm long, and exhibit well-developed twinning and concentric zoning (Fig. 3). Most of the amphibole crystals display corrosion textures and some amphibole crystals contains plagioclase inclusions within them. Apatite mainly occurs as small inclusions in the plagioclase. The granodiorite samples from the Kangding area are fine-grained, and consist of plagioclase (50–55 vol%), alkali feldspar (30–40 vol%) and minor mafic minerals (i.e., amphibole and biotite). Serious alteration of the granodiorite is observed under the microscope; the plagioclase crystals are sericitized, and the amphibole crystals are chloritized and uralitized.

3 Analytical methods

Fresh samples of granitic rocks were selected for elemental analyses. All zircon U-Pb dating, major and trace element, and whole-rock Sr-Nd-Pb isotopic analyses were conducted in the State Key Laboratory of Continental Dynamics, Northwest University in Xi'an, China.

The chips of whole-rock samples were comminuted to 200 mesh-sizes with a tungsten carbide ball mill. Major elements were analyzed by X-ray fluorescence (Rikagu RIX 2100) with a less than 2% analytical error. Trace element and REE contents were analyzed with an inductively coupled plasma mass spectrometer (ICP-MS Agilent 7500a), using USGS and Chinese national rock standards (BCR-2, GSR-1, and GSR-3). For most trace elements, analytical error is less than 10% (Liu et al., 2007).

Whole-rock Sr-Nd-Pb isotopic data were obtained using a Nu Plasma HR multi-collector

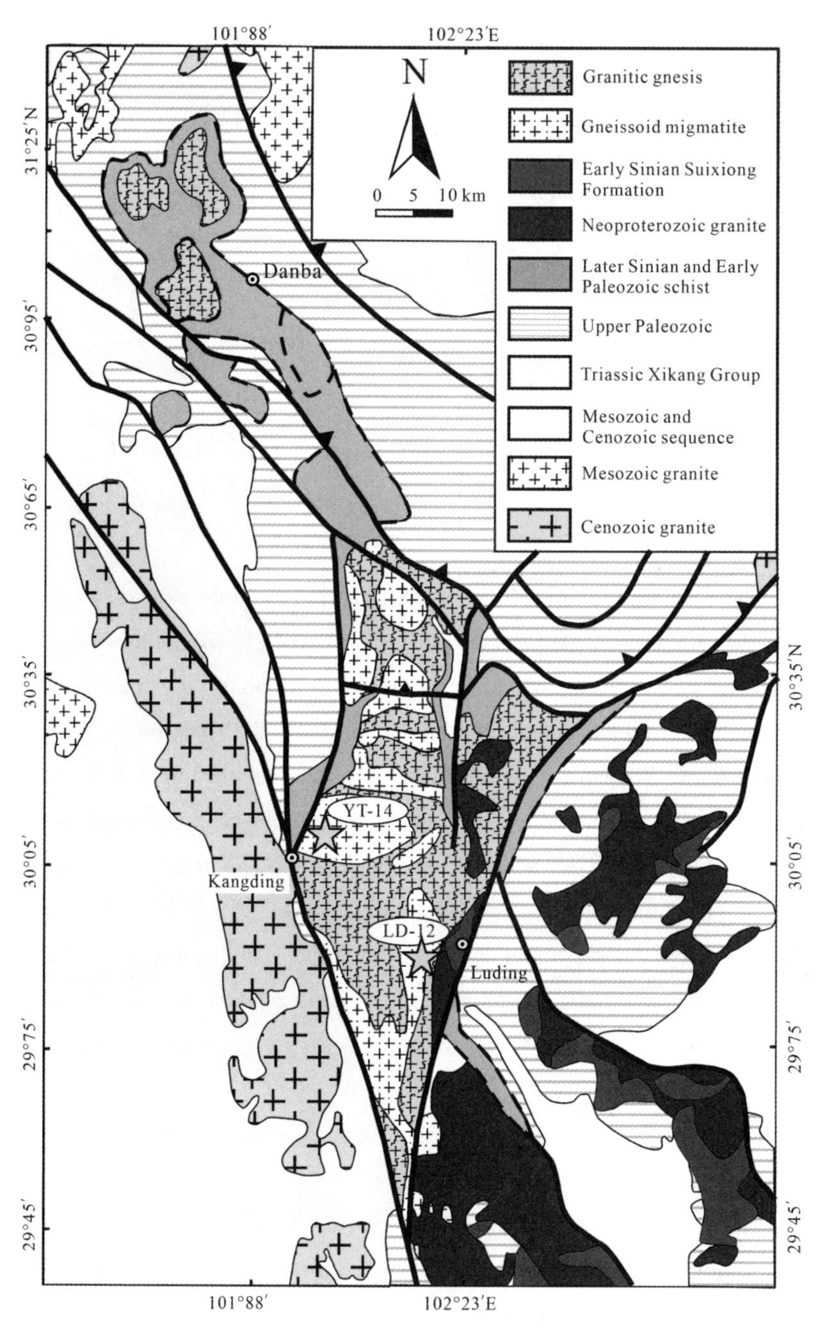

Fig. 2 Geological map of the Kangding and Danba area, showing the distribution of Neoproterozoic granites and metamorphic rocks (after Zhou et al., 2002).

mass spectrometer. Sr and Nd isotopic fractionations were corrected to $^{87}Sr/^{86}Sr = 0.1194$ and $^{146}Nd/^{144}Nd = 0.7219$, respectively. During the period of analysis, the NIST SRM 987 standard gave an average value of $^{87}Sr/^{86}Sr = 0.710250 \pm 12$ (2σ, $n = 15$) and the La Jolla standard gave an average value of $^{146}Nd/^{144}Nd = 0.511859 \pm 6$ (2σ, $n = 20$). Whole rock Pb

Fig. 3 Field photographs and microscope of the quartz monzodiorite and granodiorite from the Luding and Kangding area.

(a) field photo of the Luding quartz monzodiorite; (b) microscope of the Luding diorite; (c) field photo of the Kangding granodiorite; (d) microscope of the Kangding granodiorite. The geological hammer in (a) have 35 cm in length, the clip is 2 cm length. Plg: plagioclase; Amp: amphibole; Bi: biotite.

was separated by anion exchange in HCl-Br columns, and Pb isotopic fractionation was corrected to $^{205}Tl/^{203}Tl = 2.387\ 5$. Within the period of analysis, 30 measurements of NBS981 yielded average values of $^{206}Pb/^{204}Pb = 16.937 \pm 1\ (2\sigma)$, $^{207}Pb/^{204}Pb = 15.491 \pm 1\ (2\sigma)$, and $^{208}Pb/^{204}Pb = 36.696 \pm 1\ (2\sigma)$. BCR-2 standard yielded average values of $^{206}Pb/^{204}Pb = 18.742 \pm 1\ (2\sigma)$, $^{207}Pb/^{204}Pb = 15.620 \pm 1\ (2\sigma)$, and $^{208}Pb/^{204}Pb = 38.705 \pm 1\ (2\sigma)$. Whole rock Hf was also separated by a single anion exchange columns, and 22 measurements of JCM 475 gave an average value of $^{176}Hf/^{177}Hf = 0.282\ 161\ 3 \pm 0.000\ 001\ 3$ (Yuan et al., 2008).

Internal morphology was examined with cathodoluminescence (CL) microscopy prior to U-Pb isotopic dating. Zircon Laser Ablation Inductively Coupled Plasma Mass Spectrometry (LA-ICP-MS) U-Pb analyses were conducted on Agilent 7500a ICP-MS equipped with a 193 nm laser, following the method ofYuan et al. (2004). The $^{207}Pb/^{235}U$ and $^{206}Pb/^{238}U$ ratios were calculated using the GLITTER program, which was corrected using the Harvard zircon 91500 as external calibration. Common Pb contents were subsequently valuated using the method described in Andersen (2002). The age calculations and plotting of concordia diagrams were made using ISOPLOT (version 3.0; Ludwig, 2003). The errors quoted in tables and figures are at 2σ levels.

4 Results

4.1 Zircon LA-ICP-MS U-Pb age

Zircon U-Pb concordia diagrams and CL images of quartz monzodiorite (LD12) and granodiorite (YT14) from the Luding-Kangding area are presented in Fig. 4, and the results of the analyses are listed in Table 1.

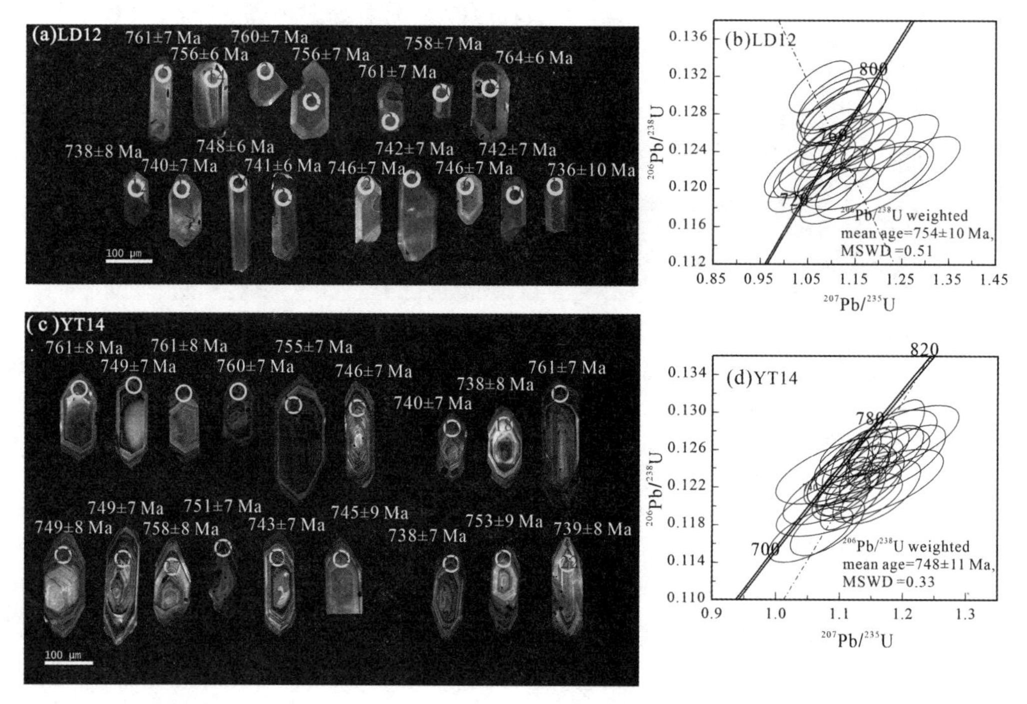

Fig. 4 Cathodoluminescence (CL) images (a, c) and LA-ICPMS U-Pb zircon concordia diagrams (b, d) of representative zircons in the Luding quartz monzodiorite and Kangding granodiorite. Ellipse dimensions are 2σ.

Zircons from quartz monzodiorite (LD12) are euhedral, colorless, long prismatic crystals, with aspect ratios ranging from 2 : 1 to 3 : 1. Most of the grains have no well-developed oscillatory zoning (Fig. 4a). One out of thirty-six spots (31#) yields discordant U-Pb age, and five spots (4#, 19#, 20#, 22# and 29#) yield older $^{206}Pb/^{238}U$ ages (811 ± 5 Ma to 884 ± 7 Ma), which leads to consider them as inherited cores. In addition, significantly younger ages (681 ± 5 Ma to 718 ± 8 Ma) were obtained from five spots, which may be caused by Pb loss in subsequent geological process. There are concordant 25 spots have U concentrations from 112 ppm to 547 ppm, Th concentrations from 86 ppm to 407 ppm, and Th/U ratios from 0.35 to 1.50. These concordant 25 spots are characterized by $^{206}Pb/^{238}U$ ages ranging from 727 ± 7 Ma to

Table 1 Results of zircon LA-ICP-MS U-Pb results for the diorite and tonalite from the Luding-Kangding area, western margin of the Yangtze Block.

Analysis	$^{207}Pb/^{206}Pb$	2δ	$^{207}Pb/^{235}U$	2δ	$^{206}Pb/^{238}U$	2δ	$^{208}Pb/^{232}Th$	2δ	Th/U	$^{207}Pb/^{206}Pb$	2δ	$^{207}Pb/^{235}U$	2δ	$^{206}Pb/^{238}U$	2δ	$^{208}Pb/^{232}Th$	2δ
Diorites																	
LD12-01	0.062 11	0.001 51	0.964 33	0.043 12	0.112 61	0.001 80	0.029 81	0.000 30	1.26	678	34	686	11	688	5	594	6
LD12-02	0.063 21	0.002 05	1.069 44	0.065 74	0.122 70	0.002 46	0.033 58	0.000 53	0.86	715	48	738	16	746	7	668	10
LD12-03	0.061 41	0.001 93	1.033 11	0.061 38	0.122 02	0.002 38	0.035 50	0.000 52	0.94	654	47	720	15	742	7	705	10
LD12-04	0.065 47	0.001 68	1.326 05	0.063 10	0.146 90	0.002 54	0.037 81	0.000 48	0.91	789	35	857	14	884	7	750	9
LD12-05	0.065 76	0.001 85	1.136 08	0.059 76	0.125 29	0.002 30	0.035 10	0.000 44	1.08	799	40	771	14	761	7	697	9
LD12-06	0.076 06	0.002 25	1.286 37	0.071 50	0.122 66	0.002 46	0.040 84	0.000 59	1.14	1 097	39	840	16	746	7	809	11
LD12-07	0.067 86	0.002 21	1.116 61	0.068 88	0.119 35	0.002 48	0.034 55	0.000 50	1.09	864	47	761	17	727	7	687	10
LD12-08	0.063 36	0.001 62	1.020 56	0.048 50	0.116 82	0.001 96	0.030 44	0.000 31	1.44	720	36	714	12	712	6	606	6
LD12-09	0.063 74	0.002 27	1.128 45	0.076 78	0.128 41	0.002 82	0.038 34	0.000 71	0.76	733	53	767	18	779	8	760	14
LD12-10	0.062 57	0.001 94	1.052 44	0.061 56	0.121 99	0.002 38	0.035 90	0.000 51	1.03	694	46	730	15	742	7	713	10
LD12-11	0.071 44	0.003 20	1.191 85	0.102 66	0.121 00	0.003 38	0.037 96	0.000 89	0.67	970	65	797	24	736	10	753	17
LD12-12	0.070 09	0.003 36	1.171 44	0.109 02	0.121 22	0.002 76	0.036 78	0.000 31	0.92	931	101	787	25	738	8	730	6
LD12-13	0.062 13	0.001 54	0.976 31	0.044 88	0.113 97	0.001 86	0.030 50	0.000 33	1.18	679	35	692	12	696	5	607	6
LD12-14	0.063 43	0.001 43	1.088 11	0.044 74	0.124 42	0.001 98	0.040 38	0.000 38	1.07	723	30	748	11	756	6	800	7
LD12-15	0.065 64	0.001 79	1.083 15	0.055 26	0.119 68	0.002 14	0.033 97	0.000 39	1.21	795	38	745	13	729	6	675	8
LD12-16	0.062 36	0.002 41	1.013 00	0.075 08	0.117 81	0.002 74	0.033 96	0.000 63	0.93	686	59	710	19	718	8	675	12
LD12-17	0.066 20	0.002 10	1.109 82	0.066 58	0.121 58	0.002 44	0.034 99	0.000 50	1.11	813	46	758	16	740	7	695	10
LD12-18	0.062 37	0.001 78	1.033 95	0.055 38	0.120 24	0.002 18	0.035 17	0.000 43	1.35	687	42	721	14	732	6	699	8
LD12-19	0.052 32	0.001 63	1.036 97	0.061 14	0.143 74	0.002 66	0.036 94	0.000 51	1.01	299	50	722	15	866	7	733	10
LD12-20	0.055 66	0.001 53	1.107 94	0.057 10	0.144 38	0.002 48	0.037 31	0.000 46	1.06	439	42	757	14	869	7	740	9
LD12-21	0.072 39	0.002 01	1.248 72	0.064 78	0.125 12	0.002 34	0.036 66	0.000 48	1.03	997	37	823	15	760	7	728	9
LD12-22	0.065 67	0.001 27	1.213 85	0.041 58	0.134 05	0.001 90	0.041 50	0.000 52	0.35	796	24	807	10	811	5	822	10
LD12-23	0.064 40	0.002 13	1.104 57	0.069 28	0.124 40	0.002 54	0.037 32	0.000 62	0.80	755	49	756	17	756	7	741	12
LD12-24	0.062 20	0.001 48	0.955 63	0.041 62	0.111 43	0.001 76	0.034 11	0.000 38	0.74	681	33	681	11	681	5	678	7

Continued

Analysis	$^{207}Pb/^{206}Pb$	2δ	$^{207}Pb/^{235}U$	2δ	$^{206}Pb/^{238}U$	2δ	$^{208}Pb/^{232}Th$	2δ	Th/U	$^{207}Pb/^{206}Pb$	2δ	$^{207}Pb/^{235}U$	2δ	$^{206}Pb/^{238}U$	2δ	$^{208}Pb/^{232}Th$	2δ
LD12-25	0. 062 51	0. 002 14	1. 117 78	0. 072 98	0. 129 69	0. 002 74	0. 038 06	0. 000 66	0. 80	692	51	762	17	786	8	755	13
LD12-26	0. 069 26	0. 001 58	1. 174 09	0. 048 94	0. 122 95	0. 001 96	0. 034 53	0. 000 32	1. 50	906	30	789	11	748	6	686	6
LD12-27	0. 067 07	0. 002 02	1. 158 00	0. 065 60	0. 125 22	0. 002 44	0. 037 30	0. 000 54	0. 94	840	43	781	15	761	7	740	11
LD12-28	0. 062 03	0. 001 86	1. 102 84	0. 062 40	0. 128 95	0. 002 44	0. 036 97	0. 000 50	1. 04	675	44	755	15	782	7	734	10
LD12-29	0. 062 40	0. 002 08	1. 219 33	0. 077 22	0. 141 71	0. 002 92	0. 043 96	0. 000 68	0. 93	688	50	809	18	854	8	870	13
LD12-30	0. 069 50	0. 001 94	1. 195 99	0. 062 40	0. 124 81	0. 002 30	0. 037 38	0. 000 47	1. 14	914	38	799	14	758	7	742	9
LD12-31	0. 078 09	0. 002 32	1. 297 51	0. 072 06	0. 120 51	0. 002 42	0. 035 86	0. 000 48	1. 05	1 149	39	845	16	733	7	712	9
LD12-32	0. 064 03	0. 002 10	1. 128 37	0. 070 36	0. 127 81	0. 002 60	0. 036 32	0. 000 62	0. 80	743	49	767	17	775	7	721	12
LD12-33	0. 065 12	0. 001 88	1. 093 52	0. 059 22	0. 121 78	0. 002 26	0. 033 12	0. 000 38	1. 41	778	41	750	14	741	6	659	7
LD12-34	0. 063 28	0. 001 68	1. 098 01	0. 054 16	0. 125 85	0. 002 18	0. 036 23	0. 000 43	1. 10	718	38	752	13	764	6	719	8
LD12-35	0. 062 83	0. 001 92	1. 105 54	0. 063 70	0. 127 62	0. 002 44	0. 038 08	0. 000 56	0. 85	702	45	756	15	774	7	755	11
LD12-36	0. 059 62	0. 001 61	1. 076 21	0. 054 22	0. 130 91	0. 002 26	0. 039 25	0. 000 49	0. 93	590	40	742	13	793	6	778	10
Tonalite																	
YT14-02	0. 068 24	0. 002 26	1. 197 54	0. 074 56	0. 127 28	0. 002 86	0. 038 73	0. 000 37	0. 61	876	70	799	17	772	8	768	7
YT14-03	0. 069 68	0. 001 98	1. 203 42	0. 052 30	0. 125 30	0. 002 76	0. 038 38	0. 000 60	0. 48	919	27	802	12	761	8	761	12
YT14-04	0. 066 40	0. 002 57	1. 100 14	0. 081 26	0. 120 16	0. 002 76	0. 036 68	0. 000 42	0. 44	819	83	753	20	732	8	728	8
YT14-05	0. 065 41	0. 001 81	1. 054 97	0. 054 20	0. 116 98	0. 002 42	0. 035 77	0. 000 34	0. 46	788	60	731	13	713	7	710	7
YT14-06	0. 065 87	0. 001 84	1. 119 47	0. 057 76	0. 123 25	0. 002 60	0. 037 66	0. 000 35	0. 52	802	60	763	14	749	7	747	7
YT14-07	0. 065 73	0. 001 91	1. 135 02	0. 061 34	0. 125 23	0. 002 66	0. 038 27	0. 000 36	0. 50	798	62	770	15	761	8	759	7
YT14-09	0. 065 22	0. 001 50	1. 125 29	0. 032 40	0. 125 17	0. 002 48	0. 036 95	0. 000 40	0. 50	781	15	766	8	760	7	733	8
YT14-11	0. 064 75	0. 001 45	1. 109 47	0. 030 08	0. 124 29	0. 002 44	0. 037 62	0. 000 41	0. 38	766	14	758	7	755	7	746	8
YT14-12	0. 066 48	0. 002 30	1. 161 31	0. 075 72	0. 126 70	0. 002 94	0. 038 67	0. 000 39	0. 47	821	74	783	18	769	8	767	7
YT14-15	0. 067 40	0. 001 65	1. 140 40	0. 050 64	0. 122 71	0. 002 56	0. 037 39	0. 000 35	0. 40	850	52	773	12	746	7	742	7
YT14-16	0. 067 23	0. 001 86	1. 127 21	0. 057 52	0. 121 60	0. 002 56	0. 037 07	0. 000 35	0. 58	845	59	766	14	740	7	736	7
YT14-17	0. 067 91	0. 001 87	1. 173 11	0. 048 72	0. 125 28	0. 002 74	0. 044 30	0. 000 65	0. 45	866	25	788	11	761	8	876	13

Continued

Analysis	$^{207}Pb/^{206}Pb$	2δ	$^{207}Pb/^{235}U$	2δ	$^{206}Pb/^{238}U$	2δ	$^{208}Pb/^{232}Th$	2δ	Th/U	$^{207}Pb/^{206}Pb$	2δ	$^{207}Pb/^{235}U$	2δ	$^{206}Pb/^{238}U$	2δ	$^{208}Pb/^{232}Th$	2δ
YT14-18	0.064 97	0.001 99	1.087 23	0.061 90	0.121 37	0.002 72	0.037 14	0.000 37	0.37	773	66	747	15	738	8	737	7
YT14-19	0.066 36	0.001 47	1.146 28	0.030 34	0.125 27	0.002 48	0.036 77	0.000 38	0.49	818	13	775	7	761	7	730	7
YT14-20	0.067 60	0.001 57	1.130 91	0.033 90	0.121 32	0.002 46	0.037 60	0.000 42	0.50	856	16	768	8	738	7	746	8
YT14-21	0.064 43	0.002 52	1.100 17	0.082 00	0.123 85	0.002 96	0.037 93	0.000 39	0.47	756	85	753	20	753	9	753	8
YT14-22	0.069 53	0.001 99	1.165 08	0.051 72	0.121 51	0.002 74	0.038 08	0.000 57	0.54	915	27	784	12	739	8	755	11
YT14-23	0.067 45	0.001 54	1.192 59	0.033 94	0.128 22	0.002 58	0.043 56	0.000 51	0.31	852	14	797	8	778	7	862	10
YT14-24	0.066 20	0.001 48	1.152 60	0.031 34	0.126 26	0.002 52	0.037 04	0.000 38	0.56	813	14	778	7	766	7	735	7
YT14-25	0.065 37	0.001 75	1.110 48	0.054 40	0.123 20	0.002 66	0.037 67	0.000 36	0.36	786	57	758	13	749	8	747	7
YT14-26	0.064 80	0.001 41	1.100 82	0.042 20	0.123 20	0.002 52	0.037 71	0.000 36	0.28	768	47	754	10	749	7	748	7
YT14-27	0.066 99	0.002 28	1.152 47	0.073 60	0.124 76	0.002 90	0.038 05	0.000 37	0.58	838	72	778	17	758	8	755	7
YT14-28	0.066 87	0.001 53	1.139 58	0.032 82	0.123 57	0.002 52	0.037 67	0.000 44	0.21	834	15	772	8	751	7	747	9
YT14-30	0.065 75	0.001 59	1.107 01	0.036 18	0.122 08	0.002 54	0.035 52	0.000 44	0.41	798	18	757	9	743	7	705	9
YT14-31	0.066 73	0.001 52	1.108 95	0.031 96	0.120 48	0.002 46	0.034 87	0.000 39	0.41	829	15	758	8	733	7	693	8
YT14-32	0.068 96	0.002 56	1.164 42	0.081 64	0.122 46	0.002 98	0.037 22	0.000 39	0.50	898	78	784	19	745	9	739	8
YT14-33	0.066 57	0.001 62	1.081 24	0.047 48	0.117 80	0.002 50	0.035 95	0.000 35	0.35	824	52	744	12	718	7	714	7

Table 2 Analytical results of major(%) and trace(ppm) element from the diorite and tonalite from the Luding-Kangding area, western margin of the Yangtze Block.

Sample	LD09	LD10	LD11	LD14	LD15	LD16	YT01	YT02	YT05	YT08	YT10	YT11	YT12	YT13	YT15
Lithology	Diorite						Tonalite								
SiO_2	63.65	61.08	63.78	60.76	63.56	60.83	66.44	67.59	67.58	67.62	66.02	66.17	65.9	65.68	65.32
TiO_2	0.45	0.52	0.41	0.56	0.44	0.51	0.26	0.27	0.27	0.26	0.32	0.30	0.30	0.30	0.24
Al_2O_3	16.29	17.16	16.74	17.01	16.52	17.08	17.56	16.74	16.76	16.7	16.87	17.08	17.43	17.17	18.32
$Fe_2O_3{}^T$	4.68	5.58	4.22	6.02	4.35	5.64	2.90	2.88	2.80	2.80	3.55	3.59	3.39	3.45	2.80
MnO	0.08	0.09	0.08	0.10	0.07	0.09	0.06	0.06	0.06	0.05	0.08	0.08	0.08	0.08	0.09

Continued

Sample	LD09	LD10	LD11	LD14	LD15	LD16	YT01	YT02	YT05	YT08	YT10	YT11	YT12	YT13	YT15
Lithology	Diorite						Tonalite								
MgO	2. 37	2. 81	2. 13	3. 15	2. 33	2. 88	0. 92	0. 93	0. 88	0. 92	1. 16	1. 17	1. 09	1. 11	0. 88
CaO	4. 64	5. 89	4. 00	5. 99	4. 10	5. 93	3. 80	3. 58	3. 62	4. 16	4. 23	4. 28	4. 13	4. 11	3. 91
Na_2O	3. 94	4. 02	4. 17	3. 92	4. 17	3. 98	4. 41	4. 22	4. 17	3. 97	4. 73	4. 28	4. 43	4. 05	4. 86
K_2O	2. 07	1. 43	2. 55	1. 49	2. 21	1. 42	2. 25	2. 25	2. 22	1. 89	1. 67	1. 66	1. 96	1. 91	2. 34
P_2O_5	0. 12	0. 16	0. 12	0. 18	0. 12	0. 17	0. 14	0. 14	0. 14	0. 11	0. 15	0. 16	0. 15	0. 15	0. 11
LOI	1. 86	1. 15	1. 59	1. 11	2. 00	1. 06	1. 43	1. 64	1. 46	1. 59	1. 32	1. 30	1. 43	1. 51	1. 12
Total	100. 15	99. 89	99. 79	100. 29	99. 87	99. 59	100. 17	100. 3	99. 96	100. 07	100. 10	100. 07	100. 29	99. 52	99. 99
Li	14. 4	14. 3	12. 2	14. 4	16. 3	15. 8	20. 5	20. 9	20. 4	20. 4	22. 1	22. 1	15. 9	15. 7	31. 7
Be	1. 31	1. 15	1. 32	1. 12	1. 31	1. 15	1. 74	1. 67	1. 66	1. 71	1. 59	1. 58	1. 56	1. 57	2. 07
Sc	12. 0	12. 4	10. 9	14. 3	11. 6	14. 5	5. 19	5. 26	5. 13	4. 92	5. 18	5. 11	5. 89	6. 21	5. 55
V	95. 2	110	86. 2	118	92. 1	110	26. 3	27. 1	26. 0	26. 5	37. 1	37. 6	35. 1	36. 0	30. 0
Cr	25. 3	28. 1	49. 2	33. 0	24. 1	30. 4	3. 51	44. 8	3. 03	3. 58	11. 3	7. 07	5. 78	12. 0	4. 38
Co	78. 0	57. 8	43. 0	71. 8	58. 5	45. 5	49. 7	35. 7	58. 2	54. 7	71. 6	57. 3	56. 0	55. 6	75. 5
Ni	15. 2	17. 1	29. 1	19. 2	14. 6	16. 5	2. 99	35. 2	2. 78	3. 08	8. 85	5. 41	4. 66	8. 23	3. 64
Cu	22. 9	22. 0	17. 3	20. 8	19. 1	23. 7	3. 41	4. 37	3. 79	2. 67	7. 69	7. 18	22. 9	21. 3	5. 55
Zn	53. 6	62. 1	49. 3	65. 1	51. 3	64. 1	53. 7	55. 3	53. 7	52. 9	62. 4	62. 7	51. 9	52. 3	48. 8
Ga	16. 8	17. 2	16. 8	17. 3	17. 0	18. 4	19. 2	18. 7	18. 6	18. 7	18. 2	17. 9	17. 9	18. 3	18. 4
Ge	1. 19	1. 16	1. 16	1. 22	1. 23	1. 17	1. 09	1. 07	1. 10	1. 08	1. 19	1. 13	1. 14	1. 18	1. 30
Rb	62. 2	37. 4	74. 3	39. 9	65. 8	38. 7	55. 6	61. 3	58. 7	52. 8	45. 5	44. 5	41. 8	45. 8	45. 5
Sr	381	488	414	472	382	493	446	426	425	469	530	533	531	537	486
Y	18. 6	16. 0	17. 2	17. 8	17. 7	16. 7	16. 4	14. 7	14. 7	9. 01	10. 3	9. 71	9. 75	10. 3	15. 0
Zr	126	145	133	135	116	121	159	149	136	174	124	119	127	107	107
Nb	5. 13	4. 06	4. 60	4. 34	4. 90	4. 11	5. 41	5. 50	5. 40	4. 27	4. 87	4. 78	4. 72	4. 62	5. 81
Cs	0. 90	1. 15	0. 84	1. 20	0. 95	1. 14	1. 64	1. 63	1. 59	1. 51	1. 89	1. 75	1. 64	1. 67	2. 42
Ba	498	439	669	453	535	445	1 006	1 074	1 069	764	705	719	821	809	893

Continued

Sample	LD09	LD10	LD11	LD14	LD15	LD16	YT01	YT02	YT05	YT08	YT10	YT11	YT12	YT13	YT15
Lithology	Diorite						Tonalite								
La	10.4	12.5	16.9	18.9	14.0	19.0	21.1	25.1	21.1	31.5	38.9	14.2	20.6	34.6	15.1
Ce	25.4	28.4	35.6	40.0	30.7	40.1	41.4	48.5	41.6	60.1	73.9	28.5	39.9	66.0	30.5
Pr	3.50	3.60	4.21	4.75	3.86	4.65	4.72	5.45	4.73	6.75	8.14	3.23	4.42	7.43	3.66
Nd	15.2	15.2	16.4	18.9	15.7	18.9	17.8	20.6	17.8	25.0	29.4	12.6	16.4	27.8	14.8
Sm	3.39	3.24	3.32	3.72	3.28	3.58	3.30	3.69	3.24	4.08	4.20	2.40	2.70	4.29	3.07
Eu	0.85	0.94	0.86	0.99	0.88	1.01	1.00	1.03	0.98	1.06	1.09	0.99	1.06	1.12	1.03
Gd	3.15	2.98	3.06	3.44	3.05	3.25	2.93	3.12	2.83	3.13	3.28	2.12	2.34	3.30	2.80
Tb	0.49	0.45	0.46	0.50	0.46	0.48	0.42	0.42	0.39	0.35	0.38	0.29	0.31	0.38	0.41
Dy	2.98	2.69	2.81	3.04	2.89	2.89	2.49	2.41	2.31	1.75	1.99	1.69	1.73	2.00	2.50
Ho	0.62	0.54	0.57	0.61	0.58	0.57	0.50	0.46	0.46	0.31	0.36	0.33	0.34	0.36	0.49
Er	1.83	1.56	1.68	1.74	1.72	1.65	1.54	1.37	1.39	0.85	1.01	0.94	0.95	0.99	1.47
Tm	0.28	0.24	0.25	0.26	0.26	0.24	0.23	0.20	0.20	0.12	0.15	0.14	0.13	0.14	0.23
Yb	1.80	1.50	1.68	1.67	1.72	1.60	1.38	1.21	1.24	0.81	0.93	0.91	0.89	0.88	1.53
Lu	0.27	0.23	0.25	0.25	0.26	0.24	0.20	0.18	0.18	0.14	0.15	0.14	0.14	0.14	0.24
Hf	3.24	3.45	3.43	3.19	2.92	2.99	3.81	3.52	3.25	4.24	2.98	2.83	2.98	2.56	2.78
Ta	0.47	0.37	0.40	0.39	0.43	0.34	0.39	0.38	0.41	0.31	0.47	0.42	0.40	0.35	0.61
Pb	10.9	8.89	10.0	8.94	11.5	9.22	16.6	16.9	16.9	13.2	12.0	12.6	12.7	12.5	22.6
Th	2.79	2.66	4.84	4.60	3.64	4.76	4.62	5.28	4.50	6.22	5.33	2.25	2.69	5.06	4.06
U	0.86	0.49	0.84	0.53	0.97	0.58	0.73	0.80	0.69	0.67	0.66	0.52	0.45	0.45	0.96
$Mg^{\#}$	54.1	54.0	54.1	54.9	55.5	54.3	42.5	42.9	42.3	43.4	43.2	43.2	42.8	42.9	42.3
Sr/Y	20.50	30.44	24.08	26.42	21.59	29.52	27.20	29.00	28.95	52.07	51.28	54.94	54.52	52.09	32.44
Rb/Sr	0.16	0.08	0.18	0.08	0.17	0.08	0.12	0.14	0.14	0.11	0.09	0.08	0.08	0.09	0.09
Eu^*/Eu	0.79	0.92	0.83	0.84	0.85	0.90	0.99	0.93	0.99	0.91	0.90	1.35	1.29	0.91	1.08
$(La/Yb)_N$	4.14	5.94	7.23	8.13	5.86	8.51	11.01	14.85	12.20	27.85	29.93	11.24	16.53	28.05	7.07

793 ± 6 Ma (weighted mean age of 754 ± 10 Ma, MSWD = 0.51, 2σ), which represent the crystallization age of the quartz monzodiorite in the Luding area. Zircons from the granodiorite (YT-14) are euhedral, fawn to colorless, 100–150 μm long, and with aspect ratios of 1∶1 to 3∶1. Most of the grains are gray and display well-developed oscillatory zoning in CL images (Fig. 4b). One (spot 13#) out of thirty-four spots yields discordant U-Pb ages. Spots 1# and 8# yield significantly younger $^{206}Pb/^{238}U$ ages (681 ± 7 Ma to 683 ± 7 Ma), which are considered to be caused by Pb loss. Four spots (29#, 34#, 35#, and 36#) yield $^{206}Pb/^{238}U$ ages from 803 ± 8 Ma to 970 ± 11 Ma, which are clearly older than the other grains. Other 27 spots have U contents from 101 ppm to 463 ppm, Th contents from 47 to 186 ppm, and Th/U ratios from 0.21 to 0.60. These 27 spots have concordant $^{206}Pb/^{238}U$ ages from 713 ± 7 Ma to 778 ± 7 Ma (weighted mean age of 748 ± 11 Ma, MSWD = 0.33, 2σ), which represent the crystallization age of the granodiorite in the Kangding area.

4.2 Major and trace element chemistry

Major and trace element analyses of the quartz monzodiorite from the Kangding area are listed in Table 2. According to field observations, quartz monzodiorite is mainly present in the northern part of the pluton, whereas the granodiorite is mainly present in the southern part of the pluton (Fig. 1b).

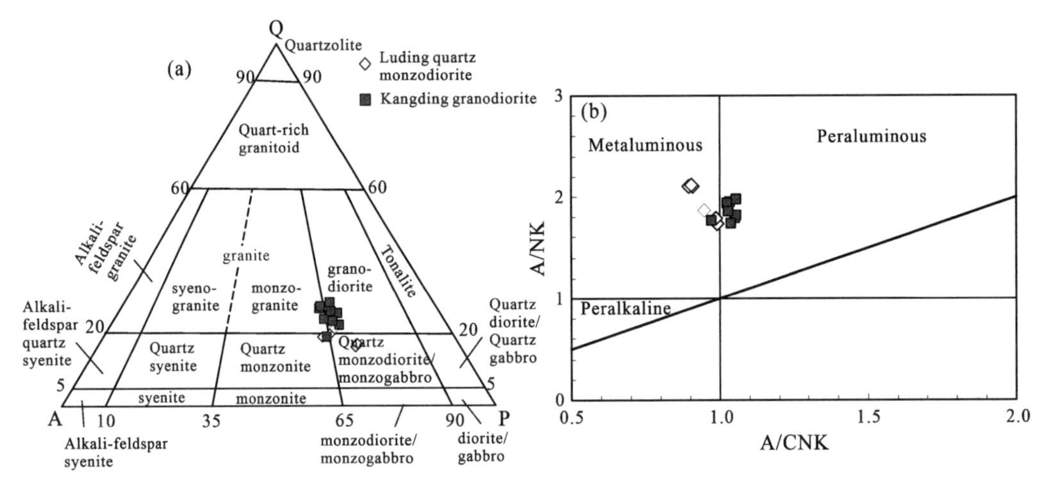

Fig. 5 Q-A-P (a) and A/NK [$Al_2O_3/(Na_2O+K_2O)$] vs. A/CNK [molar ratio $Al_2O_3/(CaO+Na_2O+K_2O)$] (b) diagrams for the Luding quartz monzodiorite and Kangding granodiorite.

4.3 High-Mg# quartz monzodiorite

Six quartz monzodiorite samples from the Kangding pluton are characterized by low SiO_2 contents (60.76–63.78 wt%), TiO_2 = 0.41–0.56 wt%, Al_2O_3 contents from 16.29 wt% to 17.16 wt%, and A/CNK values of 0.90–0.99. In the Q-A-P diagrams, the six samples plotted in the field of quartz monzodiorite, all the samples are sodic, with Na_2O = 3.92–4.17 wt%, K_2O = 1.42–2.55 wt%, and Na_2O/K_2O ratios of 1.6–2.8. All the samples plot in the calc-alkaline field in the SiO_2 vs. K_2O diagram, in the R_1-R_2 diagram, the samples plotted in the field of diorite to tonalite (Fig. 6). The quartz monzodiorite samples have MgO contents from 2.13 wt% to 3.15 wt%, $Fe_2O_3^T$ = 4.22–6.02 wt%, and high $Mg^{\#}$ values of 54.1–54.5, suggesting the involvement of mantle components in their source regions (Fig. 6). The quartz monzodiorite samples yield right steep slope LREE patterns with $(La/Yb)_N$ ratios from 4.14 to 8.51, flat HREE patterns with $(Dy/Yb)_N$ ratios from 1.1 to 1.2, and insignificant Eu anomalies (Eu^*/Eu = 0.79–0.92). The primitive mantle normalized spider diagrams (Fig. 7) show that the quartz monzodiorite samples are enriched in Rb, Ba, Sr, Th and depleted in Nb,

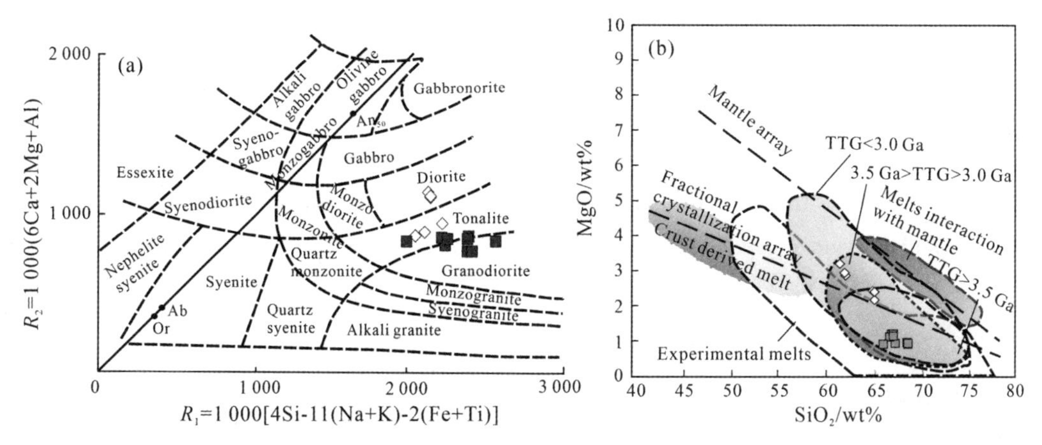

Fig. 6 Plots of R_1 vs. R_2 diagrams for classification of plutonic rocks (after de la Roche et al., 1980) (a) and SiO_2-MgO diagrams (after Moyen and Martin, 2012) (b) for the Luding quartz monzodiorite and Kangding granodiorite.

Symbols as in Fig. 5.

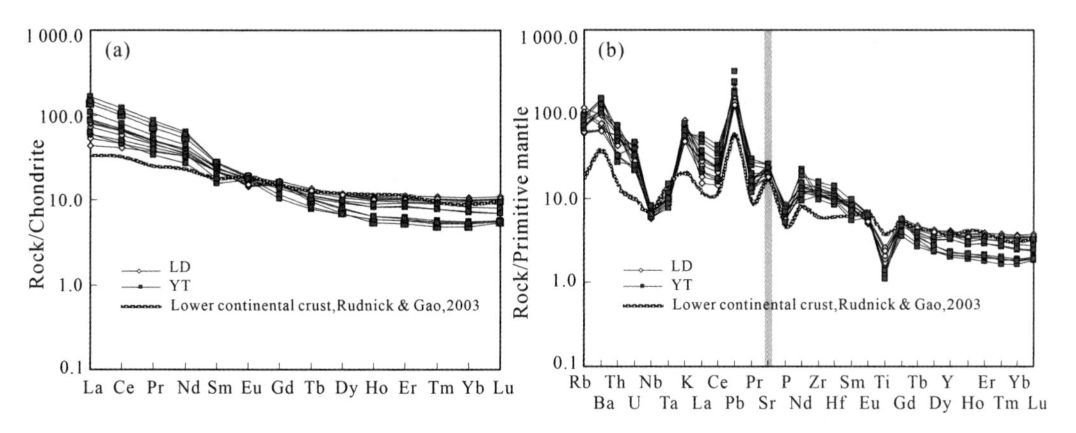

Fig. 7 Chondrite-normalized REE patterns (a) and primitive mantle (PM) normalized trace element spider diagrams (b) for the Luding quartz monzodiorite and Kangding granodiorite.

The normalized values are from Sun and McDonough (1989).

Ta and Ti, and that they have low Nb/Ta (10.8–12.1) ratios. These features are similar to those of the crustal-derived rocks or island arc volcanic rocks (Wilson, 1989), which have Sr contents from 381 to 493 ppm, Rb contents from 37.4 ppm to 74.3 ppm, Y contents from 16.0 ppm to 18.6 ppm, and Sr/Y ratios from 20.5 to 30.4.

4.4 Granodiorite

Compared to the quartz monzodiorite samples, the granodiorite samples have higher SiO_2 contents (65.32–67.62 wt%), lower TiO_2 (0.24–0.32 wt%), MgO (0.88–1.17 wt%) and $Fe_2O_3^T$ (2.80–3.59 wt%) contents, and lower $Mg^{\#}$ values (42.3–43.4), as shown in the Q-A-P diagram (Fig. 5), the samples plotted in the granodiorite field. Like the quartz monzodiorite, the granodiorite are also sodic (Na_2O/K_2O = 1.9 – 2.8) and metaluminous (A/CNK=0.98–1.06). The granodiorite samples are also enriched in LREE and show high $(La/Yb)_N$ ratios (7.1 – 29.9), but they show positive Eu anomalies (Eu^*/Eu = 0.93 – 1.35), indicating a plagioclase-rich source region. These samples also are also characterized by flat HREE patterns with $(Dy/Yb)_N$ ratios from 1.09 to 1.51. The samples are enriched in Rb, Ba, Th, and U, and depleted in Nb, Ta, and Ti in the primitive mantle normalized spider diagrams (Fig. 7). The granodiorite samples have relatively high Sr content (425–537 ppm) but low Y content (9.01–16.4 ppm), which results in high Sr/Y ratios (27.2–54.5).

4.5 Sr-Nd-Pb isotopic compositions

Whole rock Sr-Nd-Pb isotopic compositions are given in Tables 3 and 4. Initial isotopic values were calculated according to the LA-ICP-MS zircon U-Pb dates for the quartz monzodiorite and the granodiorite. Whole-rock Nd model age was calculated using the model of DePaolo and Wasserburg (1976).

The three quartz monzodiorite samples (LD09, LD10, and LD15) have Sr concentrations of 381 – 488 ppm, Rb concentrations of 37.4 – 65.8 ppm, $(^{87}Sr/^{86}Sr)_i$ of 0.703 513 – 0.704 519, $^{143}Nd/^{144}Nd$ ratios of 0.512 352–0.512 489, $\varepsilon_{Nd}(t)$ values from +2.4 to +4.8, and Nd model ages from 1.08 Ga to 1.31 Ga. The granodiorite samples (YT02, YT10 and YT15) are characterized by similar Rb (45.5–61.3 ppm) and Sr (426–536 ppm) contents, and they have $(^{87}Sr/^{86}Sr)_i$ of 0.704 071 – 0.705 414, $^{143}Nd/^{144}Nd$ ratios of 0.512 596 – 0.512 373, $\varepsilon_{Nd}(t)$ values from +0.8 to +1.3, and Nd model ages from 1.21 Ga to 1.41 Ga (Table 3). As shown in the $\varepsilon_{Nd}(t)$ vs. $(^{87}Sr/^{86}Sr)_i$ plot (Fig. 8), Sr-Nd isotopic compositions of the quartz monzodiorite and granodiorite are similar to the Neoproterozoic mafic intrusion and the diorites of the Hannan region, in the northwestern margin of the Yangtze Block (Zhao and Zhou, 2009; Zhao et al., 2010a), which are clearly different from the MORB (Zimmerer and McIntosh, 2012), suggesting significant crustal components in their source region.

Table 3 Whole-rock Sr-Nd isotopic composition for the diorite and tonalite from the Luding-Kangding area, western margin of the Yangtze Block.

Rock	Sample No	Sr /ppm	Rb /ppm	$^{87}Sr/^{86}Sr$	2SE	$^{87}Rb/^{86}Sr$ [a]	$^{87}Sr/^{86}Sr$ (750 Ma)	Nd /ppm	Sm /ppm	$^{143}Nd/^{144}Nd$	2SE	$^{147}Sm/^{144}Nd$ [a]	T_{DM} /Ga [b]	$\varepsilon_{Nd}(t)$ [c]	$^{143}Nd/^{144}Nd$ (750 Ma)
Diorite	LD09	381	62. 2	0. 709 137	0. 000 009	0. 472	0. 704 079	15. 2	3. 39	0. 512 475	0. 000 007	0. 135	1. 14	2. 74	0. 512 409
Diorite	LD10	488	37. 4	0. 706 265	0. 000 010	0. 221	0. 703 895	15. 2	3. 24	0. 512 552	0. 000 012	0. 129	0. 99	4. 83	0. 512 489
Diorite	LD15	382	65. 8	0. 708 848	0. 000 009	0. 498	0. 703 513	15. 7	3. 28	0. 512 414	0. 000 007	0. 126	1. 16	2. 37	0. 512 352
Tonalite	YT02	426	61. 3	0. 709 874	0. 000 006	0. 417	0. 705 414	20. 6	3. 69	0. 512 271	0. 000 004	0. 108	1. 24	1. 32	0. 512 218
Tonalite	YT10	530	45. 5	0. 707 179	0. 000 014	0. 248	0. 704 522	29. 4	4. 20	0. 512 138	0. 000 003	0. 086	1. 27	0. 83	0. 512 096
Tonalite	YT15	486	45. 5	0. 706 970	0. 000 020	0. 271	0. 704 071	14. 8	3. 07	0. 512 335	0. 000 005	0. 126	1. 27	0. 92	0. 512 273

[a] $^{87}Rb/^{86}Sr$ and $^{147}Sm/^{144}Nd$ ratios were calculated using Rb, Sr, Sm and Nd contents, measured by ICP-MS.

[b] T_{DM} values were calculated using present-day $(^{147}Sm/^{144}Nd)_{DM}=0.213\ 7$ and $(^{143}Nd/^{144}Nd)$ DM = 0. 513 15.

[c] $\varepsilon_{Nd}(t)$ values were calculated using present-day $(^{147}Sm/^{144}Nd)_{CHUR}=0.196\ 7$ and $(^{143}Nd/^{144}Nd)_{CHUR}=0.512\ 638$.

Table 4 Whole-rock Pb isotopic composition for the diorite and tonalite from the Luding-Kangding area, western margin of the Yangtze Block.

Rock	Sample	U	Th	Pb	$^{206}Pb/^{204}Pb$	2σ	$^{207}Pb/^{204}Pb$	2σ	$^{208}Pb/^{204}Pb$	2σ	$^{238}U/^{204}Pb$ [a]	$^{232}Th/^{204}Pb$ [a]	$^{206}Pb/^{204}Pb$ (750 Ma) [b]	$^{207}Pb/^{204}Pb$ (750 Ma) [b]	$^{208}Pb/^{204}Pb$ (750 Ma) [b]
Diorite	LD09	0. 86	2. 79	10. 92	18. 148 6	0. 000 4	15. 581 2	0. 000 4	37. 996 6	0. 001 1	4. 888	16. 369	18. 091 4	15. 578 5	37. 935 4
Diorite	LD10	0. 49	2. 66	8. 89	17. 949 2	0. 000 3	15. 567 0	0. 000 3	38. 129 7	0. 000 7	3. 433	19. 125	17. 909 1	15. 565 1	38. 058 2
Diorite	LD15	0. 97	3. 64	11. 45	18. 190 7	0. 000 4	15. 584 4	0. 000 3	38. 149 1	0. 000 9	5. 276	20. 364	18. 129 0	15. 581 5	38. 073 0
Tonalite	YT02	0. 80	5. 28	16. 92	17. 798 8	0. 000 5	15. 566 2	0. 000 5	37. 898 3	0. 001 3	2. 941	19. 815	17. 764 4	15. 564 6	37. 824 3
tonalite	YT10	0. 66	5. 33	11. 98	17. 809 7	0. 000 4	15. 567 3	0. 000 4	38. 383 9	0. 001 2	3. 424	28. 475	17. 769 6	15. 565 4	38. 277 5
tonalite	YT15	0. 96	4. 06	22. 63	18. 031 5	0. 000 5	15. 581 2	0. 000 5	37. 734 3	0. 001 3	2. 620	11. 413	18. 000 9	15. 579 7	37. 691 6

[a] Calculated by measured whole-rock U, Th and Pb contents (Table 1) and present-day whole-rock Pb isotopic ratios.

[b] Initial Pb isotopic ratio at $t=750$ Ma, calculated using single-stage model.

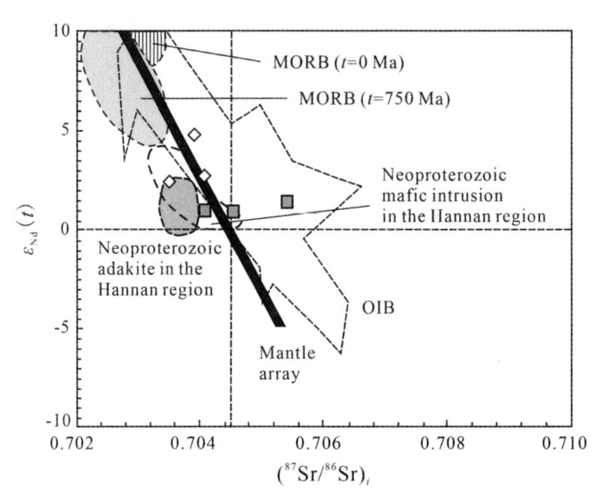

Fig. 8 $(^{87}Sr/^{86}Sr)_i$ vs. $\varepsilon_{Nd}(t)$ plot for the Luding quartz monzodiorite and Kangding granodiorite. Data are from Table 3; Symbols as in Fig. 5.

In the $^{206}Pb/^{204}Pb$ - $^{207}Pb/^{204}Pb$ and $^{206}Pb/^{204}Pb$ - $^{207}Pb/^{204}Pb$ diagrams (Fig. 9), the quartz monzodiorite and the granodiorite are plotted in the transitional zone between lower continental crust and MORB, indicating also a depleted source region. The quartz monzodiorite have $(^{206}Pb/^{204}Pb)_i$ of 17.909–18.128, $(^{207}Pb/^{204}Pb)_i$ of 15.565–15.581 and $(^{208}Pb/^{204}Pb)$ of 37.935–38.073 (Table 4). The granodiorite have $(^{206}Pb/^{204}Pb)_i$ of 17.764–18.001, $(^{207}Pb/^{204}Pb)_i$ of 15.564–15.579, and $(^{208}Pb/^{204}Pb)$ of 37.691–38.277.

5 Discussion

5.1 Causes of the geochemical variations between the quartz monzodiorite and the granodiorite

The quartz monzodiorite and the granodiorite from the Luding-Kangding area are characterized by identical formation ages but different geochemical features. The quartz monzodiorite from the Luding area have lower SiO_2 contents and higher TiO_2, $Fe_2O_3^T$, MgO contents and $Mg^{\#}$ values, as shown in the $\varepsilon_{Nd}(t)$ vs. $(^{87}Sr/^{86}Sr)_i$ diagram (Fig. 8). The quartz monzodiorite have relatively depleted Sr-Nd isotopic compositions, whereas the granodiorite have more evolved Sr-Nd isotopic compositions. These geochemical variations may be caused by: ① different partial melting conditions, resulting in different melting reactions from a homogeneous source region; ② two distinct source regions; and ③ fractional crystallization and assimilation of wall rocks.

The clear differences in Sr-Nd isotopic compositions between the quartz monzodiorite and the granodiorite preclude the first possibility, because melts that derive from a homogeneous

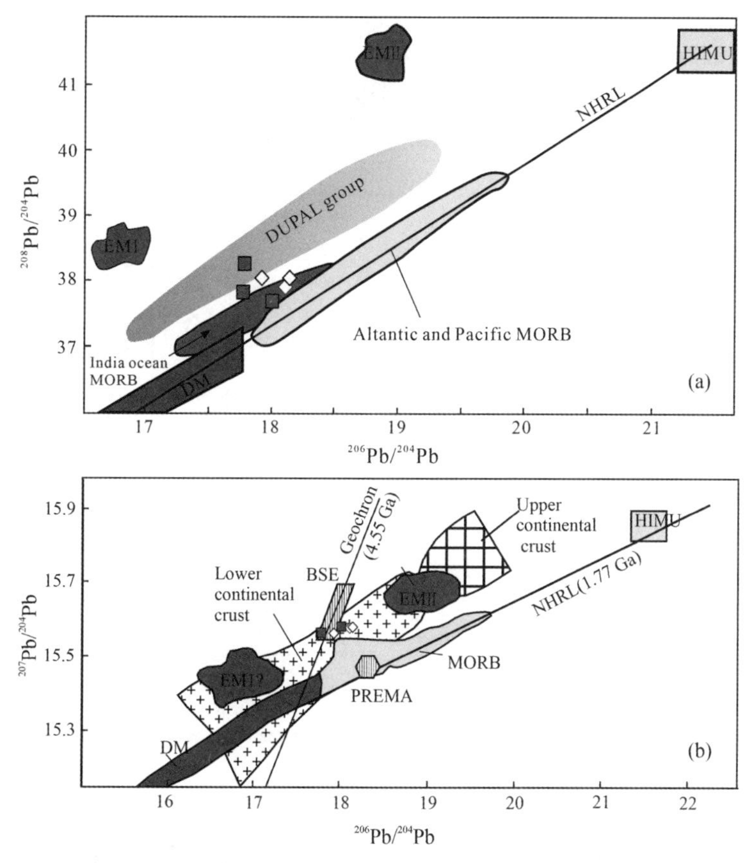

Fig. 9 $^{207}Pb/^{204}Pb$ vs. $^{206}Pb/^{204}Pb$ (a) $^{208}Pb/^{204}Pb$ vs. $^{206}Pb/^{204}Pb$ (b) diagrams for the Luding quartz monzodiorite and Kangding granodiorite (revised from Rollinson, 1993). NHRL, northern hemisphere reference line (Th/U = 0.4). Symbols as in Fig. 5.

source region have similar isotopic ratios.

Fractional crystallization of hornblende and plagioclase from the primitive dioritic melt would have caused a decrease in TiO_2, $Fe_2O_3^T$, and MgO contents, and an increase in K_2O and Na_2O contents, which is consistent with the evolution trend from the quartz monzodiorite to the granodiorite. Moreover, the assimilation of some crustal rocks could have resulted in the enrichment of isotopic compositions. However, the evolved granodiorite yield more positive Eu anomalies ($Eu^*/Eu = 0.90-1.35$) and higher Ba contents (705-1 069 ppm), which indicate that the assimilation and fractional crystallization (AFC) model is not the most plausible explanation for the geochemical variations between the quartz monzodiorite and the granodiorite.

The high Al_2O_3/TiO_2 ratios and low CaO/Na_2O ratios shown by the studied quartz monzodiorite are typical from granitic melts derived from plagioclase-poor (<5%) pelitic rocks (Sylvester, 1994, 1998). On the other hand, the granodiorite from the Kangding area are characterized by lower Rb/Ba and Rb/Sr ratios than the quartz monzodiorite from the Luding area, suggesting a basaltic source region (Fig. 10). Furthermore, the relatively evolved Sr-Nd isotopic composition of the granodiorite (Fig. 8) may be caused by contamination with upper crustal components. Regarding all these considerations, we propose that the quartz monzodiorite and the granodiorite probably derived from two distinct source regions.

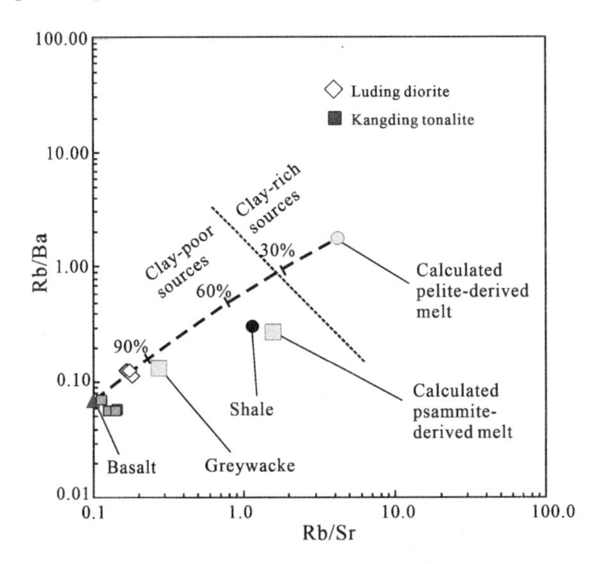

Fig. 10 Rb/Sr vs. Rb/Ba diagrams for the Luding quartz monzodiorite and Kangding granodiorite. Symbols as in Fig. 5.

5.2 Origin of the high-Mg# quartz monzodiorite: Differentiation of mantle-derived mafic magma or dehydration melting of mafic lower crust?

Quartz monzodiorite from the Luding area are sodic, with Na_2O/K_2O ratios ranging from 1.9 to 2.8, and high Mg# values (54.0–55.5), which indicate that mantle components were present in their source region (Rapp et al., 1999; Martin et al., 2005; Moyen and Martin, 2012). Previous experimental works indicate that sodic intermediate melts may be formed by: ① differentiation of mantle-derived mafic magma (Stern et al., 1989), or ② dehydration melting of mafic lower crust under high temperature conditions (Rapp and Watson, 1995). Compared to the typical mantle-derived high-Mg quartz monzodiorite or high-Mg andesite (Stern et al., 1989; Qian and Hermann, 2010), the quartz monzodiorite from the Luding area have low MgO (2.13%–3.15%), Cr (24.1–49.2 ppm), and Ni (14.6–29.1 ppm) contents, and low Mg# values (54.0–55.5), which precludes that the quartz monzodiorite in the Luding area were formed by the differentiation of a mantle-derived mafic melt. Moreover, the fact that the studied high-Mg# quartz monzodiorite have high Th contents (2.79–4.84 ppm) suggests that the primary melt was derived from the partial melting of lower crust.

Melting of newly formed mafic lower crust under high temperature can produce silicic magma with the same initial isotopic values as its mafic source (Petford and Gallagher, 2001). Neoproterozoic (760–780 Ma) mafic dykes from the Kangding area consist of depleted Nd-Hf isotopic compositions [$\varepsilon_{Nd}(t)$ = +1.73 to +8.62], which suggests significant Neoproterozoic crustal growth in the western margin of the Yangtze Block (Lin et al., 2007). Melting of these mafic rocks under high temperature conditions (>1 200 ℃) would have produced dioritic to tonalitic melts (Rapp and Watson, 1995). The quartz monzodiorite from the Luding area have low Sr contents (381–483 ppm) and moderate HREE contents (e.g., Y contents of 16.0–18.6 ppm, Yb contents of 1.50–1.80 ppm), resulting in low Sr/Y ratios (20.5–30.4) in the Y vs. Sr/Y diagram. All the quartz monzodiorite samples are plotted in the field of typical island arc rocks, suggesting the absence of garnet in their source region (Defant and Drummond, 1990).

The studied quartz monzodiorite have higher $Mg^{\#}$ values (54.0–55.5) and MgO contents than those of experimental melts of mafic lower crust (Fig. 11), suggesting significant mantle components in their source region (Moyen and Martin, 2012). The intermediate to felsic igneous rocks that have high $Mg^{\#}$ values (>45) are considered to result from: ① interaction with mantle wedge (Rapp et al., 1999; Martin, 1999; Martin et al., 2005); ② assimilation of mantle peridotite at middle-lower crust level (Qian and Hermann, 2010); ③ melts derived from delaminated lower continental crust that has interacted with the asthenosphere (Gao et al., 2004); and ④ mixing of a primitive felsic melt and mantle-derived mafic melts (Qin et al., 2010). Melts that were derived from the delaminated lower crust typically display adakitic affinity (Gao et al., 2004), which contrasts with the high-$Mg^{\#}$ quartz monzodiorite from the Luding area. The absence of mafic enclaves in the high-$Mg^{\#}$ quartz monzodiorite indicates that mixing of a primitive felsic melt and mantle-derived mafic melts is not plausible either. Then

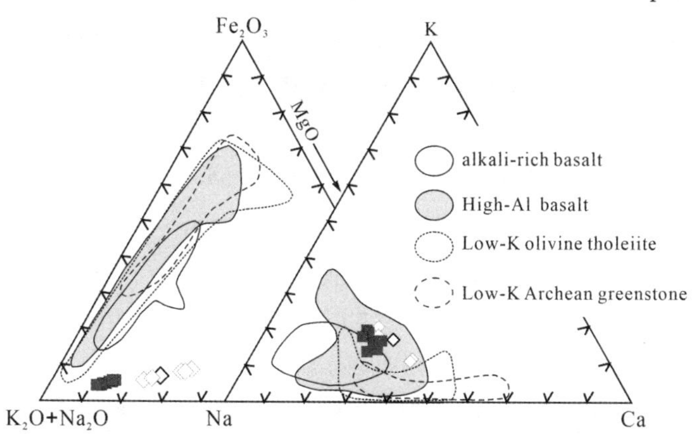

Fig. 11 Ternary A-F-M and molar Na-K-Ca diagrams for the rocks from the Luding quartz monzodiorite and Kangding granodiorite (revised from Zhao et al., 2010a).

The experimental melts for different protoliths are from Rapp and Watson (1995). The Luding quartz monzodiorite and Kangding granodiorite plot within the field for dehydration melts of high-Al basalts, but are slightly richer in MgO.

the most plausible model that can account for the high MgO, Cr and Ni contents of the studied quartz monzodiorite may be the crust-derived felsic melts incorporated some mafic residues in their source regions (Zhao et al., 2010b). The absence of Mg-rich orthopyroxene or clinopyroxene in the quartz monzodiorite seemly do not support the above model. García-Moreno et al. (2006) proposed that the mafic enclaves can dissolute in hydrous granitic magmas, this mechanism can account for the high CaO, FeO and MgO contents in the granites. Then we argued that the high MgO, Cr and Ni contents of the Luding quartz monzodiorite may be resulted from the assimilation some mafic minerals in their source region, the mafic minerals may dissolute in the host granitic to dioritic magma during the ascent process.

5.3 Dehydration melting of middle-lower crust to form the granodiorite

As mentioned above, the granodiorite did not result from the fractional crystallization of the dioritic melt. The granodiorite are characterized by higher SiO_2 contents (65.32% – 67.59%) and lower TiO_2 contents (0.24%–0.32%) than the quartz monzodiorite, and they are also sodic, with Na_2O/K_2O ratios ranging from 1.9 to 2.6. All the granodiorite samples are plotted in the calc-alkaline field in the SiO_2-K_2O diagram (Fig. 5), and are characterized by more evolved Sr-Nd isotopic compositions than the quartz monzodiorite (Fig. 8), evolved $(^{87}Sr/^{86}Sr)_i$ values (0.704 0–0.705 4), $\varepsilon_{Nd}(t)$ values from 0.8 to 1.3, Nd model ages from 1.2 Ga to 1.4 Ga, and low Rb/Ba and Rb/Sr ratios (Fig. 10). All these data suggest that the granodiorite mainly derived from partial melting of Neoproterozoic basaltic lower crust. Additionally, their high Sr (425–537 ppm) and Ba (705–1 074 ppm) contents indicate that these granodiorite derived from a plagioclase-rich source region. Their extremely low Rb/Sr ratios (0.08–0.14) indicate that sedimentary rocks were scarce in their source region.

Experimental works suggest that high degree (20% – 40%) partial melting of basaltic amphibolites under high temperature condition (1 050–1 100 ℃) can produce high-Al, sodic felsic melts (high-Al trondhjemitic-tonalitic, granodioritic, quartz dioritic, dioritic melts), leaving a granulite (plagioclase + clinopyroxene ± orthopyroxene ± olivine) residue at 0.8 GPa, and garnet granulite to eclogite (garnet + clinopyroxene) residues at 1.2–3.2 GPa (Rapp and Watson, 1995). The fact that the Kangding granodiorite have high Sr contents (425 – 537 ppm) and low Y contents (9.01 – 16.4 ppm), which results in high and variable Sr/Y ratios (27.2–54.9), suggests that a garnet residue was present in their source region (Defant and Drummond, 1990; Foley et al., 2002). In summary, the Kangding granodiorite are considered to be the result of high degree (20% – 40%) partial melting of juvenile basaltic crust (probably from the upper part of the lower arc crust, because it has more plagioclase components) at relatively high pressure (>1.0 GPa) and high temperature (1 050–1 100 ℃) conditions.

5.4 Geodynamic implications

Various geodynamic models have been proposed to explain the genesis of the Neoproterozoic mafic-felsic magmatism along the western and northern margins of the Yangtze Block (Li et al., 2002, 2003; Zhou et al., 2002, 2006; Zhao et al., 2010a; Liu and Zhao, 2012). The number of geochemical and geo-chronological works that support that this Neoproterozoic mafic-felsic magmatism was formed at subduction-related environments is progressively increasing (Zhou et al., 2006; Zhao et al., 2010a; Liu and Zhao, 2012). Zhou et al. (2006) studied the 750 Ma Xuelongbao adakitic complex from the southern part of the Longmenshan fault, in the western margin of the Yangtze Block. These granitoids are Na-rich quartz monzodiorite and granodiorite, enriched in Sr (320 – 780 ppm) and depleted in Y (< 10 ppm), which suggests that a garnet residue occurred in their source region. Moreover, these granitoids showed relatively depleted Sr-Nd isotopic compositions (+ 0.36 to + 2.88) and La/Ce ratios. All these features led these authors to propose that this adakitic complex derived from partial melting of a subducted oceanic slab. According to the similarities between the Neoproterozoic Xuelongbao adakitic complex and the Cenozoic adakitic Na-rich tonalites in the Andes, the western margin of the Yangtze Block was considered to be an Andean-type continental margin in Neoproterozoic times (Fig. 12). The Kangding granodiorite, the Luding quartz monzodiorite, and the 750 Ma Xuelongbao adakitic complex have similar geochemical features and zircon ages, suggesting that they were formed in the same tectonic setting. In an oceanic slab subduction setting, melting of mafic continental crust by fluids or mafic melts that

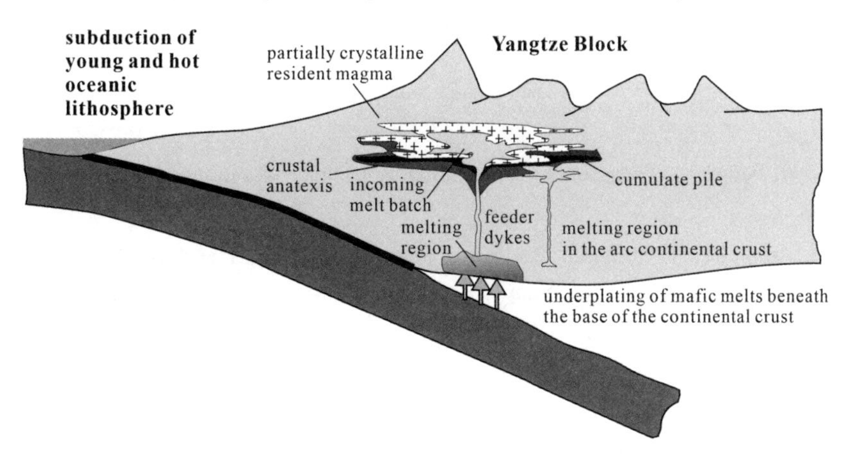

Fig. 12 Schematic model for the formation of the Neoproterozoic high-Mg diorites and tonalities from the Luding-Kangding area, western part of the Yangtze Block (revised from Stern, 2002 and Kemp et al., 2007). The eastward (present day orientation) subduction of the young and hot oceanic lithosphere beneath the Yangtze Block, fluids and mafic melts that underplate beneath the base of the Yangtze continental crust, high degree partial melting of the mafic continental crust can produce the dioritic melts in the Luding area. Melting of relatively felsic continental crust can produce the granodiorite in the Kangding area.

derive from metasomatized mantle wedge underplated beneath the base of the arc continental crust could have produced Na-rich dioritic melts. In such a context, the high-Mg# values of the Luding quartz monzodiorite were probably caused by the assimilation of some mafic minerals in the source region, and the granodiorite in the Kangding area were probably produced by melting of relatively felsic continental crust.

6 Concluding remarks

The high-Mg# quartz monzodiorite and the adakitic granodiorite of the western margin of the Yangtze Block have zircon LA-ICP-MS U-Pb ages of 754 ± 10 Ma (MSWD = 0.51, 2σ) and 748 ± 11 Ma (MSWD = 0.33, 2σ), respectively. These ages are identical to that of the 750 Ma adakitic Xuelongbao complex, suggesting that the Neoproterozoic magmatism along the western margin of the Yangtze Block was formed in an Andean-type active continental margin. Furthermore, the high-Mg# quartz monzodiorite and the granodiorite derived from two distinct source regions. The high-Mg# quartz monzodiorite were formed by high degree partial melting of newly formed mafic lower crust, and they assimilated some residual mafic minerals in their source region. The granodiorite derived from a plagioclase-rich source region.

Acknowledgements Financial support for this study was jointly provided by National Natural Science Foundation of China (Grant Nos. 41421002, 41190072 and 41102037), Program for Changjiang Scholars and Innovative Research Team in University (Grant IRT1281) and MOST Special Fund from the State Key Laboratory of Continental Dynamics.

References

Andersen, T., 2002. Correction of common lead in U-Pb analyses that do not report ^{204}Pb. Chemical Geology, 192, 59-79.

Defant, M.J., 1990. Derivation of some modern arc magmas by melting of young subducted lithosphere. Nature 347, 662-665.

de la Roche, H., Leterrier, J., Grandclaude, P., Marchal, M., 1980. A classification of volcanic and plutonic rocks using R_1-R_2 diagram and major-element analyses: Its relationships with current nomenclature. Chemical Geology, 29, 183-210.

DePaolo, D., Wasserburg, G., 1976. Nd isotopic variations and petrogenetic models. Geophysical Research Letters, 3, 249-252.

Foley, S., Tiepolo, M., Vannucci, R., 2002. Growth of early continental crust controlled by melting of amphibolite in subduction zones. Nature, 417, 837-840.

Gao, S., Rudnick, R.L., Yuan, H.L., Liu, X.M., Liu, Y.S., Xu, W.L., Ling, W.L., Ayers, J., Wang, X. C., Wang, Q.H., 2004. Recycling lower continental crust in the North China craton. Nature, 432, 892-897.

García-Moreno, O., Castro, A., Corretgé, L.G., El-Hmidi, H., 2006. Dissolution of tonalitic enclaves in ascending hydrous granitic magmas: An experimental study. Lithos, 89(3-4),245-258.

Kemp, A.I.S., Hawkesworth, C.J., Foster, G.L., Paterson, B.A., Woodhead, J.D., Hergt, J.M., Gray, C. M., Whitehouse, M.J., 2007. Magmatic and Crustal Differentiation History of Granitic Rocks from Hf-O Isotopes in Zircon.Science, 315, 980-983.

Li, X.H., Li, Z.X., Ge, W., Zhou, H., Li, W., Liu, Y., Wingate, M. T. D., 2003. Neoproterozoic granitoids in South China: Crustal melting above a mantle plume at ca. 825 Ma? Precambrian Research, 122, 45-83.

Li, X.H., Li, Z.X., Zhou, H., Liu, Y., Kinny, P.D., 2002. U-Pb zircon geochronology, geochemistry and Nd isotopic study of Neoproterozoic bimodal volcanic rocks in the Kangdian Rift of South China: Implications for the initial rifting of Rodinia. Precambrian Research, 113, 135-154.

Lin, G., Li, X., Li, W., 2007. SHRIMP U-Pb zircon age, geochemistry and Nd-Hf isotope of Neoproterozoic mafic dyke swarms in western Sichuan: Petrogenesis and tectonic significance. Science in China: Series D, Earth Sciences, 50, 1-16.

Liu, Z.R., Zhao, J.H., 2012. Mineralogical constraints on the origin of Neoproterozoic felsic intrusions, NW margin of the Yangtze Block, South China. International Geology Review, 55, 590-607.

Liu, S., Yan, Q., Li, Q., Wang, Z., 2009. Petrogenesis of granitoid rocks in the Kangding Complex, western margin of the Yangtze Craton and its tectonic significance.Acta Petrologica Sinica, 25, 1883-1896.

Liu, Y., Liu, X., Hu, Z., Diwu, C., Yuan, H., Gao, S., 2007. Evaluation of accuracy and long-term stability of determination of 37 trace elements in geological samples by ICP-MS. Acta Petrologica Sinica, 23, 1203-1210.

Ludwig, K.R., 2003. ISOPLOT 3. 0: A geochronological toolkit for Microsoft Excel. Berkeley Geochronology Center, Special publication, 4.

Martin, H., 1999. Adakitic magmas: modern analogues of Archaean granitoids. Lithos, 46, 411-429.

Martin, H., Smithies, R.H., Rapp, R., Moyen, J.F., Champion, D., 2005. An overview of adakite, tonalite-trondhjemite-granodiorite (TTG), and sanukitoid: Relationships and some implications for crustal evolution. Lithos, 79, 1-24.

Moyen, J.F., Martin, H., 2012. Forty years of TTG research. Lithos, 148, 312-336.

Petford, N., Gallagher, K., 2001. Partial melting of mafic (amphibolitic) lower crust by periodic influx of basaltic magma. Earth & Planetary Science Letters, 193, 483-499.

Qian, Q., Hermann, J., 2010. Formation of High-Mg Diorites through Assimilation of Peridotite by Monzodiorite Magma at Crustal Depths. Journal of Petrology, 51, 1381-1416.

Qin, J.F., Lai, S.C., Diwu, C.R., Ju, Y.J., Li, Y.F., 2010. Magma mixing origin for the post-collisional adakitic monzogranite of the Triassic Yangba pluton, Northwestern margin of the South China block: Geochemistry, Sr-Nd isotopic, zircon U-Pb dating and Hf isotopic evidences. Contributions to Mineralogy and Petrology, 159, 389-409.

Rapp, R.P., Shimizu, N., Norman, M.D., Applegate, G.S., 1999. Reaction between slab-derived melts and peridotite in the mantle wedge: Experimental constraints at 3. 8 GPa. Chemical Geology, 160, 335-356.

Rapp, R.P., Watson, E.B., 1995. Dehydration Melting of Metabasalt at 8 – 32 kbar: Implications for Continental Growth and Crust-Mantle Recycling. Journal of Petrology, 36, 891-931.

Rollinson, H.R., 1993. Using geochemical data: Evaluation, presentation, interpretation. Routledge.

Stern, R.A., Hanson, G.N., Shirey, S.B., 1989. Petrogenesis of mantle-derived, LILE-enriched Archean monzodiorites and trachyandesites (sanukitoids) in southwestern Superior Province. Canadian Journal of Earth Sciences, 26, 1688-1712.

Stern, R.J., 2002.Subduction Zones. Rev. Geophys., 40, 1012.

Sylvester, P.J., 1994. Archean granite plutons. Archean Crustal Evolution, 11, 261-314.

Sylvester, P.J., 1998. Post-collisional strongly peraluminous granites. Lithos, 45, 29-44.

Sun, S.S., McDonough, W.F., 1989. Chemical and isotopic systematics of oceanic basalts: Implications for mantle composition and processes. Geological Society, London, Special Publications, 42, 313-345.

Wang, X.L., Zhou, J.C., Griffin, W.L., Zhao, G., Yu, J.H., Qiu, J.S., Zhang, Y.J., Xing, G.F., 2014. Geochemical zonation across a Neoproterozoic orogenic belt: Isotopic evidence from granitoids and metasedimentary rocks of the Jiangnan orogen, China. Precambrian Research, 242, 154-171.

Wilson, B.M., 1989. Igneous petrogenesis a global tectonic approach. Springer.

Yuan, H.L., Gao, S., Dai, M.N., Zong, C.L., Gunther, D., Fontaine, G.H., Liu, X.M., Diwu, C.R., 2008. Simultaneous determinations of U-Pb age, Hf isotopes and trace element compositions of zircon by excimer laser-ablation quadrupole and multiple-collector ICP-MS. Chemical Geology, 247, 100-118.

Yuan, H.L., Gao, S., Liu, X.M., Li, H.M., Gunther, D., Wu, F.Y., 2004. Accurate U-Pb age and trace element determinations of zircon by laser ablation inductively coupled plasma mass spectrometry. Geostandard Newsletters, 28, 353-370.

Zhao, J.H., Zhou, M.F., 2008. Neoproterozoic adakitic plutons in the northern margin of the Yangtze Block, China: Partial melting of a thickened lower crust and implications for secular crustal evolution. Lithos, 104, 231-248.

Zhao, J.H., Zhou, M.F., 2009. Secular evolution of the Neoproterozoic lithospheric mantle underneath the northern margin of the Yangtze Block, South China. Lithos, 107, 152-168.

Zhao, J.H., Zhou, M.F., Zheng, J.P., Fang, S.M., 2010a. Neoproterozoic crustal growth and reworking of the Northwestern Yangtze Block: Constraints from the Xixiang dioritic intrusion, South China. Lithos, 120, 439-452.

Zhao, J.H., Zhou, M.F., Jian-Ping, Z., 2010b. Metasomatic mantle source and crustal contamination for the formation of the Neoproterozoic mafic dike swarm in the northern Yangtze Block, South China. Lithos, 115, 177-189.

Zhou, M.F., Yan, D.P., Kennedy, A.K., Li, Y., Ding, J., 2002. SHRIMP U-Pb zircon geochronological and geochemical evidence for Neoproterozoic arc-magmatism along the western margin of the Yangtze Block, South China. Earth & Planetary Science Letters, 196, 51-67.

Zhou, M.F., Yan, D.P., Wang, C.L., Qi, L., Kennedy, A., 2006.Subduction-related origin of the 750 Ma Xuelongbao adakitic complex (Sichuan Province, China): Implications for the tectonic setting of the giant Neoproterozoic magmatic event in South China. Earth & Planetary Science Letters, 248, 286-300.

Zimmerer, M.J., McIntosh, W.C., 2012. The geochronology of volcanic and plutonic rocks at the Questa caldera: Constraints on the origin of caldera-related silicic magmas. Geological Society of America Bulletin, 124, 1394-1408.

Petrogenesis of Early Cretaceous high-$Mg^{\#}$ granodiorites in the northeastern Lhasa terrane, SE Tibet: Evidence for mantle-deep crustal interaction①

Lai Shaocong Zhu Renzhi

Abstract: High-$Mg^{\#}$ [molar Mg/(Mg + Fe)] intermediate to felsic rocks, such as high-Mg andesitic rocks (HMA) and Mg andesites (MA), are usually produced by melting of mantle-derived peridotites and/or interaction with mantle-derived peridotites. Both are crucial in tracking mantle-crust interaction processes in the deep crust. We present LA-ICP-MS in-situ zircon U-Pb isotopic data, whole-rock geochemistry and whole-rock Sr-Nd-Pb-Hf isotopic data for the Basu granodiorites which are exposed between the Jiali fault and the Bangong-Nujiang suture, SE Tibet. The LA-ICP-MS zircon U-Pb data show that the granodiorites were formed at ca. 122. 1 ± 1. 0 Ma, coeval with other granitoids in this region. The granodiorites belong to the high-K calc-alkaline and high-$Mg^{\#}$(49. 7–53. 7) series, similar to typical MA. The decreasing A/CNK from 0. 97 to 0. 87 and increasing FM (molar Fe+Mg) from 0. 08 to 0. 16 indicate a clinopyroxene entrainment and I-type trend. The rocks also display enriched Rb, Th, U and Pb, but depleted P, Ti and obvious negative Eu anomalies. Evolved whole-rock isotopic compositions of Basu granodiorites include high initial $^{87}Sr/^{86}Sr$ (0. 709 7–0. 711 4), negative $\varepsilon_{Nd}(t)$ (−12. 0 to −7. 2) and $\varepsilon_{Hf}(t)$ values (−9. 6 to −7. 9), and lower to upper crustal affinity of Pb isotopes which imply that they were derived from an ancient and evolved lower crust source. Therefore, we suggest that these granodiorites were products of interaction between partial melting of ancient basaltic lower crust and mantle-derived peridotites. Together with a geochronology dataset from the literature, we suggest that there was intensive granitic magmatism occurred in the Early Cretaceous range from 133 Ma to 113 Ma in the northeastern Lhasa terrane, similar to those in the central and northern Lhasa terrane. Furthermore, we have compared with the dataset of in-situ zircon Lu-Hf isotopes, whole-rock Sr-Nd from the literatures and the relationship with zircon U-Pb ages in the northeastern Lhasa terrane, all of which indicate the importance of evolved and ancient continental crust source in the Early Cretaceous magmatism in the northeastern Lhasa terrane, but also that the mantle-derived materials play a necessary role in the generation. In addition, the amount of enriched and evolved isotopic component gradually decreased from ca.140 Ma to 100 Ma.

① Published in *Journal of Asian Earth Sciences*, 2019, 177.

1 Introduction

High-K calc-alkaline I-type granitoids are usually generated by partial melting of pre-existing meta-igneous rocks in the crust (Roberts and Clemens, 1993), however, some display high-$Mg^{\#}$ [molar Mg/(Mg + Fe)] including HMA and MA, with enrichment in large ion lithophile elements and depletion in high field strength elements, which have been attributed to melts derived from an interplay of crust-mantle interactions (Qian and Hermann, 2010). Therefore, the high-K calc-alkaline and high-$Mg^{\#}$ I-type granitoids have been regarded as an important tracer of the interaction between mantle- and crust-derived materials (Qian and Hermann, 2010; Wang et al., 2011; Chen et al., 2015).

In the northeastern Lhasa terrane, southern Tibet, there is an elongated belt of granitoid batholiths between the Jiali fault and Bangong-Nujiang suture (Searle et al., 1987; Chiu et al., 2009; Lin et al., 2012), which was considered as a crucial component to understand the magmatism and geodynamic processes in the Andean-type continental margin before the collision between Asia and India (Searle et al., 1987; Harris et al., 1990; Yin and Harrison, 2000; Mo et al., 2005; Chiu et al., 2009; Zhu et al., 2009a; Lin et al., 2012; Pan et al., 2014). They share a similar role to Early Cretaceous magmatism in the central and northern Lhasa terrane (Li et al., 2016, 2018; Chen et al., 2017; Hu et al., 2017; Fan et al., 2017). Chiu et al. (2009) suggested that these granitoids were abundant in the Early Cretaceous, and derived from a predominantly crustal source due to the Late Jurassic to Early Cretaceous continental collision of Lhasa-Qiangtang terranes. However, Zhu et al. (2009a) proposed that these Early Cretaceous granitoids were likely generated in a setting associated with southward subduction of Bangong-Nujiang ocean floor, where mantle wedge-derived magmas may provide the materials and heat. Therefore, a two-stage petrogenetic model was proposed (Lin et al., 2012): ① mantle-wedge-derived mafic magmas associated with assimilation of the lower continental crust; and ② additional differentiation and contamination which occurred in upper crust. The interaction of crust and mantle could have involved subducted oceanic slab (Yogodzinski et al., 1995), subducted continental crust (Qin et al., 2010), subducted sediments (Behn et al., 2011; Zhu et al., 2019) and delamination of lower continental crust (Gao et al., 2004). However, exact petrogenetic mechanisms of the associated high-$Mg^{\#}$ rocks were still poorly constrained.

In recent field studies, we investigated the high-$Mg^{\#}$ granodiorites near the Basu county in the Zhaxize batholith, which may provide key evidence to clarify the interaction between mantle- and crust-derived materials in the deep crust. Hence we present the new LA-ICP-MS in-situ zircon U-Pb data and whole-rock geochemistry and Sr-Nd-Pb-Hf isotopic data and compile a comprehensive dataset of Early Cretaceous magmatism in the central and northern

Lhasa terrane (Table 1) to constrain the petrogenesis of these high-$Mg^{\#}$ and high-K calc-alkaline granodiorites and offer new evidence to understand the interaction processes between mantle and crust during the Early Cretaceous in the central and northern Lhasa terrane.

2 Geological setting and petrography

The Tibet plateau is generally composed of the Songpan-Garze flysch complex, and several tectonic terranes including Northern Qiangtang, Southern Qiangtang, Lhasa, and Himalaya, which were separated by the Jinshajiang, Longmucuo-Shuanghu-Changning-Menglian, Bangong-Nujiang, and Yarlung-Zangbo sutures from north to south. The Lhasa terrane is bordered by the Bangong-Nujiang suture in the north and the Yarlung-Zangbo suture in the south (Fig. 1a), and dispersed from Gondwana during Permian to Triassic times to drift northward and finally collide with the Qiangtang terrane during Late Trassic to Early Cretaceous (Kapp et al., 2005a,b; Zhu et al., 2011; Chen et al., 2017; Liu et al., 2017; Zeng et al., 2017; Fan et al., 2017; Li et al., 2018). Three typical magmatic belts in the Lhasa terranes can be identified (Fig. 1a) (Zhu et al., 2009b, 2011), as follows: the widespread Cretaceous-early Tertiary Gangdese batholiths and Linzizong volcanic successions that have been known for decades in the southern Lhasa subterrane (Coulon et al., 1986; Harris et al., 1990; Mo et al., 2005, 2007; Wen et al., 2008; Ji et al., 2009; Lee et al., 2009), and were regarded as the products of northward subduction of Neo-Tethyan oceanic lithosphere. Abundant Mesozoic magmatic rocks are present in the central and northern Lhasa subterrane (Coulon et al., 1986; Harris et al., 1990; Guynn et al., 2006; Chu et al., 2006; Zhu et al., 2009a; Chiu et al., 2009), which were the so-called northern magmatic belt related to the closure and subduction of the Bangong-Nujiang Tethyan ocean (Zhu et al., 2009a,b, 2011, 2013, 2016; Chiu et al., 2009; Qu et al., 2012; Sui et al., 2013; Chen et al., 2014; Li et al., 2016, 2018).

The central and northern Lhasa subterrane was covered with widespread Carboniferous metasedimentary rocks and Jurassic-Early Cretaceous volcano-sedimentary rocks, plus minor Ordovician, Silurian, and Triassic limestone (Pan et al., 2004). Voluminous Mesozoic volcanic rocks that consist of andesites, dacite, rhyolite, and associated volcaniclastic rocks are exposed within the subterrane (Kapp et al., 2005a,b; Zhu et al., 2009a,b, 2011; Sui et al., 2013; Chen et al., 2014). In addition, the associated Mesozoic plutonic rocks mainly intruded the pre-Ordovician and Carboniferous-Permian metasedimentary successions, mainly occurred in the Early Cretaceous (Table 1) (Xu et al., 1985; Harris et al., 1990; Guynn et al., 2006; Zhu et al., 2009a, 2011, 2016; Qu et al., 2012; Sui et al., 2013). There are abundant dioritic enclaves within the Early Cretaceous granitoids (Zhu et al., 2009a, 2011, 2016; Sui et al., 2013).

Table 1 Summary of Early Cretaceous magmatism in the central and northern Lhasa terrane.

Sample	GPS Position	Lithology	SiO_2 /wt%	A/CNK	$Mg^{\#}$	Age /Ma	zircon $\varepsilon_{Hf}(t)$	Bulk-rock $(^{87}Sr/^{86}Sr)_i$	Bulk-rock $\varepsilon_{Nd}(t)$	Mineral assemblage	References
TransHimalayan (Bomi-Chayu) batholith											
ET103A	29°95′42″N, 95°38′45″E	Granite	70.69	1.10	34.8	118.6±2.0	−15.06 to −9.51	0.711 8	−10.6		Chiu et al., 2009; Lin et al., 2012
ET104B	29°50′75″N, 96°60′44″E	Granite	73.04	1.12	33.2	114.8±1.7	−5.36 to 0	0.708 9	−7.14	Kf(40%)+Pl(30%)+Qtz(20%)+Bi(5%−10%)+Hbl(1%−3%)	Chiu et al., 2009; Lin et al., 2012
ET105A	29°50′03″N, 96°60′66″E	Granite	67.35	1.07	40.9	122.5±1.5	−12.78 to −7.79	0.710 6	−7.85	Kf(40%)+Pl(30%)+Qtz(20%)+Bi(5%−10%)+Hbl(1%−4%)	Chiu et al., 2009; Lin et al., 2012
ET105G	29°50′03″N, 96°60′66″E	Enclave	52.23	0.91	59.5	122.5±1.5		0.705 2	−4.32		Chiu et al., 2009; Lin et al., 2012
ET105B	29°50′03″N, 96°60′66″E	Granite	73.49	1.04	42.4	122.5±1.5		0.702 2	−7.36	Kf(40%)+Pl(30%)+Qtz(20%)+Bi(5%−10%)+Hbl(1%−3%)	Chiu et al., 2009; Lin et al., 2012
ET107A	29°37′21″N, 96°91′19″E	Granite	73.05			122.5±1.5		0.715 6	−6.05	Kf(40%)+Pl(30%)+Qtz(20%)+Bi(5%−10%)+Hbl(1%−3%)	Chiu et al., 2009; Lin et al., 2012
ET117A	29°32′13″N, 97°13′43″E	Granite	62.72	0.98	60.7	116.9±1.8	−7.49 to +1.62	0.710 1	−7.11	Kf(40%)+Pl(30%)+Qtz(20%)+Bi(5%−10%)+Hbl(1%−3%)	Chiu et al., 2009; Lin et al., 2012
ET119A	29°50′74″N, 96°75′40″E	Enclave	64.14	1.04	55.5	116.9±1.8		0.705 5			Chiu et al., 2009; Lin et al., 2012
ET120A	29°74′17″N, 96°02′09″E	Granite	70.43	1.10	42.3	109.0±1.1	−13.28 to +3.08	0.702 9	−3.61	Kf(40%)+Pl(30%)+Qtz(20%)+Bi(5%−10%)+Hbl(1%−5%)	Chiu et al., 2009; Lin et al., 2012
ET120C	29°74′17″N, 96°02′09″E	Enclave	55.23	0.93	49.5	109.0±1.1		0.705 8	−2.63		Chiu et al., 2009; Lin et al., 2012
ET120D	29°74′17″N, 96°02′09″E	Enclave	45.03	0.73	47.2	109.0±1.1		0.705 8	−1.74		Chiu et al., 2009; Lin et al., 2012

Continued

Sample	GPS Position	Lithology	SiO_2 /wt%	A/CNK	Mg#	Age /Ma	zircon $\varepsilon_{Hf}(t)$	Bulk-rock $(^{87}Sr/^{86}Sr)_i$	Bulk-rock $\varepsilon_{Nd}(t)$	Mineral assemblage	References
ET120E	29°74′17″N, 96°02′09″E	Enclave	51.13	0.78	64.4	109.0±1.1		0.705 3	−2.76		Chiu et al., 2009; Lin et al., 2012
ET125A	29°75′65″N, 95°71′63″E	Granite	68.32	1.09	41.7	125.1±1.5	−15.00 to −9.38	0.717 9	−12.6		Chiu et al., 2009; Lin et al., 2012
ET106A2	29°38′55″N, 96°86′90″E	Granite	78.30	1.15		122.5±1.5	−12.78 to −7.79		−11.3	Kf(40%)+Pl(30%)+Qtz(20%)+Bi(5%−10%)+Hbl(1%−3%)	Chiu et al., 2009; Lin et al., 2012
ET219B2	29°39′22″N, 96°85′15″E	Granite	75.61	1.11		125.0±0.6	−10.93 to −7.51		−11.4	Kf(40%)+Pl(30%)+Qtz(20%)+Bi(5%−10%)+Hbl(1%−3%)	Chiu et al., 2009; Lin et al., 2012
ET220B	29°39′22″N, 96°85′15″E	Granite	74.17	1.10		125.0±0.6	−8.88 to −3.29		−10.6	Kf(40%)+Pl(30%)+Qtz(20%)+Bi(5%−10%)+Hbl(1%−3%)	Chiu et al., 2009; Lin et al., 2012
ET221B	29°39′22″N, 96°85′15″E	Granite	75.27	1.10		125.0±0.6			−10.7	Kf(40%)+Pl(30%)+Qtz(20%)+Bi(5%−10%)+Hbl(1%−3%)	Chiu et al., 2009; Lin et al., 2012
ET222B	29°39′22″N, 96°85′15″E	Granite	76.14	1.13		125.0±0.6			−10.6	Kf(40%)+Pl(30%)+Qtz(20%)+Bi(5%−10%)+Hbl(1%−3%)	Chiu et al., 2009; Lin et al., 2012
ET115F1	28°59′91″N, 97°24′96″E	Granite	74.05	1.14	20.0	133.1±1.3	−9.15 to −5.16	0.718 1	−9.52	Kf(40%)+Pl(30%)+Qtz(20%)+Bi(5%−10%)+Hbl(1%−3%)	Chiu et al., 2009; Lin et al., 2012
ET116B	28°67′24″N, 97°46′98″E	Granite	73.17	1.23	31.0	132.6±0.9	−27.28 to −6.84	0.727 4	−7.86	Kf(40%)+Pl(30%)+Qtz(20%)+Bi(5%−10%)+Hbl(1%−3%)	Chiu et al., 2009; Lin et al., 2012
73-73		Granite	73.33	1.21	28.0	128±2		0.744 9	−12.7	Kf(40%)+Pl(30%)+Qtz(20%)+Bi(5%−10%)+Hbl(1%−3%)	Chiu et al., 2009; Lin et al., 2012
T1066	29°30′00″N, 95°30′00″E	Granite	64.37−68.48	0.95−1.05	39−62	125.2±1.0		0.710 4−0.715 5	−13.8 to −8.6	Kf(25%−30%)+Pl(35%−40%)+Qtz(20%−25%)+Bi(5%−10%)	Pan et al., 2014
CY1-01	28°35′00″N, 97°10′00″E	Monzo-granite	69.92−70.78	1.00−1.03	33.9−38.7	129.1±1.6	−12.5 to −2.9	0.717 3−0.717 9	−10.9 to −9.2	Kf(35%)+Pl(25%)+Qtz(30%)+Bi(5%−10%)	Zhu et al., 2009a

Continued

Sample	GPS Position	Lithology	SiO_2 /wt%	A/CNK	$Mg^{\#}$	Age /Ma	zircon $\varepsilon_{Hf}(t)$	Bulk-rock $(^{87}Sr/^{86}Sr)_i$	Bulk-rock $\varepsilon_{Nd}(t)$	Mineral assemblage	References
CY6-01		Monzo-granite	74.47–76.77	1.01–1.05	21.8–30.7	128.5±3.1		0.712 0–0.716 7	−9.5 to −7.6	Kf(35%)+Pl(25%)+Qtz(30%)+Bi(5%–10%)	Zhu et al.,2009a
ABQ-13	30°02′36″N, 96°46′29″E	Granodiorite	60.9–68.2	0.87–0.99	49.7–53.7	122.2±1.0	−29.6 to −7.9	0.709 7–0.711 4	−12.0 to −7.2	Kf(20%)+Pl(40%)+Qtz(20%)+Bi(3%–5%)+Hbl(5%–10%)	This study
Northern Lhasa subterrane											
YH06-3	32°17.970′N, 82°33.165′E	Andesite	55.52	0.78		131.2±1.4	+12.3 to +18.8			Pl(20%)+Bi(5%)+Cpx(5%)	Zhu et al.,2011
SB01-2	30°59.600′N, 92°33.340′E	Monzo-granite	71.65	1.11		118.4±0.5	−6.0 to +5.7			Pl(40%)+Kf(35%)+Qtz(20%)+Bi(3%)	Zhu et al.,2011
NR18-1	31°56.751′N, 92°09.328′E	Granodiorite porphyry	66.40	0.95		110.4±0.7	−3.5 to +0.6			Pl(15%)+Bi(10%)+Qtz(5%)	Zhu et al.,2011
NQ16-1	31°47.360′N, 92°04.097′E	Monzo-granite	71.64	1.05		117.5±0.7	−10.2 to −5.0				Zhu et al.,2011
YH15-1	32°30.058′N, 82°26.722′E	Monzo-granite				110.1±0.7	+5.0 to +8.9			Pl(35%)+Kf(40%)+Qtz(15%)+Bi(8%)	Zhu et al.,2011
YH22-4	32°21.175′N, 82°26.862′E	Rhyolite	75.54	1.03		110.6±0.6	+4.7 to +9.2			Pl(15%)+Kf(10%)+Qtz(5%)+Bi(5%)	Zhu et al.,2011
YH04-2	32°17.084′N, 82°32.858′E	Andesite	55.61	0.81		116.7±1.2	+13.5 to +18.4			Pl(30%)	Zhu et al.,2011
YH01-2	32°16.318′N, 82°31.392′E	Rhyolite	73.98	1.15	37.0	109.0±1.0	+6.8 to +11.9			Pl(10%)+Qtz(15%)	Zhu et al.,2011
NM01-1	31°50.630′N, 87°05.250′E	Dacite	62.88	1.02		111.9±0.9	+7.7 to +10.1			Pl(25%)	Zhu et al.,2011

Continued

Sample	GPS Position	Lithology	SiO_2 /wt%	A/CNK	Mg#	Age /Ma	zircon $\varepsilon_{Hf}(t)$	Bulk-rock $(^{87}Sr/^{86}Sr)_i$	Bulk-rock $\varepsilon_{Nd}(t)$	Mineral assemblage	References
DG05-1	31°19.700′N, 88°54.890′E	Dacite	67.06	1.04		114.3±0.6	−0.9 to +1.9			Pl(25%)+Bi(5%)+Qtz(3%)	Zhu et al.,2011
DG01-1	31°21.270′N, 88°55.800′E	Rhyolite	74.36	1.20		115.7±0.6	−1.5 to +4.2			Kf(20%)+Pl(15%)+Qtz(3%)	Zhu et al.,2011
NQ12-10	31°28.803′N, 92°06.433′E	Andesite	59.51	0.90		110.8±0.6	−10.0 to −7.7			Pl(20%)+Bi(10%)	Zhu et al.,2011
NQ09-1	31°45.971′N, 92°37.403′E	Monzogr anite	69.29	0.98		110.7±0.8	−6.8 to +0.9			Pl(40%)+Kf(35%)+Qtz(15%)+Bi(8%)	Zhu et al.,2011
08DX21	31°02.859′N, 91°41.445′E	Syenogranite	75.64	1.07		110.7±0.6	−5.8 to −4.5			Pl(20%)+Kf(50%)+Qtz(20%)+Bi(8%)	Zhu et al.,2011
ZGP06-1	31°32.290′N, 87°29.970′E	Dacite	65.77	1.02		91.0±0.8	+5.2 to +8.2			Pl(40%)+Cpx(10%)	Wang et al.,2014
YH22-1	32°21.175′N, 82°26.862′E	Basalt	48.82		61.0	111.8±3.2	−2.3 to +5.3	0.706 2	−0.6	Cpx(15%)+Opx(5%)+Pl(25%)+Om(15%)+Chl(5%−8%)	Sui et al.,2013
YH01-1	32°16.318′N, 82°31.392′E	Basalt	50.90		54.0	110.0±0.7	+3.8 to +6.1	0.707 2	0.9	Cpx (10%) + Pl (70%) + Om (10%)+Cb(10%)	Sui et al.,2013
YH02-1	32°16.318′N, 82°31.392′E	Basalt	55.11		66.0	108.9±1.1	+2.8 to +9.7	0.706 0	2.1	Pl(40%)+Cpx(5%)+Om(10%)+volcanic glass(25%)+Chl(5%)	Sui et al.,2013
YH04-3		Basalt	48.86–51.38			116.2±0.6					Zhu et al.,2015
YH07-4		Trachy andesite	56.01			121.1±1.4					Zhu et al.,2015
YH10-2	32°30.058′N, 82°26.722′E	Quartz diorite porphyrite	63.06		45.0	109.7±0.8	+12.5 to +16.9	0.704 2	3.6	Pl(20%)+Hbl(15%)+Qtz(<5%)	Sui et al.,2013

Continued

Sample	GPS Position	Lithology	SiO_2 /wt%	A/CNK	$Mg^{\#}$	Age /Ma	zircon $\varepsilon_{Hf}(t)$	Bulk-rock		Mineral assemblage	References
								$(^{87}Sr/^{86}Sr)_i$	$\varepsilon_{Nd}(t)$		
YH10-6	32°30. 058′N, 82°26. 722′E	Dioritic enclave	55. 45		47. 0	110. 4±1. 4	+12. 8 to +14. 7	0. 704 3	3. 4	Pl(20%)+Hbl(15%)+Qtz(<5%)	Sui et al.,2013
YH0921-5	32°32. 416′N, 81°51. 034′E	Diabasic dyke				120. 8±1. 2					Zhu et al.,2016
YH06-1		Andesite	54. 81			131. 2±1. 4					Zhu et al.,2016
BG01-1		Granitoid				110. 7	−3. 0 to −0. 8				Zhu et al.,2016
RBG01-1		Granitoid				111. 9	−13. 5 to +2. 7				Zhu et al.,2016
R08DX21		Granitoid				111. 3	−7. 0 to −5. 4				Zhu et al.,2016
RBG22-1		Granitoid				114. 8	−7. 1 to −3. 1				Zhu et al.,2016
RBG08		Granitoid				115. 6	−0. 6 to 6. 4				Zhu et al.,2016
BG21-1		Granitoid				115. 9	−7. 1 to −2. 2				Zhu et al.,2016
BG01-3		Granitoid				110. 3	−3. 7 to +0. 2				Zhu et al.,2016
BG02-1		Granitoid				113. 6	−3. 6 to +2. 9				Zhu et al.,2016
BG02-6		Granitoid				114. 6	−1. 7 to +2. 3				Zhu et al.,2016

Continued

Sample	GPS Position	Lithology	SiO_2 /wt%	A/CNK	$Mg^{\#}$	Age /Ma	zircon $\varepsilon_{Hf}(t)$	Bulk-rock		Mineral assemblage	References
								$(^{87}Sr/^{86}Sr)_i$	$\varepsilon_{Nd}(t)$		
BG13-1		Granitoid				129. 9	+2. 7 to +5. 6				Zhu et al., 2016
BG16-1		Granitoid				134	+2. 2 to +8. 4				Zhu et al., 2016
RBG16-1		Granitoid				132. 7	+5. 0 to +11. 0				Zhu et al., 2016
RBG11-1		Granitoid				130. 7	+4. 4 to +7. 7				Zhu et al., 2016
BG03-1		Granitoid				124	-7. 4 to +1. 3				Zhu et al., 2016
BG07-2		Granitoid				126. 8	-1. 7 to +4. 2				Zhu et al., 2016
BG19-1		Granitoid				120. 9	-5. 5 to -2. 1				Zhu et al., 2016
BG24-1		Granitoid				117. 8	-8. 6 to -1. 7				Zhu et al., 2016
BG28-1		Granitoid				120. 3	-7. 7 to -2. 6				Zhu et al., 2016
RBG03-3		Granitoid				118. 1	-6. 8 to -5. 7				Zhu et al., 2016
RBG09-1		Granitoid				118. 6	-3. 2 to +5. 9				Zhu et al., 2016
RBG08-1		Granitoid				130. 3	-8. 2 to -1. 9				Zhu et al., 2016

Continued

Sample	GPS Position	Lithology	SiO_2 /wt%	A/CNK	$Mg^{\#}$	Age /Ma	zircon $\varepsilon_{Hf}(t)$	Bulk-rock $(^{87}Sr/^{86}Sr)_i$	Bulk-rock $\varepsilon_{Nd}(t)$	Mineral assemblage	References
RBG03-3-1		Granitoid				137.3	−9.3 to −8.6				Zhu et al., 2016
RBG22-1		Granitoid				124	−5.8 to +0.1				Zhu et al., 2016
RBG16-1-0		Granitoid				132.7	5.6				Zhu et al., 2016
BG21-1		Granitoid				122.9	−5.7 to −2.5				Zhu et al., 2016
BG17-1		Granitoid				122.1	−5.0 to −0.8				Zhu et al., 2016
BD08-18	31°25.00′N, 88°50.000′E	Granite	74.03–75.36	1.08–1.09	48.9–59.7	109.6±1.4		0.7100–0.7123	−10.34 to −8.65	Kf(35%–40%) + Pl(15%–20%) + Qtz(30%) + Bi(5%–10%) + Hbl(2%)	Qu et al., 2012
BD08-21	31°26.00′N, 88°50.001′E	Granite	73.55–73.86	0.97–0.98	3.1–5.8	112.2±0.9		0.7086–0.7090	−7.51 to −4.27	Kf(35%–40%) + Pl(15%–20%) + Qtz(30%) + Bi(5%–10%) + Hbl(2%)	Qu et al., 2012
BD08-24	31°27.00′N, 88°50.002′E	Granite	72.44–73.25	0.97–1.01	30.8–36.5	113.7±0.5		0.7088–0.7090	−7.33 to −4.90	Kf(35%–40%) + Pl(15%–20%) + Qtz(30%) + Bi(5%–10%) + Hbl(2%)	Qu et al., 2012
NQ1307	31°30′00″N, 92°30′00″E	Rhyolite	71.04–75.18	1.08–1.22	14.1–29.6	110.5±0.7	+3.7 to −2.4	0.7042–0.7096	−5.03 to −4.74	(Pl+Qtz)(30%) + Groundmass	Chen et al., 2017
NQ1306	31°30′00″N, 92°30′00″E	Andesite	58.45–64.46	0.76–1.26	27.4–41.3	105.5±1.3	−11.3 to −3.4	0.7041–0.7218	−5.17 to −4.88	Pl(30%) + Hbl(10%) + Bi(2%) + Groundmass	Chen et al., 2017

Continued

Sample	GPS Position	Lithology	SiO_2 /wt%	A/CNK	$Mg^{\#}$	Age /Ma	zircon $\varepsilon_{Hf}(t)$	Bulk-rock $(^{87}Sr/^{86}Sr)_i$	Bulk-rock $\varepsilon_{Nd}(t)$	Mineral assemblage	References
12T036	31°40.019′N, 90°37.020′E	Andesite	58.21–59.98	0.86–0.96	54–58	112.9±0.8	−9.1 to +0.5			Pl+Qtz+Groundmass	Hu et al., 2017
13T034	31°43.010′N, 90°54.030′E	Andesite				116.8±0.7	−6.7 to −1.4			Pl+Qtz+Groundmass	Hu et al., 2017
13T046	31°55.080′N, 90°51.993′E	Rhyolite	57.72–78.17	0.87–0.98	51–57	116.0±1.2	−6.4 to +2.3			Qtz+Groundmass	Hu et al., 2017
13T091	32°03.798′N, 90°43.418′E	Andesite	59.25–78.05	0.88–0.96	41–59	114.3±0.8	−6.4 to +6.4			Pl+Groundmass	Hu et al., 2017
13T109	32°06.164′N, 90°41.098′E	Granodiorite	62.86–67.95	0.91–1.03	32–60	118.4±0.8	+0.9 to +6.5			Pl+Hbl+Qtz+minor Bi	Hu et al., 2017
13T118	32°06.045′N, 90°44.725′E	Granodiorite				116.7±0.6	+2.5 to +6.8			Pl+Hbl+Qtz+minor Bi	Hu et al., 2017
ZT1	32°24.45′N, 84°23.30′E	Rhyolite	74.9–77.3	1.16–1.29	9.08–10.8	116.9±1.0	+4.3 to +6.1			Pl+Qtz+Groundmass	Fan et al., 2018
ZT2	32°24.45′N, 84°23.30′E	Rhyolite	75.7–78.0	1.03–1.34	7.58–9.20	111.8±1.0	+4.3 to +6.1			Pl+Qtz+Groundmass	Fan et al., 2018
Central Lhasa subterrane											
GJ0611	32°02.420′N, 82°12.040′E	Rhyolite	74.39	1.08		143±2	−11.5 to −4.9			Qtz(5%–25%)+Kf(5%–30%)+Pl(10%–30%)+Bi(3%–5%)	Zhu et al., 2009b
GJ0612	32°02.420′N, 82°12.040′E	Rhyolite	77.91	1.19		129±1	−7.7 to −5.2			Qtz(5%–25%)+Kf(5%–30%)+Pl(10%–30%)+Bi(3%–5%)	Zhu et al., 2009b
08YR11	31°43.812′N, 82°09.182′E	Tonalite	63.59	0.93		134.3±1.7	−9.9 to −7.7			Pl(60%)+Kf(5%)+Qtz(15%)+Hbl(12%)+Bi(5%)	Zhu et al., 2011

Continued

Sample	GPS Position	Lithology	SiO_2 /wt%	A/CNK	$Mg^{\#}$	Age /Ma	zircon $\varepsilon_{Hf}(t)$	Bulk-rock $(^{87}Sr/^{86}Sr)_i$	Bulk-rock $\varepsilon_{Nd}(t)$	Mineral assemblage	References
08YR14	31°41.533′N, 82°09.996′E	Rhyolitic breccia				133.8±1.1	−11.5 to −4.3			Qtz(10%)+Pl(5%)	
08YR16	31°40.816′N, 82°10.550′E	Rhyolite	73.35	1.19		142.9±1.0	−7.1 to −1.6			Pl(10%)+Qtz(5%)	
08CQ35	30°56.343′N, 84°34.281′E	Monzogranite	74.33	1.05		122.6±0.8	−13.8 to −12.6			Pl(30%)+Kf(40%)+Qtz(20%)+Bi(8%)	
DX13-1	31°30.080′N, E85°10.720′	Dacite	68.05	1.04	32.5	121±1	−10.2 to −6.6	0.714 6	−10.7	Pl(15%)+Kf(5%)+Qtz(10%)+Bi(3%)	Zhu et al., 2009b
DX2-1	31°27.510′N, 84°56.080′E	Dacite	66.48	1.03	34.1	130±1	−10.7 to −5.2	0.714 1	−10.5	Pl(20%)+Kf(5%)+Qtz(5%)+Bi(5%)	Zhu et al., 2009b
MB05-7	30°06.067′N, 92°19.557′E	Monzogranite	71.64	1.06		129.9±1.0	−7.0 to −5.6			Pl(35%)+Kf(40%)+Qtz(20%)+Bi(3%)	Zhu et al., 2011
DXL1-3	31°39.310′N, 85°11.700′E	Rhyolitic tuff	73.95	0.90	54.0	112±1	−0.6 to +2.9	0.707 3	−5.9		Zhu et al., 2009b
DX19-1	31°16.910′N, 85°07.570′E	Granodiorite	71.10	1.01	38.9	107±1	−1.4 to +2.6	0.707 7	−4.6	Pl(50%)+Qtz(25%)+Kf(10%)+Bi(8%)+Hbl(3%)	Zhu et al., 2009b
DX21-1	31°00.000′N, 85°07.020′E	Rhyolite	75.30	1.16	43.5	111±1	−5.8 to −0.2	0.708 2	−8.3	Qtz(25%)+Pl(12%)+Kf(8%)	Zhu et al., 2009b
NX5-2	30°45.660′N, 85°31.850′E	Granodiorite	67.52	1.00	39.4	109±1	−6.7 to −2.6	0.709 5	−8.1	Pl(45%)+Kf(28%)+Qtz(20%)+Hbl(5%–15%)	Zhu et al., 2009b
NX5-3	30°45.660′N, 85°31.850′E	Diorite	57.41	0.82	47.5	108±1	−7.2 to −0.7	0.708 8	−7.8	Pl(50%)+Qtz(20%)+Hbl(18%)+Bi(10%)	Zhu et al., 2009b
GB-8	30°45.560′N, 85°32.250′E	Granodiorite	68.43	0.95	31.1	116±1	−7.5 to −3.5	0.716 4	−7.9	Pl(40%)+Kf(30%)+Qtz(20%)+Bi(5%)+Hbl(3%–15%)	Zhu et al., 2009b

Continued

Sample	GPS Position	Lithology	SiO_2 /wt%	A/CNK	$Mg^{\#}$	Age /Ma	zircon $\varepsilon_{Hf}(t)$	Bulk-rock $(^{87}Sr/^{86}Sr)_i$	Bulk-rock $\varepsilon_{Nd}(t)$	Mineral assemblage	References
GRC02-1	31°03. 980′N, 88°12. 410′E	Dacite	64. 61	1. 13	50. 0	114. 0±0. 7	−14. 4 to −4. 1			Pl(30%)+Qtz(10%)+Bi(3%)	Zhu et al., 2011
GRC03-2	31°13. 570′N, 88°07. 610′E	Dacite	65. 28	1. 03	42. 0	113. 8±0. 5	−11. 8 to −4. 5			Pl(25%)+Qtz(10%)	Zhu et al., 2011
GRC03-1	31°13. 570′N, 88°07. 610′E	Basalt	50. 84	0. 73	44. 0				−3. 5		
SZ10-1	30°53. 400′N, 88°39. 740′E	Dacite	64. 45	1. 35	45. 0	112. 1±0. 4	−7. 8 to −4. 7	0. 714 7	−9. 1	Qtz(25%)+Pl(20%)+Bi(10%)	Zhu et al., 2011
SZ09-2	30°53. 400′N, 88°39. 740′E	Dacite	47. 66–50. 03	0. 65–0. 87	55–62			0. 707 1–0. 718 2			
SZ01-1*	30°45. 380′N, 88°55. 340′E	Dacite	64. 81	0. 82		116. 7±0. 6	−8. 6 to −4. 7			Pl(30%)+Qtz(20%)+Bi(5%)	Zhu et al., 2015
SZ07-1	30°45. 960′N, 88°54. 080′E	Dacite	64. 91–66. 21	0. 97–1. 05	35–56	110. 9±0. 5	−9. 0 to −4. 6	0. 712 3–0. 7141	−9. 3 to −9. 4	Pl(20%)+Bi(15%)	Zhu et al., 2011
MB12-1	30°04. 515′N, 92°09. 248′E	Granodiorite	68. 48	0. 90		88. 3±0. 5	−3. 8 to −1. 7		−2. 3 to +0. 6	Pl(50%)+Kf(20%)+Qtz(25%)+Bi(3%)	Meng et al., 2010
SQ0666	32°20. 60′N, 80°01. 58′E	Andesite	58. 70	0. 94		102±1	+1. 2 to +6. 0			Pl (25% – 30%) + Cpx (20%–25%)	Zhu et al., 2009b
CMN04-2	31°21. 05′N, 85°54. 49′E	Rhyolite	70. 44	1. 14		125±1	−14. 2 to −11. 2			Qtz(5%–25%)+Kf(5%–30%)+Pl(10%–30%)+Bi(3%–5%)	Zhu et al., 2009b
SZ39	30°49. 71′N, 87°55. 84′E	Granite porphyry	76. 58	1. 13	7. 2	125±1	−13. 9 to −11. 1	0. 718 0	−13. 7		Zhu et al., 2009b
SZ43	30°49. 94′N, 87°55. 82′E	Rhyolite	72. 66	1. 23	20. 9	129±1	−13. 2 to −10. 3	0. 720 9	−13. 1	Qtz(5%–25%)+Kf(5%–30%)+Pl(10%–30%)+Bi(3%–5%)	Zhu et al., 2009b

Continued

Sample	GPS Position	Lithology	SiO_2 /wt%	A/CNK	$Mg^{\#}$	Age /Ma	zircon $\varepsilon_{Hf}(t)$	Bulk-rock $(^{87}Sr/^{86}Sr)_i$	Bulk-rock $\varepsilon_{Nd}(t)$	Mineral assemblage	References
SZ48	30°50. 46′N, 87°55. 82′E	Andesite	60. 47	0. 99	42. 2	111±1	−9. 5 to −3. 0	0. 714 8	−9. 9	Pl (25% − 30%) + Cpx (20% −25%)	Zhu et al., 2009b
SZ52	30°50. 46′N, 87°55. 82′E	Dacite	64. 97	1. 12	31. 9	107±1	−9. 9 to +0. 5	0. 713 8	−10. 4	Qtz(5%−25%) + Kf(5%−30%) + Pl(10%−30%) + Bi(3%−5%)	Zhu et al., 2009b
SZ08-1	30°46. 37′N, 88°50. 63′E	Diorite	57. 62	0. 92		111±1	−12. 6 to −4. 5				Zhu et al., 2009b
11BH-28	32°13. 217′N, 81°14. 496′E	Granite porphyry	77. 01	1. 25	26. 8					Kf(25%) + Qtz(15%)	Cao et al., 2016
11BH-29-1	32°13. 203′ N, 81°14. 488′ E	Granite host	77. 25	1. 12	19. 8					Pl(15%) + Kf(50%) + Qtz(30%) + Bi(3%)	Cao et al., 2016
11BH-29-2	32°13. 203′ N, 81°14. 488′ E	Dioritic enclave	61. 66	0. 89	51. 3					Pl (60%) + Hbl (10%) + Kfs (20%) + Qtz(5%) + Bi(5%)	Cao et al., 2016
11BH-30	32°12. 217′N, 81°15. 002′E	Granite	72. 17	1. 03	44. 8					Pl(20%) + Kf(45%) + Qtz(25%) + Bi(8%)	Cao et al., 2016
11BH-31-1	32°11. 064′N, 81°16. 378′E	Granite host	71. 45	1. 02	43. 1	130. 5±1. 2	−2. 5 to −1. 2			Pl(25%) + Kf(45%) + Qtz(20%) + Bi(6%) + Hbl(3%)	Cao et al., 2016
11BH-31-2	32°11. 064′N, 81°16. 378′E	Dioritic enclave	60. 02	0. 87	49. 1	141. 1±1. 5	−2. 9 to +0. 9			Pl(58%) + Hbl(12%) + Kf(20%) + Qtz(4%) + Bi(5%)	Cao et al., 2016
11BH-32	32°07. 558′N, 81°16. 215′E	Granite	71. 67	1. 07	46. 1					Pl(25%) + Kf(45%) + Qtz(22%) + Bi(8%)	Cao et al., 2016
11BH-102	31°51. 469′N, 84°02. 499′E	Granite porphyry	72. 20	1. 10	41. 6	141. 6±1. 4	−10. 7 to −6. 6			Kfs(5%) + Qtz(15%)	Cao et al., 2016
11BH-103	31°51. 115′N, 84°00. 493′E	Granodiorite	62. 18	0. 87	50. 1					Pl(50%) + Hbl(15%) + Kf(25%) + Qtz(15%) + Bi(5%)	Cao et al., 2016

Continued

Sample	GPS Position	Lithology	SiO_2 /wt%	A/CNK	$Mg^{\#}$	Age /Ma	zircon $\varepsilon_{Hf}(t)$	Bulk-rock $(^{87}Sr/^{86}Sr)_i$	Bulk-rock $\varepsilon_{Nd}(t)$	Mineral assemblage	References
11BH-104	31°47.319′N, 83°58.283′E	Granodiorite	65.24	0.93	54.8	136.9±1.4	−6.8 to −3.3			Pl(50%)+Hbl(10%)+Kf(25%)+Qtz(15%)	Cao et al., 2016
11BH-104-1	31°47.319′N, 83°58.283′E	Dioritic enclave	55.65	0.85	58.2	137.7±1.6	−6.0 to −2.7			Pl(50%)+Hbl(25%)+Kf(15%)+Qtz(5%)+Bi(5%)	Cao et al., 2016
11BH-105	31°43.577′N, 83°57.335′E	Granodiorite	64.03	0.91	53.3					Pl(45%)+Hbl(10%)+Kf(25%)+Qtz(15%)+Bi(5%)	Cao et al., 2016
11BH-106	31°41.128′N, 83°56.217′E	Granodiorite	68.41	1.01	54.2	130.4±1.3	−6.2 to −2.2			Pl(40%)+Hbl(6%)+Kf(30%)+Qtz(15%)+Bi(6%)	Cao et al., 2016
SZ01-1	30°45.380′N, 88°55.340′E	Dacite	66.49–67.38	0.82–1.04	37–39	116.9±1.3	−8.2 to −4.4	0.7132	−9.4	Pl(20%–25%)+Qtz(20%)+Bi(10%–15%)	Chen et al., 2014
SZ02-1	30°45.380′N, 88°55.340′E	Andesite	60.51–61.11	0.93–1,02	46–51			0.7209–0.7218	−13.0 to −13.6		Chen et al., 2014
DG01-1	31°20.000′N, 88°54.000′E	Rhyolite	75.08	1.20	8.0	115.7±0.6	−1.1 to +4.6			Pl(20%–25%)+Qtz(20%)+Bi(10%–15%)	Chen et al., 2014
DG05-1	31°20.000′N, 88°54.001′E	Rhyolite	68.74	1.04	25.0	114.3					Chen et al., 2014

Qt: Quartz; Pl: Plagioclase; Kf: Akaline Feldspar; Hbl: Hornblende; Bi: Biotite; Cpx: Clinopyroxene; Opx: Orthopyroxene; Chl: Chlorite; Prh: Prehnite; Om: Opaque mineral; Cb: Carbonate mineral; Grt: Garnet.

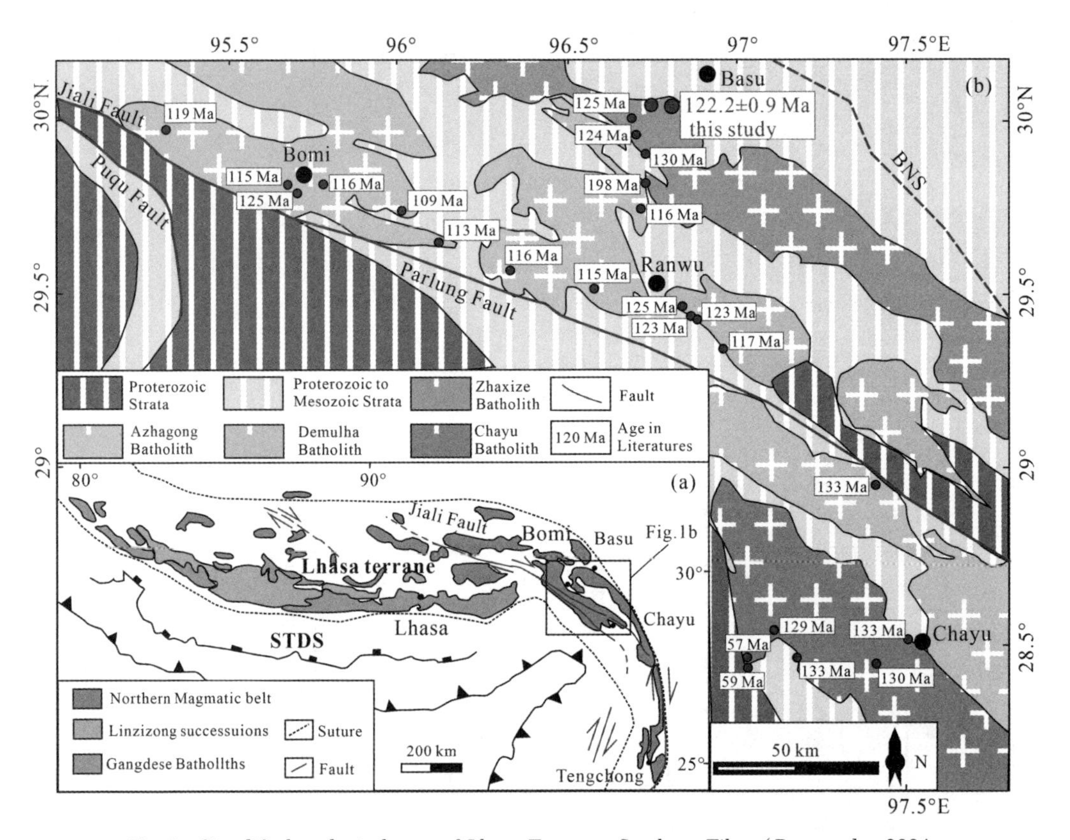

Fig. 1 Simplified geological map of Lhasa Terrane, Southern Tibet (Pan et al., 2004; Chiu et al., 2009; Lin et al., 2012) (a); Distribution of major intrusive rocks and batholiths in the northeastern Lhasa terrane (Chiu et al., 2009; Lin et al., 2012) (b).

The geochronologic data in the map was collected from Chiu et al. (2009), Zhu et al. (2009a) and Pan et al. (2014). STDS: South Tibet Detachment System; BNS: Bangong-Nujiang suture.

In the eastern part of the Lhasa terrane, extensive Cretaceous granitoids and minor Ordivician, Jurassic, and Cenozoic granitoids are exposed in a NW-SE belt to the southwest of the Bangong-Nujiang suture zone (Pan et al., 2004; Zhu et al., 2009a; Chiu et al., 2009; Lin et al., 2012), cropping out in Carboniferous-Permian, Devonian and Proterozoic metamorphic rocks. These Cretaceous granitoids mostly occurred as batholiths which were mainly distributed in the Bomi, Basu, Ranwu and Chayu area (Fig. 1b), including Zhaxuze batholith near Basu, Azhagong batholith near Bomi and Chayu, Demulha batholith near Ranwu, and Chayu batholith between Chayu and Shama. These mainly consist of monzogranites and granodiorites with minor dioritic veins and dioritic enclaves (Table 1) (Pan et al., 2004; Chiu et al., 2009; Zhu et al., 2009b; Lin et al., 2012; Pan et al., 2014).

This study focuses on the Zhaxize batholith near Basu county (Fig. 1b), which are rounded and elongated ~NW-SE, parallel to the long axis of the southeastern part of Bomi-Chayu batholith that is located in between the Jiali fault and the Bangong-Nujiang suture (Chiu et al., 2009; Lin et al., 2012). These granitoids are distributed in the southwest of Basu county, and for the most part consist of granodiorites (Fig. 2a,b). Here, these granodiorites are fine- to medium-grained (Fig. 2b), contain quartz (18% – 28%), alkaline feldspar (18%–24%), sodic plagioclase (35%–45%), hornblende (~5%–10%), biotite (~3%–5%) with accessory minerals including titanite, apatite, zircon, magnetite and other Fe-Ti oxides (Fig. 2c,d). Some hornblendes are euhedral to subhedral but partly chloritized, many plagioclases are strongly zoned (Fig. 2c,d).

Fig. 2 Field and microscope petrological features of Basu granodiorites from Zhaxize batholith in the northeastern Lhasa terrane.

Bi: biotite; Pl: plagioclase; Hbl: hornblende; Kf: K-feldspar; Qz: quartz; Sp: titanite.

3 Analytical methods

3.1 Whole-rock geochemical and Sr-Nd-Pb-Hf isotopic analyses

Representative whole rock analyses (samples ABQ03, 04, 05, 08, 10, 11, 12, 14, 17 and 20) were performed at the State Key Laboratory of Continental Dynamics, Northwest University, Xi'an, China. Fresh chips of whole-rock samples (ca 1 kg) were powdered to pass 200 mesh using a tungsten carbide ball mill. Major elements were analyzed using X-ray fluorescence (Rikagu RIX 2100). Analyses of U.S. Geological Survey and Chinese national rock standards BCR-2, GSR-1, and GSR-3, which indicate that both analytical precision and accuracy for major elements are generally better than 5%. Trace elements were determined by (Bruker Aurora M90) inductively coupled plasma mass spectrometry (ICP-MS) following the method of Qi et al. (2000). Sample powders were dissolved using an HF + HNO_3 mixture in a high-pressure PTFE bomb at 190 ℃ for 48 h. The analytical precision was better than ± 5%−10% (relative) for most of the trace elements.

Meanwhile, the representative whole-rock Sr-Nd-Pb-Hf isotopic data (samples ABQ-4, -5 and -17) were obtained using a Nu Plasma HR multi-collector (MC) mass spectrometer. The Sr and Nd isotopic fractionation was corrected to $^{87}Sr/^{86}Sr$ = 0. 119 4 and $^{146}Nd/^{144}Nd$ = 0. 721 9, respectively (Chu et al., 2009). During the sample runs, the La Jolla standard yielded an average of $^{143}Nd/^{144}Nd$ = 0. 511 862 ± 5 (2σ), and the NBS987 standard yielded an average of $^{87}Sr/^{86}Sr$ ratio = 0. 710 236 ± 16 (2σ). The total procedural Sr and Nd blanks are b1 ng and b50 pg, and NIST SRM-987 and JMC-Nd were used as certified reference standard solutions for $^{87}Sr/^{86}Sr$ and $^{143}Nd/^{144}Nd$ isotopic ratios, respectively. The BCR-1 and BHVO-1 standards yielded an average of $^{87}Sr/^{86}Sr$ ratio are 0. 705 014 ± 3 (2σ) and 0. 703 477 ± 20 (2σ). The BCR-1 and BHVO-1 standards yielded an average of $^{146}Nd/^{144}Nd$ ratio are 0. 512 615 ± 12 (2σ) and 0. 512 987 ± 23 (2σ). Whole-rock Pb was separated by an anion exchange in HCl-Br columns, Pb isotopic fractionation was corrected to $^{205}Tl/^{203}Tl$ = 2. 387 5. Within the analytical period, 30 measurements of NBS981 gave average values of $^{206}Pb/^{204}Pb$ = 16. 937 ± 1 (2σ), $^{207}Pb/^{204}Pb$ = 15. 491 ± 1 (2σ), $^{208}Pb/^{204}Pb$ = 36. 696 ± 1 (2σ). BCR-2 standard gave $^{206}Pb/^{204}Pb$ = 18. 742 ± 1 (2σ), $^{207}Pb/^{204}Pb$ = 15. 620 ± 1 (2σ), and $^{208}Pb/^{204}Pb$ = 38. 705 ± 1 (2σ). Total procedural Pb blanks were in the range of 0. 1−0. 3 ng. Whole-rock Hf was also separated by single anion exchange columns. In the course of analysis, 22 measurements of the JCM 475 standard yielded an average of $^{176}Hf/^{177}Hf$ = 0. 282 161 3 ± 0. 000 001 3 (2σ) (Yuan et al., 2007).

3. 2 Zircon U-Pb isotopic analyses

Zircon grains were separated using conventional heavy liquid and magnetic techniques. Representative grains (~5 kg) from Basu were hand-picked and mounted on epoxy resin discs, polished and carbon coated. Internal morphology was examined using cathodoluminescence (CL) prior to U-Pb isotopic analyses. Laser ablation (LA)-ICP-MS zircon U-Pb analyses were conducted on an Agilent 7500a ICP-MS equipped with a 193-nm laser, following method of Bao et al. (2017) at the State Key Laboratory of Continental Dynamics, Northwest University, Xi'an, China. The $^{207}Pb/^{206}Pb$ and $^{206}Pb/^{238}U$ ratios were calculated using the GLITTER data reduction software program (http://www. glitter-gemoc. com/), and calibrated using the Harvard zircon 91500 as external standard. These correction factors were then applied to each sample to correct for both instrumental mass bias and depth-dependent elemental and isotopic fractionation. Common Pb contents were evaluated using the method described by Andersen (2002). The age calculations and plotting diagrams were performed using ISOPLOT (version 3.0; Ludwig, 2003). The errors quoted in tables and figures are at the 2σ level.

These analytical data of representative samples from the Basu pluton, SE Tibet, are listed in Table 2 (Zircon U-Pb data), Table 3 (major and trace elements), and Table 4 (Bulk-rock Sr-Nd-Pb-Hf isotopic compositions).

4 Results

4.1 Zircon U-Pb age

We chose typical granodioritic samples from the Basu area (Fig. 3), SE Tibet, SW China for zircon U-Pb geochronology. Analytical results are listed in Table 2 and in Figs. 1-3.

CL images of zircon grains show notably euhedral, prismatic with clear oscillatory zoning (Fig. 3a). The seventeen reliable analytical spots from ABQ07 have Th (145-502 ppm) and U (228-468 ppm) contents with Th/U ratios of 0.54-1.02, and yield $^{206}Pb/^{238}U$ ages from 121±2.0 Ma to 125±2.0 Ma, with a weighted mean age of 122.2±1.0 Ma (MSWD=0.63, n =17, 1σ) (Fig. 3b,c), representing the magma crystallization age of the granodiorites.

4.2 Major and trace elemental geochemistry

In the SiO_2 vs. K_2O+Na_2O and SiO_2 vs. K_2O (Fig. 4a,b), A/CNK vs. FM (Fig. 4c) and Q-A-P (Fig. 4d) diagram, these samples are plotted as high-K, calc-alkaline, intermediate and granodiorite area, and also show the clinopyroxene entrainment and I-type trend (Zhu et al., 2017).

The granodiorites from the Zhaxize batholith in the Basu area have moderate SiO_2 range from 60.9 wt% to 68.2 wt%, K_2O = 2.90-3.90 wt%, Al_2O_3 contents = 14.8-16.1 wt%, A/CNK ratios range from 0.87 to 0.99, CaO = 3.80-5.45 wt%, $Fe_2O_3{}^T$ = 3.42-6.35 wt%, and MgO = 1.45-3.16 wt% with $Mg^\#$ = 49.7-53.7. The primitive mantle-normalized trace element diagrams (Fig. 5a) shows clear positive Rb, Th, U, and Pb anomaly, enrichment of large ion lithophile element and sharp depletions in HFSE including Nb, P, and Ti. The chondrite-normalized rare-earth element (REE) diagrams (Fig. 5b) show enrichment of LREE with high $(La/Yb)_N$ = 5.97-11.0, flat HREE, and moderate negative Eu anomalies (δEu = 0.59-0.72), which display a similar differentiation trend to granitoids in Bomi (Pan et al., 2014) and fractionated I-type granites in Chayu (Zhu et al., 2009a). However, they are different from granitoids from Demulha batholith in Ranwu, which show strongly fractionated trends with notable depletion of Ba, Sr, P, Eu and Ti (Lin et al., 2012).

Table 2 Results of zircon LA-ICP-MS U-Pb data for the Basu granodiorites in the northeastern Lhasa terrane.

Analysis	Content /ppm		Ratios									Age /Ma							
	Th	U	Th/U	$^{207}Pb/^{206}Pb$	1σ	$^{207}Pb/^{235}U$	1σ	$^{206}Pb/^{238}U$	1σ	$^{208}Pb/^{232}Th$	1σ	$^{207}Pb/^{206}Pb$	1σ	$^{207}Pb/^{235}U$	1σ	$^{206}Pb/^{238}U$	1σ	$^{208}Pb/^{232}Th$	1σ
ABQ-07 from Zhaxize batholith N 30°02. 356′, E 96°46. 292′																			
ABQ-07-01	315	406	0. 78	0. 057 02	0. 002 99	0. 148 91	0. 007 54	0. 018 94	0. 000 25	0. 005 88	0. 000 06	492	119	141	7	121	2	119	1
ABQ-07-02	267	378	0. 71	0. 052 98	0. 002 62	0. 140 60	0. 006 72	0. 019 25	0. 000 25	0. 006 03	0. 000 06	328	115	134	6	123	2	122	1
ABQ-07-03	252	382	0. 66	0. 049 94	0. 001 78	0. 129 95	0. 003 86	0. 018 87	0. 000 24	0. 006 11	0. 000 10	192	45	124	3	121	2	123	2
ABQ-07-04	259	384	0. 67	0. 057 09	0. 003 87	0. 151 62	0. 010 04	0. 019 26	0. 000 27	0. 005 98	0. 000 07	495	154	143	9	123	2	121	2
ABQ-07-05	320	339	0. 94	0. 057 89	0. 002 02	0. 154 95	0. 004 46	0. 019 41	0. 000 25	0. 006 09	0. 000 09	526	41	146	4	124	2	123	2
ABQ-07-06	360	468	0. 77	0. 048 33	0. 001 71	0. 130 78	0. 003 82	0. 019 62	0. 000 24	0. 006 14	0. 000 09	115	46	125	3	125	2	124	2
ABQ-07-07	223	334	0. 67	0. 050 22	0. 002 61	0. 130 11	0. 006 54	0. 018 79	0. 000 25	0. 005 93	0. 000 06	205	121	124	6	120	2	119	1
ABQ-07-08	237	309	0. 77	0. 048 59	0. 002 82	0. 125 72	0. 007 10	0. 018 77	0. 000 26	0. 005 94	0. 000 06	128	131	120	6	120	2	120	1
ABQ-07-09	273	288	0. 95	0. 051 01	0. 003 57	0. 132 89	0. 009 11	0. 018 89	0. 000 28	0. 005 95	0. 000 06	241	161	127	8	121	2	120	1
ABQ-07-10	286	377	0. 76	0. 053 68	0. 002 95	0. 140 70	0. 007 48	0. 019 01	0. 000 26	0. 005 95	0. 000 06	358	127	134	7	121	2	120	1
ABQ-07-11	187	280	0. 67	0. 047 31	0. 002 54	0. 125 66	0. 006 52	0. 019 26	0. 000 26	0. 006 12	0. 000 08	65	118	120	6	123	2	123	2
ABQ-07-12	145	266	0. 54	0. 048 07	0. 002 05	0. 130 13	0. 004 86	0. 019 64	0. 000 26	0. 006 85	0. 000 13	103	62	124	4	125	2	138	3
ABQ-07-13	150	228	0. 66	0. 052 15	0. 003 35	0. 138 56	0. 008 66	0. 019 27	0. 000 28	0. 006 05	0. 000 07	292	149	132	8	123	2	122	1
ABQ-07-14	223	313	0. 71	0. 057 87	0. 002 31	0. 151 07	0. 005 17	0. 018 94	0. 000 26	0. 005 35	0. 000 12	525	51	143	5	121	2	108	2
ABQ-07-15	253	355	0. 71	0. 049 32	0. 001 85	0. 129 65	0. 004 1	0. 019 07	0. 000 24	0. 006 44	0. 000 10	163	50	124	4	122	2	130	2
ABQ-07-16	502	492	1. 02	0. 050 07	0. 001 77	0. 133 15	0. 003 86	0. 019 29	0. 000 24	0. 006 30	0. 000 09	198	44	127	3	123	2	127	2
ABQ-07-17	312	338	0. 92	0. 058 93	0. 002 07	0. 153 26	0. 004 40	0. 018 87	0. 000 24	0. 006 52	0. 000 10	565	40	145	4	121	2	131	2

Table 3 Major and trace element of the Basu high-Mg# granodiorites in the northeastern Lhasa terrane.

Sample	ABQ03	ABQ04	ABQ05	ABQ08	ABQ10	ABQ10R	ABQ11	ABQ12	ABQ14	ABQ17	ABQ20
SiO_2	65.84	61.23	64.05	60.90	64.33	64.44	67.38	67.64	65.77	68.22	61.53
TiO_2	0.41	0.55	0.48	0.55	0.44	0.45	0.38	0.37	0.44	0.32	0.56
Al_2O_3	15.22	15.63	15.53	15.65	15.73	15.79	15.01	14.82	15.51	15.31	16.12
$Fe_2O_3^T$	4.49	6.11	5.52	6.35	4.85	4.87	4.19	4.02	4.85	3.42	6.18
MnO	0.08	0.11	0.10	0.10	0.08	0.08	0.08	0.08	0.08	0.06	0.10
MgO	2.07	3.04	2.58	3.16	2.29	2.32	1.81	1.79	2.20	1.45	2.67
CaO	4.08	5.26	5.17	5.32	4.64	4.67	4.04	4.00	4.39	3.80	5.45
Na_2O	2.74	2.53	2.42	2.74	2.42	2.45	2.59	2.52	2.50	2.67	3.35
K_2O	3.84	3.28	3.16	3.06	3.51	3.53	3.74	3.85	3.51	3.90	2.90
P_2O_5	0.09	0.12	0.11	0.12	0.09	0.10	0.09	0.08	0.10	0.07	0.13
LOI	1.24	1.67	1.14	1.95	1.41	1.39	0.94	0.91	1.03	0.89	1.16
Total	100.10	99.53	100.26	99.90	99.79	100.09	100.25	100.08	100.38	100.11	100.15
Li	25.6	18.9	22.2	24.7	20.1	21.6	19.1	38.6	47.9	24.4	35.2
Be	2.29	1.70	1.93	1.91	1.80	1.89	2.19	2.15	1.91	2.43	2.02
Sc	12.8	18.6	15.9	19.2	12.6	13.0	12.1	11.4	13.9	9.60	17.1
V	98.6	142.0	123.3	147.4	106.6	109.6	88.1	85.4	101.3	69.3	131.9
Cr	20.1	28.9	24.3	28.4	21.2	21.8	15.3	16.4	17.0	12.5	20.0
Co	91.2	100.4	81.6	78.0	86.5	90.4	78.7	88.7	82.9	113.4	72.7
Ni	7.84	10.65	9.24	10.27	8.43	9.25	6.57	6.35	7.47	6.09	7.88
Cu	7.11	6.63	6.04	26.48	9.80	10.2	3.67	4.57	5.89	3.05	13.32
Zn	45.1	67.8	54.2	61.7	46.1	47.7	47.8	44.0	47.2	33.5	59.6
Ga	15.8	16.4	16.4	16.9	15.7	16.1	15.6	15.3	15.6	15.3	17.3
Ge	1.50	1.49	1.50	1.54	1.35	1.41	1.55	1.48	1.45	1.51	1.52
Rb	189.2	145.1	153.3	144.5	156.3	164.2	178.3	189.6	167.9	200.8	157.0
Sr	264.2	322.0	299.1	345.8	296.9	307.6	246.3	246.2	274.7	237.3	323.3
Y	26.30	25.61	26.31	26.80	19.94	20.52	27.01	24.52	24.04	24.13	28.12
Zr	144.4	157.1	135.4	132.3	127.6	125.9	148.1	145.0	147.1	122.4	144.4
Nb	12.9	10.5	11.7	10.8	9.6	10.3	13.0	13.0	10.9	13.0	11.9
Cs	12.01	5.38	6.69	7.62	7.99	8.27	6.03	10.84	13.82	11.99	10.41
Ba	417.1	475.0	417.0	372.4	444.2	457.6	388.8	405.7	467.5	370.1	411.5
Hf	4.23	4.29	3.75	3.76	3.60	3.56	4.26	4.21	4.02	3.78	3.91
Ta	1.93	1.21	1.33	1.26	1.07	1.22	1.65	1.81	1.40	1.90	1.40
Pb	25.8	26.9	23.4	20.4	22.0	22.1	21.3	24.1	21.1	22.8	20.5
Th	32.0	21.7	23.1	23.2	22.7	22.1	26.4	31.5	25.9	27.8	15.7
U	7.07	3.89	3.45	4.56	4.11	4.21	5.67	6.73	6.12	7.83	3.75
La	35.9	30.0	24.5	34.2	29.5	31.1	22.9	27.3	21.9	21.7	31.6
Ce	70.2	60.1	56.0	67.2	57.1	59.7	48.9	53.8	45.0	44.0	62.7
Pr	7.51	6.73	6.61	7.33	5.97	6.26	5.57	5.86	5.05	4.83	6.88
Nd	27.0	25.3	25.6	27.0	21.4	22.3	21.5	21.6	19.5	18.3	26.0

Continued

Sample	ABQ03	ABQ04	ABQ05	ABQ08	ABQ10	ABQ10R	ABQ11	ABQ12	ABQ14	ABQ17	ABQ20
Sm	4.93	4.90	4.97	5.17	3.91	4.07	4.45	4.18	3.97	3.79	5.05
Eu	0.90	1.00	0.97	0.97	0.89	0.92	0.85	0.85	0.91	0.82	1.04
Gd	4.44	4.49	4.49	4.70	3.58	3.67	4.08	3.80	3.73	3.60	4.66
Tb	0.67	0.67	0.68	0.71	0.54	0.55	0.65	0.59	0.58	0.57	0.72
Dy	4.04	4.12	4.17	4.29	3.25	3.30	4.03	3.66	3.66	3.65	4.45
Ho	0.83	0.84	0.86	0.87	0.66	0.68	0.85	0.76	0.76	0.75	0.90
Er	2.52	2.48	2.55	2.60	1.98	2.03	2.56	2.34	2.31	2.33	2.70
Tm	0.38	0.36	0.38	0.39	0.30	0.30	0.40	0.36	0.35	0.36	0.40
Yb	2.64	2.48	2.58	2.59	2.04	2.04	2.76	2.55	2.47	2.59	2.79
Lu	0.40	0.36	0.38	0.39	0.31	0.31	0.42	0.38	0.37	0.40	0.42
$Mg^{\#}$	51.8	53.7	52.1	53.7	52.4	52.6	50.2	50.9	51.4	49.7	50.2
A/CNK	0.95	0.90	0.92	0.89	0.97	0.97	0.96	0.95	0.98	0.99	0.87
FM	0.11	0.15	0.13	0.16	0.12	0.12	0.10	0.09	0.11	0.08	0.14
FeO^T/MgO	1.95	1.81	1.93	1.81	1.91	1.89	2.08	2.02	1.98	2.12	2.08
$Fe_2O_3^T+MgO+TiO_2$	6.97	9.70	8.58	10.06	7.58	7.64	6.38	6.18	7.49	5.19	9.41

Table 4 Whole-rock Sr-Nd-Pb-Hf for the Basu granodiorite in the northeastern Lhasa terrane.

Sample	ABQ-4	ABQ-5	ABQ-5R	ABQ-17
t/Ma	122			
Rb/ppm	145	153	153	201
Sr/ppm	322	299	299	237
$^{87}Sr/^{86}Sr$	0.712 446	0.713 007	0.712 273	0.715 551
2σ	0.000 005	0.000 006	0.000 006	0.000 008
$^{87}Sr/^{86}Sr(t)$	0.710 221	0.710 476	0.709 742	0.711 372
Sm/ppm	4.90	4.97	4.97	3.79
Nd/ppm	25.3	25.6	25.6	18.3
$^{143}Nd/^{144}Nd$	0.512 205	0.512 119	0.511 999	0.511 968
2σ	6	5	12	4
T_{DM2}/Ga	1.30	1.41	1.57	1.62
$\varepsilon_{Nd}(t)$	−7.22	−8.91	−11.26	−11.98
U/ppm	3.89	3.45		7.83
Th/ppm	21.73	23.12		27.79
Pb/ppm	26.88	23.37		22.84
$^{206}Pb/^{204}Pb$	18.925	18.957		19.183
2σ	2	3		4
$^{207}Pb/^{204}Pb$	15.759	15.764		15.781
2σ	2	3		3
$^{208}Pb/^{204}Pb$	39.535	39.616		39.750
2σ	6	8		9
$^{206}Pb/^{204}Pb(t)$	18.749	18.777		18.765

Continued

Sample	ABQ-4	ABQ-5	ABQ-5R	ABQ-17
$^{207}Pb/^{204}Pb(t)$	15.750	15.756		15.761
$^{208}Pb/^{204}Pb(t)$	39.215	39.224		39.265
Lu/ppm	0.36	0.38	0.38	0.40
Hf/ppm	4.3	3.8	3.8	3.8
$^{176}Hf/^{177}Hf$	0.282 500	0.282 458	0.282 467	0.282 465
2σ	5	3	4	3
T_{DM2}/Ga	2.37	2.77	2.74	2.80
$\varepsilon_{Nd}(t)$	−7.93	−9.61	−9.30	−9.42

Rb, Sr, Sm, Nd, U, Th, Pb, Lu and Hf concentrations were analyzed by ICP-MS. T_{DM2} represent the two-stage model age and were calculated using present-day $(^{147}Sm/^{144}Nd)_{DM} = 0.213\ 7$ and $(^{147}Sm/^{144}Nd)_{CHUR} = 0.513\ 15$. Initial Pb isotopic ratios were calculated for 120 Ma using single-stage model. T_{DM2} were calculated using present-day $(^{176}Lu/^{177}Hf)_{DM} = 0.038\ 4$ and $(^{176}Hf/^{177}Hf)_{DM} = 0.283\ 25$, $f_{CC} = -0.55$ and $f_{DM} = 0.16$. $\varepsilon_{Hf}(t)$ values were calculated using present-day $(^{176}Lu/^{177}Hf)_{CHUR} = 0.033\ 2$ and $(^{176}Hf/^{177}Hf)_{CHUR} = 0.282\ 772$.

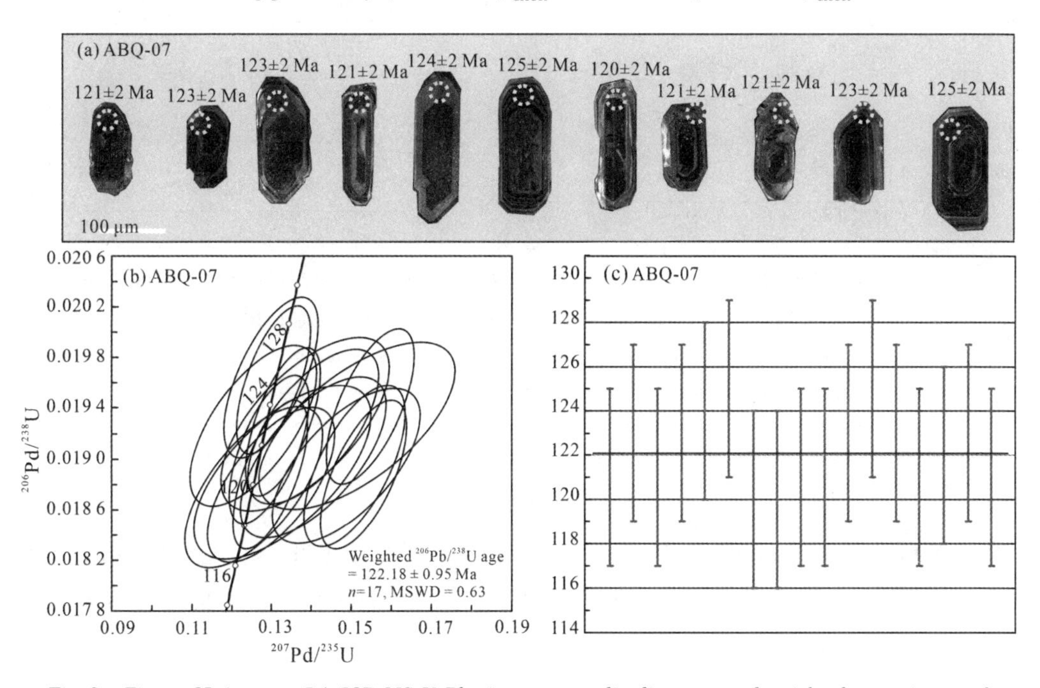

Fig. 3　Zircon CL images, LA-ICP-MS U-Pb zircon concordia diagram, and weighted mean image of representative zircon grains for the Basu granodiorites in the northeastern Lhasa terrane.

4.3　Whole-rock Sr-Nd-Pb-Hf isotopes

Whole-rock Sr-Nd-Pb-Hf isotopic data for the granodiorites in the Basu are listed in Table 4. The initial $^{87}Sr/^{86}Sr$ isotopic ratios (I_{Sr}) and $\varepsilon_{Nd}(t)$ values are calculated at the time of magma crystallization. These samples from Basu area have high I_{Sr} ratios ranging from 0.709 742−0.711 372 and $\varepsilon_{Nd}(t)$ values ranging from −12.0 to −7.2 (Fig. 6a), T_{DM2} values of

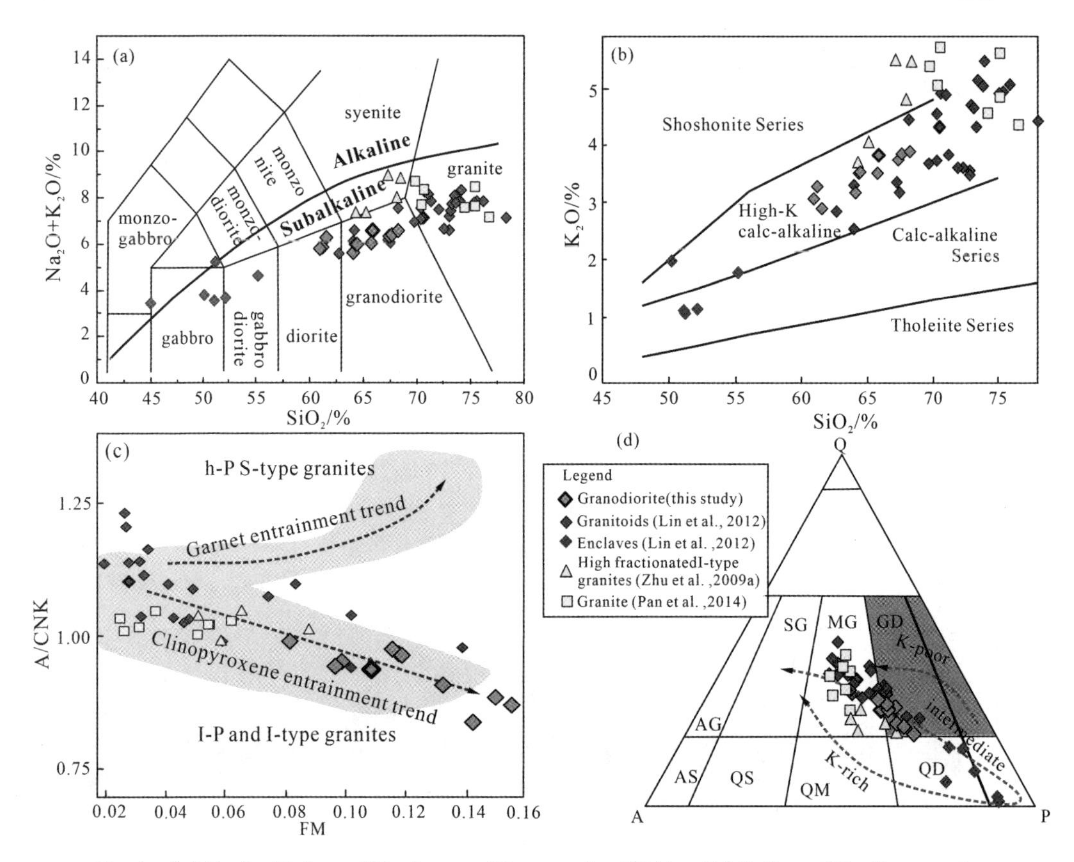

Fig. 4 (a) Na_2O+K_2O vs. SiO_2 diagram (Frost et al., 2001); (b) K_2O vs. SiO_2 diagram after (Roberts and Clemens 1993); (c) Peraluminosity vs. "maficity" relationships (Moyen and Martin, 2012). A/CNK: molar Al/(2Ca+Na+K); FM: molar Fe+Mg. (d) Q-A-P modal classification of Streicksen (1975) for the Basu graodiorites, showing similar with TTG (Grey field=tonalite+granodiorite). These samples are also plotted along the K-poor and intermediate calc-alkaline differentiation trend of Lameyre and Bowden (1982) (T=tholeiitic suite, A=alkaline suite; the calc-alkaline suites are subdivided into: a=K-poor; b=intermediate; c=K-rich).

1.30–1.62 Ga, and initial $^{206}Pb/^{204}Pb$, $^{207}Pb/^{204}Pb$, and $^{208}Pb/^{204}Pb$ ratios of 18.749–18.777, 15.750–15.761, and 39.215–39.265 (Fig. 7), respectively. Initial $^{176}Hf/^{177}Hf$ ratios and $\varepsilon_{Hf}(t)$ values of the early Cretaceous zircons were calculated according their crystallization age. The granodiorites show evolved Hf isotopic compositions, with $\varepsilon_{Hf}(t)$ values of −9.6 to −7.9 (Fig. 6b). These data show the notably evolved Hf isotopic components, which may imply various ancient crustal sources involving in during the evolutional process of magma.

5 Discussion

5.1 Intensive Early Cretaceous magmatism in the northeastern Lhasa terrane

In southern Tibet, the elongated belt of granitoids that crops out in the northeastern Lhasa

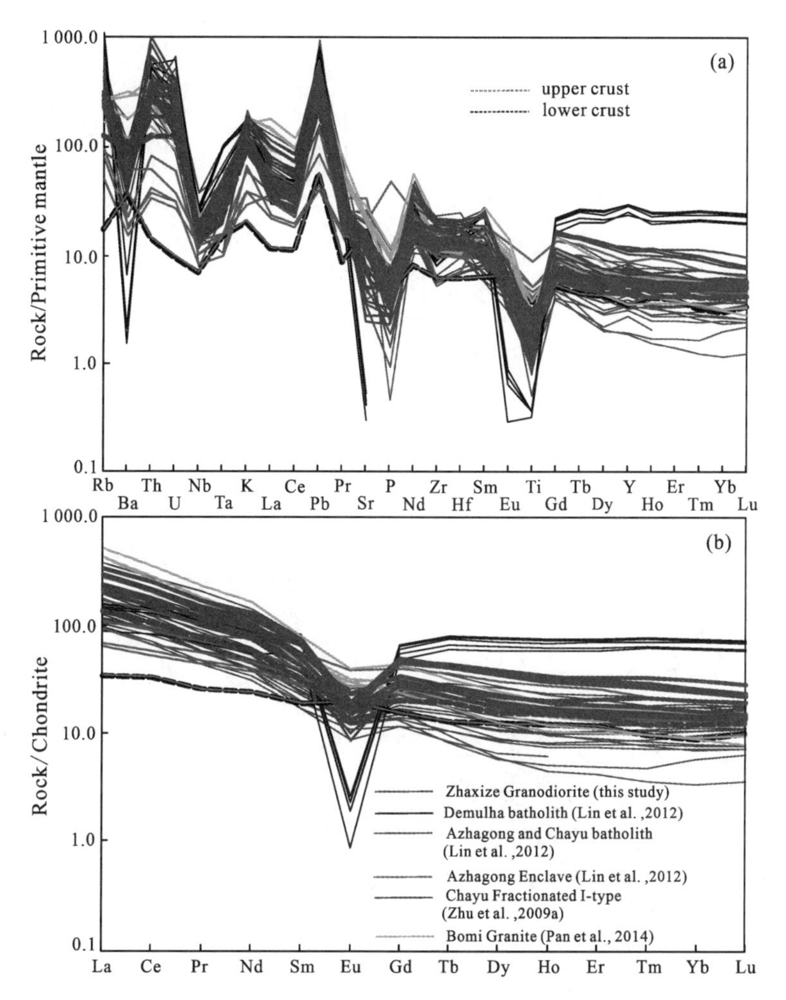

Fig. 5 Primitive-mantle-normalized trace element spider and chondrite-normalized REE patterns and diagram [(a) and (b)] for the Basu granodiorites.

The primitive mantle and chondrite values are from Sun and McDonough (1989).

The upper and lower averages reference from Rudnick and Gao (2003).

terrane between the Jiali fault and Bangong-Nujiang suture zone, has been known as the "Trans-Himalayan batholiths" (Searle et al., 1987; Booth et al., 2004; Chiu et al., 2009). Sensitive high-resolution ion microprobe (SHRIMP) and LA-ICP-MS zircon U-Pb ages and LA-MC-ICP-MS zircon in-situ Lu-Hf isotopic data have suggested that these granitoids ($n=24$) were emplaced principally in the Early Cretaceous (133−110 Ma), sub-ordinately in the Paleocene ($n=4$), and one in the earliest Jurassic (Chiu et al., 2009). Later, Zhu et al. (2009a) reported ca. 130 Ma high fractionated I-type granites in the Chayu area and Pan et al. (2014) also reported ca. 125 Ma granites in the Bomi area. In the present study, LA-ICP-MS zircon U-Pb data show that the granodiorites in the Basu area are also Early Cretaceous (ca. 122 Ma). All of these results indicate that there was intensive granitic magmatism in the early

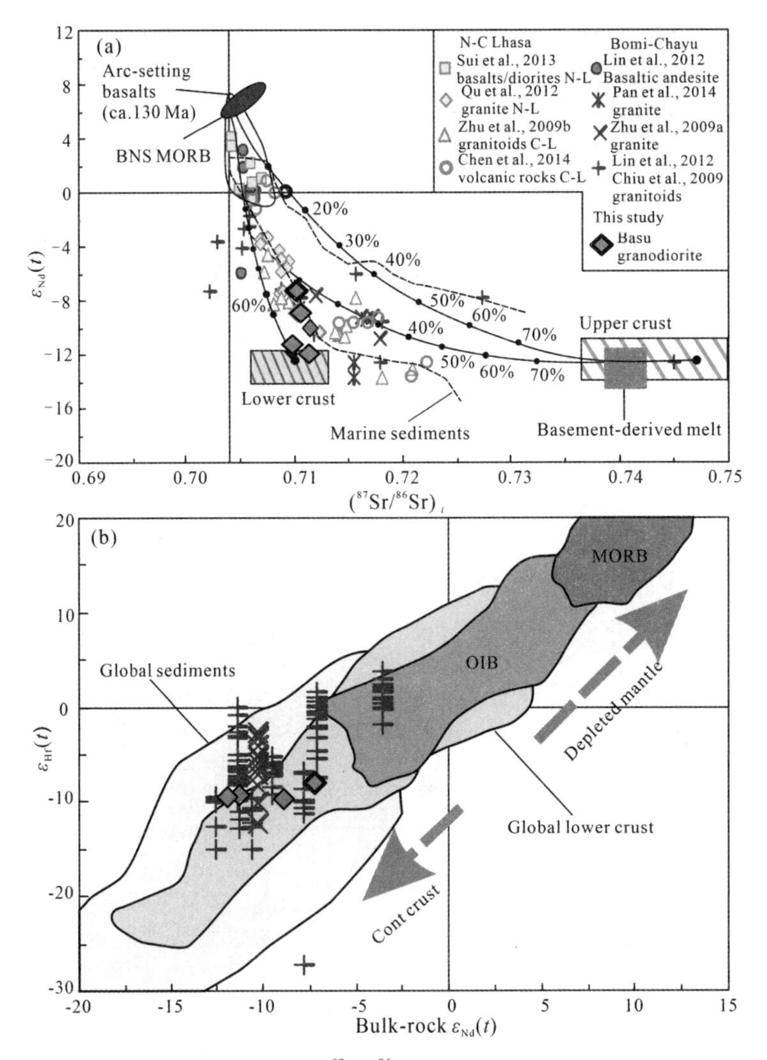

Fig. 6 The $\varepsilon_{Nd}(t)$ values vs. initial $^{87}Sr/^{86}Sr$ (a) and $\varepsilon_{Nd}(t)$ values vs. $\varepsilon_{Hf}(t)$ (b).

The data of Juvenile lower crust and Back-arc basalts from Chen et al. (2014; referenced Zhu et al. and Wang et al. unpublished data). Basement-derived melts (Zhu et al., 2011). Lower continental crust (Miller et al., 1999) and upper continental crust (Harris et al., 1990). Global sediments referenced from Vervoort and Blichert-Toft (1999), Global lower crust referenced from Dobosi et al. (2003), and OIB and MORB referenced from Kempton and McGill (2002). Symbols as Fig. 4.

Cretaceous range from 133 Ma to 113 Ma in the northeastern Lhasa terrane between the Jiali fault and Bangong-Nujiang suture zone (Table 1; Figs. 1 and 8). In addition, as shown in Table 1 and Fig. 8 ($N = 140$), both central and northern Lhasa terrane and Trans-Himalayan batholiths show similarly intensive magmatism during the early Cretaceous.

5.2 Petrogenesis of the high-K calc-alkaline granodiorites

The bulk-rock geochemistry shows that the early Cretaceous granodiorites in the Zhaxize

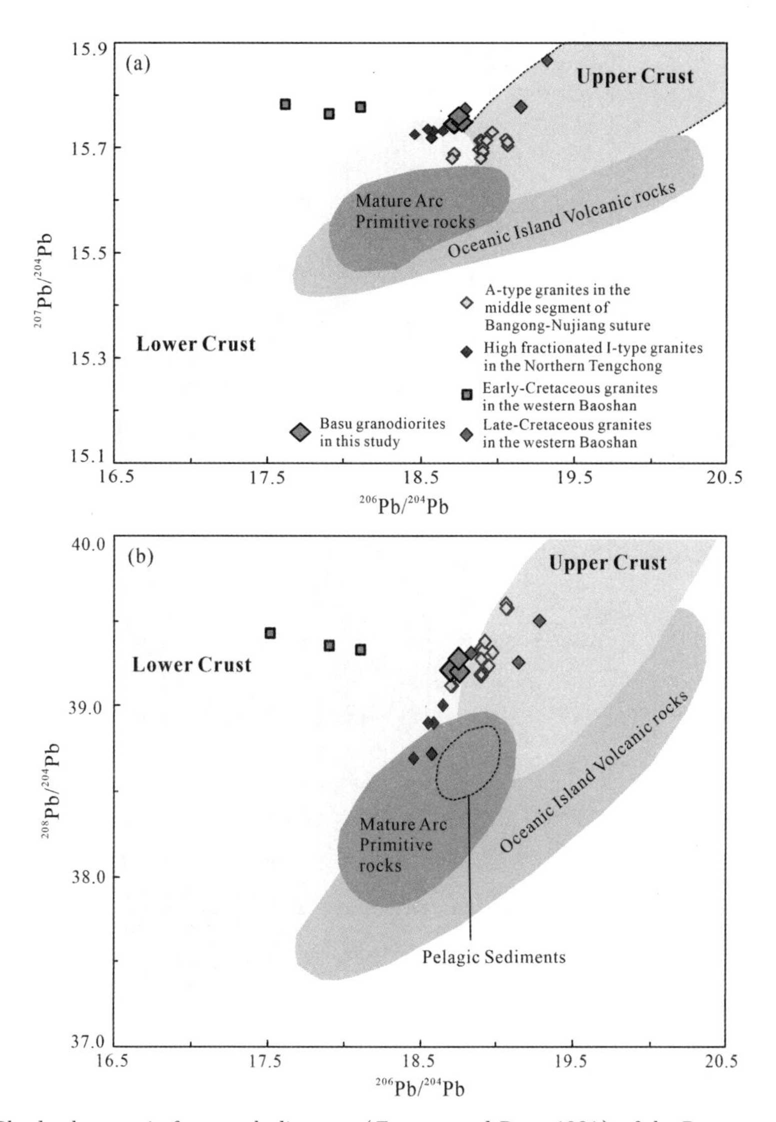

Fig. 7 Pb plumbotectonic framework diagrams (Zartman and Doe, 1981) of the Basu granodiorites. A-type granites data are from Qu et al. (2012), I-type granites data are from Zhu et al. (2015), Cretaceous granites in Baoshan Block from Zhu et al. (2018). Symbols as Fig. 4.

batholith are high-K calc-alkaline and display a notable clinopyroxene entrainment and I-type trend (Fig. 4), similar to other high-K calc alkaline I-type granitoids in the Bomi-Chayu area (Zhu et al., 2009a; Lin et al., 2012). Roberts and Clemens (1993) have proposed that most high-K, calc-alkaline I-type rocks are granodiorites and tonalites, which suggested derivation by partial fusion of pre-existing meta-igneous rocks, in contrast with the inferred metasedimentary protolith S-type granitoids (Chappell and White, 2001). In addition, experimental data suggested that high-K, calc-alkaline I-type granitoids could be only derived from the partial melting of hydrous, calc-alkaline to high-K calc-alkaline, mafic to intermediate

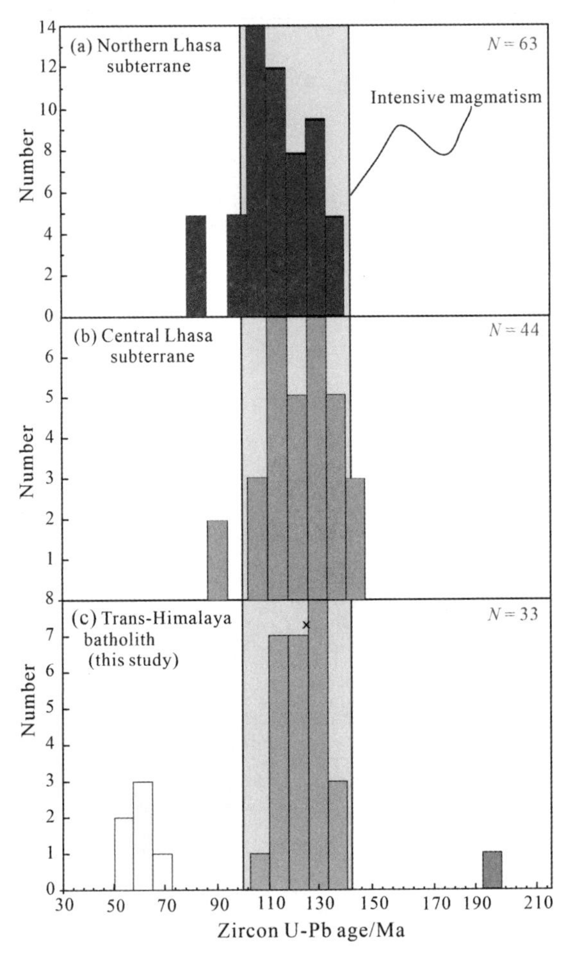

Fig. 8 Age data histograms and relative probability graphs of zircon U-Pb data for granitoids in the northeastern Lhasa terrane.

Date from Chiu et al. (2009), Zhu et al. (2009a), Pan et al. (2014) and this study.

metamorphic rocks in the crust (Roberts and Clemens, 1993). Similarly, high total $Fe_2O_3{}^T$ + MgO + TiO_2 contents (5.19 - 10.1 wt%) are consistent with those from tonalitic and granodioritic melts that was derived by partial melting of meta-igneous rocks, but contrasted with those from experimental studies of S-type granitoid melts, which have low total Fe+Mg+Ti (<3%-4%) (Stevens et al., 2007; Champion and Bultitude, 2013; Zhu et al., 2018). As mentioned above, these samples show a trend of clinopyroxene entrainment and I-type character due to negative relationship between A/CNK and FM (Moyen and Martin, 2012) (Fig. 4c). The whole-rock Sr-Nd isotopic compositions have moderate initial $^{87}Sr/^{86}Sr$ (0.709 742-0.711 372) and negative $\varepsilon_{Nd}(t)$ values (-12.0 to -7.2) (Fig. 6a), similar to granitoids in the Bomi-Chayu area (Zhu et al., 2009a; Lin et al., 2012) and some granitic samples from central and northern Lhasa terrane (Zhu et al., 2009b; Qu et al., 2012). Their proximity to the lower crust domain suggests derivation from an evolved lower crust source. Also, negative and consistent $\varepsilon_{Nd}(t)$ and $\varepsilon_{Hf}(t)$ (−9.6 to −7.9) values (Fig. 6b), together with their old model ages from 2.80 Ga to 1.30 Ga, also indicate an ancient and enriched continental source. Similarly, the rocks show strongly affinity to continental source, closer to lower crust (Fig. 7), with respect to their moderate initial $^{206}Pb/^{204}Pb$ (18.749–18.777) and high initial $^{207}Pb/^{204}Pb$ (15.750 – 15.761) and $^{208}Pb/^{204}Pb$ (39.215 – 39.265) isotopic compositions (Table 4). Therefore, we can conclude that these granodiorites were probably derived from an ancient lower crust source. However, a notably characteristic which deserve serious consideration is the rocks show high-$Mg^{\#}$ (range from 49.7 to 53.7).

Within the high-$Mg^{\#}$ intermediate to felsic rocks, it is important to define whether they are HMA, MA or more "normal" granitic rocks, because HMA are thought to represent the products of partial melting of hydrous mantle wedge peridotites and MA represent the products of melts from lower crust, subducted oceanic crust or sediments which have interacted with mantle peridotites (Qin et al., 2010; Deng et al., 2009, 2010; Wang et al., 2011; Moyen and Martin, 2012; Zhu et al., 2018, 2019). Experimental data have suggested that high-$Mg^{\#}$ (49.7–53.7), are higher than common rocks ($Mg^{\#} < 45$) that were derived from partial melting of metabasaltic rocks at crustal condition (temperature < 1 000 ℃, pressures = 1–4.0 GPa) (Fig. 9a, b) (Rapp and Watson, 1995; Rapp et al., 1999; Moyen and Martin, 2012). On the SiO_2 vs. $Mg^{\#}$ diagram (Fig. 9b, c), such samples plot in the field of interaction trend from crustal melts to peridotites-related melts and/or boninite, sanukitoids and HMA from Piip (Shimoda et al., 1998; Wang et al., 2011). Meanwhile, both on the SiO_2 vs. MgO and SiO_2 vs. FeO^T/MgO diagrams (Fig. 9c, d), the Basu granodiorites in Zhaxize batholith plot in the field of low-FeO^T series and the intersection between MA melts and partial melting of basalt from experimental results, with similar or even higher than MgO content than MA melts from experimental data and MA in Paunamn (Carroll and Wyllie, 1989; Rapp et al., 1999; Prouteau et al., 2001; Zhu et al., 2017), although they are lower than those typical HMA around the world (Stern et al., 1989; Lopez and Cameron, 1997; Qian and Hermann, 2010). Therefore, we consider that the Basu high-$Mg^{\#}$ granodiorites in Zhaxize batholith are different from HMA and common granitic rocks, but similar to those typical MA melts which have been regarded as the products of the interaction between the melts from oceanic crust or basaltic lower crust and mantle-wedge peridotites. However, the melts from partial melting of subducted oceanic crust are usually adakitic, with high Sr/Y ratios (Martin et al., 2005), with depleted and positive isotopic compositions indicating affinity to MORB and/or depleted mantle sources. These features contrast with the samples in this study. For all these reasons, we suggest that the Basu high-$Mg^{\#}$ granodiorites were derived from partial melting of ancient basaltic lower crust source and subsequently interaction with mantle-derived peridotites, similar to the typical MA.

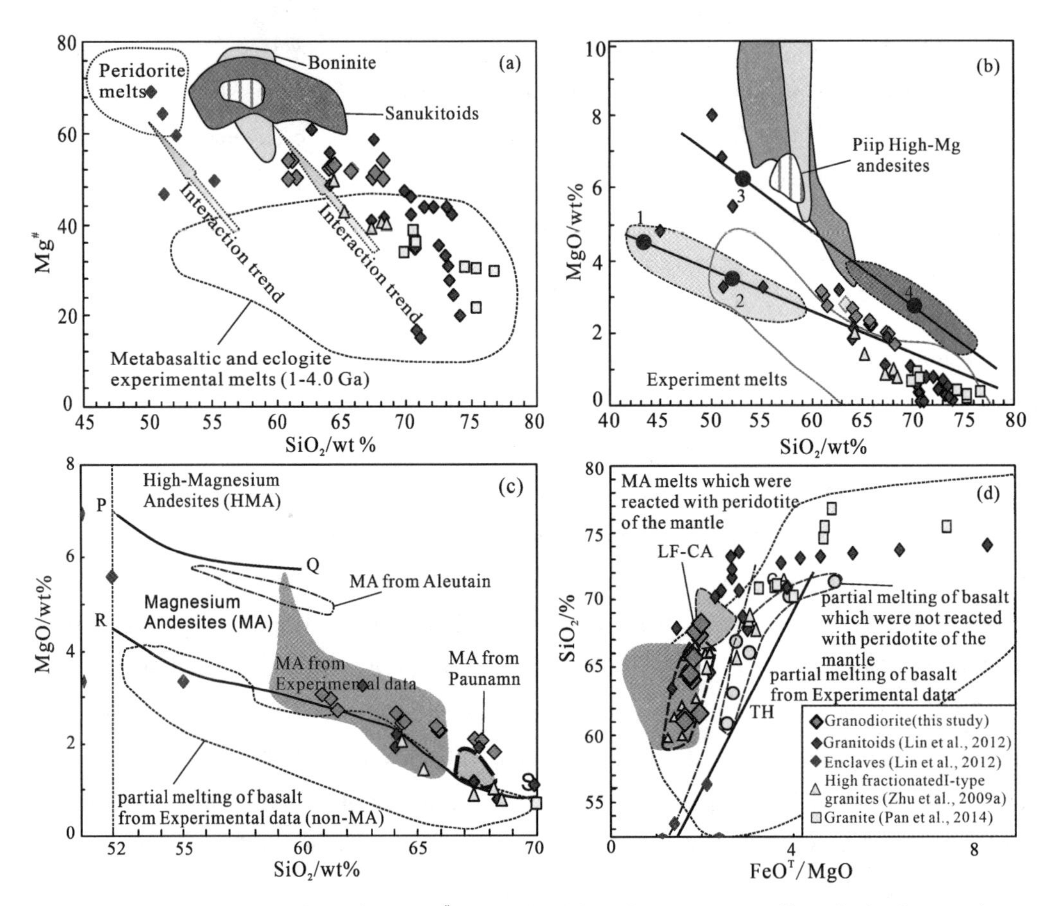

Fig. 9　(a) Plots of SiO_2 (wt%) vs. $Mg^{\#}$. Marked fields outline experimentally melts by Rapp and Watson (1995, 1999). (b) MgO (wt%) vs. SiO_2 (wt%) systematic diagram after Moyen and Martin (2012). Trend (A) is a differentiation trend (irrespective of the actual process generating it). Sample (1) has higher MgO than sample (2), Interactions of magma (2) with the mantle will shift it to position (3); from this point onwards, the magma will define a new trend (B). Yet samples (3) and (4) have experienced interactions with the mantle. If the two yellow and green fields represent two sample sets, it is not immediately intuitive that the mantle-contaminated set is group (B); Experimental Liquids after Martin and Moyen (2002). (c) SiO_2-MgO and SiO_2-FeO^T/MgO. (d) for the granitoids in the northeastern Lhasa terrane.

The lines PQ and RS mark the boundary of HMA/MA and MA/non-MA respectively (Deng et al., 2009, 2010). The dashed line means SiO_2 = 52 wt%. The area of MA, partial melting of basalt from experimental data and MA melts which were reacted with peridotite of mantle referenced to (Carroll and Wyllie, 1989; Rapp et al., 1999; Prouteau et al., 2001). Symbols as Fig. 4.

5.3　Early Cretaceous magmatic source in the northeastern Lhasa terrane

Primarily, we need to know sources and petrogenetic mechanisms of the Early Cretaceous intensive magmatic event which was in the central and northern Lhasa terrane, which was closely associated with the closure and southward subduction of Bangong-Nujiang Tethyan ocean (Zhu et al., 2009b, 2011, 2016, 2019; Li et al., 2018). As shown in Table 1: in the northern Lhasa case, there are abundant granitoids, associated mafic enclaves, basalt-andesite-dacite-rhyolites, and trachyandesites. These volcanic rocks show significant positive zircon $\varepsilon_{Hf}(t)$ values (+2.8 to +18.8, but a few was slightly negative), low whole-rock initial $^{87}Sr/^{86}Sr$ ratios (0.704 2–0.706 2) and associated whole-rock $\varepsilon_{Hf}(t)$ values range from −0.6 to +3.6, which indicate depleted mantle-derived sources and little assimilation from evolved crustal materials. Meanwhile, the zircon Hf isotopes of granitoids are highly variable (−13.5 to +16.9), which implies significant interaction between mantle and crustal components (Zhu et al., 2011, 2016; Sui et al., 2013; Wang et al., 2014; Chen et al., 2017; Hu et al., 2017; Fan et al., 2017). In the central Lhasa case, most of the granitoids and andesite-dacite-rhyolite associations display evolved zircon Hf [$\varepsilon_{Hf}(t)$ values = −14.4 to +4.6] and whole-rock Sr-Nd isotopic characteristics [high initial $^{87}Sr/^{86}Sr$ = 0.707 1–0.721 8; $\varepsilon_{Nd}(t)$ values = −13.7 to +0.6], with a lack of mantle-derived basalts, and have therefore been regarded as the products of melting of ancient continental crustal materials with mantle-derived magmas providing enough heat and limited materials (Meng et al., 2010; Zhu et al., 2009b, 2011; Chen et al., 2014; Cao et al., 2016). Chiu et al. (2009) combined in-situ zircon U-Pb and Lu-Hf isotopic compositions for the so-called eastern Trans-Himalayan batholiths, which were emplaced in the northeastern Lhasa terrane between Jiali fault and Bangong-Nujiang suture. The results suggested that (Fig. 10a,b): ① Zircon Hf compositions of these Early Cretaceous granitoids are also dominated by negative but highly variable $\varepsilon_{Hf}(t)$ values range from +3.72 to −27.3. These may indicate a multi-component source with inefficient mixing of components rather than a closed and single source (Kemp et al., 2007); And ② zircon Hf isotopic composition of granitoids near the Bomi-Ranwu area display positive $\varepsilon_{Hf}(t)$ values (up to +3.72). Therefore Chiu et al. (2009) suggested that these granitoids are different from those in the Gangdese batholith in the southern Lhasa terrane, which are dominated by depleted, juvenile material, but are similar to the granitoids in the northern Lhasa terrane of the northern magmatic belt, which have significant contributions from ancient continental crust (Chu et al., 2006; Zhu et al., 2009a, 2011). Abundant whole-rock Sr-Nd isotopic data of Early Cretaceous granitoids have also been reported in this area (Fig. 6a and Fig. 10c,d) (Zhu et al., 2009a; Lin et al., 2012; Pan et al., 2014), with results that indicate their source was evolved and ancient continental crust, close to lower crust, rather than depleted and juvenile. However, we cannot ignore the coeval mafic enclaves and basaltic to dacitic volcanic rocks that have higher and even to positive $\varepsilon_{Nd}(t)$ values (up to +3) and lower initial $^{87}Sr/^{86}Sr$ values (0.705–0.707). These were regarded as the products of deep level differentiation of the mantle-wedge-derived mafic magmas associated with assimilation by the lower continental crust of the Lhasa terrane (Lin et al., 2012). All the above whole-rock Sr-Nd data is consistent with

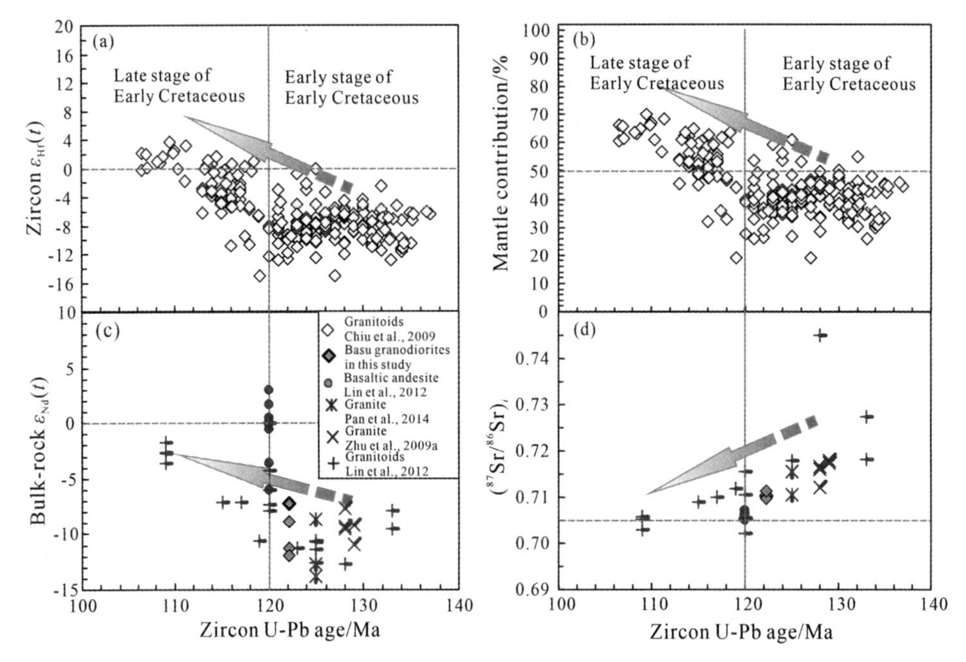

Fig. 10 Zircon $\varepsilon_{Hf}(t)$, Mantle contribution (%), Bulk-rock $\varepsilon_{Nd}(t)$, and Bulk-rock initial $^{87}Sr/^{86}Sr$ vs. Zircon $^{206}Pb/^{238}U$ data (Ma) for the granitoids in northeastern Lhasa terrane, SW China. The data for the Early Cretaceous magmatism in the northeastern Lhasa terrane was referenced in Table 1. Mantle contribution (%) calculated by Zhu et al. (2011). The yellow symbols are samples from Chiu et al. (2009) and other symbols as Figs. 4 and 6. (For interpretation of the references to color in this figure legend, the reader in referred to the web version of this article.)

the conclusion from whole-rock Pb and Hf isotopic data in this study (Figs. 6b and 7). In the view of these datasets of zircon Hf and whole-rock Sr-Nd, together with emplacement age, we propose an isotopic evolutionary trend along the age from ca. 140 Ma to 100 Ma (Fig. 10), as follows: ① The contribution from ancient and evolved continental source dominated in the source of Early Cretaceous intensive magmatism in the northeastern Lhasa terrane. A few positive zircon $\varepsilon_{Hf}(t)$ and whole-rock $\varepsilon_{Nd}(t)$ values indicate that depleted or juvenile materials also play a role. ②From 140 Ma to 110 Ma, the zircon $\varepsilon_{Hf}(t)$ values changed from completely negative to more positive components. While whole-rock $\varepsilon_{Nd}(t)$ values were changing from −14 to −1.5 (except coeval volcanic rocks can up to +3), the initial $^{87}Sr/^{86}Sr$ was also changing from higher (ca. 0.744 9) to lower (ca. 0.702 2). And ③ from ca. 120 Ma, some positive zircon $\varepsilon_{Hf}(t)$ and whole-rock $\varepsilon_{Nd}(t)$ values occurred, and the initial $^{87}Sr/^{86}Sr$ values fell to lower than 0.705. The above discussion implies that the degree of enrichment and evolved isotopic components were gradually decreasing. This evolutionary trend may be associate with the development of basaltic to dacitic volcanic rocks and mafic enclaves, which represent materials from depleted and juvenile mantle sources (Lin et al., 2012). In addition, Zhu et al. (2009a) suggested that heterogeneous zircon Hf isotopic data and bulk-rock zircon saturation temperature indicated that mantle-derived materials likely played a role in the generation. In summary, we conclude that an evolved and ancient continental crust source dominated the genesis of Early Cretaceous magmatism in the northeastern Lhasa terrane, but mantle-derived materials also played a necessary role. Both of these Early Cretaceous intensively magmatism in the central and northern Lhasa and Trans-Himalayan batholith were products of closure and southward subduction of Bangong-Nujiang Tethyan ocean (Fig. 11), those magmatism in the central Lhasa were formed by partial melting of ancient continental crust (Zhu et al., 2009b; Cao et al., 2016), those in the northern Lhasa were represented the highly variable interaction of mantle-crustal components (Zhu et al., 2011) with the Basu high-$Mg^{\#}$ granodiorites are products of interaction of mantle-crustal materials at depth (in this study).

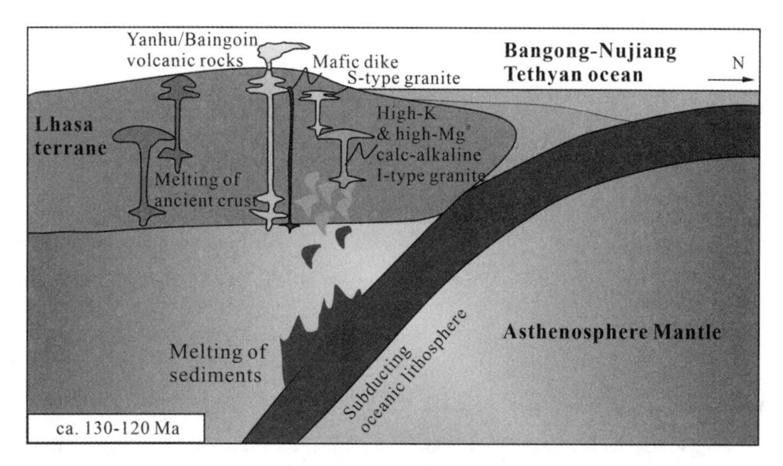

Fig. 11 Geodynamic and petrogenetic models for the Early Cretaceous magmatism in the central and northern Lhasa terrane (modified after Zhu et al., 2019).

6 Conclusions

(1) LA-ICP-MS zircon U-Pb data indicate that the Basu granodiorite in Zhaxize batholith was emplaced ca. 122.1 ± 1.0 Ma. Together with dataset from literature, we suggest that there was intensive granitic magmatism in the Early Cretaceous range from 133 Ma to 113 Ma in the northeastern Lhasa terrane between the Jiali fault and Bangong-Nujiang suture zone, similar to those in the central and northern Lhasa terrane.

(2) The Early Cretaceous Basu granodiorites are high-K calc-alkaline, metaluminous and high-$Mg^{\#}$(49.7–53.7), similar to typical MA. Combining their evolved whole-rock Sr-Nd-Pb-Hf isotopic components, we infer that then were products of interaction between partial melting of ancient basaltic lower crust and mantle-derived peridotites.

(3) Based on the dataset of in-situ zircon Lu-Hf isotopes, whole-rock Sr-Nd in the literature and relationship with zircon U-Pb age, we also conclude that: ① The degree of

enrichment and evolved isotopic component gradually decreased from strongly evolved to relative moderate, in the temporal range from ca.140 Ma to 100 Ma. And ② the evolved and ancient continental crust source dominated in the Early Cretaceous magmatism in the northeastern Lhasa terrane, but the mantle-derived materials also play a necessary role in the generation.

Acknowledgements Thanks so much for the help and constructive comments from Chief editor Prof. Mei-Fu Zhou and handled editor Prof. Derek Wyman. Thanks for the English polishing from Prof. Mike Fowler. This work was jointly supported by the National Natural Science Foundation of China (Grant Nos. 41802054, 41421002, and 41190072) and China Postdoctoral Science Foundation Grant. 2018M643713. Support was also provided by the MOST Special Fund from the State Key Laboratory of Continental Dynamics, Northwest University.

References

Andersen, T., 2002. Correction of common lead in U-Pb analyses that do not report ^{204}Pb. Chemical Geology, 192, 59-79.

Bao, Z., Chen, L., Zong, C., Yuan, H., Chen, K., Dai, M., 2017. Development of pressed sulfide powder tablets for in situ, Sulfur and Lead isotope measurement using LA-MC-ICP-MS. International Journal of Mass Spectrometry, 421.

Behn, M.D., Kelemen, P.B., Hirth, G., Hacker, B.R., Massonne, H.J., 2011. Diapirs as the source of the sediment signature in arc lavas. Nature Geosci, 4 (9), 641-646.

Booth, A.L., Zeitler, P.K., Kidd, W.S.F., Wooden, J., Liu, Y., Idleman, B., Hren, M., Chamberlain, C. P., 2004. U-Pb zircon constraints on tectonic evolution of southeastern Tibet, Namche Barwa area. Am. J. Sci., 304, 889-929.

Carroll, M.R., Wyllie, P.J., 1989. Experimental Phase Relations in the System Tonalite-Peridotite-H_2O at 15kb: Implications for Assimilation and Differentiation Processes near the Crust-Mantle Boundary. Journal of Petrology, 30(6), 1351-1382.

Cao, M.J., Qin, K.Z., Li, G.M., Li, J.X., Zhao, J.X., Evans, N.J., et al., 2016. Tectonomagmatic evolution of Late Jurassic to Early Cretaceous granitoids in the west Central Lhasa subterrane, Tibet. Gondwana Res., 39, 386-400.

Chu, M.F., Chung, S.L., Song, B., Liu, D.Y., O'Reilly, S.Y., Pearson, N.J., Ji, J.Q., Wen, D.J., 2006. Zircon U-Pb and Hf isotope constraints on the Mesozoic tectonics and crustal evolution of southern Tibet. Geology, 34, 745-748.

Champion, D.C., Bultitude, R.J., 2013. The geochemical and Sr, Nd isotopic characteristics of Paleozoic fractionated S-types granites of north Queensland: Implications for S-type granite petrogenesis. Lithos, 162-163(2), 37-56.

Chappell, B.W., White, A.J.R., 2001. Two contrasting granite types: 25 years later. Journal of the Geological Society of Australia, 48(4), 489-499.

Chen, J.L., Xua, J.F., Yua, H.X., et al., 2015. Late Cretaceous high-$Mg^{\#}$ granitoids in southern Tibet:

Implications for the early crustal thickening and tectonic evolution of the Tibetan Plateau? Lithos, 232, 12-22.

Chen, S.S., Shi, R.D., Gong, X.H., Liu, D.L., Huang, Q.S., Yi, G.D., Wu, K., Zou, H.B., 2017. A syn-collisional model for Early Cretaceous magmatism in the northern and central Lhasa subterranes. Gondwana Research, 41, 93-109.

Chen, Y., Zhu, D. C., Zhao, Z. D., Meng, F.Y., Wang, Q., Santosh, M., Wang, L.Q., Dong, G.C., Mo, X.X., 2014. Slab breakoff triggered ca. 113Ma magmatism around Xainza area of the Lhasa terrane, Tibet. Gondwana Research, 26(2), 449-463.

Chiu, H.Y., Chung, S.L., Wu, F.Y., Liu, D.Y., Liang, Y.H., Lin, I.J., Lizuka, Y., Xie, L.W., Wang, Y.B., Chu, M.F., 2009. Zircon U-Pb and Hf isotopic constraints from eastern Transhimalayan batholiths on the pre-collisional magmatic and tectonic evolution in southern Tibet. Tectonophysics, 477, 3-19.

Chu, Z.Y., Chen, F.K., Yang, Y.H., Guo, J.H., 2009. Precise determination of Sm, Nd concentrations and Nd isotopic compositions at the nanogram level in geological samples by thermal ionization mass spectrometry. Journal of Analytical Atomic Spectrometry, 24, 1534-1544.

Coulon, C., Maluski, H., Bollinger, C., Wang, S., 1986. Mesozoic and Cenozoic volcanic rocks from central and southern Tibet: ^{39}Ar-^{40}Ar dating, petrological characteristics and geodynamical significance. Earth Planetary Sci. Lett., 79, 281-302.

Deng, J. F., Cui, L., Feng, Y. F., et al., 2010. High magnesian andesitic/dioritic rocks (HMA) and magnesian andesitic/dioritic rocks(MA): Two igneous rock types related to oceanic subduction. Geology in China, 37(4), 1112-1118.

Deng, J.F., Flower, M.F.J., Liu, C., et al., 2009. Nomeuclature, diagnosis and origin of high-magnesian andesites (HMA) and magnesian andesites (MA): A review from petrographic and experimental data. Geochimica et Cosmochimica Acta, 73(13), A279.

Dobosi, G., Kempton, P.D., Downes, H., Embey-Isztin, A., Thirlwall, M., Greenwood, P., 2003. Lower crustal granulite xenoliths from the Pannonian Basin, Hungary, Part 2: Sr-Nd-Pb-Hf and O isotope evidence for formation of continental lower crust by tectonic emplacement of oceanic crust. Contributions to Mineralogy and Petrology, 144(6), 671-683.

Fan, J.J., Li, C., Sun, Z.M., Xu, W., Wang, M., Xie, C.M., 2017. Early cretaceous MORB-type basalt and A-type rhyolite in northern Tibet: evidence for ridge subduction in the Bangong-Nujiang Tethyan ocean. Journal of Asian Earth Sciences, 154, 187-201.

Frost, B.R., Barnes, C.G., Collins, W.J., Arculus, R.J., Ellis, D.J., Frost, C.D., 2001. A geochemical classification for granitic rocks. Journal of Petrology, 42(11), 2033-2048.

Gao, S., Rudnick, R.L., Yuan, H.L., Liu, X.M., Liu, Y.S., Xu, W.L., Ling, W.L., Ayers, J., Wang, X.C., Wang, Q.H., 2004. Recycling lower continental crust in the North China craton. Nature, 432, 892-897.

Guynn, J.H, Kapp, P., Pullen, A., et al., 2006. Tibetan basement rocks near Amdo reveal "missing" Mesozoic tectonism along the Bangong suture, central Tibet. Geology, 34, 505-508.

Harris, N.B.W., Inger, S., Xu, R., 1990. Cretaceous plutonism in Central Tibet: An example of post-collision magmatism? J. Volcanol. Geotherm. Res., 44, 21-32.

Hu, P.Y., Zhai, Q.G., Jahn, B.M., Wang, J., Li, C., Chung, S.L., et al., 2017. Late Early Cretaceous magmatic rocks (118 − 113 Ma) in the middle segment of the Bangong-Nujiang suture zone, Tibetan Plateau: Evidence of lithospheric delamination. Gondwana Research, 44, 116-138.

Ji, W.Q., Wu, F.Y., Chung, S.L., Li, J. X., Liu, C.Z., 2009. Zircon U-Pb geochronology and Hf isotopic constraints on petrogenesis of the Gangdese batholith, southern Tibet. Chemical Geology, 262 (3), 229-245.

Kapp, J.L.D., Harrison, T. M., Grove, M., Lovera, O.M., Lin, D., 2005a. Nyainqentanglha Shan: A window into the tectonic, thermal, and geochemical evolution of the Lhasa block, southern Tibet. Journal of Geophysical Research: Solid Earth (1978—2012), 110(B8).

Kapp, P., Yin, A., Harrison, T.M., Ding, L., 2005b. Cretaceous-Tertiary shortening, basin development, and volcanism in central Tibet. Geological Society of America Bulletin, 117, 865-878.

Kemp, A.I.S., Hawkesworth, C.J., Foster, G.L., Paterson, B.A., Woodhead, J.D., Hergt, J.M., Whitehouse, M.J., 2007. Magmatic and crustal differentiation history of granitic rocks from Hf-O isotopes in zircon. Science, 315(5814), 980-983.

Kempton, P.D., McGill, R., 2002. Procedures for the analysis of common lead at the NERC Isotope Geosciences Laboratory and an assessment of data quality. NIGL Rep Ser, 178.

Lameyre, J., Bowden, P., 1982. Plutonic rock types series: Discrimination of various granitoid series and related rocks. Journal of Volcanology and Geothermal Research, 14(1-2), 169-186.

Lee, H.Y., Chung, S.L., Lo, C.H., Ji, J., Lee, T.Y., Qian, Q., Zhang, Q., 2009. Eocene Neo-Tethyan slab breakoff in southern Tibet inferred from the Linzizong volcanic record. Tectonophysics, 477, 20-35.

Li, S.M., Zhu, D.C., Wang, Q., Zhao, Z., Zhang, L.L., Liu, S.A., et al., 2016. Slab-derived adakites and subslab asthenosphere-derived OIB-type rocks at 156 ± 2 Ma from the north of Gerze, central Tibet: Records of the Bangong-Nujiang oceanic ridge subduction during the Late Jurassic. Lithos, 262, 456-469, 831.

Li, S.M., Wang, Q., Zhu, D.C., Stern, R. J., Cawood, P. A., Sui, Q.L., Zhao, Z., 2018. One or two Early Cretaceous arc systems in the Lhasa Terrane, southernTibet. Journal of Geophysical Research: Solid Earth, 123, 3391-3413.

Liu, D., Shi, R., Ding, L., Huang, Q., Zhang, X., Yue, Y., et al., 2017. Zircon U-Pb age and Hf isotopic compositions of Mesozoic granitoids in southern Qiangtang, Tibet: Implications for the subduction of the Bangong-Nujiang Tethyan ocean. Gondwana Research, 41, 157-172.

Lin, I.J., Chung, S.L., Chu, C.H., Lee, H.Y., Gallet, S., Wu, G., Ji, J., Zhang, Y., 2012. Geochemical and Sr-Nd isotopic characteristics of Cretaceous to Paleocene granitoids and volcanic rocks, SE Tibet: Petrogenesis and tectonic implications. Journal of Asian Earth Sciences, 53, 131-150.

Lopez, R., Cameron, K.L., 1997. High-Mg andesites from the Gila Bend Mountains, southwestern Arizona: Evidence for hydrous melting of lithosphere during Miocene extension. Geological Society of America Bulletin, 109(7), 900-914.

Ludwig, K.R., 2003. ISOPLOT 3.0: A geochronological toolkit for Microsoft Excel. Berkeley Geochronology Center, Special Publication, 4.

Martin, H., Moyen, J.F., 2002. Secular changes in TTG composition as markers of the progressive cooling of the Earth. Geology, 30, 319-322.

Martin, H., Smithies, R.H., Rapp, R., Moyen, J.F., Champion, D., 2005. An overview of adakite, tonalite-trondhjemite-granodiorite (TTG) and sanukitoid: Relationships and some implications for crustal evolution. Lithos, 79, 1-24.

Meng, F.Y., Zhao, Z.D., Zhu, D.C., Zhang, L.L., Guan, Q., Liu, M., Yu, F., Mo, X.X., 2010. Petrogenesis of Late Cretaceous adakite-like rocks in Mamba from the eastern Gangdese, Tibet. Acta Petrol. Sinica, 26, 2180-2192 (in Chinese with English abstract).

Miller, C., Schuster, R., Kotzli, U., et al., 1999. Post-collisional potassic and ultrapotassic magmatism in SW Tibet: Geochemical and Sr-Nd-Pb-O isotopic constraints for mantle source characteristics and petrogenesis. J. Petrol., 40, 1399-1424.

Mo, X.X., Dong, G., Zhao, Z., Zhou, S., Wang, L., Qiu, R., Zhang, F., 2005. Spatial and temporal distribution and characteristics of granitoids in the Gangdese, Tibet and implication for crustal growth and evolution. Geology Journal of Chinese University, 11, 281-190.

Mo, X.X., Hou, Z., Niu, Y., Dong, G., Qu, X., Zhao, Z., Yang, Z., 2007. Mantle contributions to crustal thickening during continental collision: evidence from Cenozoic igneous rocks in southern Tibet. Lithos, 96, 225-242.

Moyen, J.F., Martin, H., 2012. Forty years of TTG research. Lithos, 148(148), 312-336.

Pan, F.B., Zhang, H.F., Xu, W.C., Guo, L., Wang, S., Luo, B.J., 2014. U-Pb zircon chronology, geochemical and Sr-Nd isotopic composition of Mesozoic-Cenozoic granitoids in the SE Lhasa terrane: Petrogenesis and tectonic implications. Lithos, 192-195(4), 142-157.

Pan, G., Ding, J., Yao, D., Wang, L., chief compilers, 2004. Guidebook of 1∶1 500 000 geologic map of the Qinghai-Xizang (Tibet) plateau and adjacent areas. Chengdu: Chengdu Cartographic Publishing House, 48.

Prouteau, G., Scaillet, B., Pichavant, M., Maury, R., 2001. Evidence for mantle metasomatism by hydrous silicic melts derived from subducted oceanic crust. Nature, 410(6825), 197-200.

Qi, L., Hu J., Gregoire, D.C., 2000. Determination of trace elements in granites by inductively coupled plasma mass spectrometry. Talanta, 51, 507-513.

Qian, Q., Hermann, J., 2010. Formation of high-Mg diorites through assimilation of peridotite by monzodiorite magma at crustal depths. Journal of Petrology, 51(7), 1381-1416(36).

Qin, J.F., Lai, S.C., Grapes, R., Diwu, C.R., Ju, Y.J., Li, Y.F., 2010. Origin of late Triassic high-Mg adakitic granitoid rocks from the Dongjiangkou area, Qinling orogen, central China: Implications for subduction of continental crust. Lithos, 120(3-4), 347-367.

Qu, X.M., Wang, R.J., Xin, H.B., Jiang, J.H., Chen, H., 2012.Age and petrogenesis of A-type granites in the middle segment of the Bangonghu-Nujiang suture, Tibetan plateau. Lithos, 146, 264-275.

Rapp, R.P., Shimizu, N., Norman, M.D., Applegate, G.S., 1999. Reaction between slab-derived melts and peridotite in the mantle wedge: Experimental constraints at 3.8GPa. Chemical Geology, 160, 335-356.

Rapp, R.P., Watson, E.B., 1995. Dehydration melting of metabasalt at 8 – 32 kbar: Implications for continental growth and crust-mantle recycling. Journal of Petrology, 36, 891-931.

Roberts, M.P., Clemens, J.D., 1993. Origin of high-potassium, talc-alkaline, I-type granitoids. Geology, 21,

825-828.

Rudnick, R.L., Gao, S., 2003. Composition of the continental crust. Treatise Geochem., 3, 1-64.

Searle, M.P., Windley, B.F., Coward, M.P., Cooper, D.J.W., Rex, D., Li, T., Xiao, X., Jan, M.Q., Thakur, V.C., Kumar, S., 1987. The closing of Tethys and the tectonics of Himalaya. Geol. Soc. Amer. Bull., 98, 678-701.

Shimoda, G., Tatsumi, Y., Nohda, S., Ishizaka, K., Jahn, B.M., 1998. Setouchi high-Mg andesites revisited: Geochemical evidence for melting of subducting sediments. Earth & Planetary Science Letters, 160 (3-4), 479-492.

Stern, R.A., Hanson, G.N., Shirey, S.B., 1989. Petrogenesis of mantle-derived, LILE-enriched Archean monzodiorites and trachyandesites (sanukitoids) in southwestern Superior Province. Canadian Journal of Earth Science, 26, 1688-1712.

Stevens, G., Villaros, A., Moyen, J.F., 2007. Selective peritectic garnet entrainment as the origin of geochemical diversity in S-type granites. Geology, 35, 9-12.

Streckeisen, A., 1975. To each plutonic rock its proper name. Earth-Science Reviews, 12, 1-33.

Sui, Q.L., Wang, Q., Zhu, D.C., Zhao, Z.D., Chen, Y., Santosh, M., Hu, Z.C., Yuan, H.L., Mo, X.X., 2013. Compositional diversity of ca. 110 Ma magmatism in the northern Lhasa Terrane, Tibet: Implications for the magmatic origin and crustal growth in a continent-continent collision zone. Lithos, 168-169(3), 144-159.

Sun, S.S., McDonough, W.F., 1989. Chemical and isotopic systematics of oceanic basalts: Implications or mantle composition and processes. Geological Society, London, Special Publications, 42(1), 313-345.

Vervoort, J.D., Blichert-Toft, J., 1999. Evolution of the depleted mantle: Hf isotope evidence from juvenile rocks through time. Geochimica et Cosmochimica Acta, 63(3-4), 533-556.

Wang, Q., Li, Z.X., Chung, S.L., Wyman, D.A., Sun, Y.L., Zhao, Z.H., Zhu, Y.T., Qiu, H.N., 2011. Late triassic high-Mg andesite/dacite suites from northern Hohxil, north Tibet: Geochronology, geochemical characteristics, petrogenetic processes and tectonic implications. Lithos, 126(1-2), 54-67.

Wang, Q., Zhu, D.C., Zhao, Z.D., Liu, S.A., Chung, S.L., Li, S.M., et al., 2014. Origin of the ca. 90 Ma magnesia-rich volcanic rocks in se nyima, central Tibet: Products of lithospheric delamination beneath the Lhasa-Qiangtang collision zone. Lithos, 198-199, 24-37.

Wen, D.R., Liu, D., Chung, S.L., Chu, M.F., Ji, J., Zhang, Q., Song, B., Lee, T.Y., Yeh, M.W., Lo, C.H., 2008. Zircon SHRIMP U-Pb ages of the Gangdese Batholith and implications for Neotethyan subduction in southern Tibet. Chemical Geology, 252, 191-201.

Xu, R., Scharer, U., Allègre, C.J., 1985. Magmatism and metamorphism in the Lhasa Block (Tibet): A geochronological study. Journal of Geology, 93, 41-57.

Yin, A., Harrison, T.M., 2000. Geologic evolution of the Himalayan-Tibetan orogen. Ann. Rev. Earth Plan. Sci., 28(1), 221-280.

Yuan, H.L., Gao, S., Luo, Y., Zong, C.L., Dai, M.N., Liu, X.M., Diwu, C.R., 2007. Study of Lu-Hf geochronology: A case study of eclogite from Dabie UHP Belt. Acta Petrologica Sinica, 23 (2), 233-239 (in Chinese with English abstract).

Zartman, R.E., Doe, B.R., 1981. Plumbotectonics: The model. Tectonophysics, 75, 135-162.

Zeng, Y.C., Xu, J.F., Chen, J.L., Wang, B.D., Kang, Z.Q., Huang, F., 2017. Geochronological and geochemical constraints on the origin of the Yunzhug ophiolite in the Shiquanhe-Yunzhug-Namu Tso ophiolite belt, Lhasa terrane, Tibetan plateau. Lithos, 300-301, 250-260.

Zhu, D., Mo, X., Wang, L., Zhao, Z., Niu, Y., Zhou, C., Yang, Y., 2009a. Petrogenesis of highly fractionated I-type granites in the Zayu area of eastern Gangdese, Tibet: Constraints from zircon U-Pb geochronology, geochemistry and Sr-Nd-Hf isotopes. Science in China: Series D, Earth Sciences, 52(9), 1223-1239.

Zhu, D.C., Li, S.M., Cawood, P.A., Wang, Q., Zhao, Z.D., Liu, S.A., et al., 2016. Assembly of the Lhasa and Qiangtang terranes in central tibet by divergent double subduction. Lithos, 245, 7-17.

Zhu, D.C., Mo, X.X., Niu, Y.L., Zhao, Z.D., Wang, L.Q., Liu, Y.S., Wu, F.Y., 2009b. Geochemical investigation of Early Cretaceous igneous rocks along an east-west traverse throughout the central Lhasa Terrane, Tibet. Chemical Geology, 268, 298-312.

Zhu, D.C., Zhao, Z.D., Niu, Y.L., Dilek, Y., Hou, Z.Q., Mo, X.X., 2013. The origin and pre-Cenozoic evolution of the Tibetan Plateau. Gondwana Research, 23, 1429-1454.

Zhu, D.C., Zhao, Z.D., Niu, Y.L., Mo, X.X., Chung, S.L., Hou, Z.Q., Wang, L.Q., Wu, F.Y., 2011. The Lhasa Terrane: Record of a microcontinent and its histories of drift and growth. Earth & Planetary Science Letters, 301, 241-255.

Zhu, R.Z., Lai, S.C., Santosh, M., Qin, J.F., Zhao, S.W., 2017. Early Cretaceous Na-rich granitoids and their enclaves in the Tengchong Block, SW China: Magmatism in relation to subduction of the Bangong-Nujiang Tethys ocean. Lithos, 286-287, 175-190.

Zhu, R.Z., Lai, S.C., Qin, J.F., Zhao, S.W., Santosh, M., 2019. Petrogenesis of high-K calc-alkaline granodiorite and its enclaves from the SE Lhasa block, Tibet (SW China): Implications for recycled subducted sediments. Geological Society of America Bulletin in press. https://doi.org/10.1130/B1841.1.

Zhu, R.Z., Lai, S.C., Qin, J.F., Zhao, S.W., Santosh, M., 2018. Strongly peraluminous fractionated S-type granites in the Baoshan Block, SW China: Implications for two-stage melting of fertile continental materials following the closure of Bangong-Nujiang Tethys. Lithos, 316-317, 178-198.

Zhu, R.Z., Lai, S.C., Qin, J.F., Zhao, S.W., 2015. Early Cretaceous highly fractionated I-type granites from the northern Tengchong block, western Yunnan, SW China: Petrogenesis and tectonic implications. Journal of Asian Earth Sciences, 145-163.

Three stages of Early Paleozoic magmatism in the Tibetan-Himalayan orogen: New insights into the final Gondwana assembly①

Lai Shaocong　Zhu Renzhi

Abstract: Earth's evolution involves deep, hot rocks rising upward by convection to near-surface environments, as well as the unique scenarios such as the formation and breakup of Gondwana and Pangea. These processes have been well recorded by the widely distributed magmatism. To understand the tectonic evolution and Gondwana assembly, we compiled Early Paleozoic zircon U-Pb ages (n = 67), in-situ zircon Hf isotopes (n = 1 011), and bulk-rock elemental compositions (n = 293) and Sr-Nd isotopes of magmatic rocks in the Tibetan-Himalayan orogen. Three stages of Paleozoic magmatism were identified here, including early (>490 Ma) to middle (490–470 Ma) to late (<470 Ma) stages, in response to the final assembly of the Gondwana supercontinent. Early-stage magmatic rocks are characterized by highly variable SiO_2 (48.0–80.0 wt%) and MgO (0.02–9.65 wt%) contents with $Mg^{\#}$ range from 5.4 to 78, K_2O/Na_2O ratios ranging from 0.14 to 2.81, and whole-rock Sr-Nd ($(^{87}Sr/^{86}Sr)_i$ (0.703 5–0.734 0), $\varepsilon_{Nd}(t)$ (−9.5 to +1.0) and zircon Hf [$\varepsilon_{Hf}(t)$ = −15 to +8.0] isotopic compositions with significant mantle contributions; they could be generated in an Andean-type arc setting along the active northern continental Gondwana margin. Middle-stage magmas were dominated by fertile continental crustal signatures (SiO_2 > 70 wt%, A/CNK > 1.10, MgO < 2.69 wt%) in a crustal thickening setting. Late-stage magmas also have highly variable Si, Mg and isotopic components, with coeval mantle-derived magmas developing in an extensive setting. Importantly, the widespread presence of early Cambrian to late Ordovician peraluminous high-K calc-alkaline magmatism in the Tibetan-Himalayan orogen indicates that the reworking of ancient continental crustal materials played a key role in reconstructing and stabilizing the final Gondwana assembly.

1 Introduction

A fascinating aspect of Earth's evolution is not only that deep, hot rocks rise upward, by convection to near-surface environments but also the unique spectacle of the multistage assembly of supercontinents (e.g., Gondwana) and associated Wilson cycle-type convergence

① Published in *Journal of Asian Earth Sciences*, 2021, 221.

(Veevers, 2004; Cawood et al., 2007). In the Tibetan-Himalayan orogen, a series of microcontinents and/or terranes and blocks south of the Shuanghu suture were inferred to have dispersed from the northern Gondwana margin, drifting northward and successively accreted to the Eurasia continent since the Paleozoic (Zhang et al., 2007, 2012; Zhang and Tang, 2009; Zhu et al., 2012; Metcalfe, 2013). These processes well recorded the underlying cause of a wide range of associated magmatic processes. They have, in turn, promoted the overall differentiation of the planet, the growth of oceanic crust, reworking of continental crust, and even the formation of an atmosphere. Therefore, Early Paleozoic magmatism in the Tibetan-Himalayan orogen has been regarded as a key to understanding the final assembly of the Gondwana supercontinent and the initial subduction along the margin of the *peri*-Gondwana assembly (Gehrels et al., 2006; Cawood and Buchan, 2007; Cawood et al., 2007; Zhu et al., 2012; Wang et al., 2013; Li et al., 2016; Gao et al., 2019; Liu et al., 2019). However, the nature of Early Paleozoic magmatism from the early Cambrian to late Ordovician has hotly debated, with proposals ranging from an Andean-type active continental margin (Cawood et al., 2007; Zhu et al., 2012; Hu et al., 2015), to the collision and subsequently post-collision and extension (Miller et al., 2001; Visonà et al., 2010; Gao et al., 2019), to lithospheric thickening and delamination (Li et al., 2016; Zhao et al., 2017), and even to the rift events (Murphy and Nance, 1991; Liu et al., 2019). Most importantly, if there are different phases of Early Paleozoic magmatism, what contribution and mechanism could generate and stabilize the final Gondwana assembly? Understanding can be achieved by recoding the relationships between geological, geochronological, geochemical and tectonic data and the Gondwana assembly process (Cawood et al., 2007; Liu et al., 2019).

In this study, to help us better understand the temporal process of the final assembly of Gondwana, we documented and compiled zircon U-Pb geochronological, whole-rock geochemical and zircon Hf isotopic data from the early Cambrian to late Ordovician magmatism in the Tibetan-Himalayan orogen, including Himalaya, Lhasa, South Qiangtang and its southeastern extension of Tengchong-Baoshan terrane. Three stages of distinctive early Paleozoic magmatism were identified: $a > 490$ Ma calc-alkaline to high-K calc-alkaline suite with significant mantle contribution, $a = 490 - 470$ Ma strongly peraluminous high-K calc-alkaline suite without any mantle inputs, and $a < 470$ Ma low-K calc-alkaline to high-K calc-alkaline suite associated with upwelling mantle inputs (Fig. 1). These stages are consistent with geological and tectonic records and evidence the response to the final evolutional processes from an Andean arc setting along the active continental margin, to collision and crustal thickening, and to slab break-off in an extensive setting.

2 Geological background

The Tibetan-Himalayan orogen consists of a series of allochthonous Gondwanan-derived

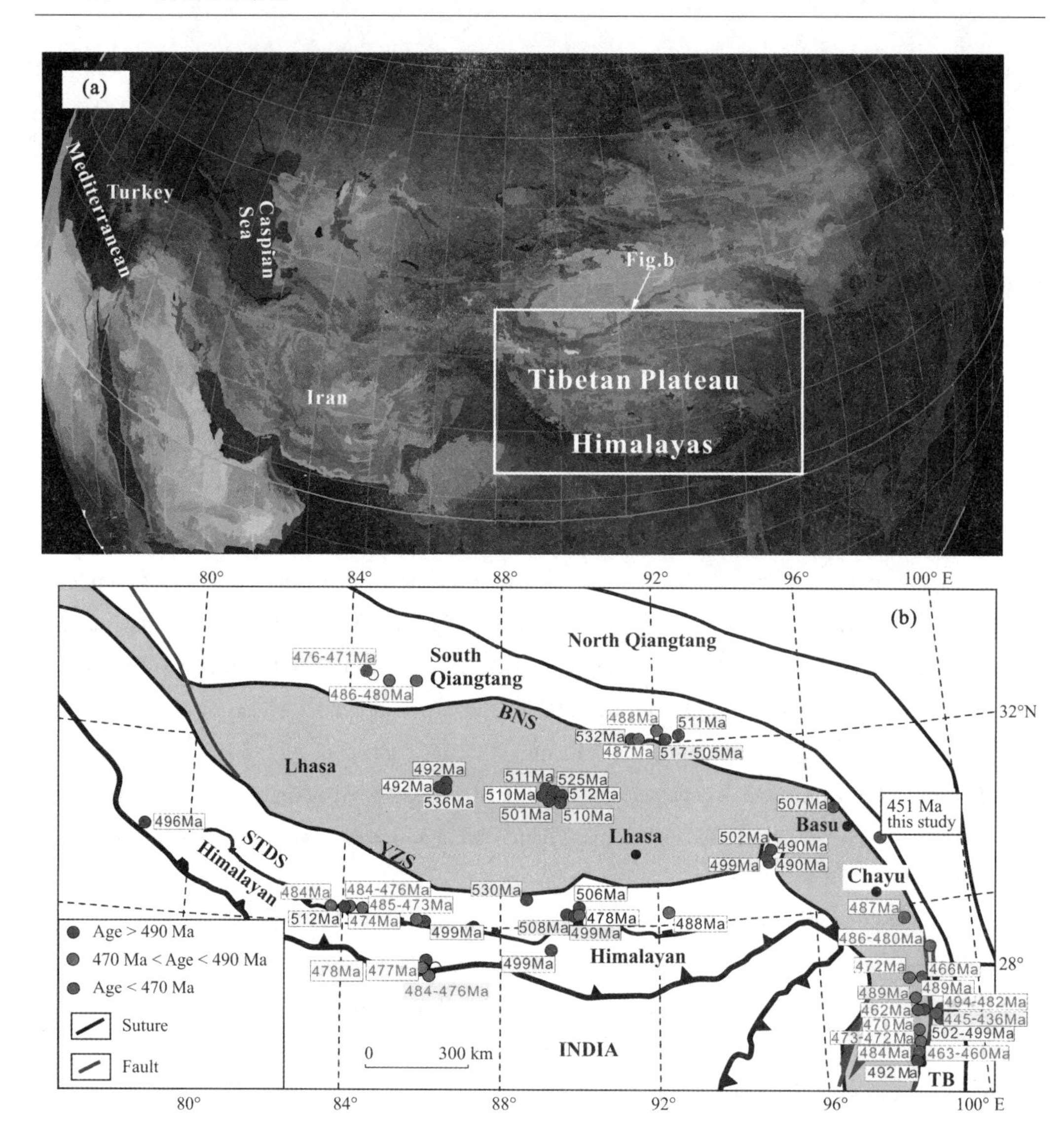

Fig. 1 (a) Tethyan orogens from mid-terrane to SE Asia (downloaded from Google Earth). (b) Distribution of the main continental blocks in the Tibetan-Himalayan orogenic system (Zhang and Tang, 2009; Zhang et al., 2012; Hu et al., 2015; Zhang et al., 2018). The data are from Supplementary Tables 1 and 2.

continental fragments that were accreted to Asia during the Mesozoic (Allégre et al., 1984; Zhang et al., 2007, 2012, 2018; Zhang and Tang, 2009), until the collision of India with Asia during the beginning of Early Cenozoic (Yin and Harrison, 2000; Zhang et al., 2012). In fact, those terranes, such as southern Qiangtang, Lhasa and Himalaya, are located at southern to central Himalayan-Tibetan orogen have been proposed to be derived from northern Gondwana, and those terranes, such as Songpan-Ganzi and northern Qiangtang, in the northern part were derived from Asia (Zhang and Tang, 2009). The continental fragments include the Songpan-Ganzi, Qiangtang, Lhasa and Himalayan terranes, separated by the

Jinsha suture, Bangong-Nujiang suture, and Indus-Yarlung suture, respectively (Yin and Harrison, 2000; Zhang et al., 2007, 2012, 2014, 2018; Zhang and Tang, 2009).

The stratigraphic units in the south Qiangtang terrane are mainly pre-Ordovician, Carboniferous, and Permian sedimentary sequences (Zhang et al., 2006, 2018). The pre-Ordovician consists of *meta*-sandstone and mica-quartz schist (Pan et al., 2004; Zhang et al., 2006). Glaciomarine deposits and cold-water biota fossils were recognized from the Carboniferous to Permian sedimentary sequences, with a Gondwana affinity (Li and Zheng, 1993; Zhang et al., 2006). The Lhasa terrane is considered to have been dispersed from the Gondwana supercontinent during the Permian to Triassic, and later drifted northward to finally collide with the Qiangtang terrane at the mid-Cretaceous (Zhang et al., 2012, 2014; Lu et al., 2019; Yan and Zhang, 2020; Ji et al., 2021). The Lhasa terrane represented a microcontinent possibly underlain by a Proterozoic basement with local Archean basement (Zhang et al., 2007; Zhu et al., 2011) and its southern part is characterized by the existence of juvenile crust (Zhu et al., 2011). The Greater Himalaya succession of the Main Central thrust sheet in the Kathmandu region comprises a succession of Neoproterozoic to Lower Paleozoic silica-clastics with some carbonates (Cawood et al., 2007). Temporally, it is facies equivalent of the Greater Himalayan and Tethyan sequences of the High Himalaya (Myrow et al., 2003), but DeCelles et al. (2000) proposed a rival accretion model. The southeastern extensions of the Baoshan-Tengchong and Shan-Thai terranes share stratigraphic and paleontological similarities to the Gondwana supercontinent (Metcalfe, 2013). The dominant stratigraphic sequences of these terranes contain pre-Mesozoic high-grade metamorphic rocks and Mesozoic-Cenozoic sedimentary and igneous rocks (YNBGMR, 1991; Zhong, 1998).

3 Data and results

Magmatic data regarding Early Paleozoic magmatic rocks in the Tibetan-Himalayan orogen have been compiled, including geochronological ($n = 67$), in situ zircon Hf isotopic ($n = 1\,011$) and geochemical ($n = 297$) data, which are summarized in Figs. 2 – 4 and Supplementary Table 2. Zircon U-Pb and Lu-Hf isotopic, whole-rock major and trace elemental and Sr-Nd isotopic data of granitoids and volcanic rocks have been used for interpretation. Owing to the focused nature of the current study, the major compilation has been restricted to the time intervals from ~530 Ma to 430 Ma, including three intervals, older than ~490 Ma, between ~490 Ma and 470 Ma, and younger than ~470 Ma, which well responded to the final magmatic activity on the northern margin of the Gondwana assembly. The data have been labelled in Fig. 1 and plotted on temporal evolution diagrams (Figs. 2–4) to illustrate early Paleozoic magmatic events across the Tibetan-Himalayan orogen located in the northern area and along the margin of the final Gondwana assembly.

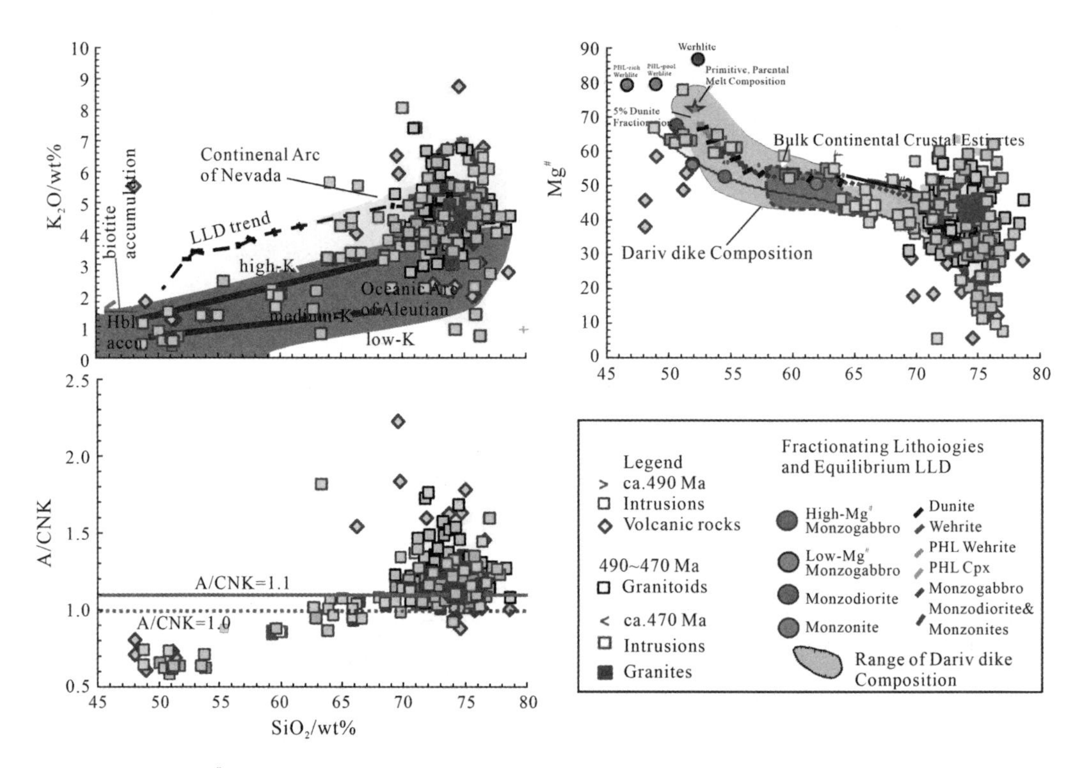

Fig. 2 K_2O, $Mg^{\#}$, and A/CNK vs. SiO_2 for the early Paleozoic magmatism (from Bucholz et al., 2014).
The Nevada and Aleutian arc data are from http://georoc. mpch-mainz.gwdg.de/georoc/.

For magmas older than 490 Ma in the Tibetan-Himalayan orogen, the zircon U-Pb ages range from 530 Ma to 490 Ma, zircon $\varepsilon_{Hf}(t)$ values range from −15 to +8.0, SiO_2 contents range from 48.0 wt% to 80.0 wt%, MgO contents range from 0.02 wt% to 9.65 wt% with $Mg^{\#}$ = 5.4−77.9, K_2O/Na_2O ratios range from 0.14 to 2.81, A/CNK values range from 0.61 to 2.22, La/Yb ratios range from 1.00 to 30.0, and Eu/Eu* values range from 0.06 to 1.08.

For magmas between 490 Ma and 470 Ma in the Tibetan-Himalayan orogen, zircon U-Pb ages range from 490 Ma to 470 Ma, zircon $\varepsilon_{Hf}(t)$ values are range from −16 to +1.0 (with only one up to 0), SiO_2 contents range from 69.3 wt% to 78.7 wt%, MgO contents range from 0.09 wt% to 2.69 wt% with $Mg^{\#}$ = 23−64 (most <50), most K_2O/Na_2O ratios range from 0.9 to 3.8, most A/CNK values are >1.0, La/Yb ratios can be up to 50, and most Eu/Eu* values are <0.64.

For magmas younger than ~470 Ma in the Tibetan-Himalayan orogen, zircon U-Pb ages range from 470 Ma to 430 Ma, zircon $\varepsilon_{Hf}(t)$ values range from −15 to +7.0, SiO_2 contents range from 50.0 wt% to 77.8 wt%, MgO contents range from 0.07 wt% to 7.72 wt% with $Mg^{\#}$ = 14.4 to 68.1, K_2O/Na_2O ratios range from 0.1 to 3.2, most A/CNK values range from 0.60 to 1.80, most La/Yb ratios are <26, and most Eu/Eu* values range from 0.04 to 1.04.

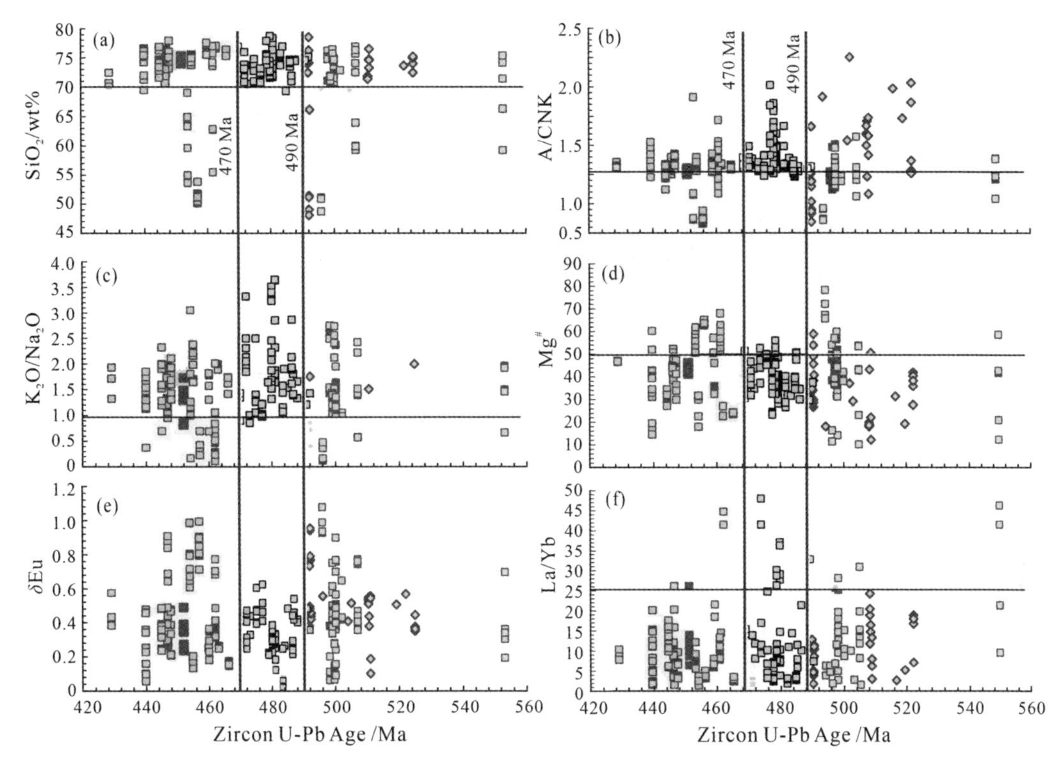

Fig. 3 Temporal evolution of geochemical indicators with the zircon U-Pb data.
The data are from Supplementary Table 2.

4 Discussion

4.1 Three stages of Early Paleozoic magmatism in the Tibetan-Himalayan orogen

For the first time, we identify three stages of early Paleozoic magmatism in the Tibetan-Himalayan orogen: 530–490 Ma, 490–470 Ma, and 470–430 Ma (Figs. 2 and 3). Most magmatic rocks forming during the early to late-stages belong to the calc-alkaline to high-K calc-alkaline, metaluminous to peraluminous, and high- to low-$Mg^{\#}$ series (Fig. 1); however, the middle-stage magmatism is characterized by high-Si contents (with most $SiO_2 > 70$ wt%), and high-K and is strongly peraluminous, with low MgO and $Mg^{\#}$ and significantly negative Eu anomalies. These features imply that the source of the middle-stage magmatism was possibly derived from highly-evolved continental crustal material with significant plagioclase residue but no other mantle-derived input. Whole-rock Sr-Nd isotopes of the middle-stage magmatism also indicate the evolved and enriched continental crustal source without any depleted components. Even the zircon Hf isotopic data show that there are no any depleted mantle-derived magma and/or juvenile crustal components that have contributed significant materials in the source

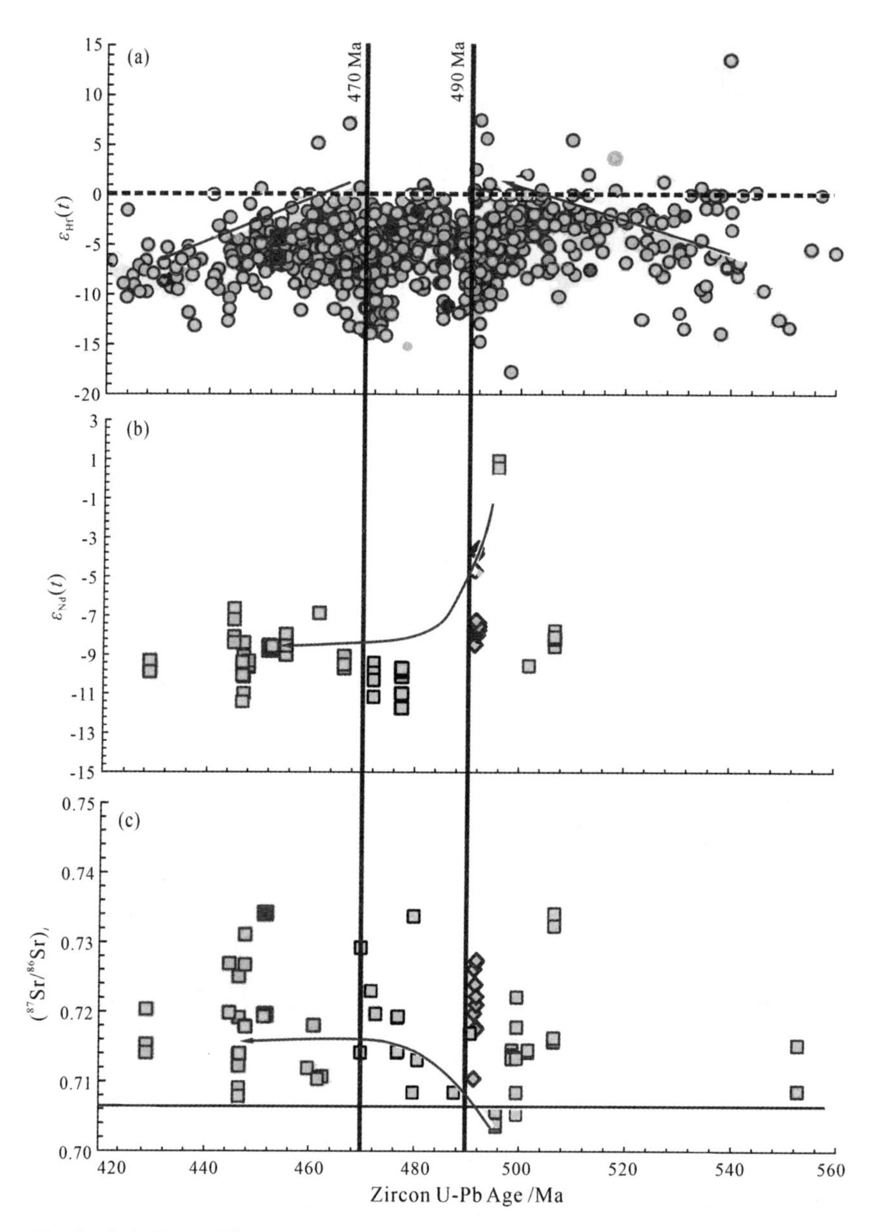

Fig. 4 (a) Zircon Hf isotopic composition and (b and c) whole-rock Nd and Sr isotopic components vs. zircon U-Pb data.

The data are from Supplemen-tary Table 2.

region of 490 – 470 Ma magmatism from the Tibetan-Himalayan orogen along the northern margin of the Gondwana assembly. In contrast, the early- and late-stages are quite different.

In the early stage, most samples belong to the typical high-K calc-alkaline peraluminous series and are associated with a few calc-alkaline and metaluminous series (Fig. 1), with highly variable SiO_2 contents, K/ Na ratios, Eu/Eu* values, A/CNK values, $Mg^{\#}$ and low La/Yb ratios (Supplemental Table 1 and Figs. 2 and 3). Meanwhile, the whole-rock initial

$^{87}Sr/^{86}Sr$ ratios are as low as 0.704 and $\varepsilon_{Nd}(t)$ values are up to +1.0, and zircon $\varepsilon_{Hf}(t)$ values can be up to +8.0 (Fig. 4); In the late-stage, most samples also plot into the field of high-K calc-alkaline series but a significant number of samples belong to the tholeiitic to calc-alkaline series (Fig. 1). The other geochemical and isotopic indicators exhibit characteristics similar to those from the early stage (Figs. 2-4). Therefore, we conclude that there are three stages of distinctive magmatism in the Tibetan-Himalayan Tethyan orogen from 530 Ma to 430 Ma, responding to the final evolution along the northern margin of the Gondwana assembly. This finding is consistent with various geological shreds of evidence, although there are widespread high-K calc-alkaline peraluminous granites ranging from the late Neoproterozoic to the early Ordovician that represented significant crustal melting (Cawood et al., 2007): ① ~491 Ma monazite inclusions in the metamorphic garnet in the Bhimphedi group indicate the date of garnet growth during regional metamorphism (Gehrels et al., 2006). ② The widespread Cambrian volcanic rocks and turbidites in the western Tethyan Himalaya and Lhasa terranes, including acid and basic tuffs, andesite and felsic volcanic rocks and bimodal volcanic suite, lasted until 490 Ma (Cawood et al., 2007; Zhu et al., 2012). ③ The Palung granite crosscuts the undeformed dyke at ~473 Ma. ④ The early Ordovician (~470 Ma) strata unconformably overlie the calc-alkaline granite in the Karakorum in the northern area of the Indus-Tsangpo suture zone (Le Fort et al., 1994). ⑤ There are widespread Cambrian-Ordovician angular unconformities and basal conglomerates in the Tethyan Himalaya in the northern Indian continent and southern Qiangtang, central Lhasa and Baoshan terranes (Li et al., 2010; Ji et al., 2009; Myrow et al., 2006; Huang et al., 2009, 2012). And ⑥ the lack of 490-470 Ma strata in the southeastern Tibet (Baoshan terrane) (Huang et al., 2009; Cai et al., 2013) is consistent with the crustal and lithospheric mantle thickening and shortening at 490-470 Ma (Gehrel et al., 2006, Cawood et al., 2007). Consequently, we suggest that the distinctive three-stage magmatism was probably closely related to different three-stage tectonic processes in response to the final evolution of the Gondwana assembly from the early Cambrian to the late Ordovician

4.2 Three-stage tectono-magmatic evolution along the northern margin of Gondwana during the early Paleozoic

The petrotectonic nature of the three-stage magmatisms must be clarified to help us better understand the tectonic processes along the northern margin of the Gondwana assembly during the early Paleozoic.

In the early stage, the petrological suite is mainly composed metabasalt, metarhyolites, andesitic, basaltic and rhyolitic volcanic rocks, mafic pillow, diorite, granodiorite and granites (Miller et al., 2001; Cawood, 2007; Zhu et al., 2012; Hu et al., 2013; Gao et al., 2019;

Dong et al., 2013; Liu et al., 2009), which is a typical calc-alkaline series and similar to the continental arc series. Geochemical data show the presence of low-Si and high-Mg magmatism with some positive zircon $\varepsilon_{Hf}(t)$ and whole-rock $\varepsilon_{Nd}(t)$ and low initial $^{87}Sr/^{86}Sr$, implying significant mantle-derived contributions in the generation of early-stage magmatism (530–490 Ma). Temporally, it consistent with the conclusion of Cawood et al. (2007), who suggested that an Andean-type magmatic arc in the northern margin of the Indian continent was active from ~530 Ma to 490 Ma, following the Gondwana assembly. The coeval mafic and granitic rocks also suggested that the early Paleozoic magmatism in the northwest Himalaya could be related to the formation of the Gondwana supercontinent (Miller et al., 2001). The ~492 Ma bimodal volcanism in the Lhasa terrane has been interpreted as resulting from partial melting of an enriched lithospheric mantle metasomatized by subduction-related components and those rhyolites have been considered to be derived from anatexis of the ancient basement that induced by mantle-derived basaltic melts (Zhu et al., 2012). In addition, both the early-stage granites and rhyolites exhibit features similar to those typical arc volcanic rocks (Hu et al., 2013), which not only inherited mantle melt signatures but are also associated with various degrees of inputs from mantle-derived materials (Zhu et al., 2012; Gao et al., 2019). Therefore, we conclude that the ~530–490 Ma magmatism in the Tibetan-Himalayan is typical Andean-type arc magmatism in which mantle-derived magmas played the necessary role in the generation of these continental arc magmatic activities and signatures.

In the middle stage, the 490–470 Ma magmas have notably continental signatures, not only because of their high-K calc-alkaline, strongly peraluminous features, and high Si and low Mg contents, but also because the whole-rock Sr-Nd and zircon Hf isotopic composition exhibit typical "crustal signatures". There is widespread early Paleozoic high-K calc-alkaline strongly peraluminous magmatism distributed from NW-SE Turkey, Central Iran, North Pakistan, NW India, Nepal, Tibet, to Tengchong-Baoshan of southeastern Tibetan-Himalayan orogen along the northern margin of Gondwana, which belongs to S-type granites that were produced by melting of ancient continental metapelite-dominated sources (Wang et al., 2013). Crustal melting to produce the strongly peraluminous and high-K calc-alkaline granite magmas requires three main mechanisms of heating that can extensively melt a fertile crust (Bea, 2012): accumulation of radiogenic heat after crustal thickening; increased heat flux from the mantle, mostly as a result of asthenospheric upwelling or mantle wedge convection; and advection by hot mantle magmas emplaced into the crust.

Both the older (>490 Ma) and younger (<470 Ma) suites are generally associated with mantle-derived rocks that have high-Mg contents and juvenile isotopic signatures (Figs. 2 and 3), indicating that the mantle-derived magmas could support the heat and/or materials in the generation of the early- and late-stage high-K calc-alkaline peraluminous magmatic activity. In contrast, however, we cannot identify any contribution from the mantle for the features of the ~490–470 Ma magmatism, in which case anatexis of the ancient continental crustal materials by thickening the crust should be considered. During ~490–470 Ma, the abundant deformed and metamorphic events occurred at ~490–480 Ma in the Kathmandu region (Cawood et al., 2007). Cawood et al. (2007) also thought that the absence of any arc activity along the northern margin of India at ~470 Ma also suggested that subduction had terminated by this time. In southeastern Tibetan Plateau, the absence of the early Ordovician strata (ca. 490–470 Ma) in the Baoshan terrane (Huang et al., 2009, 2012; Cai et al., 2013) indicates a continental crustal thickening setting during that time (Zhao et al., 2017). Various pieces of evidence have suggested that the continental and lithospheric thickening and shortening occurred at ~490–470 Ma because of the collision event, which was caused by an amalgamation of the East Gondwana margin and outward micro-continent (Gehrel et al., 2006; Cawood et al., 2007). In the continental-crustal-thickening case, enough heat (from radiogenic decay of K, U and Th) could be produced at the notably shortened and thickened crust and lithosphere (Sylvester, 1998), where the anatexis of various continental materials can easily develop. Those pure-crustal melts could then exhibit high-K calc-alkaline, strongly peraluminous and highly evolved isotopic signatures.

In the late stage, the ~457 Ma Shao La mafic rocks were identified in the Greater Himalaya, which were contemporaneous with the granitic plutons in the Everest Kharta area (Visonà et al., 2010). Similar to the Cambrian bimodal gabbro-granite in the Mandi Kaplas area, the Everest Kharta basic and acidic plutonism implies that the Paleozoic Bhimphedian orogeny was active until the late Ordovician (Visonà et al., 2010), When the Shao La basalt and basaltic andesite developed in an extensive supra-subduction zone with the geochemical fingerprint of a volcanic-arc (or back-arc) setting. Meanwhile, the coeval acidic magmatism was crustal melt that was triggered by the upwelling of the metasomatized mantle in an extensive setting (Visonà et al., 2010). From the data in this study, the late Ordovician magmatism (< 470 Ma) show characteristics similar to those of the > 490 Ma magmatism, with highly variable Si and Mg contents, K/Na ratios, Eu anomalies, A/CNK values, and zircon $\varepsilon_{Hf}(t)$ values and low La/Yb ratios (Figs. 3 and 4). The presence of Shao La basalt, high-Mg andesitic rocks, and positive zircon $\varepsilon_{Hf}(t)$ values that can be up to +7.0 also imply significant contribution and input from upwelling mantle-derived materials (Visonà et al., 2010; Xing et al., 2017; Li et al., 2016). The associated acidic plutonism was also attributed to the melting of a crustal source with various contributions from mantle-derived components (Visonà et al., 2010; Xing et al., 2017; Li et al., 2016; Zhao et al., 2014, 2017). However, two scenarios have been proposed to interpret the late Ordovician magmatism (< 470 Ma): delamination of the thickened crust and lithospheric mantle (Li et al., 2016; Zhao et al., 2017) and the presence of an extensive setting where upwelling of mantle magmas and arc-related magmatism occurred, maybe including slab breakoff (Visonà et al., 2010; Xing et al., 2017). In the delamination model, the Sr/Y ratio is high and/or adakitic rocks were abundantly developed, but none of this type of rock has yet been identified. If an extensive setting is present: the sub-arc mantle wedge could contain melts from both the sub-slab asthenosphere enriched in the high-field strength elements (HFSEs) and the normal sub-arc mantle depleted in HFSEs (Ferrari, 2004). A hybrid basaltic magma that had affinity with arc-type and within-plate suites would be produced (Zhu et al., 2017) (e.g. the Late Ordovician Shao La basalt and high-Mg andesitic rocks, Visonà et al., 2010; Xing et al., 2017). The hybrid magma would then provide heat to induce partial melting of the ancient and mature continental basement, where various continent-derived granitic melts have been produced (Wang et al., 2013; Dong et al., 2013; Li et al., 2016; Zhao et al., 2014, 2017).

4.3 Widespread reworking of ancient continental crustal materials to reconstruct Gondwana during the early Paleozoic

Although there is significant mantle-derived contribution before 490 Ma and after 470 Ma at the margin of Gondwana assembly, the three stages of strongly peraluminous, high-K calc-alkaline magmatism occurred during the early Paleozoic along the northern margin of the Gondwana assembly in the present-day Himalaya, Lhasa, Southern Qiangtang, Tengchong and Baoshan terranes (Fig. 1b). The Northern Himalayan granite belt in the Tethyan Himalayan Sequences is a 900 km curvilinear (Cawood et al., 2007). The Greater Himalayan granite belt lies between the Northern- and Lesser- Himalayan belt and extends for 1 800 km along a strike from Zanskar in the west to Bhutan in the east (Cawood et al., 2007). The Lesser Himalayan granite belt extends for some 1 900 km along a strike from southeast of Kathmandu to Pakistan in the west (Cawood et al., 2007). The Cambrian (>490 Ma) granites extend for 600 km in central Lhasa. The 490 – 470 Ma granites extend for > 1 000 km in southern Qiangtang (Fig. 1b). The Cambrian to late Ordovician granites extend for 600 km in the Tengchong and Baoshan terranes of southeastern Tibetan Plateau.

Strongly peraluminous granites are generally regarded as important indicators of the compositional maturity of the continental crust (Wu et al., 2017; Zhu et al., 2018), such as those typical S-type granites from eastern Australia (Jeon et al., 2014), the Variscan orogenic belt and Erzgebirge in Europe (Breiter, 2012), South China (Huang and Jiang, 2014) and southwestern China (Zhu et al., 2018). In this study, we find these strongly peraluminous, high-K calc-alkaline granites with high SiO_2 (> 70 wt%) and low $Mg^{\#}$ (< 45) to be widely distributed through the northern margin of the Gondwana assembly, from the early Cambrian to late Ordovician (Figs. 2 and 3). These peraluminous and high-K magmas are typical products of melting of the ancient and evolved continental crust (Cawood et al., 2007; Wang et al., 2013). Although all were derived from the reworking of the ancient and evolved continental crust materials, those from the early (> 490 Ma), middle (490 – 470 Ma) and the late (<470 Ma) stages display three distinctive features (Fig. 5): The peraluminous granites in the early stage (530–490 Ma) were closely correlated to the directly mantle-derived products, including metabasalts, andesitic to basaltic volcanic rocks, mafic pillow, and diorites (Miller et al., 2001; Cawood, 2007; Zhu et al., 2012) in an Andean-type setting along with the active continental margin (Fig. 5b). The middle-stage peraluminous magmatism was dominated by pure evolved continent crustal materials in a crustal thickening setting (Gehrel et al., 2006; Cawood et al., 2007; Zhao et al., 2017) (Fig. 5c). The late-stage peraluminous events were also closely related to the mantle-derived magmas, for example, the late Ordovician Shao La basalt, high-Mg andesitic rocks (Visonà et al., 2010; Xing et al., 2017), in an extensive setting (Fig. 5d). Notably, the strongly peraluminous high-K calc-alkaline magmatism was also indisputably dominant and became the most typical feature of the early Cambrian to late Ordovician magmatism in response to the final Gondwana assembly. It was characterized by low Sr/Y (< 20) and La/Yb ratios (< 20). Its Eu/Eu* values suggest an amalgamation of relatively thin arcs with primarily granulitic residues (<30 km) (Chapman et al., 2015; Zhao et al., 2017) rather than the thicker compressional arc (> 30 – 40 km) with the residue of eclogitic facies (Ducea, 2002). This implies that there is no residue of eclogite facies but just

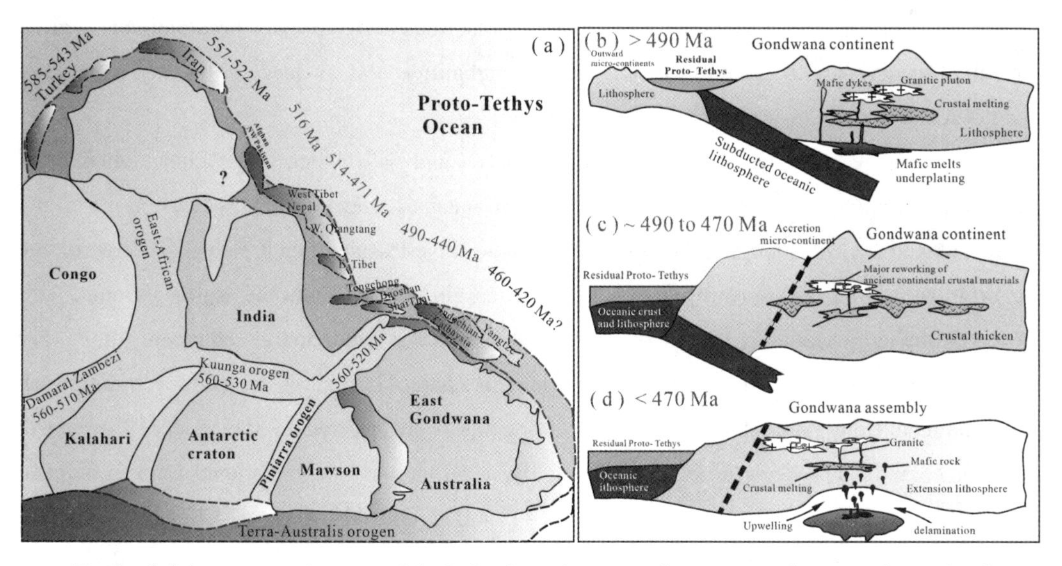

Fig. 5 (a) A reconstruction map of the India-Australia prototethyan margin showing paleographical locations of the microcontinents and associated Early Paleozoic magmatism (modified after Cawood et al., 2007 and Wang et al., 2010, 2013). (b-d) A model for final Gondwana assembly process from Andeans-type continental margin setting, to collision and crustal thicken, and to post-collision extension setting (modified after Cawood et al., 2007; Wang et al., 2013; Zhao et al., 2017).

the granulite facies in the deeper parts of the batholith in the northern margin of the Gondwana assembly. Ducea (2002) suggested a thickness of at least 20−25 km and that a convectively removable root could be developed. Meanwhile, the residues of thinner arcs must be mostly in the granulite facies, which is not denser than peridotite and thus would likely escape foundering and has not been gravitationally unstable with respect to the underlying mantle. Where the lower crust would not easily delaminate into the mantle, the crustal growth in this continental arc was efficiently. Therefore, we can conclude that the widespread strongly peraluminous high-K calc-alkaline granitic magmatism played a dominant role in reconstructing the final northern margin of the Gondwana assembly, which stabilize the margin of the Gondwana supercontinent during evolutionary process.

5 Conclusions

Distinctive three-stage magmatism occurred in the Tibetan-Himalayan orogen, from early-(~530−490 Ma) to middle (~490−470 Ma) to late stages (~ <470 Ma), in response to the final assembly of the Gondwana supercontinent. Both the early- and late-stage magmas were calc-alkaline to high-K calc-alkaline, metaluminous to peraluminous, with highly variable Si and Mg contents and whole-rock Sr-Nd and zircon Hf isotopic compositions. These mantle-derived magmas played a key role in the generation of early- and late-stage magmas, which generated at an Andean-type arc setting along the active continental margin in an extensive setting. In contrast, the middle-stage magmas are dominated by strongly peraluminous high-K calc-alkaline series with notably isotopic crustal signatures and an absence of any significant mantle-derived input in the crustal thicken setting.

Most notably, the presence of widespread early Cambrian to late Ordovician high-K calc-alkaline peraluminous magmatism in the Tibetan-Himalayan orogen indicates that the reworking of ancient and evolved continental crustal materials played a significant role in reconstructing and stabilizing the final assembly of the northern margin of the Gondwana supercontinent.

Acknowledgements Thank you so, very much for constructive comments and very carefully polish and improvement from Prof. Kai-jun Zhang. Thank you for editor handing work and comments from Prof. Mei-fu Zhou. This work has been inspired by the work of Cawood et al. (2007) and Cawood and Buchan (2007). It was supported by the National Natural Science Foundation of China (Grant Nos. 41802054, 41190072 and 41772050), China Postdoctoral Science Special Foundation (Grant. 2019 T120937) and Foundation (Grant. 2018 M643713, Natural Science Foundation of Shannxi Grant. 2019JQ-719) and Shannxi Postdoctoral Science Foundation.

Appendix A. Supplementary material Supplementary data to this article can be found online at https://doi.org/10.1016/j.jseaes.2021.104949.

References

Allégre, C.J., Courtillot, V., Tapponnier, P., Hirn, A., Mattauer, M., Coulon, C., Jaeger, J.J., Achache, J., Schärer, U., Marcoux, J., Burg, J.P., Girardeau, J., Armijo, R., Gariépy, C., Göpel, C., et al., 1984. Structure and evolution of the Himalaya-Tibet orogenic belt. Nature, 307 (5946), 17-22.

Bea, F., 2012. The sources of energy for crustal melting and the geochemistry of heat producing elements. Lithos, 153 (8), 278-291.

Breiter, K., 2012. Nearly contemporaneous evolution of the A- and S-type fractionated granites in the Krušné Hory/Erzgebirge mts. central Europe. Lithos, 151, 105-121.

Bucholz, C.E., Jagoutz, O., Schmidt, M.W., Sambuu, O., 2014. Fractional crystallization of high-K arc magmas: Biotite versus amphibole-dominated fractionation series in the Dariv Igneous Complex, Western Mongolia. Contributions to Mineralogy and Petrology, 168, 1072-1100.

Cai, Z.H., Xu, Z.Q., Duan, X.D., Li, H.Q., Cao, H., Huang, X.M., 2013. Early stage of early Palaeozoic orogenic event in western Yunnan Province, southeastern margin of Tibet Plateau. Acta Petrologica Sinica 29, 2123-2140 (in Chinese with English abstract).

Cawood, P.A., Johnson, M.R.W., Nemchin, A.A., 2007. Early palaeozoic orogenesis along the Indian margin of Gondwana: Tectonic response to Gondwana assembly. Earth & Planetary Science Letters, 255(1-2), 70-84.

Cawood, P.A., Buchan, C., 2007. Linking accretionary orogenesis with supercontinent assembly. Earth-Science Reviews, 82 (3-4), 217-256.

Chapman, J.B., Ducea, M.N., DeCelles, P.G., Profeta, L., 2015. Tracking changes in crustal thickness during orogenic evolution with Sr/Y: An example from the North American Cordillera. Geology, 43 (10), 919-922.

DeCelles, P.G., Gehrels, G.E., Quade, J., Lareau, B., Spurlin, M., 2000. Tectonic implications of U-Pb zircon ages of the Himalayan orogenic belt in Nepal. Science, 288, 497-499.

Dong, M., Dong, G., Mo, X., Santosh, M., Zhu, D., Yu, J., Nie, F., Hu, Z., 2013. Geochemistry, zircon U-Pb geochronology and Hf isotopes of granites in the Baoshan block, Western Yunnan: Implications for early Paleozoic evolution along the Gondwana margin. Lithos, 179, 36-47.

Ducea, M.N., 2002. Constraints on the bulk composition and root foundering rates of continental arcs: A California arc perspective. Journal of Geophysical Research: Solid Earth, 107 (B11), ECV 15-1-ECV 15-13.

Ferrari, L., 2004. Slab detachment control on mafic volcanic pulse and mantle heterogeneity in central Mexico. Geology, 32 (1), 77.

Gao, L.E., Zeng, L., Hu, G., Wang, Y., Wang, Q., Guo, C., Hou, K., 2019. Early Paleozoic magmatism along the northern margin of East Gondwana. Lithos, 334-335, 25-41.

Gehrels, G.E., DeCelles, P.G., Ojha, T.P., Upreti, B.N., 2006. Geologic and U-Th-Pb geochronologic evidence for early Paleozoic tectonism in the Kathmandu thrust sheet, central Nepal Himalaya. GSA Bulletin, 118 (1-2), 185-198.

Hu, P., Li, C., Wang, M., Xie, C., Wu, Y., 2013. Cambrian volcanism in the Lhasa terrane, southern Tibet: Record of an early Paleozoic Andean-type magmatic arc along the Gondwana proto-Tethyan margin. Journal of Asian Earth Sciences, 77, 91-107.

Hu, P.Y., Zhai, Q.G., Jahn, B.M., Wang, J., Li, C., Lee, H.Y., Tang, S.H., 2015. Early Ordovician granites from the south Qiangtang terrane, northern Tibet: Implications for the early Paleozoic tectonic evolution along the Gondwanan proto-Tethyan margin. Lithos, 220-223, 318-338.

Huang, L.C., Jiang, S.Y., 2014. Highly fractionated S-type granites from the giant Dahutang Tungsten deposit in Jiangnan orogen, southeast China: Geochronology, petrogenesis and their relationship with W-mineralization. Lithos, 202-203 (4), 207-226.

Huang, Y., Deng, G.B., Peng, C.L., Hao, J.X., Zhang, G.X., 2009. The discovery and significance of absence in earlymiddle Ordovician in southern Baoshan, western Yunnan. Guizhou Geology, 26 (98), 1-6 (in Chinese with English abstract).

Huang, Y., Hao, J.X., Bai, L., Deng, G.B., Zhang, G.X., Huang, W.J., 2012. Stratigraphic and petrologic response to Late Pan-African movement in Shidian area, western Yunnan Province. Geological Bulletin of China, 31, 306-313 (in Chinese with English abstract).

Jeon, H., Williams, I.S., Bennett, V.C., 2014. Uncoupled O and Hf isotopic systems in zircon fromthe contrasting granite suites of the New England orogen, eastern Australia: Implications for studies of Phanerozoic magma genesis. Geochim. Cosmochim. Acta, 146, 132-149.

Ji, C., Yan, L.L., Lu, Lu., Jin, X., Huang, Q., Zhang, K.J., 2021. Anduo Late Cretaceous high-K calc-alkaline and shoshonitic volcanic rocks in central Tibet, western China: Relamination of the subducted Meso-Tethyan oceanic plateau. Lithos, 400-401, 106345.

Ji, W.H., Chen, S.J., Zhao, Z.M., Li, R.S., He, S.P., Wang, C., 2009. Discovery of the Cambrian volcanic rocks in the Xainza area, Gangdese orogenic belt, Tibet, China and its significance. Geological Bulletin of China, 9, 1350-1354 (in Chinese with English abstract).

Le Fort, P., Tongiorgi, M., Gaetani, M., 1994. Discovery of a crystalline basement and Early Ordovician marine transgression in the Karakorum mountain range, Pakistan. Geology, 22, 941-944.

Li, C., Zheng, A., 1993. Paleozoic stratigraphy in the Qiangtang region of Tibet: Relations of the Gondwana and Yangtze continents and ocean closure near the end of the Carboniferous. International Geology Review, 35 (9), 797-804.

Li, C., Wu, Y.W., Wang, M., Yang, H.T., 2010. Significant progress on Pan-African and Early Paleozoic orogenic events in Qinghai-Tibet Plateau: Discovery of Pan-African orogenic unconformity and Cambrian System in the Gangdese area, Tibet, China. Geological Bulletin of China, 29, 1733-1736 (in Chinese with English Abstract).

Li, G.J., Wang, Q.F., Huang, Y.H., Gao, L., Yu, L. i., 2016. Petrogenesis of middle Ordovician peraluminous granites in the Baoshan block: Implications for the early Paleozoic tectonic evolution along east Gondwana. Lithos, 245, 76-92.

Liu, Y., Li, S., Santosh, M., Cao, H., Yu, S., Wang, Y., Zhou, J., Zhou, Z., 2019. The generation and reworking of continental crust during early Paleozoic in Gondwanan affinity terranes from the Tibet Plateau. Earth Sci. Rev.,190, 486-497.

Liu, S., Hu, R.Z., Gao, S., Feng, C.X., Huang, Z.L., Lai, S.C., Yuan, H.L., Liu, X.M., Coulson, I. M., Feng, G.Y., Wang, T., Qi, Y.Q., 2009. U-Pb zircon, geochemical and Sr-Nd-Hf isotopic constraints on the age and origin of early Paleozoic I-type granite from the Tengchong-Baoshan Block, Western Yunnan

Province, SW China. Journal of Asian Earth Sciences, 36, 168-182.

Lu, L., Zhang, K.J., Jin, X., Zeng, L., Yan, L.L., Santosh, M., 2019. Crustal thickening of central Tibet prior to the Indo-Asian collision: Evidence from petrology, geochronology, geochemistry and Sr-Nd-Hf isotopes on K-rich charnockite-granite suite in eastern Qiangtang. Journal of Petrology, 60, 827-854.

Metcalfe, I., 2013. Gondwana dispersion and Asian accretion: Tectonic and palaeogeographic evolution of eastern Tethys. Journal of Asian Earth Sciences, 66, 1-33.

Miller, C., Thoni, M., Frank, W., Grasemann, B., Klotzli, U., Guntli, P., Draganits, E., 2001. The early Paleozoic magmatic event in the Northwest Himalaya, India: Source, tectonic setting and age of emplacement. Geological Magazine, 138, 237-251.

Murphy, J.B., Nance, R.D., 1991. Supercontinent model for the contrasting character of Late Proterozoic orogenic belts. Geology, 19 (5), 469.

Myrow, P.M., Hughes, N.C., Paulsen, T.S., Williams, I.S., Parcha, S.K., Thompson, K.R., Bowring, S. A., Peng, S.C., Ahluwalia, A.D., 2003. Integrated tectonostratigraphic analysis of the Himalaya and implications for its tectonic reconstruction. Earth Planet. Sci. Lett., 212 (3-4), 433-441.

Myrow, P.M., Thompson, K.R., Hughes, N.C., Paulsen, T.S., Sell, B.K., Parcha, S.K., 2006. Cambrian stratigraphy and depositional history of the northern Indian Himalaya, Spiti Valley, north-central India. Geol. Soc. Am. Bull., 118 (3-4), 491-510.

Pan, G.T., Ding, J., Yao, D.S., Wang, L.Q., 2004. Guidebook of 1:1 500 000 geologic map of the Qinghai-Xizang (Tibet) plateau and adjacent areas. Chengdu: Chengdu Cartographic Publishing House, 1-48.

Sylvester, P.J., 1998. Postcollisional strongly peraluminous granites. Lithos, 45 (1-4), 29-44.

Veevers, J.J., 2004. Gondwanaland from 650 to 500 Ma assembly through 320 Ma merger in Pangea to 185–100 Ma breakup: Supercontiental tectonics via stratigraphy and radiometric dating. Earth-Science Reviews, 68.

Visonà, D., Rubatto, D., Villa, I.M., 2010. The mafic rocks of Shao La (Kharta, S. Tibet): Ordovician basaltic magmatism in the greater Himalayan crystallines of central-eastern Himalaya. Journal of Asian Earth Sciences, 38 (1-2), 14-25.

Wang, Y.J., Zhang, F.F., Fan, W.M., Zhang, G.W., Chen, S.Y., Cawood, P.A., Zhang, A.M., 2010. Tectonic setting of the South China Block in the early Paleozoic: Resolving intracontinental and ocean closure models from detrital zircon U-Pb geochronology. Tectonics, 29, TC6020.

Wang, Y., Xing, X., Cawood, P.A., Lai, S., Xia, X., Fan, W., Liu, H., Zhang, F., 2013. Petrogenesis of early Paleozoic peraluminous granite in the Sibumasu Block of SW Yunnan and diachronous accretionary orogenesis along the northern margin of Gondwana. Lithos, 182-183, 67-85.

Wu, Fuyuan, Liu, Xiaochi, Ji, Weiqiang, Wang, Jiamin, Yang, L., 2017. Highly fractionated granites: Recognition and research. Sci. China: Ser. D, Earth Sci., 60 (7), 1201-1219.

Xing, X., Wang, Y., Cawood, P.A., Zhang, Y., 2017. Early paleozoic accretionary orogenesis along northern margin of Gondwana constrained by high-Mg metaigneous rocks, SW Yunnan. International Journal of Earth Sciences, 106 (5), 1469-1486.

Yan, L.L., Zhang, K.J., 2020. Infant intra-oceanic arc magmatism due to initial subduction induced by oceanic plateau accretion: A case study of the Bangong Meso- Tethys, central Tibet, western China. Gondwana Research, 79, 110-124.

YBGMR (Yunnan Bureau Geological Mineral Resource), 1991. Regional geology of Yunnan Province. Beijing: Geological Publishing House, 1-729 (in Chinese with English abstract).

Yin, A., Harrison, T.M., 2000. Geologic evolution of the Tibetan Plateau. Annual Review of Earth and Planetary Sciences, 28 (1), 211-280.

Zhang, Kaijun, Tang, Xianchun, 2009. Eclogites in the interior of the Tibetan plateau and their geodynamic implications. Chinese Science Bulletin, 54 (15), 2556-2567.

Zhang, K.J., Zhang, Y.X., Xia, B.D., He, Y.B., 2006. Temporal variations of the Mesozoic sandstone composition in the Qiangtang block, northern Tibet (China): Implications for provenance and tectonic setting. Journal of Sedimentary Research, 76, 1035-1048.

Zhang, K.J., Zhang, Y.X., Li, B., Zhong, L.F., 2007. Nd isotopes of siliciclastic rocks from Tibet, western China: Constraints on the pre-Cenozoic tectonic evolution. Earth & Planetary Science Letters, 256, 604-616.

Zhang, K.J., Zhang, Y.X., Tang, X.C., Xia, B., 2012. Late Mesozoic tectonic evolution and growth of the Tibetan plateau prior to the Indo-Asian collision. Earth-Science Reviews, 114 (3-4), 236-249.

Zhang, K.J., Xia, B., Zhang, Y.X., Liu, W.L., Zeng, L.u., Li, J.F., Xu, L.F., 2014. Central Tibetan Meso-Tethyan oceanic plateau. Lithos, 210-211, 278-288.

Zhang, Y.X., Jin, X., Zhang, K.J., Sun, W.D., Liu, J.M., Zhou, X.Y., Yan, L.L., 2018. Newly discovered Late Triassic Baqing eclogite in central Tibet indicates an anticlockwise West-East Qiangtang collision. Scientific Reports, 8 (1).

Zhao, S.W., Lai, S.C., Gao, L., Qin, J.F., Zhu, R.Z., 2017. Evolution of the proto-Tethys in the Baoshan block along the east Gondwana margin: Constraints from early Palaeozoic magmatism. International Geology Review, 59 (1), 1-15.

Zhao, S.W., Lai, S.C., Qin, J.F., Zhu, R.Z., 2014. Zircon U-Pb ages, geochemistry, and Sr-Nd-Pb-Hf isotopic compositions of the Pinghe pluton, Southwest China: Implications for the evolution of the early Palaeozoic Proto-Tethys in Southeast Asia. International Geology Review, 56 (7), 885-904.

Zhong, D.L., 1998. Paleo-Tethyan orogenic belt in the Western Part of the Sichuan and Yunnan provinces. Beijing: Science Press, 231 (in Chinese).

Zhu, D.C., Zhao, Z.D., Niu, Y., Dilek, Y., Wang, Q., Ji, W.H., Dong, G.C., Sui, Q.L., Liu, Y.S., Yuan, H.L., Mo, X.X., 2012. Cambrian bimodal volcanism in the Lhasa terrane, southern Tibet: Record of an early Paleozoic Andean-type magmatic arc in the Australian proto-Tethyan margin. Chemical Geology, 328, 290-308.

Zhu, D.C., Zhao, Z.D., Niu, Y., Mo, X.X., Chung, S.L., Hou, Z.Q., Wang, L.Q., Wu, F. Y., 2011. The Lhasa Terrane: Record of a microcontinent and its histories of drift and growth. Earth & Planetary Science Letters, 301 (1-2), 241-255.

Zhu, R.Z., Lai, S.C., Qin, J.F., Zhao, S.W., Santosh, M., 2018. Strongly peraluminous fractionated S-type granites in the Baoshan Block, SW China: Implications for two-stage melting of fertile continental materials following the closure of Bangong- Nujiang Tethys. Lithos, 316-317, 178-198.

Zhu, R.Z., Lai, S.C., Qin, J.F., Zhao, S.W., Wang, J.B., 2017. Late Early-Cretaceous quartz diorite-granodiorite-monzogranite association from the Gaoligong belt, southeastern Tibet Plateau: Chemical variations and geodynamic implications. Lithos, 288-289, 311-325.

Petrogenesis of the Zheduoshan Cenozoic granites in the eastern margin of Tibet: Constraints on the initial activity of the Xianshuihe Fault①

Lai Shaocong Zhao Shaowei

Abstract: The Zheduoshan Miocene granitic pluton is exposed at the eastern margin of Tibet and along the strike-slip Xianshuihe Fault, and is the product of syn-tectonic magmatism closely related to this fault. This paper is focused on the petrogenesis of different granitic lithological units in the Zheduoshan composite intrusion, and the results of geochronology and lithology show that the Zheduoshan Miocene granitic pluton is incremental assembly by three stages of granitic magma influx and growth, represented by fine-grain biotite granite at 18.0 Ma, corase-grain and porphyraceous biotite monzogranite at 16.0 Ma and medium-grain two-mica monzogranite at 14.0 Ma. Combining with the geochemical signatures, these granitic rocks have high intial $^{87}Sr/^{86}Sr$ ratios, enriched Nd and Hf isotopic compositions, revealing that the sources of these granitic rocks are metabasatic rocks for fine-grain biotite granite, greywackes for coarse-grain biotite monzogranite and medium-grain monzogranite. These granites have high Sr/Y ratios, revealing that these granitic magma form at high pressure condition. The Sr/Y ratios and calculated crystallization pressure gradually decreased, implying the pressure gradually decreasing with the formation of these three stages of granites, which is probably caused by the tectonic mechanism transition from compression to strike-slip extension during the generation of these granites at 18.0–14.4 Ma. This tectonic mechanism change implied the initial activity of Xianshuihe Fault at least before 14.4 Ma.

1 Introduction

Tibetan Plateau is the Earth's largest and highest orogenic plateau resulted from the Asia indentation by India (Zhang et al., 2017). The eastern margin of Tibetan Plateau is a key region to test the intra-continental tectonic evolution (Wang et al., 2012). There are series of large-scale strike-slip shear fault zones developed at the eastern margin of Tibetan Plateau, such as Xianshuihe Fault, Ailaoshan-Red River Fault (Fig. 1a). The Xianshuihe Fault is a lithospheric-scale left-lateral strike-slip fault (Roger et al., 1995), and cut across all the geological units between eastern margin of Tibetan Plateau and the Yangtze Block, which is

① Published in *Journal of Geodynamics*, 2018, 117.

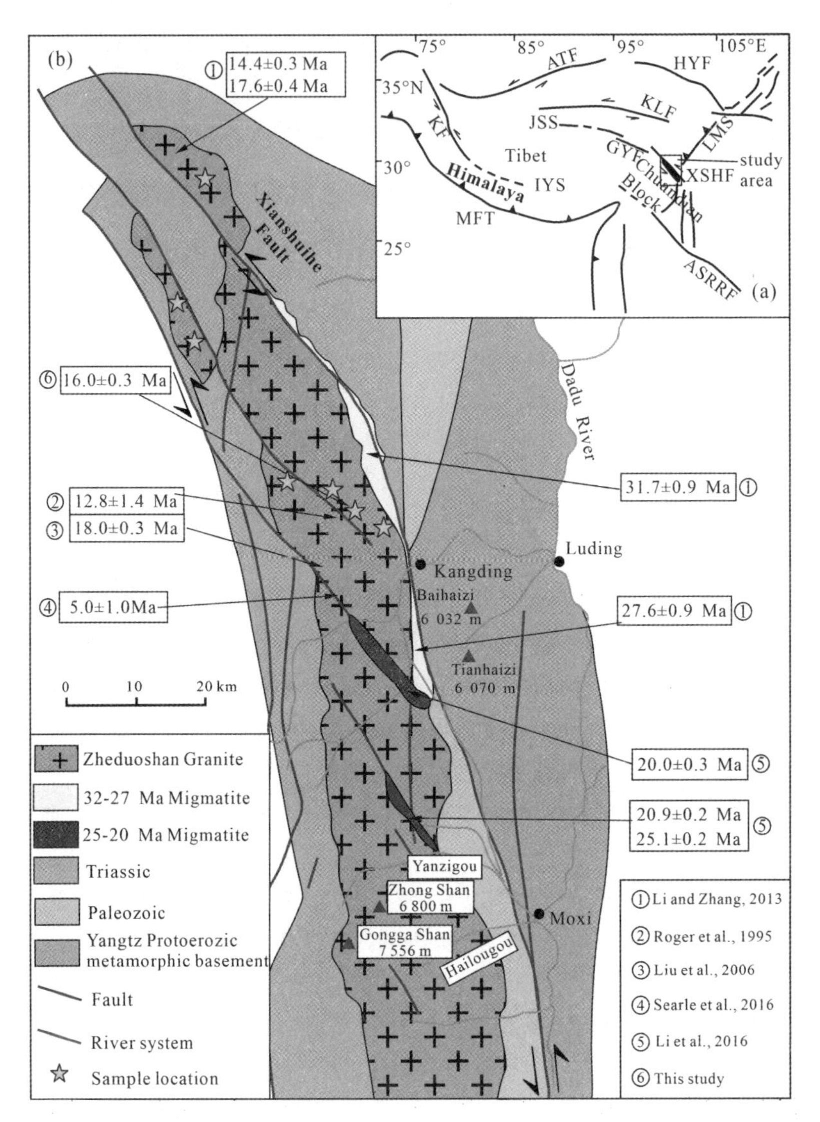

Fig. 1 Tectonic sketches of east margin of Tibetan Plateau (a) and Zheduoshan granitic pluton (b) after Li and Zhang (2013) and Searle et al. (2016).

ATF: Altyn Tagh Fault; HYF: Haiyuan Fault; KLF: Kunlun Fault; JSS: Jinsha suture; GYF: Ganze-Yushu Fault; XSHF: Xianshuihe Fault; LMS: Longmenshan Thrust Belt; KF: Karakoram Fault; IYS: Indus-Yalu suture; MFT: Main Frontal Thrust; ASRRF: Ailaoshan Red River Fault (For interpretation of the references to colour in this figure legend, the reader is referred to the web version of this article.)

formed in the process of eastward escape and internal accommodation of thickened crust in Tibetan Plateau after the collision between Indian and Asian continents (Royden et al., 1997; Clark and Royden, 2000). The Xianshuihe Fault and Longmenshan Thrust Belt is a seismically active area, where several large earthquakes in history of China are documented, for example, the Wenchuan earthquake ($M=7.9$; Hubbard and Shaw, 2009). The Xianshuihe Fault and Longmenshan Thrust Belt are also key tectonic belts of crust thickness transition from 60 km to

80 km in Tibetan Plateau to 35−40 km in Sichuan Basin (Zhang et al., 2010). Consequently, the research on initial active timing of Xianshuihe Fault and deep dynamic process in this area is very important to understanding the rising and internal accommodation of crustal materials in eastern margin of Tibetan Plateau. The formation and emplacement of Cenozoic Zheduoshan granitic pluton is closely related to the activity of the Xianshuihe Fault (Roger et al., 1995). This pluton is controlled by the early ductile shear zone and recent brittle deformation (Allen et al., 1991). The eastern part of the pluton displays an obvious ductile deformation, which is directly related to emplacement of the granitic intrusion (Roger et al., 1995; Wang et al., 1997). Thus the Zheduoshan granitic pluton is regarded as the syn-tectonic origin associated with the activity of Xianshuihe Fault (Roger et al., 1995; Wang et al., 1997). This paper selects the Zheduoshan granitic pluton, and presents the new and detail lithological and geochemical, geochronologic and isotopic data to constrain the initial activity of Xianshuihe Fault.

2 Geological setting and lithology

The Zheduoshan granitic pluton is located in the western margin of Yangtz Block, expanding NNW-SSE along the Xianshuihe Fault (Fig. 1; Liu et al., 2006; Zhang et al., 2017). This pluton covers more than 2 000 km^2, and ca. 120 km in length from Maanshan in south to Danba Baitukan in north, and is also known as Gongga pluton (Roger et al., 1995; Zhang et al., 2004). To the south, north, and west, the pluton are intrusive contact with the Triassic strata, and to the east, it contacts with the Paleozoic and Proterozoic metamorphic basement of western margin of the Yangtze Block (Zhou et al., 2002; Liu et al., 2006). In this region, there are two stages of migmatization along the northeastern margin of the Zheduoshan pluton and paralleling the Xianshuihe Fault, with ages of 32−27 Ma and 25−20 Ma, representatively (Fig. 1; Li et al., 2015, 2016).

The Zheduoshan granitic pluton is composite intrusion, comprised of Triassic-Jurassic granitic rocks, granodiorite, diorite and Cenozoic granitic rocks (Li et al., 2015; Searle et al., 2016). The Cenozoic granitic rocks in the pluton are mainly coarse-grain porphyraceous biotite monzogranite and medium-grain two-mica monzogranite that is characterized by alignment of elongated minerals. The fine-grain biotite granite is enclosed in coarse-grain biotite monzogranite, and the boundary is obvious, indicating that the biotite granite is older than the biotite monzogranite. The pegmatite occurs in the coarse-grain biotite monzogranite and medium-grain two-mica monzogranite, and as irregular vein distribution in these rocks, revealing that pegmatite is latest formation.

The types of rocks in the Zheduoshan pluton are various, including coarse-grain biotite monzogranite, medium-grain two-mica monzogranite, fine-grain biotite granite and pegmatite. They have similar mineral assemblages, but different volume percentages. The coarse-grain biotite monzogranites and fine-grain biotite granites are mainly comprised of plagioclase, K-feldspar, quartz and biotite, and the perthite is the phenocryst in the coarse-grain biotite monzogranite. The medium-grain two-mica monzogranites are mainly comprised of feldspar, K-feldspar, quartz, biotite and muscovite. The accessory minerals in these rocks are similar, such as ilmenite, zircon and apatite. The pegmatite is mainly comprised of quartz, plagioclase and biotite (Fig. 2).

Fig. 2 Field photos of Zheduoshan pluton (a) and petrographic charateristics of representative medium-grain two-mica monzogranite (b-d).
Af: alkaline-feldspar; Pl: plagioclase; Bi: biotite; Mus: muscovite; Q: quartz.

3 Analytic methods

Samples of the coarse-grain biotite monzogranite, medium-grain two-mica monzogranite, fine-grain biotite granite and pegmatite from the Zheduoshan pluton were analyzed at the State Key Laboratory of Continental Dynamics, Northwest University in Xi'an, China.

For the major and trace element analyses, weathered surfaces of samples were removed and the fresh parts were chipped and powdered to 200 mesh using a tungsten carbide ball mill. Major and trace elements were analyzed respectively by X-ray fluorescence (XRF; Rikagu RIX 2100) and inductively coupled plasma-mass spectrometry (ICP-MS; Agilent 7500a). Analyses of USGS and Chinese national rock standards (BCR-2, GSR-1, and GSR-3) revealed that the analytical precision and accuracy for the major elements were generally better than 5%. For the trace element analyses, sample powders were digested using an HF+HNO_3 mixture in high-pressure Teflon bombs at 190 ℃ for 48 h (Liu et al., 2007). For most trace elements, the analytical error was less than 2% and the precision better than 10%.

Whole rock Sr-Nd-Pb-Hf isotopic data were obtained using a Nu Plasma HR multi-collector mass spectrometer. Sr and Nd isotopic fractionations were corrected to $^{87}Sr/^{86}Sr$ = 0.119 4 and $^{146}Nd/^{144}Nd$ = 0.721 9, respectively. During the period of analysis, the NIST SRM 987 standard yielded a mean value of $^{87}Sr/^{86}Sr$ = 0.710 250±12 (2σ, n=15) and the La Jolla standard gave a mean of $^{146}Nd/^{144}Nd$ = 0.511 859±6 (2σ, n=20). Whole rock Pb was separated by anion exchange in HCl-Br columns, and Pb isotopic fractionation was corrected to $^{205}Tl/^{203}Tl$ = 2.387 5. Within the period of analysis, 30 measurements of NBS 981 gave average values of $^{206}Pb/^{204}Pb$ = 16.937 ± 1 (2σ), $^{207}Pb/^{204}Pb$ = 15.491 ± 1 (2σ), $^{208}Pb/^{204}Pb$ = 36.696±1 (2σ). The BCR-2 standard gave $^{206}Pb/^{204}Pb$ = 18.742 ± 1 (2σ), $^{207}Pb/^{204}Pb$ = 15.620±1 (2σ), and $^{208}Pb/^{204}Pb$ = 38.705±1 (2σ). Total procedural Pb blanks were in the range of 0.1–0.3 ng. Whole rock Hf was also separated by single anion exchange columns. In the course of analysis, 22 measurements of JCM 475 yielded an average of $^{176}Hf/^{177}Hf$ = 0.282 161 3±0.000 001 3 (2σ) (Yuan et al., 2007).

For U-Pb dating, zircon grains from the coarse-grain biotite monzogranite were separated using conventional heavy liquid and magnetic techniques. Representative zircon grains were handpicked and mounted in epoxy resin disks, then polished and carbon coated. The internal morphology was examined by cathodoluminescence (CL) prior to U-Pb isotope analyses. Laser ablation (LA) ICP-MS zircon U-Pb analyses were conducted on an Agilent 7500a ICP-MS equipped with a 193 nm laser, following the methods of Yuan et al. (2004). The $^{207}Pb/^{235}U$ and $^{206}Pb/^{238}U$ ratios were calculated using the GLITTER program (Macquarie University), and corrections were applied using Harvard zircon 91500 as an external calibration standard. The correction factors were applied to each sample to correct for both instrumental mass bias and depth-dependent elemental and isotopic fractionation. Details of the technique are described in Yuan et al. (2004). Common Pb contents were corrected following Andersen (2002). Age calculations and concordia diagrams were completed using ISOPLOT version 3.0 software (Ludwig, 2003). Errors quoted in tables and figures are at the 1σ level.

4 Results

4.1 Major and trace elements

The major and trace elements data are listed in Table 1. The compositions of granitic rocks in Zheduoshan pluton fall into the fields of granite-granodiorite on a An-Ab-Or diagram

(Fig. 3a). The coarse-grain biotite monzogranites have high SiO_2 and K_2O, with contents of 72.08–74.02 wt% and 3.50–5.61 wt% respectively, belonging to the high-K calc-alkaline series (Fig. 3b), Na_2O contents of 3.13–4.22 wt%, alkaline contents of 7.72–8.74 wt%, K_2O/Na_2O ratios of 0.83–1.79, Al_2O_3 contents of 14.15–14.66 wt% and A/CNK ratios of 1.04–1.08, belonging to peraluminous granite (Fig. 3c). They have low MgO contents of 0.29–0.41 wt%, high $Fe_2O_3^T$ contents of 1.31–2.07 wt%, TiO_2 contents of 0.15–0.40 wt%, CaO contents of 1.18–1.82 wt%, $Mg^\#$ values of 26–33. The fine-grain biotite granites have SiO_2 contents of 70.63–72.07 wt%, K_2O contents of 3.79–4.30 wt%, belonging to high-K calc-alkaline series (Fig. 3b), Na_2O contents of 4.61–4.82 wt%, alkaline contents of 8.61–8.91 wt%, K_2O/Na_2O ratios of 0.79–0.93, Al_2O_3 contents of 15.79–16.27 wt%, A/CNK ratios of 1.29–1.35, belonging to peraluminous granite. They also have MgO contents of 0.16–0.51 wt%, $Fe_2O_3^T$ contents of 0.88–1.89 wt%, CaO contents of 1.72–1.86 wt%, $Mg^\#$ values of 26–34. The medium-grain two-mica monzogranites have SiO_2 contents of 72.59–73.14 wt%, K_2O contents of 4.91–5.10 wt%, belonging to high-K calc-alkaline series (Fig. 3b), Na_2O contents of 3.44–3.57 wt%, K_2O/Na_2O ratios of 13.9–1.48, Al_2O_3 contents

Table 1 Analytical results of major(wt%) and trace(ppm) element of the Zheduoshan Miocene granites.

Sample	KKD07	KKD14	KKD01	KKD08	KKD09	KKD17	KKD21	MNG04	MNG05	MNG06	MNG09	KKD04
Lithology	fine-grain biotite granites		coarse-grain biotite monzogranites					medium-grain two-mica monzogranites				pegmatite
SiO_2	72.07	70.63	72.71	72.97	74.02	72.08	72.92	72.68	73.14	72.59	73.04	79.34
TiO_2	0.1	0.28	0.24	0.18	0.15	0.4	0.31	0.22	0.23	0.24	0.27	0.18
Al_2O_3	16.27	15.79	14.66	14.26	14.15	14.47	14.43	14.37	14.7	14.41	14.45	10.62
$Fe_2O_3^T$	0.88	1.89	1.82	1.41	1.31	2.07	1.83	1.5	1.48	1.54	1.66	1.48
MnO	0.01	0.02	0.02	0.01	0.01	0.02	0.02	0.02	0.02	0.02	0.03	0.01
MgO	0.16	0.51	0.39	0.31	0.29	0.41	0.33	0.26	0.27	0.29	0.32	0.27
CaO	1.86	1.72	1.82	1.21	1.18	1.34	1.2	1.31	1.26	1.39	1.42	0.51
Na_2O	4.82	4.61	4.22	3.32	3.43	3.13	3.16	3.57	3.44	3.54	3.53	1.98
K_2O	3.79	4.3	3.5	5.4	5.18	5.61	5.57	5.05	5.1	5.03	4.91	5.26
P_2O_5	0.05	0.09	0.07	0.08	0.06	0.1	0.07	0.06	0.06	0.07	0.07	0.04
LOI	0.34	0.37	0.41	0.38	0.4	0.38	0.5	0.5	0.62	0.59	0.62	0.35
Total	100.35	100.21	99.86	99.53	100.18	100.01	100.34	99.54	100.32	99.71	100.32	100.04
$Mg^\#$	26	34	29	30	30	28	26	25	26	27	27	26
A/CNK	1.06	1.03	1.04	1.06	1.06	1.06	1.08	1.05	1.09	1.04	1.05	1.08
Li	15.4	42.3	36.5	29.8	29.1	29.1	59.5	51.8	57.7	50.1	56.1	31.4
Be	3.17	2.26	2.99	2.56	2.56	2.77	3.99	4.65	7.18	5.25	5.4	0.97
Sc	1.23	2.16	2.44	2	1.83	2.07	2.47	2.3	2.33	2.24	2.38	1.9

Continued

Sample	KKD07	KKD14	KKD01	KKD08	KKD09	KKD17	KKD21	MNG04	MNG05	MNG06	MNG09	KKD04
Lithology	fine-grain biotite granites		coarse-grain biotite monzogranites					medium-grain two-mica monzogranites				pegmatite
V	5.98	16.6	11.6	9.59	9.62	11.8	9.33	8.42	8.67	9.18	10.7	7.71
Cr	1.25	3.19	3.22	1.41	2.64	2.64	2.39	4.15	2.16	2.31	5.24	3.99
Co	108	104	124	125	106	78	142	123	182	120	99.6	133
Ni	1.95	2.92	4.42	2.33	3.25	2.59	3.89	3.33	2.76	2.04	9.04	3.59
Cu	0.86	1.36	1.07	0.93	0.93	1.53	1.28	1.18	1.16	1.25	1.58	1.55
Zn	25.2	54.4	47	35.6	32.8	58.8	58.2	57.5	57.1	55.5	60.7	46.2
Ga	22.8	20.5	24.8	21.4	20.7	23.8	24.8	24.1	24.4	24	24.5	16.4
Ge	0.75	0.7	0.97	0.94	0.88	1.11	1.15	1.18	1.21	1.17	1.2	0.91
Rb	129	164	144	216	205	214	255	266	274	257	255	216
Sr	389	811	395	352	348	345	226	225	225	247	249	258
Y	1.79	5.14	3.88	4.8	2.82	5.83	7.8	9.33	9.92	10.4	11.6	2.35
Zr	107	174	173	160	119	240	222	159	178	181	182	80.9
Nb	6	5.2	12.3	10.6	9.5	15.4	16.6	13.3	13.5	14	15.1	10.5
Cs	1.18	2.5	2.84	3.7	3.52	4.04	6.62	5.85	6.98	5.64	5.54	2.55
Ba	810	1761	782	1107	1081	1167	926	895	902	1027	1014	911
Hf	2.94	4.1	4.33	4.46	3.44	5.58	5.52	4.35	4.81	4.85	4.8	2.16
Ta	0.38	0.47	0.77	0.76	0.72	1.04	1.64	1.51	1.59	1.89	2.07	0.61
Pb	38.8	38.2	29.9	40.1	38.5	30	35.1	42.6	41.4	40.7	38.5	41.5
Th	6.78	16.5	90.2	50.8	37.6	46.6	54	46.8	39.5	41.5	42.8	57.8
U	1.68	6.57	2.3	7.03	3.23	1.96	4.22	10.9	8.28	11.2	8.24	2.25
La	12.6	57.7	110	69.8	50.5	159	129	83.2	85.6	82.1	94.5	76.4
Ce	19.3	91.6	185	118	84.2	263	212	139	145	135	157	131
Pr	1.98	8.99	17.9	11.9	8.29	26.6	21.5	13.8	14.2	13.4	15.4	13.3
Nd	6.67	28.7	59.4	39.8	27.4	78.4	63.8	44.2	44.4	42	48.2	44.1
Sm	1.11	3.73	7.25	5.75	3.89	8.89	7.93	6.15	5.74	5.21	6.01	6.21
Eu	0.78	1.05	0.66	0.72	0.7	1.21	0.95	0.75	0.77	0.81	0.83	0.48
Gd	0.85	2.71	4.42	3.93	2.59	5.76	5.33	4.3	4.04	3.6	4.19	3.97
Tb	0.09	0.25	0.34	0.34	0.21	0.44	0.47	0.43	0.41	0.36	0.41	0.29
Dy	0.39	1.12	1.21	1.31	0.8	1.71	2.01	2	1.97	1.87	2.09	0.94
Ho	0.06	0.18	0.14	0.17	0.1	0.2	0.28	0.3	0.32	0.32	0.35	0.09
Er	0.18	0.49	0.37	0.43	0.26	0.57	0.71	0.84	0.92	0.97	1.08	0.23
Tm	0.03	0.06	0.04	0.05	0.03	0.06	0.08	0.11	0.12	0.14	0.16	0.02
Yb	0.16	0.38	0.22	0.27	0.16	0.33	0.43	0.66	0.74	0.89	0.98	0.1
Lu	0.03	0.06	0.03	0.04	0.02	0.05	0.06	0.09	0.1	0.12	0.13	0.02
∑REE	46	202	391	257	182	552	452	305	314	298	343	280

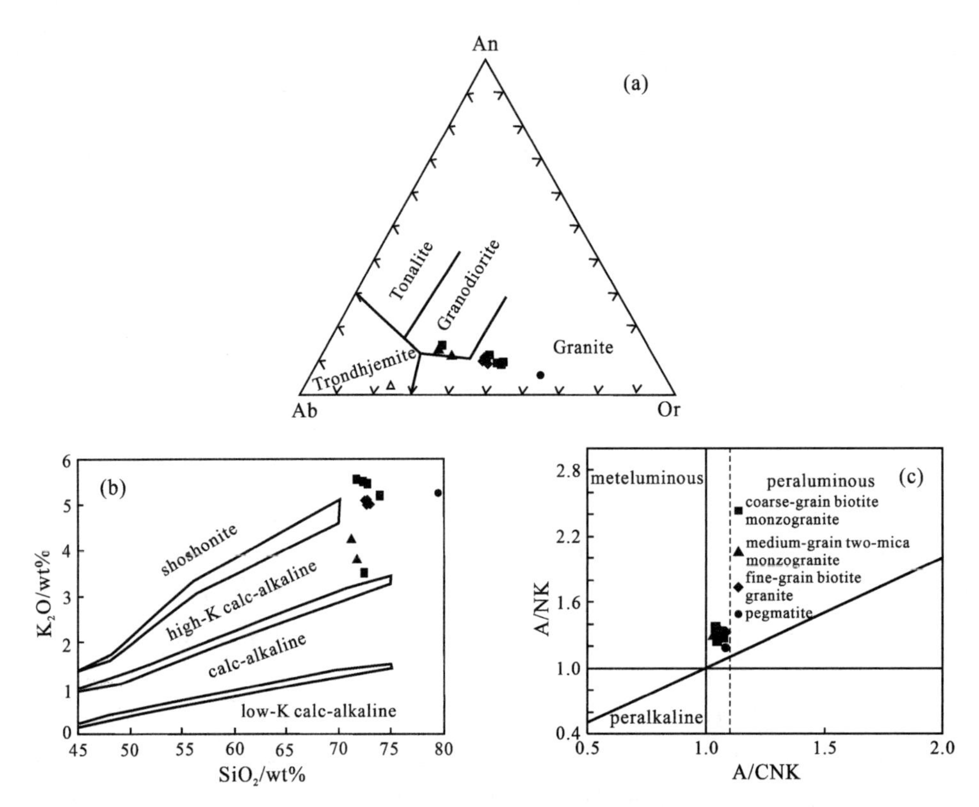

Fig. 3 Diagrams of An-Ab-Or (Barker, 1979), SiO_2-K_2O (Rollinson, 1993) and A/CNK-A/NK (Maniar and Piccoli, 1989) for granitic rocks in Zheduoshan pluton.

of 14.37 – 14.70 wt%, A/CNK ratios of 1.04 – 1.09, belong to peraluminous granite (Fig. 3c), MgO contents of 0.26 – 0.32 wt%, $F_2O_3{}^T$ contents of 1.48 – 1.66 wt%, CaO contents of 1.26–1.42 wt%, $Mg^{\#}$ ratios of 25–26. The pegmatite have SiO_2 content of 79.34 wt%, K_2O content of 5.26 wt%, Na_2O contents of 1.98 wt%, K_2O/Na_2O ratios of 2.66, Al_2O_3 contents of 10.62 wt%, CaO content of 0.51 wt%, A/CNK ratio of 1.19, belonging to high-K calc-alkaline and peraluminous granite, MgO content of 0.27 wt%, $Fe_2O_3{}^T$ content of 1.48 wt%, $Mg^{\#}$ value of 26.

On the primitive mantle normalized spider diagram, all of the coarse-grain biotite monzogranites, fine-grain biotite granites, medium-grain two-mica monzogranites and pegmatite are enriched in large ion lithophile element (LILE), and depleted in Nb, Ta, Ti, and P (Fig. 4). In the chondrite normalized rare earth element (REE) pattern diagrams, they are all enriched in LREE and depleted in HREE (Fig. 4). The fine-grain biotite granites have positive Eu anomalies with δEu values of 1.01–2.45, REE contents of 46–202 ppm, and the medium-grain two-mica monzogranites, coarse-grain biotite monzogranites and pegmatite show negative Eu anomalies with δEu values of 0.45 – 0.57, 0.36 – 0.67, and 0.30, respectively, REE contents of 298–343 ppm, 182–552 ppm, and 280 ppm, respectively.

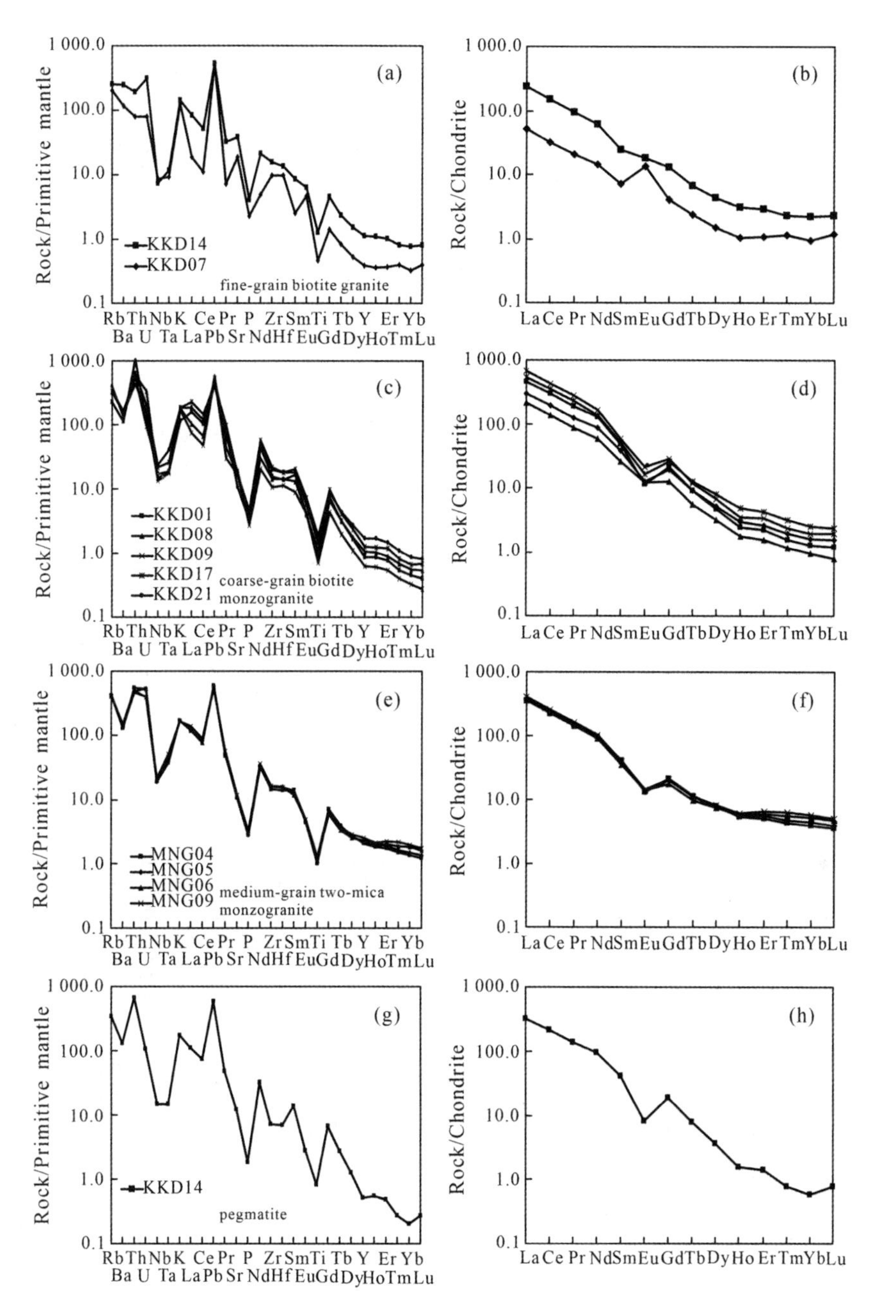

Fig. 4 Primitive mantle-normalized trace element spider and chondrite-normalized REE pattern diagrams of fine-grain biotite granite, coarse-grain and porphyraceous biotite monzogranite, medium-grain two mica monzogranite and pegmatite.

The primitive mantle and Chondrite values are from Sun and McDonough (1989).

4.2 Zircon U-Pb results

In this paper, we selected zircons from the one sample (KKD 20) of coarse-grain biotite monzogranite for zircon U-Pb isotopic measurement. The data are listed in Table 2 and Fig. 5.

Table 2　Analytical results of zircon U-Pb isotpic compositions of coarse-grain biotite monzogranite in Zheduoshan pluton.

Analysis	Contents/ppm			Ratios						Age/Ma					
	Th	U	Th/U	$^{207}Pb/^{235}U$	1σ	$^{206}Pb/^{238}U$	1σ	$^{208}Pb/^{232}Th$	1σ	$^{207}Pb/^{235}U$	1σ	$^{206}Pb/^{238}U$	1σ	$^{208}Pb/^{232}Th$	1σ
KKD20-01	532	1 586	0. 34	0. 015 11	0. 000 49	0. 002 38	0. 000 04	0. 000 91	0. 000 09	15. 2	0. 5	15. 3	0. 3	18. 0	2. 0
KKD20-02	332	703	0. 47	0. 017 35	0. 002 49	0. 002 68	0. 000 06	0. 001 46	0. 000 06	17. 0	2. 0	17. 3	0. 4	29. 0	1. 0
KKD20-03	462	1 228	0. 38	0. 017 84	0. 001 84	0. 002 44	0. 000 05	0. 000 77	0. 000 02	18. 0	2. 0	15. 7	0. 3	15. 5	0. 3
KKD20-04	1 661	1 836	0. 90	0. 018 33	0. 001 12	0. 002 40	0. 000 04	0. 000 76	0. 000 02	18. 0	1. 0	15. 5	0. 3	15. 4	0. 4
KKD20-05	1 799	9 998	0. 18	0. 013 13	0. 000 33	0. 002 06	0. 000 02	0. 000 68	0. 000 01	13. 2	0. 3	13. 3	0. 1	13. 7	0. 2
KKD20-06	577	952	0. 61	0. 014 90	0. 001 41	0. 002 35	0. 000 04	0. 000 76	0. 000 04	15. 0	1. 0	15. 1	0. 3	15. 4	0. 8
KKD20-07	630	1 442	0. 44	0. 016 43	0. 000 92	0. 002 50	0. 000 04	0. 000 81	0. 000 03	16. 5	0. 9	16. 1	0. 3	16. 4	0. 6
KKD20-08	340	1 142	0. 30	0. 015 98	0. 002 09	0. 002 40	0. 000 05	0. 000 76	0. 000 08	16. 0	2. 0	15. 4	0. 3	15. 0	2. 0
KKD20-09	81	291	0. 28	0. 982 52	0. 021 72	0. 102 19	0. 001 23	0. 031 91	0. 000 60	695	11	627	7	635	12
KKD20-10	953	2 013	0. 47	0. 018 33	0. 001 12	0. 002 67	0. 000 04	0. 000 84	0. 000 01	18. 0	1. 0	17. 2	0. 3	17. 1	0. 2
KKD20-11	320	1 533	0. 21	0. 018 24	0. 001 23	0. 002 64	0. 000 04	0. 000 83	0. 000 02	18. 0	1. 0	17. 0	0. 2	16. 9	0. 4
KKD20-12	1 246	3 088	0. 40	0. 016 00	0. 000 86	0. 002 47	0. 000 04	0. 000 73	0. 000 02	16. 1	0. 9	15. 9	0. 3	14. 7	0. 4
KKD20-13	2 268	4 015	0. 56	0. 015 75	0. 000 75	0. 002 39	0. 000 03	0. 000 71	0. 000 02	15. 9	0. 7	15. 4	0. 2	14. 3	0. 4
KKD20-14	331	2 651	0. 13	0. 017 76	0. 001 42	0. 002 50	0. 000 05	0. 000 78	0. 000 03	18. 0	1. 0	16. 1	0. 3	15. 9	0. 5
KKD20-15	676	1 833	0. 37	0. 015 88	0. 000 52	0. 002 50	0. 000 04	0. 000 82	0. 000 02	16. 0	0. 5	16. 1	0. 2	16. 5	0. 5
KKD20-16	755	1 854	0. 41	0. 015 68	0. 001 21	0. 002 35	0. 000 05	0. 000 83	0. 000 04	16. 0	1. 0	15. 1	0. 3	16. 8	0. 8
KKD20-17	161	462	0. 35	0. 016 59	0. 001 38	0. 002 61	0. 000 06	0. 000 95	0. 000 12	17. 0	1. 0	16. 8	0. 4	19. 0	2. 0
KKD20-18	1 092	2 145	0. 51	0. 016 34	0. 000 70	0. 002 49	0. 000 03	0. 000 74	0. 000 02	16. 5	0. 7	16. 0	0. 2	15. 0	0. 4
KKD20-19	508	1 854	0. 27	0. 017 43	0. 000 96	0. 002 49	0. 000 04	0. 000 75	0. 000 03	17. 5	1. 0	16. 0	0. 3	15. 2	0. 6
KKD20-20	396	1 551	0. 26	0. 015 71	0. 000 39	0. 002 47	0. 000 03	0. 000 89	0. 000 07	15. 8	0. 4	15. 9	0. 2	18. 0	1. 0
KKD20-21	385	2 099	0. 18	0. 018 48	0. 001 15	0. 002 58	0. 000 04	0. 000 89	0. 000 05	19. 0	1. 0	16. 6	0. 3	18. 0	1. 0
KKD20-22	327	849	0. 38	0. 017 99	0. 002 35	0. 002 51	0. 000 06	0. 000 79	0. 000 03	18. 0	2. 0	16. 2	0. 4	15. 9	0. 7
KKD20-23	1 303	1 294	1. 01	0. 017 15	0. 001 33	0. 002 53	0. 000 05	0. 000 73	0. 000 02	17. 0	1. 0	16. 3	0. 3	14. 7	0. 4
KKD20-24	113	2 378	0. 05	0. 198 69	0. 003 39	0. 028 97	0. 000 32	0. 008 57	0. 000 22	184	3	184	2	172	4
KKD20-25	399	1 096	0. 36	0. 015 76	0. 000 25	0. 002 48	0. 000 03	0. 001 14	0. 000 08	15. 9	0. 2	16. 0	0. 2	23. 0	2. 0
KKD20-26	338	881	0. 38	0. 022 97	0. 001 78	0. 003 44	0. 000 07	0. 000 95	0. 000 05	23. 0	2. 0	22. 1	0. 4	19. 0	1. 0

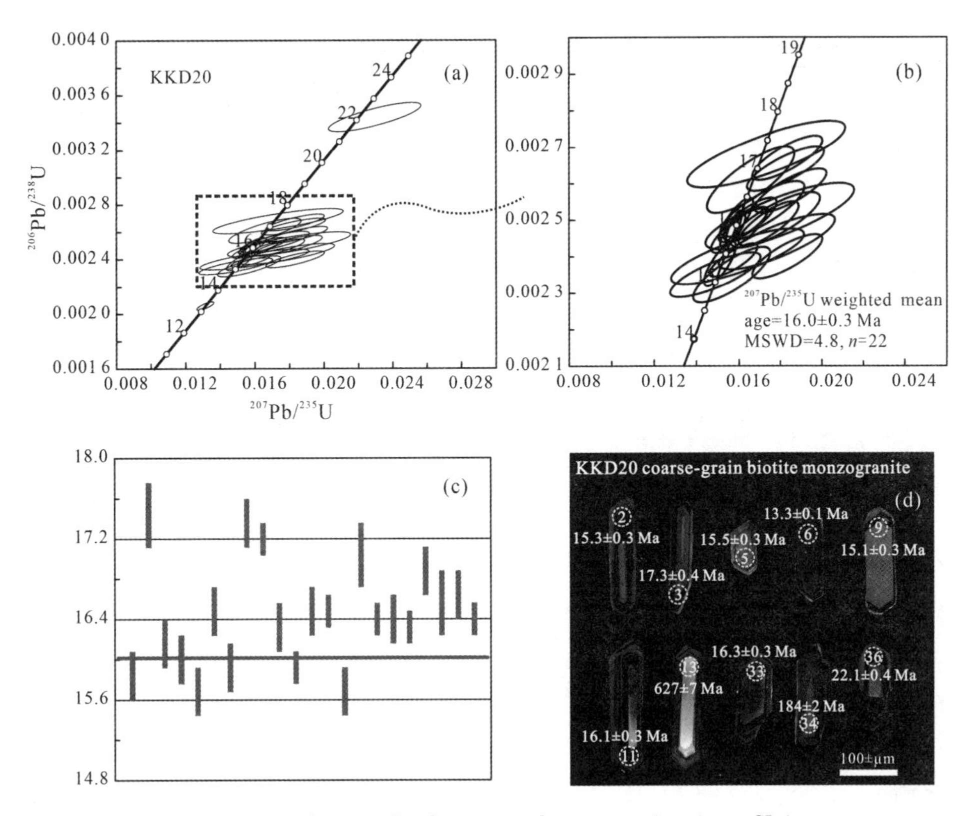

Fig. 5 Zircon U-Pb concordia diagrams and representative zircon CL images for coarse-grain biotite monzogranite.

Zircons selected from the coarse-grain biotite monzogranite are colorless and transparent, stubby to elongate prisms (100–350 μm long), with aspect ratios of 1 : 1–3 : 1. On the CL images, most of the zircons have well-defined oscillatory zoning, representing the magmatic origin, and several zircons show light cores, representing a wall-rock xenocryst or inherent origin (Fig. 5). Twenty-six spots have been analyzed and the spot 9 and 24 have $^{206}Pb/^{238}U$ ages of 627 ± 7 Ma and 184 ± 2 Ma, respectively, which are the ages of xenocrysts or inherent zircons. The other 24 spots have concordant $^{206}Pb/^{238}U$ ages of 13.3 – 22.1 Ma, giving a weighted mean age of 16.0 ± 0.3 Ma (MSWD = 4.8, n = 22), representing the peak crystallization age of coarse-grain biotite monzogranite. The age of 22.1 Ma is the starting crystallization time of magma and the age of 13.1 Ma is terminal crystallization time of magma. These Miocene zircons have variable Th contents of 81–2 268 ppm, and U contents of 291–9 998 ppm, with Th/U ratios of 0.18–1.01.

4.3 Sr-Nd-Pb-Hf isotopic data

We selected the fine-grain biotite granites, coarse-grain biotite monzogranites and medium-grain two-mica monzogranites to analyze whole-rock Sr-Nd-Pb-Hf isotopic compositions and the results are listed in Table 3 and Figs. 6 and 7.

Table 3 Whole rock Sr-Nd-Pb-Hf isotopic data for samples from Zheduoshan Miocene granites.

Sample	KKD07	KKD08	MNG04	MNG05
Lithology	fine-grain biotite granite	coarse-grain biotite monzogranite	medium-grain two-mica monzogranite	
t/Ma	18	16	14	14
Rb/ppm	129	216	266	274
Sr/ppm	389	352	225	225
$^{87}Rb/^{86}Sr$	0.958	1.775	3.419	3.522
$^{87}Sr/^{86}Sr$	0.707 266	0.707 414	0.708 795	0.708 498
2σ	10	21	12	5
$^{87}Sr/^{86}Sr(t)$	0.707 021	0.707 010	0.708 115	0.707 798
Sm/ppm	1.11	5.75	6.15	5.74
Nd/ppm	6.67	39.8	44.2	44.4
$^{147}Sm/^{144}Nd$	0.100 2	0.087 2	0.084 2	0.078 2
$^{143}Nd/^{144}Nd$	0.512 346	0.512 620	0.512 248	0.512 186
2σ	5	21	3	4
T_{DM2}/Ga	1.09	0.72	1.22	1.30
$\varepsilon_{Nd}(t)$	−5.5	−0.1	−7.4	−8.6
U/ppm	1.68	7.03	10.9	8.28
Th/ppm	6.78	50.8	46.8	39.5
Pb/ppm	38.8	40.1	42.6	41.4
$^{206}Pb/^{204}Pb$	18.539	18.586	18.493	18.493
2σ	1	1	1	1
$^{207}Pb/^{204}Pb$	15.667	15.675	15.658	15.658
2σ	1	1	1	1
$^{208}Pb/^{204}Pb$	38.970	39.059	38.948	38.941
2σ	1	1	1	1
$^{206}Pb/^{204}Pb(t)$	18.531	18.558	18.451	18.465
$^{207}Pb/^{204}Pb(t)$	15.667	15.674	15.656	15.656
$^{208}Pb/^{204}Pb(t)$	38.960	38.993	38.898	38.898
Lu/ppm	0.03	0.04	0.09	0.10
Hf/ppm	2.94	4.46	4.35	4.81
$^{176}Hf/^{177}Hf$	0.282 750	0.282 676	0.282 618	0.282 599
2σ	4	4	5	5
$\varepsilon_{Hf}(t)$	−0.4	−3.1	−5.2	−5.8
T_{DM2}/Ga	1.16	1.33	1.54	1.58

Rb, Sr, Sm, Nd, U, Th, Pb, Lu and Hf concentrations were analyzed by ICP-MS. T_{DM2} represent the two-stage model age and were calculated using present-day $(^{147}Sm/^{144}Nd)_{DM} = 0.213\ 7$, $(^{147}Sm/^{144}Nd)_{DM} = 0.513\ 15$, $(^{147}Sm/^{144}Nd)_{crust} = 0.101\ 2$, $(^{147}Sm/^{144}Nd)_{CHUR} = 0.196\ 7$, $(^{143}Nd/^{144}Nd)_{CHUR} = 0.512\ 638$. $\varepsilon_{Nd}(t) = [(^{143}Nd/^{144}Nd)_S(t)/(^{143}Nd/^{144}Nd)_{CHUR}(t) - 1] \times 10^4$, $T_{DM2} = \frac{1}{\lambda}\{1 + [(^{143}Nd/^{144}Nd)_S - ((^{147}Sm/^{144}Nd)_S - (^{147}Sm/^{144}Nd)_{crust})(e^{\lambda t} - 1) - (^{143}Nd/^{144}Nd)_{DM}]/[(^{147}Sm/^{144}Nd)_{crust} - (^{147}Sm/^{144}Nd)_{DM}]\}$, $\lambda = 6.54 \times 10^{-12}$/a.

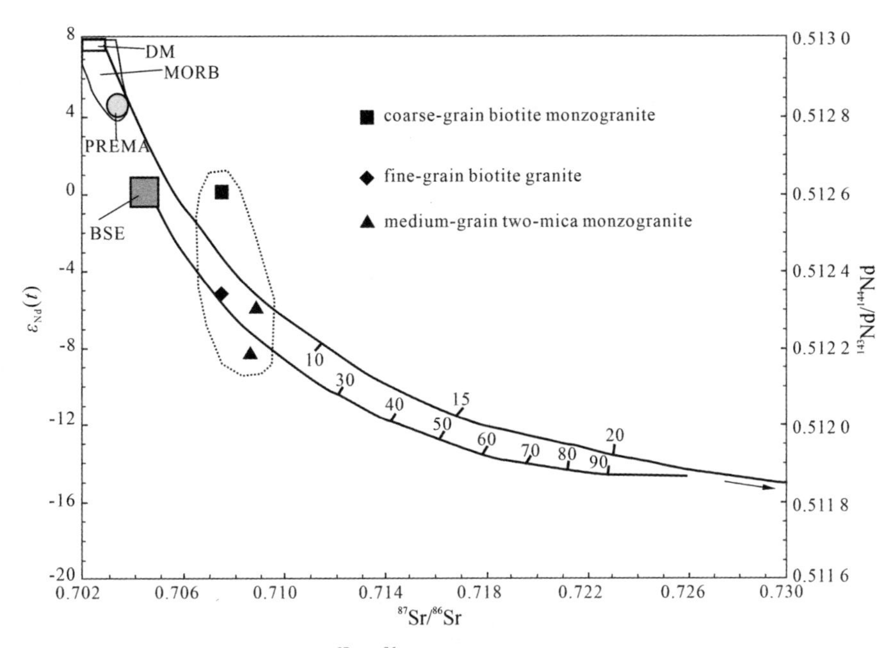

Fig. 6 Diagram of initial $^{87}Sr/^{86}Sr$-$\varepsilon_{Nd}(t)$ for Zheduoshan Miocene granites.

The fine-grain biotite granite has initial $^{87}Sr/^{86}Sr(t)$ ratio of 0.707 021, negative $\varepsilon_{Nd}(t)$ value of −5.5 with Nd two-stage model age of 1.09 Ga, slightly lower evolved Pb isotopic compositions with $^{206}Pb/^{204}Pb(t)$ = 18.531, $^{207}Pb/^{204}Pb(t)$ = 15.667, $^{208}Pb/^{204}Pb(t)$ = 38.960, $\varepsilon_{Hf}(t)$ value of −0.4 with Hf two-stage model age of 1.16Ga. The corse-grain monzogranite has initial $^{87}Sr/^{86}Sr(t)$ ratio of 0.707 010, $\varepsilon_{Nd}(t)$ value of −0.1 with Nd two-stage model age of 0.72 Ga, slightly lower evolved Pb isotopic compositions with $^{206}Pb/^{204}Pb(t)$= 18.558, $^{207}Pb/^{204}Pb(t)$= 15.674, $^{208}Pb/^{204}Pb(t)$= 38.993, $\varepsilon_{Hf}(t)$ value of −3.1 with Hf two-stage model age of 1.33 Ga. The medium-grain two-mica monzogranites have initial $^{87}Sr/^{86}Sr(t)$ ratio of 0.707 798−0.708 115, $\varepsilon_{Nd}(t)$ value of −8.6 to −7.4 with Nd two-stage model ages of 1.22−1.30 Ga, slightly lower evolved Pb isotopic compositions with $^{206}Pb/^{204}Pb(t)$= 18.458−18.465, $^{207}Pb/^{204}Pb(t)$= 15.656, $^{208}Pb/^{204}Pb(t)$= 38.898, $\varepsilon_{Hf}(t)$ value of −5.8 to −5.2 with Hf two-stage model ages of 1.54−1.58 Ga.

5 Discussion

5.1 Petrogenesis of the Zheduoshan Miocene granites

Granites generally form large batholiths or plutons at the active continental margins and collision orogens and their petrogenesis remains among the most debated topics in geology (Hawkesworth and Kemp 2006; Castro et al., 2010). Some geologists believe that the granite is crystallization and evolution from one magma chamber or magma mixing (e.g., Castro et al., 2013), and others consider that the granitic large batholith is incremental assembly according

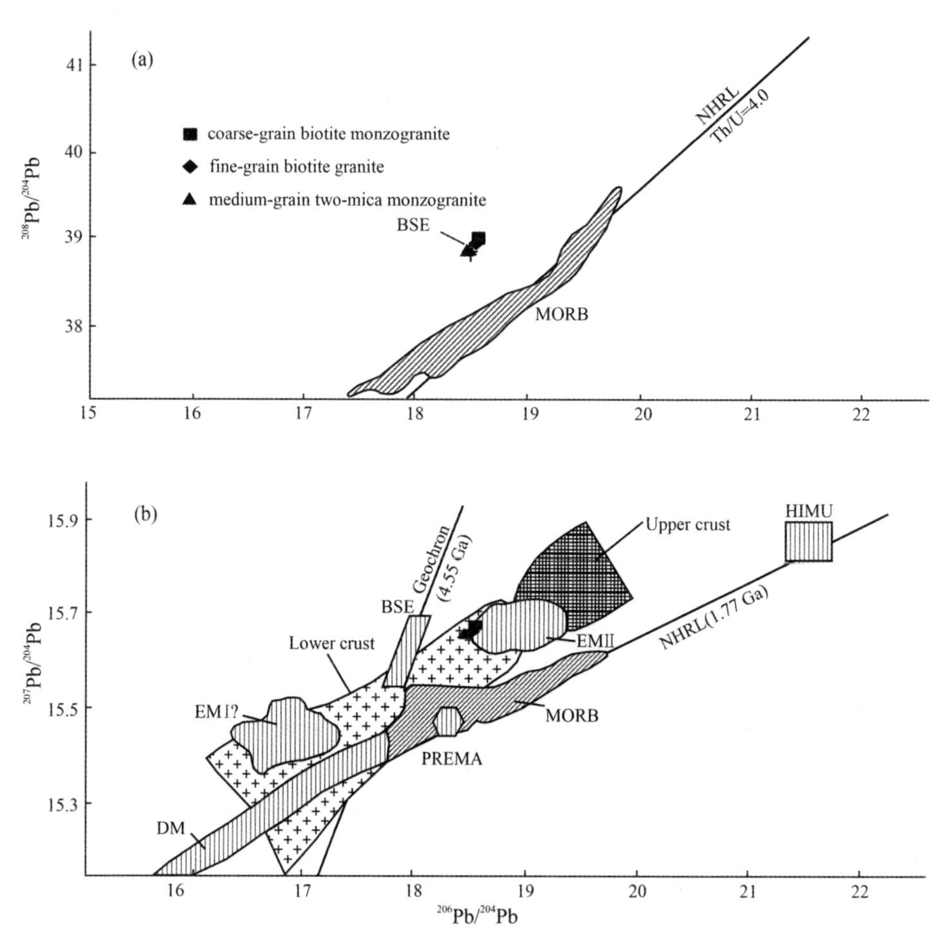

Fig. 7 Diagrams of $^{206}Pb/^{204}Pb$-$^{207}Pb/^{204}Pb$ and $^{206}Pb/^{204}Pb$-$^{208}Pb/^{204}Pb$ for Zheduoshan Miocene granites.

NHRL: northern hemisphere reference line with Th/U = 0.4; BSE: bulk silicate Earth vaule; MORB: mid-ocean ridge basalt; HIMU: mantle with high U/Pb ratios; EM: enriched mantle; DM: depleted mantle; PREMA: frequently observed prevalent mantle composition (Zinder and Hart, 1986).

to multiple magma emplacement (e.g., Coleman et al., 2004; Paterson et al., 2011). The Zheduoshan Miocene granitic pluton exposes more than 2 000 km^2 and is comprised of different granitic rocks including fine-grain biotite granite, coarse-grain biotite monzogranite, medium-grain two-mica monzogranite and pegmatite. These granitic rocks have similar geochemical signatures but their textures and mineral grain sizes are different and their geochemical data do not show obvious linear relationship. Thus, these granitic rocks can't be crystallized and evolved from one magma chamber but formed by different granitic magma emplacement and assembly. The different granitic rocks represent their different sources. The fine-grain biotite granites are enclosed in the coarse-grain biotite monzogranite as enclaves. The field contact relationship of other lithologic units is unclear according to our field investigation. The pegmatite is a product of post-magmatic hydrothermalism and it cut cross the granitic body with an obvious boundary. Liu et al. (2006) reported the fine-grain biotite granite with SHRIMP zircon U-Pb age of 18.0 ± 0.3 Ma, representing the crytallizaiton age of early granitic magmatism in this region. This paper show the coarse-grainte biotite monzogranite has zircon U-Pb age of 16.0 ± 0.3 Ma, which is younger than the age of fine-grain biotite granite and consistent with the field contact relationship between them. The medium-grain monzogranite occurs at the margin of the Zheduoshan pluton and the minerals are orientational alignment. Li and Zhang (2013) reveal that the meidium-grain monzogrnaite have a crystallizaiton age of 14.4±0.3 Ma (SHRIMP) and 17.4±0.4 Ma (LA-ICP-MS), representing the lastest granitic magmatism. Roger et al. (1995) showed that the pegmatite has crystallization age of 10.4 ± 1.2 Ma according to Rb-Sr age of plagioclase and 12.02±0.80 Ma–10.38±0.01 Ma according to $^{40}Ar/^{39}Ar$ age of K-feldspar and biotite, which probably represent the active time of post-magmatic hydrothermal. Therefore, we consider that the Zheduoshan Miocene pluton was formed by gradually assembly of three stages of granitic magmatism, early fine-grain biotite granite, interim coarse-grain biotite monzogranite and late medium-grain monzogranite. This could reconfirm that the large granitic pluton or batholith is formed by incremental assembly.

All the fine-grain biotite granites, coarse-grain biotite monzogranites and medium monzogranites are high-K calc-alkaline peraluminous granite, and have high initial $^{87}Sr/^{86}Sr(t)$ ratios (> 0.706), enriched Nd and Hf isotopes [$\varepsilon_{Nd}(t) < 0$, $\varepsilon_{Hf}(t) < 0$] (Fig. 6), resembling crustal initial Pb isotopic ratios (Fig. 7). These geochemical and isotopic signatures reveal that the sources of these granitic rocks are crustal materials.

The fine-grain biotite granites have relativley high Na_2O contents with K_2O/Na_2O ratios of 0.79–0.93, high CaO/Na_2O and low Al_2O_3/TiO_2, Rb/Ba and Rb/Sr ratios, suggesting that the sources are enriched in CaO, Sr and Ba, depleted in Rb. Combining with relatively high $Mg^{\#}$ values of 26–34, the sources of the fine-grain biotite granites could be metabasite (Rapp et al., 1995; Rushmer, 1991). The fine-grain biotite granites have positive Eu anomalies with δEu values of 1.01–2.45, high Sr contents of 389–811 ppm, low Y contents of 1.79–5.14 ppm, high Sr/Y ratios of 158–257, which imply that the granites were produced in relatively high pressure condition enough to stabilize garnet and destabilize plagioclase (Kay and Mpodozis, 2001; Chapman, 2015; Chiaradia, 2015). The coarse-grain biotite monzogranites have high K_2O/Na_2O ratios of 0.83 – 1.79, high CaO/Na_2O ratios and low Rb/Ba, Rb/Sr ratios and $Mg^{\#}$ values of (26–30), suggesting that the sources of biotite monzogranites have more felsic composition than biotite granite and could be felsic lithology, such as metagreywacke (Fig. 8). The coarse-grain biotite monzogranites are enriched in LILE, depleted in Nb, Ta, P and Zr, and have high Sr/Y ratios of 29 – 123, revealing that the coarse-grain biotite monzogranites were also produced in the relatively high pressure condition.

The geochemical characteristics of medium-grain two-mica monzogranites are similar to the coarse-grain biotite monzogranites, high K_2O/Na_2O ratios of 1.39 – 1.48, high CaO/Na_2O ratios and low Rb/Ba and Rb/Sr ratios, $Mg^{\#}$ values of 25–27, suggesting the source is also metagreywacke. The medium-grain two-mica monzogranites have Sr contents of 225–249 ppm, Y contents of 9.33–11.6 ppm, Sr/Y ratios of 21–24, indicating that medium-grain two-mica monzogranites is also produced in the relatively high pressure condition. Thus, according to the geochemical signatures, the granitic magmatisms are all active at high pressure condition.

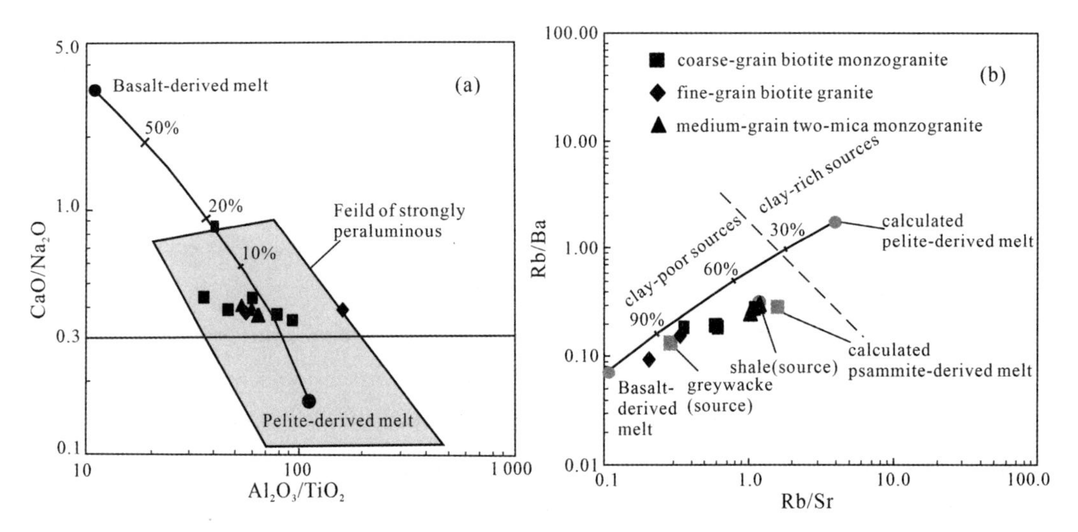

Fig. 8 Diagrams of Al_2O_3/TiO_2 vs. CaO/Na_2O and Rb/Sr vs. Rb/Ba for Zheduoshan Miocene granites after Sylvester (1998).

5.2 Tectonic implications

Most deformation of Songpan-Ganzi tectonic belt and crustal thickening in eastern margin of Tibetan Plateau are Traissic-Jurassic in age, including coeval magmatism, folded flysch sedimentary rocks and a complete Barrovian-type metamorphic sequence (Xu et al., 1991; Horrowfield and Wilson, 2005; Wilson et al., 2006; Roger et al., 2003, 2004, 2010; Weller et al., 2013; Searle et al., 2016). However, the Cenozoic structures in eastern margin of Tibetan Plateau are dominated by a series of large-scale strike-slip faluts, such as the Ganzi-Yushu and Xianshuihe Faults (Searle et al., 2016). The Zheduoshan granitic rocks are Middle-Late Miocene in age, considered to be related to the early actively of Xianshuihe Fault and syn-tectonic products (Roger et al., 1995; Wang et al., 1997). The timing of Zheduoshan Cenozoic granitic magmatism could reveal the initial activity of Xianshuihe Fault (Roger et al., 1995; Wang et al., 1997; Zhang et al., 2004). Searle et al. (2016) showed that the lecuograinte cutted by Xianshuihe Fault have crystallization age of 5 Ma, and the initial activity of Xianshuihe Fault could be after 5 Ma. However, the Xianshuihe Fault is a sustainable strike-slip fault in Cenozoic with offset at ca. 60 km (Roger et al., 1995; Yan and Lin, 2015, 2017;

Zhang et al., 2017; Bai et al., 2018). The lecuo-granite cutted by Xianshuihe Fault could only indicate the activity of this fault after the formation of lecuo-granite, but cannot reveal the true timing of initial activity.

The Zheduoshan Miocene granitic pluton is incremental assembly by three stages granitic magmatism, fine-grain biotite granites with age of 18.0±0.3 Ma (Liu et al., 2006), coarse-grain biotite monzogranites with age of 16.0±0.3 Ma, medium-grain two-mica monzogranites with age of 14.4±0.3 Ma (Li and Zhang, 2013). The Sr/Y ratios of these granitic rocks are gradually decreasing with the crystallization age (Fig. 9), indicating that the formation pressure of these graintic rocks is gradually decreasing. This could be further confirmed by the crystallization pressure of these granitic intrusions. According to the polynomial equation between crystallization pressure of quartz-oversaturated granitic system and normative quartz contents (Yang, 2017), we calculated the crystallizaiton of the Zheduoshan granites through the equation as following:

$$P=-0.2426(Qtz)^3+26.392(Qtz)^2-980.74(Qtz)+12563 \quad (R^2=0.9943)$$

where P is pressure in MPa, and R denotes correlation coefficient. The results indicate the crystallization pressure for fine-graine biotitie granite at 500 MPa, coarse-grain biotite monzogranite at 194 MPa, and medium-grain two-mica monzogranite at 204 MPa. Therefore, the formation pressures of these tectonic granitic rocks is gradually decreasing. It is implication that the tectonic machanism transition from compression to strike-slip extension, and the initial activity of Xianshuihe Fault could be at least before 14.4 Ma. This is consistent with the results gotten by Wang et al. (2012), who reported the two-phase growth of high topography in eastern margin of Tibetan Plateau during 30–25 Ma and 10–15 Ma ago respectively, using thermochronology of zircon and apitite in Pengguan Massif.

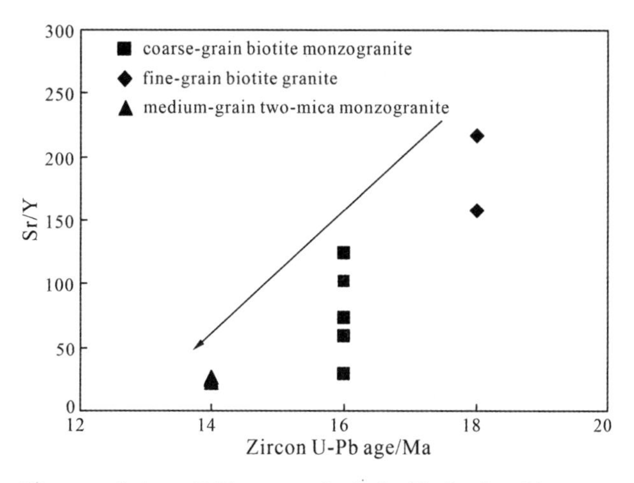

Fig. 9 Diagram of zircon U-Pb age vs. Sr/Y for Zheduoshan Miocene granites.

6 Conclusions

(1) The Zheduoshan Miocene granitic rocks are incremental assembly by three stages of granitic magmatism, the early fine-grain biotite granites derived from partial melting of metabasite, both of interim coarse-grain biotite monzogranites and last medium-grain two-mica monzogranites derived from partial melting of metagreywacke.

(2) The Sr/Y ratios and calculated crystallization pressure of Zheduoshan Miocene granitic rocks decrease with crystallization age, revealing the tectonic mechanism transition from compression to strike-slip extension. The initial activity of Xianshuihe Fault is at least before 14.4 Ma.

Acknowledgements This work was jointly supported by the National Natural Science Foundation of China (Grant No. 41190072 and 41372067), and the Fundamental Research Funds for the Central Universities (No. 300102278112).

References

Allen, C.R., Luo, Z.L., Qian, H., Wen, Y.Z., Zhou, H.W., Huang, W.S., 1991. Field study of a highly active fault zone: The Xian Shui He fault of southwestern China. Geological Society of America Bulletin, 103(9), 1178-1199.

Andersen, T., 2002. Correction of common lead in U-Pb analyses that do not report ^{204}Pb. Chemical Geology, 192, 59-79.

Bai, M.K., Chevalier, M.L., Pan, J.W., Replumaz, A., Leloup, P.H., Métois, M., Li, H.B., 2018. Southeastward increase of the late Quaternary slip-rate of the Xianshuihe fault, eastern Tibet. Geodynamic and seismic hazard implications. Earth & Planetary Science Letters, 485, 19-31.

Barker, F., 1979. Trondhjemites: Definition, environment and hypotheses of origin. In: Barker, F. Trondhjemites, Dacites and related rocks. Amsterdam: Elsevier.

Castro, A., Gerya, T., García-Casco, A., Fernández, C., Díaz Alvarado, J., Moreno-Ventas, L., Loew, I., 2010. Melting relations of MORB-sediment mélanges in underplated mantle wedge plumes: Implications for the origin of cordilleran-type batholiths. Journal of Petrology, 51, 1267-1295.

Castro, A., 2013. Tonalite-granodiorite suites as cotectic systems: A review of experimental studies with applications to granitoid petrogenesis.Earth-Science Reviews, 124, 68-95.

Chapman, J.B., Ducea, M.N., DeCelles, P.G., Profeta, L., 2015. Tracking changes in crustal thickness during orogenic evolution with Sr/Y: An example from the North American Cordillera. Geology, 43, 919-922.

Chiaradia, M., 2015. Crustal thickness control on Sr/Y signatures of recent arc magmas: An Earth scale perspective. Sicentific Reports, 5.

Clark, M.K., Royden, L.H., 2000. Topographic ooze: Building the eastern margin of Tibet by lower crustal flow. Geology, 28, 703-706.

Coleman, D.S., Gray, W., Glazner, A.F., 2004.Rethinking the emplacement and evolution of zoned plutons:

Geochronologic evidence for incremental assembly of the Tuolumne Intrusive Suite, California. Geology, 32 (5), 433-436.

Harrowfield, M.J., Wilson, C.J.L., 2005. Indosinian deformation of the Songpan-Garze fold belt, northeastern Tibetan plateau. Journal of Structural Geology, 27, 101-117.

Hawkesworth, C.J., Kemp, A.I.S., 2006. Evolution of the Continental crust. Nature, 443, 811-817.

Hubbard, J., Shaw, J.H., 2009. Uplift of the Longmen Shan and Tibetan Plateau and the 2008 Wenchuan earthquake. Nature, 458, 194-197.

Kay, S.M., Mpodozis, C., 2001. Central Andean ore deposits linked to evolving shallow subduction systems and thickening crust. GSA Today, 11, 4-9.

Li, H.L., Zhang, Y.Q., 2013. Zircon U-Pb geochronology of the Konggar granitoid and migmatite: Contraints on the Oligo-Miocene tectono-thermal evolution of the Xianshuihe fault zone, East Tibet. Tectonophysics, 606, 127-139.

Li, H.L., Zhang, Y.Q., Zhang, C.H., Dong, S.W., Zhu, F.S., 2015. Middle Jurassic syn-kinematic magmatism, anatexis and metamorphism in the Zheduo-Gonggar massif, implication for the deformation of the Xianshuihe fault zone, East Tibet. Journal of Asian Earth Sciences, 107, 35-52.

Li, H.L., Zhang, Y.Q., Zhang, C.H., Wang, J.C., 2016. Zircon U-Pb study of two-staged Oligo-Miocene migmatization along the Xianshuihe fault zone, East Tibet Plateau. Earth Science Frontiers, 23(2), 222-237 (in Chinese with English abstract).

Liu, S.W., Wang, Z.Q., Yan, Q.R., Li, Q.G., Zhang, D.H., Wang, J.G., 2006. Timing, petrogenesis and geodynamica significance of Zheduoshan Granitoids. Acta Petrologica Sinica, 22(2), 343-352.

Liu, Y., Liu, X.M., Hu, Z.C., Diwu, C.R., Yuan, H.L., Gao, S., 2007. Evaluation of accuracy and long-term stability of determination of 37 trace elements in geological samples by ICP-MS. Acta Petrologica Sinica, 23(5), 1203-1210(in Chinese with English abstract).

Ludwig, K.R., 2003. ISOPLOT 3.0: A Geochronological Toolkit for Microsoft Excel. Berkeley Geochronology Center, Special Publication, 4, 71.

Maniar, P.D., Piccoli, P.M., 1989. Tectonic discrimination of granitoids. Geological Society of American Bulletin, 101, 635-643.

Paterson, S.R., Okaya, D., Memeti, V., Economos, R., Miller, R.B., 2011. Magma addition and flux calculations of incrementally constructed magma chambers in continental margin arcs: Combined field, geochronologic, and thermal modeling studies. Geosphere, 7(6), 1439-1468.

Rapp, R.P., Watson, E.B., 1995. Dehydration melting of metabasalt at 8-32 kbar: Implications for continental growth and crust-mantle recycling. Journal of Petrology, 36, 891-931.

Roger, F., Arnaud, N., Gilder, S., Tapponnier, P., Jolivet, M., Brunel, M., Malavielle, J., Xu, Z., 2003. Geochronological and geochemical constraints on Mesozoic suturing in East Central Tibet. Tectonics, 22, 1037.

Roger, F., Calassou, S., Lancelot, J., Malavielle, J., Mattauer, M., Zhiqin, X., Ziwen, H., Liwei, H., 1995. Miocene emplacement and deformation of the Konga Shan granite (Xianshui He fault zone, west Sichuan, China): Geodynamic implications. Earth & Planetary Science Letters, 130, 201-216.

Roger, F., Jolivet, M., Malavielle, J., 2010. The tectonic evolution of the Songpan-Garze (north Tibet) and adjacent areas from Proterozoic to present: A synthesis. Journal of Asian Earth Sciences, 39, 254-269.

Roger, F., Malavielle, J., Leloup, H., Calassou, S., Xu, Z., 2004. Timing of granite emplacement and cooling in the Songpan-Garze fold belt (eastern Tibetan plateau) with tectonic implications. Journal of Asian Earth Sciences, 22, 465-481.

Rollinson, H., 1993. Using Geochemical Data: Evaluation, Presentation, Interpretation. London: Longman Scientific & Technical.

Royden, L.H., Burchfiel, B.C., King, R.W., Wang, E., Chen, Z., Shen, F., Liu, Y., 1997. Surface deformation and lower crust flow in the eastern Tibet. Science, 276, 788-790.

Rushmer, T., 1991. Partial melting of two amphibolites: Contrasting results under fluid-absent conditions. Contributions to Mineralogy and Petrology, 107, 41-59.

Searle, M.P., Roberts, N.M.W., Chung, S.L., Lee, Y.H., Cook, K.L., Elliott, J.R., Weller, O.M., St-Onge, M.R., Xu, X.W., Tan, X.B., Li, K., 2016. Age and anatomy of the Gongga Shan batholith, eastern Tibetan plateau, and its relationship to the active Xianshui-he fault. Geosphere, 12(3), 1-23.

Sun, S.S., McDonough, W.F., 1989. Chemical and isotopic systematics of oceanic basalts: Implications for mantle composition and processes. In: Saunders, A.D., Norry, M.J. Magmatism in the Ocean Basins. Geological Society Special Publication, London, 42, 313-345.

Sylvester, P.J., 1998. Post-collisional strongly peraluminous granites. Lithos, 45, 29-44.

Wang, E., Kirby, E., Furlong, K.P., van Soest, M., Xu, G., Shi, X., Kamp, P.J.J., Hodges, K.V., 2012. Two-phase growth of high topography in eastern Tibet during the Cenozoic. Nature Geoscience, 640-645.

Wang, Z.X., Xu, Z.Q., Yang, T.N., 1997. Origin of the Zheduoshan granite and its tectonic setting. Journal of Chendu University of Technology, 24(1), 48-55 (in Chinese with English abstract).

Weller, O.M., St-Onge, M.R., Waters, D.J., Rayner, N., Searle, M.P., Chung, S.L., Palin, R.M., Lee, Y.H., Xu, X., 2013. Quantifying Barrovian metamorphism in the Danba structural culmination of eastern Tibet. Journal of Metamorphic Geology, 31, 909-935.

Wilson, C.J.L., Harrowfield, M.J., Reid, A.J., 2006. Brittle modification of Triassic architecture in eastern Tibet: Implications for the construction of the Cenozoic plateau. Journal of Asian Earth Sciences, 27, 341-357.

Xu, Z.Q., Hou, L.W., Wang, D.K., Wang, Z.X., 1991. "Xikang-type" folds and their deformation mechanism-a new fold type in orogenic belts. Regional Geology of China, 1, 1-9.

Yan, B., Lin, A.M., 2015. Systematic deflection and offset of the Yangtze River drainage system along the strike-slip Ganzi-Yushu-Xianshuihe Fault Zone. Journal of Geodynamics, 87, 13-25.

Yan, B., Lin, A.M., 2017. Holocene activity and paleoseismicity of the Selaha Fault, southeastern segment of the strike-slip Xianshuihe Fault Zone, Tibetan Plateau. Tectonophysics, 694, 302-318.

Yang, X.M., 2017. Estimation of crystallization pressure of granite intrusions. Lithos, 286-287, 324-329.

Yuan, H.L., Gao, S., Liu, X.M., Li, H.M., Gunther, D., Wu, F.Y., 2004. Accurate U-Pb age and trace element determinations of zircon by laser ablation-inductively coupled plasma mass spectrometry. Geostandard Newsletters, 28, 353-370.

Yuan, H.L., Gao, S., Luo, Y., Zong, C.L., Dai, M.N., Liu, X.M., Diwu, C.R., 2007. Study of Lu-Hf geochronology: A case study of eclogite from Dabie UHP Belt. Acta Petrologica Sinica, 23(2), 233-239 (in Chinese with English abstract).

Zhang, Y.Q., Chen, W., Yang, N., 2004. $^{40}Ar/^{39}Ar$ dating of shear deformation of the Xianshuihe fault zone in west Sichuan and its tectonic significance. Science in China: Series D, Earth Sciences, 47(9), 794-803.

Zhang, Y.Z., Replumaz, A., Leloup, P.H., Wang, G.C., Bernet, M., van der Beek, P., Paquette, J.L., Chevalier, M.L., 2017. Cooling history of the Gongga batholith: Implications for the Xianshuihe Fault and Miocene Kinematics of SE Tibet. Earth & Planetary Science Letters, 465, 1-15.

Zhang, Z., Yuan, X., Chen, Y., Tian, X., Kind, R., Li, X., Teng, J., 2010. Seismic signature of the collision between the eastern Tibet escape flow and the Sichuan basin. Earth & Planetary Science Letters, 292, 254-264.

Zhou, M.F., Yan, D.P., Kennedy, A.K., Li, Y.Q., Ding, J., 2002. SHRIMP U-Pb zircon geochronological and geochemical evidence for Neoproterozoic arc-magmatism along the western margin of Yangtze Block, South China. Earth & Planetary Science Letters, 196, 51-67.

Neoproterozoic gabbro-granite association from the Micangshan area, northern Yangtze Block: Implication for crustal growth in an active continental margin[①]

Lai Shaocong Qin Jiangfeng Long Xiaoping Li Yongfei Ju Yinjuan
Zhu Renzhi Zhao Shaowei Zhang Zezhong Zhu Yu Wang Jiangbo

Abstract: Neoproterozoic igneous rocks along the northern Yangtze Block are closely related to the evolution of the Rodinia supercontinent. This paper presents zircon U-Pb dating, whole-rock geochemistry and Sr-Nd isotopes of the gabbro-granite association from the Micangshan area, northern Yangtze Block. Zircon LA-ICP-MS U-Pb dating indicates that the tonalite and monzogranite from the Pinghe display identical ages of 869±4 Ma (MSWD=1.3, 2σ) and 859±8 Ma (MSWD=1.9, 2σ), respectively. According to the regional geology, the gabbro may be younger. The monzogranite and tonalite display almost identical geochemical features, both of them have high Na_2O/K_2O ratios, with insignificant negative Eu anomalies, relatively juvenile Sr-Nd isotopic compositions. The monzogranite have initial $(^{87}Sr/^{86}Sr)_i$ ratios of 0.703 403–0.705 218, and positive $\varepsilon_{Nd}(t)$ values of +1.23 to +2.98. The tonalite have $(^{87}Sr/^{86}Sr)_i$ ratios of 0.700 387–0.706 222, positive $\varepsilon_{Nd}(t)$ values of +1.98 to +7.79, indicating that they were derived from juvenile source. The 860 Ma minzogranite-tonalite association suggests crustal growth event in an active continental margin. The gabbro has low $(La/Yb)_N$ ratios (1.80–7.61) and total rare earth element contents ($\sum$REE=44.95–83.92 ppm), they have initial $(^{87}Sr/^{86}Sr)_i$ ratios of 0.701 913–0.704 815, positive $\varepsilon_{Nd}(t)$ values of +0.65 to +5.73, indicting an island-arc affinity. In combination with the other Neoproterozoic igneous rocks along the western and northern margin of the Yangtze Block, we propose that the gabbro and granitoid association from the Micangshan area represent two stage crustal growth events in the active continental margin setting, and these results imply that the Neoproterozoic crustal growth events in the Yangtze Block may be controlled by subduction.

1 Introduction

The tectonic evolution of the Rodinia supercontinent in the Yangtze Block is hotly debated (X.H. Li et al., 2003; Zhou, Yan, Kennedy, Li et al., 2002; Zhou, Yan, Wang, Qi, and

① Published in *Geological Journal*, 2018, 53.

Kennedy, 2006). The mantle plume model (Li, Li, Zhou, Liu, and Kinny, 2002; Z.X. Li et al., 2003), continental rifting model (Wu et al., 2006), and active continental margin model (Zhao, Zhou, and Jian-Ping, 2010) have been proposed to explain the genesis of these Neoproterozoic igneous rocks. Petrogenesis and geodynamic implication for the Neoproterozoic mafic and calc-alkaline granitoids remains controversial. This paper reports the Neoproterozoic gabbro, tonalite, and monzogranite (860 Ma) from the Pinghe area, Micangshan, northern Yangtze Block. Detailed petrology and geochemistry works reveal that the gabbro, tonalite, and monzogranite association was formed in an active continental margin, these results provide detailed petrological constraints on the Neoproterozoic tectonic evolution of the northern Yangtze Block.

2 Geological setting and field geology

Micangshan is located in the convergent region among the Yangtze, Southern Qinling and Northeastern Tibetan Plateau (Fig. 1a). The crystalline basement in this area comprises the Late-Archean Houhe Group and the Early Paleoproterozoic Huodiya Group. The Houhe Group consists of the Hekou migmatites, Bajiaoshu granitic genesis, and the Wangjiaping granulites. The Huodiya Group is composed of cordierite-bearing argillaceous schists, the fold basement contains low-grade metamorphic rocks of Shangliang Formation and Mawozi Formation (Qiu, 1993). Proterozoic igneous rocks in this region include Neoproterozoic ultramafic rocks, intermediate to alkaline rocks. The overlying strata include marine and continental sediments. The Devonian and Carboniferous strata are absent in this area (He, 2010). Neoproterozoic alkaline igneous rocks occur in the Pinghe, Zhongziyuan and Shuimo pluton (Fig. 1b).

The Pinghe pluton locates in the boundary between the Nanjiang and Wangcang county, alkaline igneous rocks in this area cover an area of 9.2 km^2. The primary intrusions include the Chadizi (NhC), Bailinghe (NhB), Wangjiaping (NhW), and Dashuwan (NhD) unites (He, 2010). According to the previous geological survey (Qiu, 1993), these small intrusions have a different lithology, mainly including jacupirangite, ijolite, urtite, and alkaline syenites. The Pinghe pluton was formed by mutistage intrusions. The jacupirangite, ijolite, and urtite in the center, the nepheline syenite, alkaline syenite, and some carbonatite in the margin. Some diorite and granite intrusions occurred in the eastern margin (Qiu, 1993). These alkaline rocks intruded into the Mesoproterozoic Shangliang Formation, Jinjiahe Formation, and Mawozi Formation, which were mainly consisted of sandstone, shale, limestone, and dolomite.

The studied samples collected from the Pinghe complex include gabbro, tonalite, and monzogranite.

Gabbro (PH18 to 32): These samples display dark-grey, medium- to coarse-grained, subeuhedral to anhedral, with hypidiomorphic and gabbroic textures (Fig. 2a,b). It mainly

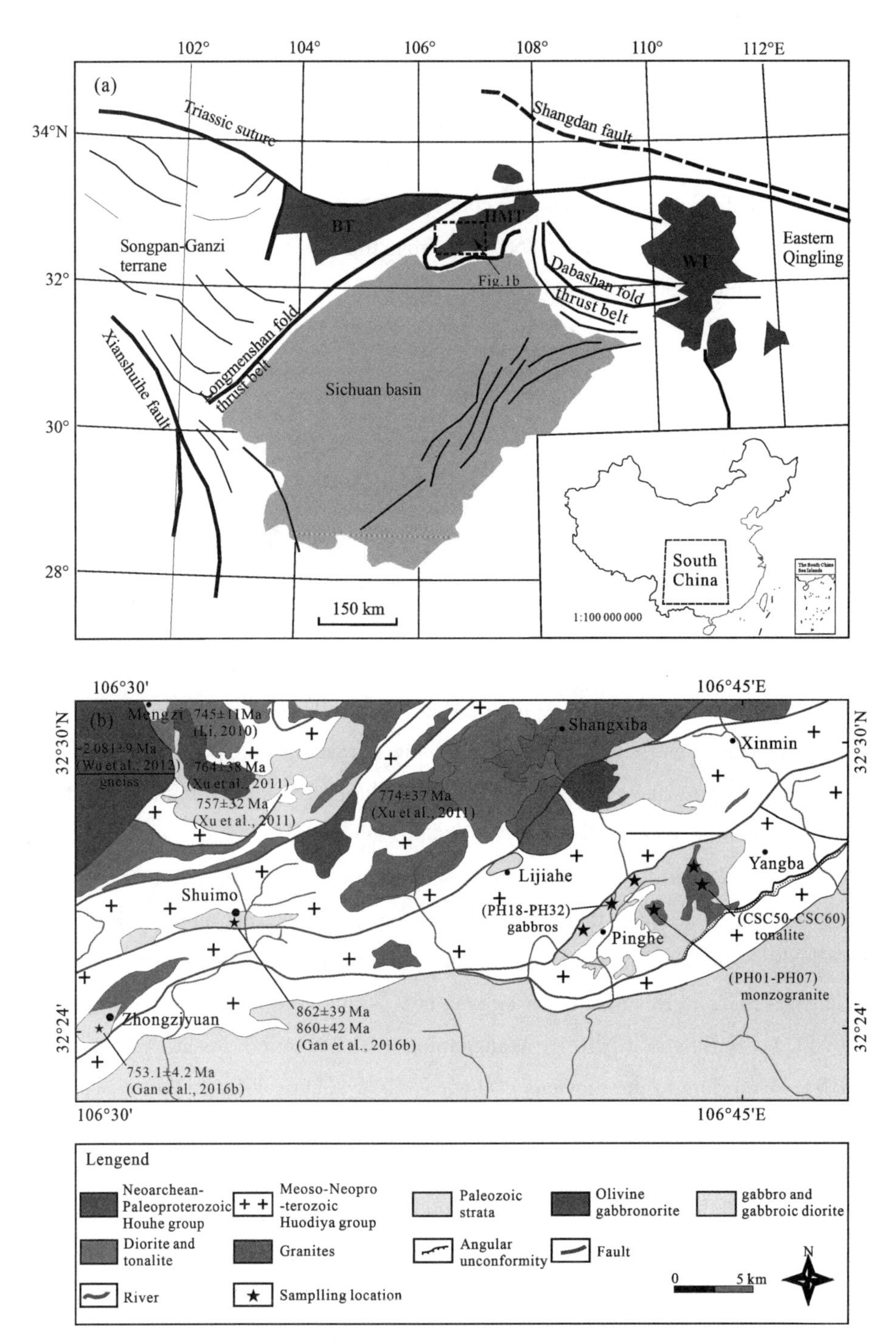

Fig. 1　(a) Sketch geological map of the Yangtze Block (after Wang et al., 2014) and the (b) geological map of the Pinghe complex (after Gan et al., 2017).

BT: Bikou Terrane; HMT: Hannan-Michangshan Terrane; WT: Wudang Terrane.

(Colour figure can be viewed at wileyonlinelibrary.com)

contains clinopyroxene (25%-30%), plagioclase (An_{50-70}, 55%-65%), amphibole (20%-25%), and minor olivine (<3%). Accessory minerals include magnetite, apatite, and zircon. The subeuhedral clinopyroxene grains are 0.2-0.5 mm in size, with corrosion texture. The plagiclase display sericitization.

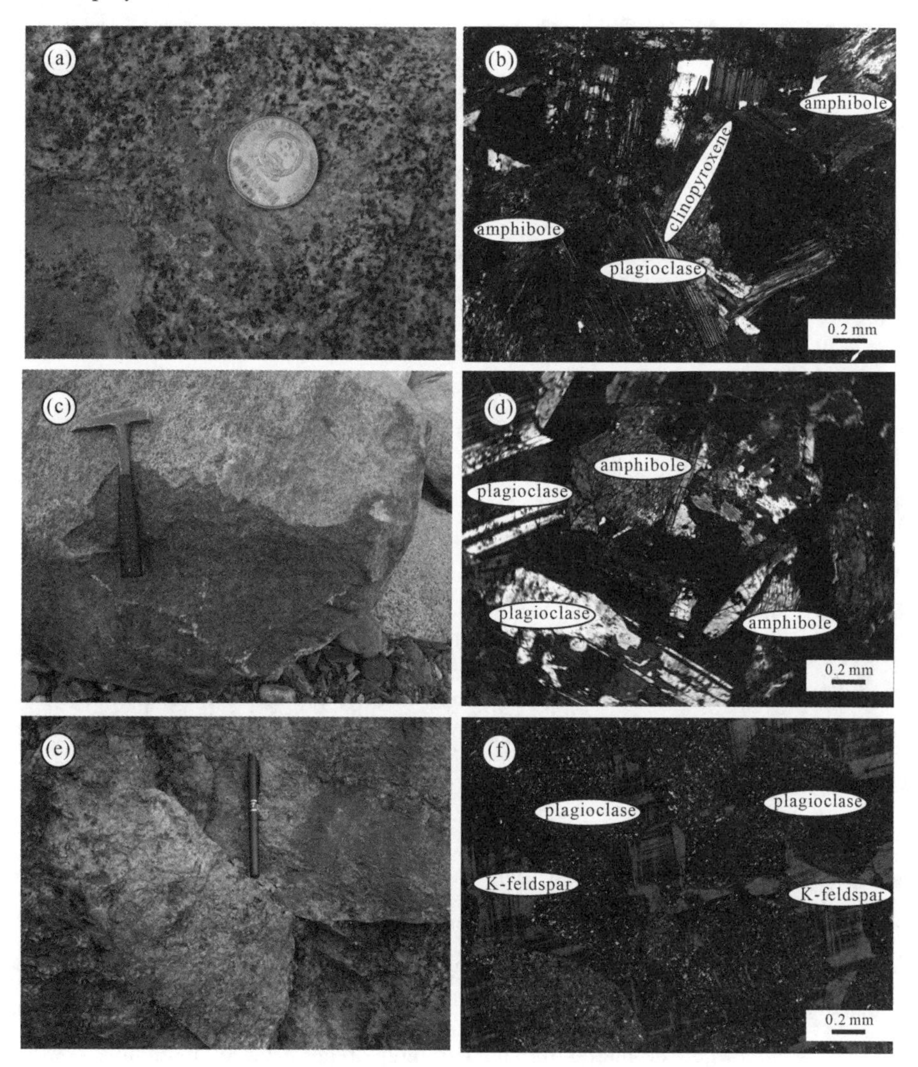

Fig. 2 Field photographs and microscope of the gabbro (a,b), tonalite (c,d) and monzogranite (e,f).

Tonalite (*CSC50 to 60*): Dark grey to green dark, medium- to coarse-grained (Fig. 2c,d). It was a serious alteration which may be metasomatized by alkaline fluid. The rocks contain alkali-feldspar (10%), plagioclase (An_{25-30}, 40%), amphibole 30%-35%, quartz (15%-20%). The aegirine-augite and iron impregnation 5%-10% surround the pyroxene and amphibole. Accessory minerals include magnetite, hematite, limonite, Fe-Ti oxides, apatite, and zircon.

Monzogranite (*PH-01 to 08*): It mainly contains microcline (40%-45%), sericitized plagioclase (30%–35%), and quartz (20%–25%). Accessory minerals (~3%) include magnetite, apatite, and zircon (Fig. 2e,f).

3 Analytical methods

3.1 Zircon LA-ICP-MS U-Pb analyses

The zircon grains were separated by using conventional heavy liquid and magnetic techniques. Representative zircon grains were handpicked and mounted in epoxy resin disks and then polished and coated with carbon. Internal morphology was examined using cathodoluminescence (CL) prior to U-Pb analyses.

Laser ablation ICP-MS zircon U-Pb analyses were conducted on an Agilent 7500a ICP-MS equipped with a 193-nm laser, which is housed at the State Key Laboratory of Continental Dynamics, Northwest University, Xi'an, China, following the method of Yuan et al. (2004). The $^{207}Pb/^{206}Pb$ and $^{206}Pb/^{238}U$ ratios were calculated by using the GLITTER program and corrected using the Harvard zircon 91500 as external calibration. The detailed analytical technique is described in Yuan et al. (2004). Common Pb contents were therefore evaluated by using the method described in Andersen (2002). The age calculations and plotting of concordia diagrams were made using ISOPLOT (version 3.0; Ludwig, 2003). The errors quoted in tables and figures are at the 2σ level.

3.2 Major and trace elements

Whole-rock samples were trimmed to remove weathered surfaces, cleaned with deionized water, crushed, and then powdered through a 200 mesh screen using a tungsten carbide ball mill. Major elements were analyzed using X-ray fluorescence (XRF) spectrometer (Rikagu RIX 2100) at the State Key Laboratory of Continental Dynamics, Northwest University, Xi'an, China. Analyses of USGS and Chinese national rock standards (BCR-2, GSR-1, and GSR-3) indicate that both analytical precision and accuracy for major elements are generally better than 5%.

Trace elements were determined by using a Bruker Aurora M90 inductively coupled plasma mass spectrometry (ICP-MS) at the Guizhou Tuopu Resource and Environmental Analysis Center, following the method of Qi, Hu, and Cregoire (2000). Sample powders were dissolved using an $HF+HNO_3$ mixture in a high-pressure PTFE bomb at 185 ℃ for 36 h. The accuracies of the ICP-MS analyses are estimated to be better than ±5%–10% (relative) for most elements.

3.3 Whole-rock Sr-Nd isotopes

Whole-rock Sr-Nd isotopic data were obtained by using a Nu Plasma HR multicollector mass spectrometer at the Guizhou Tuopu Resource and Environmental Analysis Center. The Sr and Nd isotopes were determined by using a method similar to that of Chu, Chen, Yang, and Guo (2009). Sr and Nd isotopic fractionation was corrected to $^{87}Sr/^{86}Sr = 0.1194$ and

$^{146}Nd/^{144}Nd = 0.7219$, respectively. During the period of analysis, a Neptune multicollector ICP-MS was used to measure the $^{87}Sr/^{86}Sr$ and $^{143}Nd/^{144}Nd$ isotope ratios. NIST SRM-987 and JMC-Nd were used as certified reference standard solutions for $^{87}Sr/^{86}Sr$ and $^{143}Nd/^{144}Nd$ isotopic ratios, respectively. BCR-1 and BHVO-1 were used as the reference materials.

4 Results

4.1 Zircon LA-ICP-MS U-Pb dating

Zircons from the tonalite (CSC-59) and monzogranite (PH-11) are collected for CL imaging and LA-ICP-MS U-Pb dating. The U-Pb dating results are listed in the Table. 1. The sampling locations, lithology, and dating results are summarized in Figs. 3 and 4.

The zircons are generally euhedral, with length of 100–200 μm. Most grains are colorless or light gray, prismatic, and subtransparent. Most grains exhibit oscillatory zoning in the CL images, suggesting a magmatic origin (Hoskin, Kinny, Wyborn, and Chappell, 2000; Hoskin and Schaltegger, 2003).

4.1.1 Tonalite (CSC-59)

Zircons from the tonalite are tiny crystals, they have crystals length minor than 100 μm, most grains display corrosion structure (Fig. 3a). Two spots display discordant ages, other 34 concordant spots display variable $^{206}Pb/^{238}U$ ages of 673±7 Ma to 916±12 Ma. In combination to the CL images and Th/U ratios, 28 out of 36 spots have been selected for calculating the weighted mean age. These zircons have $^{206}Pb/^{238}U$ ages of 849.2±8.1 Ma to 887.3±11.3 Ma (Fig. 4a), yielding a weighted mean age of 869 ± 4 Ma (MSWD = 1.3, n = 28, 2σ), representing the crystallization age of the tonalite.

4.1.2 Monzogranite (PH-11)

Zircons from the monzogranite are euhedral. They have crystals length of 200–300 μm, with aspect ratios of 2 : 1 to 3 : 1. Most grains display oscillatory zoning with dark rounded inherited cores (Fig. 3b). Five out of 36 spots display discordant ages, the other 31 spots can be subdivided into two groups: Group 1: 17 spots have $^{206}Pb/^{238}U$ ages of 301.9±2.8 Ma to 618.2±5.7 Ma, these spots have Th = 75–151 ppm, U = 455–3 435 ppm, with low Th/U ratios of 0.03–0.15 (with an exception of 0.27), these values are significantly lower than those of the magmatic zircons (Wu and Zheng, 2004), suggesting Pb-loss or representing the Paleozoic metamorphism age. Group 2 contains 14 spots, these spots have Th = 33.2–280.0 ppm, U = 221 – 1 489 ppm, with relative higher Th/U ratios of 0.10 – 0.96 (with two exceptions lower than 0.1), suggesting magmatic origin. These spots have $^{206}Pb/^{238}U$ ages of 842.4 ± 8.6 Ma to 884.5 ± 8.8 Ma (Fig. 4b), yielding a weighted mean age of 859 ± 8 Ma (MSWD = 1.9, n = 12, 2σ), representing the crystallization age of the monzogranite.

Table 1 Zircon LA-ICP MS U-Th-Pb isotopic analysis results for the Neoproterozoic Pinghe complex from the Micangshan region, Northern Yangtze Block.

Spot no.	Content/ppm			Isotopic ratios								Age/Ma							
	U	Th	Th/U	$^{207}Pb/^{206}Pb$	2σ	$^{207}Pb/^{235}U$	2σ	$^{206}Pb/^{238}U$	2σ	$^{208}Pb/^{232}Th$	2σ	$^{207}Pb/^{206}Pb$	2σ	$^{207}Pb/^{235}U$	2σ	$^{206}Pb/^{238}U$	2σ	$^{208}Pb/^{232}Th$	2σ
Tonalite																			
CSC-59-01	69	16	0.23	0.069 64	0.003 94	1.463 00	0.076 86	0.152 36	0.002 72	0.042 88	0.002 34	918	112	915	32	914	15	849	45
CSC-59-02	232	109	0.47	0.070 34	0.002 30	1.422 61	0.037 48	0.146 68	0.001 71	0.045 20	0.000 98	938	66	899	16	882	10	894	19
CSC-59-03	829	635	0.77	0.070 18	0.001 77	1.316 05	0.022 02	0.136 00	0.001 35	0.039 30	0.000 46	934	51	853	10	822	8	779	9
CSC-59-04	440	153	0.35	0.068 96	0.001 96	1.367 08	0.028 87	0.143 77	0.001 53	0.042 03	0.000 72	897	58	875	12	866	9	832	14
CSC-59-05	285	113	0.40	0.072 32	0.002 16	1.278 86	0.029 21	0.128 23	0.001 41	0.034 83	0.000 67	995	59	836	13	778	8	692	13
CSC-59-06	310	112	0.36	0.068 20	0.002 23	1.364 21	0.036 13	0.145 06	0.001 70	0.043 22	0.000 94	875	66	874	16	873	10	855	18
CSC-59-07	323	104	0.32	0.072 77	0.002 58	1.421 96	0.042 18	0.141 70	0.001 77	0.043 19	0.001 40	1 008	70	898	18	854	10	855	27
CSC-59-08	671	206	0.31	0.069 88	0.002 28	1.060 41	0.027 88	0.110 04	0.001 27	0.029 47	0.001 05	925	66	734	14	673	7	587	21
CSC-59-10	492	214	0.44	0.068 79	0.002 26	1.380 95	0.036 77	0.145 56	0.001 71	0.041 51	0.000 82	893	66	881	16	876	10	822	16
CSC-59-11	304	139	0.46	0.071 4	0.002 79	1.388 31	0.046 91	0.141 00	0.001 89	0.043 81	0.001 10	969	78	884	20	850	11	867	21
CSC-59-12	210	60	0.29	0.063 78	0.002 27	1.267 81	0.037 88	0.144 15	0.001 76	0.039 65	0.001 04	734	74	831	17	868	10	786	20
CSC-59-13	166	51	0.31	0.068 53	0.002 28	1.390 03	0.037 87	0.147 08	0.001 75	0.042 46	0.001 07	885	67	885	16	885	10	841	21
CSC-59-14	454	229	0.51	0.069 43	0.001 78	1.348 18	0.023 58	0.140 80	0.001 43	0.040 36	0.000 53	912	52	867	10	849	8	800	10
CSC-59-15	227	68	0.30	0.069 66	0.002 06	1.359 76	0.030 79	0.141 53	0.001 56	0.040 77	0.000 84	919	60	872	13	853	9	808	16
CSC-59-16	199	59	0.30	0.068 47	0.002 09	1.358 86	0.032 38	0.143 90	0.001 62	0.042 37	0.000 91	883	62	871	14	867	9	839	18
CSC-59-17	116	30	0.26	0.070 48	0.002 87	1.397 46	0.049 93	0.143 78	0.001 98	0.043 29	0.001 50	942	81	888	21	866	11	857	29
CSC-59-18	111	42	0.38	0.068 84	0.002 65	1.344 82	0.044 66	0.141 66	0.001 86	0.042 68	0.001 15	894	78	865	19	854	11	845	22
CSC-59-19	101	38	0.37	0.067 99	0.002 91	1.366 32	0.051 88	0.145 72	0.002 08	0.042 22	0.001 32	868	86	875	22	877	12	836	26
CSC-59-20	195	62	0.32	0.069 60	0.002 1	1.401 71	0.032 91	0.146 04	0.001 64	0.044 01	0.000 90	917	61	890	14	879	9	871	17
CSC-59-21	245	104	0.43	0.068 32	0.002 14	1.371 57	0.034 19	0.145 58	0.001 68	0.042 26	0.000 85	878	64	877	15	876	9	837	16
CSC-59-22	201	71	0.35	0.068 81	0.002 36	1.343 47	0.038 27	0.141 57	0.001 73	0.039 60	0.000 95	893	69	865	17	854	10	785	19
CSC-59-23	133	42	0.32	0.069 53	0.003 36	1.456 92	0.064 05	0.151 95	0.002 42	0.046 89	0.001 79	915	96	913	26	912	14	926	35

Continued

Spot no.	Content/ppm			Isotopic ratios								Age/Ma							
	U	Th	Th/U	$^{207}Pb/^{206}Pb$	2σ	$^{207}Pb/^{235}U$	2σ	$^{206}Pb/^{238}U$	2σ	$^{208}Pb/^{232}Th$	2σ	$^{207}Pb/^{206}Pb$	2σ	$^{207}Pb/^{235}U$	2σ	$^{206}Pb/^{238}U$	2σ	$^{208}Pb/^{232}Th$	2σ
CSC-59-24	152	60	0. 40	0. 070 05	0. 002 76	1. 425 54	0. 048 73	0. 147 56	0. 002 01	0. 046 16	0. 001 21	930	79	900	20	887	11	912	23
CSC-59-25	293	78	0. 27	0. 069 21	0. 001 93	1. 395 78	0. 028 77	0. 146 25	0. 001 58	0. 041 35	0. 000 84	905	56	887	12	880	9	819	16
CSC-59-27	117	29	0. 24	0. 068 4	0. 002 95	1. 364 47	0. 052 41	0. 144 66	0. 002 08	0. 040 88	0. 001 43	881	87	874	23	871	12	810	28
CSC-59-28	212	109	0. 51	0. 067 23	0. 002 15	1. 354 79	0. 034 89	0. 146 15	0. 001 71	0. 043 28	0. 000 79	845	65	870	15	879	10	856	15
CSC-59-29	331	132	0. 40	0. 066 65	0. 002 09	1. 312 5	0. 032 98	0. 142 82	0. 001 65	0. 041 86	0. 000 84	827	64	851	14	861	9	829	16
CSC-59-30	127	56	0. 44	0. 069 6	0. 002 47	1. 398 47	0. 041 96	0. 145 73	0. 001 85	0. 042 57	0. 001 01	917	71	888	18	877	10	843	19
CSC-59-31	130	46	0. 36	0. 070 21	0. 003 74	1. 382 73	0. 067 97	0. 142 84	0. 002 47	0. 043 27	0. 001 83	934	106	882	29	861	14	856	36
CSC-59-32	177	35	0. 20	0. 069 53	0. 002 75	1. 463 69	0. 050 58	0. 152 68	0. 002 11	0. 045 15	0. 001 85	915	79	916	21	916	12	893	36
CSC-59-33	226	73	0. 32	0. 068 17	0. 002 02	1. 361 85	0. 031 38	0. 144 88	0. 001 64	0. 043 01	0. 000 87	874	60	873	13	872	9	851	17
CSC-59-34	561	250	0. 44	0. 068 09	0. 001 80	1. 362 03	0. 025 70	0. 145 08	0. 001 53	0. 041 65	0. 000 60	871	54	873	11	873	9	825	12
CSC-59-35	196	40	0. 20	0. 069 12	0. 002 27	1. 394 01	0. 037 61	0. 146 28	0. 001 77	0. 045 62	0. 001 32	902	66	886	16	880	10	902	25
CSC-59-36	326	125	0. 38	0. 068 10	0. 002 39	1. 351 54	0. 039 97	0. 143 95	0. 001 81	0. 041 29	0. 001 12	872	71	868	17	867	10	818	22
Monzogranite																			
PH-11-01	1 554	107	0. 07	0. 102 66	0. 002 35	1. 215 48	0. 016 71	0. 085 87	0. 000 84	0. 150 86	0. 002 05	1 673	42	808	8	531	5	2 840	36
PH-11-02	3 436	156	0. 05	0. 065 69	0. 001 45	0. 574 46	0. 007 23	0. 063 43	0. 000 60	0. 032 08	0. 000 54	796	46	461	5	396	4	638	11
PH-11-03	2 749	89	0. 03	0. 065 25	0. 001 61	0. 592 47	0. 009 87	0. 065 85	0. 000 65	0. 046 57	0. 001 20	783	51	472	6	411	4	920	23
PH-11-04	204	33	0. 16	0. 068 47	0. 001 94	1. 318 00	0. 028 31	0. 139 60	0. 001 52	0. 048 47	0. 001 17	883	57	854	12	842	9	957	23
PH-11-05	2 030	101	0. 05	0. 063 38	0. 001 42	0. 685 64	0. 009 00	0. 078 45	0. 000 75	0. 026 65	0. 000 53	721	47	530	5	487	4	532	10
PH-11-06	261	35	0. 13	0. 068 53	0. 001 85	1. 389 65	0. 027 53	0. 147 07	0. 001 56	0. 048 33	0. 001 19	885	55	885	12	885	9	954	23
PH-11-07	2 249	171	0. 08	0. 065 15	0. 001 55	0. 634 71	0. 009 71	0. 070 66	0. 000 69	0. 028 66	0. 000 58	779	49	499	6	440	4	571	11
PH-11-08	1 055	73	0. 07	0. 067 43	0. 001 65	0. 687 86	0. 011 22	0. 073 98	0. 000 73	0. 047 49	0. 001 02	851	50	532	7	460	4	938	20
PH-11-09	2 191	83	0. 04	0. 063 96	0. 001 44	0. 557 66	0. 007 47	0. 063 23	0. 000 60	0. 034 45	0. 000 68	740	47	450	5	395	4	685	13
PH-11-10	1 606	69	0. 04	0. 066 53	0. 001 53	0. 837 61	0. 011 85	0. 091 31	0. 000 88	0. 036 90	0. 000 87	823	47	618	7	563	5	732	17
PH-11-11	384	60	0. 16	0. 066 79	0. 001 61	1. 330 01	0. 021 00	0. 144 41	0. 001 43	0. 044 43	0. 000 80	831	49	859	9	870	8	879	16

Continued

Spot no.	Content/ppm			Isotopic ratios								Age/Ma							
	U	Th	Th/U	$^{207}Pb/^{206}Pb$	2σ	$^{207}Pb/^{235}U$	2σ	$^{206}Pb/^{238}U$	2σ	$^{208}Pb/^{232}Th$	2σ	$^{207}Pb/^{206}Pb$	2σ	$^{207}Pb/^{235}U$	2σ	$^{206}Pb/^{238}U$	2σ	$^{208}Pb/^{232}Th$	2σ
PH-11-12	2 561	115	0. 04	0. 062 52	0. 001 43	0. 552 40	0. 007 75	0. 064 08	0. 000 61	0. 035 06	0. 000 74	692	48	447	5	400	4	697	14
PH-11-13	2 026	149	0. 07	0. 061 92	0. 001 57	0. 471 55	0. 008 25	0. 055 23	0. 000 55	0. 019 35	0. 000 50	671	53	392	6	347	3	387	10
PH-11-14	289	56	0. 19	0. 068 62	0. 002 03	1. 350 69	0. 031 23	0. 142 74	0. 001 60	0. 035 71	0. 000 99	888	60	868	13	860	9	709	19
PH-11-15	1 059	103	0. 10	0. 066 26	0. 001 64	1. 302 53	0. 022 01	0. 142 56	0. 001 44	0. 040 81	0. 000 96	815	51	847	10	859	8	809	19
PH-11-16	443	64	0. 15	0. 068 11	0. 001 90	1. 318 48	0. 027 72	0. 140 39	0. 001 52	0. 043 84	0. 001 18	872	57	854	12	847	9	867	23
PH-11-17	456	122	0. 27	0. 066 69	0. 002 09	0. 678 63	0. 017 02	0. 073 80	0. 000 83	0. 045 24	0. 000 77	828	64	526	10	459	5	894	15
PH-11-18	2 281	90	0. 04	0. 063 61	0. 001 50	0. 655 59	0. 009 87	0. 074 75	0. 000 73	0. 026 84	0. 000 71	729	49	512	6	465	4	535	14
PH-11-19	3 062	186	0. 06	0. 081 84	0. 001 88	0. 641 50	0. 009 05	0. 056 85	0. 000 55	0. 065 67	0. 000 99	1 242	44	503	6	356	3	1 286	19
PH-11-20	4 616	77	0. 02	0. 059 93	0. 001 40	0. 425 75	0. 006 24	0. 051 52	0. 000 50	0. 045 74	0. 001 65	601	50	360	4	324	3	904	32
PH-11-21	1 489	38	0. 03	0. 064 35	0. 002 11	1. 240 31	0. 033 56	0. 139 80	0. 001 65	0. 075 46	0. 003 21	753	68	819	15	844	9	1 471	60
PH-11-22	1 071	157	0. 15	0. 067 05	0. 001 76	0. 929 55	0. 017 49	0. 100 55	0. 001 04	0. 033 28	0. 000 71	839	54	667	9	618	6	662	14
PH-11-23	2 111	84	0. 04	0. 062 64	0. 001 42	0. 497 37	0. 006 76	0. 057 58	0. 000 55	0. 024 22	0. 000 53	696	47	410	5	361	3	484	10
PH-11-24	516	42	0. 08	0. 085 90	0. 002 14	1. 000 70	0. 016 94	0. 084 49	0. 000 86	0. 099 90	0. 001 72	1 336	47	704	9	523	5	1 925	32
PH-11-25	1 758	75	0. 04	0. 062 32	0. 001 41	0. 525 12	0. 007 21	0. 061 11	0. 000 58	0. 025 03	0. 000 54	685	48	429	5	382	4	500	11
PH-11-26	399	55	0. 14	0. 067 54	0. 001 80	1. 330 69	0. 025 93	0. 142 88	0. 001 50	0. 046 41	0. 001 04	855	54	859	11	861	8	917	20
PH-11-27	323	311	0. 96	0. 067 89	0. 001 81	1. 346 69	0. 026 19	0. 143 86	0. 001 51	0. 045 69	0. 000 55	865	54	866	11	867	9	903	11
PH-11-28	308	280	0. 91	0. 067 75	0. 001 81	1. 336 08	0. 026 16	0. 143 03	0. 001 51	0. 044 82	0. 000 55	861	54	862	11	862	9	886	11
PH-11-29	1 190	85	0. 07	0. 066 68	0. 001 56	1. 300 49	0. 019 37	0. 141 44	0. 001 39	0. 043 94	0. 001 03	828	48	846	9	853	8	869	20
PH-11-30	1 471	116	0. 08	0. 066 07	0. 001 56	0. 916 87	0. 013 86	0. 100 64	0. 000 98	0. 033 18	0. 000 73	809	48	661	7	618	6	660	14
PH-11-31	4 331	147	0. 03	0. 062 42	0. 001 44	0. 412 69	0. 005 93	0. 047 95	0. 000 46	0. 038 54	0. 000 80	689	48	351	4	302	3	764	16
PH-11-32	236	41	0. 17	0. 068 20	0. 001 91	1. 362 20	0. 029 06	0. 144 86	0. 001 57	0. 042 50	0. 001 07	875	57	873	12	872	9	841	21
PH-11-33	326	34	0. 10	0. 069 25	0. 001 89	1. 170 85	0. 023 84	0. 122 63	0. 001 31	0. 053 05	0. 001 38	906	55	787	11	746	8	1 045	26
PH-11-34	2 166	95	0. 04	0. 069 58	0. 001 63	0. 616 87	0. 009 19	0. 064 30	0. 000 63	0. 051 53	0. 001 02	916	47	488	6	402	4	1 016	20
PH-11-35	222	33	0. 15	0. 067 46	0. 002 77	1. 308 04	0. 047 55	0. 140 63	0. 001 97	0. 057 08	0. 002 36	852	83	849	21	848	11	1 122	45

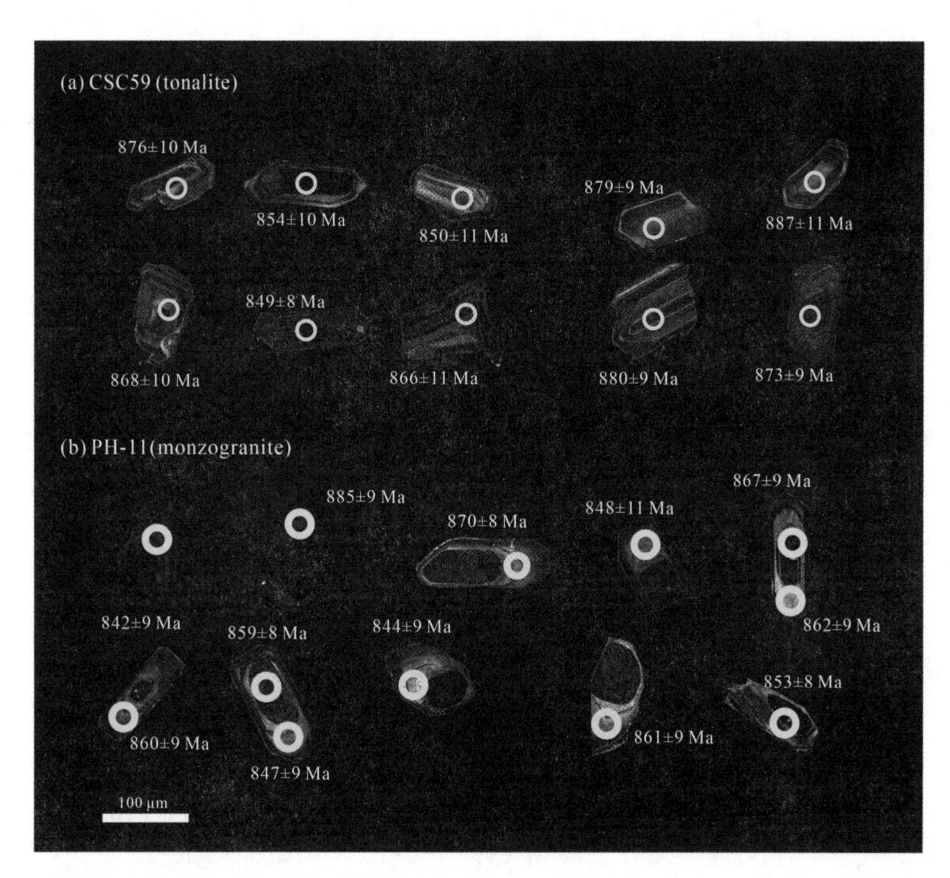

Fig. 3 Cathodoluminescence (CL) images of representative zircons in the Pinghe complex.
(a) tonalite; (b) monzogranite.

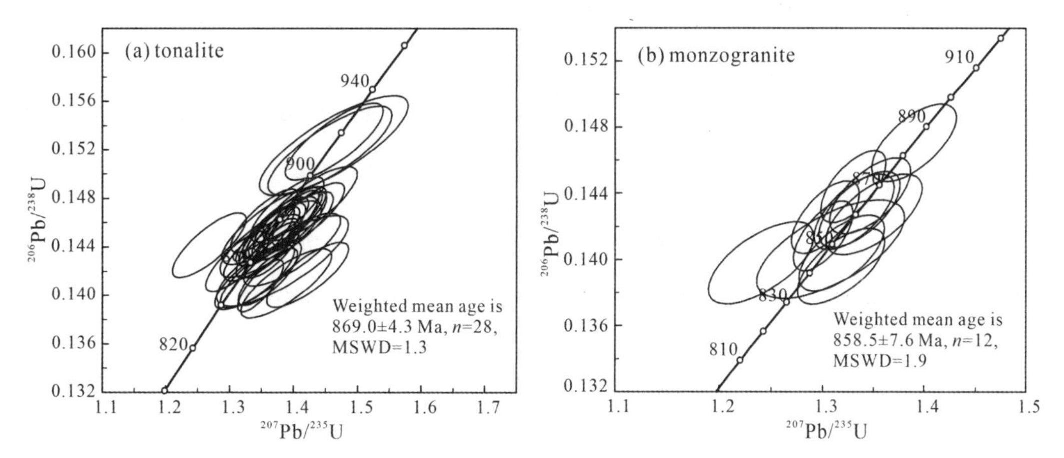

Fig. 4 LA-ICP-MS U-Pb zircon concordia diagrams of representative zircons in in the Pinghe complex.
(a) tonalite; (b) monzogranite. Ellipse dimensions are 2σ.

4. 2 Major and trace elements

Whole rock major- and trace-element data and corresponding CIPW normative calculations from 49 samples are listed in Table 2. Most samples display high loss on ignition (LOI) ranging from 0. 32 to 6. 56. We recalculated these data after deducting the LOI. According to the total alkali-silica (TAS) geochemical classification diagram (Middlemost, 1994), these rocks fell into three fields: granite to syenite, gabbroic diorite to gabbro, foid syenite to foid monzodiorite.

4. 2. 1 Gabbro

The gabbro has SiO_2 = 47. 16–53. 90 wt%, Al_2O_3 = 13. 11 – 19. 29 wt%, TiO_2 = 0. 71 – 1. 30 wt%. They have high MgO (4. 15–9. 29 wt%) and $Fe_2O_3^T$(6. 31–10. 24 wt%) contents with high $Mg^{\#}$ values of 61–69. These rocks have Na_2O = 1. 81–3. 20 wt%, K_2O = 0. 52–1. 42 wt%, with high Na_2O/K_2O ratios of 2. 2–3. 8. As shown in the TAS diagram (Fig. 5a), all the samples plot in the field of gabbro. These rocks display right-sloped REE patterns, they have variable $(La/Yb)_N$ ratios of 1. 80–7. 61, positive Eu anomalies of 1. 00–1. 32. They have ΣREE contents of 44. 95–83. 92 ppm. In the primitive mantle-normalized trace element spider diagrams, these rocks display enrichment in Rb, Ba, K, and La, toughs in Th, U, Nb, Ta, Ti, and P, similar to those of the island-arc basalts.

The gabbroic diorite has SiO_2 = 51. 50–58. 87 wt%, high Al_2O_3 contents of 16. 08–24. 79 wt%, TiO_2 = 0. 69–1. 16 wt%, low MgO (1. 90–3. 19 wt%), and $Fe_2O_3^T$(4. 10–8. 27 wt%) contents with $Mg^{\#}$ values of 47. 3 – 52. 1. These rocks have Na_2O = 3. 01 – 3. 84 wt%, K_2O = 0. 12–1. 65 wt%, high Na_2O/K_2O ratios of 2. 3–2. 5. These rocks have extremely low ΣREE contents of 7. 23 – 8. 89 ppm (with an exception of 120. 3 ppm), they have low $(La/Yb)_N$ ratios of 3. 92 – 6. 60 and extremely high Eu anomalies of 3. 53 – 5. 09 (with an exception of 1. 07).

4. 2. 2 Tonalite

The tonalite has SiO_2 = 65. 84–68. 71 wt%, Al_2O_3 = 14. 05–16. 76 wt%, MgO = 0. 75–1. 45 wt%, $Fe_2O_3^T$ = 2. 08–4. 04 wt%, $Mg^{\#}$ values of 44. 1–45. 6. These rocks have CaO = 1. 57–5. 77 wt%, Na_2O = 7. 67 – 8. 40 wt%, K_2O = 1. 58 – 1. 84 wt%, high Na_2O/K_2O ratios of 4. 3 – 5. 0, A/CNK values of 0. 56–0. 97. All the samples plot in the dacite field (Fig. 5a), suggesting they are intermediate sodic rocks. These rocks display right-sloped REE patterns with insignificant Eu anomalies, they have variable $(La/Yb)_N$ ratios of 18. 5 – 42. 6, Eu^*/Eu = 0. 85 – 0. 95, high ΣREE contents of 74. 8–139. 5 ppm. These rocks display enrichment in Rb, Ba, K, and La, depletion in Th, U, Nb, Ta, Ti, and P. They have low Rb (17. 7–25. 4 ppm), high Sr (165–187 ppm), and Ba (585–696 ppm), with low Rb/Sr ratios of 0. 09–0. 14.

Table 2 Major(wt%) and trace(ppm) element for the Pinghe Neoproterozoic igneous rocks, Micangshan region, northern Yangtze Block.

Sample	PH01	PH02	PH03	PH06	PH07	CSC50	CSC51	CSC52	CSC53	CSC54	CSC55	CSC56	CSC60	PH18	PH19	PH20	PH21	PH22	PH23	PH24	PH25	PH26	PH27	PH28	PH29	PH30	PH31	PH32
	Monzogranite					Tonalite								Gabbro														
SiO_2	68.76	74.80	71.03	70.93	71.63	66.58	68.58	67.53	68.71	67.80	66.24	65.84	68.50	48.57	49.86	50.00	47.16	48.54	54.37	50.06	55.24	49.31	54.27	53.37	52.00	57.03	56.05	59.35
TiO_2	0.06	0.05	0.14	0.11	0.10	0.41	0.46	0.32	0.46	0.41	0.43	0.37	0.41	1.21	0.94	0.93	1.30	0.93	0.74	0.92	0.71	1.09	0.89	0.74	0.69	0.71	0.91	1.16
Al_2O_3	16.56	13.20	15.19	14.35	13.76	14.46	14.75	14.89	14.05	14.62	16.76	15.82	14.24	13.45	14.83	14.10	13.27	14.57	19.89	14.13	19.59	15.07	21.23	23.70	25.03	20.54	20.32	16.21
$Fe_2O_3^T$	1.59	0.86	2.11	2.01	1.84	2.08	3.13	2.97	2.42	2.64	4.04	2.80	3.01	8.13	9.92	10.03	8.72	10.46	6.97	10.14	6.47	9.77	5.17	4.99	4.77	4.14	4.80	8.34
MnO	0.03	0.01	0.05	0.04	0.04	0.04	0.03	0.03	0.03	0.03	0.04	0.04	0.03	0.14	0.18	0.18	0.15	0.19	0.12	0.18	0.11	0.17	0.08	0.07	0.07	0.07	0.07	0.18
MgO	2.59	0.84	0.48	0.46	0.43	0.75	1.12	1.06	0.82	0.94	1.45	0.96	0.97	7.53	8.60	9.48	7.69	9.30	4.50	9.23	4.25	8.47	2.33	1.93	2.04	1.93	1.92	3.22
CaO	1.39	0.83	3.55	3.45	3.36	5.77	2.37	3.59	3.81	3.36	1.57	4.40	3.24	16.37	12.88	12.65	17.58	13.14	8.67	12.74	8.63	13.08	12.82	12.00	12.17	12.27	12.64	5.66
Na_2O	6.67	5.88	5.48	5.49	5.33	7.97	7.84	7.88	8.01	8.40	7.67	7.93	7.78	3.21	2.01	1.85	2.83	2.08	3.17	1.91	3.28	2.12	3.05	3.03	3.06	3.14	3.13	3.87
K_2O	2.31	3.53	1.93	3.12	3.47	1.84	1.61	1.63	1.58	1.70	1.68	1.74	1.71	1.37	0.56	0.56	1.30	0.55	1.35	0.53	1.46	0.63	0.12	0.13	0.14	0.13	0.13	1.66
P_2O_5	0.03	0.01	0.04	0.04	0.04	0.10	0.11	0.09	0.10	0.09	0.12	0.10	0.11	0.01	0.22	0.21	0.01	0.23	0.21	0.17	0.27	0.29	0.04	0.03	0.03	0.03	0.03	0.36
$Mg^{\#}$	79.1	69.5	34.5	34.6	35.1	45.6	45.5	45.5	44.1	45.3	45.5	44.4	42.9	68.3	66.9	68.8	67.3	67.5	60.0	68.0	60.5	66.9	51.3	47.4	49.9	52.1	48.2	47.3
A/CNK	1.04	0.88	0.86	0.77	0.74	0.56	0.78	0.70	0.64	0.67	0.97	0.69	0.69	0.37	0.54	0.53	0.35	0.52	0.89	0.53	0.86	0.54	0.75	0.88	0.92	0.74	0.72	0.88
δ	3.13	2.78	1.96	2.65	2.71	4.08	3.50	3.68	3.58	4.12	3.76	4.09	3.53	3.78	0.96	0.83	4.10	1.25	1.79	0.84	1.83	1.20	0.89	0.96	1.14	0.76	0.82	1.87
Li	48	14.6	11.2	13.1	10.7	6.21	7.7	6.85	7.09	7.14	8.93	7.1	7.37	2.31	9.68	12.7	2.1	9.42	10.4	11	10.5	10.6	0.15	0.26	0.2	0.11	0.13	13.7
Be	6.21	2.69	2.26	2.51	2.1	1	1.03	1.02	1.03	1.12	1.16	1.1	1.42	0.51	0.41	0.4	0.5	0.34	1.07	0.37	1.06	0.37	0.19	0.17	0.2	0.21	0.17	1.22
Sc	10.1	11.5	9.52	10.7	11.8	10.4	12.8	11.1	11.8	12.1	14.5	12.9	12.6	27.8	19.8	21.1	28.8	21.1	19.9	22.1	19.1	21.5	13.3	9.07	8.65	8.74	12.1	21.4
V	2.62	1.04	5.14	3.62	3.82	12.2	18.2	16.3	17	14.1	22.8	15.5	17.6	143	152	162	149	156	109	163	101	163	166	137	125	134	173	141
Cr	0.59	1.15	0.50	3.76	3.01	3.44	7.90	4.57	3.62	7.17	7.65	7.03	6.80	27.72	47.25	52.20	29.97	52.11	62.46	54.18	59.31	51.84	5.17	5.19	8.07	7.52	9.36	5.02
Co	8.69	14.9	23.4	15	16	5.17	8.49	7.33	5.88	6.3	8.36	8.63	6.8	39.3	45.2	43.8	40.7	46.5	28.8	46.7	28.3	49.2	26.9	22	25.2	26.3	32.5	31.8
Ni				5.14	5.3	7.2	9.7	6.32	4.69	8.17	8.6	6.82	5.98	56.6	66.9	57.2	56.7	61.2	58.1	56.3	52.7	59.6	12.1	10.5	12	13.1	14.8	13.9
Cu	8.51	24.48	7.97	18.99	19.89	18.27	12.87	12.06	13.59	16.65	15.39	16.20	11.70	13.77	39.60	30.33	14.13	31.50	21.24	26.37	15.21	29.61	18.54	17.73	17.10	16.83	20.88	29.25
Zn	38.1	24.9	75.5	42.4	37.8	41.3	35	36.3	44.6	26.6	45.8	30	19.4	90.4	102	118	82.6	127	101	131	110	118	79.1	82.7	63.8	78.3	83	137
Ga	26.7	19.2	21	20.7	21.2	15	19	17.8	18	18.1	22.1	18.5	19.1	14.2	12.9	12.7	14.6	12.8	18.5	12.5	18.1	13.3	15	15.9	15.5	14.2	14.7	17.9
As	0.75	0.1	2.73	0.88	0.82	0.74	0.92	0.71	2.17	0.61	0.85	0.69	0.74	0.89	1.34	1.56	1.13	1.24	1.88	1.24	3.78	1.2	1.03	0.75	0.69	0.73	0.96	0.99
Rb	47.5	92.2	54.2	69.6	66.4	20.6	19.2	17.7	18.2	19	25.4	21.7	20.3	30.9	10.7	11.2	28.8	10.8	38	10.6	43.1	12.6	1.49	1.22	1.4	1.3	1.56	30.5
Sr	113	283	443	466	436	166	181	167	182	165	185	187	184	111	561	518	111	528	437	504	441	574	747	797	811	774	759	333
Y	6.3	2.32	11.3	10.1	9.47	10.5	8.83	9.07	9.75	8.9	11.6	10.2	11.4	18.2	20.1	21.8	18.6	21.2	16.2	21	16.3	22.3	3.11	1.75	1.89	1.89	2.46	29.7
Zr	122	89.2	96.9	116	130	230	261	213	245	209	305	225	234	134	62.8	71.9	139	69.4	78.5	68.9	107	70.2	7.48	5.19	6.25	6.06	5.09	113
Nb	7.13	4.97	10.8	10.1	9.78	8.46	9.76	6.68	10.7	9.12	9.92	7.77	9.17	1.94	4	4.39	1.91	4.02	5.69	3.96	6.13	5.5	0.48	0.37	0.35	0.38	0.5	12.1

Continued

Sample	PH01	PH02	PH03	PH06	PH07	CSC50	CSC51	CSC52	CSC53	CSC54	CSC55	CSC56	CSC60	PH18	PH19	PH20	PH21	PH22	PH23	PH24	PH25	PH26	PH27	PH28	PH29	PH30	PH31	PH32
	Monzogranite					Tonalite								Gabbro														
Cs	1.17	1.79	1.28	1.52	1.19	0.13	0.15	0.12	0.14	0.14	0.16	0.14	0.16	0.36	1.7	1.58	0.34	1.87	1.51	1.6	1.4	1.73	0.3	0.29	0.29	0.31	0.28	0.63
Ba	233	386	391	620	505	696	611	585	655	640	587	653	682	22.4	290	259	23.2	263	316	247	303	291	69	72.6	73	67.4	68.2	623
La	8.85	14.8	8.82	8.89	25.2	23.3	42.8	35.3	33.9	18.9	28.3	26.1	29.3	4.71	9.61	9.58	4.71	10.3	15.9	9.37	17.5	10.7	1.44	1.35	1.36	1.3	1.38	21.4
Ce	12.1	19.8	17	16.2	35.8	33.4	59.3	50.2	45.9	29.9	45.9	40.2	43.3	13.2	21.3	22.2	13.5	23	31.4	21.3	33.7	24.2	2.79	2.59	2.62	2.51	2.52	45.1
Pr	1.15	1.73	2.29	2.08	3.45	3.85	6.46	5.27	5.36	3.62	4.99	4.17	4.82	2.07	2.79	3.03	2.17	3.02	3.68	2.77	3.84	3.15	0.36	0.31	0.3	0.3	0.31	5.55
Nd	3.64	4.67	8.92	8.27	11	14.1	20.9	17.2	18.6	13.1	16.9	14.2	17.1	10.3	13	14.1	10.7	13.2	14.8	12.6	14.8	14	1.61	1.3	1.29	1.28	1.39	23.1
Sm	0.6	0.54	1.76	1.68	1.73	2.46	3.15	2.62	3	2.58	3.27	2.45	3.16	2.74	3.14	3.51	2.91	3.35	3.16	3.29	3.14	3.6	0.43	0.31	0.32	0.3	0.38	5.1
Eu	0.600	0.677	1.111	1.226	1.180	0.667	0.728	0.636	0.769	0.689	0.756	0.683	0.855	0.891	1.337	1.366	0.920	1.326	1.039	1.307	1.009	1.435	0.498	0.494	0.502	0.472	0.517	1.733
Gd	0.65	0.46	1.54	1.45	1.51	1.89	1.98	1.81	2.06	2.02	2.25	2.02	2.55	2.70	3.03	3.31	2.82	3.28	2.72	3.18	2.73	3.46	0.43	0.28	0.29	0.28	0.37	4.83
Tb	0.11	0.07	0.29	0.26	0.25	0.33	0.33	0.32	0.34	0.32	0.40	0.33	0.41	0.50	0.55	0.60	0.52	0.60	0.48	0.58	0.47	0.61	0.08	0.05	0.05	0.05	0.06	0.86
Dy	0.70	0.30	1.62	1.54	1.46	1.81	1.70	1.65	1.80	1.60	2.12	1.79	2.14	3.10	3.38	3.72	3.17	3.59	2.76	3.53	2.68	3.73	0.51	0.28	0.29	0.30	0.41	5.02
Ho	0.16	0.07	0.34	0.32	0.30	0.34	0.31	0.31	0.34	0.30	0.39	0.34	0.39	0.61	0.69	0.74	0.64	0.72	0.53	0.70	0.53	0.74	0.10	0.06	0.06	0.06	0.08	1.01
Er	0.53	0.23	1.05	1.00	0.94	0.94	0.88	0.86	0.94	0.87	1.07	0.93	1.05	1.80	1.98	2.15	1.87	2.07	1.57	2.04	1.59	2.12	0.29	0.16	0.18	0.17	0.23	2.96
Tm	0.09	0.03	0.16	0.15	0.14	0.12	0.11	0.11	0.12	0.11	0.14	0.12	0.13	0.26	0.27	0.30	0.26	0.30	0.23	0.29	0.23	0.30	0.04	0.02	0.02	0.02	0.03	0.42
Yb	0.64	0.26	1.04	1.01	0.97	0.79	0.72	0.72	0.77	0.73	0.84	0.76	0.79	1.79	1.75	1.93	1.87	1.91	1.50	1.82	1.48	1.92	0.26	0.15	0.15	0.16	0.20	2.83
Lu	0.10	0.04	0.15	0.15	0.14	0.12	0.10	0.10	0.11	0.10	0.12	0.10	0.11	0.29	0.25	0.28	0.29	0.28	0.21	0.29	0.22	0.28	0.04	0.02	0.02	0.02	0.03	0.42
Hf	3.36	2.58	2.54	3.11	3.39	5.09	5.34	4.46	5.12	4.17	6.15	4.47	5.02	3.37	1.67	1.94	3.42	1.98	2.04	2.06	2.72	1.98	0.22	0.14	0.16	0.16	0.15	2.85
Ta	0.76	0.60	0.46	0.60	0.57	0.49	0.58	0.41	0.90	0.53	0.66	0.75	0.61	0.57	0.26	0.27	0.46	0.29	0.40	0.27	0.57	0.37	0.11	0.06	0.04	0.06	0.13	0.77
W	59.92	104.68	54.85	109.19	98.36	18.45	31.12	18.98	12.50	20.17	24.87	19.39	19.32	107.17	28.90	20.25	97.12	22.93	48.02	20.57	20.17	45.93	61.40	30.55	54.42	65.90	106.02	82.94
Tl	0.18	0.31	0.20	0.26	0.24	0.11	0.07	0.05	0.06	0.06	0.06	0.06	0.06	0.07	0.05	0.05	0.06	0.05	0.09	0.05	0.10	0.05	0.01	0.01	0.00	0.01	0.00	0.11
Pb	4.27	4.5	7.91	10.5	9.97	3.41	4.61	3.59	4.09	3.93	6.19	3.57	3.99	0.78	2.19	3.71	1.14	2.78	6.59	2.1	5.77	2.33	1.26	10.8	1.16	1.48	1.82	7.17
Bi	0.128	0.085	-0.002	0.051	0.051	0.006	0.058	0.029	0.048	0.07	0.045	0.135	0.082	0.123	0.071	0.148	0.131	0.03	0.033	0.034	0.013	0.052	0.041	0.047	0.047	0.039	0.032	0.042
Th	4.59	5.7	3.23	3.9	12.1	5.29	5.52	5.42	4.95	5.16	6.64	5.98	5.57	0.39	0.8	0.76	0.35	0.82	2.82	0.74	3.28	0.97	0.1	0.05	0.1	0.08	0.08	1.25
U	0.82	1.2	0.91	1.39	1.25	1.1	1.11	1.08	1.05	0.97	1.27	1.1	1.1	0.16	0.32	0.32	0.21	0.32	0.55	0.29	0.65	0.33	0.07	0.06	0.07	0.06	0.07	0.4
ΣREE	30	44	46	44	84	84	139	117	114	75	107	94	106	45	63	67	46	67	80	63	84	70	9	7	7	7	8	120
Eu^*/Eu	2.93	4.15	2.06	2.40	2.24	0.95	0.89	0.89	0.95	0.92	0.85	0.94	0.92	1.00	1.32	1.22	0.98	1.22	1.08	1.24	1.05	1.24	3.53	5.09	5.05	5.00	4.22	1.07
Nb/La	0.81	0.34	1.22	1.14	0.39	0.36	0.23	0.19	0.32	0.48	0.35	0.30	0.31	0.41	0.42	0.46	0.41	0.39	0.36	0.42	0.35	0.51	0.33	0.27	0.26	0.29	0.36	0.57

A/CNK: $(Al_2O_3/102)/(CaO/56+Na_2O/62+K_2O/94)$; $Mg^{\#}=(MgO/40)/(Fe_2O_3^T/80+MgO/40)$; δ: Rittmann index, $(Na_2O+K_2O)^2/(SiO_2-43)$.

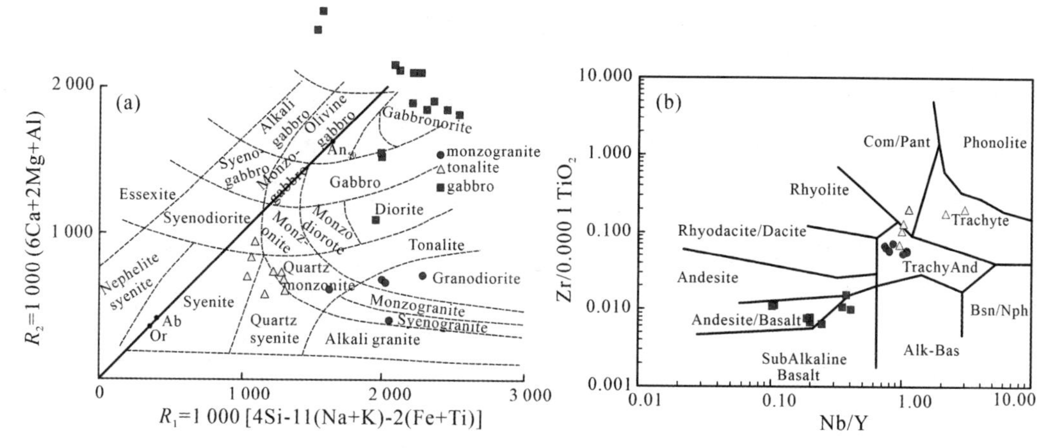

Fig. 5 R_1-R_2 discrimination (a) and Nb/Y vs. Zr/0. 000 1TiO_2(b) diagrams for the Pinghe complex, northern Yangtze Block.

4. 2. 3 Monzogranite

These rocks display high SiO_2 (68. 76 – 74. 80 wt%), moderate Al_2O_3 (13. 20 – 16. 56 wt%) contents, low MgO (0. 43–2. 59 wt%) and $Fe_2O_3{}^T$(0. 86–2. 11 wt%) contents, with variable $Mg^{\#}$ values of 35–79. They have Na_2O = 5. 33–6. 67 wt%, K_2O = 1. 93–3. 54 wt%, Na_2O/K_2O ratios of 1. 5 – 2. 9, with A/CNK ratios of 0. 74 – 1. 04. In the primitive mantle-normalized trace-element spider diagram (Fig. 6), these monzogranite display enrichment in large ion lithophile elements (LILE, e.g. Rb, K, and Pb), depletion in Nb, Ta, Ti, P. The rocks display high Rb (47. 5–92. 2 ppm) contents, low Sr (113–466 ppm), and Ba (233–620 ppm) contents, resulting in low Rb/Sr ratios of 0. 12–0. 42. They have variable Nb/Ta ratios of 8. 01–23. 32. In the chondrite-normalized REE diagram (Fig. 6), the monzogranite is enriched in LREE, display positive Eu anomalies (Eu^*/Eu = 0. 98 – 1. 32). They have low total REE contents ($\sum$REE) of 16–84 ppm, $(La/Yb)_N$ ratios of 1. 80–8. 49.

4. 3 Whole-rock Sr-Nd isotopic compositions

Representative samples from the Pinghe area were selected for Sr-Nd isotopic analyses, the results are listed in Table 3. The initial values were calculated on the basis of crystallization age = 860 Ma.

The gabbro have initial $(^{87}Sr/^{86}Sr)_i$ ratios of 0. 703 241 – 0. 704 541, positive $\varepsilon_{Nd}(t)$ values of +0. 65 to +2. 05, with single stage model ages of 1. 48 – 1. 87 Ga. The tonalite has the initial $(^{87}Sr/^{86}Sr)_i$ ratios of 0. 705 927–0. 706 222, and positive $\varepsilon_{Nd}(t)$ values of +2. 92 to +3. 66, with two-stage model ages of 1. 18 – 1. 23 Ga. The monzogranite have low initial $(^{87}Sr/^{86}Sr)_i$ ratios of 0. 703 403–0. 705 218, positive $\varepsilon_{Nd}(t)$ values of +1. 23 to +2. 98, with two-stage model ages of 1. 22 – 1. 34 Ga. As shown in the $(^{87}Sr/^{86}Sr)_i$ vs. $\varepsilon_{Nd}(t)$ diagram

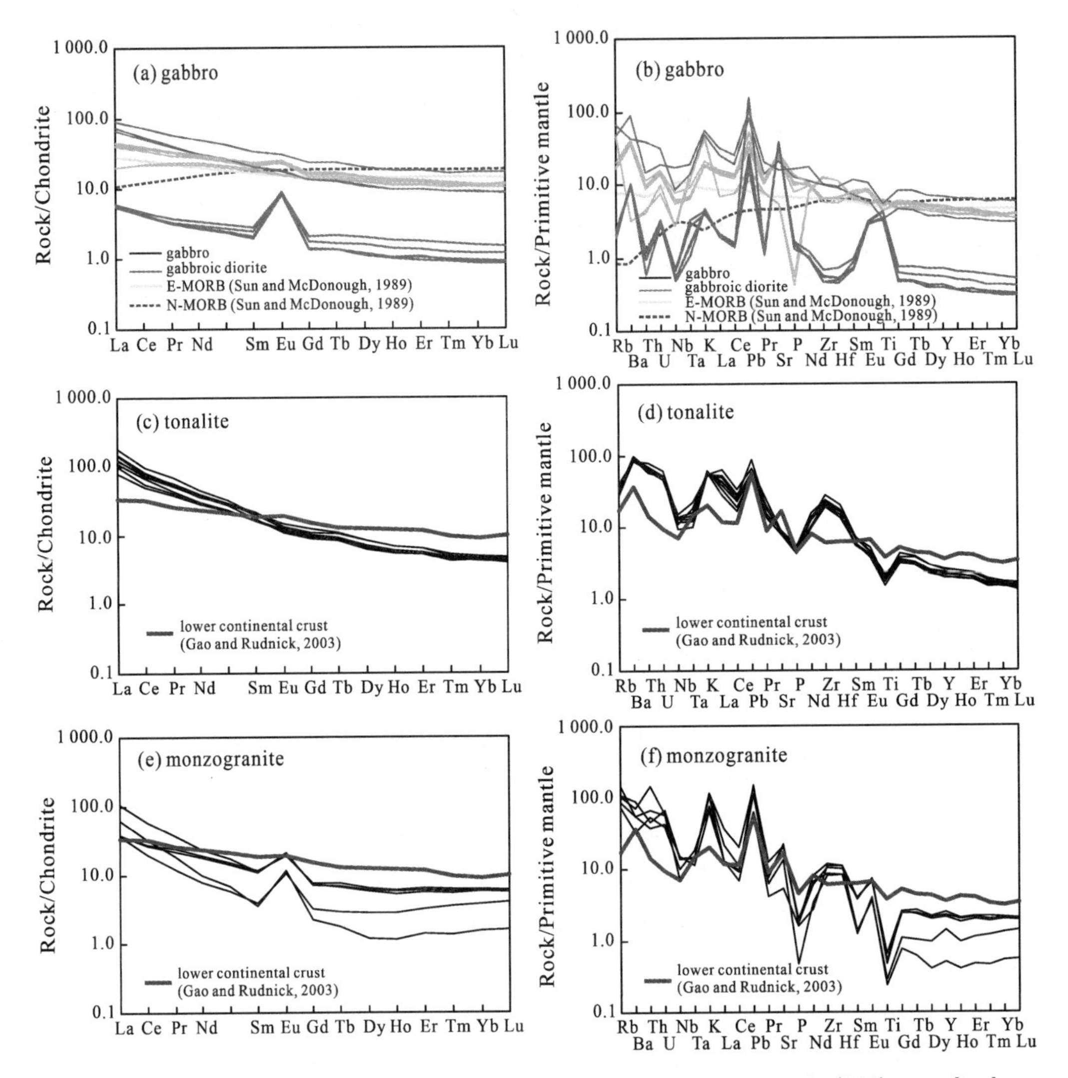

Fig. 6 Chondrite-normalized REE patterns (a, c, and e) and primitive mantle (PM) normalized trace element spider diagrams (b, d, and f) for the Pinghe complex, northern Yangtze Block. The normalized values are from Sun and McDonough (1989).

(Fig. 7), all the samples display similar Sr-Nd isotopic compositions with the Neoproterozoic adakitic rocks (Y.P. Dong et al., 2012) and mafic intrusion from the Hannan region.

Table 3 Whole-rock Sr-Nd isotopic analysis results for the Neoproterozoic Pinghe complex from the Micangshan region, Northern Yangtze Block.

Rock type	Samples	Sr /ppm	Rb /ppm	$^{87}Sr/^{86}Sr$	2*SE*	$^{87}Rb/^{86}Sr$	$^{87}Sr/^{86}Sr(t)$ (t=870 Ma)	Nd /ppm	Sm /ppm
Gabbro	PH-18	111	31	0.713 344	0.000 032	0.804 8	0.703 34	10.3	2.74
	PH-20	518	11	0.705 319	0.000 014	0.062 6	0.704 541	14.1	3.51
	PH-23	437	38	0.706 368	0.000 009	0.251 6	0.703 241	14.8	3.16
	PH-26	574	13	0.705 300	0.000 021	0.063 5	0.704 51	14	3.6

Continued

Rock type	Samples	Sr /ppm	Rb /ppm	$^{87}Sr/^{86}Sr$	2*SE*	$^{87}Rb/^{86}Sr$	$^{87}Sr/^{86}Sr(t)$ (t=870 Ma)	Nd /ppm	Sm /ppm
Tonalite	CSC-52	167	18	0.710 030	0.000 026	0.306 4	0.706 222	17.2	2.62
	CSC-60	187	22	0.710 098	0.000 018	0.335 5	0.705 927	14.2	2.45
Monzogranite	PH-02	283	92	0.715 125	0.000 015	0.943 0	0.703 403	4.67	0.54
	PH-07	436	66	0.710 696	0.000 019	0.440 7	0.705 218	11	1.73

Rock type	Samples	$^{143}Nd/^{144}Nd$	2sm	$^{147}Sm/^{144}Nd$	T_{DM} Ga	T_{DM2} Ga	$_{Nd}(t)$ t=870 Ma	$^{143}Nd/^{144}Nd(t)$ t=870 Ma
Gabbro	PH-18	0.512 498	0.000 050	0.160 8	1.87	1.34	1.27	0.511 581
	PH-20	0.512 479	0.000 054	0.150 5	1.61	1.29	2.05	0.511 621
	PH-23	0.512 324	0.000 042	0.129 1	1.48	1.33	1.41	0.511 588
	PH-26	0.512 436	0.000 046	0.155 5	1.86	1.38	0.65	0.511 549
Tonalite	CSC-52	0.512 191	0.000 019	0.092 1	1.2	1.23	2.92	0.511 665
	CSC-60	0.512 298	0.000 011	0.104 3	1.19	1.18	3.66	0.511 703
Monzogranite	PH-02	0.512 067	0.000 055	0.069 9	1.15	1.22	2.98	0.511 668
	PH-07	0.512 121	0.000 06	0.095 1	1.32	1.34	1.23	0.511 579

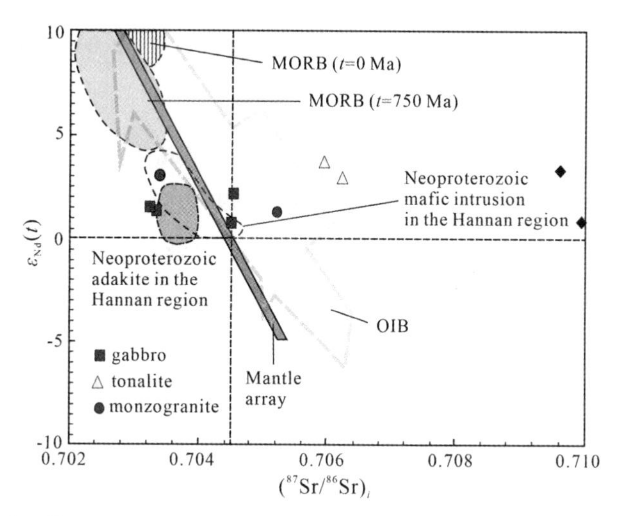

Fig. 7 $(^{87}Sr/^{86}Sr)_i$ vs. $\varepsilon_{Nd}(t)$ plot for the Pinghe complex, northern Yangtze Block.

Data are from Table 3; symbols as in Fig. 5.

5 Discussion

5.1 Neoproterozoic magmatic activity in the northern Yangtze Block

Neoproterozoic igneous rocks along the western and northern margin of the Yangtze Block mainly occurred in the Panxi, Bikou and Hannan area, which lasted for more than 200 million years (Zhao et al., 2010; Zhao and Zhou, 2009; Zhao, Zhou, and Zheng, 2013; Zhou et al., 2006). They are considered to be closely related to the evolution of the Rodinia supercontinent (Li et al., 2002; Z.X. Li et al., 2003; Zhao et al., 2013; Zhao and Cawood, 2012; Zhou, Yan, Kennedy, et al., 2002; Zhou et al., 2006) and Neoproterozoic crustal growth events in the Yangtze Block (Zhao et al., 2013). Based on detailed geochronology works (Y.P. Dong et al., 2012) on the mafic and intermediate rocks from the Zhenyuan gabbro (857±46 Ma), Xihe gabbro (824±4 Ma), Daheba granodiorite (871±77 Ma), and Shatan diorite (840±6 Ma), Y.P. Dong et al. (2012) proposed the 870–830 Ma Micangshan early arc. Zhou, Yan, Kennedy, et al. (2002) argued that the Neoproterozoic (860–750 Ma) igneous rocks along the western Yangtze Block display an island-arc affinity, they proposed a long-term active continental margin model in the Yangtze Block. In combination with the recent works on the Neoproterozoic igneous rocks from the Yangtze Block (Y.P. Dong et al., 2012; Y. Dong et al., 2011; Wang et al., 2014; Zhao et al., 2013), we summarize the Neoproterozoic magmatism along the northern Yangtze Block (Table 4), it can be seen that mafic rocks in this area display younger ages than those of the granitoids. The tonalite and monzogranite have ages of 860–870 Ma, it can be inferred that the gabbro also has a younger age.

5.2 The monzogranite-tonalite association: Melting of juvenile mafic crust in an active continental margin

Na-rich intermediate to felsic rocks were considered to have resulted from dehydration melting of basaltic lower crust (Rapp and Watson, 1995) or hot oceanic crust in subduction zone (Peacock, Rushmer and Thompson, 1994). In the condition of mafic melts underplating beneath the lower crust, heat and fluids from the hydrous mafic melts would induce partial melting of lower crust and produce Na-rich granitic melts (Annen, Blundy, and Sparks, 2006; Petford and Atherton, 1996; Petford and Gallagher, 2001). As shown in the $(^{87}Sr/^{86}Sr)_i$ vs. $\varepsilon_{Nd}(t)$ diagram (Fig. 7), both the monzogranite and tonalite display low $(^{87}Sr/^{86}Sr)_i$ ratios, positive $\varepsilon_{Nd}(t)$ values of +1.23 to +1.34, these features are identical with the juvenile basaltic lower crust (Rollinson, 1993), suggesting that partial melting of juvenile basaltic crust is the most plausible way to form the monzogranite and tonalite.

The monzogranite display high Rb (47.5–92.2 ppm) and K (1.93–3.54 wt%) contents, low Sr (113–466 ppm) and Ba (233–620 ppm) contents. In the Rb/Sr vs. Rb/Ba diagram (Fig. 8), the monzogranite samples mainly plot in the transitional zone between the basalt-derived melt and the greywacke-derived melt, the tonalite display lower Rb/Sr and Rb/Ba ratios and mainly plotted in the field that derived from juvenile basaltic rocks (Fig. 8). The above geochemical features indicate the tonalite and monzogranite were originated from melting of juvenile basaltic crust, which imply the crustal growth events in the northern Yangtze Block.

Table 4 The summary of the Neoproterozoic magmatism in the Hannan-Micangshan along the northern margin of the Yangtze Block, South China.

Tectonic regime	Location	Sample no	Lithology	Analytical methods	Age interpreation	Age /Ma	Isotopic compositions	Model age /Ga	References
Micangshan area	Xihe	MC01	Granite	LA-ICP-MS	Mean $^{206}Pb/^{238}U$ age	829 ± 5.0			Y.P. Dong et al., 2012
	Xihe	MC02	Gabbro			824 ± 4.0			
	Shatan	ZNJ-025	Diorite			840 ± 6.0			
	Guanwushan	ZNJ-027	Granite			838 ± 17			
	Zhengyuan	ZWC-022	Norite gabbro		Upper intercept age	857 ± 46			
	Wangcang	ZWC-012	Granodiorite		Upper intercept age	871 ± 77			
	Beiba	BB39	Gabbro	SHRIMP	Upper intercept age	814 ± 9.0	$^{87}Sr/^{86}Sr$ = 0.703 77 –0.704 59; $\varepsilon_{Nd}(t)$ +0.23 to +1.98; $\varepsilon_{Hf}(t)$ –0.7 to +17.7	T_{DMHf} 0.66 –1.40	Zhao and Zhou, 2009
	Tianpinghe	MCTP-01	Granite	ELA-ICP-MS	Mean $^{206}Pb/^{238}U$ age	863 ± 10			Ling, Cheng, and Ren, 2006
	Mnegzi	08-95	Norite gabbro	LA-ICP-MS	Mean $^{206}Pb/^{238}U$ age	764 ± 38	$\varepsilon_{Hf}(t)$ +6.92 to +13.99	T_{DMHf} 0.74–1.1	Xu et al., 2011
		08-104TW	Amphibole Aabbro			757 ± 32	$\varepsilon_{Hf}(t)$ +4.74 to +14.97	T_{DM2Hf} 0.7–1.2	
		08-109TW	Quartz diorite			774 ± 34	$\varepsilon_{Hf}(t)$ –4.46 to +13.1	T_{DM2Hf} 0.8–1.7	
		08-99TW	K-rich granite			745 ± 11	$\varepsilon_{Hf}(t)$ –16.49 to +8.33	T_{DM2Hf} 0.9–2.0	
	Pinghe	PH2-6	Monzogranite			742 ± 6	$\varepsilon_{Nd}(t)$ +3.6 to +5.2	T_{DM2Nd} 0.96–1.07	Gan, Lai and Qin, 2016a
		CSC-59	Quartz diorite			869 ± 4			Gan et al., 2017
	Shuimo	SM02	Syenite			869 ± 4		T_{DMNd} 0.89–1.42;	
		SM26	Syenite			860 ± 5		T_{DM2Nd} 0.89–1.29	
	Zhongziyuan	SM03	Gabbro			753 ± 4	$^{87}Sr/^{86}Sr$ = 0.703 56 –0.704 93; $\varepsilon_{Nd}(t)$ +1.6 to +4.5	T_{DMNd} 1.11–2.08	
	Xinmin	Yb06	Amphibole gabbro			776.5	$^{87}Sr/^{86}Sr$ = 0.703 47; $\varepsilon_{Nd}(t)$ +3.1	T_{DMNd} 1.55	Gan, Lai, and Qin, 2016b

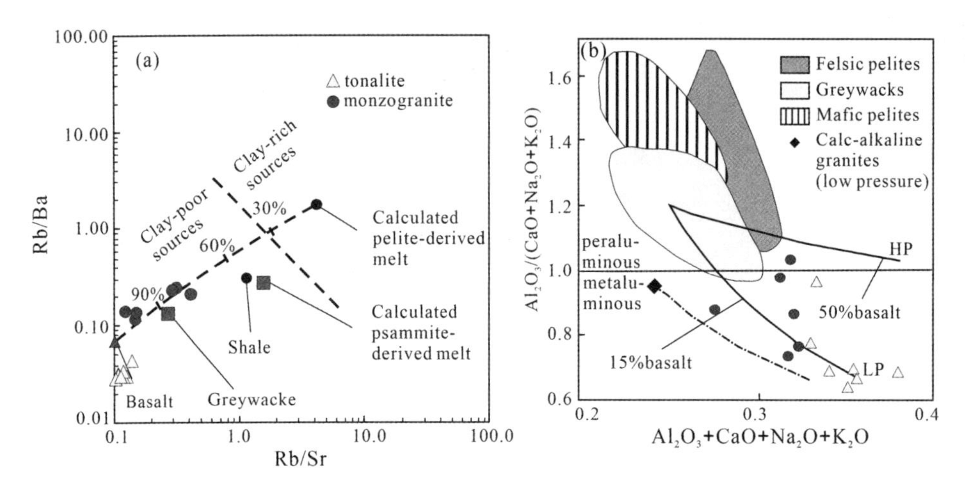

Fig. 8 (a) Rb/Sr vs. Rb/Ba diagram (after Liégeois, Navez, Hertogen, and Black, 1998) and (b) $Al_2O_3+CaO+Na_2O+K_2O$ vs. $Al_2O_3/(CaO+Na_2O+K_2O)$ diagrams (after Patiño Douce, 1999) for the Neoproterozoic tonalite and monzogranite from the Pinghe area, Micangshan region.

5.3 Genesis of the gabbro: Mafic melts derived from metasomatized mantle wedge

Basaltic rocks can provide important information about the melting mechanism of the mantle (Xia, 2014; Xu et al., 2002). According to the mineral assemblage and geochemical features of the basalt, we can deduce the composition and melting conditions of the mantle lithosphere. In addition the middle ocean ridge basalt (MORB), which formed by decompression melting of depleted asthenosphere. Other basaltic rocks may be ascribed to: ① melting of the mantle wedge in subduction zone (Wilson, 1989); ② mantle plume (Li et al., 2002; Wang et al., 2008); or ③ continental rift in an extensional setting (Zheng et al., 2007). The gabbro from the Pinghe complex displays variable SiO_2 of 47.16–59.35 wt%, Al_2O_3 = 13.27–25.03 wt%. According to the CIPW calculation results, most samples have variable normative quartz contents of 3.6% – 11.3%. These gabbro display similar REE patterns with the island-arc basalts (Pearce, 1982; Xia, 2014), they are enriched in LREE, with positive Eu anomalies of 1.00 – 1.32. Four samples have extremely high Eu^*/Eu anomalies of 3.53–5.09, suggesting the accumulation of plagioclase. Compared with the N-MORB (Sun and McDonough, 1989), these gabbros display enrichment in Sr, K, Rb, Ba, and Th, depletion in Nb, Ta, Ti, Zr, and Hf, these features are identical with the basalts that formed in island arc or active continental margin, suggesting the involvement of subduction-related fluids or contamination by continental crust (Wilson, 1989). Compared with the normal MORB, these gabbros display higher Nb/Yb and Th/Yb ratios, similar to those of the E-MORB, suggesting enrichment components in their source region (Fig. 9a). The Nb/Yb vs.

TiO_2/Yb diagram (Fig. 9b) also reveal the involvement of enrichment components. These gabbro rocks display low and variable Nb/La ratios of 0. 06–0. 76, the gabbro has Zr = 5. 09–139 ppm, variable Zr/Y ratios of 2. 06–7. 47. In the Hf/3-Th-Ta diagram (Fig. 8d), all the gabbro samples plotted in the field of island-arc tholeiitic basalts (IAT) to EMORB. The gabbro samples display relative juvenile Sr-Nd isotopic compositions, $(^{87}Sr/^{86}Sr)_i$ = 0. 703 241–0. 704 541, $\varepsilon_{Nd}(t)$ values of +0. 65 to +2. 05. In summary, the gabbros from the Pinghe complex display geochemical affinities with the island-arc basalts, they were formed by melting of the metasomatized juvenile mantle wedge in subduction zone.

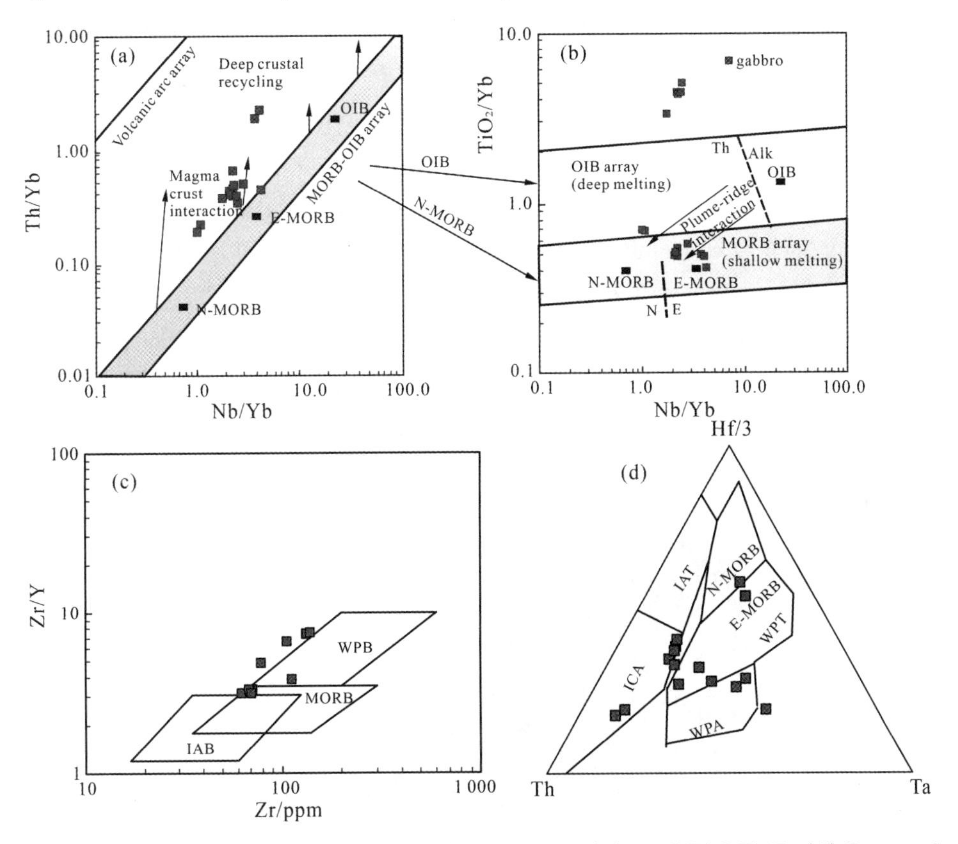

Fig. 9 Nb/Yb-Th/Yb (a), Nb/Yb-TiO_2/Yb (b), Zr-Zr/Y (c), and Hf/3-Th-Ta (d) diagrams for the Neoproterozoic gabbro from the Pinghe area, northernYangtze Block (after Pearce, 2008).

IAB: island-arc basalts; WPB: within-plate basalts; MORB: mid-ocean ridge basalts; IAT: island-arc tholeiites basalts; ICA: island-arc calc-alkaline basalts; WPA: alkaline series in within-plate setting; WPT: tholeiitic series in within-plate setting.

5. 4 Implications for the Neoproterozoic crustal growth in the northern Yangtze Block

Neoproterozoic igneous rocks along the northern and western margin of the Yangtze Block is considered to be closely related with the evolution of Rodinia supercontinent (Li et al.,

2002; X.H. Li et al., 2003; Zhao et al., 2013; Zhao and Cawood, 2012; Zhou et al., 2002; Zhou et al., 2006). X.H. Li et al. (2003) proposed that the 825–780 Ma magmatic event in the Yangtze Block were caused by mantle plume, the igneous rocks that have ages >850 Ma were considered to have resulted from the collision between the Yangtze and Cathaysia blocks. The break-up event in the Yangtze Block mainly occurred in the age of 850–830 Ma, this is supported by the bimodal volcanic rocks in the Jiangnan orogenic belt (Wang et al., 2014) and the Bikou terrane (Wang et al., 2008). The tonalite and monzogranite have ages of 860–870 Ma, which is prior to the extension events in the Yangtze Block (Z.X. Li et al., 2003; Wang et al., 2014), suggesting that they may formed in a convergent setting, that is, the subduction zone. In combination with their juvenile Sr-Nd isotopic compositions, it can be inferred that the tonalite and monzogranite represent the crust growth events in the northern Yangtze Block. In the case of the oceanic subduction, dehydration fluids or melts from the subducted oceanic crust may induce melting of juvenile arc crust, this melting process would produce granitic melt with juvenile isotopic compositions (Fig. 10). This crustal growth event is also recorded by the igneous in the Huangling complex (Zhao et al., 2013) and Xuelongbao adakitic rocks (750 Ma) along the western Yangtze Block (Zhou et al., 2006). The gabbro in this study may record the latter crust growth event. Finally, it can be concluded that the Neoproterozoic crust growth events along the northern and western Yangtze Block lasted for more than 100 million years, which may be controlled by oceanic subduction.

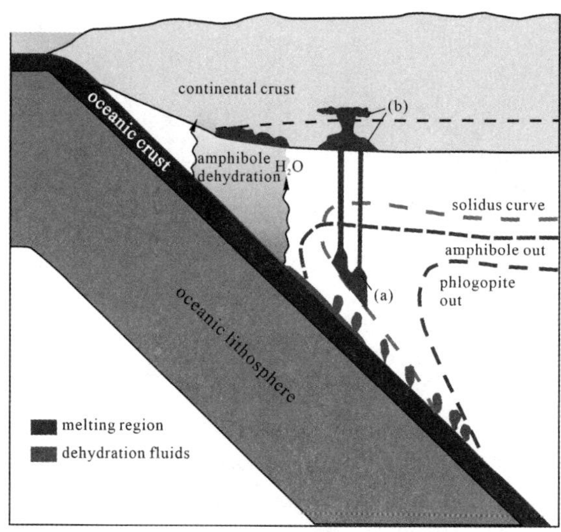

(a) melting of metasomatized mantle wedge to generate the gabbro
(b) melting of continental crust in variable depth to generate the tonalite and monzogranite

Fig. 10 A cartoon model for the genesis of the Neoproterozoic gabbro and granitoids from the Micangshan region, northern Yangtze Block (after Wyllie, Osmaston, and Morrison, 1984; Deng et al., 2015).

6 Concluding remarks

The Pinghe gabbro-granitoids association from the Micangshan region, northern Yangtze Block have zircon LA-ICP-MS U-Pb ages of 860−870 Ma, which was prior to the collision age between the Yangtze and Cathaysia blocks. Detailed geochemistry works indicate that the tonalite-monzogranite were derived from juvenile basaltic crust in a convergent setting. The younger gabbro displays geochemical affinity with island-arc basalts, indicating the melts were derived from a juvenile metasomatized mantle wedge. In summary, the Neoproterozoic Pinghe gabbro-granitoids association represent two stage crustal growth events in the Yangtze Block. These results suggest a long-term convergent setting along the northern Yangtze Block in Neoproterozoic time, which have great contribution to the crustal growth in the Yangtze Block.

Acknowledgements Financial support for this study was jointly provided by National Natural Science Foundation of China (Grant Nos. 41370267, 41421002, and 41190072, 41102037), the Foundation for the Author of National Excellent Doctoral Dissertation of PR China (Grant 201324), Program for Changjiang Scholars and Innovative Research Team in University (Grant IRT1281) and MOST Special Fund from the State Key Laboratory of Continental Dynamics.

References

Andersen, T., 2002. Correction of common lead in U-Pb analyses that do not report ^{204}Pb. Chemical Geology, 192, 59-79.

Annen, C., Blundy, J.D., Sparks, R.S.J., 2006. The genesis of intermediate and silicic magmas in deep crustal hot zones. Journal of Petrology, 47(3), 505-539.

Chu, Z.Y., Chen, F.K., Yang, Y.H., Guo, J.H., 2009. Precise determination of Sm, Nd concentrations and Nd isotopic compositions at the nanogram level in geological samples by thermal ionization mass spectrometry. Journal of Analytical Atomic Spectrometry, 24(11), 1534-1544.

Deng, J.F., Feng, Y.F., Di, Y.J., Liu, C., Xiao, Q.H., Su, S.G., ... Yao, T., 2015. Magmatic arc and ocean-Continent transition: Discussion. Geological Review, 61(3), 473-485 (in Chinese with English abstract).

Dong, Y., Liu, X., Santosh, M., Zhang, X., Chen, Q., Yang, C., Yang, Z., 2011. Neoproterozoic subduction tectonics of the northwestern Yangtze Block in South China: Constrains from zircon U-Pb geochronology and geochemistry of mafic intrusions in the Hannan massif. Precambrian Research, 189, 66-90.

Dong, Y.P., Liu, X.M., Santosh, M., Chen, Q., Zhang, X.N., Li, W., ... Zhang, G.W., 2012. Neoproterozoic accretionary tectonics along the northwestern margin of the Yangtze Block, China: Constraints from zircon U-Pb geochronology and geochemistry. Precambrian Research, 196-197, 247-274.

Gan, B.P., Lai, S.C., Qin, J.F., 2016a. Geochemistry of the bojite from the Xinmin area, Micang mountain.

Acta Petrologica et Mineralogica, 35(2), 213-228 (in Chinese with English abstract).

Gan, B.P., Lai, S.C., Qin, J.F., 2016b. Petrogenesis and implication for the Neoproterozoic monzogranite in Pinghe, Micang Mountain. Geological Review, 62(4), 929-944 (in Chinese with English abstract).

Gan, B.P., Lai, S.C., Qin, J.F., Zhu, R.Z., Zhao, S.W., Li, T., 2016. Neoproterozoic alkaline intrusive complex in the northwestern Yangtze Block, Micang Mountains region, south China: Petrogenesis and tectonic significance. International Geology Review, 59(3), 311-332.

He, L., 2010. Characteristics and structural setting of the Pinghe alkaline complex in northern Sichuan (Thesis for Master degree). Chengdu: Chengdu University of Technology.

Hoskin, P.W.O., Kinny, P.D., Wyborn, D., Chappell, B. W., 2000. Identifying accessory mineral saturation during differentiation in granitoid magmas: An integrated approach. Journal of Petrology, 41, 1365-1396.

Hoskin, P.W.O., Schaltegger, U., 2003. The composition of zircon and igneous and metamorphic petrogenesis. Reviews in Mineralogy and Geochemistry, 53, 27-62.

Li, X.H., Li, Z.X., Ge, W., Zhou, H., Li, W., Liu, Y., Wingate, M.T.D., 2003. Neoproterozoic granitoids in South China: Crustal melting above a mantle plume at ca. 825 Ma? Precambrian Research, 122(1-4), 45-83.

Li, X.H., Li, Z.X., Zhou, H., Liu, Y., Kinny, P.D., 2002. U-Pb zircon geochronology, geochemistry and Nd isotopic study of Neoproterozoic bimodal volcanic rocks in the Kangdian rift of South China: Implications for the initial rifting of Rodinia. Precambrian Research, 113(1-2), 135-154.

Li, Z.X., Li, X.H., Kinny, P.D., Wang, J., Zhang, S., Zhou, H., 2003. Geochronology of Neoproterozoic syn-rift magmatism in the Yangtze craton, South China and correlations with other continents: Evidence for a mantle super plume that broke up Rodinia. Precambrian Research, 122(1-4), 85-109.

Liégeois, J.P., Navez, J., Hertogen, J., Black, R., 1998. Contrasting origin of post-collisional high-K calc-alkaline and shoshonitic versus alkaline and peralkaline granitoids: The use of sliding normalization. Lithos, 45, 1-28.

Ling, W.L., Cheng, J.P., Ren, B.F., 2006. Neoproterozoic dioritic-granitic complexes of the Yangtze craton, petrogenesis and its tectonic significances. Geochimica et Cosmochimica Acta, 70(18): A361-A361.

Ludwig, K.R., 2003. ISOPLOT 3.0: A geochronological toolkit for Microsoft Excel. Berkeley Geochronology Center, Special Publication, 4.

Middlemost, E.A.K., 1994. Naming materials in the magma/igneous rock system. Earth-Science Reviews, 37, 215-224.

Patiño Douce, A.E., 1999. What do experiments tell us about the relative contributions of crust and mantle to the origin of granitic magmas? Geological Society, London, Special Publications, 168, 55-75.

Peacock, S.M., Rushmer, T., Thompson, A.B., 1994. Partial melting of subducting oceanic crust. Earth & Planetary Science Letters, 121(1-2), 227-244.

Pearce, J.A., 1982. Trace element characteristics of lavas from destructive plate boundaries. In: Thorpe, R. S. Andesites. Chichester: Wiley, 525-548.

Pearce, J.A., 2008. Geochemical finger printing of oceanic basalts with applications to ophiolite classification and the search for Archean oceanic crust. Lithos, 100(1-4), 14-48.

Petford, N., Atherton, M., 1996. Na-rich partial melts from newly under plated basaltic crust: The Cordillera

Blanca Batholith, Peru.Journal of Petrology, 37(6), 1491-1521.

Petford, N., Gallagher, K., 2001. Partial melting of mafic (amphibolitic) lower crust by periodic influx of basaltic magma. Earth & Planetary Science Letters, 193(3-4), 483-499.

Qi, L., Hu, J., Cregoire, D.C., 2000. Determination of trace elements in granites by inductively coupled plasma mass spectrometry.Talanta, 51(3), 507-513.

Qiu, J.X., 1993. The alkaline rocks in Qinling Orogenic belt. Beijing: Geololical Press, 1-179(in Chinese).

Rapp, R.P., Watson, E.B., 1995. Dehydration melting of meta basalt at 8 – 32 kbar: Implications for continental growth and crust-mantle recycling. Journal of Petrology, 36, 891-931.

Rollinson, H. R., 1993. Using geochemical data: Evaluation, presentation, interpretation. London: Pearson Education Limited, 1-284.

Sun, S.S., Mc Donough, W.F., 1989.Chemical and isotopic systematics of oceanic basalts: Implications for mantle composition and processes. In: Saunders, A. D., Norry, M. J. Magmatism in the Ocean Basins. Geological Society, London, Special Publications, 42, 313-345.

Wang, X.C., Li, X.H., Li, W.X., Li, Z.X., Liu, Y., Yang, Y.H.,..., Tu, X.L., 2008. The Bikou basalts in the northwestern Yangtze Block, South China: Remnants of 820 – 810 Ma continental flood basalts? Geological Society of America Bulletin, 120, 1478-1492.

Wang, X. L., Zhou, J. C., Griffin, W. L., Zhao, G., Yu, J. H., Qiu, J. S., ..., Xing, G. F., 2014. Geochemical zonation across a Neoproterozoic orogenic belt: Isotopic evidence from granitoids and metasedimentary rocks of the Jiangnan orogen, China. Precambrian Research, 242, 154-171.

Wilson, B.M., 1989. Igneous petrogenesis a global tectonic approach. Springer.

Wu, R.X., Zheng, Y.F., Wu, Y.B., Zhao, Z.F., Zhang, S.B., Liu, X., Wu, F.Y. 2006. Reworking of juvenile crust: Element and isotope evidence from Neoproterozoic granodiorite in South China.Precambrian Research, 146, 179-212.

Wu, Y., Zheng, Y., 2004. Genesis of zircon and its constraints on interpretation of U-Pb age. Chinese Science Bulletin, 49, 1554-1569.

Wyllie, P.J., Osmaston, M.F., Morrison, M.A., 1984. Constraints imposed by experimental petrology on possible and impossible magma sources and products (and discussion). Philosophical Transactions of the Royal Society of London: Series A, Mathematical and Physical Sciences, 310, 439-456.

Xia, L.Q., 2014. The geochemical criteria to distinguish continental basalts from arc related ones. Earth-Science Reviews, 139(1), 195-212.

Xu, J.F., Castillo, P.R., Li, X.H., Yu, X.Y., Zhang, B.R., Han, Y.W., 2002. MORB-type rocks from the paleo-Tethyan Mian-Lueyang northern ophiolite in the Qinling mountains, central China: Implications for the source of the low $^{206}Pb/^{204}Pb$ and high $^{143}Nd/^{144}Nd$ mantle component in the Indian Ocean. Earth & Planetary Science Letters, 198, 323-337.

Xu, X.Y., Li, T., Chen, J.L., Li, P., Wang, H.L., Li, Z.P., 2011. Zircon U-Pb age and petrogenesis of intrusions from Mengzi area in the northern margin of Yangtze plate: Actor. Petrologica Sinica, 27(3), 699-720 (in Chinese with English abstract).

Yuan, H.L., Gao, S., Liu, X.M., Li, H.M., Gunther, D., Wu, F.Y., 2004. Accurate U-Pb age and trace element determinations of zircon by laser ablation-inductively coupled plasma mass spectrometry.

Geostandards Newsletter, 28, 353-370.

Zhao, G.C., Cawood, P.A., 2012. Precambrian geology of China. Precambrian Research, 222-223, 13-54.

Zhao, J.H., Zhou, M.F., 2009. Secular evolution of the Neoproterozoic lithospheric mantle underneath the northern margin of the Yangtze Block, South China.Lithos, 107, 152-168.

Zhao, J.H., Zhou, M.F., Zheng, J.P., 2010. Metasomatic mantle source and crustal contamination for the formation of the Neoproterozoic mafic dike swarm in the northern Yangtze Block, South China. Lithos, 115, 177-189.

Zhao, J.H., Zhou, M.F., Zheng, J.P., 2013. Neoproterozoic high-K granites produced by melting of newly formed mafic crust in the Huangling region, South China. Precambrian Research, 233, 93-107.

Zheng, Y.F., Zhang, S.B., Zhao, Z.F., Wu, Y.B., Li, X.H., Li, Z.X., Wu, F.Y., 2007. Contrasting zircon Hf and o isotopes in the two episodes of Neoproterozoic granitoids in South China: Implications for growth and reworking of continental crust.Lithos, 96(1-2), 127-150.

Zhou, M.F., Yan, D.P., Kennedy, A., Li, Y.Q., D.J., 2002. SHRIMP U-Pb zircon geochronological and geochemical evidence for Neoproterozoic arc-magmatism along the western margin of the Yangtze Block, South China. Earth & Planetary Science Letters,196(1-2), 51-67.

Zhou, M.F., Yan, D.P., Wang, C.L., Qi, L., Kennedy, A., 2006.Subductio related origin of the 750 Ma Xuelongbao adakitic complex (Sichuan Province, China): Implications for the tectonic setting of the giant Neoproterozoic magmatic event in South China. Earth & Planetary Science Letters, 248, 286-300.

Granitic magmatism in eastern Tethys domain (western China) and their geodynamic implications[①]

Lai Shaocong Zhu Renzhi Qin Jiangfeng Zhao Shaowei Zhu Yu

Abstract: Western China locates in the eastern section of the Tethys domain, granitic rocks in this region with variable formation ages and geochemistry record key information about the crust-mantle structure and thermal evolution during the convergent process of Tethys. In this study, we focus on some crucial granitic magmatism in the western Yangtze, Qinling orogen, and western Sanjiang tectonic belt, where magma sequence in the convergent orogenic belt can provide important information about the crust-mantle structure, thermal condition and melting regime that related to the evolution processes from Pre- to Neo-Tethys. At first, we show some features of Pre-Tethyan magmatism, such as Neoproterozoic magmatism (ca. 870–740 Ma) in the western margin of the Yangtze Block were induced by the assembly and breakup of the Rodinia supercontinent. The complication of voluminous Neoproterozoic igneous rocks indicated that the western Yangtze Block underwent the thermodynamic evolution from hot mantle-cold crust stage (ca. 870–850 Ma) to hot mantle and crust stage (ca. 850–740 Ma). The Neoproterozoic mantle sources beneath the western Yangtze Block were progressively metasomatized by subduction-related compositions from slab fluids (initial at ca. 870 Ma), sediment melts (initial at ca. 850 Ma), to oceanic slab melts (initial at ca. 825 – 820 Ma) during the persistent subduction process. Secondly, the early Paleozoic magmatism can be well related to three distinctive stages (variable interaction of mantle-crust to crustal melting to variable sources) from an Andeans-type continental margin to collision to extension in response to the evolution of Proto-Tethys and final assembly of Gondwana continent. Thirdly, the Paleo-Tethys magmatism, Triassic granites in the Qinling orogenic display identical formation ages and Lu-Hf isotopic compositions with the related mafic enclaves, indicate a coeval melting event of lower continental crust and mantle lithosphere in the Triassic convergent process and a continued hot mantle and crust thermal condition through the interaction of subducted continental crust and upwelling asthenosphere. Finally, the Meso- and Neo-Tethyan magmatism: Early Cretaceous magmatism in the Tengchong Block are well responding to the subduction and closure of Bangong-Nujiang Meso-Tethys, recycled sediments metasomatized mantle by subduction since 130 Ma and subsequently upwelling asthenosphere since ca. 122 Ma that causes melting of heterogeneous continental crust until the final convergence, this process well recorded the changing

① Published in *Acta Geologica Sinica* (*English Edition*) ,2022,96(2).

thermal condition from hot mantle-cold crust to hot mantle and crust; The Late Cretaceous to Early Cenozoic magmatism well recorded the processes from Neo-Tethyan ocean slab flat subduction, steep subduction, to initial collision of India-Asia, it resulted in a series of continental arc magmatism with enriched mantle to crustal materials at late Cretaceous, increasing depleted and/or juvenile materials at the beginning of Early Cenozoic, and increasing evolved crustal materials in the final stage, implying a continued hot mantle and crust condition during that time. Then we can better understand the magmatic processes and variable melting from the mantle to crust during the evolution of Tethys, from Pre-, Paleo-, Meso-, to Neo-, both they show notably intensive interaction of crust-mantle and extensive melting of the heterogeneous continent during the final closure of Tethys and convergence of blocks, and thermal perturbation by a dynamic process in the depth could be the first mechanism to control the thermal condition of mantle and crust and associated composition of magmatism.

1 Introduction

Since Suess (1893) proposed "Tethys" to study of the past ocean in the Earth history, and later it has been regarded as ocean ancestral "Mediterranean" to the Alpine-Himalayan orogen systems (Laubscher and Bernoulli, 1977; Jenkyns, 1980), the study of Tethys has been ongoing for more than a century, until the theory of plate tectonics provides valuable insight into the past Tethyan ocean has evolved (Ricon, 1995; Condie, 1997). Almost all geologists used this concept to understand the Phanerozoic ocean-continental evolution on the Earth (Berra and Angiolini, 2014). As for present-day Asia, or eastern Tethys domain, a multi-stage collage of continental blocks that derived from the eastern Gondwana (Metcalfe, 2013), various subduction-related igneous rocks occurred at different blocks assembled by the multiple Tethyan oceans and associated collision-related magmatism. There are mainly four phases of Tethys linked to the Rodinia and Gondwana supercontinent break-up and Asian amalgamation (Sciunnach and Garzanti, 2012; Metcalfe, 2013; Deng et al., 2014; Li S C et al., 2018; Lai and Zhu, 2021): ① the Proto-Tethys opened from the rifting of assembly Rodinia and mainly closed at the end of the Early Paleozoic; ② Paleo-Tethys was initial from Devonian and end at the Triassic, mainly included the North China, South China, Tarim, and Indochina blocks dispersed from Gondwana and merged into Asia; ③ Meso-Tethys opened in the late Early Permian between these separated blocks from Cimmerian and Condwana, including Southern Qiangtang, Lhasa and Sibumasu (including Tengchong-Baoshan Block), until its final closure at the late Early-Cretaceous (Zhu D C et al., 2016; Zhu R Z et al., 2017a,b); ④ Neo-Tethys represented the final collage of micro-continent blocks that drifted from Gondwana since Late Triassic and closed in Late Cretaceous to Cenozoic (Zhu et al., 2011), it has been a final accreted event to the Eurasian continent. During these processes, many crucial magmatism along the main convergent belts were remains debated and needed further study.

Abundant heterogeneous continent pieces and orogenies were distributed in this region and recorded the multi-stage history of Tethys (Fig. 1) (Zhang et al., 1995, 2001a, 2015; Xu et al., 2015), including long-term subduction-accretion, arc-continent collisions, and continent-continent collisions resulted in multiple orogenies and related magmatism (e.g., granitoid), uplift and deformation (Metcalfe, 2013; Cao et al., 2017). In a series of merging processes, widespread plutonic rocks during the Tethys were closely associated with the convergent mechanisms (Zhang et al., 1995, 2010, 2015). We focused on the Pre-Tethyan stage through Proterozoic plutonic magmatism in the western margin of Yangtze, Proto-Tethyan magmatism of early Paleozoic three stage magmatism in the Tibetan-Himalyan orogen, Paleo-Tethyan magmatism of Triassic plutonic rocks in Qinling orogen, Meso- and Neo-Tethyan magmatism of Mesozoic to Cenozoic plutonic rocks in the SE Tibetan-Himalayan orogen, aimed to provide some crucial constrains to help us better understand the ocean-continent evolution and associated magmatic process (e.g., thermal condition, crust-mantle interaction) during the amalgamation of a various microcontinent.

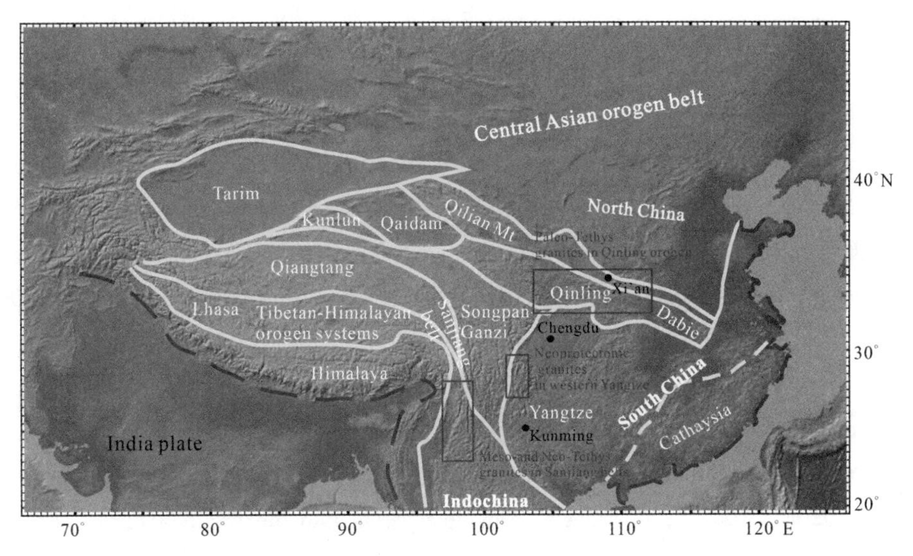

Fig. 1 The geological distribution of major units in the Asia downloaded from GEOMAP APP (http:// www. geomapapp. org./).

Red line area represents the study region. Yellow lines are major tectonic boundary. Purple line represents part National boundary.

Through the data of petrology, zircon U-Pb and Lu-Hf isotope, bulk-rock Sr-Nd-Pb-Hf isotope, bulk-rock major-trace element and associated mineral chemistry on these regions, we emphasized that these key magmatism mainly represented the multi-stage crust-mantle interaction and associated crustal melting in different convergent process, including the persistent Neoproterozoic subduction beneath the western Yangtze plate (e.g., interaction of slab-derived fluids, subducted sediments, and slab melts with mantle wedge) (Zhu Y et al., 2019a-c, 2020a,b, 2021a,b); the Triassic subducted continental crust beneath the northern China induced the interaction between subducted continent and upper mantle (Qin et al., 2008, 2009, 2010a,b, 2013); upwelling mantle-derived mafic magmas induced the melting of continental crust caused by Bangong-Nujiang Meso-Tethys slab rollback to slab break-off (Zhu R Z et al., 2015, 2017a,b, 2018a,b, 2019); The absence of Late Cretaceous to Early Paleocene mafic magma indicates cold mantle, abundant Early Eocene mafic rocks reveal a hot mantle, suggesting transition from Neo-Tethyan subduction to Indian-Asian collision (Zhao S W et al., 2016, 2017a, 2019, 2020; Zhu R Z et al., 2021). These results well record key information of various magmatism during the evolution of Tethys, it will help us different convergent mechanism during the history of western China.

2 Pre-Tethys: A case from Neoproterozoic magmatism in the western Yangtze Block, South China

The western Yangtze Block is characterized by voluminous late Mesoproterozoic to Neoproterozoic granitic rocks associated with mafic-ultramafic rocks (Zhao and Cawood, 2012; Lai et al., 2015a,b; Zhao J H et al., 2018, 2019; Zhu Y et al., 2019a-c, 2020a, b, 2021a, b), which are magmatic responses during the assembly and breakup of Rodinia supercontinent (Zhao and Cawood, 2012; Zhao J H et al., 2018, 2019; Lai and Zhu, 2020). Although the mantle plume is a dominated controversial model for the Neoproterozoic magmatism (Li et al., 2003, 2006, 2008), an increasing number of igneous, metamorphic, sedimentary, geophysical evidence strengthen the long-lived subduction setting (e.g., Sun et al., 2009; Gao R et al., 2016; Zhao J H et al., 2017, 2018, 2019, 2021; Li J Y et al., 2018, 2021). Therefore, numerous Neoproterozoic granitoids and mafic rocks are windows for probing the crustal growth and reworking as well as mantle metasomatism. Here we focused on Neoproterozoic igneous rocks in the western Yangtze Block (Lai and Zhu, 2020; Zhu Y et al., 2019a-c, 2020a, 2021a), in order to provide insights into the thermodynamic conditions of crust and mantle.

The Neoproterozoic magmatic rocks in the western Yangtze Block include gabbros, diorites, tonalites, granodiorites, monzogranites, and granites (e.g., Zhao J H et al., 2018, 2019, 2021). The detailed compilation of these igneous rocks, which were sourced from different magmatic depth (metasomatized mantle source-juvenile mafic crust source-mature continental crust source), suggested that which were predominately formed at ca. 870–740 Ma (Lai and Zhu, 2020; Zhao et al., 2018). The melting of the Neoproterozoic metasomatized lithospheric mantle source in the western Yangtze Block began at ca. 870 Ma (Fig. 2) during the early stage of subduction, which can be demonstrated by the positive zircon $\varepsilon_{Hf}(t)$ values (up to +20.6) and whole-rock $\varepsilon_{Nd}(t)$ values of ca. 870–850 Ma hornblende gabbros, gabbros, gabbro-diorites, and diorites (e.g., Zhao J H et al., 2019; Zhu Y et al., 2021b), this correlates to the cold crust (Fig. 2). Therefore, there is a stage of hot mantle and cold crust at ca. 870–850 Ma. After ca. 850 Ma, the western Yangtze Block entered into the stage of hot mantle and hot crust, which can be evidenced by hetergeneous crust sources were heated and melted by mantle-derived magmas, resulted in the generation of both metasomatized mantle-derived mafic-intermediate rocks, juvenile mafic lower crust (meta-igneous)-derived granodiorite, and mature crust-derived granites (e.g., Zhao J H et al., 2008, 2019; Zhu Y et al., 2019, 2020, 2021). In summary, the western Yangtze Block underwent the thermodynamic evolution from hot mantle-cold crust to hot mantle and crust.

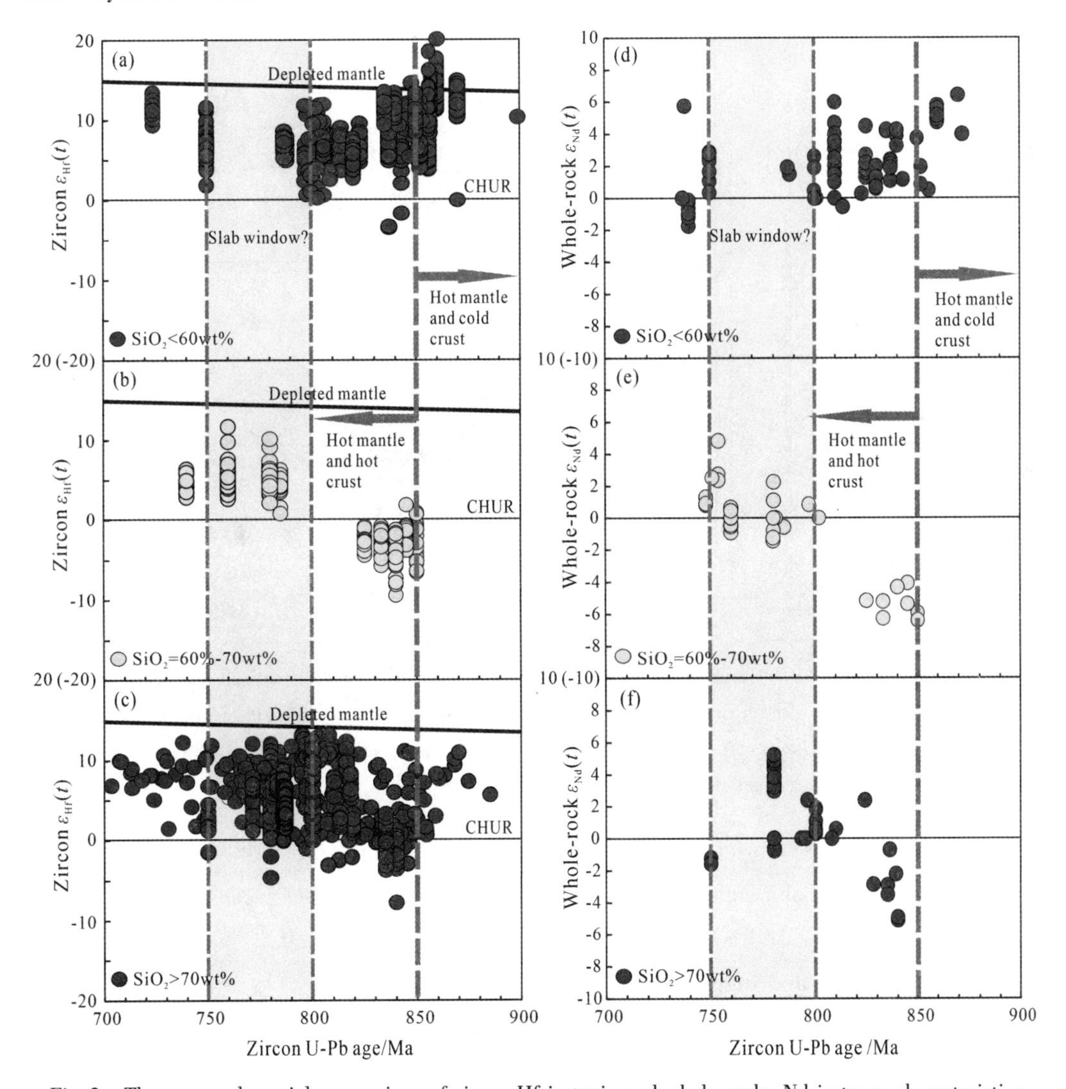

Fig. 2 The temporal-spatial comparison of zircon Hf isotopic and whole-rocks Nd isotopes characteristics for the Neoproterozoic magmatism in the western Yangtze Block, South China.

For the Neoproterozoic metasomatized mantle-derived magmatism (from hot mantle-cold crust stage to hot mantle and crust stage) , in one word, which suggested the Neoproterozoic mantle sources beneath the western Yangtze Block were gradually metasomatized by the subduction slab fluids, sediment melts, and oceanic slab melts (Fig. 3) . The subducted slab fluids were firstly introduced into the mantle source at ca. 870 Ma (Zhao J H et al., 2019) . The subducted sediment melts-related mantle metasomatism began at ca. 850 Ma (Zhao J H et al., 2019a; Zhu et al., 2020b) , and the slab melts-related metasomatized mantle-derived magmatism occurred at ca. 825–820 Ma (Munteanu et al., 2010; Zhao J H et al., 2019) .

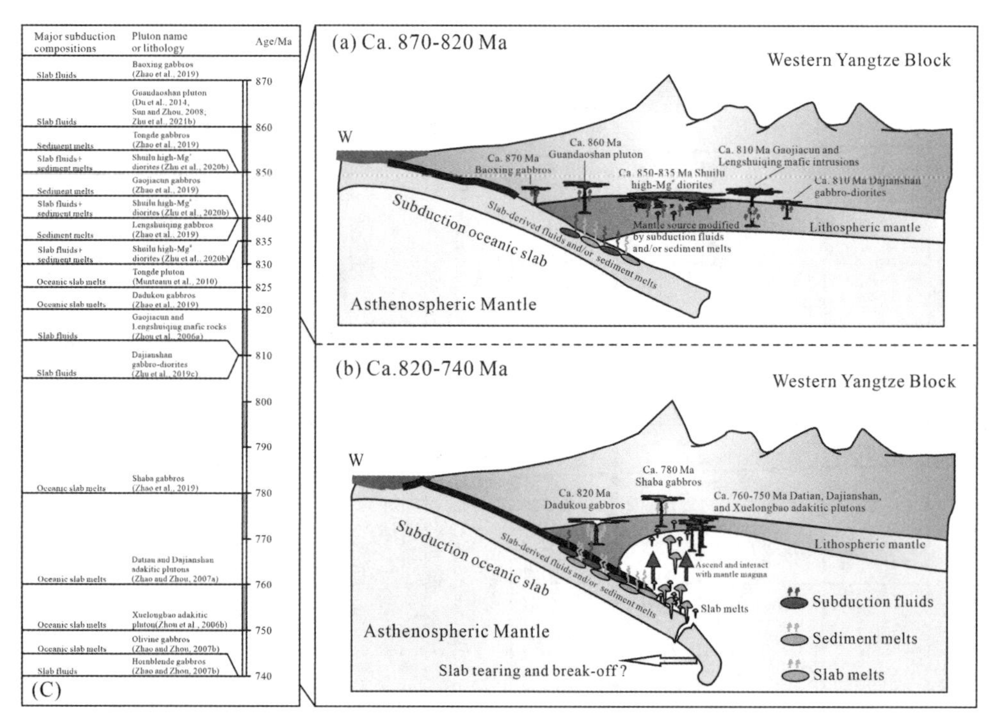

Fig. 3 A sketch map (a, b) and a summary (c) for the Neoproterozoic metasomatized mantle magmatism in the subduction zone, which display that Neoproterozoic mantle source was progressively metasomatized by the subduction-related compositions from slab fluids, sediment melts, to oceanic slab melts during persistent subduction process (modified after Zhu Y et al., 2020b) .

As for the Neoproterozoic intra-crustal magmatism (hot mantle and crust stage) , the firstly partial melting of dominantmature crust sources indicated by negative zircon $\varepsilon_{\mathrm{Hf}}(t)$ values (up to −9. 36) and whole-rock $\varepsilon_{\mathrm{Nd}}(t)$ values (up to −6. 38) of ca. 850 – 835 Ma pearluminous granites and high-maficity I-type granites (Fig. 2b, c, e, f) , which were induced by the mantle-derived magma during the early subduction process (e.g., Zhu Y et al., 2019a, 2021a) . The voluminous ca. 860–810 Ma mantle-derived mafic-ultramafic intrusion (e.g., the Tongde, Gaojiacun and Lengshuiqing intrusions) (Zhou et al., 2006a; Sun and Zhou, 2008;

Munteanu et al., 2010) leading to the generation of juvenile mafic lower crust (Zhao et al., 2018, 2021; Zhu Y et al., 2019b). Therefore, the extensive partial melting of (thickened) juvenile mafic crust occurred after ca. 835 Ma (Fig. 2b,c,e,f). Notably, the ca. 800–750 Ma (i.e., the late stage of Neoproterozoic subduction process) witnessed the widespread remelting of juvenile curst, but not the significant melting of metasomatized mantle wedge source (Fig. 2). We speculated that this phenomenon was caused by the oceanic slab tear and breakoff during the late stage of subduction process (Although the accurate time of slab breakoff have not been constrained in detail) (Zhao et al., 2018), which may lead to the asthenosphere upwelling through the slab window, and resulting the large-scale crust melting, including the juvenile mafic crust formed by the underplating of metasomatized mantle-derived magma.

3 Proto-Tethys: Three-stage of early Paleozoic magmatism in the Tibetan-Himalayan orogen

Early Paleozoic magmatism in the Tibetan-Himalayan orogen has been considered as a key information to understand the evolution of Proto-Tethys including: ①final assembly of the Gondwana supercontinent and ②the initial subduction along the margin of the peri-Gondwana assembly (Cawood and Buchan, 2007; Cawood et al., 2007; Zhu et al., 2012; Wang Y J et al., 2013; Li et al., 2016; Gao et al., 2019; Liu et al., 2019; Lai and Zhu, 2021, and reference therein). The related tectonic setting with Proto-Tethys of early Paleozoic magmatism from the early Cambrian to Late Ordovician has hotly debated, with proposals ranging from an Andean-type active continental margin (Cawood et al., 2007; Zhu et al., 2012), to the collision-related and then extension setting (Visonà et al., 2010; Gao et al., 2019), to lithospheric thickening and delamination (Zhao S W et al., 2014, 2017b; Li et al., 2016), and even the rift events (Liu et al., 2019). What contribution and mechanism could generate during the evolution of Proto-Tethys and stabilize the final Gondwana assembly? Understanding has been achieved by recoding the relationships of geological, geochronological, and geochemical data (Cawood et al., 2007; Liu et al., 2019; Lai and Zhu, 2021).

Three stages of Paleozoic magmatism were identified through compiled early Paleozoic zircon U-Pb-Hf isotopes and bulk-rock geochemistry of magmatic rocks in the Tibetan-Himalayan orogen in Lai and Zhu (2021), detailed evidences and processes are as follows (Fig. 4): ①Early to middle to late stages were in response to the final assembly of the Gondwana supercontinent and closure of Proto-Tethys. Early-stage magmatic rocks (>490 Ma) have highly variable SiO_2 (48.0–80.0 wt%) and MgO (0.02–9.65 wt%), K_2O/Na_2O ratios (0.14–2.81), and whole-rock Sr-Nd $(^{87}Sr/^{86}Sr)_i$ (0.7035–0.7340), $\varepsilon_{Nd}(t)$ (−9.5 to +1.0) and $\varepsilon_{Hf}(t)$ (−15 to +8.0) isotopic compositions with significant mantle contributions, they could be generated in an Andean-type arc setting along the active northern continental Gondwana margin. ② Middle-stage magmas (490 – 470 Ma) were dominated by fertile continental crustal signatures (SiO_2>70 wt%, A/CNK >1.10, MgO<2.69 wt%) in a crustal thickening setting. ③Late-stage magmas (<470 Ma) have highly variable Si, Mg and isotopic components again that similar with early-stage one, also with coeval mantle-derived magmas developing in an extensive setting. The widespread presence of early Cambrian to late Ordovician peraluminous high-K calc-alkaline magmatism in the Tibetan-Himalayan orogen, as we emphasized that has indicated the reworking of ancient continental crustal materials have played significant role in reconstructing and stabilizing the final Gondwana assembly and the closure of Proto-Tethys.

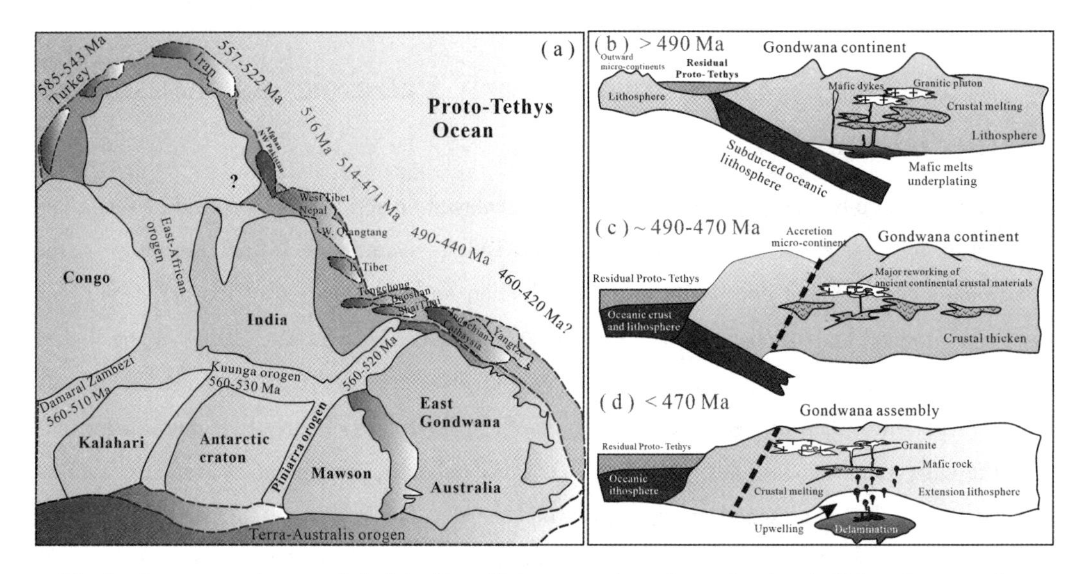

Fig. 4 A reconstruction map (a) of the Proto-Tethyan margin and associated early Paleozoic magmatism (modified after Cawood et al., 2007; Wang Y J et al., 2013; Lai and Zhu, 2021); (b,d) a model for final Gondwana assembly and Proto-Tethyan process from Andeans-type continental margin setting, to collision and crustal thicken, and to post-collision extension setting (modified after Cawood et al., 2007; Wang Y J et al., 2013; Zhao S W et al., 2014, 2017; Lai and Zhu, 2021).

4 Paleo-Tethys: Triassic magmatism in the Qinling orogen in response to the final convergence of Paleo-Tethys

4.1 Magmatism, metamorphism, and orogeny in the Qinling orogenic belts

The Triassic Qinling-Dabie-Sulu orogen is the most prominent tectonic feature in central China. It originated from continent-continent collision between the North China Block (NCB) and the Yangtze Block (YZB) (e.g., Meng and Zhang, 2000; Ratschbacher et al., 2003; Zheng et al., 2011) and plays a key role in the understanding of the evolution of the intervening Paleo-Tethys ocean and the terminal assembling of the eastern Asian continents (Fig. 1). It is the archetype for high (HP) and ultrahigh-pressure (UHP) metamorphism in cratonal continental lithosphere (e.g., Zhang et al., 2001a; Hacker et al., 2004; Ernst et al, 2007). However, there are striking geologic differences between the Qinling (the Qinling Orogenic Belts, the western section of the orogen) and the Dabie Shan (the Dabie Mountains, the central section of the orogen) and Sulu (the northeastern part of the orogen) sections of the orogen. In "Dabie" and "Sulu", regional exposure of Early to Middle Triassic HP and UHP metamorphic rocks indicates deep continental subduction (> 100 km, metamorphic diamond and coesite) and subsequent rapid exhumation of YZB continental lithosphere from beneath the NCB (Hacker et al., 1998, 2000, 2004, 2006, 2009; Ratschbacher et al., 2000, 2006; Zheng, 2008, 2012). In the Qinling, distributed and voluminous high-K calc-alkaline plutonism accompanied the Triassic collision; the most widespread felsic granitoids contain mafic enclaves and dykes (Sun et al., 2002; Wang X X et al., 2007, 2011, 2013; Qin et al., 2008, 2009, 2010a,b, 2013), Triassic UHP metamorphism is unknown in the Qinling. This Triassic granitic magmatism was previously considered the result of a signal and short-lived (220–205 Ma) collision event (Sun et al., 2002).

Recent detailed zircon dating works revealed that the Triassic granitic magmatism in the Qinling area was a long-term process (248 – 194 Ma, Li et al., 2013; Qin et al., 2013), although it remains unclear whether this long-term process is continuous or multi-staged. Geochronologic studies of the mafic enclaves revealed that they have zircon U-Pb ages identical to their host granitoids. These mafic enclaves (mainly diorite with some gabbros) have intermediate to mafic composition, their evolved Sr-Nd-Pb isotopic compositions suggest that they were derived from enriched mantle lithosphere (Qin et al., 2009, 2010a,b, 2013; Wang Y J et al., 2013), and the zircon Lu-Hf isotopic compositions in the mafic enclaves indicate that there is some asthenospheric components in their source region (Qin et al., 2010b, 2013). Why is there extensive partial melting of mantle lithosphere in the Triassic Qinling orogen? What is the role of mantle-derived melts in the genesis of the Triassic granitoids in the Qinling? A better understanding of the ages and modeling of the geochemical characteristics of the mafic enclaves will provide pivotal information on the type and extent of melting of the mantle lithosphere in the Qinling and thus on its composition and temporal evolution. Furthermore, the genetic links between the partial melting processes of the orogenic crust, i.e., the formation of the different granitoids, the intrusion of mantle-derived mafic melts (mafic rocks that are associated with the granites) into the crust, and the associated metamorphism is still poorly understood. Consequently, single geodynamic model scenarios—as currently invoked—appear inappropriate for the genesis of the extensive granitic magmatism in the Qinling Orogenic Belts.

Understanding these issues has significance for investigating the tectonic evolution of the Triassic Qinling-Dabie-Sulu orogen and for continental subduction orogens in general.

4.2 Granitic magmatism sequence in the Triassic Qinling Orogenic Belt

The classic work of Sun et al. (2002) considered the Triassic—overwhelmingly granitic—magmatism in the Qinling to have formed in a single and short period (220 – 205 Ma). However, my own detailed zircon U-Pb dating in the Wulong plutonic complex near the central Qinling Foping dome revealed a 235–200 Ma formation period (Qin et al., 2013). Based on petrologic and geochemical features, this period encompassed three stages: the first stage (235–225 Ma) comprises minor high-$Mg^{\#}$ adakitic quartzdiorite; the second stage (225 – 210 Ma) involves voluminous high-K, calc-alkaline granites and associated mafic enclaves (mainly diorite with minor gabbro and hornblendite); the third stage (~207 Ma) includes K-feldspar bearing monzogranite without mafic enclaves (Qin et al., 2013). Furthermore, Zhu et al. (2010) reported ~242 Ma granites in the northern Qinling, and Jin et al. (2005) reported 245–238 Ma adakitic granites in the western Qinling. These observations suggest that the Triassic Qinling magmatism formed in stages on a regional scale, and that the spatial distribution of these stages can be used to trace the Triassic collisional process between the NCB and YZB.

Comprehensive data revealed that Late Triassic granites are mainly high-K calc-alkaline, and display adakitic affinity, e.g., enriched in Sr, Ba, extremely depleted in Y and HREE, without significant negative Eu anomalies. According their geochemistry and formation age, three main types can be recognized: ① Quartz diorite formed at 235 Ma to 225 Ma were characterized by low Si, high K, Mg contents, which was considered to be resulted from high degree (>40%) partial melting of mafic lower crust and undergone subsequent interaction with enriched mantle wedge during their ascent. ② High-K calc-alkaline granodiorites, tonalite and monzogranite formed at 220~205 Ma also display adakitic geochemical features, the high $Mg^{\#}$ adakitic granodiorite and tonalite from the Dongjiangkou (Qin et al., 2010a,b) provide exact evidences for partial melting of subducted continental crust: exhumation of subducted continental crust that resulted from slab break off would melt in condition of aqueous fluids derived from the decomposition of hydrous minerals and heat from the upwelling asthenosphere, producing adakitic melts at relatively high pressure (amphibole ± garnet stable field). During their ascent, the adakitic magma bodies extensively react with ambient peridotite to form pyroxenite and orthopyroxene-rich zone, this is evidenced by the outward increasing trend of MgO (3.13–3.76 wt%), $Mg^{\#}$(~66), Cr (166–199 ppm), Ni (74.4–84.4 ppm), P_2O_5(0.19–2.21 wt%) and Y (16.7–17.4 ppm) contents, as well as Nb/Ta, Rb/Sr and Th/U ratios, indicating more mantle components in the pluton margin. Extensive melt-mantle interactions would make the Cr, Ni, Mg, Y, Rb and K transfer into the granitic melt, and produce the Mg-rich hybrid magmas. ③ Biotite granite formed at ~200 Ma should be resulted from partial melting of middle orogenic crust, which was induced by fluids from lower crust or mantle lithosphere.

4.3 Coeval mafic magmatism and implication for role of mantle lithosphere

The mafic enclaves and dykes that are hosted by the overwhelmingly felsic complexes are a key feature of the Triassic Qinling plutonism. Their detailed zircon LA-ICPMS U-Pb dating (Wang et al., 2007, 2011; Qin et al., 2009, 2010a,b, 2013) demonstrated that their ages (220 - 210 Ma) are identical to the second stage granite (i. e., high-K calc-alkaline granodiorite). These mafic enclaves represent mafic mafic magma that derived from melting of metasonatized enriched mantle lithosphere (Qin et al., 2009a, b, 2013; Wang X X et al., 2013). It remains unclear, however, whether these enclaves originated from an ancient continental mantle lithosphere or an enriched mantle wedge, metasomatized by subduction-related fluids or melts. Furthermore, it is unclear whether juvenile mantle material (asthenosphere-derived melt) was involved in the genesis of the mafic enclaves. This is a key for the understanding of the partial melting process of the mantle lithosphere in the Triassic. The variable zircon Lu-Hf isotopic compositions indicate that these mafic enclaves may be formed by reworking of Neo-Proterozoic sub-continental lithospheric mantle (Fig. 5). The widespread Late Triassic mafic enclaves in the granites suggest a regional partial melting of sub-continental lithospheric mantle during Late Triassic time.

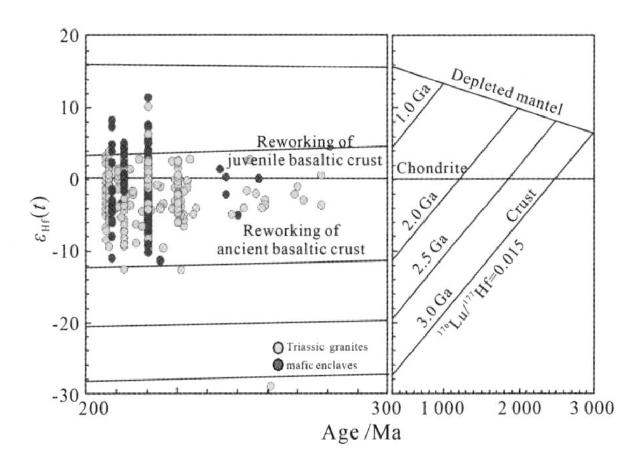

Fig. 5 $\varepsilon_{Hf}(t)$ vs. time diagram for the Triassic granites and mafic enclaves from the Qinling orogenic belt. Yellow circle: granites; blue circle: mafic enclaves.

4.4 Possible geodynamic process for the Qinling Triassic granites

The Late Triassic granites in the Qinling orogenic belt may be formed by by multi-stage partial melting of subducted Yangtze continental crust. The partial melting of continental crust may be caused by: ① the clockwise rotation of the Yangtze Block cause extension setting in the Dabie area, which led to the rapid exhumation of HP-UHP metamorphic rocks, while compression setting in the Qinling area and cause extensive partial melting of subducted continental lithosphere, led to the widespread Late-Triassic granites magmatism and associated mafic magmatiam; ② there is higher convergence rate in the Qinling area than those of the Dabie-Sulu area during the Triassic collision between the Yangtze and North China blocks; ③ continental lithosphere in the western and eastern part of the Yangtze Block has different composition and thermal structure.

In combination with regional tectonic setting, this thesis proposed a new petrogentic model (Fig. 6) to explain the genesis of these Late Triassic granites: ① the sub-continental lithospheric mantle was metasomatized by Late Paleozoic to Early Mesozoic northward subduction of the Mianlue oceanic crust beneath the south Qinling terrane; ② dense and refractory mafic lower crust that was trapped in mantle depth by continental subduction will melt at high temperature to produce the early (235 – 225 Ma) quartz diorite, subsequent interaction with mantle peridotite would elevate and their $Mg^{\#}$ and metasomatized the overriding mantle wedge, in combination with the gravity of high-density oceanic lithosphere, this would led to the slab break-off in the Qinling area; ③ the slab break-off cause asthenosphere upwelling and exhumation of the subducted continental crust, this would led to the extensive partial melting of subducted continental crust and overlying enriched wedge, produce large-scale high-K calc-alkaline granites and associated mafic enclaves.

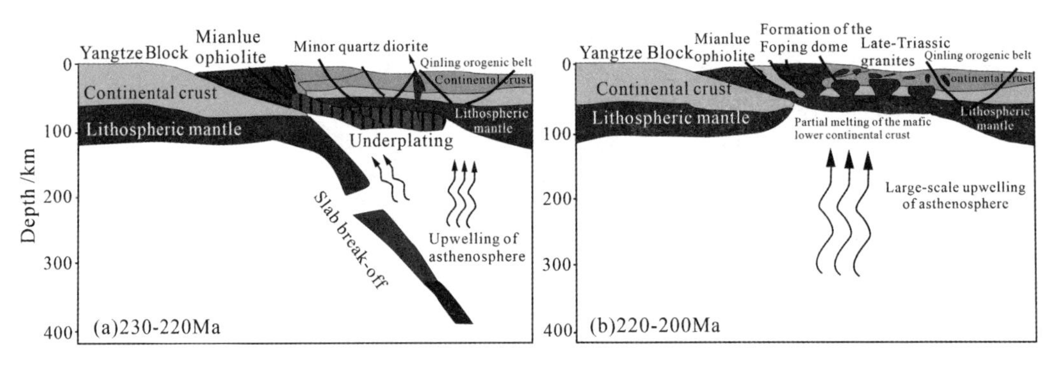

Fig. 6 The possible model for Triassic granites and the continental subduction (revised from Qin et al., 2013)

5 Meso-Tethys: Early Cretaceous magmatism in response to the final closure of Bangong-Nujiang Meso-Tethys

The northern magmatic belt, as the products of the subducting of Bangong-Nujiang Meso-Tethyan ocean, which was distributed from the western part in the central and northern Lhasa Block of Tibet to southeastern part in the Tengchong Block of its southeastern extension (Wang Y

J et al., 2015a,b). However, the subducting and closure processes caused different magmatic response: abundant Early Cretaceous basalts-andesites-dacites-rhyolites and mafic dykes-granitoids in the central and northern Lhasa Block (Zhu et al., 2016), which formed by subducting, detach, roll-back and slab break-off of Bangong-Nujiang Meso-Tethyan oceanic lithosphere (Zhu et al., 2016); In contrast to, the abundantly Early Cretaceous intermediate to felsic granitoids and minor dacites and rhyolites in the Tengchong Block (Xu et al., 2012; Xie et al., 2016; Zhu et al., 2017a,b). Whether both of them were triggered by the similarly subducting, detach, roll-back and slab break-off of Bangong-Nujiang Meso-Tethyan oceanic lithosphere or not? If so, what's condition and composition in response to the geodynamic processes?

Then several results would be proposed (Zhu R Z et al., 2018b):

(1) The Early Cretaceous magmatism in the Tengchong Block was mainly composed by the intermediate to felsic granitoids. To the west, which were seperated by the Gudong-Tengchong fault, and to the east, which were separated by the Nujiang-Lushui-Ruili fault. Their zircon U-Pb age was mainly developed from ca. 130 Ma to ca. 110 Ma, and three stages should be focused on: ①ca. 130–122 Ma, the rocks were mainly developed in the Donghe area of the northeastern Tengchong and Dulongjiang area of the northern Gaoligong, these granodiorites-monzogranites-syenogranites are weakly peraluminous and high-K calc-alkaline series (Zhu et al., 2015, 2017a); ②ca. 122–116 Ma, the rocks were developed in the Xiaotang-Mangdong and Menglian area in the eastern Tengchong Block, these quartz monzodiorite-granodiorites are Na-rich calc-alkaline (Zhu et al., 2017a); ③ca. 114–110 Ma, the magmatism in this stage was mainly developed in the Lushui-Pianma area of the south-middle part of Gaoligong (Zhu et al., 2017b). From east to west, the Na-rich calc-alkaline quartz diorites were gradually changed to high-K calc alkaline granodiorites and monzogranites.

(2) The Early Cretaceous amphiboles are mainly magnesian-hornblende (Zhu R Z et al., 2017a, 2018b), a few was ferro-tschermakite, and their plagioclases were mainly labradorite-andesine but a few was bytownite. Dulongjiang granodiorites were derived from recycled subducting sediments-derived melts by the mantle-like materials in the mantle wedge (Zhu R Z et al., 2019); Donghe granites were typically high fractionated I-type granites (Zhu et al., 2015), which derived from partial melting of lower crustal materials; The Xiaotang-Mangdong and Menglian quartz monzodiorites-granodiorites were derived from melting of various basaltic lower crustal components (Zhu et al., 2017a), triggered by the underplate mantle-deirved magmas; The Pianma quartz diorites-granodiorites-monzogranites were derived from various crustal materials in the depth by mantle-like magmas (Zhu et al., 2017a). The abundant mafic enclaves and variable zircon Hf isotope indicate that the mantle-crust interaction play the crucial role (Zhu R Z et al., 2018b).

(3) In contrast to the Early Cretaceous magmatism in the central and northern Lhasa Block,

the absence of basaltic-andesitic volcanic rocks in the Tengchong Block. The Early-Cretaceous granitic rocks show more evolved geochemical features than those in the central and northern Lhasa Block. From ca. 122 Ma, the Na-rich calc-alkaline rocks have depleted zircon Hf components, and their enclaves are mantle-derived. These features indicate that the mantle components have significant roles, from ca. 122 Ma, although the crust-derived materials were also pre-dominantly.

The Early Cretaceous intensive magmatic events have notably increased zircon $\varepsilon_{Hf}(t)$ values until to +10.0 since 122 Ma imply the increasing asthenosphere mantle contributions during this stage (Fig. 7). But before it, the basaltic rocks were of very enriched zircon Hf components that a major enriched mantle materials contributions because of significant metasomatized by recycled subducted sediments since ca. 130 Ma (Zhu R Z et al., 2019), thus there could be a upwelling mantle materials induce the developing of this intensive

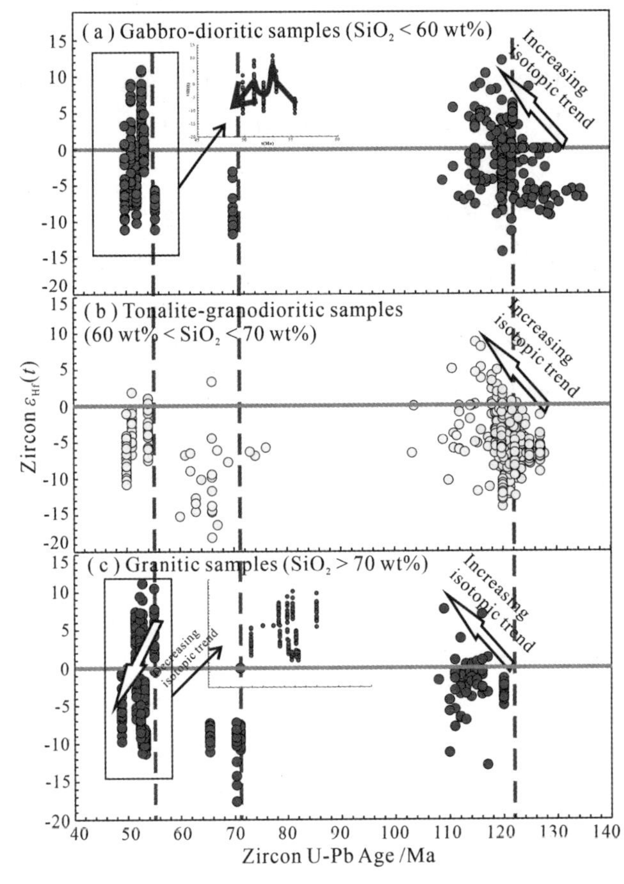

Fig. 7 The temporal-spatial comparison of zircon Hf isotopic characteristics for Early Cretaceous to Early Cenozoic magmatism in the Tengchong terrane in response to Meso- and Neo-Tethys evolution.

Data from Cong et al., 2011; Qi et al., 2011; Chen et al., 2014; Zhu R Z et al., 2015, 2017a,b, 2018a-c; 2019, 2020a, 2021; Cao et al., 2016, 2017; Xie et al., 2016, 2020; Zhao S W et al., 2016, 2017a, 2019, 2020; Ma et al., 2021b.

magmatism from 130 Ma to 110 Ma, from metasomatized- to asthenosphere- mantle, from dominated fertile crustal isotopic signatures (through recycling) to significant mantle isotopic controls (through upwelling asthenosphere). This interpretation could also be supported by increasing depleted isotopic signatures not only basaltic rocks but also intermediated to felsic rocks (Fig. 7). This could be interpreted as closely related to the tectonic processes on closure of Qiangtang-Baoshan and Lhasa-Tengchong Blocks, from soft-collision to slab break-off of Bangong-Nujiang Tethyan ocean (Zhu et al., 2016), where subduction results the mantle metasomatized by recycling crustal materials and slab break-off induce the upwelling materials from asthenosphere (Fig. 8). This observation will couple with the geophysical data that the isotopic shift and change of convergence rate has closely relationship (Fig. 7). This slab break-off has been confirmed by variable geochemical and isotopic signatures of diorite-granodiorite-monzogranite in eastern Tengchong (Zhu R Z et al., 2017b).

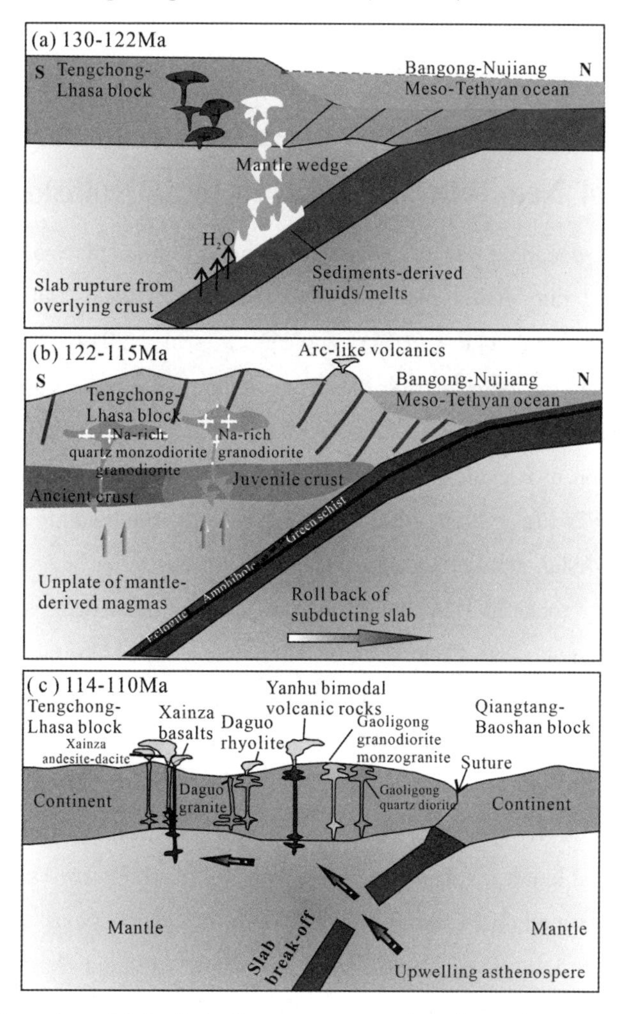

Fig. 8 The geodynamic model for Early Cretaceous magmatism of Tengchong terrane in response to the closure of Bangong-Nujiang Meso-Tethys (modified from Zhu, 2018).

Therefore, we may propose that the Early Cretaceous intensive magmatic event in eastern Tengchong Block could be resulted from the processes from final steepen subduction, collision between Lhasa-Tengchong and Qiangtang-Baoshan blocks to slab break-off of Bangong-Nujiang Tethys (Fig. 8). In the final stage of Bangong-Nujiang Meso-Tethys, the enriched mantle materials were produced by subduction. Subsequently, in the slab break-off setting, a slab window may result in the partial melting of various sources region including upwelling asthenospheric mantle, enriched lithospheric mantle, and overlying crust (Ferrari, 2004). The sub-arc mantle wedge could possibly contain melts from both sub-slab asthenosphere that not only enriched in HFSE and normal sub-arc mantle that depleted in HFSE but also supported more primitive isotopic composition into the magmatism (Ferrari, 2004), then arc magmatism shifts to more contribution of mantle and/or juvenile source. Therefore, the thermal condition of Tengchong micro-continent during early Cretaceous in response to final closure of Bangong-Nujiang Meso-Tethys could be transferred from hot mantle and cold crust to hot mantle and crust, from 130 Ma, 122 Ma, to 110 Ma.

6 Neo-Tethys: Late Cretaceous to early Eocene magmatism: From subduction of Neo-Tethyan Ocean to initial collision of India-Asia

Over the past a decade, we still focused on the evolution of Neo-Tethys and collision between Indian and Indian-Asian plates in the Tengchong Block, an Andean-type active continental margin arc (Ma et al., 2014; Wang Y J et al., 2014, 2015b). Although the Gaoligong Formation is considered as the basement of the Tengchong Block (Zhong, 1998), the abundant Late Cretaceous to early Eocene mafic to granitic intrusions are distinguished, which are the extension of Gangdese magmatic belt (Xu et al., 2015; Xie et al., 2016; Zhao S W et al., 2016, 2017a, 2019, 2020), and records the evolution from Neo-Tethyan subduction to Indian-Asian syn-collision process. The crystallization age of Late Cretaceous to early Eocene magmatic rocks in Tengchong Block are concentrated in 76–50 Ma, and could be subdivided into two stages, Late Cretaceous to early Paleocene at 76–64 Ma and early Eocene at 55–50 Ma, respectively (Fig. 1). The Late Cretaceous magmatic rocks are mainly exposed at Guyong, Husa, Jiucheng and Sudian area, and they are shoshonitic diorite and high-K calc-alkaline granite-granodiorite in compositions (Xu et al., 2012; Cao et al., 2016; Zhao S W et al., 2017a, 2020; Zhu R Z et al., 2021). The early Eocene magmatic rocks are mainly occurred at Xima-Tongbiguan, Nabang and Bangwan area, which are calc-alkaline to high-K calc-alkaline granite-granodiorite-quartz diorite-gabbro in compositions (Zhao S W et al., 2016, 2019). These two stages of granitic magmatism have different characteristics not only in lithology but also in geochemistry, isotope and the petrogenesis of enclaves enclosed in their corresponding host rocks.

The Late Cretaceous to early Paleocene granite-granodiorite in Tengchong Block are mainly characterized by high and variable SiO_2(51.5–76.5 wt%), K_2O (3.10–5.08 wt%) and Na_2O (3.05–3.85 wt%), low and variable MgO (0.06–5.59 wt%) and CaO (0.37–4.07 wt%) contents, enriched Nd and zircon Hf isotopic compositions, indicating they were mainly derived from partial melting of crustal metapelite or metagreywacke materials. In addition, the plagioclases compositions and texture in the Late Cretaceous to Paleocene granites and their enclaves show normal zoning and lack of a resorption zoning, indicating the Late Cretaceous to Paleocene granites are not involved in more mafic magmatism during their generation. They are products of almost pure crustal materials melting during the long-time heat accumulation (Moyen, 2019) and the enclave in Late Cretaceous to Early Paleocene granites is an "autolith" (Zhao et al., 2020). It's could be further confirmed that the Late Cretaceous to Paleocene mafic rocks are lack or rare in the Tengchong Block.

The early Eocene (55–50 Ma) magmatism in Tengchong Block has an obvious flare-up, similar to coeval magmatism in Gangdese magmatic belt (Ma et al., 2014; Wang Y J et al., 2014, 2015; Zhao S W et al., 2016, 2019). The Eocene intrusions are diversity, comprising of grabbro, hornblendite, diorite, quart-diorite, granodiorite and granite. Overall, the mafic rocks mainly expose in the western part but the granitic rocks occur in all Tengchong Block. The petrogenesis of Eocene mafic and granitic magmatic rocks is closely related. Both of mafic and granitic rocks have similar compositional variations trend in space with increasing values of enriched compositions (e.g., K_2O, LREE, and Th), $^{87}Sr/^{86}Sr$, $\varepsilon_{Nd}(t)$ and $\varepsilon_{Hf}(t)$, which could be resulted from the interaction between them (Zhao S W et al., 2019). In addition, the mafic rocks generally occur around the granitic plutons as a satellite and mafic microgranular enclaves (MME) are abundant enclosed in the granitic body. These field characteristics indicate the mafic rocks could play an important role in the formation of granitic rocks. The early Eocene mafic rocks in the Tengchong Block are mainly hornblende-bearing gabbro and hornblendite. They could be as a water donor and provide the fluid and heat to promote the partial melting of lower crustal materials (Castro, 2019). Then magma mixing occurred in the granitic rocks is prevalent. The plagioclase in Eocene granite and MMEs show textures with resorption zoning and "M" type chemistry compositional variation, which is interpreted as products of magma mixing (Zhao et al., 2020). The high basalt flux supported and promoted the successively lower crustal melting and magma mixing, even the differentiation of these granitic mush.

The different lithology assemblages between Late Cretaceous-Early Paleocene to Early Eocene magmatism could reveal different original mechanism and tectonic-thermal evolution in Tengchong Block during the Neo-Tethyan subduction and Indian-Asian collision (Fig. 9). The basic difference of magmatic activity is whether melting of lithospheric mantle or not. The lack of mafic magmatism in Late Cretaceous to Early Paleocene indicates a cold thermal condition but the abundant mafic rocks in early Eocene reveal a hot thermal condition. The change of thermal condition could be resulted by the tectonic realm transition from Neo-Tethyan subduction to Indian-Asian collision during the earliest occurring mafic rocks at ca. 55 Ma (Fig. 9).

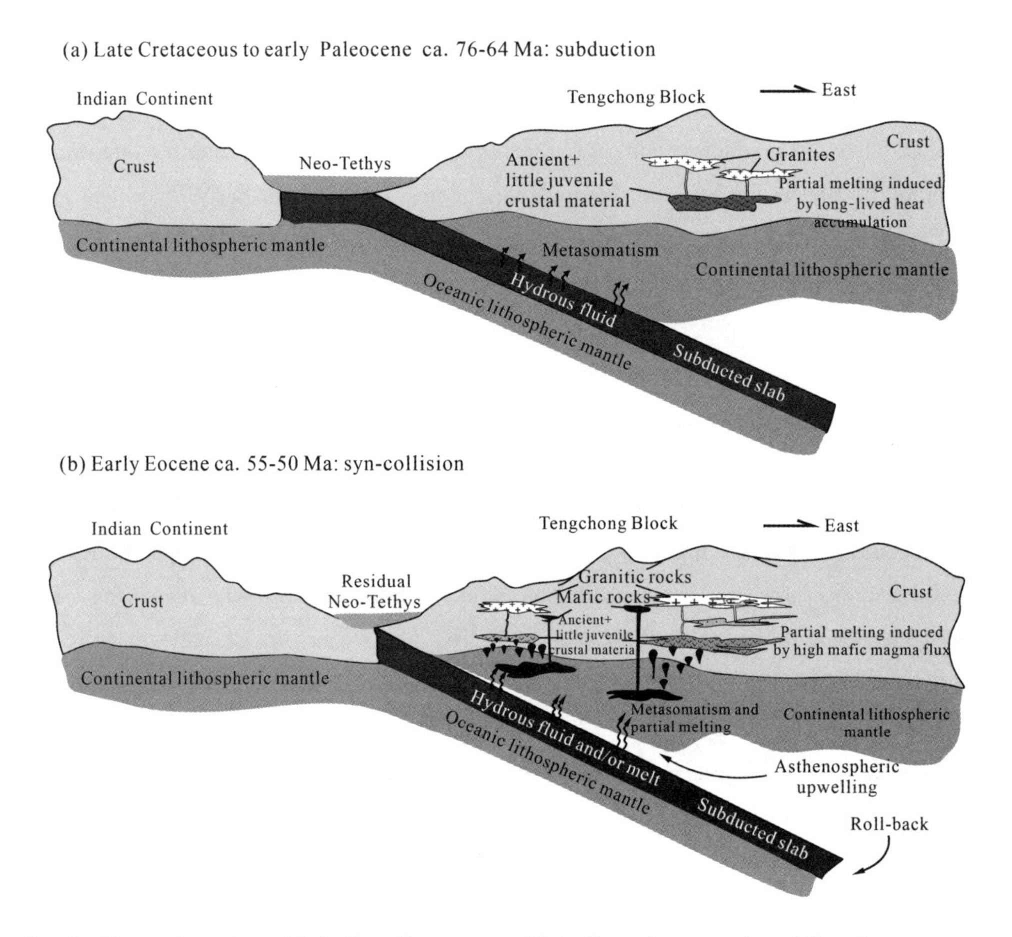

Fig. 9 The geodynamic model for Late Cretaceous to Early Cenozoic magmatism of Tengchong terrane in response to final closure of Neo-Tethys (Zhao S W et al., 2020).

It is interesting in a signal granitic pluton. They show different emplacement level according to the amphibole barometers. In the western of Tengchong Block, the Xima-Tongbiguan pluton could emplace at lower-middle crust with ca. 5–7 kbar, but the Bangwan pluton emplaced at upper crust with 2–3 kbar (our unpublished data). The metamorphic degree is also decrease from west to east, granulite-facies or high amphibole facies metamorphism in the west (Ma L Y et al., 2021a) but low-grade metamorphism in the east. These signatures could show that the Tengchong Block could be a tilted lower-middle to upper arc crust profile and exposed a transcrust magmatic system. Therefore, the Eocene granitic rocks in the Tengchong Block could provide the evolved information of granitic magmatic reservoirs at different depth.

For the Late Cretaceous to early Cenozoic intensive magmatism in the Tengchong arc segment has notably increasing zircon $\varepsilon_{Hf}(t)$ values and later shows decreasing with temporal evolution (Fig. 7). But the detailed processes should be figured out from ca. 70 Ma to 50 Ma. In the late Cretaceous stage ca. 70－60 Ma, both bulk-rock and zircon isotopic components show dominated enriched features, with heavy O isotopic composition, evolved both Sr, Nd, Pb and Hf isotopic components (Qi et al., 2011; Chen et al., 2014; Cao et al., 2017; Zhao S W et al., 2017a; Zhu R Z et al., 2021). This suggests a very enriched source. New identified shoshonitic dioritic rocks also has very evolved isotopic components with high Mg and high K content that confirms that there is an enriched mantle sources beneath the continental lithosphere. This enriched mantle could be resulted from recycled subducted sediments due to abrupted change of angle and other geophysical parameters of Neo-Tethyan oceanic slab, from flatten to steepen subduction since ca. 70 Ma (Zhu R Z et al., 2021). The coeval and contacted shoshonitic diorite and high K and high silica granites imply the shoshonitic mafic magmas could provide enough heat and materials to the overlying extensive melting of ancient continental crust, inducing the products of high-volume fertile granites with both high silic, alkaline, LILEs and REE content (Zhu R Z et al., 2021). This process indicates the trans-crustal system from metasomatized mantle to evolved crust pathway. Until to the initial collision of India-Asia, the initial magmatism was induced by the upwelling asthenosphere that generated by slab break-off of Neo-Tethyan ocean, resulting the basaltic magmatism and its associated differentiated felsic rocks. This hot mantle-derived provide heat to melt the ancient mafic lower crust that produced intermediated rocks. As the collision following, more and more continental crustal materials involved into the intensive melting that produce not only the enriched mantle- and crustal intermediated to felsic rocks. Then we can infer that the Late Cretaceous to Early Cenozoic magmatism in the Tengchong Block well recorded the continued melting in response to the process from subduction of Neo-Tethys to initial collision of India-Asia.

7 Conclusions

The heterogeneous continents and/or micro-continents have occurred various drifting, motion, and convergence along the eastern Tethyan regime, including the Proto-, Paleo-, Meso-, and Neo-Tethyan ocean. The convergent belts in western China recorded much crucial information to trace these processes. Based on a comprehensive interpretation of granite petrology and geochemical data, we try to answer some questions on crustal melting, thermal condition and interaction of crust-mantle in multi-stage convergent process. Several conclusions are as follows: ①For the Pre-Tethyan stage, Neoproterozoic granitoid in the western margin of

Yangtze underwent the thermodynamic evolution from hot-mantle and cold-crust to hot-mantle and -crust under persistent subduction process. The Neoproterozoic mantle sources were progressively metasomatized by the subduction-related compositions from slab fluids, sediment melts, to oceanic slab melts. ②Proto-Tethys: three stage different early Paleozoic magmatism from variable crust-mantle interaction to purely crustal melting to variable interaction again that indicate the tectonic setting evolution from an Andeans-type continental margin to collision to extension setting after the final closure of Proto-Tethys and assembly of Gangdwana continent. ③Paleo-Tethys stage: Triassic granites and coeval mafic rocks recorded the deep interaction of subducted continental crust and upwelling asthenosphere in response to the final convergence of South China and North China after the final closure of Paleo-Tethys between them, where hot mantle induces the subsequently melting of subducted continental crust. ④Meso- to Neo-Tethys stage: Early Cretaceous to Early Cenozoic plutonic magmatism in the Tengchong Block consistents with the final closure of Bangong-Nujiang Meso-Tethys and Neo-Tethyan ocean, metasomatized mantle resulted from an abrupted change of slab and/or convergence geophysical parameters initially reactive the intensive magmatism within ca. 30 – 10 Ma, where thermal condition changed from hot-mantle and cold-crust to hot mantle and crust. In these cases, hot mantle means the upwelling of metasomatized and/or asthenosphere mantle-derived materials and heat, hot-crust means that there is heterogeneous continental crust melting through heat and/or materials were provided from depth. We, therefore, infer those granites and associated mafic rocks could be the effective cases to help us better understand the sources, thermal condition and crust-mantle interaction in response to the complex convergent processes during the evolution of Tethys.

Acknowledgements This work was supported by the National Natural Science Foundation of China (Grant Nos. 40872060, 41102307, 41372067, 41772052, 41802054, 41190072, and 41421002), China Postdoctoral Science Special Foundation (Grant Nos. 2019T120937 and 2018M643713), Natural Science Foundation of Shaanxi (Grant. No. 2019JQ-719) and Shaanxi Postdoctoral Science Foundation.

References

Berra, F., and Angiolini, L., 2014. The evolution of the Tethys region throughout the Phanerozoic: A brief tectonic reconstruction. In: Marlow L., Kendall C., and Yose L. Petroleum Systems of the Tethyan Region: AAPG Memoir 106:1-27.

Cao, H.W., Zou, H., Zhang, Y.H., Zhang, S.T., Zheng, L., Zhang, L.K., Tang, L., and Pei, Q.M., 2016. Late Cretaceous magmatism and related metallogeny in the Tengchong area: Evidence from geochronological, isotopic and geochemical data from the Xiaolonghe Sn deposit, western Yunnan, China. Ore Geology Reviews, 78:196-212.

Cao, W.R., Lee, C.T., and StarLackey, J., 2017. Episodic nature of continental arc activity since 750 Ma: A global compilation. Earth & Planetary Science Letters, 461:85-95.

Castro, A., 2019. The dual origin of I-type granites: The contribution from laboratory experiments. Geological Society London Special Publications, 491: 101-145.

Cawood, P.A., and Buchan, C., 2007. Linking accretionary orogenesis with supercontinent assembly. Earth-Science Reviews, 82:217-256.

Cawood, P.A., Johnson, M.R.W., and Nemchin, A.A., 2007. Early palaeozoic orogenesis along the Indian margin of Gondwana: Tectonic response to Gondwana assembly. Earth & Planetary Science Letters, 255(1-2):1-84.

Chen, X., Wang, D., Wang, X.L., Gao, J.F., Shu, X.J., Zhou, J.C., and Qi, L., 2014. Neoproterozoic chromite-bearing high-Mg diorites in the western part of the Jiangnan orogen, southern China: Geochemistry, petrogenesis and tectonic implications. Lithos, 200-201:35-48.

Condie, K.C., 1997. Contrasting sources for upper and lower continental aust: The greenstone connection. The Journal of Geology, 105(6):729-736.

Cong, F., Lin, S.L., Zou, G.F., Li, Z.H., Xie, T., Peng, Z.M., and Liang, T., 2011. Magma mixing of granites at Lianghe: In-situ zircon analysis for trace elements, U-Pb ages and Hf isotopes. Science China: Earth Sciences, 54:1346-1359.

Deng, J., Wang, Q., Li, G., Li, C., and Wang, C., 2014. Tethys tectonic evolution and its bearing on the distribution of important mineral deposits in the Sanjiang region, SW China. Gondwana Research, 26(2): 419-437.

Ernst, W.G., Tsujimori, T., Zhang, R., and Liou, J.G., 2007. Permo-Triassic collision, subduction-zone metamorphism, and tectonic exhumation along the East Asian Continental Margin. Annual Review of Earth and Planetary Sciences, 35(1): 73-110.

Ferrari, L., 2004. Slab detachment control on mafic volcanic pulse and mantle heterogeneity in central Mexico. Geology, 32(1): 77-80.

Gao, L.E, Zeng, L., Hu, G., Wang, Y., Wang, Q., Guo, C., and Hou, K., 2019. Early Paleozoic magmatism along the northernmargin of East Gondwana. Lithos, 334-335: 25-41.

Gao, P., Zheng, Y.F., and Zhao, Z.F., 2016. Experimental melts from crustal rocks: A lithochemical constraint on granite petrogenesis. Lithos, 266-267: 133-157.

Gao, R., Chen, C., Wang, H.Y., Lu, Z.W., Brown, L., Dong, S.W., Feng, S.Y., Li, Q.S., Li, W.H., Wen, Z.P., and Li, F., 2016. SINOPROBE deep reflection profile reveals a Neo-Proterozoic subduction zone beneath Sichuan basin. Earth & Planetary Science Letters, 454: 86-91.

Hacker, B.R., Ratschbacher, L., and Liou, J.G., 2004. Subduction, collision, and exhumation in the ultrahigh-pressure Qinling-Dabie orogen. Geological Society, London, Special Publications, 226(1): 157-175.

Hacker, B.R., Ratschbacher, L., Webb, L., Ireland, T., Walker, D., and Dong, S., 1998. U/Pb zircon ages constrain the architecture of the ultrahigh-pressure Qinling-Dabie Orogen, China. Earth & Planetary Science Letters, 161: 215-230.

Hacker, B.R., Ratschbacher, L., Webb, L., McWilliams, M.O., Ireland, T., Calvert, A., Dong, S., Wenk, H.R., and Chateigner, D., 2000. Exhumation of ultrahigh-pressure continental crust in east central

China: Late Triassic-Early Jurassic tectonic unroofing. Journal of Geophysics Research, 105 (B6): 13339-13364.

Hacker, B.R., Wallis, S.R., McWilliams, M.O., and Gan, P.B., 2009. $^{40}Ar/^{39}Ar$ Constraints on the tectonic history and architecture of the ultrahigh-pressure Sulu orogen. Journal of Metamorphic Geology, 27(9): 827-844.

Hacker, B. R., Wallis, S. R., Ratschbacher, L., Grove, M., and Gehrels, G., 2006. High-temperature geochronology constraints on the tectonic history and architecture of the ultrahigh-pressure Dabie-Sulu Orogen. Tectonics, 25(5):TC5006.

Jenkyns, H.C., 1980. Tethys: Past and present. Proceedings of the Geologists Association, 91(1-2):107-118.

Jin, W.J., Zhang, Q., He, D.F., and Jia, X.Q., 2005. SHRIMP dating of adakites in western Qinling and their implications. Acta Petrologica Sinica, 21(3):959-966 (in Chinese with English abstract).

Lai, S.C., and Zhu, R.Z., 2021. Three stages of early Paleozoic magmatism in the Tibetan-Himalayan orogen: New insights into the final Gondwana assembly. Journal of Asian Earth Sciences, 221:104949.

Lai, S.C, and Zhu, Y., 2020. Petrogenesis and geodynamic implications of Neoproterozoic typical intermediate-felsic magmatism in the western margin of the Yangtze Block, South China. Journal of Geomechanics, 26 (5):759-790 (in Chinese with English abstract).

Lai, S.C., Qin, J.F., Zhu, R.Z., and Zhao, S.W., 2015a. Neoproterozoic quartz monzodiorite-granodiorite association from the Luding-Kangding area: Implications for the interpretation of an active continental margin along the Yangtze Block (South China Block). Precambrian Research, 267(3-4):196-208.

Lai, S.C., Qin, J.F., Zhu, R.Z., and Zhao, S.W., 2015b. Petrogenesis and tectonic implication of Neoproterozoic peraluminous granitoids from the Tianquan area, western Yangtze Block, South China. Acta Petrologica Sinica, 31(8):2245-2258 (in Chinese with English abstract).

Laubscher, H., and Bernoulli, D., 1977. The Ocean Basins and Margins. New York: Plenum Press: Chapter 1.

Li, G.J., Wang, Q.F., Huang, Y.H., Gao, L., and Yu, L., 2016. Petrogenesis of middle Ordovician peraluminous granites in the Baoshan Block: Implications for the early Paleozoic tectonic evolution along east Gondwana. Lithos, 245:76-92.

Li, J.Y., Tang, M., Lee, C., Wang, X.L., Gu, Z.D., Xia, X.P., Wang, D., Du, D.H., and Li, L.S., 2021a. Rapid endogenic rock recycling in magmatic arcs. Nature Communications, 12(1):3533.

Li, J.Y., Wang, X.L., and Gu, Z.D., 2018. Petrogenesis of the Jiaoziding granitoids and associated basaltic porphyries: Implications for extensive early Neoproterozoic arc magmatism in western Yangtze Block. Lithos, s296-299:547-562.

Li, N., Chen, Y.J., Santosh, M., and Pirajno, F., 2015. Compositional polarity of Triassic granitoids in the Qinling Orogen, China: Implication for termination of the northernmost Paleo-Tethys. Gondwana Research, 27(1):244-257.

Li, S.Z., Zhao, S., Liu, X., Cao, H., Yu, S., Li, X., and Suo, Y., 2018. Closure of the Proto-Tethys Ocean and early Paleozoic amalgamation of microcontinental blocks in East Asia. Earth-Science Reviews, 186:37-75.

Li, X.H., Li, Z.X., Ge, W.C., Zhou, H.W., Li, W.X., Liu, Y., and Wingate, M.T.D., 2003.

Neoproterozoic granitoids in South China: Crustal melting above a mantle plume at ca. 825 Ma? Precambrian Research, 122(1-4):45-83.

Li, X.H., Li, Z.X., Sinclair, J.A., Li, W.X., and Carter, G., 2006. Revisiting the "Yanbian Terrane": Implications for Neoproterozoic tectonic evolution of the western Yangtze Block, South China. Precambrian Research, 151(1-2):14-30.

Liu, Y.M., Li, S.Z., Santosh, M., Cao, H.H., Yu, S.Y., Wang, Y.H., Zhou, J., Zhou, Z.Z., 2019. The generation and reworking of continental crust during early Paleozoic in Gondwanan affinity terranes from the Tibet Plateau. Earth-Science Reviews, 190:486-497.

Li, Z.X., Bogdanova, S.V., Collins, A.S., Davidson, A., De Waele, B., Ernst, R.E., Fitzsi-mons, I.C. W., Fuck, R.A., Gladkochub, D.P., Jacobs, J., Karlstrom, K.E., Lu, S., Natapov, L.M., Pease, V., Pisarevsky, S.A., Thrane, K., and Vernikovsky, V., 2008. Assembly, configuration, and break-up history of Rodinia: A synthesis. Precambrian Research, 160:179-210.

Ma, L.Y., Cawood, P.A., Ding, W., Cai, Y.F., Liu, X.J., Zhang, A.M., Li, H.R., and Ren, M., 2021. Cenozoic retrogression and exhumation of the amphibolites in the eastern Gangdese Belt, SW China. Journal of Asian Earth Sciences, 205:104574.

Ma, L.Y., Wang, Y.J., Fan, W.M., Geng, H.Y., Cai, Y.F., Zhong, H., Liu, H.C., and Xing, X.W., 2014. Petrogenesis of the early Eocene I-type granites in west Yingjiang (SW Yunnan) and its implications for the eastern extension of the Gangdese batholiths. Gondwana Research, 25:401-419.

Ma, P.F., Xia, X.P., Lai, C. K., Cai, K.D., and Yang, Q., 2021. Evolution of the Tethyan Bangong-Nujiang ocean and its se Asian connection: perspective from the early cretaceous high-mg granitoids in SW China. Lithos, 388-389(5):106074.

Meng, Q.R., and Zheng, G.W., 2000. Geologic framework and tatonic evolution of Qinling orogen, central China. Tectonophysics, 323(3-4):183-196.

Metcalfe, I., 2013. Gondwana dispersion and Asian accretion: Tectonic and palaeogeographic evolution of eastern Tethys. Journal of Asian Earth Sciences, 66: 1-33.

Moyen, J.F., 2019. Granites and crustal heat budget. Geological Society, London, Special Publications, 491.

Munteanu, M., Wilson, A., Yao, Y., Harris, C., Chunnett, G., and Luo, Y., 2010. The Tongde dioritic pluton (Sichuan, SW China) and its geotectonic setting: Regional implications of a local-scale study. Gondwana Research, 18(2):455-465.

Qi, X.X., Zhu, L.H., Hu, Z.C., and Li, Z.Q., 2011. Zircon SHRIMP U-Pb dating and Lu-Hf isotopic composition for Early Cretaceous plutonic rocks in Tengchong Block, southeastern Tibet, and its tectonic implications. Acta Petrologica Sinica, 27:3409-3421 (in Chinese with English abstract).

Qin, J.F., Lai, S.C., and Li, Y.F., 2013. Multi-stage granitic magmatism during exhumation of subducted continental lithosphere: Evidence from the Wulong pluton, South Qinling. Gondwana Research, 24: 1108-1126.

Qin, J.F., Lai, S.C., Diwu, C.R., Ju, Y.J., and Li, Y.F., 2010a. Magma mixing origin for the post-collisional adakitic monzogranite of the Triassic Yangba pluton, Northwestern margin of the South China Block: Geochemistry, Sr-Nd isotopic, zircon U-Pb dating and Hf isotopic evidences. Contributions to Mineralogy and Petrology, 159(3):389-409.

Qin, J.F., Lai, S.C., Grapes, R., Diwu, C.r., Ju, Y.J., and Li, Y.F., 2010b. Origin of Late Triassic high-Mg adakitic granitoid rocks from the Dongjiangkou area, Qinling orogen, central China: Implications for subduction of continental crust. Lithos, 120(3-4):347-367.

Qin, J.F., Lai, S.C., Grapes, R., Diwu, C.R., Ju, Y.J., and Li, Y.F., 2009. Geochemical evidence for origin of magma mixing for the Triassic monzonitic granite and its enclaves at Mishuling in the Qinling orogen (central China). Lithos, 112(3-4):259-276.

Qin, J.F., Lai, S.C., and Li, Y.F., 2008. Slab break-off model for the Triassic post-collisional adakitic granitoids in the Qinling Orogen, oentral China: Zircon U-Pb ages, geochemistry, and Sr-Nd-Pb isotopic constraints. International Geology Review, 50(12):1080-1104.

Ratschbacher, L., Franz, L., Enkelmann, E., Jonckheere, R., Pörschke, A., Hacker, B.R., Dong, S., and Zhang, Y., 2006. The Sino-Korean-Yangtze suture, the Huwan detachment, and the Paleozoic-Tertiary exhumation of (ultra) high-pressure rocks along the Tongbai-Xinxian-Dabie Mountains. Geological Society of America Special Papers, 403: 45-75.

Ratschbacher, L., Hacker, B.R., Calvert, A., Webb, L.E., Grimmer, J.C., McWilliams, M.O., Ireland, T., Dong, S., and Hu, J., 2003. Tectonics of the Qinling (oentral China): Tectonostratigraphy, geochronology, and deformation history. Tectonophysics, 366:1-53.

Ratschbacher, L., Hacker, B.R., Webb, L.E., McWilliams, M., Ireland, T., Dong, S., Calvert, A., Chateigner, D., and Wenk, H.R., 2000. Exhumation of the ultrahigh-pressure continental crust in east central China: Cretaceous and Cenozoic unroofing and the Tan-Lu fault. Journal of Geophysical Research: Solid Earth, 105(B6):13303-13338.

Ricou, L.E., 1995. The plate tectonic history of the past Tethys Ocean. The Ocean Basins and Margins, Volume 8 (Chapter 1A): The Tethys Ocean. New York Press.

Sciunnach, D., and Garzanti, E., 2012. Subsidence history of the Tethys Himalaya. Earth-Science Reviews, 111(1-2):179-198.

Suess, E., 1893. Are great oceans depths permanent? Nature Science, 2:180-187.

Sun, W.D., Li, S.G., Chen, Y.D., and Li, Y.J., 2002. Timing of syn-orogenic granitoids in the south Qinling, central China: Constraints on the evolution of the Qinling-Dabie orogenic belt. The Journal of Geology, 110:457-468.

Sun, W.H., and Zhou, M.F., 2008. The ~860 Ma, Cordilleran-type Guandaoshan dioritic pluton in the Yangtze Block, SW China: Implications for the origin of Neoproterozoic magmatism. Journal of Geology, 116(3):238-253.

Sun, W.H., Zhou, M.F., Gao, J.F., Yang, Y.H., Zhao, X.F., and Zhao, J.H., 2009. Detrital zircon U-Pb geochronological and Lu-Hf isotopic constraints on the Precambrian magmatic and crustal evolution of the western Yangtze Block, SW China. Precambrian Research, 172(1):99-126.

Visonà, D., Rubatto, D., Villa, I.M., 2010. The mafic rocks of Shao La (Kharta, S. Tibet): Ordovician basaltic magmatism in the greater Himalayan crystallines of central-eastern Himalaya. Journal of Asian Earth Sciences, 38:14-25.

Wang, Y.J, Li, S., Ma, L., Fan, W., Cai, Y., and Zhang, Y., 2015a. Geochronological and geochemical constraints on the petrogenesis of early Eocene metagabbroic rocks in Nabang (SW Yunnan) and its implications on the NeoTethyan slab subduction. Gondwana Research, 27(4):1474-1486.

Wang, X.L., Zhou, J.C., Wan, Y.S., Kitajima, K., Wang, D., Bonamici, C., Qiu, J.S., and Sun, T., 2013. Magmatic evolution and crustal recycling for Neoproterozoic strongly peraluminous granitoids from Southern China: Hf and O isotopes in zircon. Earth & Planetary Science Letters, 366(2):71-82.

Wang, X.X., Wang, T., Jahn, B.M., Hu, N.G., and Chen, W., 2007. Tectonic significance of Late Triassic post-collisional lamprophyre dykes from the Qinling Mountains (China). Geological Magazine, 144:1-12.

Wang, X.X., Wang, T., and Zhang, C.L., 2013. Neoproterozoic, Paleozoic, and Mesozoic granitoid magmatism in the Qinling Orogen, China: Constraints on orogenic process. Journal of Asian Earth Sciences, 72:129-151.

Wang, X.X., Wang, T., Castro, A., Pedreira, R., Lu, X.X., and Xiao, Q.H., 2011. Triassic granitoids of the Qinling orogen, central China: Genetic relationship of enclaves and rapakivi-textured rocks. Lithos, 126:369-387.

Wang, Y.J., Li, S.B., Ma, L.Y., Fan, W.M., Cai, Y.F., Zhang, Y.H., and Zhang, F.F., 2015. Geochronological and geochemical constraints on the petrogenesis of Early Eocene metagabbroic rocks in Nabang (SW Yunnan) and its implications on the Neotethyan slab subduction. Gondwana Research, 27: 1474-1486.

Wang, Y.J., Xing, X.W., Cawood, P.A., Lai, S.C., Xia, X.P., Fan, W.M., Liu, H.C., and Zhang, F.F., 2013. Petrogenesis of early Paleozoic peraluminous granite in the Sibumasu Block of SW Yunnan and diachronous accretionary orogenesis along the northern margin of Gondwana. Lithos, 182:67-85.

Wang, Y.J., Zhang, L.M., Cawood, P.A., Ma, L.Y., Fan, W.M., Zhang, A.M., Zhang, Y.Z., and Bi, X. W., 2014. Eocene supra-subduction zone mafic magmatism in the Sibumasu Block of SW Yunnan: Implications for Neotethyan subduction and India-Asia collision. Lithos, 206-207:384-399.

Xie, J.C., Zhu, D.C., Dong, G.C., Zhao, Z.D., Wang, Q., and Mo, X.X., 2016. Linking the Tengchong Terrane in SW Yunnan with the Lhasa Terrane in southern Tibet through magmatic correlation. Gondwana Research, 39:217-229.

Xu, Y.G., Yang, Q.J., Lan, J.B., Luo, Z.Y., Huang, X.L., Shi, Y.R., and Xie, L.W., 2012. Temporal-spatial distribution and tectonic implications of the batholiths in the Gaoligong-Tengliang-Yingjiang area, western Yunnan: Constraints from zircon U-Pb ages and Hf isotopes. Journal of Asian Earth Sciences, 53: 151-175.

Xu, Z.Q., Wang, Q., Cai, Z.H., Dong, H.W., Li, H.Q., Chen, X.J., Duan, X.D., Cao, H., Li, J., and Burg, J.P., 2015. Kinematics of the Tengchong Terrane in SE Tibet from the late Eocene to early Miocene: Insights from coeval mid-crustal detachments and strike-slip shear zones. Tectonophysics, 665:127-148.

Zhang, G.W., Dong, Y.P., and Li, S.Z., 2015a. The Mianlue Tectonic Zone of the Qinling Orogen and China Continental Tectonics. Beijing: Science Press:1-501 (in Chinese with English Abstract).

Zhang, G.W., Zhang, B.R., Yuan, X.C., and Chen, J.Y., 2001. Qinling Orogenic Belt and Continental Dynamics. Beijing: Science Press:1-855 (in Chinese).

Zhang, G.W., Zhang, B.R., Yuan, X.C., and Xiao, Q.H., 2001a. Qinling Mountains Orogenic Belt and Continental Dynamics. Beijing: Science Press:1-854 (in Chinese).

Zhang, G.W., Zhang, Z.Q., and Dong, Y.P., 1995. Nature of main tectonic-lithostratigraphic units of the

Qinling Orogen: Implications for the tectonic evolution. Acta Petrologica Sinica, 11(2):101-114 (in Chinese with English abstract).

Zhao, G.C., and Cawood, P.A., 2012. Precambrian geology of China. Precambrian Research, s222-223: 13-54.

Zhao, J.H., and Zhou, M.F., 2007a. Neoproterozoic Adakitic Plutons and Arc Magmatism along the western Margin of the Yangtze Block, South China. Journal of Geology, 115(6):675-689.

Zhao, J.H., and Zhou, M.F., 2007b. Geochemistry of Neoproterozoic mafic intrusions in the Panzhihua district (Sichuan Province, SW China): Implications for subduction-related metasomatism in the upper mantle. Precambrian Research, 152(1):27-47.

Zhao, J.H., Zhou, M.F., Yan, D.P., Yang, Y.H., and Sun, M., 2008. Zircon Lu-Hf isotopic constraints on Neoproterozoic subduction-related crustal growth along the western margin of the Yangtze Block, South China. Precambrian Research, 163(3):189-209.

Zhao, J.H., Asimow, P.D., Zhou, M.F., Zhang, J., Yan, D.P., and Zheng, J.P., 2017a. An Andean type arc system in Rodinia constrained by the Neoproterozoic Shimian ophiolite in South China. Precambrian Research, 296:93-111.

Zhao, J.H., Li, Q.W., Liu, H., and Wang, W., 2018. Neoproterozoic magmatism in the western and northern margins of the Yangtze Block (South China) controlled by slab subduction and subduction-transform-edge-propagator. Earth-Science Reviews, 187:1-18.

Zhao, J.H., Zhou, M.F., Wu, Y.B., Zheng, J.P., and Wang, W., 2019a. Coupled evolution of Neoproterozoic arc mafic magmatism and mantle wedge in the western margin of the South China Craton. Contributions to Mineralogy and Petrology, 174(4).

Zhao, J.H., Nebel, O., and Johnson, T.E., 2021. Formation and Evolution of a Neoproterozoic Continental Magmatic Arc. Journal of Petrology, 62 (8):1-53.

Zhao, S.W., Lai, S.C., Pei, X.Z., Li, Z.C., Li, R.B., Qin, J.F., Zhu, R.Z., Chen, Y.X., Wang, M., Pei, L., Liu, C.J., and Gao, F., 2020. Neo-Tethyan evolution in southeastern extension of Tibet: Constraints from Early Paleocene to Early Eocene granitic rocks with associated enclaves in Tengchong Block. Lithos, 364-365:105551.

Zhao, S.W., Lai, S.C., Pei, X.Z., Qin, J.F., Zhu, R.Z., Tao, N., and Gao, L., 2019b. Compositional variations of granitic rocks in continental margin arc: Constraints from the petrogenesis of Eocene granitic rocks in the Tengchong Block, SW China. Lithos, 326-327:125-143.

Zhao, S.W., Lai, S.C., Qin, J.F., and Zhu, R.Z., 2016. Petrogenesis of Eocene granitoids and microgranular enclaves in the western Tengchong Block: Constraints on eastward subduction of the Neo-Tethys. Lithos, 264:96-107.

Zhao, S.W., Lai, S.C., Qin, J.F., Zhu, R.Z., and Wang, J.B., 2017a. Geochemical and geochronological characteristics of Late Cretaceous to Early Paleocene granitoids in the Tengchong Block, Southwestern China: Implications for crust anatexis and thickness variations along eastern Neo-Tethys subduction zone. Tectonophysics, 694:87-100.

Zhao, S.W., Lai, S.C., Gao, L., Qin, J.F., and Zhu, R.Z., 2017b. Evolution of the proto-Tethys in the Baoshan block along the east Gondwana margin: Constraints from early Palaeozoic magmatism. International

Geology Review, 59(1):1-15.

Zhao, S.W., Lai, S.C., Qin, J.F., and Zhu, R.Z., 2014. Zircon U-Pb ages, geochemistry, and Sr-Nd-Pb-Hf isotopic compositions of the Pinghe pluton, Southwest China: Implications for the evolution of the early Palaeozoic Proto-Tethys in Southeast Asia. International Geology Review, 56:885-904

Zheng, Y.F., 2008. A perspective view on ultrahigh-pressure metamorphism and continental collision in the Dabie-Sulu orogenic belt. Chinese Science Bulletin, 53:3081-3104.

Zheng, Y.F., 2012. Metamorphic chemical geodynamics in continental subduction zones. Chemical Geology, 328:5-48.

Zheng, Y.F., Xia, Q.X., Chen, R.X. and Gao, X.Y., 2011. Partial melting, fluid super-criticality and element mobility in ultrahigh-pressure metamorphic rocks during continental collision. Earth-Science Reviews, 107 (3-4):342-374.

Zhong, D.L., 1998. Paleo-Tethyan orogenic belt in the western parts of the Sichuan and Yunnan provinces. Beijing: Science Press:1-231.

Zhou, M.F., Ma, Y.X., Yan, D.P., Xia, X.P., Zhao, J.H., and Sun, M., 2006a. The Yanbian Terrane (Southern Sichuan Province, SW China): A Neoproterozoic arc assemblage in the western margin of the Yangtze Block. Precambrian Research, 144:19-38.

Zhou, M.F., Yan, D.P., Wang, C.L., Qi, L., and Kennedy, A., 2006b. Subduction-related origin of the 750 Ma Xuelongbao adakitic complex (Sichuan Province, China): Implications for the tectonic setting of the giant Neoproterozoic magmatic event in South China. Earth & Planetary Science Letters, 248(1-2): 286-300.

Zhu, D.C., Zhao, Z.D., Niu, Y., Dilek, Y., Wang, Q., and Ji, W.H., 2012. Cambrian bimodal volcanism in the Lhasa terrane, southern Tibet: Record of an early Paleozoic Andean-type magmatic arc in the Australian proto-Tethyan margin. Chemical Geology, 328:290-308.

Zhu, D.C., Li, S.M., and Cawood, P.A., 2016. Assembly of the Lhasa and Qiangtang terranes in central Tibet by divergent double subduction. Lithos, 245:7-17.

Zhu, D.C., Zhao, Z.D., Niu, Y., Mo, X.X., Chung, S.L., Hou, Z.Q., and Wu, F. Y., 2011. The Lhasa Terrane: Record of a microcontinent and its histories of drift and growth. Earth & Planetary Science Letters, 301(1-2):241-255.

Zhu, L., Zhang, G., Lee, B., Guo, B., Gong, H., Kang, L. and Lü, S., 2010. Zircon U-Pb dating and geochemical study of the Xianggou granite in the Ma'anqiao gold deposit and its relationship with gold mineralization. Science China: Earth Sciences, 53(2):220-240.

Zhu, R.Z., Lai, S.C., Qin, J.F., Zhao, S.W., and Santosh, M., 2019. Petrogenesis of high-K calc-alkaline granodiorite and its enclaves from the SE Lhasa block, Tibet (SW China): Implications for recycled subducted sediments. GSA Bulletin, 131(7-8):1224-1238.

Zhu, R. Z., 2018. Petrogenesis and Geodynamic Implications of Early Cretaceous Granitic Rocks in the Tengchong Block, SW China (Ph.D. thesis). Xi'an: Northwest University.

Zhu, R.Z., Lai, S.C., Qin, J.F., and Zhao, S.W., 2015. Early-Cretaceous highly fractionated I-type granites from the northern Tengchong block, western Yunnan, SW China: Petrogenesis and tectonic implications. Journal of Asian Earth Sciences, 100:145-163.

Zhu, R.Z., Lai, S.C., Qin, J.F., and Zhao, S.W. 2018a. Petrogenesis of late Paleozoic-to-early Mesozoic granitoids and metagabbroic rocks of the Tengchong Block, SW China: Implications for the evolution of the eastern Paleo-Tethys. International Journal of Earth Sciences, 107(2):431-457.

Zhu, R.Z., Lai, S.C., Qin, J.F., Zhao, S.W., and Santosh, M. 2018b. Strongly peraluminous fractionated S-type granites in the Baoshan Block, SW China: Implications for two-stage melting of fertile continental materials following the closure of Bangong-Nujiang Tethys. Lithos, 316:178-198.

Zhu, R.Z., Lai, S.C., Qin, J.F., Zhao, S.W., and Wang, J.B., 2017b. Late Early-Cretaceous quartz diorite-granodiorite-monzogranite association from the Gaoligong belt, southeastern Tibet Plateau: Chemical variations and geodynamic implications. Lithos, 288-289:311-325.

Zhu, R.Z., Lai, S.C., Santosh, M., Qin, J.F., and Zhao, S.W., 2017a. Early Cretaceous Na-rich granitoids and their enclaves in the Tengchong Block, SW China: Magmatismin relation to subduction of the Bangong-Nujiang Tethys ocean. Lithos, 286(287):175-190.

Zhu, R.Z.,Słaby, E., Lai, S.C., et al., 2021. High-K calc-alkaline to shoshonitic intrusions in SE Tibet: implications for metasomatized lithospheric mantle beneath an active continental margin. Contributions to Mineralogy and Petrology, 176(10):1-19.

Zhu, Y., Lai, S.C., Qin, J.F., Zhu, R.Z., Zhang, F.Y., Zhang, Z.Z., and Zhao, S.W., 2019a. Neoproterozoic peraluminous granites in the western margin of the Yangtze Block, South China: Implications for the reworking of mature continental crust. Precambrian Research, 333:105443.

Zhu, Y., Lai, S.C., Qin, J.F., Zhu, R.Z., Zhang, F.Y., Zhang, Z.Z., and Gan, B.P., 2019b. Petrogenesis and geodynamic implications of Neoproterozoic gabbro-diorites, adakitic granites, and A-type granites in the southwestern margin of the Yangtze Block, South China. Journal of Asian Earth Sciences, 183:103977.

Zhu, Y., Lai, S.C., Qin, J.F., Zhu, R.Z., Zhang, F.Y., and Zhang, Z.Z., 2019c. Geochemistry and zircon U-Pb-Hf isotopes of the 780 Ma I-type granites in the western Yangtze Block: Petrogenesis and crustal evolution. International Geology Review, 61(10):1222-1243.

Zhu, Y., Lai, S.C., Qin, J.F., Zhu, R.Z., Zhang, F.Y., and Zhang, Z.Z., 2020a. Petrogenesis and geochemical diversity of late Mesoproterozoic S-type granites in the western Yangtze Block, South China: Co-entrainment of peritectic selective phases and accessory minerals. Lithos, 352-353:105326.

Zhu, Y., Lai, S.C., Qin, J.F., Zhu, R.Z., Liu, M., Zhang, F.Y., Zhang, Z.Z., and Yang, H., 2020b. Genesis of ca. 850−835 Ma high-Mg# diorites in the western Yangtze Block, South China: Implications for mantle metasomatism under the subduction process. Precambrian Research, 343:105738.

Zhu, Y., Lai, S.C., Qin, J.F., Zhu, R.Z., Zhao, S.W., Liu, M., Zhang, F.Y., Zhang, Z.Z., and Yang, H., 2021a. Peritectic assemblage entrainment (PAE) model for the petrogenesis of Neoproterozoic high-maficity I-type granitoids in the western Yangtze Block, South China. Lithos, 402-403:106247.

Zhu, Y., Lai, S.C., Qin, J.F., Zhu, R.Z., Liu, M., Zhang, F.Y., Zhang, Z.Z., and Yang, H., 2021b. Neoproterozoic metasomatized mantle beneath the western Yangtze Block, South China: Evidence from whole-rock geochemistry and zircon U-Pb-Hf isotopes of mafic rocks. Journal of Asian Earth Sciences, 206:104616.

Genesis of the Cenozoic sodic alkaline basalt in the Xiahe-Tongren area of the northeastern Tibetan Plateau and its continental dynamic implications①

Lai Shaocong

Objective

The Cenozoic Indo-Asian collision caused significant crustal shortening and plateau uplift in the central Tibet. The extrusion tectonics model has been widely accepted to explain the strike-slip faults around the Tibetan Plateau. Previous studies indicate that lower crust flow is the main drive force of the extrusion tectonics. Whether mantle extrusion process occurred during the Cenozoic uplift is a major problem to be addressed, which is significant for understanding the uplift mechanism and tectonic evolution of the Tibetan Plateau.

Method

Based on the comprehensive research of the petrology and geochemistry of volcanic rocks, we selected the Na-rich alkaline basalt in the Xiahe-Tongren area of the northeastern Tibetan Plateau to analyze the major and trace elements of phenocryst, glass matrix and microcrystalline mineral in the basalt, using the lithoprobe deep inversion technique, electron microprobe, in-situ laser microprobe and ICP-MS. Coupled with the theoretical petrology and phase equilibrium theory between minerals and melts, we effectively revealed the characteristics, composition, thermal state of the mantle, the mechanism of mafic magmatic origin, the interaction between the mantle and crust, and geodynamic processes in the northeastern margin of Tibetan Plateau. This study provides important dynamic constraints on the tectonic setting and the mechanism of the uplift of Tibetan Plateau, and the eastward extrusion model for the materials in the plateau during the Cenozoic.

Results

This paper reports the petrology and geochemistry of Miocene mafic and ultra-mafic volcanic rocks in the Xiahe and Lixian areas (Fig. 1). Our detailed work indicates that there is

① Published in *Acta Geologica Sinica* (*English Edition*), 2016, 90(3).

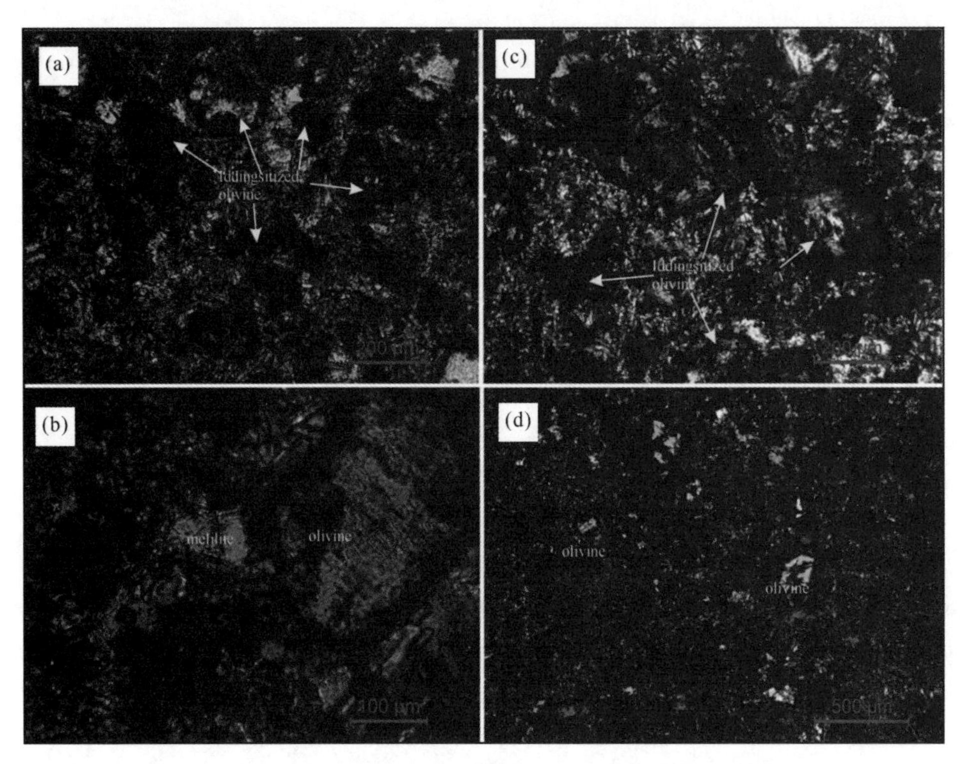

Fig. 1 Petrographic images of the Cenozoic sodic alkaline basalt in Xiahe and Lixian area, northeastern margin of the Tibetan Plateau.

(a) and (b), Porphyritic texture, iddingsitized olivine crystals in the fine-grained matrix; (c) and (d), Porphyritic texture, olivine crystal in the fine-grained matrix.

a regular trend in partial melting condition and source region of the volcanic rocks from west to east, revealing a Cenozoic eastward mantle extrusion in the northeastern margin of Tibetan Plateau. The Xiahe volcanic rocks display ocean island basalts (OIB) affinity with negative K anomalies and positive Nb and Ta anomalies, and they are alkaline with extremely high ratios of Na_2O/K_2O. In combination with relatively enriched Sr-Nd-Pb isotopic compositions, it can be considered that these basalts were derived from partial melting of carbonated pyroxenite in the western Qinling area. The Lixian tephrites are closely associated with carbonatites, strongly indicative of a role of CO_2 in their sources. They have relatively high TiO_2 (3.47%–4.66%), MgO (11.24%–18.88%) contents and low SiO_2 (41.14%–44.82%) and Al_2O_3 (5.84%–9.18%) contents. Coupled with their Sr-Nd-Pb isotopic compositions and previous experimental works, we proposed that the Lixian tephrites may be derived from partial melting of carbonatized peridotite of Paleozoic oceanic lithosphere mantle at relatively high pressures (>3 GPa), with minor hornblendite in their source region, which can account for their higher TiO_2 and $Na_2O + K_2O$ contents. This lateral variation of source region and partial melting process with spatial distribution pattern of Cenozoic volcanic rocks may reflect the eastward extrusion of western mantle material in the northeastern margin of the Tibetan Plateau.

Conclusion

Combining the temporal-spatial distribution characteristics of Cenozoic volcanic rocks in northern Tibet with the regional geology, we propose that the Cenozoic volcanic magmatism in the eastern margin of Tibetan Plateau was induced by eastward extrusion of mantle sources during the uplift of the Tibetan Plateau, which was accompanied with the gradually decreasing geothermal gradient and increasing depth of partial melting from west to east. This process may form the Na-rich alkaline basalt in the Xiahe and Lixian-Tongren area. This model is obviously different from the previous view that they were formed in a continental-rift setting. Furthermore, the partial melting of mantle sources may be closely related to the local activity of strike-slip faults during the eastward extrusion of mantle material in the Tibetan Plateau, which suggests that the mantle sources melting could be controlled by the episodic activity of the surface structure.

Acknowledgment This work was supported by the National Natural Science Foundation of China (Grant No. 41072052).

Discovery of the granulite xenoliths in Cenozoic volcanic rocks from Hoh Xil, Tibetan Plateau①

Lai Shaocong Yi Haisheng Lin Jinhui

Abstract: Two-pyroxene granulite and clinopyroxene granulite xenoliths have been recently discovered in the Late Paleogene to Neogene volcanic rocks (with ages in the range of 4.27–44.60 Ma) that outcropped in Hoh Xil, central Tibetan Plateau. Based on the electron microprobe analysis data, the xenoliths provide constraints for the formation equilibrium temperatures of the two-pyroxene granulite being about 783–818℃ as determined by two-pyroxene thermometry and the forming pressure of the clinopyroxene granulite being about 0.845–0.858 GPa that equal to 27.9–28.3 km depth respectively. It indicates that these granulite xenoliths represent the samples from the middle part of the thickened Tibetan crust. This discovery is important and significant to making further discussion on the component and thermal regime of the deep crust of the Tibetan Plateau.

Granulite is generally regarded as the metamorphic result at high temperature in lower crust. In fact, most identified granulites outcropping at surface belong to Precambrian granulite, which had withstood a long time complicated tectonic movement and recorded the metamorphic processes of different geological period[1-2]. Comparatively, the granulite xenoliths included in the Mesozoic-Cenozoic volcanic rocks could more effectively reveal the metamorphism and thermal regime at lower crust during Mesozoic-Cenozoic period. Therefore, the Mesozoic-Cenozoic granulite xenoliths have been regarded as the important rock-probe and direct evidence for discussing the material components and thermal conditions of the lower crust[1-5].

In recent decades, granulites and granulite-phase rocks have been identified one after another in the eastern and western Himalayan tectonic nodes[6-10]. This is significant to the study of deep crust of Himalayan tectonic region. However, it is very difficult to identify and collect the deep crust granulite xenoliths from the north Tibetan Cenozoic volcanic rocks. This is because of the limitation of physicogeographical circumstances at north Tibet. The only report about the granulite xenoliths identified from the Cenozoic volcanic rocks comes from the Taipinghu area of north Qiangtang[3]. This paper documents for the first time the presence of granulite xenoliths in the Late Paleogene to Neogene volcanic rocks from south Ulan Ul Lake,

① Published in *Progress in Natural Science*, 2003, 13(9).

Hol Xil and presents initial electron microprobe analysis data for these granulite xenoliths which come from the middle part of the thickened Tibetan crust. It then discusses their mineral assemblage, equilibrium temperature, forming pressure and geological significance.

1 General geological setting

The studied Late Paleogene to Neogene volcanic rocks are distributed in the Zhentouya area, south to Ulan Ul Lake, Hoh Xil (Fig. 1). They belong to Shipingding group (N_2s) stratigraphical classification, and overlie the Tuotuo River group (Et) of Paleogene red sandstone, gravel-bearing sandstone. The volcanic rocks occur as lava sheet, long flows that range in thickness from 50 m to 200 m. The main rock type is andesite and dacite, belonging to a series of typical intermediate-silicic magma rocks. Basaltic rock type has not been identified in this group of rocks. The volcanic rocks can be summarized as two eruption cycles separated by a layer of volcanic breccia with width range from 20 m to 30 m. The first eruption cycle is consisted of andesite dominantly, but the second is dacite.

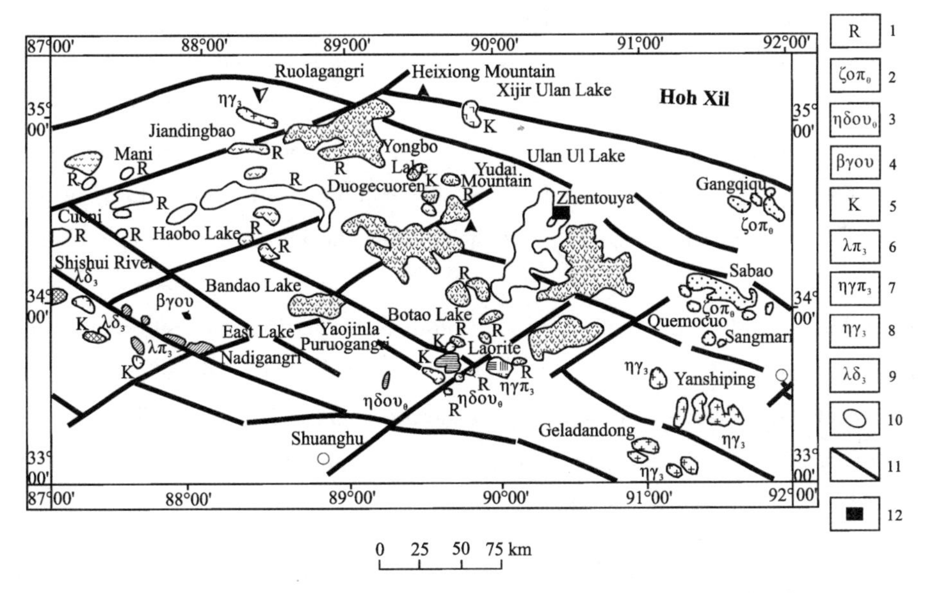

Fig. 1 Regional distribution of the Mesozoic-Cenozoic magma rocks in north Tibet area.

1. Late Paleogene to Neogene volcanic rocks; 2. Paleogene quartz-syenite porphyry; 3. Paleogene quartz-monzodiorite porphyry; 4. Paleogene basaltic-trachyandesite porphyry; 5. Cretaceous volcanic rocks; 6. Upper Cretaceous rhyolite porphyry; 7. Upper Cretaceous adamellite porphyry; 8. Upper Cretaceous adamellite; 9. Upper Cretaceous granodiorite; 10. Hypabyssal intrusive; 11. fault; 12. outcropping region of the granulite xenoliths.

The andesites are characterized by phenocrysts of plagioclase ± amphibole ± biotite. Pyroxene phenocryst can be found in some andesite. The groundmass is finely grained and sometimes with volcanic glass. Minor apatite and magnetite are restricted to the groundmass. The andesites are occasionally slightly altered, with kaolinization of feldspar, chloritization of dark-colored minerals, carbonates invading vacuoles and fractures, but the effect is limited. The dacites are mostly fresh. The vesicular and amygdaloidal structures, aphyric and porphyro-aphanitic textures have been identified. The porphyro-aphanitic dacites commonly have phenocrysts of quartz, amphibole, biotite and plagioclase. The groundmass which often exhibits interlocking texture and hyalopilitic texture is always finely grained with abundant volcanic glass. All the andesitic and dacitic rocks belong to sub-alkaline series on (K_2O+Na_2O) vs. SiO_2 diagram and they are plotted in a calc-alkaline trend on the Alk-FeO-MgO diagram. The andesitic rocks exhibit SiO_2 = 56.59% – 60.74% (mean value 58.21%), Al_2O_3 = 13.78% – 17.66% (mean value 15.37%), MgO = 2.01% – 6.10% (mean value 3.97%), Na_2O > 2.50% (2.50% – 3.90%, mean value 3.03%) and Na_2O/K_2O = 0.58 – 1.39 (mean value 0.96). The dacitic rocks exhibit SiO_2 = 61.91% – 65.70% (mean value 63.38%), Al_2O_3 = 14.87% – 17.28% (mean value 15.99%), MgO = 1.26% – 1.82% (mean value 1.57%), Na_2O > 2.40% (2.40% – 4.00%, mean value 3.40%), Na_2O/K_2O = 0.86 – 1.18 (mean value 0.95).

The stratigraphical constraints on the forming age of the volcanic rocks in the studied area had been obtained by field investigations. In fact, they overlie discordantly on the Tuotuo River group (Et) of the Paleogene system. On the other hand, some of the rock radiochronological results concerning this area that were obtained using K-Ar method can be arranged between 4.27 – 44.66 Ma [11-12]. Therefore, it is believable that the main part of the volcanic rocks developed in the studied area should be formed during the period of from Late Paleogene to Neogene.

Granulite xenoliths have been identified from both of the first (andesitic rock series) and the second (dacitic rock series) eruption cycles. Especially, some sanidine phenocrysts (0.5 – 5.0 cm) have been identified from both andesitic and dacitic rocks. The sanidines are surrounded by acicular and re-crystallized rims of potassium-oligoclase. It is thus clear that the phenocrysts of sanidine are xenocrystic.

2 Petrographic features and forming conditions of the granulite xenoliths

The granulites concerned occur as xenolith in the Cenozoic volcanic rocks from Hoh Xil, which appear fresh in drab-red color and/or greyish black color (for mafic granulite). The size generally ranges from 2 cm to 6 cm, the biggest is 12 cm. A massive structure (some of the xenoliths showing slight gneissic structure) and medium-fine grained granoblastic texture have been identified (Fig. 2). According to the mineral assemblage features the granulite xenoliths can be divided into two categories: two-pyroxene granulite and clinopyroxene granulite. The two-pyroxene granulite is composed of orthopyroxene, clinopyroxene, plagioclase, alkali feldspar and biotite. The clinopyroxene granulite is composed of clinopyroxene, plagioclase and quartz.

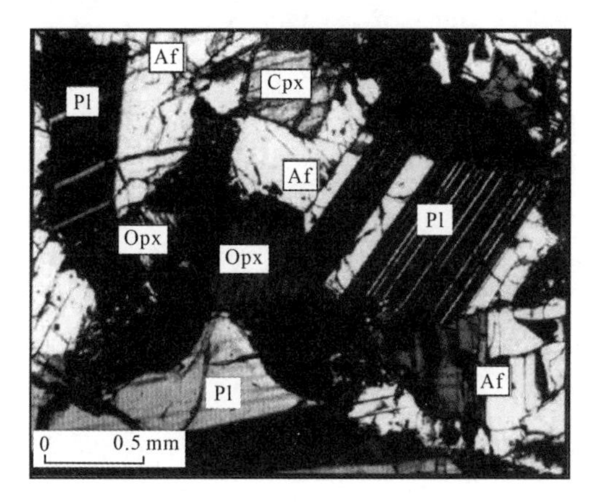

Fig. 2 Structure and mineral assemblage of the two-pyroxene granulite in thin section under a microscope. Cpx: clinopyroxene; Opx: orthopyroxene; Pl: plagioclase; Af: alkali feldspare.

Two granulite xenoliths included in Sample P2126H10 and P2106H have been studied by electron microprobe. The analysis results are shown in Table 1. According to the analysis data, the xenolith included in Sample P2126H10 belongs to typical two-pyroxene granulite, being composed of Cpx + Opx + Pl + Af + Q. The orthopyroxene shows colorless-light red color and obviously pleochroism in thin section under the microscope. Its component exhibits Wo (wollastonite) being 1.27–1.34, En (enstatite) being 64.92–66.64, Fs (ferrosilite) being 32.09–33.73, belonging to typical hypersthene. The clinopyroxene shows light green color in thin section under the microscope. Its component exhibits Wo being 44.76, En being 27.84, Fs being 27.40, belonging to salite. The biotite shows brown-red color in thin section under the microscope. Its component exhibits Fe, Ti enrichment, $Mg/(Mg+Fe^T)$ (MF) >0.5 (0.73–0.75), Ti >0.3 (0.770 0–0.773 4), belonging to titanium-ferrous biotite. The plagioclase shows allotriomorphic-granular, An = 51.03 – 61.49, belonging to labradorite. The alkali feldspar exhibits Or = 63.62, Ab = 30.77, belonging to soda-sanidine. Based on the electron microprobe analysis data (Table 1), the forming temperature of the granulite are between 783–818℃ as determined by two-pyroxene thermometry [13]. This temperature range represents the eutectic temperature at which the clinopyroxene and orthopyroxene reached the equilibrium.

The xenolith included in Sample P2106H belongs to typical clinopyroxene granulite, being composed of Cpx+Pl+Q. The clinopyroxene exhibits Wo=44.80–44.99, En=33.19–33.34, Fs=21.82–21.86, belonging to salite. The plagioclase exhibits An=50.88–65.90, belonging to labradorite.

Table 1 Electron microprobe analysis result (wt%) for the granulites from Cenozoic volcanic rocks, Hoh Xil.

Sample	P2126H10	P2126H10	P2126H10	P2126H10	P2126H10	P2126H10	P2126H10	P2126H10	P2106H	P2106H	P2106H	P2106H	P2106H
Point	1	2	3	4	5	6	7	8	9	10	11	12	13
Mineral	Cpx	Opx	Opx	Pl	Pl	Af	Bi	Bi	Cpx	Cpx	Pl	Pl	Q
SiO_2	51. 20	49. 26	51. 30	54. 26	54. 91	64. 82	36. 51	35. 33	47. 63	48. 15	57. 87	57. 04	99. 53
TiO_2	0. 15	0. 11	0. 11	0. 00	0. 04	0. 19	7. 11	7. 09	0. 42	0. 38	0. 00	0. 08	0. 00
Al_2O_3	1. 01	2. 30	2. 81	30. 61	29. 45	19. 24	15. 56	16. 50	4. 12	4. 59	26. 01	26. 09	0. 03
Cr_2O_3	0. 03	0. 08	0. 00	0. 00	0. 00	0. 05	0. 00	0. 00	0. 03	0. 00	0. 00	0. 00	0. 18
FeO	16. 40	21. 64	21. 55	0. 27	0. 13	0. 88	10. 50	11. 47	13. 31	13. 23	0. 51	0. 55	0. 00
MnO	0. 70	0. 80	0. 94	0. 00	0. 01	0. 00	0. 06	0. 16	0. 51	0. 56	0. 00	0. 05	0. 06
MgO	9. 35	25. 22	23. 27	0. 14	0. 09	0. 14	17. 32	17. 40	11. 36	11. 32	0. 13	0. 08	0. 19
CaO	20. 91	0. 67	0. 67	11. 61	10. 89	1. 13	0. 06	0. 00	21. 42	21. 16	11. 93	10. 79	0. 00
Na_2O	0. 45	0. 07	0. 04	2. 89	3. 19	3. 42	0. 46	0. 44	1. 17	1. 00	3. 09	5. 34	0. 00
K_2O	0. 00	0. 00	0. 00	0. 76	0. 88	10. 75	8. 97	8. 98	0. 00	0. 00	0. 49	0. 66	0. 02
NiO	0. 00	0. 02	0. 00	0. 07	0. 18	0. 00	0. 00	0. 03	0. 06	0. 00	0. 01	0. 00	0. 00
Total	100. 20	100. 17	100. 68	100. 61	99. 79	100. 63	96. 55	97. 40	100. 04	100. 38	100. 04	100. 68	100. 01
[O] =	6	6	6	8	8	8	22	22	6	6	8	8	2
Si	1. 975 2	1. 847 1	1. 900 1	2. 427 0	2. 472 6	2. 940 9	5. 280 8	5. 102 3	1. 833 8	1. 839 7	2. 594 5	2. 562 1	0. 996 9
Al	0. 045 9	0. 101 6	0. 122 7	1. 613 7	1. 563 0	1. 028 8	2. 652 6	2. 808 5	0. 187	0. 206 7	1. 374 4	1. 381 2	0. 000 4
Ti	0. 004 4	0. 003 1	0. 003 1	0. 000 0	0. 001 4	0. 006 5	0. 773 4	0. 770 0	0. 012 2	0. 010 9	0. 000 0	0. 002 7	0. 000 0
Cr	0. 000 9	0. 002 4	0. 000 0	0. 000 0	0. 000 0	0. 001 8	0. 000 0	0. 000 0	0. 000 9	0. 000 0	0. 000 0	0. 000 0	0. 001 4
Fe	0. 529 1	0. 678 7	0. 667 5	0. 010 1	0. 004 9	0. 033 4	1. 270 1	1. 385 3	0. 428 6	0. 422 7	0. 019 1	0. 020 7	0. 000 0
Mn	0. 022 9	0. 025 4	0. 029 5	0. 000 0	0. 000 4	0. 000 0	0. 007 4	0. 019 6	0. 016 6	0. 018 1	0. 000 0	0. 001 9	0. 000 5
Mg	0. 537 6	1. 409 5	1. 284 7	0. 009 3	0. 006	0. 009 5	3. 734 0	3. 745 5	0. 651 9	0. 644 7	0. 008 7	0. 005 4	0. 002 8
Ca	0. 864 3	0. 026 9	0. 026 6	0. 556 4	0. 525 4	0. 054 9	0. 009 3	0. 000 0	0. 883 6	0. 866 2	0. 573 1	0. 519 3	0. 000 0

Continued

Sample	P2126H10	P2126H10	P2126H10	P2126H10	P2126H10	P2126H10	P2126H10	P2126H10	P2106H	P2106H	P2106H	P2106H	P2106H
Na	0. 033 7	0. 005 1	0. 002 9	0. 250 6	0. 278 5	0. 300 9	0. 129 0	0. 123 2	0. 087 3	0. 074 1	0. 268 6	0. 465 1	0. 000 0
K	0. 000 0	0. 000 0	0. 000 0	0. 043 4	0. 050 6	0. 622 2	1. 655 1	1. 654 4	0. 000 0	0. 000 0	0. 028	0. 037 8	0. 000 3
Ni	0. 000 0	0. 000 6	0. 000 0	0. 002 5	0. 006 5	0. 000 0	0. 000 0	0. 003 5	0. 001 9	0. 000 0	0. 000 4	0. 000 0	0. 000 0
Sum	4. 013 9	4. 100 4	4. 037 0	4. 913 1	4. 909 2	4. 998 8	15. 512 0	15. 612 0	4. 103 8	4. 083 1	4. 866 7	4. 996 1	1. 002 3
Wo	44. 76	1. 27	1. 34						44. 99	44. 80			
En	27. 84	66. 64	64. 92						33. 19	33. 34			
Fs	27. 40	32. 09	33. 73						21. 82	21. 86			
An				51. 03	61. 49	5. 61					65. 9	50. 8	
Ab				29. 47	32. 59	30. 77					30. 88	45. 5	
Or				5. 10	5. 92	63. 62					3. 20	3. 70	
Ca							1. 86	0. 00					
Mg							74. 48	73. 00					
Fe							25. 33	27. 00					

Analyzed by Electron Microprobe Section, Xi'an Institute of Geology and Mineral Resources (2001).

According to the clinopyroxene-plagioclase-quartz mineral assemblage geological manometer suggested by Ellis[14], the forming pressure of the clinopyroxene granulite xenolith included in Sample P2106H has been calculated. The calculation result is summarized as below:

(1) Plagioclase (average composition of the analysis point 11 and 12 shown in table 1): Si^{4+}(0.956 1), Ti^{4+}(0.000 5), Al^{3+}(0.511 0), Fe^{3+}(0.007 4), Mg^{2+}(0.002 6), Mn^{2+}(0.000 4), Ca^{2+}(0.202 6), Na^{+}(0.136 0), K^{+}(0.012 2). Molecular formula: $(Na_{0.367}K_{0.033})_{0.4}Ca_{0.546}Fe^{3+}_{0.02}Al_{1.378}Si_{2.578}O_8$, end member component: Ab = 38.80, An = 57.70, Or = 3.5.

(2) Clinopyroxene (average composition of the analysis point 9 and 10 shown in Table 1): Si^{4+}(1.837), Ti^{4+}(0.012), $Al^{3+(IV)}$(0.163), $Al^{3+(VI)}$(0.034), Fe^{3+}(0.080), Fe^{2+}(0.346), Mg^{2+}(0.647), Mn^{2+}(0.017), Ca^{2+}(0.782), Na^{+}(0.080), K^{+}(0.000). Molecular formula: $T_{Sch16.3}(Ac+Jd)_8Hd_{36.6}Di_{39.2}$.

(3) Forming pressure: Based on the temperature (783 – 818℃) obtained from two-pyroxene granulite included in Sample P2126H10 determined by two-pyroxene thermometry, we take the equilibrium temperature of the granulite as T= 1 073 K = 800 ℃. Then the lnKd = −2.727 and the forming pressure of the clinopyroxene granulite should be 0.842 GPa. According to the pressure correction method suggested by Ellis[14], we can get the corrected pressure range as below:

$$P_{max}(\text{corrected}) = 0.858 \text{ GPa (equalent to 28.31 km depth)}$$

$$P_{min}(\text{corrected}) = 0.845 \text{ GPa (equalent to 27.90 km depth)}$$

Therefore, the granulite xenoliths included in Cenozoic volcanic rocks from Hoh Xil should come from the depth ranging from 27 km to 28 km.

3 Discussion about the geological significance

Above research indicates that the xenolith included in Sample P2126H10 belongs to typical two-pyroxene granulite, and the xenolith included in Sample P2106H belongs to typical clinopyroxene granulite, which comes from the depth ranging from 27 km to 28 km.

Up to now, despite numerous studies of the region, the geochemical features of the thickened crust, especially the material components, thermal regime, partial melting characteristics of the middle-lower part of the Tibetan doubled crust are still controversial. However, it is significant for scientists to discuss about the processes responsible for the formation/evolution of the plateau, the uplift mechanism and the magma system generation of the ultrapotassium-potassium-high potassium calc alkaline Cenozoic volcanic rocks outcropped in the Tibetan Plateau.

Recent decades, the rock association of the metamorphic basement outcropping at surface have been used to infer the component of the upper crust of the plateau[15]. In addition, crust-derived magma system, geophysical data and geothermal studies have been used to infer indirectly the characteristics of the middle-lower crust and upper mantle of the Tibetan Plateau [16,17]. However, few studies quantify the petrological structure profile about the middle-lower crust and upper mantle of the plateau, which should be set up upon the detailed petrographic and micro-structure observations. This is because, it is difficult to collect the deep-seated rock samples from Tibetan Plateau. Therefore, the discovery of the typical granulite xenoliths included in Cenozoic volcanic rocks from Hoh Xil is significant. This discovery is important to further study on the petrological structure, component, mineral phase, temperature, pressure, partial melting mechanism of the special thickened Tibetan Plateau.

However, the formation age of the granulite xenoliths has not been obtained. The definite geological significance of the granulite xenoliths included in the Cenozoic volcanic rocks from Hoh Xil needs to be determined by further research work.

Acknowledgements Thanks are due to Prof. Gao Shan (Northwest University) for his help in the thin section observation, Prof. Li Sanzhong (Ocean University of Qingdao) and Prof. Luo Zhaohua (China University of Geosciences) for their help in calculating the forming pressure of the clinopyroxene granulite.

References

[1] Shao, J.A., et al. Discovery of the Early Mesozoic granulite xenoliths in North China Craton. Science in China: Series D, 2000, 43(Supp.):245.

[2] Han, Q.J., et al. Mineral chemistry and metamorphic *p-t* conditions of granulute xenoliths in Early Mesozoic diorite in Harkin region, eastern Inner Mongolia Autonomous Region, China. Earth Science, 2000, 25(1):21 (in Chinese).

[3] Hacker, B.R., et al. Hot and dry deep crustal xenoliths from Tibet. Science, 2000, 287:2463.

[4] Vander, H.R.D., et al. Composition, Deep Structure and Evolution of Continents. Amsterdam-London-New York: Elsevier, 1999:342.

[5] Gao, S., et al. Chemical composition of the continental crust as revealed by studies in East China. Geochimica et Cosmochimica Acta, 1998, 62:1959.

[6] Li, D.W., et al. Discovery of the mafic granulite from core-complex of the middle segment, Himalayan and its tectonic significance. Earth Science, 2002, 27(1):80 (in Chinese).

[7] Zhong, D., et al. Discovered high-pressure granulite from Namjagabarwa area, Tibet. Chinese Science Bulletin, 1995, 14:1343.

[8] Liu, Y., et al. Petrology of high-pressure granulites from the eastern Himalayan syntaxis. J. Metamorphic Geol., 1997, 15:451.

[9] Yamamoto, H. Contrasting metamorphic *P-T*-time paths of the Kohistan granulites and tectonics of the western Himalayas. J. Geol. Soc., London, 1993, 150:843.

[10] Ding, L., et al. Cenozoic structural and metamorphic evolution of the eastern Himalayan syntaxis (Namche Barwa). Earth & Planetary Science Letters, 2001, 192:423.

[11] Zhang, Y.F., et al. An Introduction to the Geological Evolution of Hoh Xil and Its Adjacent Region. Beijing: Seismological Press, 1994:1 (in Chinese).

[12] Deng, W.M. Cenozoic Intraplate Volcanic Rocks in the Northern Qinghai-Xizang Plateau. Beijing: Geological Publishing House, 1998:139 (in Chinese).

[13] Wood, B.J., et al. Garnet-orthopyroxene relationships in simple and complex systems. Contrib. Mineral. Petrol., 1973, 42:109.

[14] Ellis, D.J. Osumilite-sapphirine-quartz granulites from Enderby Land, Antarctica: *P-T* conditions of metamorphism, implications for garnet-cordierite equilibrium and the evolution of the deep crust. Contrib. Mineral Petrol., 1980, 74:201.

[15] Kapp, P., et al. Blueschist-bearing metamorphic core complexes in the Qiangtang block reveal deep crustal structure of northern Tibet. Geology, 2000, 28:19.

[16] Lai, S.C., et al. Enriched upper mantle and eclogitic lower crust in north Qiangtang, Qinghai-Tibet plateau: Petrological and geochemical evidence from the Cenozoic volcanic rocks. Acta Petrologica Sinica, 2001, 17(3):459 (in Chinese).

[17] Lai, S.C., et al. Petrogenesis and its significance to continental dynamics of the Neogene high-potassium calc-alkaline volcanic rock association from north Qiangtang, Tibetan plateau. Science in China: Series D, 2001, 44(Supp.):45.

Further study on geochemical characteristics and genesis of the boninitic rocks from Bikou Group, Northern Yangtze Plate①

Li Yongfei Lai Shaocong② Qin Jiangfeng

Abstract: Compared with the major and trace elements of typical boninite, the metabasalts collected from the Nanfanba (南范坝)-Miaowanli(庙湾里) region in the Bikou(碧口) Block could be treated as boninitic characterized by low-Si, low-Ti, low-P, high-$Mg^{\#}$ and high Al_2O_3/TiO_2, consistent with geochemical features of boninite. The normal mid-ocean ridge basalt (N-MORB) normalized spider diagram displays fairly depleted high field strong elements (HFSE) (Zr, Y, Ti). Enriched refractory elements (Cr, Co, Ni) as well as light rare earth elements (LREE)-depleted chondrite-normalized REE distribution patterns suggest the boninitic magmas are derived from an extremely depleted mantle wedge in the presence of a hydrous fluid, meanwhile signifying the source region had previously undergone a high degree partial melting process yielding primary magmas with enriched large ion lithophile elements (LILE). In addition, almost all the samples in the Nb-Zr-Y and Ti-Zr-Y discrimination diagrams were plotted in the island arc basalt (IAB) field. Coupled with the island arc tholeiitic (IAT) basalt in the study region, therefore, the geochemical characteristics of the studied rocks indicate the meta-basalts probably occurred in a fore-arc subduction setting. This conclusion may be of great significance for the further study of the tectonic background of the Bikou volcanism.

Introduction

Compositionally, the term "boninite" refers to a large variety of primary or near-primary ($Mg^{\#}=0.60-0.85$) magmas with a wide variation of CaO/Al_2O_3 (most range from 0.55 to 0.75) (Crawford et al., 1989). With regard to its definition as recommended by the IUGS (Deng et al., 1993) and Hickey and Frey (1982), the classification of boninitic rocks is still a matter of petrological debate (Qiu et al., 2004). In general, rock with some typical geochemical features of boninite could be treated as a boninitic series in an intensive study. Being a tectonic indicator of fore-arc setting in a subduction zone (Sobolev and Danyushevsky,

---

① Published in *Journal of China University of Geosciences*, 2006, 17(2).

② Corresponding author.

1994; Falloon and Crawford, 1991; Crawford et al., 1989; Meijer, 1980), boninitic rocks have played an important role in geodynamic environment research over the past few years.

The tectonic nature of the Bikou volcanic group has drawn a great deal of attention, and yet still remained controversial. In the 1970s some people proposed they should be assigned to part of an ophiolitic belt developed in different settings such as IAB, MORB or ocean island basalt (OIB) (Zhao et al., 1990). Later, the suggestion that they belonged to an ophiolitic mélange was also made, and challenged (Xia et al., 1991). Recently, a new study by Yan et al. (2004) suggested the Bikou Group developed in a fore-arc setting, on the grounds that some boninitic rocks had been discovered in the northern margin of the Yangtze Block. On the basis of the previous study, and focusing on the boninitic rocks, this study further discusses their possible tectonic setting and source region, and places a geochemical constraint on the original geodynamic environment in the Bikou Group.

Geological setting

With a large scale of about 10 000 km^2, extending from the east via Bikou, west to Pingwu in Sichuan, the Bikou volcanic group is exposed in the Bikou Block, in the northern margin of the Yangtze Block and southwest to Qinling orogenic belt (Fig. 1). It has been assigned a metamorphic-sedimentary formation with an age of 846–766 Ma (Yan et al., 2004). The present volcanic rocks consist mainly of metabasalts, metaandesites and metabasalt-andesite with a minor amount of basalt, andesite, rhyolite, K-rhyolite and volcaniclastic rocks. Although the rocks have been subject to low-grade greenschist facies metamorphism, they have well preserved igneous textures and structures, such as the fumarolic, amygdaloidal, pillow structure and intergrowth, and porphyritic texture. Under the microscope, the boninitic rocks are generally shown to contain phenocrysts or microphenocrysts of clinopyroxenes, hornblendes and plagioclases. The reaction rim texture could be observed in the clinopyroxenes with a few fresh microphenocrysts of the hornblendes. The groundmass is comprised of glass and microphenocrysts of hornblende and plagioclase. The present rocks from basic volcanic rocks in the lower 1st and 2nd cycles of this group (Xu et al., 2002), and located in the Nanfanba-Miaowanli region, are mainly composed of metabasalt and tuff, both with strong carbonation.

Major elements

In terms of major elements, the rocks are compositionally uniform, SiO_2 spans 48.21% to 50.44%, TiO_2 varies from 0.15% to 0.36%, lower than that of typical boninite, and $Mg^{\#}$ ranges from 0.62 to 0.66, consistent with boninite (0.55–0.83) (Falloon and Crawford, 1991). On the whole, the results compare well with the major elements of typical boninite (Table 1), except for the lower SiO_2, and so the rocks can be regarded collectively as boninitic rocks.

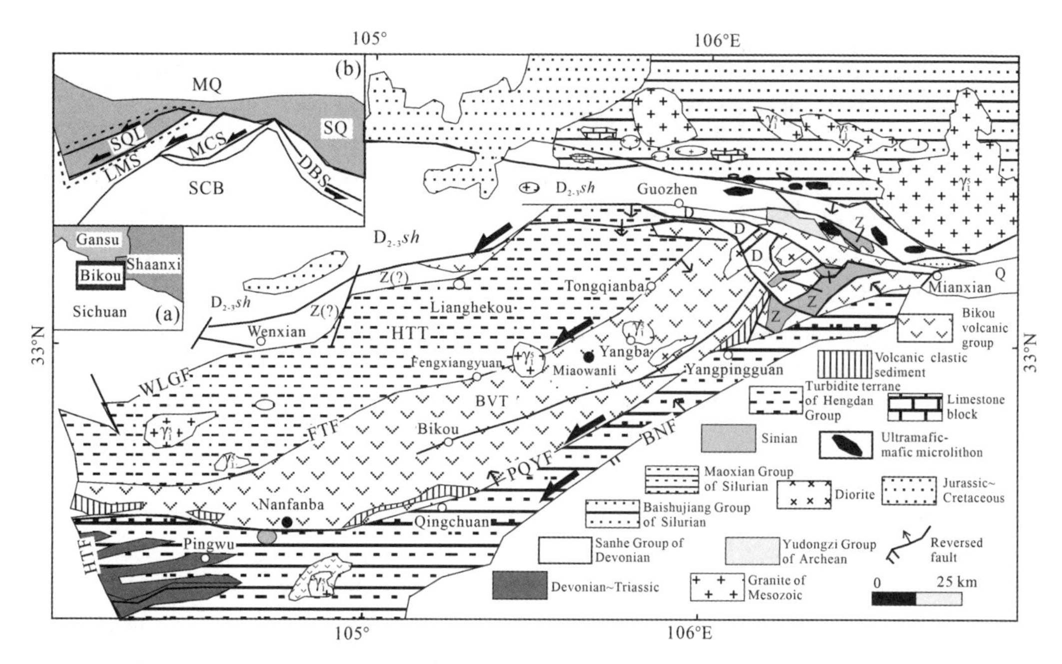

Fig. 1 Sketch geological map of Bikou area in the Southern Qinling Mountain (after Qin et al., 2005). MQ: Middle Qinling; SQ: South Qinling; SCB: South China Block; SQL: Southwest Qinling; LMS: Longmen Moutain; MCS: Micang Moutain; DBS: Daba Moutain; HTT: Hendan Terrane; BVT: Bikou volcanic group; HTF: Huya-Tucheng fault; WLGF: Wenxian-Lianghekou-Guozhen fault; FTF: Fengxiangyuan-Tongqianba fault; PQYF: Pingwu-Qingchuan-Yangpingguan fault; BNF: Beichuan-Nanba fault.

Table 1 Major(wt%) and trace($\times10^{-6}$) element contents of boninites from the study area and other typical areas.

Rock	Bikou metabasalt			Boninite			Bikou metabasalt			
Sample	BK226	BK228	BK230	Marayala	Boni island	Gape vogel	Sample	BK226	BK228	BK230
SiO_2	48.21	50.42	50.44	55.00	58.00	57.42	La	0.32	0.36	0.18
TiO_2	0.15	0.36	0.24	0.32	0.16	0.25	Ce	1.08	1.74	0.95
Al_2O_3	22.69	11.65	18.10	15.2	12.07	8.34	Pr	0.22	0.43	0.25
Fe_2O_3	-	-	-	9.31	8.46	9.61	Nd	1.52	2.57	1.65
FeO^T	4.11	6.48	5.73	-	-	-	Sm	0.59	0.90	0.57
MnO	0.06	0.14	0.10	0.12	0.12	0.21	Eu	0.19	0.32	0.18
MgO	8.38	14.16	10.33	6.87	10.00	17.80	Gd	0.80	1.71	1.10
CaO	13.18	15.09	12.26	9.59	8.27	4.91	Tb	0.11	0.28	0.16
Na_2O	3.19	1.67	2.77	2.60	1.60	0.86	Dy	0.71	1.77	1.07
K_2O	0.02	0.01	0.01	0.98	0.49	0.33	Ho	0.13	0.36	0.19
P_2O_5	0.01	0.02	0.02	0.05	0.02	0.03	Er	0.37	0.87	0.50
$Mg^{\#}$	0.64	0.66	0.62	0.57	0.69	0.77	Tm	0.06	0.14	0.08
V	69.85	203.09	143.49	-	166.00	156.00	Yb	0.40	0.87	0.55
Cr	724.41	1 581.16	956.11	262.00	632.00	1 774.00	Lu	0.07	1.36	0.09
Ni	131.94	225.35	160.01	85.00	162.00	778.00	Y	3.89	9.77	5.48
Rb	3.95	3.87	3.73	13.00	10.00	6.00	δEu	0.86	0.77	0.68

Continued

Rock	Bikou metabasalt			Boninite			Bikou metabasalt			
Sample	BK226	BK228	BK230	Marayala	Boni island	Gape vogel	Sample	BK226	BK228	BK230
Sr	41.58	18.15	39.70	102.00	74.00	100.00	$(La/Yb)_N$	0.58	0.30	0.23
Y	3.89	9.77	5.48	8.00	5.00	5.00	$(La/Sm)_N$	0.35	0.26	0.20
Sc	24.54	60.24	45.18	36.00	38.00	30.00	$(Gd/Yb)_N$	1.68	1.63	1.65
Zr	29.85	34.02	30.71	34.00	21.00	37.00	$(Ce/Yb)_N$	0.51	0.37	0.32
Ba	15.19	4.78	5.22	33.00	27.00	49.00				

The major and trace element contents of Bikou metabasalt are from Yan et al. (2004), the major and trace element contents of boninite in Marayala are from Bougault et al. (1982), the contents of boninite in Boni island are from Hickey and Frey (1982), and Cameron et al. (1983), and the contents of boninite in Cape vogel are from Jenner (1981).

Trace and rare earth elements

The total REE (rare earth elements) content in rocks is fairly low, varying from 6.57×10^{-6} to 13.67×10^{-6}. The feature of remarkably depleted LREE may be readily detected by the $(La/Yb)_N = 0.23-0.58$ as well as $\sum LREE/\sum HREE = 0.37-0.60$. In the rock-chondrite distribution pattern (Fig. 2), the rocks display a left sloping pattern with a slight trough in Eu, which strikingly distinguishes it from the average U-shaped pattern of boninite rather than closer to the boninite eastern of Tonga ridge determined by Falloon and Crawford (1991).

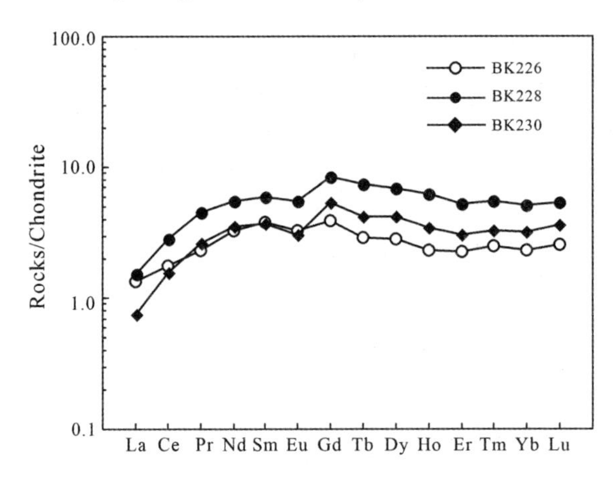

Fig. 2 Chondrite-normalized REE distribution patterns of the rocks in the study area (after Sun and McDonough, 1989).

As Table 1 shows, the content of refractory elements Co, Cr, and Ni is rather higher than those of typical boninite, whereas the Zr and Y content is the same. The ratios of Zr/chondrite (5-6), Y/chondrite (1.8-4.6) and Yb/chondrite (1.8-4.0) are much less than the corresponding values from MORB (Sun, 1980). The rocks show quite a depletion (much lower than the N-MORB reference line) in HFSE (Zr, Ti, Y, P) in the N-MORB spider diagram

(Fig. 3), except for mobile elements. This signifies the source of these rocks is possibly more depleted as a result of residual mantle after experiencing a partial melting to give birth to MORB magma (Falloon and Crawford, 1991).

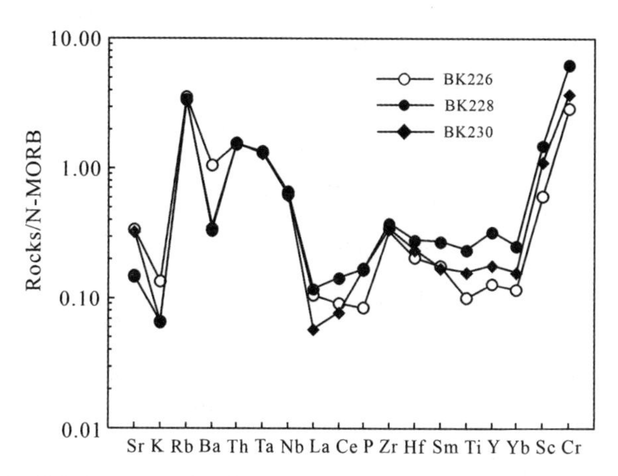

Fig. 3 N-MORB normalized trace element patterns of the rocks in study area (after Pearce, 1996, 1980).

The tectonic setting

Recently, regardless of the time of volcanic rock, the trace elements, especially, the immobile elements that are more stable than mobile ones in the alternation and metamorphism, have been a valuable geochemical fingerprint to trace the original geodynamic processes and magmatic genesis by means of their abundance, combination, ratios and soon (Lai et al. 2001, 2000a, b). Accordingly, the HFSE (Zr, Ti, Y, Nb, Ta) and their ratios appear not to be greatly affected by alteration, so have been taken as an effective tool for determining the original tectonic region/setting. The Ti/100-Zr-3Y and 2Nb-Zr/4-Y discrimination diagrams show the samples fall in the volcanic-arc basalt field (Figs. 4 and 5), which accords well with the fore-arc setting capable of generating the boninite in many cases (Sobolev and Danyushevsky, 1994; Falloon and Crawford, 1991; Crawford et al., 1989; Meijer, 1980).

Discussion and conclusion

Commonly, the boninite derived from the subduction setting is accompanied by calc-alkaline basalt (CAB) (Zhang, 1990). Significantly, the IAT (SiO_2 = 50. 28%–55. 84%) in the study area has a partial melting trend with the studied rocks in the La-La/Sm diagram (Yan et al., 2004). It is generally believed that the boninitic rocks magmas are produced by a two-stage process: extreme depletion of a mantle source by melt extraction, followed by a second stage of melting induced by enriched subduction components. The present rocks in the study area are characterized by a strong depletion in Ti, Y, and Yb as well as a slight depletion in

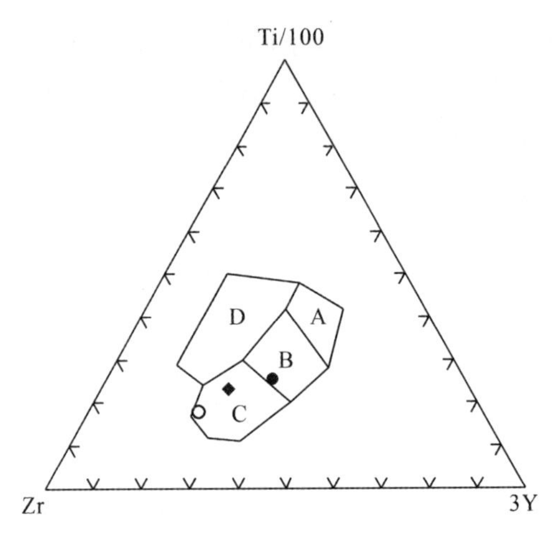

Fig. 4 Ti-Zr-Y diagram for basalts formed in different tectonic settings (after Pearce and Cann, 1973).
A: OIT; B: IAT, calc-alkali basalt (CAB) and MORB; C: CAB; D: within plate basalt (WPB).

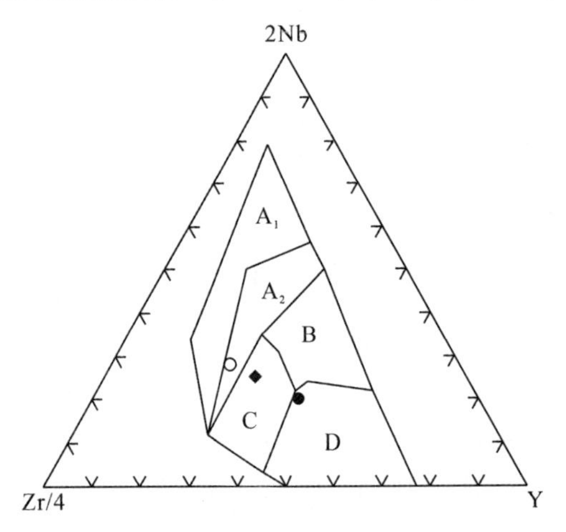

Fig. 5 Nb-Zr-Y diagram for basalts formed in different tectonic settings (after Meschede, 1986).
A_1, A_2: within plate alkaline (WPA); B: plume MORB (P-MORB); A_2, C: within plate tholeiite (WPT);
D: N-MORB; C, D: arc volcanic (ARC).

LILE, Nb, Ta, and LREE, which suggests the source region underwent an extreme partial melting process to yield the IAT, and then the residual mantle remelted to give rise to the boninitic magmas in the presence of a hydrous fluid from the subduction slab. Given a little continental material involved in the oceanic crust, the boninitic rocks with the left sloping rock chondrite-normalized pattern would be a reasonable result of the subduction slab melting.

As we know an ophiolitic belt is effective evidence of the continental breakoff and oceanic basin development in orogenic study. However, the rock associations or assemblages from the arc setting are also important in discussions of the ocean development (Lai et al., 2000a, b).

The volcanic nature of the Bikou Group has long attracted attention, and yet debate remains concerning whether it is an ophiolitic belt or not. The arc volcanic series (IAB) from Nanfanba-Miaowanli region in the Bikou Group was reported previously (Yan et al., 2004).

The MORB, ocean island tholeiite (OIT) and ocean island alkaline (OIA) were determined in the Guangpinghe and Tongqianba-Tuohe regions, respectively, in new studies. These significant rocks could offer strong petrological evidence to confirm the Bikou Group as an ophiolitic mélange, which implies a perished oceanic basin developed in the Bikou area during the Neoproterozoic. As long as boninite was found in an ophiolitic belt, those rocks were bound to occur in the fore-arc subduction setting (Zhang and Zhou, 2001). Accordingly, the present boninitic rocks would be of great significance for the further study on the original geodynamic process in the arc-basin system as well as on the tectonic nature of the study area in the Neoproterozoic.

Acknowledgment This study is supported by the National Natural Science Foundation of China (No. 40234041).

References

Bougault, H., Joron,J.L., Treuil, M., et al., 1982.Tholeiites, Basaltic Andesitic, and Andesites from Leg 60 Sites: Geochemistry, Mineralogy and Low Partition Coefficient Elements. Init. Repts. DSDP, 60:657-678.

Cameron, W.E., McCulloch, M.T., Walker, D.A., 1983. Boninite Petrogenesis: Chemical and Nd-Sr Isotopic Constraints. Earth & Planetary Science Letters, 65:75-89.

Crawford, A.J., Falloon, T.J., Green, D.H., 1989. Classification, Petrogenesis and Tectonic Setting of Boninites. In: Crawford, A J. Boninite and related rocks. London: Unwin Hyman:1-49.

Deng, J.F., Mo, X.X., Wei, Q.R., et al., 1993.The Volcanism and Tethyan Evolution in Sanjiang Region. In: Mo, X.X. The Tethyan Volcanism and Mineralization in Sanjiang Region. Beijing: Geological Publishing House: 224-233 (in Chinese).

Falloon, T.J., Crawford, A.J., 1991. The Petrogenesis of High-Calcium Boninite Lavas Dredged from the Northern Tonga Ridge. Earth & Planetary Science Letters, 102:375-394.

Hickey, R.L., Frey, F.A., 1982. Geochemical Characteristics of Boninite Series Volcanics: Implication for Their Source. Geochim. Cosmochim. Acta, 46:2099-2115.

Jenner,G.A.,1981. Geochemistry of High-Mg Andesites from Cape Vegol, Papua New Guinea. Chem. Geol., 33:307-332.

Lai, S.C., Yang, R.Y., Zhang, G.W., 2001. Tectonic Setting and Implication of the Sunjiahe Volcanic Rocks, Xixiang Group, in South Qinling. Chinese Journal of Geology, 36(3): 295-303 (in Chinese with English Abstract).

Lai, S.C., Zhang, G.W., Yang, R.Y., 2000a. Identification of the Island-Arc Magmatic Zone in the Lianghe-Raofeng-Wuliba Area, South Qinling and Its Tectonic Significance. Science in China: Series D, 30:53-63 (in Chinese).

Lai, S.C., Zhang, G.W., Yang, R.Y., 2000b. Geochemistry of the Volcanic Rock Association from Lianghe Area in Mianlue Suture Zone, Southern Qinling and Its Tectonic Significance. Acta Petrologica Sinica, 16 (3): 317-326 (in Chinese with English abstract).

Meijer, A., 1980. Primitive Arc Volcanism and a Boninite Series: Examples from Western Pacific Island Arcs.

In: The Tectonic and Geologic Evolution of Eastsouthern Asia Seas and Islands. Am. Geophys . Union Washington, 23: 269-282.

Meschede, M. A., 1986. Method of Discriminating between Different Types of Mid-Ocean Basalts and Continental Tholeiites with the Nb-Zr-Y Diagram. Chem. Geol., 56: 207-218.

Pearce, J.A., Cann, J.R., 1973. Tectonic Setting of Basaltic Volcanic Rocks Determined Using Trace Element Analysis. Earth & Planetary Science Letters, 19:290-300.

Pearce, J. A., 1980. Geochemical Evidence for the Genesis and Eruptive Setting of Lavas from Tethyan Ophiolites. In: Panayiotou, A. Proc. Internat. Ophiolite Symp. Cyprus, Geol. Surv. Dept., Nicosia, Cyprus: 261-272.

Pearce, J.A., 1996. A Users Guide to Basalt Discrimination Diagrams. In: Wyman, D.A. Trace Element Geochemistry of Volcanic Rocks: Applications for Massive Sulphide Exploration. Geochem. Short Course Notes-Geol. Assoc. Can., 12: 79-113.

Qin, J.F., Lai, S.C., Li, Y.F., 2005. Petrogenesis and Geological Significance of Yangba Granodiorites from Bikou Area, Northern Margin of Yangtze Plate. Acta Petrologica Sinica, (3):697-710 (in Chinese with English abstract).

Qiu, R.Z., Cai, Z.Y., Li, J.F., 2004.Boninite of Ophiolite Belts in Western Qinghai-Tibet Plateau and Its Geological Implication. Geoscience, 18(3):305-308 (in Chinese with English abstract).

Sobolev, A.V., Danyushevsky, L.V., 1994. Petrology and Geochemistry of Boninites from the North Termination of the Tonga Trench: Constraints on the Generation Conditions of Primary High-Ca Boninite Magmas. J. Petrol., 35:1183-1208.

Sun, S.S., Nesbitt, R.W., McCulloch, M.T., 1989. Geochemistry and Petrogenesis of Archean and Early Proterozoic Siliceous High-Magnesian Basalts. In: Crawford, A.J. Boninites. London: Academic Division of Unwin Hvman: 149-173.

Sun, S.S., McDonough, W.F., 1989. Chemical and Isotopic Systematics of Oceanic Basalts: Implications for Mantle Composition and Processes. In: Saunders, A.D., Norry, M. Magmatism in the Ocean Basin. J. Geol. Soc. Special Publ., 42:313-345.

Xia, L.Q., Xia, Z.C., Ren, Y.X., et al., 1991. The Marine Volcanic Rocks in the Qinling-Qilian Mountain. Beijing: Geological Publishing House (in Chinese with English abstract).

Xu, X.Y., Xia, Z.C., Xia, L.Q., 2002. Volcanic Cycles of the Bikou Group and Their Tectonic Implications. Geological Bulletin of China, 21:478-485 (in Chinese with English abstract).

Yan, Q.R., Andrew, D.H., Wang, Z.Q., et al., 2004. Geochemistry and Tectonic Setting of the Bikou Volcanic Terrane on the Northern Margin of the Yangtze Plate. Acta Petrologica et Mineralogica, 23(1):1-11 (in Chinese with English Abstract).

Zhang, Q., Zhou, G.Q., 2001. The Ophiolite of China. Beijing: Science Press (in Chinese).

Zhang, Q., 1990. A Preliminary Study on the Geochemistry and Origin of Boninite and Boni-Basalts. Geochimica, 19 (3):207-215 (in Chinese with English abstract).

Zhao, X.S., Ma, S.L., Zou, X,H., et al., 1990. The Study of the Age, Sequence, Volcanism and Mineralization of Bikou Group in Qinling-Dabashan. Bulletin of the Xi'an Institute of Geology and Mineral Resources, Chinese Academy of Geological Sciences, 29:1-128 (in Chinese with English abstract).

High-Mg# adakitic tonalite from the Xichahe area, South Qinling orogenic belt (central China): Petrogenesis and geological implications[①]

Qin Jiangfeng Lai Shaocong[②] Wang Juan Li Yongfei

Abstract: The Xichahe tonalite emplaced into the Proterozoic Foping metamorphic complex, south Qinling Mountains, central China, has been dated at 213.6±2.2 Ma using the LA-ICP-MS zircon U-Pb method, indicating that it is post-orogenic ($\leqslant 242\pm21$ Ma). With the exception of higher Mg# (52.94-67.59) values, the tonalite has compositional similarities to high-silica adakites from supra-subduction-zone tectonic settings. Chondrite-normalized rare-earth element patterns of the tonalite are characterized by high $(La/Yb)_N$, ratios and concave-upward shapes of HREE, and without a significant Eu anomaly ($\delta Eu = 0.65-1.10$). In conjuction with high Ba (877.13-1 963.52 ppm) and Sr (820.76-1 252.75 ppm), low Y (10.58-18.30 ppm) and HREE (e.g., Yb=0.79-1.51 ppm), trace elements, and REE patterns suggest that the tonalite magma protolith was most likely a feldsparpoor, garnet ± amphibole-rich eclogite assemblage. Isotopic values of $^{87}Sr/^{86}Sr = 0.706\,5-0.706\,9$, $I_{Sr} = 0.705\,4-0.706\,4$, $^{143}Nd/^{144}Nd = 0.512\,3-0.512\,4$, $\varepsilon_{Nd}(t) = -3.62$ to -1.52, $^{206}Pb/^{204}Pb = 18.294-18.349$, $^{207}Pb/^{204}Pb = 15.993-15.997$, and $^{208}Pb/^{204}Pb = 39.502-39.526$, are similar to those of the Proterozoic Yaolinghe Group mafic volcanics in the south Qinling Mountains; the Nd isotopic model age of 1.15-1.38 Ga is close to the formation age (~1.1Ga) for the Yaolinghe metabasalts. High Mg# and Nb/Ta ratios, and Cr and Ni contents of the Xichahe adakitic tonalitie indicate that the parental magma mixed with a mantle melt. Based on the geological setting, it is concluded that the Xichahe tonalite was unlikely to have been produced by partial melting of recycling lower continental crust, and that mixing of the parental adakite magma with mantle melts resulted from Triassic asthenospheric upwelling caused by breakoff of the subducting Mianlue oceanic slab during the postorogenic stage of the South Qinling orogenic belt.

Introduction

NA-RICH SILICIC IGNEOUS ROCKS characterized by low HREE, Y, and Sc, high Ba and Sr, and resulting high Sr/Y and La/Yb ratios include members of Archean TTG suits, Cenozoic high-silica adakites, and Phanerozoic Na-rich granitoids (Topuz et al., 2005).

① Published in *International Geology Review*, 2007, 49.

② Corresponding author.

Compositional similarties and differences among these rock types have been discussed by Martin (1999), Simithes(2000), Condie(2005), and Martin et al(2005). The genesis of adakite magma, however, is still a matter of debate(e.g., Garrison and Davidson, 2003; Prouteau and Scaillet, 2003). Although it is generally accepted that the presence of significant amounts of garnet ± amphibole are required at some stage in the petrogenesis of these magmas, either as a residual or an early crystallizing assemblage(Topuz et al., 2005), a variety of petrogenetic models have been proposed. These include: ① partial melting of subducting oceanic crust (e.g., Defant and Drummond, 1990,1993;Drummond and Defant, 1990; Kay et al., 1993) and sunsequent variable interaction of salb-derived melts with the peridotitic mantle wedge (e.g., Stern and Kilian, 1996; Rapp et al., 1999; Simithies, 2000; Prouteau et al., 2001); ② high-pressure fractional crystallization of garnet and amphibole from hydrous basaltic magma (Prouteau and Scaillet, 2003); ③ partial melting of hot, mafic lower-arc crust, triggered by underplating of basaltic magmas(e.g., Smith and Leeman, 1987; Atherton and Petford, 1993; Wareham et al., 1997; Chung et al., 2003) and/or release of fluids from underthrust continental rocks(Wang et al., 2005); and ④ derivation from ancient mafic lower crust (eclogite) that foundered into the convective mantle, subsequently melted and interacted with peridotites (Xu et al., 2002; Gao et al., 2004).

The present study focuses on the genesis of a post-orogenic high-Sr/Y, high-$Mg^{\#}$ granodiorite/tonalite pluton emplaced at ~213 Ma into the Proterozoic Foping metamorphic complex, South Qinling Mountains, central China (Fig. 1). The pluton is considered to be a product of collision between Yangtze and North China terranes that resulted from closure of the Mianlue paleoocean. Based on field relations, petrological and geochemical characteristics documented in this study, the high-$Mg^{\#}$ adakite magma that formed the pluton is considered to have been generated by mixing of thickened crustal partial melts and lithospheric mantle-derived magma.

Geological setting and samples

The Qinling orogenic belt is a multistage orogenic belt composed of two mountain chains and two sutures. The north suture lies mainly along the Shangdan tectonic belt, whereas the south suture consists of the Mianlue ophiolite complex belt in the west and possibly the Bashan fault in the east (Meng and Zhang, 1999; Sun et al., 2002;Zhang et al., 2002,2004;Lai et al., 2003, 2004a,b). The mid-Paleozoic North Qinling terrane lies north of the Shangdan fault (Li et al., 1989; Sun and Li, 1998; Zhai et al., 1998; Li and Sun., 1996). Between the Shangdan and Mianlue belts is the Late Paleozoic-Triassic South Qinling orogen (Mattauer et al.,1985) with aboudant Trassic granitoids and late Paleozoic metamorphic rocks (Li and Sun, 1996; Li et al., 1996;Meng and Zhang, 1999). South of the Mianlue belt is the Yangtze Block(Fig. 1a).

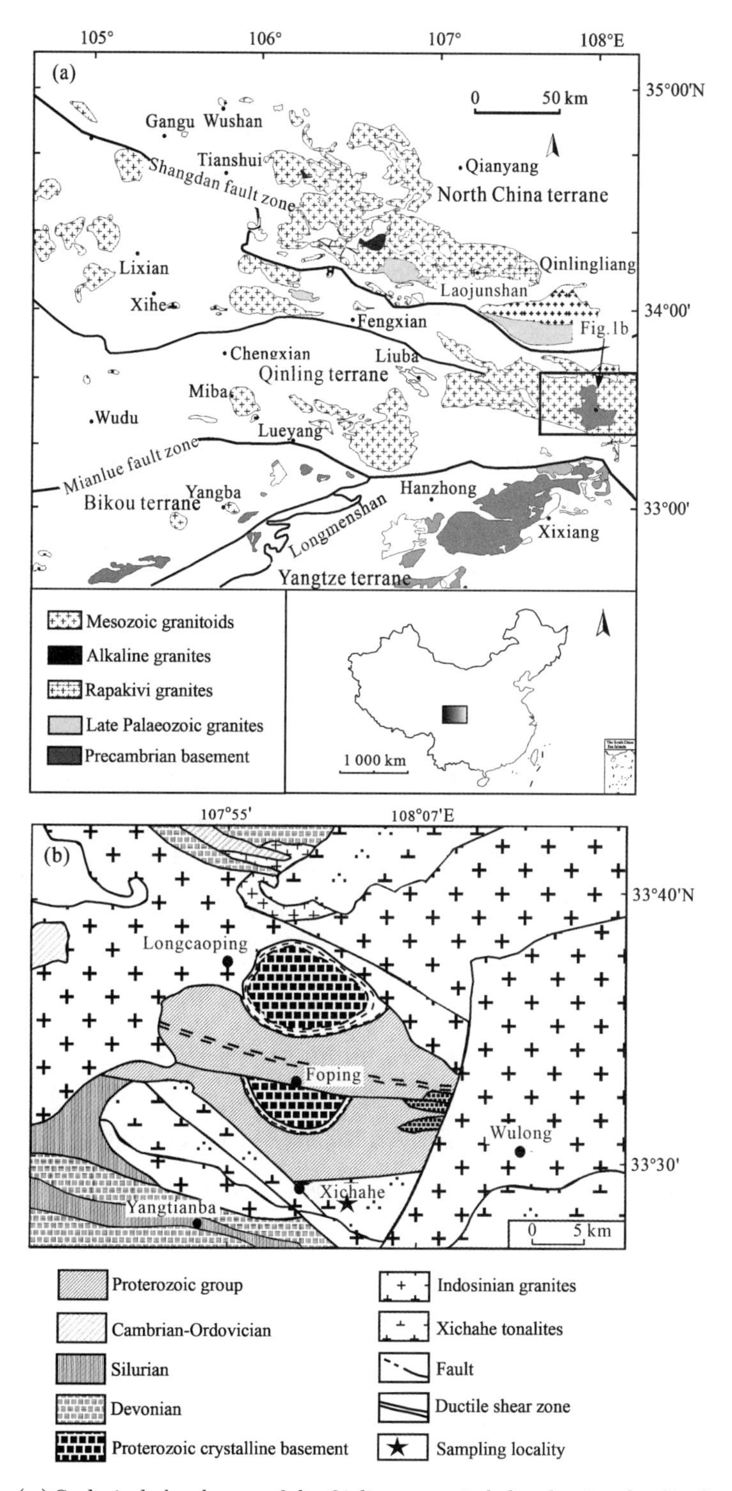

Fig. 1 (a) Geological sketch map of the Qinling orogenic belt, showing the distribution of Indosinian granites (after Zhang et al., 2001). (b) Geological sketch map of the Foping area in the South Qinling Mountains (modified from Zhang et al., 2002).

The discovery of the Mianlue suture in the South Qinling Mountains has been one of the most important advances in Qinling-Dabie orogenic research (Meng and Zhang, 1999; Li et al., 1996; Lai et al, 2003, 2004a,b; Zhang et al., 2004). The discovery has led to a new tectonic model of the Qinling orogenic belt as a multi-collision belt(e.g., Zhang et al., 2002, 2004; Meng and Zhang, 1999; Li et al., 1996; Lai et al., 2003, 2004a,b) instead of the previously accepted simple model of collision between two continents.

The Mianlue suture is separated from the Neo-Proterozoic Bikou Group and Silurian Baishuijiang Group by the Mianlue fault in the south and the Zhuangyuanbei fault to the north. It mainly consists of Sinian-Cambrian and Devonian-Carboniferous sheared thrust slabs (Sun et al., 2002). The Neo-Proterozoic-Cambrain rocks consist mainly of congomerate-bearing argillite, argilltic, polyclastic rocks, and carbonates. The Devonian system consists of turbidite, carbonates, argillite, quartzite, and quartzose sandstone, whereas the Carboniferous rocks comprise mainly carbonates and shales. In constrast, there are virtually no Ordovician and Silurian sediments within the Mianlue suture (e.g., Meng et al., 1996; Meng and Zhang, 1999).

Ophiolites and volcanics that occur in the Mianlue suture are allochthonous. The ophiolites are represented by strongly sheared metabasalt, gabbro, ultramafic rocks, and radiolarian chertff(Meng and Zhang, 1999). Some of the metabasalts display normal mid-ocean ridge basalt(N-MORB) geochemistry (Lai and Zhang, 1996; Lai et al., 2003, 2004a,b) and were metamorphosed in the Early Trassic(242–221 Ma; Li et al., 1996).

The ~400 km long South Qinling granite belt trends almost parallel to the Mianlue suture in the west and crops over an area of ~6 000 km^2 between the Daheba-Ningshan-Shanyang fault and the Shangdan suture in the east (Fig. 1a). The age of the granitoids is between 205 Ma and 220 Ma, implying a syncollisional tectonic origin(e.g., Sun et al., 2002) and they have been classified on the basis of petrography and geochemistry into three suites—the Guangtoushan, Wulong, and Dongjiangkou suites(Zhang et al., 1994; Sun et al., 2002).

The Xichahe tonalite forms part of the Wulong suite, covers an area of ~24km^2, and is located north of the Mianlue suture (Fig. 1b), where it is associated with Neoproterozoic-Silurian metasediments mainly composed of Neoproterozoic-Ordovician quartzite, graphite-marble, Cambrian-Ordovician graphite-quartzite, and Silurian micaceous quartzite. The tonalite is medium grained, massive, and composed of plagioclase (30%~45%; An_{25-35}), hornblende (25%–30%), K-feldspar (10%), quartze (10%–15%), and biotite (5%–10%), with accessory zircon, apatite, allanite, clinozoisite, and magnetite. Apatite forms rod-like or euhedral prismatic crystals and mainly occurs as inclusions in biotite, hornblende, and plagioclase. Zircon typically forms long, euhedral crystals.

Analytical methods

A sample of the Xichahe tonalite (Sample XH-03; Table 1) was selected for laser ablation inductively coupled plasma mass spectrometry (LA-ICP-MS) zircon U-Pb dating. Zircon was separated using conventional heavy liquid and magnetic techniques. Representative zircon grains were handpicked and mounted on an epoxy resin disc, then polished and coated with carbon film. The internal morphology of the grains was examined using cathodoluminescent (CATH) prior to U-Pb isotopic analyses. Zircon U-Pb dating was carried out at the State Key Laboratory of Continental Dynamics, Northwest University. The ICP-MS used in this study is ELAN6100DRC from Perkin Elmer/SCIEX (Ontario, Canada) with a dynamic reaction cell (DRC). Details of the analytical procedures of zircon analysis using LA-ICP-MS techniques are given by Yuan et al. (2003) and Gao et al. (2002).

Fresh samples of tonalites were analysed for major elements by wet chemistry at the Institute of Geochemistry, Chinese Academy of Sciences. Trace element and rare earth element (REE) concentrations were determined by ICP-MS of aqueous solutions using a VG Plasma-Quad Excell ICP-MS at the University of Hong Kong after a two-day closed beaker digestion using a mixture of HF and HNO_3 acids in high-pressure bombs. Pure element standard solutions were used for external calibration, and BHVO-1 (basalt) and SY-4 (syenite) were used as a reference materials. The procedure for the ICP-MS analysis is described by Qi et al. (2000). Accuracy of the ICP-MS analyses are estimated to be better than ± 5% (relative) for most elements.

Sr-Nd-Pb isotopic data were obtained using a Nu Plasma HR mutli-collector mass spectrometer at State Key Laboratory of Continental Dynamics, Northwest University. Sr and Nd isotopic fractionations were corrected to $^{87}Sr/^{86}Sr = 0.119\ 4$ and $^{146}Nd/^{144}Nd = 0.721\ 9$, respectively. During the analysis, the NIST SRM 987 standard yield an average value of $^{87}Sr/^{86}Sr = 0.710\ 250 \pm 12$ (2σ, $n = 15$) and the La Jolla standard give an average of $^{143}Nd/^{144}Nd = 0.511\ 859 \pm 6$ (2σ, $n = 20$). Whole-rock Pb was separated by an anion exchange in HCl-Br columns, and Pb isotopic fractionation was corrected to $^{205}Tl/^{203}Tl = 2.387\ 5$. During analysis, 30 measurements and NBS981 gave average values of $^{206}Pb/^{204}Pb = 16.937 \pm 1$ (2σ), $^{207}Pb/^{204}Pb = 15.491 \pm 1$ (2σ), and $^{208}Pb/^{204}Pb = 36.696 \pm 1$ (2σ); the BCR-2 standard gave $^{206}Pb/^{204}Pb = 18.742 \pm 1$ (2σ), $^{207}Pb/^{204}Pb = 15.620 \pm 1$ (2σ), $^{208}Pb/^{204}Pb = 38.705 \pm 1$ (2σ). Total procedural Pb blanks were in the range of 0.1–0.3ng.

Table 1 Results of zircon LA-ICP-MS U-Pb dating for Xichahe tonalites.

No.	Contents		Ratios							Age/Ma					
	^{238}U /(μg/g)	^{232}Th /(μg/g)	Th/U	$^{207}Pb^*/^{206}Pb^*$	1σ	$^{207}Pb^*/^{235}U$	1σ	$^{206}Pb^*/^{238}U$	1σ	$^{207}Pb^*/^{206}Pb^*$	1σ	$^{207}Pb^*/^{235}U$	1σ	$^{206}Pb^*/^{238}U$	1σ
XH-02	226. 49	125. 96	0. 556 1	0. 055 5	0. 001 20	0. 253 4	0. 004 9	0. 033 1	0. 000 2	287	68	216	5	209	1
XH-04	215. 78	118. 72	0. 550 2	0. 051 0	0. 001 01	0. 236 4	0. 004 2	0. 033 6	0. 000 2	240	29	215	3	213	1
XH-09	246. 25	147. 62	0. 599 5	0. 050 7	0. 001 39	0. 235 3	0. 006 0	0. 033 7	0. 000 3	228	45	215	5	213	2
XH-13	231. 91	127. 50	0. 549 8	0. 051 8	0. 001 35	0. 238 4	0. 005 7	0. 033 4	0. 000 2	278	42	217	5	212	1
XH-14	197. 54	99. 97	0. 506 1	0. 051 4	0. 001 99	0. 239 8	0. 008 9	0. 033 8	0. 000 3	261	67	218	7	214	2
XH-20	193. 85	100. 49	0. 518 4	0. 056 5	0. 001 38	0. 256 9	0. 005 8	0. 033 0	0. 000 2	362	72	221	6	208	2
XH-23	409. 67	216. 28	0. 527 9	0. 053 4	0. 000 93	0. 244 2	0. 003 6	0. 033 2	0. 000 2	347	23	222	3	210	1
XH-24	185. 82	93. 37	0. 502 5	0. 055 8	0. 001 68	0. 259 0	0. 007 3	0. 033 7	0. 000 3	297	90	220	7	213	2

Results

Crystallization age of the Xichahe tonalites

Euhedral zircon from the Xichahe tonalite has typical microscale magmatic oscillatory zoning (Fig. 2a). As shown in Table 1, eight zircons have a wide range of U(193.85–409.67 ppm) and Th(93.37–216.68 ppm) with a high Th/U=0.50–0.59>0.4, indicating that they are magmatic (Hoskin et al., 2003). The measured $^{206}Pb/^{238}U$ ratios are in good agreement with analytical precision, yielding a weighted mean age of 213.6±2.2 Ma(MSWD=0.68, 2σ) (Fig. 2b), which is regarded as the crystallization age of the tonalite.

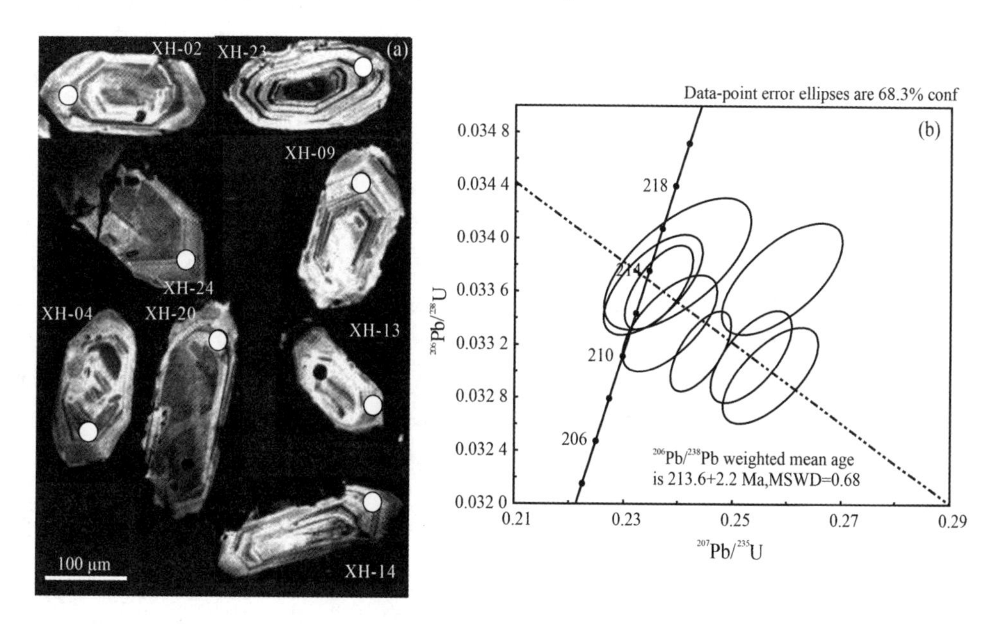

Fig. 2 Cathodoluminescence electron images(a) and zircon LA-ICP-MS U-Pb concordia diagram (b) with for Xichahe tonalites.

Major and trace element geochemistry

Major and trace element compositions of representive samples of the Xichahe tonalites are listed in Table 2. The rocks have (wt%): 56.98%–62.30% SiO_2, 0.48%–0.80% TiO_2, high Al_2O_3(14.87%–18.37%) with A/CNK[=molar $Al_2O_3/(CaO+K_2O+Na_2O)$] ranging from 0.72 to 1.10, indicating that they are metaluminous-aluminous with Na_2O = 3.10%–4.40%, K_2O=2.00%–3.20%, $Na_2O/K_2O>1.3$, high MgO with $Mg^{\#}[=Mg^{2+}/(Mg^{2+}+Fe^{2+T})]$ of ~53 to ~67, implying a mantle component in their source region(e.g., Rapp, 1997,; Rapp et al., 1999). In terms of MgO vs. SiO_2, the tonalites plot within the field of high-Mg adakite (Fig. 3a).

Table 2 Major(wt%) and trace(μ,ppm) element analyses of the Xichahe tonalite[1].

Sample	XH-03	XH-04	XH-05	XH-06	XH-07	XH-09	XH-10	XH-12	XH-14	XH-15
SiO_2	62.30	61.75	64.4	60.91	60.03	56.98	60.83	60.23	58.65	59.89
TiO_2	0.65	0.60	0.48	0.80	0.72	0.57	0.55	0.53	0.65	0.52
Al_2O_3	16.4	18.37	17.50	15.09	18.09	14.87	15.96	15.09	15.40	14.87
$Fe_2O_3^T$	1.60	1.05	0.60	1.40	1.80	1.80	1.00	1.50	2.25	1.00
FeO	3.60	3.00	2.70	3.70	3.10	4.00	3.90	4.10	4.05	4.09
MnO	0.10	0.07	0.11	0.09	0.10	0.16	0.13	0.14	0.18	0.09
MgO	3.15	3.05	2.55	3.99	3.42	6.51	4.18	5.99	6.00	4.02
CaO	3.80	3.96	3.49	4.61	4.19	6.25	4.61	6.06	5.83	4.47
Na_2O	4.00	4.40	4.0	4.20	4.10	3.80	4.00	3.40	3.80	3.10
K_2O	2.20	2.21	2.70	2.60	2.70	2.71	2.90	2.00	2.00	3.20
P_2O_5	0.30	0.23	0.20	0.37	0.36	0.53	0.31	0.30	0.37	0.35
LOI	1.47	1.05	0.92	1.67	1.10	1.31	1.04	0.23	0.37	3.87
Total	99.57	99.74	99.65	99.43	99.71	99.79	99.41	99.85	99.67	99.47
A/CNK	1.03	1.09	1.10	0.83	1.05	0.72	0.88	0.80	0.81	0.89
σ	1.99	2.33	2.10	2.58	2.72	3.03	2.67	1.69	2.15	2.35
$Mg^{\#1}$	52.94	58.19	58.62	59.15	56.60	67.59	61.05	66.42	64.00	59.19
Ba	1 243.05	1 085.55	1 963.52	1 362.98	1 365.35	877.13	887.71	904.22	1 002.00	1 297.04
Rb	89.83	86.37	84.30	104.11	103.67	121.04	140.08	78.59	66.90	132.47
Sr	906.56	899.61	902.62	837.34	865.37	956.89	820.76	1 029.29	1 252.75	941.79
Y	13.43	10.58	10.25	16.05	14.24	16.42	18.30	22.11	22.65	16.59
Zr	86.37	69.91	52.94	69.56	78.72	92.84	106.60	87.69	110.23	224.40
Nb	11.71	10.55	8.92	14.56	14.00	9.05	15.12	9.72	9.77	14.42
Th	5.58	4.00	3.88	3.55	2.14	4.04	3.90	3.07	2.50	16.10
Pb	18.06	17.31	21.47	19.46	18.36	14.52	17.49	17.06	14.52	22.31
Ga	15.49	14.93	13.82	15.70	16.00	13.04	13.69	11.51	13.67	13.31
Zn	68.10	62.53	49.86	85.18	76.95	84.04	86.28	57.03	72.08	61.38
Cu	21.79	17.40	14.59	28.03	27.36	28.54	3.54	25.53	56.04	28.40
Ni	39.78	37.84	33.03	51.88	45.84	140.32	65.56	123.34	107.67	67.34
V	97.27	91.38	78.04	117.43	111.36	138.77	104.49	121.13	153.53	119.11
Cr	97.02	88.83	74.82	121.87	109.73	365.69	192.21	347.16	294.52	198.70
Hf	3.11	2.83	2.56	2.34	2.72	3.14	3.49	3.26	3.50	6.92
Cs	4.96	5.16	4.34	6.08	6.08	6.47	2.64	3.27	2.72	5.03
Sc	11.73	10.09	9.12	13.35	12.93	16.81	17.26	19.61	19.84	14.63
Ta	0.74	0.54	0.53	0.90	0.88	0.47	1.00	0.95	0.66	0.87
Co	50.66	62.92	71.80	43.26	44.17	51.88	41.23	54.74	43.15	46.58
Li	43.42	39.64	33.284	50.692	51.3	48.35	37.61	39.07	38.10	37.73
Be	1.36	1.97	1.70	1.49	2.07	1.86	3.28	1.90	1.80	1.96
U	0.90	0.75	0.68	2.23	0.88	1.19	0.95	2.15	1.29	3.25
La	30.78	24.78	24.24	21.62	17.11	27.92	31.45	20.14	27.38	38.9
Ce	59.83	49.95	47.65	44.31	37.89	58.81	63.05	46.22	62.75	76.32

Continued

Sample	XH-03	XH-04	XH-05	XH-06	XH-07	XH-09	XH-10	XH-12	XH-14	XH-15
Pr	6.49	5.51	5.16	5.16	4.60	6.80	6.86	5.62	7.46	8.03
Nd	25.63	22.22	21.05	23.44	20.68	27.95	27.15	25.68	32.83	30.8
Sm	5.01	4.14	4.01	5.36	4.80	5.65	5.14	5.87	6.83	5.57
Eu	1.28	1.21	1.32	1.21	1.23	1.25	1.03	1.36	1.47	1.22
Gd	4.05	3.30	3.10	4.50	3.95	4.23	4.31	4.65	5.31	4.38
Tb	0.54	0.44	0.39	0.65	0.57	0.59	0.64	0.73	0.77	0.60
Dy	2.97	2.40	2.21	3.59	3.19	3.33	3.66	4.42	4.40	3.22
Ho	0.50	0.42	0.38	0.63	0.53	0.57	0.66	0.81	0.82	0.59
Er	1.32	1.05	0.99	1.6	1.44	1.64	1.83	2.19	2.40	1.62
Tm	0.17	0.15	0.13	0.22	0.19	0.23	0.24	0.34	0.33	0.24
Yb	1.04	0.79	0.83	1.27	1.17	1.41	1.49	2.15	2.13	1.51
Lu	0.16	0.13	0.12	0.20	0.17	0.22	0.22	0.33	0.35	0.26
Sr/Y	67.52	85.03	88.09	52.17	60.79	58.27	44.86	46.56	55.30	56.78
Y/Yb	12.91	13.39	12.34	12.64	12.17	11.65	12.28	10.28	10.64	10.99
La/Yb	29.60	31.37	29.20	17.02	14.62	19.80	21.11	9.37	12.85	25.76
δEu	0.84	0.97	1.10	0.73	0.84	0.75	0.65	0.77	0.72	0.73

[1] $Mg^{\#}=100Mg^{2+}/(Mg^{2+}+Fe^{2+T})$; $Fe^{2+T}=(Fe_2O_3{}^{T}\ wt\%\times 0.8998)/72$.

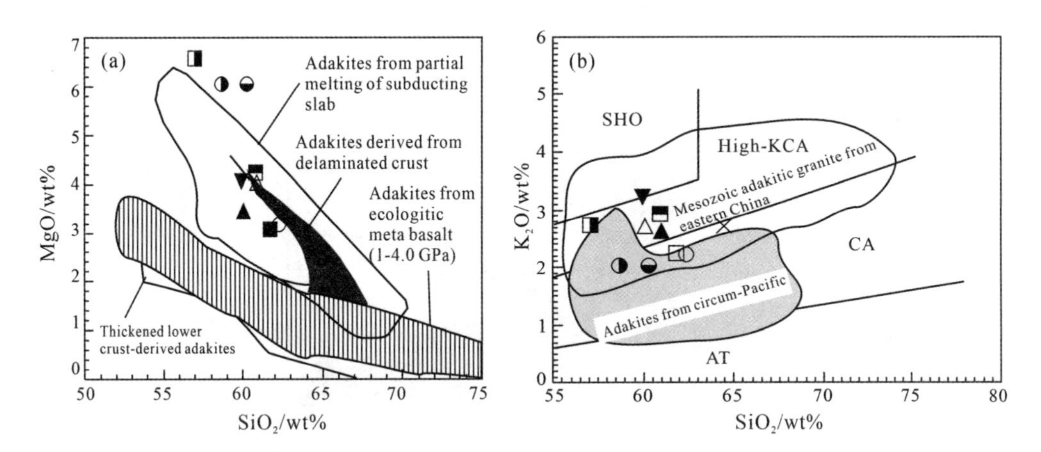

Fig. 3 Plots of (a) MgO vs. SiO_2 and (b) K_2O vs. SiO_2 (after Peccerillo and Taylor, 1976) for the Xichahe tonalite.

Adakites field from ecologitic metabasalt (1–4 GPa) is from Sen and Dunn (1994), Rapp and Watson (1995), Springer and Seck (1997). Adakites field from partial melting of subducting slab is from Defant and Drummond (1990), Stern and Kilian (1996), Martin (1999), Smithies (2000), and Defant et al. (2002). Adakites field derived from delaminated crust is from Xu et al. (2002); Field of thickened lower crust-derived adakites is from Atherton and Petford (1993), Muir et al. (1995), Petford and Atherton (1996), and Smithies (2000). Data of Mesozoic adakites from eastern China and addkites from the circum-Pacific belt after Wu et al. (2002).

The tonalite is characterized by high Sr (820.76–1 252.75 ppm), and Ba (877.13–1 963.52 ppm), lower Y (10.58–18.30 ppm) and Yb (Yb = 0.79–1.51 ppm), with the exception of two samples (XH-12, XH-14) that have higher Y (22.11–22.65 ppm) and HREE (e.g., Yb = 2.13–2.15 ppm). Furthermore, chondrite-normalized REE patterns of the tonalite have HREE concave-upward patterns and lack significant Eu anomaly with δEu [$\delta Eu = 2Eu_N/(Sm_N + Gd_N)$] ranging from 0.77 to 1.10, with high $(La/Yb)_N$ ratios of (~6.3 to ~21.2) (Fig. 4). Primitive mantle-normalized element concentration patterns show notable negative anomalies in Nb, Ta, P, K, Th, and Ti and positive Ba and Sr anomalies (Fig. 5).

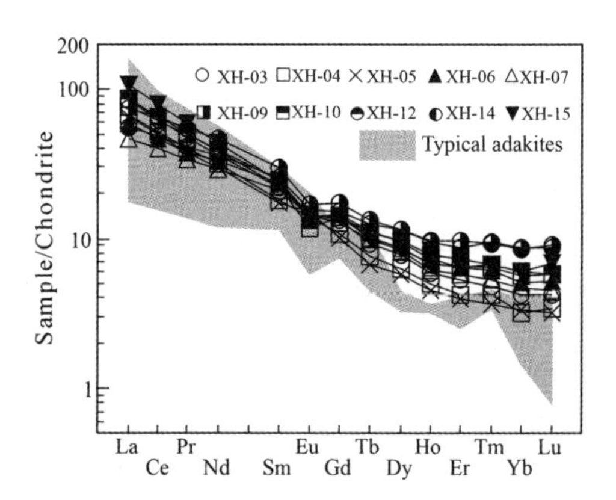

Fig. 4 Chondrite-normalized REE patterns for the Xichahe tonalites.
Normalized value are from Sun and McDonough (1989);
Patterns of the typical adakites is from Zhang et al. (2001).

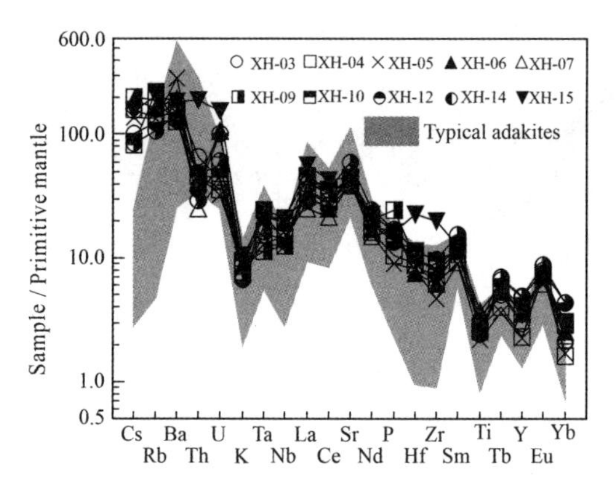

Fig. 5 Primitive mantle-normalized trace element patterns for the Xichahe tonalites.
Normalized values are from Sun and McDonough (1989);
Patterns of the typical adakites are from Zhang et al. (2001).

Geochemically, the Xichahe tonalites exhibits a distinct "adakitic affinity"—i.e., low concentration of garnet-compatible elements (Y, HREE), a high aboundance of plagioclase-compatible Sr and Ba, and high ratios of Sr/Y (44.86–88.09) and La/Yb (Table 2). These characteristics can be inferred to indicate a garnet ± amphibole-rich, feldspar-poor source. The

$Mg^{\#}$ values of the tonalite are higher than that of experimental melts of metabasalts and eclogites, and may imply that the parental magma of the tonalites are unlikely to have been produced simply by partial melting of thickened lower continental crust (Atherton and Petford, 1993; Green, 1994).

Sr-Nd-Pb isotopic compositions

Sr-Nd isotope analysis is listed in Table 3. The calcuated initial Nd and Sr isotopic composition for samples XH-12 and XH-14 yields T = 213 Ma. The Xichahe tonalite is characterized by high radiogenetic Sr with $^{87}Sr/^{86}Sr$ = 0.706 5-0.706 9, I_{Sr} = 0.705 4-0.706 4 and high non-radiogenetic Nd with $^{143}Nd/^{144}Nd$ = 0.512 3-0.512 4, $\varepsilon_{Nd}(t)$ = -3.6 to -1.5. Nd isotopic model ages (T_{DM}) range from 1.15 Ga to 1.38 Ga. We have already noted that Sr and Nd isotopic compositions of the tonalite are clearly different from MORB-like isotopic compositions of modern arc-related adakites (e.g., Kay and Kay, 1993; Sternand and Kilian, 1996). Whole-rock Pb isotopic data (Table 4) show that the Xichahe tonalites are characterized by high radiogenic Pb with present-day whole-rock Pb isotopic ratios of $^{206}Pb/^{204}Pb$ = 18.294-18.349, $^{207}Pb/^{204}Pb$ = 15.993-15.997, and $^{208}Pb/^{204}Pb$ = 39.502-39.526. Due to their young age (~213 Ma) and low U (0.75-3.25 ppm) and Th (2.14-5.58 ppm), present-day Pb isotopic ratios can be considered to represent the initial Pb isotopic ratios (e.g., Zhang et al., 2006).

Table 3 Whole-rock Sr and Nd isotopes of Xichahe tonalite.

Sample	Rb /ppm	Sr /ppm	$^{87}Rb/^{86}Sr$	$^{87}Sr/^{86}Sr$ ±1σ	I_{Sr} (210 Ma)	Sm /ppm	Nd /ppm	$^{147}Sm/^{144}Nd$[1]	$^{143}Nd/^{144}Nd$ ±1σ	$\varepsilon_{Nd}(t)$ (210 Ma)[2]	T_{DM} /Ga[3]
XH09	121.04	956.89	0.365 93	0.706 545±8	0.705 43	5.65	27.95	0.122 15	0.512 454±9	-1.52	1 15
XH14	66.90	1 252.75	0.154 50	0.706 922±42	0.706 45	6.83	32.83	0.125 80	0.512 352±6	-3.62	1 38

[1]Rations of $^{87}Rb/^{86}Sr$ and $^{147}Sm/^{144}Nd$ are calculated using Rb, Sr, Sm and Nd contents, measured by ICP-MS.

[2]Values of $\varepsilon_{Nd}(t)$ are calculated using present-day $(^{147}Sm/^{144}Nd)_{CHUR}$ = 0.196 7 and $(^{143}Nd/^{144}Nd)_{CHUR}$ = 0.512 638.

[3]T_{DM} values are calculated using present-day $(^{147}Sm/^{144}Nd)_{DM}$ = 0.213 7 and $(^{143}Nd/^{144}Nd)_{DM}$ = 0.513 15.

Table 4 Whole-rock Pb isotopes of Xichahe tonalites.

Sample	U /ppm	Th /ppm	Pb /ppm	$^{206}Pb/^{204}Pb$	1σ	$^{207}Pb/^{204}Pb$	1σ	$^{208}Pb/^{204}Pb$	1σ
XH09	1.186	4.042	14.522	18.349 39	0.001	15.997 37	0.000 339	39.526 23	0.001
XH14	1.294	2.502	14.519	18.294 17	0.001	15.993 15	0.000 384	39.501 69	0.001

Discussion

Petrogenesis of the Xichahe tonalites

The geochemistry of adakitic rocks indicates that they have a high-pressure origin within the stability field of garnet, most probably under eclogite-facies conditions. Various hypotheses have been advanced to explain the origin of adakitic magma (Wang, X.C. et al., 2006). These include partial melting of subducted oceanic crust (Defant and Drummond, 1990; Peacock et al., 1994; Gutscher et al., 1999; Sajona et al., 2000; Beate et al. 2001); partial melting of thickened lower continental crust (Atherton and Petford, 1993; Muir et al., 1995; Barnes et al., 1996; Petford and Atherton, 1996; Chung et al., 2003; Hou et al., 2004; Wang et al., 2005); or partial melting of delaminated lower crust (Xu et al., 2002; Gao et al., 2004; Wang X.C. et al., 2006), and/or assimilation- fractionation-crystallization (AFC) (e.g., Feeley and Hacker, 1995).

The Xichahe tonalite is characterized by low Y(10.25–16.59 ppm<18 ppm) and HREE (e.g., 0.79–1.51 ppm<1.8 ppm Yb), high Sr(>850 ppm) with a weak positive Sr anomaly in primitive mantle-normalized trace element patterns (Fig. 4), and high La/Yb (9.37–21.6) and Sr/Y (44.86–88.09). The geochemical signatures are comparable to those of adakites as defined by Defant and Drummond (1990). On Sr/Y vs. Y (Fig. 6a) and $(La/Yb)_N$ vs. Yb_N diagrams (Fig. 6b) diagrams, the Xichahe tonalites mainly fall within the adakite field, and on an SiO_2 vs. MgO plot (Fig. 3a), in the overlapping region between adakites derived from delaminated crust melting (Xu et al., 2002) and those from the partial melting of subducted oceanic crust (Defant and Drummond, 1990; Stern and Kilian, 1996; Martin, 1999; Smithies, 2000; Defant et al., 2002).

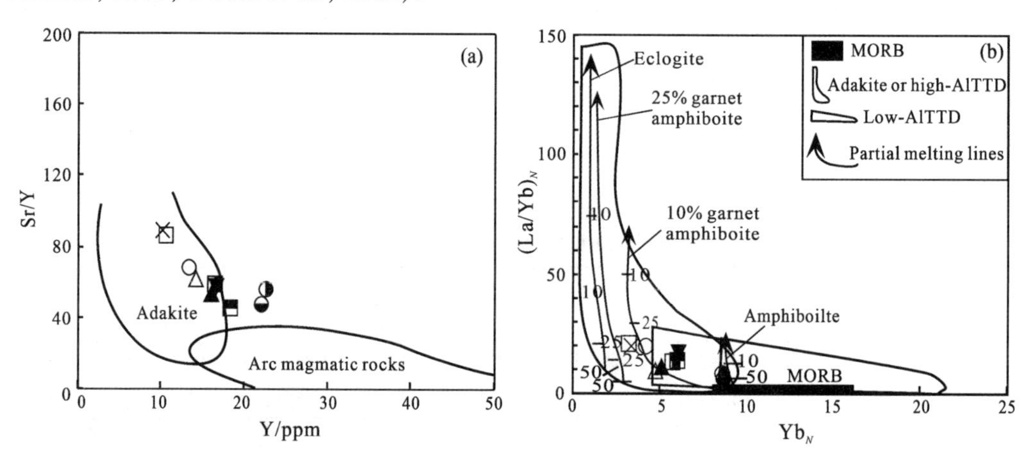

Fig. 6 Sr/Y vs. Y(a) and $(La/Yb)_N$ vs. Yb_N(b) diagrams for the Xichahe tonalites.
Field of adakite is from Defant and Drummond (1990).

However, zircon U-Pb dating indicates that the Xichahe adakitic tonalite formed at about 213 Ma (Fig. 2b), and the Mianlue Ocean closure—i.e., the Indosinian collision of the Yangtze and the North China blocks, occurred at ~242 Ma (e.g., Li et al., 1996), at which time subduction of the Mianlue oceanic crust beneath North China had ceased. Furthermore, the tonalite has higher K_2O than adakites from the circum-Pacific belt (Wu et al., 2002) and mainly plots within the high-K calc-alkaline series field on an SiO_2 versus K_2O diagram (Fig. 3b). Due to lower $\varepsilon_{Nd}(t)$ and higher initial $^{87}Sr/^{86}Sr$ than Cenozoic adakites, it can be considered that the Xichahe adakitic tonalite is unlikely to have been produced by partial melting of subducting oceanic slab during the Triassic. Additionally, the Xichahe tonalite has higher $Mg^{\#}$ values than those of experimental melts of metabasalts and eclogites (1–4 GPa; Fig. 3b), and even has higher Nb/Ta ratios (14. 80–19. 54) than that of typical continental crust (~12; Sun and McDonough, 1989), implying that the parental magma could not have originated from partial melting of thickened lower crust (cf. Atherton and Petford, 1993). Thus, it appears likely that the Xichahe adakitic tonalite was derived from partial melting of delaminated lower crust, and resulted in an eclogitic residue (e.g., Kay and Kay, 1993). Such an origin can readily account for the elevated $Mg^{\#}$ values of the Xichahe tonalite (Fig. 3a; e.g., Rapp et al., 1999). However, another mechanism could also account for the high $Mg^{\#}$ values of adakitic rocks. Ducea and Saleeby (1998) have suggested that when hot melts from the lithospheric mantle intrude the lower crust, partial melting of the amphibole-bearing eclogitic crust occurs. Mixing between these magmas and mantle peridotite could significantly increase the MgO content of adakitic magmas. Thus, it remains to resolve the mechanism that was responsible for the mantle signatures of the Xichahe tonalite.

Geodynamic process

Given that the Xichahe adakitic tonalite was produced by partial melting of recycled lower crust, as proposed for the Mesozoic high-$Mg^{\#}$ adakitic volcanics from Xinglonggou (Gao et al., 2004), what were the parent rocks, and could the Qinling-Dabie orogenic lithosphere have undergone delamination at ~210 Ma? The Sr-Nd isotopic compositions of the Xichahe tonalite are closely similar to those of the Proterozoic Yaolinghe metabasalts (Fig. 7), and their Nd model ages (1. 15–1. 38 Ga) are consistent with the eruption age of the Yaolinghe metabasalts (~1. 1 Ga; Zhang et al., 2002), indicating that the Xichahe tonalite could be genetically related to the Proterozoic Yaolinghe metabasalts.

Previous studies have shown that collision between the North China and Yangtze terranes occurred earlier in the eastern part of the orogenic belt (Yin and Nie, 1996; Zhang et al., 2002). Li et al. (2002) have proposed that the initial exhumation age of the ultra-high pressure metamorphic rocks in the Dabie orogenic belt corresponding to slab breakoff of the

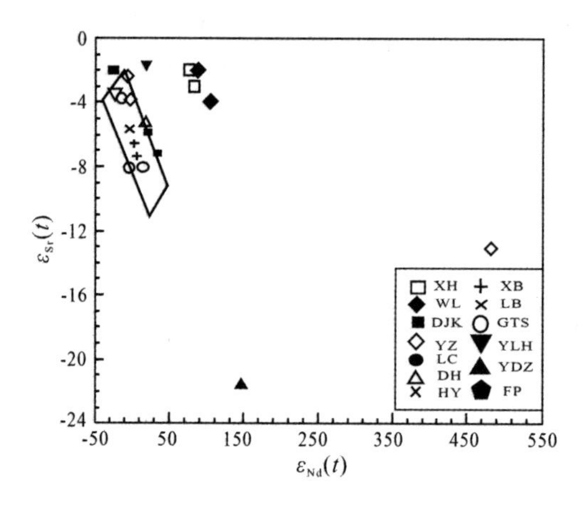

Fig. 7 $\varepsilon_{Nd}(t)$ vs. $\varepsilon_{Sr}(t)$ diagra.

XH: Xichahe tonalites; WL: Wulong granites; DJK: Dongjiangkou granites; YZ: Yanzhiba granites; LC: Laocheng; DH: Donghetaizi; HY: Huayang; XB: Xiba; LB: Liuba. GTS: Guangtoushan granites; YLH: Yaolignhe group; YDZ: Yudongzi group; FP: Foping group. Besides Xichahe and Wulong, other data are from Zhang et al(2002). Data of Wulong are from Wang (unpubl.).

subducted Mianlue oceanic crust was ~230 Ma, whereas the post-orogenic magmatism caused by delamination of the lower crust in the Dabie Orogenic belt occurred at ~130 Ma (Hacker et al., 1998). According to the tectonic setting, the possibility that lithospheric delamination in the Qinling belt occurred in the Triassic can be precluded. Furthermore, U-Pb zircon ages of 220–205 Ma for granitoids in South Qinling (Sun et al., 2002), coupled with regional geology, suggest that breakoff of the subduction slab occurred at shallow depth, causing upward movement of the asthenosphere; this process led to crustal melting and formation of the Indosinian synorogenic granitoids in the South Qinling Mountains. Thus, we conclude that the Xichahe adakitic tonalite is unlikely to have been produced by partial melting of recycling lower crust, and that its mantle signature must be attributed to other mechanisms.

With respect to the above discussion, a new model is proposed as shown in Fig. 8. Triassic collision between the Yangtze and North China terranes caused crustal thickening in the Qinling-Dabie orogen to >50 km; assuming that subducting oceanic slab breakoff occurred at shallow depth, upwelling of the asthenosphere took place along the Mianlue suture, causing partial melting of enriched lithospheric mantle. The mantle-derived melt initially underplated the thickened lower crust and lithospheric mantle so that Triassic temperatures at the boundary between mantle and amphibole-bearing eclogite were high enough to trigger dehydration melting of garnet-bearing rocks, which have Yaolinghe-like isotopic compositions; at a pressure of >1.5 GPa, this resulted in the generation of adakitic magma (e.g., Rapp et al., 1999, 2002), leaving an eclogite restite. Given that the lithospheric mantle-derived magmas intruded the adakitic magma chamber, mixing between adakitic magmas and mantle-derived melts produced the mantle signature (i.e., high $Mg^{\#}$, Nb/Ta ratios, and Cr, Ni contents) of the Xichahe adakitic tonalite.

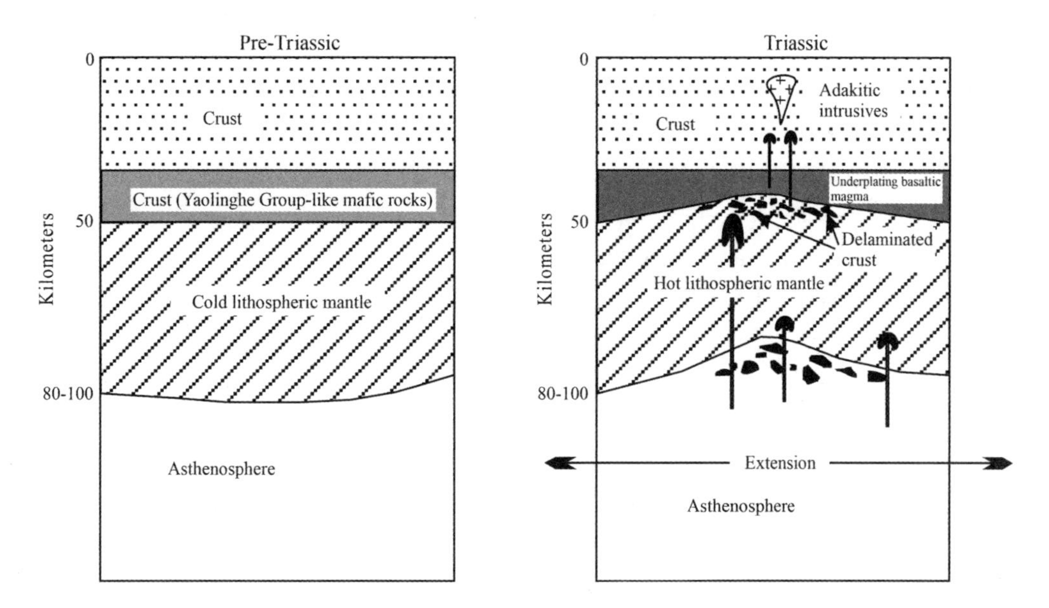

Fig. 8 Suggested model to explain formation of the Triassic Xichahe high $Mg^{\#}$ adakitic tonalites (revised from Wang et al., 2006).

(a) Relatively thick pre-Triassic lithosphere and crust. The lower portion of the thickened crust inferred to be amphibole-bearing eclogites. (b) Mianlue oceanic subducting slab breakoff occurs at a shallow depth, resulting in upwelling of asthenosphere causing partial melting of enriched lithospheric mantle. Resulting mantle-derived mafic melt was initially underplated at the boundary of thickened lower crust (> 50 km) and lithospheric mantle. As a consequence, Triassic temperatures at the boundary would have been high enough to cause dehydration melting of eclogite than had Yaolinghe-like isotopic compositions, at pressures >1.5 GPa, resulting in the generation of adakitic magma and garnet-rich eclogitic restite. Given that the lithospheric mantle-derived liquids intruded the adakitic magmas chamber, mixing between adakitic magmas and mantle-derived melts produced the mantle signature (i.e., high $Mg^{\#}$, Nb/Ta ratios, and Cr, Ni contents) of the Xichahe tonalite.

Acknowledgements This work was jointly supported by the National Natural Science Foundation of China (Grant Nos. 40572050, 40272042, and 40234041) and the Teaching and Research Award Program for Outstanding Young Teachers in Higher Education Institutions of MOE, P. R. China. Comments by Prof. Rodney Grapes helped to improve the presentation of the manuscript.

References

Atherton, M.P., Petford, N., 1993. Generation of sodium-rich magmas from newly underplated basaltic crust. Nature, 362, 144-146.

Barnes, C.G., Petersen, S.W., et al., 1996. Source and tectonic implication of tonalite-trondhjemite magmatism in the Klamath Mountains. Contributions to Mineralogy and Petrology, 123, 40-60.

Beate, B., Monzier, M., Spikings, R., Cotten, J., Silva, J., Bourdon, E., Eissen, J.P., 2001. Mio-Pliocene adakite generation related to fiat subduction in Southern Ecuador: The Quimsaeoeha-oceanic center. Earth & Planetary Science Letters, 192, 499-508.

Chung, S.L., Liu, D.Y., Ji, J., Chu, M.F., Lee, H.Y., Wen, D.J., Lo, C.H., Lee, T.Y, Qian Qing, Zhang Qi., 2003. Adakites from continental collision zones: Melting of thickened lower crust beneath southern Tibet. Geology, 31, 1021-1024.

Condie, K.C., 2005. TTGs and adakites: Are they both slab melts? Lithos, 80, 33-44.

Defant, M.J., Drummond, M.S., 1990. Derivation of some modern arc magmas by partial melting of young subducted lithosphere. Nature, 347, 662-665.

Defant, M.J., Drummond, M.S., 1993. Mount St. Helens: Potential example of the partial melting of the subducted lithosphere in a volcanic arc. Geology, 21, 547-550.

Defant, M.J., Xu, J.F., Kepezhinskas, P., 2002. Adakites: Some variations on a theme. Acta Petrologica Sinica, 18(2), 129-142.

Drummond, M.S., Defant, M.J., 1990. A model for trondhjemite-tonalite-dacite genesis and crustal growth via slab melting: Archean to modern comparisons. Journal of Geophysical Research, 95, 21503-21521.

Ducea, M., Saleeby, J., 1998. Crustal recycling beneath continental arcs: Silica-rich glass inclusions in ultramafic xenoliths from the Sierra Nevada, California. Earth & Planetary Science Letters, 158, 101-116.

Feeley, T. C., Hacker, M. D., 1995. Intracrustal derivation of Na-rich andesite and dacite magmas: An example from Volcan Ollague Andean Central Volcanic Zone. Journal of Geology, 103, 213-225.

Gao, S., Liu, X. M., Yuan, H.L., 2002. Determination of forty-two major and trace elements in USGS and NIST SRM glasses by laser ablation inducitly coupled plasma-mass spectrometry. Geostandards Newsletter, 26(2), 181-195.

Gao, S., Rudnick, R.L., Yuan, H.L., Liu, X.M., Liu, Y.S., Xu, W.L., Ling, W.L., Ayers, J., Wang, X.C., Wang, Q.H., 2004. Recycling of lower continental crust in the North China craton. Nature, 432, 892-897.

Garrison, J.M., Davidson, J.P., 2003. Dubious case for slab melting in the northern volcanic zone of the Andes. Geology, 31, 565-568.

Green, T.H., 1994. Experimental studies of trace element partitioning applicable to igneous petrogenesis: Sedona 16 years later. Chemical Geology, 117, 1-36.

Gutscher, M.A., Maury, R.C., Eissen, J.P., Bourdon, E., 1999. Can slab melting be caused by flat subduction? Geology, 28, 535-538.

Hacker, B.R., Ratschbacher, L, Webb, L., et al., 1998. U/Pb zircon ages constrain the architecture of the ultrahigh-pressure Qinling-Dabie Orogen, China. Earth & Planetary Science Letters, 161, 215-230.

Hoskin, P.W.O., Schaltegger, U., 2003. The composition of zircon and igneous and metamorphic petrogenesis. In: Hanchar, J.M., Hoskin, P.W.O. Reviews in Mineralogy and Geochemistry (Zircon), 53, 27-62.

Hou, Z.Q., Gao, Y.F., Qu, X.M., Qiu, R.Z., Mo, X.X., 2004. Origin of adakitic intrusives generated during mid-Miocene east-west extension in southern Tibet. Earth & Planetary Science Letters, 220, 139-155.

Kay, R.W., Kay, S.M., 1993, Delamination and delamination magmatism. Tectonophysics, 219, 177-189.

Kay, S.M, Ramos, V.A., Marquez, M., 1993. Evidence in Cerro Pampa volcanic rocks for slab melting prior to ridge-trench collision in southern South America. Journal of Geology, 101, 703-714.

Lai, S.C., Zhang, G.W., 1996. Geochemical features of ophiolites in Mianxian-Lueyang suture zone, Qinling Orogenic Belt. Journal of China University of Geosciences, 7, 165-172.

Lai, S.C., Liu, C.Y., Yin, H.S., 2003. Geochemistry and petrogenesis of Cenozoic andesite-dacite associations from the Hoh Xil Region, Tibetan Plateau. International Geology Review, 45, 998-1019.

Lai, S.C., Zhang, G.W., Dong, Y.P., Pei, X.Z., Chen, L., 2004a. Geochemistry and regional distribution of ophiolite and associated volcanics in Mianlüe suture, Qinling-Dabie Mountains. Science in China: Series D, 47, 289-299.

Lai, S.C., Zhang, G.W., Li, S.Z., 2004b. Ophiolites from the Mianlue suture in the Southern Qingling and their relationship with eastern paleotethys evolution. Acta Geologica Sinica, 78 (1), 107-117.

Li, S.G., Hart, S.R., Zheng, S., Liou, D., Zhang, G.W., Guo, A.L., 1989. Timing of collision between the North and South China Blocks: Sm-Nd isotopic age evidence. Science in China: Series B, 32, 1391-1400.

Li, S.G., Huang, F., Li, H., 2002. Post-collisional lithosphere delamination of Dabie-Sulu orogen. Chinese Science Bulletin, 47 (3), 259-263.

Li, S.G., Sun, W.D., 1996. A middle Silurian-early Devonian magmatic arc in the Qinling Mountains of central China: A discussion. Journal of Geology, 104, 501-503.

Li, S. G., Sun, W.D., Zhang, G.W., Chen, J.Y., Yang, Y.C., 1996. Chronology and geochemistry of metavolcanic rocks from Heigouxia Valley in Mianlue tectonic arc, South Qinling: Observation for a Paleozoic oceanic basin and its close time. Science in China: Series D, 39, 300-310.

Martin, H., 1999. The adakitic magma: Modern analogue of Archean granotoids. Lithos, 46, 411-429.

Martin, H., Smithies, R.H., Rapp, R., Moyen, J.F., Champion, D., 2005. An overview of adakite, tonalitetrondhjemite-granodiorite (TTG), and sanukitoid: Relationships and some implications for crustal evolution. Lithos, 79, 1-24.

Mattauer, M., Matte, P., Malaveile, J., Tapponier, P., Maluski, H., Xu, Zh. Q., Han, Y.L., Tang, Y. Q., 1985. Tectonics of Qinling belt: Build-up and evolution of Western Asia. Nature, 1085 (317), 496-500.

Meng, Q.R., Zhang, G.W., 1999. Timing of the collision of the North and South China blocks: Controversy and reconciliation. Geology, 27, 123-126.

Meng, Q.R., Zhang, G.W., Yu, Z.P., Mei, Z.C., 1996. Late Paleozoic sedimentation and tectonics of rift and limited ocean basin at the southern margin of the Qinling. Science in China: Series D, 39 (Suppl.), 24-32.

Muir, R.J., Weaver, S.D., Bradshaw, J.D., Eby, G.N., Evans, J.A., 1995. The Cretaceous separation point batholith, New Zealand: Granitiod magmas formed by melting of mafic lithosphere. Journal of the Geological Society of London, 152, 689-701.

Peacock, S.M., Rushmer, T., Thompson, A.B., 1994. Partial melting of subducting oceanic crust. Earth & Planetary Science Letters, 121, 227-244.

Peccerillo, R., Taylor, S.R., 1976. Sr geochemistry of Eocene calc-alkaline volcanic rocks from the Kastamonu area, northern Turkey. Contributions to Mineralogy and Petrology, 58, 63-81.

Petford, N., Atherton, M., 1996. Na-rich partial melt from newly underplated basaltic crust: The Cordmera

Blanca Batholith, Peru. Joumal of Petrology, 37, 491-521.

Prouteau, G., Scaillet, B., 2003. Experimental constraints on the origin of the 1991 Pinatubo dacite. Journal of Petrology, 44, 2203-2241.

Prouteau, G., Scaillet, B., Pichavant, M., Maury, R., 2001. Evidence for mantle metasomatism by hydrous silicic melts derived from subducted oceanic crust. Nature, 410, 197-200.

Qi, L., Hu, J., Gregoire, D.C., 2000. Determination of trace elements in granites by inductively coupled plasma-mass spectrometry. Talanta, 51, 507-513.

Rapp, R.P., 1997. Heterogeneous source regions for Archean granitoids. In: Wit, M.J., Ashwal, L.D. Greenstone belts. Oxford: Oxford University Press, 35-37.

Rapp, R.P., Shimizu, N., Norman, M.D., Applegate, G. S., 1999. Reaction between slab-derived melts and peridotite in the mantle wedge: Experimental constraints at 3.8 GPa. Chemical Geology, 160, 335-356.

Rapp, R.P., Watson, E.B., 1995. Dehydration melting of metabasalt at 8-32 kbar: Implications for continental growth and crust-mantle recycling. Journal of Petrology, 36, 891-931.

Rapp, R.P., Xiao, L., Shimizu, N., 2002. Experimental constraints on the origin of potassium-rich adakites in eastern China. Acta Petrologica Sinica, 18, 293-302.

Sajona, F.G., Maury, R.C., Pubellier, M., 2000. Magmatic source enrichment by slab-derived melts in a young post-collision setting, central Mindanao (Philippines). Lithos, 54, 173-206.

Sen, C., Dunn T., 1994. Dehydration melting of a basaltic composition amphibolite at 1.5 and 2.0 Gpa: Implication for the origin of adakites. Contributions to Mineralogy and Petrology, 117, 394-409.

Smith, D.R., Leeman, W. P., 1987. Petrogenesis of Mount St. Helens dacitic magmas. Journal of Geophysical Research, 92, 10313-10334.

Smithies, R.H., 2000. The Archaean tonalitetrondhjemite-granodiorite (TTG) series is not an analogue of Cenozoic adakite. Earth & Planetary Science Letters, 182, 115-125.

Springer, W., Seck, H.A., 1997. Partial fusion of basaltic granulites at 5−15 kbar: Implication for the origin of TTG magmas. Contributions to Mineralogy and Petrology, 123, 263-281.

Stern, C.R., Kilian, R., 1996. Role of the subducted slab, mantle wedge, and continental crust in the generation of adakites from the Andean Austral Volcanic Zone. Contributions to Mineralogy and Petrology, 123, 263-281.

Sun, S.S., McDonough, W.F., 1989. Chemical and isotopic systematics of oceanic basalts: Implication for the mantle composition and process. In: Saunders, A.D., Norry, M.J. Magmatism in the ocean basins. Geological Society of London, Special Publications, 42, 313-345.

Sun, W.D., Li, S.G., 1998. Pb isotopes of granitoids suggest Devonian accretion of Yangtze (South China) craton to North China craton: Comment. Geology, 26, 859-860.

Sun, W.D., Li, S.G., Chen, Y.D., Li, Y.J., 2002. Timing of synorogenic granotoids in the south Qinling, central China: Constraints on the evolution of the Qinling-Dabie orogenic belt. Journal of Geology, 110, 457-468.

Topuz, G., Altherr, R., Schwarz, W.H., 2005. Postcollisional plutonism with adakite-like signatures: The Eocene Saraycik granodiorite (Eastern Pontides, Turkey). Contributions to Mineralogy and Petrology, 150, 441-455.

Wareham, C.D, Millar, I.L., Vaughan, A.P.M., 1997. The generation of sodic granitic magmas, western Palmer Land, Antarctic Peninsula. Contributions to Mineralalogy and Petrology, 128, 81-96.

Wang, Q., McDermott, F., Xu, J.F., Bellon, H., Zhu, Y.T., 2005. Cenozoic K-rich adakitic volcanic rocks in the Hohxil area, northern Tibet: Lower-crustal melting in an intracontinental setting. Geology, 33, 465-468.

Wang, Q., Wyman, D.A., Xu, J.F., Zhao, Z.H., Jian, P., Xiong, X.L., Bao, Z.W., Li, C.F., Bai, Z. H., 2006. Petrogenesis of Cretaceous adakitic and shoshonitic igneous rocks in the Luzong area, Anhui Province (eastern China): Implications for geodynamics and CuAu mineralization. Lithos, 89, 424-446.

Wang, X.C., Liu, Y.S., Liu, X.M., 2006. Mesozoic adakites in the Lingqiu Basin of the central North China craton: Partial melting of underplated basaltic lower crust. Geochemical Journal, 40, 447-461.

Wu, F.Y., Ge, W.C., Sun, D.Y., 2002. The definition, discrimination of adakites and their geological role. In: Xiao, Q.H., Deng, J.F., Ma, D.Q., et al. Ways of investigation of granotoids. Beijing: Geological Publishing House, 172-191 (in Chinese with English abstract).

Xu, J.F., Shinjio, R., Defant, M.J., Wang, Q., Rapp, R.P., 2002. Origin of Mesozoic adakitic intrusive rocks in the Ningzhen area of east China: Partial melting of delaminated lower continental crust? Geology, 32, 1111-1114.

Yin, A., Nie, S., 1993. An indentation model for North and South China collision and the development of the Tanlu and Honam fault systems, eastern Asia. Tectonics, 12, 801-813.

Yuan, H.L., Wu, F.Y., Gao, S., Liu, X.M., Xu, P., Sun, D.Y., 2003. Determination of zircons from Cenozoic intrusions in Northeastern China by laser ablation ICP-MS. Chinese Science Bulletin, 48(4), 1511-1515.

Zhai, X., Day, H.W., Hacker, B.R., You, Z., 1998. Paleozoic metamorphism in the Qinling orogen, Tongbai Mountains, central China. Geology, 26, 371-374.

Zhang, B.R., Luo, T., Gao, S., Ouyang, J.P., Chen, D., Ma, Z., Han, Y., Gu, X., 1994. Geochemical study of the lithosphere, tectonism, and metallogenesis in the Qinling-Dabieshan region. Wuhan: China University of Geoscience Press, 110-122 (in Chinese with English abstract).

Zhang, G.W., Cheng, S.Y., Guo, A.L., Dong, Y.P., Lai, S.C., Yao, A.P., 2004. Mianlue paleo-suture on the southern margin of central orogenic system in Qingling-Dabie: With a discussion of the assembly of the main part of the continent of China. Geological Bulletin of China, 23(9-10), 846-853 (in Chinese with English abstract).

Zhang, G.W., Zhang, B.R., Yuan, X.C., Chen, J.Y., 2002. Qinling orogenic belt and continental dynamics. Beijing: Science Press, 855 (in Chinese with English abstract).

Zhang, H.F., Zhang, L., Harris, N., Jin, L.L., Yuan, H.L., 2006. U-Pb zircon ages, geochemical and isotopic compositions of granitoids in Songpan-Graze fold belt, eastern Tibetan Palteau: Constraints on petrogenesis and tectonic evolution of the basement. Contributions to Mineralogy and Petrology, 152, 75-88.

Zhang, Q., Wang, Y., Qian, Q., Yang, J., Wang, Y.L., Zhao, T.P., Guo, G.J., 2001. The characteristics and tectonic-metaliogenic significances of the adakites in Yanshan period from eastern China. Acta Petrologica Sinica, 17(2), 236-244.

Zircon LA-ICP-MS U-Pb age, Sr-Nd-Pb isotopic compositions and geochemistry of the Triassic post-collisional Wulong adakitic granodiorite in the South Qinling, central China, and its petrogenesis[①]

Qin Jiangfeng　Lai Shaocong[②]　Wang Juan　Li Yongfei

Abstract: The Indosinian post-collisional Wulong pluton intruded into the Mesoproterozoic Foping Group, South Qinling, central China. In the southern part of the pluton, some mafic enclaves have sharp or gradational contact relationships with the host biotite granodiorite. Geochemistry, zircon LA-ICP MS (laser ablation inductively-coupled plasma mass spectrometry) U-Pb chronology and Sr-Nd-Pb isotope geochemistry of the pluton are reported in this paper. The biotite granodiorites shows close compositional similarities to high-silica adakites. Its chondrite-normalized REE patterns are characterized by strong HREE depletion ($Yb = 0.33\times10^{-6} - 0.96\times10^{-6}$ and $Y = 4.77\times10^{-6} - 11.19\times10^{-6}$), enrichment of Ba ($774\times10^{-6} - 1\,385\times10^{-6}$) and Sr ($642\times10^{-6} - 1\,115\times10^{-6}$) and high Sr/Y (57.83 – 159.99) and Y/Yb (10.99 – 14.32) ratios, as well as insignificant Eu anomalies ($\delta Eu = 0.70 - 0.83$), suggesting a feldspar-poor, garnet ± amohibole-rich residual mineral assemblage. The mafic enclaves have higher MgO (4.15%–8.13%), Cr ($14.79\times10^{-6} - 371.31\times10^{-6}$), Ni ($20.00\times10^{-6} - 224.24\times10^{-6}$), and Nb/Ta (15.42–21.91) than the host granodiorite, implying that they are mantle-derived and might represent underplated mafic magma. Zircon LA-ICP-MS dating of the granodiorite yield a $^{206}Pb/^{238}U$ weighted mean age of 208 ± 2 Ma (MSWD = 0.50, 1σ), which is the age of the emplacement of the host biotite granodiorites. This age indicate that the Wulong pluton formed during the late-orogenic or post-collisional stage (≤ 242 ± 21 Ma) of the south Qinling Belt. The host granodiorites display $^{87}Sr/^{86}Sr = 0.705\,9 - 0.706\,2$, $I_{Sr} = 0.704\,4 - 0.705\,0$, $^{143}Nd/^{144}Nd = 0.512\,36 - 0.512\,38$, $\varepsilon_{Nd}(t) = -2.66$ to -2.26, $^{206}Pb/^{204}Pb = 18.099 - 18.209$, $^{207}Pb/^{204}Pb = 15.873 - 15.979$ and $^{208}Pb/^{204}Pb = 38.973 - 39.430$. Those ratios are similar to those of the mesoproterozoic Yaolinghe Group in South Qinling. Futhermore, its Nd isotopic model age (~1.02 Ga) is consitent with the age (~1.1 Ga) of the Yaolinghe Group. Based on the integrated geological and geochemical studies, coupled with previous studies, the authors suggest that the Wulong adakite biotite granodiorite was probably generated by dehydration melting of the Yaolinghe Group-like thickened mafic crust, triggered by underplating of mafic magma at the

---

① Published in *Acta Geologica Sinica*, 2008, 82(2).

② Corresponding author.

boundary of the thickened mafic crust and hot lithospheric mantle, and that the Wulong adakite biotite granodiorites may have resulted from thinning and delamination of lower crust or breakoff of the subducting slab of the Mianlue ocean during the Indosinian post-collisional orogenic stage of the Qinling orogenic belt.

1 Introduction

The Qinling-Dabie orogenic belt is an important orogenic belt dividing continental China into the North and South China blocks and records important information of the tectonic evolution of continental China. Accurate konwledge of the tectonics and evolution of the Qinling-Dabie orogenic belt is important to placing firm constraints on the evolution of this continent (Sun et al., 2002; Zhang et al., 2002; Lai et al., 2004a, b). In the western segment of this orogenic belt, there is a ~400 km long Indosinian granitoid belt along the north side of the Mianlue suture (Fig. 1a). Petrological, geochemical and zircon U-Pb chronological studies indicate that this suite of granitoidis syncollision granitoidis(Zhang et al., 1994; Sun et al., 2002).

Adakite are relatively low-K and high-Al, intermediate to felsic volcanic rocks. It generally enriched in Sr and strongly depleted in Y and HREE (Martin et al., 2005; Rollinson and Martin, 2005). Such rocks form in arc settings related to subduction of a hot oceanic slab (Defant and Drummond, 1990, 1993; Martin, 1999) or in the lower part of a thickened crust (Smithies, 2000; Defant et al., 2002; Kay and Kay, 2002; Toupoz et al., 2005). Plutonic granitoids originating from adakitic magma have also been reported in several localities (Bourgois et al., 1996; Zhang et al., 2001; Xu et al., 2002; Chung et al., 2003; Hou et al., 2004; Zhang et al., 2006). This study focuses on the genesis of the Wulong biotite granodiorite pluton in the Qinling Orogenic Belt, central China. The pluton was emplaced into the Proterozoic Foping Group ~225 Ma ago (Fig. 1b) and was generated during the collision between the Yangtze and North-China blocks, which followed the northward subduction of the Mianlue ocean crust. Based on field observation and detailed studies of petrology, petrography, geochemistry, zircon LA-ICP MS (laser ablation inductively-coupled plasma mass spectrometry) U-Pb chronology and Sr-Nd-Pb isotopic geochemistry, we think that the Wulong biotite granodiorite has the chemical properties of adakite and might be the product of partial melting of the thickened mafic lower crust at >45 km depth which reached eclogite facies or garnet granulite facies. The establishment of the Triassic adakitic granitoids provides convincing evidence for the lithospheric thinning during the Indosinian late-orogenic or post-collisional stage in the Qingling orogenic belt.

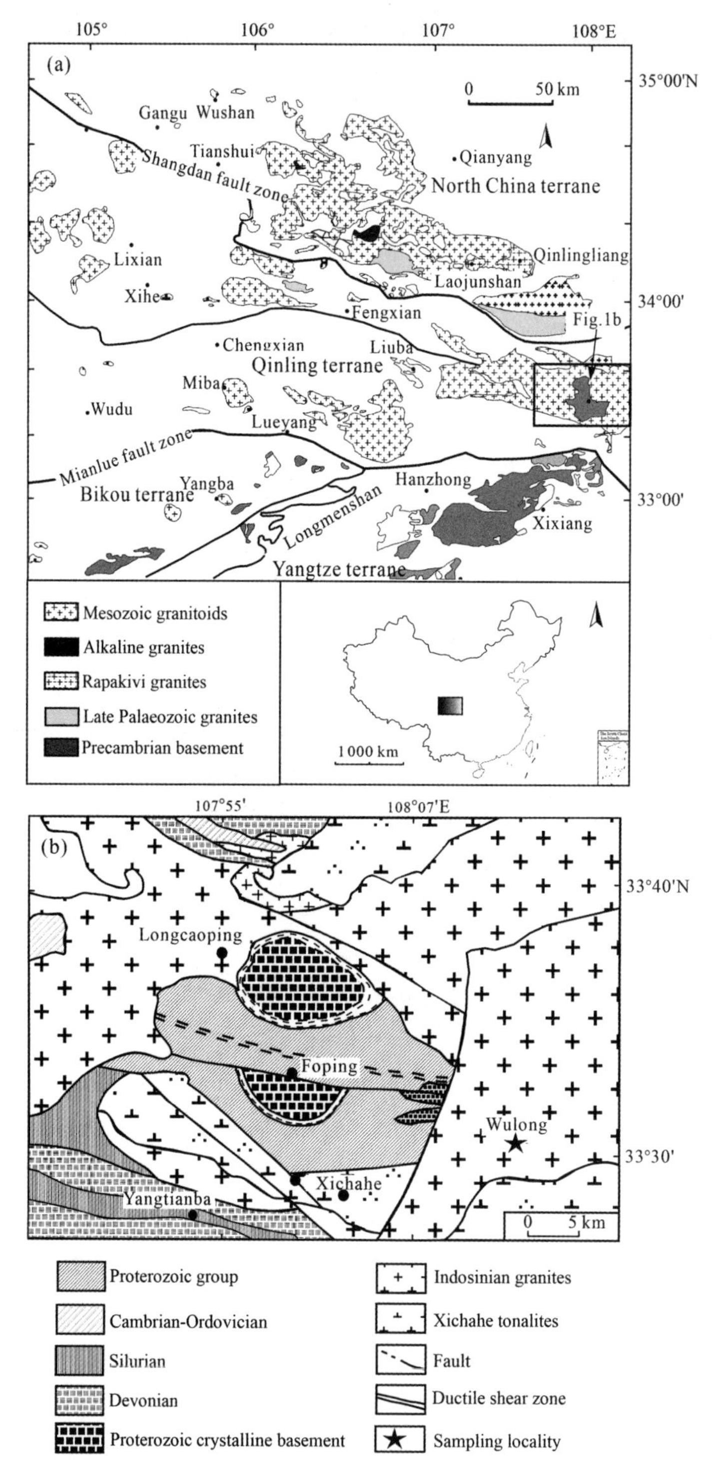

Fig. 1 (a) Geological sketch map of the Qinling orogenic belt showing distribution of Indosinian granites (after Zhang Guowei et al., 2002); (b) Geological sketch map of the Foping area in the South Qinling Mountains (modified from Zhang Guowei et al., 2002).

2 Geological setting

The Qinling orogenic belt is acomposite orogenic belt with two mountain chains and two sutures (Sun et al., 2002; Zhang et al., 2002). The north suture lies mainly along the Shangdan tectonic belt, whereas the south suture consists of the Mianlue ophiolite mélange belt in the west and presumably the Bashan fault in the east (Meng and Zhang, 1999; Zhang Guowei et al., 2002, 2004; Lai et al., 2004a,b). North to the Shangdan fault lies the North Qinling blocks, which was generated during Middle Paleozoic magmatism and metamorphism (Li et al., 1989; Li and Sun, 1996; Sun and Li, 1998; Zhai et al., 1998). Between the Shangdan belt and Mianlue belt is the South Qinling blocks, a Late Paleozoic to Triassic orogen (Mattauer et al., 1985), which is characterized by extensive Triassic granitoids and Late Paleozoic metamorphic rocks (Li and Sun, 1996; Li et al.1996; Meng and Zhang, 1999; Sun et al., 2002). South of the Mianlue suture belt is the Yangtze Block(Fig. 1a).

The discovery of the Mianlue suture in the South Qinling area has been one of the most important advances in the study of the Qinling-Dabie orogenic belt (Li et al., 1996; Meng and Zhang, 1999; Lai et al., 2004a,b; Zhang et al., 2004). This discovery led to the presentation of the view that the Qinling orogenic belt is a belt composed of two suture belts, formed by multiple collision between the North China and Yangtze plates (Li et al., 1996; Meng and Zhang, 1999; Zhang et al., 2002, 2004; Lai et al., 2004a,b) rather than simple collision between the two plates as proposed previously.

The Mianlue tectonic melangebelt is separated from the Neoproterozoic Bikou Group by the Mianlue fault on the south and from the Silurian Baishuijiang Group by the Zhuangyuanbei fault on the north. It consists predominantly of Sinian-Cambrain and Devonian-Carboniferous strongly sheared thrust slabs, forming south-vergent, imbricate thrust nappe structure. The Sinian-Cambrain strata consist mainly of gravel-bearing argillaceous rocks, argillaceous clastic rocks, pyroclastic rocks, carbonate rocks and magnesian carbonates; the Devonian System consists of turbidite, argillaceous carbonate rocks and argillaceous rocks; the Carboniferous rocks consist principally of carbonate rocks; and the Ordovician and Silurian strata are lacking(Meng et al., 1996; Meng and Zhang 1999; Sun et al., 2000).

Ophiolite mélanges and metavolcanic rocks in the Mianlue belt also occur as tectonic slabs involved in this tectonic belt. The ophiolite mélanges consists of strongly sheared metabasalts, cumulate gabbros, ultramafic rocks (ophiolite and talc schist), diabase dike swarms and radiolarian cherts (Meng and Zhang, 1999), in which some metabasalts display the geochemical features of normal mid-ocean ridge baslts(N-MORB)(Lai and Zhang, 1996; Lai et al., 2004a, b) and were subjected to metamorphism. Intensive study shows that this sequence of volcanic rocks has peak metamorphic ages of 242–221 Ma, indicating an Early

Triassic age (Li shuguang et al., 1996).

Mesozoic granitoids are widespread in the Mianlue suture. The granitoid belt is distributed roughly parallelly to the Mianlue suture in the west and mainly between the Daheba-Ningshan-Shanyang fault and Shangdan tectonic belt in the east. The granitoids belt is ~400 km long and cover an area of ~6 000 km^2. The emplacement ages of the granitoids cluster at 205–220 Ma, suggesting a syn-collisional origin (Sun et al., 2002). Granitoids in the belt were previously calssified into the Guangtoushan, Wulong, and Dongjiangkou pluton swarms (Sun et al., 2002; Zhang et al., 1994).

The Wulongpluton is exposed north of Mianlue tectonic belt over an area of ~1 800 km^2, belonging to the Wulong pluton swarm (Fig. lb). The surrounding rocks of the pluton are a Paleoproterozoic metamorphic complex known as the Fuping Group and Neopetrozoic-Late Palaeozoic strata. The pluton shows intrusive contact with the Fuping Group (Zhang et al., 2002). Three rock facies may be recognized in the pluton: porphyritic moyite in the center, medium-granied biotite granodiorite in the rims and a transition facies of granodiorite and amphibole monzogranite in between. Generally, the pluton was emplaced passively and no deformation took place.

The biotite granodiorite is grayish white, has a medium-grained texture and massive structure and is made up of plagioclase (45%–50% with An = 27–34), hornblende (5%–10%), K-feldspar (10%), quartze (~15%) and biotite (8%–12%). The monzogranite has a coarse-grained texture and composed dominantly of quartze (25%–30%), K-feldspar (30%–40%), plagioclase (30%–35%) and biotite (~5%) with accessory zircon, apatite, allanite, clinozoisite and magnetite. Apatite is euhedral rod-like and usually occurs as inclusions in biotite, hornblende and plagioclase. Zircon is mainly euhedral long-prismatic.

In the southern part of the pluton, some microgranular mafic enclaves are well developed. They commonly have an ovoid shape (up to 20 cm in diameter) and medium-and fine-grained textures and show sharp contacts with the host biotite granodiorite, but some enclaves have gradational relationships with their hosts. Petrographic observation suggest they have a typical magmatic texture, and in all the observed enclaves, no quench signals, nor feldspar or quartz xenoliths from the host granodiorite are found.

3 Analytical methods

Intensive petrographic observation was made of the samples taken from the Wulong pluton. A fresh sample without veins was chosen for the zircon chronological, major element, trace element and Sr-Nd-Pb isotopic analyses. Zircon grains were separated using the conventional heavy liquid and magnetic techniques (Di et al., 2005; Liu et al., 2006; Peng et al., 2006). Representative zircon grains were handpicked under the binocular microscope, mounted in an epoxy resin disc, and then polished to expose the centers of the zircon grains and coated with carbon film. The internal morphology was examined using cathodoluminescent (CL) analysis prior to U-Pb isotopic analyses. The CL analysis was performed with a microprobe at the Institute of Geology and Geophysics, Chinese Academy of Sciences. Zircon U-Pb dating was carried out at the State Key Laboratory of Continental Dynamics, Northwest University. The ICP-MS (inductively-coupled plasma mass spectrometer) used in this study was ELAN6100DRC from Perkin Elmer/SCIEX (ON, Canada) with a dynamic reaction cell (DRC). Details of the procedures for the analysis of zircons using LA-ICP MS have been given by Yuan et al. (2003) and Gao et al. (2002).

Major elements were analysed by means of wet chemistry at the Institute of Geochemistry, Chinese Academy of Science. Trace element and rare earth element (REE) analyses were performed by ICP-MS of bebuized solutions using a VG Plasma-Quad Excel ICP-MS at the University of Hong Kong after a 2-day closed beaker digestion using a mixture of HF and HNO_3 acids in high-pressure bombs. Pure element standard solutions were used for external calibration and BHVO-1(basalt) BCR-2 (basalt) and AGV-1 (andesite) were used as reference materials. The procedure for the ICP-MS analysis was described by Qi et al. (2000). The accuracies of the ICP-MS analyses are estimated to be better than ±5% (relative) for most trace elements.

Whole-rock Sr-Nd and Pb isotopic data were obtained using a Nu Plasma HR mutli-collector mass spectrometer at State Key Laboratory of Continental Dynamics, Northwest University. Sr and Nd were separated using AG50W-X8 (200–400 mesh), HDEHP (self-made) and AG1-X8 (200–400 mesh) ion exchange resin. Sr and Nd isotopic fractionations were corrected to $^{87}Sr/^{86}Sr = 0.1194$ and $^{146}Nd/^{144}Nd = 0.7219$ respectively. In the analysis, the NIST SRM 987 standard yielded an average value of $^{87}Sr/^{86}Sr = 0.710250 \pm 12$ (2σ, $n = 15$) and the La Jolla standard gave an average of $^{143}Nd/^{144}Nd = 0.511859 \pm 6$ (2σ, $n = 20$). Whole-rock Pb was separated by anion exchange in HCl-Br columns, and Pb isotopic fractionation was corrected to $^{205}Tl/^{203}Tl = 2.3875$. In the analytical period, 30 measurements of NBS981 gave average values of $^{206}Pb/^{204}Pb = 16.937 \pm 1$ (2σ), $^{207}Pb/^{204}Pb = 15.491 \pm 1$ (2σ), $^{208}Pb/^{204}Pb = 36.696 \pm 1$ (2σ). The BCR-2 standard gave $^{206}Pb/^{204}Pb = 18.742 \pm 1$ (2σ), $^{207}Pb/^{204}Pb = 15.620 \pm 1$ (2σ), and $^{208}Pb/^{204}Pb = 38.705 \pm 1$ (2σ). Total procedural Pb blanks were in the range of 0.1–0.3 ng.

4 Results

4.1 LA-ICP MS U-Pb zircon dating

CL images of zircons (Fig. 2) show that most zircon grains from the Wulong biotite granodiorite (sample WL-01) are pink to colorless, euhedral prismatic, with the grain size ranging from 100 μm to 200 μm and the length/ width ranging from 2∶1 to 3∶1. They display rhythmic oscillatory zoning, indicating a magma origin (Chen et al., 2007; Liu et al., 2007). Some zircons contain cores of inherited zircons. Eight analyses of zircons (not including inherited zircons) gave $U = 452\times10^{-6} - 788\times10^{-6}$, $Th = 210\times10^{-6} - 644\times10^{-6}$ and Th/U = 0.37–0.82. Their $^{206}Pb/^{238}U$ ages cluster at 206–213 Ma with a $^{206}Pb/^{238}U$ weighted mean age of 208.5 ± 2 Ma (MSWD = 0.50, 1σ), which is the best estimate of the emplacement age of the Wulong biotite granodiorite (Fig. 3). The analytical results are shown in Table 1.

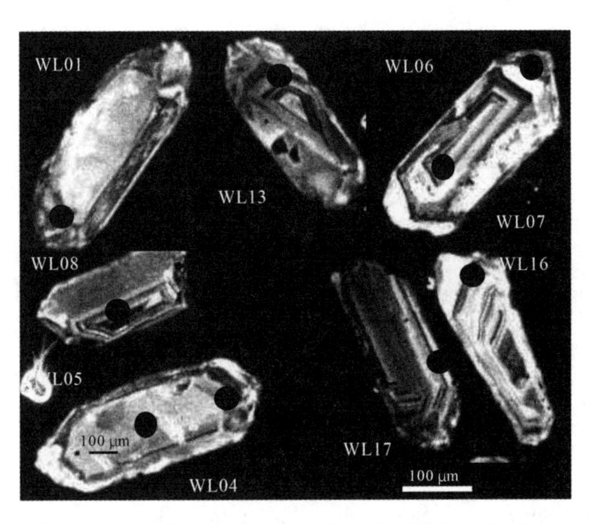

Fig. 2 Cathodoluminescence image of zircons from the Wulong biotite granodiorite.

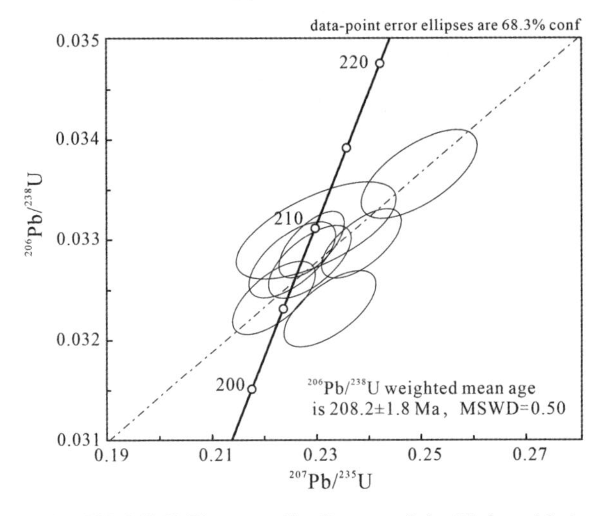

Fig. 3 Zircon LA-ICP-MS U-Pb concordia diagram of the Wulong biotite granodiorite.

4.2 Major and trace element and REE geochemistry

The results of major and trace element analyses of representive samples from biotite granodiorites and mafic microgranular enclave are given in Table 2.

Table 1 LA-ICP-MS U-Pb zircon dating of the Wulong biotite granodiorites.

Sample	Content/(μg/g)		Ratios							Age/Ma					
No.	^{238}U	^{232}Th	$^{232}Th/^{238}U$	$^{207}Pb^*/^{206}Pb^*$	1σ	$^{207}Pb^*/^{235}U$	1σ	$^{206}Pb^*/^{238}U$	1σ	$^{207}Pb^*/^{206}Pb^*$	1σ	$^{207}Pb^*/^{235}U$	1σ	$^{206}Pb^*/^{238}U$	1σ
WL01-02	788	644	0. 816 6	0. 049 7	0. 001 26	0. 222 0	0. 005 2	0. 032 4	0. 000 2	180	41	204	4	206	1
WL01-04	736	459	0. 624 6	0. 050 5	0. 001 01	0. 229 2	0. 004 1	0. 033 0	0. 000 2	217	29	210	3	209	1
WL01-05	486	215	0. 442 0	0. 049 9	0. 001 30	0. 225 5	0. 005 4	0. 032 8	0. 000 3	189	42	206	5	208	2
WL01-06	612	230	0. 375 5	0. 052 5	0. 001 21	0. 238 6	0. 005 0	0. 033 0	0. 000 2	308	35	217	4	209	1
WL01-07	452	210	0. 463 3	0. 052 3	0. 001 39	0. 232 6	0. 005 8	0. 032 3	0. 000 3	296	42	212	5	205	2
WL01-13	661	300	0. 453 2	0. 057 6	0. 001 94	0. 264 9	0. 008 5	0. 033 4	0. 000 3	215	106	210	8	210	2
WL01-15	466	255	0. 550 0	0. 053 8	0. 001 67	0. 249 7	0. 007 3	0. 033 7	0. 000 3	364	50	226	6	213	2
WL01-22	616	402	0. 652 3	0. 050 7	0. 001 24	0. 228 9	0. 005 2	0. 032 8	0. 000 2	226	38	209	4	208	1

Table 2 Major element(%) and trace element($\times 10^{-6}$) analysis of the Wulong biotite granodiorite.

Rock type	Biotite granodiorites								Mafic enclaves						
	WL02	WL04	WL05	WL10	WL13	WL15	WL18	WL21	WLB08	WLB17	WLB25	WLB26	WLB31	WLB33	WLB34
SiO_2	68. 44	65. 29	68. 92	71. 30	70. 89	70. 54	66. 04	63. 84	53. 76	53. 97	55. 78	54. 99	52. 86	53. 93	53. 41
TiO_2	0. 20	0. 26	0. 12	0. 16	0. 19	0. 21	0. 33	0. 31	0. 87	0. 82	0. 98	0. 85	0. 97	1. 22	1. 15
Al_2O_3	17. 71	17. 73	16. 28	13. 96	14. 87	15. 31	17. 06	19. 12	16. 62	15. 31	13. 56	15. 53	16. 62	17. 93	18. 59
Fe_2O_3	0. 70	0. 54	0. 60	0. 90	1. 00	1. 05	1. 40	1. 57	3. 01	2. 20	2. 68	2. 60	3. 40	2. 86	3. 30
FeO	1. 40	1. 96	1. 50	1. 00	1. 10	1. 05	1. 50	1. 43	3. 89	3. 80	5. 02	4. 10	4. 00	4. 84	4. 10
MnO	0. 14	0. 06	0. 13	0. 08	0. 09	0. 08	0. 11	0. 08	0. 18	0. 19	0. 14	0. 19	0. 20	0. 21	0. 17
MgO	0. 71	1. 31	0. 85	0. 76	0. 86	0. 85	1. 24	1. 26	6. 09	6. 26	8. 13	6. 32	6. 58	4. 26	4. 15
CaO	2. 09	2. 86	1. 81	1. 77	1. 81	1. 67	2. 60	2. 70	7. 20	8. 58	5. 61	7. 78	6. 72	4. 48	4. 73
Na_2O	4. 50	5. 00	4. 30	4. 30	4. 30	4. 00	4. 80	4. 70	3. 80	3. 60	2. 70	3. 50	4. 00	3. 90	4. 00
K_2O	2. 51	2. 30	3. 90	3. 30	3. 10	3. 30	2. 30	2. 60	1. 30	1. 90	2. 30	1. 50	1. 60	3. 00	3. 10
P_2O_5	0. 10	0. 25	0. 22	0. 10	0. 20	0. 18	0. 32	0. 22	0. 63	0. 70	0. 73	0. 67	0. 93	1. 03	1. 02
LOI	1. 21	1. 78	0. 70	1. 84	1. 05	1. 57	2. 10	2. 04	1. 46	1. 71	1. 86	1. 50	1. 08	1. 72	1. 78

Continued

Rock type	Biotite granodiorites								Mafic enclaves						
	WL02	WL04	WL05	WL10	WL13	WL15	WL18	WL21	WLB08	WLB17	WLB25	WLB26	WLB31	WLB33	WLB34
Total	99.71	99.34	99.33	99.47	99.46	99.81	99.80	99.87	99.40	99.46	99.49	99.53	99.46	99.38	99.50
A/CNK	1.27	1.11	1.11	1.01	1.08	1.16	1.13	1.24	0.80	0.65	0.79	0.72	0.81	1.01	1.00
σ	1.93	2.39	2.59	2.04	1.96	1.94	2.19	2.56	2.42	2.76	1.96	2.09	3.18	4.36	4.84
$Mg^{\#}$	38.63	49.08	42.86	43.05	43.63	43.40	44.71	44.37	62.42	66.10	66.32	63.85	62.65	50.84	51.38
Ba	1 242.29	774.69	1 224.42	1 080.67	1 025.39	1 185.40	1 385.98	1 477.58	1 196.96	1 155.69	1 496.67	982.62	669.79	462.26	485.74
Rb	100.41	85.87	114.85	98.76	97.14	96.78	80.83	54.94	40.22	40.41	117.26	35.46	94.04	198.15	204.92
Sr	762.67	1 002.77	662.89	642.99	714.83	646.77	1 115.36	1 065.91	1 396.16	1 696.53	1 741.74	1 631.32	1 518.09	836.08	856.18
Y	4.77	10.59	7.27	6.98	7.79	11.19	8.36	8.39	19.39	18.25	18.51	20.17	23.36	34.26	31.54
Zr	150.05	196.71	125.02	132.50	136.66	143.88	188.76	198.66	110.81	43.31	195.92	112.78	178.99	226.80	212.33
Nb	6.71	12.49	9.54	6.81	7.96	11.45	7.48	7.76	12.90	14.26	15.80	15.30	33.45	34.22	31.10
Th	8.12	9.67	10.18	5.76	8.33	7.34	9.82	7.53	7.28	5.12	9.07	6.16	14.13	13.50	12.27
Pb	21.57	23.51	27.92	23.96	23.77	24.55	18.63	17.46	15.12	13.85	13.09	13.47	14.39	17.94	18.58
Ga	16.92	17.90	14.97	13.68	14.53	14.53	16.82	15.61	13.07	12.60	12.83	12.23	15.10	23.45	22.62
Zn	41.98	39.28	31.47	31.03	33.91	41.16	58.67	54.66	76.62	72.66	85.84	65.42	122.54	179.57	185.19
Cu									27.63	140.34	75.12	215.13	4.55	13.74	14.79
Ni	2.50	10.85	7.75	6.13	7.12	8.07	7.58	7.94	49.42	69.15	224.24	68.50	73.01	20.00	20.10
V	33.31	44.28	31.22	29.10	34.42	38.35	56.37	54.74	176.48	199.02	161.60	204.02	209.36	185.19	197.78
Cr	9.91	23.49	19.37	13.46	16.50	17.77	18.47	22.56	110.21	172.02	371.31	175.77	164.01	14.79	14.97
Hf	5.23	6.49	4.57	5.11	4.54	5.06	5.47	6.48	3.35	1.99	5.39	3.38	4.57	6.46	5.74
Cs	5.81	6.11	3.99	3.63	4.32	3.91	5.46	3.57	1.95	1.07	10.05	0.88	7.35	13.09	13.55
Sc	5.64	5.81	4.84	4.24	4.94	7.11	6.35	5.11	26.08	26.17	18.32	26.65	25.28	30.42	19.41
Ta	0.39	1.12	0.85	0.80	0.74	1.01	0.52	0.60	0.66	0.77	0.85	0.86	1.53	2.22	1.90
Co	48.36	56.73	51.05	84.98	48.97	54.70	49.47	67.36	46.23	49.64	63.79	46.15	49.51	42.30	36.58
Li	86.83	65.34	47.62	37.39	57.75	46.21	57.25	54.52	31.44	26.57	86.31	32.44	76.97	148.03	145.98
Be	2.43	2.89	2.65	1.77	2.19	2.29	1.97	1.87	2.15	1.75	1.57	1.22	1.49	3.76	3.48

Continued

Rock type	Biotite granodiorites								Mafic enclaves						
	WL02	WL04	WL05	WL10	WL13	WL15	WL18	WL21	WLB08	WLB17	WLB25	WLB26	WLB31	WLB33	WLB34
U	0. 80	2. 16	1. 58	0. 88	0. 97	1. 27	0. 89	0. 88	2. 01	1. 86	1. 64	2. 01	3. 48	5. 59	5. 25
La	32. 22	30. 56	24. 74	20. 84	29. 02	26. 22	49. 94	39. 27	46. 00	38. 25	63. 67	48. 52	96. 12	57. 93	53. 78
Ce	62. 89	54. 67	46. 99	41. 01	52. 63	54. 78	88. 01	68. 31	104. 18	87. 25	132. 16	101. 28	181. 64	138. 66	130. 64
Pr	6. 11	5. 90	4. 74	4. 35	5. 39	6. 10	8. 23	6. 88	12. 01	10. 10	14. 53	11. 66	19. 34	17. 23	16. 25
Nd	21. 35	22. 66	16. 87	15. 80	19. 27	27. 74	28. 56	24. 37	48. 85	42. 65	57. 66	48. 70	73. 40	73. 54	70. 56
Sm	3. 29	3. 93	2. 68	2. 86	3. 19	4. 46	4. 11	3. 95	8. 54	7. 49	9. 00	8. 20	11. 10	14. 25	13. 16
Eu	0. 77	0. 92	0. 66	0. 67	0. 76	0. 93	0. 89	0. 93	2. 00	1. 88	2. 09	2. 03	2. 66	3. 00	2. 86
Gd	2. 16	2. 95	2. 18	2. 08	2. 36	3. 38	3. 25	2. 94	6. 04	5. 30	6. 45	6. 16	7. 82	10. 13	9. 02
Tb	0. 23	0. 40	0. 29	0. 26	0. 30	0. 46	0. 36	0. 38	0. 78	0. 69	0. 79	0. 77	0. 93	1. 36	1. 25
Dy	1. 05	1. 94	1. 51	1. 28	1. 49	2. 19	1. 84	1. 78	3. 95	3. 61	3. 86	4. 26	4. 77	7. 02	6. 34
Ho	0. 16	0. 36	0. 23	0. 23	0. 26	0. 39	0. 31	0. 31	0. 71	0. 68	0. 69	0. 77	0. 88	1. 26	1. 14
Er	0. 42	1. 02	0. 64	0. 58	0. 66	1. 01	0. 78	0. 81	1. 87	1. 78	1. 90	2. 11	2. 42	3. 22	3. 10
Tm	0. 05	0. 14	0. 07	0. 08	0. 09	0. 14	0. 10	0. 10	0. 25	0. 24	0. 24	0. 27	0. 31	0. 42	0. 40
Yb	0. 33	0. 96	0. 63	0. 52	0. 62	0. 82	0. 63	0. 65	1. 68	1. 59	1. 44	1. 82	2. 02	2. 66	2. 51
Lu	0. 05	0. 13	0. 09	0. 08	0. 10	0. 11	0. 09	0. 10	0. 24	0. 24	0. 24	0. 28	0. 34	0. 37	0. 35
Sr/Y	159. 99	94. 65	91. 23	92. 19	91. 76	57. 83	133. 50	127. 04	72. 00	92. 96	94. 08	80. 87	64. 98	24. 40	27. 15
Y/Yb	14. 32	10. 99	11. 63	13. 54	12. 56	13. 71	13. 37	12. 99	11. 54	11. 47	12. 87	11. 09	11. 57	12. 86	12. 55
La/Yb	96. 75	31. 70	39. 58	40. 46	46. 81	32. 13	79. 90	60. 79	27. 36	24. 04	44. 24	26. 67	47. 58	21. 74	21. 40
δEu	0. 83	0. 79	0. 81	0. 80	0. 81	0. 70	0. 72	0. 80	0. 81	0. 87	0. 80	0. 84	0. 83	0. 73	0. 76
Nb/Ta	17. 42	11. 19	11. 17	8. 48	10. 71	11. 29	14. 30	12. 98	19. 70	18. 55	18. 61	17. 79	21. 91	15. 42	16. 34

The major element geochemical characteristics of the Wulong biotite granodiorites (eight samples) are as follows: SiO_2 = 63.84% – 71.30%, Al_2O_3 = 13.96% – 19.12%, MgO = 0.71%–1.31%, CaO = 1.67%–2.86% and K_2O = 2.30%–3.90%, with K_2O/Na_2O = 0.46–0.83. A/CNK values [molar Al_2O_3/($CaO + Na_2O + K_2O$)] range from 1.01 to 1.27, indicating that the granodiorite is dominantly meta-aluminous and aluminous. Compared with the host biotite granodiorite, the mafic micro-granular enclaves show lower SiO_2 (52.86% – 55.78%), Na_2O (2.70%–4.00%), K_2O (1.30%–3.1%), K_2O/Na_2O (0.34–0.85) and A/CNK (0.65–1.01), and higher TiO_2 (0.85%–1.22%), MgO (4.15%–8.13%) and CaO (4.48%–8.58%). In the Harker diagrams (Fig. 4), the host granodiorite and microgranular mafic enclaves show different evolution trends, implying that no significant magma mixing took place between them.

Trace element data show that the Wulong biotite granodioritesis enriched in Sr (643×10^{-6}–$1\,115\times10^{-6}$ and $>400\times10^{-6}$) and Ba (up to $1\,386\times10^{-6}$) and depleted in Y (4.77×10^{-6}–11.19×10^{-6} and $<18\times10^{-6}$), Nb, Ta and Ti. The low Nb/Ta ratios (8.48 – 14.30, mostly < 12) suggest that it is continental crust-derived. The primitive mantle-normalized trace element spidergrams of the biotite granodiorites shows significant negative anomalies of Nb, Ta and Ti and weak positive anomalies of Ba and Sr (Fig. 5). Like its host granodiorite, the mafic micrognanular enclaves are enriched in large-ion lithophile elements (LILE) such as Sr and Ba, but they have weak negative anomalies of Nb and Ta and their Nb/Ta ratios (15.42 – 21.91) are notably higher than those of crust-derived magma, suggesting that they might be the product of mantle-derived magmatism subjected to contamination with crustal materials.

Chondrite-normalized REE patterns of the biotite granodiortes have HREE concave-upward shapes and are lacking in a significant Eu anomaly with δEu [$\delta Eu = 2Eu_N/(Sm_N + Gd_N)$] ranging from 0.72 to 0.83 (Fig. 6). The biotite granodiorites show wide variation in $(La/Yb)_N$ ratios (~21.7 to ~65.4). The mafic microgramular enclaves have lower $(La/Yb)_N$ ratios (14.46–32.16) than and similar Eu anomalies (δEu = 0.76 – 0.87) to their host biotite granodiorite.

4.3 Sr-Nd-Pb isotope geochemistry

The whole-rock Sr-Nd-Pb isotope data from the Wulong biotite granodiorite are presented in Table 3. The $^{87}Sr/^{86}Sr$ values of the Wulong biotite granodiorite (two sample) range between 0.705 9 and 0.706 2 with I_{Sr} = 0.704 4 – 0.705 0, exhibiting the features of crust-derived magma; the $^{143}Nd/^{144}Nd$ values of the rock range from 0.512 36 to 0.512 38 with $\varepsilon_{Nd}(t)$ = –2.66 to –2.26. Whole-rock Pb isotopic data (Table 4) show that the Xichahe tonalites is characterized by highly radiogenic Pb with the present whole-rock Pb isotopic ratios of $^{206}Pb/^{204}Pb$ = 18.099 – 18.209, $^{207}Pb/^{204}Pb$ = 15.873 – 15.979 and $^{208}Pb/^{204}Pb$ = 38.973 –

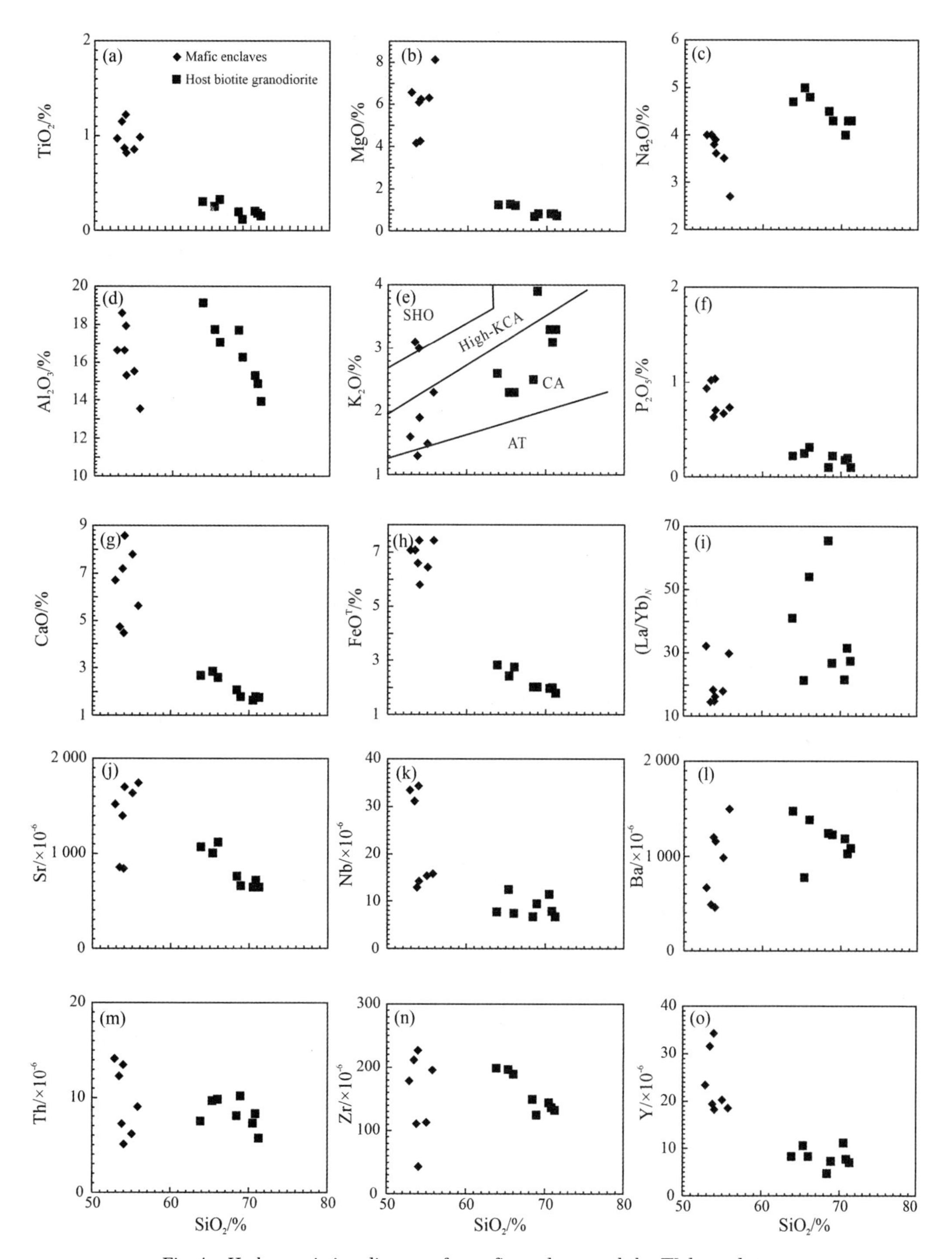

Fig. 4 Harker variation diagrams for mafic enclaves and the Wulong pluton.
The diagram of K_2O vs. SiO_2 shows the field boundaries between the medium-K (normal calc-alkaline), high-K and shoshonitic series of Peccerillo and Taylor (1976).

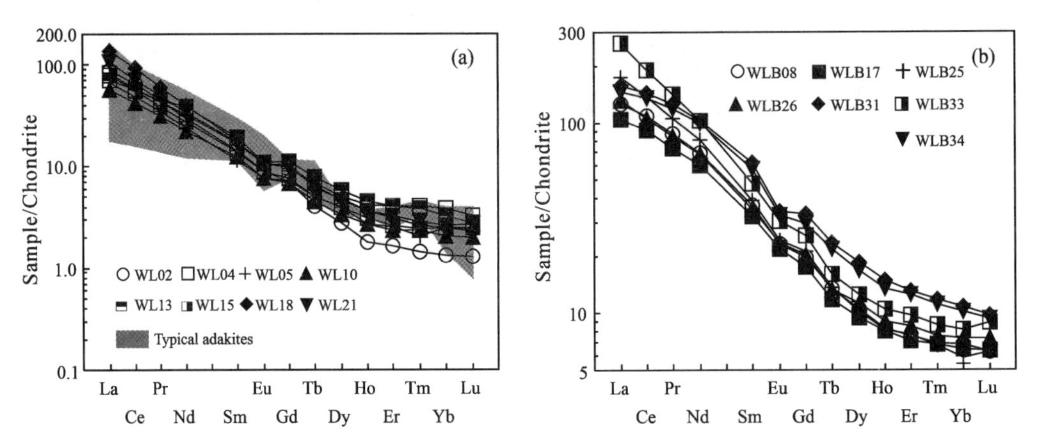

Fig. 5 Chondrite-normalized trace element patterns for the Wulong biotite granodiorites (a) and mafic enclaves (b).

Normalized values are from Sun and McDonough(1989). The field of the typical adakites are from Zhang et al.(2001).

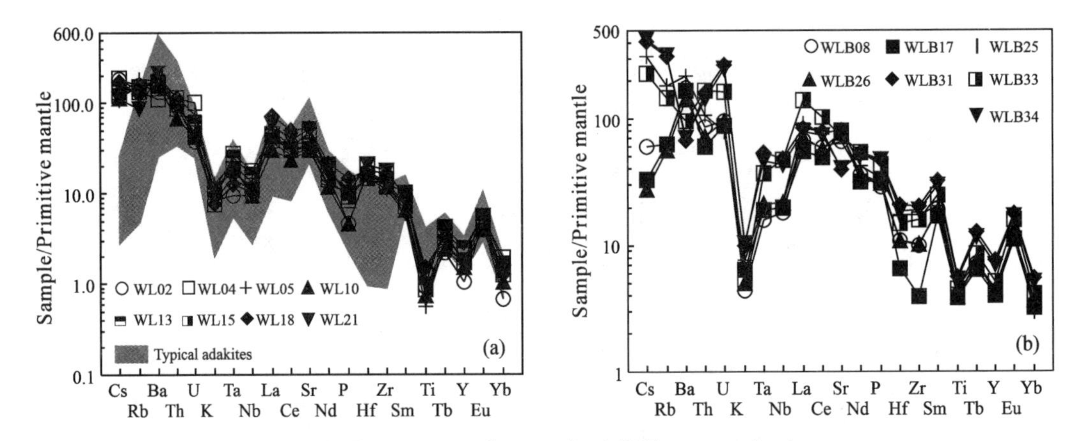

Fig. 6 Primitive mantle-normalized REE patterns for the Wulong biotite granodiorites (a) and mafic enclaves (b).

Primitive mantle-normalized valued are from Sun and McDonough(1989).

The field of the typical adakites is from Zhang Qi et al. (2001). The symbols same as in Fig. 5.

39.430. Because of its young age (~225 Ma) and very low contents of U ($0.80\times10^{-6}-2.16\times10^{-6}$) and Th ($5.76\times10^{-6}-10.18\times10^{-6}$), these isotopic ratios may be considered to represent the initial Pb isotopic ratios (e.g., Zhang et al., 2006). So no common Pb correction was made on these ratios.

5 Discussion

5.1 Petrogenesis and source region

The Wulong biotite granodiorites shows notable adakitic affinities, e.g. strong depletion in Y and HREE ($Yb=0.79\times10^{-6}-1.51\times10^{-6}$ and $<1.8\times10^{-6}$) and enriched in Ba and Sr with weak positive Sr anomalies in primitive mantle-normalized trace element patterns (Fig. 5), and high La/Yb (9.37–21.6) and Sr/Y (44.86–88.09) ratios. In the Sr/Y vs. Y and $(La/Yb)_N$ vs. Yb_N diagrams (Fig. 7a, b), all sample of the Wulong biotite granodiorites polt in the typical adakite field.

Table 3 Sr and Nd isotopic compositions of the Wulong biotite granodiorites.

Sample	$Rb/\times10^{-6}$	$Sr/\times10^{-6}$	$^{87}Rb/^{86}Sr$	$^{87}Sr/^{86}Sr\pm1\sigma$	I_{Sr}(210 Ma)	$Sm/\times10^{-6}$	$Nd/\times10^{-6}$
WL05	114.85	662.89	0.501 19	0.705 945±5	0.704 44	2.68	16.87
WL13	97.14	714.83	0.393 11	0.706 203±14	0.705 02	3.19	19.27

Sample	$^{147}Sm/^{144}Nd$[a]	$^{143}Nd/^{144}Nd\pm1\sigma$	$\varepsilon_{Nd}(t)$ (210 Ma)[b]	T_{DM}/Ga[c]
WL05	0.095 84	0.512 363±8	−2.657	1 017.221 055
WL13	0.100 12	0.512 389±9	−2.262	1 020.487 132

[a] $^{87}Rb/^{86}Sr$ and $^{147}Sm/^{144}Nd$ ratios were calculated using Rb, Sr, Sm and Nd contents, measured by ICP-MS.

[b] $\varepsilon_{Nd}(t)$ values were calculated using present-day $(^{147}Sm/^{144}Nd)_{CHUR}$ = 0.196 7 and $(^{143}Nd/^{144}Nd)_{CHUR}$ = 0.512 638.

[c] T_{DM} values were calculated using present-day $(^{147}Sm/^{144}Nd)_{DM}$ = 0.213 7 and $(^{143}Nd/^{144}Nd)_{DM}$ = 0.513 15.

Table 4 Whole-rock Pb isotopic composition of the Wulong biotite granodiorite.

Sample	$U/\times10^{-6}$	$Th/\times10^{-6}$	$Pb/\times10^{-6}$	$^{206}Pb/^{204}Pb$	1σ
WL05	1.58	10.18	27.92	18.098 13	0.001 20
WL13	0.97	8.33	23.77	18.209 38	0.000 35

Sample	$^{207}Pb/^{204}Pb$	1σ	$^{208}Pb/^{204}Pb$	1σ
WL05	15.873 38	0.001 05	38.973 46	0.002 45
WL13	15.978 73	0.000 30	39.430 48	0.001 16

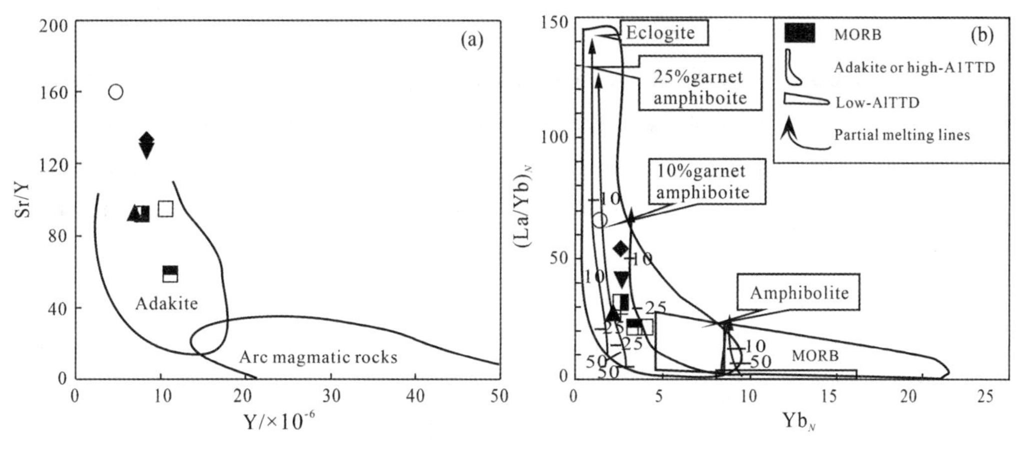

Fig. 7 Sr/Y vs. Y (a) (after Defant and Drummond, 1993; Defant et al., 2002) and $(La/Yb)_N$ vs. Yb_N (b) (after Drummond et al., 1990) diagrams for the Wulong biotite granodiorites.

Althrough the genesis of adakitic magma is still a matter of debate, it is generally accepted that they were derived from partial melting sources of basaltic compositions with significant amounts of garnet ± amohibole either as a residual phase or an early crystallizing assemblage (Martin, 1999; Rapp et al., 1999; Prouteau et al., 2001; Garrison et al., 2003; Condie, 2005; Martin et al., 2005). The petrogenetic models for these magmas include: ① partial melting of the subducted oceanic slab (Drummond, et al., 1990; Defant and Drummond, 1990; Peacock et al., 1994; Gutscher et al., 2000; Sajona et al., 2000; Beate et al., 2001), ② partial melting of the underplating basaltic lower crust (Petford and Atherton, 1996), and ③ partial melting of the delaminated lower crust (Xu et al., 2002; Gao et al., 2004; Wang et al., 2006) or thickened lower crust (Chung et al., 2003; Hou et al., 2004; Wang et al., 2005). Experimental studies show that basaltic magmas might become adakitic magmas only when the metamorphism of the basaltic rocks reaches eclogite facies or garnet-bearing granulite facies at a pressure of ~1.4 GPa (equivalent to a thickness >50 km) (Sen and Dunnt, 1994; Rapp et al., 1995). Therefore, generally speaking, partial melting is essential for the formation of adakitic rocks with the isotope geochemical characteristics of the continental crust (Zhang et al., 2006).

The Wulong biotite granodiorite has higher Na contents, with K_2O/Na_2O ratios of 0.50–0.72, and lower MgO content, with $Mg^{\#}$ ranging from ~39 to ~44, and Sr-Nd-Pb isotopic compositions have distinct features of crustal rocks, which indicates that the rock was unlikely to be the product of partial melting of subducted oceanic slab or delaminated lower continental crust. The enrichment in Sr and Ba and depletion in HREE and Y and unpronounced Eu anomalies suggest a feldspar-poor and garnet ± amphibole-rich fractionating mineral assemblage in the residual phase of the sources (Green, 1994).

The heat for melting the lower crust can be provided by underplating basaltic magmas (Petford and Gallagher, 2001; Annen and Sparks, 2002). Some experimental studies have shown that the temperatures of the lower continental crust need to be unrealistically high to produce granitoid liquids during garnet-amphibole dehydration. In the garnet-absent field, at pressure >1 GPa, amphibolites begin to melt at a relatively high temperatures of 800–900 ℃ (Wolf and Wyllie, 1991, 1994; Wyllie and Wolf, 1993).

The sparse mafic microgranular enclaves occurring in the southern part of the pluton show no significant mineralogical or geochemical evidence for their interaction with the host biotite granodiorite. Their geochemical features can place effective constraints on their source region. Geochemically, they have notably higher MgO, Cr, Ni, Ti and Nb/Ta ratios (15.42–21.91) than the crustal rocks, implying a mantle source. Futhermore, based on the above, combined with field observations, it may be inferred that they were probably derived from underplated mantle-derived magma which was syn-magmatic with the host granodioritic magmas.

In a word, the overall geochemical characteristics and Sr-Nd-Pb isotope geochemical characteristics of the Wulong adakitic biotite granodiorites suggest an origin by partial melting of mafic sources at pressure high enough to stabilize a residual mineral asseblage rich in garnet ± amphibole and poor in plagioclase. The abnormal heat was possibily provided by underplated mantle-derived mafic magma. Thus it may be inferred that the crust under the South Qinling orogenic belt was thickened to > 45 km by Triassic North China-Yangtze collisional orogeny.

5.2 Geodynamic implications

It is widely accepted that the widespread Mesozoic granitoids at the northern side of the Mianlue suture of the South Qinling terrane was generated by partial melting of the thickened (> 45 km) basaltic lower crust which resulted from underplating of asthenospheric materials along the Qinling orogenic to the crus-tmantle boundary due to delamination of the thickened mafic crust (Zhang et al., 2005) or breakoff of the subducted Mianlue oceanic slab (Davies and Von Blanckenburg, 1995; Li et al., 2002).

Our recent study of the Yangba adakitic granodiorite at the southern side of the Mianlue tectonic belt suggests that this intrusion originated by partial melting of thickened mafic crust in the post-collisional environment (Qin et al., 2005). The Wulong biotite granodiorite and Yangba granodiorite have similar geochemical characteristics and origins (Fig. 8). The spatial distribution of such adakitic rocks indicates that substantial lithospheric thinning took place in

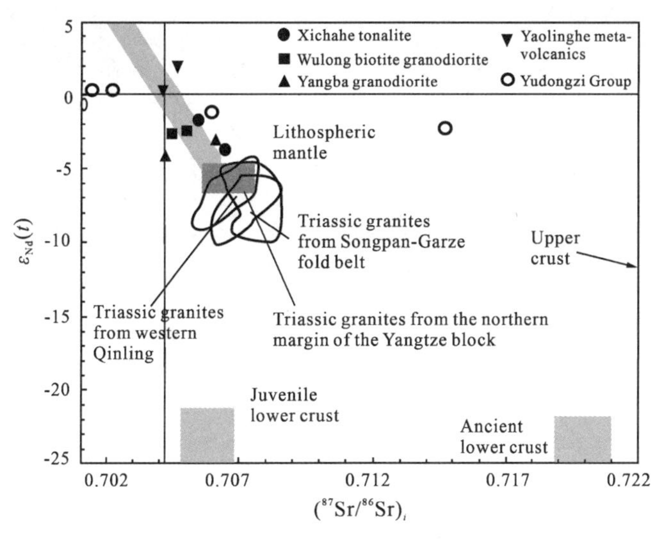

Fig. 8 $(^{87}Sr/^{86}Sr)_i$ vs. $\varepsilon_{Nd}(t)$ plot of the Wulong biotite granodiorite of the Qinling orogenic belt.

The data for Triassic granites from the Songpan-Garzê fold belt are from Zhang Hongfei et al. (2006); the data for the Triassic granites from the western Qinling are from Zhang Hongfei et al. (2005); the data for the Yangba granite are from Zhang Hongfei et al. (2007); the data for the Yudongzi Group are from Zhang Hongfei et al. (2001); the data for Yanglinghe metavolcanic rocks are from Huang and Wu (1990).

the Qinling orogenic belt in the Triassic. The isotopic composition of the Wulong biotite granodiorite is similar to that of the Proterozoic Yaolinghe Group in the South Qinling and its Nd model age (~1.02 Ga) is consistent with the that (1.1 Ga) of the latter (Zhang et al., 2002). The isotope geochemistry of Indosinian granitoids in the South Qinling by Zhang et al. (1997) suggests that the source rock of the Wulong biotite granodiorite is similar to that of the Yaolinghe Group. Therefore, it may be thought that the source rock of the biotite granodiorite is also similar to that of the Yaolinghe Group. However, the Yaolinghe Group now exposed is generally of greenschist facies and no evidence indicates that partial melting occurred (Zhang et al., 2002).

Then we propose a feasible petrogentic model for the Wulong biotite granodiorites: during the Indosinian post-collisional stage, the asthenospheric upwelling and underplating of mantle-derived magmas to the crust-mantle boundary under the Qinling orogenic belt occurred as a result of delamination or subducting slab breakoff, and tempretures at the boundary of hot mantle and amphibole-bearing eclogite may have been high enough to trigge dehydration melting of the amphibole-bearing materials at a pressure of >1.5 GPa, thus resulting in the generation of adakitic magmas (Rapp et al., 1999), and the Wulong biotite granodiorite (Fig. 9).

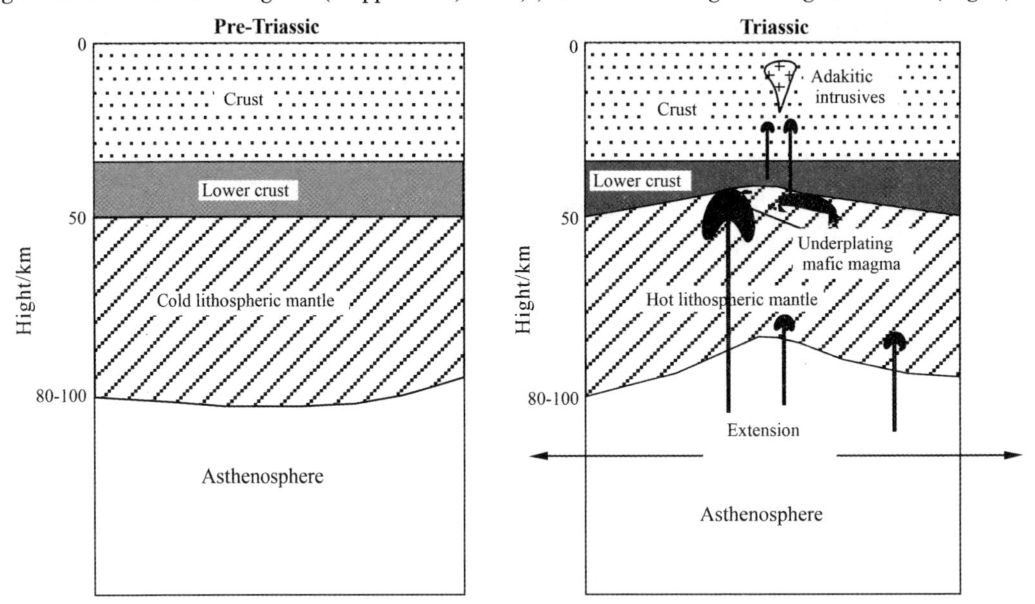

Fig. 9 Genetic model for the Triassic Wulong adakite biotite granodiorite and mafic enclaves (modified from Wang et al., 2006).

(a) The relatively thick lithosphere and thick lower crust before the Triassic. The lower portion of the thick crust is likely composed of amphibole-bearing eclogites. (b) During the Late Triassic, collision between the North China and Yangtze plates resulted in thickening and delamination of the lower crust. The lithospheric mantle was heated by the upwelling asthenosphere. At the same time, the thickened crust was also removed through delamination or foundering. Partial melting of the enriched lithospheric mantle and delaminated lower crust produced the mafic and adakitic magmas respectively.

To sum up, the geodynamic setting of the Wulong adakite biotite granodiorite may be summarized as follows: Collision between the South China and North China blocks during the Triassic resulted in delamination of the thickened lower crust (Zhang et al., 2005) or the subducting slab breakoff at shallow depth along the Qinling orogenic belt (Davies and Von Blanckenburg, 1995), which led to the upwelling of asthenosphere materials and induced partial melting of the thickened mafic lower crust at high pressures, thus forming the Wulong adakite biotite granodiorite.

6 Conclusion

(1) The Wulong biotite granodiorites in the South Qinling has an adakitic affinity and similar Sr-Nd-Pb isotopic compositions to the Proterozoic Yaolinghe Group in the South Qinling, which suggests that they were probably generated by dehydration melting of the thickened mafic crust which contained garnet and little or no feldspar residual phases and isotopically resembled the Yaolinghe Group.The mafic enclaves in the intrusion are of magmatic origin and may represent the mantle-derived underplating mafic magma.

(2) Zircon LA-ICP-MS U-Pb dating of the host biotite granodiorite yield an weighted mean age of 208. 5±2. 2 Ma (MSWD=0. 5, 1σ), which should be the crystalization age of the Wulong biotite granodiorite, which are younger than the Indosinian main orogenic phase (242±21 Ma). Therefore the Wulong pluton may have formed during the late-orogenic or post-collisional stage (≤242±21 Ma) in the South Qinling belt. The formation may be attributed to the partial melting induced by asthenospheric materials resulting from slab breakoff or thickened mafic crust delamination during the post-collisional stage of the Triassic collision between the North China and Yangtze Blocks.

Acknowledgements This work was jointly supported by the National Natural Science Foundation of China grants 40572050, 40272042 and 40234041 and the Teaching and Research Award Program for Outstanding Young Teachers in Higher Education Institutions of the Ministry of Education, China.

References

Annen, C, Sparks, R.S.J., 2002. Effects of repetitive emplacement of basaltic intrusions on thermal evolution and melt generation in the crust. Earth & Planetary Science Letters, 203:937-955.

Beate, B., Monzier, M., Spikings, R., Cotten, J., Silva, J., Bourdon, E., Eissen, J.P, 2001.Mio-Plioeene adakite generation related to flat subduction in Southern Ecuador: The Quimsaeoeha volcanic center. Earth & Planetary Science Letters, l92:561-570.

Bourgois, J., Martin, H., Lagabrielle, Y., LeMoigne, J., Jara, J.F., 1996 Subduction erosion related to spreading-ridge subduction: Taitao peninsula (Chile margin triple junction area). Geology, 24 :723-726.

Chen Weifeng, Chen Peirong, Zhou Xinmin, Huang Hongye, Ding Xing, Sun Tao, 2007. Single zircon LA-

ICP-MS dating of the Guandimiao and Wawutang granitic plutons in Hunan, South China and its petrogenetic significance. Acta Geologica Sinica (English Edition), 81(1):81-89.

Chung Sunling, Liu Dunyi, Ji, J., Chu Meifu, Lee, H.Y, Wen, D.J., Lo, C.H., Lee, T.Y., Qian Qing, Zhang Qi, 2003. Adakites from continental collision zones: Melting of thickened lower crust beneath southern Tibet. Geology, 31:1021-1024.

Condie, K.C., 2005. TTGs and adakites: Are they both slab melts? Lithos, 80:33-44.

Davies, D. W., von Blanckenburg, F., 1995. Slab breakoff: A model of lithosphere detachment and its test in the magmatism and deformation of collisional orogents. Earth & Planetary Science Letters, 129:85-102.

Defant, M.J., Drummond, M.S., 1990. Derivation of some modern arc magmas by partial melting of young subducted lithosphere. Nature, 347:662-665.

Defant, M.J., Drummond, M.S., 1993. Mount St. Helens: Potential example of the partial melting of the subducted lithosphere in a volcanic arc. Geology, 21:547-550.

Defant, M.J., Xu, J.F., Kepezhinskas, P., 2002. Adakites: Some variations on a theme. Acta Petrologica Sinica, 18(2):129-142.

Di Yongjun, Wu Ganguo, Zhang Da, Song Biao, Zang Wenshuan, Zhang Zongyi, Li Jinwen, 2005. SHRIMP U-Pb zircon geochronology of the Xiaotongguanshan and Shatanjiao intrusions and its petrological implication in the Tongling area, Anhui. Acta Geologica Sinica (English Edition), 79 (6): 795- 802.

Drummond, M.S., Defant, M.J., 1990. A model for trondhjemitetonalite-dacite genesis and crustal growth via slab melting: Archean to modern comparisons. Journal of Geophysical Research, 95:21503-21521.

Gao Shan, Rudnick, R.L., Yuan Honglin, Liu Xiaoming, Liu Yongsheng, Xu Wenliang, Ling Wenli, Ayers, J., Wang Xuance, Wang Qinghai, 2004. Recycling lower continental crust in the North China craton. Nature, 432:892-897.

Gao Shan,Liu Xiaoming, Yuan Honglin, 2002. Determination of forty two major and trace element in USGS and NIST SRM glasses by laser ablation inductively coupled plasma-mass spectrometry. Geostandards Newsletter, 26(2):181-195.

Garrison, J.M., Davidson, J.P., 2003. Dubious case for slab melting in the Northern volcanic zone of the Andes. Geology, 31:565-568.

Green, T.H., 1994. Experimental studies of traceelement partitioning applicable to igneous petrogenesis Sedona 16 years later. Chemical Geology, 117:1-36.

Gutscher, M.A., Maury, R.C., Eissen, J.P., Bourdon, E., 1999. Can slab melting be caused by flat subduction? Geology, 28(6):535-538.

Hou Zengqian, Gao Yongfeng, Qu Xiaoming, Rui Ruizhao, Mo Xuanxue, 2004. Origin of adakitic intrusives generated during mid-Miocene east-west extension in southern Tibet. Earth & Planetary Science Letters, 220:139-155.

Huang Xuan, Wu Liren. Nd-Sr isotopes of granitoids from Shaanxi Province and their significance for tectonic evolution. Acta Petroligica Sinica, 2:1-11.

Kay, R.W., Kay, S.M., 2002. Andean adakites: Three ways to make them. Acta Petrologica Sinica, 18: 303-311.

Lai Shaocong, Zhang Guowei, 1996. Geochemical features of ophiolites in Mianxian-Lueyang suture zone,

Qinling Orogenic Belt. Journal of China University of Geoscience, 7:165-172.

Lai Shaocong, Zhang Guowei, Li Sanzhong, 2004. Ophiolites from the Mianlue suture in the Southern Qingling and their relationship with eastern paleotethys evolution. Acta Geologica Sinica, 78(1):107-117.

Lai Shaocong, Zhang Guowei, Dong Yunpeng, Pei Xianzhi, Chen Liang, 2004. Geochemistry and regional distribution of ophiolite and associated volcanics in Mianlüe suture, Qinling-Dabie Mountains. Science in China: Series D, 47:289-299.

Li Shuguang, Sun Weidong, Zhang Guowei, Chen Jiayi, Yang Yongcheng, 1996. Chronology and geochemistry of metavolcanic rocks from Heigouxia Valley in Mianlue tectonic arc, South Qinling: Observation for a Paleozoic oceanic basin and its close time. Science in China: Series D, 39:300-310.

Li Shuguang, Hart, S.R., Zheng, S., Liou, D., Zhang Guowei., Guo, A.L., 1989. Timing of collision between the North and South China Blocks: Sm-Nd isotopic age evidence. Science in China: Series D, 32: 1391-1400.

Li Shuguang, Sun Weidong, 1996. A middle Silurian-early Devonian magmatic arc in the Qinling Mountains of central China: A discussion. Journal of Geology, 104:501-503.

Liu Fulai, Gerdes, Robision, A.P.T., Xue Huaimin, Ye Jianguo, 2007. Zoned zircon from ecologitic lenses in marbles from the Dabie-Sulu UHP terrane, China: A clear record of ultra-deep subduction and fast exhumation. Acta Geologic a Sinica (English Edition), 81 (2): 204-225.

Liu Shuwen, Wang Zongqi, Yan Quanren, Li Qiugen, Zhang Dehui, Wang Jianguo, Yang Bin, Li Bing, Zhang Fengshan, 2006. Indosinian tectonic setting of the Yidu Arc: Constraints from SHRIMP zircon chronology and geochemistry of the dioritic porphyrites and granites. Acta Geologic a Sinica (English edition), 80 (3):387-399.

Martin, H., 1999. The adakitic magmas: Modern analogue of Archean granotoids. Lithos, 46:411-429.

Martin, H, Smithies, R.H., Rapp, R., Moyen, J.F., Champion, D., 2005. An overview of adakite, tonalite-trondhjemite-granodiorite (TTG), and sanukitoid: Relationships and some implications for crustal evolution. Lithos, 79:1-24.

Meng Qingren, Zhang Guowei, 1999. Timing of the collision of the North and South China Blocks: Controversy and reconciliation. Geology, 27:123-126.

Mattauer, M., Matte, P., Malaveile, J., Tapponier, P., Maluslci, H., Xu, Zhang, Q., Ha, Y.L., Tang, Y. Q., 1985. Tectonics of Qinling belt: build-up and evolution of Western Asia. Nature, 1985, 1085(317): 496-500.

Meng Qingren, Zhang Guowei, Yu Zaiping, Mei Zhichao, 1996. Late Paleozoic sedimentation and tectonics of rift and limited ocean basin at southern margin of the Qinling. Science in China: Series D, 39 (Suppl.): 24-32.

Peccerillo, R., Taylor, S., 1976. Geochemistry of Eocene calc-alkaline volcanic rocks from the Kastamonu area, northern Turkey. Contributions to Mineralogy and Petrology, 58:63-81.

Peng Bingxia, Wang Yuejun, Fan Weiming, Peng Touping, Liang Xinquan, 2006. LA-ICPMS zircon U-Pb dating for three indosinian granitic plutons from central Hunan and western Guangdong provinces and its petrogenetic implication. Acta Geologica Sinica (English Edition), 80(5):660-669.

Petford, N., Atherton, M., 1996. Na-rich partial melt from newly underplated basaltic crust: The Cordmera

Blanca Batholith, Peru. Journal of Petrology, 37:491-521.

Petford, N., Gallagher, K., 2001. Partial melting of mafic (amphibolitic) lower crust by periodic influx of basaltic magma. Earth & Planetary Science Letters, 193:483-499.

Peacock, S.M., Rushmer, T., Thompson, A.B., 1994. Partial mehing of subducting oceanic crust. Earth & Planetary Science Letters, 121:227-244.

Prouteau, G., Scaillet, B., Pichavant, M., Maury, R., 2001. Evidence for mantle metasomatism by hydrous silicic melts derived from subducted oceanic crust. Nature 410:197-200.

Qi Liang, Hu, J., Gregoire, D.C., 2000. Determination of trace elements in granites by inductively coupled plasma-mass spectrometry. Talanta, 51:507-513.

Qin Jiangfeng, Lai Shaocong, Li Yongfei, 2005. Petrogenesis and geological significance of Yangba granodiorites from Bikou area, northern margin of Yangtze Block. Acta Petrologica Sinica, 21(3):697-710 (in Chinese).

Rapp, R.P., Watson, E.B., 1995. Dehydration melting of metabasalt at 8 – 32 kbar: Implications for continental growth and crustmantle recycling. Journal of Petrology, 36:891-931.

Rapp, R.P., Shimizu, N., Norman, M.D., Applegate, G.S., 1999. Reaction between slab-derived melts and peridotite in the mantle wedge: Experimental constraints at 3.8 GPa. Chemical Geology, 160:335-356.

Rollinson, H., Martin H., 2005. Geodynamic controls on adakite, TTG and sanukitoid genesis: Implications for models of crust formation. Lithos, 79:9-12.

Sajona, F.G., Maury, R.C., Pubellier, M., 2000. Magmatic source enrichment by slab-derived melts in a young post-collision setting, central Mindanao (Philippines). Lithos, 54:173-206.

Sen, C., Dunnt, T., 1994. Dehydration melting of a basaltic composition amphibolite at 1.5 GPa and 2.0 GPa: Implication for the origin of adakites. Contributions to Mineralogy and Petrology, 117:394-409.

Smithies, R.H., 2000. The Archaean tonalite-trondhjemite-granodiorite (TTG) series is not an analogue of Cenozoic adakite. Earth & Planetary Science Letters, 182:115-125.

Sun Weidong, Li Shuguang, 1998. Pb isotopes of granitoids suggest Devonian accretion of Yangtze (South China) craton to North China craton: Commemt. Geology, 26:859-860.

Sun Weidong, Li Shuguang, Chen Yadong, Li Yujing, 2002. Timing of synorogenic granotoids in the south Qinling, central China: Constraints on the evolution of the Qinling-Dabie Orogenic Belt. Journal of Geology, 110:457-468.

Sun, S.S., McDonough, W.F., 1989. Chemical and isotopic systemmatics of oceanic basalts: Implication for the mantle composition and process. In: Saunder, A.D., Norry, M.J. Magmatism in the ocean basins. Geological Society of London Special Publications, 42:313-345. London: Geological Society of London and Blackwell Scientific Publications.

Topuz, G., Altherr, R., Schwarz, W.H., 2005. Post-collisional plutonism with adakite-like signatures: The Eocene Saraycik granodiorite (Eastern Pontides, Turkey). Contributions to Mineralogy and Petrology, 150:441-455.

Wang Qiang, McDermott, F., Xu Jifeng, Bellon, H., Zhu Yitao, 2005. Cenozoic K-rich adakitic volcanic rocks in the Hohxil area, northern Tibet: Lower-crustal melting in an intracontinental setting. Geology, 33:465-468.

Wang Qiang, Wyman, D.A., Xu Jifeng, Zhao Zhenhua, Jian Ping, Xiong Xiaolin, Bao Zhiwei, Li Chaofeng, Bai Zhenghua, 2006. Petrogenesis of Cretaceous adakitic and shoshonitic igneous rocks in the Luzong area, Anhui Province (eastern China): Implications for geodynamics and Cu-Au mineralization. Lithos, 89: 424-446.

Wolf, M.B, Wyllie, P.J., 1991. Dehydration-melting of solid amphibolite at 10 kbar: Textural development, liquid interconnectivity and applications to the segregation of magmas. Mineralogy and Petrology, 44: 151-179.

Wolf, M.B., Wyllie, P.J., 1994.Dehydration-melting of amphibolite at 10 kbar: The effects of temperature and time. Contributions to Mineralogy and Petrology, 115:369-383.

Wyllie, P.J., Wolf, M.B., 1993. Amphibolite dehydration-melting: sorting out the solidus. In: Prichard, H. M., Alabaster, T., Harris, N.B.W., Neary, C.R. Magmatic processes and plate tectonics. Geological Society of London, Special Publications, 76:405-416.

Xu Jifeng, Shinjio, R., Defant, M.J., Wang Qiang, Rapp, R.P., 2002. Origin of Mesozoic adakitic intrusive rocks in the Ningzhen area of east China: Partial melting of delaminated lower continental crust? Geology, 32:1111-1114.

Yuan Honglin, Wu Fuyuan, Gao Shan, Liu Xiaoming, Xu, P., Sun Deyou, 2003.Determination of zircons from cenozoic intrusions in Northeastern China by laser ablation ICP-MS. Chinese Science Bulletin, 48 (4):1511-1520.

Zhai,X., Day,H.W., Hacker, B.R., You Zhendong, 1998. Paleozoic metamorphism in the Qinling orogen, Tongbai Mountains, central China. Geology, 26: 371-374.

Zhang Guowei,Cheng Shunyou, Guo Anlin, Dong Yunpeng, Lai Shaocong, Yao Anping, 2004. Mianlue paleo-suture on the southern margin of central orogenic system in Qingling-Dabie: With a discussion of the assembly of the main part of the continent of China. Geological Bulletin of China, 23(9-10):846-853 (in Chinese).

Zhang Guowei, Zhang Benren, Yuan Xuecheng, Chen Jiayi, 2002. Qinling Orogenic Belt and Continental Dynamics.Beijing:Science Press: 1-855 (in Chinese).

Zhang Qi, Wang Yan, Qian Qing, Yang Jinhui, Wang, Y.L., Zhao Taiping, Guo, G.J., 2001. The characteristics and tectonic-metaliogenic significances of the adakites in Yanshan period from eastern China. Acta Petrologica Sinica, 17(2):236-244.

Zhang Benren, Luo, T., Gao Shan, Ouyang Jianping, Chen, D., Ma, Z., Han, Y., Gu, X., 1994. Geochemical study of the lithosphere, tectonism and metallogenesis in the Qinling-Dabieshan region. Wuhan: Chinese University of Geoscience Press: 110-122 (in Chinese with English abstract).

Zhang Hongfei, Zhang Li, Harris N, Jin Lanlan, Yuan Honglin, 2006. U-Pb zircon ages, geochemical and isotopic compositions of granitoids in Songpan-Graze fold belt, eastern Tibetan Palteau: Constraints on petrogenesis and tectonic evolution of the basement. Contributions to Mineralogy and Petrology, 152(1): 75-88.

Zhang Chengli, Zhang Guowei, Yan Yunxiang, Wang Yu, 2005. Origin and dynamic significance of Guangyoushan granitic plutons to the north of Mianlue zone in southern Qinling. Acta Petrologica Sinica, 21 (3):711-720 (in Chinese).

Zhang Hongfei, Ouyang Jianping, Ling Wenli, Chen Yuelong, 1997. Pb, Sr, Nd isotopic composition of Ningshan granitoids, South Qinling and their deep geological information. Acta Petrologica et Mineralogica, 16(1):22-31 (in Chinese with English abstract).

Zhang Zongqing, Zhang Guowei, Tang Suohan, Wang Jinhili, 2001. On the age of metamorphic rocks of the Yudongzi Group and the Archean crystalline basement of the Qinling orogen. Acta Geologica Sinica, 75 (2):198-204.

Zhang Hongfei, Jin Lanlan, Zhang Li, Nigel Harris, Zhou Lian, Hu Shenghong, Zhang Benren, 2005. Geochemical and Pb-Sr-Nd isotopic compositions of granitoids from western Qinling belt: Constraints on basement nature and tectonic affinity. Science in China: Series D, 50 (2):184-196.

Zhang Hongfei, Xiao Long, Zhang Li, Yuan Honglin, Jin Lan lan, 2007. Geochemical and Pb-Sr-Nd isotopic compositions of Indosinian granitoids from the Bikou Block, northwest of the Yangtze plate: Constraints on petrogenesis, nature of deep crust and geodynamics. Science in China: Series D, 50 (7): 972-983.

Slab breakoff model for the Triassic post-collisional adakitic granitoids in the Qinling orogen, central China: Zircon U-Pb ages, geochemistry, and Sr-Nd-Pb isotopic constraints①

Qin Jiangfeng Lai Shaocong② Li Yongfei

Abstract: Triassic granites in the Qinling orogen of central China provide insights into regional tectono-magmatic events and tectonic evolution. Zircon LA-ICP-MS U-Pb dating show that the Yangba, Xichahe, Wulong and Guangtoushan plutons have magmatic crystallization ages of 209 ± 2 Ma, 212 ± 2 Ma and 208 ± 2 Ma, respectively, slightly younger than the time of collision between the Yangtze and North China cratons (242 Ma). Major- and trace-element geochemistry indicate that the granitoids have adakitic signatures, suggesting that they were generated by partial melting of a thickened lower crust. Geochemistry and petrology of mafic microgranular enclaves in some of the plutons indicate that they were derived from partial melting of the subcontinental lithospheric mantle (SCLM) and are genetically related to the host adakitic granites. We proposed that slab-breakoff explains the spatial, age and geochemistry characters of Triassic magamtism in the Qinling orogen. In summary, Triassic collision between the Yangtze and North China cratons caused crustal thickening in the Qinling-Dabie orogen to > 50 km, with breakoff of the subducting oceanic slab occuring at shallow depth. This caused: ① Upwelling of the asthenosphere along the Mianlue suture, with initial melting at the base of the lower crust, producing mafic melt; ② Mantle-derived melt underplated the thickened lower crust so that temperatures at the boundary between it and the mantle were high enough to trigger dehydration melting of garnet-bearing rock at a pressure of > 1.5 GPa, resulting in the generation of adakitic magma with high Sr, Ba and Sr/Y ratios, no negative Eu anomalies, and leaving eclogite as a residue; ③ Moreover, given that lithospheric mantle-derived magmas intruded the adakitic magma chamber, interaction between pristine adakitic magmas and mantle-derived magmas result in formation of the mafic enclaves, providing the adakitic magma with a mantle/crustal component, e.g. high $Mg^{\#}$ and Cr and Ni contents. We concluded that partial melting of lithospheric mantle played an important role in generation of the Triassic adakitic granites in the Qinling orogen, reflecting post-collisional slab-breakoff.

---

① Published in *International Geology Review*, 2008, 50.

② Corresponding author.

Introduction

Eastern Asian is divided into the North and South China (Yangtze) blocks by the Qinling-Dabie orogen. Accurate knowledge of the structure and tectonic history of the Qinling-Dabie orogen can place firmer constraints on the evolution of the Chinese continent (e.g., Mattauer et al., 1985). Triassic collision between the Yangtze and North China cratons (Li et al., 1996; Zhang et al., 2001) along the Qinling-Dabie orogen is generally accepted to have been due to the closure of the Mian-Lueyang ocean, which was probably an eastward branch of the Paleo-Tethys (Zhang et al., 2001). This model was advanced on the basis of clear field evidence of deformation, ophiolite geochemistry, and strata in the Mianxian-Lueyang (Mianlve) sector of the orogen (Zhang et al., 2002; Lai et al., 2004a,b). The model involved subduction of the Mianlue oceanic crust beneath the North China craton in the Late Palaeozoic. This subduction polarity continued until Early Triassic times, resulting in the closure of the Paleo-Tethys ocean, and is an integral element in the Triassic granite magmatic evolution of the South Qinling belt.

Triassic granites in the Qinling orogen comprise a ~400 km long granitoid belt along the Mianlue suture (Fig. 1). The origin of these granitoids in the South Qinling orogen is one of the most important constraints on the evolution of the Qinling-Dabie orogen. However, this origin and tectonic implications have long been debated (e.g., Zhang et al., 1994, 2002, 2005; Wang et al., 2003, 2005; Qin et al., 2005). On the basis of field observations and geochronology, Zhang et al. (2001) proposed that Triassic granite magmatism in South Qinling was caused by the northward subduction of Paleo-Tethys oceanic crust. However, new chronological data suggest that contemporaneous granite magmatism also occurred in the northern margin of the Yangtze craton and the North Qinling Block (Wang et al., 2003, 2005; Qin et al., 2005), then oceanic crust subduction is not a reasonable model for the genesis of the Triassic granites in the Qinling orogen.

Zhang et al. (2005) argued that the Triassic granites were post-collisional and resulted from the delamination of thickened orogenic crust during the post-orogenic stage of the Triassic Qinling orogen. By means of high-precision U-Pb zircon dating, Sun et al. (2002) carried out geochronological work on six granite plutons in the South Qinling. They obtained ages ranging from 220 Ma to 205 Ma, and argued that slab break-off in the South Qinling happened at a very shallow depth and disturbed the asthenosphere greatly, finally, leading to the formation of synorogenic granitoids. If the Triassic granites resulted from delamination of thickened orogenic crust or slab breakoff at a shallow depth, then evidence of such could be recorded in spatial variations in the geochemistry of Triassic granites emplaced in the Qinling orogen. We test this possibility through new geo chronology, geochemistry and Sr-Nd-Pb isotope data from Triassic

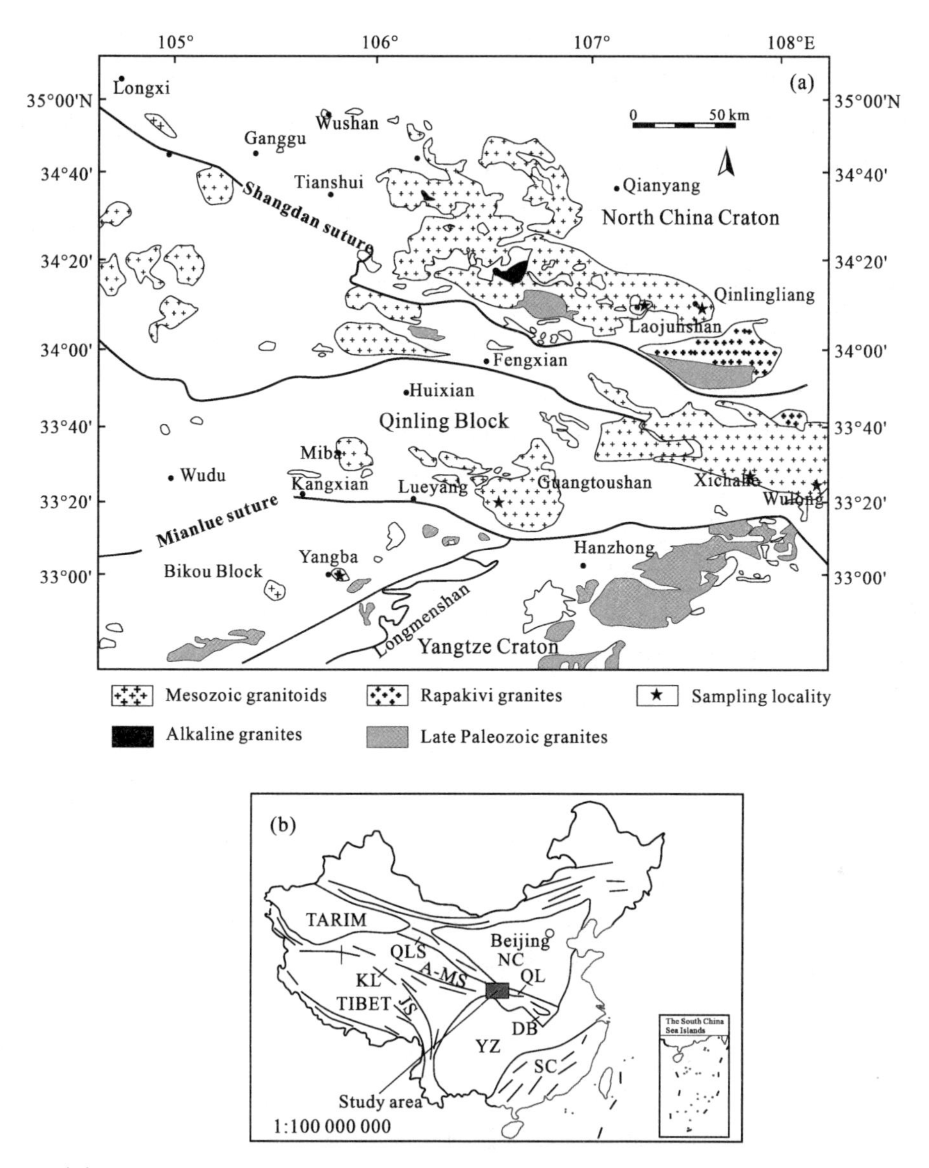

Fig. 1 (a) Geological sketch map of the Qinling orogenic belt showing distribution of Triassic granites (after Zhang et al., 2001); (b) Location of the study area.

NC: North China craton; YC: Yangtze craton; SC: South China Orogenic Belt; KL: Kunlun Mountains; QLS: Qinlian Mountains; QL: Qinling Mountains; DB: Dabie Mountains (after Zhang et al., 2006).

granitic rocks and mafic microgranular enclaves from Yangba, Xichahe, Wulong and Guangtoushan plutons in the northern margin of the Yangtze craton and the South Qinling, central China.

The results suggest that these granotoids have adakitic affinity, which may have been derived from partial melting of thickened lower crust during the Triassic post-collisional stage of the Qinling orogen, and the mafic enclaves were generated by partial melting of metasomatized sub-continental lithospheric mantle (SCLM). When combined with data from the Triassic granitoids in the North Qinling block, we propose that partial melting of the SCLM and magma mixing play important roles in the generation of the adakitic granites, which resulted from slab-breakoff during the post-collisional stage of the Qinling orogen. Our results also have important implications for the tectonic evolution of the Qinling orogen during Late Triassic time.

Geological setting and sampling location

To understand the geotectonic setting of Triassic granite magmatism, it is necessary to highlight some significant aspects of the Triassic Qinling-Dabie orogen, particularly involving collisional stages. The Qinling orogen is a multistage orogen with two mountain chains and two sutures. The north suture lies mainly along the Shangdan tectonic belt, whereas the south suture consists of the Mianlue ophiolite complex in the west and is presumably marked by the Bashan fault in the east (Zhang et al., 2001, 2004; Lai and Zhang, 1996, 2004a,b; Meng and Zhang, 1999; Sun et al., 2002). The North Qingling lies north of the Shangdan suture, and comprise Middle Paleozoic medium-grade metasedimentary and metavolcanic rocks (Li et al.,1989; Li and Sun, 1996; Sun and Li, 1998; Zhai et al., 1998). The South Qinling lies between the Shangdan and Mianlue sutures, and consists mainly of Late Paleozoic medium-grade metasedimentary and metavolcanic rocks (Mattauer et al., 1985) and Triassic granitoids (Li and Sun, 1996; Li et al.,1996; Meng and Zhang, 1999). South of the Mianlue suture lies the Yangtze craton (Fig. 1).

Triassic granites are widespread in the South Qinling. Four granite plutons have been sampled for this study (Fig. 1). The Yangba pluton occurs at the northern margin of the Yangtze craton with an outcrop area of 30–40 km^2, and is emplaced into the Precambrian Bikou Group. Mafic microgranular enclaves (MME) are widespread, and their shapes are rounded or ovoid, although some are irregular. They are commonly 20–50 cm in diameter, and the contacts of enclaves and host granotoid are sharp. The host granitoid is mainly monzogranite, which has equi-granular texture, massive structure, and is composed of plagioclase (~40% with An = 25 – 35), orthoclase (~10%), perthite (~20%), quartz (~15%), biotite (~10%), and hornblende (~10%); accessory phases include zircon, apatite, allanite, and magnetite. The mafic enclaves have hypidiomorphic-allotriomorphic textures, some with plagioclase and quartz megacrysts from the host granodiorite. The matrix is formed by laths of plagioclase, biotite, hornblende, interstitial or poikilitic quartz, rare orthoclase, apatite, zircon, ilmenite, and rutile.

The Xichahe pluton crops out over an area of ~24 km^2 and intruded a Neoproterozoic-Silurian meta-sedimentary association of the Qinling block. The main mass consists of medium-grained tonalite with a massive structure, being made up of plagioclase (~45% with An=25–

35), hornblende (~30%), orthoclase (~10%), quartz (~10%), and biotite (~5%), and dominant accessory phases including zircon, apatite, allanite, and magnetite.

The Wulong pluton is over ~1 800 km^2 and is located 60 km east of the Xichahe pluton. It is partly surrounded by a Proterozoic metamorphic complex known as the Foping Group (Zhang et al., 2001). Three phases of grantoid are present, an inequigranular porphyritic monzogranite in the center, medium-grained biotite granodiorite in the rim, and a transition belt of amphibole-bearing monzogranite. The biotite granodiorite has a massive structure and is composed of plagioclase (45%–50% with An=25–35), hornblende (10%–15%), orthoclase (20%), quartz (~10%), and biotite (~5%). The monzogranite has a coarse-grained texture and is composed of quartz (25%–30%), orthoclase (30%–40%), plagioclase (30%–35%), and biotite (5%). The dominant accessory phases are zircon, apatite, allanite, and magnetite. In the southern part of the pluton, there are a few mafic enclaves with an ovoid shape (up to 20 cm in diameter); their contacts with host granodiorite vary from sharp to gradational. Feldspar or quartz megacrysts from the host monzogranite occur in the enclaves.

The Guangtoushan pluton covers an area of 900 km^2 in the Qinling belt and consists of two phases. The inner phase is mainly biotite granodiorite and the outer phase in fine-grained trondhjemite (Sun et al., 2002). The biotite granodiorite consists of plagioclase (50%–60%), orthoclase (10%), quartz (15%–20%), and biotite (10%). Accessory phases include zircon, apatite, allanite, and magnetite.

Results

U-Pb zircon LA-ICP MS dating

Zircon U-Pb analysis results of the Yangba monzogranite (YB-01), Xichahe tonalite (XH-02), and Wulong monzogranite (WL-04) are listed in Table 1; representative zircon CL images with spot analyses are shown in Fig. 2 and Fig. 3.

Monzogranite YB-1 from the Yangba pluton was chosen for zircon U-Pb analysis. Zircons are mainly light pink to colourless euhedral crystals (Fig. 2a) with perfect pyramids and prisms. The zircons range in length from ~150 μm to ~250 μm with length/width ratios from 2:1 to 4:1. The zircon CL images (Fig. 2a) show strong oscillatory zoning, being indicative of magma zircon. Some zircons contain the inherited cores. Fifteen analyses (not including inherited cores) of zircon from the Yangba monzogranite display U = 439–1 355 ppm, Th = 210–1 388 ppm, Th/U = 0.48–0.54, $^{206}Pb/^{238}U$ ages ranging from 215–202 Ma with a $^{206}Pb/^{238}U$ weighted mean age of 209±2 Ma (MSWD=0.87, 2σ) (Fig. 3a), which should be interpreted as the crystallization age of the Yangba granodiorite pluton.

Table 1 Results of zircon LA-ICP-MS U-Pb results for the Yangba, Wulong and Xichahe Plutons[1].

Sample no.	Content/ppm			Ratios						Age/Ma					
	U	Th	Th/U	$^{207}Pb^*/^{206}Pb^*$	2σ	$^{207}Pb^*/^{235}U$	2σ	$^{206}Pb^*/^{238}U$	2σ	$^{207}Pb^*/^{206}Pb^*$	2σ	$^{207}Pb^*/^{235}U$	2σ	$^{206}Pb^*/^{238}U$	2σ
YB-01-01	960	712	0. 74	0. 045	0. 001	0. 201	0. 005	0. 033	0. 000	−30	35	186	4	207	2
YB-01-02	513	255	0. 50	0. 046	0. 002	0. 205	0. 007	0. 033	0. 000	−15	55	189	6	206	2
YB-01-03	439	210	0. 48	0. 049	0. 002	0. 226	0. 008	0. 034	0. 000	144	64	207	7	213	2
YB-01-04	866	673	0. 78	0. 045	0. 001	0. 202	0. 006	0. 033	0. 000	−16	42	187	5	207	2
YB-01-05	843	618	0. 73	0. 047	0. 002	0. 205	0. 008	0. 032	0. 000	31	64	190	7	203	2
YB-01-06	708	518	0. 73	0. 048	0. 002	0. 214	0. 009	0. 033	0. 000	86	73	197	8	207	2
YB-01-07	926	697	0. 75	0. 049	0. 001	0. 218	0. 006	0. 032	0. 000	155	45	200	5	204	2
YB-01-08	1 355	882	0. 65	0. 047	0. 001	0. 221	0. 005	0. 034	0. 000	70	41	203	4	215	2
YB-01-09	934	1388	1. 49	0. 051	0. 001	0. 235	0. 006	0. 034	0. 000	236	46	214	5	213	2
YB-01-10	1 034	555	0. 54	0. 049	0. 002	0. 224	0. 007	0. 033	0. 000	145	57	205	6	211	2
YB-01-11	826	582	0. 70	0. 048	0. 002	0. 224	0. 007	0. 034	0. 000	116	54	205	6	213	2
YB-01-12	1 015	690	0. 68	0. 051	0. 001	0. 236	0. 006	0. 034	0. 000	240	47	215	5	213	2
YB-01-13	857	737	0. 86	0. 051	0. 002	0. 227	0. 008	0. 033	0. 000	225	61	208	7	206	2
YB-01-14	817	573	0. 70	0. 053	0. 002	0. 241	0. 009	0. 033	0. 000	341	62	219	8	208	3
YB-01-15	629	340	0. 54	0. 055	0. 002	0. 241	0. 010	0. 032	0. 000	402	67	219	8	202	3
XH-02-03	213	116	0. 54	0. 049	0. 002	0. 226	0. 010	0. 034	0. 000	128	82	207	9	214	3
XH-02-04	215	107	0. 49	0. 057	0. 004	0. 252	0. 017	0. 032	0. 000	476	161	229	14	205	3
XH-02-05	194	94	0. 49	0. 055	0. 003	0. 245	0. 011	0. 033	0. 000	392	80	223	9	207	3
XH-02-06	205	101	0. 49	0. 052	0. 002	0. 244	0. 010	0. 034	0. 000	299	75	222	8	215	2
XH-02-07	339	62	0. 18	0. 059	0. 006	0. 269	0. 027	0. 033	0. 000	561	233	242	22	210	3
XH-02-09	240	124	0. 52	0. 055	0. 002	0. 251	0. 010	0. 033	0. 000	397	67	227	8	211	2
XH-02-10	406	248	0. 61	0. 060	0. 003	0. 261	0. 012	0. 032	0. 000	585	108	235	10	202	2
XH-02-11	276	195	0. 71	0. 050	0. 002	0. 240	0. 009	0. 034	0. 000	214	69	218	8	219	2
XH-02-12	227	134	0. 59	0. 051	0. 003	0. 245	0. 012	0. 035	0. 000	239	93	222	10	221	3
XH-02-13	249	145	0. 58	0. 055	0. 004	0. 242	0. 019	0. 032	0. 000	402	187	220	16	203	3

Continued

Sample no.	Content/ppm			Ratios						Age/Ma					
	U	Th	Th/U	$^{207}Pb^*/^{206}Pb^*$	2σ	$^{207}Pb^*/^{235}U$	2σ	$^{206}Pb^*/^{238}U$	2σ	$^{207}Pb^*/^{206}Pb^*$	2σ	$^{207}Pb^*/^{235}U$	2σ	$^{206}Pb^*/^{238}U$	2σ
XH-02-16	302	192	0. 63	0. 057	0. 003	0. 278	0. 015	0. 035	0. 000	503	124	249	12	223	3
XH-02-17	278	142	0. 51	0. 049	0. 002	0. 229	0. 011	0. 034	0. 000	125	88	209	9	217	3
XH-02-18	193	99	0. 51	0. 052	0. 003	0. 241	0. 015	0. 033	0. 001	302	116	219	13	212	3
XH-02-19	277	163	0. 59	0. 056	0. 003	0. 265	0. 015	0. 034	0. 000	447	134	238	12	218	3
XH-02-21	241	97	0. 40	0. 053	0. 004	0. 240	0. 019	0. 033	0. 001	326	184	219	15	209	3
XH-02-22	201	104	0. 52	0. 052	0. 004	0. 235	0. 017	0. 033	0. 001	261	129	214	14	210	4
XH-02-23	147	71	0. 48	0. 056	0. 005	0. 262	0. 023	0. 034	0. 001	444	206	237	19	216	4
XH-02-24	264	145	0. 55	0. 048	0. 003	0. 229	0. 013	0. 034	0. 001	119	104	209	11	217	3
XH-02-28	211	119	0. 56	0. 053	0. 003	0. 248	0. 014	0. 034	0. 001	330	102	225	11	215	3
XH-02-29	284	144	0. 51	0. 056	0. 003	0. 266	0. 013	0. 034	0. 000	452	82	240	10	218	3
XH-02-30	206	104	0. 50	0. 046	0. 003	0. 213	0. 015	0. 033	0. 001	9	165	196	13	212	3
XH-02-31	377	143	0. 38	0. 051	0. 003	0. 229	0. 012	0. 033	0. 000	243	130	209	10	207	2
XH-02-33	308	154	0. 50	0. 056	0. 004	0. 252	0. 019	0. 033	0. 001	438	181	228	16	208	3
XH-02-34	194	106	0. 54	0. 057	0. 004	0. 265	0. 019	0. 034	0. 001	485	170	239	16	214	3
XH-02-36	361	292	0. 81	0. 055	0. 002	0. 252	0. 011	0. 033	0. 000	422	75	228	9	210	2
XH-02-40	211	109	0. 51	0. 054	0. 004	0. 251	0. 016	0. 034	0. 001	379	115	227	13	213	3
XH-02-43	223	108	0. 48	0. 055	0. 003	0. 252	0. 015	0. 033	0. 001	421	108	228	12	210	3
WL-04-02	788	644	0. 82	0. 050	0. 001	0. 222	0. 005	0. 032	0. 000	180	41	204	4	206	1
WL-04-04	736	459	0. 62	0. 051	0. 001	0. 229	0. 004	0. 033	0. 000	217	29	210	3	209	1
WL-04-05	486	215	0. 44	0. 050	0. 001	0. 225	0. 005	0. 033	0. 000	189	42	206	5	208	2
WL-04-06	612	230	0. 38	0. 053	0. 001	0. 239	0. 005	0. 033	0. 000	308	35	217	4	209	1
WL-04-07	452	210	0. 46	0. 052	0. 001	0. 233	0. 006	0. 032	0. 000	296	42	212	5	205	2
WL-04-13	661	300	0. 45	0. 050	0. 002	0. 230	0. 010	0. 033	0. 000	215	106	210	8	210	2
WL-04-15	466	255	0. 55	0. 054	0. 002	0. 250	0. 007	0. 034	0. 000	364	50	226	6	213	2
WL-04-22	616	402	0. 65	0. 051	0. 001	0. 229	0. 005	0. 033	0. 000	226	38	209	4	208	1

[1] * = Radiogenic Pb.

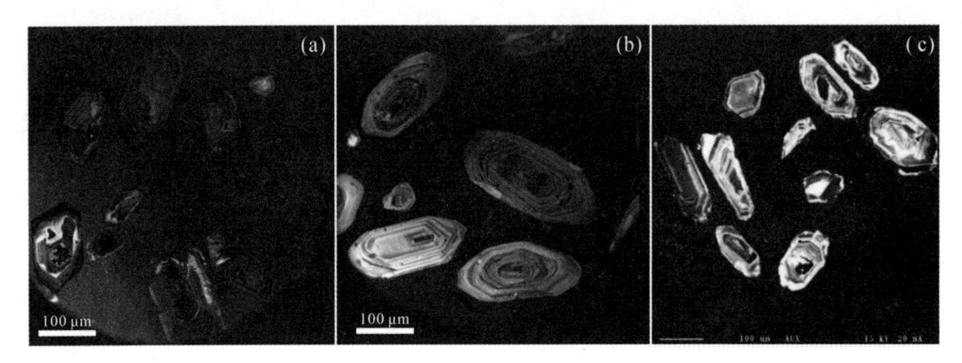

Fig. 2 Cathodoluminescence(CL) electron images of zircons from the (a)Yangba, (b)Xichahe, and (c)Wulong pluton. The black circle represent the site of the ICP analyses.

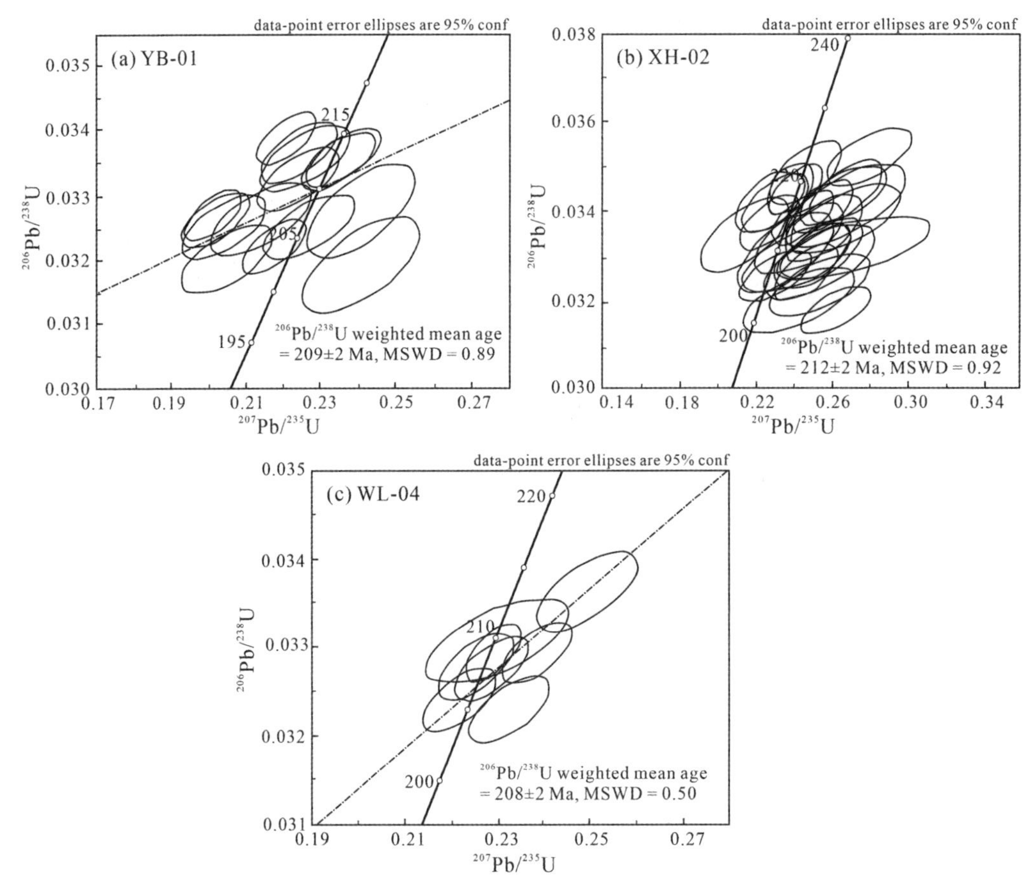

Fig. 3 LA-ICP-MS U-Pb concordia diagrams for zircon from (a)Yangba, (b)Xichahe and (c)Wulong plutons.

Tonalite XH-02 from the Xichahe pluton was chosen for zircon U-Pb analyses; the zircons are mostly euhedral crystals and display oscillatory zoning in the CL images (Fig. 2b). Twenty-seven LA-ICP-MS analysis of the zircons display U = 193–406 ppm, Th = 62–248 ppm, Th/U

$=0.38-0.63$, being indicative of magma zircons. They have $^{206}Pb/^{238}U$ ages ranging from 221–202 Ma, with a weighted mean age of 212 ± 2 Ma (MSWD=0.92, 2σ) (Fig. 3b), which should be regarded as the crystallization age of the Xichahe tonalite.

Monzogranite WL-04 from the Wulong pluton was chosen for zircon U-Pb analysis; the zircons are pink to colourless, mainly euhedral crystals with length/width ratios ranging from 2∶1 to 3∶1. Some zircons display good oscillatory zoning in the CL image (Fig. 2c), being indicative of a magmatic origin. Eight analyses of zircons from the Wulong monzogranite (not including inherited cores) display U=452–788 ppm, Th=210–644 ppm, Th/U=0.38–0.82, and have $^{206}Pb/^{238}U$ ages ranging from 213 Ma to 206 Ma, with a $^{206}Pb/^{238}U$ weighted mean age of 208 ± 2 Ma (MSWD=0.50, 2σ) (Fig. 3c), which should be the best estimate of crystallization age of the Wulong monzogranite.

Major and trace elements

Major and trace element compositions of representative samples from the Yangba, Wulong, Xichahe, and Guangtoushan plutons are given in Tables 2–4.

(1) *Yangba pluton.* According to the classification scheme of Frost et al. (2001), the Yangba monozogranite is magnesian, calc- alkaline, and metaluminous (Table 2), the $Mg^{\#}$ [$Mg^{\#}=Mg/(Mg+Fe)=50.8-54.5$] >45, implies a mantle component in the source region (Rapp et al., 1999). Chondrite-normalized REE patterns of the monozogranite have HREE concave-upward shapes and are characterized by the general absence of a significant Eu anomaly with Eu/Eu^{*} ($=\sqrt{Eux\cdot Euy/(Smx\cdot Gdy)}$) ranging from 0.84 to 0.91 (Fig. 4a). They are enriched in Sr (>900 ppm) and Ba (883–1 483 ppm), and depleted in HREE, Y and high-field-strength elements (HFSE) with Y=9.51–14.5 ppm, Yb=0.74–1.20 ppm, Y/Yb=11.12–15.10, and Sr/Y=65–95 (Table 2). Primitive mantle-normalized element concentration diagrams show significant negative anomalies of Nb-Ta, P, and Ti, in conjunction with positive anomalies in Ba and Sr (Fig. 4b).

The mafic microgranular enclaves in the Yangba pluton display some shoshonitic characteristics possessing $SiO_2<63\%$, σ (4.54–6.18) > 3.3, high K_2O content (4.22%–6.04%), $K_2O/Na_2O>1$, $Mg^{\#}=60-64$. In the primitive mantle-normalized element diagrams, they show positive anomalies in Rb, Ba, U (Th), K, La, Nd, and negative anomalies in Nb, Ta, P, and Ti; $\sum REE=357-655$ ppm, $(La/Yb)_N=32.82-51.78$; $Eu/Eu^{*}=0.76-0.85$ (Fig. 4c,d). This suggests that the mafic magmas were likely derived from metasomatized lithospheric mantle.

(2) *Xichahe pluton.* The high MgO content ($Mg^{\#}=53-67$) of the Xichahe tonalite metaluminous-aluminous series implies a mantle component in its source (e.g., Rapp et al., 1999). The Xichahe tonalite has high Sr, Ba, low Y and Yb, with the exception of two

Table 2 Analytical results of major(%) and trace(ppm) element from the Yangba pluton.

Sample	Granodiorite											Mafic enclaves					
	YB-01[a]	YB-02	YB-03	YB-04	YB-05	YB-06	YB-07	YB-08	YB-09	YB-10	YB-11	YB-17	YB-20	YB-21	YB-22	YB-23	YB-24
SiO_2	67.92	68.55	67.97	66.60	67.49	67.26	67.53	68.19	67.73	67.87	69.08	57.41	61.33	56.39	61.08	61.15	56.42
TiO_2	0.33	0.32	0.30	0.36	0.37	0.36	0.34	0.33	0.31	0.30	0.26	0.79	0.56	0.92	0.57	0.54	0.93
Al_2O_3	15.87	16.05	16.36	16.18	15.84	15.81	15.96	15.66	16.23	16.04	16.48	16.01	15.14	17.36	15.21	15.14	17.36
$Fe_2O_3^T$	2.68	2.62	2.38	2.96	2.65	2.84	2.55	2.53	2.42	2.47	2.05	5.69	4.69	5.69	4.65	4.49	5.75
MnO	0.05	0.05	0.05	0.05	0.05	0.05	0.05	0.05	0.05	0.05	0.04	0.09	0.10	0.10	0.10	0.10	0.10
MgO	1.51	1.42	1.43	1.76	1.53	1.65	1.49	1.44	1.40	1.42	1.06	4.16	3.39	3.56	3.45	3.39	3.62
CaO	2.80	2.68	2.89	2.91	2.75	2.90	2.70	2.71	2.76	2.72	2.29	4.80	4.33	5.14	4.45	4.42	5.12
Na_2O	4.80	4.56	4.92	4.73	4.59	4.66	4.53	4.56	4.76	4.64	4.98	4.07	3.41	4.90	3.31	3.35	4.89
K_2O	3.44	3.67	3.22	3.59	3.62	3.43	3.84	3.47	3.57	3.62	3.71	5.21	5.72	4.27	6.04	5.88	4.22
P_2O_5	0.17	0.18	0.17	0.21	0.18	0.18	0.18	0.17	0.16	0.17	0.12	0.82	0.52	0.71	0.53	0.52	0.71
LOI	0.36	0.36	0.34	0.38	0.44	0.38	0.35	0.41	0.45	0.41	0.40	0.63	0.34	0.48	0.58	0.53	0.58
Total	99.93	100.46	100.03	99.73	99.51	99.52	99.52	99.52	99.84	99.71	100.47	99.68	99.52	99.52	99.97	99.51	99.70
A/CNK	0.95	0.98	0.97	0.95	0.96	0.95	0.97	0.97	0.97	0.97	1.01	0.76	0.77	0.79	0.76	0.76	0.79
σ[b]	2.72	2.65	2.65	2.93	2.75	2.70	2.86	2.56	2.81	2.74	2.90	5.98	4.55	6.28	4.84	4.69	6.18
$Mg^{\#}$	53	52	55	54	54	54	54	53	54	53	51	59	59	56	60	60	56
Ba	892	1288	899	883	1 224	1 131	1 483	1 123	990	1 370	1 142	1 967	2 219	1 462	2 405	2 341	1 481
Rb	86.2	81.7	89.2	98.0	95.5	100	88.7	96.7	95.3	96.9	95.4	126	231	108	230	208	125
Sr	934	972	1 047	916	1 000	1 014	1 015	1 002	1 033	1 037	905	1 906	877	1 109	843	854	1 089
Y	12.7	11.7	11.2	13.2	14.5	13.8	13.0	12.6	12.5	12.7	9.5	27.1	25.6	22.1	25.8	24.6	22.0
Zr	154	165	170	171	150	175	159	155	150	167	151	325	276	249	236	260	253
Nb	9.53	8.50	8.17	10.30	10.60	10.40	9.28	9.09	9.27	8.83	8.44	21.1	14.6	30.1	14.9	14.2	30.7
Th	13.9	16.7	11.5	15.1	14.1	13.8	17.5	15.3	19.0	17.3	13.7	49.4	36.2	22.0	34.6	35.2	22.5
Pb	26.5	24.7	22.7	27.2	30.0	28.8	30.0	30.0	30.1	31.4	30.9	30.8	76.5	29.7	81.1	78.1	31.4

Continued

Sample	Granodiorite											Mafic enclaves					
	YB-01[a]	YB-02	YB-03	YB-04	YB-05	YB-06	YB-07	YB-08	YB-09	YB-10	YB-11	YB-17	YB-20	YB-21	YB-22	YB-23	YB-24
Ga	19.8	19.7	20.6	20.3	19.7	20.1	19.4	20.2	20.5	20.8	20.4	23.4	21.0	22.2	20.0	19.7	22.4
Ni	16.9	16.1	15.7	17.3	16.9	15.8	14.0	15.5	13.5	14.7	10.6	22.2	32.5	13.7	31.9	30.8	12.1
V	49.3	48.7	42.8	55.0	44.9	49.9	43.5	44.7	43.2	43.4	31.8	131.0	99.0	130.0	96.8	94.9	131
Cr	27.2	22.9	28.1	22.0	29.0	26.8	21.5	30.0	20.3	23.9	14.3	31.8	44.9	21.0	48.0	38.4	9.78
Hf	3.70	3.59	3.35	3.61	3.92	4.51	3.98	4.14	3.88	4.31	4.15	7.79	6.83	5.90	6.22	6.38	5.99
Cs	2.78	2.77	2.28	4.56	2.76	3.44	2.80	2.39	2.81	2.33	2.64	6.10	4.17	4.21	3.68	3.59	4.46
Sc	5.32	4.53	5.02	5.14	5.37	5.93	4.99	5.17	4.69	5.21	3.19	13.70	12.50	12.60	12.30	11.50	13.10
Ta	0.65	0.54	0.50	0.69	0.75	0.74	0.68	0.73	0.67	0.70	0.65	1.11	0.79	1.63	0.81	0.78	1.64
Co	135	128	154	140	166	170	161	197	172	171	169.0	59.5	69.5	43.5	57.6	65.0	46.80
Li	40.9	38.1	32.6	46.1	37.3	40.3	36.8	39.0	38.4	38.4	47.6	62.8	33.8	64.1	24.6	24.9	64.80
Be	2.94	2.52	2.47	3.33	2.60	2.88	2.65	2.77	2.80	2.75	3.12	4.30	4.64	3.21	4.39	4.52	3.07
U	1.95	1.74	1.79	1.56	2.13	2.57	2.11	1.62	2.10	2.02	1.58	4.98	3.67	4.71	3.37	3.58	4.34
La	34.0	37.2	31.9	36.5	41.9	40.4	47.0	42.2	37.3	41.5	30.5	151	102	84.8	102	99.8	89.7
Ce	65.8	69.9	59.2	68.5	80.3	77.2	85.6	76.7	70.5	76.8	55.2	293	192	162	195	193	166
Pr	7.13	7.15	6.19	7.34	8.76	8.28	8.88	8.13	7.68	8.14	5.96	35.2	21.0	17.1	21.6	21.0	17.4
Nd	27.2	26.0	22.8	27.0	33.2	31.8	33.1	30.5	29.2	30.4	22.5	122	79.7	62.6	81.9	78.7	64.0
Sm	4.44	3.90	3.50	4.30	5.48	5.21	5.15	4.81	4.69	4.79	3.40	19.40	13.70	9.90	14.0	13.5	10.0
Eu	1.15	1.06	0.96	1.08	1.41	1.33	1.35	1.24	1.21	1.29	0.94	4.42	3.09	2.55	3.29	3.17	2.61
Gd	3.77	3.36	2.96	3.58	4.52	4.30	4.42	4.10	3.93	4.06	2.93	15.3	11.2	8.55	11.5	11.1	8.73
Tb	0.42	0.36	0.32	0.40	0.55	0.52	0.51	0.48	0.47	0.47	0.34	1.54	1.25	0.95	1.28	1.22	0.97
Dy	2.03	1.76	1.56	1.95	2.58	2.47	2.37	2.25	2.23	2.23	1.63	6.31	5.35	4.31	5.42	5.25	4.37
Ho	0.38	0.31	0.28	0.35	0.44	0.43	0.42	0.39	0.39	0.39	0.29	0.90	0.83	0.72	0.82	0.81	0.72
Er	0.99	0.83	0.75	0.92	1.21	1.17	1.12	1.06	1.07	1.07	0.81	2.40	2.13	1.92	2.12	2.05	1.93

Continued

Sample	Granodiorite											Mafic enclaves					
	YB-01[a]	YB-02	YB-03	YB-04	YB-05	YB-06	YB-07	YB-08	YB-09	YB-10	YB-11	YB-17	YB-20	YB-21	YB-22	YB-23	YB-24
Tm	0. 14	0. 12	0. 11	0. 14	0. 17	0. 17	0. 16	0. 15	0. 16	0. 16	0. 12	0. 29	0. 28	0. 25	0. 27	0. 27	0. 26
Yb	0. 96	0. 82	0. 74	0. 92	1. 20	1. 18	1. 11	1. 06	1. 07	1. 09	0. 85	1. 89	1. 83	1. 67	1. 78	1. 76	1. 71
Lu	0. 16	0. 13	0. 12	0. 15	0. 19	0. 19	0. 18	0. 18	0. 18	0. 18	0. 15	0. 27	0. 27	0. 25	0. 26	0. 26	0. 25
Sr/Y	74	83	93	69	69	73	78	80	83	82	95	70	34	50	33	35	49
Y/Yb	13	14	15	14	12	12	12	12	12	12	11	14	14	13	15	14	13
La/Yb	35	46	43	40	35	34	42	40	35	38	36	80	55	51	58	57	52
Eu/Eu*	0. 86	0. 90	0. 91	0. 84	0. 87	0. 86	0. 87	0. 85	0. 86	0. 89	0. 91	0. 78	0. 76	0. 85	0. 79	0. 79	0. 85
Nb/Ta	15	16	16	15	14	14	14	12	14	13	13	19	19	18	18	18	19

[a] YB = Yangba pluton.

[b] Rittmann index $\sigma = (Na_2O + K_2O)^2/(SiO_2-43)$.

Table 3　Analytical results of major(%) and trace(ppm) element from the Wulong pluton.

Sample	Monzogranite								Mafic enclaves						
	WL[a]-02	WL-04	WL-05	WL-10	WL-13	WL-15	WL-18	WL-21	WLB-08	WLB-17	WLB-25	WLB-26	WLB-31	WLB-33	WLB-34
SiO_2	68. 44	65. 29	68. 92	71. 30	70. 89	70. 54	66. 04	63. 84	53. 76	53. 97	55. 78	54. 99	52. 86	53. 93	53. 41
TiO_2	0. 20	0. 26	0. 12	0. 16	0. 19	0. 21	0. 33	0. 31	0. 87	0. 82	0. 98	0. 85	0. 97	1. 22	1. 15
Al_2O_3	17. 71	17. 73	16. 28	13. 96	14. 87	15. 31	17. 06	19. 12	16. 62	15. 31	13. 56	15. 53	16. 62	17. 93	18. 59
Fe_2O_3	0. 70	0. 54	0. 60	0. 90	1. 00	1. 05	1. 40	1. 57	3. 01	2. 20	2. 68	2. 60	3. 40	2. 86	3. 30
FeO	1. 40	1. 96	1. 50	1. 00	1. 10	1. 05	1. 50	1. 43	3. 89	3. 80	5. 02	4. 10	4. 00	4. 84	4. 10
MnO	0. 14	0. 06	0. 13	0. 08	0. 09	0. 08	0. 11	0. 08	0. 18	0. 19	0. 14	0. 19	0. 20	0. 21	0. 17
MgO	0. 71	1. 31	0. 85	0. 76	0. 86	0. 85	1. 24	1. 26	6. 09	6. 26	8. 13	6. 32	6. 58	4. 26	4. 15
CaO	2. 09	2. 86	1. 81	1. 77	1. 81	1. 67	2. 60	2. 70	7. 20	8. 58	5. 61	7. 78	6. 72	4. 48	4. 73
Na_2O	4. 50	5. 00	4. 30	4. 30	4. 30	4. 00	4. 80	4. 70	3. 80	3. 60	2. 70	3. 50	4. 00	3. 90	4. 00
K_2O	2. 51	2. 30	3. 90	3. 30	3. 10	3. 30	2. 30	2. 60	1. 30	1. 90	2. 30	1. 50	1. 60	3. 00	3. 10

Continued

Sample	Monzogranite								Mafic enclaves						
	WL[a]-02	WL-04	WL-05	WL-10	WL-13	WL-15	WL-18	WL-21	WLB-08	WLB-17	WLB-25	WLB-26	WLB-31	WLB-33	WLB-34
P_2O_5	0. 10	0. 25	0. 22	0. 10	0. 20	0. 18	0. 32	0. 22	0. 63	0. 70	0. 73	0. 67	0. 93	1. 03	1. 02
LOI	1. 21	1. 78	0. 70	1. 84	1. 05	1. 57	2. 10	2. 04	1. 46	1. 71	1. 86	1. 50	1. 08	1. 72	1. 78
Total	99. 71	99. 34	99. 33	99. 47	99. 46	99. 81	99. 80	99. 87	99. 40	99. 46	99. 49	99. 53	99. 46	99. 38	99. 50
A/CNK	1. 27	1. 11	1. 11	1. 01	1. 08	1. 16	1. 13	1. 24	0. 80	0. 65	0. 79	0. 72	0. 81	1. 01	1. 00
σ	1. 93	2. 39	2. 59	2. 04	1. 96	1. 94	2. 19	2. 56	2. 42	2. 76	1. 96	2. 09	3. 18	4. 36	4. 84
$Mg^{\#}$	39	49	43	43	44	43	45	44	62	66	66	64	63	51	51
Ba	1 242	775	1 224	1 081	1 025	1 185	1 386	1 478	1 197	1 156	1 497	983	670	462	486
Rb	100	85. 9	115	98. 8	97. 1	96. 8	80. 8	54. 9	40. 2	40. 4	117	35. 5	94. 0	198	205
Sr	763	1 003	663	643	715	647	1 115	1 066	1 396	1 697	1 742	1 631	1 518	836	856
Y	4. 77	10. 6	7. 27	6. 98	7. 79	11. 2	8. 36	8. 39	19. 4	18. 3	18. 5	20. 2	23. 4	34. 3	31. 5
Zr	150	197	125	132	137	144	189	199	111	43. 3	195. 9	113	179	227	212
Nb	6. 71	12. 5	9. 54	6. 81	7. 96	11. 4	7. 48	7. 76	12. 9	14. 3	15. 8	15. 3	33. 5	34. 2	31. 1
Th	8. 12	9. 67	10. 18	5. 76	8. 33	7. 34	9. 82	7. 53	7. 28	5. 12	9. 07	6. 16	14. 13	13. 50	12. 27
Pb	21. 6	23. 5	27. 9	24. 0	23. 8	24. 6	18. 6	17. 5	15. 1	13. 8	13. 1	13. 5	14. 4	17. 9	18. 6
Ga	16. 9	17. 9	15. 0	13. 7	14. 5	14. 5	16. 8	15. 6	13. 1	12. 6	12. 8	12. 2	15. 1	23. 5	22. 6
Ni	2. 50	10. 9	7. 75	6. 13	7. 12	8. 07	7. 58	7. 94	49. 4	69. 2	224	68. 5	73. 0	20. 0	20. 1
V	33. 3	44. 3	31. 2	29. 1	34. 4	38. 3	56. 4	54. 7	176	199	162	204	209	185	198
Cr	9. 91	23. 5	19. 4	13. 5	16. 5	17. 8	18. 5	22. 6	110	172	371	176	164	14. 8	15. 0
Hf	5. 23	6. 49	4. 57	5. 11	4. 54	5. 06	5. 47	6. 48	3. 35	1. 99	5. 39	3. 38	4. 57	6. 46	5. 74
Cs	5. 81	6. 11	3. 99	3. 63	4. 32	3. 91	5. 46	3. 57	1. 95	1. 07	10. 1	0. 88	7. 35	13. 1	13. 6
Sc	5. 64	5. 81	4. 84	4. 24	4. 94	7. 11	6. 35	5. 11	26. 1	26. 2	18. 3	26. 6	25. 3	30. 4	19. 4
Ta	0. 39	1. 12	0. 85	0. 80	0. 74	1. 01	0. 52	0. 60	0. 66	0. 77	0. 85	0. 86	1. 53	2. 22	1. 90
Co	48. 4	56. 7	51. 0	85. 0	49. 0	54. 7	49. 5	67. 4	46. 2	49. 6	63. 8	46. 1	49. 5	42. 3	36. 6

Continued

Sample	Monzogranite								Mafic enclaves						
	WL[a]-02	WL-04	WL-05	WL-10	WL-13	WL-15	WL-18	WL-21	WLB-08	WLB-17	WLB-25	WLB-26	WLB-31	WLB-33	WLB-34
Li	86. 8	65. 3	47. 6	37. 4	57. 8	46. 2	57. 3	54. 5	31. 4	26. 6	86. 3	32. 4	77. 0	148	146
Be	2. 43	2. 89	2. 65	1. 77	2. 19	2. 29	1. 97	1. 87	2. 15	1. 75	1. 57	1. 22	1. 49	3. 76	3. 48
U	0. 80	2. 16	1. 58	0. 88	0. 97	1. 27	0. 89	0. 88	2. 01	1. 86	1. 64	2. 01	3. 48	5. 59	5. 25
La	32. 2	30. 6	24. 7	20. 8	29. 0	26. 2	49. 9	39. 3	46. 0	38. 3	63. 7	48. 5	96. 1	57. 9	53. 8
Ce	62. 9	54. 7	47. 0	41. 0	52. 6	54. 8	88. 0	68. 3	104	87. 2	132	101	182	139	131
Pr	6. 11	5. 90	4. 74	4. 35	5. 39	6. 10	8. 23	6. 88	12. 0	10. 1	14. 5	11. 7	19. 3	17. 2	16. 3
Nd	21. 3	22. 7	16. 9	15. 8	19. 3	27. 7	28. 6	24. 4	48. 8	42. 6	57. 7	48. 7	73. 4	73. 5	70. 6
Sm	3. 29	3. 93	2. 68	2. 86	3. 19	4. 46	4. 11	3. 95	8. 54	7. 49	9. 00	8. 20	11. 1	14. 2	13. 2
Eu	0. 77	0. 92	0. 66	0. 67	0. 76	0. 93	0. 89	0. 93	2. 00	1. 88	2. 09	2. 03	2. 66	3. 00	2. 86
Gd	2. 16	2. 95	2. 18	2. 08	2. 36	3. 38	3. 25	2. 94	6. 04	5. 30	6. 45	6. 16	7. 82	10. 13	9. 02
Tb	0. 23	0. 40	0. 29	0. 26	0. 30	0. 46	0. 36	0. 38	0. 78	0. 69	0. 79	0. 77	0. 93	1. 36	1. 25
Dy	1. 05	1. 94	1. 51	1. 28	1. 49	2. 19	1. 84	1. 78	3. 95	3. 61	3. 86	4. 26	4. 77	7. 02	6. 34
Ho	0. 16	0. 36	0. 23	0. 23	0. 26	0. 39	0. 31	0. 31	0. 71	0. 68	0. 69	0. 77	0. 88	1. 26	1. 14
Er	0. 42	1. 02	0. 64	0. 58	0. 66	1. 01	0. 78	0. 81	1. 87	1. 78	1. 90	2. 11	2. 42	3. 22	3. 10
Tm	0. 05	0. 14	0. 07	0. 08	0. 09	0. 14	0. 10	0. 10	0. 25	0. 24	0. 24	0. 27	0. 31	0. 42	0. 40
Yb	0. 33	0. 96	0. 63	0. 52	0. 62	0. 82	0. 63	0. 65	1. 68	1. 59	1. 44	1. 82	2. 02	2. 66	2. 51
Lu	0. 05	0. 13	0. 09	0. 08	0. 10	0. 11	0. 09	0. 10	0. 24	0. 24	0. 24	0. 28	0. 34	0. 37	0. 35
Sr/Y	160	95	91	92	92	58	134	127	72	93	94	81	65	24	27
Y/Yb	14	11	12	14	13	14	13	13	12	11	13	11	12	13	13
La/Yb	97	32	40	40	47	32	80	61	27	24	44	27	48	22	21
Eu/Eu*	0. 88	0. 82	0. 84	0. 84	0. 84	0. 73	0. 74	0. 84	0. 85	0. 91	0. 84	0. 87	0. 87	0. 76	0. 80
Nb/Ta	17	11	11	8	11	11	14	13	20	19	19	18	22	15	16

[a] WL = Wulong pluton.

Table 4 Analytical results of major(%) and trace(ppm) element from the Xichahe and Guangtoushan granitoids.

Sample	Tonalite										Granodiorite					
	XH[a]-03	XH-04	XH-05	XH-06	XH-07	XH-09	XH-10	XH-12	XH-14	XH-15	G[b]-02	G-03	G-04	G-06	G-10	G-11
SiO_2	62. 30	61. 75	64. 40	60. 91	60. 03	56. 98	60. 83	60. 23	58. 65	59. 89	70. 42	71. 07	71. 93	69. 99	68. 95	71. 12
TiO_2	0. 65	0. 60	0. 48	0. 80	0. 72	0. 57	0. 55	0. 53	0. 65	0. 52	0. 23	0. 12	0. 15	0. 24	0. 36	0. 20
Al_2O_3	16. 40	18. 37	17. 50	15. 09	18. 09	14. 87	15. 96	15. 09	15. 40	14. 87	15. 85	16. 72	15. 64	15. 74	16. 23	15. 68
Fe_2O_3	1. 60	1. 05	0. 60	1. 40	1. 80	1. 80	1. 00	1. 50	2. 25	1. 00	nd	nd	nd	nd	nd	nd
FeO	3. 60	3. 00	2. 70	3. 70	3. 10	4. 00	3. 90	4. 10	4. 05	4. 09	nd	nd	nd	nd	nd	nd
$Fe_2O_3{}^T$											1. 57	0. 83	1. 45	1. 76	2. 21	1. 53
MnO	0. 10	0. 07	0. 11	0. 09	0. 10	0. 16	0. 13	0. 14	0. 18	0. 09	0. 02	0. 01	0. 03	0. 03	0. 02	0. 03
MgO	3. 15	3. 05	2. 55	3. 99	3. 42	6. 51	4. 18	5. 99	6. 00	4. 02	0. 42	0. 27	0. 34	0. 42	0. 46	0. 29
CaO	3. 80	3. 96	3. 49	4. 61	4. 19	6. 25	4. 61	6. 06	5. 83	4. 47	2. 46	2. 76	2. 12	2. 24	2. 59	2. 00
Na_2O	4. 00	4. 40	4. 00	4. 20	4. 10	3. 80	4. 00	3. 40	3. 80	3. 10	4. 72	5. 70	4. 85	4. 83	4. 88	4. 67
K_2O	2. 20	2. 21	2. 70	2. 60	2. 70	2. 71	2. 90	2. 00	2. 00	3. 20	3. 11	2. 12	3. 36	3. 46	3. 21	3. 72
P_2O_5	0. 30	0. 23	0. 20	0. 37	0. 36	0. 53	0. 31	0. 30	0. 37	0. 35	0. 07	0. 05	0. 05	0. 08	0. 10	0. 06
LOI	1. 47	1. 05	0. 92	1. 67	1. 10	1. 31	1. 04	0. 23	0. 37	3. 87	0. 68	0. 31	0. 45	0. 76	0. 53	0. 58
Total	99. 57	99. 74	99. 65	99. 43	99. 71	99. 79	99. 41	99. 85	99. 67	99. 47	99. 55	99. 96	100. 37	99. 55	99. 54	99. 88
A/CNK	1. 03	1. 09	1. 10	0. 83	1. 05	0. 72	0. 88	0. 80	0. 81	0. 89	1. 01	1. 00	1. 01	1. 00	1. 00	1. 02
σ	1. 99	2. 33	2. 10	2. 58	2. 72	3. 03	2. 67	1. 69	2. 15	2. 35	2. 24	2. 18	2. 33	2. 55	2. 52	2. 50
$Mg^{\#}$	53	58	59	59	57	68	61	66	64	59	35	39	32	32	29	27
Ba	1 243	1 086	1 964	1 363	1 365	877	888	904	1 002	1 297	1 278	937	1 149	1 228	1 581	1 237
Rb	89. 8	86. 4	84. 3	104	104	121	140	78. 6	66. 9	132	79. 4	47. 3	82. 1	93. 1	85. 9	92. 7
Sr	907	900	903	837	865	957	821	1 029	1 253	942	773	1 053	805	731	904	705
Y	13. 4	10. 6	10. 3	16. 1	14. 2	16. 4	18. 3	22. 1	22. 7	16. 6	8. 91	3. 74	7. 37	9. 06	9. 68	7. 17
Zr	86. 4	69. 9	52. 9	69. 6	78. 7	92. 8	107	87. 7	110	224	143	103	122	164	212	138
Nb	11. 7	10. 6	8. 92	14. 6	14. 0	9. 05	15. 12	9. 72	9. 77	14. 4	5. 67	3. 14	4. 73	7. 60	9. 40	6. 19

Continued

Sample	Tonalite										Granodiorite					
	XH[a]-03	XH-04	XH-05	XH-06	XH-07	XH-09	XH-10	XH-12	XH-14	XH-15	G[b]-02	G-03	G-04	G-06	G-10	G-11
Th	5.58	4.00	3.88	3.55	2.14	4.04	3.90	3.07	2.50	16.1	6.37	1.52	6.25	10.7	11.0	8.39
Pb	nd	nd	nd	nd	nd	nd	nd	nd	nd	nd	12.3	11.1	20.2	15.1	16.4	15.1
Ga	nd	nd	nd	nd	nd	nd	nd	nd	nd	nd	18.7	17.1	18.4	19.0	19.0	19.3
Ni	nd	nd	nd	nd	nd	nd	nd	nd	nd	nd	2.26	2.09	2.29	2.01	2.12	2.06
V	nd	nd	nd	nd	nd	nd	nd	nd	nd	nd	14.7	9.74	13.0	14.0	23.4	12.1
Cr	nd	nd	nd	nd	nd	nd	nd	nd	nd	nd	2.51	2.15	2.41	1.92	2.10	2.05
Hf	3.11	2.83	2.56	2.34	2.72	3.14	3.49	3.26	3.50	6.92	3.49	2.60	3.39	4.18	4.95	3.58
Cs	4.96	5.16	4.34	6.08	6.08	6.47	2.64	3.27	2.72	5.03	1.17	0.69	1.06	1.32	1.43	1.32
Sc	nd	nd	nd	nd	nd	nd	nd	nd	nd	nd	2.23	1.37	1.96	2.29	2.06	1.99
Ta	0.74	0.54	0.53	0.90	0.88	0.47	1.00	0.95	0.66	0.87	0.63	0.36	0.52	0.61	0.70	0.64
Co	nd	nd	nd	nd	nd	nd	nd	nd	nd	nd	125	128	157	130	126	128
U	0.90	0.75	0.68	2.23	0.88	1.19	0.95	2.15	1.29	3.25	1.25	0.86	1.95	1.59	1.70	1.87
La	30.8	24.8	24.2	21.6	17.1	27.9	31.5	20.1	27.4	38.9	24.8	4.3	17.5	32.0	39.6	25.1
Ce	59.8	50.0	47.7	44.3	37.9	58.8	63.1	46.2	62.8	76.3	44.4	8.0	33.6	54.4	73.9	43.6
Pr	6.49	5.51	5.16	5.16	4.60	6.80	6.86	5.62	7.46	8.03	4.75	1.12	3.58	5.38	7.61	4.38
Nd	25.6	22.2	21.1	23.4	20.7	28.0	27.2	25.7	32.8	30.8	17.6	4.5	13.5	18.8	27.7	15.5
Sm	5.01	4.14	4.01	5.36	4.80	5.65	5.14	5.87	6.83	5.57	2.99	0.94	2.44	2.84	4.69	2.36
Eu	1.28	1.21	1.32	1.21	1.23	1.25	1.03	1.36	1.47	1.22	0.84	0.29	0.58	0.77	1.23	0.63
Gd	4.05	3.30	3.10	4.50	3.95	4.23	4.31	4.65	5.31	4.38	2.35	0.85	2.05	2.40	3.51	1.79
Tb	0.54	0.44	0.39	0.65	0.57	0.59	0.64	0.73	0.77	0.60	0.32	0.11	0.28	0.33	0.43	0.24
Dy	2.97	2.40	2.21	3.59	3.19	3.33	3.66	4.42	4.40	3.22	1.54	0.48	1.28	1.63	1.90	1.13
Ho	0.50	0.42	0.38	0.63	0.53	0.57	0.66	0.81	0.82	0.59	0.29	0.10	0.23	0.30	0.30	0.23
Er	1.32	1.05	0.99	1.60	1.44	1.64	1.83	2.19	2.40	1.62	0.74	0.23	0.63	0.78	0.70	0.59

Continued

Sample	Tonalite										Granodiorite					
	XH[a]-03	XH-04	XH-05	XH-06	XH-07	XH-09	XH-10	XH-12	XH-14	XH-15	G[b]-02	G-03	G-04	G-06	G-10	G-11
Tm	0.17	0.15	0.13	0.22	0.19	0.23	0.24	0.34	0.33	0.24	0.11	0.04	0.10	0.12	0.10	0.10
Yb	1.04	0.79	0.83	1.27	1.17	1.41	1.49	2.15	2.13	1.51	0.63	0.20	0.58	0.65	0.47	0.51
Lu	0.16	0.13	0.12	0.20	0.17	0.22	0.22	0.33	0.35	0.26	0.10	0.04	0.10	0.10	0.07	0.09
Sr/Y	68	85	88	52	61	58	45	47	55	57	87	281	109	81	93	98
Y/Yb	13	13	12	13	12	12	12	10	11	11	14	19	13	14	21	14
La/Yb	30	31	29	17	15	20	21	9	13	26	40	22	30	49	84	49
Eu/Eu*	0.87	1.00	1.14	0.75	0.86	0.78	0.67	0.80	0.75	0.76	0.97	0.99	0.79	0.90	0.93	0.94
Nb/Ta	16	19	17	16	16	19	15	10	15	17	9	9	9	12	13	10

[a] XH = Xichahe pluton.

[b] G = Guangtoushan pluton.

Table 5　Data of whole-rock Sr and Nd isotopes for the Wulong and Xichahe granitoids.

Sample	Rb /ppm	Sr /ppm	$^{87}Rb/^{86}Sr$[a]	$^{87}Sr/^{86}Sr \pm 2\sigma$	$(^{87}Sr/^{86}Sr)_i$ (210 Ma)	Sm /ppm	Nd /ppm	$^{147}Sm/^{144}Nd$[a]	$^{143}Nd/^{144}Nd \pm 2\sigma$	$\varepsilon_{Nd}(t)$ (210 Ma)[b]	T_{DM}[c] /Ga
WL-05	114	663	0.501	0.705 94±5	0.704 44	2.68	16.9	0.096	0.512 363±8	−2.7	1.02
WL-13	97.1	715	0.393	0.706 20±14	0.705 02	3.19	19.3	0.100	0.512 389±9	−2.3	1.02
XH-09	121	957	0.365	0.706 55±8	0.705 43	5.65	28.0	0.122	0.5124 54±9	−1.5	1.15
XH-14	66.9	125 3	0.154	0.706 92±42	0.706 45	6.83	32.8	0.126	0.5123 52±6	−3.6	1.38

[a] $^{87}Rb/^{86}Sr$ and $^{147}Sm/^{144}Nd$ ratios were calculated using Rb, Sr, Sm and Nd contents, measured by ICP-MS.

[b] $\varepsilon_{Nd}(t)$ values were calculated using present-day $(^{147}Sm/^{144}Nd)_{CHUR} = 0.196\ 7$ and $(^{143}Nd/^{144}Nd)_{CHUR} = 0.512\ 638$.

[c] T_{DM} values were calculated using present-day $(^{147}Sm/^{144}Nd)_{DM} = 0.213\ 7$ and $(^{143}Nd/^{144}Nd)_{DM} = 0.513\ 15$.

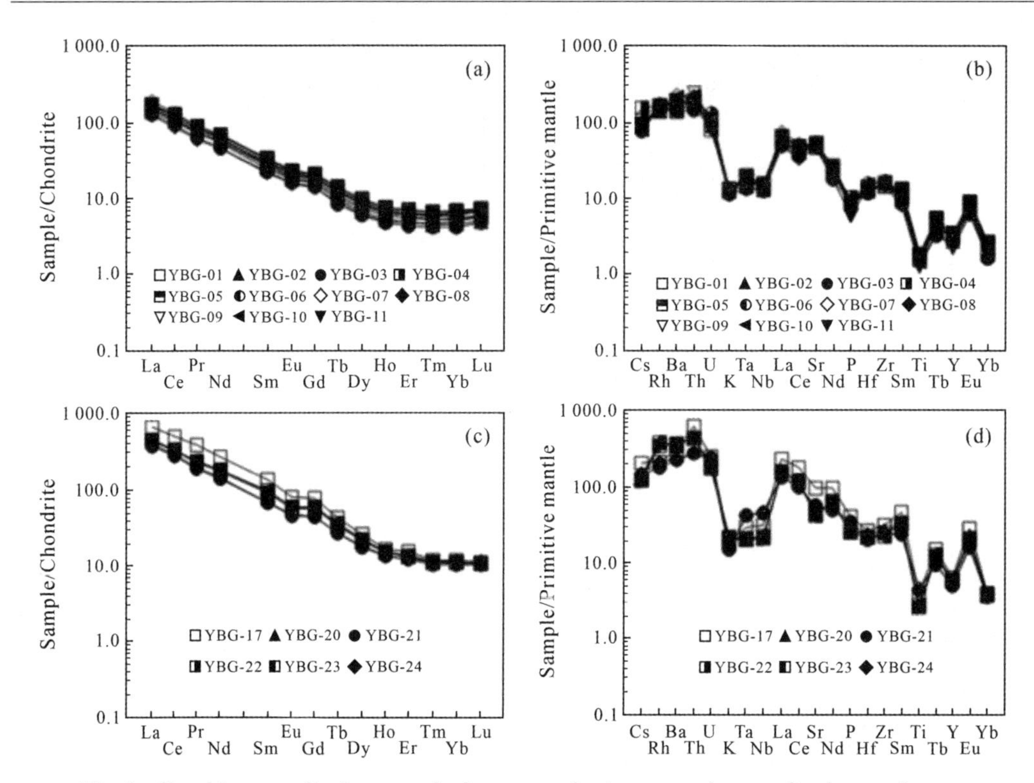

Fig. 4 Chondrite-normalized rare earth elements amd primary mantle-normalized trace element distributions for the granites from the northern margin of the Yangtze craton.

(a) and (b) Yangba granodiorites. (c) and (d) Yangba mafic microgranular enclaves. Chondrite-normalized values are from Taylor and McLennan (1985). Primitive mantle-normalized values are from Sun and McDonough (1989).

samples (sample XH-12 and XH-14) showing relatively high Y (22.11–22.65 ppm) and HREE (e.g., Yb=2.13–2.15 ppm) (Fig. 5e). Chondrite-normalized REE patterns have flat HREE patterns and are characterized by the general absence of significant Eu anomalies, with Eu/Eu^* ranging from 0.76 to 1.14 (Fig 5a). In the primitive mantle-normalized diagram significant negative spikes in Nb-Ta, P, K, Th, and Ti are evident, in conjunction with positive anomalies in Ba and Sr (Fig. 5b). Overall, the tonalite from Xichahe has a close geochemical affinity to those of adakites and TTG (Martin, 1999), such as low content of garnet-compatible elements (Y, HREE), high abundance of plagioclase-compatible Sr and Ba, and high ratios of Sr/Y (44.86–88.09) and La/Yb (Table 4).

(3) *Wulong pluton.* The Wulong monzogranite (eight samples) is dominantly metaluminous. Compared with the host monzogranite, the mafic enclaves show lower SiO_2, Na_2O, K_2O, K_2O/Na_2O, and A/CNK (0.65–1.01) and higher TiO_2, MgO, and CaO. As shown in Table 4, the Wulong monzogranite is enriched in Sr and Ba, and depleted in Y, Nb, Ta, and Ti. The low Nb/Ta ratios (8.48–14.30, most < 12) suggest that they were likely derived from continental crust. Chondrite-normalized REE patterns have steep HREE shapes

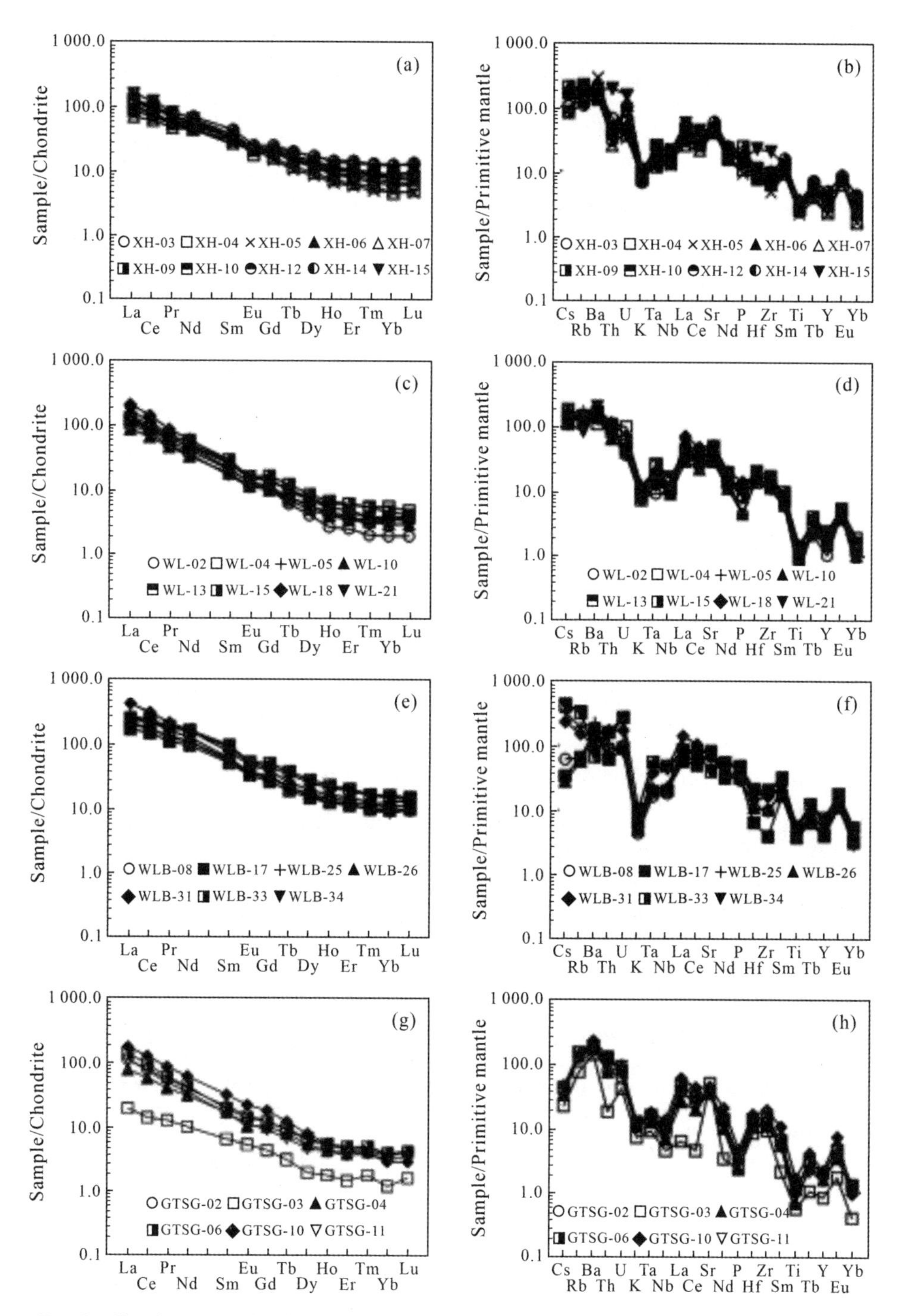

Fig. 5 Chondrite-normalized rare-earth elements and primary mantle-normalized trace element distributions for the Triassic granites in the South Qinling Mountains.

(a) and (b), Xichahe tonalite. (c) and (b), Wulong granites. (e) and (f), Wulong mafic enclaves. (g) and (h), Guangtoushan trondhjemites. Chondrite-normalized values are from Sun and McDonough (1989). Primitive mantle-normalized values are from Taylor and McLennan (1985).

and lack significant Eu anomalies, with Eu/Eu* ranging from 0.73 to 0.88 (Fig. 5c). The monzogranite shows large variation in $(La/Yb)_N$ ratios (21.7–65.4) and the mafic enclaves have lower $(La/Yb)_N$ ratios (14.46–32.16) and similar Eu anomalies (Eu/Eu* = 0.76–0.91) to the host monzogranite (Fig. 5e). Primitive mantle-normalized element patterns show significant negative anomalies of Nb-Ta, P, and Ti in conjunction with positive anomalies in Ba and Sr (Fig. 5f). The mafic enclaves have higher Nb/Ta ratios (15.42–21.91) and weaker Nb-Ta anomalies, implying that they are mantle-derived and contaminated by crustal material.

(4) *Guangtoushan pluton*. The Guangtoushan granodiorite is peraluminous. Trace-element data have high contents of total REE (ΣREE = 242–355 ppm). Chondrite-normalized REE patterns have steep HREE patterns and are characterized by the general absence of significant Eu anomalies, with Eu/Eu* ranging from 0.79 to 0.97 (Fig. 5g). In the primitive mantle-normalized element diagrams they show significant negative anomalies in Nb-Ta, P, K, Th, and Ti in conjunction with positive anomalies in Ba and Sr (Fig. 5h).

Sr-Nd-Pb isotopes

Whole-rock Sr-Nd-Pb isotope analyses are listed in Tables 5 and 6. Initial Nd and Sr isotopic compositions for the different samples are calculated at t = 210 Ma. The Xichahe tonalities (two samples) display similar characteristics, with $^{87}Sr/^{86}Sr$ = 0.706 5–0.706 9, $(^{87}Sr/^{86}Sr)_i$ = 0.705 4–0.706 4 and non-radiogenetic Nd isotope with $^{143}Nd/^{144}Nd$ = 0.512 3–0.512 4 and $\varepsilon_{Nd}(t)$ = −3.6 to −1.5. T_{DM} ranges from 1.15 Ga to 1.38 Ga. The Wulong monzogranite (two samples) has $^{87}Sr/^{86}Sr$ = 0.705 9–0.706 2, $(^{87}Sr/^{86}Sr)_i$ = 0.704 4–0.705 0, $^{143}Nd/^{144}Nd$ = 0.512 36–0.512 38, $\varepsilon_{Nd}(t)$ = −2.66 to −2.26 and Nd isotopic model ages (T_{DM}) = ~1.02 Ga. In the $(^{87}Sr/^{86}Sr)_i$-$\varepsilon_{Nd}(t)$ diagram (Fig. 6), the Wulong and Xichahe granites show similar Sr-Nd isotopic compositions to the Yangba granites and the meta-volcanic rocks from the Proterozoic Yaolinghe Group (Zhang et al., 2007a).

Table 6 Whole-rock Pb isotopes for the Wulong and Xichahe granitoids.

Sample	U/ppm	Th/ppm	Pb/ppm	$^{206}Pb/^{204}Pb$	2σ	$^{207}Pb/^{204}Pb$	2σ	$^{208}Pb/^{204}Pb$	2σ
WL-05	1.58	10.2	28.0	18.098	0.001 2	15.873	0.001 0	38.973	0.002 5
WL-13	0.97	8.33	23.8	18.209	0.000 4	15.979	0.000 3	39.430	0.001 2
XH-09	1.19	4.04	14.5	18.349	0.000 4	15.997	0.000 4	39.526	0.001 0
XH-14	1.29	2.50	14.5	18.294	0.000 4	15.993	0.000 4	39.502	0.001 0

Whole-rock Pb isotopic data (Table 6) show that both the Xichahe tonalite and Wulong monzogranite and are characterized by high radio-genetic Pb isotopic compositions with present-day whole-rock Pb isotopic ratios of $^{206}Pb/^{204}Pb$ = 18.099–18.349, $^{207}Pb/^{204}Pb$ = 15.873–15.997, and $^{208}Pb/^{204}Pb$ = 38.973–39.526. Due to their young age (~210 Ma) and low contents of U (0.75–3.25 ppm) and Th (2.14–5.58 ppm), it can be considered that the present-day Pb

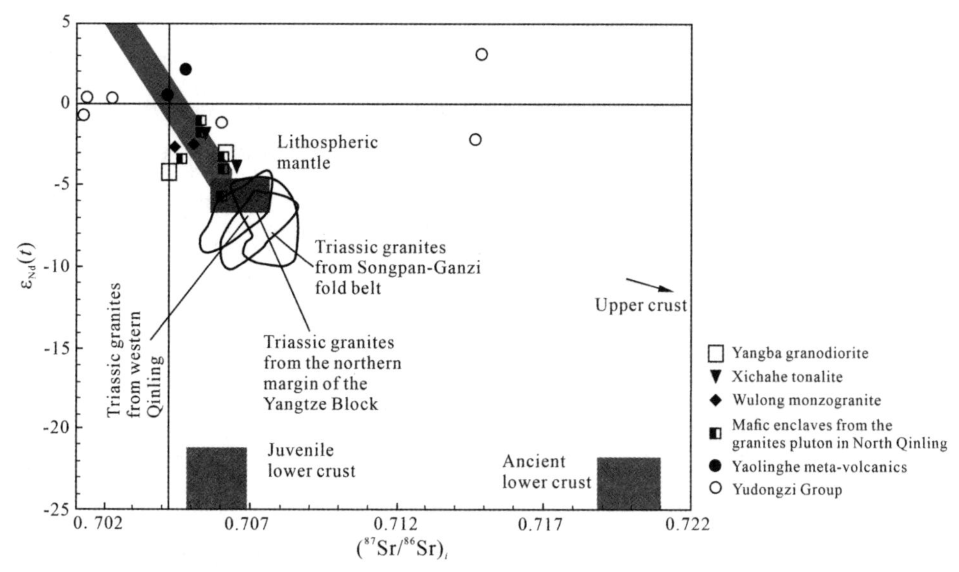

Fig. 6 $(^{87}Sr/^{86}Sr)_i$ vs. $\varepsilon_{Nd}(t)$ plot of the Wulong and Xichahe granitoids, Qinling orogenic belt.
Data for Triassic granites from the Songpan-Ganzi fold belt are from Zhang et al. (2006); data for the Triassic granites from the western Qinling are from Zhang et al. (2007a); data for Yangba granite and Triassic granites from the northern margin of the Yangtze craton are from Zhang et al.(2007b); data for the Yudongzi Group are from Zhang et al.(2001); data for the Yanglinghe metavolcanic rocks are from Huang and Wu (1990); data for mafic enclaves from Triassic granite plutons in the northern Qinling are from Wang et al.(2005).

isotopic ratios represent the initial Pb isotopic ratios (e.g., Zhang et al., 2006).

Discussion

Timing of Triassic granite magmatism in the Qinling orogen

The many geochronological studies of HP (high pressure) and UHP (ultra-high pressure) metamorphic rocks from the Qinling-Dabie-Sulu orogen suggest that the collision between the Yangtze and North China cratons occurred in the Early Triassic (~230 Ma) (e.g., Ames et al., 1996; Chavagnac and Jahn, 1996; Hacker et al., 1998; Li et al., 2000, 2002; Wan et al., 2005). Detailed geochronology and cooling history study of UHP metamorphic rocks from the Dabie Mountains reveals two rapid cooling events during 221–218 Ma and 180–170 Ma, respectively, corresponding to two stages of rapid uplift (Hacker et al., 1998; Li et al., 2002; Wang et al., 2005). The first stage is considered to be related to break-off of the subducted slab (Davis and von Blanckenburg, 1995; Li et al., 2002). Sun et al. (2002) carried out chronological studies on six granite plutons in the South Qinling obtaining U-Pb ages ranging from 220 Ma to 205 Ma, which is approximately 10 Ma later than the UHP peak metamorphic age and yet consistent with the first rapid cooling time of the UHP metamorphic rocks in the

Dabie and Sulu orogens (Li et al., 2002; Wang et al., 2005).

Li et al. (1996) obtained a Sm-Nd whole rock isochron age of 242 ± 21 Ma from metavolcanic rocks in the Heigouxia area, Mianxian country, in the western section of the Qinling-Dabie orogen. Based on geological observation and geochemical analysis, they proposed that this age represents metamorphism related to the Mianlue paleo-ocean closure. Accordingly, it was concluded that the Mianlue paleo-oceanic basin had been consumed by Late Triassic time. Detailed geochronological work on the Triassic granites in the North Qinling by Lu et al. (1999) suggested that their formation ages ranged from 214 Ma to 217 Ma. Our new geochronological study indicates that the Triassic granites as the northern margin of the Yangtze craton and in the South Qinling have zircon LA-ICPMS U-Pb age of 208 ± 2 Ma (Yangba pluton), 208±2 Ma (Wulong pluton), and 212±2 Ma (Xichahe pluton), consistent with the first rapid cooling time of the UHP metamorphic rocks in the Dabie-Sulu orogen. These ages may indicate that the extensive Triassic granite magmatism took place in the entire Qinling orogen, from the northern margin of the Yangtze craton, via the South Qinling to the North Qinling, regionally. Meanwhile, the intrusive age of these granite plutons was 10–20 m.y. younger than the metamorphic peak age of collision between the Yangtze and North China cratons. Hence, Triassic granite magmatism in the Qinling orogen should be considered to be post-collisional.

Petrogenesis of the Triassic granites in the South Qinling

Compared with the normal I-type granitoids (White and Chappell 1977), the Qinling granites display depleted Y and HREE, and are enriched in Sr and Ba with weak positive Sr anomalies in primitive mantle-normalized trace-element patterns (Figs. 4 and 5), high La/Yb and Sr/Y, and concave-upward patterns in the HREE without significant negative Eu anomalies. These geochemical signatures are comparable to adakites as defined by Defant and Drummond (1990), and shown in Sr/Y-Y and $(La/Yb)_N$-Yb_N diagrams (Fig. 7a,b).

Although the genesis of adakitic magma is still a matter of debate, it is generally accepted that it was derived from sources of basaltic composition, and that the presence of significant amounts of garnet±amphibole is required at some stage in the petrogenesis, either as a residual or an early crystallizing assemblage (Martin, 1999; Rapp et al., 1999; Prouteau et al., 2000; Garrison and Davidson, 2003; Condie, 2005; Martin et al., 2005). Petrogenetic models for the possible origin of adakite include: ① partial melting of a subducted oceanic slab (Defant and Drummond, 1990, 1993; Peacock et al., 1994; Gutscher et al., 1999; Sajona et al., 2000; Beate et al., 2001); ② partial melting of thickened lower continental crust (Atherton and Petford, 1993; Muir et al., 1995; Barnes et al., 1996; Petford and Atherton, 1996; Chung et al., 2003; Wang et al., 2005); ③ partial melting of delaminated lower crust (Xu et

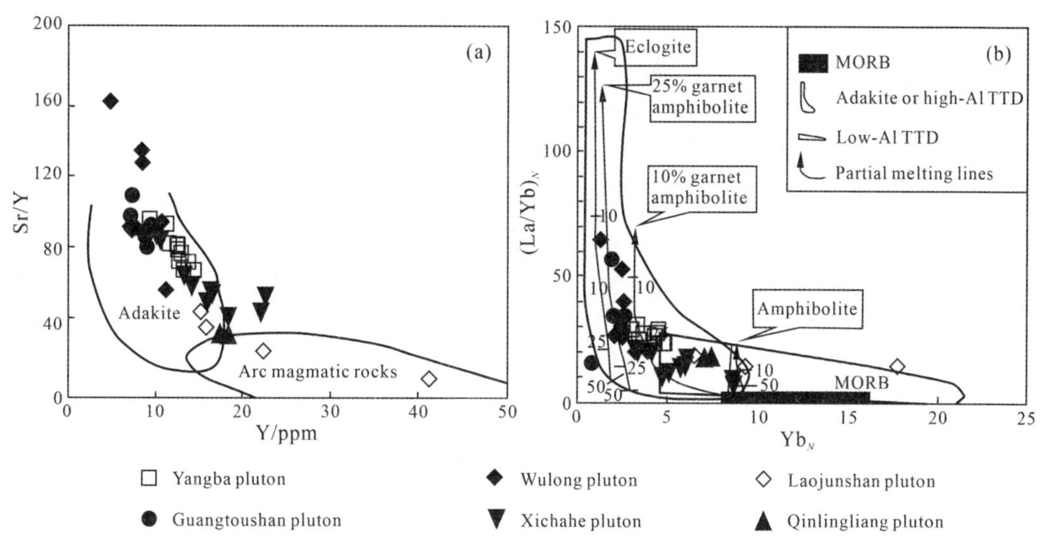

Fig. 7 (a) Sr/Y vs. Y (a) and $(La/Yb)_N$ vs. Yb_N (b) diagrams for the granites from the Qinling orogenic belt.

Field of adakite is from Defant and Drummond (1990).

al., 2002; Gao et al., 2004; Wang et al., 2006); and ④ assimilation-fractional crystallization (AFC) of basaltic magmas (e.g., Feeley and Hacker, 1995). Experimental studies show that the magma source for adakite is basaltic, transformed to garnet amphibolite and/or amphibole eclogite at pressures equivalent to a thickness > 50 km (Sen and Dunn, 1994; Rapp and Watson, 1995). Therefore, crustal thickening is essential for petrogenesis of adakitic rocks with specified crustal isotopic characteristics (Zhang et al., 2006). However, either partial melting of subducted oceanic crust or AFC of a basaltic magma appears unlikely for the studied adakitic granites in South Qinling based on the following observations.

First, zircon U-Pb dating indicates that the four granite plutons in this study were emplaced between 205 Ma and 220 Ma. The closure of the Mianlue ocean and collision of the Yangtze and the North China cratons occurred at ~242 Ma (e.g. Li et al., 1996); evidently, subduction of Mianlue oceanic crust beneath the North China craton had ceased by this time. The adakitic granites show a wide variation in MgO and somewhat higher K_2O relative to adakites from circum-Pacific regions (Wu et al., 2002). Thus, the adakitic granites in the Qinling orogen are unlikely to have been produced by partial melting of oceanic crust. Second, fractional crystallization of basaltic magma is an inefficient mechanism of producing felsic rocks (Green and Ringwood 1968), inasmuch as it requires > 70% fractional crystallization to produce trondhjemitic liquids (Spulber and Rutherford, 1983) or rhyolite (Meijer, 1983). Moreover, fractional crystallization of garnet and plagioclase could produce liquids not only with low Sr/Y and La/Yb ratios, but also significant negative Eu anomalies, inconsistent with the geochemical features of the adakitic granites in the Qinling orogen. In the North Qinling, Wang et al. (2003) reported that post-collisional granites of the Laojunshan and Qinlingliang plutons also showed some adakitic features: e.g., strong enrichment in Sr, and Ba and depletion in Y and HREE, with high Sr/Y and Y/Yb ratios and no significant Eu anomalies.

Therefore, the petrogenesis of the Triassic adakitic granitoids in the Qinling orogen can be attributed to either a thickened lower crust model (Chung et al., 2003; Wang et al., 2005) or delamination of the lower crust (Xu et al., 2002; Gao et al., 2004). This is in agreement with convergence between the North China and Yangtze cratons during the Triassic (Zhang et al., 2002). According to the depth of generation of adakitic magma (Sen and Dunn, 1994; Rapp and Watson, 1995), we suggest that the crust of the Qinling orogenic belt was likely thickened to >50 km in the Late Triassic. Partial melting of thickened lower crust in the Qinling orogen could have produced these adakitic magmas such as the Yangba, Wulong, Xichahe and Guangtoushan granitoids.

Mantle-derived magmatism in the Qinling orogen and its role in the generation of Triassic adakitic granites

It is noteworthy that most of the Triassic granite plutons in the Qinling orogen contain some mafic enclaves. These mafic enclaves might have played a fundamental role in the genesis of the adakitic magmas (Fig. 8). In the Yangba pluton, the host monozogranite has a high $Mg^{\#}$ (50.8–54.5>45), indicative of mantle contamination (Rapp et al., 1999). Petrographically, the enclaves are fine-grained, rounded or ovoid, and contain acicular apatite. Megacrysts cross the boundary of the host and enclaves, strongly suggesting that the enclaves were incorporated as magma globules into the host granodioritic magma. Generally, the mafic enclaves have higher $\sum$REE (357–654 ppm), $(La/Yb)_N$ and significantly greater negative Eu anomalies than those of the host monozogranite. In the Xichahe pluton, the tonalite displays significant mantle signatures with higher $Mg^{\#}$ (53–67), Nb/Ta ratios (14.8–19.5), and Cr and Ni contents. In the Wulong pluton, ovoid mafic enclaves (up to 20 cm in diameter) display mantle-like signatures, with high Nb/Ta ratios (15.42–21.91) and MgO (4.15%–8.13%). Sr and Nd isotopic geochemistry (Table 5) also suggests that the Wulong and Xichahe granites have crust/mantle compositions: $\varepsilon_{Nd}(t) = -3.6$ to -1.5, $(^{87}Sr/^{86}Sr)_i = 0.7044$–$0.7064$. In the North Qinling, the Qinlingliang and Laojunshan plutons contain mafic enclaves that display significant mantle signatures, with $SiO_2 = 50\%–62\%$, $K_2O + Na_2O = 7.01\%–9.4\%$, $\sigma = 5–9$, and high contents of REE, Cr, Ni, V, and Ga. Combined with their low $(^{87}Sr/^{86}Sr)_i$ (0.70514–0.70624) and high $\varepsilon_{Nd}(t)$ (−3.3 to −0.95), Wang et al. (2005) proposed that the magma of the enclave may have been derived from the subcontinental lithospheric mantle as a result of the mixing of basic and granitic magma. It seems that partial melting of the SCLM and crust/mantle interaction played an important role in the evolution of the Triassic adakitic granites in the Qinling orogen.

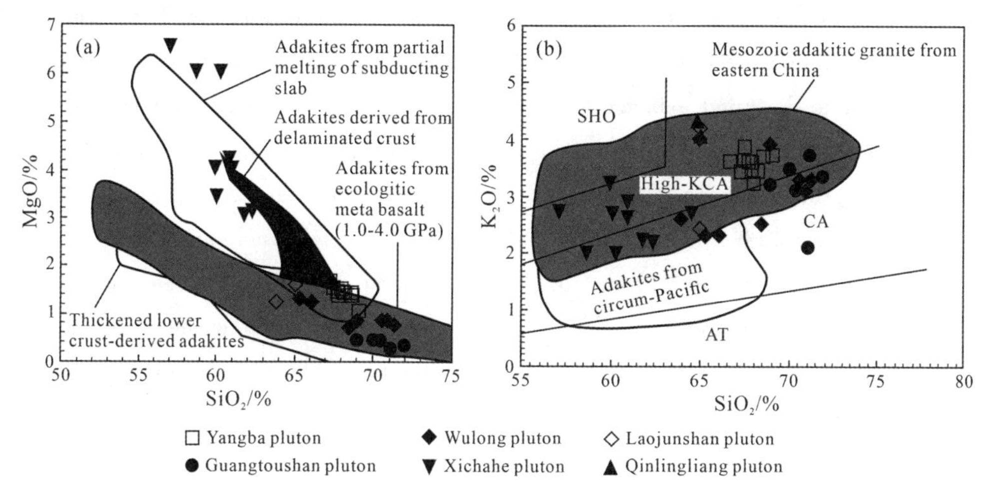

Fig. 8 Plots of MgO vs. SiO_2(a) and K_2O vs. SiO_2(b) (after Peccerillo and Taylor, 1976) for adakitic granites in the Qinling orogenic belt.

The adakite field from eclogitic metabasalt (1–4 GPa) is from Sen and Dunn (1994), and Rapp (1995), and Rapp and Watson (1995). The adakite field from partial melting of a subducting slab is from Defant and Drummond (1990), Stern and Kilian (1996), Martin (1999), Smithies (2000), and Defant et al. (2002). The adakite field derived from delaminated crust is from Xu et al. (2002). The field of thickened lower crust-derived adakites is from Atherton and Petford (1993), Muir et al. (1995), Petford and Atherton (1996), and Smithies (2000). The data for Mesozoic adakites from eastern China and adakites from the circum-Pacific belt after Wu et al. (2002).

A slab break-off model, and genesis of Triassic adakitic granites in the Qinling orogen

A self-consistent model for granitic magmatism in the Qinling orogenic belt must explain the following features: ① timing of Triassic granite emplacement across the north margin of the Yangtze craton in both the South and North Qinling, with a climax about 20 m.y. after profound collision of the Yangtze and North China cratons; ② the adakitic characteristics of the granites, particularly the high-$Mg^{\#}$ adakitic granites from the Yangba and Xichahe plutons which indicate crustal-derived melt that interacted with mantle material; and ③ the nature of mafic magmas that produced the mafic enclaves in the adakitic granites.

In the ongoing debate regarding the evolution of the Alps, Davies and von Blanckenburg (1995) proposed a "slab breakoff" model to explain syncollisional magmatism preceded by subduction of continental lithosphere to mantle depths. They proposed that slab breakoff would result in asthenospheric upwelling into a narrow rift, and, following breakoff, it would impinge on the overriding mantle lithosphere. The hypothesized thermal perturbation would lead to melting of the metasomatized overriding mantle lithosphere, producing basaltic magamtism that ultimately would result in enhanced flow and melting of lower crust, producing granitic magmatism. Breakoff would remove the force at the down-dip side of the continental crust, while enhanced heating would lead to a reduction in the strength of the underlying crust. Both effects would facilitate the freeing of a buoyant crustal sheet that could then rise toward the surface, leading to the rapid exhumation of eclogite-facies continental crust. The cessation of subduction and replacement of cold oceanic lithosphere by asthenosphere would lead to rapid uplift of the orogen. According to their conclusions, orogens that have undergone slab-breakoff have the following features:① magmatism and metamorphism are synchronous near the suture and the center of the orogen, demonstrating the general steepening of the suture during collision; ② an inverse correlation exists between maximum depth of mteamorphism and volume of synorogenic magmatism; and ③ the melts may be emplaced in both compressive and extensional environments.

Some aspects of slab breakoff-type magmatism and metamorphism can be applied to Triassic Qinling orogen, including: ① mafic magmatism followed after a short time gap by felsic magmatism, including Triassic granite plutons containing mafic enclaves. The petrology and geochemistry of the mafic enclaves suggest that they were derived from subcontinental lithospheric mantle and have been incorporated as magma globules into granodiorite magma (Qin et al., 2005; Zhang et al., 2005; Wang et al., 2005); ② the timing of uplift relative to mafic/granitic magma genesis. Emplying $^{39}Ar/^{40}Ar$ dating of hornblende, biotite, and alkali feldspar, and multi-domain diffusion modeling in the Caoping, Shahewan, Laojunshan, and Qinlingliang plutons in the Qinling orogen, Wang et al. (2004) proposed that these granites underwent rapid cooling during the Late Triassic. Furthermore, the first stage (221–218 Ma) of rapid uplift of metamorphic rocks in the Dabie orogen were considered to be related to breakoff following closure of the Mianlue ocean (Li et al., 2002; Wang et al., 2005); and ③ occurrence of mafic rocks, e.g. the mafic enclaves in some granite plutons, suggests that some mantle magma stalled and crystallized at the base of the lower crust. Subsequent partial melting of the lower crust formed granitic magma. Asthenospheric upwelling increased and slab breakoff evolved from narrow rifting to complete separation. The asthenospheric upwelling resulting from slab breakoff would impinge on the base of the lower crust, in turn triggering partial melting, resulting in granitic magmatism.

Granitic magmatism in the Qinling orogen can be explained by slab breakoff (Fig. 9). The Triassic collision between the Yangtze and North China cratons caused crustal thickening to >50 km along the Qinling-Dabie orogen. In such circumstances, slab breakoff is a natural consequence (Davies and von Blanckenburg, 1995). Assuming that breakoff occurred at shallow depth, a large thermal anomaly would result in upwelling of the asthenosphere along the Mianlue suture, and initial melting would occur at the base of the enriched underplate. This

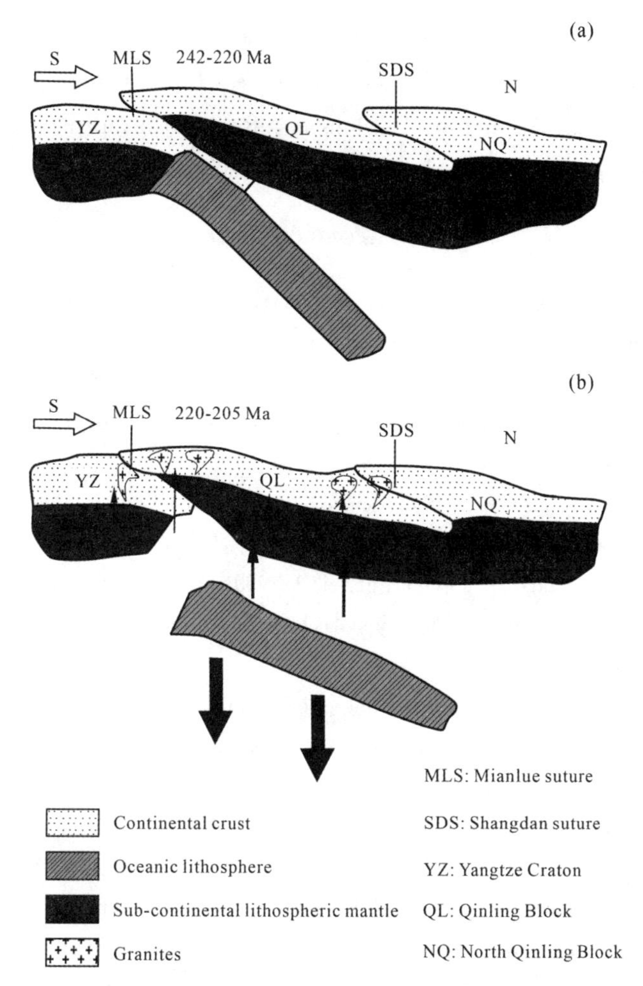

Fig. 9 Suggested model to explain formation of the Triassic post-collisional adakitic granites in the Qinling orogenic belt.

(a) Triassic collision between the Yangtze and North China craton caused crustal thickening in the Qinling-Dabie orogen to >50 km. (b) Subducting oceanic slab breakoff occurred at a shallow depth, and caused widespread granite magmatism.

would result a in melt with SiO_2 of ~60%; high Ba, Sr, and Zr; no significant Eu anomalies; and with lower crustal components (mafic enclaves). Later, mantle-derived melt underplated the thickened lower crust so that temperatures at the boundary between mantle and lower crust were high enough to trigger dehydration melting of garnet-bearing rock at a pressure of >1.5 GPa, resulting in the generation of adakitic magma with high Sr, Ba, and Sr/Y ratios and no negative Eu anomalies (e.g. Rapp et al., 1999), leaving residual eclogite. Given that the lithospheric mantle-derived magmas intruded the adakitic magma chamber (e.g., Ducea and Saleeby, 1998), interaction between pristine adakitic magmas and mantle-derived magmas would result in the formation of the mafic enclaves, and the adakitic magma would have a mantle/crustal component, e.g. high $Mg^{\#}$, Nb/Ta ratios, and Cr and Ni contents.

Acknowledgements This work are supported by the National Natural Science Foundation of China (Grant No. 40872060) , Northwest University Graduate Innovation and Creativity Funds (07YZZ26) , and the Teaching and Research Award Program for Outstanding Young Teachers in Higher Education Institutions of the Ministry of Education, P.R. China.

References

Ames, L., Tilton, G.R., Zhou, G., 1996. Geochronology and isotopic character of ultrahigh-pressure metamorphism with implications for collision of the Sino-Korean and Yangtze cratons, central China. Tectonics, 15, 472.

Atherton, M.P., Petford, N., 1993. Generation of sodium-rich magmas from newly underplated basaltic crust. Nature, 362,144-146.

Branes, C.G., Peterson, S.W., Kistler., 1996. Source and tectonic implication of quartz diorite-trondhjemite magmatism in the Klamath Mountains. Contributions to Mineralogy and Petrology, 123, 40-60.

Beate, B., Monzier, M., Spikings, R., Cotton, J., et al., 2001. Mio-Plioeene adakite generation related to fiat subduction in Southern Ecuador: The Quimsaeoeha-oleanic center. Earth & Planetary Science Letters, 192, 499-508.

Chavagnac, V., Jahn, B.M., 1996. Coesite-bearing eclogites from the Bixiling complex, Dabie Mountains, China: Sm-Nd ages, geochemical characteristics and tectonics implications. Chemical Geology, 133,29.

Chung, S.L., Liu, D.Y., Ji, J., Chu, M.F., et al., 2003. Adakites from continental collision zones: Melting of thickened lower crust beneath southern Tibet. Geology, 31,1021-1024.

Condie, K. C., 2005. TTGs and adakites: Are they both slab melts? Lithos, 80,33-44.

Davies, D.W., von Blanckenburg, F., 1995. Slab breakoff: A model of lithosphere detachment and its test in the magmatism and deformation of collisional orogents. Earth & Planetary Science Letters, 129,85-102.

Defant, M.J., Drummond, M.S., 1990. Derivation of some modern arc magmas by partial melting of young subducted lithosphere. Nature, 347,662-665.

Defant, M.J., Drummond, M.S., 1993. Mount St. Helens: Potential example of the partial melting of the subducted lithosphere in a volcanic arc. Geology, 21, 547-550.

Defant, M.J., Xu, J.F., Kepezhinskas, P., 2002, Adakites: Some variations on a theme. Acta Petrologica Sinica, 18(2), 129-142.

Drummond, M.S., Defant, M.J., 1990. A modelfor trondhjemite-quartz diorite-dacite genesis and crustal growth via slab melting: Archean to modern comparisons. Journal of Geophysical Research, 95, 21503-21521.

Ducea, M., Saleeby, J., 1998. Crustal recycling beneath continental arcs: Silica-rich glass inclusions in ultramafic xenoliths from the Sierra Nevada, California. Earth & Planetary Science Letters, 158, 101-116.

Feeley, T.C., Hacker, M.D., 1995. Intracrustal derivation of Na-rich andesite and dacite magmas: An example from Volcan Ollague Andean Central Volcanic Zone. Journal of Geology, 103, 213-225.

Frost, B.R., Branes, C.C., Collins, W.J., Arculus, R.J., et al., 2001. A geochemical classification for granite rocks. Journal of Petrology, 42, 2033-2048.

Gao, S., Rudnick, R.L., Yuan, H.L., Liu, X.M., et al., 2004. Recycling of lower continental crust in the

North China craton. Nature, 432, 892-897.

Garrison, J.M., Davidson, J.P., 2003. Dubious case for slab melting in the Northern volcanic zone of the Andes. Geology, 31,565-568.

Green, T. G., Ringwood, A. E., 1968. Genesis of the calc-alkaline ingeous rocks suit. Contributions to Mineralogy and Petrology, 18, 105-162.

Gutscher, M. A., Maury, R. C., Eissen, J. P, Bourdon, E., 1999. Can slab melting be caused by flat subduction? Geology, 28(6),535-538.

Hacker, R. B., Ratschbacher, L, Webb, L., 1998. U-Pb zircon ages constrain the architecture of the ultrahigh-pressure Qinling-Dabie Orogen, China. Earth & Planetary Science Letters, 161,215-230.

Huang, X., Wu, L.R., 1990. Nd-Srisotopes of granitoids from Shaanxi Province and their significance for tectonic evolution. Acta Petrologica Sinica, 2, 1-11.

Lai, S.C., Zhang, G.W., 1996. Geochemical features of ophiolites in Mianxian-Lueyang suture zone, Qinling Orogenic Belt. Journal of Chinese University of Geoscience, 7, 165-172.

Lai, S.C., Zhang, G.W., Dong, Y.P., Pei, X.Z., et al., 2004a. Geochemistry and regional distribution of ophiolite and associated volcanics in Mianlüe suture, Qinling-Dabie Mountains. Science in China: Series D, 47,289-299.

Lai, S.C., Zhang, G.W., Li, S.Z., 2004b. Ophiolites from the Mianlue suture in the Southern Qingling and their relationship with eastern paleotethys evolution. Acta Geologica Sinica, 78, 107-117.

Li, S.G., Sun, W.D., 1996. A middle Silurian-Early Devonian magmatic arc in the Qinling Mountains of central China: A discussion. Journal of Geology, 104, 501-503.

Li, S.G., Sun, W.D., Zhang, G.W., Chen, J.Y., et al., 1996. Chronology and geochemistry of metavolcanic rocks from Heigouxia Valley in Mianlue tectonic arc, South Qinling: Observation for a Paleozoic oceanic basin and its close time. Science in China: Series D, 39, 300-310.

Li, S.G., Hart, S.R., Zheng, S., Liou, D., et al., 1989. Timing of collision between the North and South China Blocks: Sm-Nd isotopic age evidence. Science in China: Series B, 32,1391-1400.

Li, S.G., Jagoutz, E., Chen, Y.Z., 2000. Sm-Nd and Rb-Sr isotopic chronology and cooling history of ultrahigh pressure metamorphicrocks and their country rocks at Shuanghe in the Dabie Mountains, central China. Geochim. Cosmochim. Acta, 64, 1077.

Li, S.G., Huang, F., Li, H., 2002. Post-collisional lithosphere delamination of Dabie-Sulu orogen. Chinese Science Bulletin, 47 (3), 259-263.

Lu, X.X., Wei, X.D., Xiao, Q.H., Zhang, Z.Q., et al., 1999. Geochronological studies of rapakivi granites in Qinling and its geological implication. Geological Journal of China Universities, 15(4), 372-377.

Martin, H., 1999. The adakitic magma: Modern analogue of Archean granotoids. Lithos, 46, 411-429.

Martin, H., Smithies, R.H., Rapp, R., Moyen, J.F., et al., 2005. An overview of adakite, quartz diorite-trondhjemite- granodiorite (TTG), and sanukitoid: Relationships and some implications for crustal evolution. Lithos, 79:1-24.

Mattauer, M., Matte, P., Malaveile, J., Tapponier, P., et al., 1985. Tectonics of Qinling belt: Build-up and evolution of Western Asia. Nature, 317, 496-500.

Meijer, A., 1983. The orogin of low-K rhyolites from the Mariana frontal arc. Contributions to Mineralogy and

Petrology, 83, 45-51.

Meng, Q.R., Zhang, G.W., 1999. Timing of the collision of the North and South China Blocks: Controversy and reconciliation. Geology, 27, 123-126.

Muir, R.J., Weaver, S.D., Bradshaw, J.D., Eby, G.N., et al., 1995. The Cretaceous separation point batholith, New Zealand: Granitiod magmas formed by melting of mafic lithosphere. Journal of the Geological Society of London, 152, 689-701.

Peccerillo, R., Taylor, S. R., 1976. Sr geochemistry of Eocene calc-alkaline volcanic rocks from the Kastamonu area, northern Turkey. Contributions to Mineralogy and Petrology, 58, 63-81.

Peacock, S.M., Rushmer, T., Thompson, A.B., 1994. Partial melting of subducting oceanic crust. Earth & Planetary Science Letters, 121, 227-244.

Petford, N., Atherton, M., 1996. Na-rich partial melt from newly underplated basaltic crust: The Cordmera Blanca Batholith, Peru. Journal of Petrology, 37, 491-521.

Prouteau, G., Scaillet, B., Pichavant, M., Maury, R., 2000. Evidence for mantle metasomatism by hydrous silicic melts derived from subducted oceanic crust. Nature, 410, 197-200.

Qi, L., Hu, J., Gregoire, D.C., 2000. Determination of trace elements in granites by inductively coupled plasma-mass spectrometry. Talanta, 51, 507-513.

Qin, J.F., Lai, S.C., Li, Y.F., 2005. Petrogenesis and geological significance of the Yangba granodiorites from Bikou area, northern margin of the Yangtze Block. Acta Petrologica Sinica, 21(3), 697-710.

Rapp, R. P., Watson, E. B., 1995. Dehydration melting of metabasalt at 8 – 32 kbar: Implications for continental growth and crust-mantle recycling. Journal of Petrology, 36, 891-931.

Rapp, R.P., Shimizu, N., Norman, M.D., Applegate, G.S., 1999. Reaction between slab-derived melts and peridotite in the mantle wedge: Experimental constraints at 3. 8 GPa. Chemical Geology, 160, 335-356

Sajona, F.G., Maury, R.C., Pubellier, M., 2000. Magmatic source enrichment by slab-derived melts in a young post-collision setting, central Mindanao (Philippines). Lithos, 54, 173-206.

Sen, C., Dunn, T., 1994. Dehydration melting of a basaltic composition amphibolite at 1. 5 GPa and 2. 0 GPa: Implication for the origin of adakites. Contributions to Mineralogy and Petrology, 117, 394-409.

Smithies, R.H., 2000. The Archaean quartz diorite-trondhjemite-granodiorite (TTG) series is not an analogue of Cenozoic adakite. Earth & Planetary Science Letters, 182, 115-125.

Spulber, S.D., Rutherford, M.J., 1983. The orogin of rhyolite and plagiogranite in oceanic crust: An experimental study. Journal of Petrology, 24, 1-25.

Stern, C.R., Kilian, R., 1996. Role of the subducted slab, mantle wedge and continental crust in the generation of adakites from the Andean Austral Volcanic Zone. Contributions to Mineralogy and Petrology, 123, 263-281.

Sun, W.D., Li, S.G., 1998. Pb isotopes of granitoids suggest Devonian accretion of Yangtze (South China) craton to North China craton: Comment. Geology, 26, 859-860.

Sun, W.D., Li, S.G., Chen, Y.D., Li, Y.J., 2002. Timing of synorogenic granotoids in the south Qinling, central China: Constraints on the evolution of the Qinling-Dabie Orogenic Belt. Journal of Geology, 110, 457-468.

Sun, S.S., McDonough, W.F., 1989. Chemical and isotopic systemmatics of oceanic basalts: Implication for the mantle composition and process. In: Saunder, A.D., Norry, M.J. Magmatism in the ocean basins. Geological Society of London Special Publications, 42:313-345. London: Geological Society of London and Blackwell Scientific Publications.

Taylor, S.R., McLennan, S.M., 1985. The Continental Crust: Its Composition and Evolution. Blackwell Scientific Pub., Palo Alto, CA, 1-328.

Wang, Q., McDermott, F., Xu, J.F., Bellon, H., et al., 2005. Cenozoic K-rich adakitic volcanic rocks in the Hohxil area, northern Tibet: Lower-crustal melting in an intracontinental setting. Geology, 33, 465-468.

Wang, Q., Wyman, D.A., Xu, J.F., Zhao, Z.H., et al., 2006. Petrogenesis of Cretaceous adakitic and shoshonitic igneous rocks in the Luzong area, Anhui Province (eastern China): Implications for geodynamics and Cu-Au mineralization. Lithos, 89, 424-446.

Wang, X.X., Wang, T., Lu, X.X., Xiao, Q.H., 2003. Laojunshan and Qinlingliaag rapakiv-textured granitoids in North Qinling and their tectonic setting: A possible orogenic-type rapakivi gramtoids. Acta Petrologica Sinica, 19(4), 650-660.

Wang, X.X., Wang, T., Happala, I., Lu, X.X., 2005. Genesis of mafic enclaves from rapakivi-textured granites in the Qinling and its petrological significance: Evidence of elements and Nd, Sr isotopes. Acta Petrologica Sinica, 21(3), 935-946.

Wang, F., Zhu, R.X., Li, Q., He, H.Y., et al., 2004. A differential uplifting of Qinling orogeny belt evidences from $^{40}Ar/^{39}Ar$ thermochronology of granites. Earth Science Frontiers, 11(4), 445-459.

White, A.J.R., Cappell, B.W., 1977. Ultrametamorphism and granitoid genesis. Tectonophysics, 43:7-22

Wu, F.Y., Ge, W.C., Sun, D.Y., 2002. The definition, discrimination of adakites and their geological role. In: Xiao, Q.H., Deng, J.F., Ma, D.Q., et al. The ways of investigation on granotoids. Beijing: Geological Publishing House, 172-191.

Xu, J.F., Shinjio, R., Defant, M.J., Wang, Q., et al., 2002. Origin of Mesozoic adakitic intrusive rocks in the Ningzhen area of east China: Partial melting of delaminated lower continental crust? Geology, 32, 1111-1114.

Yuan, H.L., Wu, F.Y., Gao, S., Liu, X.M., et al., 2004. Determination of zircons from Cenozoic intrusions in Northeastern China by laser ablation ICP-MS. Chinese Science Bulletin, 48, 2411-2421.

Zhai, X., Day, H.W., Hacker, B.R., You, Z.D., 1998. Paleozoic metamorphism in the Qinling orogen, Tongbai Mountains, central China. Geology, 26, 371-374.

Zhang, B.R., Luo, T., Gao, S., Ouyang, J.P., et al., 1994. Geochemical study of the lithosphere, tectonism and metallogenesis in the Qinling-Dabieshan region. Wuhan: Chinese University of Geoscience Press, 110-122.

Zhang, C.L., Zhang, G.W., Yan, Y.X., Wang, Y., 2005. Origin and dynamic significance of Guangyoushan granitic plutons to the north of Mianlue zone in southern Qinling. Acta Petrologica Sinica, 21(3), 711-720.

Zhang, H.F., Ouyang, J.P., Ling, W.L., Chen, Y.L., 1997. Pb, Sr, Nd isotopic composition of Ningshan granitoids, South Qinling and their deep geological information. Acta Petrologica et Mineralogica, 16(1), 22-31.

Zhang, H.F., Zhang, L., Harris, N., Jin, L.L., et al., 2006. U-Pb zircon ages, geochemical and isotopic compositions of granitoids in Songpan-Graze fold belt, eastern Tibetan Palteau: Constraints on petrogenesis and tectonic evolution of the basement. Contributions to Mineralogy and Petrology, 152(1), 75-88.

Zhang, H.F., Jin, L.L., Zhang, L., Nigel, Harris., et al., 2007a. Geochemical and Pb-Sr-Nd isotopic compositions of granitoids from western Qinling belt: Constraints on basement nature and tectonic affinity. Science in China: Series D, 50(2),184-196.

Zhang, H.F., Xiao, L., Zhang, L., Yuan, H.L., et al., 2007b. Geochemical and Pb-Sr-Nd isotopic compositions of Indosinian granitoids from the Bikou block, northwest of the Yangtze plate: Constraints on petrogenesis, nature of deep crust and geodynamics. Science in China: Series D, 50 (7), 972-983.

Zhang, G.W., Zhang, B.R., Yuan, X.C., Chen, J.Y., 2002. Qinling Orogenic Belt and Continental Dynamics. Beijing: Science Press, 1-855.

Zhang, G.W., Cheng, S.Y., Guo, A.L., Dong, Y.P., et al., 2004. Mianlue paleo-suture on the southern margin of central orogenic system in Qingling-Dabie-with a discussion of the assembly of the main part of the continent of China. Geological Bulletin of China, 23(9-10), 846-853.

Zhang, Z.Q., Zhang, G.W., Tang, S.H., Wang, J.H., 2001. On the age of metamorphic rocks of the Yudongzi Group and the Archean crystalline basement of the Qinling orogen. Acta Geologica Sinica, 75 (2), 198-204.

Zircon U-Pb ages, geochemistry, and Sr-Nd-Pb-Hf isotopic compositions of the Pinghe pluton, Southwest China: Implications for the evolution of the early Paleozoic Proto-Tethys in Southeast Asia①

Zhao Shaowei Lai Shaocong② Qin Jiangfeng Zhu Renzhi

Abstract: The geological record of the Neoproterozoic to early Paleozoic Proto-Tethyan Ocean in Southeast Asia is not clear. To better constrain the evolution of the Proto-Tethys, we present new geochronology, geochemistry, and petrology of the late Cambrian to Ordovician Pinghe pluton monzogranite from the Baoshan Block, western Yunnan, southwest China. Laser ablation inductively coupled plasma mass spectrometry (LA-ICP-MS) analyses of four zircon samples yield ages of 482–494 Ma and 439–445 Ma for the pluton, interpreted as two episodes within one magmatic event accompanying the whole process of subduction-collision-orogrny between buoyant blocks and oceanic crust of the Proto-Tethys. The monzogranite belongs to the strong peraluminous, high-K, calc-alkaline series and shows characteristics of both I-type and S-type granitic rocks. It is characterized by extremely high Rb/Sr and Rb/Ba but low TiO_2, MgO, FeO^T and CaO/Na_2O ratios. The monzogranite is also moderately enriched in light rare earth elements (LREEs), depleted in heavy rare earth elements (HREEs), lacks HREE fractionation, and has strongly negative Eu ($Eu/Eu^* = 0.06–0.49$), Ba, Nb, Ta, Sr, and Ti anomalies. Whole-rock $\varepsilon_{Nd}(t)$ and $\varepsilon_{Hf}(t)$ values range from −11.6 to −8.7 and −9.58 to −5.55, respectively. Nd and Hf two-stage model ages range from 1.66 Ga to 2.06 Ga and 2.14 Ga to 3.00 Ga, respectively, with variable radiogenic $^{206}Pb/^{204}Pb(t)$ (16.547–18.705), $^{207}Pb/^{204}Pb(t)$ (15.645 – 15.765), and $^{208}Pb/^{204}Pb(t)$ (38.273–38.830). These signatures suggest that the monzogranite magma was derived from partial melting of heterogeneous metapelite, which was generated from Neoarchean to Paleoproterozoic materials mixed with basaltic magma. The monzogranite magma underwent crystallization differentiation of plagioclase, K-feldspar and ilmenite. Magmatism to form the Pinghe pluton occurred in a post-collisional setting. Based on the comparison of coeval granites throughout adjacent regions (e.g., Himalayan orogen, Lhasa Terrane, and parts of Gondwana supercontinent), we propose that the Baoshan Block was derived from the northern Australian Proto-Tethyan Andean-type active continental margin of Gondwana and experienced subduction of the Proto-Tethys oceanic crust and accretion of an outboard micro-continent. The Pinghe pluton could have formed when a

① Published in *International Geology Review*, 2014,56(7).

② Corresponding author.

subducting oceanic slab broke off during collision.

Introduction

The "Tethys" was proposed by Suess (1893) as the body of water separating Gondwana in the south from Laurasia in the north. Researchers now divide the Tethys into at least two successive ocean basins, the Paleo-Tethys and the Neo-Tethys (Sengör et al., 1988). However, even the Meso-Tethys (Metcalfe, 1996a, b, 2002) and Proto-Tethys (Li et al., 1990; Chen, 1994; Pan and Chen, 1997; Guo, 2001; Lu, 2001; Xiao et al., 2003, 2009; von Raumer and Stampfli, 2008) have also been proposed. The Proto-Tethys, considered as a transmeridional ocean, opened in Neoproterozoic time and closed in late early Paleozoic time, is closely related with the break-up of Rodinia and the amalgamation of Gondwana. The record of the Proto-Tethys is not always globally considered—for instance, Scotese (2004) did not mention the Proto-Tethys in the global reconstruction project, but the geological record of the Proto-Tethyan Ocean is found throughout East and Southeast Asia. The opening and closing of successive Tethyan basins and the subsequent collision and accretion of micro-continental blocks, such as North and South China, Indochina, Tarim, Qaidam, Sibumasu, Qiangtang, Lhasa, West Burma and Woyla Terranes (Fig. 1a) (Metcalfe, 1996a, b, 2002, 2006), resulted in the formation of East and Southeast Asia (Sengör et al., 1988; Metcalfe, 1996a, b; Wopfner, 1996; Ueno, 2000). However, the location of these micro-continental blocks in Early Palaeozoic time are poor known and the relationships among them are also undefined.

The Baoshan Block in Western Yunnan plays an important role in the eastern Tethyan tectonic belt, and its evolution is significant for understanding the development of the Tethys within Southeast Asia (Chen et al., 2005, 2006, 2007; Liu et al., 2009). In addition, there may be some genetic relation between the Lhasa and Baoshan blocks (Dong et al., 2012). Information preserved in the Early Paleozoic orogenic belt was masked by intense Cenozoic orogenesis in many areas of Himalaya (Valdiya, 1995; Gehrels et al., 2003, 2006a) and western Yunnan, and these complicated tectonic environments gave rise to debate about the development of the Tethys. However, abundant unconformities between the Cambrian and Ordovician strata have been reported in many regions of the Lhasa Block and Himalayan orogen (Stöcklin and Bhattarai, 1977; Stöcklin, 1980; Hughes, 2002; Gehrels et al., 2003; Zhou et al., 2004; Li et al., 2010) and Baoshan Block (Huang et al., 2009, 2012). Corresponding magmatism and metamorphism have also been found in these areas, particularly in the Baoshan Block and Gaoligong belt in western Yunnan (Chen et al., 2006, 2007; Liu et al., 2009; Li et al., 2012). However, the geochronology and geochemistry of these magmatic and metamorphic rocks are shown in these researches rather than the defined tectonic setting and evolution of the Baoshan Block, and the affinity of Baoshan Block also remains poorly

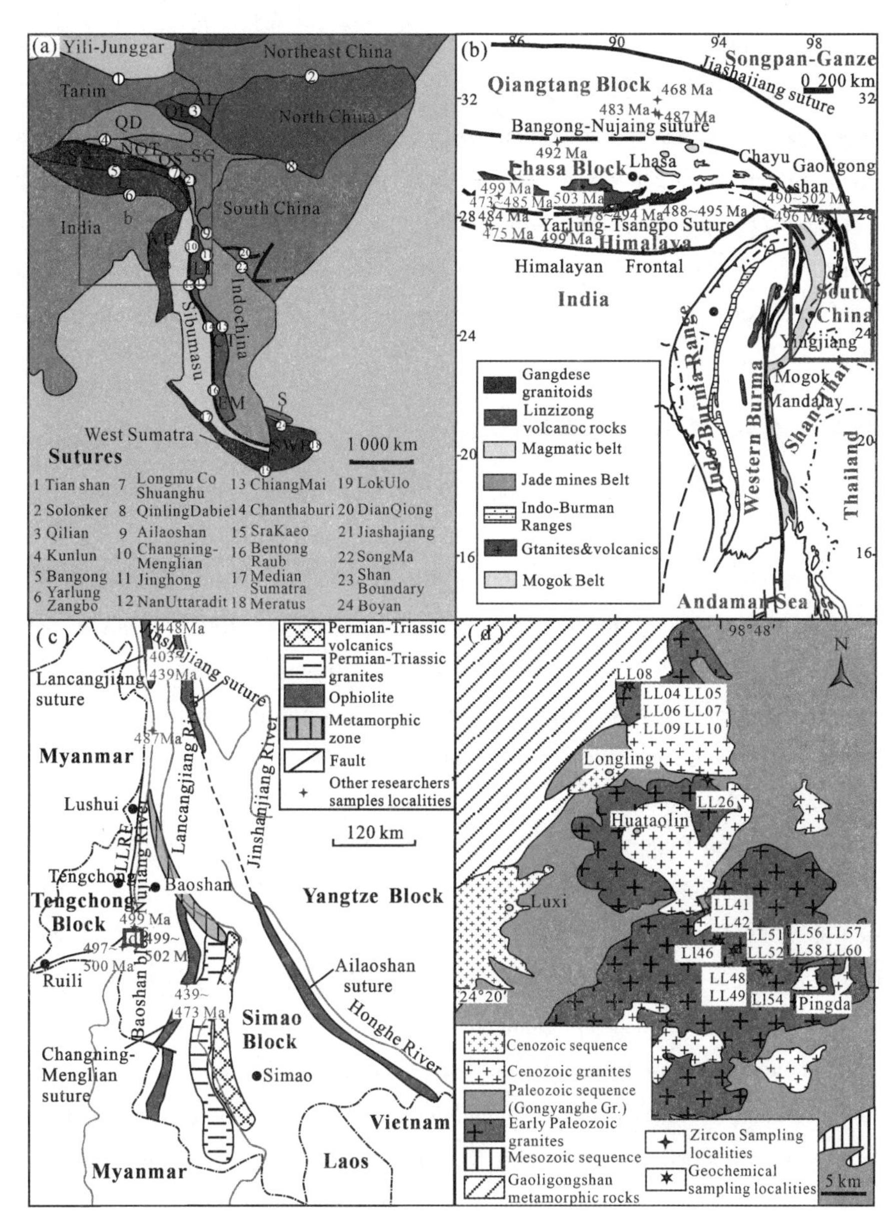

Fig. 1 (a) Distribution of principal continental blocks, arc terranes, and sutures of eastern Asian (modified after Metcalfe, 2013). (b) Geological sketch of the Tibetan Plateau (modified after Xu et al., 2012). The geochronological ages after Cawood et al. (2007), Gehrels et al. (2006b), Lee and Whitehouse (2007), Guynn et al. (2012), Wang et al. (2012), Zhang et al. (2012), Zhu et al. (2012b). (c) Geological sketch of the Tethyan belt in western Yunnan (modified after Zhong, 1998; Chen et al., 2007) and the geochronological ages after Li et al. (2012), Liu et al. (2009), Yang et al. (2012), Song et al. (2007), Jian et al. (2009), Jian and Liu (2002), Wang et al. (2013a). (d) Geological of study area (modified after Zhong, 1998; Chen et al., 2007) with the sampling localities.

Fig. 1a is the distribution map of major tectonic units in East Asia, not a geographic map of countries. AL: Ala Shan; QL: Qilian; QD: Qaidam; SG: Songpan Ganzi accretionary complex; NQT: North Qiangtang; SQT: South Qiangtang; QS: Qadom-Simao; L: Lhasa; WB: West Burma; SWB: South West Borneo; S: Semitau; LT: Lincang Arc Terrane; CT: Chanthaburi Arc Terrane; EM: East Malaya.

constrained. Moreover, the development of Proto-Tethys in western Yunnan and the location of the Baoshan Block in Proto-Tethys are unclear.

In this paper, we present new zircon U-Pb geochronology, whole rock major and trace element compositions, and whole-rock Sr-Nd-Pb-Hf isotopic characteristics of Early Paleozoic granites for the Pinghe pluton in the Baoshan Block, western Yunnan. These lines of evidences constrain not only the formation and tectonic environment of Early Paleozoic granites, but evolution of the Neoproterozoic to Early Paleozoic Proto-Tethys in western Yunnan, as preserved in the rocks of the Baoshan Block.

Geological setting and petrology

Western Yunnan is an important location for studying the Tethyan belt. This region consists of several micro-continental blocks and tectonic sutures (Fig. 1a-c): the Tengchong-Baoshan Block of Gondwana affinity, the Simao Block of Yangtze Block affinity, the Gaoligong orogenic belt, and Changning-Menglian suture zone (Wu et al., 1995; Metcalfe, 1996a,b; Wang, 1996; Wopfner, 1996; Zhang et al., 1997; Zhong, 1998; Ueno, 2000). Some researchers have proposed that the Changning-Menglian suture, a branch of the Paleo-Tethys, closed during the Triassic connecting the Baoshan Block to the west with the Simao Block to the east (e.g., Fang et al., 1994; Wu et al. 1995; Feng et al., 2002; Chen et al., 2005, 2007). The Tengchong-Baoshan Block is separated by the Gaoligong orogenic belt, and consists of the Tengchong Block, the Baoshan Block and the Luxi geosyncline (Chen et al., 2005), as well as the two blocks belonging to the northern tip of the Sibumasu terrane (Metcalfe, 1988, 1996a,b, 1998; Wu et al., 1995). Moreover, it is thought that the Tengchong Block and Baoshan Block were separated by the Bangong-Nujiang ocean basin during the development of the Neo-Tethys before it closed in early Jurassic time. Later, the two blocks were joined and led to the formation of the Gaoligong orogenic belt (Zhong, 1998). Subsequently, the Gaoligong orogen developed the strike-slip fault in Miocene time. However, the relationship between the Tengchong and Baoshan Blocks is contentious. Some researchers argue that the Tengchong and Baoshan Blocks may not have been connected with each other in Palaeozoic time: ① the basement rocks in these blocks are different—the Gongyanghe Group in the Baoshan Block and the Gaoligong Group in the Tengchong Block; ② differences in Permian stratigraphy between the Tengchong and Baoshan blocks, with some volcanics and red beds presenting on Baoshan but not on Tengchong (Jin, 1994; Wopfner, 1996; Wang et al., 2002); and ③ Li et al. (2014) suggested that the Tengchong Block was located along the Indian margin of Gondwana during the early Palaeozoic, in light of detrital zircon ages and Hf isotopic data. In addition, Jin (2002) concluded that both blocks were derived from Gondwana, but could not have been adjacent to each other during the Permo-Carboniferous, and that the juxtaposition of the Baoshan and Tengchong blocks as seen today is due later tectonic movements. On the contrary, Metcalfe (2002) argued that both the Tengchong and Baoshan blocks should be considered as part of the Sibumasu Terrane. Wang et al. (2002) considered that the Tengchong Block has a generally similar Permian sedimentary history to the Baoshan Block, although it lacks volcanic rocks.

The Pinghe pluton (Fig. 1d) is located in the western Baoshan Block and is a granitoid batholith of Late Cambrian to Early Ordovician age (Bureau of Geology and Mineral Resources of Yunnan Province, 1990). The area is bounded by the Lushui-LuXi-Ruili fault (LLRF) to the west and the Nujiang River to the east. The outcrop covers about 800 km^2 along a NNE-trending tectonic line. The granite intruded into the Gongyanghe group, a Neoproterozoic to Cambrian low-grade metamorphic, terrestrial sedimentary sequence, in the Palaeozoic time. The Pinghe pluton was also intruded by several small late Mesozoic granitoid stocks (Chen, 1987; Jin and Zhuang, 1988; Bureau of Geology and Mineral Resources of Yunnan Province, 1990; Chen et al., 2007; Liu et al., 2009).

Most of the granitoid rocks are fresh, grey to off-white, with equigranular or porphyritic textures (Fig. 2a-c) in the Pinghe pluton. The batholith mainly consists of medium- to coarse-grained biotite or two-mica monzogranite (Fig. 2d-f) with no deformation or metamorphism, and is comprised of plagioclase (~40%, An = 27), orthoclase (~15%), microcline or perthite (~15%), quartz (~20%), biotite plus muscovite (~10%), minor amphibole, and accessory minerals including zircon, apatite, garnet, magnetite, and sphene. The batholith composition ranges from acidic at the centre to neutral at the edge, and porphyritic granodiorite occurs at the margin of the batholith. From this, we collected four zircon samples and carried

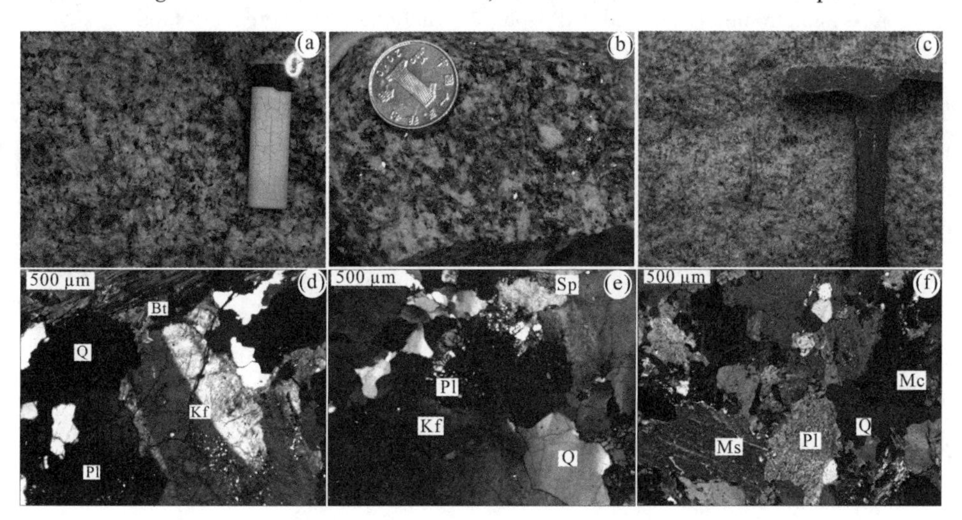

Fig. 2 Field and microscope photographs of the monzogranite from Pinghe pluton, Baoshan Block, western Yunnan, SW China.

out 16 geochemical analyses. The locations of the four samples collected for zircon dating from the Pinghe pluton were as follows: N24°39.360′, E98°47.726′ for LL08; N24°35.789′, E98°48.476′ for LL26; N24° 31.086′, E98° 47.789′ for LL46; and N24° 22.888′, E98°47.436′ for LL54.

Analytical methods

Fresh samples of granitoid rocks were selected for elemental analyses. All U-Pb dating, major and trace element, and Sr-Nd-Pb-Hf isotopic analyses were conducted in the State Key Laboratory of Continental Dynamics, Northwest University in Xi'an, China.

Major and trace element analyses

Chips of whole-rock samples were pulverized to 200 mesh using a tungsten carbide ball mill. Major elements were analysed by X-ray fluorescence (XRF; Rikagu RIX 2100) with an analytical error of less than 2%. Trace element and rare earth element (REE) contents were analysed by inductively coupled plasma mass spectrometry (ICP-MS; Agilent 7500a) using United States Geological Survey (USGS) and Chinese national rock standards (BCR-2, GSR-1 and GSR-3). For the trace element analysis, sample powders were digested using an HF + HNO_3 mixture in high-pressure Teflon bombs at 190 ℃ for 48 h. For most trace elements, analytical error is less than 2% and precision is greater than 10% (Liu et al., 2007).

Sr-Nd-Pb-Hf isotopic analyses

Whole-rock Sr-Nd-Pb-Hf isotopic data were obtained using a Nu Plasma HR multi-collector mass spectrometer. Sr and Nd isotopic fractionations were corrected to $^{87}Sr/^{86}Sr$ = 0.119 4 and $^{146}Nd/^{144}Nd$ = 0.721 9, respectively. During the analysis, the NIST SRM 987 standard yielded a mean value of $^{87}Sr/^{86}Sr$ = 0.710 250 ± 12 (2σ, n = 15) and the La Jolla standard gave a mean of $^{146}Nd/^{144}Nd$ = 0.511 859 ± 6 (2σ, n = 20). Whole-rock Pb was separated by anion exchange in HCl-Br columns, and Pb isotopic fractionation was corrected to $^{205}Tl/^{203}Tl$ = 2.387 5. Over the period of analysis, 30 measurements of NBS 981 gave average values of $^{206}Pb/^{204}Pb$ = 16.937 ± 1 (2σ), $^{207}Pb/^{204}Pb$ = 15.491 ± 1 (2σ), and $^{208}Pb/^{204}Pb$ = 36.696 ± 1 (2σ). The BCR-2 standard gave $^{206}Pb/^{204}Pb$ = 18.742 ± 1 (2σ), $^{207}Pb/^{204}Pb$ = 15.620 ± 1 (2σ), and $^{208}Pb/^{204}Pb$ = 38.705 ± 1 (2σ). Total procedural Pb blanks were in the range of 0.1–0.3 ng. Whole-rock Hf was also separated by single-anion exchange columns. In the course of analysis, 22 measurements of JCM 475 yielded an average of $^{176}Hf/^{177}Hf$ = 0.282 161 3 ± 0.000 001 3 (2σ) (Yuan et al., 2007).

Zircon U-Pb analyses

Zircon grains from the granitoid rocks were separated using conventional heavy liquid and magnetic techniques. Representative zircon grains were handpicked and mounted in epoxy resin disks, then polished and carbon coated. Internal morphology was examined by cathodoluminescence (CL) prior to U-Pb isotopic dating.

U-Pb isotopes of the zircon were analysed by laser ablation ICP-MS (LA-ICP-MS) on an Agilent 7500a equipped with a 193 nm laser, following the method of Yuan et al. (2004). The $^{207}Pb/^{235}U$ and $^{206}Pb/^{238}U$ ratios were calculated using the GLITTER program, which was corrected using the Harvard zircon 91500 as external calibration. These correction factors were applied to each sample to correct for both instrumental mass bias and depth-dependent elemental and isotopic fractionation. The detailed technique is described in Yuan et al. (2004). Common Pb contents were evaluated using the method described in Andersen (2002). Age calculations and concordia diagrams were completed using ISOPLOT version 3.0 (Ludwig, 2003). The errors quoted in tables and figures are at 2σ levels.

Results

Zircon U-Pb ages

Zircon U-Pb isotopic results and CL images of four samples from the Pinghe pluton are presented in Figs. 3 and 4 and the isotope data are listed in Supplementary Dataset 1.

The analysed zircon grains are light brown to colourless subhedral to euhedral crystals. The lengths range from 150 μm to 350 μm with aspect ratios of 3∶1 to 2∶1. Most of the grains show oscillatory zoning in the CL images indicating igneous crystallization. Parts of zircon grains display strong luminescence with inconspicuous oscillatory zoning in the core and the opposite in the margin, while other grains have contrary phenomena. However, a few zircon grains show the discontinuous oscillatory zoning between the core and margin in the CL images, which may represent the inherited or trapped zircon in the core and the crystallized zircon in the margin or the embodying of the different crystallization stages.

Thirty-six analyses were conducted for sample LL08. Ten out of 36 spots have concordant $^{206}Pb/^{238}U$ ages of 473–507 Ma, giving a weighted mean age of 482.8±6.4 Ma, MSWD=4.1, and with variable U (383–4354 ppm) and Th (54–807 ppm) contents with Th/U ratios ranging from 0.07 to 0.75. Spots 3, 5, and 12 have $^{206}Pb/^{238}U$ ages of 507 Ma, 519 Ma, and 528 Ma, respectively. According to their morphology, the ages are interpreted as the magmatism starting in the late Cambrian and continuing to the Early Ordovician. Seventeen spots yield Late Ordovician ages, and may represent another magmatic episode that subsequently occurred, with U (251–3 863 ppm), Th (119–997 ppm), and Th/U ratios of 0.05–1.57. These analyses show concordant $^{206}Pb/^{238}U$ ages of 431–451 Ma, with a weighted mean age of 441.8 ± 3.8 Ma, and MSWD = 3.3. Other spots display discordant ages, because the laser

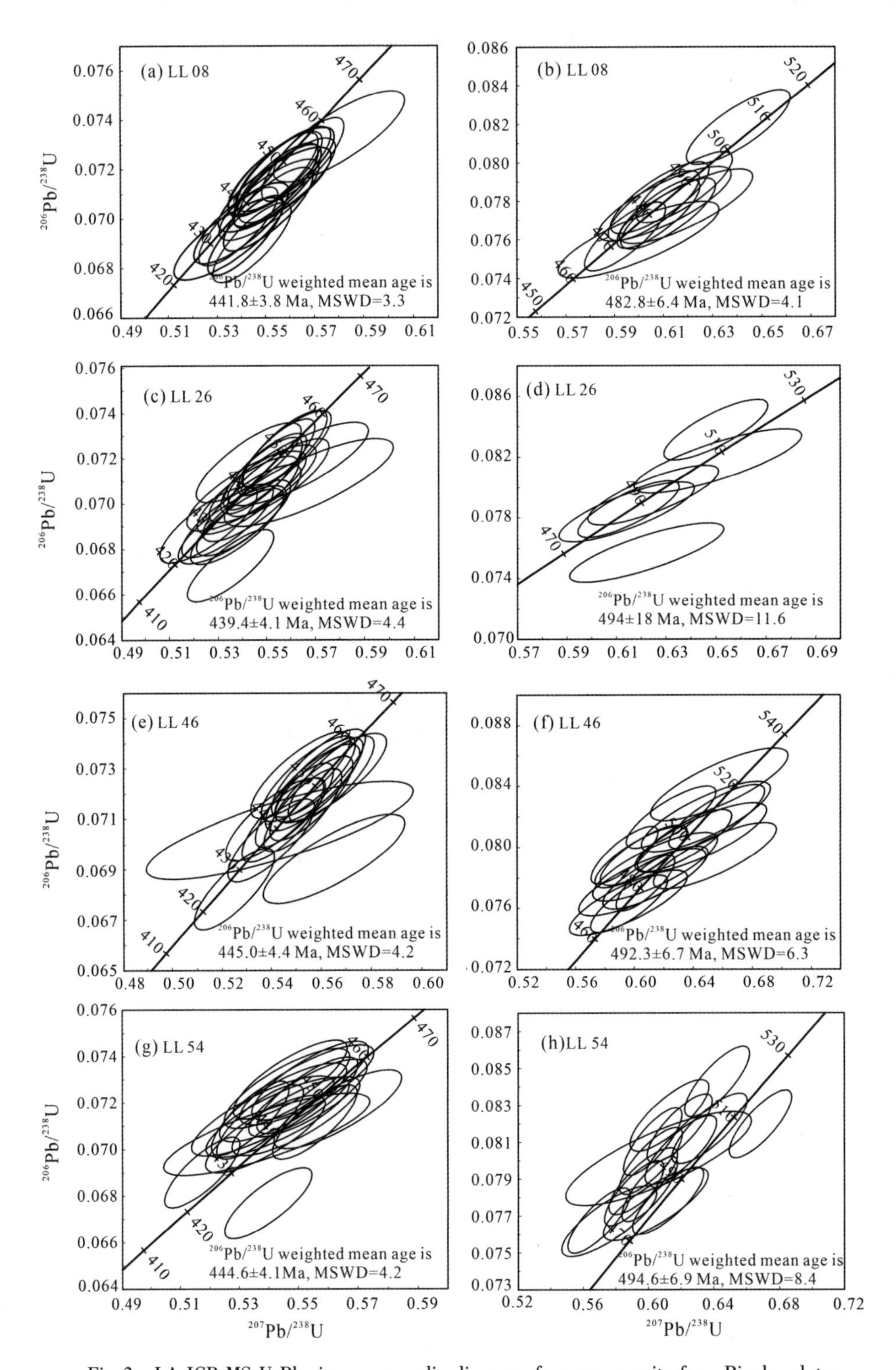

Fig. 3 LA-ICP-MS U-Pb zircon concordia diagrams for monzogranite from Pinghe pluton.

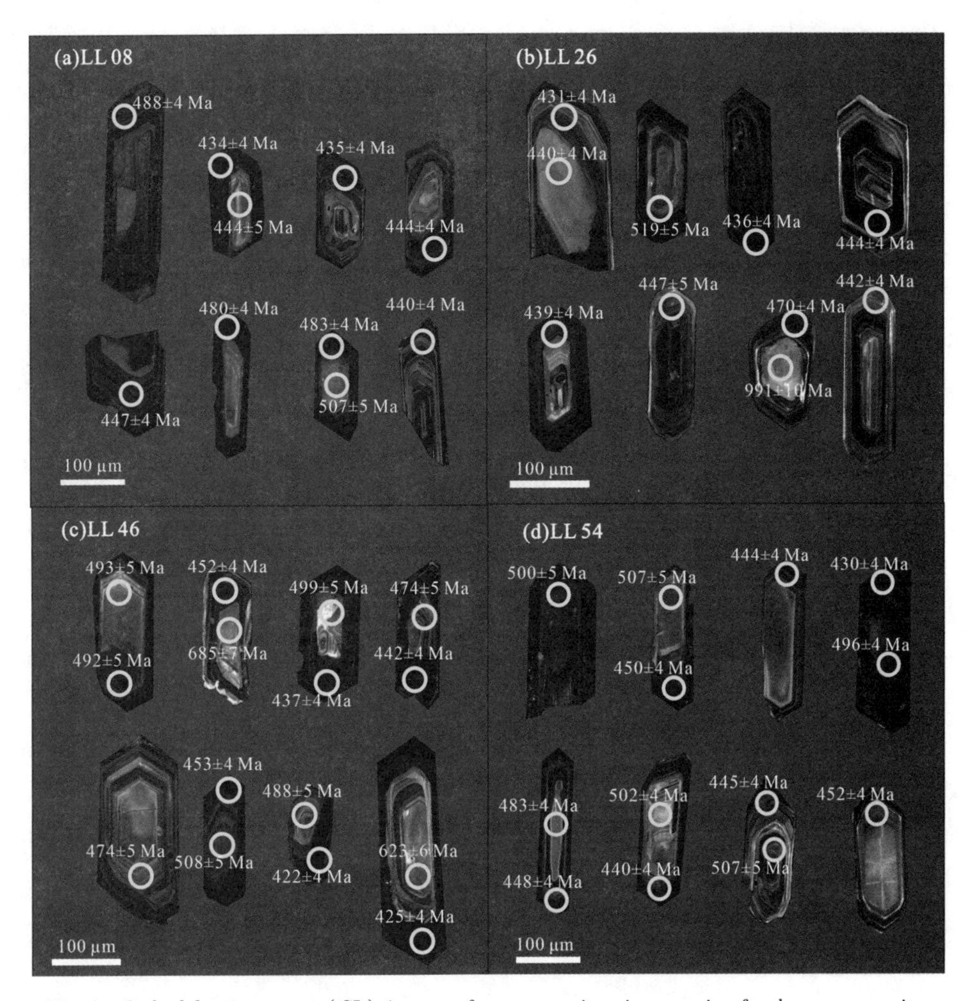

Fig. 4 Cathodoluminescence (CL) images of representative zircon grains for the monzogranite from Pinghe pluton, western Yunnan, SW China.

locations have fallen between the core and margin in the zircon or the signal yielded by the zircon was inadequate.

Thirty-six analyses were also conducted for sample LL26. Spots 16, 17, 28, 32, 34, and 35 have $^{206}Pb/^{238}U$ ages of 658–991 Ma, with variable U (109–1 574 ppm) and Th (38–1 394 ppm) contents, and their morphology suggests that they are xenocrysts. Spots 7, 13, 24, 29, 30, and 33 have concordant $^{206}Pb/^{238}U$ ages of 470–519 Ma, with a weighted mean age of 494 ± 18 Ma, MSWD = 11. 6, and have less variable U (507–1 409 ppm) and Th (98–259 ppm) contents with Th/U ratios of 0. 12 – 0. 29. Nineteen out of 36 spots display concordant $^{206}Pb/^{238}U$ ages of 420–451 Ma, giving a weighted mean age of 439. 4 ± 4. 1 Ma, MSWD = 4. 4, and more variable U (513–6 278 ppm) and Th (99–2 530 ppm) contents, with Th/U ratios of 0. 05–1. 11. Spots 10, 12, 14, 18, and 19 have unreliable ages due to laser spots located in the mixing zone between the core and margin of the zircon.

Thirty-six analyses were conducted for sample LL46. Spots 6 and 36 have $^{206}Pb/^{238}U$ ages of 623 Ma and 685 Ma, respectively, with U (286–572 ppm) and Th (67–243 ppm). Their morphology suggests that they are xenocrysts. Sixteen out of 36 spots have concordant $^{206}Pb/^{238}U$ ages of 474–521 Ma, giving a weighted mean age of 493.3±6.7 Ma, MSWD=6.3, and have U (139–1 128 ppm) and Th (45–245 ppm) contents with Th/U ratios of 0.09–0.51. The other 16 spots yield concordant $^{206}Pb/^{238}U$ ages of 425–453 Ma, with a weighted mean age of 445.0±4.0 Ma, MSWD=4.2, and contained U (397–2 764 ppm) and Th (139–502 ppm) contents with Th/U ratios of 0.08–0.40. Spots 10 and 22 have unreliable ages, possibly for the same reason as the above.

Thirty-six spots were conducted for sample LL54. Spot 4 has a $^{206}Pb/^{238}U$ age of 659 Ma, U and Th contents of 1998 ppm and 387 ppm, respectively, and is a xenocryst. Sixteen spots have concordant $^{206}Pb/^{238}U$ ages of 477–511 Ma, with a weighted mean age of 494.6±6.9 Ma, MSWD=8.4, and have U (98–2 751 ppm) and Th (41–1 471 ppm) contents with Th/U ratios of 0.07–0.56. Eighteen spots yield concordant $^{206}Pb/^{238}U$ ages of 422–454 Ma, with a weighted mean age of 444.6±4.1 Ma, MSWD=4.2, and have U (743–4 057 ppm) and Th (282–1 599 ppm) contents with Th/U ratios of 0.25–0.84. Spot 7 has an unreliable age.

In summary, the zircon U-Pb ages reveal a complex multi-episode emplacement of the Pinghe pluton, of which the granitoid rocks can be separated into two episodes in one magmatic event accompanied by the entire process of the subduction-collision-orogeny system between circumambient blocks and oceanic crust of the Proto-Tethys: ① 473–528 Ma, with peak magmatism ages of 482–494 Ma for the former episode; and ② 420–454 Ma, with peak ages of 439–445 Ma for the latter.

Major and trace element geochemistry

Analytical data for major (wt%) and trace element (ppm) analyses of 16 whole rock samples from the Pinghe pluton are listed in Table 1.

The whole-rock geochemical results show SiO_2=69.36–76.41 wt%, TiO_2=0.03–0.23 wt%, Al_2O_3=13.16–16.08 wt%, Na_2O=2.63–5.53 wt%, K_2O=2.23–5.52 wt%, Na_2O+K_2O=7.14–8.73 wt%, MgO=0.07–1.52 wt%, and $Mg^{\#}$ [$Mg^{\#}$=100Mg/(Mg+Fe)]=14.4–60.0. The A/CNK values [molar Al_2O_3/($CaO+Na_2O+K_2O$) ratio] vary from 1.04 to 1.39, indicating peraluminous composition using the A/CNK vs. A/NK diagram (Fig. 5a). All samples fall within the high-K, calc-alkaline series in the K-Na-Ca diagram (Fig. 5b). Furthermore, these samples display similar chondrite-normalized REE patterns, with slightly to moderately enriched light rare earth elements (LREEs), depleted heavy rare earth elements (HREEs), and lacked fractionation among HREEs (Fig. 6b). The $(La/Yb)_N$ and $(Tb/Yb)_N$ ratios range from 1.04 to 14.44 and 0.78 to 2.31, respectively, and negative Eu anomaly

Table 1 Major(wt%) and trace(ppm) element analysis results for the monzogranite from Pinghe pluton.

Lithology	Two-mica monzogranite						Two-mica monzogranite		Monzogranite		Biotite monzogranite		Biotite monzogranite			
Sample	LL04	LL05	LL06	LL07	LL09	LL10	LL41	LL42	LL48	LL49	LL51	LL52	LL56	LL57	LL58	LL60
SiO_2	76.41	75.52	75.36	75.26	76.39	75.95	75.98	75.93	74.04	75.03	75.29	75.27	71.19	71.12	71.79	69.36
TiO_2	0.05	0.05	0.06	0.03	0.03	0.04	0.09	0.16	0.12	0.11	0.12	0.12	0.22	0.22	0.20	0.23
Al_2O_3	13.44	13.81	13.95	14.16	13.60	13.67	13.72	13.40	14.73	14.37	13.16	13.39	14.81	14.78	14.86	16.08
$Fe_2O_3^T$	1.02	1.15	1.38	0.97	0.96	0.97	0.88	1.52	1.25	1.05	1.25	1.30	2.75	3.06	2.36	2.59
MnO	0.05	0.08	0.07	0.07	0.06	0.05	0.07	0.06	0.02	0.02	0.06	0.05	0.10	0.08	0.08	0.12
MgO	0.09	0.10	0.10	0.10	0.08	0.07	0.17	0.32	0.40	0.30	0.28	0.41	0.79	1.41	1.52	0.72
CaO	0.33	0.33	0.34	0.31	0.37	0.41	0.59	0.67	0.44	0.53	0.51	0.43	1.05	0.38	0.47	2.00
Na_2O	3.31	3.11	3.21	3.09	3.40	3.66	3.51	3.29	3.28	3.36	3.94	3.26	2.63	3.21	5.53	3.50
K_2O	4.69	5.15	4.83	5.52	4.70	4.74	4.46	3.85	4.52	4.43	4.79	5.15	4.99	4.30	2.23	4.22
P_2O_5	0.08	0.09	0.06	0.10	0.08	0.09	0.30	0.25	0.17	0.18	0.04	0.05	0.08	0.08	0.07	0.09
LOI	0.77	0.69	0.90	0.63	0.72	0.56	0.64	0.82	1.13	0.93	0.68	0.75	1.09	1.43	1.06	1.15
Total	100.24	100.08	100.26	100.24	100.39	100.21	100.41	100.27	100.10	100.31	100.12	100.18	99.70	100.07	100.17	100.06
$Mg^{\#}$	17.1	16.9	14.4	19.4	16.3	14.4	31.0	32.9	42.7	40.0	34.3	42.4	40.1	51.8	60.0	39.3
Li	41.6	52.2	51.0	46.6	40.5	38.5	208	283	111	98.6	1.0	20.4	56.0	54.0	42.8	40.5
Be	7.65	15.5	11.1	16.5	8.90	8.09	35.1	18.8	12.2	13.5	4.54	5.09	5.73	4.92	3.05	7.29
Sc	3.16	2.78	3.59	2.49	3.04	3.05	2.06	2.70	3.83	3.10	4.31	4.38	5.22	5.31	4.68	5.85
V	2.62	2.02	3.40	2.02	2.95	1.89	3.61	6.25	9.35	6.95	11.5	11.2	24.1	24.1	22.5	25.3
Cr	1.55	7.08	10.7	1.36	3.25	1.40	8.38	4.98	4.14	8.96	71.9	5.00	3.01	3.04	2.51	4.69
Co	116	98.1	108	93.9	146	109	118	75.4	92.9	90.0	133	167	76.3	110	70.9	111
Ni	2.84	4.95	9.46	2.34	4.53	1.97	6.88	5.53	3.50	7.38	39.4	5.93	2.56	2.52	2.52	4.15
Cu	0.50	1.61	1.78	1.66	0.83	0.43	2.16	1.06	14.9	8.28	5.69	4.49	191	175	4.12	15.7
Zn	22.7	19.0	24.8	15.9	20.0	20.1	39.1	64.1	53.2	47.7	23.7	45.8	779	87.8	97.9	89.8
Ga	19.3	19.7	20.0	19.6	19.0	19.4	23.0	25.0	18.9	16.7	15.6	15.6	17.8	19.4	16.8	20.7
Ge	2.46	2.61	2.30	2.55	2.41	2.43	2.65	2.13	2.20	2.17	2.01	2.03	2.22	2.01	1.50	2.12
Rb	493	423	414	447	488	500	538	446	275	255	242	269	335	340	175	222
Sr	5.96	5.53	7.65	6.01	5.85	5.33	27.0	25.9	45.7	53.6	47.7	48.6	67.8	42.9	55.3	170
Y	33.5	23.7	26.6	24.6	38.6	33.8	8.40	12.0	18.0	19.4	29.1	34.5	22.1	26.0	23.9	27.1
Zr	43.8	31.9	52.3	29.7	42.2	43.9	56.6	81.0	79.1	72.7	81.8	72.2	108	105	98.7	123

Continued

Lithology	Two-mica monzogranite						Two-mica monzogranite		Monzogranite		Biotite monzogranite		Biotite monzogranite			
Sample	LL04	LL05	LL06	LL07	LL09	LL10	LL41	LL42	LL48	LL49	LL51	LL52	LL56	LL57	LL58	LL60
Nb	12. 3	9. 51	13. 2	7. 77	11. 9	12. 1	28. 8	35. 0	13. 0	12. 3	17. 6	18. 5	13. 6	13. 4	11. 7	17. 2
Cs	28. 8	37. 9	37. 3	32. 4	26. 7	25. 2	53. 7	55. 5	31. 2	24. 5	9. 34	11. 9	21. 8	33. 3	25. 4	15. 0
Ba	22. 1	16. 7	31. 5	18. 8	22. 8	21. 5	66. 2	61. 6	256	273	227	214	518	395	170	510
La	7. 18	6. 22	7. 88	6. 22	6. 21	7. 06	10. 6	18. 7	18. 3	17. 9	15. 3	15. 9	17. 3	16. 6	25. 3	37. 9
Ce	14. 6	14. 7	19. 0	15. 3	14. 1	14. 5	22. 3	40. 9	39. 5	38. 2	33. 8	34. 3	36. 1	36. 1	52. 9	74. 8
Pr	2. 38	1. 90	2. 37	1. 93	2. 10	2. 29	2. 70	4. 81	4. 67	4. 62	4. 20	4. 38	4. 23	4. 23	6. 02	8. 39
Nd	8. 79	6. 80	8. 48	7. 04	7. 98	8. 51	9. 66	17. 3	17. 3	17. 1	16. 2	17. 0	15. 9	15. 9	21. 9	30. 9
Sm	2. 92	2. 37	2. 74	2. 41	2. 71	2. 82	2. 25	3. 90	4. 20	4. 18	4. 42	4. 83	3. 70	3. 87	4. 65	6. 59
Eu	0. 097	0. 050	0. 067	0. 049	0. 100	0. 099	0. 17	0. 19	0. 47	0. 49	0. 37	0. 37	0. 58	0. 59	0. 51	0. 93
Gd	3. 07	2. 38	2. 67	2. 46	3. 13	3. 13	1. 89	3. 23	3. 92	4. 00	4. 25	4. 80	3. 56	3. 81	4. 21	5. 61
Tb	0. 71	0. 56	0. 62	0. 58	0. 73	0. 73	0. 30	0. 47	0. 63	0. 66	0. 75	0. 87	0. 59	0. 63	0. 67	0. 86
Dy	4. 88	3. 75	4. 14	3. 89	5. 12	4. 99	1. 57	2. 38	3. 43	3. 64	4. 69	5. 48	3. 64	3. 88	3. 95	4. 71
Ho	1. 01	0. 74	0. 83	0. 75	1. 08	1. 01	0. 26	0. 38	0. 59	0. 62	0. 93	1. 11	0. 73	0. 77	0. 77	0. 89
Er	3. 16	2. 27	2. 61	2. 33	3. 44	3. 19	0. 69	0. 98	1. 55	1. 67	2. 72	3. 23	2. 10	2. 24	2. 25	2. 57
Tm	0. 55	0. 41	0. 45	0. 42	0. 60	0. 56	0. 10	0. 14	0. 23	0. 25	0. 44	0. 52	0. 32	0. 34	0. 35	0. 39
Yb	3. 94	2. 98	3. 23	3. 04	4. 27	4. 05	0. 69	0. 93	1. 45	1. 63	2. 92	3. 42	2. 08	2. 26	2. 25	2. 57
Lu	0. 56	0. 42	0. 46	0. 43	0. 62	0. 58	0. 090	0. 13	0. 21	0. 23	0. 42	0. 49	0. 31	0. 33	0. 33	0. 38
Hf	2. 14	1. 61	2. 42	1. 62	2. 15	2. 26	2. 06	2. 77	2. 49	2. 33	3. 40	2. 95	3. 52	3. 40	3. 04	4. 12
Ta	4. 67	2. 95	3. 75	2. 45	4. 52	4. 33	14. 4	9. 29	1. 81	2. 30	2. 83	2. 79	1. 84	2. 21	1. 61	2. 13
Pb	23. 3	25. 2	22. 7	27. 8	23. 7	24. 4	25. 5	22. 6	24. 5	30. 5	38. 8	54. 1	253	38. 2	7. 93	50. 9
Th	12. 3	12. 1	15. 1	10. 3	11. 7	12. 1	9. 89	18. 5	12. 5	12. 7	22. 6	24. 4	16. 3	17. 8	16. 3	22. 0
U	7. 28	8. 91	8. 36	10. 1	9. 43	10. 7	15. 5	38. 2	7. 25	13. 0	12. 9	12. 3	10. 9	11. 9	7. 41	12. 4
A/CNK	1. 21	1. 22	1. 25	1. 22	1. 20	1. 15	1. 18	1. 24	1. 33	1. 27	1. 04	1. 14	1. 27	1. 39	1. 20	1. 15
ΣREE	275. 8	219. 4	257. 3	224. 7	275. 8	276. 2	191. 7	329. 0	358. 6	359. 8	377. 1	410. 5	354. 8	360. 8	463. 6	640. 9
Rb/Sr	82. 72	76. 42	54. 09	74. 33	83. 43	93. 83	19. 92	17. 19	6. 03	4. 76	5. 08	5. 53	4. 94	7. 91	3. 16	1. 31
$(La/Yb)_N$	1. 31	1. 50	1. 75	1. 47	1. 04	1. 25	11. 01	14. 44	9. 04	7. 89	3. 74	3. 33	5. 96	5. 27	8. 05	10. 60
$(Tb/Yb)_N$	0. 82	0. 86	0. 87	0. 87	0. 78	0. 82	2. 01	2. 31	1. 97	1. 84	1. 17	1. 15	1. 30	1. 27	1. 34	1. 51
Eu/Eu^*	0. 10	0. 06	0. 08	0. 06	0. 11	0. 10	0. 25	0. 16	0. 35	0. 37	0. 26	0. 24	0. 49	0. 47	0. 35	0. 47

(Eu/Eu*) varies from 0.06 to 0.49. In the N-MORB normalized trace element spider diagram (Fig. 6a), the monzogranite patterns also show negative Ba, Nb, Ta, Sr, and Ti anomalies.

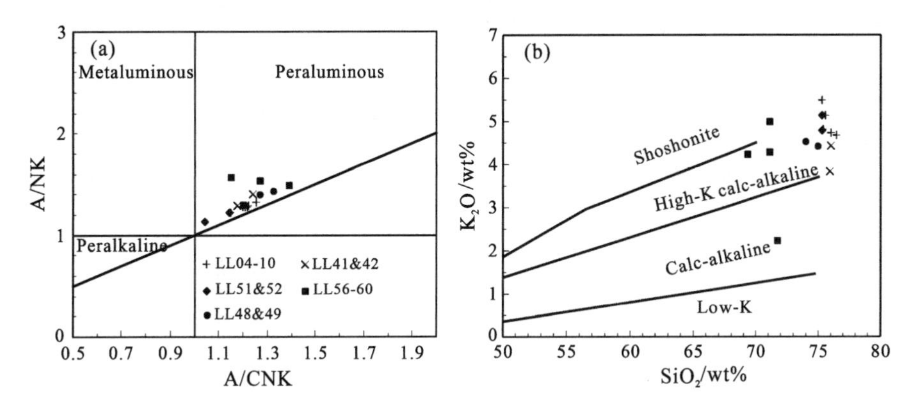

Fig. 5 (a) A/CNK ratio vs. A/NK ratio diagram (b) SiO_2 vs. K_2O diagram for the monzogranite from Pinghe pluton, western Yunnan, SW China, after Roberts and Clemens (1993).

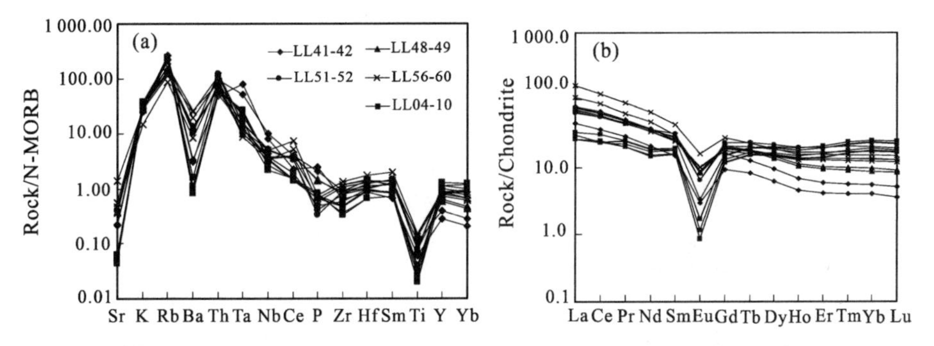

Fig. 6 N-MORB-normalized trace element spider (a) and chondrite-normalized REE pattern (b) diagrams for the monzogranite from Pinghe pluton, western Yunnan, SW China.
The N-MORB and chondrite values are from Pearce (1983) and Sun and McDonough (1989), respectively.

Whole-rock Sr-Nd-Pb-Hf isotopic composition

Analytical results of whole-rock Sr-Nd-Pb-Hf isotopes for the Pinghe pluton are given in Tables 2–4. The initial values have been calculated at $t = 440$ Ma on the basis of the LA-ICP-MS U-Pb zircon ages for Pinghe pluton.

The monzogranite has high $^{87}Rb/^{86}Sr$ ratios that result in inaccurate initial $^{87}Sr/^{86}Sr(t)$ ratios displaying a large variation (0.613 931 – 0.773 897), and initial $^{143}Nd/^{144}Nd(t)$ = 0.511 329–0.511 628. The $\varepsilon_{Nd}(t)$ values are −11.6 to −8.7 with the two-stage model ages of 1.66–2.06 Ga (Table 2), and $\varepsilon_{Hf}(t)$ values are −9.58 to −5.55 with the two-stage model ages of 2.14–3.00 Ga (Table 4). Highly various radiogenic Pb isotopic ratios are $^{206}Pb/^{204}Pb(t)$ = 16.547–18.705, $^{207}Pb/^{204}Pb(t)$ = 15.645 – 15.765, and $^{208}Pb/^{204}Pb(t)$ = 38.273 – 38.830 (Table 3).

Table 2 Whole-rock Rb-Sr and Sm-Nd isotopic data for the monzogranite from Pinghe pluton.

Sample	Rb /ppm	Sr /ppm	$^{87}Rb/^{86}Sr$	$^{87}Sr/^{86}Sr$	2σ	$^{87}Sr/^{86}Sr(t)$ (t=440 Ma)	Sm /ppm	Nd /ppm	$^{147}Sm/^{144}Nd$	$^{143}Nd/^{144}Nd$	2σ	$^{143}Nd/^{144}Nd(t)$ (t=440 Ma)	T_{DM2} /Ga	$\varepsilon_{Nd}(t)$ (t=440 Ma)
LL41	27.0	538	58.073	0.787 363	63	0.613 931	2.25	9.66	0.140 6	0.511 881	5	0.511 476	1.87	−11.6
LL48	45.7	275	17.638	0.826 573	53	0.773 897	4.20	17.3	0.147 1	0.511 753	4	0.511 329	2.06	−14.5
LL51	47.7	242	14.843	0.807 906	19	0.763 577	4.42	16.2	0.165 0	0.512 052	5	0.511 577	1.73	−9.7
LL56	67.8	335	14.423	0.804 410	36	0.761 338	3.70	15.9	0.140 5	0.512 033	5	0.511 628	1.66	−8.7

$^{87}Rb/^{86}Sr$ and $^{147}Sm/^{144}Nd$ ratios were calculated using Rb, Sr, Sm, and Nd contents analyzed by ICP-MS.

T_{DM2} represent the two-stage model age and were calculated using present-day $(^{147}Sm/^{144}Nd)_{DM}=0.2137$, $(^{147}Sm/^{144}Nd)_{DM}=0.51315$, and $(^{147}Sm/^{144}Nd)_{crust}=0.1012$. $\varepsilon_{Nd}(t)$ values were calculated using present-day $(^{147}Sm/^{144}Nd)_{CHUR}=0.1967$ and $(^{147}Sm/^{144}Nd)_{CHUR}=0.512638$.

$\varepsilon_{Nd}(t)=[(^{143}Nd/^{144}Nd)_S(t)/(^{143}Nd/^{144}Nd)_{CHUR}(t)-1]\times10^4$.

$T_{DM2}=\frac{1}{\lambda}\{1+[(^{143}Nd/_{144}Nd)_S-((^{147}Sm/^{144}Nd)_S-(^{147}Sm/^{144}Nd)_{crust})(e^{\lambda t}-1)-(^{143}Nd/^{144}Nd)_{DM}]/[(^{147}Sm/^{144}Nd)_{crust}-(^{147}Sm/^{144}Nd)_{DM}]\}$, $\lambda=6.54\times10^{-12}/a$.

Table 3 Whole-rock Pb isotopic data for the monzogranite from Pinghe pluton.

Sample	U /ppm	Th /ppm	Pb /ppm	$^{206}Pb/^{204}Pb$	2σ	$^{207}Pb/^{204}Pb$	2σ	$^{208}Pb/^{204}Pb$	2σ	$^{238}U/^{204}Pb$	$^{232}Th/^{204}Pb$	$^{206}Pb/^{204}Pb(t)$ (t=440 Ma)	$^{207}Pb/^{204}Pb(t)$ (t=440 Ma)	$^{208}Pb/^{204}Pb(t)$ (t=440 Ma)
LL41	15.5	9.9	25.5	19.339	0.000 3	15.800	0.000 3	39.401	0.000 9	39.53	25.80	16.547	15.645	38.830
LL48	7.3	12.5	24.5	19.499	0.000 5	15.809	0.000 4	39.154	0.001 1	19.12	33.83	18.148	15.734	38.406
LL51	12.9	22.6	38.8	20.189	0.000 5	15.850	0.000 4	39.134	0.000 9	21.66	38.93	18.659	15.765	38.273
LL56	10.9	16.3	252.7	18.900	0.000 5	15.776	0.000 5	38.897	0.001 3	2.76	4.23	18.705	15.765	38.804

U, Th, Pb concentrations were analyzed by ICP-MS. Initial Pb isotopic ratios were calculated for 440 Ma using single-stage model.

Table 4 Whole-rock Lu-Hf isotopic data for the monzogranite from Pinghe pluton.

Sample	Lu /ppm	Hf /ppm	$^{176}Lu/^{177}Hf$	$^{176}Hf/^{177}Hf$	2σ	$\varepsilon_{Hf}(t)$ (t=440 Ma)	$f_{Lu/Hf}$	T_{DM2} /Ga
LL41	0. 09	2. 06	0. 006 180	0. 282 348	8	−7. 13	−0. 81	2. 14
LL48	0. 21	2. 49	0. 011 972	0. 282 326	6	−9. 58	−0. 64	2. 73
LL51	0. 42	3. 40	0. 017 585	0. 282 455	6	−6. 67	−0. 47	3. 00
LL56	0. 31	3. 52	0. 012 334	0. 282 443	3	−5. 55	−0. 63	2. 39

Lu and Hf concentrations were analyzed by ICP-MS.

T_{DM2} were calculated using present-day $(^{176}Lu/^{177}Hf)_{DM}=0.0384$ and $(^{176}Hf/^{177}Hf)_{DM}=0.28325$.

$f_{CC}=-0.55$, and $f_{DM}=0.16$. f_S is the sample's $f_{Lu/Hf}$.

$\varepsilon_{Hf}(t)$ values were calculated using present-day $(^{176}Lu/^{177}Hf)_{CHUR}=0.0332$ and $(^{176}Hf/^{177}Hf)_{CHUR}=0.282772$.

$$T_{DM1}=\frac{1}{\lambda}\ln\{1+[(^{176}Hf/^{177}Hf)_S-(^{176}Hf/^{177}Hf)_{DM}]/[(^{176}Lu/^{177}Hf)_S-(^{176}Lu/^{177}Hf)_{DM}]\}.$$

$$T_{DM2}=T_{DM1}-(T_{DM1}-t)[(f_{CC}-f_S)/(f_{cc}-f_{DM})],\ \lambda=1.876\times10^{-11}/a.$$

Discussion

Early Paleozoic magmatism in the Pinghe Pluton

The four samples from Pinghe pluton yield two ages of 490 Ma and 440 Ma, which are interpreted as two episodes in one magmatic event accompanied by the whole process of the subduction-collision-orogeny system between circumambient blocks and oceanic crust of the Proto-Tethys. There are three factors accounting for the interpretation: ① two groups of zircons with different age-stages have no differences in their morphology and internal structure; ② as seen from macroscopical exposure, the batholith has no deformation and metarmophism; and ③no traces of fluid metasomatism have been found in the field. Geochronological evidence shows that the early magmatic activity began at 528 Ma and ended at 470 Ma, with late crystallization age representing the period from 454 Ma to 420 Ma, with a break of about 20 Ma between the two episodes of the magmatism. Contemporaneous magmatic events were found in other areas: Metcalfe (1998) showed a U-Pb age of 486±5 Ma from a granite body in the late Paleozoic sedimentary sequence within the Nan-Uttraadit suture zone between the Sibumasu and Indochina blocks; Li et al. (2012) showed crystallization ages of 497. 8 ± 7. 2 Ma to 500 ± 14 Ma and a metamorphic age of 459 ± 5 Ma from the biotite monzogranite gneiss in the Gaoligongshan Group; and Yang et al. (2012) reported an age of 499. 2 ± 2. 1 Ma from the metamorphosed mafic volcanics of the Gongyanghe Group; Zhang et al., (2008) dated at 490–500 Ma an inherited zircon from a gneiss having granite and granodiorite as protoliths, and found an age of 505 Ma for the metamorphic growth of zircons from calc-silicates in the Namche Barwa Complex located in the eastern Himalaya syntaxis, which is part of the High Himalayan

Crystalline Complex. In addition, several researchers have found evidences of similar Paleozoic magmatism and metamorphism from the granites in northern India, Nepal, Chinese Himalaya, and western Yunnan (Stöcklin and Bhattarai, 1977; Stöcklin, 1980; Valdiya, 1995; Argles et al., 1999; Decelles et al., 2000; Wang, 2000; Gatlos et al., 2002; Gehrels et al., 2003, 2006a, b; Xu et al., 2005; Xu et al., 2010; Dong et al., 2009; Qi et al., 2010; Zhu et al., 2012b) (Fig. 1b, c). The abundance of geochronological evidence indicates a strong magmatic activity in northern Gondwana in early Palaeozoic time, which is interpreted as reflecting Pan-African (Kennedy, 1964) and early Paleozoic orogenic events. These igneous and metamorphic rocks with the similar ages of approximately late Cambrian to Ordovician are located along northwestern India, Nepal, western and eastern Tibet, and turn southward with the Tengchong and Baoshan blocks, which make up the Neoproterozoic to early Palaeozoic Proto-Tethyan tectonic belt. The results of Wang et al. (2013b) represent a progression along-strike from old ages in the west to young in the east, ranging from NW Turkey, Central Iran, NW Pakistan, NW India, and Nepal to South Tibet and SW Yunnan. It can be inferred from their conclusion that Proto-Tethys closed from west to east was gradual.

Characteristics of magma evolution and sources

A-, I-, or S-type?

Geochemical analyses show that the monzogranite samples belong to the high-K calc-alkaline series, with peraluminous characteristics, moderate Na_2O+K_2O values (7.14%–8.13%), and 10 000 Ga/Al ratios of 2.14–3.52. Geochemical indicators are not obvious between A- and I & S-type granite (Collins et al., 1982; Whalen et al., 1987), and parts of samples fall within the I & S-type granite area while others fall within the boundary between A and I & S zones in the 10 000 Ga/Al vs. Zr diagram (Fig. 7a). However, I & S-type characteristics are more common in the samples than A-type characteristics. Meanwhile, the monzogranite also exhibits not only I-type but also a S-type evolutionary trend in the Th and Y vs. Rb and P_2O_5 vs. SiO_2 diagrams (Fig. 7b-d). Thus, we can not simply define the monzogranite as I- or S-type. These data indicate that the monzogranite underwent a complex evolution rather than a single process and was produced in a complicated tectonic setting, which is consistent with the geochronological interpretation, indicating two episodes in one magmatic event.

The evolution of magma

With increasing SiO_2 contents, the samples show decreasing major elements (e.g., TiO_2, Al_2O_3, MgO, CaO, and FeO; Fig. 8a-e), decreasing trace elements (e.g. Zr, Ba and Sr; Fig. 8f-h), and increasing Rb (Fig. 8i) and Rb/Sr ratios. This illustrates the evolution of magma differentiation. The pronounced negative correlation of SiO_2 vs. Sr and Ba (Fig. 8g-h)

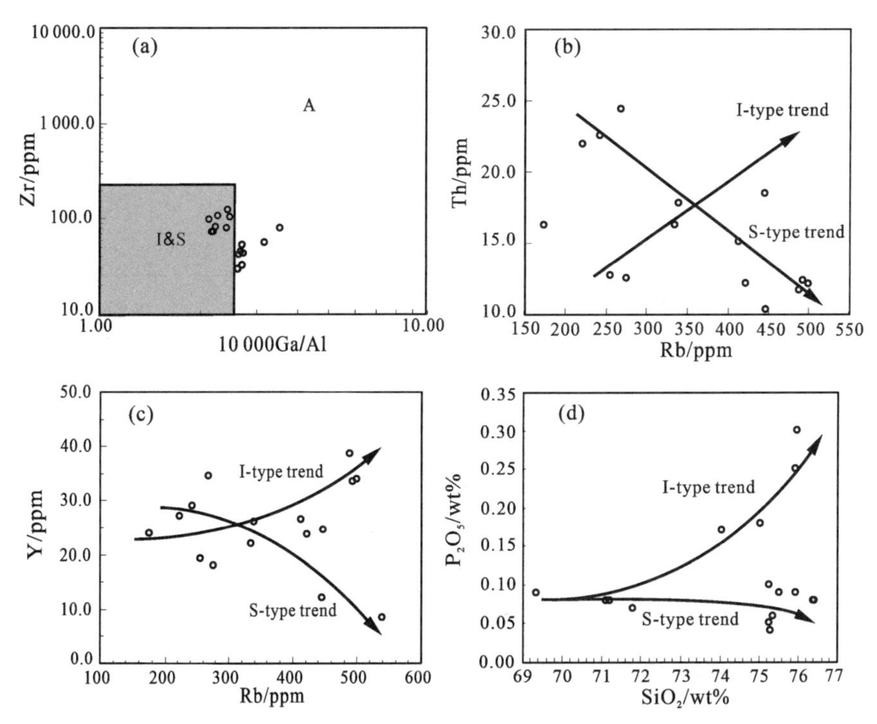

Fig. 7 The monzogranite plotted in terms of the Zr vs. 10 000 Ga/Al diagram of Whalen et al.(a) and Th & Y vs. Rb(b-c) and P_2O_5 vs. SiO_2 diagrams for the Pinghe Pluton, Baoshan Block, western Yunnan, SW China.

The trend line is from Li et al. (2007).

and depleted Eu (Fig. 6b) indicate the crystallization differentiation of plagioclase, K-feldspar, and biotite during magmatic evolution. It will be seen from the Sr vs. Rb and Ba, and Eu^* vs. Sr and Ba diagrams (Fig. 9a-d), that K-feldspar plays a more important role than plagioclase and biotite in controlling Sr and Ba contents during fractional crystallization. An obvious negative Ti (Fig. 6a) anomaly results from the fractionation of the Ti-bearing phase (ilmenite). An insignificant negative P anomaly (Fig. 6a) indicates the negligible fractionation of apatite. Simply put, crystallization differentiation was accompanied by the evolution of magmas.

Characteristics of sources

Several researchers reported that the strong peraluminous (SP) granites form from melts of meta-igneous rocks(Miller, 1985; Patiño Douce, 1995) or metasedimentary rocks (Le Fort et al., 1987; White and Chappell, 1988) in the lower crust. Sylvester (1998) considered that CaO/Na_2O ratios in the SP granites are controlled by the original sources by varying clay levels. Pelite-derived melts have lower CaO/Na_2O ratios (<0.3) than psammite-derived. The majority of the samples from Pinghe had low CaO/Na_2O ratios (0.08-0.57) and high Al_2O_3/TiO_2 ratios, which indicates that the monzogranite was derived from metapelite melting with slight mixing of basaltic or mafic magma (Fig. 10a). In addition, high Rb/Sr and Rb/Ba ratios

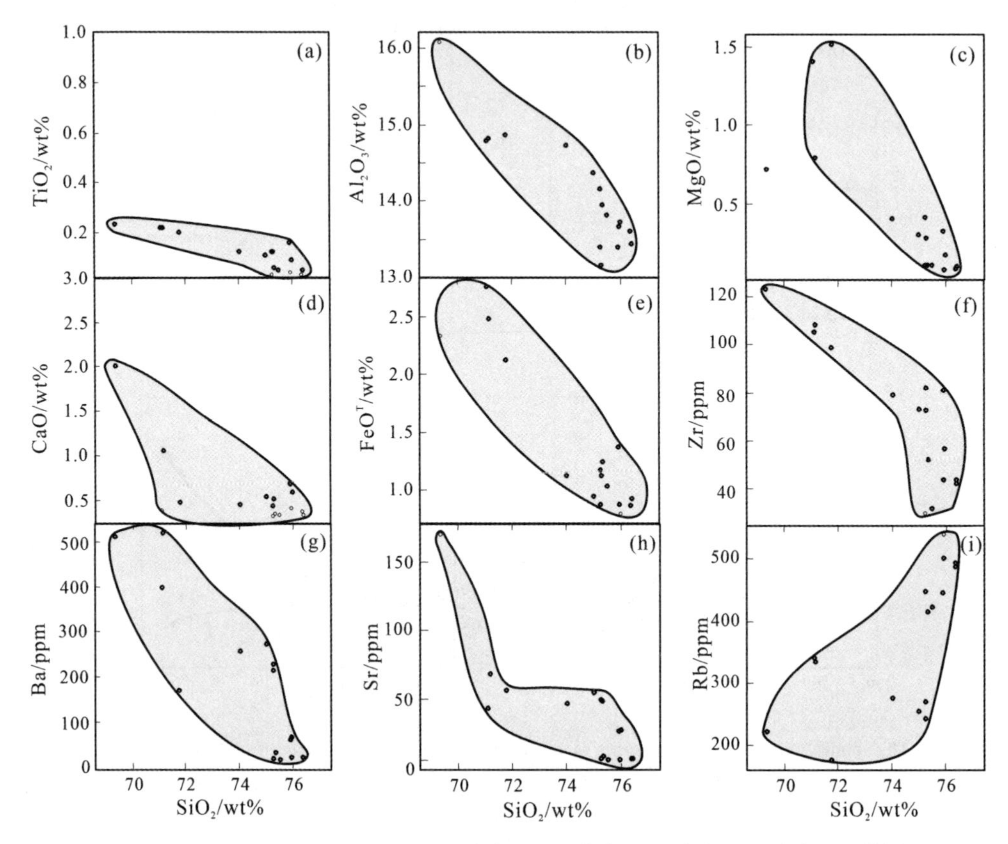

Fig. 8 SiO_2 vs. major elements: TiO_2(a), Al_2O_3(b), MgO(c), CaO(d), FeO^T(e), and trace elements: Zr(f), Ba(g), Sr(h), Rb(i) for the Pinghe pluton, Baoshan Block, western Yunnan, SW China.

The shaded zones represent the evolutive trends of the major and trace elements in the wake of SiO_2 contents.

show that the monzogranite was derived from clay-rich sources (metapelite-derived melts) (Fig. 10b). Moreover, whole-rock isotopic data from the monzogranite show negative $\varepsilon_{Nd}(t)$ values of −11.6 to −8.7 with the two-stage model ages of 1.66–2.06 Ga (Table 2), and corresponding negative $\varepsilon_{Hf}(t)$ values of −9.58 to −5.55 with the two-stage model ages of 2.14–3.00 Ga (Table 4), which seems to imply that the monzogranite was derived from crust rocks. In the discrimination diagram of $\varepsilon_{Nd}(t)$ vs. $\varepsilon_{Hf}(t)$, all data points fall within the fields of global sediments and global lower crust (Fig. 11), which also suggests that the monzogranite was sourced from the Neoarchean and Proterozoic lower crust rocks. In addition, $^{206}Pb/^{204}Pb$, $^{207}Pb/^{204}Pb$, and $^{208}Pb/^{204}Pb$ ratios have large values and spans (16.547–18.705, 15.645–15.765, and 38.273–38.830, respectively) (Table 3), which shows that the monzogranite probably had a heterogeneous source. Therefore, these samples suggest that the Pinghe pluton was derived from inhomogeneous melts of metapelite shaped by Neoarchean to Paleoproterozoic basement rocks, recording with some slight mixing of basaltic magma.

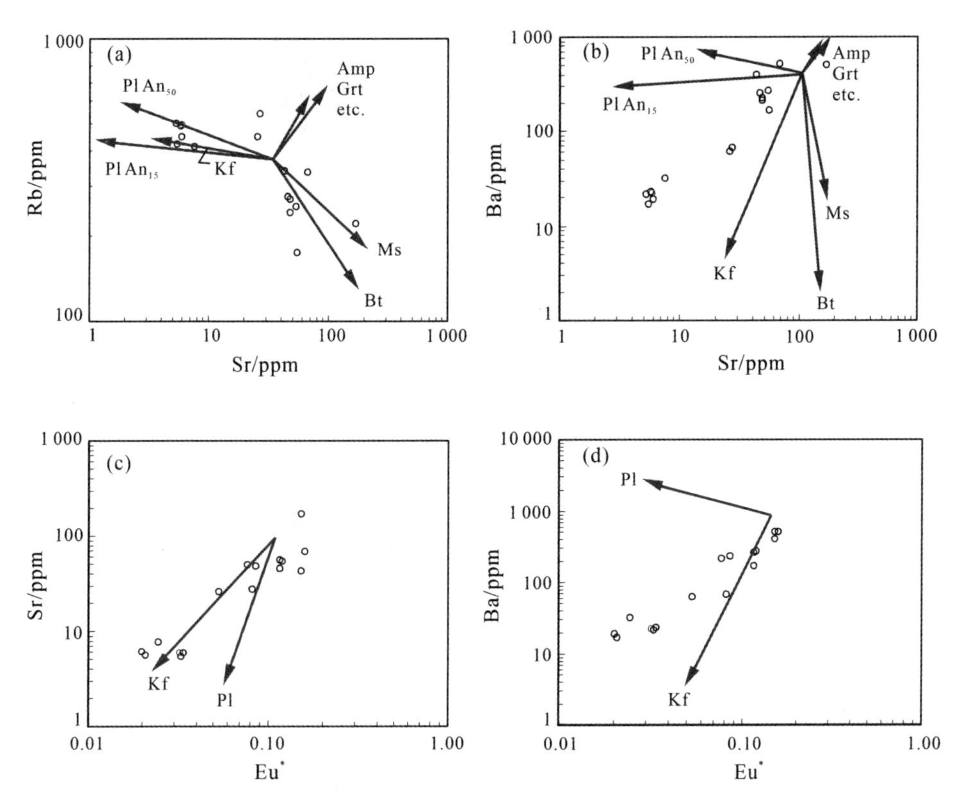

Fig. 9 Diagrams of Sr vs. Rb and Ba, Eu* vs. Sr and Ba for the monzogranite from Pinghe pluton, Baoshan Block, western Yunnan, SW China, according to Janoušek et al. (2004).

Pl: plagioclase; Kf: K-feldspar; Bt: biotite; Ms: muscovite; Grt: garnet; Amp: amphibole.

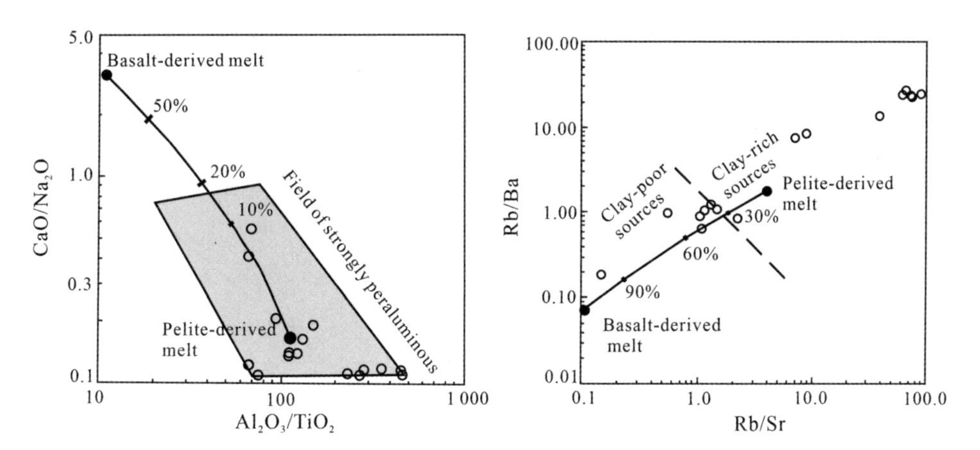

Fig. 10 Plots of Al_2O_3/TiO_2 vs. CaO/Na_2O and Rb/Sr vs. Rb/Ba discrimination diagrams for the monzogranite from Pinghe pluton, Baoshan Block, west Yunnan, SW China, after Sylvester (1998).

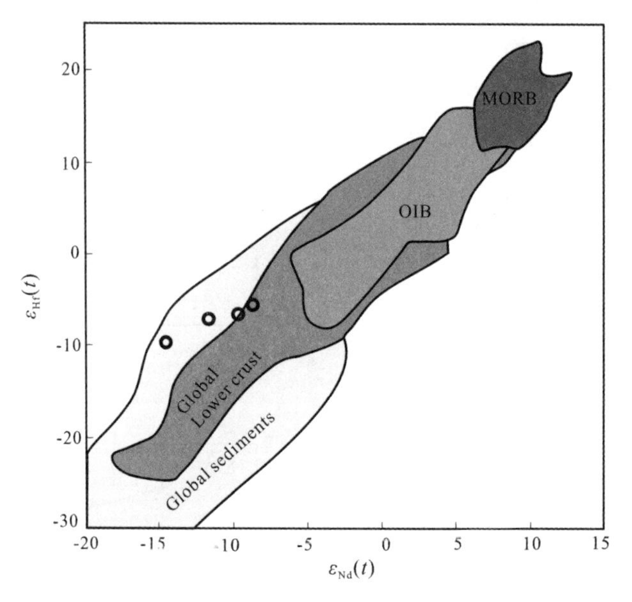

Fig. 11 Geochemical discrimination diagram of $\varepsilon_{Nd}(t)$ vs. $\varepsilon_{Hf}(t)$ for the monzogranite from Pinghe pluton, Baoshan Block, western Yunnan, SW China, after Dobosi et al. (2003).

Tectonic setting and significance

The results show that the samples from Pinghe pluton are strongly peraluminous monzogranite, which formed as a consequence of post-collisional processes in various orogens (Sylvester, 1998). Pinghe samples mostly fall within the collision field in the discrimination diagrams (Fig. 12a,b) of Yb+Nb vs. Rb (Pearce et al., 1984; Pearce, 1996) and R_1 vs. R_2 (Harris et al., 1986). In addition, the diagram of Yb+Nb vs. Rb shows that the monzogranite has the trend of arc-continent collision, and seems to indicate an active continental margin in the Baoshan Block in Cambro-Ordovician time. Cawood and Buchan (2007) argued that final

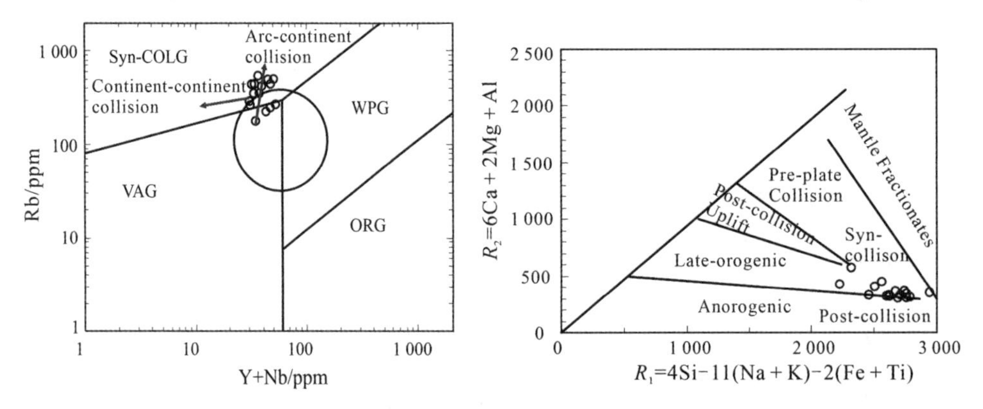

Fig. 12 Geochemical discrimination diagrams of Rb vs. Y+Nb(a) and R_1[Si−11(Na+K)−2(Fe+Ti)] vs. R_2(6Ca+2Mg+Al) for the monzogranite from Pinghe pluton, Baoshan Block, western Yunnan, SW China, after Pearce et al. (1984), Pearce(1996), and Harris et al. (1986), respectively.

assembly of Gondwana occurred between 570 Ma and 510 Ma, amalgamating the various components of East and West Gondwana. Moreover, they reported that supra-subduction zone ophiolites are preserved in greenstone successions in eastern Australia, which formed in association with orogenesis between 520 Ma and 490 Ma. This further corroborates the inference that the regional tectonic setting was post-collision in late Cambrian to Early Ordovician time, when the Pinghe pluton was produced between 490 Ma and 440 Ma.

Our results point to a similarity with the late Cambrian to Ordovician igneous rocks between the Baoshan Block, Himalayan orogeny, and Lhasa block (Stöcklin and Bhattarai, 1977; Stöcklin, 1980; Valdiya, 1995; Metcalfe, 1998; Argles et al., 1999; Decelles et al., 2000; Wang, 2000; Gatlos et al., 2002; Gehrels et al., 2003, 2006a,b; Xu et al., 2005, 2010; Zhang et al., 2008; Dong et al., 2009; Qi et al., 2010; Li et al., 2010; Yang et al., 2012; Zhu et al., 2012b), as well as similar unconformities, which were represented as either angular uncomformity or basal conglomerate ranging from the Cambrian to Ordovician sequences in these areas (Stöcklin and Bhattarari, 1977; Stöcklin, 1980; Hughes, 2002; Gehrels et al., 2003; Zhou et al., 2004; Huang et al., 2009, 2012; Li et al., 2010). Based on these sources, we propose that the Baoshan Block was in contact with Himalayan orogen and Lhasa block in late Cambrian to Ordovician time, and experienced similar tectonic events during that time. Cawood et al. (2007) interpreted the Himalayan orogen as being related to Andean-type orogenic activity on the northern margin of the Indian subcontinent in Cambro-Ordovician time. Meanwhile, Zhu et al. (2011, 2012a, 2012b) reported that the Lhasa Terrane came from Gondwana, near what is now northern or northwestern Australia, and represented an early Paleozoic Andean-type magmatic arc in Gondwana and the Proto-Tethyan ocean margin, where exposed Archean to Proterozoic cratonic blocks (Mazumder et al., 2000; Wingate and Evans, 2003; Griffin et al., 2004) appear to indicate the sources of the igneous rocks. If the Himalayan orogen and Lhasa Terrane as part of a Proto-Tethyan margin underwent Andean-type subduction and an associated orogenic event, the adjoining Baoshan Block must have experienced comparable tectonomagmatic events. Moreover, Metcalfe (2011, 2013) argued that there are three factors accounting for the Baoshan Block derived from NW Australian Gondwana in Palaeozoic time: ① the similarity of Cambro-Early Permian faunas and late Carboniferous and early Permian *Glossopteris* between these two continents; ② the distribution of Permian glacial-marine diamictites; and ③ the identical paleomagnetic data from Devonian to early Permian. Taking into account all available geochronological and petrological information, the Baoshan Block may have originated from northern Australia and played a part in the Proto-Tethyan Andean-type active continental margin of northern Gondwana. In addition, Cawood et al. (2007) and Zhu et al. (2012b) used the slab break-off model to interpret the origin of the early Palaeozoic tectonomagmatism along the Indian-Australia Proto-Tethyan margin. Considering the global geodynamic perspective, the slab break-off model is also suitable for the Pinghe pluton of the adjoining Baoshan Block. It can be concluded that the Baoshan Block underwent subduction with an unknown accretion micro-continental block in Neoproterozoic to early Cambrian time (Fig. 13a), and collision-orogenesis in late Cambrian to Ordovician time along the margin of northern Australian Gondwana (Fig. 13b). Due to insufficient evidence, the accretion micro-continental block cannot be named. However, Metcalfe (2002, 2011, 2013) reported that South China has been in contact with the NE Gondwana Himalayan-West Australian region in the early Palaeozoic and separated from Gondwana in Devonian time, as proved by endemic faunas and paleomagnetic data. From the Cambro-Ordovician to Silurian, palaeobiological data were identical between Sibumasu and South China, which is consistent with paleomagentic evidence after Li et al. (2004). These indicators suggest that South China is potentially the unknown accretion micro-continental block and collided with Sibumasu rerrane including Baoshan Block, but we do not have enough evidence to support the inference.

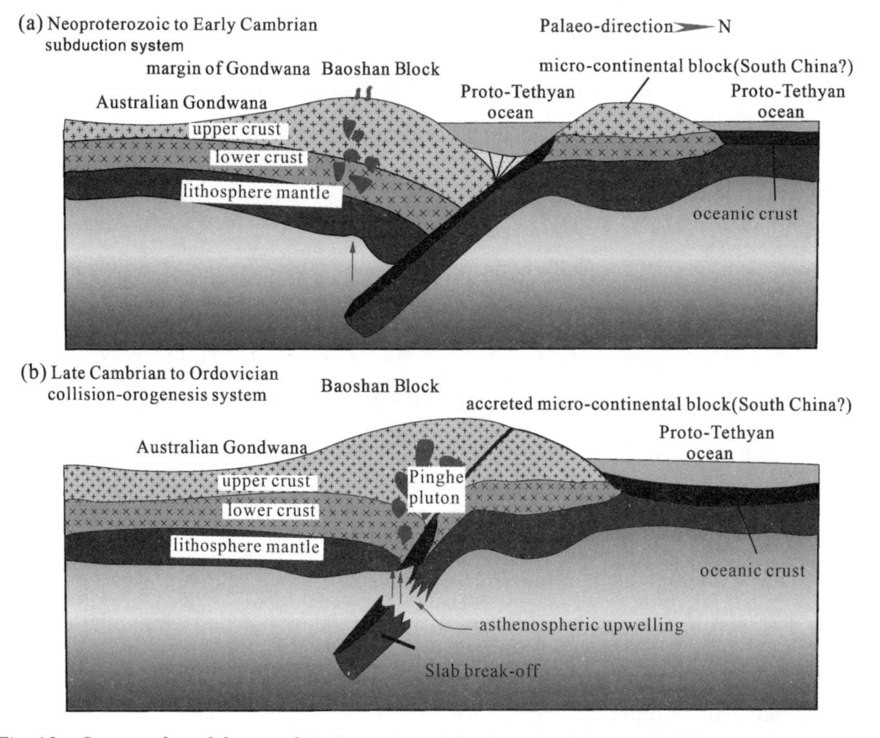

Fig. 13 Suggested model to explain formation of the late Cambrian to Ordovician post-collision monzogranite in Pinghe pluton of Baoshan Block, SW China.

(a) The Proto-Tethyan oceanic crust subducted under the Australian Gondwana margin in the Neoproterozoic to early Cambrian time; (b) during the late Cambrian to Ordovician time, post-collision between Baoshan (the margin of northern Australian Gondwana) and accretion micro-continental block (South China?), subducting oceanic slab break-off induced the widespread granitic magmatism.

Nevertheless, evidence from geological records of late Cambro-Ordovician time may have been obscured by Mesozoic and Cenozoic tectonic movements. Moreover, this model is proposed mostly based on the limited data currently available for the Cambrian to Ordovician geology in the Baoshan Block and northern Australia. Therefore, further investigations and further geological data on the Cambrian to Ordovician are needed to test the interpretation of geodynamics, which was responsible for generation of the Pinghe pluton.

Conclusion

(1) The Pinghe pluton is made up of equigranular or porphyritic two-mica or biotite monzogranite and occurs as a batholith. LA-ICP-MS U-Pb zircon dating yield emplacement ages of 473–528 Ma, with peak ages of 482–494 Ma; and 420–454 Ma, with peak ages of 439–445 Ma. This is interpreted as the result of two episodes of one magmatic event.

(2) The samples belong to the SP, high-K, calc-alkaline series and are enriched in LREEs and depleted in HREEs, with negative Eu, Ba, Sr, Nb, Ta, and Ti anomalies. They also show whole-rock Sr-Nd-Pb-Hf isotopic compositions with abnormally high initial $^{87}Sr/^{86}Sr$ and $\varepsilon_{Nd}(t)$ values of −11.4 to −8.7 and $\varepsilon_{Hf}(t)$ values of −9.58 to −5.55. The Nd and Hf isotopic two-stage model age values were 1.66–2.06 Ga and 2.14–3.00 Ga, respectively, with variable initial Pb isotopic compositions. The monzogranite had low CaO/Na_2O and high Al_2O_3/TiO_2. Taking into account the geochemical signatures, the monzogranite was probably derived from partial melting of heterogeneous metapelite shaped by Neoarchean to Paleoproterozoic rocks with slight mixing of basaltic magma, and underwent fractional crystallization of K-feldspar, plagioclase, and ilmenite during evolution of the magma.

(3) According to the comparison of the adjacent regional lithology, geochronology, and tectonics, the SP monzogranite was produced in an active continental margin arc-continent post-collisional setting and the Baoshan Block was likely derived from the NW Australian Proto-Tethyan Andean-type active continental margin of the Gondwana supercontinent. The formation of Pinghe pluton was interpreted using the model of slab break-off, and the potential block that collided with Baoshan Block was South China.

Acknowledgement We are grateful to Prof. Robert J. Stern for his kind help.

Funding This work was jointly supported by the National Natural Science Foundation of China (41190072) and the Teaching and Research Award Program for Outstanding Young Teachers in Higher Education Institutions of MOE, P.R. China. Support was also provided by the MOST Special Fund from the State Key Laboratory of Continental Dynamics, Northwest University and Province Key Laboratory Construction Item (08JZ62).

Supplemental data Supplemental data for this article can be accessed at https://www.dx.doi.org/10.1080/00206814.2014.905998.

References

Andersen, T., 2002. Correction of common lead in U-Pb analyses that do not report ^{204}Pb. Chemical Geology, 192,59-79.

Argles, T.W., Prin, C.I., Foster, G.L., Vance, D., 1999. New garnet for old? Cautionary tales from young mountain belt. Earth & Planetary Science Letters, 172, 301-309.

Bureau of Geology and Mineral Resources of Yunnan Province, 1990. Regional Geology of Yunnan Province. Beijing: Geological Publishing House, 729.

Cawood, P.A., Buchan, C., 2007. Linking accretionary orogenesis with supercontinent assembly. Earth-Science Reviews, 82, 217-256.

Cawood, P.A., Johnson, M.R.W., Nemchin, A.A., 2007. Early Palaeozoic orogenesis along the Indian margin of Gondwana: Tectonic response to Gondwana assembly. Earth & Planetary Science Letters, 255, 70-84.

Chen, F.K., Li, Q.L., Wang, X.L., Li, X.H., 2005. Early Paleozoic magmatism in Baoshan-Tengchong Block of the Tethyan Belt, Yunnan Province. Acta Geoscience Sinica, 26(Sup), 93.

Chen, F.K., Li, Q.L., Wang, X.L., Li, X.H., 2006. Zircon age and Sr-Nd-Hf isotopic composition of migmatite in the eastern Tengchong block, western Yunnan. Acta Petrologica Sinica, 22, 439-448.

Chen, F.K., Li, X.H., Wang, X.L., Li, Q.L., Siebel, W., 2007. Zircon age and Nd-Hf isotopic composition of the Yunnan Tethyan belt, southwestern China. International Journal of Earth Science, 96, 1179-1194.

Chen, J.C., 1987. Discussion on the age division and the selects of isotopic age determination for granitic rock in western Yunnan. Yunnan Geology, 6, 101-112 (in Chinese with English abstract).

Chen, Z.L., 1994. Tethyan Geology for 100 years. Tethyan Geology, 18, 1-22 (in Chinese with English abstract).

Collins, W.J., Beams, S.D., White, A.J.R., Chappell, B.W., 1982. Nature and origin A-type granites with particular reference to southeastern Australia. Contributions to Mineralogy and Petrology, 80, 189-200.

DeCelles, P.G., Gehrels, G.E., Quade, J., LaReau, B., Spurlin, M., 2000. Tectonic implications of U-Pb zircon ages of the Himalayan orogenic belt in Nepal. Science, 288, 497-499.

Dobosi, G., Kempton, P.D., Downes, H., Embey-Isztin, A., Thirlwall, M., Greenwood, P., 2003. Lower crustal granulite xenoliths from the Pannonian Basin, Hungrary, Part 2: Sr-Nd-Pb-Hf and O isotope evidence for formation of continental lower crust by tectonic emplacement of oceanic crust. Contributions to Mineralogy and Petrology, 144, 671-683.

Dong, M.L., Dong, G.C., Mo, X.X., Zhu, D.C., Nie, F., Xie, X.F., Wang, X., Hu, Z.C., 2012. Geochronology and geochemistry of the Early Paleozoic granitoid in Baoshan block, western Yunnan and their implications. Acta Petrologica Sinica, 28(5), 1453-1464 (in Chinese with English abstract).

Dong, X., Zhang, Z.M., Wang, J.L., Zhao, G.C., Liu, F., Wang, W., Yu, F., 2009. Provenance and formation age of the Nyingchi Group in the southern Lhasa terrane, Tibetan Plateau: Petrology and zircon U-Pb geochronology. Acta Petrologica Sinica, 25(7), 1678-1694 (in Chinese with English abstract).

Fang, N.Q., Liu, B.P., Feng, Q.L., Jia, J.H., 1994. Late Palaeozoic and Triassic deep-water deposits and tectonic evolution of the Palaeotethys in the Changning-Menglian and Lancangjiang belts, southwestern

Yunnan. Journal of Southeast Asian Earth Sciences, 9, 363-374.

Feng, Q.L., 2002. Stratigraphy of volcanic rocks in the Changning-Menglian Belt in southwestern Yunnan, China. Journal of Asian Earth Sciences, 20, 657-664.

Gatlos, E.J., Harrison, T.M., Manning, C.E., Grove, M., Rai, S.M., Hubbard, M.S., Uprete, B.N., 2002. Records of the evolution of the Himalayan orogen from in situ Th-Pb ion microprobe dating of monazite: Eastern Nepal and western Garhwal. Journal of Asian Earth Science, 20, 459-479.

Gehrels, G.E., Decelles, P.G., Martin, A., Ojha, T.P., Pinhassi, G., Upreti, B.N., 2003. Initiation of the Himalayan orogen as an early Paleozoic thin-skinned thrust belt. GSA Today, 13, 4-9.

Gehrels, G.E., Decelles, P.G., Ojha, T.P., Upreti, B. N., 2006a. Geologic and U-Pb geochronologic evidence for early Paleozoic tectonism in the Dadeldhura thrust sheet, far-west Nepal Himalaya. Journal of Asian Earth Sciences, 28, 385-408.

Gehrels, G.E., Decelles, P.G., Ojha, T.P., Upreti, B.N., 2006b. Geologic and U-Pb geochronologic evidence for early Paleozoic tectonism in the Kathmandu thrust sheet, central Nepal Himalaya. Geological Society of America Bulletin, 118, 185-198.

Griffin, W.L., Belousova, E.A., Shee, S.R., Pearson, N.J., O'Reilly, S.Y., 2004. Archean crustal evolution in the northern Yilgarn Craton: U-Pb and Hf-isotope evidence from detrital zircons. Precambrian Research, 131, 231-282.

Guo, F.X., 2001. Paleozoic tectono-paleobiogeologyphy of Xinjiang, China. Xinjiang Geology, 19(1), 20-27 (in Chinese with English abstract).

Guynn, J., Kapp, P., Gehrels, G.E., Ding, L., 2012. U-Pb geochronology of basement rocks in central Tibet and paleogeographic implications. Journal of Asian Earth Sciences, 43, 23-50.

Harris, N.B.W., Marzouki, F.M.H., Ali, S., 1986. The Jabel Sayid complex, Arabian shield: Geochemical constraints on the origin of peralkaline and related granites. Journal of the Geological Society, 143, 287-295.

Huang, Y., Deng, G.B., Peng, C.L., Hao, J.X., Zhang, G.X., 2009. The discovery and significance of absence in early-middle Ordovician in southern Baoshan, western Yunnan. Guizhou Geology, 26(98), 1-6 (in Chinese with English abstract).

Huang, Y., Hao, J.X., Bai, L., Deng, G.B., Zhang, G.X., Huang, W.J., 2012. Stratigraphic and petrologic response to Late Pan-African movement in Shidian area, western Yunnan Province. Geological Bulletin of China, 31, 306-313 (in Chinese with English abstract).

Hughes, N.C., 2002. Late Middle Cambrian trace fossils from the LejopygeArmata horizon, Zanskar Valley, India, and the use of Precambrian/Cambrian ichnostratigraphy in the Indian subcontinent. Special Papers in Palaeontology, 67, 135-151.

Janoušek, V., Finger, F., Roberts, M., Frýda, J., Pin, C., Dolejš, D., 2004. Deciphering the petrogenesis of deeply buried granites: Whole-rock geochemical constraints on the origin of largely undepleted felsic granulites from the Moldanubian Zone of the Bohhemian Massif. Transactions of the Royal Society of Edinburgh: Earth Sciences, 95, 141-159.

Jian, P., Liu, D.Y., 2002. U-Pb zircon dating of the Caledonian Gongbo gabbro from the Mid-Jinshajiang area, Sichuan Province. Geological Review, 48 (Suppl), 17-21 (in Chinese with English abstract).

Jian, P., Liu, D.Y., Kröner, A., Zhang, Q., Wang, Y.Z., Sun, X.M., Zhang, W., 2009. Devonian to Permian plate tectonic cycle of the Paleo-Tethys Orogen in southwest China (Ⅱ): Insights from zircon ages of ophiolites, arc/back-arc assemblages and within-plate igneous rocks and generation of the Emeishan CFB province. Lithos, 113, 767-784.

Jin, S.C., Zhuang, F.L., 1988. Study of melting inclusion in granites at Longling in Luxi Area. Journal of Kunming Institute of Technology, 13, 1-15 (in Chinese with English abstract).

Jin, X.C., 1994. Sedimentary and paleogeographic significance of Permo-Carboniferous sequences in Western Yunnan China. Geologisches Institut der Universität zu Köln Sonderveröffentlichungen, 99,136.

Jin, X. C., 2002. Permo-Carboniferous sequences of Gondwana affinity in southwest China and their paleogeographic implications. Journal of Asian Earth Sciences, 20, 633-646.

Kennedy, W.Q., 1964. The structural differentiation of Africa in the Pan-African (± 500 m. y.) tectonic episode. Annual Report of the Research Institute of African Geology, University of Leeds, 8, 48-49.

Le Fort, P., Cuney, M., Deniel, C., France-Lanord, C., Sheppard, S.M.F., Upreti, B.N., Vidal., P., 1987. Crustal generation of the Himalayan leucogranites. Tectonophysics, 134, 39-57.

Lee, J., Whitehouse, M.J., 2007. Onset of mid-crustal extensional flow in southern Tibet: Evidence from U-Pb zircon ages. Geology, 35, 45-48.

Li, C., Wu, Y.W., Wang, M., Yang, H.T., 2010. Significant progress on Pan-African and Early Paleozoic orogenic events in Qinghai-Tibet Plateau-discovery of Pan-African orogenic unconformity and Cambrian System in the Gangdise area, Tibet, China. Geological Bulletin of China, 29, 1733-1736.

Li, D.P., Luo, Z.H., Chen, Y.L., Liu, J.Q., Jin, Y., 2014. Deciphering the origin of the Tengchong block, west Yunnan: Evidence from detrital zircon U-Pb ages and Hf isotopes of Carboniferous strata. Tectonophysics, 614, 66-77.

Li, P.W., Rui, G., Junwen, C., Ye, G., 2004. Paleomagnetic analysis of eastern Tibet: Implications for the collisional and amalgamation history of the Three Rivers Region, SW China. Journal of Asian Earth Sciences, 24, 291-310.

Li, X.H., Li, Z.X., Li, W.X., Liu,Y., Yuan, C., Wei, G.J., Qi, C.S., 2007. U-Pb zircon geochemical and Sr-Nd-Hf isotopic constraints on age and origin of Jurassic I-type and A-type granites from central Guangdong, SE China: A major igneous event in response to foundering of zircon using the laser ablation-MC-ICPMS technique. Chemical Geology, 220, 121-137.

Li, X.Z., Pan, G.T., Luo, J.N., 1990. A boundary between Gondwanaland and Laurasia Continents in Sanjiang region. Contribution to the Geology of the Qinghai-Xizang (Tibet) Plateau, 20, 217-233 (in Chinese with English abstract).

Li, Z.H., Lin, S.L., Cong, F., Xie, T., Zou, G.F., 2012. U-Pb ages of zircon from metamorphic rocks of the Gaoligongshan Group in western Yunnan and its tectonic significance. Acta Petrologica Sinica, 28(5), 1529-1541 (in Chinese with English abstract).

Liu, S., Hu, R.Z., Gao, S., Feng, C.X., Huang, Z.L., Lai, S.C., Yuan, H.L.,Liu, X.M., Coulson, I. M., Feng, G.Y., Wang, T., Qi, Y.Q., 2009. U-Pb zircon, geochemical and Sr-Nd-Hf isotopic constraints on the age and origin of Early Palaeozoic I-type granite from the Tengchong-Baoshan block, western Yunnan Province, SW China. Journal of Asian Earth Sciences, 36, 168-182.

Liu, Y., Liu, X.M., Hu, Z.C., Diwu, C.R., Yuan, H.L., Gao, S., 2007. Evaluation of accuracy and long-term stability of determination of 37 trace elements in geological samples by ICP-MS. Acta Petrologica Sinica, 23 (5), 1203-1210 (in Chinese with English abstract).

Lu, S. N., 2001. From Rodinia to Gondwanaland Supercontinents-thinking about problems of researching Neoproterozoic supercontinents. Earth Science Frontiers, 8 (4), 441-448 (in Chinese with English abstract).

Ludwig, K.R., 2003. ISOPLOT 3. 0: A geochronological toolkit for Microsoft Excel. Berkeley Geochronology Center, Special Publication, 4.

Mazumader, R., Bose, P.K., Sarkar, S., 2000. A commentary on the tectono-sedimentary record of the pre-2. 0Ga continental growth of India vis-à-vis possible pre-Gondwana Afro-Indian supercontinent. Journal of African Earth Sciences, 30(5), 201-217.

Metcalfe, I., 1988. Origin and assembly of Southeast Asian continental terranes. Geological Society of London Special Publication, 37, 101-118.

Metcalfe, I., 1996a. Pre-Cretaceous evolution of SE Asian terranes. In: Hall, R., Blundell, D. Tectonic evolution of Southeast Asian. Geology Society of London, Special Publication, 37, 101-118.

Metcalfe, I., 1996b. Gondwanaland dispersion, Asian accretion and evolution of eastern Tethys. Australian Journal of Earth Science, 43, 605-623.

Metcalfe, I., 1998. Palaeozoic and Mesozoic geological evolution of the SE Asian region: Multidisciplinary constraints and implications for biogeography. In: Hall, R., Holloway, J.D. Biogeography and geological evolution of SE Asia. Amsterdam: Backbuys Publishers, 25-44.

Metcalfe, I., 2002. Permian tectonic framework and palaeogeography of SE Asia. Journal of Asian Earth Sciences, 20, 551-566.

Metcalfe, I., 2006. Paleozoic and Mesozoic tectonic evolution and palaeogeography of East Asian crustal fragments: The Korean Peninsula in context. Gondwana Research, 9, 24-46.

Metcalfe, I., 2011. Palaeozoic-Mesozoic history of SE Asia. Geological Society, London, Special Publications, 355, 7-35.

Metcalfe, I., 2013. Gondwana dispersion and Asian accretion: Tectonic and palaeogeographic evolution of eastern Tethys. Journal of Asian Earth Sciences, 66, 1-33.

Miller, C.F., 1985. Are strongly peraluminous magmas derived from pelitic sedimentary sources? The Journal of Geology, 93, 673-689.

Pan, G.T., Chen, Z.L., 1997. Geological-tectonic evolution in the eastern Tethys. Beijing: Geological Publishing House, 218 (in Chinese).

Patiño Douce, A.E., 1995. Experimental generation of hybrid silicic melts by reaction of high Al basalt with metamorphic rocks. The Journal of Geology, 100, 623-639.

Pearce, J.A., 1983. Role of the subcontinental lithosphere in magma genesis at active continental margins. In: Hawkesworth, C.J., Norry, M.J. Continental basalts and mantle xenoliths. Nantwich, U.K., Cheshire, Shiva, 230-249.

Pearce, J.A., 1996. Sources and settings of granitic rocks. Episodes, 19, 120-125.

Pearce, J.A., Harris, N.B.W., Tindle, A.G., 1984. Trace element discrimination diagrams for the tectonic interpretation of granitic rocks. Journal of Petrology, 25, 956-983.

Qi, X.X., Li, H.Q., Li, T.F., Cai, Z.H., Yu, C.L., 2010. Zircon SHRMP U-Pb dating for garnet-rich granite veins in high-pressure granulites from the Namche Barwa Complex, eastern syntaxis of the Himalayas and the relationship with exhumation. Acta Petrologica Sinica, 26(3), 975-984 (in Chinese with English abstract).

Roberts, M.P., Clemens, J.D., 1993. Origin of high-potassium, calc-alkaline, I-type granitoids. Geology, 21, 835-828.

Scotese, C.R., 2004. A continental drift flipbook. The Journal of Geology, 112, 729-741.

Sengör, A.M.C., Altiner, D., Cin, A., Hsü, K.J., 1988. Origin and assembly of the Tethyside orogenic collage at the expense of Gondwana Land. Geological Society, London, Special Publications, 37, 119-181.

Song, S.G., Ji, J.Q., Wei, C.J., Su, L., Zheng, Y.D., Song, B., Zhang, L.F., 2007. Early Paleozoic granite in Nujiang River of northwest Yunnan in SW China and its tectonic implications. Chinese Science Bulletin, 52, 2402-2406.

Stöcklin, J., 1980. Geology of Nepal and regional frame: Thirty-third William Smith Lecture. Journal of the Geological Society, 137, 1-34.

Sun, S.S., McDonough, W.F., 1989. Chemical and isotopic systematics of oceanic basalts: Implications for mantle composition and processes. In: Saunders, A.D., Norry, M.J. Magmatism in the Ocean Basins. London, Geological Society Special Publication, 42, 313-345.

Sylvester, P.J., 1998. Post-collisional strongly peraluminous granites. Lithos, 45, 29-44.

Ueno, K., 2000. Permian fusulinacean faunas of the Sibumasu and Baoshan block, implication for the paleogeographic reconstruction of the Cimmerian continent. Geoscience Journal, 4, 160-163.

Valdiya, K.S., 1995. Proterozoic sedimentation and Pan-African geodynamic development in the Himalaya. Precambrian Research, 74, 35-55.

Von Raumer, J.F., Stampfli, G.M., 2008. The birth of the Rheic Ocean-Early Paleozoic subsidence patterns and subsequent tectonic plate scenarios. Tectonophysics, 461, 9-20.

Wang, B.D., Wang, L.Q., Pan, G.T., Yin, F.G., Wang, D.B., Tang, Y., 2013a. U-Pb zircon dating of Early Paleozoic gabbro from the Nantinghe ophiolite in the Changning-Menglian suture zone and its geological implication. Chinese Science Bulletin, 58, 920-930.

Wang, K.Y., 1996. The tectonic belt in the Sanjiang area of the Yunnan Province and rock units of the Precambrian basements and tectonic evolution along the western margin of the Yangtze Block. Geology of Yunnan, 15, 138-148 (in Chinese with English abstract).

Wang, X.D., Shi, G.R., Sugiyama, T., 2002. Permian of West Yunnan, Southwest China: A biostratigraphic synthesis. Journal of Asian Earth Sciences, 20, 647-656.

Wang, X.X., Zhang, J.J., Santosh, M., Liu, J., Yan, S.Y., Guo, L., 2012. Andean-type orogeny in the Himalayas of south Tibet: Implications for early Paleozoic tectonics along the Indian margin of Gondwana. Lithos, 154, 248-262.

Wang, Y.J., Xing, X.X., Cawood, P.A., Lai, S.C., Xia, X.P., Fan, W.M., Liu, H.C., Zhang, F.F., 2013b. Petrogenesis of early Paleozoic peraluminous granite in the Sibumasu Block of SW Yunnan and diachronous accretionary orogenesis along the northern margin of Gondwana. Lithos, 182-183, 67-85.

Wang, Y.Z., 2000. Tectonics and mineralization of Southern Sanjiang area. Beijing: Geological Publishing House, 45-49 (in Chinese).

Whalen, J.B., Currie, K.L., Chappell, B.W., 1987. A-type granites: Geochemical characteristics, discrimination and petrogenesis. Contributions to Mineralogy and Petrology, 95, 407-419.

White, A.J.R., Chappell, B.W., 1988. Some supracrustal (S-type) granites of the Lachlan Fold Belt. Transactions of the Royal Society of Edinburgh: Earth Sciences, 79, 169-181.

Wingate, M.T.D., Evans, D.A.D., 2003. Palaeomagnetic constraints on the Proterozoic tectonic evolution of Australia. Geological Society, London, Special Publications, 206, 77-91.

Wopfner, H., 1996. Gondwana origin of the Baoshan and Tengchong terrenes of west Yunnan. In: Hall, R., Blundell, D., eds. Tectonic evolution of Southeast Asian. Journal of Geology Society,106, 539-547.

Wu, H., Boulter, C.A., Ke, B., Stow, D.A.V., Wang, Z., 1995. The Changning-Menglian suture zone: A segment of the major Cathaysian-Gondwana divide in Southern Asian. Tectonophysics, 242, 267-280.

Xiao,W.J., Windley, B.F., Hao, J., Zhai, M.G., 2003. Accretion leading to collision and the Permian Solonker suture, Inner Mongolia, China: Termination of the central Asian orogenic belt. Tectonics, 22 (6), 1069.

Xiao, W.J., Windley, B.F., Yong, Y., Yan, Z., Yuan, C., Liu, C., Li, J., 2009. Early Paleozoic to Devonian multi-accretionary model for the Qilian shan, NW China. Journal of Asian Earth Sciences, 35, 323-333.

Xu, W.C., Zhang, H.F., Parrish R., Harris, N., Guo, L., Yuan, H.L., 2010. Timing of granulite-facies metamorphism in the eastern Himalayan syntaxis and its tectonic implications. Tectonophysics, 485(1-4), 234-244.

Xu, Y.G., Yang, Q.J., Lan, J.B., Luo, Z.Y., Huang, X.L., Shi, Y.R., Xie, L.W., 2012. Temporal-spatial distribution and tectonic implications of the batholiths in the Gaoligong-Tengliang-area, western Yunnan: Constraints from zircon U-Pb ages and Hf isotopes. Journal of Asian Earth Sciences, 53, 151-175.

Xu, Z.Q., Yang, J.S., Liang., F.H., Qi, X.X., Zeng, L.S., Liu, D.Y., Li, H.B., Wu, C.L., Shi, R.D., Chen, S.Y., 2005. Pan-African and Early Paleozoic orogenic events in the Himalaya terrane: Inference from SHRIMP U-Pb zircon ages. Acta Petrologica Sinica, 21 (1), 1-12 (in Chinese with English abstract).

Yang, X.J., Jia, X.C., Xiong, C.L., Bai, X.Z., Huang, B.X., Luo, G., Yang, C.B., 2012. LA-ICP-MS zircon U-Pb age of metamorphic basic volcanic rock in Gongyanghe Group of southern Gaoligong Mountain, western Yunnan Province, and its geological significance. Geological Bulletin of China, 31(2/3), 264-276 (in Chinese with English abstract).

Yuan, H.L., Gao, S., Liu, X.M., Li, H.M., Gunther, D., Wu, F.Y., 2004. Accurate U-Pb age and trace element determinations of zircon by laser ablation-inductively coupled plasma mass spectrometry. Geostandard Newsletters, 28, 353-370.

Yuan, H.L., Gao, S., Luo, Y., Zong, C.L., Dai, M.N., Liu, X.M., Diwu, C.R., 2007. Study of Lu-Hf geochronology: A case study of eclogite from Dabie UHP Belt. Acta Petrologica Sinica, 23(2), 233-239 (in Chinese with English abstract).

Zhang, C.H., Wang, Z.Q., Li, J.P., Song, M.S., 1997. Tectonic frame for deformation of the metamorphic rocks in the Ximeng area of the western Yunnan. Regional Geology of China, 16, 171-179 (in Chinese with English abstract).

Zhang, Z.M., Dong, X., Santosh, M., Liu, F., Wang, W., Yiu, F., He, Z.Y., Shen, K., 2012. Petrology and geochronology of the Namche Barwa Complex in the eastern Himalayan syntaxis, Tibet: Constraints on the origin and evolution of the north-eastern margin of the Indian Craton. Gondwana Research, 21, 123-137.

Zhang, Z.M., Wang, J.L., Shen, K., Shi, C., 2008. Petrology and geochronology of the Namche Barwa Complex in the eastern Himalayan syntaxis, Tibet. Atca Petrologica Sinica, 24(7), 1627-1637 (in Chinese with English abstract).

Zhong, D.L., 1998. Paleo-Tethyan Orogenic Belt in the Western Part of the Sichuan and Yunnan Provinces. Beijing: Science Press, 231 (in Chinese).

Zhou, Z.G., Liu, W.C., Liang, D.Y., 2004. Discovery of the Ordovician and its basal conglomerate in the Kangmar area, southern Tibet-with a discussion of the relation of the sedimentary cover and unifying basement in the Himalayas. Geological Bulletin of China, 23, 655-663 (in Chinese with English abstract).

Zhu, D.C., Zhao, Z.D., Niu, Y.L., Dilek, Y., Wang, Q., Ji, W., Dong, G.C., Sui, Q.L., Liu, Y.S., Yuan, H.L., Mo, X.X., 2012b. Cambrian bimodal volcanism in the Lhasa Terrane, southern Tibet: Record of an early Paleozoic Andean-typemagmatic arc in the Australian proto-Tethyan margin. Chemical Geology, 328, 290-308.

Zhu, D.C., Zhao, Z.D., Niu, Y.L., Mo, X.X., Chung, S.L., Hou, Z.Q., Wang, L.Q., Wu, F.Y., 2011. The Lhasa Terrane: Record of a micro-continent and its histories of drift and growth. Earth & Planetary Science Letters, 301, 241-255.

Zhu, D.C., Zhao, Z.D., Niu, Y.L., Wang, Q., Yildirim, D., Dong, G.C., Mo, X.X., 2012. Origin and Paleozoic tectonic evolution of the Lhasa Terrane. Geological Journal of China Universities, 18(1), 1-15 (in Chinese with English abstract).

Early-Cretaceous highly fractionated I-type granites from the northern Tengchong Block, western Yunnan, SW China: Petrogenesis and tectonic implications[①]

Zhu Renzhi Lai Shaocong[②] Qin Jiangfeng Zhao Shaowei

Abstract: Western Yunnan, an important constituent of the southeastern segment of the East Tethyan tectonic domain, lies along a transformed orientation from the NWW trending Himalayan-Tethyan segment to the northerly trending Southeast Asian segment. However, the geodynamical setting of the Early Cretaceous tectonothermal magmatism along the Bangong-Nujiang-Lushui-Luxi-Ruili belt as the Tethyan branch in western Yunnan (SW China) remains controversial. The Donghe granitoid, which is located between the Gaoligong and Tengliang belts in the northern Tengchong Block, reveals its petrogenesis and its tectonics, both of which play a vital role in resolving previous disputes. Our zircon laser ablation inductively coupled plasma mass spectrometry U-Pb dating of granites from the Donghe batholith yields ages of 119.9±0.9 Ma to 130.6±2.5 Ma. These granites display features typical of highly fractionated I-type granites: high SiO_2 contents(>71 wt%), high K contents (K_2O = 3.88–5.66 wt%), calc-alkaline character, slight peraluminosity (A/CNK = 1.02–1.16), and a highly differentiated index ranging from 83.6 to 95.6. In addition, as SiO_2 contents increase, the rare earth element (REE) abundances, especially heavy REE abundances, and REE pattern slopes change gradually, but the negative Eu anomalies increase sharply, while the degree of enrichment in Rb, Th, U, and Pb and depletion in Ba, Nb, Sr, P, and Ti are enhanced. These features indicate that K-feldspar, ± plagioclase, ± biotite, ± amphibole, ± apatite, ± sphene/garnet, and ± Fe-Ti oxides such as ilmenite play the major role in the fractional crystallization process. The high initial $^{87}Sr/^{86}Sr$ (0.7067 and 0.7079) and negative $\varepsilon_{Nd}(t)$ values (−10.1 to −8.6), with T_{DM2} ranging from 1.39 Ga to 1.49 Ga, indicate that the sources were mainly derived from the mature ancient middle to lower crust and minor mantle-derived materials. The initial $^{206}Pb/^{204}Pb$, $^{207}Pb/^{204}Pb$, and $^{208}Pb/^{204}Pb$ ratios of 18.462–18.646, 15.717–15.735, and 38.699–39.007, respectively, signify that some subduction-related material such as ocean island volcanic rocks and mature arc primitive rocks may be involved as sources. Based on an analysis of similar zircon saturated temperature and geochemical characteristics of typical highly fractionated I-type granites in SE and SW China, and consideration of the regional

① Published in *Journal of Asian Earth Sciences*, 2015, 100.

② Corresponding author.

geological setting, we suggest that the parent magma may be derived from the ancient middle to lower continental crust and mantle-derived basaltic magma.These were generated in the setting of a westward subducted Lushui-Luxi-Ruili fault (LLR) Tethyan oceanic slab, where mantle-wedge-derived sources provided enough heat and material to melt the middle to lower ancient crust. Taking into account the temporal-spatial distributions of Early Cretaceous magmatic rocks in the region, we further suggest the existence of an Andean-type active continental margin from the Lhasa Block to the Tengchong Block along the Bangong-Nujiang-LLR Tethys Ocean during the Early Cretaceous.

1 Introduction

The study of Tethys has been ongoing for more than a century since Suess (1893) proposed it as an equatorial ocean ancestral to the Alpine-Himalayan mountain ranges. The theory of plate tectonics provides valuable insight into the mechanisms by which Tethys has evolved (Condie, 1997). Western Yunnan, an important constituent of the southeastern segment of the East Tethyan tectonic domain, lies along a transformed orientation from the NWW trending Himalayan-Tethyan segment to the northerly trending Southeast Asian segment (Hutchison, 1989; Metcalfe, 1996a,b, 2002, 2013; Zhang et al., 2008, 2012a; Wang et al.,2013, 2014; Deng et al., 2014), so it well records the evolutionary history of Tethys. Two stages are primarily involved: ①the accretion of some continents and microcontinents of Gondwana biotic affinity to continents and/or microcontinents of Cathaysian biotic affinity from the Late Paleozoic to the Early Mesozoic (Metcalfe, 2011a,b, 2013) and ② the accretion of Gondwana to the Asian continent during the Mesozoic to Cenozoic (Hall, 1997, 2002, 2011, 2012; Metcalfe, 2013).

The Tengchong Block,a crucial constituent of the Neo-Tethys domain in SW China, being located between the Banggong-Nujiang-Lushui-Longling-Ruili (LLR) suture zone and the Yarlung-Zangbo-Myitkyina suture zone, exposes abundant Mesozoic granitoids (Cao et al., 2014; Chen et al., 1991; Cong et al., 2009, 2010a, 2011b; Li et al., 2012a,b; Luo et al., 2012; Qi et al., 2011; Xie et al., 2010; Xu et al., 2012; Yang et al., 2006,2009; Yang and Xu, 2011; Zou et al., 2011a) and Cenozoic granitoid rocks (Ma et al., 2014). These ribbon pattern magmatic belts were dominated by Early Cretaceous granitoids, forming the southeastern extension of the northern magmatic belt including the central and northern Lhasa block (Cao et al., 2014; Li et al., 2012a,b; Qi et al., 2011; Xie et al., 2010; Xu et al., 2012; Yang et al., 2006; Zou et al., 2011a) along the Bangong-Nujiang-LLR suture zone. However, the geodynamical setting of the Early Cretaceous tectonothermal magmatism along the Bangong-Nujiang-LLR belt in the Tengchong Block has remained controversial because of the following pieces of conflicting evidence: ① the active continental margin is related to the eastward subduction of the Putao-Myitkyina paleo-oceanic slab as a branch of the Neo-Tethys

Ocean (Cong et al., 2010a, 2011a,b); ② the active continental margin is related to the southward subduction of the Bangong-Nujiang Tethyan oceanic slab (Li et al., 2012a,b); ③ the southward subduction of the Bangong-Nujiang Tethyan oceanic slab occurred during the collision between the Lhasa-Tengchong and the Qiangtang-Baoshan blocks (Qi et al., 2011; Zou et al., 2011a); ④ there is a post-collision setting after the southward subduction of the Bangong-Nujiang Tethyan oceanic slab (Cao et al., 2014; Luo et al., 2012; Yang et al., 2006; Yang and Xu, 2011); and ⑤ there is a post-collisional setting between the Lhasa and Qiangtang blocks being partly influenced by the far field of the Neo-Tethyan oceanic slab (Xu et al., 2012). In other words, a post-collision-related setting is supported by the peraluminous S-type granites in the northern Gaoligong belt, and a subduction-related setting is supported by the metaluminous to peraluminous I-type granites, granodiorites, and diorites in the southern Tengliang belts. The Donghe granitoid located between the Gaoligong and Tengliang belts, therefore, reveals its petrogenesis and tectonics, both of which play a pivotal role in understanding the geodynamical setting of the Mesozoic magmatism along the Bangong-Nujiang-LLR suture as the branch of the Neo-Tethyan domain from Tibet to western Yunnan, SW China.

We first present zircon U-Pb age and major and trace element and Sr-Nd-Pb isotopic composition data for the Donghe granitoid in the northern Tengchong Block. Our main aims are twofold: ① to constrain their petrogenesis and tectonic implications in the northern Tengchong Block and ② to explore the geodynamical setting of the Early Cretaceous magmatism along the Bangong-Nujiang-LLR suture from the Lhasa terrane in the Tibetan Plateau to the Tengchong Block in western Yunnan and further understand the temporal-spatial evolution of Tethys in western Yunnan, SW China. The highly fractionated I-type granites in the Tengchong Block as the southeast extension of the Tibetan Plateau are reported for the first time. Coupled with consideration of the other published evidence in the literature, our results may imply an Andean-type active continental margin along the Bangong-Nujiang-LLR belt triggered by the southwestern subducted Bangong-Nujiang-LLR Tethyan oceanic slab during the Early Cretaceous.

2 Geological setting and petrography

Western Yunnan is an important constituent of the southeastern segment of the East Tethyan tectonic domain (Kou et al., 2012) (Fig. 1a). From east to west, this region comprises the Simao-Indochina, Baoshan-Shan-Thai, and Tengchong blocks (Wang et al., 2006a,b; Chen et al., 2007; Cong et al., 2011a,b; Xu et al., 2012; Wang et al., 2013, 2014; Deng et al., 2013; Metcalfe, 2013), separated by the Changning-Menglian suture and the Lushui-Luxi-Ruili fault (LLRF), respectively.

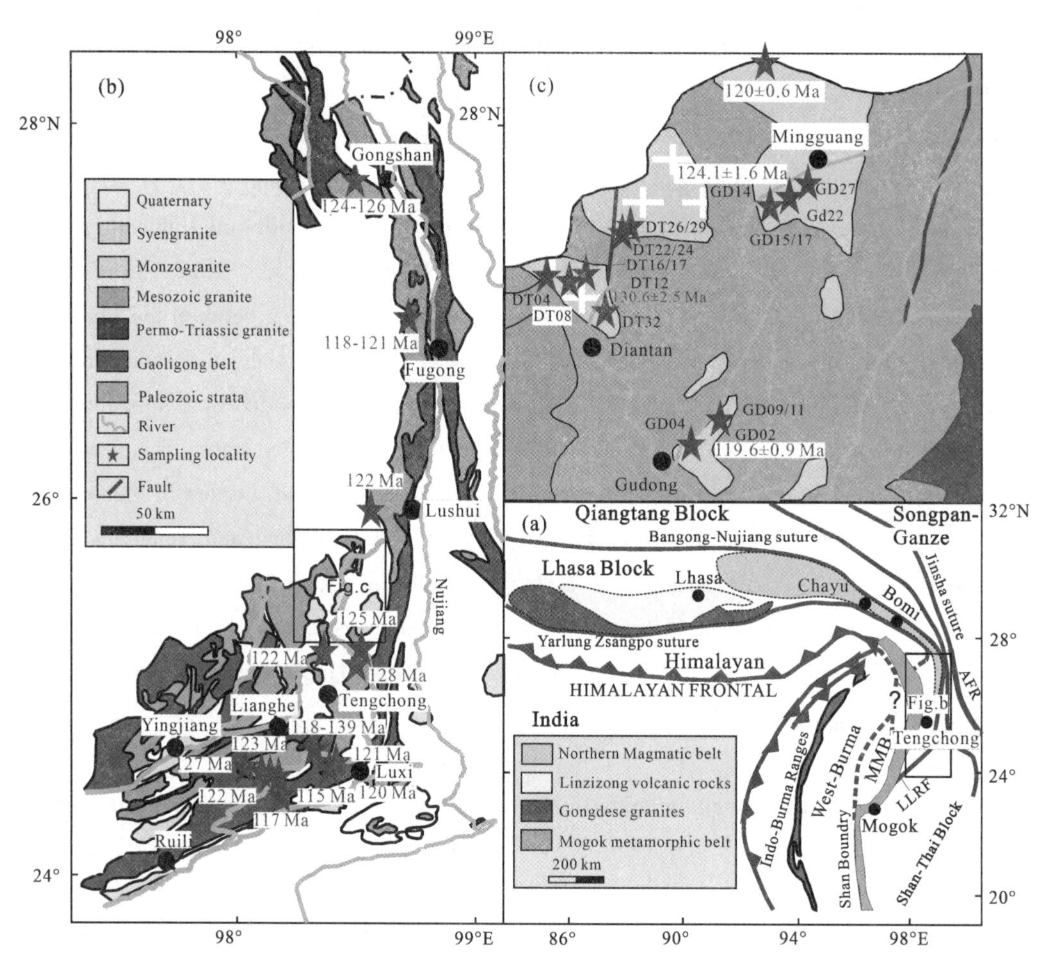

Fig. 1 (a) Distribution of main continental blocks of SE Asia (modified after Xu et al., 2012 and Ma et al., 2014). (b) Distribution of principal magma and strata of western Yunnan (modified after Xu et al., 2008). (c) Study area in the Tengchong Block (modified after YNBGMR, 1991). All the age data in the diagram are from Table 4.

The Simao-Indochina Block of South China affinity (Metcalfe, 1996a,b; Wopfner, 1996; Zhong, 1998; Ueno, 2000) comprises a Proterozoic metamorphic succession of volcano-clastic rocks and carbonates (Zhong, 1998), unconformably overlain by a Paleozoic package of carbonate and siliciclastic rocks, with typical Cathaysia flora and fauna (Feng, 2002; YNBGMR, Yunnan Bureau of Geology and Mineral Resources, 1991; Zhong, 1998). The Baoshan, Tengchong, and Shan-Thai blocks (Metcalfe, 1988, 1996a,b, 1998, 2002, 2013; Shi and Archbold, 1995, 1998) have stratigraphic and paleontological affinities to Gondwana (Fan and Zhang, 1994; Feng, 2002; Fontaine, 2002; Metcalfe, 1996a,b, 2002, 2013; Zhong, 1998). The main stratigraphic package includes pre-Mesozoic high-grade metamorphic rocks and Mesozoic-Cenozoic sedimentary and igneous rocks (YNBGMR, 1991; Zhong, 1998).

The Tengchong Block, the northern part of Sibumasu, is bounded to the eastern Baoshan Block by the LLRF and to the western Burma Block by the Putao-Myitkyina suture zone (Li et al., 2004) (Fig. 1b). Based on the occurrence of Permo-Carboniferous glacio-marine deposits and overlying post-glacial black mudstones, as well as the Gondwana-like fossil assemblages in the blocks, one can conclude that it was derived from the margins of western Australian on the eastern Gondwana supercontinent (Jin, 1996). Its characteristic strata has a Mesoproterozoic metamorphic basement (Gaoligong Mountain Group), which is overlain by Late Paleozoic clastic sedimentary rocks and carbonates, Mesozoic-Tertiary granitoids, and Tertiary-Quaternary volcano-sedimentary sequences (YNBGMR, 1991). The Gaoligong Mountain Group is composed of quartzites, two mica quartz schists, feldspathic gneisses, migmatites, amphibolites, and marble. Zircons from a paragneiss sample and an orthogneiss sample give ages of 1 053–635 Ma and 490–470 Ma, respectively (Song et al., 2010). The Paleozoic sedimentary strata are dominated by Carboniferous clastic rocks, Upper Triassic to Jurassic turbidites, Cretaceous red beds, and Cenozoic sandstones (YNBGMR, 1991; Zhong, 2000).

There are abundant granitic gneiss, migmatite, and leucogranite that mainly outcrop in the Tengchong Block and most of them have been previously considered to be Proterozoic in age (YNBGMR, 1991). According to recent geochronology, however, there are also massive granitoids from the Paleozoic to the Cenozoic in the Tengchong Block (with zircon U-Pb ages of 232–206 Ma) present south of Tengchong county and north of Lianghe county (Li et al., 2011; Zou et al., 2011b). Younger granitoids with zircon U-Pb ages of 139–115 Ma are common in the eastern part of the Tengchong Block near Tengchong and Lianghe counties (Chen et al., 1991; Yang et al., 2006; Yang and Xu, 2011; Xie et al., 2010; Cong et al., 2010a, 2011a,b; Qi et al., 2011; Zou et al., 2011a; Li et al., 2012a,b; Luo et al., 2012; Xu et al., 2012; Cao et al., 2014). Late Cretaceous S-type and A-type granites with zircon U-Pb ages of 76–68 Ma are also present in the mid-western part of the Tengchong Block (Jiang et al., 2012; Xu et al., 2012; Ma et al., 2013). In addition, granites with zircon ages as young as 66–52 Ma have been found near the border between China and Myanmar in the Tengchong Block (Booth et al., 2004; Liang et al., 2008; Chiu et al., 2009; Xie et al., 2010; Xu et al., 2012), and all of these granitoids intruded into the Paleozoic and Mesozoic strata. The granites at Gaoligong lie to the west of the Lushui-Luxi-Ruili fault and have undergone strong shearing with a northerly trending foliation and a subhorizontally plunging mineral lineation (Wang and Burchfiel, 1997; Wang et al., 2016a,b; Zhang et al., 2011, 2012c).

The Donghe granitoid (Fig. 1c) can be divided into coarse-grain porphyry granite (Fig. 2a) distributed near the town of Gudong; mid- to coarse-grain monzogranites (Fig. 2b) distributed near the town of Mingguang, which contain some mafic enclaves in the parent pluton (Fig. 2c), the first two of which are both called monzogranite; and mid- to coarse-grain

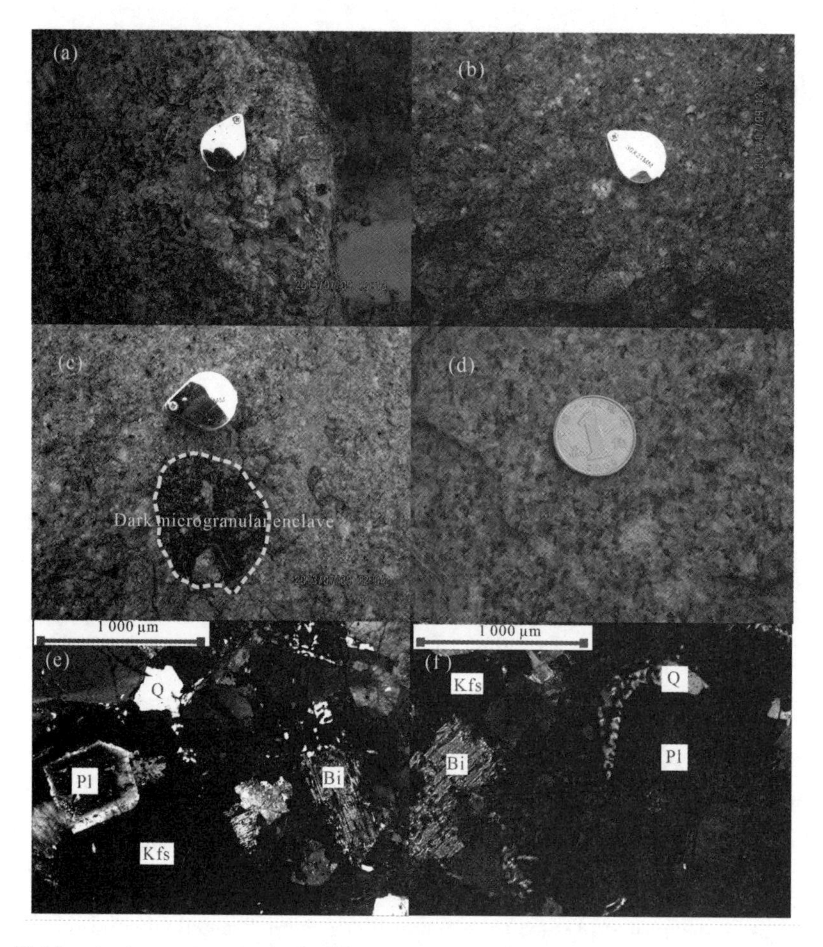

Fig. 2 Field and microscope petrography of monzogranite and syengranite from the Donghe granitoid in the Tengchong Block, western Yunnan, SW China.

syenogranites outcropped near the town of Diantan(Fig. 2d). Combining field with microscopic observations (Fig. 2e,f), we find that the granites near Gudong have commonly porphyaceous texture; both of these monzogranites show feldspar phenocrysts mosaicing with other main minerals, such as quartz and plagioclase, containing 30% – 40% quartz, 40% – 55% K-feldspar, 5% – 10% plagioclase, 3% – 7% biotite, and 0% – 3% hornblende. Accessory minerals include zircon, apatite, titanite, magnetite, and ilmenite. The grain sizes for K-feldspar, plagioclase, and quartz grange from 300 μm to 500 μm, from 200 μm to 300 μm, and from 100 μm to 300 μm, respectively. The syenogranites are mid-coarse grained, with 35%–40% quartz, 45%–60% K-feldspar, 3%–5% plagioclase, 5%–8% biotite, and minor amounts (< 3%) of chlorite. Accessory minerals include zircon, apatite, magnetite, and ilmenite. Grain sizes for K-feldspar, plagioclase, and quartz range from 500 μm to 1 000 μm, from 300 μm to 500 μm, and from 100 μm to 300 μm, respectively. In total, almost all of the feldspars are K-feldspar in syenogranites. For both, K-feldspar megacrysts comprise the dominant mineral type, including amazonite with grid bicrystals and orthoclase with Carlsbad twin, which shows hypidiomorphic granular texture. Some xenomorphic quartz and a smaller amounts of hexagonal quartz are present in the monzogranite and syenogranite, biotite is inlaid between the feldspar and the quartz, and a few acicular crystals of apatite are also present.

3 Analytical methods

3.1 Major and trace elements

Whole-rock samples were trimmed to remove weathered surfaces, cleaned with deionized water, crushed, and then powdered through a 200 mesh screen using a tungsten carbide ball mill. Major elements were analyzed using an X-ray fluorescence (XRF) spectrometer (Rikagu RIX 2100) at the State Key Laboratory of Continental Dynamics, Northwest University, Xi'an, China. Analyses of USGS and Chinese national rock standards (BCR-2, GSR-1, and GSR-3) indicate that both analytical precision and accuracy for major elements are generally better than 5%.

Trace elements were determined by using a Bruker Aurora M90 inductively coupled plasma mass spectrometry (ICP-MS) at the Guizhou Tuopu Resource and Environmental Analysis Center, following the method of Qi et al. (2000). Sample powders were dissolved using an $HF + HNO_3$ mixture in a high-pressure PTFE bomb at 185℃ for 36 h. The accuracies of the ICP-MS analyses are estimated to be better than ± 5% – 10% (relative) for most elements.

3.2 Sr-Nd-Pb isotopic analyses

Whole-rock Sr-Nd-Pb isotopic data were obtained by using a Nu Plasma HR multi-collector mass spectrometer at the Guizhou Tuopu Resource and Environmental Analysis Center.

The Sr and Nd isotopes were determined by using a method similar to that of Chu et al. (2009). Sr and Nd isotopic fractionation was corrected to $^{87}Sr/^{86}Sr = 0.1194$ and $^{146}Nd/^{144}Nd = 0.7219$, respectively. During the period of analysis, a Neptune multi-collector ICP-MS was used to measure the $^{87}Sr/^{86}Sr$ and $^{143}Nd/^{144}Nd$ isotope ratios. NIST SRM-987 and JMC-Nd were used as certified reference standard solutions for $^{87}Sr/^{86}Sr$ and $^{143}Nd/^{144}Nd$ isotopic ratio, respectively. BCR-1 and BHVO-1 were used as the reference materials.

Whole-rock Pb was separated by an anion exchange in HCl-Br columns, with Pb isotopic fractionation corrected to $^{205}Tl/^{203}Tl = 2.3875$. Within the analytical period, 30 measurements of NBS981 gave average values of $^{206}Pb/^{204}Pb = 16.937 \pm 1\ (2\sigma)$, $^{207}Pb/^{204}Pb = 15.491 \pm 1\ (2\sigma)$, and $^{208}Pb/^{204}Pb = 36.696 \pm 1\ (2\sigma)$. The BCR-2 standard gave $^{206}Pb/^{204}Pb = 18.742 \pm 1\ (2\sigma)$, $^{207}Pb/^{204}Pb = 15.620 \pm 1\ (2\sigma)$, and $^{208}Pb/^{204}Pb = 38.705 \pm 1\ (2\sigma)$. Total procedural

Pb blanks were in the range of 0. 1–0. 3 ng.

3. 3 Zircon U-Pd analyses

Zircon was separated from three ~4 kg samples (GD02, GD14, and DT12) taken from various sampling locations within the Donghe granitoid. The zircon grains were separated by using conventional heavy liquid and magnetic techniques. Representative zircon grains were handpicked and mounted in epoxy resin disks and then polished and coated with carbon. Internal morphology was examined using cathodoluminescent (CL) prior to U-Pb analyses.

Laser ablation ICP-MS zircon U-Pb analyses were conducted on an Agilent 7500a ICP-MS equipped with a 193-nm laser, which is housed at the State Key Laboratory of Continental Dynamics, Northwest University, Xi'an, China, following the method of Yuan et al. (2004). The $^{207}Pb/^{206}Pb$ and $^{206}Pb/^{238}U$ ratios were calculated by using the GLITTER program, and corrected using the Harvard zircon 91500 as external calibration. These correction factors were then applied to each sample to correct for both instrumental mass bias and depth-dependent elemental and isotopic fractionation. The detailed analytical technique is described in Yuan et al. (2004). Common Pb contents were therefore evaluated by using the method described in Andersen (2002). The age calculations and plotting of concordia diagrams were made using ISOPLOT (version 3. 0; Ludwig, 2003). The errors quoted in tables and figures are at the 2σ level.

4 Results

The Donghe granitoid can be divided into Qipanshi granites, Mingguang granites, and Diantan granites (Wang et al., 2014). All the analytical data are listed in Table 1 (major and trace elements), Table 2 (Sr-Nd isotopes), and Table 3 (Pb isotopes).

Table 1 Major(wt%) and trace(ppm) element analysis result for the Donghe granitoid.

Sample	Qipanshi monzogranite			Mingguang monzogranite					Diantan syenogranite								
	GD04	GD09	GD11	GD15	GD17	GD22	GD27	GD28	DT04	DT08	DT16	DT17	DT22	DT24	DT26	DT29	DT32
SiO_2	72. 7	71. 7	73. 4	76. 0	76. 6	76. 8	74. 3	74. 7	76. 5	76. 0	77. 4	77. 4	77. 1	77. 9	76. 4	76. 3	76. 3
TiO_2	0. 32	0. 33	0. 30	0. 15	0. 14	0. 15	0. 21	0. 20	0. 03	0. 04	0. 03	0. 03	0. 04	0. 04	0. 04	0. 04	0. 13
Al_2O_3	14. 4	14. 3	13. 5	12. 7	12. 6	12. 9	13. 2	13. 1	12. 5	13. 1	12. 8	12. 7	12. 2	11. 5	12. 7	12. 7	12. 8
$Fe_2O_3{}^T$	1. 90	2. 26	1. 98	1. 24	1. 06	0. 81	1. 62	1. 57	1. 05	0. 75	0. 72	0. 80	1. 06	0. 81	1. 07	0. 93	0. 59
MnO	0. 04	0. 05	0. 05	0. 03	0. 03	0. 03	0. 05	0. 05	0. 04	0. 01	0. 01	0. 01	0. 06	0. 04	0. 05	0. 03	0. 03
MgO	0. 60	0. 76	0. 59	0. 23	0. 23	0. 22	0. 41	0. 40	0. 04	0. 03	0. 04	0. 03	0. 08	0. 03	0. 06	0. 04	0. 14
CaO	1. 84	1. 62	1. 71	0. 94	0. 89	0. 81	1. 35	1. 29	0. 56	0. 46	0. 47	0. 30	0. 28	0. 99	0. 19	0. 51	0. 50
Na_2O	3. 69	3. 57	3. 39	3. 18	3. 05	2. 87	3. 38	3. 28	3. 54	3. 79	3. 87	3. 76	2. 41	2. 27	3. 13	3. 32	2. 89
K_2O	3. 88	4. 11	4. 12	4. 98	5. 01	5. 13	4. 51	4. 57	4. 71	4. 87	4. 42	4. 58	5. 66	5. 32	5. 04	5. 12	5. 27
P_2O_5	0. 08	0. 08	0. 07	0. 03	0. 03	0. 03	0. 05	0. 04	0. 02	0. 01	0. 02	0. 02	0. 02	0. 01	0. 02	0. 01	0. 02

Continued

Sample	Qipanshi monzogranite			Mingguang monzogranite					Diantan syenogranite								
	GD04	GD09	GD11	GD15	GD17	GD22	GD27	GD28	DT04	DT08	DT16	DT17	DT22	DT24	DT26	DT29	DT32
LOI	0.77	0.80	0.59	0.36	0.35	0.48	0.46	0.40	0.58	0.52	0.51	0.52	0.73	0.59	0.99	0.60	0.88
Total	100	100	100	100	100	100	100	100	100	100	100	100	100	100	100	100	100
$Mg^{\#}$	42.4	43.9	41.0	30.2	33.6	38.8	37.1	37.3	8.15	8.53	11.5	8.04	15.0	7.95	11.6	9.11	35.6
A/NK	1.40	1.38	1.34	1.20	1.20	1.25	1.27	1.27	1.14	1.14	1.15	1.14	1.21	1.21	1.20	1.15	1.23
A/CNK	1.06	1.08	1.02	1.03	1.04	1.09	1.03	1.03	1.04	1.06	1.07	1.08	1.15	1.02	1.16	1.06	1.13
Li	26.6	33.8	29.2	42.6	42.1	25.5	66.4	70.1	6.0	1.1	5.8	3.0	4.1	1.2	3.8	2.6	23.6
Be	2.58	2.53	2.24	4.36	4.11	4.55	5.13	4.72	2.87	2.26	3.20	2.49	1.94	1.97	2.80	2.83	3.43
Sc	7.24	7.69	6.98	5.99	5.82	5.38	7.22	6.08	6.03	5.11	6.87	4.94	4.92	4.75	5.83	5.57	4.69
V	36.1	38.1	33.5	14.0	11.5	12.8	22.2	21.6	2.6	1.8	2.6	2.3	2.6	2.4	1.9	2.4	7.1
Cr	8.93	7.23	7.53	8.98	6.43	4.11	6.95	5.57	2.74	4.98	2.62	2.49	4.52	2.91	4.76	2.31	3.12
Co	43.0	37.9	40.9	58.9	61.3	37.5	98.0	81.6	57.5	114.0	61.6	52.2	53.5	65.7	51.1	59.8	51.1
Ni	10.1	5.1	5.8	6.7	4.1	2.9	3.8	7.5	2.7	2.6	2.9	2.8	5.3	3.2	5.1	2.6	3.1
Cu	6.89	5.22	5.37	1.51	1.44	11.69	1.66	2.34	0.76	0.47	0.44	0.25	0.10	0.22	1.12	0.19	0.45
Zn	60.4	46.7	48.2	23.1	30.9	16.8	29.5	34.6	48.8	29.7	42.4	27.5	41.5	42.7	39.5	24.9	77.2
Ga	16.5	16.6	15.5	15.0	14.3	14.4	16.2	15.6	12.0	11.3	11.9	10.9	12.4	13.0	12.4	10.7	12.6
Ge	1.21	1.19	1.18	1.62	1.61	1.58	1.65	1.59	1.19	1.23	1.52	1.32	0.90	1.31	1.20	1.23	1.59
As	1.83	1.62	1.46	1.60	1.94	1.30	1.45	1.57	0.90	0.70	0.91	0.63	0.90	1.50	0.75	0.74	1.27
Rb	270	284	269	488	472	517	489	504	348	356	442	455	444	387	375	373	557
Sr	164	164	154	56	49	56	103	103	18	20	11	10	28	56	31	36	27
Y	24.2	24.5	23.8	54.2	47.4	54.0	84.2	63.2	25.4	19.0	33.6	18.6	25.7	26.5	28.6	29.0	28.2
Zr	166	156	159	109	117	99	130	102	86	87	82	84	86	75	113	88	105
Nb	13.3	13.6	12.6	19.6	19.6	17.5	21.6	18.3	15.4	14.3	14.7	14.9	13.3	13.1	17.2	14.1	15.4
Sb	0.16	0.16	0.17	0.41	0.31	0.47	0.37	0.33	0.37	0.13	0.23	0.16	0.21	0.27	0.15	0.24	0.43
Cs	3.60	4.23	3.65	9.41	9.75	13.90	8.63	8.90	2.52	1.52	2.87	2.88	2.64	2.12	1.82	2.15	12.10
Ba	403	493	421	141	132	177	214	219	170	176	81	72	259	282	229	182	75
La	35.8	33.9	36.9	35.1	35.1	52.0	34.6	33.3	20.9	6.1	12.4	8.5	20.3	15.9	11.8	15.6	21.5
Ce	65.0	55.6	70.7	69.5	71.2	94.2	66.9	60.8	42.2	19.8	35.9	23.9	38.2	37.9	36.4	32.9	51.4
Pr	7.07	6.65	7.41	7.87	8.03	12.70	8.07	6.98	4.18	1.26	2.89	1.71	4.18	3.22	2.71	3.08	4.39
Nd	24.0	22.9	24.7	26.8	28.2	43.0	30.2	25.2	16.3	5.0	12.0	6.7	15.4	11.9	10.6	12.3	15.9
Sm	4.40	4.30	4.40	5.76	5.65	8.28	7.47	5.83	4.03	1.46	3.84	1.89	3.45	3.07	3.28	3.28	3.41
Eu	0.73	0.75	0.69	0.35	0.29	0.40	0.46	0.44	0.17	0.13	0.10	0.05	0.18	0.19	0.17	0.17	0.23
Gd	3.69	3.62	3.77	5.77	5.05	6.76	7.94	6.15	4.25	1.82	4.76	2.29	3.57	3.22	3.85	3.87	3.15
Tb	0.61	0.61	0.60	1.07	0.91	1.22	1.54	1.13	0.71	0.38	0.83	0.38	0.61	0.56	0.73	0.68	0.60
Dy	3.60	3.69	3.52	7.17	6.02	7.60	10.50	7.67	4.40	2.85	5.37	2.63	3.98	3.60	4.89	4.56	3.97
Ho	0.71	0.73	0.78	1.52	1.33	1.59	2.34	1.70	0.89	0.62	1.10	0.53	0.81	0.77	1.03	0.96	0.85
Er	2.25	2.31	2.20	5.09	4.39	5.18	7.78	5.62	2.63	2.02	3.36	1.64	2.48	2.31	3.07	2.79	2.81
Tm	0.34	0.35	0.32	0.79	0.73	0.83	1.31	0.93	0.39	0.33	0.50	0.24	0.36	0.33	0.46	0.40	0.45
Yb	2.42	2.53	2.32	5.78	5.01	5.56	8.91	6.40	2.46	2.17	3.15	1.55	2.34	2.13	3.09	2.66	3.50

Continued

Sample	Qipanshi monzogranite			Mingguang monzogranite					Diantan syenogranite								
	GD04	GD09	GD11	GD15	GD17	GD22	GD27	GD28	DT04	DT08	DT16	DT17	DT22	DT24	DT26	DT29	DT32
Lu	0.37	0.37	0.34	0.83	0.77	0.79	1.33	0.94	0.35	0.31	0.43	0.22	0.34	0.32	0.43	0.39	0.52
Hf	4.18	4.64	4.01	3.61	4.21	3.40	4.29	3.48	3.53	3.95	4.14	3.68	3.33	2.95	4.56	3.54	3.62
Ta	1.25	1.28	1.44	3.09	2.99	3.06	6.12	4.27	1.62	2.37	2.65	2.11	1.57	1.43	2.43	2.11	3.23
W	186	187	199	228	234	160	311	236	221	380	230	183	216	229	213	208	160
Tl	0.70	0.68	0.65	1.25	1.27	1.78	1.24	1.21	1.27	1.41	1.31	1.36	1.91	1.81	1.47	1.33	1.38
Pb	19.6	18.6	19.2	21.4	21.0	23.2	22.5	22.5	29.8	23.9	34.5	21.2	19.4	16.9	29.1	30.0	26.8
Bi	0.05	0.08	0.08	0.06	0.05	0.10	0.10	0.10	0.23	0.04	0.09	0.22	0.13	0.17	0.12	0.09	0.08
Th	24.9	25.6	25.7	44.2	39.9	52.3	38.4	35.7	24.7	20.4	26.1	18.3	26.7	19.5	27.1	24.6	49.5
U	2.42	3.62	3.28	8.33	7.71	8.48	8.61	7.44	3.67	2.88	4.54	3.45	3.29	3.56	4.44	4.36	8.06
Eu^*	0.18	0.19	0.17	0.06	0.05	0.05	0.06	0.07	0.04	0.08	0.02	0.02	0.05	0.06	0.05	0.05	0.07
Sr/Y	6.78	6.69	6.47	1.03	1.04	1.04	1.22	1.63	0.69	1.05	0.33	0.55	1.08	2.11	1.08	1.25	0.97
Y/Yb	10.0	9.68	10.3	9.38	9.46	9.71	9.45	9.88	10.3	8.76	10.7	12.0	11.0	12.4	9.26	10.9	8.06
Gd/Yb	1.52	1.43	1.63	1.00	1.01	1.22	0.89	0.96	1.73	0.84	1.51	1.47	1.53	1.51	1.25	1.46	0.90
∑REE	151	138	159	173	173	240	189	163	104	44.3	86.6	52.3	96.2	85.4	82.5	83.7	113

Table 2　Whole-rock Rb-Sr and Sm-Nd isotopic data from the Donghe granitoid.

Sample	$^{87}Sr/^{86}Sr$	2*SE*	Sr /ppm	Rb /ppm	$^{143}Nd/^{144}Nd$	2*SE*	Nd /ppm	Sm /ppm	T_{DM2} /Ga	$\varepsilon_{Nd}(t)$	I_{Sr}
GD11	0.715 292	12	154	269	0.512 125	8	25	4	1.39	−8.6	0.706 667
GD17	0.749 311	13	49	472	0.512 063	4	28	6	1.49	−10.1	0.701 781
GD27	0.731 371	16	103	489	0.512 107	11	30	7	1.46	−9.6	0.707 891
DT16	0.878 851	67	11	442	0.512 169	14	12	4	1.43	−9.1	0.660 448
DT26	0.763 816	26	31	375	0.512 16	7	11	3	1.44	−9.2	0.698 583

$^{87}Rb/^{86}Sr$ and $^{147}Sm/^{144}Nd$ ratios were calculated using Rb, Sr, Sm, and Nd contents analyzed by ICP-MS.

T_{DM2} values represent the two-stage model age and were calculated using present-day $(^{147}Sm/^{144}Nd)_{DM}$ = 0.213 7, $(^{147}Sm/^{144}Nd)_{DM}$ = 0.513 15, and $(^{147}Sm/^{144}Nd)_{crust}$ = 0.101 2.

$\varepsilon_{Nd}(t)$ values were calculated using present-day $(^{147}Sm/_{144}Nd)_{CHUR}$ = 0.196 7 and $(^{147}Sm/^{144}Nd)_{CHUR}$ = 0.512 638.

$$\varepsilon_{Nd}(t)=[(^{143}Nd/^{144}Nd)_S(t)/(^{143}Nd/^{144}Nd)_{CHUR}(t)-1]\times10^4.$$

$$T_{DM2}=\frac{1}{\lambda}\{1+[(^{143}Nd/^{144}Nd)_S-((^{147}Sm/^{144}Nd)_S-(^{147}Sm/^{144}Nd)_{crust})(e^{\lambda t}-1)-(^{143}Nd/^{144}Nd)_{DM}]/[(^{147}Sm/^{144}Nd)_{crust}-(^{147}Sm/^{144}Nd)_{DM})]\}.$$

Table 3　Whole-rock Pb isotopic data from the Donghe granitoid.

Sample	U	Th	Pb	$^{206}Pb/^{204}Pb$	2*SE*	$^{207}Pb/^{204}Pb$	2*SE*	$^{208}Pb/^{204}Pb$	2*SE*	$^{238}U/^{204}Pb$	$^{232}Th/^{204}Pb$	$(^{206}Pb/^{204}Pb)_i$	$(^{207}Pb/^{204}Pb)_i$	$(^{208}Pb/^{204}Pb)_i$
GD11	3.3	25.7	19.2	18.961	7	15.736	5	39.700	10	11.040	88.742	18.578	15.717	38.723
GD17	7.7	39.9	21.0	19.295	7	15.768	7	40.099	22	23.968	127.242	18.462	15.726	38.699

Continued

Sample	U	Th	Pb	$^{206}Pb/^{204}Pb$	2*SE*	$^{207}Pb/^{204}Pb$	2*SE*	$^{208}Pb/^{204}Pb$	2*SE*	$^{238}U/^{204}Pb$	$^{232}Th/^{204}Pb$	$(^{206}Pb/^{204}Pb)_i$	$(^{207}Pb/^{204}Pb)_i$	$(^{208}Pb/^{204}Pb)_i$
GD27	8. 6	38. 4	22. 5	19. 456	9	15. 773	11	40. 161	26	25. 056	114. 637	18. 586	15. 729	38. 899
DT16	4. 5	26. 1	34. 5	18. 940	10	15. 747	7	39. 558	14	8. 487	50. 054	18. 646	15. 732	39. 007
DT26	4. 4	27. 1	29. 1	18. 892	11	15. 752	4	39. 578	10	9. 838	61. 597	18. 550	15. 735	38. 900

U, Th, and Pb concentrations were analyzed by ICP-MS. Initial Pb isotopic ratios were calculated for 120 Ma(GD11–27) and 130 Ma(DT16–26) using a single-stage model.

4. 1 Zircon U-Pb age

Zircons from three representative samples are analyzed from the Qipanshi(sample GD02, Fig. 3a), Mingguang(sample GD14, Fig. 3b), and Diantan(sample DT12, Fig. 3c) granites from the Tengchong area, respectively, in western Yunnan (SW China). Analytical results are listed in the Supplementary Dataset Table. The sampling locations, lithology, and dating results are summarized in the Supplementary Dataset Table and Figs. 1 and 3. The separated zircons are generally euhedral, measuring up to 100–400 μm with length: width ratios of 2:1 to 4:1. Most crystals are colorless or light brown, prismatic, and transparent to subtransparent; they exhibit clear oscillatory zoning in CL images and have Th/U ratios >0. 4, which is typical of magmatic origin (Hoskin and Black, 2000; Hoskin and Schaltegger, 2003).

4. 1. 1 Diantan syenogranite

Seventeen reliable analytical data are obtained from sample DT12. All of the analyzed grains have high U (210–3 317 ppm) and Th (196–2 898 ppm) contents with Th/U ratios of 0. 58–1. 41, except one with an extremely low Th/U ratio of 0. 28; the normal Th/U ratios of magmatic zircon are usually >0. 4. The eleven analyses on zircon crystals yield $^{206}Pb/^{238}U$ ages from 126±2. 0 Ma to 137±2. 0 Ma, with a weighted mean age of 130. 6±2. 5 Ma (MSWD= 1. 0, $n=11$, 2σ), representing the magma crystallization age of the monzogranites. Spots 12, 13, 14 and 15 give a $^{206}Pb/^{238}U$ ages from 142±2. 0 Ma to 155±2. 0 Ma; according to their morphology, these zircons show inconspicuous nuclear structure(Fig. 3c). Spot 12's Th (485 ppm), U (582 ppm), and Pb (55. 9 ppm) contents and Th/U ratio of 0. 83, integrated with its morphology, leads us to infer it as the inherited age. Spot 13 and 14 have abnormally high Th (2 786 ppm and 2 898 ppm) and U (3 317 Ma and 2 644 ppm), Th/U=0. 84 and 1. 10, and abnormally high Pb (284. 6 ppm and 239. 8 ppm), suggesting its old age may be unreliable as it is influenced by abnormally high Th, U, and Pb contents. Spot 15 has relatively low Th (196 ppm) and U (299 ppm), Th/U=0. 83, and abnormally low Pb (25. 9 ppm); combined with its inconspicuous CL image, this leads us to suggest that the age results may be influenced by metamorphic fluid. Spots 17 and 16 have $^{206}Pb/^{238}U$ ages from 65. 4± 1. 0 Ma to 75. 5 ± 1. 0 Ma; their morphologies suggest that may have interacted with

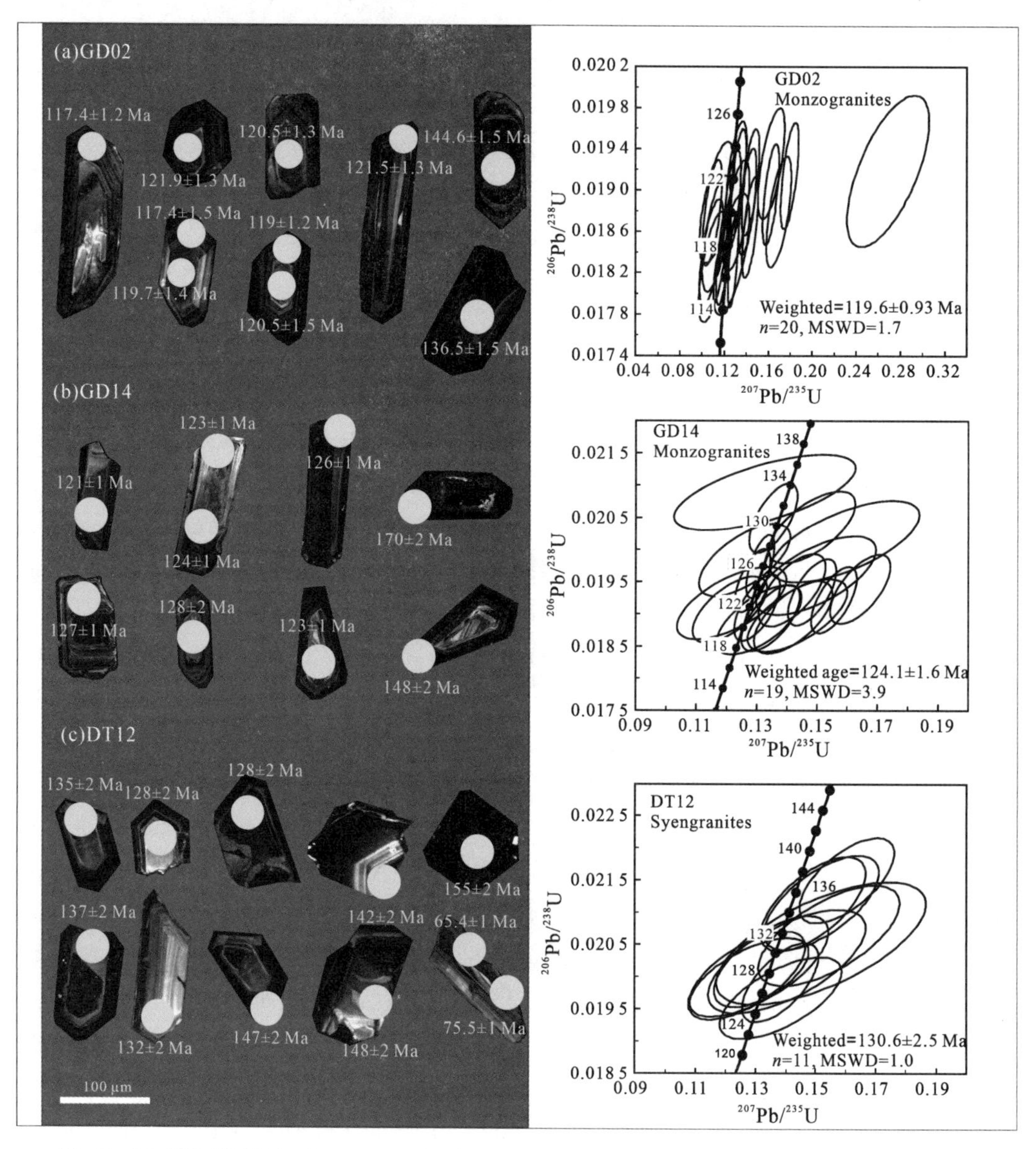

Fig. 3 LA-ICP-MS U-Pb zircon concordia diagram and CL images of representative zircon grains for the Donghe granitoid in the Tengchong Block, western Yunnan, SW China.

metamorphic fluid, and CL images show inconspicuous oscillatory zoning and an abnormal Th/U ratio of 0. 28. Herein, we consider the weighted mean age of 130. 6±2. 5 Ma as the main crystallization stage age of the Diantan granites.

4. 1. 2 Mingguang monzogranites

Twenty-five reliable analytical data are obtained from sample GD14. The analyzed grains have widely varying U (147–2 410 ppm) and Th (131–2 444 ppm) contents with high Th/U ratios of 0. 40–5. 74. Nineteen analyses on zircon crystals yield $^{206}Pb/^{238}U$ ages between 121±1. 0 Ma and 133±2. 0 Ma, with a weighted mean age of 124. 1±1. 6 Ma (MSWD=3. 9, $n=$

19, 2σ), representing the magma crystallization age of the monzogranites. Spots 25 and 24 give $^{206}Pb/^{238}U$ ages of 148 ± 2.0 Ma and 170 ± 2.0 Ma, respectively. Spot 24 has extremely high U (4 700 ppm) and Th (3 405 ppm), Th/U = 0.72, and abnormally high Pb (481.5 ppm); spot 25 also has high U (2 057 ppm) and Th (1 305 ppm), Th/U = 0.63, and high Pb (182.8 ppm); both their morphologies show very dark color and inconspicuous oscillatory zoning(Fig. 3b). We suggest these age results may be influenced by abnormal Th, U, and Pb contents. Spots 20, 21, 22, and 23 have $^{206}Pb/^{238}U$ ages from 60.5 ± 0.6 Ma to 103 ± 1.0 Ma. According to CL images, they are colorless and show inconspicuous oscillatory zoning. These spots have U contents of 361 – 2 410 ppm and Th contents of 343 – 1 209 ppm, with Th/U = 0.45 – 1.02, and their younger ages may be interpreted as unreliable because of radiogenic lead loss via metamorphic fluid. We suggest a weighted mean age of 124.1 ± 1.6 Ma as the main crystallization stage age.

4.1.3 Qipanshi monzogranites

Twenty-nine reliable analytical data are obtained from sample GD02. The analyzed grains have widely varying U (85 – 1 778 ppm) and Th (135 – 2 452 ppm) contents, with high Th/U ratios of 0.40 – 2.37. Twenty analyses on zircon crystals yield $^{206}Pb/^{238}U$ ages are between 116.6 ± 1.4 Ma and 122.5 ± 1.9 Ma, with a weighted mean age of 119.9 ± 0.9 Ma (MSWD = 1.7, n = 20, 2σ), representing the magma crystallization age of the porphyritic granites. Spots 21, 22, 23, 24, 25, and 26 give a weighted mean age of 126.5 ± 1.8 Ma (MSWD = 1.5). Spots 29 and 28 have concordant $^{206}Pb/^{238}U$ ages of 136.5 Ma to 144.6 Ma; the CL images show these two spots being located in the zircon core and having obviously nuclear structure (Fig. 3a). Spot 29 has U and Th contents of 2 017 Ma and 838 ppm, respectively, so Th/U = 0.42; spot 30 has U and Th contents of 638 ppm and 1 170 ppm, respectively, so Th/U = 1.84; thus these two spots may represent inherited cores. One additional spot (27) gives a concordant $^{206}Pb/^{238}U$ age of 113.6 ± 1.3 Ma, which, by considering aspective factors, may be interpreted as caused by radiogenic lead loss via metamorphic fluid. We suggest a weighted mean age of 119.9 ± 0.9 Ma as the main crystallization stage age.

4.2 Major and trace elemental geochemistry

In the Q-ANOR diagram (Fig. 4), these granitoids plot from alkali granite to monzogranite. In the A/CNK vs. A/NK diagram (Fig. 5a) and SiO_2 vs. ($Na_2O + K_2O$) – CaO diagram (Fig. 5b), we may define these granitoids as high-K, calc-alkaline, and slightly peraluminous granites. In the Harker, Rb vs. Th, and Rb vs. Y diagrams (Fig. 6), they show typical regularities with the increasing SiO_2 and Rb content.

4.2.1 Diantan syenogranite

The samples from the Diantan granitic pluton have the following contents: SiO_2 = 76.0 –

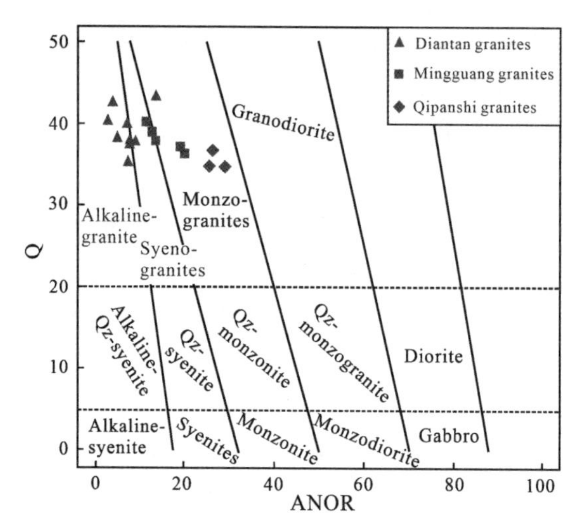

Fig. 4 Q-ANOR normative composition diagram (Streckeisen and Le Maitre, 1979) for classification of the Donghe granitoid.

$Q' = Q/(Q + Or + Ab + An) \times 100$; $ANOR = An/(Or + An) \times 100$.

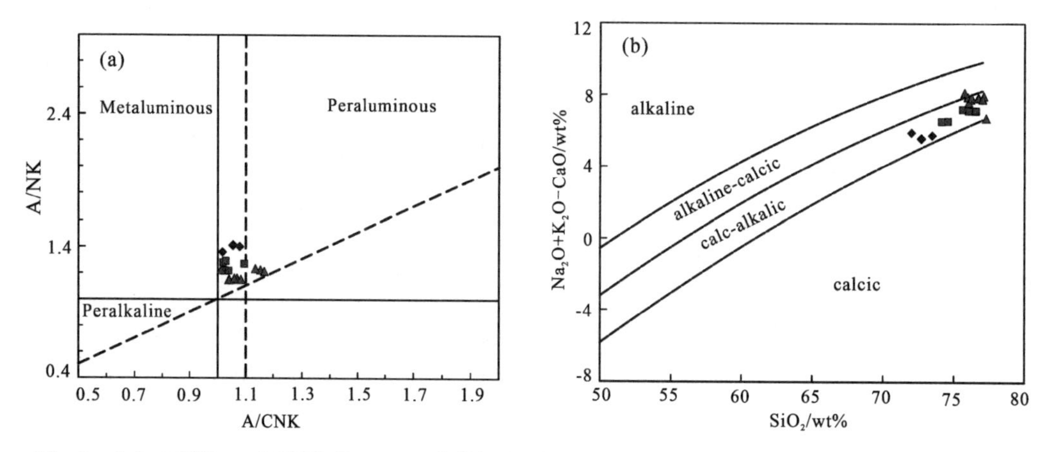

Fig. 5 (a) A/NK vs. A/CNK diagram and (b) Na_2O+K_2O-CaO vs. SiO_2 diagram (Frostet al., 2001) for the Donghe granitoid in the Tengchong Block, western Yunnan, SW China.

Symbols are as in Fig. 4.

77.9 wt%, K_2O = 4.42–5.66 wt%, K_2O/Na_2O = 1.14–2.35, K_2O+Na_2O = 7.59–8.66 wt%. TiO_2 = 0.03–0.04 wt% (except one at 0.13), CaO = 0.19–0.99 wt%, P_2O_5 = 0.01–0.02 wt%, $Fe_2O_3{}^T$ = 0.59–1.07 wt%, MgO = 0.03–0.04 wt% (except one at 0.14), and $Mg^\#$ = 7.95–15.0 (except one at 35.6). They also exhibit Al_2O_3 contents of 11.5–13.1 wt%, A/NK (molar Al_2O_3/Na_2O+K_2O) ratios of 1.1–1.2, and A/CNK (molar $Al_2O_3/CaO+Na_2O+K_2O$) ratios of 1.02–1.16, but most of them are <1.1.

The chondrite-normalized rare-earth element (REE) diagram (Fig. 7c) shows total REE contents of 44.3–112.7 ppm and enrichment in light REEs (LREEs) [with LREE to heavy REE (HREE) ratios of 3.22–6.11], with HREE values of Yb_N = 9.11–20.6, $(La/Yb)_N$ =

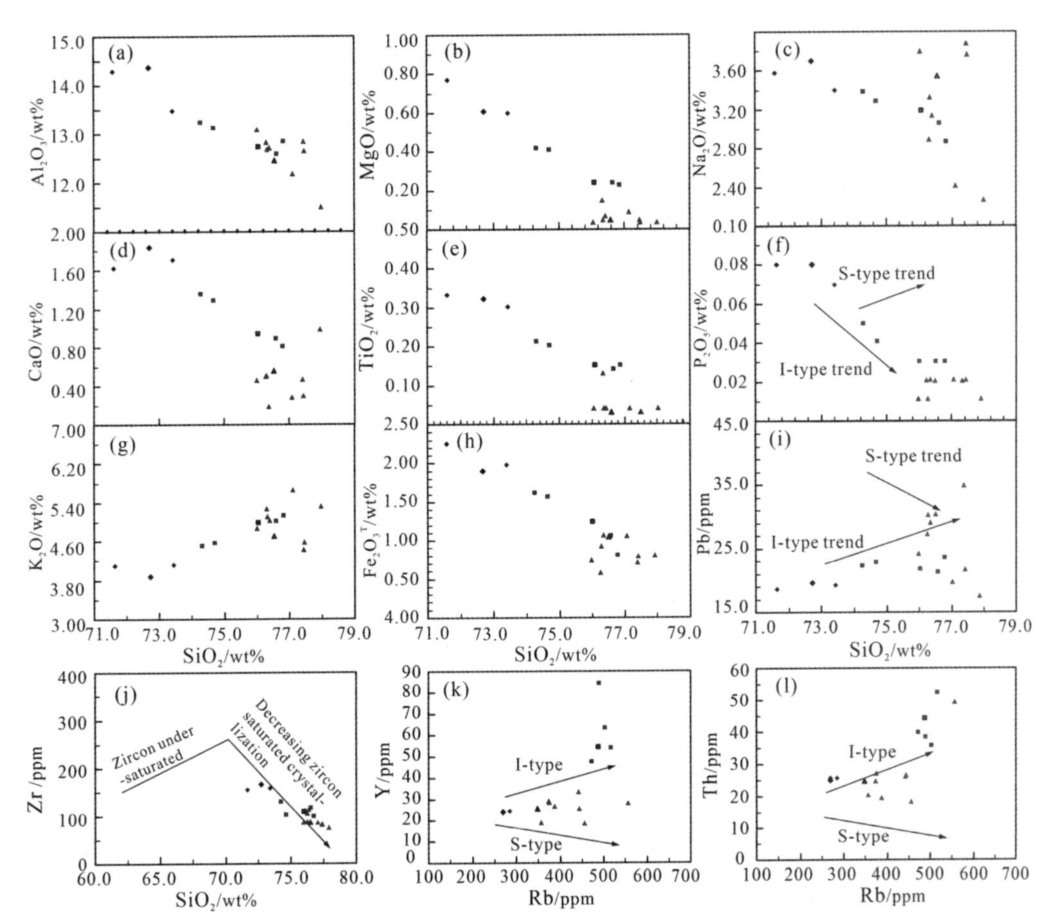

Fig. 6 SiO_2 vs. Al_2O_3(a), MgO (b), Na_2O (c), CaO (d), TiO_2(e), P_2O_5(f), K_2O (g), $Fe_2O_3^T$(h), and Pb (i) variation diagrams showing that the granites follow the I-type trend proposed by Chappell and White (1992) and Zr vs. SiO_2 diagram (j) and Y vs. Rb (k) and Th vs. Rb (l) variation diagrams (from Li et al., 2007) for the Donghe granitoid in the northern Tengchong Block, western Yunnan, SW China.

Symbols are as in Fig. 4.

2.03–6.22, Sr/Y = 0.55 – 2.11, Y/Yb = 8.06 – 12.44, and Gd/Yb = 0.84 – 1.73. In addition, the samples exhibit sharply negative Eu anomalies (δEu = 0.16–0.22).

The primitive mantle-normalized trace element spider diagram (Fig. 7f) shows enrichment in large ion lithophile elements (LILE) such as Rb, Th, U, and K, an especially positive Pb anomaly, and enrichment in high field strength elements (HFSE), with relatively marked depletions in Ba, Nb, Sr, P, Ti, and Eu, with Sr = 10.2–55.9 ppm, Ba = 72.3–282 ppm, and Y = 18.6–33.6 ppm.

4.2.2 Mingguang monzogranite

The samples from the Mingguang granitic pluton have the following contents: SiO_2 = 74.3–76.8 wt%, K_2O = 4.51–5.13 wt%, K_2O/Na_2O = 1.33–1.79, K_2O+Na_2O = 7.85–8.16 wt%,

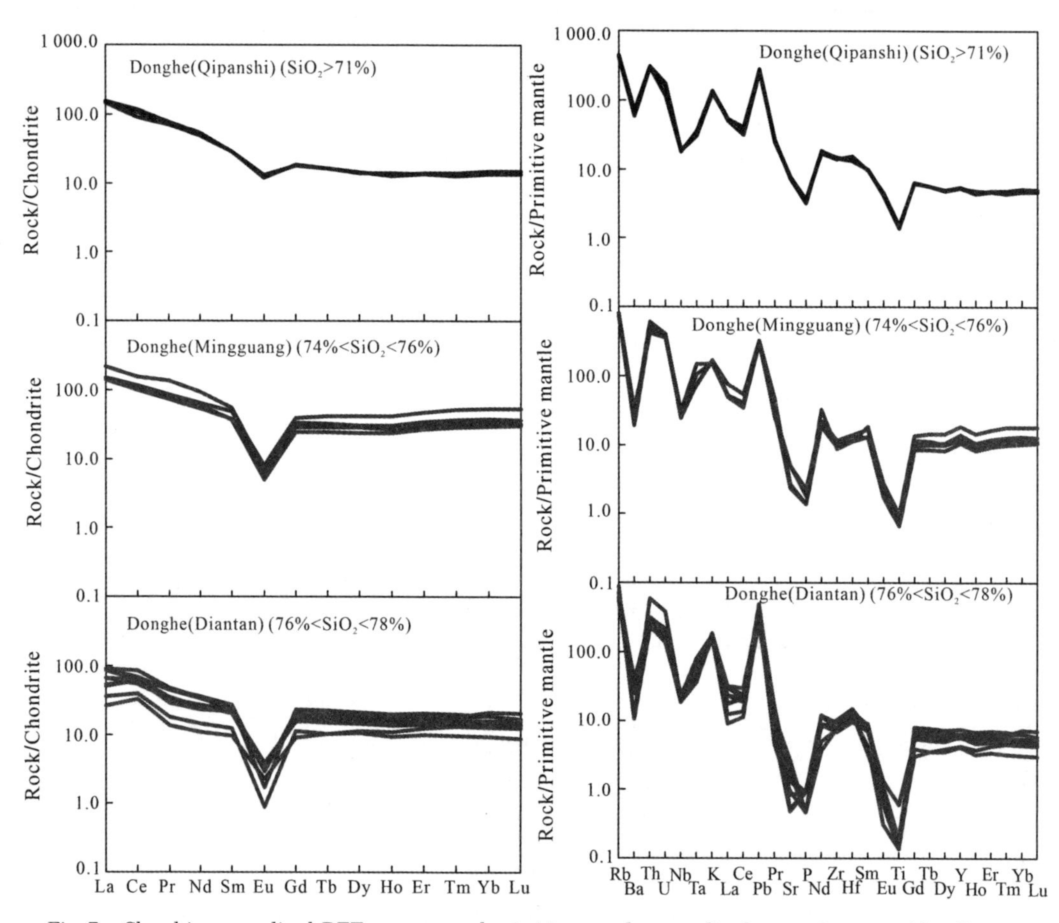

Fig. 7 Chondrite-normalized REE patterns and primitive-mantle-normalized trace element spider diagram for the Donghe granitoid, western Yunnan.

The primitive mantle and chondrite values are from Sun and McDonough (1989).

TiO_2 = 0. 14–0. 21 wt%, CaO = 0. 81 – 1. 35 wt%, P_2O_5 = 0. 03–0. 05 wt%, $Fe_2O_3{}^T$ = 1. 06–1. 62 wt% (except one at 0. 81), MgO = 0. 22–0. 41 wt%, and $Mg^{\#}$ = 30. 2–38. 8. They also exhibit Al_2O_3 contents of 12. 6 – 13. 2 wt%, A/NK [molar Al_2O_3/(Na_2O + K_2O)] ratios of 1. 2–1. 3, and A/CNK [molar Al_2O_3/(CaO + Na_2O + K_2O)] ratios of 1. 03–1. 09.

The chondrite-normalized REE diagram (Fig. 7b) shows total REE contents of 163–240 ppm and enrichment in LREE (LREE/HREE = 3. 55 – 7. 13), with HREE values of Yb_N = 29. 5–52. 4, $(La/Yb)_N$ = 2. 79 – 5. 03, and Gd/Yb = 0. 89 – 1. 22. In addition, the samples exhibit sharply negative Eu anomalies (δEu = 0. 16–0. 22).

The primitive mantle-normalized trace element spider diagram (Fig. 7e) also shows enrichment in Rb, Th, U, and K, a positive Pb anomaly, and enrichment in HFSE, with relatively slight depletion in Ba, Nb, Sr, P, Ti, and Eu, with Sr = 49. 2–103 ppm and Ba = 132–219 ppm.

4.2.3 Qipanshi monzogranite

The samples from the Qipanshi granitic pluton have the following contents: SiO_2 = 71.7–73.4 wt%, K_2O = 3.88–4.12 wt%, K_2O/Na_2O = 1.05–1.22, K_2O+Na_2O = 7.51–7.68 wt%, TiO_2 = 0.30–0.33 wt%, CaO = 1.62–1.84 wt%, P_2O_5 = 0.07–0.08 wt%, $Fe_2O_3^T$ = 1.90–2.26 wt%, MgO = 0.59–0.76 wt%, and $Mg^{\#}$ = 41.0–43.9. They also exhibit Al_2O_3 contents of 13.5–14.4 wt%, A/NK [molar $Al_2O_3/(Na_2O+K_2O)$] ratios of 1.3–1.4, and A/CNK [molar $Al_2O_3/(CaO+Na_2O+K_2O)$] ratios of 1.02–1.08.

The chondrite-normalized REE diagram (Fig. 7a) shows total REE contents of 138–159 ppm and enrichment in LREE (LREE/HREE = 8.73–10.5), with HREE values of Yb_N = 13.6–14.9, $(La/Yb)_N$ = 9.61–11.4, Sr/Y = 6.47–6.78, Y/Yb = 9.68–10.3, and Gd/Yb = 1.43–1.63. In addition, the samples exhibit sharply negative Eu anomalies (δEu = 0.52–0.58).

The primitive mantle-normalized trace element spider diagram (Fig. 7d) displays features similar to those of Mingguang granites, with Sr = 154–164 ppm, Ba = 403–493 ppm, and Y = 23.8–24.5 ppm.

Overall, the samples from the Qipanshi granites, to the Mingguang granites, to the Diantan granites display the following typical regularities (Fig. 6): increasing SiO_2, TiO_2, Al_2O_3, $Fe_2O_3^T$, MgO, CaO, P_2O_5, $Mg^{\#}$, and A/NK, decreasing Sr and Ba, and increasing K_2O, K_2O/Na_2O, Na_2O+K_2O, and Rb, with Na_2O and MnO being nearly constant. In addition, the REE abundances (Fig. 7a-c), especially the HREE abundances, display obvious differences from Qipanshi, to Mingguang, to Diantan granites. The REE pattern slopes change slightly but negative Eu anomalies increasing sharply, with the positive Rb, Th, U, and Pb anomalies and negative Ba, Nb, Sr, P, and Ti anomalies changing gradually and the positive Y anomaly distinctively gradually decreasing with increasing SiO_2 of fractional crystallization (Fig. 7d-f). These uncommon chemical variations, REE patterns, and multi-element variation patterns can be seen in some typical highly differentiated I-type granites in NE China (Wu et al., 2003 a), SE China (Chen et al., 2000; Li et al., 2007; Qiu et al., 2008; Tao et al., 2013), and SW China (Zhu et al., 2009a) (in Table 5). Li et al. (2007) interpreted the chemical variations as the fractionated crystallization of ferromagnesian minerals, plagioclase, apatite, and Fe-Ti oxides, attributing the uncommon REE pattern to separated accessory minerals including apatite, allanite, titanite, and monazite (Wu et al., 2003a; Li et al., 2007, and references therein).

4.3 Whole-rock Sr-Nd-Pb isotopes

Whole-rock Sr-Nd-Pb isotopic data for the granitoids from the Tengchong Block are listed in Tables 2 and 3. All the initial $^{87}Sr/^{86}Sr$ isotopic ratios (I_{Sr}) and $\varepsilon_{Nd}(t)$ values are calculated at the time of magma crystallization. These samples can be divided three parts: the Qipanshi granitoids sample (GD11), the Mingguang granitoids samples (GD17 and GD27), and the Diantan granitoids samples (DT16 and DT26).

4.3.1 Diantan syenogranite

The Diantan granites (DT16 and DT26) have low I_{Sr} ratios ranging from 0.660 5 to 0.700 0, but stable $\varepsilon_{Nd}(t)$ values ranging from −9.2 to −9.1, and T_{DM2} values of 1.43–1.44 Ga. According to the extremely low Sr content (11 ppm), the test result may be unreliable, casting suspicion on the abnormal I_{Sr} ratio of 0.660 5, and so we eliminated it from the dataset. The low result of 0.700 0 may be caused by abnormally low Sr contents (30.9 ppm and 49.2 ppm) from the influence of fluid, weathering, or disequilibrium differentiation, so these low values should be used cautiously.

The initial $^{206}Pb/^{204}Pb$, $^{207}Pb/^{204}Pb$, and $^{208}Pb/^{204}Pb$ ratios were 18.550–18.646, 15.732–15.735, and 38.900–39.007, respectively.

4.3.2 Mingguang monzogranite

The Mingguang granites (GD17 and GD27) display higher I_{Sr} = 0.701 8–0.707 9, $\varepsilon_{Nd}(t)$ = −10.1 to −9.6, and T_{DM2} = 1.46–1.49 Ga.

The initial $^{206}Pb/^{204}Pb$, $^{207}Pb/^{204}Pb$ and $^{208}Pb/^{204}Pb$ ratios were 18.462–18.586, 15.726–15.729, and 38.699–39.899, respectively.

4.3.3 Qipanshi monzogranite

The sample (GD11) from the Qipanshi granites displays I_{Sr} ratios of 0.706 7 and negative $\varepsilon_{Nd}(t)$ values of −8.6, with T_{DM2} values of 1.39 Ga.

The initial $^{206}Pb/^{204}Pb$, $^{207}Pb/^{204}Pb$, and $^{208}Pb/^{204}Pb$ ratios were 18.578, 15.717, and 38.723, respectively.

4.3.4 Summary of sample findings

The three sets of samples share similar results of reliable initial $^{87}Sr/^{86}Sr$ ratios (0.706 7 and 0.707 9) and negative $\varepsilon_{Nd}(t)$ values (−10.1 to −8.6), with T_{DM2} ranging from 1.39 Ga to 1.49 Ga. The initial $^{206}Pb/^{204}Pb$, $^{207}Pb/^{204}Pb$, and $^{208}Pb/^{204}Pb$ ratios of 18.462–18.646, 15.717–15.735, and 38.699–39.007, respectively, plot into the low crust and upper crust near the primitive arc regions in the Zartman and Doe (1981) Pb tectonic-setting framework diagrams (Fig. 12a,b), indicating unusually highly enriched components.

5 Discussion

5.1 Petrogenesis of the Early Cretaceous granites from the northern Tengchong Block

5.1.1 The highly fractionated I-type granites

The Early Cretaceous granites from the Donghe share similar features of high SiO_2, high

K, calc-alkalinity, peraluminosity (A/CNK = 1.02 − 1.16, but most being < 1.1), and relatively high sodium contents (Na_2O = 2.27−3.87 wt%, with an average value of 3.27 wt%, and most of them being > 3.20 wt%). Their CIPW normative corundum < 1% and diopside indicate that the high-Si felsic monzogranites and syenogranites are I-type granites (Chappell and White, 1974), an identification that is also supported by the presence of hornblende in the absence of cordierite and corundum. Moreover, in the ($Na_2O + K_2O$)/Cao vs. 10 000 Ga/Al and ($Na_2O + K_2O$)/CaO vs. Zr + Nb + Ce + Y diagrams (Fig. 8e, f), all samples plot into the field of I- and S-types.

Geochemically, from the Qipanshi monzogranite, to the Mingguang monzogranite, to the Diantan syenogranite, as SiO_2 and total alkalinity increase, the FeO^T/MgO ratio [from 2.68−3.02, to 3.31−4.85, to 11.9−24.3 (except one at 3.79)] and ($Na_2O + K_2O$)/CaO ratio [from 4.11−4.74, to 5.84−9.88, to 14.7−28.8 (except for points at 7.67 and 43.0)] display distinctively different degrees of increases among the three plutons (Fig. 8e, f), indicating partly different degrees of significantly intensive magmatic fractionation from Qipanshi and Mingguang monzogranites to Diantan syenogranites. This conclusion is also supported by the petrography (Fig. 2), which gradually changes from coarse grain to mid-coarse grain in the field. Moreover, all samples show obvious P_2O_5 decreases with increasing SiO_2 (Fig. 6f); this is a crucial criterion for distinguishing I-type granites from S-type granites owing to the high apatite solubility in strongly peraluminous melts (Wolf and London, 1994). Chappell and White (1992) also consider Pb moving in the opposite direction as an important criterion for distinguishing I-type from S-type granite (Fig. 6i). Similarly to the findings of Li et al. (2007), the Donghe granites also display increasing Y and Th with increasing Rb, indicative of the typical trend for I-type granites (Fig. 6k, l). In addition, we cannot attribute them to A-type granites owing to the lack of mafic alkaline minerals such as arfvedsonite and riebeckite, relatively low Nb, La, Ce, Zr, Zn, and Ga contents (Wu et al., 2003a), and Ga/Al ratio of < 2.6. Furthermore, in the Zr + Nb + Ce + Y and ($Na_2O + K_2O$)/CaO vs. 10 000 Ga/Al discrimination diagrams (Fig. 8e, f) and Rb-Ba-Sr diagram (Fig. 9), all samples plot into the field of I- and S-types and are slightly differentiated to strongly differentiated granites, indicating an intensively fractionated trend from Qipanshi monzogranites, to Mingguang monzogranites, to Diantan syenogranites.

Clemens (2003) proposed that crystal fractionation probably plays the major role in the differentiation of many granitic magmas, especially those emplaced at high levels or in volcanic environments. Along with the formation of the Donghe granitoid, gradually advanced fractional crystallization has been taking place, as evidenced by the positive Rb and Pb anomalies and negative Ba, Nb, Sr, P, Ti, and Eu anomalies.

Taken together, the increasing Ba and varying Rb Sr contents with increasing Sr contents

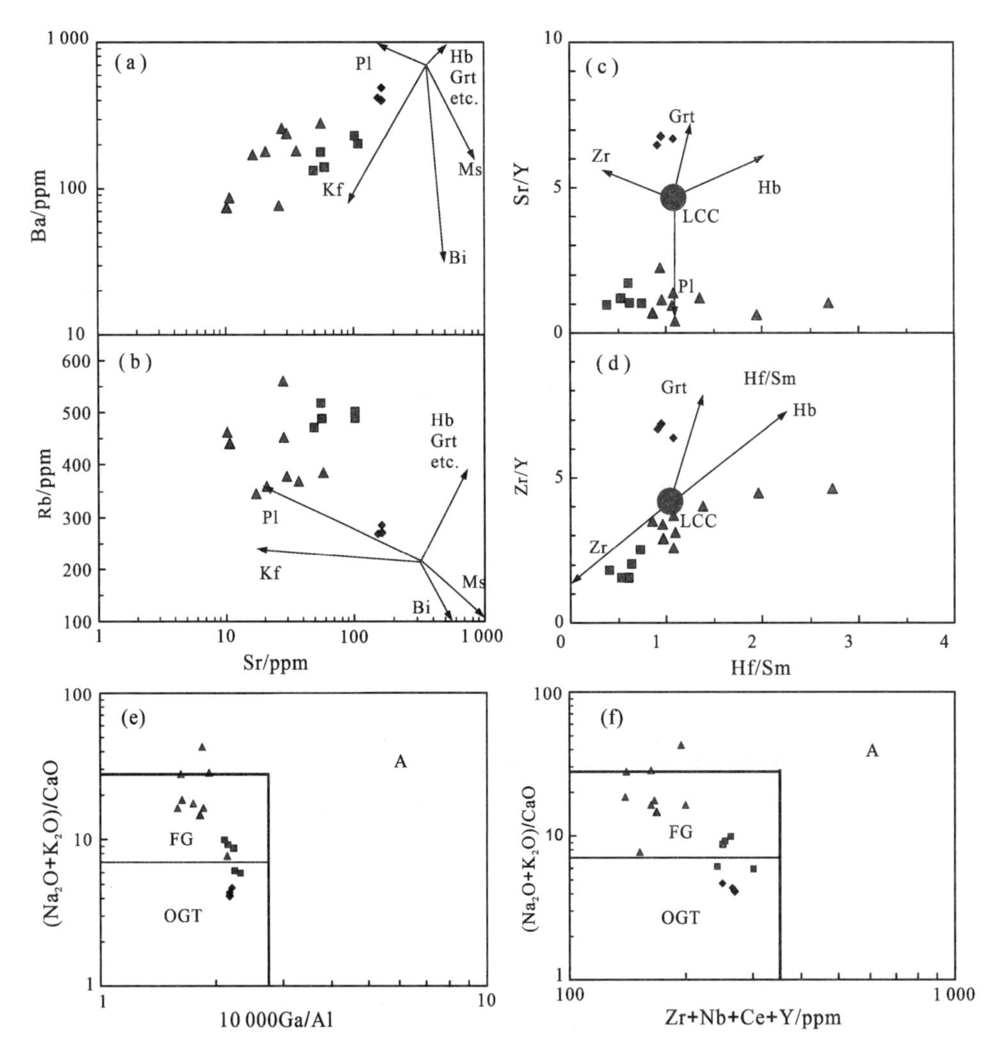

Fig. 8 (a) Ba vs. Sr and (b) Rb vs. Sr diagrams (Janoušek et al., 2004); (c) Sr/Y vs. Hf/Sm and (d) Zr/Y vs. Hf/Sm diagrams (Rudnick and Gao, 2003); (e) $(Na_2O+K_2O)/CaO$ vs. 10 000 Ga/Al and (f) $(Na_2O+K_2O)/CaO$ vs. Zr+Nb+Ce+Y diagrams (Whalen et al., 1987) for the Donghe granitoid in the northern Tengchong Block, western Yunnan, SW China.

Symbols are as in Fig. 4.

(Fig. 8a,b), along with increasing Rb/Ba ratios with increasing Rb/Sr ratios and deceasing Ba/Sr ratios with increasing Ba contents (Fig. 10a) and notably negative δEu and depletion in Ba, Sr, Eu, P, Nb, and Ti, imply the importance of fractional crystallization of K-feldspar, plagioclase, apatite, sphene, Ti-rich minerals, and some biotite during the process of magma evolution (Clemens, 2003; Healy et al., 2004; Le Fort et al., 1987; Miller, 1995; Patiño Douce et al., 1990; Sylvester, 1998; Wang et al., 2007, 2013, and references therein). The flat HREE pattern and relatively high Y content indicate that the source melting occurred at pressures below the garnet stability field (Rapp and Watson, 1995) and that fractional

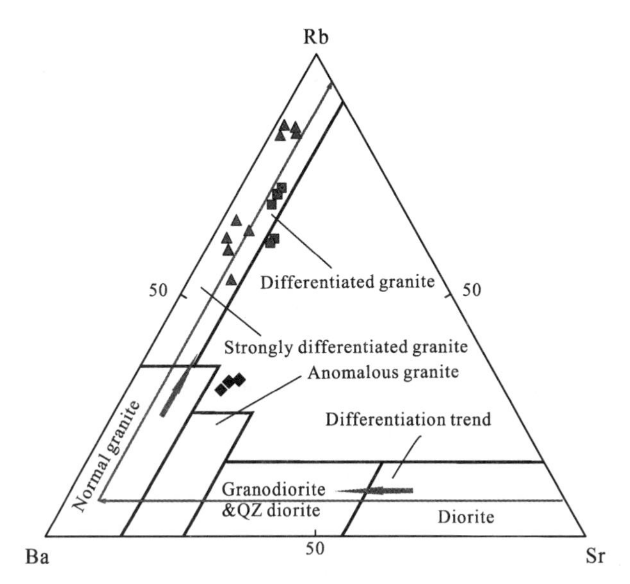

Fig. 9 Rb-Ba-Sr diagram (Wang et al., 2013) for the Donghe granitoid in the northern Tengchong Block, western Yunnan.

Symbols are as in Fig. 4.

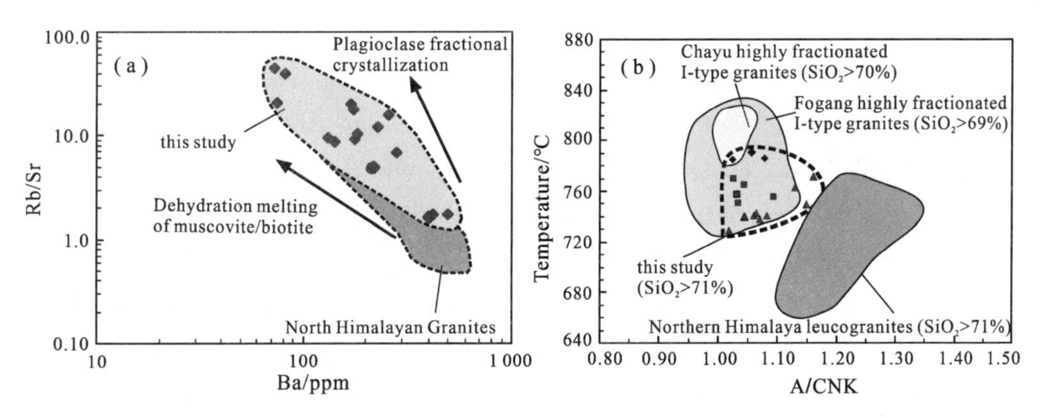

Fig. 10 (a) Rb/Sr vs. Ba diagram and (b) the zircon saturated temperatures vs. A/CNK diagram (Zhu et al., 2009a) for the Early Cretaceous Donghe granitoid in the northern Tengchong Block, western Yunnan, SW China.

The Northern Himalayan granites were modified from Inger and Harris (1993) and Zhang et al. (2004). Zircon saturated temperatures were calculated by Watson and Harrison (1983) and northern Himalayan granite data are from Zhang et al. (2004). Symbols are as in Fig. 4.

crystallization of amphibole and sphene took place (Jaques and Green, 1980; Wilson, 1989).

The Sr/Y vs. Hf/Sm and Zr/Y vs. Hf/Sm diagrams (Fig. 8c,d) indicate that the remaining plagioclase plays an important role in the majority of the granitic rocks and that residual hornblende might contribute to the high Hf/Sm ratios. The positive correlation in Fig. 10 indicates a significant role of residual zircon, with little garnet and/or hornblende during melting (Rudnick and Gao, 2003).

5. 1. 2 Petrogenesis

Generally, the highly fractionated I-type granites can be interpreted as ① derived by melting of a mixed source in the lower crust resulting from underplating or intrusion of mantle-derived magma (Wu et al., 2003b; Li et al., 2007) or ② a two-stage process including mixing of mantle-derived basaltic and crustal felsic magma in the lower crust and subsequent extensive differentiation of the parent magma (Qiu et al., 2008; Tao et al., 2013; Chen et al., 2000; Zhu et al., 2009a).

As above mentioned in the discussion of whole-rock Sr-Nd-Pb, the Donghe granitoid including Qipanshi, Mingguang, and Diantan granites displays strikingly similar features of low initial $^{87}Sr/^{86}Sr$ (0.706 7 and 0.707 9) and negative $\varepsilon_{Nd}(t)$ values (−10.1 to −8.6), with T_{DM2} ranging from 1.39 Ga to 1.49 Ga and initial $^{206}Pb/^{204}Pb$, $^{207}Pb/^{204}Pb$, and $^{208}Pb/^{204}Pb$ ratios of 18.462 − 18.646, 15.717 − 15.735, and 38.699 − 39.007, respectively. These isotopic features are similar to those of contemporaneous granites from the Gaoligong tectonic belt (Yang et al., 2006; Xu et al., 2012) and the northern magmatic belt (Chen et al., 2013; Qu et al., 2012; Pan et al., 2014; Zhu et al., 2009a, b) (Figs. 11 and 12). They have lower initial Sr contents than the Zayu highly fractionated granites (Zhu et al., 2009a) and Fogang highly fractionated granites (Li et al., 2007) but $\varepsilon_{Nd}(t)$ values that are similar to these other granites. The Sr-Nd isotopic results indicate that the major proportion of lower crustal materials (around 60% − 75%), the minority of mantle-derived materials (~12% − 18%), and little or no upper crustal materials were involved in the generation of these granitoids during the Mesoproterozoic (see Wu et al., 2003b) (Fig. 11). Furthermore, the Pb isotopic data suggest that the lower crustal material, along with some upper crustal material, played a major role in the generation of these granitoids (Fig. 12a). The characteristics of mature arc primitive and oceanic island volcanic rocks are shown in Fig. 12b. Another crucial finding is that I-type magmatism could also result from the reworking of sedimentary materials by mantle-like magmas critically involved in crustal growth and to some extent by non-mantle-like isotope ratios of bulk rocks; these new added crustal materials can reach from 50% to 85% (Kemp et al., 2007). The $\delta^{18}O$ values (from 4.99‰ to 10.59‰, < 11‰) of the Donghe granitoid (Chen et al., 1991) indicate that the source may involve some mantle-derived material (Miller, 1985), whereas the $\delta^{18}O$-$^{87}Sr/^{86}Sr$ diagram indicates that the source may be derived from altered oceanic basalt (Magaritz et al., 1978). Moreover, the zircon saturation temperature reaches 729−790 ℃, a regime similar to that of highly fractionated I-type granites in SE China (Li et al., 2007) and SW China (Zhu et al., 2009a) (Fig. 10b), Zhu et al. (2009a) conclude that this implies that mantle-derived basalts may play a role in the generation of magma.

Patiño Douce (1999) also proposed that almost all the granitic rocks were formed by

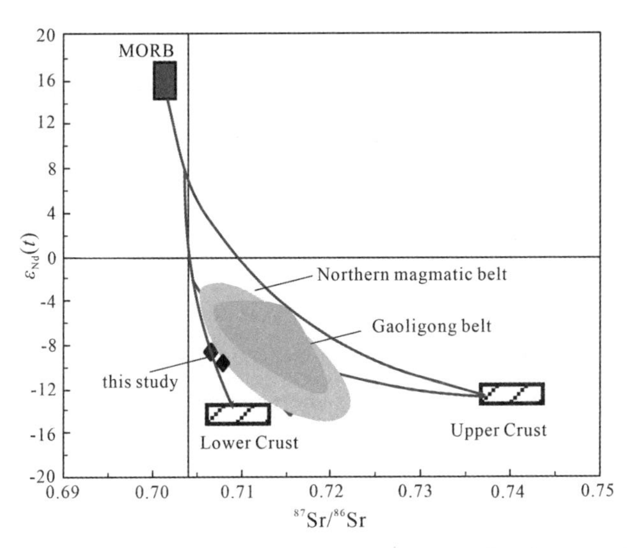

Fig. 11 Initial Sr-Nd isotopic composition for the Early Cretaceous Donghe granitoid in the northern Tengchong Block, western Yunnan, SW China.

The diagram was modified from Lin et al. (2012). The northern magmatic belt data are from Chen et al. (2013), Pan et al. (2014), Qu et al. (2012), and Zhu et al. (2009a,b) and the Gaoligong belt data are from Yang et al. (2006).

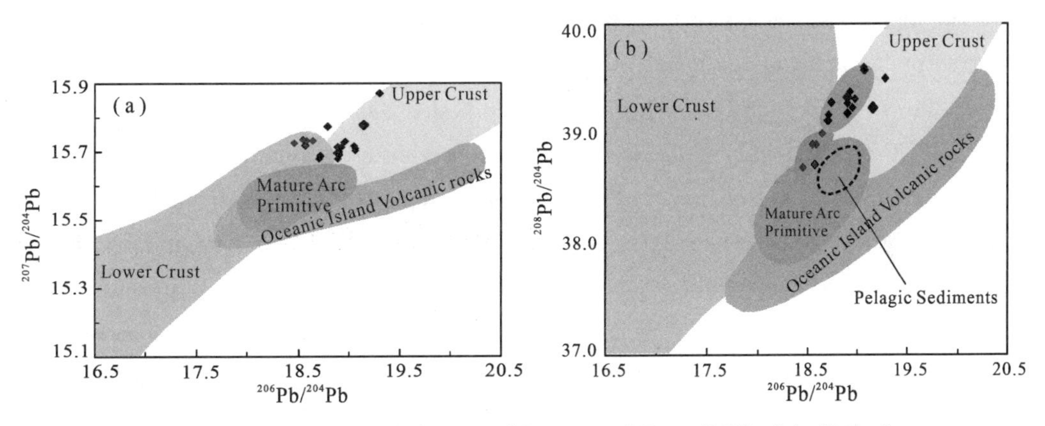

Fig. 12 Pb plumbotectonic framework diagrams (Zartman and Doe, 1981) of the Early Cretaceous Donghe granitoid in the northern Tengchong Block, western Yunnan, SW China.

Blue plots are from Qu et al. (2012) and red plots are from this study.

hybrid magmas except for the peraluminous leucogranites. Considering the petrographical, geochemical, and Sr-Nd isotopic similarities with other highly fractionated granites in NE, SE, and SW China, and the fact that all their parent magmas have involved melting from both mantle-derived and crust-derived materials, and further considering our new evidence, we conclude that the parent magma of the Donghe granitoid could be a hybrid magma generated by the mixing of mostly ancient-crust-derived material and some material from a mantle-derived source, with even some arc-related and oceanic-island-related materials probably also involved in these processes.

In addition, the features of high K, high K_2O/Na_2O, slight peraluminosity, and high SiO_2 hint that the basaltic magma has medium to high K contents (Sisson et al., 2005; Li et al., 2007 reference therein), with Sr-Nd isotopic evolution indicating an origin from partial melting of subducted ocean evolved through assimilation fractional crystallization with mature continental crust, as evidenced by the dark microgranular enclave found in the host granitoids (Castillo et al., 1999) (Fig. 2c). As mentioned above, gradually advanced fractional crystallization has been taking place, as evidenced by the positive Rb and Pb anomalies and negative Ba, Nb, Sr, P, Ti, and Eu anomalies, and played an indispensable role in the subsequent crystallization processes of the Donghe granitoid.

Therefore, we suggest that the petrogenesis of the Donghe granitoid could be interpreted as a hybrid magma produced via both mantle- and crustal-derived melting and later experiencing strong fractional crystallization. Its petrogenesis is similar to that of the Zayu highly fractionated granites in the southeastern Tibetan Plateau (Zhu et al., 2009a).

5.2 Tectonic implications of the Early Cretaceous highly fractionated granites

The highly fractionated I-type granites are generally related to: ① anorogenic magmatism triggered by the breakup and foundering of a subducted flat slab beneath continental crust (Li et al., 2007); ② post-orogenic magmatism (Wu et al., 2003b); ③ subducted oceanic floor (Zhu et al., 2009a); and ④ the underplating of basaltic magmas, triggered by lithospheric delamination during the beginning of the post-orogenic episode (Chen et al., 2000). Further, Roberts and Clemens (1993) suggest that the high-K, calc-alkaline, I-type granites could be generated in two tectonic settings: ① a continental arc setting such as of Andean type, in which the source of the hybrid magma was mantle- and crustal-derived or ② a post-collision setting caused by decompression following crustal thickening similar to that of Caledonia.

The Bangong-Nujiang suture, as the remains of the Tethys Ocean, extends for more than 2 800 km across the central Tibetan Plateau and western Yunnan and to the south it is correlated with the Shan boundary in Myanmar (Wang et al., 2014). The Bangong-Nujiang Tethyan Ocean lithosphere was subducted both southward under the Lhasa Block of the Tibetan Plateau (Hsü et al., 1995; Pan et al., 2006; Zhu et al., 2013) and southeastward under the Tengchong Block in western Yunnan (Wang et al., 2014; Zou et al., 2011b) during the Permian to the Early Cretaceous, until the closure of the Banggong-Nujiang Tethys Ocean at ~100 Ma (Baxter et al., 2009; Zhang et al., 2012b), as evidenced from the ages of ophiolites and radiolarians. However, the Neo-Tethys Ocean subducted northward under the Lhasa terrane and the Tengchong Block from around the Jurassic (Barley et al., 2003; Chu et al., 2006; Kapp et al. 2007; Searle et al., 2007; Xu et al., 2012; Wang et al., 2014) to ~50 Ma, resulting in the formation of the southern Gangdese magmatic arc. Geophysical data (Tilmann and Ni, 2003; Zhang and Santosh, 2011) indicate that the northern edge of the Neo-Tethyan oceanic lithosphere stretches across the Lhasa terrane and reaches to the southern margin of the Bangong-Nujiang suture, but it just reached the southern Lhasa subterrane during the Early Cretaceous (Zhu et al., 2011), so these granites may not be generated by a low-angle Neo-Tethys subducted flat slab. In addition, to the west, the west Burma Block, which separated from northern Gondwana during the Late Jurassic, collided with the Tengchong Block in the Late Cretaceous and this, along with the closure of the Putao-Myitkyina paleo-ocean (Metcalfe, 2002; Li et al., 2005; Cong et al., 2011a). Li et al. (2004), suggests an eastward subduction of the Putao-Myitkyina paleo-oceanic slab beneath the Tengchong Block in the Early Cretaceous (Li et al., 2004; Wang et al., 2012), but it just reached the western Tengchong Block and induced a series of I-type granites (~120 Ma) (Wang et al., 2012).

In the Rb vs. Y+Nb and Rb vs. Ta+Yb diagrams (Fig. 13a,c), almost all of the samples fall into the field of syn-COLG, but their character follows the trend of arc-continent collision.

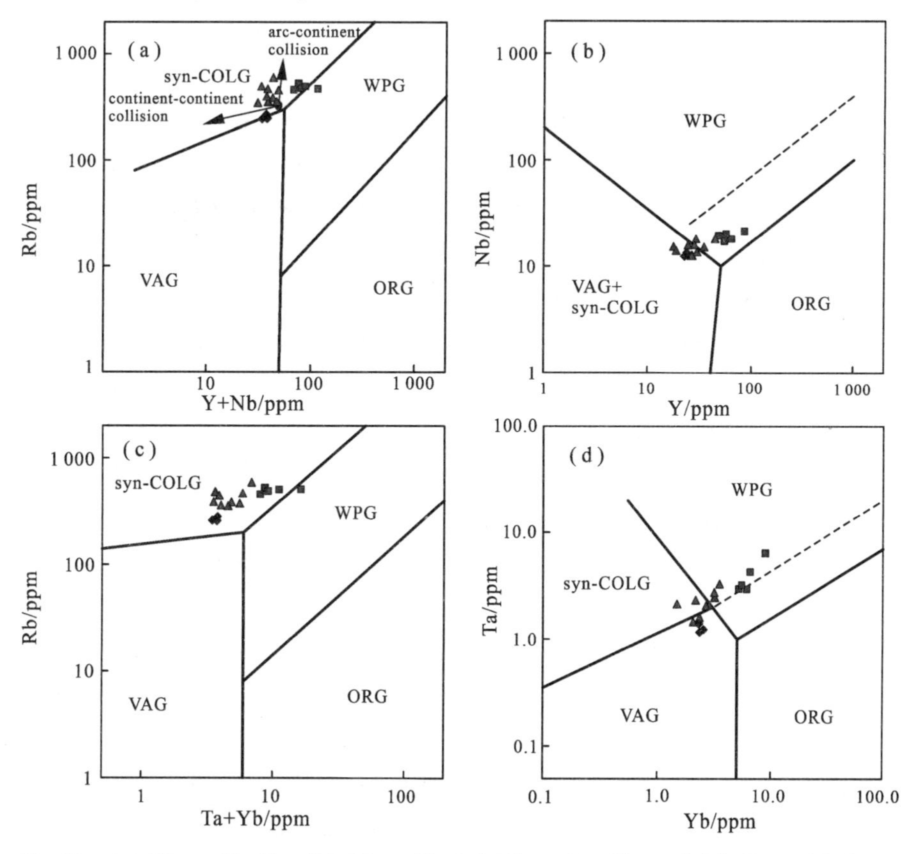

Fig. 13 (a) Rb vs. Y+Nb, (b) Nb vs. Y, (c) Rb vs. Ta+Yb, and (d) Ta vs. Yb diagrams (Pearce et al., 1984; Pearce, 1996) for the Donghe granitoid in the northern Tengchong Block, western Yunnan, SW China.

Symbols are as in Fig. 4.

In the Nb vs. Y and Ta vs. Yb diagrams (Fig. 13b,d), nearly half the samples fall into the fields of syn-COLG and VAG, with other samples falling into the WPG field. According to these granite tectonic discrimination diagrams (Pearce et al., 1984; Pearce, 1996), the Donghe granitoid is more likely dominated by an arc-continental collision setting with some partly related to volcanic arc and within-plate settings. The statistical characteristics of Early Cretaceous ribbon-pattern magmatic belt in the Tengchong Block (see Table 4) indicate that arc-continent collision and volcanic arc settings should be considered, and so the post-collision setting related to the crustal thickening model may not be reasonable. Up to now, no evidence could confirm that the Tengchong Block was a thickening crust during the Early Cretaceous. Through evidence from the eastern Himalayan syntaxis, Pan et al. (2014) suggested that the thickening crust occurred in the Late Cretaceous, and the crustal shortening and thickening of the Lhasa Block mainly occurred during the Late Cretaceous (Kapp et al., 2003). Therefore, we are inclined to accept the first model proposed by Roberts and Clemens (1993) as mentioned above, so that, in the tectonic setting, the basaltic intrusions can produce temperatures high enough to induce the partial melting of the middle to lower crust in the island arc district that was triggered by the mantle-wedge fluid (Patiño Douce, 1999).

In addition, the negative Nb, Ta, Ti, and P anomalies and enriched initial $^{87}Sr/^{86}Sr$ ratios of ~0.706 7 and ~0.707 9 are typically indicative of subduction-related magmatism (Spurlin et al., 2005). Some dark microgranular enclaves could be observed within the host granites in the field (Fig. 2c) and some homochronous Na-rich granodiorites, diorite, and gabbro have also been reported along the eastern margin of the Tengchong Block in previous work (see Table 4), which imply a setting related to the subducted oceanic slab and later triggering of an arc-continent collision. Furthermore, some ~120–130 Ma oceanic island arc-related volcanic rocks can be found along the LLRF as the southeast extension of the Bangong-Nujiang suture (Bai et al., 2002; Gao et al., 2012). Moreover, the existence of ~130 Ma andesite (Chiu et al., 2009) has also been proposed along the Bangong-Nujiang suture. The strata of the Bangong-Nujiang suture belongs to a marine setting until at least ~118 Ma, and even to 115–100 Ma (Kapp et al., 2005b, 2007), indicating that the eastward subducted Bangong-Nujiang-LLR oceanic slab was likely still active between the Tengchong and Baoshan blocks until at least ~120 Ma. Zou et al. (2011b) reported that some island arc granites northwest of Luxi were triggered by the southwestward subducted Bangong-Nujiang-Lushui-Ruili ocean slab between the Zuogong-Baoshan Block and the Bomi-Tengchong Block at ~120 Ma. Zhu et al. (2009a) also points out that the coeval Zayu highly fractionated I-type granites in almost same tectonic position were generated in a setting associated with the southward subduction of the Bangong-Nujiang Tethyan oceanic floor.

Table 4 Early Cretaceous granitoid distribution in the Tengchong Block.

Tectonic position	Location	Lithology	Method	Age /Ma	Geochemical characters	Isotope	Model age /Ga	Similatity	References
Teng-Liang belt	Lianghe	Biotite plagioclase gneiss	Zircon U-Pb	120.4±1.7	High SiO_2, Al_2O_3	Zircon $\varepsilon_{Hf}(t)$ −13.9 to −10.7	T_{DM2Hf} 1.90−2.00		Cong et al., 2011a
	Lianghe	Alkali-feldspar granites	Zircon U-Pb	127	High SiO_2, K	Zircon $\varepsilon_{Hf}(t)$ −9.1 to −5.4	T_{DM2Hf} 1.50−1.70		Cong et al., 2011b
		Granodiorites	Zircon U-Pb	115	High SiO_2	Zircon $\varepsilon_{Hf}(t)$ −4.5 to 0	T_{DM2Hf} 1.10−1.40		
		Diorite	Zircon U-Pb	122	High SiO_2, K	Zircon $\varepsilon_{Hf}(t)$ 3.6 to 6.2	T_{DMHf} 0.54 to 0.67		
	Lianghe	Granites	Zircon U-Pb	127±1.0	Peraluminous, high K				Cong et al., 2010a
	Lianghe	Granitoids	Zircon U-Pb Zircon U-Pb Zircon U-Pb	139.6±2.3 129.7±5.2 118.5±4.2	Peraluminous, High-K	Zircon $\varepsilon_{Hf}(t)$ −30.7 to −20.2	T_{DMHf} 2.46−3.09	Central and northern Lhasa terrane	Li et al., 2012a
	Tengchong	Granodiorites	Shrimp	122±1.3	Metaaluminous, high Na	Zircon $\varepsilon_{Hf}(t)$ −9.6 to −4.8	T_{DMHf} 0.98−1.19	Central Lhasa terrane	Qi et al., 2011
		Quartzmonzonite	Shrimp	125±1.3	Metaaluminous, high Na	Zircon $\varepsilon_{Hf}(t)$ −7.8 to −4.7	T_{DMHf} 0.98−1.11		
	Menglian	Granites	Zircon U-Pb	128±2.4	Peraluminous, high K calc-alkaline				Luo et al., 2012
	Bangmu	Monzogranite	Shrimp	120.5±1.7	Peraluminous, calc-alkaline			Northern Lhasa terrane	Zou et al., 2011
	Lianghe	K-feldspar granite	Zircon U-Pb	115.8±1.0	Peraluminous, high K, calc-alkaline			eastern Gangdese magmatic belt	Xie et al., 2010

Continued

Tectonic position	Location	Lithology	Method	Age /Ma	Geochemical characters	Isotope	Model age /Ga	Similatity	References
Teng-Liang belt	Lianghe	Monzogranite	Zircon U-Pb	123. 8±2. 5	Peraluminous, high K, calc-alkaline			Central and northern Lhasa terrane	Li et al., 2012b
		Quartz-diorite	Zircon U-Pb	127. 1±0. 96	Metaluminous, high K calc-alkaline	Zircon $\varepsilon_{Hf}(t)$ −7. 6 to −3. 8	T_{DM2Hf} 1. 42–1. 66		
		Gabbro-diorite	Zircon U-Pb	122. 6	Low SiO_2, high MgO				
	Luxi	Gneissic granite	Zircon U-Pb	117	Peraluminous				Cong et al., 2009
Gaoli-gong belt	Gaoligong	Granites	Zircon U-Pb	126–121	Peraluminous S-type granites	Zircon $\varepsilon_{Hf}(t)$ −12 to −2	T_{DMHf} 0. 93–1. 11	Northern magmatic belt	Xu et al., 2012
	Gaoligong	Granites	Shrimp	118–124	Peraluminous S-type granites				Yang and Xu., 2011
	Gaoligong	Monzogranites	Shrimp	118–126	Peraluminous S-type granites	I_{Sr} 0. 706 2–0. 718 6 $\varepsilon_{Hf}(t)$ −12. 3 to −5. 7	T_{DM2Nd} 1. 30–1. 40	Nothtern Lhasa terrane	Yang et al., 2006
North Teng-chong	Jingjiguan	Monzogranite	Zircon U-Pb	120±0. 6	Peralumious S-type granites	Zircon $\varepsilon_{Hf}(t)$ − 4. 9 to −1. 9		Bangong-Nujiang Magmatic belt	Cao et al., 2014
		Molybdenite	Re-Os	122±0. 7					
	Diantan	K-feldspar granite	Mineral isochron	118–127	I-type				Chen et al., 1991
		Monzogranite	Mineral isochron	138					
	Mingguang	Monzogranite	Mineral isochron	127–143	I&S-type				
		K-feldspar granite	Mineral isochron	134					
	Qipanshi	Granodiorite-porphyry	Mineral isochron	117	Peraluminous S-type				
		Quartzmonzo-diorite	Mineral isochron	122					
North Tengc-hong	Donghe	Monzogranites	Zircon U-Pb	119–124	Peraluminous, high K, calc-alkaline	I_{Sr} 0. 706 7–0. 707 9 $\varepsilon_{Hf}(t)$ −8. 6 to −10. 1	T_{DM2Nd} 1. 39–1. 49	Northern magmatic belt	This study
		Mynegranites	Zircon U-Pb	130					

Table 5 Distributions of highly fractionated in China.

Location	Age /Ma	Petrography	Geochemical features	Isotopes	Model age	Differentiated Index	Reference
NE China	122–130	Leucogranites to alkaline granites	High SiO_2, peralkalis, and peraluminous, 10–20 and 200–1 500 × chondrite abundances, tetrad effect significantly negative Eu amomailies	I_{Sr} 0.703±0.010, $\varepsilon_{Nd}(t)$ +1.9 to +2.5; I_{Sr} = 0.703, $\varepsilon_{Nd}(t)$ −1.1 to 0.1	T_{MD2} 720–760 Ma T_{MD2} 920–1 015 Ma		Jahn et al., 2001
NE China	184–137	Monzogranites, syengranites, to alkaline granites	High SiO_2, alkalis, and peraluminous, enrich in LREE and significantly negative Eu anomalies, depleted in Ba, Nb, Sr, P, Eu, and Ti. But positive Rb, Pb.	I_{Sr} 0.704 5±0.001 5, $\varepsilon_{Nd}(t)$ +1.3 to +2.8	T_{DM} 720–840 Ma	88.55–98.54	Wu et al., 2003a
NE China	184–138	Monzogranites, syengranites, to alkaline granites	High SiO_2, alkalis, and peraluminous, enrich in LREE and significantly negative Eu anomalies, depleted in Ba, Nb, Sr, P, Eu, and Ti. But positive Rb, Pb.	I_{Sr} 0.704 5±0.001 5, $\varepsilon_{Nd}(t)$ +1.3 to +2.9	T_{DM} 720–840 Ma	88.55– 98.55	Wu et al., 2003b
SW China	130	Monzogranites	High SiO_2, K_2O, A/CNK (1.00–1.05), enriched in Rb, Th, U, and Pb, depleted in Ba, Nb, Ta, Sr, P, Ti, and Eu. Significant negative Eu	$\varepsilon_{Hf}(t)$ −10.9 to −7.6, I_{Sr} 0.712 0 –0.717 9, zircon $\varepsilon_{Hf}(t)$ −12.8 to −2.9	T_{DM2} 1.4–2.0 Ga	82–92	Zhu et al., 2009
SE China	158–165	Mozogranites	High SiO_2, alkalis, A/CNK (0.89–1.13), enrich in Rb and Th and deplete in Ba, Nb (Ta), Sr, P, Eu and Ti. Significant negative Eu	I_{Sr} 0.709 8–0.713 6, $\varepsilon_{Nd}(t)$ −4.3 to −12.2, $\varepsilon_{Hf}(t)$ −11.5 to −3.1	T_{DM2} 1.57–1.88 Ga		Li et al., 2007

Continued

Location	Age /Ma	Petrography	Geochemical features	Isotopes	Model age	Differentiated Index	Reference
SE China	91–96	Syenogranites	High SiO_2, alkaline, metaluminous, enrich in Rb, Th, U, Pb, depleted in Ba, Sr, P, Ti, Eu.	$\varepsilon_{Nd}(t)$ −4.2 to −5.5, $\varepsilon_{Hf}(t)$ −11.6 to 4.5	T_{DM2} 1.32–1.34 Ga	95–97.2	Qiu et al., 2008
SE China	156	Biotite granites and two-mica granites	High SiO_2, K, A/CNK (1.05–1.33); enrich in Rb, Th, Pb, LREE, depleted in Ba, Nb, Sr, P, Ti, and sharply negative Eu anomaly.	$\varepsilon_{Nd}(t)$ −13.0 to −7.7, $\varepsilon_{Hf}(t)$ −13.8 to −7.9; $\delta^{18}O$ 7.91‰–10.08‰	T_{DM2} 1.57–2.00 Ga		Tao et al., 2013
SW China	120–130	Monzogranites to syenogranites	High SiO_2, high K, A/CNK (1.03–1.16), enriched in Rb, Th, U, and Pb, depleted in Ba, Nb, Ta, Sr, P, Ti, and Eu. Significant negative Eu	I_{Sr} 0.7067–0.7079, $\varepsilon_{Nd}(t)$ −8.6 to −10.1. initial $^{206}Pb/^{204}Pb$ 18.462–18.646, $^{207}Pb/^{204}Pb$ 15.717–15.735, $^{208}Pb/^{204}Pb$ 38.699–39.007	T_{DM2} 1.39–1.49 Ga	83.6–95.6	this study

We therefore consider that the Donghe granitoid was generated in a setting of a westward subducted LLR Tethyan oceanic slab (as the southeastern extension of Bangong-Nujiang Tethys Ocean). The setting was likely induced by an arc-continent collision, in which mantle-wedge-derived magma provided enough heat for the melting of ancient crust to generate a hybrid magma from both mantle- and crustal-derived melting and this hybrid magma later experienced strongly fractional crystallization (Fig. 14). Previously published evidence relates this to an Andean-type active continental margin from the Lhasa Block to the Tengchong Block along the Bangong-Nujiang-LLR Tethys Ocean during the Early Cretaceous.

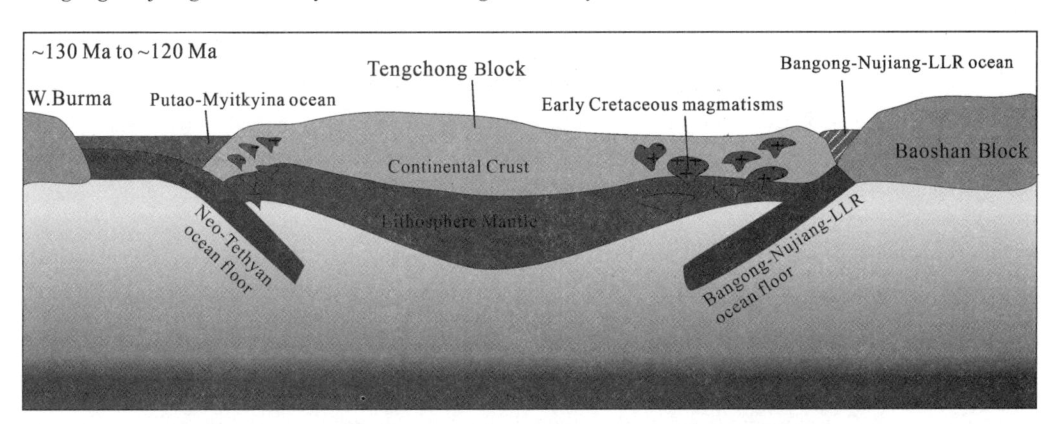

Fig. 14 Tectonic model for the Early Cretaceous Donghe granitoid in the Tengchong Block, western Yunnan, SW China.

Fig. modified after Kapp et al. (2007), Zhu et al.(2013), and Wang et al. (2012). LLR: Lushui-Luxi-Ruili.

5.3 Widespread magmatism from the Lhasa Block to the Tengchong Block during the Early Cretaceous

Our U-Pb age data show that the slightly peraluminous, high-K, calc-alkaline granites in the Tengchong Block were formed between 130 Ma and 119 Ma, which is consistent with abundant Early Cretaceous granitoids (of age range from 143 Ma to 115 Ma) outcropping near Gaoligong, Lianghe, Luxi, and Tengchong counties in the Tengchong Block (Cao et al., 2014; Chen et al., 1991; Cong et al., 2009, 2010a, 2011a,b; Li et al., 2012a,b; Luo et al., 2012; Qi et al., 2011; Xie et al., 2010; Xu et al., 2012; Yang et al., 2009; Yang and Xu, 2011; Zou et al., 2011a) (Table 4) and indicates intensive magmatic activity in the Tengchong Block during the Early Cretaceous. Most researchers considered these Early Cretaceous granites as the magmatic response to the evolution of Tethys, but some researchers regarded these granitoid rocks as the southeastern extension of the magmatic belt in the central and northern Lhasa subterrane (Cao et al., 2014; Li et al., 2012a,b; Qi et al., 2011; Xie et al., 2010; Xu et al., 2012; Yang et al., 2009; Zou et al., 2011a) (Fig. 15).

Recently, some researchers have also reported that there are some Triassic granitoid rocks

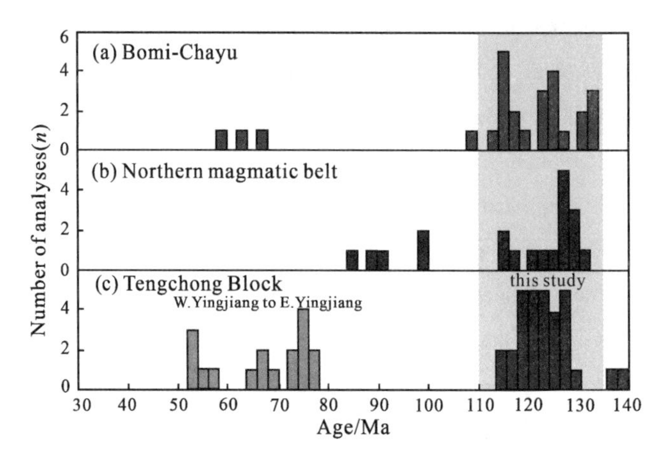

Fig. 15 Histogram of age data from the Bomi-Chayu, northern magmatic belt, and Tengchong Block. This figure is from Xu et al. (2012); the Bomi-Chayu, northern magmatic belt, and Yingjiang data come from Xu et al. (2012) and Ji et al. (2009). All other data are given in Table 4.

in the Tengchong Block from 232 Ma to 206 Ma (Cong et al., 2010b; Li et al., 2011; Zou et al., 2011b). We may conclude that the Tengchong Block was once a segment separated from the Gondwana supercontinent since the Early Permian (Sengör, 1984; Metcalfe, 1984, 1996a; Fan and Zhang, 1994; Wopfner, 1996; Zhong, 1998; Pan et al., 2006), where magmatism occurred from 232 Ma to 115 Ma in the eastern Tengchong Block, even to the Cenozoic (Ma et al., 2014; Wang et al., 2014; Yang et al., 2009). These Early Cretaceous magmatic events are characterized by negative $\varepsilon_{Hf}(t)$ [−12.3 to −5.7, except for some of age ~115 Ma, which display positive $\varepsilon_{Hf}(t)$] and a lower initial $^{87}Sr/^{86}Sr$ range from 0.7062 to 0.7186. Most two-stage model ages concentrate on the Mesoproterozoic (see Table 4).

The northern magmatic belt including the central and northern Lhasa terrane (Zhu et al., 2011) has also been undergoing a relatively long period of magmatism from 240 Ma to 110 Ma (Xu et al., 2012; Zhu et al., 2011), with intensification from 135 Ma to ~110 Ma (Xu et al., 2012). Moreover, the northern magmatic belt was dominated by peraluminous granites (Xu et al., 1985, 2012; Harris et al., 1990; Kapp et al., 2005a,b; Chung et al., 2005 and references therein) and most of these magmas are characterized by negative $\varepsilon_{Hf}(t)$ [except for some of age ~110 Ma, which display positive $\varepsilon_{Hf}(t)$] (Chu et al., 2006; Zhu et al., 2011). These features precisely share some similarities with those Early Cretaceous granitoid rocks in the Tengchong Block (see Table 4).

Consequently, we may conclude that both the Tengchong and Lhasa blocks had been undergoing a series of intense magmatism related to the processes of subduction, collision, and closure of the Bangong-Nujiang Tethys Ocean during the Early Cretaceous, and these granitoid rocks were derived from ancient Mesoproterozoic continental curst and some mantle-derived material.

6 Conclusions

(1) Zircon U-Pb geochronology indicates that the Donghe granitoid in the Tengchong Block was emplaced at ~119 Ma to 130 Ma, coeval with the Early Cretaceous granitic rocks in the Gaoligong belt in the central and northern Lhasa terrane.

(2) The Donghe granitoid contains highly fractionated granites, with high SiO_2 and high K contents, and are calc-alkaline in character and slightly peraluminous. Moreover, the negative Eu anomalies increase sharply with the positive Rb, Th, U, and Pb anomalies and the negative Ba, Nb, Sr, P, and Ti anomalies change gradually and the positive Y anomaly gradually decreases with increasing SiO_2 of fractional crystallization, as the result of the fractionation crystallization of K-feldspar, ± plagioclase, ± biotite, ± amphibole, ± apatite, ± sphene/garnet, and ± Fe-Ti oxides.

(3) The low initial $^{87}Sr/^{86}Sr$ (0.706 7 and 0.707 9) and negative $\varepsilon_{Nd}(t)$ values (−10.1 to −8.6), with T_{DM2} ranging from 1.39 Ga to 1.49 Ga, and the initial $^{206}Pb/^{204}Pb$, $^{207}Pb/^{204}Pb$, and $^{208}Pb/^{204}Pb$ ratios of 18.462−18.646, 15.717−15.735, and 38.699−39.007, respectively, suggest that the parent magma has mixed with the more ancient middle to lower crustal materials and less mantle-derived basaltic magma during the Mesoproterozoic.

(4) The Donghe granitoid, tectonically, formed in a geodynamical setting related to the westward subducted LLR Tethyan oceanic slab (as the southeastern extension of the Bangong-Nujiang Tethys Ocean), where mantle wedge-derived magma provided enough heat for the melting of ancient crust.

(5) The widespread magmatism of the Lhasa block to the Tengchong Block may imply an Andean-type active continental margin from the Lhasa block to the Tengchong Block along the Bangong-Nujiang-LLR Tethys Ocean during the Early Cretaceous.

Acknowledgements This work was jointly supported by the National Natural Science Foundation of China (Grant Nos. 41190072, 41421002, and 41372067) and the program for Changjiang Scholars and Innovative Research Team in University (Grant IRT1281). Support was also provided by the MOST Special Fund from the State Key Laboratory of Continental Dynamics, Northwest University, and Province Key Laboratory Construction Item (08JZ62). We thank the anonymous reviewers and, particularly, Editor-in-Chief Bor-Ming Jahn for providing constructive comments.

Appendix A. Supplementary material Supplementary data associated with this article can be found, in the online version, at http://dx.doi.org/10.1016/j.jseaes.2015.01.014.

References

Bai, X.Z., Jia, X.C., Yang, X.J., Xiong, C.L., Liang, B., Huang, B.X., Luo, G., 2002. LA-ICP-MS zircon U-Pb dating and geochemical characteristics of Early Cretaceous volcanic rocks in Longli-Ruili fault belt, western Yunnan Province. Geol. Bull. China, 31 (2/3), 297-305.

Barley, M.E., Pickard, A.L., Zaw, K., Rak, P., Doyle, M.G., 2003. Jurassic to Miocene magmatism and metamorphism in the Mogok metamorphic belt and the India-Eurasia collision in Myanmar. Tectonics., 22, 1019.

Baxter, A.T., Aitchison, J.C., Zyabrev, S.V., 2009. Radiolarian age constraints on Mesotethyan ocean evolution, and their implications for development of the Bangong-Nujiang suture, Tibet. J. Geol. Soc. Lond., 166, 689-694.

Booth, A.L., Zeitler, P.K., Kidd, W.S.F., Wooden, J., Liu, Y.P., Idleman, B., Hren, M., Chamberlain, C.P., 2004. U-Pb zircon constraints on the tectonic evolution of southeastern Tibet, Namche Barwa area. Am. J. Sci., 304, 889-929.

Cao, H.W., Zhang, S.T., Lin, J.Z., Zheng, L., Wu, J.D., Li, D., 2014. Geology, geochemistry and geochronology of the Jiaojiguanliangzi Fe-polymetallic Deposit, Tengchong County, western Yunnan (China): Regional tectonic implications. J. Asian Earth Sci., 81, 142-152.

Castillo, P.R., Janney, P.E., Solidum, R.U., 1999. Petrology and geochemistry of Camiguin island, southern Philippines: Insights to the source of adakites and other lavas in a complex arc setting. Contrib. Miner. Petrol., 134, 33-51.

Chappell, B.W., White, A.J.R., 1974. Two contrasting granite types. Pac. Geol., 8, 173-174.

Chappell, B.W., White, A.J.R., 1992. I- and S-type granites in the Lachlan Fold Belt. Geol. Soc. Am. Spec. Pap., 272, 1-26.

Chen, J.S., Lin, W.X., Chen, L.Z., 1991. Series and unit research on Tin-bearing granites of Tengchong-Lianghe area. Yunnan Inst. Geol. Sci., 10, 241-289 (in Chinese with English abstract).

Chen, C.H., Lin, W., Lu, H.Y., Lee, C.Y., Tien, J.L., Lai, Y.H., 2000. Cretaceous fractionated I-type granitoids and metaluminous A-type granites in SE China: The Late Yanshanian post-orogenic magmatism. Geol. Soc. Am. Spec. Pap., 350, 195-205.

Chen, F., Li, X.H., Wang, X.L., Li, Q.L., Siebel, W., 2007. Zircon age and Nd-Hf isotopic composition of the Yunnan Tethyan belt, southwestern China. Int. J. Earth Sci., 96 (6), 1179-1194.

Chen, Y., Zhu, D.C., Zhao, Z.D., Meng, F.Y., Wang, Q., Santosh, M., Mo, X.X., 2013. Slab breakoff triggered ca. 113 Ma magmatism around Xainza area of the Lhasa Terrane, Tibet. Gondwana Res.

Chiu, H.Y., Chung, S.L., Wu, F.Y., Liu, D.Y., Liang, Y.H., Lin, I.J., Lizuka, Y., Xie, L.W., Wang, Y.B., Chu, M.F., 2009. Zircon U-Pb and Hf isotopic constraints from eastern Transhimalayan batholiths on the precollisional magmatic and tectonic evolution in southern Tibet. Tectonophysics, 477, 3-19.

Chu, M.F., Chung, S.L., Song, B., Liu, D.Y., O'Reilly, S.Y., Pearson, N.J., Ji, J.Q., Wen, D.J., 2006. Zircon U-Pb and Hf isotope constraints on the Mesozoic tectonics and crustal evolution of southern Tibet. Geology, 34, 745-748.

Chu, Z.Y., Chen, F.K., Yang, Y.H., Guo, J.H., 2009. Precise determination of Sm, Nd concentrations and Nd isotopic compositions at the nanogram level in geological samples by thermal ionization mass spectrometry. J. Anal. At. Spectrom., 24, 1534-1544.

Chung, S.L., Chu, M.F., Zhang, Y., Xie, Y., Lo, C.H., Lee, T.Y., Wang, Y., 2005. Tibetan tectonic evolution inferred from spatial and temporal variations in post collisional magmatism. Earth Sci. Rev., 68 (3), 173-196.

Clemens, J.D., 2003. S-type granitic magmas-petrogenetic issues, models and evidence. Earth Sci. Rev., 61, 1-18.

Condie, K.C., 1997. Plate Tectonics and Crustal Evolution. Oxford (Chapter 1).

Cong, F., Lin, S.L., Li, Z.H., Zou, G.F., Geng, Q.R., 2009. Ziron U-Pb age of gneissic granites in the Tengchong Block, Western Yunnan. Acta Geol. Sinica, 83, 651-658 (in Chinese with English abstract).

Cong, F., Lin, S.L., Xie, T., Li, Z.H., Zou, G.F., Liang, T., 2010a. Rare earth element geochemistry and U-Pb age of zircons from granites in Tengchong-Lianghe Area, western Yunnan. J. Jilin Univ.: Earth Sci. Ed., 40, 573-580 (in Chinese with English abstract).

Cong, F., Lin, S.L., Zou, G.F., Li, Z.H., Xie, T., Peng, Z.M., Liang, T., 2010b. Trace elements and Hf isotope compositions and U-Pb age of igneous zircons from the Triassic Granite in Lianghe, western Yunnan. Acta Geol. Sinica, 84, 1155-1164.

Cong, F., Lin, S.L., Zou, G.F., Li, Z.H., Xie, T., Peng, Z.M., Liang, T., 2011a. Magma mixing of granites at Lianghe: In-situ zircon analysis for trace elements, U-Pb ages and Hf isotopes. Sci. China Earth Sci., 54, 1346-1359.

Cong, F., Lin, S.L., Zou, G.F., Xie, T., Li, Z.H., Tang, F.W., Peng, Z.M., 2011b. Geochronology and petrogenesis for the protolith of biotite plagioclase gneiss at Lianghe, western Yunnan. Acta Geol. Sinica, 85, 870-880.

Deng, J., Wang, Q.F., Li, G.J., Li, C.S., Wang, C.M., 2014. Tethys tectonic evolution and its bearing on the distribution of important mineral deposits in the Sanjiang region. SW China. Gondwana Res., 26 (2), 419-437.

Fan, C.J., Zhang, Y.F., 1994. The structure and tectonics of western Yunnan. J. SE Asian Earth Sci., 9, 355-361.

Feng, Q.L., 2002. Stratigraphy of volcanic rocks in the Changning-Menglian belt in southwestern Yunnan, China. J. Asian Earth Sci., 20, 657-664.

Fontaine, H., 2002. Permian of Southeast Asia: An overview. J. Asian Earth Sci., 20, 567-588.

Frost, B.R., Barnes, C.G., Collins, W.J., Arculus, R.J., Ellis, D.J., Frost, C.D., 2001. A geochemical classification for granitic rocks. J. Petrol., 42 (11), 2033-2048.

Gao, Y.J., Lin, S.J., Cong, F., Zou, G.F., Xie, T., Tang, F.W., Li, Z.H., Liang, T., 2012. LA-ICP-MS zircon U-Pb dating and geological implications for the Early Cretaceous volcanic rocks on the southeastern margin of the Tengchong block, western Yunnan. Sediment. Geol. Tethyan Geol., 32 (4), 59-64.

Hall, R., 1997. Cenozoic plate tectonic reconstructions of SE Asia. Geol. Soc. Lond. Spec. Publ., 126 (1), 11-23.

Hall, R., 2002. Cenozoic geological and plate tectonic evolution of SE Asia and the SW Pacific: computer-based reconstructions, model and animations. J. Asian Earth Sci., 20, 353-431.

Hall, R., 2011. Australia-SE Asia collision: Plate tectonics and crustal flow (in the SE Asian gateway; history and tectonics of the Australia-Asia collision). Geol. Soc. Spec. Publ., 355, 75-109.

Hall, R., 2012. Late Jurassic-Cenozoic reconstructions of the Indonesian region and the Indian Ocean.

Tectonophysics, 570, 1-41.

Harris, N.B.W., Inger, S., Xu, R.H., 1990. Cretaceous plutonism in Central Tibet: An example of postcollision magmatism. J. Volcanol. Geoth. Res., 44, 21-32.

Healy, B., Collins, W.J., Richards, S.W., 2004. A hybrid origin for Lachlan S-type granites: The Murrumbridgee batholith example. Lithos, 79, 197-216.

Hoskin, P.W.O., Black, L.P., 2000. Metamorphic zircon formation by solid-state recrystallization of protolith igneous zircon. J. Metamorph. Geol., 18, 423-439.

Hoskin, P.W.O., Schaltegger, U., 2003. The composition of zircon and igneous and metamorphic petrogenesis. Rev. Mineral. Geochem., 53, 27-62.

Hsü, K.J., Pan, G.T., Sengör, A.M.C., 1995. Tectonic evolution of the Tibetan Plateau: A working hypothesis based on the archipelago model of orogenesis. Int. Geol. Rev., 37, 473-508.

Hutchison, C.S., 1989. Geological Evolution of South-east Asia. Oxford and New York: Clarendon Press, 1-368.

Inger, S., Harris, N.B.W., 1993. Geochemical constraints on leucogranite magmatism in the Langtang Valley, Nepal Himalaya. J. Petrol., 34, 345-368.

Jahn, B.M., Wu, F.Y., Capdevila, R., Martineau, F., Wang, Y.X., Zhao, Z.H., 2001. Highly evolved juvenile granites with tetrad REE patterns: The Woduhe and Baerzhe granites from the Great Xing'an (Khingan) Mountains in NE China. Lithos, 59, 171-198.

Janoušek, V., Finger, F., Roberts, M., Frýda, J., Pin, C., Dolejš, D., 2004. Deciphering the petrogenesis of deeply buried granites: Whole-rock geochemical constraints on the origin of largely undepleted felsic granulites from the Moldanubian Zone of the Bohemian Massif. Geol. Soc. Am. Spec. Pap., 389, 141-159.

Jaques, A.L., Green, D.H., 1980. Anhydrous melting of peridotite at 0-15 kb pressure and the genesis of tholeiitic basalts. Contrib. Miner. Petrol., 73, 287-310.

Ji, W.Q., Wu, F.Y., Chung, S.L., Li, J.X., Liu, C.Z., 2009. Zircon U-Pb geochronology and Hf isotopic constraints on petrogenesis of the Gangdese batholith, southern Tibet. Chem. Geol., 262 (3), 229-245.

Jiang, B., Gong, Q.J., Zhang, J., Ma, N., 2012. Late Cretaceous aluminium A-type granites and its geological significance of Dasongpo Sn deposit, Tengchong, west Yunnan. Acta Petrol. Sinica, 28, 1477-1492 (in Chinese with English abstract).

Jin, X.C., 1996. Tectono-stratigraphic units in western Yunnan and their counterparts in southeast Asia. Cont. Dynam., 1, 123-133.

Kapp, P., Murphy, M.A., Yin, A., Harrison, T.M., Ding, L., Guo, J., 2003. Mesozoic and Cenozoic tectonic evolution of the Shiquanhe area of western Tibet. Tectonics, 22 (4).

Kapp, J.L.D., Harrison, T.M., Grove, M., Lovera, O.M., Lin, D., 2005a. Nyainqentanglha Shan: A window into the tectonic, thermal, and geochemical evolution of the Lhasa block, southern Tibet. J. Geophys. Res.: Solid Earth, (1978—2012) 110 (B8).

Kapp, P., Yin, A., Harrison, T.M., Ding, L., 2005b. Cretaceous-Tertiary shortening, basin development, and volcanism in central Tibet. Geol. Soc. Am. Bull., 117, 865-878.

Kapp, P., Peter, G.D., Gehrels, George E., Matthew, H., Lin, D., 2007. Geological records of the Lhasa-Qiangtang and Indo-Asian collisions in Nima area of central Tibet. Geol. Soc. Am. Bull., 9, 917-993.

Kemp, A.I.S., Hawkesworth, C.J., Foster, G.L., Paterson, B.A., Woodhead, J.D., Hergt, J.M., Whitehouse, M.J., 2007. Magmatic and crustal differentiation history of granitic rocks from Hf-O isotopes in zircon. Science, 315 (5814), 980-983.

Kou, K.H., Zhang, Z.C., Santosh, M., Huang, H., Hou, T., Liao, B.L., Li, H.B., 2012. Picritic porphyrites generated in a slab-window setting: Implications for the transition from Paleo-Tethyan to Neo-Tethyan tectonics. Lithos, 155, 375-391.

Le Fort, P., Cuney, M., Deniel, C., France-Lanord, C., Sheppard, S.M.F., Upreti, B.N., Vidal, P., 1987. Crustal generation of the Himalayan leucogranites. Tectonophysics, 134, 39-57.

Li, X.Z., Liu, C.J., Ding, J., 2004. Correlation and connection of the main suture zones in the Greater Mekong subregion. Sediment. Geol. Tethyan Geol., 4, 1.

Li, P.W., Gao, R., Cui, J.W., Guan, Y., 2005. Paleomagnetic results from the Three Rivers region, SW China: Implications for the collisional and accretionary history. Acta Geosci. Sinica, 26 (5), 387-404 (in Chinese with English abstract).

Li, X.H., Li, Z.X., Li, W.X., Liu, Y., Yuan, C., Wei, G., Qi, C., 2007. U-Pb zircon, geochemical and Sr-Nd-Hf isotopic constraints on age and origin of Jurassic I- and A-type granites from central Guangdong, SE China: A major igneous event in response to foundering of a subducted flat-slab? Lithos, 96 (1), 186-204.

Li, H.Q., Xu, Z.Q., Cai, Z.H., Tang, Z.M., Yang, M., 2011. Indosinian epoch magmatic event and geological significance in the Tengchong block, western Yunnan Province. Acta Petrol. Sinica, 27 (7), 2165-2172.

Li, Z.H., Lin, S.L., Cong, F., Zou, G.F., Xie, T., 2012a. U-Pb dating and Hf isotopic compositions of quartz diorite and monzonitic granite from the Tengchong Lianghe block, western Yunnan, and its geological implications. Acta Geol. Sinica, 86 (7), 1047-1062 (in Chinese with English abstract).

Li, Z.H., Lin, S.L., Cong, F., Zou, G.F., Xie, T., 2012b. Early Cretaceous magmatism in Tengchong-Lianghe block, western Yunnan. Bull. Mineral. Petrol. Geochem., 31, 590-598 (in Chinese with English abstract).

Liang, Y.H., Chung, S.L., Liu, D., Xu, Y., Wu, F.Y., Yang, J.H., Wang, Y., Lo, C.H., 2008. Detrital zircon evidence from Burma for reorganization of the eastern Himalayan river system. Am. J. Sci., 308, 618-638.

Lin, I.J., Chung, S.L., Chu, C.H., Lee, H.Y., Gallet, S., Wu, G., Ji, J., Zhang, Y., 2012. Geochemical and Sr-Nd isotopic characteristics of Cretaceous to Paleocene granitoids and volcanic rocks, SE Tibet: Petrogenesis and tectonic implications. J. Asian Earth Sci., 53, 131-150.

Ludwig, K.R., 2003. ISOPLOT 3.0: A Geochronological Toolkit for Microsoft Excel. Special Publication, 4. Berkeley Geochronology Center.

Luo, G., Jia, X.C., Yang, X.J., Xiong, C.L., Bai, X.Z., Huang, B.X., Tan, X.L., 2012. LA-ICP-MS zircon U-Pb dating of southern Menglian granite in Tengchong area of western Yunnan Province and its tectonic implications. Geol. Bull. China, 31, 287-296 (in Chinese with English abstract).

Ma, N., Deng, J., Wang, Q.F., Wang, C.M., Zhang, J., Li, G.J., 2013. Geochronology of the Dasongpo tin deposit, Yunnan Province: Evidence from zircon LA-ICP-MS U-Pb ages and cassiterite LA-MC-ICP-MS

U-Pb age. Acta Petrol. Sinica, 29, 1223-1235 (in Chinese with English abstract).

Ma, L.Y., Wang, Y.J., Fan, W.M., Geng, H.Y., Cai, Y.F., Zhong, H., Liu, H.C., Xing, X.W., 2014. Petrogenesis of the early Eocene I-type granites in west Yingjiang (SW Yunnan) and its implication for the eastern extension of the Gangdese batholiths. Gondwana Res., 25, 401-419.

Magaritz, M., Whitford, D.J., James, D.E., 1978. Oxygen isotopes and the origin of high $^{87}Sr/^{86}Sr$ andesites. Earth Planet. Sci. Lett., 40, 220-230.

Metcalfe, I., 1984. Stratigraphy, palaeontology and palaeogeography of the Carboniferous of Southeast Asia. Mem. Geol. Soc. Fr., 147, 107-118.

Metcalfe, I., 1988. Origin and assembly of south-east Asian continental terranes. Geol. Soc. Lond. Spec. Publ., 37 (1), 101-118.

Metcalfe, I., 1996a. Pre-Cretaceous evolution of SE Asian terranes. Geol. Soc. Lond. Spec. Publ., 106 (1), 97-122.

Metcalfe, I., 1996b. Gondwanaland dispersion, Asian accretion and evolution of eastern Tethys. Aust. J. Earth Sci., 43 (6), 605-623.

Metcalfe, I., 1998. Palaeozoic and Mesozoic geological evolution of the SE Asian region: Multidisciplinary constraints and implications for biogeography. Biogeogr. Geol. Evol. SE Asia, 25-41.

Metcalfe, I., 2002. Permian tectonic framework and palaeogeography of SE Asia. J. Asian Earth Sci., 20 (6), 551-566.

Metcalfe, I., 2011a. Palaeozoic-Mesozoic history of SE Asia. Geol. Soc. Lond. Spec. Publ., 355 (1), 7-35.

Metcalfe, I., 2011b. Tectonic framework and Phanerozoic evolution of Sundaland. Gondwana Res., 19, 3-21.

Metcalfe, I., 2013. Gondwana dispersion and Asian accretion: Tectonic and paleogeography evolution of eastern Tethys. J. Asian Earth Sci., 66, 1-33.

Miller, C.F., 1985. Are strongly peraluminous magmas derived from pelitic sedimentary sources? J. Geol., 93, 673-689.

Andersen, T., 2002. Correction of common lead in U-Pb analyses that do not report ^{204}Pb. Chem. Geol., 192, 59-79.

Pan, G.T., Mo, X.X., Hou, Z.Q., Zhu, D.C., Wang, L.Q., Li, G.M., Zhao, Z.D., Geng, Q.R., Liao, Z. L., 2006. Spatial-temporal framework of the Gangdese Orogenic Belt and its evolution. Acta Petrol. Sinica, 22, 521-533 (in Chinese with English abstract).

Pan, F.B., Zhang, H.F., Xu, W.C., Guo, L., Wang, S., Luo, B.J., 2014. U-Pb zircon chronology, geochemical and Sr-Nd isotopic composition of Mesozoic Cenozoic granitoids in the SE Lhasa terrane: Petrogenesis and tectonic implications. Lithos, 192, 142-157.

Patiño Douce, A.E., 1999. What do experiments tell us about the relative contributions of crust and mantle to the origin of the granitic magmas. Geol. Soc. Lond., 168, 55-75.

Patiño Douce, A.E., Humphreys, E.D., Johnston, A.D., 1990. Anatexis and metamorphism in tectonically thickened continental crust exemplified by the Sevier hinterland, western North America. Earth Planet. Sci. Lett., 97, 290-315.

Pearce, J.A., 1996. Sources and settings of granitic rocks. Episodes, 19, 120-125.

Pearce, J.A., Harris, N.B.W., Tindle, A.G., 1984. Trace element discrimination diagrams for the tectonic interpretation of granitic rocks. J. Petrol., 25, 956-983.

Qi, L., Hu, J., Gregoire, D.C., 2000. Determination of trace elements in granites by inductively coupled plasma mass spectrometry. Talanta, 51, 507-513.

Qi, X.X., Zhu, L.H., Hu, Z.C., Li, Z.Q., 2011. Zircon SHRIMP U-Pb dating and Lu-Hf isotopic composition for Early Cretaceous plutonic rocks in Tengchong block, southeastern Tibet, and its tectonic implications. Acta Petrol. Sinica, 27, 3409-3421 (in Chinese with English abstract).

Qiu, J.S., Xiao, E., Hu, J., Xu, X.S., Jiang, S.Y., Li, Z., 2008. Petrogenesis of high fractionated I-type granites in the coastal area of northeastern Fujian Province: Constraints from zircon U-Pb geochronology, geochemistry, and Nd-Hf isotopes. Acta Petrol. Sinica, 24 (11), 2468-2484.

Qu, X.M., Wang, R.J., Xin, H.B., Jiang, J.H., Chen, H., 2012. Age and petrogenesis of A type granites in the middle segment of the Bangonghu-Nujiang suture, Tibetan plateau. Lithos, 146, 264-275.

Rapp, R.P., Watson, E.B., 1995. Dehydration melting of metabasalt at 8 – 32 kbar: Implications for continental growth and crust-mantle recycling. J. Petrol., 36, 891-931.

Roberts, M.P., Clemens, J.D., 1993. Origin of high-potassium, talc-alkaline, I-type granitoids. Geology, 21, 825-828.

Rudnick, R.L., Gao, S., 2003. Composition of the continental crust. Treatise Geochem., 3, 1-64.

Searle, M.P., Noble, S.R., Cottle, J.M., Wales, D.J., Mitchell, A.H.G., Hlaing, T., Horstwood, M.S.A., 2007. Tectonic evolution of the Mogok metamorphic belt, Burma (Myanmar) constrained by U-Th-Pb dating of metamorphic and magmatic rocks. Tectonics, 26 (3).

Sengör, A.C., 1984. The Cimmeride orogenic system and the tectonics of Eurasia. Geol. Soc. Am. Spec. Pap., 195, 1-74.

Shi, G.R., Archbold, N.W., 1995. Permian brachiopod faunal sequence of the Shan Thai terrane: Biostratigraphy, palaeobiogeographical affinities and plate tectonic/palaeoclimatic implications. J. SE Asian Earth Sci., 11, 177-187.

Shi, G.R., Archbold, N.W., 1998. Permian marine biogeography of SE Asia. Biogeogr. Geol. Evol. SE Asia, 57-72.

Sisson, T.W., Ratajeski, K., Hankins, W.B., Glazner, A.F., 2005. Voluminous granitic magmas from common basaltic sources. Contrib. Miner. Petrol., 148 (6), 635 661.

Song, S.G., Niu, Y.L., Wei, C.J., Ji, J.Q., Su, L., 2010. Metamorphism, anatexis, zircon ages and tectonic evolution of the Gongshan block in the northern Indochina continent: An eastern extension of the Lhasa Block. Lithos, 120, 327-346.

Spurlin, M.S., Yin, A., Horton, B.K., Zhou, J., Wang, J., 2005. Structural evolution of the Yushu-Nangqian region and its relationship to syncollisional igneous activity, east-central Tibet. Geol. Soc. Am. Bull., 117 (9-10), 1293-1317.

Streckeisen, A., Le Maitre, R.W., 1979. A chemical approximation to the modal QAPF classification of the igneous rocks. Neues Jb. Mineral. Abh., 136, 169-206.

Suess, E., 1893. Are great ocean depths permanent. Nat. Sci., 2 (13), 180-187.

Sun, S.S., McDonough, W.F., 1989. Chemical and isotopic systematics of oceanic basalts: Implications or mantle composition and processes. Geol. Soc. Lond. Spec. Publ., 42 (1), 313-345.

Sylvester, P.J., 1998. Postcollisional strongly peraluminous granites. Lithos, 45, 29-44.

Tao, J., Li, W., Li, X., Cen, T., 2013. Petrogenesis of early Yanshanian highly evolved granites in the Longyuanba area, southern Jiangxi Province: Evidence from zircon U-Pb dating, Hf-O isotope and whole-rock geochemistry. Sci. China: Earth Sci., 56 (6), 922-939.

Tilmann, F., Ni, J., 2003. Seismic imaging of the downwelling Indian lithosphere beneath central Tibet. Science, 300, 1424-1427.

Ueno, K., 2000. Permian fusulinacean faunas of the Sibumasu and Baoshan blocks, implications for the paleogeographic reconstruction of the Cimmerian continent. Geosci. J., 4, 160-163.

Wang, E., Burchfiel, B., 1997. Interpretation of Cenozoic tectonics in the right-lateral accommodation zone between the Ailao Shan shear zone and the eastern Himalayan syntaxis. Int. Geol. Rev., 39, 191-219.

Wang, R., Xia, B., Zhou, G.Q., Zhang, Y.Q., Yang, Z.Q., Li, W.Q., Wei, D.L., Zhong, L.F., Xu, L. F., 2006a. SHRIMP zircon U-Pb dating for gabbro from the Tiding ophiolite in Tibet. Chin. Sci. Bull., 51, 1776-1779.

Wang, Y.J., Fan, W.M., Zhang, Y.H., Peng, T.P., Chen, X.Y., Xu, Y.G., 2006b. Kinematics and $^{40}Ar/^{39}Ar$ geochronology of the Gaoligong and Chongshan shear systems, western Yunnan, China: Implications for early Oligocene tectonic extrusion of SE Asia. Tectonophysics, 418, 235-254.

Wang, Y.J., Fan, W.M., Sun, M., Liang, X.Q., Zhang, Y.H., Peng, T.P., 2007. Geochronological, geochemical and geothermal constraints on petrogenesis of the Indosinian peraluminous granites in the South China Block: A case study in the Hunan Province. Lithos, 96, 475-502.

Wang, H., Lin, F.C., Li, X.Z., Shi, M.F., Liu, C.J., Shi, H.Z., 2012. Tectonic unit division and Neo-Tethys tectonic evolution in north-central Myanmar and its adjacent areas. Geol. China, 39, 912-922.

Wang, Y.J., Xing, X.W., Cawood, P.A., Lai, S.C., Xia, X.P., Fan, W.M., Liu, H.C., Zhang, F.F., 2013. Petrogenesis of early Paleozoic peraluminous granite in the Sibumasu Block of SW Yunnan and diachronous accretionary orogenesis along the northern margin of Gondwana. Lithos, 182, 67-85.

Wang, C.M., Deng, J., Emmanuel, J.M., Carranza, M., Santosh, 2014. Tin metallogenesis associated with granitoids in the southwestern Sanjiang Tethyan Domain: Nature, deposit types, and tectonic setting. Gondwana Res., 26 (2), 576-593.

Watson, E.B., Harrison, T.M., 1983. Zircon saturation revisited: Temperature and composition effects in a variety of crustal magma types. Earth Planet. Sci. Lett., 64, 295-304.

Whalen, J.B., Currie, K.L., Chappell, B.W., 1987. A-type granites: Geochemical characteristics, discrimination and petrogenesis. Contrib. Miner. Petrol., 95, 407-419.

Wilson, B.M., 1989. Igneous Petrogenesis. Springer (Part one).

Wolf, M.B., London, D., 1994. Apatite dissolution into peraluminous haplogranite melts: An experimental study of solubilities and mechanisms. Geochim. Cosmochim. Acta, 58 (19), 4127-4145.

Wopfner, H., 1996. Gondwana origin of the Baoshan and Tengchong terrenes of west Yunnan. Geol. Soc. Lond. Spec. Publ., 106 (1), 539-547.

Wu, F.Y., Jahn, B.M., Wilde, S.A., Lo, C.H., Yui, T.F., Lin, Q., Sun, D.Y., 2003a. Highly fractionated I-type granites in NE China (Ⅰ): Geochronology and petrogenesis. Lithos, 66 (3), 241-273.

Wu, F.Y., Jahn, B.M., Wilde, S.A., Lo, C.H., Yui, T.F., Lin, Q., Sun, D.Y., 2003b. Highly fractionated I-type granites in NE China (Ⅱ): Isotopic geochemistry and implications for crustal growth in the Phanerozoic. Lithos, 67 (3), 191-204.

Xie, T., Lin, S.L., Cong, F., Li, Z.H., Zou, G.F., Li, J.M., Liang, T., 2010. LA-ICP-MS zircon U-Pb dating for K-feldspar granites in Lianghe region, western Yunnan and its geological significance. Geotectonica Metallogenia, 34, 419-428 (in Chinese with English abstract).

Xu, R.H., Schärer, U., Allègre, C.J., 1985. Magmatism and metamorphism in the Lhasa block (Tibet): A geochronological study. J. Geol., 93, 41-57.

Xu, Y.G., Lan, J.B., Yang, Q.J., Huang, X.L., Qiu, H.N., 2008. Eocene break-off of the Neo-Tethyan slab as inferred from intraplate-type mafic dykes in the Gaoligong orogenic belt, eastern Tibet. Chem. Geol., 255, 439-453.

Xu, Y.G., Yang, Q.J., Lan, J.B., Luo, Z.Y., Huang, X.L., Shi, Y.R., Xie, L.W., 2012. Temporal-spatial distribution and tectonic implications of the batholiths in the Gaoligong-Tengliang-Yingjiang area, western Yunnan: Constraints from zircon U-Pb ages and Hf isotopes. J. Asian Earth Sci., 53, 151-175.

Yang, Q.J., Xu, Y.G., 2011. The emplacement of granites in Nujiang-Gaoligong belt, western Yunnan, and response to the evolvement of Neo-Tethy. J. Jilin Univ.: Earth Sci. Ed., 41, 1353-1361 (in Chinese with English abstract).

Yang, Q.J., Xu, Y.G., Huang, X.L., Luo, Z.Y., 2006. Geochronology and geochemistry granites in the Gaoligong tectonic belt, western Yunnan, tectonic implications. Acta Petrol. Sinica, 22, 817-834 (in Chinese with English abstract).

Yang, Q.J., Xu, Y.G., Huang, X.L., Luo, Z.Y., Shi, Y.R., 2009. Geochronology and geochemistry granites in the Tengliang area, western Yunnan, tectonic implications. Acta Petrol. Sinica, 25, 1092-1104 (in Chinese with English abstract).

YBGMR (Yunnan Bureau Geological Mineral Resource), 1991. Regional Geology of Yunnan Province. Beijing: Geological Publishing House, 1-729 (in Chinese with English abstract).

Yuan, H.L., Gao, S., Liu, X.M., Li, H.M., Gunther, D., Wu, F.Y., 2004. Accurate U-Pb age and trace element determinations of zircon by laser ablation-inductively coupled plasma mass spectrometry. Geo-stand. Newslett., 28, 353-370.

Zartman, R.E., Doe, B.R., 1981. Plumbotectonics-the model. Tectonophysics, 75, 135-162.

Zhang, Z., Santosh, M., 2011. Tectonic evolution of Tibet and surrounding regions. Gondwana Res., 21, 1-3.

Zhang, H.F., Harris, N., Parrish, R., Kelley, S., Zhang, L., Rogers, N., Argles, T., King, J., 2004. Causes and consequences of protracted melting of the mid-crust exposed in the North Himalayan antiform. Earth Planet. Sci. Lett., 228, 195-212.

Zhang, Z.M., Wang, J.L., Shen, K., Shi, C., 2008. Paleozoic circus-Gondwana orogens: Petrology and geochronology of the Namche Barwa Complex in the eastern Himalayan syntaxis, Tibet. Acta Petrol. Sinica, 24, 1627-1637 (in Chinese with English abstract).

Zhang, B., Zhang, J.J., Zhong, D.L., Wang, X.X., Qu, J.F., Guo, L., 2011. Structural feature and its significance of the northernmost segment of the Tertiary Biluoxueshan Chongshan shear zone, east of the Eastern Himalayan Syntaxis. Sci. China Earth Sci., 54, 959-974 (in Chinese with English abstract).

Zhang, Z.M., Dong, X., Santosh, M., Liu, F., Wang, W., Yiu, F., He, Z.Y., Shen, K., 2012a.

Petrology and geochronology of the Namche Barwa Complex in the eastern Himalayan syntaxis, Tibet: Constrains on the origin and evolution of the north eastern margin of the Indian craton. Gondwana Res., 21, 123-137.

Zhang, K.J., Zhang, Y.X., Tang, X.C., Xia, B., 2012b. Late Mesozoic tectonic evolution and growth of the Tibetan plateau prior to the Indo-Asian collision. Earth Sci. Rev., 114 (3), 236-249.

Zhang, B., Zhang, J., Zhong, D., Yang, L., Yue, Y., Yan, S., 2012c. Polystage deformation of the Gaoligong metamorphic zone: Structures, $^{40}Ar/^{39}Ar$ mica ages, and tectonic implications. J. Struct. Geol., 37, 1-18.

Zhong, D.L., 1998. Paleo-Tethyan Orogenic Belt in the Western Parts of the Sichuan and Yunnan provinces. Beijing: Science Press, 231.

Zhong, D.L., 2000. Paleotethys Sides in West Yunnan and Sichuan, China. Beijing: Science Press, 1-248 (in Chinese with English abstract).

Zhu, D.C., Mo, X.X., Wang, L.Q., Zhao, Z.D., Niu, Y.L., Zhou, C.Y., Yang, Y.H., 2009a. Petrogenesis of highly fractionated I-type granites in the Zayu area of eastern Gangdese, Tibet: Constraints from zircon U-Pb geochronology, geochemistry and Sr-Nd-Hf isotopes. Sci. China, Ser. D Earth Sci., 52 (9), 1223-1239.

Zhu, D.C., Mo, X.X., Niu, Y.L., Zhao, Z.D., Wang, L.Q., Liu, Y.S., Wu, F.Y., 2009b. Geochemical investigation of Early Cretaceous igneous rocks along an east-west traverse throughout the central Lhasa Terrane, Tibet. Chem. Geol., 268, 298-312.

Zhu, D.C., Zhao, Z.D., Niu, Y.L., Mo, X.X., Chung, S.L., Hou, Z.Q., Wang, L.Q., Wu, F.Y., 2011. The Lhasa Terrane: Record of a microcontinent and its histories of drift and growth. Earth Planet. Sci. Lett., 301, 241-255.

Zhu, D.C., Zhao, Z.D., Niu, Y.L., Dilek, Y., Hou, Z.Q., Mo, X.X., 2013. The origin and pre-Cenozoic evolution of the Tibetan Plateau. Gondwana Res., 23, 1429-1454.

Zou, G.F., Lin, S.L., Li, Z.H., Cong, F., Xie, T., Tang, W.Q., 2011a. SHRIMP zircon U-Pb dating of Bangmu admellite in Luxi, western Yunnan, and its tectonic implications. Geol. China, 38, 77-85 (in Chinese with English abstract).

Zou, G.F., Lin, S.L., Li, Z.H., Cong, F., Xie, T., 2011b. Geochronology and geochemistry of the Longtang Granite in the Lianghe Area, western Yunnan and its tectonic implications. Geotectonica Metallogenia, 35, 439-451.

Evolution of the Proto-Tethys in the Baoshan Block along the East Gondwana margin: Constraints from early Paleozoic magmatism[①]

Zhao Shaowei Lai Shaocong[②] Gao Liang Qin Jiangfeng Zhu Renzhi

Abstract: Zircon U-Pb ages and geochemical and isotopic data for Late Ordovician granites in the Baoshan Block reveal the early Paleozoic tectonic evolution of the margin of East Gongwana. The granites are high-K, calc-alkaline, metaluminous to strongly peraluminous rocks with A/CNK values of 0.93–1.18, are enriched in SiO_2, K_2O, and Rb, and depleted in Nb, P, Ti, Eu, and heavy rare earth elements, which indicate the crystallization fractionation of the granitic magma. Zircon U-Pb dating indicates that they formed at ca. 445 Ma. High initial $^{87}Sr/^{86}Sr$ ratios of 0.719 761–0.726 754, negative $\varepsilon_{Nd}(t)$ values of −8.3 to −6.6, and two-stage model ages of 1.52–1.64 Ga suggest a crustal origin, with the magmas derived from the partial melting of ancient metagreywacke at high temperature. A synthesis of data for the early Paleozoic igneous rocks in the Baoshan Block and adjacent Tengchong Block indicates two stages of flare-up of granitic and mafic magmatism caused by different tectonic settings along the East Gondwana margin. Late Cambrian to Early Ordovician granitic rocks (ca. 490 Ma) were produced when underplated mafic magmas induced crustal melting along the margin of East Gondwana related to the break-off of subducted Proto-Tethyan oceanic slab. In addition, the cession of the mafic magmatism between late Cambrian-Early Ordovician and Late Ordovician could have been caused by the collision of the Baoshan Block and outward micro-continent along the margin of East Gondwana and crust and lithosphere thickening. The Late Ordovician granites in Baoshan Block were produced in an extensional setting resulting from the delamination of an already thickened crust and lithospheric mantle followed by the injection of synchronous mafic magma.

1 Introduction

The Neoproterozoic to early Paleozoic oceanic basin along the margin of East Gondwana from northwest Turkey, central Iran, northwest Pakistan, northwest India, and Nepal to south Tibet and southwest Yunnan, is known as the "Proto-Tethys" (e.g. von Raumer and Stampfli,

① Published in *International Geology Review*, 2017,59(1).

② Corresponding author.

2008; Zhu et al., 2012). The Cambrian to Early Ordovician magmatism, coeval metamorphism, and geochronology of detrital zircons in these areas are consistent with the existence of this ocean (e.g., DeCells et al., 2000, 2004; Gehrels et al., 2006a,b, 2011; Cawood et al., 2007; Liu et al., 2009; Dong et al., 2012, 2013; Wang et al., 2012, 2013; Zhu et al., 2012, 2013; Hu et al., 2013; Zhao et al., 2014, 2016; Ding et al., 2015). Early Paleozoic volcanism is prevalent in the west Tethyan Himalaya (Garzanti et al., 1986; Brookfield, 1993; Valdiya, 1995), and the rocks include acidic and basic tuffs, basalts, andesites, and felsic volcanic rocks, indicative of an immature arc environment during the Cambrian to Early Ordovician along the East Gondwana margin (Garzanti et al., 1986; Myrow et al., 2006a,b). However, the subsequent Late Ordovician tectonic evolution of the Proto-Tethys and the margin of East Gondwana were poorly constrained. Li et al. (2016) proposed the Gondwana margin crustal and lithospheric thickening and subsequent delamination during the Ordovician, but Xing et al. (2015) considered the continuous subduction of Proto-Tethys at this time. Therefore, the Late Ordovician magmatism in the Baoshan Block could provide important insights into clarifying the tectonic evolution of the Proto-Tethys and the orogeny along the margin of East Gondwana.

In this paper we focus on the petrogenesis and geodynamic implications of the Late Ordovician granites in the Baoshan Block, southwest Yunnan. Our new observations and data, together with previously published information on the early Paleozoic magmatism in the Baoshan Block and the adjacent Tengchong Block (Chen et al., 2007; Dong et al., 2012, 2013; Xiong et al., 2012; Eroğlu et al., 2013; Wang et al., 2013; Zhao et al., 2014, 2016; Xing et al., 2015; Li et al., 2016), indicate two stages flare-up of granitic and mafic magmatism in the Late Cambrian to Ordovician, and these periods could represent different stages in the evolution of the Proto-Tethys and orogenesis along the margin of East Gondwana.

2 Geological background and petrology

Southwest Yunnan, located at the southeastern margin of the Tibetan Plateau, consists of several micro-continental blocks and tectonic belts such as the Simao/Indochina Block of Yangtze affinity, the Baoshan and Tengchong blocks of Gondwana affinity, the Gaoligong belt, and the Changning-Menglian suture zone (Fig. 1; Wu et al., 1995; Metcalfe, 1996a, b; Wang, 1996; Wopfner, 1996; Zhang et al., 1997; Zhong, 1998; Ueno, 2000). The closure of the Paleo-Tethys resulted in the formation of the Changning-Menglian suture zone, which separates the Simao Block in the east from the Baoshan Block in the west (Fang et al., 1994; Wu et al., 1995; Feng, 2002). The Tengchong and Baoshan Blocks are separated by the Gaoligong belt, and to the west these two blocks are bound by the Sagaing Fault and the Mogok metamorphic belt. The Tengchong and Baoshan blocks are also considered to represent the

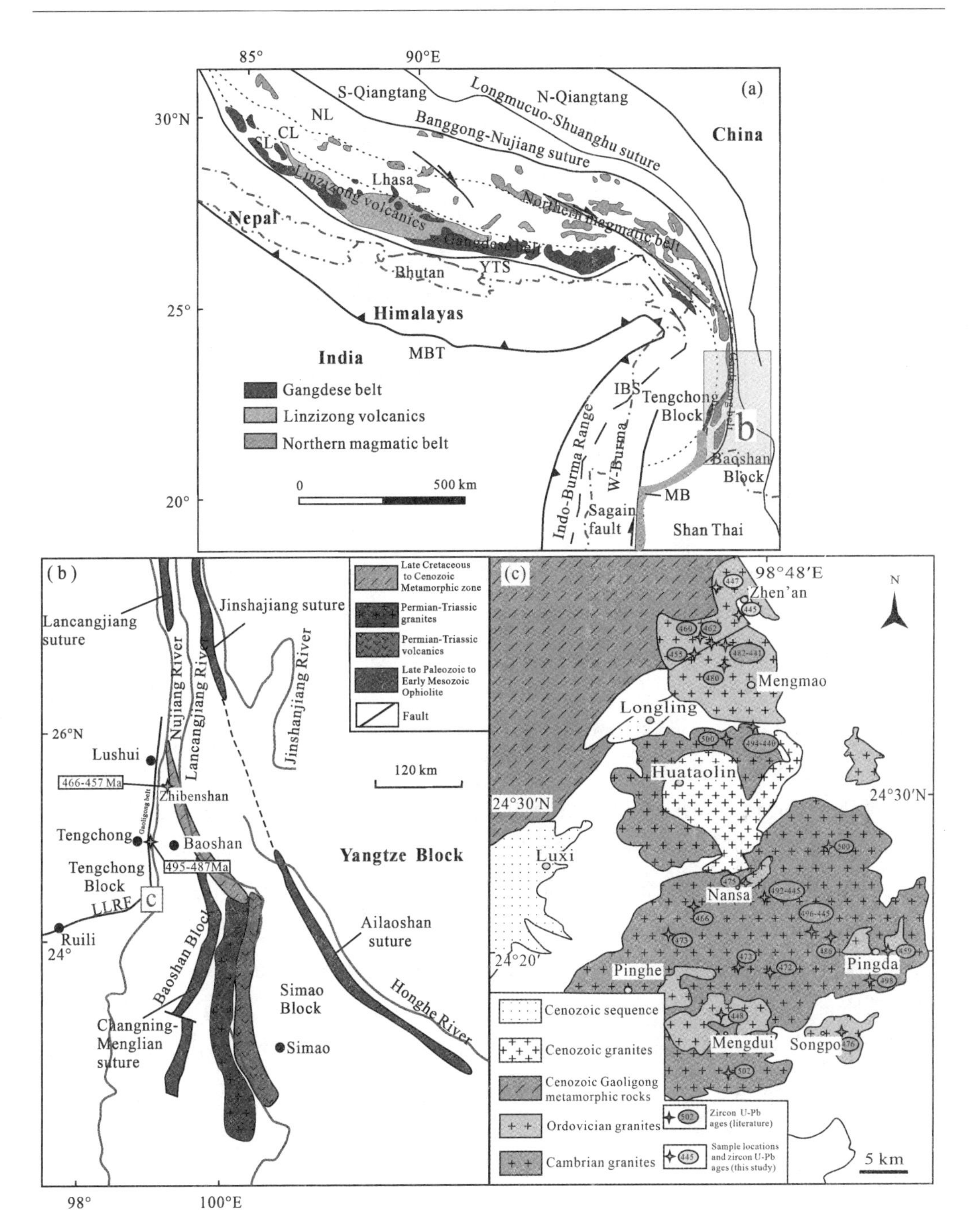

Fig. 1 (a) Simplified geological map of the Himalaya-Tibet tectonic realm (modified after Wang et al., 2013); (b) geological sketch of the Tethyan belt in western Yunnan (modified after Zhong, 1998; Chen et al., 2007); (c) simplified geological map of the early Paleozoic granitic batholith and intrusions in Baoshan Block.

Age data from Dong et al. (2012, 2013), Li et al. (2016), Wang et al. (2013, 2015b), Zhao et al. (2014, 2016), Xiong et al. (2012). MBT: Main Boundary Thrust; YTS: Yarlung-Tsangpo suture; SL, CL, NL: Southern, Central, Northern Lhasa Block; IBS: Indo-Burma suture; MB: Mogok metamorphic belt.

northern tip of the Sibumasu terrane (Metcalfe, 1996a,b, 1998, 2013), which is the eastern part of the Cimmerian continent (Sengör, 1984).

Paleozoic to Cenozoic granitic and mafic rocks are abundant in the Baoshan and Tengchong blocks, and the Gaoligong belt (Chen et al., 2007, 2015; Liu et al., 2009; Cong et al., 2011; Dong et al., 2012, 2013; Xu et al., 2012; Yang et al., 2012; Wang et al., 2013; Ma et al., 2014, 2015a; Qi et al., 2015; Xing et al., 2015; Xie et al., 2016; Li et al., 2016), and some of these had previously been considered to represent a Proterozoic basement (BGMRYP, 1990). However, recent high-precision zircon U-Pb dating and geochemical analyses have shown that these granites are closely related to the evolution of the Tethys (Xu et al., 2012; Ma et al., 2014; Wang et al., 2014, 2015b; Chen et al., 2015; Qi et al., 2015; Zhao et al., 2016). The Pinghe pluton in the western part of the Baoshan Block is the largest early Paleozoic granitic batholith in the block (Chen et al., 2007; Liu et al., 2009; Dong et al., 2012, 2013; Wang et al., 2013; Zhao et al., 2014), and these Paleozoic granites were intruded by late Mesozoic granitoid stocks (Chen et al., 2007). U-Pb dating of zircons in the monzogranites of the Pinghe batholith indicate two periods of early Paleozoic magmatism at ca. 490 Ma and 440 Ma (Zhao et al., 2014).

In this paper, we focus our attention on the Late Ordovician Zhen'an granitic intrusion that is located in the northern part of the Pinghe pluton. There is a fault that separates the Zhen'an intrusion in the north and the Mengmao intrusion in the south (Fig. 1; BGMRYP, 1990). The strike of the Zhen'an intrusion is parallel to the Gaoligong belt, and these granitic rocks have experienced strong Cenozoic shearing that resulted in a N-S trend foliation (Wang et al., 2006, 2008; Zhang et al., 2012). The rocks in this intrusion are deformed and exhibit gneissic structures, mylonitic textures, and mineral preferred orientations (Fig. 2), which was considered as the migmatite caused by regional thermal dynamo metamorphism (BGMRYP, 1990). However, recently precise zircon U-Pb dating and petrologic study indicate that the protolith of the so-termed migmatite is granitic rock and emplaced at the early Paleozoic (Dong et al., 2012; Xiong et al., 2012; Li et al., 2016). These rocks are syenogranite in composition (Fig. 3) and comprise phenocrysts of Kf (K-feldspar), Pth (perthite) and Pl (plagioclase) (15%) in a matrix of Qz (quartz, 25%), Pl (25%), Kf+Pth (27%) and Bi (biotite, 8%). The accessory minerals include zircon, apatite and titanite.

3 Analytical methods

3.1 Zircon U-Pb isotopic analyses

Zircon U-Pb isotopic analyses were undertaken at the State Key Laboratory of Continental Dynamics, Northwest University, Xi'an, China. Zircon grains were separated using

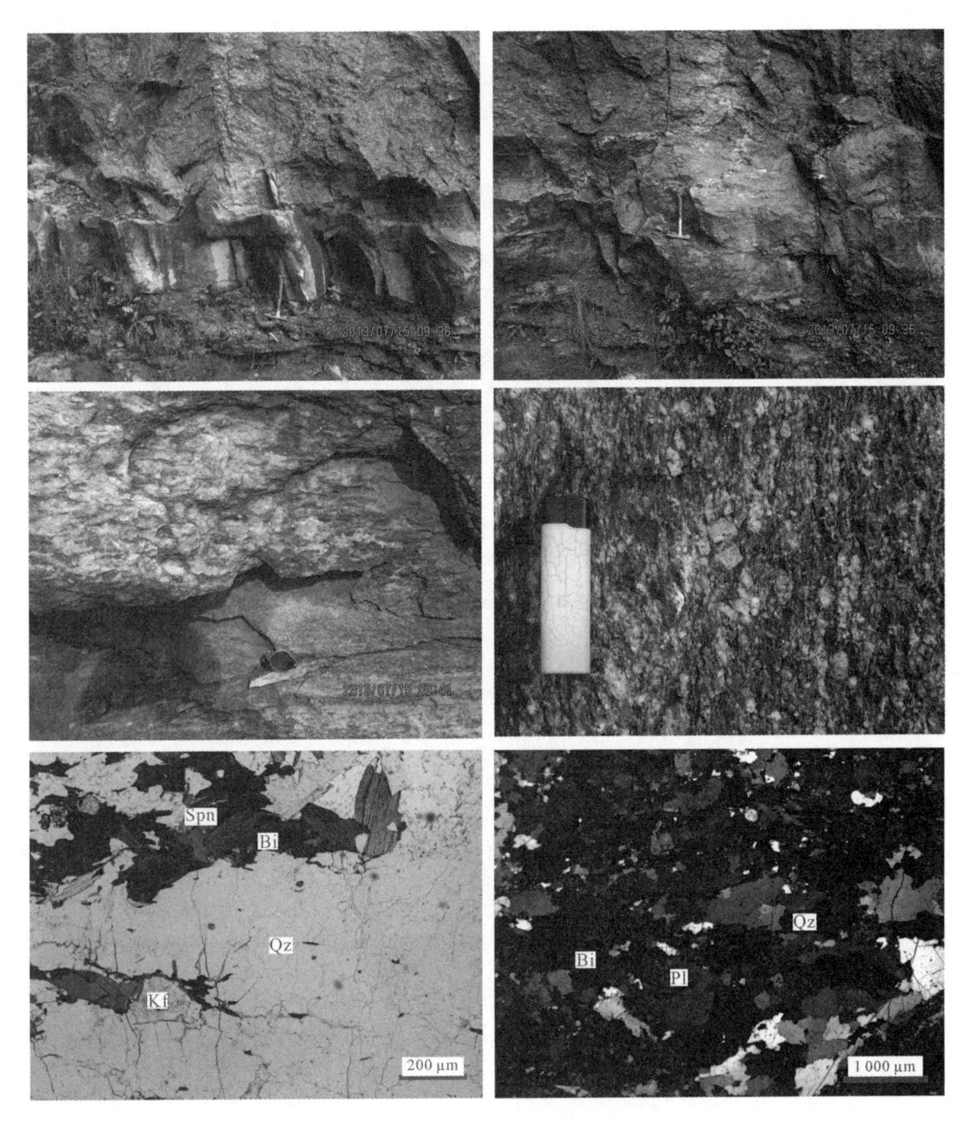

Fig. 2 Field, hand-specimen, and microscope photographs of granitic rocks from the Zhen'an intrusion, Baoshan Block.

Pl: plagioclase; Bi: biotite; Qz: quartz; Kf: K-feldspar; Spn: sphene.

conventional heavy liquid and magnetic separation techniques. Representative zircon grains were hand-picked under a binocular microscope and mounted in epoxy resin discs, then ground, polished, and coated with carbon. The morphology and internal structures of the zircons were examined using reflected light photomicrographs and cathodolu minescence (CL) images prior to isotope analysis.

Laser ablation inductively coupled plasma mass spectrometry (LA-ICP-MS) zircon U-Pb analyses were conducted using an Agilent 7500a ICP-MS equipped with a 193 nm wavelength ArF excimer laser. During analysis, the spot diameter was 30 μm, and we followed the methods

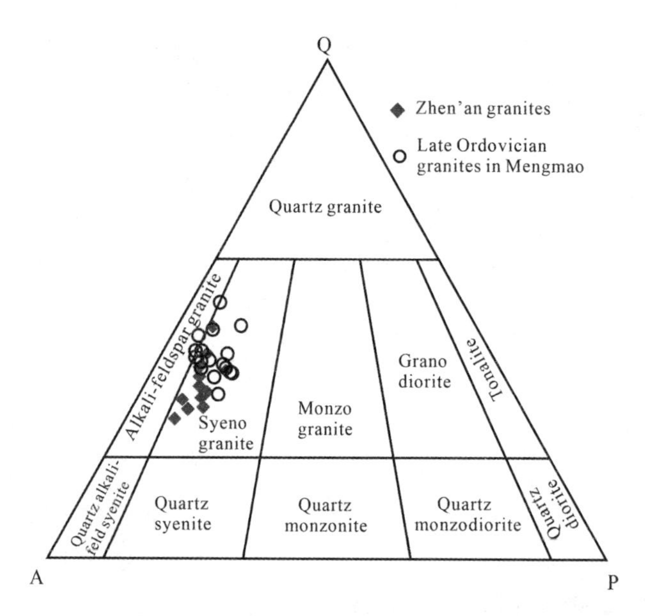

Fig. 3 Diagram of Q-A-P for granitic rocks from the Zhen'an intrusion, Baoshan Block.
Q: normative quartz; P: normative albite + anorthite; A: normative orthoclase. Data for Late Ordovician granite in Mengmao from Dong et al. (2012), Xiong et al. (2012), Li et al. (2016).

described by Yuan et al. (2004). The $^{207}Pb/^{235}U$ and $^{206}Pb/^{238}U$ ratios were calculated using the GLITTER program, which was corrected using the Harvard standard zircon 91500 for external calibration. Correction factors were applied to each sample in order to correct for both instrumental mass bias and depth-dependent elemental and isotopic fractionation. The techniques used followed those described by Yuan et al. (2004). Common Pb contents were evaluated using the method described in Andersen (2002). Age calculations and concordia diagrams were completed using ISOPLOT version 3.0 (Ludwig, 2003). The errors quoted in tables and figures are at the 2σ level.

3.2 Geochemical and isotopic analyses

Weathered surfaces of the granitoid samples were removed, and the fresh parts were then chipped and powdered to ~200 mesh using a tungsten carbide ball mill. Major and trace elements were analysed using X-ray fluorescence (XRF; Rikagu RIX 2100) and ICP-MS (Agilent 7500a), respectively, at the State Key Laboratory of Continental Dynamics, Northwest University. Analyses of USGS and Chinese national rock standards (BCR-2, GSR-1 and GSR-3) showed that the analytical precision and accuracy for the major elements were generally better than 5%. For trace element analyses, powdered samples were digested in an $HF + HNO_3$ mixture in high-pressure Teflon bombs at 190 ℃ for 48 h. For most trace elements, the analytical errors were less than 2% and the precision greater than 10% (Liu et al., 2007).

Whole-rock Sr-Nd isotopic data were obtained using a Nu Plasma HR multi-collector mass spectrometer at the Guizhou Tuopu Resource and Environmental Analysis Centre, Guiyang, China. Sr and Nd isotopic fractionations were corrected to $^{87}Sr/^{86}Sr = 0.119\,4$ and $^{146}Nd/^{144}Nd = 0.721\,9$, respectively. During the period of analysis, the NIST SRM 987 standard yielded a mean value of $^{87}Sr/^{86}Sr = 0.710\,250 \pm 12$ (2σ, $n = 15$) and the La Jolla standard gave a mean of $^{146}Nd/^{144}Nd = 0.511\,859 \pm 6$ (2σ, $n = 20$). The Nd model ages were calculated as two-stage model ages using present-day values of $(^{147}Sm/^{144}Nd)_{DM} = 0.213\,7$, $(^{147}Sm/^{144}Nd)_{DM} = 0.513\,15$, and $(^{147}Sm/^{144}Nd)_{crust} = 0.101\,2$. The values of $\varepsilon_{Nd}(t)$ were calculated present-day values of $(^{147}Sm/^{144}Nd)_{CHUR} = 0.196\,7$ and $(^{147}Sm/^{144}Nd)_{CHUR} = 0.512\,638$. Details of the formulae are given in Supplementary Table 3.

4 Results

4.1 Zircon U-Pb data

Two samples were collected from the Zhen'an intrusion for zircon U-Pb isotope analyses, and the results are shown in Fig. 4 and listed in Supplementary Table 1. The zircon grains were mainly prismatic with obvious oscillatory zoning on CL images. The crystals had lengths of 100–200 μm and aspect ratios of 1∶1.5 to 1∶3. Some zircons had light cores, indicative of zircon captured from the country rocks and/or inherited from the source region (Fig. 5).

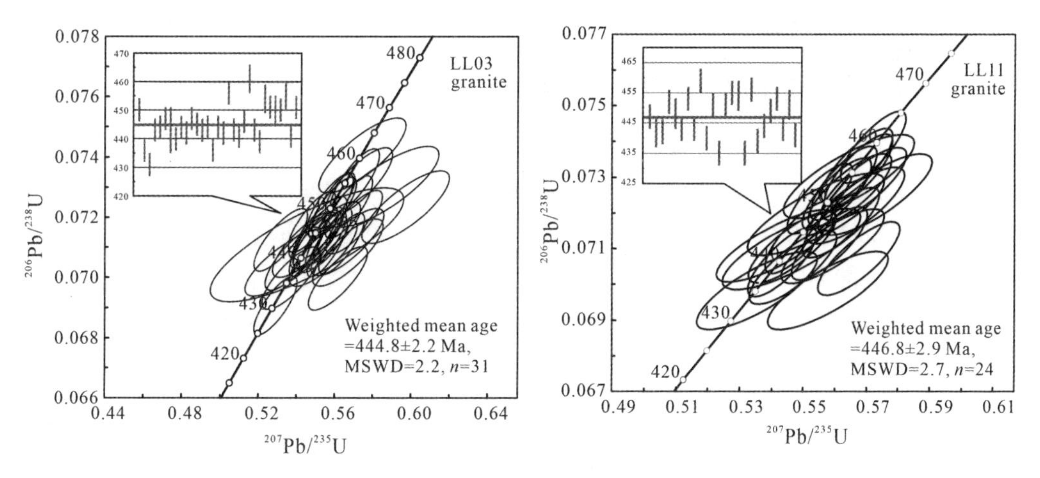

Fig. 4 U-Pb concordia diagrams and weighted mean ages of $^{206}Pb/^{238}U$ for zircons from the Zhen'an granitic intrusion, Baoshan Block.

Thirty-two spots were analysed on the zircons from sample LL03, and 31 grains yielded Late Ordovician ages with $^{206}Pb/^{238}U$ values of 431–461 Ma, a weighted mean age of 444.8 ± 2.2 Ma (MSWD = 2.2, $n = 31$), variable and high U (387–5 340 ppm) and Th (89–605 ppm) contents, and Th/U ratios of 0.08–0.95. One spot yielded a $^{206}Pb/^{238}U$ age of 864 ±

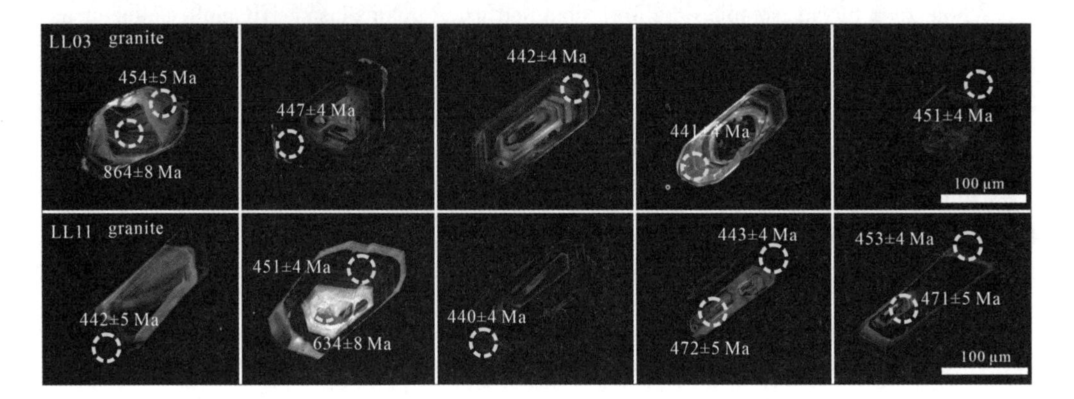

Fig. 5 CL images of representative zircons for granitic rocks from Zhen'an intrusion, Baoshan Block.

8 Ma, which we interpret as representing a captured and/or inherited zircon.

Twenty-four grains from sample LL11 yielded Late Ordovician ages. These grains had U contents of 163–3 157 ppm, Th contents of 77–1 303 ppm, Th/U ratios of 0. 11–0. 85, and $^{206}Pb/^{238}U$ ages of 435–459 Ma, which gave a weighted mean age of 446. 8 ± 2. 9 Ma (MSWD = 2. 7, n = 24). We interpret the age as the time of crystallisation of the Zhen'an granitoid intrusion. Six other grains from this sample yielded ages ranging from 471 Ma to 730 Ma, and these represent captured and/or inherited zircon cores according to the grain morphologies.

In summary, the Zhen'an intrusion emplaced at Late Ordovician, with zircon U-Pb age of ca. 445 Ma.

4. 2 Geochemical and isotopic data

The results of major and trace element analyses and whole-rock Sr-Nd isotopic data for granites from the Zhen'an intrusion are listed in Supplementary Tables 2 and 3. The granites are relatively enriched in SiO_2(71. 49–76. 68 wt%). The total-alkali contents (6. 85–9. 44 wt%) are high and variable, with relatively high K_2O contents of 3. 37–6. 33 wt% and relatively low Na_2O contents of 2. 02–4. 62 wt%. Moreover, the granites have low MgO contents (0. 22–0. 57 wt%), $Mg^{\#}$ values of 27 – 34, TiO_2 contents of 0. 13 – 0. 31 wt%, and relatively high CaO contents of 0. 88 – 1. 49 wt%. The geochemical signature shows that these are high-K, calc-alkaline and metaluminous to strongly peraluminous granites with A/CNK ratios of 0. 93–1. 18 (Fig. 6). Trace elemental data show that the granites have high Rb (199 – 271 ppm) and extremely low Sr (45. 6–79. 6 ppm) contents, which give high Rb/Sr ratios of 2. 8–6. 0. These granites also have high Y (24. 7 – 43. 7 ppm) contents and relatively low concentrations of HFSEs (e.g. P, Ti). A strong enrichment in LILEs is indicated on the N-MORB-normalised trace element spider diagram (Fig. 7a), and the granites have high REE contents of 125–210 ppm and strong negative Eu anomalies with δEu values of 0. 24–0. 5 (Fig. 7b).

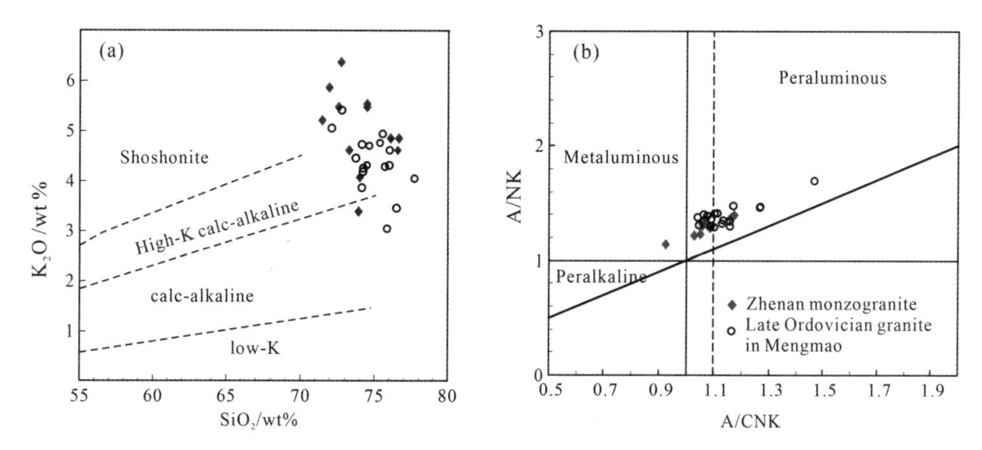

Fig. 6 Diagrams of K_2O vs. SiO_2 after Rollinson(1993) (a) and A/NK[molar $Al_2O_3/(Na_2O+K_2O)$] vs. A/CNK [molar $Al_2O_3/(CaO+Na_2O+K_2O)$] (Maniar and Piccoli, 1989) (b) for granitic rocks in Zhen'an intrusion.

Data sources as in Fig. 3.

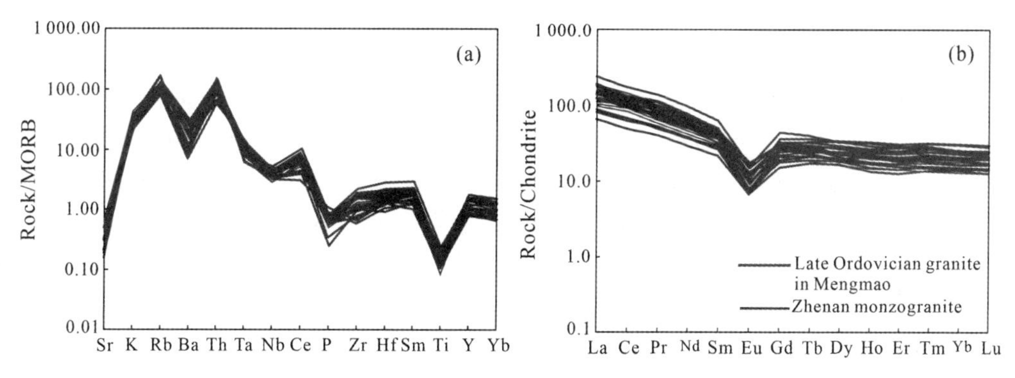

Fig. 7 N-MORB-normalized trace element spider (a) and chondrite-normalized REE pattern (b) diagrams for samples from the Zhen'an intrusion, Baoshan Block.

The N-MORB and chondrite values are from Pearce (1983) and Sun and McDonough (1989), respectively. Data sources as in Fig. 3.

All the isotopic data were calculated at $t = 440$ Ma on the basis of the LA-ICP-MS U-Pb zircon ages. Whole-rock isotopic data for the representative samples show that these granites have variable initial Sr ratios. Wu et al. (2000) reported that the high-Rb/Sr rocks commonly yield widely scattered and sometimes unreasonably low initial Sr ratios of less than 0.700, all with very large errors. Consequently, these authors chose samples with $^{87}Rb/^{86}Sr < 10$ in order to obtain more precise values of I_{Sr}. Removing our samples with $^{87}Rb/^{86}Sr > 10$, the available and precise initial Sr values for these granites are 0.719 761 – 0.726 754. They also have concordant $\varepsilon_{Nd}(t)$ values of −8.3 to −6.6, with corresponding two-model ages of 1.52 – 1.64 Ga, indicating ancient crustal sources.

5 Discussion

5.1 Petrogenesis of Late Ordovician granites

5.1.1 Magmatic type and evolution

The Late Ordovician granites in the Zhen'an intrusion have high SiO_2 contents of 72.01–76.68 wt%, and K_2O contents of 3.37–6.33 wt%. They belong to high-K calc-alkaline, and metaluminous to strong peraluminous series (Fig. 6), and normative-CIPW calculation indicate corundum contents of 0.63 – 2.33 wt%, except for sample LL25 (no corundum). These signatures reveal that the granites in the Zhen'an intrusion are likely to be S-type granites (Clemens, 2003), similar to the coeval and congenetic granites in the Mengmao intrusion, Baoshan Block (Dong et al., 2012; Xiong et al., 2012; Li et al., 2016). The granitic rocks in the Zhen'an intrusion show strong troughing in Sr, Ba, Nb, P, Ti, and Eu (Fig. 6). These geochemical signatures could have been caused by the fractionation of early crystallized mineral in the granitic magma. In combination with data for the Mengmao intrusion, fractional crystallization is supported by the geochemical trend of diagrams K_2O vs. Rb and Th vs. Th/Nd (Fig. 8a,b). The negative Nb and Ti anomalies were attributed to fractionation of the Ti-bearing phase (e.g., ilmenite). The significant negative P anomaly resulted from the fractionation of apatite. The pronounced negative Sr, Ba and Eu values indicated the crystallization differentiation of plagioclase, K-feldspar, and biotite. It is further evidenced by the diagrams of Sr vs. Rb and Ba, and Eu^* vs. Sr and Ba (Fig. 8c,d) that K-feldspar plays a more important role than plagioclase and biotite in controlling Sr and Ba contents during fractionation. Simply put, crystallization differentiation was accompanied by evolution of the Zhen'an granitic magmas.

5.1.2 Characteristics of sources

The isotopic data of Zhen'an granites show enrichments in Nd isotopes, with negative $\varepsilon_{Nd}(t)$ values of −8.0 to −6.6 and two-stage model ages of 1.52–1.62 Ga. These results, coupled with the high initial $^{87}Sr/^{86}Sr$ ratios, indicate ancient crustal sources (Fig. 9). As inferred from Figs. 3 and 8, the components of Zhen'an granites could be mostly close to the primary granitic magma. These granites have high ratios of CaO/Na_2O (0.3–0.47), Rb/Ba and Rb/Sr. Sylvester (1998) demonstrated that strongly peraluminous granites derived from pelites tend to have lower CaO/Na_2O ratios (<0.3) than their psammite-derived counterparts. Taking the isotopic data into account, the Zhen'an granites may therefore have been derived from the partial melting of metagreywacke, as indicated by geochemical diagrams (Fig. 10a,c). In the Rb/Sr vs. Rb/Ba diagram (Fig. 10b), the data plotting into the clay-rich field could have been caused by the fractionated crystallization process that altered the action of

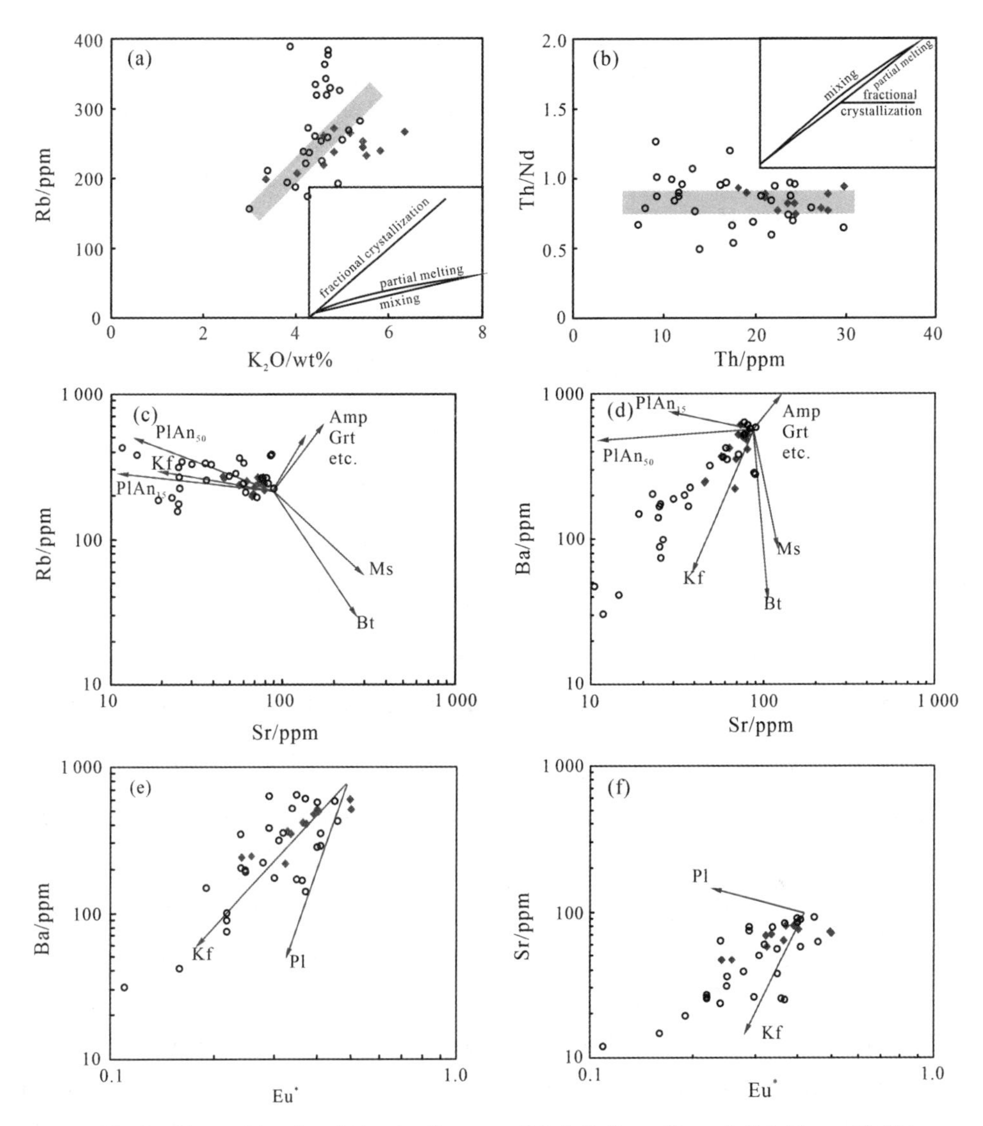

Fig. 8 Compositional variation in diagrams of (a) K_2O vs. Rb and (b) Th vs. Th/Nd after Schiano et al. (2010); (c,d) Sr vs. Rb and Ba, Eu^* vs. Sr and Ba after Janoušek et al. (2004) for the Zhen'an granites.

Pl: plagioclase; Kf: K-feldspar; Bt: biotite; Ms: muscovite; Grt: garnet; Amp: amphibole. Data sources as in Fig. 3.

mobile elements (Lee and Morton, 2015). The Al_2O_3/TiO_2 ratios of granites can be used as an indicator of the partial melting temperature (Sylvester, 1998), with high Al_2O_3/TiO_2 ratios (>100) corresponding to low-temperature granite melts and low Al_2O_3/TiO_2 ratios (<100) corresponding to high-temperature melts. The Zhen'an granites have Al_2O_3/TiO_2 ratios of 45–97, suggesting that the magma was produced at the high-temperature condition. This is further evidenced by the zircon saturation temperature (Watson and Harrison, 1983) of the granites, with values of 742–810 ℃.

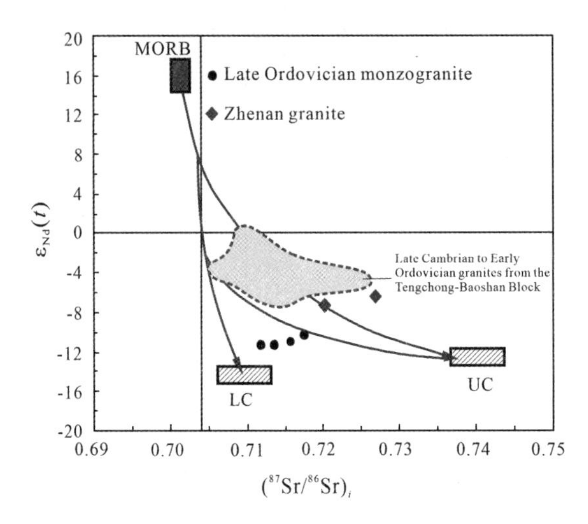

Fig. 9 $(^{87}Sr/^{86}Sr)_i$ vs. $\varepsilon_{Nd}(t)$ diagram for Zhen'an granites in the Baoshan Block.
The field of late Cambrian to Early Ordovician monzogranites is modified after Zhao et al. (2016). Data for Late Ordovician monzogranite from Wang et al. (2015a).

5.2 Early Paleozoic continuous magmatism along the Gondwana margin

Zircon U-Pb dating indicates that the Zhen'an intrusion formed at the Late Ordovician, with ages of ca. 445 Ma. Early Paleozoic magmatism in the Baoshan-Tengchong Block includes the Pinghe monzogranites at ca. 490 Ma (Chen et al., 2007; Liu et al., 2009; Dong et al., 2012; Wang et al., 2013); Nasa, Songpo, Pingda, and Mengdui two-mica granites at ca. 448–476 Ma; Zhen'an and Mengmao granites at ca. 445–480 Ma (this study; Chen et al., 2007; Dong et al., 2012; Xiong et al., 2012; Li et al., 2016); Shuangmaidi granites at ca. 460–470 Ma (Li et al., 2016); Zhibenshan monzogranites at ca. 457–466 Ma (Wang et al., 2015a); metabasalts in the Bangmai area at ca. 499 Ma (Yang et al., 2012), high-Mg mafic rocks in the Huimian and Manlai area at ca. 454–462 Ma (Xing et al., 2015); and gneissic granitic rocks in the Tengchong Block with ages from 487 Ma to 500 Ma (Song et al., 2007; Li et al., 2012; Wang et al., 2013; Zhao et al., 2016). In addition, arc-related calc-alkaline plutonism, rhyolitic-andesitic volcanism, and low- to medium-grade metamorphism (550–525 Ma) are evident in central Iran and Turkey (Ramezani and Tucker, 2003; Gessner et al., 2004). The early Paleozoic basaltic and acidic metavolcanic rocks occur in the Lhasa Block (Zhu et al., 2012, 2013; Hu et al., 2013; Ding et al., 2015), and granites in the Himalaya region (Cawood et al., 2007; Wang et al., 2012) and Qiangtang Block (Pullen et al., 2011). Coupled with early Paleozoic magmatism in central Iran, Turkey, the Himalayan area, Lhasa, Qiangtang, Baoshan-Tengchong, and Sibumasu (e.g., Stöcklin and Bhattarari, 1977; Garzanti et al., 1986; Brookfield, 1993; Valdiya, 1995; DeCelles et al., 2000; Gehrels et

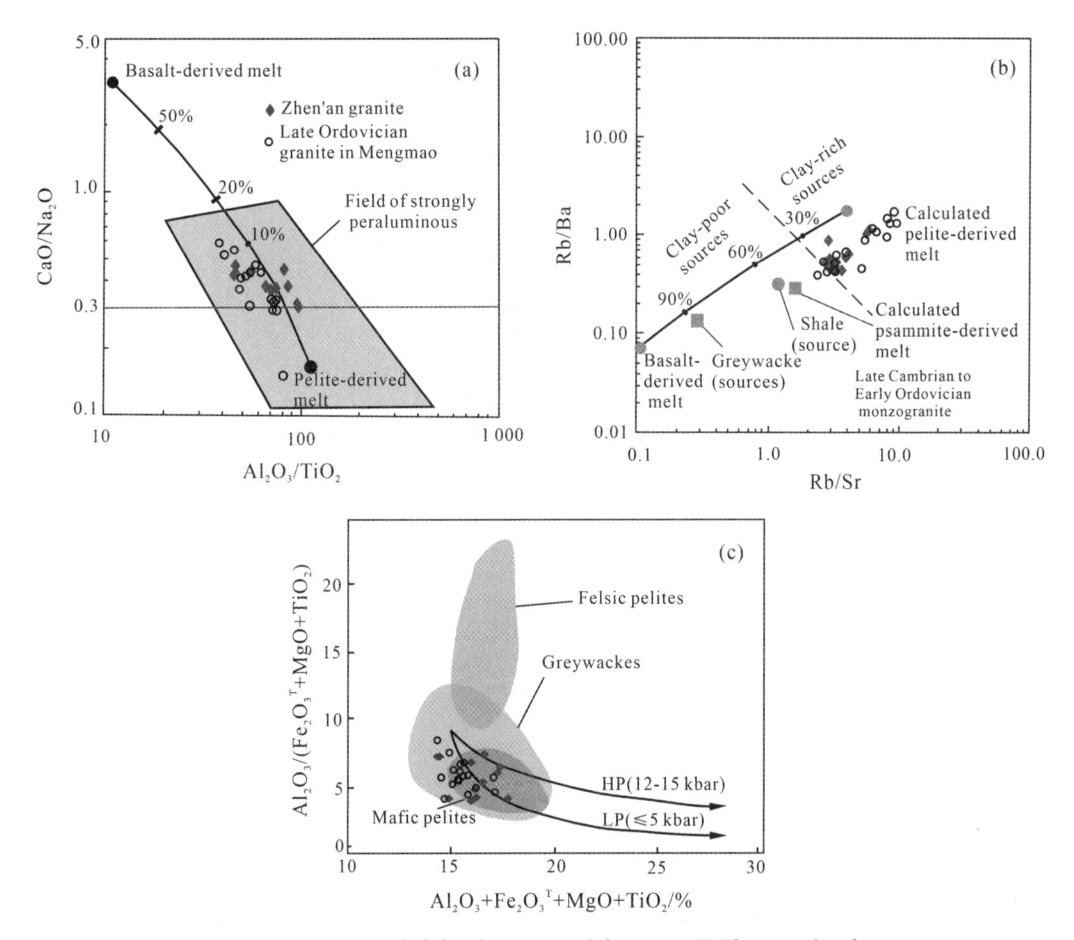

Fig. 10 Relative probability histogram of the zircon U-Pb ages of early Palaeozoic granitic rocks in the Baoshan-Tengchong Block.

Data sources from Chen et al. (2007), Liu et al. (2009), Dong et al. (2012, 2013), Song et al. (2007), Wang et al. (2013, 2015b), Zhao et al. (2014, 2015), Wang et al. (2015a), Xiong et al. (2012), Li et al. (2016).

al., 2006a,b, 2011; Myrow et al., 2006a,b; Pullen et al., 2011; Wang et al., 2012; Zhu et al., 2012, 2013; Hu et al., 2013; Lin et al., 2013; Ding et al., 2015), it is clear that there was extensive magmatism along the margin of the East Gondwana continent that was related to evolution of the Proto-Tethys and orogenesis during late Cambrian to Ordovician.

Although the early Paleozoic magmatism in the East Gondwana margin is successive, statistical data from the Baoshan-Tengchong Block show that it could have occurred in two stages of flare-up of granitic magmatism, at ca. 490 Ma and 460 Ma (Fig. 11), and Late Ordovician magmatism was limited in the Baoshan Block. It is also interesting that the corresponding two stages of mafic rocks occurred in the Baoshan Block at ca. 499 Ma and 462 Ma (Yang et al., 2012; Xing et al., 2015), and the absence of mafic magmatism in the Baoshan Block from 499 Ma to 462 Ma.

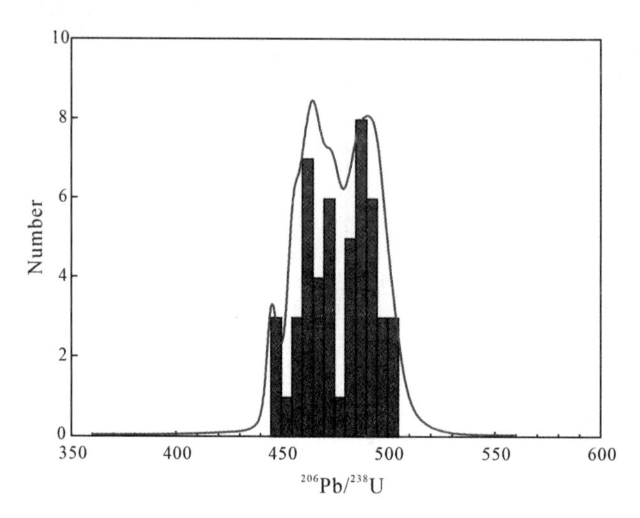

Fig. 11 Diagrams of Al_2O_3/TiO_2 vs. CaO/Na_2O (a), Rb/Sr vs. Rb/Ba (b), $Al_2O_3+Fe_2O_3{}^T+MgO+TiO_2$ vs. $Al_2O_3/(Fe_2O_3{}^T+MgO+TiO_2)$ (c) for Late Ordovician granites in Baoshan Block after Sylvester (1998) and Patiño Douce (1999), respectively. Data sources as in Fig. 3.

5.3 Tectonic setting

The late Cambrian to Early Ordovician igneous rocks produced in the margin of East Gondwana (e.g., central Iran, Turkey, Himalaya, Lhasa, Qiangtang, Tengchong, Baoshan) include not only basic to acidic volcanic and plutonic rocks, but also the corresponding metamorphic rocks, and these rocks have ages ranging from 530 Ma to 490 Ma (e.g., Stöcklin and Bhattarari, 1977; DeCelles et al., 2000; Gehrels et al., 2006a,b, 2011; Cawood et al., 2007; Chen et al., 2007; Liu et al., 2009; Dong et al., 2012, 2013; Wang et al., 2012, 2013; Zhu et al., 2012, 2013; Hu et al., 2013; Zhao et al., 2014, 2016; Ding et al., 2015). Two possible tectonic regimes have been proposed as the context for this late Cambrian to Early Ordovician magmatism: ① a late extensional stage of the long-lasting Pan-African orogenic cycle, which ended with the formation of the Gondwana supercontinent (Girard and Bussy, 1999; Miller et al., 2001); and ② an Andean-type active continental margin related to the subduction of Proto-Tethyan oceanic lithosphere and the assembly of outboard micro-continents (e.g., Cawood et al., 2007; Wang et al., 2012, 2013; Zhu et al., 2012, 2013; Hu et al., 2013; Zhao et al., 2014, 2016; Ding et al., 2015). There is much evidence in support of an Andean-type orogen following subduction of the Proto-Tethys oceanic lithosphere along the margin of East Gondwana during the late Cambrian to Early Ordovician, including arc magmatism (Cawood et al., 2007; Zhu et al., 2012, 2013; Zhao et al., 2014). Early Paleozoic angular unconformities and basal conglomerates typify the Cambrian-Early Ordovician and Middle-Late Ordovician strata in the Baoshan Block (Huang et al., 2009, 2012; Cai et al., 2013), and also the Cambrian and Ordovician strata in various areas of the Himalaya, the southern Qiangtang and Lhasa Blocks (Le Fort et al., 1994; Valdiya, 1995; Liu et al., 2002; Zhou et al., 2004; Fig. 9 in; Wang et al., 2015a). Altogether, therefore, the early Paleozoic (530–490 Ma) arc magmatism and stratigraphy for the margin of East Gondwana in the Himalaya, Lhasa, Qiangtang, Tengchong and Baoshan blocks clearly indicate the development of an orogen after closure of the Proto-Tethys during the early Paleozoic. The late Cambrian to Early Ordovician mafic and granitic rocks (ca. 500–490 Ma) in the Baoshan Block represent the continuation of an early Paleozoic granitic belt that extended along the margin of East Gondwana (Wang et al., 2013), with synchronous igneous rocks forming in the Himalaya, Qiangtang and Lhasa Blocks (Cawood et al., 2007; Wang et al., 2012; Zhu et al., 2012; Ding et al., 2015). These igneous rocks have been interpreted as the result of crustal anatexis induced by the subduction and break-off of the Proto-Tethyan oceanic slab below an Andean-type active continental margin (Fig. 12a; Zhu et al., 2012; Dong et al., 2013; Wang et al., 2013; Zhao et al., 2014, 2016).

The crust and lithospheric mantle thickening and shortening, caused by the amalgamation of the East Gondwana margin and outward micro continent, took place at 490–470 Ma (Fig. 12 b; Gehrel et al., 2006b; Cawood et al., 2007). This can be further confirmed by the absence of the Early Ordovician strata (ca. 490–470 Ma) and the presence of Late Ordovician strata in the Baoshan Block (Huang et al., 2009, 2012; Cai et al., 2013), and within-plate basalt with ages of ca. 457 Ma in the Greater Himalaya (Visonà et al., 2010).

Xing et al. (2015) show demonstrated high-Mg igneous rocks in the Baoshan Block, with ages of 454–462 Ma, and they proposed that the mafic rocks were generated in a Late Ordovician island arc setting related to subduction of the Proto-Tethyan Ocean along the margin of East Gondwana and partial melting of metasomatized mantle source involving the subducted sediment-derived components. In fact, the high-Mg igneous rocks experienced a complex evolution of primary magma, having variable SiO_2 content of 53. 57–69. 10 wt%, MgO content of 3. 21–9. 12 wt%, Cr content of 41–1 236 ppm, and Ni content of 30–597 ppm. Thus, the samples with lowest SiO_2 and highest MgO may represent the primary melt. Based on both their Sr-Nd isotopic and geochemical data, we consider that the high-$Mg^{\#}$ mafic rocks in Baoshan Block were due to partial melting of the delaminated thickened lithospheric mantle and/or lower crust and the melt can be equilibrated with the mantle materials leading to enhanced MgO contents and $Mg^{\#}$ values during ascent (Xu et al., 2008), because: ① closure of the Proto-Tethys was suggested as occurring before 490 Ma (e.g., Wang et al., 2012; Zhu et al., 2012; Li et al., 2016) and then the Baoshan Block in the East Gondwana margin could be in the orogenic stage and crust and lithospheric mantle thickening after 490 Ma; and ② a sufficiently

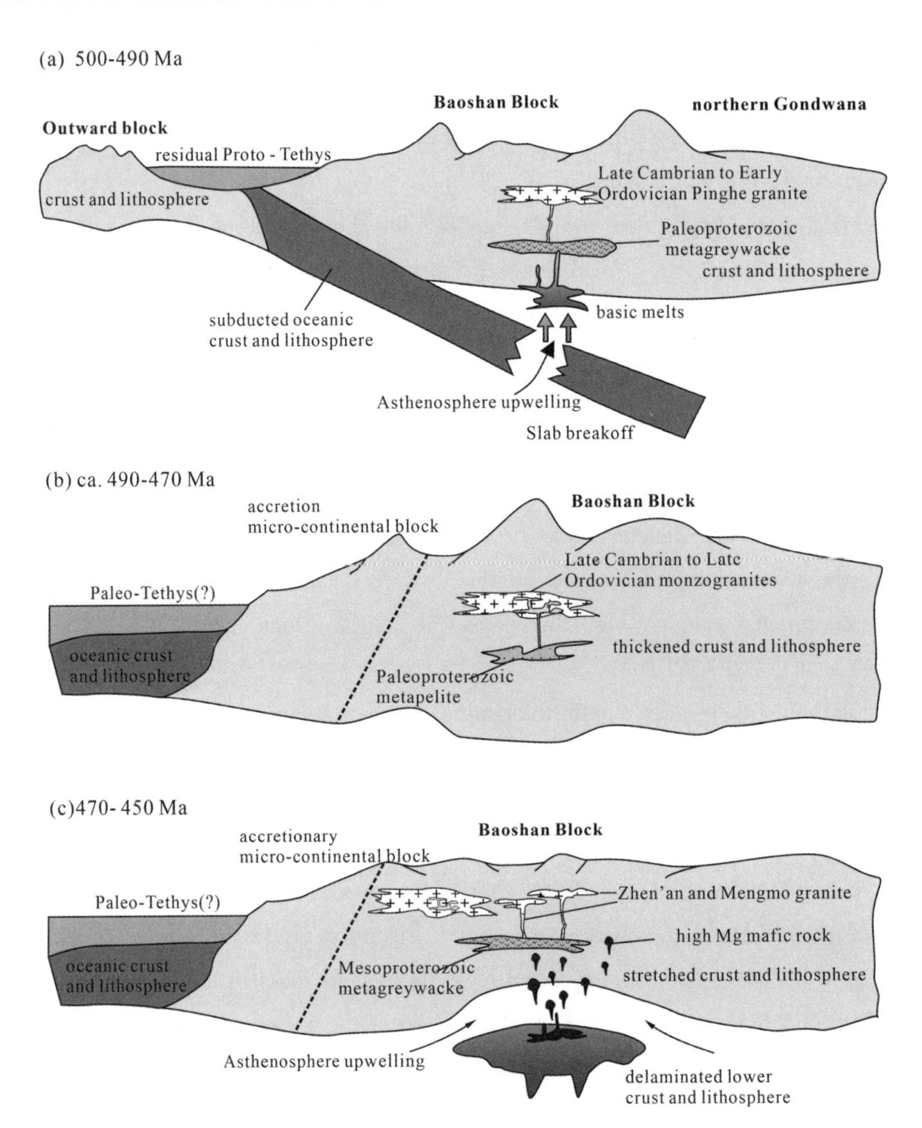

Fig. 12 Schematic illustrations showing the petrogenesis and tectonic setting for the late Cambrian to Early Ordovician magmatism along the East Gondwana margin.

high temperature is required to induce melting of the metasomatized mantle wedge (Fallon and Crawford, 1991; Macpherson and Hall, 2001), and this would have occurred by delamination of an already thickened lower crust and lithospheric mantle followed by sinking into the asthenosphere to induce rising temperature and melting. Due to the thickened crust and lithospheric mantle at 490–470 Ma (Gehrels et al., 2006b; Cawood et al., 2007; Huang et al., 2012; Cai et al., 2013) and the Late Ordovician mafic and granitic rocks at 462–445 Ma (Visonà et al., 2010; Li et al., 2016; Xiong et al. 2012; Xing et al., 2015; Zhao et al., 2014; this study), delamination of the thickened lower curst and lithospheric mantle and partial melting could have taken place at 470–450 Ma (Fig. 12c). We mentioned that the two stages of flare-up of granitic and mafic magmatism in the Baoshan Block occurred at 490 Ma and 460 Ma (Fig. 11; Yang et al., 2012; Xing et al., 2015), and could have been caused by subducted slab break-off of the Proto-Tethyan Ocean and delamination of thickened lower crust and lithospheric mantle in the Baoshan Block, respectively. The end of the mafic magmatism between 499 Ma and 462 Ma could have been caused by amalgamation of the Baoshan Block and outward micro-continent (South China?), because collision could not have provided effective heat and sufficent H_2O to induce the melting of mantle and thus an effective mechanism resulting in the decompression melting of mantle. The presence of granitic rocks at 490–470 Ma can be derived from partial melting of ancient crustal materials caused by the compression (Niu, 2013). Furthermore, the intrusive Late Ordovician mafic magma (Xing et al., 2015) could have provided the heat necessary for crusting and thus led to the melting of crustal material to produce the Late Ordovician granitic magma, such as the Zhen'an, Mengmao, and Zhibenshan granites (Xiong et al., 2012; Wang et al., 2015; Li et al., 2016; this study).

6 Conclusions

(1) The Zhen'an intrusion in the Baoshan Block was emplaced during the Late Ordovician at ca. 445 Ma, and the intrusive rocks are high-K, calc-alkaline, metaluminous to strongly peraluminous granites with crustal isotopic characteristics. The Late Ordovician granitic magma in the Baoshan Block experienced crystallization fractionation, and could be derived from the partial melting of metagreywacke.

(2) The early Paleozoic continuous granitic rocks in the Baoshan Block show two stages of flare-up, at ca. 490 Ma and 460 Ma. These two stages of flare-up of the magmatism could represent different tectonic regimes along the margin of East Gondwana.

(3) The Late Cambrian to Early Ordovician (ca. 500–490 Ma) magmas probably formed along an Andean-type active continental margin of East Gondwana in response to the subduction and break-off of a Proto-Tethyan oceanic slab. The break-off of this slab resulted in asthenospheric upwelling and the underplating of mafic magma, thus inducing partial melting of crustal materials. Then, the crust and lithosphere could have been thickened at ca. 490–470 Ma. The Late Ordovician (470–450 Ma) granitic magmas resulted from the delamination of an already thickened crust and lithospheric mantle, decompression and injection of the mafic magma that was derived from partial melting of delamination of thickened crust and lithosphere, leading to the partial melting of crustal materials.

Acknowledgements This work was jointly supported by the National Natural Science Foundation of China (Grant Nos 41190072 and 41421002) program for Changjiang Scholars and Innovative Research Team at the University (Grant IRT 1281).

References

Andersen, T., 2002. Correction of common lead in U-Pb analyses that do not report ^{204}Pb. Chemical Geology, 192, 59-79.

Brookfield, M.E., 1993. The Himalayan passive margin from Precambrian to Cretaceous times. Sedimentary Geology, 84, 1-35.

Bureau of Geology and Mineral Resources of Yunnan Province, 1990. Regional Geology of Yunnan Province. Beijing: Geological Publishing House, 729 (in Chinese).

Cai, Z.H., Xu, Z.Q., Duan, X.D., Li, H.Q., Cao, H., Huang, X.M., 2013. Early stage of early Paleozoic orogenic event in western Yunnan Province, southeastern margin of Tibet Plateau. Acta Petrologica Sinica, 29, 2123-2140 (in Chinese with English abstract).

Cawood, P.A., Johnson, M.R.W., Nemchin, A.A., 2007. Early Paleozoic orogenesis along the Indian margin of Gondwana: Tectonic response to Gondwana assembly. Earth & Planetary Science Letters, 255, 70-84.

Chen, F.K., Li, X.H., Wang, X.L., Li, Q.L., Siebel, W., 2007. Zircon age and Nd-Hf isotopic composition of the Yunnan Tethyan belt, southwestern China. International Journal of Earth Science, 96, 1179-1194.

Chen, X.C., Hu, R.Z., Bi, X.W., Zhong, H., Lan, J.B., Zhao, C.H., Zhu, J.J., 2015. Petrogenesis of metaluminous A-type granitoids in the Tengchong-Lianghe tin belt of southwestern China: Evidences from zircon U-Pb ages and Hf-O isotopes, and whole-rocks Sr-Nd isotopes. Lithos, 212-215, 93-110.

Clemens, J. D., 2003. S-type granitic magmas-Petrogenesis issues, model and evidence. Earth-Science Reviews, 61, 1-18.

Cong, F., Lin, S.L., Zou, G.F., Li, Z.H., Xie, T., Peng, Z.M., Liang, T., 2011. Magma mixing of granites at Lianghe: In-site zircon analysis for trace elements, U-Pb ages and Hf isotopes. Science China: Earth Sciences, 54, 1346-1359.

DeCelles, P.G., Gehrels, G.E., Najman, Y., Martin, A.J., Carter, A., Garzanti, E., 2004. Detrital geochronology and geochemistry of Cretaceous-Early Miocene strata of Nepal: Implications for timing and diachroneity of initial Himalayan orogenesis. Earth & Planetary Science Letters, 227, 313-330.

DeCelles, P.G., Gehrels, G.E., Quade, J., LaReau, B., Spurlin, M., 2000. Tectonic implications of U-Pb zircon ages of the Himalayan orogenic belt in Nepal. Science, 288, 497-499.

Ding, H.X., Zhang, Z.M., Dong, X., Yan, R., Lin, Y.H., Jiang, H.Y., 2015. Cambrian ultrapotassic rhyolites from the Lhasa terrane, south Tibet: Evidence for Andean-type magmatism along the northern active margin of Gondwana. Gondwana Research, 27, 1616-1629.

Dong, M.L., Dong, G.C., Mo, X.X., Santosh, M., Zhu, D.C., Yu, J.C., Nie, F., Hu, Z.C., 2013. Geochemistry, zircon U-Pb geochronology and Hf isotopes of granites in the Baoshan Block, Western Yunnan: Implications for early Paleozoic evolution along the Gondwana margin. Lithos, 179, 36-47.

Dong, M.L., Dong, G.C., Mo, X.X., Zhu, D.C., Nie, F., Xie, X.F., Wang, X., Hu, Z.C., 2012. Geochronology and geochemistry of the early Paleozoic granitoid in Baoshan block, western Yunnan and their implications. Acta Petrologica Sinica, 28(5), 1453-1464 (in Chinese with English abstract).

Eroğlu, S., Siebel, W., Danišík, M., Pfänder, J.A., Chen, F.K., 2013. Multi-system geochronological and isotopic constraints on age and evolution of the Gaoligongshan metamorphic belt and shear zone system in western Yunnan, China. Journal of Asian Earth Sciences, 73, 218-239.

Fallon, T.J., Crawford, A.J., 1991. The petrogenesis of high-calcium boninite lavas dredged from the northern Tonga ridge. Earth & Planetary Science Letters, 102(3-4), 375-394.

Fang, C.J., Zhang, Y.F., 1994. The structure and tectonics of western Yunnan. Journal of Southeast Asian Earth Sciences, 9, 355-361.

Feng, Q.L., 2002. Stratigraphy of volcanic rocks in the Changning-Menglian belt in southwestern Yunnan, China. Journal of Asian Earth Sciences, 20(6), 657-664.

Garzanti, E., Casnedi, R., Jadoul, F., 1986. Sedimentary evidence of a Cambro-Ordovician orogenic event in the northwestern Himalaya. Sedimentary Geology, 48, 237-265.

Gehrels, G.E., DeCelles, P.G., Ojha, T.P., Upreti, B.N., 2006a. Geological and U-Pb geochronological evidence for early Paleozoic tectonism in the Dadeldhura thrust sheet, far-west Nepal Himalaya. Journal of Asian Earth Sciences, 28, 385-408.

Gehrels, G.E., DeCelles, P.G., Ojha, T.P., Upreti, B.N., 2006b. Geologic and U-Th-Pb geochronologic evidence for early Paleozoic tectonism in the Kathmandu thrust sheet, central Nepal Himalaya. Geological Society of America Bulletin, 118, 185-198.

Gehrels, G.E., Kapp, P., DeCelles, P.G., Pullen, A., Blakey, R., Weislogel, A., Ding, L., Guynn, J., Martin, A., McQuarrie, N., Yin, A., 2011. Detrital zircon geochronology of pre-Tertiary strata in the Tibetan-Himalayan orogen. Tecotinics, 30, TC5016.

Gessner, K., Collins, A.S., Ring, U., Gungor, T., 2004. Structural and thermal history of poly-orogenic basement: U-Pb geochronology of granitoid rocks in the southern Menderes Massif, Western Turkey. Journal of the Geological Society, 161, 93-101.

Girard, M., Bussy, F., 1999. Late Pan-African magmatism in the Himalaya: New geochronological and geochemical data from the Ordovician Tso Morari metagranites (Ladakh, NW India). Schweizerische Mineralogische and Petroraphische Motterlungen, 79, 399-418.

Hu, P.Y., Li, C., Wang, M., Xie, C.M., Wu, Y.W., 2013, Cambrian volcanism in the Lhasa terrane, southern Tibet: Record of an early Paleozoic Andean-Type magmatic arc along the Gondwana proto-Tethyan margin. Journal of Asian Earth Sciences, 77, 91-107.

Huang, Y., Deng, G.B., Peng, C.L., Hao, J.X., Zhang, G.X., 2009. The discovery and significance of absence in early-middle Ordovician in southern Baoshan, western Yunnan. Guizhou Geology, 26(98), 1-6 (in Chinese with English abstract).

Huang, Y., Hao, J.X., Bai, L., Deng, G.B., Zhang, G.X., Huang, W.J., 2012. Stratigraphic and petrologic response to Late Pan-African movement in Shidian area, western Yunnan Province. Geological Bulletin of China, 31, 306-313 (in Chinese with English abstract).

Janoušek, V., Finger, F., Roberts, M., Fryda, J., Pin, C., Dolejš, D., 2004. Deciphering the petrogenesis of deeply buried granites: Whole-rock geochemical constraints on the origin of largely underpleted felsic granulites from the Moldanubian zone of the Bohemian Massif. Transactions of the Royal Society of Edinburgh: Earth Sciences, 95, 141-159.

Le Fort, P., Tongiorgi, M., Gaetani, M., 1994. Discovery of a crystalline basement and early Ordovician marine transgression in the Karakorum mountain range, Pakistan. Geology, 22, 941-944.

Lee, C.T.A., Morton, D.M., 2015. High silica granites: Terminal porosity and crystal setting in shallow magma chambers. Earth & Planetary Science Letters, 409, 23-31.

Li, G.J., Wang, Q.F., Huang, Y.H., Gao, L., Yu, L., 2016. Petrogenesis of middle Ordovician peraluminous granites in the Baoshan block: Implication for the early Paleozoic tectonic evolution along East Gondwana. Lithos, 245, 76-92.

Li, Z.H., Lin, S.L., Cong, F., Xie, T., Zou, G.F., 2012. U-Pb ages of zircon from metamorphic rocks of the Gaoligong Group in western Yunnan and its tectonic significance. Acta Petrologica Sinica, 28, 1529-1541 (in Chinese with English abstract).

Lin, Y.L., Yeh, M.W., Lee, T.Y., Chung, S.L., Iizuka, Y., Charusiri, P., 2013. First evidence of the Cambrian basement in Upper Peninsula of Thailand and its implication for crustal and tectonic evolution of the Sibumasu terrane. Gondwana Research, 24(3-4), 1031-1037.

Liu, S., Hu, R.Z., Gao, S., Feng, C.X., Huang, Z.L., Lai, S.C., Yuan, H.L., Liu, X.M., Coulson, I. M., Feng, G. Y., Wang, T., Qi, Y. Q., 2009. U-Pb zircon, geochemical and Sr-Nd-Hf isotopic constraints on the age and origin of Early Paleozoic I-type granite from the Tengchong-Baoshan Block, western Yunnan Province, SW China. Journal of Asian Earth Sciences, 36, 168-182.

Liu, W.C., Liang, D.Y., Wang, K.Y., Zhou, Z.G., Li, G.B., Zhang, X.X., 2002. The discovery of the Ordovician and its significance in the Kangmar area, southern Tibet. Earth Science Frontiers, 9, 247-248. (in Chinese with English abstract).

Liu, Y., Liu, X.M., Hu, Z.C., Diwu, C.R., Yuan, H.L., Gao, S., 2007. Evaluation of accuracy and long-term stability of determination of 37 trace elements in geological samples by ICP-MS. Acta Petrologica Sinica, 23(5), 1203-1210 (in Chinese with English abstract).

Ludwig, K.R., 2003. ISOPLOT 3.0: A geochronological toolkit for Microsoft Excel. Berkeley Geochronology Center, Special Publication, 4, 71.

Ma, L.Y., Wang, Y.J., Fan, W.M., Geng, H.Y., Cai, Y.F., Zhong, H., Liu, H.C., Xing, X.W., 2014. Petrogenesis of the early Eocene I-type granites in west Yingjiang (SW Yunnan) and its implications for the eastern extension of the Gangdese batholiths. Gondwana Research, 25, 401-419.

Macpherson, C.G., Hall, R., 2001. Tectonic setting of Eocene boninite magmatism in the Izu-Bonin-Mariana forearc. Earth & Planetary Science Letters, 186(2), 215-230.

Maniar, P.D., Piccoli, P.M., 1989. Tectonic discrimination of granitoids. Geological Society of American Bulletin, 101, 635-643.

Metcalfe, I., 1996a. Pre-Cretaceous evolution of SE Asian terranes. In: Hall, R., Blundell, D. Tectonic evolution of Southeast Asian, 37. Geology Society of London, Special Publication, 101-118.

Metcalfe, I., 1996b. Gondwanaland dispersion, Asian accretion and evolution of eastern Tethys. Australian Journal of Earth Sciences, 43, 605-623.

Metcalfe, I., 1998. Palaeozoic and Mesozoic geological evolution of the SE Asian region, multidisciplinary constraints and implications for biogeography. Amsterdam: Backbuys Publishers, 25-44.

Metcalfe, I., 2013. Gondwana dispersion and Asian accretion: Tectonic and palaeogeographic evolution of eastern Tethys. Journal of Asian Earth Sciences, 66, 1-33.

Miller, C., Thöni, M., Frank, W., Gramsemann, B., Klötzli, U., Guntli, P., Draganits, E., 2001. The early Paleozoic magmatic event in the Northwest Himalaya, India: Source, tectonic setting and age of emplacement. Geological Magazine, 138(3), 237-251.

Myrow, P.M., Snell, K.E., Hughes, N.C., Paulsen, T.S., Heim, N.A., Parcha, S.K., 2006a. Cambrian depositional history of the Zanskar valley region of the Indian Himalaya: Tectonic implications. Journal of Sedimentary Research, 76, 364-381.

Myrow, P.M., Thompson, K.R., Hughes, N.C., Paulsen, T.S., Sell, B.K., Parcha, S.K., 2006b. Cambrian stratigraphy and depositional history of the northern Indian Himalaya, Spiti Valley, north-central India. Geological Society of America Bulletin, 118, 491-510.

Niu, Y.L., 2013. Global tectonic and geodynamics: A petrological and geochemical approach. Beijing: Science Press, 307 (in Chinese).

Patiño Douce, A.E., 1999. What do experiments tell us about the relative contributions of crust and mantle to the origin of granitic magma? Geological Society, London, Special Publications, 168, 55-75.

Pearce, J.A., 1983. Role of the subcontinental lithosphere in magma genesis at active continental margin. In: Hawkesworth, C. J., Norry, M. J. Continental basalts and mantle xenoliths, U. K, Cheshire, Shiva, Nantwich, 230-249.

Pullen, A., Kapp, P., Gehrels, G.E., Ding, L., Zhang, Q.H., 2011. Metamorphic rocks in central Tibet: Lateral variations and implications for crustal structure. Geological Society of American Bulletin, 123, 585-600.

Qi, X.X., Zhu, L.H., Grimmer, J.C., Hu, Z.C., 2015. Tracing the Transhimalayan magmatic belt and the Lhasa block southward using zircon U-Pb, Lu-Hf isotopic and geochemical data: Cretaceous-Cenozoic granitoids in the Tengchong block, Yunnan, China. Journal of Asian Earth Sciences, 110, 170-188.

Ramezani, J., Tucker, R.D., 2003, The Saghand Region, Central Iran: U-Pb geochronology petrogenesis and implications for Gondwana Tectonics. American Journal of Science, 303, 622-665.

Rollinson, H., 1993, Using Geochemical data: Evaluation, presentation, interpretation. London: Longman Scientific & Technical.

Schiano, P., Monzier, M., Eissen, J.P., Martin, H., Koga, K.T., 2010. Simple mixing as the major control of the evolution of volcanic suites in the Ecuadorian Andes. Contributions to Mineralogy and Petrology, 160, 297-312.

Sengör, A.M.C., 1984. The Cimmeride orogenic system and the tectonic of Eurasia. Geological Society of American Special Paper, 195, 1-82.

Song, S.G., Ji, J.Q., Wei, C.J., Su, L., Zheng, Y.D., Song, B., Zhang, L.F., 2007. Early Paleozoic granite in Nujiang River of northwest Yunnan in southwestern China and its tectonic implications. Chinese Science Bulletin, 52, 2402-2406.

Stöcklin, J., Bhattarari, K.D., 1977. Geology of the Kathmandu area and central Mahabharat range. Report, 1-86.

Sun, S.S., McDonough, W.F., 1989. Chemical and isotopic systematics of oceanic basalts: Implications for mantle composition and processes. In Saunders, A.D., Norry, M.J. Magmatism in the Ocean Basins, 42. London, Geological Society Special Publication, 313-345.

Sylvester, P.J., 1998. Post-collisional strongly peraluminous granites. Lithos, 45, 29-44.

Ueno, K., 2000. Permian fusulinacean faunas of the Sibumasu and Baoshan block, implication for the paleogeographic reconstruction of the Cimmerian continent. Geoscience Journal, 4, 160-163.

Valdiya, K.S., 1995. Proterozoic sedimentation and Pan-African geodynamic development in the Himalaya. Precambrian Research, 74, 35-55.

Visonà, D., Rubatto, D., Villa, I.M., 2010. The mafic rocks of Shao La (Kharta, S. Tibet): Ordovician basaltic magmatism in the greater Himalayan crystallines of central-eastern Himalaya. Journal of Asian Earth Sciences, 38, 14-25.

von Raumer, J.F., Stampfli, G.M., 2008. The birth of the Rheic Ocean-Early Paleozoic subsidence patterns and subsequent tectonic plate scenarios. Tectonophysics, 461, 9-20.

Wang, C.M., Deng, J., Lu, Y.J., Bagas, L., Kemp, A.I.S., McCuaig, T.C., 2015a. Age, nature, and origin of Ordovician Zhibenshan granite from the Baoshan terrane in the Sanjiang region and its significance for understanding Proto-Tethys evolution. International Geology Review, 57, 1922-1939.

Wang, G., Wan, J.L., Wang, E., Zheng, D.W., Li, F., 2008. Late Cenozoic to recent transtensional deformation across the Southern part of the Gaoligong shear zone between the Indian plate and SE margin of the Tibetan plateau and its tectonic origin. Tectonophysics, 460, 1-20.

Wang, K.Y., 1996. The tectonic belt in the Sanjiang area of the Yunnan Province and rock units of the Precambrian basements and tectonic evolution along the western margin of the Yangtze block. Geology of Yunnan, 15, 138-148 (in Chinese with English abstract).

Wang, X.X., Zhang, J.J., Santosh, M., Liu, J., Yan, S.Y., Guo, L., 2012. Andean-type orogeny in the Himalayas of south Tibet: Implications for early Paleozoic tectonics along the Indian margin of Gondwana. Lithos, 154, 248-262.

Wang, Y.J., Fan, W.M., Zhang, Y.H., Peng, T.P., Chen, X.Y., Xu, Y.G., 2006. Kinematics and $^{40}Ar/^{39}Ar$ geochronology of the Gaoligong and Chongshan shear systems, western Yunnan, China: Implications for early Oligocene tectonic extrusion of SE Asia. Tectonophysics, 418, 235-254.

Wang, Y.J., Li, S.B., Ma, L.Y., Fan, W.M., Cai, Y.F., Zhang, Y.H., Zhang, F.F., 2015b. Geochronological and geochemical constraints on the petrogenesis of Early Eocene metagabbroic rocks in Nabang (SW Yunnan) and its implications on the Neotethyan slab subduction. Gondwana Research, 27, 1474-1486.

Wang, Y.J., Xing, X.X., Cawood, P.A., Lai, S.C., Xia, X.P., Fan, W.M., Liu, H.C., Zhang, F.F., 2013. Petrogenesis of early Paleozoic peraluminous granite in the Sibumasu Block of SW Yunnan and diachronous accretionary orogenesis along the northern margin of Gondwana. Lithos, 182-183, 67-85.

Wang, Y.J., Zhang, L.M., Cawood, P.A., Ma, L.Y., Fan, W.M., Zhang, A.M., Zhang, Y.Z., Bi, X.W., 2014. Eocene supra-subduction zone mafic magmatism in the Sibumasu block of SW Yunnan: Implications for Neotethyan subduction and India-Asia collision. Lithos, 206-207, 384-399.

Watson, E.B., Harrison, T.M., 1983. Zircon saturation revisited: Temperature and compositional effects in variety of crustal magma type. Earth & Planetary Science Letters, 64, 295-304.

Wopfner, H., 1996. Gondwana origin of the Baoshan and Tengchong terrenes of west Yunnan. In: Hall, R., Blundell, D. Tectonic evolution of Southeast Asian, 106:London,Journal of Geology Society, 539-547.

Wu, F.Y., Jahn, B.M., Wilde, S., Sunm D.Y., 2000. Phanerozoic crustal growth: U-Pb and Sr-Nd isotopic evidence from the granites in northeastern China. Tectonophysics, 328, 89-113.

Wu, H., Boulter, C.A., Ke, B., Stow, D.A.V., Wang, Z.C., 1995. The Changning-Menglian suture zone: A segment of the major Cathaysian-Gondwana divide in Southern Asian. Tectonophysics, 242, 267-280.

Xie, J.C., Zhu, D.C., Dong, G.C., Zhao, Z.D., Wang, Q., Mo, X.X., 2016. Linking the Tengchong Terrane in SW Yunnan with the Lhasa Terrane in southern Tibet through magmatic correlation. Gondwana Research (in press).

Xing, X.X., Wang, Y.J., Cawood, P.A., Zhang, Y.Z., 2015. Early Paleozoic accretionary orogenesis along northern margin of Gondwana constrained by high-Mg metaigneous rocks, SW Yunnan. International Journal of Earth Sciences (in press).

Xiong, C.L., Jia, X.C., Yang, X.J., Luo, G., Bai, X.Z., Huang, B.X., 2012. LA-ICP-MS zircon U-Pb dating of Ordovician Mengmao monzogranites in Longling area of western Yunnan Province and its tectonic setting. Geological Bulletin of China, 31, 277-286 (in Chinese with English abstract).

Xu, W., Hergt, J.M., Gao, S., Pei, F., Wang, W., Yang, D., 2008. Interaction of adakitic melt-peridotite: Implications for the high-$Mg^{\#}$ signature of Mesozoic adakitic rocks in the eastern North China Craton. Earth & Planetary Science Letters, 265(1-2), 123-137.

Xu, Y.G., Yang, Q.J., Lan, J.B., Luo, Z.Y., Huang, X.L., Shi, Y.B., Xie, L.W., 2012. Temporal-spatial distribution and tectonic implications of the batholiths in the Gaoligong-Tengliang-Yingjiang area, western Yunnan: Constraints from zircon U-Pb ages and Hf isotopes. Journal of Asian Earth Sciences, 53, 151-175.

Yang, X.J., Jia, X.C., Xiong, C.L., Bai, X.Z., Huang, B.X., Gai, L., Yang, C.B., 2012. LA-ICP-MS zircon U-Pb age of metamorphic basic volcanic rock in Gongyanghe Group of southern Gaoligong Mountain, western Yunnan Province, and its geological significance. Geological Bulletin of China, 31, 264-276 (in Chinese with English abstract).

Yuan, H.L., Gao, S., Liu, X.M., Li, H.M., Gunther, D., Wu, F.Y., 2004. Accurate U-Pb age and trace element determinations of zircon by laser ablation-inductively coupled plasma mass spectrometry. Geostandards and Geoanalytical Research, 28, 353-370.

Zhang, B., Zhang, J.J., Zhong, D.L., Yang, L.K., Yue, Y.H., Yan, S.Y., 2012. Polystage deformation of the Gaoligong metamorphic zone: Structures, $^{40}Ar/^{39}Ar$ mica ages, and tectonic implications. Journal of Structural Geology, 37, 1-18.

Zhang, C.H., Wang, Z.Q., Li, J.P., Song, M.S., 1997. Tectonic frame for deformation of the metamorphic rocks in the Ximeng area of the western Yunnan. Regional Geology of China, 16, 171-179 (in Chinese with English abstract).

Zhao, S.W., Lai, S.C., Qin, J.F., Zhu, R.Z., 2014. Zircon U-Pb ages, geochemistry, and Sr-Nd-Pb-Hf isotopic compositions of the Pinghe pluton, Southwest China: Implication for the evolution of the early Palaeozoic Proto-Tethys in Southeast Asia. International Geology Review, 56(7), 885-904.

Zhao, S.W., Lai, S.C., Qin, J.F., Zhu, R.Z., 2016. Tectono-magmatic evolution of the Gaoligong belt, southeastern margin of the Tibetan plateau: Constraints from granitic gneisses and granitoid intrusions. Gondwana Research, 35, 238-256.

Zhong, D.L., 1998. Paleo-Tethyan Orogenic Belt in the western Part of the Sichuan and Yunnan provinces.

Beijing: Science Press, 231 (in Chinese).

Zhou, Z.G., Liu, W.C., Liang, D.Y., 2004. Discovery of the Ordovician and its basal conglomerate in the Kangmar area, southern Tibet: With a discussion of the relation of the sedimentary cover and unifying basement in the Himalayas. Geological Bulletin of China, 23, 655-663 (in Chinese with English abstract).

Zhu, D.C., Zhao, Z.D., Niu, Y.L., Dilek, Y., Hou, Z.Q., Mo, X.X., 2013. The origin and pre-Cenozoic evolution of the Tibetan Plateau. Gondwana Research, 23, 1429-1454.

Zhu, D.C., Zhao, Z.D., Niu, Y.L., Dilek, Y., Wang, Q., Ji, W.H., Dong, G.C., Sui, Q.L., Liu, Y.S., Yuan, H.L., Mo, X.X., 2012. Cambrian bimodal volcanism in the Lhasa Terrane, southern Tibet: Record of an early Paleozoic Andean-type magmatic arc in the Australian proto-Tethyan margin. Chemical Geology, 328, 290-308.

Tectono-magmatic evolution of the Gaoligong belt, southeastern margin of the Tibetan Plateau: Constraints from granitic gneisses and granitoid intrusions①

Zhao Shaowei Lai Shaocong② Qin Jiangfeng Zhu Renzhi

Abstract: The Gaoligong belt is located in the southeastern margin of the Tibetan Plateau, and is bound by the Tengchong and Baoshan blocks. This paper presents new data from zircon geochronology, geochemistry, and whole-rock Sr-Nd-Pb-Hf isotopes to evaluate the tectonic evolution of the Gaoligong belt. The major rock types analysed in the present study are granitic gneiss, granodiorite, and granite. They are metaluminous to peraluminous and belong to high-K, calc-alkaline series. Laser ablation inductively coupled plasma mass spectrometry (LA-ICP-MS) analyses of zircons from nine granitic rocks yielded crystallization ages of 495–487 Ma, 121 Ma, 89 Ma, and 70–63 Ma. The granitoids can be subdivided into the following four groups: ① Early Paleozoic granitic gneisses with high $\varepsilon_{Nd}(t)$ and $\varepsilon_{Hf}(t)$ values of −3.45 to −1.06 and −1.16 to 2.09, and model ages of 1.16–1.33 Ga and 1.47–1.63 Ga, respectively. Their variable $^{87}Sr/^{86}Sr$ and Pb values resemble the characteristics of the Early Paleozoic Pinghe granite in the Baoshan Block. Our data suggest that the rocks were derived from the break-off of the Proto-Tethyan oceanic slab between the outboard continent and the Baoshan Block, which induced the partial melting of Mesoproterozoic pelitic sources mixed with depleted mantle materials. ② Early Cretaceous granodiorites with low $\varepsilon_{Nd}(t)$ and $\varepsilon_{Hf}(t)$ values of −8.92 and −4.91 with Nd and Hf model ages of 1.41 Ga and 1.49 Ga, respectively. These rocks have high initial $^{87}Sr/^{86}Sr$ (0.711 992) and lower crustal Pb values, suggesting that they were derived from Mesoproterozoic amphibolites with tholeiitic signature, leaving behind granulite residue at the lower crust. ③ Early Late Cretaceous granites with low $\varepsilon_{Nd}(t)$ and $\varepsilon_{Hf}(t)$ values of −9.58 and −4.61 with Nd and Hf model ages of 1.43 Ga and 1.57 Ga, respectively. These rocks have high initial $^{87}Sr/^{86}Sr$ (0.713 045) and lower crustal Pb isotopic values. These rocks were generated from the partial melting of Mesoproterozoic metapelitic sources resulting from the delamination of thickened lithosphere, following the closure of the Bangong-Nujiang Ocean and collision of the Lhasa-Qiangtang blocks. And ④ Late Cretaceous to Paleogene granitic gneisses with low $\varepsilon_{Nd}(t)$ and $\varepsilon_{Hf}(t)$ values of −10 to −4.41 and −8.71 to −5.95, Nd model ages ranging from 1.08 Ga to 1.43 Ga, and Hf model ages from 1.53 Ga to

① Published in *Gondwana Research*, 2016, 35.

② Corresponding author.

1.67 Ga, respectively. These rocks show high initial $^{87}Sr/^{86}Sr$ (0.713 201 and 0.714 662) and lower crustal Pb values. The data suggest that these rocks are likely related to the eastward subduction of the Neo-Tethyan Oceanic slab, which induced partial melting of Mesoproterozoic lower crustal metagreywacke. The results presented in this study from the Gaoligong belt offer important insights on the evolution of the Proto-Tethyan, Bangong-Nujiang, and Neo-Tethyan oceans in the southeastern margin of the Tibetan Plateau.

1 Introduction

The closure of the Tethyan Ocean and the collision between the Indian and Eurasian continents constructed the Himalayan orogen and elevated the Tibetan Plateau (e.g., Yin and Harrison, 2000; Searle et al., 2007; Gibbons et al., 2015).Several large-scale and numerous small-scale strike-slip faults and shear zones developed in the convergent belt including the Gaoligong belt (Wang et al., 2006). These shear zones are considered to have formed through intra-continental deformation by northward subduction of the Indian plate beneath the Eurasian Plate, which led to the southeastward extrusion of the Indochina block away from the convergent front (Tapponnier et al., 1982, 1986; Replumaz and Tapponnier, 2003; Wang et al., 2006; Replumaz et al., 2014). These shear zones and strike-slip faults accommodate southeastward material flow between several blocks such as the Tengchong, Baoshan, Indochina (Eroğlu et al., 2013) by crustal shortening or lateral extrusion outward from the plateau (Dewey et al., 1989; Wang and Burchfiel, 1997; Wang et al., 2006). Therefore, investigations to constrain the processes of deformation and southeastward material extrusion are crucial in understanding the geodynamics of the India-Eurasia continental collision (Socquet and Pubelier, 2005; Schoenbohm et al., 2006; Searle, 2006).

The Gaoligong belt in southwest Yunnan plays a vital role in our understanding of the subduction of the Tethyan Oceanic slab and the collision between the Indian and Eurasian continents. Previous research in this region has mainly focused on the Cenozoic deformation events (Wang et al., 2006, 2008; B. Zhang et al., 2010; Eroğlu et al., 2013) and the basement which was defined as the product of Meso- to Neoproterozoic tectonics (BGMRYP, 1990; Zhong, 1998). However, recent geochronological investigations have revealed multiple stages of magmatism and metamorphism in this belt (Eroğlu et al., 2013). The magmatism includes Early Paleozoic ca. 490 Ma (granitic gneiss, 500 Ma, Li et al., 2012a; gneissic granite, 487 Ma, Song et al., 2007) and Early Cretaceous, and the latter contains magmatic rocks including volcanics with ca. 120 Ma age (Yang et al., 2006; Cong et al., 2011; Bai et al., 2012; Li et al., 2012b, c). In order to understand the multi-stages of magmatism in the Gaoligong belt, we present new zircon U-Pb geochronology, whole-rock major and trace element geochemistry, and whole-rock Sr-Nd-Pb-Hf isotopic compositions of a suite of representative samples from the Gaoligong belt. These data reveal four stages of magmatism in the belt corresponding to 495–487 Ma, 121 Ma, 89 Ma, and 70–63 Ma. We correlate the magmatism with the evolution of the Proto-Tethyan, Bangong-Nujiang, and Neo-Tethyan oceans.

2 Geological setting

The N-S trending Gaoligong belt is located at the western part of the Nujiang valley and to the east of Longchuanjiang. The region lies northward of the Eastern Himalayan Syntaxis and southwestward of the Tengchong-Baoshan area and the Burma Block, extending over a length of 400 km in the Yunnan Province in China (Fig. 1; BGMRYP, 1990). The belt forms the

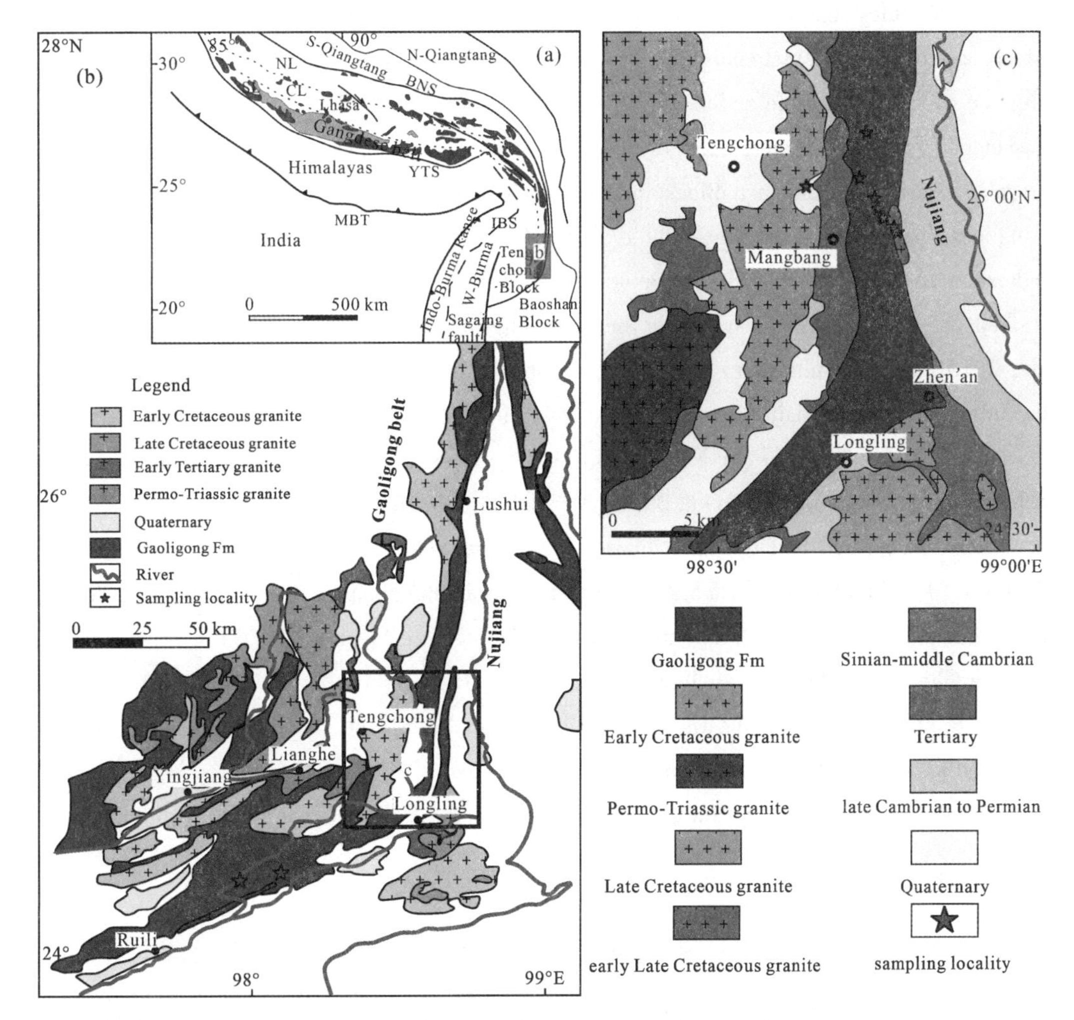

Fig. 1 Simplified geological map of the Himalaya-Tibet tectonic realm (a, modified after Wang et al., 2013), the Gaoligong belt, southeastern margin of Tibetan Plateau (b, modified after Xu et al., 2008) and the samples localities in the belt (c).

boundary between the Tengchong and Baoshan blocks and separates the Tengchong Block to the west and the Baoshan Block to the east. The Tengchong and Baoshan blocks have been considered as part of the Sibumasu continent, which broke away from eastern Gondwana in the early Permian (Metcalfe, 2002, 2011, 2013). However, the tectonic affinity of the Tengchong Block remains controversial (C. M. Wang et al., 2014). The relationship between the Tengchong and Baoshan blocks is also not clear, and several lines of evidence were evaluated by Zhao et al. (2014). Jin (2002) suggested that the earlier locations of the Tengchong and Baoshan blocks were different from their present position, although both these blocks were juxtaposed during Permo-Carboniferous. The juxtaposition of the Baoshan and Tengchong blocks as seen today was due to later tectonic movements. In addition, the Gaoligong Formation contains the oldest basement rocks (1 094–840 Ma; Zhong, 1998) in the Gaoligong belt, which are composed of metamorphosed volcano-sedimentary rocks, tonalite, and gabbro, in the absence of Early Paleozoic sediments. The late Paleozoic strata directly cover the metamorphic basement (Fang and Zhang, 1994). The Proterozoic units in the formation are similar to the Mogok rocks in the Sibumasu Block in northern Myanmar (Chen, 1991; Cong et al., 1993; Fang and Zhang, 1994; Zhong, 1998). In the Gaoligong belt, the dextral ductile shear sense indicators suggest dextral strike-slip motions (Wang et al., 2006). Several researchers have studied the strike-slip shear system in the Gaoligong belt (Wang et al., 2006, 2008; Eroğlu et al., 2013). The results show that the core of the shear zone consists of a narrow belt of mylonitic rocks with right-lateral strike-slip fabrics that were formed during the northward movement of India relative to Eurasia during Oligocene and Miocene. The main phases of deformation have been dated as ca. 35–21 Ma and ca. 19–12 Ma (Eroğlu et al., 2013). Lin et al. (2009) suggested that the main phase of deformation in the Gaoligong belt occurred from 22 Ma to 11 Ma and from 18 Ma to 13 Ma. The identical shear process revealed for the Karakorum-Jiali-Parlung fault system in western Tibet (Lee et al., 2003; Lin et al., 2009) and the Chongshan shear zone (Wang et al., 2006; B.Zhang et al., 2010) suggests that the Gaoligong belt is the southeastern continuation of this fault, and that it records the eastward extrusion of Tibet and crustal movement around the Eastern Himalayan Syntaxis (Eroğlu et al., 2013) resulting from the continuing collision between the Indian and Eurasian continents.

3 Samples and petrography

A total of 31 representative samples of granitic gneisses and granites were collected from the Gaoligong belt. Twenty-two samples were selected for whole-rock chemical analyses to determine the major and trace element concentrations, seven of them were also analyzed for whole-rock Sr-Nd-Pb-Hf isotopic compositions, and nine other samples were used for geochronologic analyses. These samples can be divided into three groups according to the lithology characteristics: ① gneiss (LL65-LL72, LL84 and LL86, LL246 and LL247, LL251-LL253, LL256-LL258, a total 16 samples were selected for geochemical analyses; LL64, LL85, LL244, LL250, LL259, LL260, a total 6 samples were taken for zircon U-Pb analyses); ②granodiorite (LL232-LL238, 4 samples were taken for geochemical analyses; LL234, this sample was used for geochronologic analyses); and ③ two-mica granite (LL264 and LL268, these samples were used for geochemical analyses; LL265 and LL266, these samples were used for zircon U-Pb analyses). As shown in Table 1 and Fig. 2, the granitic gneisses display mylonitic texture and have a gneissic structure, and they are comprised of plagioclase (Pl), K-feldspar (Kf), biotite (Bi), and quartz (Qz). Although these granitic gneisses have similar mineral assemblages, the contents of each mineral are different (Table 1). The augen structure in some gneisses indicates right-lateral shear (Fig. 2h). The granodiorites show coarse-grained texture and massive structure with a mineral assemblage of Pl+Kf+Qz+Bi+Hb; the biotite and hornblende are mostly chloritized. The two-mica granites exhibit equigranular texture and massive structure with a major mineral assemblage of Pl, Kf, Qz, Bi, and muscovite (Ms). The minerals display weak alignment. The common accessory minerals are zircon, apatite, and magnetite. Alumina-rich metamorphic minerals are absent. Therefore, we infer that the protoliths of the gneisses are igneous rocks, as also brought out in the geochemical discrimination plots (Fig. 3).

Table 1 Simplified sample description and locations.

Sample	Latitude(N)	Longitude(E)	Lithology	Mineral assemblage
LL64	24°13.758′	98°08.628′	biotite granitic gneiss	Phenocryst: Pl(10%)+Kf(5%)+Qz(3%)+Hb(2%) Matrix: Pl(25%)+Kf(20%)+Qz(20%)+Bi(15%)
LL65				
LL66				
LL68				
LL70				
LL71				
LL72				
LL84	24°14.131′	97°59.360′	granitic gneiss	Phenocryst: Pl(12%)+Kf(8%) Matrix: Pl(20%)+Kf(25%)+Qz(25%)+Bi(10%)
LL85				
LL86				
LL232	25°00.280′	98°39.064′	granodiorite	Pl(25%)+Kf(40%)+Qz(20%)+Bi(12%)+Hb(3%)
LL233				
LL234				
LL235				
LL238				
LL244	25°00.686′	98°42.061′	granitic gneiss	Phenocryst: Pl, Kf(20%)+Qz(10%) Matrix: Pl, Kf(40%)+Qz(20%)+Bi(10%)
LL246				
LL247				

Continued

Sample	Latitude(N)	Longitude(E)	Lithology	Mineral assemblage
LL250	25°09.531′	98°42.372′	granitic gneiss	Phenocryst: Pl(15%)+Kf(10%)+Qz(5%)
LL251				Matrix: Pl(15%)+Kf(20%)+Qz(30%)+Bi(5%)
LL252				
LL253				
LL256	24°59.803′	98°43.600′	granitic gneiss	Phenocryst: Pl(10%)+Kf(10%)
LL257				Matrix: Pl(20%)+Kf(25%)+Qz(25%)+Bi(10%)
LL258				
LL259				
LL260	24°58.423′	98°43.779′	biotite granitic gneiss	Phenocryst: Pl(2%)+Kf(8%) Matrix: Pl, Kf, Qz(75%)+Bi(15%)
LL264	24°55.606′	98°46.133′	two mica granite	Pl(35%)+Kf(30%)+Qz(20%)+Bi, Ms(15%)
LL265				
LL266	24°55.640′	98°46.408′		
LL268				

4 Analytical methods

Samples of gneisses, granodiorites, and granites from the Gaoligong belt were analyzed at the State Key Laboratory of Continental Dynamics, Northwest University, Xi'an, China (SKLCDNWUC). Weathered surfaces of samples were removed and the fresh parts were then chipped and powdered to about a 200 mesh size using a tungsten carbide ball mill. Major and trace elements were analyzed by X-ray fluorescence (XRF; Rikagu RIX 2100) and inductively coupled plasma mass spectrometry (ICP-MS; Agilent 7500a), respectively. Analyses of USGS and Chinese national rock standards (BCR-2, GSR-1, and GSR-3) showed that the analytical precision and accuracy for the major elements were generally better than 5%. For the trace element analyses, sample powders were digested using an $HF+HNO_3$ mixture in high-pressure Teflon bombs at 190 ℃ for 48 h. For most trace elements, the analytical error was less than 2% and the precision was greater than 10% (Liu et al., 2007).

Whole-rock Sr-Nd-Pb-Hf isotopic data were obtained using a Nu Plasma HR multi-collector mass spectrometer at the SKLCDNWUC. The Sr and Nd isotopic fractionations were corrected to $^{87}Sr/^{86}Sr=0.1194$ and $^{146}Nd/^{144}Nd=0.7219$, respectively. During the analysis period, the NIST SRM 987 standard yielded a mean value of $^{87}Sr/^{86}Sr=0.710250\pm12$ (2σ, $n=15$) and the La Jolla standard gave a mean of $^{146}Nd/^{144}Nd=0.511859\pm6$ (2σ, $n=20$). Whole-rock Pb was isolated by anion exchange in HCl-Br columns, and Pb isotopic fractionation was corrected to $^{205}Tl/^{203}Tl=2.3875$. Within the period of analysis, 30 measurements of the NBS 981 standard gave average values of $^{206}Pb/^{204}Pb=16.937\pm1$ (2σ), $^{207}Pb/^{204}Pb=15.491\pm1$ (2σ), and $^{208}Pb/^{204}Pb=36.696\pm1$ (2σ). The BCR-2 standard gave

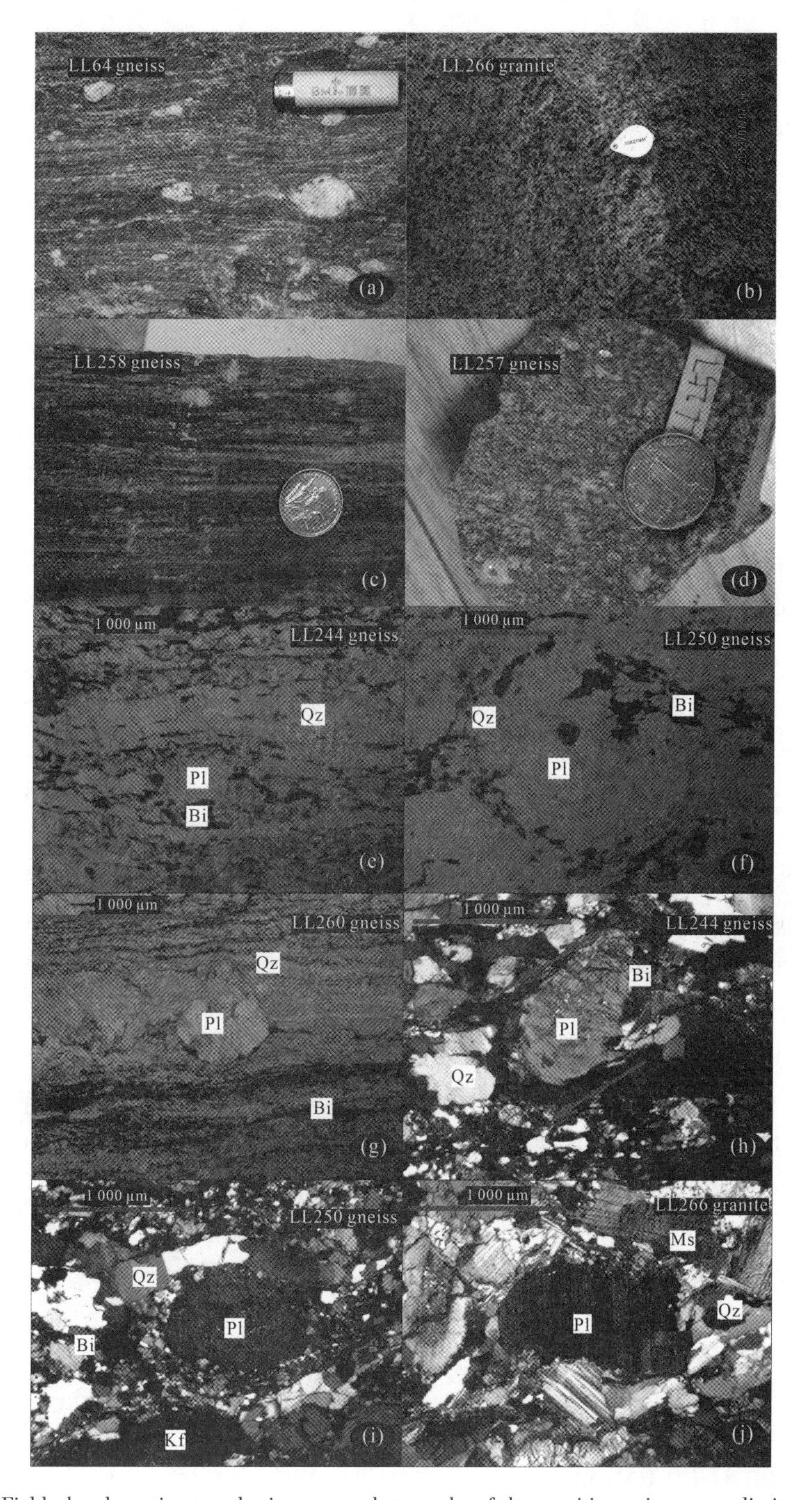

Fig. 2 Field, hand specimen and microscope photographs of the granitic gneiss, granodiorite, granite from the Gaoligong belt, southeastern margin of Tibetan Plateau.

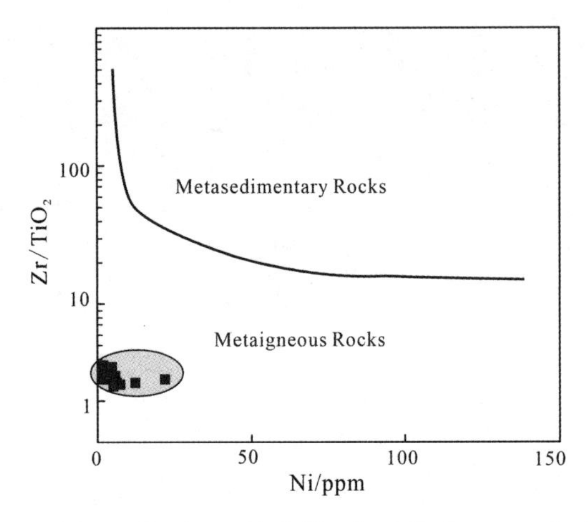

Fig. 3 Zr/TiO_2 vs. Ni plots with fields after Winchester et al. (1980).

values of $^{206}Pb/^{204}Pb = 18.742 \pm 1$ (2σ), $^{207}Pb/^{204}Pb = 15.620 \pm 1$ (2σ), and $^{208}Pb/^{204}Pb = 38.705 \pm 1$ (2σ). The total procedural Pb blanks were in the range of 0.1–0.3 ng. Whole-rock Hf was also separated by single anion exchange columns. In the course of analysis, 22 measurements of the JCM 475 standard yielded an average of $^{176}Hf/^{177}Hf = 0.282\ 161\ 3 \pm 0.000\ 001\ 3$ (2σ) (Yuan et al., 2007).

For U-Pb dating, zircon grains from the gneisses, granodiorites, and granites were separated using conventional heavy liquid and magnetic techniques. Representative zircon grains were handpicked and mounted in epoxy resin disks, then polished and carbon coated. Internal morphology was examined by cathodoluminescence (CL) prior to U-Pb isotopic analyses. Laser ablation (LA) ICP-MS zircon U-Pb analyses were conducted on an Agilent 7500a ICP-MS equipped with a 193 nm laser, which is located at the SKLCDNWUC, following the methods of Yuan et al. (2004). The $^{207}Pb/^{235}U$ and $^{206}Pb/^{238}U$ ratios were calculated using the GLITTER program, which was corrected using the Harvard zircon 91500 as an external calibration standard. These correction factors were applied to each sample to correct for both instrumental mass bias and depth-dependent elemental and isotopic fractionation. The detailed technique is described in Yuan et al. (2004). Common Pb contents were evaluated using the method described in Andersen (2002). Age calculations and concordia diagrams were completed using ISOPLOT version 3.0 software (Ludwig, 2003). The errors quoted in tables and figures are at 1σ levels.

5 Results

5.1 Zircon U-Pb dating

Results from zircon U-Pb dating of the nine samples from the Gaoligong belt are presented in the following sections. The isotopic data and rare earth element (REE) contents of zircons are listed in Supplementary Dataset.

Based on the geochronological data, the nine samples can be subdivided into four groups: Early Paleozoic granitic gneisses, Early Cretaceous granodiorites, early Late Cretaceous granites, and Late Cretaceous to Paleogene granitic gneisses.

5.1.1 Early Paleozoic granitic gneisses (LL244, LL250, and LL260)

Zircon grains in these three samples are identical. They are euhedral and columnar with lengths ranging from 100 μm to 200 μm and aspect ratios of 1:1 to 2:1. Their similar internal structures shown in the CL images (Fig. 4) with inner cores showing clear oscillatory zoning suggest magmatic protoliths. The thin, dark convoluted rims represent zircon overgrowth.

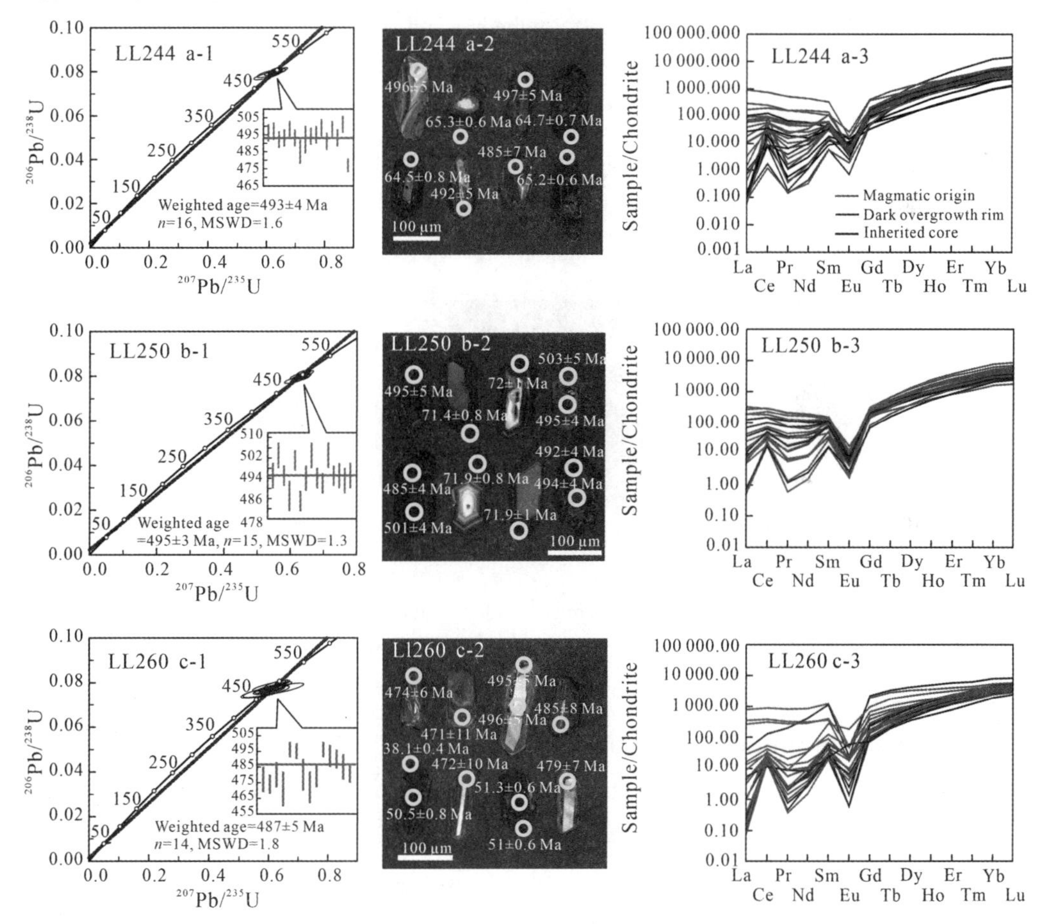

Fig. 4 LA-ICP-MS U-Pb zircon concordia diagrams, cathodoluminescence (CL) images, and chondrite-normalized REE patterns of representative zircon grains for Early Paleozoic granitic gneisses from the Gaoligong belt.

Twenty-six spots were analyzed from samples LL244, 24 from LL250, and 22 from LL260. The ages from the zoned cores in LL244 range from 477 Ma to 501 Ma with ^{206}Pb/^{238}U weighted mean age of 493.1 ± 3.5 Ma (MSWD = 1.6, n = 16). The zircons show relatively high Th/U ratios of 0.01–0.97. Another group of ages ranging from 64.5 Ma to 65.3 Ma with relatively low Th/U ratios (0. 04–0. 06), correlate with zircon overgrowth rims, although the significance of these ages with respect to geological events in the area are not clear. Five grains (#4, 10, 11, 12, and 17) yielded older ages of 909 Ma, 792 Ma, 811 Ma, 2606 Ma, and 724 Ma, indicating inherited zircon cores. In sample LL250, the crystallization ages of protolith are 495. 2 ± 2. 9 Ma (MSWD = 1. 3, n = 15) and the ages from overgrowth rims range from 70 Ma to 72 Ma (n = 9). Data from LL260 show that the crystallization ages of protolith is 486. 6 ± 5. 2 Ma (MSWD = 1. 8, n = 14) with relatively high Th/U ratios (0. 33–0. 78). The ages from overgrowth rims range from 52 Ma to 38. 1 Ma (n = 7). Spot #22 shows inherited core with an age of 572 Ma.

5. 1. 2 Early Cretaceous granodiorites (LL234)

The zircon grains from LL234 are mostly euhedral and prismatic in shape ranging in length from 100 μm to 150 μm with aspect ratios of 1∶1–1. 5∶1. The grains are colorless, transparent, and show clear oscillatory zoning in the CL images (Fig. 5), suggesting magmatic origin, as also confirmed by their high Th/U ratios of 0. 28–3. 65. Thirty-three spots were analyzed, and these yield concordant $^{206}Pb/^{238}U$ ages from 117 Ma to 126 Ma with a weighted mean age of 120. 9 ± 1. 0 Ma (MSWD = 3. 1, n = 33), which is interpreted as the crystallization age of the granodiorite.

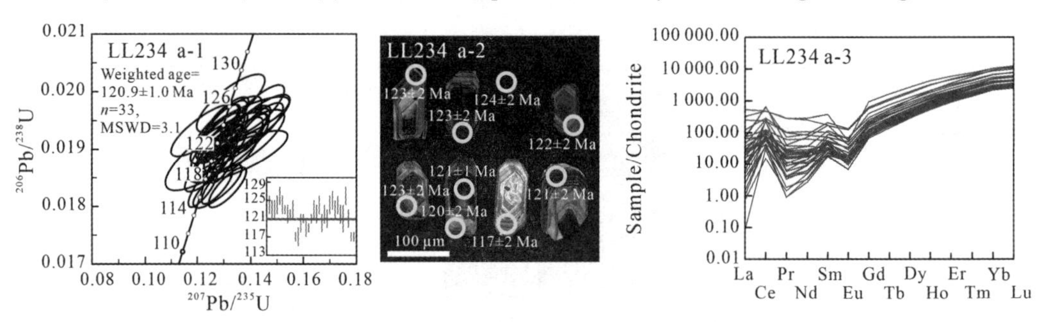

Fig. 5 LA-ICP-MS U-Pb zircon concordia diagrams, cathodoluminescence (CL) images, and chondrite-normalized REE pattern of representative zircon grains for Early Cretaceous granodiorites from the Gaoligong belt.

5. 1. 3 Early Late Cretaceous two-mica granites (LL265 and LL266)

The zircon grains separated from LL265 and LL266 are characterized by euhedral to subhedral long column crystals. Their lengths range from 150 μm to 250 μm with aspect ratios of 1. 5∶1 – 2∶1. Almost all of the grains display typical core-rim structure in the CL images (Fig. 6). Some of the cores are bright, structureless, and oval in shape, which is indicative of inherited zircon cores. Others show magmatic origin with clear oscillatory zoning. The dark, thin rims likely represent the overgrowth.

Twenty-nine spots from LL265 yielded variable ages from 72. 1 Ma to 975 Ma including inherited zircon core, magmatic zircons, and overgrowth rims. Fifteen spots in the core domains show variable ages in the range of 105–975 Ma. The oscillatory zoned grains yielded ages in the range of 84. 4–93 Ma with $^{206}Pb/^{238}U$ weighted mean age of 88. 9–1. 6 Ma (MSWD = 6. 1, n =

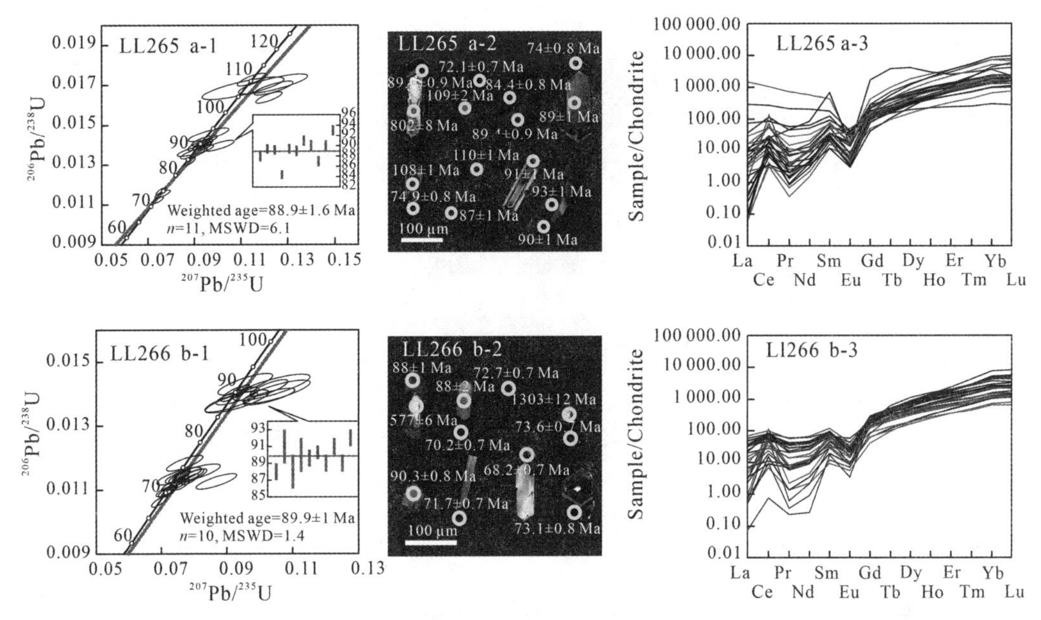

Fig. 6 LA-ICP-MS U-Pb zircon concordia diagrams, cathodoluminescence (CL) images, and chondrite-normalized REE pattern of representative zircon grains for early Late Cretaceous two-mica granites from the Gaoligong belt.

11) and relatively high Th/U ratios (0. 03–1. 18), which represents the formation age of the two-mica granite. The dark, thin rims have ages from 72. 1 Ma to 74. 9 Ma ($n=3$).

Thirty-one points from LL266 yielded ages similar to LL265, ranging from 68. 2 Ma to 1 303 Ma. Spots 4, 19, and 26 show characteristics of inherited zircon cores with ages of 577 Ma, 1 303 Ma, and 869 Ma, respectively. Ten of the 31 spots yielded ages from 88 Ma to 92 Ma with a weighted mean age of 89. 86 ± 0. 92 Ma (MSWD = 1. 4, n = 10), which is interpreted as the formation age of the two-mica granite. The other 18 spots from the dark rims yielded ages from 68. 2 Ma to 76 Ma.

5. 1. 4 Late Cretaceous to Paleogene granitic gneisses (LL64, LL85, and LL259)

(1) *Sample LL64: biotite granitic gneiss.* Zircon grains in the LL64 sample are colorless and transparent, and the columnar euhedral crystals show prismatic and double pyramid faces. The crystals range in length from 150 μm to 300 μm with length/width ratios of 3∶1–5∶1. The grains display obvious oscillatory zoning in the CL images (Fig. 7), and almost all of their Th/U ratios are higher than 0. 1 (ranging from 0. 07 to 1. 48), which is indicative of magmatic zircons. A total of 33 analyses were carried out and all the data plot on or close to concordia. Five spots (#3, 7, 16, 20, and 32) yielded ages of 937 Ma, 823 Ma, 490 Ma, 933 Ma, and 466 Ma, respectively, which were interpreted as inherited zircon cores. The other grains yielded apparent $^{206}Pb/^{238}U$ ages from 67. 1 Ma to 72. 6 Ma with a weighted age of 69. 6±0. 5 Ma (MSWD = 2. 2, n = 28), which is interpreted as the crystallization age of the protolith of the gneiss.

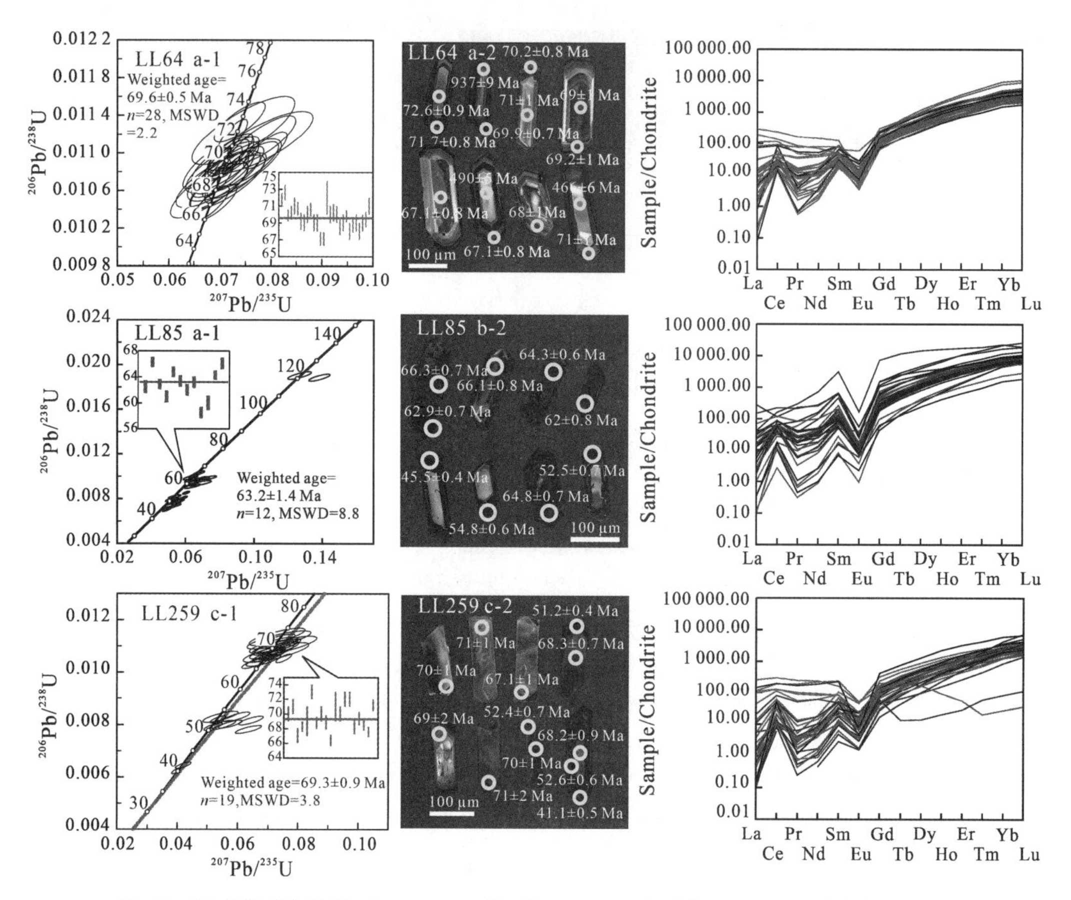

Fig. 7 LA-ICP-MS U-Pb zircon concordia diagrams, cathodoluminescence (CL) images, and chondrite-normalized REE pattern of representative zircon grains for Late Cretaceous to Paleogene granitic gneisses from the Gaoligong belt.

(2) *Sample LL85: granitic gneiss*. The zircon grains in this rock are columnar or tabular subhedral crystals ranging in length from 100 μm to 200 μm with aspect ratios of 1∶1 to 1.5∶1. Internal structures in the CL images (Fig. 7) display dark rims of overgrowth, but partial grains have preserved magmatic oscillatory zoning. Thirty-two grains were analyzed from this sample. The inherited zircon cores yielded ages of 122 Ma and 121 Ma from spots #11 and 22, respectively. The ages from 58.5 Ma to 66.3 Ma correspond to the magmatic crystallization age, with $^{206}Pb/^{238}U$ weighted average age of 63.2 ± 1.4 Ma (MSWD = 8.8, n = 12) and variable Th/U ratios. The ages ranging from 44.8 to 54.8 Ma are from the dark overgrowth rims.

(3) *Sample LL259: granitic gneiss*. In the CL images, some of the zircon grains in this rock show magmatic oscillatory zoning and long columnar or needle shapes, whereas the others are surrounded by dark structureless rims (Fig. 7). The zircon grains have lengths of 200–400 μm with aspect ratios of 2∶1–5∶1. Thirty-one analyses were performed on the zircon grains from this rock and the results can be divided into three groups: ① inherited zircon cores with

ages of 120 Ma and 121 Ma (#25 and 27); ② magmatic zircons with ages from 66.4 Ma to 73 Ma with a weighted mean age of 69.3±0.9 Ma (MSWD=3.8, $n=19$) and relatively high Th/U ratios (0.15–5.39); and ③ overgrowth rims ages with ages ranging from 40 Ma to 53.3 Ma ($n=10$).

Our data above show that the Gaoligong belt experienced multi-stage magmatism. Early Paleozoic magmatism occurred during 487–495 Ma, which is similar to the Early Paleozoic granites in the Baoshan Block (Dong et al., 2012, 2013; Wang et al., 2013; Zhao et al., 2014). Cretaceous magmatism occurred during 121–63 Ma, and this period can be subdivided into three stages: 121 Ma, 89 Ma, and 70–63 Ma. The ages yielded from the dark overgrowth rims mainly correspond to Late Cretaceous to Paleocene, with ages from ca. 76–40 Ma, although the related geologic events remain obscure. Hence these ages are not considered further in the discussion.

5.2 Geochemical data

5.2.1 Major and trace elements data

Geochemical data on 22 samples analyzed in this study are listed in Table 2 and plotted in Figs. 8 and 9. The salient geochemical features are discussed below.

(1) *Early Paleozoic granitic gneisses.* They are characterized by high SiO_2 (71.27–81.18 wt%) and low MgO (0.15–0.69 wt%) contents, together with low $Mg^{\#}$ values [$Mg^{\#}$ = 100Mg/(Mg+Fe)] ranging from 17.40 to 38.33. The Al_2O_3, K_2O, and Na_2O contents of these rocks are 9.67–14.68 wt%, 3.16–6.61 wt%, and 1.39–2.89 wt%, respectively, with high K_2O/Na_2O ratios of 1.59–4.76. The Early Paleozoic granitic gneisses plot in the granite field in the An-Ab-Or diagram (Fig. 8a). These rocks belong to the high-K, calc-alkaline series and are strongly peraluminous with high A/CNK values ranging from 1.10 to 1.21. In the chondrite-normalized REE and primitive mantle-normalized trace element diagrams (Fig. 9), the Early Paleozoic granitic gneisses show enrichment in the LREEs and depletion in the HREEs. They also show a slight fractionation with $(La/Yb)_N$ values ranging from 11.29 to 29.77, except for sample LL253 [$(La/Yb)_N$ = 103.45]. Overall, these samples have high total REE contents (96–684.2 ppm) and strongly negative Eu anomalies (Eu/Eu^* = 0.18–0.40). They also show depletion in Ba, Nb, Sr, P, and Ti, but enrichment in Rb, Th, U, and Pb.

(2) *Early Cretaceous granodiorites.* These rocks are characterized by low SiO_2 (63.15–66.00 wt%) and high MgO (1.53–1.79 wt%) contents, and the corresponding $Mg^{\#}$ values range from 45.39 to 53.74. They show high Al_2O_3 (14.95–17.56 wt%), K_2O (2.78–3.63 wt%), and Na_2O (3.53–3.76 wt%) contents, and the K_2O/Na_2O ratios are 0.74–1.03. The samples fall within the granodiorite field in the An-Ab-Or (Fig. 8a). They also show high-K, calc-alkaline, and metaluminous to peraluminous characteristics (Fig. 8b,c) with A/CNK values of 0.98–1.18. These rocks have total REE contents of 149–205.3 ppm and similar

Table 2 Analytical results of major(wt%) and trace(ppm) elements for granitic gneisses and granitoid intrusions from the Gaoligong belt.

Age	Early Paleozoic					Early Cretaceous				Early Late Cretaceous		Late Cretaceous to Paleogene										
	493 Ma		495 Ma			121 Ma				90 Ma		70 Ma						65 Ma		70 Ma		
Sample	LL246	LL247	LL251	LL252	LL253	LL232	LL233	LL235	LL238	LL264	LL268	LL65	LL66	LL68	LL70	LL71	LL72	LL84	LL86	LL256	LL257	LL258
Lithology	granitic gneiss		granitic gneiss			granodiorite				two-mica monzogranite		granitic gneiss						granitic gneiss		granitic gneiss		
SiO_2	81.18	78.14	76.39	78.00	71.27	66.00	63.15	63.46	65.87	72.83	75.24	71.69	71.16	71.69	71.46	71.30	71.17	75.96	75.21	72.75	75.14	72.10
TiO_2	0.17	0.14	0.07	0.08	0.37	0.59	0.61	0.63	0.50	0.21	0.10	0.32	0.37	0.36	0.38	0.23	0.37	0.05	0.10	0.27	0.12	0.35
Al_2O_3	9.67	11.32	13.18	11.82	14.68	16.36	17.54	17.56	14.95	14.92	14.53	14.58	14.65	14.29	14.42	14.05	14.66	13.47	13.15	14.43	13.55	14.33
$Fe_2O_3{}^T$	1.95	1.66	0.75	0.79	2.66	3.99	4.52	4.29	3.21	1.60	0.96	2.55	2.96	2.92	2.98	1.99	2.99	0.50	1.41	2.28	1.27	2.34
MnO	0.05	0.04	0.01	0.01	0.02	0.07	0.07	0.08	0.08	0.03	0.03	0.05	0.06	0.06	0.06	0.04	0.07	0.01	0.03	0.04	0.02	0.04
MgO	0.18	0.15	0.20	0.18	0.69	1.79	1.67	1.53	1.60	0.31	0.19	0.74	0.84	0.82	0.86	0.57	0.88	0.06	0.16	0.45	0.21	0.43
CaO	0.73	0.38	0.74	0.50	1.22	2.30	4.27	4.18	3.05	0.93	0.58	1.80	1.92	1.90	1.95	1.63	2.01	1.02	0.95	1.49	1.15	1.50
Na_2O	1.99	1.39	2.50	2.04	2.89	3.53	3.61	3.76	3.57	3.05	3.28	2.89	3.06	2.90	2.94	2.27	2.95	2.96	2.74	2.49	2.56	3.11
K_2O	3.16	6.61	6.04	5.92	5.02	3.63	2.78	2.80	3.55	5.18	4.43	5.15	4.52	4.70	4.64	6.43	4.76	5.33	5.50	5.31	6.01	5.70
P_2O_5	0.03	0.04	0.06	0.04	0.07	0.15	0.17	0.17	0.13	0.12	0.13	0.13	0.17	0.16	0.16	0.21	0.15	0.02	0.03	0.09	0.03	0.11
LOI	1.08	0.56	0.38	0.62	0.93	1.59	1.59	1.83	3.45	0.90	0.90	0.32	0.41	0.44	0.36	0.79	0.36	0.42	0.42	0.69	0.34	0.38
Total	100.19	100.43	100.32	100.00	99.82	100.00	99.98	100.29	99.96	100.08	100.37	100.22	100.12	100.24	100.21	99.51	100.37	99.80	99.70	100.29	100.40	100.39
A/CNK	1.21	1.12	1.10	1.11	1.18	1.18	1.05	1.04	0.98	1.21	1.29	1.07	1.09	1.07	1.08	1.03	1.07	1.08	1.08	1.15	1.06	1.02
$Mg^{\#}$	17.70	17.40	38.33	34.68	37.68	51.11	46.27	45.39	53.74	31.11	31.57	40.34	39.81	39.56	40.21	40.03	40.68	21.85	20.91	31.51	27.82	29.98
K_2O/N_2O	1.59	4.76	2.42	2.90	1.74	1.03	0.77	0.74	0.99	1.70	1.35	1.78	1.48	1.62	1.58	2.83	1.61	1.80	2.01	2.13	2.35	1.83
Li	76.9	78.1	19.6	11.4	47.5	30.4	25.9	22.8	24.6	148	153	113	117	123	134	79.1	120	14.0	27.4	82.5	26.5	38.7
Be	4.99	2.84	1.81	1.48	2.45	2.76	2.71	2.66	1.99	9.29	10.4	8.27	10.4	8.83	8.24	7.52	8.77	3.27	2.95	3.25	2.85	5.31
Sc	1.76	1.84	1.70	1.49	4.81	11.4	10.9	11.8	9.45	2.44	2.92	6.23	7.42	7.22	7.80	4.88	7.18	2.45	4.52	4.75	1.85	2.32
V	6.40	5.12	4.53	4.86	26.7	76.0	82.9	81.0	61.6	10.5	4.70	32.6	36.8	34.8	38.1	22.5	38.4	3.41	5.40	18.0	7.59	23.8

Continued

Age	Early Paleozoic					Early Cretaceous				Early Late Cretaceous		Late Cretaceous to Paleogene										
	493 Ma		495 Ma			121 Ma				90 Ma		70 Ma						65 Ma		70 Ma		
Sample	LL246	LL247	LL251	LL252	LL253	LL232	LL233	LL235	LL238	LL264	LL268	LL65	LL66	LL68	LL70	LL71	LL72	LL84	LL86	LL256	LL257	LL258
Lithology	granitic gneiss		granitic gneiss			granodiorite				two-mica monzogranite		granitic gneiss						granitic gneiss		granitic gneiss		
Cr	4.39	1.03	2.23	4.29	6.83	12.7	13.2	11.4	14.8	21.2	2.32	8.51	21.0	8.97	9.14	6.60	9.07	0.72	1.92	2.39	1.27	7.58
Co	119	181	104	146	91.4	78.4	71.7	97.0	33.6	105	144	69.8	62.5	118	79.5	117	72.2	138	114	67.2	137	83.1
Ni	4.72	2.37	3.26	5.89	4.74	5.35	5.56	5.49	7.77	22.4	2.41	5.69	12.4	6.12	6.17	5.93	6.36	2.81	2.72	1.85	2.01	5.45
Cu	0.72	0.66	0.71	0.98	1.93	4.97	5.68	5.87	2.50	2.95	30.3	4.31	1.75	6.52	10.4	16.8	5.92	0.45	0.51	1.26	1.12	2.13
Zn	55.1	47.6	23.6	21.0	79.5	50.9	63.7	64.4	38.9	72.3	34.0	44.1	59.6	53.6	53.6	68.8	60.3	12.1	32.9	46.0	28.4	49.6
Ga	16.5	16.9	13.4	10.3	21.2	19.8	21.3	21.1	16.4	26.1	25.7	17.7	19.4	18.0	18.1	17.0	18.4	15.2	16.3	17.6	17.2	20.1
Ge	1.73	2.00	1.54	1.58	1.69	1.56	1.31	1.30	1.49	1.58	1.98	1.85	1.86	1.83	1.76	1.86	1.80	1.66	1.50	1.53	1.44	1.60
Rb	259	458	266	259	287	218	133	120	170	339	331	306	317	280	283	317	295	321	345	222	251	311
Sr	28.2	35.8	55.5	46.8	87.3	295	329	328	271	103	54.3	185	139	166	171	169	182	58.0	54.8	158	113	140
Y	41.7	51.2	27.8	16.2	31.8	27.8	25.1	32.1	22.9	10.2	12.9	40.8	37.6	57.8	49.9	22.7	34.3	59.7	52.4	29.0	14.6	88.8
Zr	444	460	105	84.2	297	208	216	233	186	142	62.6	155	179	181	171	121	175	71.9	150	221	169	286
Nb	21.4	22.3	8.08	7.55	27.3	12.8	12.1	14.6	10.6	25.8	27.7	20.8	29.5	23.3	23.4	15.6	22.5	13.6	16.7	20.6	15.2	47.6
Cs	19.5	21.6	4.87	4.96	7.10	3.61	2.83	2.05	2.84	20.6	23.7	33.2	39.2	39.5	40.0	26.8	41.0	4.19	5.23	7.36	8.09	8.23
Ba	89.8	157	437	367	368	599	649	645	661	516	185	412	258	296	312	362	359	133	121	713	358	504
La	92.9	153	37.0	19.8	127	47.9	38.3	40.8	33.1	49.0	17.1	40.4	46.3	36.9	41.3	22.3	40.7	12.1	51.3	68.1	68.0	90.7
Ce	169	303	73.2	41.6	262	89.8	69.8	76.4	64.0	81.3	24.5	79.7	91.0	73.9	82.5	46.2	81.3	23.0	107	128	133	175
Pr	23.5	35.5	8.08	4.49	30.3	9.54	7.68	8.99	6.88	8.97	3.94	9.23	10.8	8.63	9.57	5.39	9.51	2.85	13.0	14.1	14.5	19.6
Nd	80.7	122	28.8	16.2	102	33.5	27.6	33.4	24.9	30.1	13.4	33.7	39.6	31.5	34.8	19.8	35.4	10.4	47.4	49.6	48.6	69.3
Sm	16.4	23.7	5.81	3.38	17.0	5.91	5.12	6.57	4.69	5.04	2.65	6.80	8.26	6.37	6.85	4.26	7.34	3.15	11.4	8.65	8.35	13.7

Continued

Age	Early Paleozoic					Early Cretaceous				Early Late Cretaceous		Late Cretaceous to Paleogene										
	493 Ma		495 Ma			121 Ma				90 Ma		70 Ma						65 Ma		70 Ma		
Sample	LL246	LL247	LL251	LL252	LL253	LL232	LL233	LL235	LL238	LL264	LL268	LL65	LL66	LL68	LL70	LL71	LL72	LL84	LL86	LL256	LL257	LL258
Lithology	granitic gneiss		granitic gneiss			granodiorite				two-mica monzogranite		granitic gneiss						granitic gneiss		granitic gneiss		
Eu	1.36	1.27	0.55	0.42	0.87	1.11	1.28	1.46	0.99	0.66	0.39	1.04	0.87	0.95	0.97	0.93	1.06	0.40	0.45	1.10	1.15	1.08
Gd	12.6	19.3	5.24	3.04	13.5	5.22	4.63	5.87	4.28	3.71	2.27	6.06	7.39	6.07	6.21	3.94	6.62	4.24	10.1	7.27	6.11	12.4
Tb	1.85	2.64	0.85	0.50	1.82	0.77	0.70	0.90	0.66	0.44	0.40	0.98	1.16	1.09	1.04	0.67	1.02	1.00	1.57	1.01	0.72	2.10
Dy	9.42	12.5	4.85	2.83	8.34	4.54	4.11	5.30	3.79	2.05	2.26	6.22	6.68	7.75	7.13	4.03	6.04	7.64	9.15	5.33	3.25	13.2
Ho	1.57	1.95	0.88	0.53	1.11	0.93	0.84	1.08	0.78	0.32	0.39	1.25	1.21	1.76	1.54	0.75	1.13	1.73	1.82	0.96	0.50	2.81
Er	4.21	4.72	2.29	1.44	2.03	2.70	2.44	3.11	2.24	0.84	1.01	3.91	3.31	5.75	4.89	2.03	3.17	5.68	5.17	2.59	1.27	8.55
Tm	0.59	0.60	0.30	0.21	0.18	0.39	0.36	0.45	0.32	0.11	0.15	0.69	0.48	0.95	0.79	0.29	0.48	0.97	0.73	0.35	0.17	1.37
Yb	3.84	3.69	1.70	1.26	0.88	2.51	2.40	2.93	2.14	0.71	0.94	4.93	3.05	6.22	5.12	1.78	3.14	6.63	4.53	2.18	1.05	8.93
Lu	0.53	0.49	0.22	0.17	0.12	0.37	0.35	0.42	0.31	0.094	0.12	0.76	0.44	0.92	0.75	0.25	0.46	0.98	0.67	0.30	0.15	1.16
Hf	13.0	13.5	3.29	2.97	7.90	5.17	5.29	5.72	4.55	4.03	2.16	4.49	5.21	5.18	4.87	3.53	4.76	2.87	5.03	5.89	5.25	7.65
Ta	0.87	2.34	1.09	1.23	1.97	1.18	0.99	1.33	0.91	5.50	6.58	4.45	5.36	4.10	4.01	3.18	4.39	1.03	0.99	1.60	1.50	9.62
Pb	34.6	59.4	54.5	51.0	56.2	7.32	19.7	20.8	9.04	38.8	34.4	49.5	51.3	48.0	47.5	50.3	44.3	71.1	68.6	35.6	54.1	55.1
Th	60.9	116	32.3	18.5	116	27.2	15.5	18.0	18.4	33.0	10.4	36.4	32.5	28.9	24.6	13.6	28.2	17.1	77.5	23.0	92.8	52.0
U	7.66	8.75	5.46	3.71	12.7	2.16	2.08	2.47	2.08	4.83	5.90	15.5	4.49	11.2	10.8	20.1	10.8	6.71	11.1	2.52	23.8	5.18
ΣREE	418.8	684.2	169.9	96	566.7	205.3	165.6	187.6	149	183.4	69.4	195.6	220.61	188.8	203.4	112.6	197.3	80.78	264.4	289.2	287	420.2
$(La/Yb)_N$	17.36	29.77	15.66	11.29	103.5	13.68	11.43	10.00	11.08	49.52	13.04	5.88	10.88	4.26	5.79	8.98	9.31	1.31	8.13	22.38	46.59	7.29
Eu^*	0.29	0.18	0.31	0.40	0.18	0.61	0.81	0.72	0.68	0.47	0.48	0.49	0.34	0.46	0.45	0.69	0.47	0.33	0.13	0.42	0.49	0.25

patterns to the REE and trace element patterns within Early Paleozoic granitic gneisses, i.e., they are enriched in LREE [$(La/Yb)_N$ = 10.00–13.68], Rb, Th, U, and Pb, but depleted in HREE, Eu (Eu/Eu^* = 0.61–0.81), Ba, Nb, Sr, P, and Ti (Fig. 9).

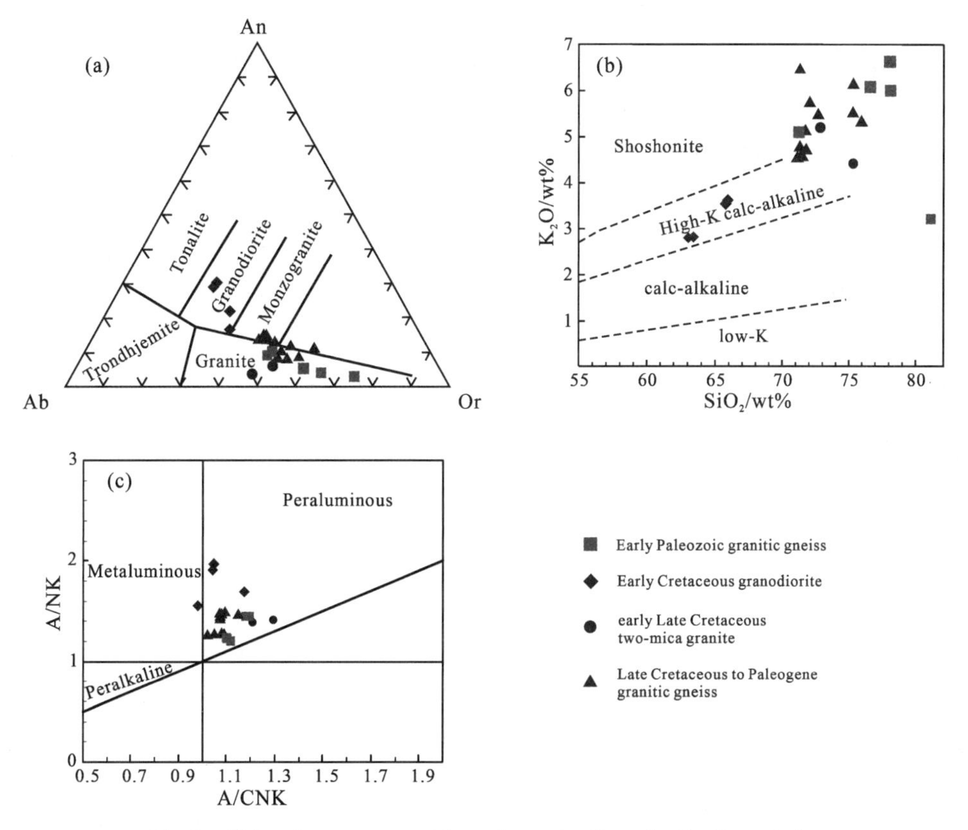

Fig. 8 Major element compositions of samples from the Gaoligong belt.

(a) An-Ab-Or diagram (Barker, 1979); (b) K_2O vs. SiO_2 classification diagram after Rollinson (1993); (c) A/NK [molar $Al_2O_3/(Na_2O+K_2O)$] vs. A/CNK [molar $Al_2O_3/(CaO+Na_2O+K_2O)$] (Maniar and Piccoli, 1989).

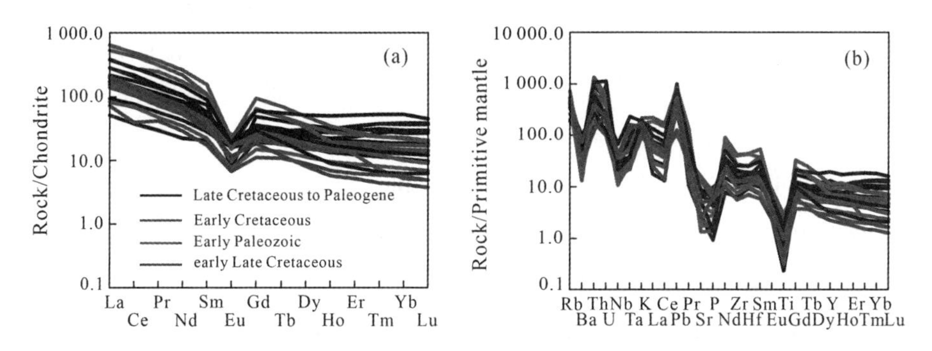

Fig. 9 Chondrite-normalized REE pattern (a) and primitive mantle (PM) normalized trace element spider (b) diagrams for the samples from Gaoligong belt.

Primitive mantle and Chondrite values are from Sun and McDonough (1989).

(3) *Early Late Cretaceous granites*. They have high SiO_2 (72.83–75.24 wt%), Al_2O_3 (14.53–14.92 wt%), and K_2O (4.43–5.18 wt%) contents, but low MgO (0.19–0.31 wt%) and Na_2O (3.05–3.28 wt%) contents. They also have high K_2O/Na_2O ratios of 1.35–1.70. The early Late Cretaceous granites fall in the granite field in the An-Ab-Or and belong to the high-K, calc-alkaline series and strongly peraluminous granites with A/CNK ratios ranging from 1.21 to 1.29 (Fig. 8). Similar REE and trace element patterns (Fig. 9) indicate that these samples are enriched in the LREEs [$(La/Yb)_N$ = 13.04–49.52], Rb, Th, U, and Pb, but depleted in the HREEs, Eu (Eu/Eu^* = 0.47–0.48), Ba, Nb, Sr, P, and Ti.

(4) *Late Cretaceous to Paleogene granitic gneisses*. They are characterized by high SiO_2 (71.16–75.96 wt%) and low MgO (0.06–0.88 wt%) contents, and the corresponding $Mg^\#$ values range from 20.91 to 40.68. They display Al_2O_3, K_2O, and Na_2O contents of 13.15–14.66 wt%, 4.52–6.43 wt%, and 2.27–3.11 wt%, respectively, and high K_2O/Na_2O ratios of 1.48–2.83. These rocks fall in the monzogranite and granite field in the An-Ab-Or (Fig. 8a). Furthermore, they belong to the high-K, calc-alkaline series and peraluminous granites with A/CNK ratios ranging from 1.03 to 1.15. Compared with the other samples, the Late Cretaceous to Paleogene granitic gneisses display similar REE and trace element patterns in the chondrite-normalized REE patterns and primitive mantle-normalized trace element pattern diagrams (Fig. 9). These granitic gneisses are enriched in the LREEs but depleted in the HREEs, except for samples LL65, LL70, and LL84, which are depleted in the middle REEs. They also have positive Rb, Th, U, and Pb anomalies and negative Eu (Eu/Eu^* = 0.13–0.69), Ba, Nb, Sr, P, and Ti anomalies.

Although field observations and petrologic data from the Gaoligong belt show that almost all of these rocks have experienced deformation, the large ion lithophile elements (LILEs) including Rb, Sr, Ba, and K do not seem to be significantly influenced, as they display linear relationship between SiO_2 and LILEs. This suggests that the LILEs were not mobile and the original concentrations of these elements have been largely preserved, and can be used for evaluating the petrogenesis of the protoliths of gneisses, granodiorites, and granites.

5.2.2 Whole-rock Sr-Nd-Pb-Hf isotopic compositions

Results of the whole-rock Sr-Nd-Pb-Hf isotopic analyses for the samples from the Gaoligong belt are given in Table 3. The initial values have been calculated based on the LA-ICP-MS U-Pb zircon ages.

Table 3 Whole-rock Sr-Nd-Pb-Hf isotopic data for samples from the Gaoligong belt.

Sample	LL65	LL84	LL235	LL246	LL252	LL257	LL264
t/Ma	70	65	120	470	470	70	90
Rb/ppm	306	321	120	259	259	251	339
Sr/ppm	185	58.0	328	28.2	46.8	113	103

Continued

Sample	LL65	LL84	LL235	LL246	LL252	LL257	LL264
$^{87}Rb/^{86}Sr$	4. 786	16. 028	1. 056	26. 615	16. 064	6. 427	9. 514
$^{87}Sr/^{86}Sr$	0. 717 961	0. 728 324	0. 713 793	0. 740 408	0. 738 023	0. 722 997	0. 725 212
2σ	15	14	6	30	9	5	16
$^{87}Sr/^{86}Sr(t)$	0. 713 201	0. 714 662	0. 711 992	0. 562 184	0. 630 451	0. 716 605	0. 713 045
Sm/ppm	6. 8	3. 15	6. 57	16. 4	3. 38	8. 35	5. 04
Nd/ppm	33. 7	10. 4	33. 4	80. 7	16. 2	48. 6	30. 1
$^{147}Sm/^{144}Nd$	0. 122 1	0. 182 6	0. 118 9	0. 123	0. 125 7	0. 104	0. 101 2
$^{143}Nd/^{144}Nd$	0. 512 378	0. 512 122	0. 512 12	0. 512 234	0. 512 365	0. 512 354	0. 512 091
2σ	27	10	11	19	23	22	4
$^{143}Nd/^{144}Nd(t)$	0. 512 322	0. 512 050	0. 511 777	0. 511 880	0. 512 003	0. 512 306	0. 512 031
T_{DM2}/Ga	1. 13	1. 43	1. 41	1. 33	1. 16	1. 08	1. 43
$\varepsilon_{Nd}(t)$	−4. 41	−10	−8. 92	−3. 45	−1. 06	−4. 71	−9. 58
U/ppm	15. 5	6. 7	2. 5	7. 7	3. 7	23. 8	4. 8
Th/ppm	36. 4	17. 1	18	60. 9	18. 5	92. 8	33
Pb/ppm	49. 5	71. 1	20. 8	34. 6	51	54. 1	38. 8
$^{206}Pb/^{204}Pb$	18. 896	18. 897	18. 935	18. 955	18. 978	19. 057	18. 922
2σ	4	5	3	5	3	4	3
$^{207}Pb/^{204}Pb$	15. 776	15. 774	15. 769	15. 777	15. 784	15. 787	15. 766
2σ	3	5	3	4	3	3	3
$^{208}Pb/^{204}Pb$	39. 510	39. 509	39. 727	39. 661	39. 463	39. 689	39. 560
2σ	9	16	7	20	7	8	8
$^{238}U/^{204}Pb$	20. 22	6. 08	7. 69	14. 32	4. 70	28. 52	8. 04
$^{232}Th/^{204}Pb$	48. 55	15. 92	57. 54	116. 72	24. 04	113. 94	56. 35
$^{206}Pb/^{204}Pb(t)$	18. 675	18. 840	18. 791	17. 872	18. 623	18. 746	18. 809
$^{207}Pb/^{204}Pb(t)$	15. 766	15. 771	15. 762	15. 716	15. 764	15. 776	15. 760
$^{208}Pb/^{204}Pb(t)$	39. 341	39. 461	39. 383	36. 900	38. 894	39. 300	39. 307
Lu/ppm	0. 76	0. 98	0. 42	0. 53	0. 17	0. 15	0. 09
Hf/ppm	4. 49	2. 87	5. 72	13	2. 97	5. 25	4. 03
$^{176}Lu/^{177}Hf$	0. 024 157	0. 048 636	0. 010 468	0. 005 764	0. 008 308	0. 004 181	0. 003 321
$^{176}Hf/^{177}Hf$	0. 282 514	0. 282 621	0. 282 582	0. 282 589	0. 282 520	0. 282 530	0. 282 591
2σ	5	5	3	4	5	4	4
$\varepsilon_{Hf}(t)$	−8. 71	−5. 95	−4. 91	2. 09	−1. 16	−7. 2	−4. 61
T_{DM2}/Ga	1. 67	1. 56	1. 49	1. 47	1. 63	1. 53	1. 57

Rb, Sr, Sm, Nd, U, Th, Pb, Lu and Hf concentrations were analyzed by ICP-MS.

T_{DM2} represent the two-stage model age and were calculated using present-day $(^{147}Sm/^{144}Nd)_{DM} = 0.213\,7$ and $(^{147}Sm/^{144}Nd)_{DM} = 0.513\,15$.

$\varepsilon_{Hf}(t)$ values were calculated using present-day $(^{147}Sm/^{144}Nd)_{CHUR} = 0.196\,7$ and $(^{147}Sm/^{144}Nd)_{CHUR} = 0.512\,638$.

Initial Pb isotopic ratios were calculated for 440 Ma using single-stage model.

T_{DM2} were calculated using present-day $(^{176}Lu/^{177}Hf)_{DM} = 0.038\,4$ and $(^{176}Hf/^{177}Hf)_{DM} = 0.283\,25$, $f_{CC} = -0.55$ and $f_{DM} = 0.16$.

$\varepsilon_{Hf}(t)$ values were calculated using present-day $(^{176}Lu/^{177}Hf)_{CHUR} = 0.033\,2$ and $(^{176}Hf/^{177}Hf)_{CHUR} = 0.282\,772$.

The Early Paleozoic granitic gneisses (LL246, LL252) have abnormally low initial $^{87}Sr/^{86}Sr(t)$ ratios of 0.562 184–0.630 451. The $\varepsilon_{Nd}(t)$ values range from −3.45 to −1.06 with Nd model ages of 1.16–1.33 Ga, and the $\varepsilon_{Hf}(t)$ values range from −1.16 to 2.09 with two-stage model ages of 1.47–1.63 Ga. The Early Paleozoic rocks also have variable radiogenic Pb compositions [$^{206}Pb/^{204}Pb(t)$ = 17.872–18.623, $^{207}Pb/^{204}Pb(t)$ = 15.716–15.764, and $^{208}Pb/^{204}Pb(t)$ = 36.900–38.894].

The Early Cretaceous granodiorite (LL235) shows $^{87}Sr/^{86}Sr(t)$ of 0.711 992 and $\varepsilon_{Nd}(t)$ value of −8.92. The Nd model age is 1.41 Ga, $\varepsilon_{Hf}(t)$ value is −4.91 with two-stage model age of 1.49 Ga, and moderately radiogenic Pb composition [$^{206}Pb/^{204}Pb(t)$ = 18.791, $^{207}Pb/^{204}Pb(t)$ = 15.762, and $^{208}Pb/^{204}Pb(t)$ = 39.383].

The early Late Cretaceous granite (LL264) has a high initial Sr ratio [$^{87}Sr/^{86}Sr(t)$ = 0.713 045], and $\varepsilon_{Nd}(t)$ and $\varepsilon_{Hf}(t)$ values of −9.58 and −4.61, respectively, with corresponding model ages of 1.43 Ga and 1.57 Ga. The Pb composition for the granite is $^{206}Pb/^{204}Pb(t)$ = 18.809, $^{207}Pb/^{204}Pb(t)$ = 15.760, and $^{208}Pb/^{204}Pb(t)$ = 39.307.

Compared with the other samples, the Late Cretaceous to Paleogene granitic gneisses (LL65, LL84, and LL257) have high initial Sr ratios (0.713 201–0.716 605), $\varepsilon_{Nd}(t)$ and $\varepsilon_{Hf}(t)$ values of −10 to −4.41 and −8.71 to −5.95, respectively, with corresponding Nd model ages of 1.08–1.43 Ga and Hf model ages of 1.53–1.67 Ga, and Pb compositions of $^{206}Pb/^{204}Pb(t)$ = 18.675–18.840, $^{207}Pb/^{204}Pb(t)$ = 15.766–15.776, and $^{208}Pb/^{204}Pb(t)$ = 39.300–39.461.

In order to characterize the isotopic features for the rocks from the Gaoligong belt, we have compiled Sr-Nd isotope data (Fig. 10) for the following: Early Paleozoic granite from the Gaoligong belt and Baoshan blocks (Wang et al., 2013); Early Cretaceous granite from the Gaoligong belt (Yang et al., 2006), North and Central Lhasa (Zhu et al., 2009a; Qu et al., 2012; Chen et al., 2014), and eastern Gangdese (Zhu et al., 2009b); Early Cretaceous granitoids from eastern Lhasa (Pan et al., 2014); and early Late Cretaceous granitoids in the Gangdese (adakites, ca. 90 Ma from Jiang et al., 2012; calc-alkaline rocks, 103–85 Ma from Wen et al., 2008a) and northern Lhasa (ca. 90 Ma magnesia-rich volcanic rocks from Q.Wang et al., 2014).

6 Discussion

The samples analyzed in this study have been subdivided into four groups on the basis of the geochronologic data. In order to clarify the multi-stage tectono-magmatism in the Gaoligong belt, we will discuss the magmatism belonging to each group in the following sections.

6.1 Early Paleozoic magmatism

The protoliths of Early Paleozoic granitic gneisses in the belt were identified as granites.

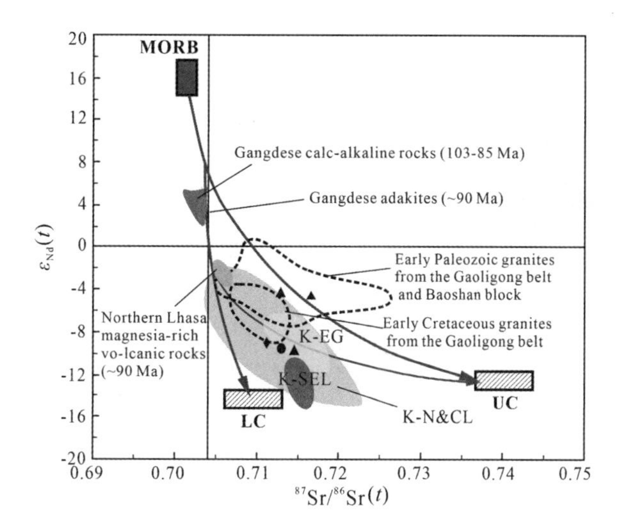

Fig. 10 $^{87}Sr/^{86}Sr(t)$ vs. $\varepsilon_{Nd}(t)$ diagram for the samples from the Gaoligong belt.

Symbols as in Fig. 8. Data for the Gaoligong belt and Baoshan Block Early Paleozoic granite are from Wang et al. (2013). Data for the Gaoligong belt Cretaceous granite are after Yang et al. (2006). Data for North-Central Lhasa Cretaceous granite are after Zhu et al. (2009a), Qu et al. (2012), Chen et al. (2014). Data for eastern Gangdese Cretaceous granites and eastern Lhasa Early Cretaceous granitoids are from Zhu et al. (2009b), Pan et al. (2014), respectively, and early Late Cretaceous granitoids in the Gangdese (calc-alkaline rocks, 103–85 Ma from Wen et al., 2008a; adakites, ca. 90 Ma from Jiang et al., 2012) and early Late Cretaceous magnesia-rich volcanic rocks in northern Lhasa (ca. 90 Ma from Q. Wang et al., 2014).

They are peraluminous to strongly peraluminous, high-K, calc-alkaline; the results of CIPW normal mineral calculations show >1% corundum and lack diopside. These signatures of the early Paleozoic rocks in the belt are similar to those of S-type granites. This is further confirmed in the diagrams of 10 000 Ga/Al vs. Zr, FeO^T/MgO, $(Na_2O+K_2O)/CaO$, and Y (Fig. 11). All the Early Paleozoic samples have obvious positive Rb anomalies and negative Eu, Ba, Sr, P, Ti, Nb, and Ta anomalies, representing fractional crystallization of K-feldspar, plagioclase, biotite, apatite, and ilmenite (Fig. 12). In addition, minerals such as monazite, apatite, K-feldspar in the residual phase may also contribute to these anomalies (Guo and Li, 2007). Generally, the petrogenesis of peraluminous granites could be summarized into two different models: ① fractionation of alumina-poor magma; and ② partial melting of crustal rocks involving metasedimentary and/or metaigneous sources that sometimes mix with materials derived from the mantle (Syvester, 1998; Clemens, 2003 and references therein). Available geochemical and experimental data show that the fractionation of alumina-poor magma is usually characterized by metaluminous and Na-rich components with low K_2O/Na_2O ratios. However, the Early Paleozoic samples show peraluminous to strongly peraluminous characteristics and high K_2O/Na_2O ratios (1.59–4.76), which are not consistent with the fractionation of alumina-poor magma. Therefore, the Early Paleozoic rocks were probably derived from partial

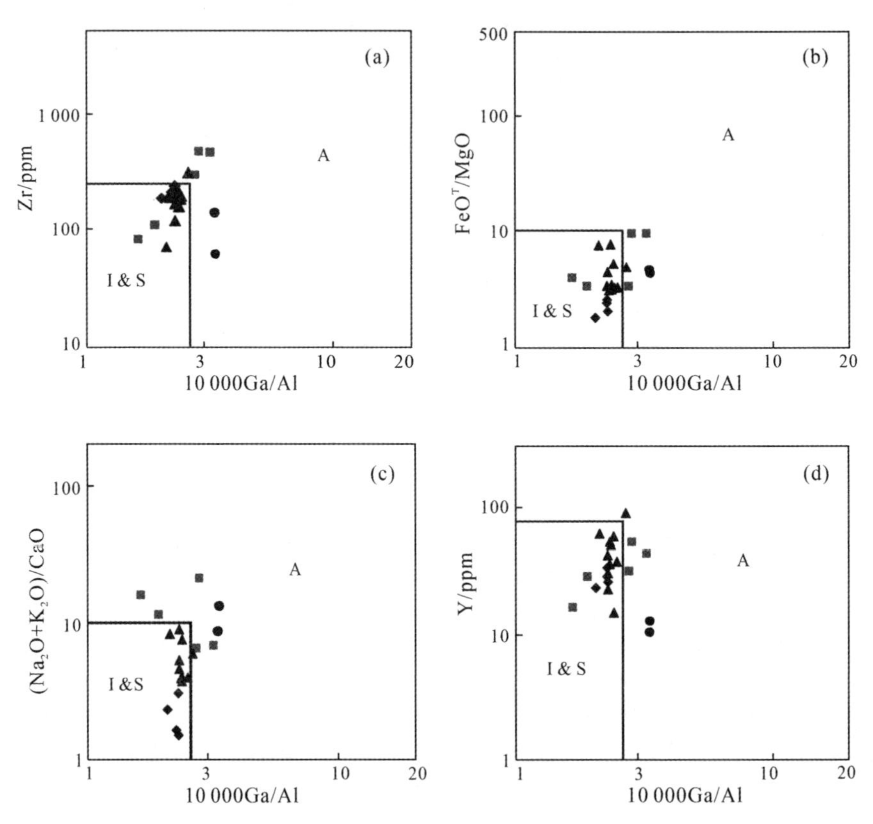

Fig. 11 The diagrams of Zr, FeO^T/MgO, $(Na_2O+K_2O)/CaO$, and Y vs. 10 000 Ga/Al, respectively, after Whalen et al. (1987).

Symbols as in Fig. 8.

melting of crustal rocks. This is further confirmed by the diagram of $\varepsilon_{Nd}(t)$ vs. $\varepsilon_{Hf}(t)$ values (Fig. 13). The samples also have high CaO/Na_2O ratios (0.25–0.42). Generally, strongly peraluminous granites with high CaO/Na_2O ratios (>0.3) can be generated by the mixing of pelite-derived melts with mafic magmas, which is consistent with the features in the diagrams of AFM vs. CFM, CaO/Na_2O vs. Al_2O_3/TiO_2, and Rb/Ba vs. Rb/Sr (Fig. 14). This is also confirmed by the isotopic data [i.e., low negative $\varepsilon_{Nd}(t)$ and positive $\varepsilon_{Hf}(t)$ values]. The rocks also display Nd and Hf model ages of 1.16–1.33 Ga and 1.47–1.63 Ga, respectively. All these features suggest that Early Paleozoic rocks in the belt were derived from partial melting of Mesoproterozoic pelitic components that interacted with depleted mantle materials.

Three samples (LL244, LL250, and LL260) of granitic gneiss yielded Paleozoic ages of ca. 487–494 Ma, which are interpreted as the crystallization ages of the protolith. A similar Early Paleozoic magmatism has been documented in the adjacent block (Baoshan Block). Zhao et al. (2014) reported ages of ca. 490 Ma and 440 Ma from monzogranites in the Pinghe pluton, and similar ages have been also reported by Chen et al. (2007), Dong et al. (2012, 2013), Liu et al. (2009), and Wang et al. (2013) in the Baoshan Block. In addition,

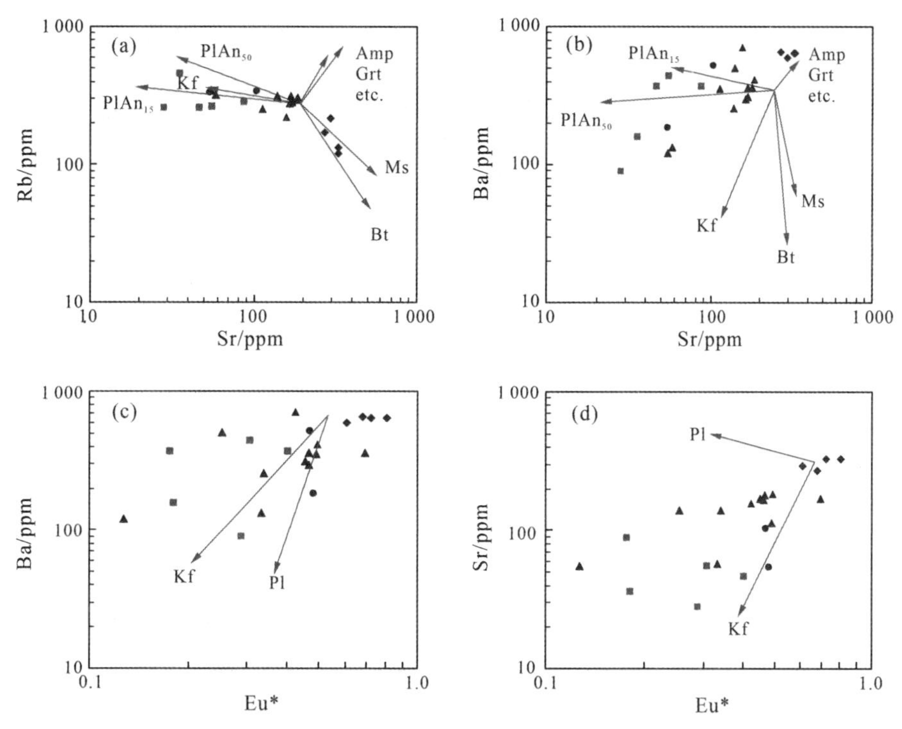

Fig. 12 Plots of Sr vs. Rb and Ba, Eu* vs. Sr and Ba for the samples from the Gaoligong belt according to Janoušek et al. (2004).

Pl: plagioclase; Kf: K-feldspar; Bt: biotite; Ms: muscovite; Grt: garnet; Amp: amphibole. Symbols as in Fig. 5.

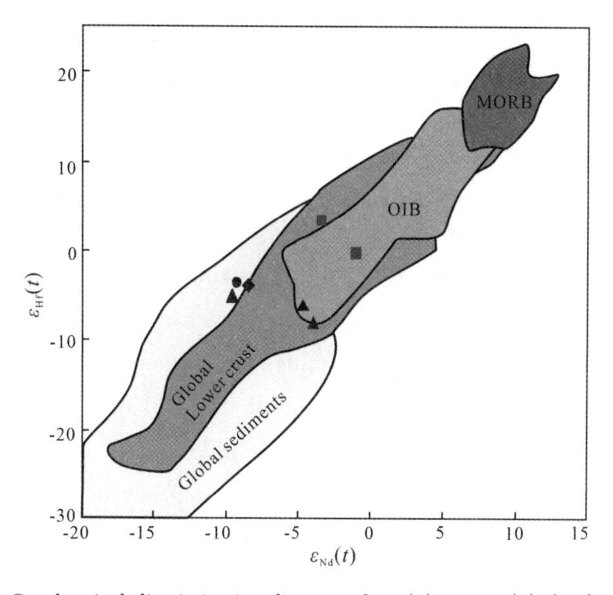

Fig. 13 Geochemical discrimination diagram of $\varepsilon_{Nd}(t)$ vs. $\varepsilon_{Hf}(t)$ for the samples from the Gaoligong belt after Dobosi et al. (2003).

Symbols as in Fig. 8.

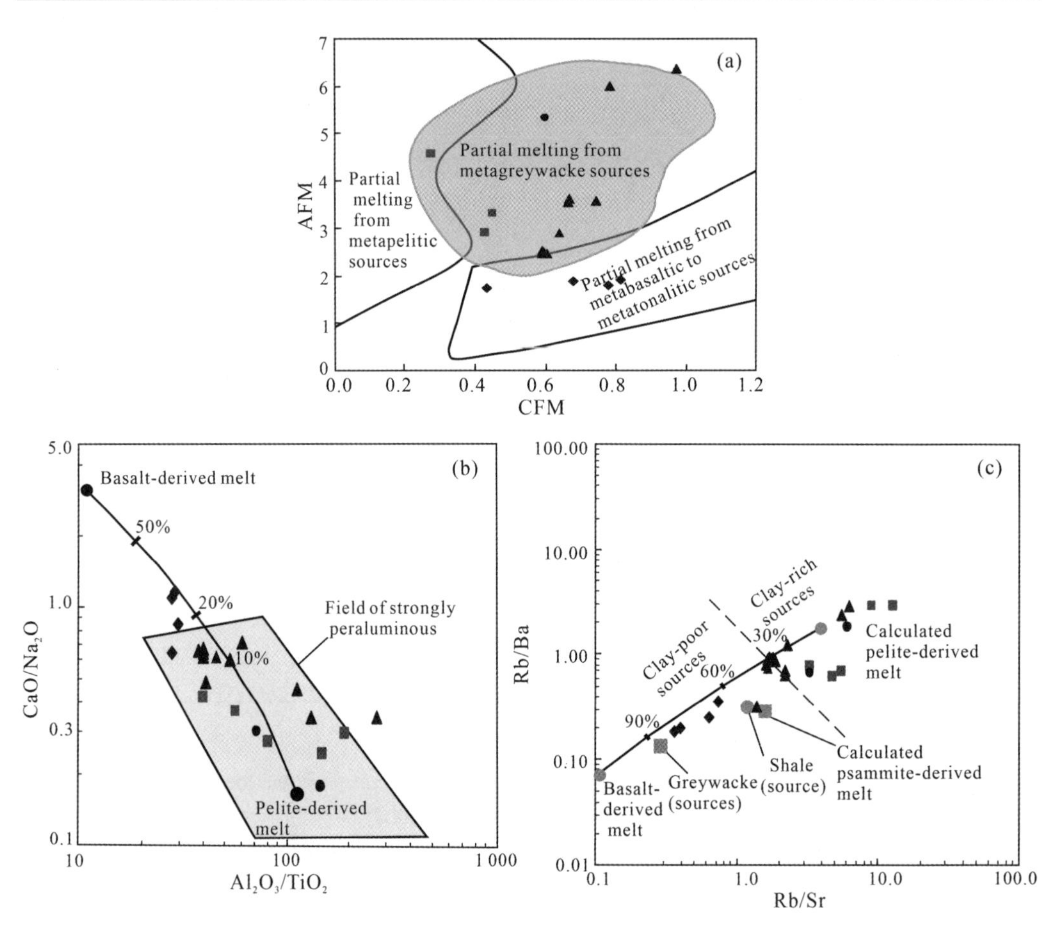

Fig. 14 Diagrams of AFM [molar $Al_2O_3/(MgO+FeO^T)$] vs. CFM [molar $CaO/(MgO+FeO^T)$] after Altherr et al. (2000) (a), Al_2O_3/TiO_2 vs. CaO/Na_2O (b) and Rb/Sr vs. Rb/Ba (c) after Sylvester (1998) for the samples from the Gaoligong belt. Symbols as in Fig. 8.

gneissic granites have been reported from the Maji area in the Gongshan Block at the northern part of the Gaoligong belt which show protolith crystallization age of 487 Ma (Song et al., 2007). Biotite monzonite gneisses have been also found in the Gaoligong belt with crystallization ages of 497–500 Ma (Li et al., 2012a). Furthermore, many gneissic rocks from the Gaoligong belt have inherited zircons with ages of 450–510 Ma (Cong et al., 2009). Thus, Early Paleozoic magmatism in the Gaoligong belt is widespread, and the rocks resemble the granites from the Pinghe pluton in the Baoshan Block, not only in terms of the crystallization ages, but also in relation to the geochemical and isotopic data, especially the abnormal low initial $^{87}Sr/^{86}Sr$ and variable Pb isotopes, the extremely low Sr compositions, and the high Rb/Sr ratios (Chen et al., 2007; Dong et al., 2013; Wang et al., 2013; Zhao et al., 2014). Thus, we regard the magmatism in the Gaolingong belt and Baoshan Block to be products of a coeval tectono-magmatic event relate with syn- to post-collision stage (Fig. 16). This might correlate with the slab break-off of the Proto-Tethyan Oceanic subduction slab between the outboard continent and the Baoshan Block on the margin of the Gondwana continent (Fig. 17a), as evaluated in detail by Zhao et al. (2014).

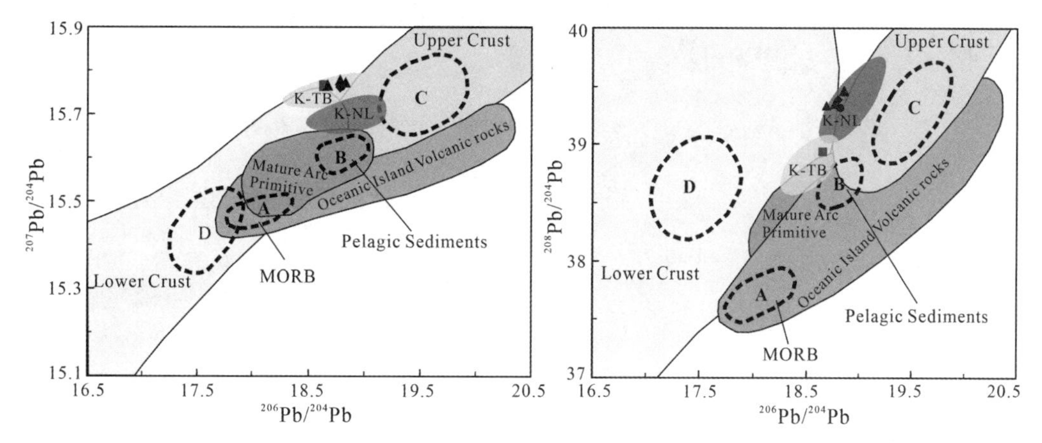

Fig. 15 $^{207}Pb/^{204}Pb$ vs. $^{206}Pb/^{204}$(a) and $^{208}Pb/^{204}Pb$ vs. $^{206}Pb/^{204}Pb$ diagrams for the samples from the Gaoligong belt after Zartman and Doe (1981).

Symbols as in Fig. 8.

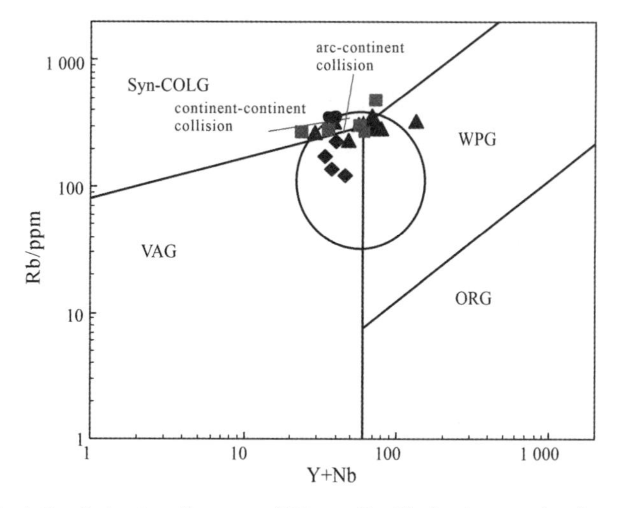

Fig. 16 Geochemical discrimination diagrams of Rb vs. Y + Nb for the samples from the Gaoligong belt after Pearce et al. (1984), and Pearce (1996).

Symbols as in Fig. 8.

However, there is no evidence for Neoproterozoic to Early Paleozoic magmatism in the Tengchong Block or in the Baoshan Block. Although several workers have mentioned the occurrence of Early Paleozoic protolith of gneiss in the Gaoligong belt (Song et al., 2007; Li et al., 2012a; Wang et al., 2013; this study), we consider that the protolith of the gneiss may be similar to the monzogranite in the Pinghe pluton in the Baoshan Block. This evidence seems to imply that the Tengchong and Baoshan blocks did not experience tectono-magmatism during

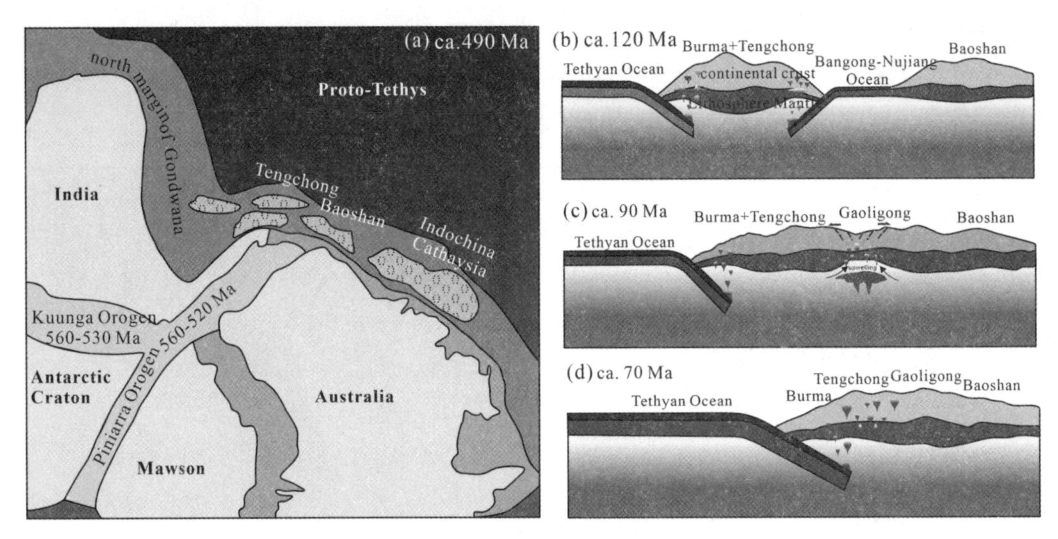

Fig. 17 (a) The tectonic framework of the India-Australia Proto-Tethyan margin. (b-d) Schematic illustrations displaying the tectonic settings for the generation of the magmatism during Early Cretaceous, early Late Cretaceous, and Late Cretaceous to Paleogene.

Early Paleozoic as compared to the outboard continents of Gondwana, although both these blocks were located along the margin of Gondwana supercontinent (Jin, 1996; Wophner, 1996; Jin, 2002; Metcalfe, 2002, 2011, 2013). In addition, the following factors might indicate that the two blocks were separate units during Paleozoic: ① The basement in the Tengchong Block is composed of Mesoproterozoic metamorphic rocks termed as the Gaoligong Formation (BGMRYP, 1990; Jin, 1996), and the Gongyanghe Group in the Baoshan Block consists of a Neoproterozoic to Cambrian low-grade metamorphic, terrestrial sedimentary sequence. ② The Neoproterozoic rocks are unknown in the Tengchong Block and the Early Paleozoic strata are poorly developed with only a few Early Devonian strata (and possibly some Silurian); however, the sedimentary sequences in the Baoshan Block are complete except for the absence of Upper Cambrian to Upper Ordovician strata (Jin, 1996; Huang et al., 2009, 2012), and the thicknesses and components of the Late Carboniferous units in both blocks are also different. This suggests that the two blocks were located in different sedimentary environments during that time (Jin, 1996; Wophner, 1996). ③ Volcanic activity (Woniusi Basalt) and red beds are developed in the Baoshan Block but not in the Tengchong Block, and there are successive carbonate formations in the Tengchong Block (Jin, 1994, 1996; Wophner, 1996; Wang et al., 2002). ④ Paleontological investigations indicate that the contemporary fusulinid faunas in the Tengchong Block share similar features with those in the Baoshan and Sibumasu blocks; however, visible differences in species compositions still exist between the faunas of the Tengchong Block and those of the other two blocks, suggesting that there are apparent regional features associated with the fauna of the Tengchong Block. And

⑤ Li et al. (2014) suggested that the Tengchong Block was located along the Indian margin of Gondwana during the Early Paleozoic based on detrital zircon ages and Hf isotopic data, but the Baoshan Block was correlated with the Australian margin of Gondwana. All these lines of evidence show that the Tengchong and Baoshan blocks were not connected, although they were apparently not far from each other during the Early Paleozoic. Thus, the Gaoligong belt should have formed subsequent to the Early Paleozoic.

6.2 Early Cretaceous magmatism

The Early Cretaceous granodiorites exhibit high-K, calc-alkaline, and metaluminous to peraluminous characteristics and contain hornblende, which is suggestive of typical I-type granite. This inference is also consistent with other geochemical features (Fig. 11). The rocks are enriched in Rb, depleted in Ba, Sr, Ti, Eu, P, Nb, and Ta, and show trends of increasing MgO and decreasing FeO^T with increasing SiO_2, which are indicative of either the separation of plagioclase, K-feldspar, biotite, apatite, and ilmenite during the magma evolution or minerals left in sources which caused these anomalies. The granodiorites have high initial Sr (0.711 992), negative $\varepsilon_{Nd}(t)$ and $\varepsilon_{Hf}(t)$ values (−8.92 and −4.91, respectively) with model ages of 1.41 Ga and 1.49 Ga, respectively, and lower crust Pb isotopic compositions, thus indicating crustal sources. Experimental studies show that intermediate to silicic magmas can be derived from dehydration melting of tholeiitic amphibolite leaving behind granulite residues at 8−12 kbar and garnet granulite to eclogite residues at 12−32 kbar (Rushmer, 1991; Rapp and Watson, 1995). The magma from this model is usually low in K_2O and high in $Na_2O/K_2O > 1$, which is consistent with the Early Cretaceous granodiorite with Na_2O/K_2O ratios (0.97−1.34). In addition, the granodiorites are not depleted in HREEs because the Ga/Yb ratios range from 1.9 to 2.1, which argue against garnet as the residual mineral. Thus, we suggest that the granodiorite may have been generated by partial melting of Mesoproterozoic tholeiitic amphibolite that left a granulite residue at the lower crust. We also obtained similar results from the geochemical diagrams of AFM vs. CFM, CaO/Na_2O vs. Al_2O_3/TiO_2, Rb/Ba vs. Rb/Sr (Fig. 14), and $\varepsilon_{Nd}(t)$ vs. $\varepsilon_{Hf}(t)$ (Fig. 13), as well as the Pb isotopic ratios (Fig. 15).

The Early Cretaceous magmatism in the Gaoligong belt is dominantly characterized by I-type granodiorites with an emplacement age of 121 Ma, and the rocks are of a lower crustal origin. Coeval magmatism took place at the Dulongjiang, Fugong, and Lushui areas in the northern Gaoligong belt, and the rocks were named Gaoligong granites by Yang et al. (2006) and Xu et al. (2012), with ages ranging from 118 Ma to 126 Ma and 121 Ma to 126 Ma, respectively. Similar magmatic rocks were also found in the Lianghe and Tengchong areas located along the margin of the Gaoligong belt (Cong et al., 2011; Qi et al., 2011; Li et al.,

2012b,c), with ages of ~120 Ma. Corresponding volcanic rocks were also found in the inner metamorphic belt with an age of 130 Ma (Bai et al., 2012) and in the Lianghe area with an age of 121 Ma (Gao et al., 2012). Therefore, the Early Cretaceous magmatism was widely distributed in the Gaoligong belt and these intermediate to silicic rocks in the area form an Early Cretaceous magmatic belt, which runs northward to the Zayu area, southeast Tibet (Zhu et al., 2009b), and southward to the Tengchong-Lianghe area, west Yunnan. In addition, the Early Cretaceous magmatism that occurred at the Northern magmatic belt in the Lhasa terrane is represented by granitoids with ages ranging from 240 Ma to 110 Ma, and it intensified during the Early Cretaceous (Xu et al., 2012). A comparison with the Early Cretaceous granites between the Gaoligong belt and the Northern magmatic belt in the Lhasa terrane (Yang et al., 2006; Chui et al., 2009, Yang and Xu, 2011; Xu et al., 2012) shows similar lithology, emplacement ages, major and trace elemental compositions, and isotopic data (Figs. 10 and 15), suggesting that the Early Cretaceous granite is the eastern extension of the Northern magmatic belt. In addition, the formation of the Early Cretaceous igneous rocks in the Northern magmatic belt remains controversial with diverse opinions (Xu et al., 2012). One of the prevailing models consider these as products of the Neo-Tethyan low-angle northward subduction (Coulon et al., 1986; Ding et al., 2003) and the other models invoke the subduction of the Bangong-Nujiang Oceanic slab or collision of the Lhasa-Qiangtang blocks (Zhu et al., 2009a, 2011; Xu et al., 2012). If the former one is valid, there must have been episodic magmatism and a complex massif from south to north in the overlying plate, but such a feature is absent in the Lhasa block. Meanwhile, the occurrence of coeval volcanic rocks shows that the formation of the Early Cretaceous rocks may have involved the subduction of the Bangong-Nujiang Oceanic slab, and not the thickening of crust resulting from the collision between the Lhasa and Qiangtang blocks. Therefore, the Early Cretaceous magmatic rocks were formed in an arc environment resulting from the southward subduction of the Bangong-Nujiang Oceanic slab (Fig. 17b). This is further confirmed by the diagram of Rb vs. Y+Nb (Fig. 16). All the Early Cretaceous samples fall in the VAG field, which is suggestive of a subduction-related setting.

6.3 Early Late Cretaceous magmatism

Early Late Cretaceous two-mica granites show high-K, calc-alkaline, and strongly peraluminous characteristics. The pronounced depletions in Ba, Sr, Nb, P, Ti, and Eu and enrichment in Rb demonstrate advanced fractional crystallization of plagioclase, K-feldspar, biotite, apatite, and ilmenite during the magma evolution or the anomalies can be attributed to the minerals in the residual phase. They also display high initial Sr ratio (0.713 045), negative $\varepsilon_{Nd}(t)$ and $\varepsilon_{Hf}(t)$ values of −9.58 and −4.61 with model ages of 1.43 Ga and

1.57 Ga, respectively, and lower crust Pb isotopic compositions, which represent the lower crustal origin (Figs. 13 and 15). High K_2O/Na_2O ratios (1.35–1.70) and low CaO/Na_2O ratios (0.18–0.3, < 0.3) (Fig. 14) indicate that the early Late Cretaceous granites were derived from partial melting of Mesoproterozoic middle to lower crustal meta-pelite sources.

The early Late Cretaceous magmatism in the Gaoligong belt show ages of 89 Ma, and coeval igneous rocks are widely distributed in the Lhasa block (Meng et al., 2014 and references therein) and Burma terrane (Mitchell et al., 2012; J.G. Wang et al., 2014). However, the lithological features of the early Late Cretaceous igneous rocks within the Gaoligong belt, Lhasa block, and Burma terrane are different, and this may have been caused by different sources and petrogenesis, even possibly by different tectonic settings. The rocks in the Gangdese batholith and Burma terrane are characterized by norite, hornblendite (Ma et al., 2013), charnockite (Z.M. Zhang et al., 2010), diorite, and granodiorite with positive zircon $\varepsilon_{Hf}(t)$ and whole-rock $\varepsilon_{Nd}(t)$ values, and low initial $^{87}Sr/^{86}Sr$ values (Fig. 10) (Wen et al., 2008a; Mitchell et al., 2012; J.G. Wang et al., 2014), which is suggestive of magma generation in arc setting. In addition, the early Late Cretaceous Gangdese intrusions show adakitic geochemical characteristics (Wen et al., 2008a; Jiang et al., 2012; Zheng et al., 2014), and these may have been derived from the partial melting of the subducted oceanic slab in the Neo-Tethyan arc. However, our samples are characterized by high SiO_2, a high initial $^{87}Sr/^{86}Sr$ value (0.7130), and negative whole-rock $\varepsilon_{Nd}(t)$ and $\varepsilon_{Hf}(t)$ values of −9.6 and −4.6, respectively. Also, the location is far from the western Myanmar arc by more than 200 km. Therefore, the petrogenesis of the early Late Cretaceous two-mica granite is not likely related to the subduction of the Neo-Tethys. In addition, Q.Wang et al. (2014) show that the ca. 90 Ma magnesia-rich volcanic rocks (andesites and dacites) in the northern Lhasa subterrane have high $Mg^{\#}$ of 54–64, low $^{87}Sr/^{86}Sr(t)$ values (0.7054–0.7065), negative $\varepsilon_{Nd}(t)$ of −3.2 to −1.7, and positive $\varepsilon_{Hf}(t)$ of +3.8 to +7.0. They proposed that the volcanic rocks were most likely derived from partial melting of a delaminated mafic lower crust following the final Lhasa-Qiangtang amalgamation. The closure of the Bangong-Nujiang Ocean and the collision of the Lhasa-Qiangtang blocks took place at the Early Cretaceous as indicated by the oldest nonmarine strata (~118 Ma, Kapp et al., 2007) and molasses deposition (Late Cretaceous, Pan et al., 2006). Therefore, we propose that the early Late Cretaceous two-mica granites in the Gaoligong belt were derived from partial melting of the middle to lower crust resulting from the delamination of the thickened lithosphere, which was caused by the closure of the Bangong-Nujiang Ocean and collision of the Lhasa-Qiangtang blocks; the upwelling asthenosphere provided the heat that resulted in the melting (Fig. 17c). This inference is consistent with the results of the diagram for Rb vs. Y+Nb (Fig. 16).

6.4 Late Cretaceous to Paleogene petrogenesis and magmatism

The Late Cretaceous to Paleogene granitic gneisses show high-K, calc-alkaline, and weakly peraluminous characteristics, and they exhibit an affinity of S-type granite (Fig. 11). Fractional crystallization occurred during the evolution of magma because of the pronounced depletion in Ba, Sr, Nb, P, Ti, and Eu and enrichment in Rb. The separation minerals are plagioclase, K-feldspar, biotite, apatite, and ilmenite. In addition, the minerals (such as monazite, apatite, K-feldspar, etc.) in the residual phase could also cause these anomalies. They show high initial Sr values (0.713 201–0.716 605), negative $\varepsilon_{Nd}(t)$ and $\varepsilon_{Hf}(t)$ values ranging from −10 to −4.41 and −8.71 to −5.95, respectively, with Nd model ages of 1.08–1.43 Ga and Hf model ages of 1.53–1.67 Ga, and lower crust Pb isotopic compositions (Figs. 13 and 15). High K_2O/Na_2O (1.48–2.83) and CaO/Na_2O (0.34–0.72, >0.3) ratios indicate that these Late Cretaceous to Paleogene rocks were derived from partial melting of Paleo-to Mesoproterozoic lower crustal metagreywacke, which is consistent with the geochemical features (Fig. 14).

The Late Cretaceous to Paleogene igneous rocks in the Gaoligong belt yielded ages of 70–65 Ma. Similar magmatism occurred in the Tengchong and Baoshan blocks (78–61 Ma, Chen et al., 2007, 2015; Qi et al., in press), Gaoligong belt (~76 Ma, Eroğlu et al., 2013; 76–68 Ma, Xu et al., 2012), southeastern Tibet (Booth et al., 2004; Guo et al., 2012; Zhang et al., 2013), and Burma (Barley et al., 2003), which were caused by subduction of the Neo-Tethys. However, Wen et al. (2008b) proposed that a Gangdese magmatic gap or quiescent period existed between ca. 80 Ma and 70 Ma, which suggests that there were differences in the Neo-Tethyan subduction between the Tibet (northward subduction) and Yunnan (eastward subduction) regions. The generation of Late Cretaceous to Paleogene igneous rocks in Yunnan took place before the collision between the Indian and Eurasian continents (~55 Ma), and this should have been related to the eastward subduction of the Neo-Tethyan Oceanic slab and lack of interactions between crustal rocks and materials from the mantle [as indicated by the evidence from the negative $\varepsilon_{Nd}(t)$ and $\varepsilon_{Hf}(t)$ values] (Fig. 17d). Heat input resulted from the thickening crust, convection of the mantle wedge, and radioactivity.

7 Conclusions

The granitic rocks analyzed in this study yielded four stages of crystallization ages as ca. 495–487 Ma, 121 Ma, 89 Ma, and 70–63 Ma. Thus, these rocks can be subdivided into four groups as follows.

(1) Early Paleozoic granitic gneisses: their geochronologic, geochemical, and isotopic

data are markedly similar to those of the Early Paleozoic granite in the Pinghe pluton, Baoshan Block. This was interpreted asthe break-off of the subducting Proto-Tethyan Oceanic slab between the outboard continent and the Baoshan Block, which induced partial melting of Mesoproterozoic pelitic sources that interacted with depleted mantle materials. Based on a comparison between the Tengchong and Baoshan blocks, it can be inferred that these blocks were not connected. Also, the Tengchong Block might not have experienced the Early Paleozoic tectono-magmatic event.

(2) Early Cretaceous granodiorites: these I-type granodiorites in the Gaoligong belt were interpreted to have formed within arc setting that resulted from the southward subduction of the Bangong-Nujiang Oceanic slab. This led to the partial melting of Mesoproterozoic tholeiitic amphibolite, leaving granulite residue in the lower crust.

(3) Early Late Cretaceous granites: the formation of these rocks likely involved the partial melting of Mesoproterozoic middle to lower crustal metapelitic sources resulting from the delamination of the thickened lithosphere, which was caused by the closure of the Bangong-Nujiang Ocean and collision of the Lhasa-Qiangtang blocks.

(4) Late Cretaceous to Paleogene granitic gneisses: these rocks were likely related to the eastward subduction of the Neo-Tethyan Oceanic slab, which resulted in the partial melting of Mesoproterozoic lower crustal metagreywacke.

Acknowledgements We are very grateful to Prof. M. Santosh for his kind help with English language.This work was jointly supported by the National Natural Science Foundation of China (Grant Nos. 41190072 and 41421002), and program for Changjiang Scholars and Innovative Research Team in University (Grant IRT 1281).

Supplementary data Supplementary data to this article can be found online at https://dx.doi.org/10.1016/j.gr.2015.05.007.

References

Altherr, R., Holl, A., Hegner, E., Langer, C., Kreuzer, H., 2000. High-potassium, calc-alkaline I-type plutonism in the European Variscides: Northern Vosges (France) and northern Schwarzwald Variscides: northern Vosges (France) and northern Schwarzwald (Germany). Lithos, 50, 51-73.

Andersen, T., 2002. Correction of common lead in U-Pb analyses that do not report ^{204}Pb. Chemical Geology, 192, 59-79.

Bai, X.Z., Jia, X.C., Yang, X.J., Xiong, C.L., Liang, B., Huang, B.X., Luo, G., 2012. LA-ICP-MS zircon U-Pb dating and geochemical characteristics of Early Cretaceous volcanic rocks in Longling-Ruili fault belt, western Yunnan Province. Geological Bulletin of China, 31, 297-305 (in Chinese with English abstract).

Barker, F., 1979. Trondhjemites: Definition, environment and hypotheses of origin. In: Barker, F. Trondhjemites, Dacites and related rocks. Amsterdam: Elsevier.

Barley, M.E., Pickard, A.L., Zaw, K., Pak, P., Doyle, M.G., 2003. Jurassic to Miocene magmatism and metamorphism in the Mogok metamorphic belt and the India-Eurasia collision in Myanmar. Tectonics, 22, 1-11.

Booth, A.L., Zeitler, P.K., Kidd, W.S.F., 2004. U-Pb zircon constraints on the tectonic evolution of southeastern Tibet, Namche Barwa area. American Journal of Science, 304, 889-929.

Bureau of Geology and Mineral Resources of Yunnan Province, 1990. Regional Geology of Yunnan Province. Beijing: Geological Publishing House, 1-729 (in Chinese).

Chen, J.C., 1991. Characteristics of Pb, Sr isotopic compositions in west Yunnan. Geology Science, 2, 174-183 (in Chinese with English abstract).

Chen, F.K., Li, X.H., Wang, X.L., Li, Q.L., Siebel, W., 2007. Zircon age and Nd-Hf isotopic composition of the Yunnan Tethyan belt, southwestern China. International Journal of Earth Science, 96, 1179-1194.

Chen, Y., Zhu, D.C., Zhao, Z.D., Meng, F.Y., Wang, Q., Santosh, M., Wang, L.Q., Dong, G.C., Mo, X.X., 2014. Slab breakoff triggered ca. 113 Ma magmatism around Xainza area of the Lhasa Terrane, Tibet. Gondwana Research, 26, 449-463.

Chen, X.C., Hu, R.Z., Bi, X.W., Zhong, H., Lan, J.B., Zhao, C.H., Zhu, J.J., 2015. Petrogenesis of metaluminous A-type granitoids in the Tengchong-Lianghe tin belt of southwestern China: Evidences from zircon U-Pb ages and Hf-O isotopes, and whole-rocks Sr-Nd isotopes. Lithos, 212-215, 93-110.

Chui, H.Y., Chung, S.L., Wu, F.Y., Liu, D.Y., Liang, Y.H., Lin, I.J., Lizuka, Y., Xie., L.W., Wang, Y.B., Chu, M.F., 2009. Zircon U-Pb and Hf isotopic constraints from eastern Transhimalayan batholiths on the precollisional magmatic and tectonic evolution in southern Tibet. Tectonophysics, 477, 3-19.

Clemens, J.D., 2003. S-type granitic magmas-petrogenetic issues, models and evidence. Earth-Science Reviews, 61, 1-18.

Cong, B.L., Wu, G.Y., Zhang, Q., Zhang, R.Y., Zhai, M.G., Zhao, D.S., Zhang, W.H., 1993. Petro-tectonic evolution of the Paleo-Tethys in the west Yunnan, China. Science in China: Series B, 23, 1201-1207 (in Chinese with English abstract).

Cong, F., Lin, S.L., Li, Z.H., Zou, G.F., Geng, Q.R., 2009. Zircon U-Pb Age of Gneissic Granitic in the Tengchong Block, Western Yunnan. Acta Geologica Sinica, 83, 651-658 (in Chinese with English abstract).

Cong, F., Lin, S.L., Zou, G.F., Li, Z.H., Xie, T., Peng, Z.M., Liang, T., 2011. Magma mixing of granites at Lianghe: In-site zircon analysis for trace elements, U-Pb ages and Hf isotopes. Science China: Earth Sciences, 54, 1346-1359.

Coulon, C., Maluski, H., Bollinger, C., Wang, S., 1986. Mesozoic and Cenozoic volcanic rocks from central and southern Tibet: ^{39}Ar-^{40}Ar dating, petrological characteristics and geodynamical significance. Earth & Planetary Science Letters, 79, 281-302.

Dewey, J.F., Cande, S., Pitman, W.C.I., 1989. Tectonic evolution of the India/Eurasia collision zone. Eclogae Geologicae Helvetiae, 82, 717-734.

Ding, L., Kapp, P., Zhong, D.L., Deng, W.M., 2003. Cenozoic Volcanism in Tibet: Evidence for a Transition from Oceanic to Continental Subduction. Journal of Petrology, 44, 1833-1865.

Dobosi, G., Kempton, P.D., Downes, H., Embey-Isztin, A., Thirlwall, M., Greenwood, P., 2003. Lower crustal granulite xenoliths from the Pannonian Basin, Hungary, Part 2: Sr-Nd-Pb-Hf and O isotope evidence for formation of continental lower crust by tectonic emplacement of oceanic crust. Contributions to Mineralogy and Petrology, 144, 671-683.

Dong, M.L., Dong, G.C., Mo, X.X., Zhu, D.C., Nie, F., Xie, X.F., Wang, X., Hu, Z.C., 2012. Geochronology and geochemistry of the Early Paleozoic granitoid in Baoshan block, western Yunnan and their implications. Acta Petrologica Sinica, 28(5), 1453-1464 (in Chinese with English abstract).

Dong, M.L., Dong, G.C., Mo, X.X., Santosh, M., Zhu, D.C., Yu, J.C., Nie, F., Hu, Z.C., 2013. Geochemistry, zircon U-Pb geochronology and Hf isotopes of granites in the Baoshan Block, Western Yunnan: Implications for Early Paleozoic evolution along the Gondwana margin. Lithos, 179, 36-47.

Eroğlu, S., Siebel, W., Danišík, M., Pfänder, J.A., Chen, F.K., 2013. Multi-system geochronological and isotopic constraints on age and evolution of the Gaoligongshan metamorphic belt and shear zone system in western Yunnan, China. Journal of Asian Earth Sciences, 73, 218-239.

Fang, C.J., Zhang, Y.F., 1994. The structure and tectonics of western Yunnan. Journal of Southeast Asian Earth Sciences. 9, 355-361.

Gao, Y.J., Lin, S.L., Cong, F., Zou, G.F., Xie, T., Tang, F.W., Li, Z.H., Liang, T., 2012. LA-ICP-MS zircon U-Pb dating and geological implications for the Early Cretaceous volcanic rocks on the southeastern margin of the Tengchong block, western Yunnan. Sedimentary Geology and Tethyan Geology, 32, 59-64 (in Chinese with English abstract).

Gibbons, A.D., Zhairovic, S., Muller, R.D., Whittaker, J.M., Yatheesh, V., 2015. A tectonic model reconciling evidence for the collisions between India, Eurasia and intra-oceanic arcs of the central-eastern Tethys. Gondwana Research, 28(2), 451-492.

Guo, S.S., Li, S.G., 2007. Petrological and geochemical constraints on the origin of leucogranites. Earth Science Frontiers, 14, 290-298 (in Chinese with English abstract).

Guo, L., Zhang, H.F., Harris, N., Randall, P., Xu, W.C., Shi, Z.L., 2012. Paleogene crustal anatexis and metamorphism in Lhasa terrane, eastern Himalayan syntaxis: Evidence from U-Pb zircons ages and Hf isotopic compositions of the Nyingchi Complex. Gondwana Research, 21, 100-111.

Huang, Y., Deng, G.B., Peng, C.L., Hao, J.X., Zhang, G.X., 2009. The discovery and significance of absence in early-middle Ordovician in southern Baoshan, western Yunnan. Guizhou Geology, 26(98), 1-6 (in Chinese with English abstract).

Huang, Y., Hao, J.X., Bai, L., Deng, G.B., Zhang, G.X., Huang, W.J., 2012. Stratigraphic and petrologic response to Late Pan-African movement in Shidian area, western Yunnan Province. Geological Bulletin of China, 31, 306-313 (in Chinese with English abstract).

Janoušek, V., Finger, F., Roberts, M., Frýda, J., Pin, C., Dolejš, D., 2004. Deciphering the petrogenesis of deeply buried granites: Whole-rock geochemical constraints on the origin of largely undepleted felsic granulites from the Moldanubian Zone of the Bohhemian Massif. Transactions of the Royal Society of Edinburgh: Earth Sciences, 95, 141-159.

Jiang, Z.Q., Wang, Q., Li, Z.X., Wyman, D.A., Tang, G.J., Jia, X.H., Yang, Y.H., 2012. Late Cretaceous (ca. 90 Ma) adakitic intrusive rocks in the Kelu area, Gangdese Belt (southern Tibet): Slab melting and implications for Cu-Au mineralization. Journal of Asian Earth Sciences, 53, 67-81.

Jin, X.C., 1994. Sedimentary and paleogeographic significance of Permo-Carboniferous sequences in west Yunnan, China. Geologisches Institut der Universităt zu Kŏln Sonderverŏffentlichungen, 99, 136.

Jin, X.C., 1996. Tectono-Stratigraphic Units of western Yunnan, China and Their Counterparts in Southeast Asia. Continental Dynamics, 1, 123-133.

Jin, X.C., 2002. Permo-Carboniferous sequences of Gondwana affinity in southwest China and their paleogeographic implications. Journal of Asian Earth Sciences, 20, 633-646.

Kapp, P., Decelles, P.G., Gehrels, G.E., Heizler, M., Ding, L., 2007. Geological records of the Lhasa-Qiangtang and Indo-Asian collisions in the Nima area of central Tibet. Geological Society of America Bulletin, 119, 917-933.

Lee, H.Y., Chung, S.L., Wang, J.R., Wen, D.J., Lo, C.H., Yang, T.F., Zhang, Y.Q., Xie, Y.W., Lee, T.Y., Wu, G.Y., Ji, J.Q., 2003. Miocene Jiali faulting and its implications for Tibetan tectonic evolution. Earth & Planetary Science Letters, 205, 185-194.

Li, Z.H., Lin, S.L., Cong, F., Xie, T., Zou, G.F., 2012a. U-Pb ages of zircon from metamorphic rocks of the Gaoligongshan Group in western Yunnan and its tectonic significance. Acta Petrologica Sinica, 28(5), 1529-1541 (in Chinese with English abstract).

Li, Z.H., Lin, S.L., Cong, F., Zou, G.F., Xie, T., 2012b. U-Pb Dating and Hf Isotopic Compositions of Quartz Diorite and Monzonitic Granite from the Tengchong-Lianghe Block, western Yunnan, and its Geological Implications. Acta Geologica Sinica, 86, 1047-1062 (in Chinese with English abstract).

Li, Z.H., Lin, S.L., Cong, F., Zou, G.F., Xie, T., 2012c. Early Cretaceous Magmatism in Tengchong-Lianghe Block, Western Yunnan. Bulletin of Mineralogy, Petrology and Geochemistry, 31, 590-598 (in Chinese with English abstract).

Li, D.P., Luo, Z.H., Chen, Y.L., Liu, J.Q., Jin, Y., 2014. Deciphering the origin of the Tengchong block, west Yunnan: Evidence from detrital zircon U-Pb ages and Hf isotopes of Carboniferous strata. Tectonophysics, 614, 66-77.

Lin, H.T., Lo, C.H., Chung, S.L., Hsu, F.J., Yeh, M.W., Lee, T.Y., Ji, J.Q., Wang, Y.Z., Liu, D.Y., 2009. $^{40}Ar/^{39}Ar$ dating of the Jiali and Gaoligong shear zones: Implications for crustal deformation around the Eastern Himalayan Syntaxis. Journal of Asian Earth Sciences, 34, 674-685.

Liu, Y., Liu, X.M., Hu, Z.C., Diwu, C.R., Yuan, H.L., Gao, S., 2007. Evaluation of accuracy and long-term stability of determination of 37 trace elements in geological samples by ICP-MS. Acta Petrologica Sinica, 23 (5), 1203-1210 (in Chinese with English abstract).

Liu, S., Hu, R.Z., Gao, S., Feng, C.X., Huang, Z.L., Lai, S.C., Yuan, H.L., Liu, X.M., Coulson, I. M., Feng, G.Y., Wang, T., Qi, Y.Q., 2009. U-Pb zircon, geochemical and Sr-Nd-Hf isotopic constraints on the age and origin of Early Paleozoic I-type granite from the Tengchong-Baoshan block, western Yunnan Province, SW China. Journal of Asian Earth Sciences, 36, 168-182.

Ludwig, K.R., 2003. ISOPLOT 3.0: A geochronological toolkit for Microsoft Excel. Berkeley Geochronology Center, Special Publication, 4, 71.

Ma, L. Wang, Q., Li, Z.X., Wyman, D.A., Jiang, Z.Q., Yang, J.H., Gou, G.N., Guo, H.F., 2013. Early Late Cretaceous (ca. 93 Ma) norites and hornblendites in the Milin area, eastern Gangdese: Lithosphere-asthenosphere interaction during slab roll-back and an insight into early Late Cretaceous (ca. 100–80 Ma) magmatic "flare-up" in southern Lhasa (Tibet). Lithos, 172-173, 17-30.

Maniar, P.D., Piccoli, P.M., 1989. Tectonic discrimination of granitoids. Geological Society of American Bulletin, 101, 635-643.

Meng, F.Y., Zhao, Z.D., Zhu, D.C., Mo, X.X., Guan, Q., Huang, Y., Dong, G.C., Zhou, S., Depaolo, D., Harrison, T.M., Zhang, Z.C., Liu, J.L., Liu, Y.S., Hu, Z.C., Yuan, H.L., 2014. Late Cretaceous magmatism in Mamba area, central Lhasa subterrane: Products of back-arc extension of Neo-Tethyan Ocean? Gondwana Research, 26, 505-520.

Metcalfe, I., 2002. Permian tectonic framework and palaeogeography of SE Asia. Journal of Asian Earth Sciences, 20, 551-566.

Metcalfe, I., 2011. Palaeozoic-Mesozoic history of SE Asia. Geological Society, London, Special Publications, 355, 7-35.

Metcalfe, I., 2013. Gondwana dispersion and Asian accretion: Tectonic and palaeogeographic evolution of eastern Tethys. Journal of Asian Earth Sciences, 66,1-33.

Mitchell, A., Chung, S.L., Oo, T., Lin, T.H., Hung, C.H., 2012. Zircon U-Pb ages in Myanmar: Magmatic-metamorphic events and the closure of a neo-Tethyan ocean? Journal of Asian Earth Sciences, 56, 1-23.

Pan, G.T., Mo, X.X., Hou, Z.Q., Zhu, D.C., Wang, L.Q., Li, G.M., Zhao, Z.D., Geng, Q.R., Liao, Z.L., 2006. Spatial-temporal framework of the Gangdese Orogenic Belt and its evolution. Acta Petrologica Sinica, 22, 521-533 (in Chinese with English abstract).

Pan, F.B., Zhang, H.F., Xu, W.C., Guo, L., Wang, S., Luo, B.J., 2014. U-Pb zircon chronology, geochemical and Sr-Nd isotopic composition of Mesozoic-Cenozoic granitoids in the SE Lhasa terrane: Petrogenesis and tectonic implications. Lithos, 192-195, 142-157.

Pearce, J.A., 1996. Sources and settings of granitic rocks. Episodes, 19, 120-125.

Pearce, J.A., Harris, N.B.W., Tindle, A.G., 1984. Trace element discrimination diagrams for the tectonic interpretation of granitic rocks. Journal of Petrology, 25, 956-983.

Qi, X.X., Zhu, L.H., Hu, Z.C., Li, Z.Q., 2011. Zircon SHRIMP U-Pb dating and Lu-Hf isotopic composition for Early Cretaceous plutonic rocks in Tengchong block, southeastern Tibet, and its tectonic implications. Acta Petrologica Sinica, 27, 3409-3421 (in Chinese with English abstract).

Qi, X.X., Zhu, L.H., Grimmer, J.C., Hu, Z.C., 2014. Tracing the Transhimalayan magmatic belt and the Lhasa block southward using zircon U-Pb, Lu-Hf isotopic and geochemical data: Cretaceous-Cenozoic granitoids in the Tengchong block, Yunnan, China. Journal of Asian Earth Sciences (in press).

Qu, X.M., Wang, R.J., Xin, H.B., Jiang, J.H., Chen, H., 2012. Age and petrogenesis of A-type granites in the middle segment of the Bangonghu-Nujiang suture, Tibetan plateau. Lithos, 146-147, 264-275.

Rapp, R.P., Watson, E.B., 1995. Dehydration Melting of Metabasalt at 8-32 kbar: Implication for Continental Growth and Crust-Mantle Recycling. Journal of Petrology, 36, 891-931.

Replumaz, A., Tapponnier, P., 2003. Reconstruction of deformed collision zone Between India and Asia by backward motion of lithospheric blocks. Journal of Geophysical Research, 108 (B6),2285.

Replumaz, A., Capitanio, A., Guillot, S., Negredo, A.N., Villasenor, A., 2014. The coupling of Indian subduction and Asian continental tectonics. Gondwana Research, 26, 608-626.

Rollinson, H., 1993. Using geochemical data: Evaluation, presentation, interpretation. London: Longman Scientific & Technical.

Rushmer, T., 1991. Partial melting of two amphibolites: contrasting experimental results under fluid-absent conditions. Contributions to Mineralogy and Petrology, 107, 41-59.

Schoenbohm, L.M., Burchfield, B.C., Chen, L.Z., Yin, J.Y., 2006. Miocene to present activity along the Red River fault, China, in the context of continental extrusion, upper-crustal rotation and lower-crustal flow. Geological Society of America Bulletin, 118, 672-688.

Searle, M.P., 2006. Role of the Red River Shear zone, Yunnan and Vietnam, in the continental extrusion of SE Asia. Journal of the Geological Society, 163, 1025-1036.

Searle, M.P., Noble, S.R., Cottle, J.M., Waters, D.J., Mitchell, A.H.G., Hlaing, T., 2007. Tentonic evolution of the Mogok metamorphic belt, Burma (Myanmar) constrained by U-Th-Pb dating of metamorphic and magmatic rocks. Tectonics, 26, 1-24.

Socquet, A., Pubelier, M., 2005. Cenozoic deformation in western Yunnan (China-Myanmar border). Journal of Asian Earth Sciences, 24, 495-515.

Song, S.G., Ji, J.Q., Wei, C.J., Su, L., Zheng, Y.D., Song, B., Zhang, L.F., 2007. Early Paleozoic granite in Nujiang River of northwest Yunnan in SW China and its tectonic implications. Chinese Science Bulletin, 52, 2402-2406.

Sun, S.S., McDonough, W.F., 1989. Chemical and isotopic systematics of oceanic basalts: Implications for mantle composition and processes. In: Saunders, A.D., Norry, M.J. Magmatism in the Ocean Basins. Geological Society, London, Special Publications, 42, 313-345.

Sylvester, P.J., 1998. Post-collisional strongly peraluminous granites. Lithos, 45, 29-44.

Tapponnier, P., Peltzer, G., Armijo, R., Le Dain, A.Y., Cobbold, P., 1982. Propagating extrusion tectonics in Asia: new insights from simple experiments with plasticine. Geology, 10, 611-616.

Tapponnier, P., Peltzer, G., Armijo, R., 1986. On the mechanics of the collision between India and Asia. Geological Society, London, Special Publications, 19, 113-157.

Wang, E.C., Burchfiel, B.C., 1997. Interpretation of Cenozoic tectonics in the right-lateral accommodation zone between the Ailaoshan shear zone and the eastern Himalayan syntaxis. International Geology Review, 39, 191-219.

Wang, X.D., Shi, G.R., Sugiyama, T., 2002. Permian of west Yunnan, Southwest China: A biostratigraphic synthesis. Journal of Asian Earth Science, 20, 647-656.

Wang, Y.J., Fan, W.M., Zhang, Y.H., Peng, T.P., Chen, X.Y., Xu, Y.G., 2006. Kinematics and $^{40}Ar/^{39}Ar$ geochronology of the Gaoligong and Chongshan shear systems, western Yunnan, China: Implications for early Oligocene tectonic extrusion of SE Asia. Tectonophysics, 418, 235-254.

Wang, G., Wan, J.L., Wang, E., Zheng, D.W., Li, F., 2008. Late Cenozoic to recent transtensional deformation across the Southern part of the Gaoligong shear zone between the Indian plate and SE margin of the Tibetan plateau and its tectonic origin. Tectonophysics, 460, 1-20.

Wang, Y.J., Xing, X.X., Cawood, P.A., Lai, S.C., Xia, X.P., Fan, W.M., Liu, H.C., Zhang, F.F., 2013. Petrogenesis of early Paleozoic peraluminous granite in the Sibumasu Block of SW Yunnan and diachronous accretionary orogenesis along the northern margin of Gondwana. Lithos, 182-183, 67-85.

Wang, C.M., Deng, J., Carranza, E.J.M., Santosh, M., 2014. Tin metallogenesis associated with granitoids in the southwestern Sanjiang Tethyan Domain: Nature, deposit types, and tectonic setting. Gondwana Research, 26, 576-593.

Wang, J.G., Wu, F.Y., Tan, X.C., Liu, C.Z., 2014. Magmatic evolution of the Western Myanmar Arc documented by U-Pb and Hf isotopes in detrital zircon. Tectonophysics, 612-613, 97-105.

Wang, Q., Zhu, D.C., Zhao, Z.D., Liu, S.A., Chung, S.L., Li, S.M., Liu, D., Dai, J.G., Wang, L.Q., Mo, X.X., 2014c. Origin of the ca. 90 Ma magnesia-rich volcanic rocks in SE Nyima, central Tibet: Products of lithospheric delamination beneath the Lhasa-Qiangtang collision zone. Lithos, 198-199, 24-37.

Wen, D.R., Chung, S.L., Song, B., Iizuka, Y., Yang, H.J., Ji, J.Q., Liu, D.Y., Gallet, S., 2008a. Late Cretaceous Gangdese intrusions of adakitic geochemical characteristics, SE Tibet: Petrogenesis and tectonic implications. Lithos, 105, 1-11.

Wen, D.R., Liu, D.Y., Chung, S.L., Chu, M.F., Ji, J.Q., Zhang, Q., Song, B., Lee, T.Y., Yeh, M.W., Lo, C.H., 2008b. Zircon SHRIMP U-Pb ages of the Gangdese Batholith and implications for Neotethyan subduction in southern Tibet. Chemical Geology, 252, 191-201.

Whalen, J.B., Currie, K.L., Chappell, B.W., 1987. A-type granites: Geochemical characteristics, discrimination and petrogenesis. Contributions to Mineralogy and Petrology, 95, 407-419.

Winchester, J.A., Park, R.G., Holland, J.G., 1980. The geochemistry of Lewisian semipelitic schists from the Gairloch district, North-West Scotland. Scottish Journal of Geology, 16, 165-179.

Wopfner, H., 1996. Gondwana origin of the Baoshan and Tengchong terrenes of west Yunnan. In: Hall, R., Blundell, D. Tectonic evolution of Southeast Asian. Journal of Geology Society, London, 106, 539-547.

Xu, Y.G., Lan, J.B., Yang, Q.J., Huang, X.L., Niu, H.N., 2008. Eocene break-off of the Neo-Tethyan slab as inferred from intraplate-type mafic dykes in the Gaoligong orogenic belt, eastern Tibet. Chemical Geology, 255, 439-453.

Xu, Y.G., Yang, Q.J., Lan, J.B., Luo, Z.Y., Huang, X.L., Shi, Y.R., Xie, L.W., 2012. Temporal-spatial distribution and tectonic implications of the batholiths in the Gaoligong-Tengliang-Yingjiang area, western Yunnan: Constraints from zircon U-Pb ages and Hf isotopes. Journal of Asian Earth Sciences, 53, 151-175.

Yang, Q.J., Xu, Y.G., 2011. The Emplacement of Granites in Nujiang-Gaoligong Belt, Western Yunnan, and Response to the Evolvement of Neo-Tethys. Journal of Jilin University: Earth Science Edition, 41, 1353-1361 (in Chinese with English abstract).

Yang, Q.J., Xu, Y.G., Huang, X.L., Luo, Z.Y., 2006. Geochronology and geochemistry of granites in the Gaoligong tectonic belt, western Yunnan: Tectonic implications. Acta Petrologica Sinica, 22, 817-834 (in Chinese with English abstract).

Yin, A., Harrison, T.M., 2000. Geological evolution of the Himalayan-Tibetan Orogen. Annual Review of Earth and Planetary Sciences, 28, 211-280.

Yuan, H.L., Gao, S., Liu, X.M., Li, H.M., Gunther, D., Wu, F.Y., 2004. Accurate U-Pb age and trace element determinations of zircon by laser ablation-inductively coupled plasma mass spectrometry. Geostandard Newsletters, 28, 353-370.

Yuan, H.L., Gao, S., Luo, Y., Zong, C.L., Dai, M.N., Liu, X.M., Diwu, C.R., 2007. Study of Lu-Hf geochronology: A case study of eclogite from Dabie UHP Belt. Acta Petrologica Sinica, 23(2), 233-239 (in Chinese with English abstract).

Zartman, R.E., Doe, B.R., 1981. Plumbotectonics: The Model. Tectonophysics, 75, 135-162.

Zhang, B., Zhang, J.J., Zhong, D.L., 2010. Structure kinematics and ages of transpression during strain-partitioning in the Chongshan shear zone, western Yunnan, China. Journal of Structural Geology, 32, 445-463.

Zhang, Z.M., Zhao, G.C., Santosh, M., Wang, J.L., Dong, X., Shen, K., 2010. Late Cretacecous charnockite with adakitic affinities from the Gangdese batholith, southeastern Tibet: Evidence for Neo-Tethyan mid-ocean ridge subduction? Gondwana Research, 17, 615-631.

Zhang, Z.M., Dong, X., Xiang, H., Liou, J.G., Santosh, M., 2013. Building of the Deep Gangdese Arc, South Tibet: Paleocene Plutonism and Granulite-Facies Metamorphism. Journal of Petrology, 54, 2547-2580.

Zhao, S.W., Lai, S.C., Qin, J.F., Zhu, R.Z., 2014. Zircon U-Pb ages, geochemistry and Sr-Nd-Pb-Hf isotopic compositions of the Pinghe pluton, Southwest China: Implications for the evolution of the Early Paleozoic Proto-Tethys in Southeast Asia. International Geology Review, 56, 885-904.

Zheng, Y.C., Hou, Z.Q., Gong, Y.L., Liang, W., Sun, Q.Z., Zhang, S., Fu, Q., Huang, K.X., Li, Q.Y., Li, W., 2014. Petrogenesis of Cretaceous adakite-like intrusions of the Gangdese Plutonic Belt, southern Tibet: Implications for mid-ocean ridge subduction and crustal growth. Lithos, 190-191, 240-263.

Zhong, D.L., 1998. Paleo-Tethyan Orogenic Belt in the Western Part of the Sichuan and Yunnan Provinces. Beijing: Science Press, 1-230 (in Chinese).

Zhu, D.C., Mo, X.X., Niu, Y.L., Zhao, Z.D., Wang, L.Q., Liu, Y.S., Wu, F.Y., 2009a. Geochemical investigation of Early Cretaceous igneous rocks along an east-west traverse throughout the central Lhasa Terrane, Tibet. Chemical Geology, 268, 298-312.

Zhu, D.C., Mo, X.X., Wang, L.Q., Zhao, Z.D., Niu, Y.L., Zhou, C.Y., Yang, Y.H., 2009b. Petrogenesis of highly fractionated I-type granites in the Zayu area of eastern Gangdese, Tibet: Constraints from zircon U-Pb geochronology, geochemistry and Sr-Nd-Hf isotopes. Science in China: Series D, Earth Sciences, 52, 1223-1239.

Zhu, D.C., Zhao, Z.D., Niu, Y.L., Mo, X.X., Chung, S.L., Hou, Z.Q., Wang, L.Q., Wu, F.Y., 2011. The Lhasa Terrane: Record of a microcontinent and its histories of drift and growth. Earth & Planetary Science Letters, 301, 241-255.

Petrogenesis of Eocene granitoids and microgranular enclaves in the western Tengchong Block: Constraints on eastward subduction of the Neo-Tethys①

Zhao Shaowei Lai Shaocong② Qin Jiangfeng Zhu Renzhi

Abstract: Eocene granitic and related igneous rocks in the western Tengchong Block are considered to be the result of eastward subduction of Neo-Tethyan oceanic lithosphere beneath the Tengchong Block.In this paper we show that the granitic and mafic rocks in the western Tengchong Block exhibit a systematic compositional variation from west to east, with Na-rich granodiorites in the Nabang area (west) that differ from coeval high-K calc-alkaline granodiorites in the Bangwan area (east), and with tholeiitic mafic rocks in the Nabang area that differ from shoshonitic mafic microgranular enclaves (MMEs) in granodiorites of the Bangwan area. In addition, high-silica biotite granites were intruded into the granodiorites in the Bangwan area. The host granodiorites, MMEs, and biotite granites in the Bangwan area yield zircon U-Pb ages of ca. 50 Ma. The MMEs have relatively low SiO_2 contents (53.1–64.95 wt%) and $Mg^{\#}$ values (37–45), and high K_2O (4.14–5.02 wt%) and $\sum$REE contents (331–509 ppm); the MMEs contain acicular apatites that indicate quenching. The host granodiorites also have high K_2O (4.48–5.95 wt%) and $\sum$REE compositions (320–459 ppm), and together with the MMEs they are enriched in Th but depleted in Nb and Ti. The Sr-Nd-Pb isotopic compositions of the host granodiorites and the MMEs are similar, with $\varepsilon_{Hf}(t)$ values of −10.8 to −1.0 and −11.1 to 3.3, respectively. The geochemical data and igneous textures suggest that the MMEs represent a mafic magma that was derived from the partial melting of mantle pyroxenite, with the melting induced by the influx of fluids/melts from the recycling of sediments in the subducted slab. The mafic melts then caused the partial melting of lower crustal tonalitic rocks to produce granodioritic magma that was subsequently mixed with mafic magma. The biotite granites have relatively high SiO_2 contents and low $Mg^{\#}$ values that indicate a purely crustal origin and derivation from the partial melting of upper crustal metagraywacke. The similar isotopic compositions of the biotite granites and the granodiorites can be interpreted to reflect the influence of fluids released during crystallization of the granodiorites. A comparison of contemporaneous Eocene granitic and mafic rocks in Nabang, Tongbiguan, Longchuan, and Bangwan in the Tengchong Block reveals that the mafic rocks vary from tholeiitic through high-K

---

① Published in *Lithos*, 2016, 264.

② Corresponding author.

calc-alkaline to shoshonitic in composition from west to east, while the granitic rocks vary from Na- to K-rich. These systematic compositional variations indicate that the arc magmatism in the western Tengchong Block was the result of the eastward subduction of the Neo-Tethyan oceanic lithosphere. Fluids or/and melts from the subducted slab (including sediments) are likely to have played an important role in producing these compositional variations.

1 Introduction

The flare-up in Late Cretaceous-Eocene magmatism, represented by the Gangdese Belt and Linzizong volcanic successions, provides information on the northward subduction of the Neo-Tethyan oceanic lithosphere and the subsequent India-Asia collision (e.g., Chung et al., 2005; Guo et al., 2007, 2013; Wang et al.,2015; Wen et al., 2008a,b; Zhu et al., 2011; Zhang et al., 2013). Recent seismic tomographic results indicate eastward subduction of the Indian continental slab beneath the Burma-Tengchong Blocks (e.g., Li et al., 2008; He et al., 2010; Zhao et al., 2011a), from which it might be inferred that the Neo-Tethyan slab should also have been subducted eastward beneath the Burma-Tengchong Blocks prior to subduction of the continental slab. The Eocene magmatic rocks in the Lhasa Block are dominated by the calc-alkaline Linzizong volcanic rocks (Mo et al., 2007, 2008) and plutonic equivalents (the Gangdese plutonics), which include gabbros and granites in the Yangbajing and Dangxung areas (Gao et al., 2010; Zhao et al., 2011). In addition, the enrichment components of Eocene magmatic rocks in the Lhasa Block are progressively increasing from south to north (Zhao et al., 2011). Contemporaneous intrusive rocks are also widespread in the Nabang area (westmost Tengchong Block) near the Myanmar-China boundary, and represent the products of eastward subduction of Neo-Tethyan oceanic crust. The intrusives comprise mafic to acid rocks with ages of ca. 50–55 Ma (Ji et al., 2000; Ma et al., 2014; Wang et al., 2014, 2015;Xu et al.,2012). Ji et al. (2000) interpreted the mafic rocks as representing the subducted Neo-Tethys upper-oceanic crust. Wang et al. (2014, 2015) suggested that rollback of the subducted oceanic slab took place, inducing mafic magmatism that was influenced by components derived from the subducted slab. They also identified a systematic compositional variation in the Eocene mafic rocks from west to east in the western Tengchong Block. Ma et al. (2014) suggested that the Eocene Na-rich granitic rocks in the Nabang area formed from melting induced by underplated basaltic magma. However, high-K calc-alkaline granodiorites with shoshonitic mafic microgranular enclaves (MMEs) and high-silica biotite granites occur in the Bangwan area, east of Nabang (east of the western Tengchong Block). Compared with the mafic and granitic rocks in the Nabang area, the granodiorites and MMEs in the Bangwan area have higher contents of K_2O, Rb, Th, and the rare earth elements (REEs), and clearly differ in composition from the coeval mafic and granitic rocks in the Nabang area. The Eocene rocks in the western Tengchong Block also appear to be increasing enrichment composition from west to east, similar to the situation in the Lhasa Block (Gao et al., 2010; Zhao et al., 2011). Generally, magmatism that is related to a subduction zone occurs mainly within a magmatic arc at a distance from the trench of 150–300 km, and there is usually an obvious spatial trend in the composition of extrusive and intrusive rocks, from near-trench tholeiitic rocks through calc-alkaline rocks to shoshonitic rocks. The Early Eocene mafic and granitic magmatism in the western Tengchong Block shows a spatial variation from west to east, which is closely related to subduction of the Neo-Tethyan oceanic lithosphere that existed between India and Asia, but the detailed reasons for the systematic compositional variations in the Eocene magmatism have not been clarified. Consequently, our study has been aimed at clarifying ① the petrogenesis of the high-K calc-alkaline granodiorites, its MMEs, and the associated peraluminous biotite granites; and ② the relationship between the mafic and granitic magmas in the western Tengchong Block and the eastward subduction of the Neo-Tethyan oceanic slab.

2 Geological setting

The study area is located in southeastern Yingjiang County, which tectonically belongs to the Tengchong Block (Fig. 1). This block is bound by the Gaoligong Belt to the east, the Sagaing Fault to the west (Replumaz and Tapponnier, 2003), and the Ruili Fault to the southeast (Fig. 1). Voluminous Mesozoic to Cenozoic magmatism took place in the Tengchong Block (Chen et al., 2015; Guo et al., 2015; Ma et al., 2014; Qi et al., 2015; Wang et al., 2007; Wang et al., 2014, 2015; Xu et al., 2008, 2012; Xie et al., in press; Zhao et al., 2016), mainly granitic plutons and basaltic volcanics. Xu et al. (2012) proposed that the granites in the Gaoligong-Tengliang-Yingjiang area are the eastern extension of the Gangdese and Northern magmatic belts in the Lhasa Block. The Early Cretaceous Gaoligong batholith (126–121 Ma) is composed of S-type granites (Yang et al., 2009; Zhu et al., 2015), the Late Cretaceous Tengliang granites (76–64 Ma) are strongly peraluminous and have an ancient crustal origin (Chen et al., 2015; Qi et al., 2015; Xu et al., 2012; Zhao et al., 2015), and the Early Eocene magmatic rocks (55–50 Ma) in the Yingjiang region near the China-Myanmar border are dominated by granites and mafic rocks that are related to juvenile mantle-derived rocks (e.g., Ma et al., 2014; Wang et al., 2014, 2015). Xu et al. (2012) and Zhu et al. (2015) considered that the Gaoligong batholith resulted from subduction of the Bangong-Nujiang oceanic lithosphere and the Baoshan-Tengchong collision. On the other hand, the Cenozoic granites may be related to the evolution of the Neo-Tethys (Ma et al., 2014; Wang et al., 2014, 2015; Zhao et al., 2016), and Eocene basaltic dikes with ages of 42–40 Ma, with intraplate characteristics, represent magmatism induced by the break-off of the subducting Neo-

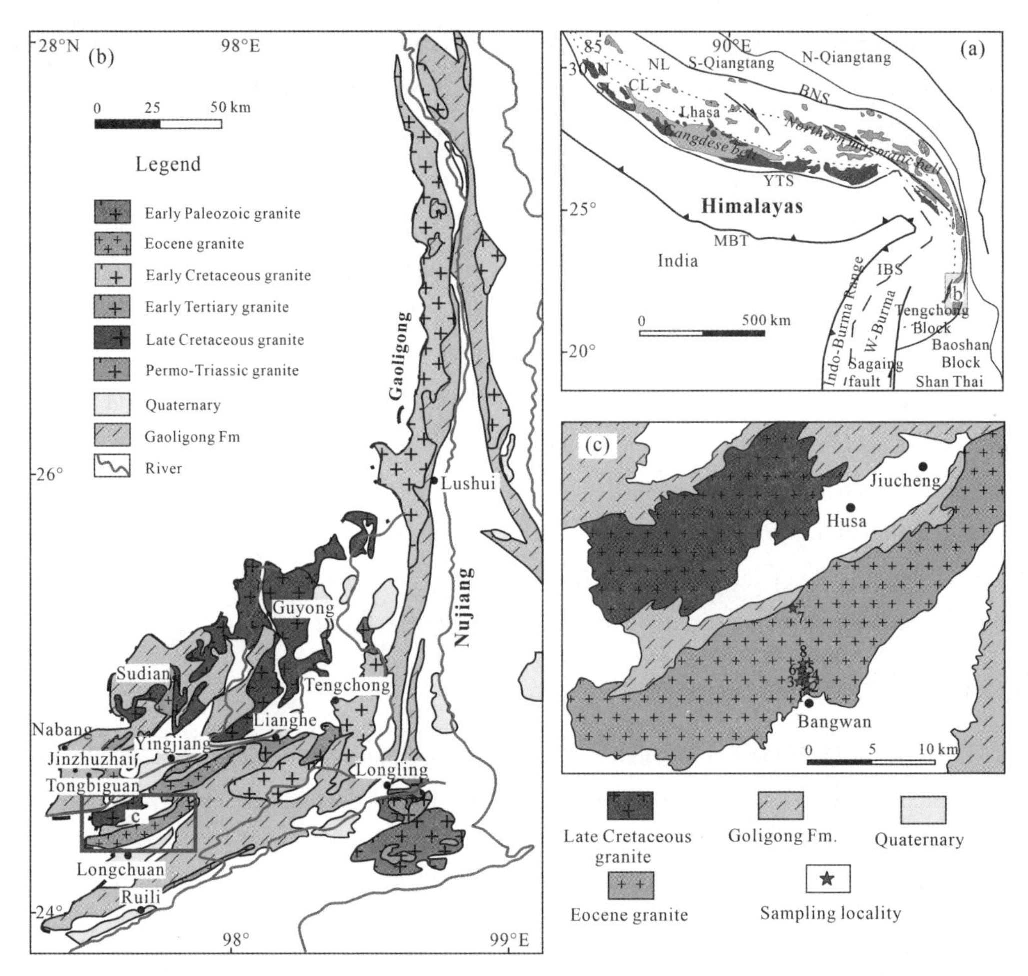

Fig. 1 Simplified geological map of Himalaya-Tibet tectonic realm (a), Tengchong Block and adjacent area in southeastern margin of Tibetan Plateau (b), and the samples localities in the Bangwan batholith(c).

Tethyan slab (Xu et al., 2008). Multiple stages of eruptive basaltic and basaltic-andesitic volcanism have taken place in the Tengchong Block since the Late Miocene, including during the periods 5.5–4.0 Ma, 3.9–0.9 Ma, 0.8–0.01 Ma, and younger than 0.01 Ma (Wang et al., 2007; Guo et al., 2015).

3 Sample descriptions and petrography

The Bangwan batholith, located near Bangwan village, Yingjiang County, is a compound massif that trends ENE-WSW, approximately parallel to the Ruili Fault. The rocks are mainly porphyraceous granodiorites that have been intruded by fine-grained biotite granites. The biotite granites form stocks that are several square meters in size in the outcrops, and the boundaries between the biotite granites and the granodiorites are sharp. Round or ovoid MMEs occur in the granodiorites but are absent from the biotite granites. The MMEs are mostly 5 – 50 cm in diameter, and have fine-grained textures with clusters of mafic minerals (Fig. 2).

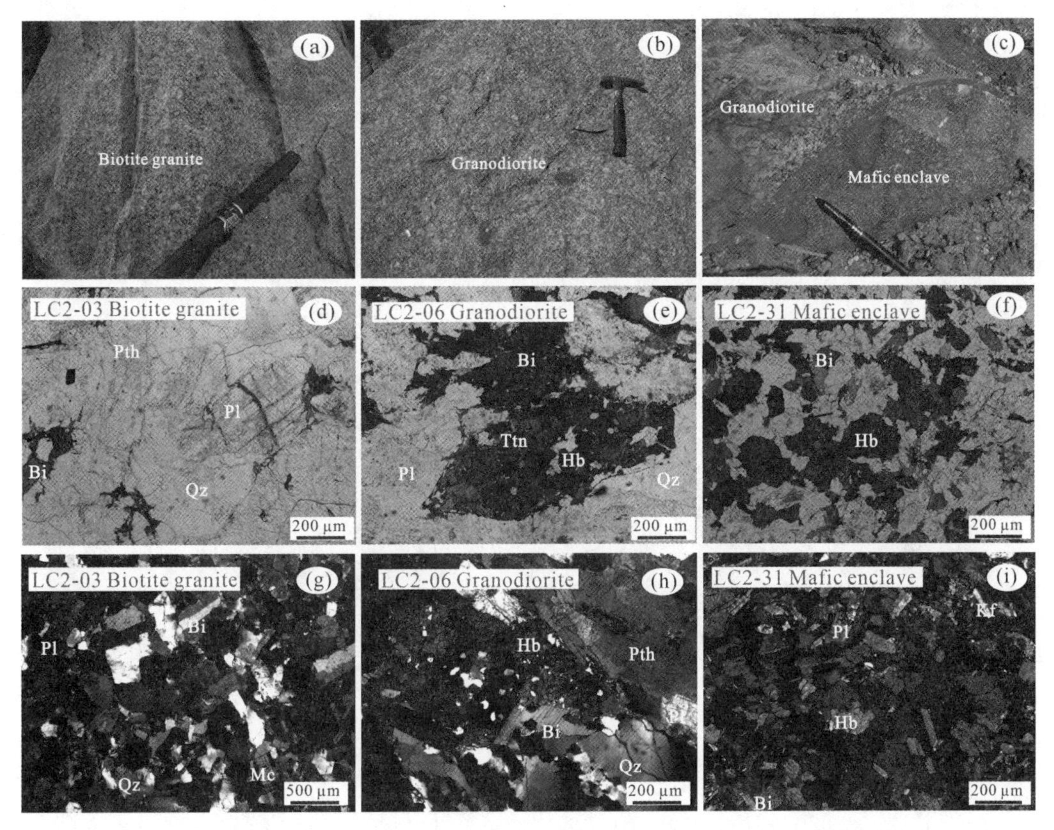

Fig. 2 Field, and microscope photographs of the granodiorites, MMEs, and biotite granites from Bangwan batholith, western Tengchong Block.

Pl: plagioclase; Bi: biotite; Hb: hornblende; Qz: quartz; Pth: perthite; Kf: K-feldspar; Mc: microcline; Ttn: titanite.

The granodiorites are composed of 5% phenocrysts of perthite (Pth) up to 20–30 mm in length set in a matrix of Pth (20%–25%), microcline (Mc, 10%–15%), plagioclase (Pl, 30%), quartz (Qz, 13%–15%), biotite (Bi, 9%–12%), and hornblende (Hb, 3%–5%). The accessory minerals in the granodiorites are mainly titanite, zircon, apatite, and magnetite. The MMEs in the granodiorites contain Hb (20%–25%), Bi (20%–25%), Pl (30%), K-feldspar (Kf, 20%–23%), and Qz (2%–5%). The presence of abundant acicular apatites and K-feldspar megacrysts in the MMEs suggests interaction between mafic and felsic magma. The biotite granites consist of Kf (15%–25%), Pth (20%–25%), Mc (0–15%), Pl (20%–28%), Qz (20%–25%), and Bi (5%–7%), and accessory minerals include zircon, apatite, and magnetite. Details of the mineral assemblages are given in Supplementary Table 1.

4 Results

4.1 Zircon U-Pb dating

Four samples were collected for zircon dating: LL96 and LC2-17 from the fine-grained biotite granites, LC2-09 from the porphyraceous granodiorites, and LC2-30 from the mafic enclaves. Zircon CL images and U-Pb isotopic results are presented in Fig. 3, and U-Pb isotopic data are listed in Supplementary Table 2.

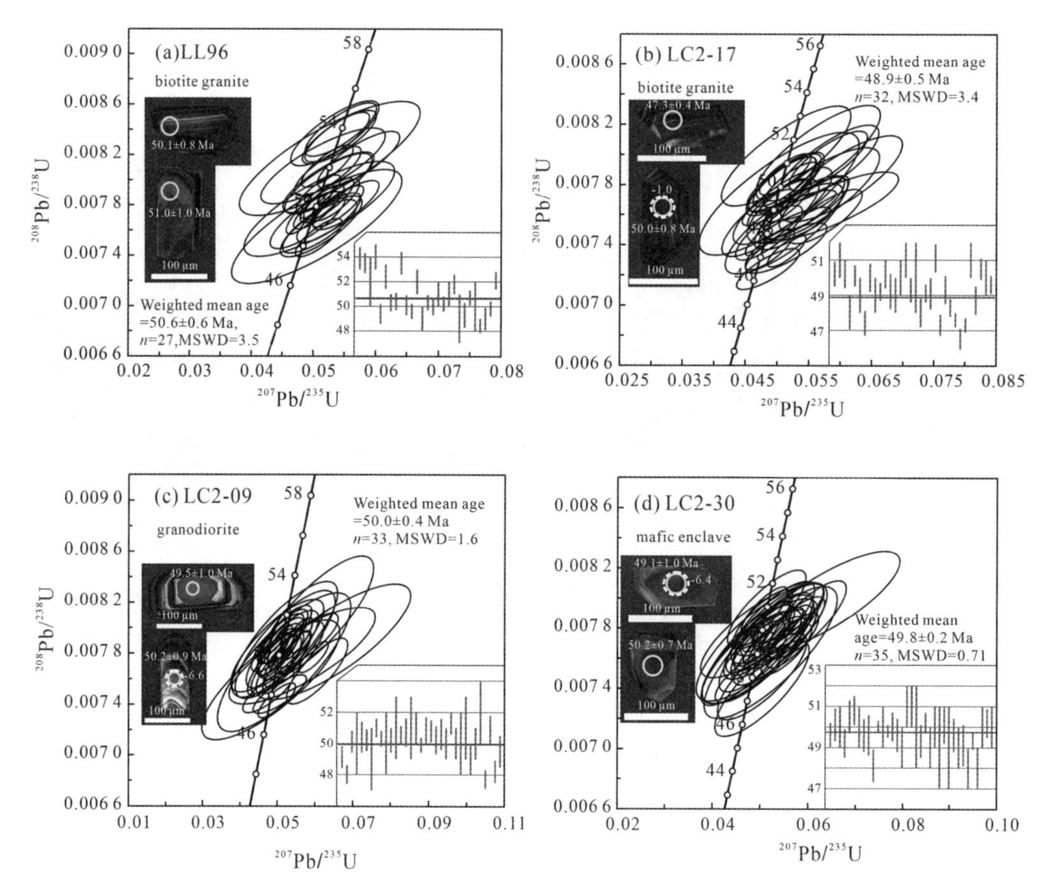

Fig. 3 U-Pb concordia diagrams and weighted mean ages of $^{206}Pb/^{238}U$ for zircons, and CL images of representative zircons from Bangwan granitic batholith, western Tengchong Block.

Zircon grains from the biotite granites are clear, stubby to elongate prisms (80–250 μm long) with well-defined oscillatory zoning. Twenty-eight spots were analyzed on LL96 zircons. Spot 16 yielded a $^{206}Pb/^{238}U$ age of 62.7 Ma and is probably a wall-rock xenocryst according to the CL image. The other 27 spots have U contents of 132–2 087 ppm, Th contents of 133–1 490 ppm, Th/U ratios of 0.22–1.04, and $^{206}Pb/^{238}U$ ages of 48.4–54.0 Ma, with a weighted mean age of 50.6 ± 0.6 Ma (MSWD = 3.5, n = 27). Thirty-two spots on LC2-17 zircons were analyzed, and have U contents of 470–4 358 ppm, Th contents of 366–3 077 ppm, Th/U ratios of 0.41–1.37, and a $^{206}Pb/^{238}U$ weighted mean age of 48.9 ± 0.5 Ma (MSWD = 3.4, n = 32). Thus, the age of crystallization of the biotite granite is 48.9–50.6 Ma.

Zircon grains from the granodiorites are elongate prismatic crystals (150–250 μm long) with aspect ratios of 1∶2 to 1∶3. In general, they show well-defined oscillatory zoning. Two spots (10 and 31) gave $^{206}Pb/^{238}U$ ages of 457–1 056 Ma, which can be interpreted as the ages of xenocrysts, as inferred from the zircon morphology. Thirty-three spots yielded concordant $^{206}Pb/^{238}U$ ages of 47.7–52.0 Ma, giving a weighted mean age of 50.0 ± 0.4 Ma (MSWD = 1.6, n = 33), which can be considered the crystallization age of the granodiorites.

Zircon grains from the MMEs are subhedral and 50–100 μm long. They are mainly dark and unzoned in CL images, and a few grains show weak zoning. Thirty-five spots yielded concordant $^{206}Pb/^{238}U$ ages of 48.0–51.0 Ma and a weighted mean age of 49.8 ± 0.2 Ma (MSWD = 0.71, n = 35), which can be considered the age of crystallization of the MMEs. They have variable U contents of 150–8 636 ppm, Th contents of 126–5 260 ppm, and Th/U ratios of 0.22–1.65.

In summary, the biotite granites, granodiorites, and MMEs were contemporaneous, with all crystallizing during the Early Eocene at ca. 50 Ma.

4.2 Major and trace elements

The results of major and trace elemental analyses of the biotite granites, granodiorites, and mafic enclaves from the Bangwan batholith are listed in Supplementary Table 3.

The biotite granites are characterized by high SiO_2 (74.33–75.41 wt%) and K_2O (5.57–6.23 wt%) contents, K_2O/Na_2O ratios of 1.93–2.48, A/CNK ratios of 1.00–1.09 (Fig. 4c), relatively low MgO (0.14–0.22 wt%) and TiO_2 (0.09–0.13 wt%) contents, and $Mg^{\#}$ values of 18–31. They belong to the high-K calc-alkaline series (Fig. 4b). The biotite granites are enriched in highly incompatible trace elements such as the light rare earth elements [LREEs; e.g., $(La/Yb)_N$ = 9.1–16.8] and large-ion lithophile elements (LILEs; Rb, K, and Th), but depleted in Nb, Ta, P, and Ti (Fig. 5a,b). They show very strong negative Eu anomalies (δEu = 0.19–0.22) and nearly flat HREE patterns with uniform $(Dy/Yb)_N$ ratios of 1.3–1.9 (Fig. 5a,b). It is noteworthy that the biotite granites have high total REE compositions of 266–435 ppm.

The granodiorites are mainly quartz monzonite in composition (Fig. 4a). Compared with the biotite granites, the granodiorites have relatively low SiO_2 contents of 67.21–69.57 wt%, but high MgO (0.74–1.06 wt%) and TiO_2 (0.42–0.52 wt%) contents, and $Mg^{\#}$ values of 36–38. They are transitional from metaluminous to weakly peraluminous and belong to the high-

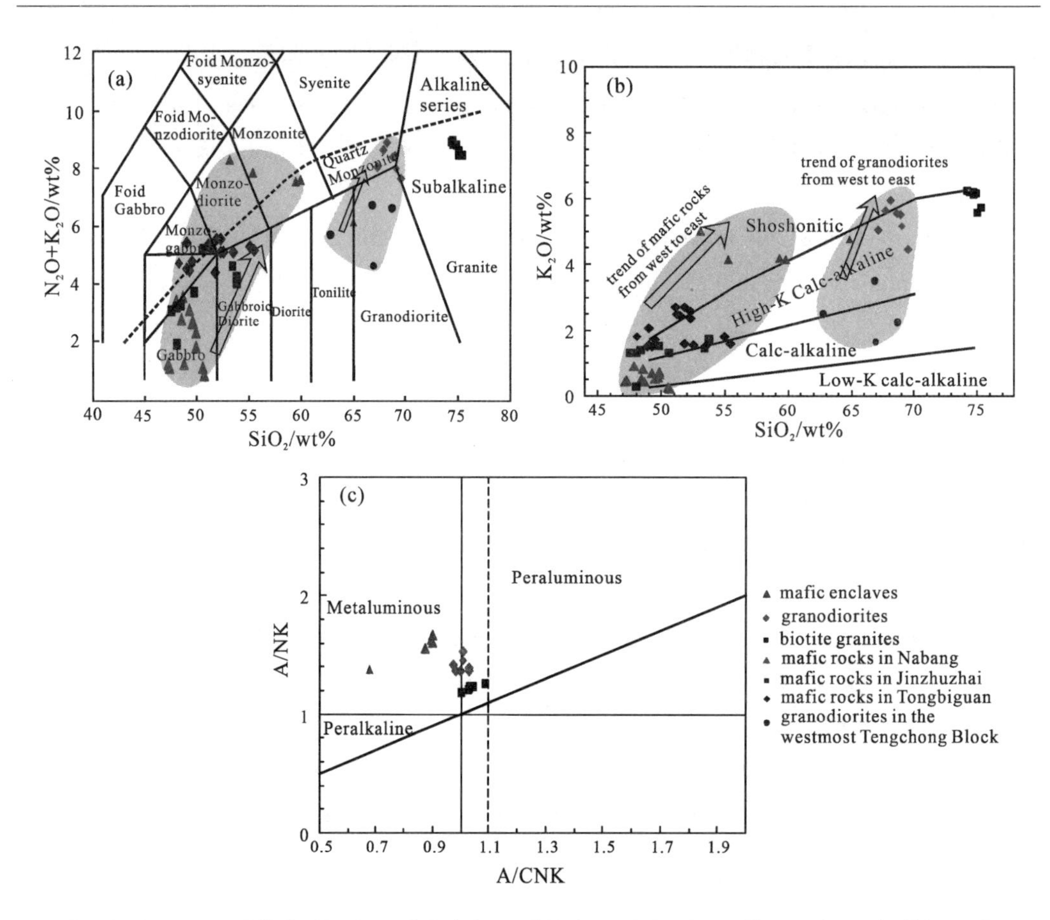

Fig. 4 Diagrams of K_2O+Na_2O vs. SiO_2(a), and K_2O vs. SiO_2 after Rollinson (1993) (b), and A/NK [molar Al_2O_3/(Na_2O+K_2O)] vs. A/CNK [molar Al_2O_3/($CaO+Na_2O+K_2O$)] (Maniar and Piccoli, 1989) (c) for granodiorites, MMEs, and biotite granites.
Data for coeval mafic and granitic rocks in Nabang, Jinzhuzhai and Tongbiguan from Wang et al. (2014, 2015), and Ma et al. (2014).

K calc-alkaline series, with high K_2O contents of 4.48–5.95 wt%, K_2O/Na_2O ratios of 1.40–2.02, and A/CNK ratios of 0.97 – 1.03. The granodiorites are enriched in LREEs and depleted in HREEs; they have flat HREE patterns, extremely high total REE compositions of 320–459 ppm, negative Eu anomalies (δEu = 0.31 – 0.55), high $(La/Yb)_N$ ratios of 9.5–20.4, and uniform $(Gd/Yb)_N$ ratios of 1.8 – 2.3 (Fig. 5c,d). In the primitive mantle normalized spider diagram, the granodiorites show spikes in Rb, Sr, and Th, and troughs in Ba, Nb, Ta, and Ti, similar to the upper continental crust (Fig. 5c).

The MMEs are mainly monzodiorite to monzonite in composition (Fig. 4a). Compared with the host granodiorites, the MMEs have lower and variable SiO_2 contents of 53.1–64.95 wt%, MgO contents of 2.31 – 2.76 wt%, $Mg^{\#}$ values of 37 – 45, and TiO_2 contents of 0.47 – 1.23 wt%. They also have high K_2O contents of 4.14–5.02 wt%, K_2O/Na_2O ratios of 1.1–3.41,

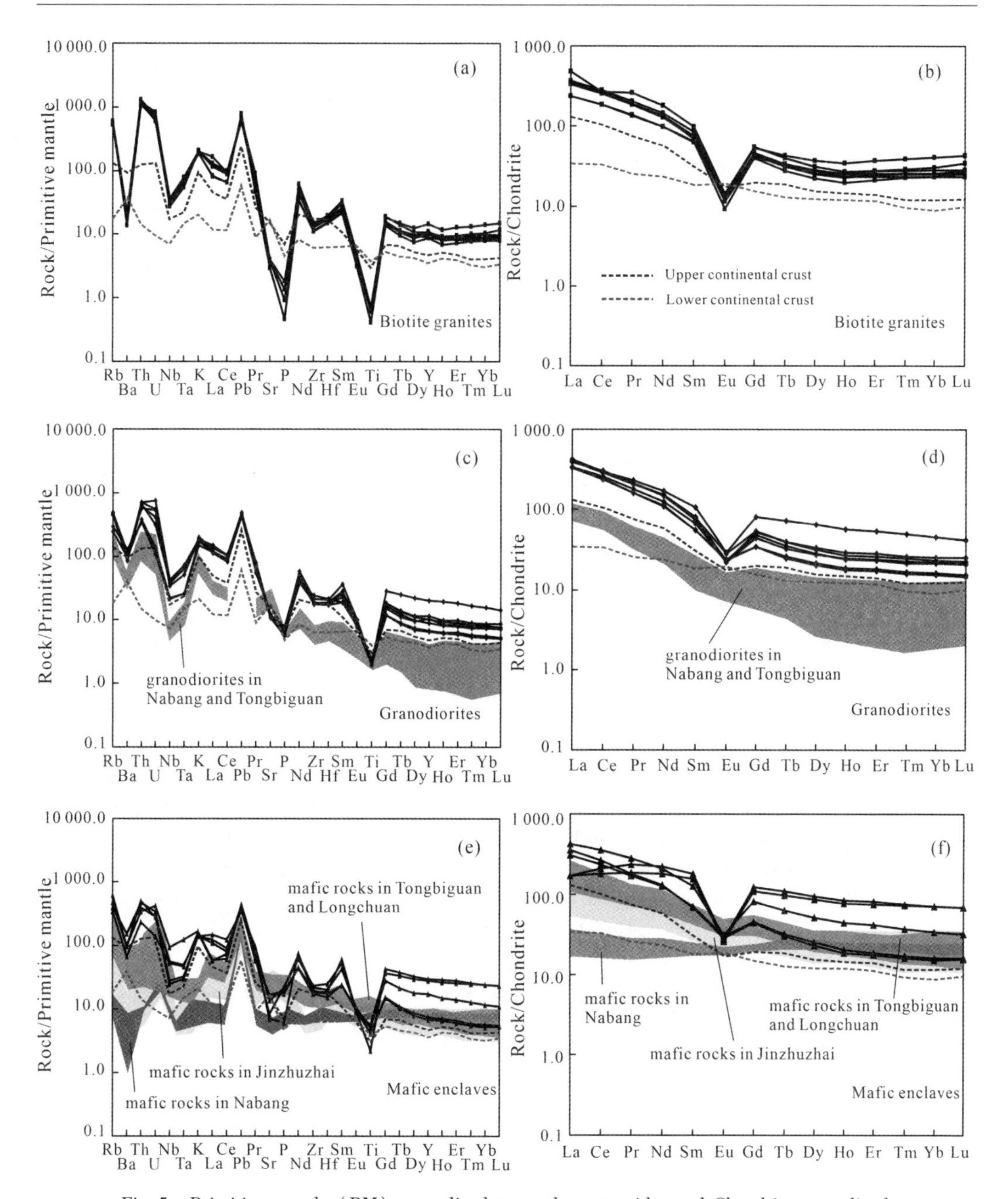

Fig. 5 Primitive mantle (PM) normalized trace element spider and Chondrite-normalized REE pattern diagrams for granodiorites, MMEs, and biotite granites.

The Primitive mantle and Chondrite values are from Sun and McDonough (1989). Data for coeval mafic and granitic rocks in Nabang, Jinzhuzhai, Tongbiguan and Longchuan from Wang et al. (2014, 2015), and Ma et al. (2014).

and they are shoshonitic, metaluminous, and transitional from subalkaline to alkaline, with A/CNK ratios of 0.67 – 0.9. The MMEs also have high total REE compositions of 331 – 509 ppm, are enriched in LREEs, depleted in HREEs, have flat HREE patterns and negative Eu anomalies (δEu = 0.18 – 0.53), variable $(La/Yb)_N$ ratios of 2.4 – 23.3, and uniform

(Gd/Yb)$_N$ ratios of 1. 5–2. 9 (Fig. 5e ,f). They show spikes in Rb, Th(U) , K, and Nd, and troughs in Ba, Nb, Ta, and Ti, similar to the upper continental crust (Fig. 5e ,f).

4. 3 Sr-Nd-Pb isotopic data

Whole-rock Sr-Nd-Pb isotopic compositions are given in Supplementary Table 4 and Figs. 6 and 7. Initial isotopic values were calculated at $t = 50$ Ma on the basis of LA-ICP-MS zircon U-Pb ages for the biotite granites, granodiorites, and MMEs.

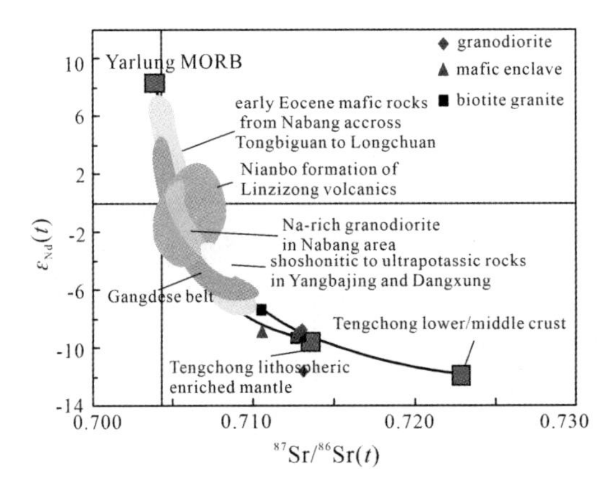

Fig. 6 $^{87}Sr/^{86}Sr(t)$ vs. $\varepsilon_{Nd}(t)$ diagram for granodiorites, MMEs, and biotite granites.

Data for Nianbo volcanic rocks from Mo et al. (2008) and Lee et al. (2012), for Gangdese belt granitic rocks from Wen et al. (2008a, 2008b) and Huang et al. (2010), for early Eocene mafic rocks in Nabang, Tongbiguan and Longchuan from Wang et al. (2014), for Na-rich granodiorite in Nabang from Ma et al. (2014), for shoshonitic to ultrapotassic rocks in Yangbajing and Dangxung from Zhao et al. (2011b) and Gao et al. (2010). The data for end member of Yarlung MORB from Xu and Castillo. (2004), and for Tengchong lithospheric enriched mantle and lower/ middle crust from Wang et al. (2014).

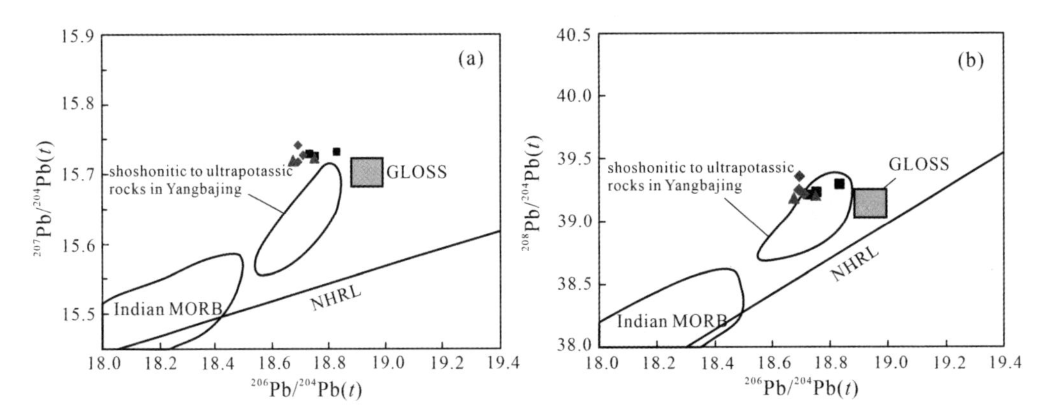

Fig. 7 $^{207}Pb/^{204}Pb$ vs. $^{206}Pb/^{204}$(a) and $^{208}Pb/^{204}Pb$ vs. $^{206}Pb/^{204}Pb$ (b) diagrams for granodiorites, MMEs, and biotite granites.

Data for Indian MORB from Hofmann (1988), for shoshonitic to ultrapotassic rocks in Yangbajing from Gao et al. (2010).

The biotite granites have high initial $^{87}Sr/^{86}Sr(t)$ ratios of 0. 710 526 – 0. 713 002, negative $\varepsilon_{Nd}(t)$ values of −9. 2 to −7. 4 with unified model ages of 1. 25 – 1. 39 Ga, and evolved Pb isotopic compositions with $^{206}Pb/^{204}Pb(t)$ = 18. 735 2–18. 832 9, $^{207}Pb/^{204}Pb(t)$ = 15. 725 0–15. 731 5, and $^{208}Pb/^{204}Pb(t)$ = 39. 211 6–39. 292 8.

The granodiorites have similar isotopic compositions to the biotite granites, with high initial $^{87}Sr/^{86}Sr(t)$ ratios of 0. 712 751–0. 713 073, negative $\varepsilon_{Nd}(t)$ values of −11. 6 to −8. 8 with unified model ages of 1. 34–1. 54 Ga, and slightly lower evolved Pb isotopic compositions with $^{206}Pb/^{204}Pb(t)$ = 18. 695 0 – 18. 711 4, $^{207}Pb/^{204}Pb(t)$ = 15. 718 3 – 15. 741 9, and $^{208}Pb/^{204}Pb(t)$ = 39. 220 7–39. 350 7.

The MMEs have initial $^{87}Sr/^{86}Sr(t)$ ratios of 0. 710 545 – 0. 712 570, negative $\varepsilon_{Nd}(t)$ values of −9. 1 to −8. 8 with unified model ages from 1. 35 Ga to 1. 37 Ga, and evolved Pb isotopic compositions with $^{206}Pb/^{204}Pb(t)$ = 18. 675 4–18. 752 4, $^{207}Pb/^{204}Pb(t)$ = 15. 718 8–15. 724 2, and $^{208}Pb/^{204}Pb(t)$ = 39. 184 4–39. 212 3.

4. 4 Zircon Hf isotope compositions

The zircons from three samples (granodiorite LC2-09, mafic enclave LC2-30, and biotite granite LC2-17) were analyzed for in situ Lu-Hf isotopes, and the results are listed in Supplementary Table 5 and Fig. 8. The initial $^{176}Hf/^{177}Hf$ ratios and $\varepsilon_{Hf}(t)$ values of the Eocene zircons were calculated using their crystallization ages. The Hf isotopic compositions are similar among the granodiorites, mafic enclaves, and biotite granites. Zircons from the granodiorites show mostly negative $\varepsilon_{Hf}(t)$ values of −10. 8 to −1. 0, with two-stage model ages

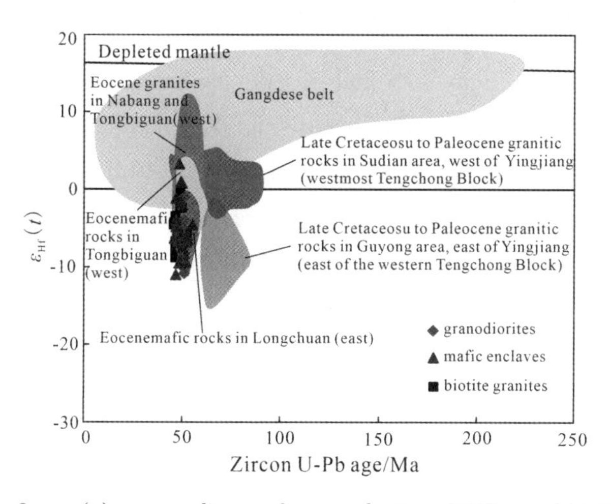

Fig. 8 $\varepsilon_{Hf}(t)$ vs. age diagram for granodiorites, MMEs, and biotite granites.

Data for Gangdese belt from Ji et al. (2009), for mafic and granitic rocks in Nabang, Tongbiguan and Longchuan from Wang et al. (2014) and Ma et al. (2014), for Late Cretaceosus to Paleocene granitic rocks in westmost Tengchong Block from Xu et al. (2012), for Late Cretaceosu to Paleocene granitic rocks in east of western Tengchong Block from Qi et al. (2015) and Chen et al. (2015).

of 1 186–1 809 Ma. Zircons from the mafic enclaves also show mostly negative $\varepsilon_{Hf}(t)$ values of −11. 1 to −1. 2, with two-stage model ages of 1 197–1 825 Ma, and the other three grains have positive $\varepsilon_{Hf}(t)$ values of 0. 6 to 3. 3 with model ages of 1 911–1 084 Ma. Zircons from the biotite granites have negative $\varepsilon_{Hf}(t)$ values of −9. 7 to −0. 7 and two-stage model ages of 1 166–1 735 Ma.

4. 5 Mineral chemistry

The mineral chemical data for representative amphiboles and biotites from the granodiorites and MMEs are listed in Supplementary Table 6 .

Amphiboles occur in both the granodiorites and MMEs. Formulae are calculated on the basis of 23 (O) and 13 cations (excluding Ca, Na, and K). The amphiboles in the MMEs have high Ca_B values of 1. 79–2. 06, low Na_B values of 0. 07–0. 26, and low $Mg^{\#}$ values of 38–52, indicating these are mainly ferrohornblendes or ferrotschermakites, characterized by TiO_2 contents of 0. 81 – 1. 74 wt% (nomenclature after Leake et al., 1997). Amphiboles in the granodiorites are ferrohornblendes that are characterized by low $Mg^{\#}$ values of 32–42 and TiO_2 contents of 0. 23–2. 04 wt%.

The biotites are subhedral and characterized by low $Mg^{\#}$ values of 34–43 in the MMEs and 32–37 in the granodiorites.

5 Discussion

5. 1 Petrogenesis of the host granodiorites

The granodiorites have an affinity with high-K calc-alkaline I-type granites, as indicated by their metaluminous to weakly peraluminous nature and their hornblende contents. They have high K_2O contents and are enriched in incompatible elements and LILEs (Fig. 5c,d; e.g., U, Th, Rb, and REEs). These geochemical features suggest that significant amounts of crustal material contributed to the magma. The host granodiorites also have similar trace element and isotopic compositions to the MMEs, but the granodiorites are unlikely to have been derived from the partial melting of the MMEs, due to both the granodiorites and MMEs being contemporaneous at ca. 50 Ma. The granodiorites possess almost invariable SiO_2 contents in the narrow range of 67. 81–69. 09 wt%, and there is no obvious linear relationship between SiO_2 and K_2O, $Mg^{\#}$, Rb, and LREEs (not shown), which suggests negligible contamination of the parent magma by crustal material (wall rocks). Roberts and Clemens (1993) suggested that the most suitable source materials for high-K calc-alkaline I-type granites are hydrated, calc-alkaline, and high-K calc-alkaline andesites and basaltic andesites. This would require an appropriate volume of andesitic source rocks in the lower crust before melting. However, the east of the western Tengchong Block is made of ancient crustal material, and the lower-crustal components of ancient blocks are usually granulitic and tonalitic (Rudnick and Gao, 2003). This possibility is confirmed by the Late Cretaceous to Paleogene lower-crust-derived magmatic rocks in the Guyong area, east of Yingjiang (east of the western Tengchong Block) which have high initial Sr ratios (0.718 2–0.745 7) and negative $\varepsilon_{Nd}(t)$ (−12.4 to −11.2) and $\varepsilon_{Hf}(t)$ values (−15.5 to −2.7) (Fig. 8, Chen et al., 2015; Qi et al., 2015). This differ from the coeval granitic rocks in Sudian area, west of Yingjiang (westmost Tengchong Block), that have positive $\varepsilon_{Hf}(t)$ values (0 to 4.09) (Fig. 8, Xu et al., 2012). The implication is that mantle-derived mafic magma or juvenile crustal rocks involved in the granitic magma in the westmost Tengchong Block (Nabang-Tongbiguan-Sudian area), but the partial melting of ancient crustal material are mainly occurs in the east of western Tengchong Block (Guyong-Bangwan area). In addition, the SiO_2 contents and other geochemical signatures indicate that the granodiorites were possibly derived from the partial melting of lower-crustal granulite facies rocks (Rapp and Watson, 1995; Rushmer, 1991). Patiño Douce (2005) conducted experiments on vapor-absent partial melting of tonalite at pressures of 15 – 32 kbar, and produced leucogranitic melts. Generally, without additional aqueous fluids, melts from the lower crust are the result of fluid-absent dehydration melting, and the melts become progressively drier as all the H_2O preserved in the solids is consumed. However, the host granodiorites in our study area contain abundant biotite and hornblende, and are more likely to be the product of fluid-present melting (Milord et al., 2001). Yardley and Valley (1997) have shown that rocks in the granulite facies are essentially "dry". Thus, it would be essential to introduce exotic aqueous fluids if granodioritic magmas were to be produced. Lóperz et al. (2005) conducted an experiment where mafic and tonalitic rocks were interbedded, to simulate the effects of fluids released from a crystallizing hydrous magma emplaced within a tonalitic continental crust. The results showed that the fluids in the mafic magma could return to the continental crust through andesite magmatism due to the high solubility of water (up to 15 wt%) in andesitic melts (Carmichael, 2002). The fluids from the mafic magma could carry H_2O, K_2O, and LREEs to the tonalitic rocks in the lower crust, inducing melting of the tonalitic rocks to form granodioritic melt, and the volume of the granodioritic melt could be twice that of the basic magma (Lóperz et al., 2005). The MMEs in the granodiorites of the present study could represent such a mafic magma, and the geochemical data show that the MMEs are also enriched in K_2O, LILEs, and LREEs, which means that the fluids released from the MMEs should have similar geochemical features. The isotopic data for the granodiorites resemble those for the MMEs, with high initial Sr ratios of 0.713 001–0.713 073, negative $\varepsilon_{Nd}(t)$ and $\varepsilon_{Hf}(t)$ values of −11.6 to −8.8 and −10.8 to −1.0, respectively, and evolved Pb compositions close to those of global subducting sediments (GLOSS), which are dominated by terrigenous materials and are similar in composition to the upper continental crust (Plank and Langmuir, 1998; Figs. 6 – 8). The H_2O-rich mafic magmas ascended through the crust, exsolved aqueous fluids that rose faster than the mafic magma, and exchanged trace elements and isotopic compositions with subsolidus tonalitic rocks in the lower crust while inducing melting and making the trace element and isotopic compositions homogeneous (Holk and Taylor, 2000). The amphiboles and biotites in the host granodiorites have $Mg^{\#}$ values of 32–42 and 32–37, respectively, which are lower than the values obtained for the same minerals in the MMEs, suggesting a more silicic protolith for the granodiorites. Therefore, the granodiorites are likely to have been derived from the partial melting of lower-crustal tonalitic rocks with an involvement of K-, REE-, and Th-rich fluids released from the mafic magmas.

5. 2 Petrogenesis of the MMEs

MMEs are widespread in granitoids worldwide (Didier and Barbarin, 1991), and there are generally three types: ① refractory residues after partial melting of crust that was the source of the granitic magma (Chappell et al., 1987, 2000; Collins, 1998); ② xenoliths of country rock or schlieren/accumulations of early crystallized minerals during the evolution of the granitic magma (Shellnutt et al., 2010); and ③ samples of mafic magma from the mantle that was mixing with felsic magma from the crust (Bonin, 2004; Yang et al., 2007). The MMEs enclosed in the granodiorites of the present study are rounded or ovoid, and characterized by fine-granular igneous textures (Fig. 2). The zircon U-Pb data and CL images indicate a magmatic origin and crystallization age of ca. 49. 4 Ma, clearly different from the wall rocks but coeval with the host granodiorites. In addition, no cumulate textures were observed in the enclaves, suggesting that the enclaves are not cognate fragments of cumulates or early crystallized concretionary bodies. The MMEs contain K-feldspar megacrysts and acicular apatites, and have low and variable SiO_2 contents, which suggests some degree of hybridization between the host granodiorite and the mafic magma. As SiO_2 contents in the MMEs increase from 53. 10 wt% to 64. 95 wt%, Rb compositions decrease from 394 ppm to 214 ppm, which could indicate that the mafic magma had a more incompatible composition than the host granodiorite, and the incompatible components could have diffused from the mafic enclaves to the host granodiorite during the mixing of the mafic and granitic magmas. Thus, we assume that the most basic MMEs represent the original mafic magma derived from the mantle, although most MMEs must be hybrids containing a proportion of felsic material. Therefore, in the following discussion we use the MMEs samples with the lowest SiO_2 and highest MgO contents to discuss their petrogenesis.

The MMEs are shoshonitic and are enriched in LILEs (e.g., Rb, K, and Pb), LREEs, and Th, but depleted in Nb, Ta, and Ti (Fig. 5e,f). The widely accepted notion is that mafic magmas are derived from mantle-related sources, where melts more silicic than andesitic compositions cannot form (i.e., the melts must have less than 57 wt% SiO_2; Lloyd et al., 1985; Baker et al., 1995). Thus, the most reasonable petrogenesis for the MMEs involves partial melting of the sub-continental lithospheric mantle at a destructive plate margin, where significant metasomatism by subduction-related fluids and melts is likely to have occurred (e.g., Cousens et al., 2001; Leslie et al., 2009). Subduction-related fluids are generally enriched in elements such as Ba, Pb, Rb, and U (e.g., Brenan et al., 1995; Elliott et al., 1997; Hawkesworth et al., 1990), and depleted in Th, Nb, Ta, Hf, Ti, and REEs (e.g., Ayers et al., 1997; Keppler, 1996; Kessel et al., 2005), and melts derived from subducted oceanic slabs are marked by low Th/La, Th/Nb, and Th/Yb ratios, and they become gradually enriched in Na_2O from 20 kbar to 30 kbar; the magmas range from tonalitic to trondhjemitic, and from metaluminous to peralkaline (Prouteau et al., 2001). However, the MMEs of the present study have extremely high Th and REE contents, indicating recycled sediments (Fig. 9a,b; e.g., Conticelli et al., 2009; Plank, 2005). The incompatible trace element patterns of these MMEs closely resemble the patterns of trace elements in upper crustal

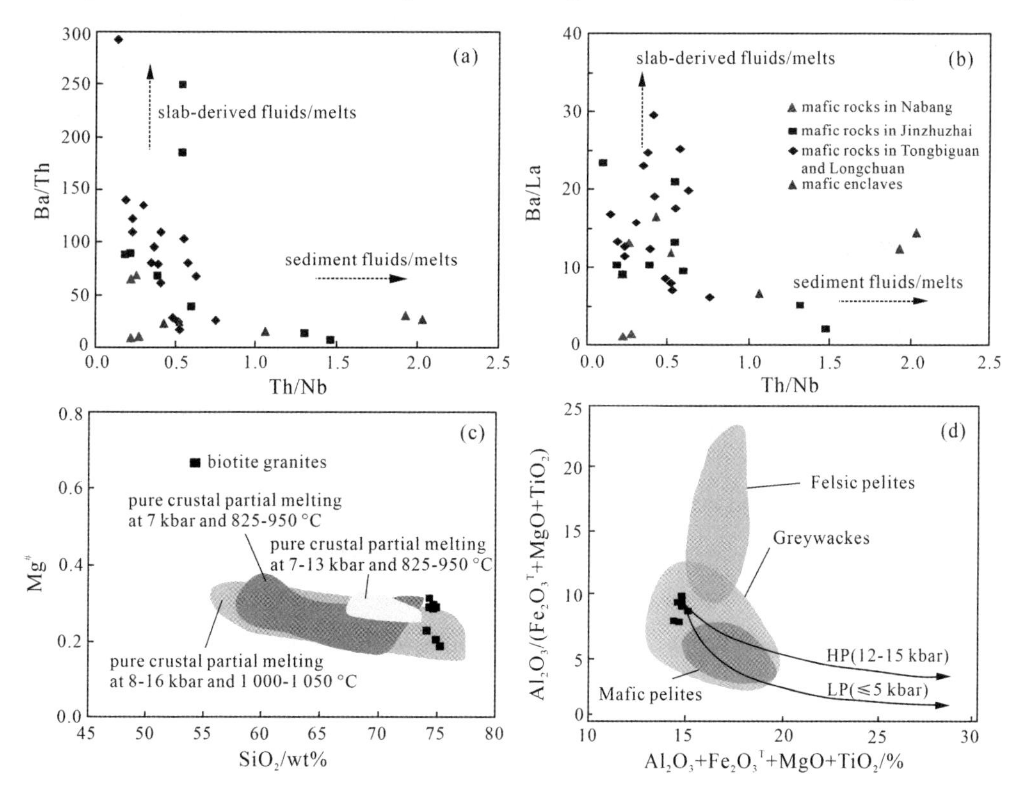

Fig. 9 Th/Nb vs. Ba/Th (a) and Th/Nb vs. Ba/La (b) diagrams for MMEs, and SiO_2 vs. $Mg^{\#}$ ratios (c) and $Al_2O_3+Fe_2O_3{}^T+MgO+TiO_2$ vs. $Al_2O_3/(Fe_2O_3{}^T+MgO+TiO_2)$ (d) diagrams for biotite granite.

rocks, also suggesting the involvement of sediment recycling (Avanzinelli et al., 2009). In general, the fluids/melts related to a subducted oceanic slab would not change the Nd isotopic compositions in the mantle source that was overlying the subducted slab, which commonly has positive $\varepsilon_{Nd}(t)$ values (e.g., Gertisser and Keller, 2003; Guo et al., 2005; Shimoda et al., 1998). Nevertheless, the MMEs in our study area have low $\varepsilon_{Nd}(t)$ values of −8.8, resembling the supposed values of Tengchong enriched lithospheric mantle (Wang et al., 2014, 2015), as well as mostly negative $\varepsilon_{Hf}(t)$ values of −11.1 to −1.2 and several positive $\varepsilon_{Hf}(t)$ values of 0.6 to 3.3, which suggest that the negative $\varepsilon_{Nd}(t)$ and $\varepsilon_{Hf}(t)$ values result from the effects of subducted sediments, while some of the positive $\varepsilon_{Hf}(t)$ values could be indicative of a residual mantle source. The high $^{87}Sr/^{86}Sr(t)$ ratios and evolved Pb compositions show an affinity with GLOSS (Fig. 7). The isotopic and trace elemental compositions of the MMEs in the Bangwan area are clearly different from those of the mafic rocks in the Nabang and Tongbiguan areas (Figs. 6 and 8), which could indicate different origins and sources of the mafic magmatism in the two areas. We suggest, therefore, that the MMEs were derived from partial melting of sub-continental lithospheric mantle that had been metasomatized by fluids and melts from recycled subducted sediments rather than from subducted oceanic slab. These mafic magmas were subsequently mixed with felsic magmas.

Klimm et al. (2008) demonstrated that the trace element budget of metasomatic agents at 2.5 GPa depends on the solubility of accessory phases such as allanite and monazite (controlling REE, Th, and U), and rutile (controlling Ti and Nb). At normal slab/mantle interface temperatures, these accessory phases are stable in residues. However, with sufficiently high temperatures and pressures, these phases become oversaturated in fluids/melts, and they release Th and REEs into the fluids/melts (Hermann, 2002; Kessel et al., 2005; Klimm et al., 2008). Thus, with regard to subduction of the Neo-Tethyan ocean lithosphere beneath the Tengchong Block, the temperatures and pressures during melting of the subducted sediments must have been high enough to destabilize the allanite and monazite, and according to Hermann (2002) and Klimm et al. (2008) this means temperatures greater than 1 000 ℃ in the pressure range of 2.5–4.0 GPa, which is much hotter than the estimates for temperatures at the slab/mantle interface for both "cold" and "warm" subduction (e.g., Kelemen et al., 2003; Peacock, 2003). Thus, a higher-than-normal geothermal gradient is needed if allanite and monazite are to melt and release Th and the REEs. Wang et al. (2014) suggested that the prevalent Eocene mafic magmatism in the western Tengchong Block was caused by rollback of the subducted slab, with the slab becoming steeper than normal. Under such conditions we can infer that the slab/mantle interface experienced an increase in temperature and pressure sufficient to induce the melting of allanite and monazite, thereby releasing their high Th and REE contents into the fluids/melts. Moreover, the highly fractionated REE patterns could result from two scenarios: ① garnet occurring as a residue when fluids/melts are derived from recycled sediments (Avanzinelli et al., 2009; Kerrick and Connolly, 2001), with a depletion in HFSEs indicating that rutile is part of the residue; and ② the sources already having fractionated RRE patterns and being depleted in HFSEs before partial melting occurs.

Generally, the rocks and minerals that are derived from mantle peridotites that have been metasomatized by fluids/melts from a subducting slab should have high MgO contents and $Mg^{\#}$ values (e.g., Kelemen, 1995; Prouteau et al., 2001). The MMEs in our study area actually have low MgO contents of 2.31–2.76 wt%, and $Mg^{\#}$ values of 37–45. Moreover, the calcic amphiboles in the MMEs have low $Mg^{\#}$ values of 38–52, and the $Mg^{\#}$ values of the biotites are in the range 34–43. We infer, therefore, that the metasomatized lithospheric mantle rocks from which the MMEs were derived were pyroxenites rather than peridotites.

5.3 Petrogenesis of the biotite granites

The biotite granites are high-K, calc-alkaline, and peraluminous, and they were coeval with the granodiorites and MMEs. Their trace element patterns are similar to those of the upper crust. Compared with the granodiorites and MMEs, the biotite granites have higher SiO_2, K_2O, Rb, Th, and U contents, and lower Sr and Ba contents, which suggest that upper crustal materials were involved in their formation. The low $Mg^{\#}$ values also show that the biotite granite magmas were probably the products of partial melting of pure crustal material (Fig. 9c). The biotite granites have high CaO/Na_2O ratios of 0.33–0.41 and Al_2O/TiO_2 ratios of 89–144, indicating that the magma was derived from metagraywacke (Sylvester, 1998), and on the $Al_2O_3+Fe_2O_3{}^T+MgO+TiO_2$ vs. $Al_2O_3/(Fe_2O_3{}^T+MgO+TiO_2)$ diagram the granites fall in the graywacke field (Fig. 9d; Patiño Douce, 1999). However, the biotite granites have much higher Rb, Th, and REE contents than melts produced by dehydration-related partial melting of metagraywacke. These elements are easily dissolved in fluids, and therefore if we interpret the biotite granites to be the result of metagraywacke melting induced by the emplacement of granodiorites, the concomitant release of fluids from the granodiorite magma may have imparted the biotite granites with the extra K_2O, Rb, Th, and REE. The isotopic data also support this inference, as the high initial $^{87}Sr/^{86}Sr$ ratios, negative $\varepsilon_{Nd}(t)$ and $\varepsilon_{Hf}(t)$ values, and evolved Pb compositions of the biotite granites are identical to the host granodiorites (Figs. 6–8). We propose, therefore, that fluids played an important role in the petrogenesis of the biotite granites, as they were derived from partial melting of upper crustal metagraywacke, induced by the emplacement of granodioritic magma, but an influx of fluids from the granodioritic magma enriched the composition of the biotite granite magma so that it now represents both upper crustal material and the introduced fluid.

5.4 Regime for the formation of the granodiorites, MMEs, and biotite granites

The zircon analyses reveal that the host granodiorites, MMEs, and biotite granites all formed at ca. 50 Ma, coeval with the widespread acidic-basic rocks in the western Tengchong Block (e.g., Ma et al., 2014; Wang et al., 2014, 2015; Xu et al., 2012). The obvious negative Eu anomalies (0.18–0.55) indicate a preferential Eu fractionation into a co-existing fluid phase rather than into feldspar (Irber, 1999; Muecke and Clarke, 1981), suggesting fluids/melts played an important role in the formation of the mafic, granodioritic, and biotite granitic magmas. Consequently, we propose the following regime for the Bangwan granitic rocks and MMEs. The mafic magma was derived from the partial melting of sub-continental lithospheric mantle pyroxenite that had been metasomatized by fluids/melts derived from the subducted sediments. Subsequently, the mafic magmas ascended and crystallized under or within the lower crust, and released K-, Th-, and REE-rich fluids into the tonalitic rocks of the lower crust. This process induced partial melting of the tonalitic rocks, and produced the granodioritic magma, which then mixed with some of the intruded mafic magma to generate the MMEs. In addition, the ascending granodioritic magmas caused partial melting of upper crustal metagraywacke, at the same time releasing K-, Th-, and REE-rich fluids to produce the fluid-rich conditions under which the biotite granitic magmas were formed (Fig. 10). The geochemical and isotopic signatures of the MMEs, granodiorites, and biotite granites have a certain similarity, due to the fluids/melts exchanged during these processes of melting (Holk and Taylor, 2000).

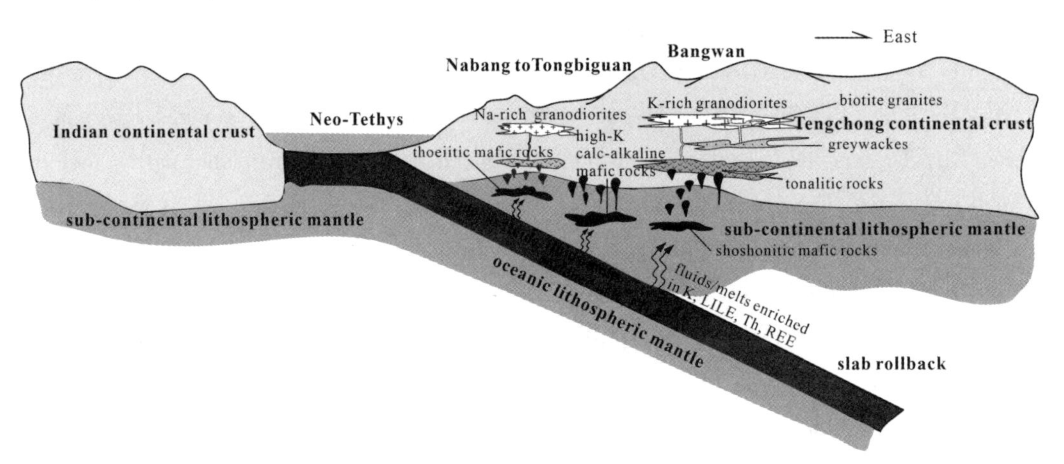

Fig. 10 Schematic diagram for the petrogenesis of granodiorites, MMEs and biotite granites in Bangwan area, and the relationship between systematic compositional variation in the mafic rocks and granitic rocks in the western Tengchong Block and eastward subduction of the Neo-Tethys.

5.5 Geochemical and isotopic trends of Eocene magmatic rocks from west to east in the western Tengchong Block

The flare-up of Eocene magmatism in the western Tengchong Block produced mafic to acid rocks with ages of 50–55 Ma (e.g., Ma et al., 2014; Wang et al., 2014, 2015; Xu et al., 2012; this study). For comparable SiO_2 contents (ca. 62–69 wt%), the granitic rocks in the Nabang area (westmost Tengchong Block) are enriched in Na_2O and relatively depleted in K_2O, with Na_2O/K_2O ratios of 0.9–2.0 (Ma et al., 2014), whereas the granitic rocks in the Bangwan area (east of Nabang, east of the western Tengchong Block) have higher K_2O contents, Na_2O/K_2O ratios of 0.5–0.7, and belong to the high-K calc-alkaline series. The contemporaneous mafic rocks show a similar variation, with K_2O, Rb, and REE contents increasing from west to east in the western Tengchong Block (Wang et al., 2014). In addition, the $\varepsilon_{Nd}(t)$ and $\varepsilon_{Hf}(t)$ values of the granitic and mafic rocks also decrease from west to east (Figs. 6 and 8; e.g., Ma et al., 2014; Wang et al., 2014, 2015). In summary, from west to east the mafic rocks show a transition from tholeiitic through high-K calc-alkaline to shoshonitic rocks, and the granitic rocks show a range from Na-rich to K-rich granitoids (Fig. 1; Ma et al., 2014; Wang et al., 2014; this study). This systematic compositional variation is likely to be an imprint of the arc magmatism in the Neo-Tethyan subduction zone, and it shows that the direction of subduction was from west to east. The trends are identical to those in the coeval igneous rocks of the Gangdese magmatic arc belt, and in that region a northward subduction of the Neo-Tethyan oceanic lithosphere is inferred (Zhao et al., 2011). We propose that such spatial variations in geochemistry were controlled by the expected variations in the fluids and melts derived from the subducted slab (including its sediments), which were released into the sub-continental lithospheric mantle above the subduction zone and which would correlate with the gradual increase in the depth of the subduction zone from west to east. The fluids/melts in the shallow zone (west) would be mainly hydrous, and in the middle zone (middle) they carry more K and LILEs (Rb, Ba, and Sr) and would lack Th and REEs during the dehydration of serpentine, and in the deep zone (east) they would be enriched in K, LILEs, Th, and REEs, especially when accessory minerals such as allanite and monazite became unstable. These fluids/melts would have metasomatized the overlying mantle material and induced its partial melting to produce the mafic rocks. The variations in the chemistry of the metasomatizing fluids/melts would have produced the observed geochemical variations in the mafic magmas from west to east in the western Tengchong Block (Fig. 10). These mafic magmas then ascended to induce melting of crustal material, thus forming the granodioritic magma that then became mixed with the mafic magma.

6 Conclusions

(1) The granodiorites, MMEs, and biotite granites in the Bangwan area of thewestern Tengchong Block are contemporaneous, with zircon U-Pb ages of ca. 50 Ma.

(2) The geochemical and evolved Sr-Nd-Pb isotopic compositions show that the mafic magma of the MMEs was shoshonitic, and was probably derived from the partial melting of lithospheric mantle pyroxenite that had been metasomatized by fluids/melts from the subducted sediments. The granodiorites were formed by partial melting of lower crustal tonalitic rocks during the intrusion of mafic magmas at the base of the lower crust, assisted by the K-, REE-, and Th-rich fluids that were released from those mafic magmas. The biotite granites were formed by the partial melting of upper crustal metagraywackes during emplacement of the granodiorites, assisted by the release of fluids from the granodioritic magma.

(3) Comparisons of the coeval magmatic rocks in the western Tengchong Block show that the mafic and granitic rocks have similar geochemical and isotopic trends, with transitions from west to east in the mafic rocks from tholeiitic through high-K calc-alkaline to shoshonitic, and in the granitic rocks from Na-rich to K-rich. These geochemical variations show that the Eocene magmatic rocks in the western Tengchong Block were related to a magmatic arc produced by eastward subduction of the Neo-Tethyan oceanic lithosphere.

Acknowledgements This work was jointly supported by the National Natural Science Foundation of China (Grant Nos. 41190072, 41421002 and 41102037), program for Changjiang Scholars and Innovative Research Team in University (Grant IRT 1281), the Foundation for the Author of National Excellent Doctoral Dissertation of PR China (201324).

Appendix A. Supplementary data Supplementary data to this article can be found online at https://dx.doi.org/10.1016/j.lithos.2016.08.25.

References

Avanzinelli, R., Lustrino, M., Mattei, M., Melluso, L., Conticelli, S., 2009. Potassic and ultrapotassic magmatism in the circum-Tyrrhenian region: Significance of carbonated pelitic vs. pelitic sediment recycling at destructive plate margins. Lithos, 113, 213-227.

Ayers, J.C., Dittmer, S.K., Layne, G.D., 1997. Partitioning of elements between peridotite and H_2O at 2.0–3.0 GPa and 900–1 000 ℃, and application to models of subduction processes. Earth & Planetary Science Letters, 150, 381-398.

Barker, M.B., Hirschmann, M.M., Ghiorso, M.S., Stolper, E.M., 1995. Compositions of near-solidus peridotite melts from experiments and thermodynamic calculations. Nature, 375, 308-311.

Bernan, J.M., Shaw, H.F., Ryerson, F.J., Phinney, D.L., 1995. Mineral-aqueous fluids partitioning of trace elements at 900 ℃ and 2.0 Gpa: Constrains on the trace element chemistry of mantle and deep crustal fluids. Geochimica et Cosmochimica Acta, 59, 3331-3350.

Bonin, B., 2004. Do coeval mafic and felsic magmas in post-collisional to within-plate regimes necessarily imply two contrasting mantle and crustal, sources? A review. Lithos, 17, 1-24.

Carmichael, I., 2002. The andesite aqueduct: Perspectives on the evolution of intermediate magmatism in west-central (105–99°W) Mexico. Contributions to Mineralogy and Petrology, 143, 641-663.

Chappell, B., White, A., Wyborn, D., 1987. The importance of residual source material (restite) in granite petrogenesis. Journal of Petrology, 28, 1111-1138.

Chappell, B.W., White, A.J.R., Williams, I.S., Wyborn, D., Wyborn, L.A.I., 2000. Lachlan Fold Belt granites revisited: High- and low-temperature granites and their implications. Australian Journal of Earth Sciences, 47, 123-138.

Chen, X.C., Hu, R.Z., Bi, X.W., Zhong, H., Lan, J.B., Zhao, C.H., Zhu, J.J., 2015. Petrogenesis of metaluminous A-type granitoids in the Tengchong-Lianghe tin belt of southwestern China: Evidences from zircon U-Pb ages and Hf-O isotopes, and whole-rocks Sr-Nd isotopes. Lithos, 212-215, 93-110.

Chung, S.L., Chu, M.F., Zhang, Y.Q., 2005. Tibet tectonic evolution inferred from spatial and temporal variations in post-collisional magmatism. Earth-Science Reviews, 68, 173-196.

Collins, W.J., 1998. Evaluation of petrogenetic models for Lachlan Fold Belt granitoids: Implications for crustal architecture and tectonic models. Australian Journal of Earth Sciences, 45, 483-500.

Conticelli, S., Guarnieri, L., Farinelli, A., Mattei, M., Avanzinelli, R., Bianchini, G., Boari, E., Tommasini, S., Tiepolo, M., Prelevic, D., Venturelli, G., 2009. Trace elements and Sr-Nd-Pb isotopes of K-rich, shoshonitic, and calc-alkaline magmatism of the Western Mediterranean Region: Genesis of ultrapotassic to calc-alkaline magmatic associations in a post-collisional geodynamic setting. Lithos, 107, 68-92.

Cousens, B.L., Aspler, L.B., Chiarenzelli, J.R., Donaldson, J.A., Sandeman, H., Peterson, T.D., Lecheminant, A.N., 2001. Enriched Archean lithopheric mantle beneath western Churchill Province tapped during Paleoproterozoic orogenesis. Geology, 29, 827-830.

Didier, J., Barbarin, B., 1991. Enclaves and Granite Petrology. Amsterdam: Elsevier, 625.

Elliott, T., Plank, T., Zindler, A., White, W., Bourdon, B., 1997. Element transport from slab to volcanic front at the Mariana arc. Journal of Geophysical Research, 102, 14991-15019.

Gao, Y.F., Yang, Z.S., Hou, Z.Q., Wei, R.H., Meng, X.J., Tian, S.H., 2010. Eocene potassic and ultrapotassic volcanism in south Tibet: New constraints on mantle source characteristics and geodynamic processes. Lithos, 117, 20-32.

Gertisser, R., Keller, J., 2003. Trace element and Sr, Nd, Pb, and O isotope variations in medium-K and high-K volcanic rocks from Merapi volcano, central Java. Indonesia: Evidence for the involvement of subducted sediments in Sunda arc magma genesis. Journal of Petrology, 44, 457-489.

Guo, Z.F., Hertogen, J., Liu, J.Q., Pasteels, P., Boven, A., Pumzalan, L., He, H.Y., Luo, X.J., Zhang, W.H., 2005. Potassic magmatism in Western Sichuan and Yunnan provinces, SE Tibet, China: Petrological and geochemical constraints on petrogenesis. Journal of Petrology, 46, 33-78.

Guo, Z.F., Wilson, M., Liu, J.Q., 2007. Post-collisional, adakites in south Tibet: Products of partial melting of subduction-modified lower crust. Lithos, 96, 205-224.

Guo, Z.F., Wilson, M., Zhang, M.L., Cheng, Z.H., Zhang, L.H., 2013. Post-collisional, K-rich, mafic magmatism in south Tibet: Constraints on Indian slab-to-wedge transport processes and plateau uplift. Contributions to Mineralogy and Petrology, 165, 1311-1340.

Guo, Z.F., Cheng, Z.H., Zhang, M.L., Zhang, L.H., Li, X.H., Liu, J.Q., 2015. Post-collisional high-K calc-alkaline volcanism in Tengchong volcanic field, SE Tibet: Constraints on Indian eastward subduction and slab detachment. Journal of the Geological Society, 172, 624-640.

Hawkesworth, C.J., Turner, S.P., Peate, D.W., McDermott, F., van Carlsteren, P.W., 1990. Continental mantle lithosphere and shallow level enrichments processes in Earth's mantle. Earth & Planetary Science Letters, 96, 256-268.

He, R.Z., Zhao, D.P., Gao, R., Zheng, H.W., 2010. Tracing the Indian lithospheric mantle beneath central Tibetan Plateau using teleseismic tomography. Tectonophysics, 491, 230-243.

Hermann, J., 2002. Allanite: Thorium and light rare earth element carrier in subducted crust. Chemical Geology, 192, 289-306.

Hofmann, A.W., 1988. Chemical differentiation of the Earth: The relationship between mantle, continental crust, and oceanic crust. Earth & Planetary Science Letters, 90, 279-314.

Holk, G.J., Taylor, H.P., 2000.Water as a petrologic catalyst driving $^{18}O/^{16}O$ homogenization and anatexis of the middle crust in the metamorphic core complexes of British Columbia. International Geology Review, 42, 97-130.

Huang, Y., Zhao, Z. D., Zhang, F. D., Zhu, D. C., Dong, G. C., Mo, X.X., 2010. Geochemistry and implication of the Gangdese batholiths from Renbu and Lhasa areas in southern Gangdese, Tibet. Acta Petrologica Sinica, 26, 3131-3142 (in Chinese with English abstract).

Irber, W., 1999. The lanthanide tetrad effect and its correlation with K/Rb, Eu/Eu*, Sr/Eu, Y/Ho, and Zr/Hf of evolving peraluminous granite suites. Geochimica et Cosmochimica Acta, 63, 489-508.

Ji, J.Q., Zhong, D.L., Chen, C.Y., 2000. Geochemistry and genesis of Nabang metamorphic basalt, southwest Yunnan, China: Implications for the subducted slab break-off. Acta Petrologica Sinica, 16, 433-442 (in Chinese with English abstract).

Ji, W.Q., Wu, F.Y., Chung, S.L., Li, J.X., Liu, C.Z., 2009. Zircon U-Pb geochronology and Hf isotopic constraints on petrogenesis of the Gangdese batholith, southern Tibet. Chemical Geology, 262, 229-245.

Kelemen, P.B., 1995. Genesis of high Mg# andesites and the continental crust. Contributions to Mineralogy and Petrology, 120, 1-19.

Kelemen, P.B., Rilling, J.L., Parmentier, E.M., Mehl, L., Hacker, B.R., 2003. Thermal Structure due to Solid-State Flow in the Mantle Wedge Beneath Arcs. Inside the Subduction Factory, 138, 293-311.

Keppler, H., 1996. Constraints from partitioning experiments on the composition of subduction-zone fluids. Nature, 380, 237-240.

Kerrick, D.M., Connolly, J.A.D., 2001. Metamorphic devolatilization of subducted marine sediments and the transport of volatiles into the Earth's mantle. Nature, 411, 293-296.

Kessel, R., Schmidt, M.W., Ulmer, P., Pettke, T., 2005. Trace element signature of subduction-zone fluids, melts and supercritical liquids at 120–180 km depth. Nature, 437, 724-727.

Klimm, K., Blundy, J.D., Green, T.H., 2008. Trace element partitioning and accessory phase saturation during H_2O-saturated melting of basalt with implications for subduction zone chemical fluxes. Journal of

Petrology, 49, 523-553.

Leake, B.E., Woolley, A.R., Arps, C.E.S., Birch, W.D., Gilbert, M.C., Grice, J.D., Hawthorne, F.C., Kato, A., Kisch, H.J., Krivovichev, V.G., Linthout, K., Laird, J., Mandarino, J., Maresch, W.V., Nickel, E.H., Rock, N.M.S., Schumacher, J.C., Smith, D.C., Stephenson, N.C.N., Ungaretti, L., Whittaker, E.J.W., Youzhi, G., 1997. Nomenclature of amphiboles: Report of the Subcommittee on amphiboles of the International Mineralogical Association Commission on New Minerals and Mineral Names. The Canadian Mineralogist, 35, 219-246.

Lee, H.Y., Chung, S.L., Ji, J., Qian, Q., Gallet, S., Lo, C.H., Lee, T.Y., Zhang, Q., 2012. Geochemical and Sr-Nd isotopic constraints on the genesis of Cenozoic Linzizong volcanic successions, southern Tibet. Journal of Asian Earth Sciences, 53, 96-114.

Leslie, R.A.J., Danyushevsky, L.V., Crawford, A.J., Verbeeten, A.C., 2009. Primitive shoshonites from Fiji: Geochemistry and source components. Geochemistry, Geophysics and Geosystems, 10, 131-147.

Li, C., van Der Hilst, R.D., Meltzer, A.S., Engdahl, E.R., 2008. Subduction of the Indian lithosphere beneath the Tibetan plateau and Myanmar. Earth & Planetary Science Letters, 274, 157-168.

Lloyd, F.E., Arima, M., Edgar, A.D., 1985. Partial melting of a phlogopite-clinopyroxenite nodule from south-west Uganda: An experimental study bearing on the origin of highly potassic continental rift volcanics. Contributions to Mineralogy and Petrology, 91, 321-329.

Lóperz, S., Castro, A., García-Casco, A., 2005. Production of granodiorite melt by interaction between hydrous magma and tonalitic crust: Experimental constraints and implication for the generation of Archaean TTG complexes. Lithos, 79, 229-250.

Ma, L.Y., Wang, Y.J., Fan, W.M., Geng, H.Y., Cai, Y.F., Zhong, H., Liu, H.C., Xing, X.W., 2014. Petrogenesis of the early Eocene I-type granites in west Yingjiang (SW Yunnan) and its implications for the eastern extension of the Gangdese batholiths. Gondwana Research, 25, 401-419.

Maniar, P.D., Piccoli, P.M., 1989. Tectonic discrimination of granitoids. Geological Society of American Bulletin, 101, 635-643.

Milord, I., Sawyer, E.W., Brown, M., 2001. Formation of diatexite migmatite and granite magma during anatexis of semipelitic metasedimentary rocks: An example from St. Malo, France. Journal of Petrology, 42, 487-505.

Mo, X.X., Hou, Z.Q., Niu, Y.L., Dong, G.C., Qu, X.M., Zhao, Z.D., Yang, Z.M., 2007. Mantle contributions to crustal thickening during continental collision: Evidence from Cenozoic igneous rocks in southern Tibet. Lithos, 96, 225-242.

Mo, X.X., Niu, Y.L., Dong, G.C., Zhao, Z.D., Hou, Z.Q., Zhou, S., Ke, S., 2008. Contribution of syncollisional felsic magmatism to continental crust growth: A case study of the Paleogene Linzizong volcanic Succession in southern Tibet. Chemical Geology, 250, 49-67.

Muecke, G.K., Clarke, D.B, 1981. Geochemical evolution of the South Mountain Batholith, Nova Scotia: Rare-earth element evidence. The Canadian Mineralogist, 19, 133-145.

Patiño Douce, A.E., 1999. What do experiments tell us about the relative contributions of crust and mantle to the origin of granitic magma? Geological Society, London, Special Publications, 168, 55-75.

Patiño Douce, A.E., 2005. Vapor-Absent Melting of Tonalite at 15 – 32 kbar. Journal of Petrology, 46,

275-290.

Peacock, S.M., 2003. Thermal structure and metamorphic evolution of subducting slabs. In: Eiler, J.M. Inside the Subduction Factory. Geophysics Monograph Series, 138, AGU, Washington, D.C., 7-22.

Plank, T., 2005. Constraints from Thorium/Lanthanum on Sediment Recycling at Subduction Zones and the Evolution of the Continents. Journal of Petrology, 46, 921-944

Plank, T., Langmuir, C.H., 1998. The chemical composition of subducting sediment and its consequences for the crust and mantle. Chemical Geology, 145, 325-394.

Prouteau, G., Scaillet, B., Pichavant, M., Maury, R., 2001. Evidence for mantle metasomatism by hydrous silicic melts derived from subducted oceanic crust. Nature, 410, 197-200.

Qi, X.X., Zhu, L.H., Grimmer, J.C., Hu, Z.C., 2015. Tracing the Transhimalayan magmatic belt and the Lhasa block southward using zircon U-Pb, Lu-Hf isotopic and geochemical data: Cretaceous-Cenozoic granitoids in the Tengchong block, Yunnan, China. Journal of Asian Earth Sciences, 110, 170-188.

Rapp, R.P., Watson, E.B., 1995. Dehydration melting of metabasalt at 8－32 kbar: Implications for continental growth and crust-mantle recycling. Journal of Petrology, 36, 891-931.

Replumaz, A., Tapponnier, P., 2003. Reconstruction of deformed collision zone Between India and Asia by backward motion of lithospheric blocks. Journal of Geophysical Research, 108 (B6), 2285.

Roberts, M.P., Clemens, J.D., 1993. Origin of high-potassium, calc-alkaline, I-type granitoids. Geology, 21, 825-828.

Rollinson, H., 1993. Using Geochemical Data: Evaluation, Presentation, Interpretation. London: Longman Scientific & Technical.

Rudnick, R.L., Gao, S., 2003. Composition of the Continental Crust. In: Rudnick, R.L., Holland, H.D., Turekian, K.T. Treatise on Geochemistry, 3, Elsevier, 1-64.

Rushmer, T., 1991. Partial melting of two amphibolites: Contrasting experimental results under fluid-absent conditions. Contributions to Mineralogy and Petrology, 107, 41-59.

Shellnutt, J.G., Jahn, B.M., Dostal, J., 2010. Elemental and Sr-Nd isotope geochemistry of microgranular enclaves from peralkaline A-type granitic plutons of the Emieshan large igneous province, SW China. Lithos, 119, 34-46.

Shimoda, G., Tatsumi, Y., Nohda, S., Ishizaha, K., Jahn, B. M., 1998. Setouchi high-Mg andesites revisited: Geochemical evidence for melting of subducting sediments. Earth & Planetary Science Letters, 160, 479-492.

Sun, S.S., McDonough, W.F., 1989. Chemical and isotopic systematics of oceanic basalts: Implications for mantle composition and processes. In: Saunders, A.D., Norry, M.J. Magmatism in the Ocean Basins. Geological Society Special Publication, London, 42, 313-345.

Sylvester, P.J., 1998. Post-collisional strongly peraluminous granites. Lithos, 45, 29-44.

Wang, Y., Zhang, X.M., Jiang, C.S., Wei, H.Q., Wan, J.L., 2007. Tectonic controls on the late Miocene-Holocene volcanic eruptions of the Tengchong volcanic field along the southeastern margin of Tibetan plateau. Journal of Asian Earth Sciences, 30, 375-389.

Wang, Y.J., Zhang, L.M., Cawood, P.A., Ma, L.Y., Fan, W.M., Zhang, A.M., Zhang, Y.Z., Bi, X.W., 2014. Eocene supra-subduction zone mafic magmatism in the Sibumasu block of SW Yunnan: Implications for Neotethyan subduction and India-Asia collision. Lithos, 206-207, 384-399.

Wang, R., Richards, J.P., Hou, Z.Q., An, F., Creaser, R.A., 2015a. Zircon U-Pb age and Sr-Nd-Hf-O isotope geochemistry of the Paleocene-Eocene igneous rocks in western Gangdese: Evidence for the timing of Neo-Tethyan slab breakoff. Lithos, 224-225, 179-194.

Wang, Y.J., Li, S.B., Ma, L.Y., Fan, W.M., Cai, Y.F., Zhang, Y.H., Zhang, F.F., 2015b. Geochronological and geochemical constraints on the petrogenesis of Early Eocene metagabbroic rocks in Nabang (SW Yunnan) and its implications on the Neotethyan slab subduction. Gondwana Research, 27, 1474-1486.

Wang, R., Richards,J.P,.Hou,Z.Q.,An,F.,Creaser,R.A.,2015a.Zircon U-Pb age and Sr-Nd-Hf-O isotope geochemistry of the Paleocene-Eocene igneous rocks in western Gangdese: Evidence for the timing of Neo-Tethyan slab breakoff. Lithos, 224-225.

Wen, D.R., Liu, D.Y., Chung, S.L., Chu, M.F., Ji, J.Q., Zhang, Q., Song, B., Lee, T.Y., Yeh, M.W., Lo, C.H., 2008a. Zircon SHRIMP U-Pb ages of the Gangdese Batholith and implications for Neotethyan subduction in southern Tibet. Chemical Geology, 252, 191-201.

Wen, D.R., Chung, S.L., Song, B., Iizuka, Y., Yang, H.J., Ji, J.Q., Liu, D.Y., Gallet, S., 2008b. Late Cretaceous Gangdese intrusions of adakitic geochemical characteristics, SE Tibet: Petrogenesis and tectonic implications. Lithos, 105, 1-11.

Xie, J.C., Zhu, D.C., Dong, G.C., Zhao, Z.D., Wang, Q., Mo, X.X., 2016. Linking the Tengchong Terrane in SW Yunnan with the Lhasa Terrane in southern Tibet through magmatic correlation. Gondwana Research (in press).

Xu, J.F., Castillo, P.R., 2004. Geochemical and Nd-Pb isotopic characteristics of the Tethyan asthenosphere: Implications for the origin of the Indian Ocean mantle domain. Tectonophysics, 393, 9-27.

Xu, Y.G., Lan, J.B., Yang, Q.J., Huang, X.L., Niu, H.N., 2008. Eocene break-off of the Neo-Tethyan slab as inferred from intraplate-type mafic dykes in the Gaoligong orogenic belt, eastern Tibet. Chemical Geology, 255, 439-453.

Xu, Y.G., Yang, Q.J., Lan, J.B., Luo, Z.Y., Huang, X.L., Shi, Y.B., Xie, L.W., 2012. Temporal-spatial distribution and tectonic implications of the batholiths in the Gaoligong-Tengliang-Yingjiang area, western Yunnan: Constraints from zircon U-Pb ages and Hf isotopes. Journal of Asian Earth Sciences, 53, 151-175.

Yang, J.H., Wu, F.Y., Wilde, S.A., Xie, L.W., Yang, Y.H., Liu, X.M., 2007. Tracing magma mixing in granite genesis: In situ U-Pb dating and Hf-isotope analysis of zircons. Contributions to Mineralogy and Petrology, 153, 177-190.

Yang, Q.J., Xu, Y.G., Huang, X.L., Luo, Z.Y., Shi, Y.R., 2009. Geochronology and geochemistry of granites in the Tengliang area, western Yunnan: Tectonic implication. Acta Petrologica Sinica, 25, 1092-1104 (in Chinese with English abstract).

Yardley, B.W.D., Valley, J.W., 1997.The petrologic case for a dry lower crust. Journal of Geophysical Research Solid Earth, 102, 12173-12185.

Zhang, Z.M., Dong, X., Xiang, H., Liou, J.G., Santosh, M., 2013. Building of the Deep Gangdese Arc, South Tibet: Paleocence Plutonism and Granulite-Facies Metamorphism. Journal of Petrology, 54, 2547-2580.

Zhao, D.P., Yu, S., Ohtani, E., 2011a. East Asia: Seismotectonics, magmatism and mantle dynamics. Journal of Asian Earth Sciences, 40, 689-709.

Zhao, Z.D., Zhu, D.C., Dong, G.C., Mo, X.X., Depaolo, D., Jia, L.L., Hu, Z.C., Yuan, H.L., 2011b. The ca. 54 Ma gabbro-granite intrusive in southern Dangxung area, Tibet: Petrogenesis and implications. Acta Petrologica Sinicam, 27, 3513-3524 (in Chinese with English abstract).

Zhao, S.W., Lai, S.C., Qin, J.F., Zhu, R.Z., 2016. Tectono-magmatic evolution of the Gaoligong belt, southeastern margin of the Tibetan plateau: Constraints from granitic gneisses and granitoid intrusions. Gondwana Research, 35, 238-325.

Zhu, D.C., Zhao, Z.D., Niu, Y.L., Mo, X.X., Chung, S.L., Hou, Z.Q., Wang, L.Q., Wu, F.Y., 2011. The Lhasa Terrane: Record of a microcontinent and its histories of drift and growth. Earth & Planetary Sciences Letters, 301, 241-255.

Zhu, R.Z., Lai, S.C., Qin, J.F., Zhao, S.W., 2015. Early-Cretaceous highly fractionated I-type granites from the northern Tengchong block, western Yunnan, SW China: Petrogenesis and tectonic implications. Journal of Asian Earth Sciences, 100, 145-163.

Neoproterozoic alkaline intrusive complex in the northwestern Yangtze Block, Micang Mountains region, South China: Petrogenesis and tectonic significance[①]

Gan Baoping　Lai Shaocong[②]　Qin Jiangfeng　Zhu Renzhi　Zhao Shaowei　Li Tong

Abstract: Voluminous Neoproterozoic mafic-ultramafic, felsic, and alkaline intrusions are found in the northern Yangtze Block, South China. Here, we present whole-rock major and trace element, and Sr-Nd isotopic compositions, together with zircon U-Pb ages, for syenite and gabbro samples from the Shuimo-Zhongziyuan alkaline intrusive complex in the Micang Mountains region at the northwestern margin of the Yangtze Block. Zircon U-Pb dating yields crystallization ages for the Na- and K-rich Shuimo syenites of 869 ± 4 Ma (MSWD = 0.85, 2σ) and 860 ± 5 Ma (MSWD = 0.47, 2σ), respectively, and for the Zhongziyuan gabbros of 753 ± 4 Ma (MSWD = 0.23, 2σ), indicating that the syenites and gabbros represent different stages of magmatism. The syenites include both Na-and K-rich types and have high values of the Rittman index (σ), and high SiO_2 and Na_2O+K_2O contents. These syenites are enriched in light rare earth elements (LREE) and large-ion lithophile elements (LILE), but depleted in high-field-strength elements (HFSE), with high $(La/Yb)_N$ values and small negative and positive Eu anomalies ($Eu/Eu^* = 0.74-1.17$). In contrast, the gabbros have lower SiO_2 and $Na_2O + K_2O$ contents, are only slightly enriched in LREEs, are enriched in LILE but depleted in HFSEs, and have small negative and positive Eu anomalies ($Eu/Eu^* = 0.86-1.37$). The syenites have low initial $^{87}Sr/^{86}Sr$ (0.703 340) and $\varepsilon_{Nd}(t)$ values (+1.9 to +7.7). The gabbros have relatively high initial $^{87}Sr/^{86}Sr$ (0.703 562-0.704 933) and positive $\varepsilon_{Nd}(t)$ values (+1.6 to +4.5). These data suggest that the syenites and gabbros are isotopically similar and were largely derived from melts of depleted mantle. The syenites underwent significant fractional crystallization and small amounts of crustal contamination during magma evolution. In contrast, the gabbros were formed by partial melting (> 15%) of a garnet lherzolite source and might also have experienced crustal assimilation. Taking into account the geochemical signatures and magmatic events, we propose that the Shuimo syenites formed in an intra-arc rifting setting, however, the Zhongziyuan gabbros were most likely produced in a subduction-related, continental margin arc setting during the Neoproterozoic, thus suggesting that the alkaline intrusive complexs were formed by the arc-related magmatism in the Micang Mountains.

---

① Published in *International Geology Review*, 2017, 59(3).

② Corresponding author.

1 Introduction

The Yangtze Block is one of the largest and most complex Precambrian blocks in South China (Wang et al., 2010). The Neoproterozoic magmatic rocks in the Hannan-Micangshan area of the northwestern periphery of the Yangtze Block are predominantly characterized by felsic and minor mafic-ultramafic intrusions (Zhou et al., 2002a; Ling et al., 2003; Xiao et al., 2007; Dong et al., 2012; Cai et al., 2015; Zhang et al., 2015). In the past decade, the characteristics of the Neoproterozoic igneous rocks in the northern and western parts of the Yangtze Block have been reexamined in the context of the tectonic evolution of the whole block. Three broad types of tectonic model have been proposed for the tectonic setting of the Yangtze Block from the late Mesoproterozoic to early Neoproterozoic: ① the mantle-plume model (Li, 1999; Li et al., 2002a, b; Ling et al., 2003; Wang and Li, 2003; Wang et al., 2011); ② the island arc model (Zhou et al., 2002a, b, 2006; Wanget al., 2006; Dong et al., 2011a,b, 2012; Zhao and Zhou, 2007, 2008, 2009, 2012); and ③ the plate-rift model (Wu et al., 2006, 2007; Zheng et al., 2006, 2007, 2008). These models are diverse, and the tectonic evolution of the Yangtze Block remains unclear. As shown in Fig. 1, the Hannan and Micangshan massifs are located in the northwestern corner of the Yangtze Block at the

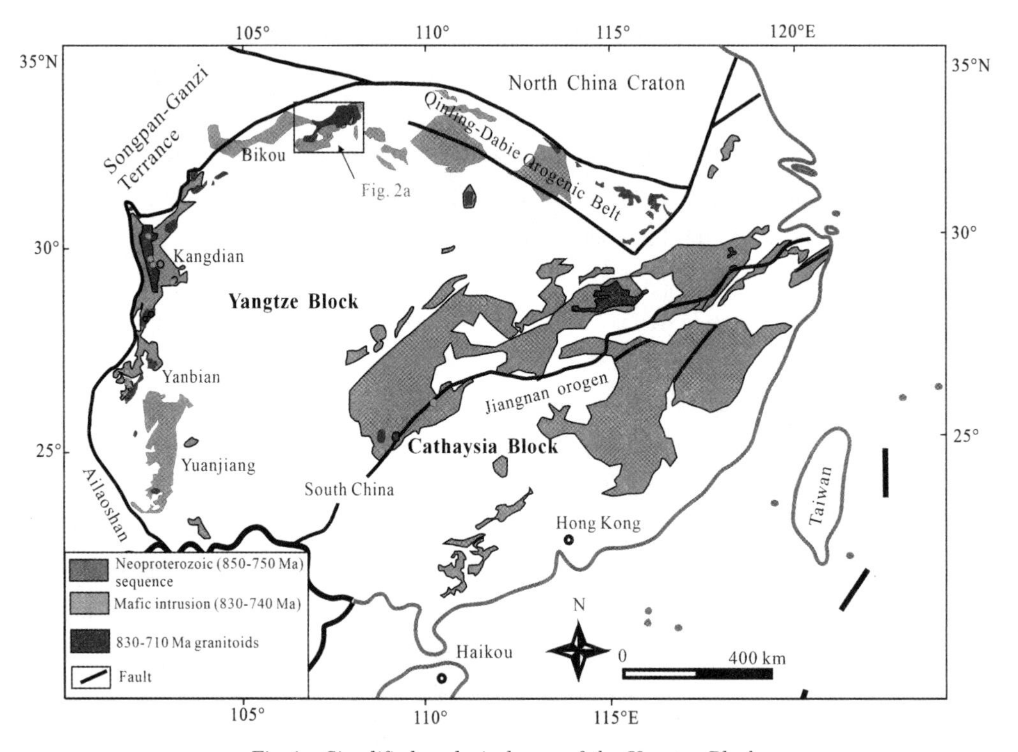

Fig. 1 Simplified geological map of the Yangtze Block (modified after Wang et al., 2010a; and Zhou et al., 2014).

intersection of the western and northern margins of the block (Dong et al., 2011a, 2012). Previous investigations into the Neoproterozoic alkaline complex rocks of the Micang Mountains region of the northwestern margin of the Yangtze Block have focused mainly on constraining the petrology and geochemistry of the rocks. Although progress has been made in understanding the magmatic evolution of these alkaline rocks, there is still a need to provide tighter constraints on the Neoproterozoic tectonic environment and associated evolution of magmatism (Xu, 1993; Li and Zhang, 1994; Xiao et al., 1997, 1998; Ma et al., 1997a; Zhou and Zhou, 1998; He, 2010; Yu and Xiao, 2011).

In this study, we present for the first time, zircon U-Pb ages, detailed petrological descriptions, and geochemical and Sr-Nd isotopic composition data for the Shuimo-Zhongziyuan alkaline intrusive complex in the Micang Mountains area on the northwestern margin of the Yangtze Block, South China. Our main objectives are to constrain the petrogenesis of the alkaline intrusive complex (syenite and gabbro) and the nature of their source, to infer the tectonic setting of this intrusive complex, and to explore the geodynamic implications of Neoproterozoic magmatism in the Hannan-Micangshan area (Fig. 2). Together with other published data, our geochemical and age data provide better temporal and spatial constraints on the tectonic evolution of the region during the Neoproterozoic.

2 Geological setting

The South China Block is composed of the Yangtze Block in the west and the Cathaysia Block in the east (Li et al., 2009; Zhao and Zhou, 2009; Zhao et al., 2010; Zhang and Zheng, 2013; Meng et al., 2015) (Fig. 1). The Yangtze and Cathaysia blocks are generally regarded as having amalgamated along the Jiangnan orogen during the Neoproterozoic (Li et al., 2003a; Wang et al., 2008; Zheng et al., 2008). The Yangtze Block is separated from the North China Block by the Qinling-Dabie-Sulu high-ultrahigh-pressure metamorphic belt to the north, and from the Tibetan Plateau by the Longmenshan fault to the west (Gao et al., 1999; Zhang et al., 2005). The rocks of the Yangtze Block comprise highly metamorphosed Neoarchean-Palaeoproterozoic crystalline basement, greenschist-facies metamorphosed Meso-to Neoproterozoic transitional basement, and unmetamorphosed Sinian-Mesozoic clastic and carbonate rock cover (Dong et al., 2012; Peng et al., 2012).

The Micang Mountains are located in the Yangtze Block and Qinling orogenic belt transition area. The Micang Mountains nappe structural belt is located on the northwestern margin of the Yangtze Block. The Micang Mountains are bounded to the east by the western end of the Dabashan arc-shaped tectonic belt, to the west by the northern part of the Longmenshan tectonic belt, to the north by the South Qinling orogenic belt, and to the south by the Sichuan basin. The Micang Mountains region contains exposures of the Yangtze Block crystalline

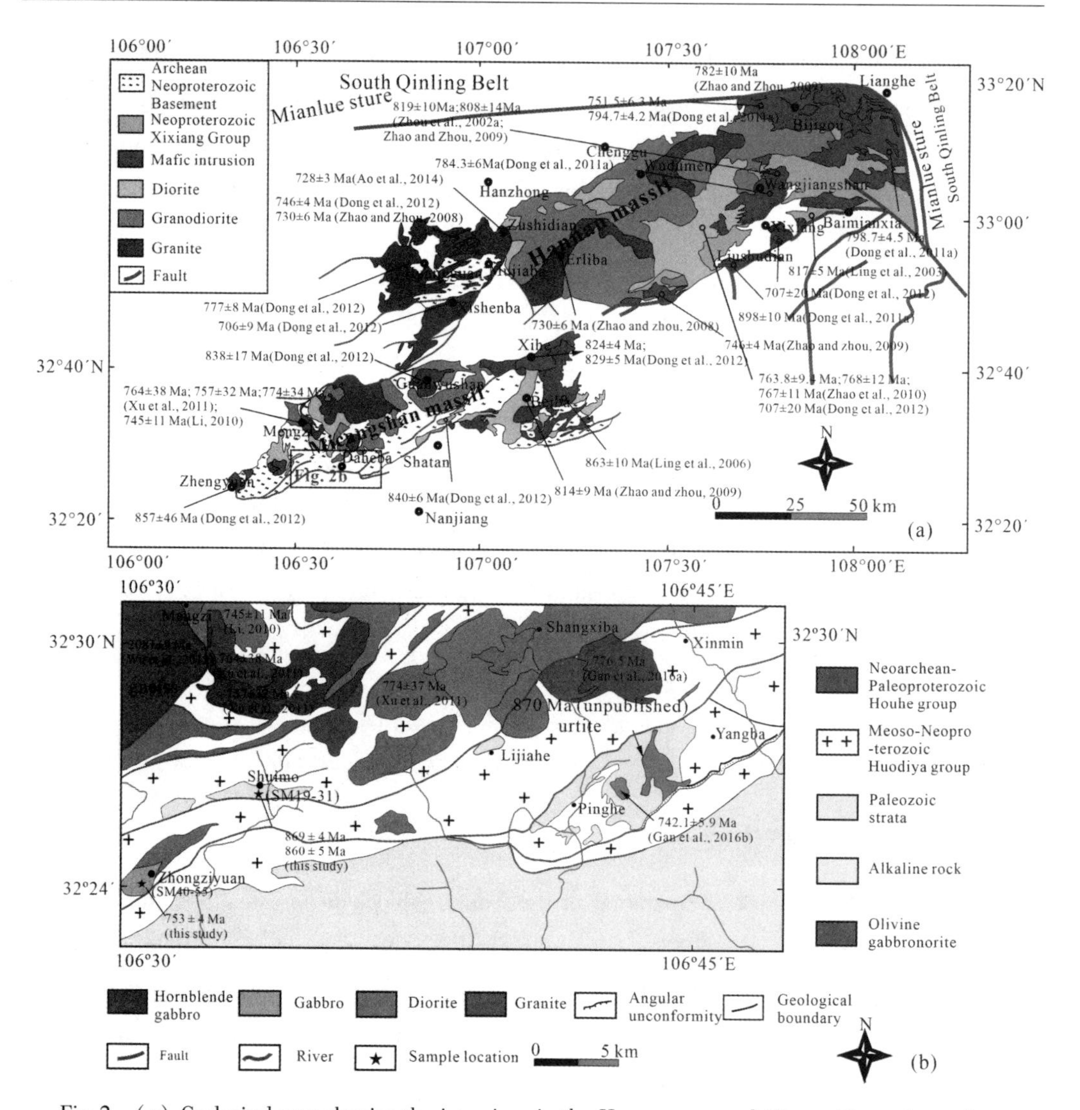

Fig. 2 (a) Geological map showing the intrusions in the Hannan area and Micang Mountains region (modified after Dong et al., 2011a, 2012). (b) Geological map of the Shuimo-Zhongziyuan area, Micang Mountains.

basement, folded basement, and sedimentary cover. The basement consists predominantly of the Neoarchean-Palaeoproterozoic Houhe Group and Meso-to Neoproterozoic Huodiya Group, which are unconformably overlain by Sinian and Phanerozoic strata by fault contact. The Houhe Group consists mainly of migmatites, trondhjemitic gneisses, amphibolites, and migmatites with minor marble (He et al., 1995; Zhao and Zhou, 2008). The Houhe Group rocks are similar to the rock units in the Kongling Group (Gao et al., 1999; Ling et al., 2006) to the east and those in the Kangding Group (Zhao and Zhou, 2007) to the southwest. One grey gneiss sample from the Houhe Group has yielded a zircon weighted mean $^{207}Pb/^{206}Pb$ age of 2 081 ± 9 Ma, interpreted as the formation age of the Houhe complex (Wu et al., 2012)

(Fig. 2b). The Houhe Group is unconformably overlain by greenschist-facies metamorphosed Meso- to Neoproterozoic Huodiya Group. The Huodiya Group consists of (from bottom to top) the Mawozi Formation, which is mainly marble, Shangliang Formation, which is aleuvitic slates with marble, and Tiechuanshan Formation, which is volcanic and clastic rocks (Ling et al., 1998). Sedimentary cover includes Sinian-Jurassic rocks but lacks a Devonian and Carboniferous record. Sinian-Triassic strata typical of Yangtze Block sedimentary rocks consist mainly of shallow-marine carbonate rocks, including Upper Triassic to Middle Jurassic continental-facies clastic rocks (Ma et al., 1997b; He, 2010; Xu et al., 2011; Tian et al., 2010).

3 Petrography and sample descriptions

The Micang Mountains massif is composed mainly of Neoproterozoic intrusion of granite, diorite, and gabbro, with minor alkaline rocks (Fig. 2a). The typical alkaline complex of the Micang Mountains region is located in Nanjiang and Wangcang counties in Sichuan. In this study, we mainly focus on the alkaline syenites and gabbros of the Shuimo-Zhongziyuan area. Ten samples of alkaline syenite and 16 samples of gabbro were selected from Shuimo and Zhongziyuan in Guangyuan City, Sichuan Province, respectively. The syenite and gabbro samples were collected from 32°27.563′ N, 106°33.856′ E and 32°24.325′ N, 106°31.526′ E, respectively (Fig. 2b).

The syenites are grey, medium to coarse grained, pegmatoidal, and massive (Fig. 3a). The syenites are composed of 30%–45% subhedral perthite, 25%–35% plagioclase, 5%–10% K-feldspar, 5%–8% quartz, and minor nepheline. The secondary mineral mainly include: the subhedral-anhedral amphibole, pyroxene, and biotite is irregular, and these minerals alteration phenomenon are obvious; the apatite, magnetite, titanite, and zircon are the most common accessory minerals (Fig. 3c-f). In the K-rich syenite, the dark minerals show evidence of substantial alteration (Fig. 3d).

The gabbros are dark grey to dark green, and medium to coarse-grained (Fig. 3b). They are composed of 30%–35% plagioclase, 20%–30% pyroxene, 10%–20% K-feldspar, ~10% amphibole, and ~5% biotite, with accessory apatite, magnetite, titanite, and zircon (Fig. 3g, h). Some minerals also show evidence of alteration.

4 Analytical methods for geochemistry

4.1 Major and trace elements

Whole-rock samples were trimmed to remove weathered surfaces, cleaned with deionized water, crushed, and then powdered through a 200 mesh screen using a tungsten carbide ball

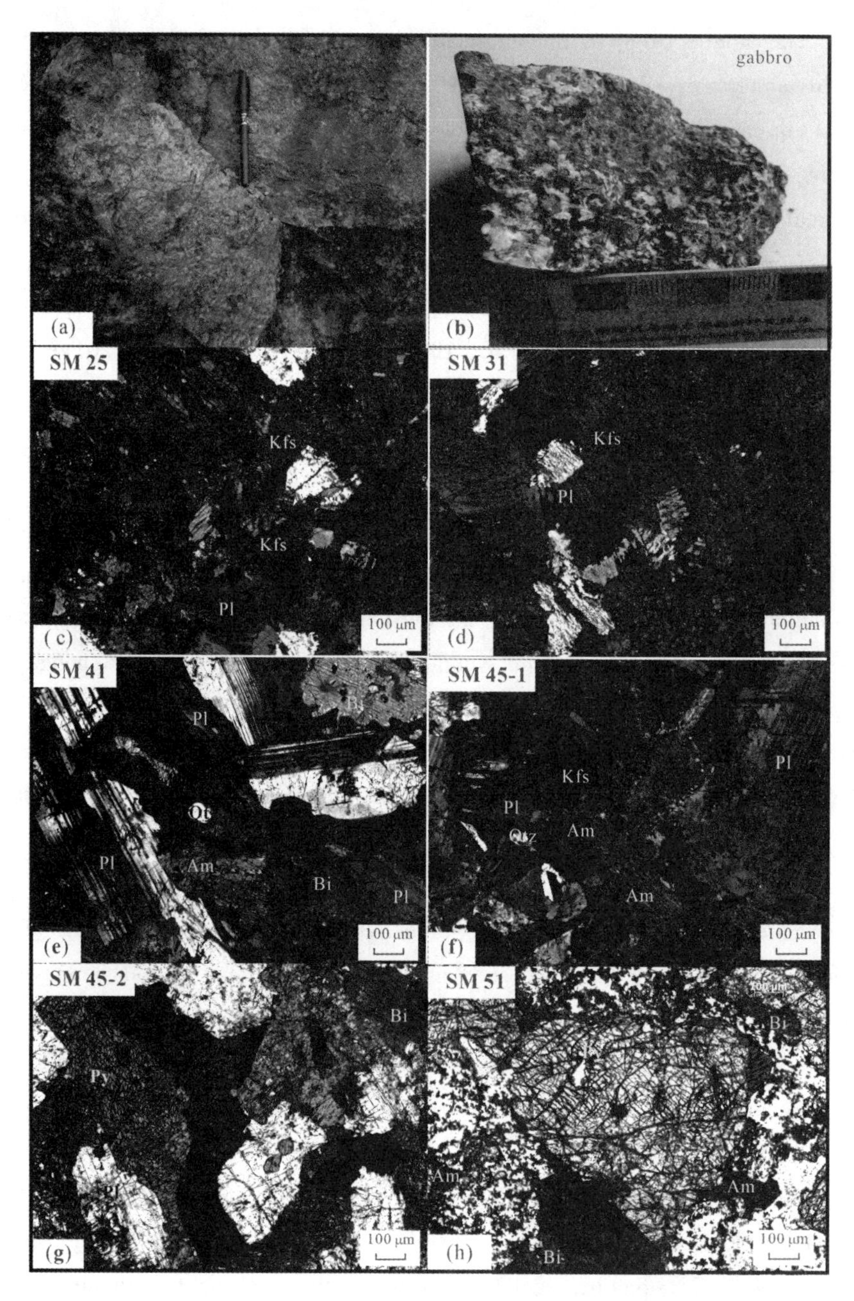

Fig. 3 Representative hand specimens (a,b) and microscope photographs (c-h) of syenites and gabbros from the Micang Mountains region at the northern margin of the Yangtze Block.

Syenite (c,d) and gabbro (e-h), with c-f in cross-polarized light and (g and h) in plane-polarized light. Py: pyroxene; Am: amphibole; Bi: biotite; Pe: perthite; Pl: plagioclase; Kfs: K-feldspar; Qtz: quartz (Shen, 2009).

mill. Major elements wereanalysed using an X-ray fluorescence (XRF) spectrometer (Rikagu RIX 2100) at the Guizhou Tuopu Resource and Environmental Analysis Center, the major element data had analytical uncertainties of less than ±5%.

Trace elements were determined by using a Bruker Aurora M90 inductively coupled plasma mass spectrometry (ICP-MS) at the Guizhou Tuopu Resource and Environmental Analysis Center, following the method of Qi et al. (2000). Sample powders were dissolved using an HF + HNO_3 mixture in a high-pressure PTFE bomb at 185℃ for 36 h. The accuracies of the ICP-MS analyses are estimated to be better than ± 5% – 10% (relative) for most elements.

4.2 Sr-Nd isotopic analyses

Whole-rock Sr-Nd isotopic data were obtained by using a Nu Plasma HR multi-collector mass spectrometer at the Guizhou Tuopu Resource and Environmental Analysis Center, the Sr, Nd isotopic analysis reference Chu et al. (2009). Sr and Nd isotopic fractionation was corrected to $^{87}Sr/^{86}Sr$ = 0.119 4 and $^{146}Nd/^{144}Nd$ = 0.721 9, respectively. During the period of analysis, a Neptune multi-collector ICP-MS was used to measure the $^{87}Sr/^{86}Sr$ and $^{143}Nd/^{144}Nd$ isotope ratios. NIST SRM-987 and JMC-Nd were used as certified reference standard solutions for $^{87}Sr/^{86}Sr$ and $^{143}Nd/^{144}Nd$ isotopic ratios, respectively. BCR-1 and BHVO-1 were used as the reference materials.

4.3 Zircon U-Pb analyses

Zircon was separated from three ~4 kg samples (SM02, SM03, and SM26) taken from various sampling locations along the margin northern of the Yangtze Block, Micang Mountains region. The zircon grains were separated by using conventional heavy liquid and magnetic techniques. Representative zircon grains were handpicked and mounted in epoxy resin disks and then polished and coated with carbon. Internal morphology was examined using cathodoluminescent (CL) prior to U-Pb analyses.

Laser ablation (LA) ICP-MS zircon U-Pb analyses were conducted on an Agilent 7500a ICP-MS equipped with a 193-nm laser, which is housed at the State Key Laboratory of Continental Dynamics, Northwest University, Xi'an, China, following the method of Yuan et al. (2004). The $^{207}Pb/^{206}Pb$ and $^{206}Pb/^{238}U$ ratios were calculated by using the GLITTER programme, which was corrected using the Harvard zircon 91500 as external calibration. These correction factors were then applied to each sample to correct for both instrumental mass bias and depth-dependent elemental and isotopic fractionation. The detailed analytical technique is described in Yuan et al. (2004). Common Pb contents were, therefore, evaluated by using the method described in Andersen (2002). The age calculations and plotting of concordia diagrams were made using ISOPLOT version 3.0 (Ludwig, 2003). The errors quoted in tables and figures are at 2σ level.

5 Analytical results

5.1 Zircon LA-ICP-MS U-Pb ages

The zircon LA-ICP-MS U-Pb concordia diagrams and cathodoluminescence (CL) images from syenite (SM02 and SM26) and gabbro (SM03) of the Shuimo-Zhongziyuan area in the Micang Mountains region are presented in Fig. 4, along with the results in Supplementary Table 1. The separated zircons are broken, generally subhedral to xenomorphic, irregularly shaped, colourless or light grey, prismatic, and translucent to transparent, and have lengths ranging from 50 μm to 100 μm with length-to-width ratios of 1∶2 to 1∶1. The CL images show that most of the grains have oscillatory zoning (Fig. 4), indicating igneous crystallization. In the CL images, some parts of zircon grains (syenite and gabbro) display discontinuous oscillatory zoning between the core and margin, which may represent inherited or trapped zircon in the core and crystallized zircon in the margin, or may reflect different stages of crystallization. The Th/U ratios of zircons are generally considered to reflect their origin (Maas et al., 1992). Magmatic zircons have Th/U ratios mostly ranging from 0.2 to 1.0, whereas metamorphic zircons have lower Th/U ratios (<0.1) (Williams et al., 2009). Almost all the zircons from our samples almost display Th/U ratios >0.2, apart from three samples with a Th/U ratio of 0.1, suggesting most have a magmatic origin.

5.1.1 Shuimo syenite (SM02 and SM26)

Thirty-six reliable analytical data (spots) were obtained from sample SM02, an Na-rich syenite. Seven spots yield discordant U-Pb ages, and spots 30−35 give significantly younger $^{206}Pb/^{238}U$ ages from 664 ± 7 Ma to 792 ± 8 Ma, which may be due to Pb loss from subsequent geological processes. Spots 36 yield older $^{206}Pb/^{238}U$ ages ($2\ 149 \pm 21$ Ma), which suggests they are inherited. The remaining 22 concordant spots have U contents ranging from 42.96 ppm to 1 414.81 ppm, Th contents ranging from 58.81 ppm to 1 637.35 ppm, and Th/U ratios ranging from 0.11 to 32.96. These 22 concordant spots are characterized by $^{206}Pb/^{238}U$ ages ranging from 855 ± 9 Ma to 884 ± 9 Ma, with a weighted mean age of 869 ± 4 Ma (MSWD = 0.85; $n=22$; 2σ) (Fig. 4a). This age is interpreted as the age of syenite emplacement in the Shuimo area.

Twenty-four reliable analytical data were obtained from sample SM26, a K-rich syenite. Seven spots yield discordant U-Pb ages, and spots 12−15 give significantly younger $^{206}Pb/^{238}U$ ages ranging from 582 ± 4 Ma to 810 ± 10 Ma, which are considered to be caused by Pb loss. Two spots (16 and 17) yield $^{206}Pb/^{238}U$ ages ranging from 939 ± 10 Ma to 978 ± 7 Ma and are clearly older than the other grains. The remaining 11 spots have U contents ranging from 43.90 ppm to 669.20 ppm, Th contents ranging from 64.90 ppm to 1 311.40 ppm, and Th/U ratios from 0. 13 to 13. 44. These 11 concordant spots are characterized by $^{206}Pb/^{238}U$ ages ranging from 846 ± 11 Ma to 866 ± 5 Ma, with a weighted mean age of 860 ± 5 Ma (MSWD = 0. 47; n = 11; 2σ) (Fig. 4c). This is taken to be the syenite emplacement age in the Shuimo area.

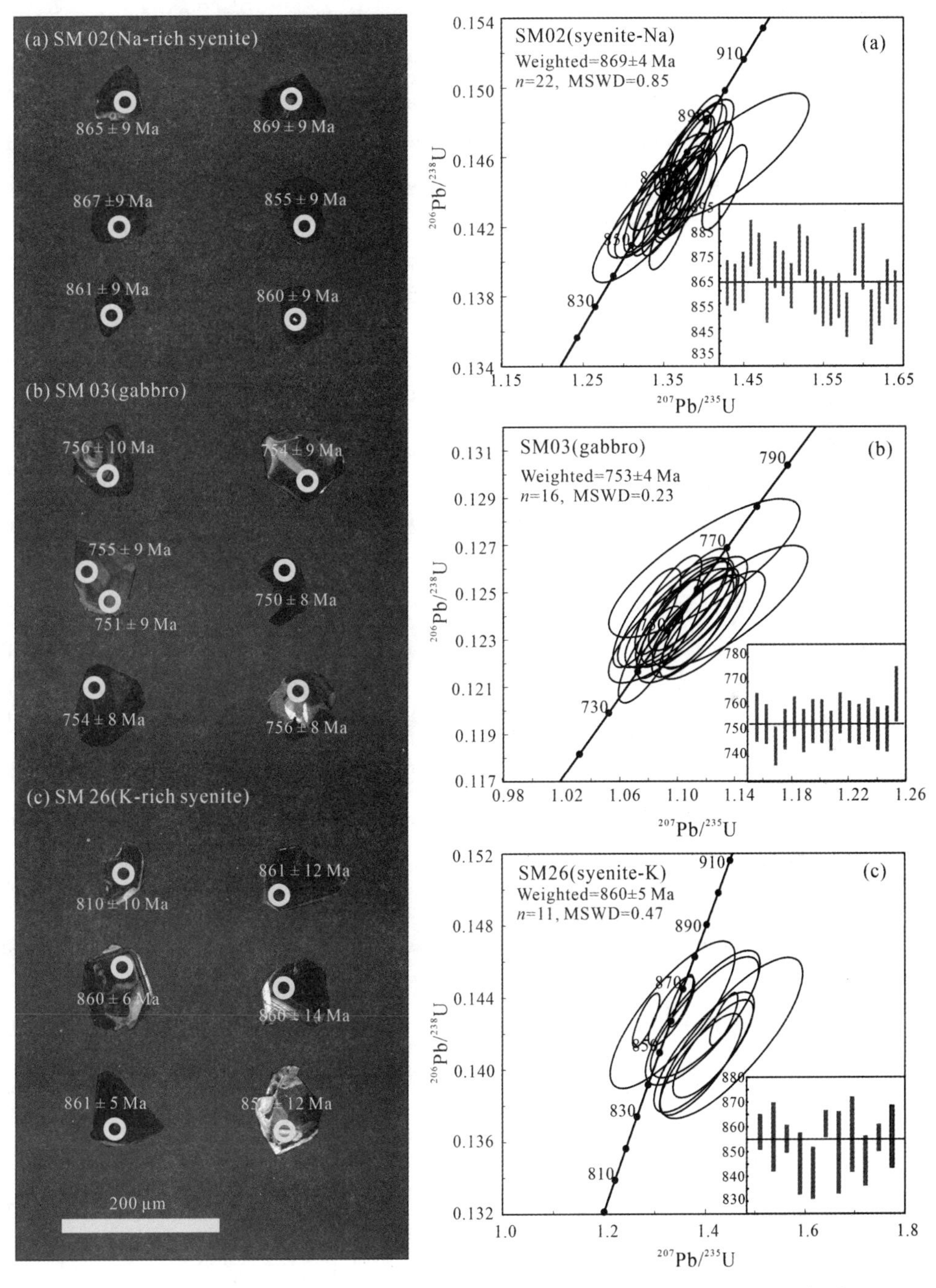

Fig. 4 Cathodoluminescence (CL) images (left) and LA-ICP-MS U-Pb zircon concordia diagrams (right), with the positions of zircon age analyses on the CL images shown by the yellow circles.

5.1.2 Zhongziyuan gabbro (SM03)

Twenty-eight reliable analyses were obtained from sample SM03. Four spots yield discordant U-Pb ages. Spots 17–23 give $^{206}Pb/^{238}U$ ages ranging from 595±6 Ma to 686±8 Ma. Regarding their morphology, these zircons show inconspicuous internal structures. Spot 24 yields older $^{206}Pb/^{238}U$ ages (969±10 Ma), which indicates this zircon is inherited. There are 16 concordant spots, which have U contents ranging from 105.10 ppm to 807.40 ppm, Th contents ranging from 64.90 ppm to 980.70 ppm, and Th/U ratios ranging from 0.62 to 1.38. These 16 concordant spots are characterized by $^{206}Pb/^{238}U$ ages ranging from 744±8 Ma to 765±11 Ma, with a weighted mean age of 753±4 Ma (MSWD=0.23; n=16; 2σ) (Fig. 4b). This is interpreted as the emplacement age of the gabbro in the Zhongziyuan area.

In summary, the compositional and U-Pb age data for zircons from the three samples define magmatic events at ca. 860 Ma and 750 Ma at the northwestern margin of the Yangtze Block, South China, in the Micang Mountains region.

5.2 Major and trace element compositions

The results of the major and trace element analyses of 10 syenite and 16 gabbro samples from the Shuimo-Zhongziyuan alkaline intrusive complex are listed in Supplementary Table 2. In a Q-A-P-F diagram (Fig. 5a), data for the syenite samples fall into the field of alkali feldspar (A), indicating that the syenites belong to the alkaline rock series, whereas the gabbros fall within the fields of gabbro and diorite (P), implying that the samples are mafic intrusive rocks. These samples are further distinguished in Fig. 5b, in the total alkalis (Na_2O+K_2O) vs. silica (SiO_2) (TAS) classification diagram of Middlemost (1994), the Shuimo-Zhongziyuan rock samples fall in the fields of gabbro and syenite and are classified as sub-alkaline and alkaline series, respectively.

The σ [$(K_2O+Na_2O)^2/(SiO_2-43)$] values for the syenite samples range from 4.21 to 6.34, meaning that these rocks are classified as alkaline series. In the SiO_2 vs. A.R. diagram (Fig. 5c), all the syenite samples fall within the alkaline field. However, the gabbros trend from calc-alkaline to alkaline. The major elements differ between the alkaline syenites and gabbros. Compared with the gabbros, the 10 syenite samples (SM19–31) from the Shuimo area, including the Na-rich SM19–25 (Na_2O=5.73–6.69 wt%, K_2O=3.23–4.6 wt%) and K-rich SM26–31 (Na_2O = 4.01 – 4.68 wt%, K_2O = 7.72 – 8.79 wt%) samples, are characterized by higher SiO_2 (62.52–68.80 wt%) and Na_2O + K_2O (9.52–13.47 wt%) contents, as well as lower Al_2O_3 (11.15–13.85 wt%), TiO_2 (0.05–0.21 wt%), $Fe_2O_3^T$ (1.01–2.92 wt%), MgO (0.38–1.89 wt%), and CaO (1.71–5.83 wt%) contents. The 16 gabbro samples (SM40–55) have lower SiO_2 (46.0–51.90 wt%) and Na_2O + K_2O (1.26–5.45 wt%) contents and Na_2O >K_2O, as well as higher Al_2O_3 (12.40–19.50 wt%), TiO_2

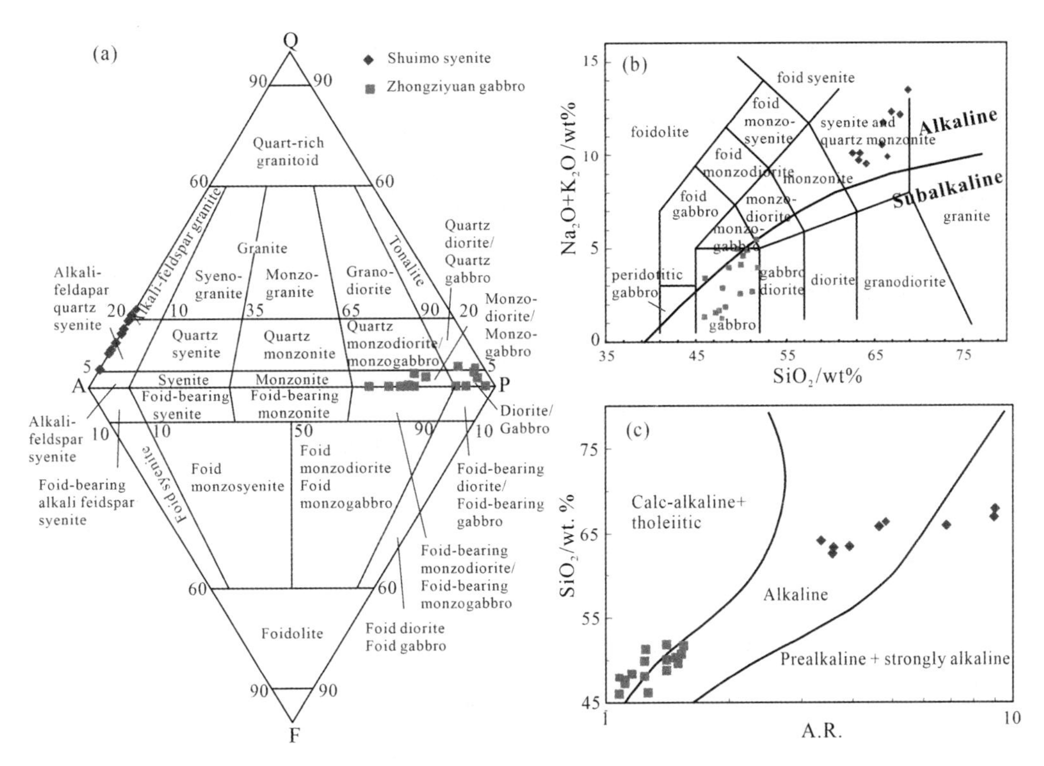

Fig. 5 (a) Q-A-P-F diagram; (b) total alkali vs. silica (TAS) diagram (after Middlemost, 1994); (c) SiO_2 vs. A.R. [$(Al_2O_3+CaO+Na_2O+K_2O)/(Al_2O_3+CaO-Na_2O-K_2O)$] diagram (modified after Wright, 1969).

(0.87–2.29 wt%), $Fe_2O_3^T$ (9.30–16 wt%), MgO (4.20–11 wt%), and CaO (8.21–14.10 wt%) contents. The $Mg^{\#}$ values of the syenites and gabbros have a mean of 50. It is important to note that the alkaline syenites have higher loss on ignition values as compared with the gabbros. In the major element Harker diagrams (Fig. 6), most elements of the two groups of rocks show good linear correlations with SiO_2, although Al_2O_3 and TiO_2 contents of the gabbros and K_2O contents of the syenites show some scatter (Fig. 6b, e, g). Moreover, the major elements also show good linear correlations with SiO_2 in the Mengzi and Hannan area mafic intrusive rocks.

The syenites have lower total rare earth element (REE) concentrations than the gabbros, with an average of 60 ppm. The syenite samples exhibit strong enrichment in light REE and significantly fractionated REE [relatively high $(La/Yb)_N$ = 5.54–17.4], with small negative and positive Eu anomalies (Eu/Eu^* = 0.74–1.17) (Fig. 7a). The negative Eu anomalies may be associated with residual plagioclase in the magma source. In contrast to the alkaline syenites, the gabbro samples exhibit flat heavy REE patterns, slightly enriched light REE patterns, and small positive or negative Eu anomalies (Eu/Eu^* = 0.86 – 1.37). The $(La/Yb)_N$ ratios for the gabbro samples range from 1.71 to 4.34, and indicated that the

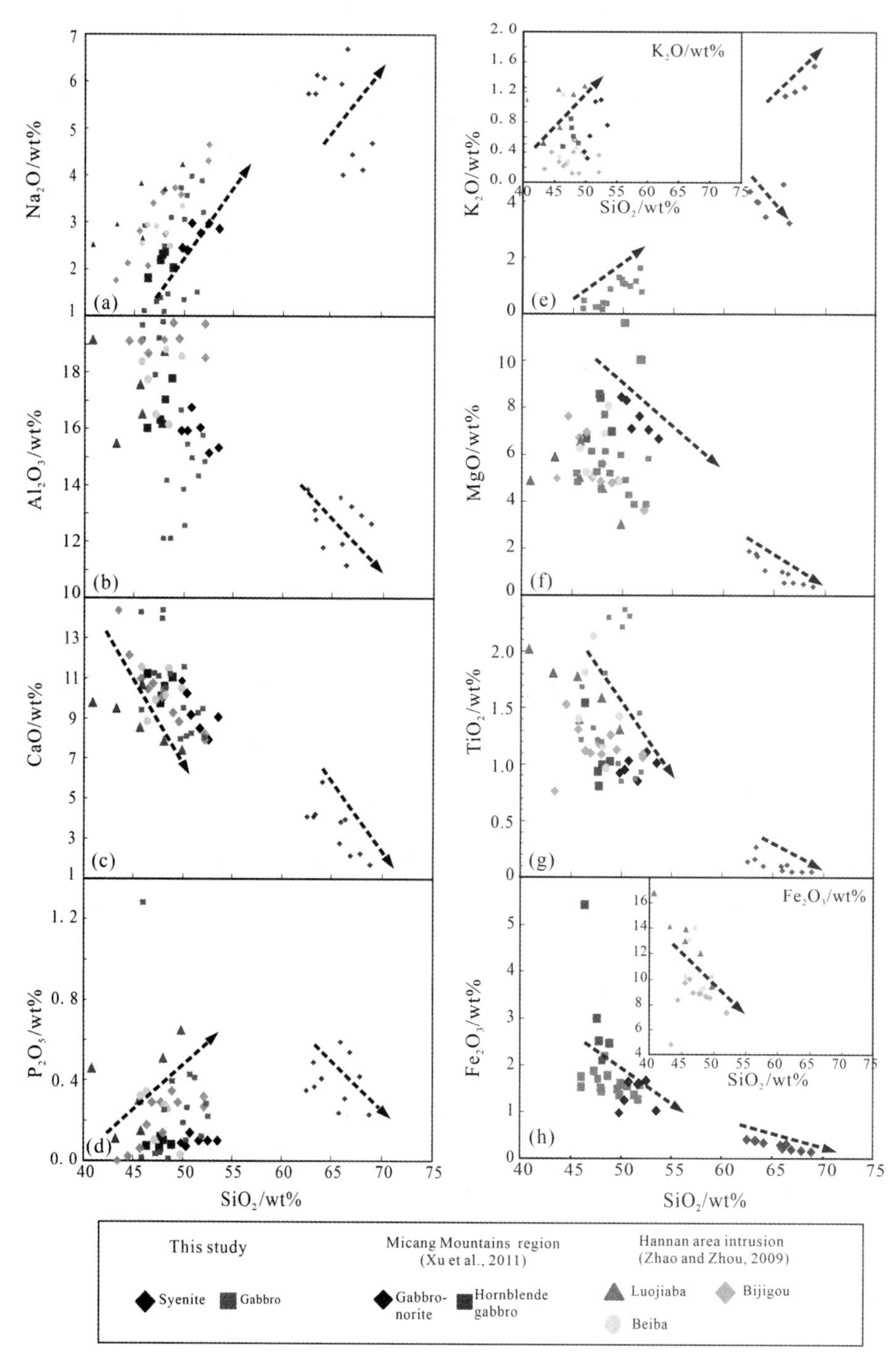

Fig. 6 Plots of SiO_2 vs. various major oxides (a-h) for the Shuimo-Zhongziyuan alkaline intrusive complex in the Micang Mountains region.

The arrows signify a fractionation trend.

fractionation between light and heavy REE is not marked. Although the trace elements of the mafic rocks in Hannan-Micangshan region in the spider diagram are some different to the OIB, the REE of the mafic rocks show some similarity feature with OIB (Fig. 7c), indicating that they may be derived from the same source region.

In N-MORB-normalized trace element diagrams (Fig. 7b), the alkaline complex samples are generally characterized by enrichment in large-ion lithophile elements (LILE) (e.g., Rb, Ba, Th, and U), positive Pb anomalies, and depletion in high-field-strength elements (HFSEs) (e.g., Nb and Ta) (Fig. 7b). The gabbros are also enriched in LILEs and depleted in HFSEs, the Hannan area mafic intrusions also have similar characteristics (Fig. 7d).

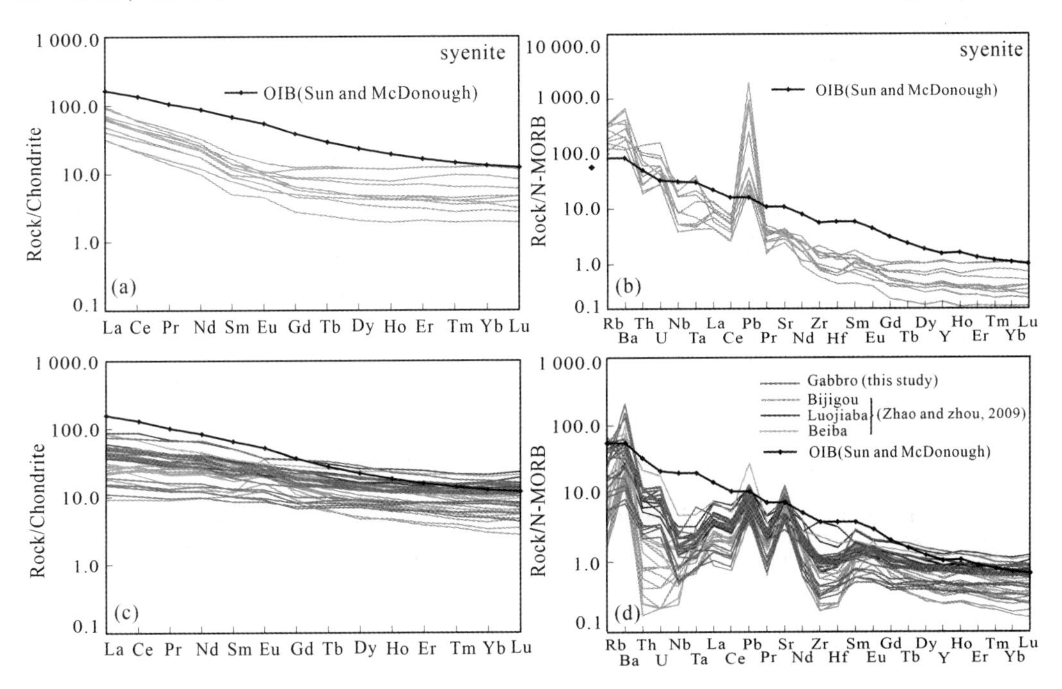

Fig. 7 Diagrams of chondrite-normalized REE (a-c) and N-MORB-normalized trace element patterns (b-d) of the Shuimo syenites and Zhongziyuan gabbros from the Micang Mountains region. The chondrite and N-MORB values were taken from Sun and McDonough (1989).

5.3 Sr-Nd isotope compositions

Whole-rock Sr-Nd isotopic compositions are presented in Supplementary Table 3. The initial $^{87}Sr/^{86}Sr$ ratios and $\varepsilon_{Nd}(t)$ values are calculated at t(syenite) = 860 Ma and t(gabbro) = 750 Ma.

The Shuimo alkaline syenite samples have Sr concentrations of 191 − 221 ppm, Rb concentrations of 47−134 ppm, low initial $^{87}Sr/^{86}Sr$ ratios of 0.700 549−0.703 340, positive $\varepsilon_{Nd}(t)$ values of +1.9 to +7.7, and Nd isotopic T_{DM} model ages of 0.89−1.42 Ga. The Zhongziyuan gabbro samples are characterized by a wide range of Sr concentrations from 339 ppm to 1 197 ppm, Rb concentrations from 6 ppm to 28 ppm, initial $^{87}Sr/^{86}Sr$ ratios from 0.703 562 to 0.704 933, positive $\varepsilon_{Nd}(t)$ values from +1.6 to +4.5, and Nd isotopic T_{DM} model ages of 1.32–2.08 Ga.

6 Discussion

6.1 Formation age of the intrusive complex

Although previous studies have reported geochronological data for the alkaline rocks of the Micang Mountains region, there has been no clear conclusion concerning the formation ages of these rocks. The northwest Sichuan geological team in 1992 measured hornblende K-Ar ages of xenoliths in ultrabasic alkaline rocks (Lijiahe carbonate rock) and obtained ages ranging from 746.6 Ma to 775.2 Ma. Qiu (1993) measured an isochron age of 687 Ma using whole-rock Rb-Sr methods. He (2010) provided a whole-rock Rb-Sr isotopic age for Shaheba alkaline rocks of 779.4 Ma. These previous ages for the alkaline intrusive rocks range from 687 Ma to 779 Ma, which span a significant range. Several studies have investigated igneous rocks ranging from mafic to felsic in composition distributed across the Hannan-Micangshan massif at the northwestern margin of the Yangtze Block. The formation ages of these rocks range from 700 Ma to 900 Ma (Table 1), and include ages of 740–770 Ma for magmatic rocks in the Mengzi area (Li, 2010; Xu et al., 2011), 863±10 Ma for the Tianpinghe granite (Ling et al., 2006), 852 Ma for the Xihe mafic-acidic rocks (Dong et al., 2012), 814 Ma for the Beiba gabbro (Zhao and Zhou, 2009), 742.1±5.9 Ma for the Pinghe monzogranite (Gan et al., 2016b), and 865 Ma for the Pinghe alkaline rocks (our unpublished data).

Our U-Pb age data show that the formation ages of the Shuimo syenite are 869±4 Ma (for the Na-rich samples) and 860±5 Ma (for the K-rich samples), and that of the Zhongziyuan gabbro is 753±4 Ma. The ages suggest that these rocks likely formed during multiple magmatic events. On the basis of these ages and associated uncertainties, we infer that the mafic-intermediate-felsic rocks in the Micang Mountains region are the products of Neoproterozoic magmatic activity (Fig. 2) and that the alkaline complex formed during the mid-Neoproterozoic ("Jingningian").

6.2 Characteristics and petrogenesis of the intrusive complex

The syenite samples from the Shuimo area, compared with the Zhongziyuan gabbros, have higher values of σ, SiO_2 and Na_2O+K_2O contents, and lower Al_2O_3, TiO_2, $Fe_2O_3^T$, MgO, and CaO contents. The Zhongziyuan gabbro samples have lower SiO_2 and Na_2O+K_2O contents and $Na_2O>K_2O$, and higher Al_2O_3, $Fe_2O_3^T$, MgO, and CaO contents. Furthermore, both the syenites and gabbros have low Rb and high Sr contents (Supplementary Table 2), suggesting that the major and trace element compositions of these rocks can be used to infer the characteristics of their source.

Table 1 Isotopic ages of the Neoproterozoic magmatism in the Hannan-Micang Mountains along the northern margin of the Yangtze Block, South China.

Tectonic position	Location	Sample	Lithology	Method	Interpretation	Age /Ma	Isotopic	Model age /Ga	Reference
Micangshan massif	Xihe	MC01	Granite	LA-ICP-MS	Mean $^{206}Pb/^{238}U$ age	829 ± 5.0			Dong et al., 2012
	Xihe	MC02	Gabbro			824 ± 4.0			
	Shatan	ZNJ-025	Diorite			840 ± 6.0			
	Guanwushan	ZNJ-027	Granite			838 ± 17			
	Zhengyuan	ZWC-022	Gabbronorite		upper intercept age	857 ± 46			
	Wangcang	ZWC-012	Granodiorite		upper intercept age	871 ± 77			
	Beiba	BB39	Gabbro	SHRIMP	upper intercept age	814 ± 9.0	I_{Sr} 0.703 77–0.704 59; $\varepsilon_{Nd}(t)$ +0.23 to +1.98; $\varepsilon_{Hf}(t)$ −0.7 to +17.7	T_{DMHf} 0.66–1.40	Zhao and Zhou, 2009
	Tianpinghe	MCTP-01	Granite	ELA-ICP-MS	Mean $^{206}Pb/^{238}U$ age	863 ± 10			Ling et al., 2006
	Mengzi	08-95	Gabbronorite	LA-ICP-MS	Mean $^{206}Pb/^{238}U$ age	764 ± 38	$\varepsilon_{Hf}(t)$ +6.92 to +13.99	T_{DMHf} 0.74–1.1	Xu et al., 2011
		08-104TW	Hornblende gabbro			757 ± 32	$\varepsilon_{Hf}(t)$ +4.74 to +14.97	T_{DM2Hf} 0.7–1.2	
		08-109TW	Quartz diorite			774 ± 34	$\varepsilon_{Hf}(t)$ −4.46 to +13.1	T_{DM2Hf} 0.8–1.7	
		08-99TW	Moyite			745 ± 11	$\varepsilon_{Hf}(t)$ −16.49 to +8.33	T_{DM2Hf} 0.9–2.0	Li, 2010
	Pinghe	PH2-6	Monzogranite			742.1 ± 5.9	$\varepsilon_{Nd}(t)$ +3.6 to +5.2	T_{DM2Nd} 0.96–1.07	Gan et al., 2016b
		PH11	Andesitic basalt			858.5 ± 7.6			unpublished data
		CSC-1	Urtite			871.1 ± 4.1			
		CSC-59	Tonalite			869 ± 4.3			

Continued

Tectonic position	Location	Sample	Lithology	Method	Interpretation	Age /Ma	Isotopic	Model age /Ga	Reference
Micangshan massif	Shuimo	SM02	Syenite (Na-rich)	LA-ICP-MS	Mean $^{206}Pb/^{238}U$ age	869±4	I_{Sr} 0.700 55–0.703 34; $\varepsilon_{Nd}(t)$ +1.9 to +7.7	T_{DMNd} 0.89–1.42; T_{DM2Nd} 0.89–1.29	this study
		SM26	Syenite (K-rich)			860±5			
	Zhongziyuan	SM03	Gabbro			753±4	I_{Sr} 0.703 56–0.704 93; $\varepsilon_{Nd}(t)$ +1.6 to +4.5;	T_{DMNd} 1.11–2.08	
	Xinmin	Yb06	Hornblende gabbro			776.5	I_{Sr} 0.703 47 $\varepsilon_{Nd}(t)$ +3.1	T_{DMNd} 1.55	Gan et al., 2016a
Hannan massif	Mujiaba	ZNZ-030	Gabbronorite	LA-ICP-MS	Mean $^{206}Pb/^{238}U$ age	746±4.0			Dong et al., 2012
	Wudumen	HN-03	Tonalite			718±9.0			
	Xishenba	ZNZ-009	Monzogranite			706±9.0			
	Huangguan	ZNZ-032	Monzogranite			777±8.0			
	Xixiang	ZXX-005	Syenogranite			707±20			
	Wangjiangshan	ZXX-001	Gabbro			784.3±6			Dong et al., 2011a
	Liushudian	ZXX-007	Gabbro			898±10			
	Xijiaba	10BJG-01	Gabbro			798.7±4.5			
	Youshui	10BJG-12	Diabase			794.7±5.2			
	Youshui	10BJG-23	Gabbro			751.5±6.3			
	Wudumen	MCWD-1	Tonalite	ELA-ICP-MS		789±10			Ling et al., 2006
	Wangjiangshan	WBZ1	Diorite	SHRIMP	Mean $^{206}Pb/^{238}U$ age	819±10	I_{Sr} 0.703 27–0.703 73 $\varepsilon_{Nd}(t)$ +3.37 to +3.83 $\varepsilon_{Hf}(t)$ +6 to +15.3	T_{DMHf} 0.77–1.14	Zhou et al., 2002a; Zhao and Zhou et al., 2009
	Wangjiangshan	WQZ1	Gabbro			808±14			
	Bijigou	BQZ4	Gabbro			782±10	I_{Sr} 0.703 309–0.703 615 $\varepsilon_{Nd}(t)$ +2.95 to +3.87 $\varepsilon_{Hf}(t)$ +4.9 to +10	T_{DMHf} 0.95–1.13	

Continued

Tectonic position	Location	Sample	Lithology	Method	Interpretation	Age /Ma	Isotopic	Model age /Ga	Reference
Hannan massif	Luojiba	LJB16	Gabbro	SHRIMP	Mean $^{206}Pb/^{238}U$ age	746±4.0	I_{Sr} 0.703 69–0.703 91; $\varepsilon_{Nd}(t)$ +0.93 to +1.34; $\varepsilon_{Hf}(t)$ +0.28 to +5.73	T_{DMHf} 0.99–1.27	Zhao and Zhou, 2009
	Xixiang	XX16	Diorite intrusion	SHRIMP	Mean $^{206}Pb/^{238}U$ age	763.8±9.4	I_{Sr} 0.703 31–0.703 52	T_{DM} 0.97–1.09;	Zhao et al.,2010
		XX05		LA-ICP-MS		768±12	$\varepsilon_{Nd}(t)$ +1.94 to +3.16;	T_{DM2} 1.09–1.3	
		XX10				767±11	$\varepsilon_{Hf}(t)$ +5.5 to +9.8		
	Wudumen	WDM25	Granodiorite	SHRIMP	Mean $^{206}Pb/^{238}U$ age	735±8.0	I_{Sr} 0.703 65–0.703 83; $\varepsilon_{Nd}(t)$ +0.16 to +1.16	T_{DMHf} 0.94–1.27; T_{DM2Hf} 1.07–1.58	Zhao and Zhou,2008
	Erliba	RL23	Granodiorite			730±6.0	I_{Sr} 0.703 54–0.703 79; $\varepsilon_{Nd}(t)$ +0.32 to +2.06		
	lower Xixiang	BMX-Zir	Dacite	TIMS	a concordant U-Pb age	950±4.0			Ling et al., 2003
	upper Xixiang	SJH-02	Rhyolite			897±3.0			
	Tiechuanshan	XG-02	Rhyolite			817±5.0			
	Chenggu	Q92102-6	Plagiogranite (biotite)	$^{40}Ar/^{39}Ar$		796±20			Zhang et al.,2002
	Sunjiahe (Xixiang)	SJH06-4	Pyroxene-basalt	LA-ICP-MS	Mean $^{206}Pb/^{238}U$ age	845±17			Xu et al., 2009
		SJH06-3	Rhyolite			832.9±4.9			
	Sunjiahe		Rhyolite	Zircon U-Pb		839±2.0			Zhao et al.,2006
	Wudumen		Tonalite			764±2.0			
	Huangjiaying (Nanzheng)		Moyite			778±5.0			
	Zushidian	EL-1	Trondhjemite	LA-ICP-MS		728±3.0	$\varepsilon_{Hf}(t)$ +5 to +12.2	T_{DMHf} 0.81–1.10	Ao et al., 2014

In Harker diagrams (Fig. 6), the syenites exhibit well-developed negative correlations of Na_2O, Al_2O_3, MgO, CaO, TiO_2, P_2O_5, and Fe_2O_3 with SiO_2, and positive correlations of K_2O with SiO_2. On the same diagrams, the gabbros display negative correlations of CaO, Fe_2O_3, and MgO with SiO_2, and positive correlations of Na_2O, K_2O, and P_2O_5 with SiO_2. These features indicate that fractional crystallization may have played a major role in magma evolution. The intrusion complex samples are generally characterized by enrichment in LILEs and depletion in HFSEs, and the HFSEs depletion may have resulted from mantle metasomatism related to subducted fluids or sedimentary materials (Sun and McDonough, 1989).

In the SiO_2 vs. A.R. diagram (Fig. 5c), the samples lie in the calc-alkaline-alkaline-peralkaline range and fall mainly within the field of the alkalic series. A. R. is positively correlated with SiO_2, which suggests that the sampled rocks may have ultimately originated from similar magmas (Wilson, 1989). In a Harker diagram (Fig. 6), some samples display discrete distributions, although the major elements have fairly high correlations with SiO_2, and the samples have moderate $Mg^{\#}$ values (43% - 72%), which together indicates that these intrusive rocks underwent normal magmatic evolution. In addition, the A. R. values of the syenites are higher than those of the gabbros, and the A.R. values of the K-rich syenites are higher than those of the Na-rich syenites, implying that the syenite may have had a complex source and tectonic setting (Huang, 2013). Furthermore, the alkali contents of the samples may reflect compositional changes in the magma during its evolution. For example, sample SM26 has a high A.R. value of 31. 6 and may identify the existence of a silica-rich alkali fluid.

An La/Sm-La correlation diagram (Fig. 8a) shows fairly constant values of La/Sm in the process of magma differentiation and fractional crystallization, revealing that the intrusive rocks may be composed of mantle-derived primitive basaltic magma. In a chondrite-normalized REE diagram (Fig. 7a), the Shuimo syenite has a small negative and positive Eu anomaly, and also

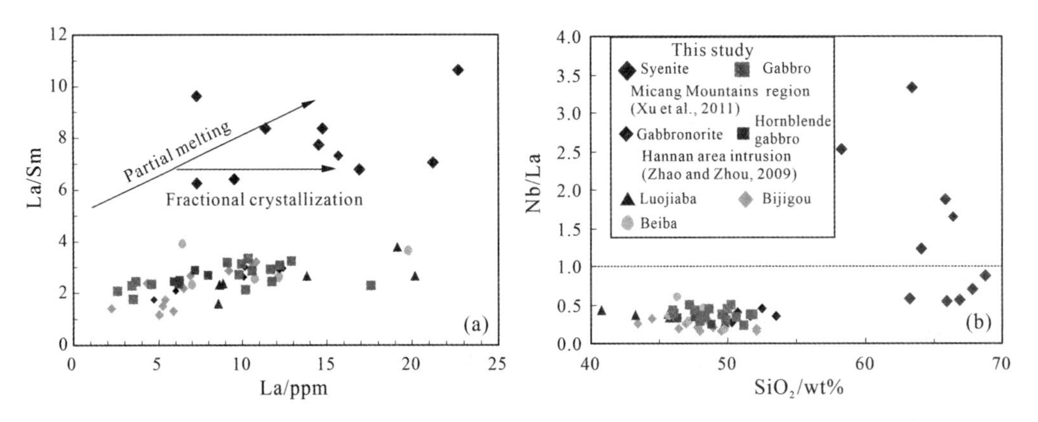

Fig. 8 (a) La vs. La/Sm diagram (Allegre and Minster, 1978); (b) SiO_2 vs. Nb/La diagram.

have some positive Ba, Sr anomalies (Fig. 7b), because of the different level of the fractional crystallization. In a SiO_2 vs. Nb/La diagram (Fig. 8b), all gabbros, including the Hannan and Micang Mountains region mafic intrusion, have Nb/La ratios < 1, which again indicates that the magmatic source of these rocks was affected by crustal contamination, however, the syenite show relatively scattered. In a diagram of SiO_2 vs. $\varepsilon_{Nd}(t)$ (Fig. 9), the gabbro data show a moderate linear correlation, suggesting that the gabbros underwent crustal contamination during their formation, but it is unclear whether the syenites were similarly affected.

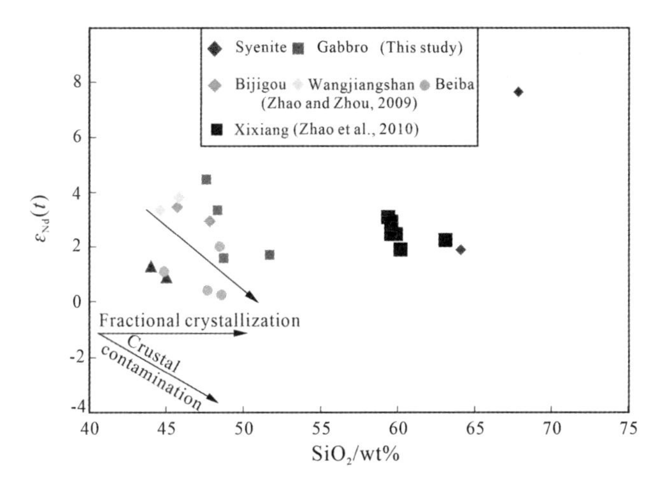

Fig. 9 SiO_2 vs. $\varepsilon_{Nd}(t)$ diagram.

The trace element characteristics and Sr-Nd isotopic compositions of the Shuimo-Zhongziyuan intrusive complex in the Micang Mountains region provide a new constraint on the nature of their mantle source, magmatic evolution, and geodynamic implication. The Shuimo alkaline syenite samples have low initial $^{87}Sr/^{86}Sr$ ratios of less than 0. 703 340 and high Rb/Sr, including SM23 I_{Sr} = 0. 700 549, and high positive $\varepsilon_{Nd}(t)$ values (+1. 9 to +7. 7) that have a rather wide range (Supplementary Table 3). Wu et al. (2000) reported that the high $^{87}Rb/^{86}Sr$ rocks commonly yield widely scattered and sometimes unreasonably low initial Sr ratios of less than 0. 700, all with very large errors. Therefore, initial Sr values for such high Rb/Sr rocks cannot be used in petrogenetic discussions (Wu et al., 2000; Jahn et al., 2001). Compared with the alkaline syenites that the Zhongziyuan gabbro samples are char-acterized by the relatively high initial $^{87}Sr/^{86}Sr$ ratios (0. 703 562–0. 704 815) and positive $\varepsilon_{Nd}(t)$ values (+1. 6 to +4. 5). As shown in a diagram of $\varepsilon_{Nd}(t)$ vs. $(^{87}Sr/^{86}Sr)_i$ (Fig. 10), the syenites and gabbros have relatively low and constant initial $^{87}Sr/^{86}Sr$ ratios and positive $\varepsilon_{Nd}(t)$ values, suggesting that crustal contamination was insignificant in these rocks, and the magmas were derived from depleted mantle. Comprehensive the above information, and the syenite samples Nb/La (< 1 or > 1) values have widely scattered (Fig. 8b), however, the Nb/La average value of the continental crust is 0.7 (Taylor and McCulloch, 1989), we believe that the syenites may have also been affected by crustal contamination during the latter stages of magma evolution.

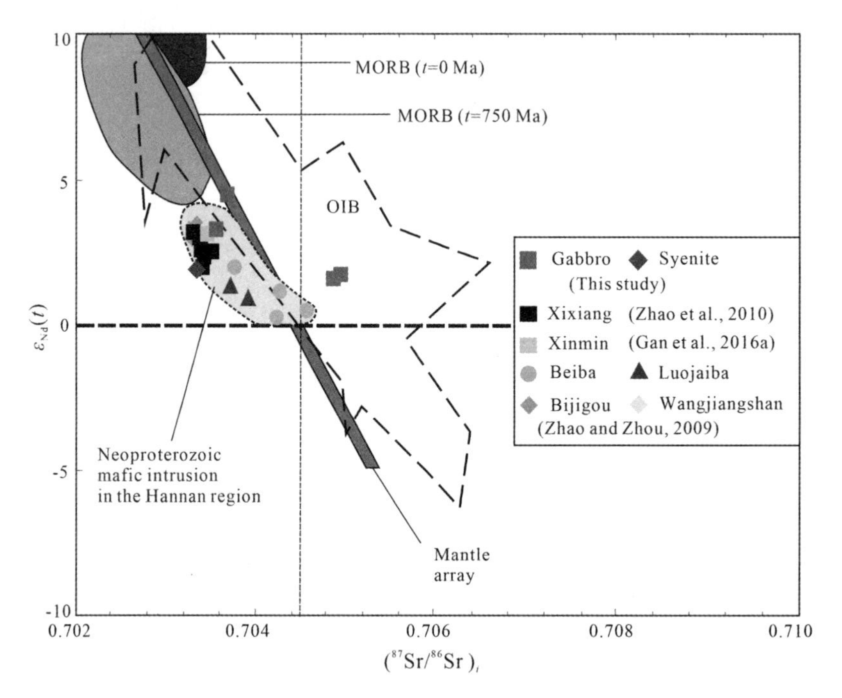

Fig. 10 Initial $^{87}Sr/^{86}Sr$ ratios vs. $\varepsilon_{Nd}(t)$ plot for the alkaline intrusive complex in the Micang Mountains region, South China.

OIB field after White and Duncan (1996), MORB and corrected MORB (t=750 Ma) fields after Zimmer et al. (1995).

The Zhongziyuan gabbros have silica contents (SiO_2 = 46.0–51.9 wt%) and the MgO values vary from 4.2 to 11, which display flat HREE patterns and weakly enriched LREEs in chondrite-normalized REE patterns (Fig. 7c), and show the LILEs enrichment and HFSEs depletion in an N-MORB-normalized trace element diagram (Fig. 7d), indicating they may also have been derived from a basaltic mantle source. In the La/Sm vs. Sm/Yb diagram (Fig. 11a), data for gabbro samples fall near the OIB field, suggesting that the source may be originated from the similar mantle source to OIB with the enriched mantle source. In Fig. 11b, the gabbro samples, including the Mengzi and Hannan area mafic rocks, lie to the lower right of the OIB field, showing that there is a certain degree of contamination of lithosphere mantle during the rock-formed process. A partial melting model based on the Sm/Yb vs. La/Sm diagram data and published partition coefficients (Salters and Andreas, 2004) indicates that the gabbro samples lie on a garnet lherzolite batch melting line, suggesting a garnet lherzolite source with a degree of partial melting of >15% (Fig. 11a), and the Mengzi area mafic rocks (gabbronorite and gabbro) also have the same characteristics, Micang Mountain region (Xu et al., 2011). The Bijigou intrusion was generated by partial melting of a depleted garnet lherzolite to garnet-spinel lherzolite mantle source, the Beiba and Luojiaba intrusions were generated at shallower depth of garnet-spinel lherzolite to spinel lherzolite, and followed the sequence of decreasing degree of melting (Zhao and Zhou, 2009).

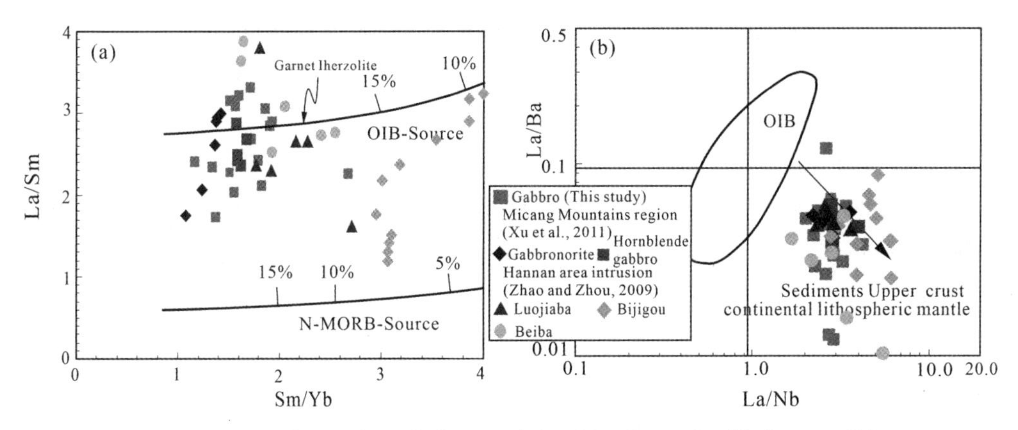

Fig. 11 La/Sm vs. Sm/Yb diagram (a) and La/Ba vs. La/Nb diagram (b) for the gabbro samples (after Xu et al., 2011).

Several genetic models have been proposed to interpret the formation of syenites. The main models include: ① the partial melting of crustal rocks (Whalen et al., 1987; Wang et al., 2002); ② magma mixing of either basic and silicic melts with subsequent differentiation of the hybrid magma (Zhao et al., 1995; Mingram et al., 2000; Vernikovsky et al., 2003), or the mixing of mantle-derived, silica-undersaturated alkaline magmas with lower crust derived granitic magmas (Dorais, 1990; Yang et al., 2008); and ③ extensive fractional crystallization of mantle-derived basaltic magmas (Litvinovsky et al., 2002). Our results show that formation of the Shuimo syenites cannot be explained by the first or second models, because the present petrographical, geochemical, and zircon U-Pb data suggest that the syenites have relatively high silica contents (silica-saturated syenites), and alkaline magmas were derived from depleted mantle, and experienced fractional crystallization in the process of magma evolution. Besides, our conclusions are consistent with the findings of Yu (1992).

In summary, we consider that the alkaline syenites originated by melting of depleted mantle and subsequently formed by extensive fractional crystallization and small amounts of crustal contamination during magma evolution and ascent. The gabbros also originated from ca. 15% melting of depleted garnet lherzolite mantle and subsequent contamination by crustal assimilation during the magma ascent.

6.3 Tectonic setting and implications

6.3.1 The tectonic setting of the alkaline intrusive complex

Alkaline igneous rocks are typical products of continental magmatism, and the association of alkaline magmatism with extensional tectonic settings (Eby, 1990, 1992; Black and Liegeois, 1993) means that such rocks can provide significant information on post-collisional and/or intraplate extensional magmatic processes within the continental lithosphere (Yang et al., 2005, 2012; Peng et al., 2008; Cai et al., 2011; Zhang et al., 2015). Yu (1992) argued that the alkaline rocks of the Qinba area were formed as a result of the fracturing of an orogenic belt, because the rocks have transitional characteristics between continental magmatism and orogenic belt magmatism.

Th, Ta, and Hf have similar geochemical characteristics, therefore, their ratios are unaffected by magmatic processes and fingerprint their magma source (Zhang et al., 1999a). Therefore, we now use the ratios between these three elements to distinguish the tectonic setting of magmatism.

In a Th/Yb-Ta/Yb diagram (Fig. 12a), almost all the gabbro samples fall within the VAB field, indicating that the gabbros formed in a volcanic arc tectonic environment. However, the syenites lie outside the VAB field and instead fall in the alkaline rock field (ALK), and therefore have similar characteristics to intra-continental basalts.

According to the Th/Hf-Ta/Hf tectonic discrimination diagram (Fig. 12b), all but three of the syenite samples have Ta/Hf values >1 (SM21, 23, 25) and are concentrated mainly in the continental plate inner-mantle field (rift zone). Xiao et al. (1997) proposed that the Pinghe alkaline complex was the result of nephelinic alkaline magmatism involving the partial fusion of upper mantle material in the initial stages of intracontinental rifting and its subsequent intrusion and dispersal through faults in a passive setting during the Neoproterozoic. Our results mainly support this interpretation and show that the alkaline complex of the Shuimo and Pinghe areas formed in an intra-arc rifting environment during the Neoproterozoic. The complex lithology and different types alkaline rocks formed as a result of mantle melting and the subsequent rise of an alkaline magmatic diapir and its entry into and emplacement in the crust. The gabbro samples have the characteristics of intracontinental basalt (Fig. 12b), which we further distinguish in the Hf/3-Th-Ta and 2Nb-Zr/4-Y triangular tectonic discrimination diagrams (Fig. 12c,d). In these diagrams, gabbro samples fall within the calc-alkaline volcanic arc field, reflecting the characteristics of continental margin arc basalts, which supports the findings of our associated research on hornblende gabbro from the Xinmin of the Micang Mountains area (Gan et al., 2016a). Besides, the Hannan area (Beiba, Luojiaba, and Bijigou) and Micangshan region (Mengzi) mafic intrusions have island-arc features, they were formed in subduction setting (Zhao and Zhou, 2009; Xu et al., 2011).

Taking into account the above petrographic, geochemical, and Sr-Nd isotopic data, and zircon U-Pb ages, combined with the geological history of the Micang Mountains region during the Neoproterozoic, we consider that the Shuimo-Zhongziyuan alkaline intrusive complex is a

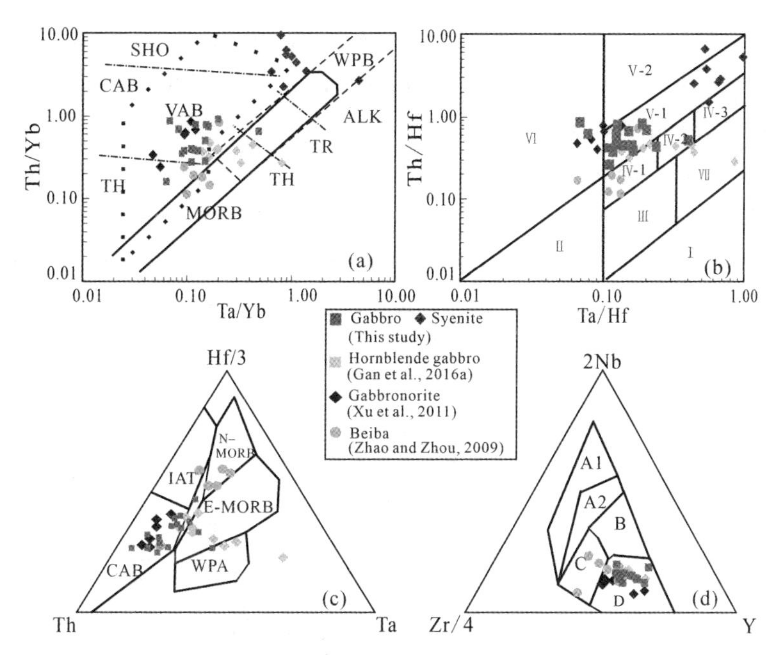

Fig. 12 Tectonic setting discrimination diagrams for the alkaline intrusion complex in the Micang Mountains region.

(a) Th/Yb vs. Ta/Yb (Pearce, 1983); (b) Ta/Hf vs. Th/Hf (Zhang et al., 1999a); (c) Hf/3-Th-Ta (Wood, 1979); (d) 2Nb-Zr/4-Y (Meschede, 1986). SHO: shoshonite; TH: tholeiite; TR: transitional; ALK: alkaline; CAB: Calc-alkaline basalt; SHO; TH; TR; ALK; Ⅰ: depleted mantle zone; Ⅱ: N-MORB; Ⅲ: oceanic intraplate tholeiite; OIB and transitional and enrichment type of mid-ocean ridge basalt area; Ⅳ: margin of separated continental plate (rift zone); Ⅴ: continental intraplate Ⅴ-1: Continental plate inner-mantle zone, Ⅴ-2: continental intraplate crust zone; Ⅵ: subduction zones and island arc zone; Ⅶ: mantle plume (hot) zone; WPA: within-plate basalt; E-MORB: enriched mid-ocean ridge basalt; N-MORB: depleted mid-ocean ridge basalt; IAT: island arc tholeiite; CAB: calc-alkaline basalt; A1, A2: within-plate alkaline basalts; A2, C: intraplate tholeiite; P-MORB: enriched, hot mid-ocean ridge basalt; C, D: volcanic arc basalt.

product of different stages of Neoproterozoic magmatism. At ca. 0.86 Ga, mantle melting led to the generation of alkaline magma, and this was followed by significant fractional crystallization, eventually causing diapiric rise and emplacement of the magmatic mass into the crust in an intra-arc rifting environment, forming the observed complex lithology and diverse types of alkaline syenites (K-and Na-rich). At ca. 0.75 Ga, mafic magmas produced by a depleted mantle source and possibly affected by crustal assimilation during magma ascent formed the mafic intrusive rocks (gabbros) in a continental margin arc tectonic environment. This arc-like environment was most prob-ably related to subduction system. Combined with the above information, we can see that the Hannan area and Micang Mountains region mafic rocks have similar geological characteristics in the northwestern Yangtze Block in Neoproterozoic time.

6.3.2 Geodynamic implications

Various studies have been made of the petrology, geo-chemistry, and zircon U-Pb geochronology of the Hannan-Micangshan intrusive complex at the north-western margin of the

Yangtze Block. These studies show that the Hannan-Micangshan area was still situated in a convergent margin setting during the Neoproterozoic (870 – 706 Ma), characterized by magmatic activity, and the formation of a multi-stage volcanic arc. These Neoproterozoic magmatic rocks were distributed continuously in time but irregularly in space, suggesting that the Hannan-Micangshan area underwent long-lived tectonic events from 900 Ma to 700 Ma (Figs. 2 and 13). Subsequently, these basement complexes were overlain unconformably by Neoproterozoic Sinian strata, such as the Doushantuo and Dengying formations (Zhang et al., 2015).

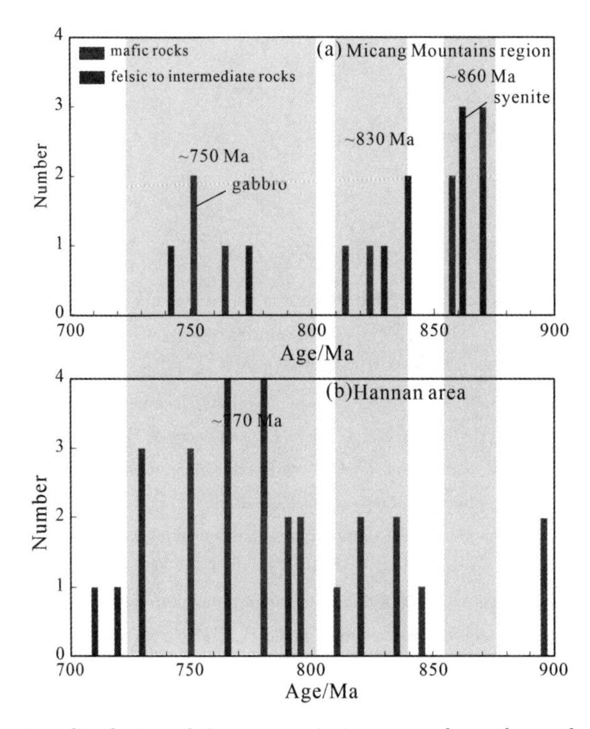

Fig. 13 Age distribution of Neoproterozoic igneous rocks at the northern margin of the Yangtze Block between 900 Ma and 700 Ma.

(a) Micang Mountains region; (b) Hannan area. Data from this study and previous studies (Table 1).

Neoproterozoic magmatic activity in the Micang Mountains region can be divided into three periods at ca. 860 Ma, 830 Ma, and 750 Ma. The alkaline syenites and gabbros formed in two different stages (Fig. 13a). Although magmatism in the Hannan area from 850 Ma to 700 Ma occurred continuously, the age distribution of rocks shows a peak at ca. 770 Ma (Fig. 13b). These data suggest that magmatism in the Micang Mountains region occurred earlier (>800 Ma) (Dong et al., 2012) than in the Hannan area (< 800 Ma) during the Neoproterozoic (Fig. 14a). Our age data, therefore, constrain the Neoproterozoic tectonic evolution of the Micang Mountains region and Hannan area at the northern margin of the Yangtze Block.

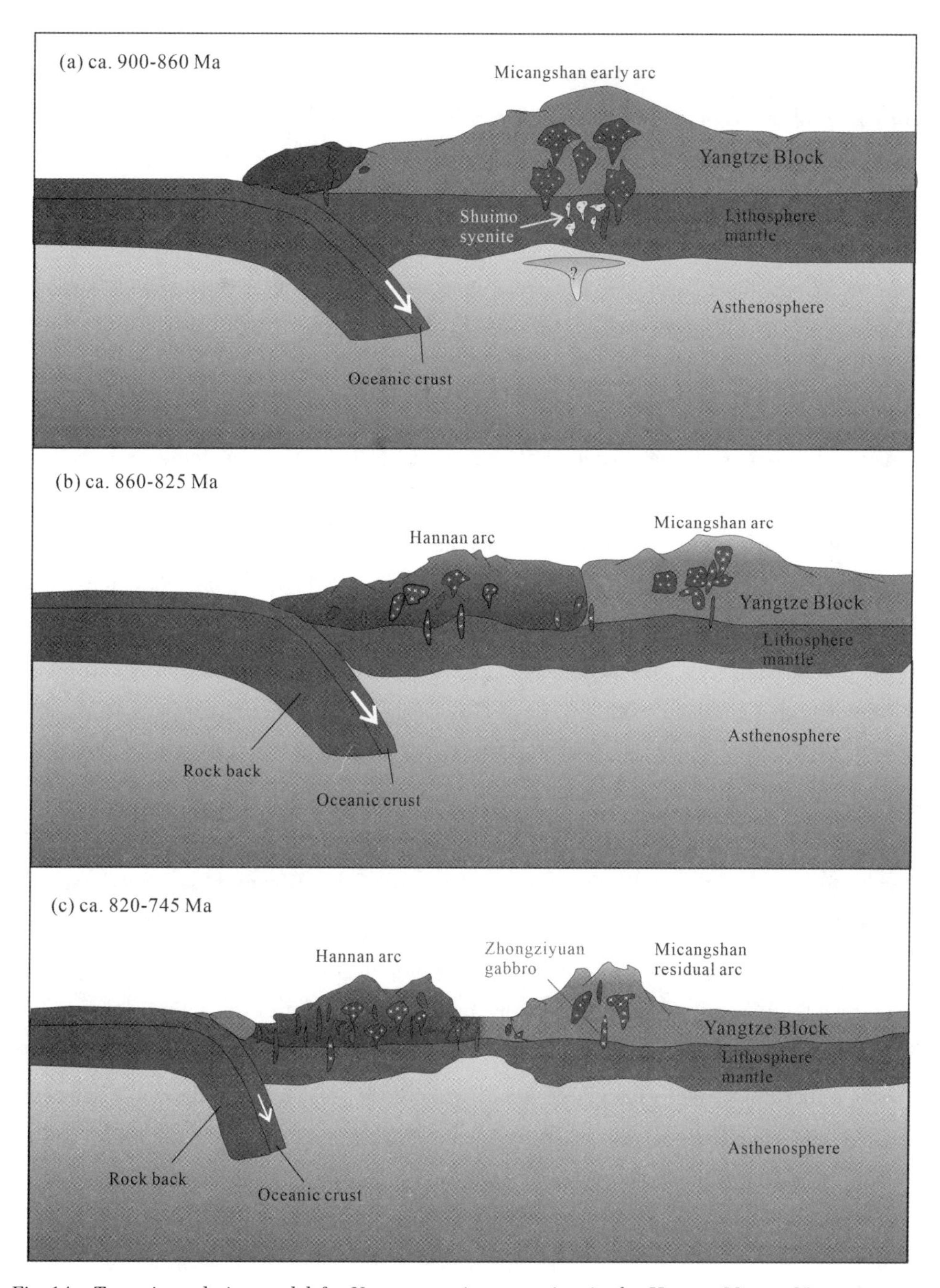

Fig. 14 Tectonic evolution model for Neoproterozoic magmatism in the Hannan-Micang Mountains area at the northern margin of the Yangtze Block (revised from Dong et al., 2012; Dong and Santosh, 2015; and Zhang et al., 2015).

Defining the temporal and spatial patterns of Neoproterozoic magmatism has implications for reconstructing the Rodinia supercontinent. The Neoproterozoic extrusive and intrusive rocks on the margins of the Yangtze Block are key to understanding the amalgamation and break-up of this supercontinent (Li et al., 2003b; Du et al., 2014; Zhang et al., 2015). Magmatic events ranging in age from 900 Ma to 700 Ma occurred in the Hannan-Micangshan area in the north-western corner of the Yangtze Block. However, previous studies of the petrogenesis and tectonic settings of these intrusive and volcanic rocks have produced equivocal results. Li et al.(2003b) proposed a mantle plume model to explain the break-up of Rodinia between 830 Ma and 745 Ma, and Zheng et al. (2006, 2007, 2008) proposed a similar rifting model. However, other recent studies have suggested that these igneous rocks are related to oceanic subduction (Zhou et al., 2002a, b, 2006; Zhao and Zhou, 2007, 2008, 2009; Dong et al., 2011a, 2012; Wang et al., 2012).

Although the Neoproterozoic magmatic events took place in the Hannan-Micangshan area at the northern margin of the Yangtze Block, most of the intrusions indicate in an active continental margin arc setting from 900 Ma to 745 Ma (Ling et al., 2003). The rocks in the Hannan-Micangshan area are composed mainly of subduction-related plutons (Zhou et al., 2002b; Zhao et al., 2010; Xu et al., 2011; Dong et al., 2011b, 2012; Dong and Santosh, 2015) that intruded Mesoproterozoic strata and are unconformably overlain by Sinian strata. These Neoproterozoic rocks have been considered to be linked with the assembly and break-up of the Rodinia supercontinent (Xie et al., 1999; Wang et al., 2010; Cawood et al., 2013; Liu et al., 2015). The Hannan and Micang Mountains massifs in the northwestern corner of the Yangtze Block are characterized by well-preserved mafic-ultramafic plutons and granitoids that record a detailed Neoproterozoic tectonic history, including the stages of continental margin arc, back-arc basin, and intracontinental rift. Therefore, the Micang Mountains region is an ideal location for studying Neoproterozoic continental crust convergence and break-up, the results of which have implications for the tectonic evolution Rodinia (Cheng and Ling, 2000).

Our results allow us to propose a model for the Neoproterozoic tectonic evolution of the area around the northwestern margin of the Yangtze Block. Oceanic crust was subducted southwards under the northern margin of the Yangtze Block, and tectonic thermal events caused pronounced arc-related magmatic activity in the Micangshan area at ca. 900 – 860 Ma (Fig. 14a). Although the alkaline syenite investigated here is inferred to have formed at depth in an intra-arc rifting environment, this is not inconsistent with the Micangshan area having been located in an earlier volcanic arc, after which the magmatic events may have been associated with the break-up of Rodinia during the early Neoproterozoic. Subsequently, the magmatic events gradually migrated to the Hannan area, and extensive mafic and granitic intrusions were emplaced into the Hannan massif (Fig. 14b). The roll back of the subducted oceanic lithosphere generated the Hannan arc during ca. 820–745 Ma and produced voluminous Neoproterozoic felsic intrusions associated with many mafic-ultramafic bodies (Li et al., 2003b; Zhao and Zhou, 2007, 2009; Zhao et al., 2010) (Fig. 14c), and these igneous rocks in the Hannan area also record the subduction of an oceanic slab and the growth of continental crust during the Neoproterozoic (Zhao et al., 2010).

7 Conclusions

We conducted a detailed petrological, geochemical, Sr-Nd isotopic, and zircon U-Pb dating investigation of the Micang Mountains intrusive complex at the northern margin of the Yangtze Block, southern China. The major conclusions are as follows:

(1) U-Pb dating yielded ages for the Shuimo syenite of 869 ± 4 Ma (Na-rich samples) and 860 ± 5 Ma (K-rich samples), and for the Zhongziyuan gabbro of 753 ± 4 Ma. These rocks are the products of different stages of magmatism that occurred in the Micang Mountains region at the northern edge of the Yangtze Block during the Neoproterozoic.

(2) The syenite and gabbro in the Micang Mountains region originated from melting depleted mantle during the Neoproterozoic. The syenite underwent substantial fractional crystallization and small amounts of crustal contamination during magma evolution and ascent. In contrast, the gabbro was formed by the partial melting (>15%) of garnet lherzolite and may have undergone crustal assimilation during magma ascent.

(3) The Shuimo syenite formed in an intra-arc rifting tectonic environment, the gabbro in the Zhongziyuan area was most likely formed in an active continental margin arc setting that was related to continuous subduction along the northern margin of the Yangtze Block during the Neoproterozoic.

Acknowledgements We are grateful to Professor Robert J. Stern for our help and also thank the three anonymous reviewers and Professor Shengyao Yu of for providing constructive comments. This work was supported by the National Natural Science Foundation of China (Grant No. 41372067); National Natural Science Fund Project (Grant No. 41190072); Natural Science Foundation Innovation Group (Grant No. 41421002); Ministry of Education Innovation Team (Grant No. IRT1281); State Key Laboratory of Continental Dynamics.

References

Allegre, C.J., Minster, J.F., 1978. Quantitative method of trace element behavior in magmatic processes. Earth & Planetary Science Letters, 38, 1-25.

Andersen, T.,2002. Correction of common lead in U-Pb analyses that do not report ^{204}Pb. Chemical Geology, 192(1-2), 59-79.

Ao, W.H., Zhang, Y.K., Zhang, R.Y., Zhao, Y., Sun, Y., 2014. Neoproterozoic crustal accretion of the northern margin of Yangtze Plate constrains from geochemical characteristics, LA-ICP-MS zircon U-Pb chronology and Hf isotopic compositions of trondhjemite from Zushidian area, Hannan region. Geological Review, 60(6), 1393-1408 (in Chinese with English abstract).

Black, R., Liegeois, J.P., 1993. Cratons, mobile belts, alkaline rocks and continental lithospheric mantle;

the Pan-African testimony. Journal of the Geological Society of London, 150(1), 89-98.

Cai, J.H., Yan, G.H., Mu, B.L.,Reng, K.X., Li, F.T., Yang, B., 2011. Geochronology and geochemistry of Sungshuang alkaline complex in Jixian County, Tianjin. Journal of Jilin University: Earth Science Edition, 41(6), 1901-1913 (in Chinese).

Cai, Y.F., Wang, Y.J.,Cawood, P.A., Zhang, Y.Z., Zhang, A. M., 2015. Neoproterozoic crustal growth of the Southern Yangtze Block: Geochemical and zircon U-Pb Geochronological and Lu-Hf isotopic evidence of Neoproterozoicdiorite from the Ailaoshan zone. Precambrian Research, 266, 137-149.

Cawood, P.A., Wang, Y., Xu, Y., Zhao, G., 2013. Locating South China in Rodinia and Gondwana: A fragment of greater India lithosphere? Geology, 41, 903-906.

Cheng, J.P., Ling, W.L., 2000. Micangshan Neoproterozoic igneous province in northwest Yangtze Block, South China: A potential testee in Rodinia supercontinent reconstruction. Geological Science and Technology Information, 19(3), 12-16 (in Chinese with English abstract).

Chu, Z.Y., Chen, F.K., Yang, Y.H., Guo, J.H.,2009. Precise determination of Sm, Nd concentrations and Nd isotopic compositions at the nanogram level in geological samples by thermal ionization mass spectrometry. Journal of Analytical Atomic Spectrometry, 24, 1534-1544.

Dong, Y.P., Liu, X.M., Santosh, M., Chen, Q., Zhang, X.N., Li, W., He, D.F., Zhang, G.W.,2012. Neoproterozoic accretionary tectonics along the northwestern margin of the Yangtze Block, China: Constraints from zircon U-Pb geochronology and geochemistry. Precambrian Res, 196, 247-274.

Dong, Y.P., Liu, X.M., Santosh, M., Zhang, X.N., Chen, Q.,Yang, C., Yang, Z., 2011a. Neoproterozoic subduction tectonics of the northwestern Yangtze Block in South China: Constrains from zircon U-Pb geochronology and geochemistry of mafic intrusions in the Hannan Massif. Precambrian Researcher, 189, 66-90.

Dong, Y.P., Santosh, M., 2015. Tectonic architecture and multiple orogeny of the Qinling Orogenic Belt, central China. Gondwana Research, 1-40.

Dong, Y.P., Zhang, G.W., Hauzenberger, C., Neubauer, F., Yang, Z., Liu, X.M., 2011b. Palaeozoic tectonics and evolutionary history of the Qinling orogen: Evidence from geochemistry and geochronology of ophiolite and related volcanic rocks. Lithos, 122, 39-56.

Dorais, M.J., 1990. Compositional variations in pyroxenes and amphiboles of the Belknap Mountain complex, New Hampshire: Evidence for origin of silica-saturated alkaline rocks. American Mineralogist, 75, 1092-1105.

Du, L.L., Guo, J.H.,Nutman, A.P., Wyman, D., Geng, Y., Yang, C., Liu, F., Ren, L., Zhou, X., 2014. Implications for Rodinia reconstructions for the initiation of Neoproterozoic subduction at ~860 Ma on the western margin of the Yangtze Block: Evidence from the Guandaoshan Pluton. Lithos, 196-197, 67-82.

Eby, G.N., 1990. A-type granitoids: A review of their occurrence and chemical characteristics and speculations on their petrogenesis. Lithos, 26, 115-134.

Eby, G.N., 1992. Chemical subdivision of the A-type granitoids: Petrogenetic and tectonic implications. Geology, 20, 641-644.

Gan, B.P., Lai, S.C., Qin, J.F.,2016a. Geochemistry of the bojite from the Xinmin area, Micang Mountain. Acta Petrologica et Mineralogica, 35(2), 213-228 (in Chinese with English abstract).

Gan, B.P., Lai, S.C., Qin, J.F.,2016b. Petrogenesis and implication for the Neoproterozoic monzogranite in Pinghe, Micang Mountains. Geological Review, 62(4), 929-944 (in Chinese with English abstract).

Gao, S., Ling, W.L.,Qiu, Y., Zhou, L., Hartmann, G., Simon, K., 1999. Contrasting geochemical and Sm-Nd isotopic compositions of Archean metasediments from the Kongling high-grade terrain of the Yangtze craton: Evidence for cratonic evolution and redistribution of REE during crustal anatexis. Geochimica et Cosmochimica Acta, 63, 2071-2088.

He, D.L., Liu, D.Z., Deng, M.S., Zhou, Y.,1995. Age of crystalline basement of Yangtze platform in the Michang Mountains, Sichuan. Acta Geologica Sichuan, 15(3), 176-183 (in Chinese without English abstract).

He, L., 2010. Characteristics and structural setting of the Pinghe alkalic complex in northern Sichuan. Chengdu: Master degree paper of Chengdu University of Technology, 1-61 (in Chinese with English abstract).

Huang, Y.P.,2013. The petrogeochemical characteristics of Bengge alkaline complex igneous rock and its deep metal-logenic geological significance, Northwest Yunnan. Chengdu: Chengdu University of Technology, 1-74 (in Chinese with English abstract).

Jahn, B.M., Wu, F.Y.,Capdevila, R., Martineau, F., Wang, Y.X., Zhao, Z.H., 2001. Highly evolved juvenile granites with tetrad REE patterns: The Woduhe and Baerzhe granites from the Great Xing'an (Khingan) Mountains in NE China. Lithos, 59, 171-198.

Li, T.,2010. The study of Neoproterozoic tectonic-magmatic events in the Northern margin of the Yangtze continental. Xi'an: Master degree paper of Chang'an University, 1-55 (in Chinese with English abstract).

Li, T.Z., Zhang, J.X.,1994. On the composition of alkali-granite in Micangshan area. Journal of Southwest Institute of Technology, 9(4), 39-50 (in Chinese with English abstract).

Li, X.H.,1999. U-Pb zircon ages of granites from the southern margin of the Yangtze Block: Timing of the Neoproterozoic Jinning Orogeny in SE China and implications for Rodinia assembly. Precambrian Research, 97, 43-57.

Li, X.H., Li, W.X., Li, Z.X., Lo, C.H., Wang, J., Ye, M.F., Yang, Y.H.,2009. Amalgamation between the Yangtze and Cathaysia Blocks in South China: Constraints from SHRIMP U-Pb zircon ages, geochemistry and Nd-Hf isotopes of the Shuangxiwu volcanic rocks. Precambrian Research, 174, 117-128.

Li, X.H., Li, Z.X., Ge, W.C., Zhou, H.W., Li, W.X., Liu, Y., Wingate, M.T.D.,2003a. Neoproterozoic granitoids in South China: Crustal melting above a mantle plume at ca. 825 Ma? Precambrian Research, 122 (s1-4), 45-83.

Li, X.H., Li, Z.X., Zhou, H., Liu, Y., Kinny, P.D., 2002a. U-Pb zircon geochronology, geochemistry and Nd isotopic study of Neoproterozoic bimodal volcanic rocks in the Kangdian Rift of South China: Implications for the initial rifting of Rodinia. Precambrian Research, 113, 135-154.

Li, X.H., Li, Z.X., Zhou, H., Liu, Y., Liang, X., Li, W.,2002b. SHRIMP zircon U-Pb age, geochemistry and Nd isotope of the Guandaoshan granite in western Sichuan: Petrogenesis and tectonic implications. Sciences China, 32, 60-68 (in Chinese with English abstract).

Li, Z.X., Li, X.H.,Kinny, P.D., Wang, J., Zhang, S., Zhou, H., 2003b. Geochronology of Neoproterozoic syn-rift magmatism in the Yangtze Craton, South China and correlations with other continents: Evidence for a mantle superplume that broke up Rodinia. Precambrian Research, 122, 85-109.

Ling, W.L., Gao, S., Cheng, J.P., Jiang, L.S., Yuan, H.L., Hu,Z. C., 2006. Neoproterozoic magmatic events within the Yangtze continental interior and along its northern margin and their tectonic implication: Constraint from the ELA-ICP-MS U-Pb geochronology of zircons from the Huangling and Hannan complexes. Acta Petrologica Sinica, 22, 387-396 (in Chinese with English abstract).

Ling, W.L., Gao, S., Zhang, B.R., Li, H.M., Liu, Y., Cheng, J.P., 2003. Neoproterozoic tectonic evolution of the northwestern Yangtze craton, South China: Implications for amalgamation and break-up of the Rodinia Supercontinent. Precambrian Research, 122, 111-140.

Ling, W.L., Gao, S., Zheng, H.F., Zhou, L., Zhao, Z.B., 1998. An Sm/Nd istopic dating study of the Archen Kongling complex in the Huangling area of the Yangtze craton. Chinese Science Bulletin, 43(1), 86-89.

Litvinovsky, B.A., Jahn, B.M., Zanvilevich, A.N., Saunders, A., Poulain, S., Kuzmin, D.V., Reichow, M.K., Titov, A.V., 2002. Petrogenesis of syenite-granite suites from the Bryansky Complex (Transbaikalia, Russia): Implications for the origin of A-type granitois magmas. Chemical Geology, 189, 105-133.

Liu, L.,Yang, X., Santosh, M., Aulbach, S., Zhou, H., Geng, J.Z., Sun, W.D., 2015. Neoproterozoic intraplate crustal accretion on the northern margin of the Yangtze Block: Evidence from geochemistry, zircon SHRIMP U-Pb dating and Hf isotopes from the Fuchashan complex. Precambrian Research, 268, 97-114.

Ludwig, K.R., 2003. ISOPLOT 3.0: A geochronological toolkit for Microsoft Excel. Berkeley Geochronology Center, Special Publication, 4.

Ma, R.Z., Xiao, Y.F., Wei, X.G., He, Z.W., Li, Y.G., 1997a. The magmatic activity and tectonic evolution in the Micangshan area, China. Journal of Mineralogy and Petrology, 17(Suppl.), 76-82 (in Chinese with English abstract).

Ma, R.Z., Xiao, Y.P., Wei, X.G., He, Z.W., Li, Y.G., 1997b. Research on the geochemical property and genesis of basic and ultrabaisc rocks of Jinning period in the Micangshan area, Sichuan Province. Journal of Mineralogy and Petrology, 17(Suppl.), 34-47 (in Chinese with English abstract).

Maas, R.,Kinny, P.D., Williams, I.S., Froude, D.O., Compston, W., 1992. The Earth's oldest known crust: A geochronological and geochemical study of 3 900−4 200 Ma old detrital zircons from Mt. Narryer and Jack Hills, Western Australia. Geochimica et Cosmochimica Acta, 56, 1281-1300.

Meng, E., Liu, F.L., Du, L.L., Liu, P.H., Liu, J.H., 2015. Petrogenesis and tectonic significance of the Baoxing granitic and mafic intrusions, southwestern China: Evidence from zircon U-Pb dating and Lu-Hf isotopes, and whole-rock geochemistry. Gondwana Research, 28, 800-815.

Meschede, M., 1986. A method of discriminating between different type of mid-ocean ridge basalts and continental tholeiites with the Nb-Zr-Y diagram. Chemical Geology, 56, 207-218.

Middlemost, E.A.K., 1994. Naming materials in the magma/igneous rock system. Earth-Science Reviews, 37 (3-4), 215-224.

Mingram, B., Trumbull, R.B., Littman, S., Gerstenberger, H., 2000. A petrogenetic study of androgenic felsic magmatism in the Cretaceous Paresis ring complex, Namibia: Evidence for mixing of crust and mantle-derived components. Lithos, 54, 1-22.

Pearce, J.A., 1983. Role of the subcontinental lithosphere in magma genesis at active continental margins. In:

Hawkesworth, C.J., Norry, M.J. Continental basalts and mantle xenoliths, Nantwich, U.K, Cheshire, Shiva, 230-249.

Peng, M., Wu, Y., Gao, S., Zhang, H., Wang, J., Liu, X., Gong, H., Zhou, L., Hu, Z., Liu, Y., Yuan, H., 2012. Geochemistry, zircon U-Pb age and Hf isotope compositions of Paleoproterozoic aluminous A-type granites from the Kongling terrain, Yangtze Block: Constraints on petrogenesis and geologic implications. Gondwana Research, 22(1), 140-151.

Peng, P., Zhai, M.G., Guo, J.H., Zhang, H.F., Zhang, Y.B., 2008. Petrogenesis of Triassic post-collisional syenite plutons in the Sino-Korean Craton: An example from DPRK. Geological Magazine, 145, 637-647.

Qi, L., Hu, J., Gregoire, D.C., 2000. Determination of trace elements in granites by inductively coupled plasma mass spectrometry. Talanta, 51, 507-513.

Qiu, J.X., 1993. Alkalic rocks in Qingling-Dabashan area. Beijing: Geological Publishing House, 1-179.

Salters, V.J.M., Andreas, S., 2004. Composition of the depleted mantle. Geochemistry, Geophysics, Geosystems, 5(5), 469-484.

Shen, Q.H., 2009. The recommendation of a systematic list of mineral abbreviations. Acta Petrologica et Mineralogica, 28(5), 497-501 (in Chinese with English abstract).

Sun, S.S., McDonough, W.F., 1989. Chemical and isotopic systematics of oceanic basalts: Implications for mantle com-position and processes. In: Saunders, A.D., Norry, M.J. Magmatism in the Ocean Basins. Geological Society, London, Special Publications, 42, 313-345.

Taylor, S.R., McCulloch, M.T., 1989. The Geochemical evolution of the continental crust. Earth Planet. Sci. Lett., 94, 257-273.

Tian, Y.T., Zhu, C.Q., Xu, M., Rao, S., Kohn, B.P., Hu, S.B., 2010. Exhumation history of the Micangshan-Hannan Dome since Cretaceous and its tectonic significance: Evidences from Apatite Fission Track analysis. Chinese Journal Geophysics, 53(4), 920-930 (in Chinese with English abstract).

Vernikovsky, V.A., Pease, V.L., Vernikovskaya, A.E., Romanov, A.P., Gee, D.G., Travin, A.V., 2003. First report of early Triassic A-type granite and syenite intrusions from Taimyr: Product of the northern Eurasian superplume. Lithos, 66, 23-66.

Wang, J., Li, Z.X., 2003. History of Neoproterozoic rift basins in South China: Implications for Rodinia break-up. Precambrian Research, 122, 141-158.

Wang, L.J., Griffin, W.L., Yu, J.H., O'Reilly, S.Y., 2010. Precambrian crustal evolution of the Yangtze Block tracked by detrital zircons from Neoproterozoic sedimentary rocks. Precambrian Research., 177, 131-144.

Wang, L.J., Yu, J.H., Griffin, W.L., O'Reilly, S.Y., 2012. Early crustal evolution in the western Yangtze Block: Evidence from U-Pb and Lu-Hf isotopes on detrital zircons from sedimentary rocks. Precambrian Research, 222(223), 368-385.

Wang, Q., Wyman, D.A., Li, Z.X., Bao, Z.W., Zhao, Z.H., Wang, Y.X., Jian, P., Yang, Y.H., Chen, L.L., 2010a. Petrology, geochronology and geochemistry of ca. 780 Ma A-type granites in South China: Petrogenesis and implications for crustal growth during the breakup of the supercontinent Rodinia. Precambrian Research, 178, 185-208.

Wang, X.C., Li, Z.X., Li, X.H., Li, Q.L., Zhang, Q.R., 2011. Geochemical and Hf-Nd isotope data of

Nanhua rift sedimentary and volcaniclastic rocks indicate a Neoproterozoic continental flood basalt provenance. Lithos, 127, 427-440.

Wang, X.L., Zhao, G.C., Zhou, J.C., Liu, Y.S., Hua, J.,2008. Geochronology and Hf isotopes of zircon from volcanic rocks of the Shuangqiaoshan Group, South China: Implications for the Neoproterozoic tectonic evolution of the eastern Jiangnan orogen. Gondwana Research, 14(3), 355-367.

Wang, X.L., Zhou, J.C.,Qiu, J.S., Zhang, W.L., Liu, X.M., Zhang, G.L., 2006. LA-ICP-MS U-Pb zircon geochronology of the Neoproterozoic igneous rocks from Northern Guangxi, South China: Implications for tectonic evolution. Precambrian Research, 145, 111-130.

Wang, Y.J., Zhang, Y.H., Fan, W.M., Xi, X.W., Guo, F., Lin, K., 2002. Petrogenesis of the Indosinian peraluminous granites in Hunan Province: Numerical simulation of magmatic underplating and heat effect of crustal thickening. Science China: Series D, Earth Sciences, 32, 491-499 (in Chinese).

Whalen, J.B., Currie, K.L., Chappell, B.W., 1987. A-type granites: Geochemical characteristics, discrimination and petrogenesis. Contributions Mineral. Petrol., 95, 407-419.

White, W.M., Duncan, R.A.,1996. Geochemistry and geo-chronology of the Society Islands: New evidence for deep mantle recycling. In: Basu, A., Hart, S. Earth Processes: Reading the Isotopic Code. Geophysicalmonograph, 95, AGU, Washington, DC, 183-206.

Williams, I., Cho, D., Kim, S.,2009. Geochronology, and geochemical and Nd-Sr isotopic characteristics, of Triassic plutonic rocks in the Gyeonggi Massif, South Korea: Constraints on Triassic post-collisional magmatism. Lithos, 107, 239-256.

Wilson, M.,1989. Igneous petrogenesis. London: Unwin Hyman, 466.

Wood, D.A.,1979. A variably veined suboceanic upper mantle-genetic significance for mid-ocean ridge basalts from geochemical evidence. Geology, 7, 499-503.

Wright, J.B.,1969. A simple alkalinity ratio and its application to questions of non-orogenic granite genesis. Geological Magazine, 106, 370-384.

Wu, F.Y., Jahn, B.M., Wilde, S., Sun, D.Y.,2000. Phanerozoic continental crustal growth: Sr-Nd isotopic evidence from the granites in northeastern China. Tectonophysics, 328, 89-113.

Wu, R.X., Zheng, Y.F., Wu, Y.B., Zhao, Z.F., Zhang, S.B., Liu, X.M., Wu, F.Y.,2006. Reworking of juvenile crust: Element and isotope evidence from Neoproterozoic granodiorite in South China. Precambrian Research, 146(s3-4), 179-212.

Wu, Y.B., Gao, S., Zhang, H.F., Zheng, J.P., Liu, X.C., Wang, H., Gong, H.J., Zhou, L., Yuan, H. L.,2012. Geochemistry and zircon U-Pb geochronology of Paleoproterozoic arc related granitoid in the Northwestern Yangtze Block and its geological implications. Precambrian Research, 200-203, 26-37.

Wu, Y.B., Zheng, Y.F., Tang, J., Gong, B., Zhao, Z.F., Liu, X. M.,2007. Zircon U-Pb dating of water-rock interaction during Neoproterozoic rift magmatism in South China. Chemical Geology, 246(1-2), 65-86.

Xiao, L., Zhang, H.F., Ni, P.Z., Xiang, H., Liu, X.M.,2007. LA-ICP-MS U-Pb zircon geochronology of early Neoproterozoic mafic-intermediat intrusions from NW margin of the Yangtze Block, South China: Implication for tectonic evolution. Precambrian Research, 154, 221-235.

Xiao, Y.F., Ma, R.Z., He, Z.W., Wei, X.G.,1997. The units characteristics and tectonic settings of the alkalic complex in Micangshan. Journal of Petrology, 17 (Suppl.), 59-66 (in Chinese with English abstract).

Xiao, Y.F., Ma, R.Z., Wei, X.G., He, Z.W., Li, Y.G.,1998. The characteristics and genesis of the basic intrusive complex in Chengjiang period, Micangshan, Sichuan. Journal of Chengdu University of Technology, 25(4), 537-542 (in Chinese with English abstract).

Xie, C.F., Fu, J.M., Xiong, C.Y., Zhang, Y.M., Hu, N., Huang, Z.X., 1999. Precambrian crustal evolution in the central Northern margin of the Yangtze Block. Gondwana Research, 2, 515-518.

Xu, J.F.,1993. Studies of essential minerals in alkaline rocks of Micangshan area and their genetic information. Acta Petrologica et Mineralogica, 12(3), 269-278 (in Chinese with English abstract).

Xu, X.Y., Li, T., Chen, J.L., Li, P., Wang, H.L., Li, Z.P.,2011. Zircon U-Pb age and petrogenesis of intrusions from Mengzi area in the northern margin of Yangtze plate. Acta Petrologica Sinica, 27(3), 699-720 (in Chinese with English abstract).

Xu, X.Y., Xia, L.Q., Chen, J.L., Ma, Z.P., Li, X.M., Xia, Z.C., Wang, H.L.,2009. Zircon U-Pb dating and geochemical study of volcanic rocks from Sunjiahe Formation of Xixiang Group in northern margin of Yangtze Plate. Acta Petrological Sinica, 25(12), 3309-3326 (in Chinese with English abstract).

Yang, J.H., Chung, S.L., Wilde, S.A., Wu, F.Y., Chu, M.F., Lo, C.H., Fan, H.R.,2005. Petrogenesis of post-orogenic syenites in the Sulu Orogenic Belt, East China: Geochronological, geochemical and Nd-Sr isotopic evidence. Chemical Geology, 214, 99-125.

Yang, J.H., Sun, J.F., Zhang, M., Wu, F.Y., Wilde, S.A.,2012. Petrogenesis of silica-saturated and silica-undersaturated syenites in the northern North China Craton related to post-collisional and intraplate extension. Chemical Geology, 328, 149-167.

Yang, J.H., Wu, F.Y., Wilde, S.A., Chen, F., Liu, X.M., Xie, L. W., 2008. Petrogenesis of an Alkali syenite-granite-rhyolite suite in the Yanshan Fold and Thrust Belt, Eastern North China Craton: Geochronological, geochemical and Nd-Sr-Hf isotopic evidence for lithospheric thinning. Journal of Petrology, 49, 315-351.

Yu, H.J., Xiao, Y.F.,2011. Characteristics and genetic analysis of the Pinghe alkaline rock, Nanjiang, Sichuan. Journal of Minerals, S1(Suppl.), 186-187 (in Chinese without abstract).

Yu, X.H.,1992. The relation of alkaline rocks in the Qinling-Daba mountains region and the tectonic evolution of the orogen and their features. Regional Geology of China, 3, 233-240 (in Chinese with English abstract).

Yuan, H.L., Gao, S., Liu, X.M., Li, H.M., Gunther, D., Wu, F.Y.,2004. Accurate U-Pb age and trace element determinations of zircon by laser ablation-inductively coupled plasma mass spectrometry. Geostandards & Geoanalytical Research, 28(3), 353-370.

Zhang, C.J., Wang, Y.L., Hou, Z.Q., 1999a. Th, Ta and Hf characteristics and the tectonic setting of magmatic source region of Emeishan basalts. Geological Review, 45(Suppl.), 858-860(in Chinese with English abstract).

Zhang, R.Y., Sun, Y., Zhang, X., Ao, W., Santosh, M., 2015. Neoproterozoic magmatic events in the South Qinling Belt, China: Implications for amalgamation and breakup of the Rodinia supercontinent. Gondwana Research, 30, 6-23.

Zhang, S.B., Zheng, Y.F.,2013. Formation and evolution of Precambrian continental lithosphere in South China. Gondwana Research, 23, 1241-1260.

Zhang, Z.J., Badal, B., Li, Y.K., Chen, Y., Yang, L.P., Teng, J. W., 2005. Crust-upper mantle seismic velocity structure across Southeastern China. Tectonophysics, 395, 137-157.

Zhang, Z.Q., Zhang, G.W., Tang, S.H., Zhang, Q.D., Wang, J.H., 2002. Age and rapid condensation reason of the Hannan intrusion complex. Chinese Science Bulletin, 45(23), 2567-2572 (in Chinese without English abstract).

Zhao, F.Q., Zhao, W.P., Zuo, Y.C., Li, Z.H., Xue, K.Q., 2006. U-Pb geochronology of Neoproterozoic magmatic rocks in Hanzhong, Southern Shaanxi, China. Geological Bulletin of China, 25(5), 383-388 (in Chinese with English abstract).

Zhao, J.H., Zhou, M.F., 2007. Geochemistry of Neoproterozoic mafic intrusions in the Panzhihua district (Sichuan Province, SW China): Implications for subduction related metasomatism in the upper mantle. Precambrian Research, 152(1), 27-47.

Zhao, J.H., Zhou, M.F., 2008. Neoproterozoic adakitic plutons in the northern margin of the Yangtze Block, China: Partial melting of a thickened lower crust and implications for secular crustal evolution. Lithos, 104, 231-248.

Zhao, J.H., Zhou, M.F., 2009. Secular evolution of the Neoproterozoic lithospheric mantle underneath the northern margin of the Yangtze Block, South China. Lithos, 107(3), 152-168.

Zhao, J.H., Zhou, M.F., Zheng, J.P., Fang, S.M., 2010. Neoproterozoic crustal growth and reworking of the Northwestern Yangtze Block: Constraints from the Xixiang dioritic intrusion, South China. Lithos, 120(3-4), 439-452.

Zhao, J.X., Shiraishi, K., Ellis, D.J., Sheraton, J.W., 1995. Geochemical and isotopic studies of syenites from the Yamoto Mountains, East Antarctica: Implication for the origin of syenitic magmas. Geochimica et Cosmochimica Acta, 59, 1363-1385.

Zheng, Y.F., Wu, R.X., Wu, Y.B., Zhang, S.B., Yuan, H.L., Wu, F.Y., 2008. Rift melting of juvenile arc-derived crust: Geochemical evidence from Neoproterozoic volcanic and granitic rocks in the Jiangnan orogen, South China. Precambrian Research, 163(3-4), 351-383.

Zheng, Y.F., Zhang, S.B., Zhao, Z.F., Wu, Y.B., Li, X.H., Li, Z.X., Wu, F.Y., 2007. Contrasting zircon Hf and O isotopes in the two episodes of Neoproterozoic granitoids in South China: Implications for growth and reworking of continental crust. Lithos, 96(12), 127-150.

Zheng, Y.F., Zhao, Z.F., Wu, Y.B., Zhang, S.B., Liu, X.M., Wu, F.Y., 2006. Zircon U-Pb age, Hf and O isotope constraints on protolith origin of ultrahigh-pressure eclogite and gneiss in the Dabie orogen. Chemical Geology, 231(1-2), 135-158.

Zhou, L.D., Zhou, G.F., 1998. Study on mineralogy and geochemistry of the Pinghe ultrabasic-alkaline rock body, Nanjiang, Sichuan. Journal of Chengdu University of Technology, 25(2), 246-256 (in Chinese with English abstract).

Zhou, M.F., Kennedy, A.K., Sun, M., Malpas, J., Lesher, C.M., 2002a. Neoproterozoic arc-related mafic intrusions in the northern margin of South China: Implications for accretion of Rodinia. Journal of Geology, 110, 611-618.

Zhou, M.F., Yan, D., Kennedy, A.K., Li, Y.Q., Ding, J., 2002b. SHRIMP U-Pb zircon geochronological and geochemical evidence for Neoproterozoic arc-magmatism along the western margin of the Yangtze Block,

South China. Earth & Planetary Science Letters, 196(1-2), 51-67.

Zhou, M.F., Yan, D.P., Wang, C.L., Qi, L., Kennedy, A., 2006. Subduction-related origin of the 750 Ma Xuelongbao adakitic complex (Sichuan Province, China): Implications for the tectonic setting of the giant Neoproterozoic magmatic event in South China. Earth & Planetary Science Letters, 248(1-2), 286-300.

Zhou, M.F., Zhao, X.F., Chen, W.T., Li, X.C., Wang, W., Yan, D.P., Qiu, H.N., 2014. Proterozoic Fe-Cu metallogeny and supercontinental cycles of the southwestern Yangtze Block, southern China and northern Vietnam. Earth-Science Reviews, 139, 59-82.

Zimmer, M., Kroner, A., Jochum, K.P., Reischmann, T., Todt, W., 1995. The Gabal Gerf complex: A Precambrian N-MORB ophiolite in the Nubian Shield, NE Africa. Chemical Geology, 123, 29-51.

Geochemical and geochronological characteristics of Late Cretaceous to Early Paleocene granitoids in the Tengchong Block, Southwestern China: Implications for crustal anatexis and thickness variations along the eastern Neo-Tethys subduction zone[①]

Zhao Shaowei Lai Shaocong[②] Qin Jiangfeng Zhu Renzhi Wang Jiangbo

Abstract: The Tengchong Block of Southwestern China is key to tracing the eastward subduction of Neo-Tethys and collision between Indian and Asian continents. The block contains a magmatic belt that represents the southeastward continuation of the Gangdese belt, produced by the eastward subduction of eastern Neo-Tethyan oceanic lithosphere. In this paper we present geochemical and geochronological data of Late Cretaceous to Early Paleocene (~64 Ma) granitic rocks of the Guyong and Husa batholiths in the Tengchong Block. These can be subdivided into high-silica peraluminous granites and low-silica metaluminous granodiorites, and all belong to the high-K calc-alkaline series, are enriched in LILE, and depleted in HFSE. The Guyong granitoids have high initial Sr ratios of 0.706 511–0.711 753, negative $\varepsilon_{Nd}(t)$ values of −11.6 to −9.2, two-stage model ages of 1.39–1.55 Ga, and Pb isotopic compositions that indicate a crustal affinity. The Husa granodiorites also have high initial Sr ratios of 0.716 496, negative $\varepsilon_{Nd}(t)$ value of −16.5, two-stage model age of 1.89 Ga, variable $\varepsilon_{Hf}(t)$ values of −18.1 to 3.4 and Pb isotopic compositions similar to lower crustal values. These geochemical and isotopic data indicate that the Guyong granitoids were likely derived from partial melting of ancient crustal metapelite or mixed pelite-greywacke sources, while the Husa granodiorites were derived from the partial melting of lower crustal mixed sources involving metasedimentary and metaigneous rocks. To understand the thermal state and architecture of the Late Cretaceous to Early Paleocene magmatic arc crust, the crust-derived intermediate to acidic igneous rocks of the southern-central Lhasa and Tengchong blocks and eastern Himalayan syntaxis are compared. We infer that partial melting of crust occurred at great depth in the southern Lhasa Block, intermediate depths in the eastern Himalayan syntaxis, and shallow depths in the central Lhasa and Tengchong Block. Sr/Y ratios indicate that the Tengchong Block was characterized by thin crust, while the eastern Himalayan syntaxis and northernmost part of southern Lhasa Block

---

① Published in *Tectonophysics*, 2017, 694.

② Corresponding author.

contained crust of normal thickness, and the southernmost part of the southern Lhasa Block contained thickened crust.

1 Introduction

The Tibetan Plateau and adjacent blocks were important regions during the subduction of Neo-Tethyan oceanic crust and India-Asia collision and orogenesis. Mesozoi-Cenozoic magmatism is widespread throughout the Lhasa Block, which is subdivided into southern, central and northern Lhasa Block (Zhu et al., 2011a; Qi et al., 2015; Fig. 1a). The Gangdese belt and Linzizong volcanics mainly occur in the southern and central Lhasa Block, and Northern magmatic belt in the central and northern Lhasa Block (Chung et al., 2005; Chu et al., 2006; Zhu et al., 2011a). The Northern magmatic belt was produced by the subduction of Bangong-Nujiang oceanic lithosphere and subsequent collision of the Lhasa and Qiangtang blocks, and the igneous rocks emplaced largely during the Jurassic to Early Cretaceous, with ages of 240–110 Ma (Zhu et al., 2009, 2011a) and comprise peraluminous or S-type granitic plutons (Chung et al., 2005; Chu et al., 2006). Arc igneous rocks in the Gangdese belt were emplaced between 190 Ma and 42 Ma (Wen et al., 2008; Ji et al., 2009; Wu et al., 2010). The two magmatic belts recorded the gradually northward amalgamation of India-Lhasa-Qiangtang along the Asian continental margin. Mitchell et al. (2012) suggested that the Neo-Tethyan subduction gave rise to the Wuntho-Popa magmatic arc before the Late Cretaceous (>105 Ma) in west Burma Block, and this could northward prolong to Gangdese magmatic arc. The Tengchong Block is apparently located between the Wuntho-Popa and Gangdese magmatic arc (Fig. 1a). Recent seismic tomographic results indicate the Indian continental slab has been subducted to the east beneath the Burma-Tengchong Blocks (e.g., Li et al., 2008; He et al., 2010; Zhao et al., 2011). This result indicates that the Neo-Tethyan oceanic lithosphere was subducted beneath the Burma-Tengchong Blocks prior to continental collision. Late Cretaceous to Eocene magmatism of the Tengchong Block is coeval with, and similar to that of Gangdese magmatic arc belt and Linzizong volcanics, west Burma Block and Mogok metamorphic belt(e.g., Mitchell et al., 2012; Ma et al., 2014; Wang et al., 2014, 2015; Xu et al., 2015;Zhao et al., 2016a,b). Early Cretaceous magmatic rocks related to subduction of Bangong-Nujiang Ocean are exposed in the eastern part of Tengchong Block, and represent the southeastward continuation of the Northern magmatic belt in central and northern Lhasa Block (Xu et al., 2012; Zhu et al., 2015), which extends southward into the Mogok metamorphic belt (Mitchell et al., 2012). Thus, the Lhasa, Tengchong, and west Burma blocks, and Mogok metamorphic belt have been structurally linked since Early Cretaceous (Mitchell et al., 2012; Ma et al., 2014; Wang et al., 2014, 2015; Xu et al., 2012, 2015; Xie et al., 2016).

In addition, Kornfeld et al. (2014) show ca. 87° clockwise rotation of Tengchong Block since ca. 40 Ma with respect to stable Eurasia, suggesting that the Tengchong Block was sub-parallel to Lhasa Block, and that the Neo-Tethyan slab was subducted northward beneath Tengchong Block prior to 40 Ma. However, the subduction of Indian continent was northward to the Lhasa Block, through eastern Himalayan syntaxis, and eastward to the Burma-Tengchong blocks in the present tectonic framework (e.g., Li et al., 2008; He et al., 2010; Zhao et al., 2011). Thus, we use the eastward subduction of Neo-Tethys beneath the Tengchong Block in our paper.

The initial age of India-Asia collision has been extensively studied (Aitchison et al., 2007; Cai et al., 2011; Chu, et al., 2011; Ding et al., 2003, 2005; Hu el al., 2012; Mo et al., 2007; Najman et al., 2010; Sun et al., 2012; van Hinsbergen et al., 2012; Yi et al., 2011), with most placing the event in the Late Cretaceous to Paleocene (ca. 65–55 Ma). The recent studies show that the precise age of initial collision in the Lhasa Block was 59 ± 1 Ma (Hu et al., 2015; Wu et al., 2014), and suturing gradually propagated eastward until the collision of eastern Himalayan syntaxis and Tengchong Block at ca. 52–57 Ma (Xu et al., 2008; Ding et al., 2001; Zhang et al., 2010). Thus, the Late Cretaceous to Early Paleocene crustal architecture and anatexis along the magmatic arc are important compositions of the evolution of Neo-Tethys prior to continental collision. However, the genesis of Late Cretaceous to Paleocene granites and their tectonic setting in Tengchong Block remain debated: ① the granitic rocks are dual I-type and S-type in composition, related to the eastward subduction of the Neo-Tethys beneath the Asian continent and representing the melting products of thickened crust in the hinterland (Xu et al., 2012; Yang et al., 2009; Ma et al., 2013; Qi et al., 2015); and ② the granitic rocks are associated with an back-arc extension related to the subduction of Neo-Tethys (Chen et al., 2015; Wang et al., 2014). In this study we constrain the Late Cretaceous to early Paleocene thermal state and architecture of the magmatic arc crust related to the subduction of Neo-Tethys by comparing granitic rocks of the Tengchong Block with crust-derived intermediate to acidic rocks in the southern-central Lhasa Block and eastern Himalayan syntaxis.

2 Geological setting

The Tengchong Block is bound from Baoshan Block by the Gaoligong dextral strike-slip fault to the east and the Ruili fault to the southeast, and from the West Burma Block by the Sagaing fault to the west (Replumaz and Tapponnier, 2003; Fig. 1). Detrital zircon U-Pb age and Hf data from Carboniferous strata indicate that the Tengchong Block was located along the Indian margin of Gondwana during the Early Paleozoic (Li et al., 2014). Actually, the ca. 1. 17 Ga zircon age group occurs in the Carboniferous strata of Tengchong Block, similar to that in Lhasa Block, which is interpreted as an origin of southwest Australian margin of Gondwana (Zhu et al., 2011b). In contrast, the 0. 95 Ga and 0. 5 Ga detrital zircon age populations in

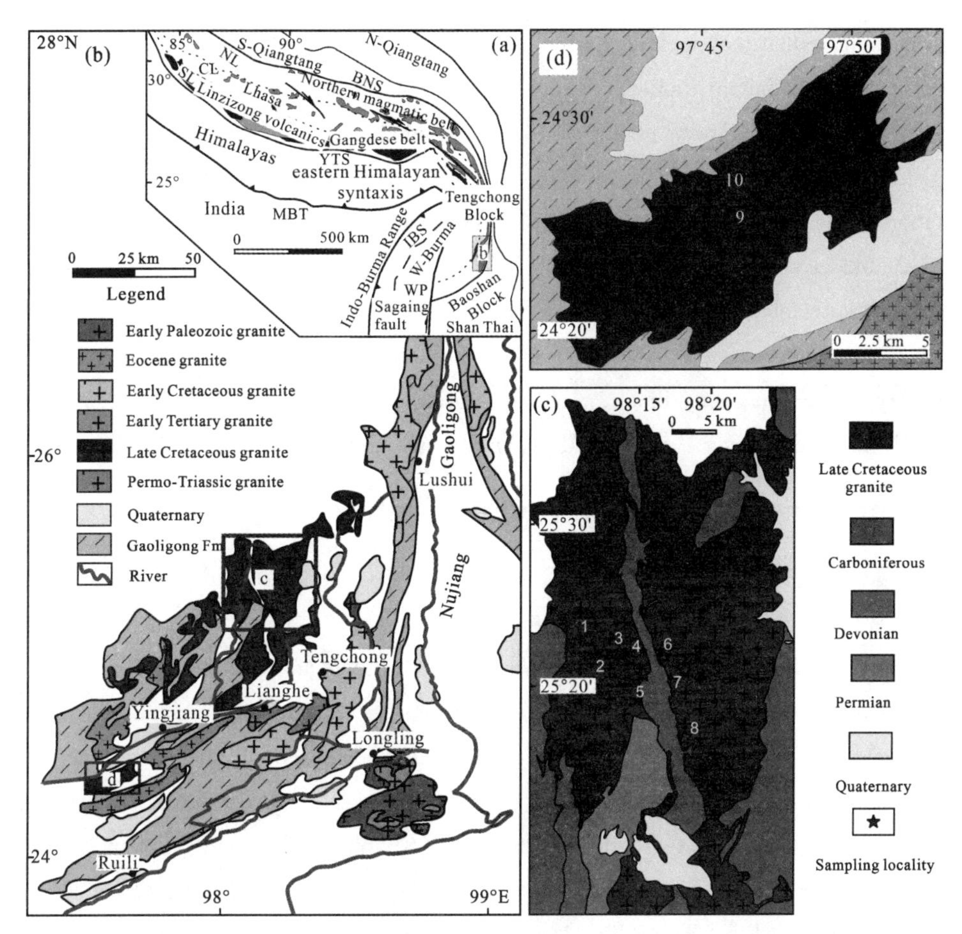

Fig. 1 Simplified geological map of Himalaya-Tibet tectonic realm after Qi et al. (2015) (a), Tengchong Block and adjacent area in southeast of Tibetan Plateau after Xu et al. (2008) (b) and the samples localities in the Guyong batholith (c) and Husa batholith (d).

BNS: Banggong-Nujiang suture; NL, CL, SL: southern, central, northern Lhasa Block; YTS: Yarlung-Tsangpo suture; MBT: Main Boundary Thrust; IBS: Indo-Burma suture; WP: Wuntho-Popa magmatic arc.

the Carboniferous strata of Tengchong Block are more compared with those in west Qiangtang and Tethyan Himalaya terranes and match those defined by zircons from Neoproterozoic metasedimentary rocks in High Himalaya (Li et al., 2014; Zhu et al., 2011b; Gehrels et al., 2003, 2006a,b). Thus the origin of Tengchong Block is still controversial. The basement of Tengchong Block is summarized as Gaoligong formation (Zhong, 1998), and is composed of Paleoproterozoic metamorphosed volcano-sedimentary rocks, tonalite, and gabbro, similar to the Mogok metamorphic rocks in the eastern Burma Highland (Chen, 1991; Cong et al., 1993; Fang and Zhang, 1994; Mitchell, 1993). Early Paleozoic strata have not been identified in the Tengchong Block, and Late Paleozoic-Mesozoic sedimentary rocks directly overlie the basement, consisting of Devonian-Triassic carbonate and low- to medium-grade metamorphic clastic sedimentary rocks (Li et al., 2014). Zircon U-Pb ages from the Gaoligong formation indicate that they include Early Paleozoic (547–454 Ma) and Cretaceous to Paleogene (ca. 120–50 Ma) granitoid intrusions (Cao et al., 2014, 2016; Eroğlu et al., 2013; Li et al., 2016; Xu et al., 2012; Wang et al., 2014, 2015; Ma et al., 2013; Zhao et al., 2016a, b, 2017). Extensive Late Mesozoic and Cenozoic igneous rocks (ca. 76–50 Ma) occur in a N-S orientation in the Tengchong-Lianghe-Yingjiang-Longchuan area (e.g., Cao et al., 2016; Xu et al., 2012; Ma et al., 2014; Wang et al., 2014, 2015; Zhao et al., 2016a, b), representing magmatic activity related to subduction of Neo-Tethys and collision of Indian-Asian continent. Since Late Miocene, multiple stages of eruptive basaltic and basaltic-andesitic volcanism have taken place in the Tengchong Block, including during the periods 5.5–4.0 Ma, 3.9–0.9 Ma, 0.8–0.01 Ma, and younger than 0.01 Ma (Wang et al., 2007; Guo et al., 2015).

3 Sample descriptions and petrography

Samples were collected from the Guyong and Husa batholiths in northern Tengchong County and southwestern Yingjiang County, respectively (Fig. 1). Sample localities and detailed mineralogical descriptions are shown in Fig. 1c, d and listed in Table 1.

The Guyong batholith consists of porphyraceous or equigranular granites and biotite granites (Fig. 2). The porphyritic biotite granites are composed of phenocrysts of Kf (K-feldspar, 5%) in a matrix of Qz (quartz, 25%–30%), Pl (plagioclase, 20%–23%), Mc (microcline) + Pth (perthite) + Kf (30%), Bi (biotite, 12% – 15%), and minor Hb (hornblende, 0%–3%). The equigranular granites are comprised of Qz (22%–30%), Pl (25%–40%), Mc + Pth + Kf (27%–42%), and Bi (2%–8%). Accessory minerals in the granites include zircon, apatite, titanite, and magnetite.

Rocks in the Husa batholith are mainly granodiorite in composition, and exhibit porphyritic texture (Fig. 2). The granodiorites are composed of phenocrysts of Mc (10%) in a matrix of Qz (20%), Pl (30%), Mc (10%), Kf (15%), Bi (7%), Hb (8%). Accessory minerals include zircon, apatite, allanite, and titanite.

4 Analytic methods

4.1 Zircon U-Pb and Lu-Hf isotopic analyses

Zircon U-Pb isotopic analyses were conducted at the State Key Laboratory of Continental Dynamics, Northwest University, Xi'an, China. Zircon grains were separated by the conventional heavy liquid and magnetic separation techniques. Representative zircon grains were handpicked under a binocular microscope, and mounted in epoxy resin discs, then ground and polished, and coated with carbon. Morphology and internal structures were examined using reflected light photograph and cathodoluminescence (CL) prior to isotope analyses.

Table 1 Simplified samples descriptions and sampling locations for the Guyong and Husa batholith, Tengchong Block.

Group	Sample	Latitude(N)	Longitude(E)	Lithology	Texture	Mineral assemblage
Guyong batholith						
1	GY02	25°22.695′	98°12.251′	Biotite granite	Porphyraceous	Phenocryst: Kf(5%)
	GY04					Matrix: Qz(25%)+Pl(20%)+Mc(15%)+Kf(20%)+Bi(12%)+Hb(3%)
	GY06					
	GY07					
2	GY10	25°21.476′	98°13.647′	Granite	Fine grained equigranular	Qz(30%)+Pl(25%)+Mc(20%)+Kf(22%)+Bi(3%)
	GY11					
	GY16					
	GY17					
3	GY23	25°21.564′	98°14.037′	Granite	Medium grained equigranular	Qz(30%)+Pl(40%)+Kf(27%)+Bi(3%)
	GY28					
4	GY31	25°21.269′	98°14.657′	Granite	Fine grained equigranular	Qz(28%)+Pl(35%)+Mc(15%)+Kf(20%)+Bi(2%)
5	GY36	25°20.715′	98°15.840′	Granite	Coarse grained equigranular	Qz(22%)+Pl(40%)+Kf(30%)+Bi(8%)
	GY37					
	GY42					
	GY46					
6	GY47	25°22.806′	98°16.401′	Granite	Medium grained equigranular	Qz(25%)+Pl(30%)+Mc(15%)+Kf(25%)+Bi(5%)
	GY48					
7	GY53	25°20.830′	98°17.673′	Biotite granite	Porphyraceous	Phenocryst: Kf(5%)
	GY55					Matrix: Qz(30%)+Pl(23%)+Mc(10%)+Kf(20%)+Bi(12%)
8	GY63	25°12.876′	98°17.127′	Granite	Medium grained equigranular	Qz(25%)+Pl(35%)+Mc(10%)+Kf(25%)+Bi(5%)
Husa batholith						
9	LL120	24°27.454′	97°45.244′	Granodiorite	Porphyraceous	Phenocryst: Mc(10%)
	LL121					Matrix: Qz(20%)+Pl(30%)+Mc(10%)+Kf(15%)+Bi(7%)+Hb(8%)
10	LC20	24°27.855′	97°45.040′	Granodiorite	Porphyraceous	Phenocryst: Mc(10%)
	LC21					Matrix: Qz(20%)+Pl(30%)+Mc(10%)+Kf(15%)+Bi(7%)+Hb(8%)

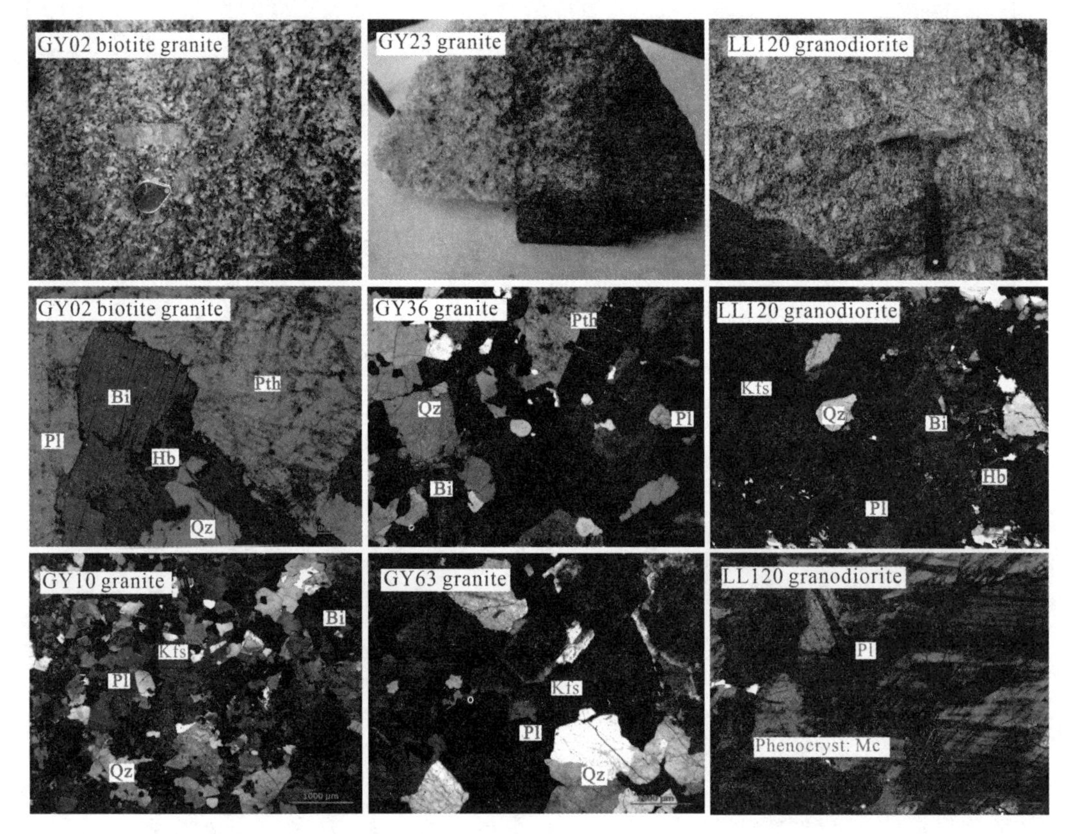

Fig. 2 Field, hand specimen and microscope photographs of the granitic rocks from Guyong and Husa batholith, Tengchong Block.

Pl: plagioclase; Bi: biotite; Hb: hornblende; Qz: quartz; Pth: perthite; Kf: K-feldspar; Mc: microcline.

Laser ablation ICP-MS zircon U-Pb analyses were conducted on an Agilent 7500a ICP-MS equipped with a 193-nm wavelength ArF excimer laser. During analyses, the spot diameter was 30 μm and the methods of the zircon U-Pb analyses were in the light of Yuan et al. (2004). The $^{207}Pb/^{235}U$ and $^{206}Pb/^{238}U$ ratios were calculated using the GLITTER program, which was corrected using the Harvard zircon 91500 as external calibration. These correction factors were applied to each sample to correct for both instrumental mass bias and depth-dependent elemental and isotopic fractionation. The detailed technique is described in Yuan et al. (2004). Common Pb contents were evaluated using the method described in Andersen (2002). Age calculations and concordia diagrams were completed using ISOPLOT version 3. 0 (Ludwig, 2003). The errors quoted in tables and figures are at 2σ levels.

The in situ zircon Hf isotopic analyses were made using a Neptune MC-ICPS. The laser repetition rate was 6 Hz at 100 mJ and the spot sizes were 30 μm. The detailed analytical technique is depicted by Yuan et al. (2008). During analyses, the $^{176}Hf/^{177}Hf$ and $^{176}Lu/^{177}Hf$ ratios of the standard zircon (91500) were 0. 282 294±15 (2σ) and 0. 000 31, respectively. The notations of $\varepsilon_{Hf}(t)$ value, $f_{Lu/Hf}$, two-stage model ages are defined as in Wu et al. (2007).

4.2 Whole rock geochemical and Sr-Nd-Pb isotopic analyses

Weathered surfaces of the granitoid samples were removed and the fresh parts were chipped and powdered to about 200 mesh size using a tungsten carbide ball mill. Major and trace element were analyzed by X-ray fluorescence (XRF; Rikagu RIX 2100) and inductively coupled plasma mass spectrometry (ICP-MS; Agilent 7500a), respectively, at the State Key Laboratory of Continental Dynamics, Northwest University, Xi'an, China. Analyses of USGS and Chinese national rock standards (BCR-2, GSR-1, and GSR-3) showed that the analytical precision and accuracy for the major elements were generally better than 5%. For the trace element analyses, sample powders were digested using an $HF+HNO_3$ mixture in high-pressure Teflon bombs at 190 ℃ for 48 h. For most trace elements, the analytical error was <2% and the precision was >10% (Liu et al., 2007).

Whole-rock Sr-Nd-Pb isotopic data were obtained using a Nu Plasma HR multi-collector mass spectrometer at both the Guizhou Tuopu Resource and Environmental Analysis Center and the State Key Laboratory of Continental Dynamics, Northwest University, Xi'an China. The Sr and Nd isotopic fractionations were corrected to $^{87}Sr/^{86}Sr = 0.1194$ and $^{146}Nd/^{144}Nd = 0.7219$, respectively. During the analysis period, the NIST SRM 987 standard yielded a mean value of $^{87}Sr/^{86}Sr = 0.710250 \pm 12$ (2σ, $n=15$) and the La Jolla standard gave a mean of $^{146}Nd/^{144}Nd = 0.511859 \pm 6$ (2σ, $n=20$). Whole-rock Pb was isolated by anion exchange in HCl-Br columns, and Pb isotopic fractionation was corrected to $^{205}Tl/^{203}Tl = 2.3875$. Within the period of analysis, 30 measurements of the NBS 981 standard gave average values of $^{206}Pb/^{204}Pb = 16.937 \pm 1$ (2σ), $^{207}Pb/^{204}Pb = 15.491 \pm 1$ (2σ), and $^{208}Pb/^{204}Pb = 36.696 \pm 1$ (2σ). The BCR-2 standard gave values of $^{206}Pb/^{204}Pb = 18.742 \pm 1$ (2σ), $^{207}Pb/^{204}Pb = 15.620 \pm 1$ (2σ), and $^{208}Pb/^{204}Pb = 38.705 \pm 1$ (2σ). The total procedural Pb blanks were in the range of 0.1–0.3 ng.

5 Results

5.1 Zircon U-Pb results

Four samples (GY06, GY17, GY46, and LC28) from Guyong and Husa batholith were selected for zircon U-Pb isotopic measurement.

Zircon grains from the Guyong batholith are mostly transparent, light-brown or colorless, subhedral to euhedral crystal, 100–300 μm long, with aspect ratios of 2∶1 to 3∶1. On the representative zircon cathodoluminescence (CL) images, mostly grains display clear concentric oscillatory zoning core with thin and dark rim, indicating the magmatic origin and affected by late-stage metamorphism. Individual grains possess light-colored and oval cores, that are interpreted to be inherited. Zircon grains from the Husa batholith are transparent, light-brown or colorless, euhedral to subhedral crystal, 150–300 μm long with aspect ratios of 1.5 : 1 to 3 : 1. On the CL images, they show clear concentric oscillatory zoning. Zircon U-Pb isotopic results and CL images are presented in Fig. 3 and Fig. 4, respectively, and the U-Pb isotopic data are listed in a Supplementary Table 1.

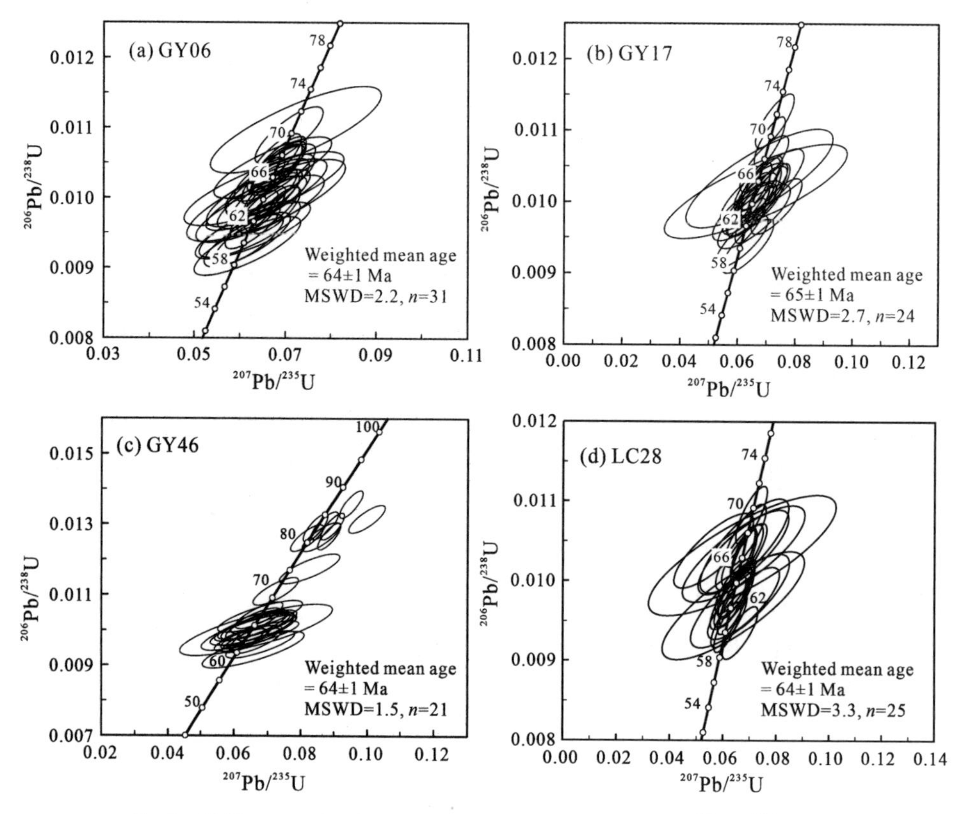

Fig. 3 U-Pb concordia diagrams and weighted mean ages of $^{206}Pb/^{238}U$ for zircons from Guyong and Husa batholith, Tengchong Block.

Analyses of 33 zircon grains from sample GY06 yielded ages of 60–1 789 Ma, while two spots from the light cores yielded $^{206}Pb/^{238}U$ ages of 1 789 ± 32 Ma and 1 080 ± 33 Ma, which can be interpreted as the ages of xenocrysts, as inferred from zircon morphology. All others spots yielded the Early Tertiary ages. The zircons have variable Th (121–1 079 ppm) and U (313–3 067 ppm) contents, with Th/U ratios of 0.27–0.89, and $^{206}Pb/^{238}U$ ages from 60 ± 2 Ma to 70 ± 2 Ma, with a weighted mean age of 64 ± 1 Ma (MSWD = 2.2, n = 31) (Fig. 3a).

The zircon grains from sample GY17 mostly show magmatic zoning with a dark metamorphic rim. Three core analytical spots yielded $^{206}Pb/^{238}U$ ages of 899 Ma, 634 Ma, and

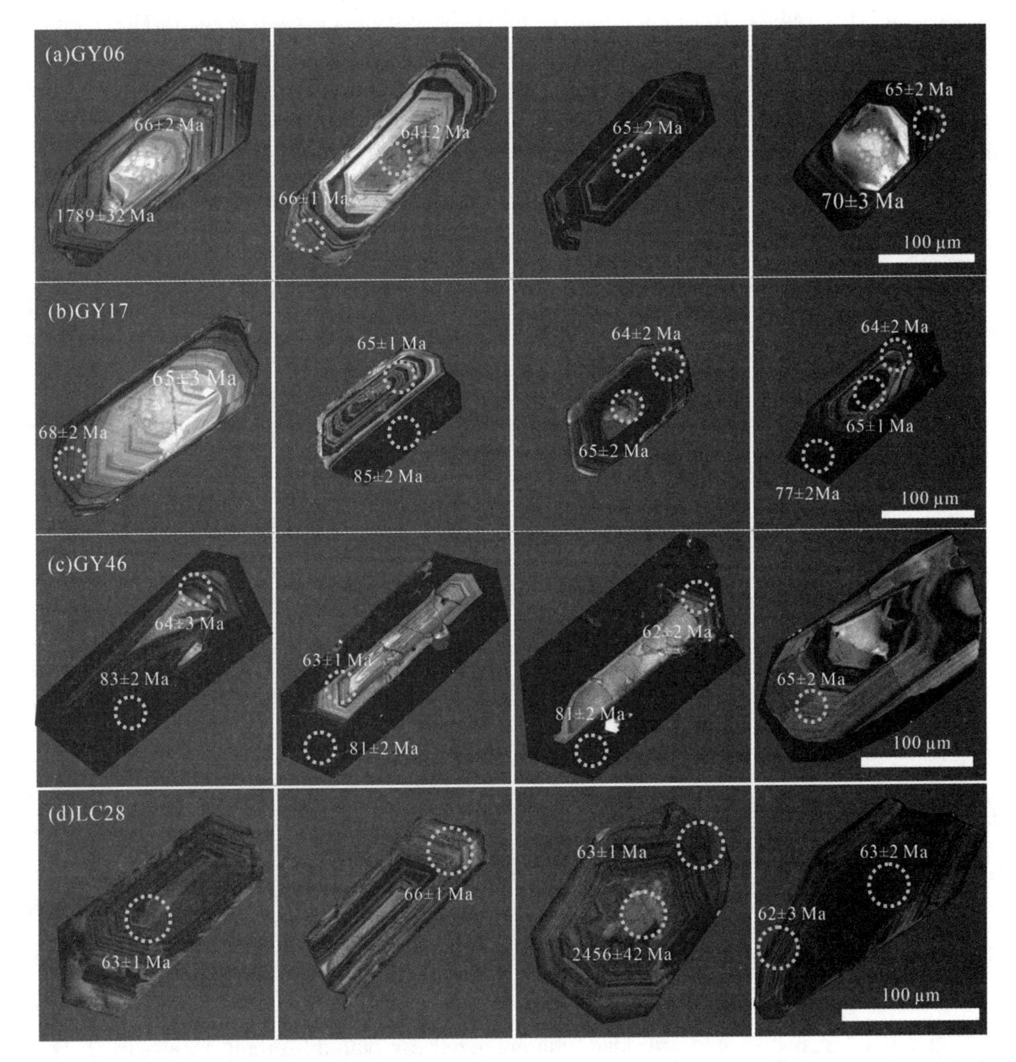

Fig. 4 CL images of representative zircons for the rocks from Guyong and Husa batholiths, Tengchong Block.

89 Ma, which are interpreted as the ages of xenocrysts. The dark rims were apparently affected by the late metamorphism and yielded $^{206}Pb/^{238}U$ ages of 77 Ma and 85 Ma, however, these ages have no geological implications, as the dark rim could be crystallized from a Pb-rich fluid and the ages could not represent this actually geological event. The other 24 spots have Early Tertiary ages with variable Th (66–3 205 ppm) and U (144–6 396 ppm) contents, Th/U ratios of 0.26 to 1.36. They yielded $^{206}Pb/^{238}U$ ages of 60±2 Ma to 71±2 Ma, with a weighted mean age of 65±1 Ma (MSWD=2.7, n=24) (Fig. 3b).

Thirty spots were analyzed in zircons from GY46. The nine zircon dark rims yielded $^{206}Pb/^{238}U$ ages of 75 Ma to 87 Ma, however, these ages cannot be applied to interpret the geological event. The remaining 21 spots yielded $^{206}Pb/^{238}U$ ages of 61±2 Ma to 71±2 Ma, giving a weighted mean age of 64±1 Ma (MSWD=1.5, n=21) (Fig. 3c), and the zircons have variable Th (144–1 705 ppm) and U (331–3 899 ppm) contents and Th/U ratios of 0.29–1.31.

Thirty-one spots were analyzed for LC28, and 25 spots yielded Late Cretaceous to Early Paleocene age, and have Th and U contents of 83 – 2 382 ppm and 129 – 3 568 ppm respectively, and Th/U ratios of 0.26 – 1.66, indicating a magmatic origin. Representative grains yielded ages from 60 Ma to 69 Ma, with a weighted mean age of 64 ± 1 Ma (MSWD = 3.3, n=25; Fig. 3d), which can be interpreted as the crystallization age of the granodiorites. Another six spots yielded ages from 72 Ma to 2 456 Ma, which could be interpreted as the ages of xenocrysts.

In summary, the Guyong and Husa granitic rocks were emplaced during Late Cretaceous to Early Paleocene, at ca. 64 Ma.

5.2 Major and trace elements

The results of major and trace element analyses of the granites from Guyong and the granodiorites from Husa batholith are listed in Table 2.

Rocks of the Guyong batholith are high-K calc-alkaline (Fig. 5a), weakly peraluminous (Fig. 5b), with A/CNK ratios of 1.01 – 1.11. The granites contain relatively high SiO_2 (70.64–75.77 wt%) and K_2O (4.14–5.08 wt%) contents, and low contents of total Fe_2O_3 (0.80–3.55 wt%), CaO (0.38–2.28 wt%), MgO (0.06–0.65 wt%), and Na_2O (3.05–3.52 wt%). They also have low $Mg^{\#}$ [molar 100Mg/(Mg + Total Fe^{2+})] values of 11 – 40. Trace element results indicate that the Guyong granitic rocks have high contents of Th (24.9–75.9 ppm), Pb (25.4–43.7 ppm), and Rb (283–497 ppm), low Ni (2.8–15.1 ppm) and Cr (2.8–46.4 ppm) contents, and are depleted in the Ti, Nb, and Eu, with δEu values of 0.12 – 0.58 (Fig. 6). They are also characterized by variable total rare earth element (ΣREE) values (66–266 ppm), low Sr/Y (0.5–6.2) and $(La/Yb)_N$ ratios (1.5–12.8).

In contrast, the granodiorites of Husa batholith contain relatively low SiO_2(64.50–66.99 wt%) and K_2O (3.10 – 3.78 wt%) contents, and high MgO contents of 1.19 – 1.53 wt%, Na_2O contents of 3.44–3.85 wt% and TiO_2 contents of 0.52–0.66 wt%, with $Mg^{\#}$ values of 38–40, and A/CNK ratios of 0.96–0.97. They are high-K, calc-alkaline and metaluminous. The Husa granodiorites share similar trace element and REE characteristics to the Guyong granites (Fig. 6). However, they have relatively high Sr/Y ratios of 6.8 – 11.5 and low Th, Pb, Rb contents, and negative Eu anomalies with δEu values of 0.60–0.71.

5.3 Sr-Nd-Pb isotopic compositions

Whole rock Sr-Nd-Pb isotopic compositions are given in Table 3 and Figs. 7–8. Initial isotopic values were calculated at t=64 Ma on the basis of LA-ICP-MS zircon U-Pb ages for the samples.

Table 2 Analytical results of major (wt%) and trace (ppm) elements for granitic rocks from Guyong and Husa batholith, Tengchong Block.

Sample	GY02	GY04	GY07	GY10	GY11	GY16	GY23	GY31	GY36	GY37	GY42	GY47	GY48	GY53	GY55	GY63	LL120	LL121	LC21	LC22
SiO_2	72.33	72.33	70.64	75.20	75.58	75.04	75.62	75.22	70.44	73.97	73.21	75.77	75.51	76.45	73.75	75.72	66.99	65.37	65.89	64.50
TiO_2	0.22	0.26	0.28	0.11	0.08	0.10	0.07	0.06	0.33	0.16	0.22	0.13	0.12	0.30	0.32	0.05	0.52	0.54	0.59	0.66
Al_2O_3	14.03	13.80	14.55	12.92	12.87	12.95	12.88	13.23	14.47	13.59	13.48	12.78	12.85	12.13	13.08	13.18	15.75	15.98	15.82	16.15
$Fe_2O_3{}^T$	2.15	2.34	2.71	1.30	1.02	1.23	1.11	0.80	3.55	1.80	2.48	1.20	1.30	2.05	2.26	1.10	4.38	4.92	4.69	5.44
MnO	0.06	0.06	0.08	0.05	0.04	0.05	0.07	0.04	0.10	0.05	0.07	0.01	0.01	0.03	0.05	0.05	0.08	0.09	0.08	0.10
MgO	0.49	0.55	0.64	0.17	0.13	0.16	0.10	0.07	0.61	0.28	0.40	0.12	0.12	0.36	0.65	0.06	1.19	1.34	1.21	1.53
CaO	2.17	1.97	2.28	1.02	0.94	1.05	0.86	0.86	1.92	1.46	1.48	0.54	0.48	0.38	1.56	0.54	3.84	4.07	3.49	3.83
Na_2O	3.36	3.05	3.28	3.29	3.27	3.35	3.45	3.47	3.52	3.19	3.16	3.25	3.17	3.26	3.12	3.35	3.45	3.56	3.85	3.44
K_2O	4.14	4.53	4.38	4.75	4.85	4.79	4.66	4.93	4.14	4.65	4.62	5.03	5.08	4.46	4.14	4.85	3.29	3.10	3.48	3.78
P_2O_5	0.06	0.06	0.08	0.03	0.02	0.02	0.02	0.02	0.08	0.04	0.06	0.03	0.03	0.08	0.10	0.01	0.15	0.16	0.17	0.17
LOI	0.73	0.58	0.75	0.75	0.77	0.81	0.66	0.84	0.44	0.37	0.32	0.67	0.84	0.82	0.55	0.78	0.33	0.38	0.32	0.66
Total	99.74	99.53	99.67	99.59	99.57	99.55	99.50	99.54	99.60	99.56	99.50	99.53	99.51	100.32	99.58	99.69	99.97	99.51	99.59	100.26
$Mg^{\#}$	35	35	35	23	23	23	17	17	29	27	27	19	18	29	40	11	39	39	38	40
K_2O/Na_2O	1.23	1.49	1.34	1.44	1.48	1.43	1.35	1.42	1.18	1.46	1.46	1.55	1.60	1.37	1.33	1.45	0.95	0.87	0.90	1.10
A/CNK	1.01	1.02	1.02	1.04	1.04	1.03	1.05	1.05	1.05	1.05	1.05	1.09	1.11	1.11	1.05	1.12	0.97	0.96	0.96	0.97
Li	42.3	44.3	49.9	34.6	25.9	34.8	64.3	15.5	76.2	45.8	58.8	10.3	10.0	36.1	79.4	9.82	19.6	21.3	37.3	30.9
Be	3.47	3.12	3.16	3.4	3.29	3.45	5	4.72	4.47	3.69	3.75	3.85	3.86	5.08	8.24	3.92	2.69	2.84	3.49	2.84
Sc	6.63	6.36	8.06	4.29	4.22	4.56	5.17	5.22	6.91	5.22	5.32	5.10	3.97	3.42	7.33	5.66	8.35	9.58	9.11	9.94
V	28.5	28.5	34.0	10.1	9.81	9.01	6.48	4.4	26.5	13.9	19.7	7.87	8.64	17.7	30.5	2.34	59.4	68.6	57.6	74.2
Cr	4.24	5.60	3.67	2.79	3.97	5.91	3.73	3.04	5.51	4.74	8.62	3.07	3.1	7.54	46.4	4.96	2.77	4.69	4.42	4.57
Co	38.7	127	36.1	44.1	42.7	43.9	53	60.2	39.9	62.4	60	67.9	75.2	55.8	54.9	44.6	107	109	69.7	74.2
Ni	5.10	5.22	3.58	3.27	2.78	4.02	3.02	2.93	4.28	4.33	14.7	3.10	3.42	7.11	15.1	3.23	2.90	5.27	2.47	2.67
Cu	1.14	2.30	1.77	0.62	0.72	1.27	0.11	0.69	2.64	1.90	5.89	1.41	1.81	3.24	2.73	1.10	4.44	4.54	3.87	5.81
Zn	38.7	62.7	52.6	34.9	23.9	41.2	40.8	46.6	124	65	100	51	29.8	73.7	54.6	62.6	52.8	59.0	61.3	66.6
Ga	16.1	15.1	16.7	13.9	13.7	14.5	14.8	14.7	19.7	16.3	16.9	18.7	18.7	16.3	18.8	14.6	19.0	19.9	20.7	20.5
Ge	1.25	1.2	1.25	1.45	1.41	1.44	1.75	1.63	1.64	1.48	1.51	1.51	1.47	1.49	1.47	1.38	1.23	1.29	1.39	1.35
Rb	283	303	309	404	388	401	497	475	440	406	420	384	391	487	494	433	109	111	156	153
Sr	158	148	159	50.7	40.9	49.3	25.1	36.4	118	102	96.8	46.2	46.5	89.9	140	13.4	320	324	270	309

Continued

Sample	GY02	GY04	GY07	GY10	GY11	GY16	GY23	GY31	GY36	GY37	GY42	GY47	GY48	GY53	GY55	GY63	LL120	LL121	LC21	LC22
Y	26.8	27.2	25.7	34.4	28.8	29.3	42.6	35.1	46.2	32.3	39.8	26.2	29.4	83.4	40.5	26.9	29.6	28.3	39.5	32.3
Zr	96.1	116	144	68.2	52	73.2	84.7	66.5	154	103	144	115	122	20.7	142	96.2	185	207	193	202
Nb	14.1	15.1	15.6	16.5	16.3	16.5	17.3	15.9	21.4	16.5	20.6	13.2	13.4	29.6	22.3	18.3	12.6	12.0	14.4	13.3
Cs	6.15	6.46	6.77	7.93	6.5	7.87	8.47	4.82	10	6.73	7.77	4.84	4.85	11.9	12.3	5.52	1.73	1.89	2.81	3.92
Ba	355	415	375	50.4	38.9	53.4	30.1	116	167	174	170	154	155	299	307	13.3	972	872	733	1031
La	37.1	33.3	38.4	16.9	11	14.5	11.3	9.88	45.8	31.5	37	38.9	33.3	87.2	58.4	14.2	38.9	44.1	43.3	64.2
Ce	65.7	56.3	72.9	38.2	26	33.2	27.3	26.1	95.7	56.8	73.8	83	55.1	184	122	26.5	73.2	80.7	81.3	112
Pr	6.78	5.65	7.48	3.71	2.53	3.08	2.92	2.67	9.78	6.25	7.85	8.7	6.74	20.5	12.6	3.79	7.98	8.55	9.33	11.8
Nd	23.8	19.9	27	14.1	9.81	12.2	12.3	11	34.4	21.9	27.6	30.4	26.1	72.4	42.4	15.4	29.3	30.6	34.1	39.8
Sm	4.4	3.73	4.63	3.61	2.49	2.95	3.95	3.57	5.97	4.25	4.91	5.5	5.89	13.9	7.4	4.67	5.72	5.65	6.78	6.55
Eu	0.72	0.67	0.73	0.29	0.23	0.27	0.16	0.20	0.53	0.46	0.48	0.27	0.34	0.81	0.73	0.19	1.28	1.24	1.30	1.34
Gd	3.73	3.45	4.04	3.65	2.66	2.92	4.49	3.83	5.15	3.76	4.25	5.02	5.68	11.79	6.07	4.58	5.20	5.18	6.40	6.05
Tb	0.66	0.61	0.69	0.66	0.51	0.54	0.82	0.68	0.85	0.63	0.72	0.8	0.87	2.02	0.98	0.73	0.81	0.79	1.01	0.88
Dy	3.96	3.99	3.96	4.4	3.63	3.67	5.65	4.5	5.25	3.9	4.42	4.47	5.07	12	5.65	4.41	4.84	4.63	6.27	5.24
Ho	0.82	0.85	0.81	0.96	0.79	0.82	1.21	0.96	1.12	0.83	0.99	0.88	0.99	2.47	1.18	0.88	0.99	0.95	1.29	1.05
Er	2.53	2.64	2.46	3.07	2.58	2.68	3.9	3.11	3.78	2.73	3.19	2.62	2.81	7.27	3.77	2.59	2.88	2.74	3.83	3.11
Tm	0.38	0.42	0.38	0.51	0.45	0.47	0.68	0.53	0.6	0.45	0.52	0.34	0.39	1.08	0.57	0.38	0.42	0.40	0.57	0.46
Yb	2.78	2.93	2.78	3.9	3.29	3.58	5.29	3.92	4.57	3.39	3.98	2.18	2.47	6.79	4.08	2.57	2.73	2.62	3.78	3.09
Lu	0.42	0.44	0.42	0.6	0.51	0.56	0.8	0.62	0.71	0.54	0.62	0.31	0.34	0.91	0.63	0.38	0.38	0.37	0.56	0.47
Hf	2.92	3.34	3.99	2.84	2.07	3	3.97	2.92	5.71	3.26	4.31	4	4.09	0.82	4.47	3.74	4.44	4.93	4.95	5.05
Ta	2.12	2.39	2.11	3.65	4.4	3.18	4.92	4.32	2.65	2.41	2.96	1.53	1.62	4.16	3.26	1.53	1.12	1.11	1.27	1.20
Pb	25.9	26.2	26.6	32.6	33.2	33.3	42.9	40.5	25.4	28.4	28.5	36.9	37.2	32.9	32.3	43.7	19.2	19.3	21.8	21.0
Th	31.6	28.4	39.9	30.1	24.9	30.1	27.9	25.1	63.2	49	49.8	31.3	37.7	75.9	51.5	44	17.4	18.2	20.6	25.4
U	7.7	7.37	9.62	9.73	10.5	7.88	16.6	20.1	13.3	12.9	12.5	4.91	6.32	13.2	9.88	9.81	3.06	2.87	3.60	3.54
Sr/Y	5.9	5.4	6.2	1.5	1.4	1.7	0.6	1.0	2.6	3.2	2.4	1.8	1.6	1.1	3.5	0.5	10.8	11.5	6.8	9.5
$(La/Yb)_N$	10	8	10	3	2	3	2	2	7	7	7	13	10	9	10	4	10	12	8	15
δEu	0.54	0.58	0.52	0.24	0.28	0.28	0.12	0.16	0.29	0.35	0.32	0.16	0.18	0.19	0.33	0.13	0.71	0.70	0.60	0.65
REE	154	135	167	95	66	81	81	72	214	137	170	183	146	423	266	81	175	189	200	256

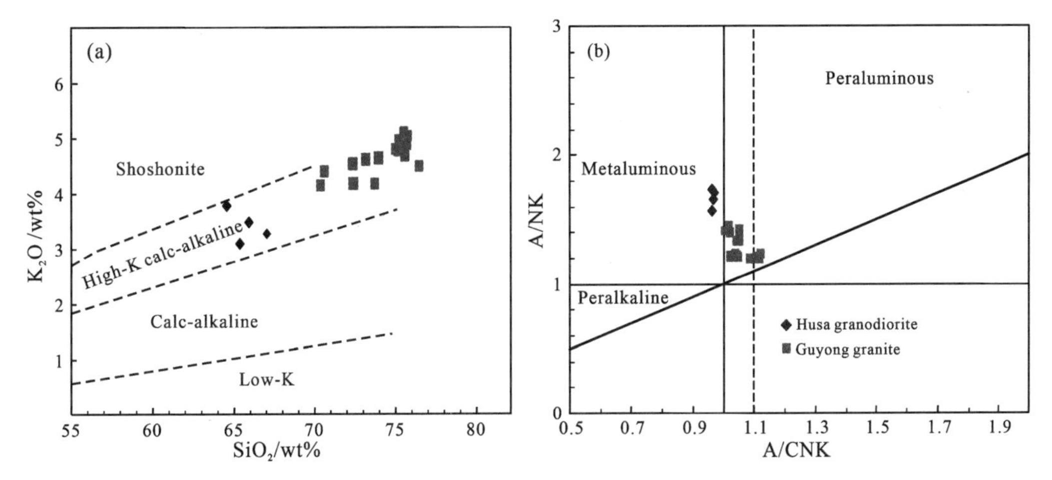

Fig. 5 Diagrams of K_2O vs. SiO_2 after Rollinson (1993) (a) and A/NK [molar $Al_2O_3/(Na_2O+K_2O)$] vs. A/CNK [molar $Al_2O_3/(CaO+Na_2O+K_2O)$] (Maniar and Piccoli, 1989) (b) for granitic rocks in Guyong and Husa batholiths.

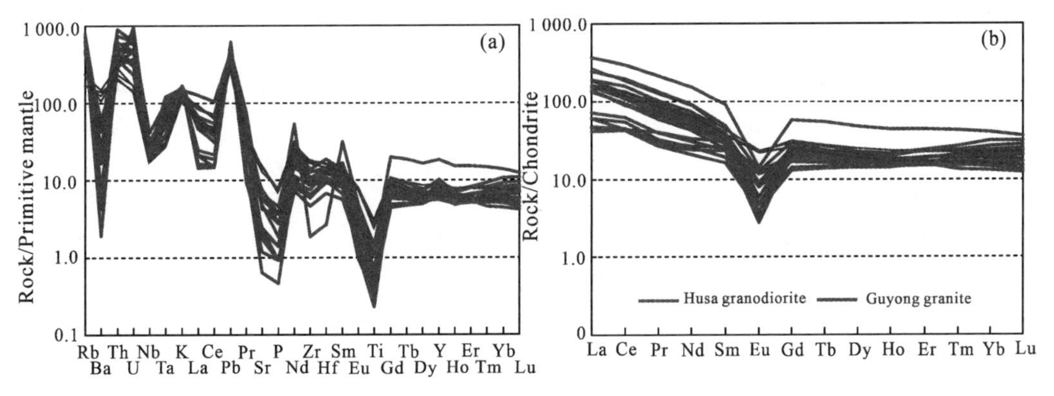

Fig. 6 Primitive mantle (PM) normalized trace element spider (a) and Chondrite-normalized REE pattern (b) diagrams for the samples from Guyong and Husa batholiths, Tengchong Block.
The Primitive mantle and Chondrite values are from Sun and McDonough (1989).

Table 3 Whole rocks Sr-Nd-Pb isotopic data for samples from Guyong and Husa batholith, Tengchong Block.

Sample	GY07	GY16	GY42	GY55	LL120
t/Ma	64	64	64	64	64
Rb/ppm	309	401	420	494	109
Sr/ppm	159	49.3	96.8	140	320
$^{87}Rb/^{86}Sr$	5.6	23.6	12.6	10.2	0.983
$^{87}Sr/^{86}Sr$	0.714 969	0.727 950	0.723 184	0.717 268	0.717 389
2σ	22	25	24	18	7
$^{87}Sr/^{86}Sr(t)$	0.709 853	0.706 511	0.711 753	0.707 977	0.716 496
Sm/ppm	4.63	2.95	4.91	7.4	5.72
Nd/ppm	27	12.2	27.6	42.4	29.3

Continued

Sample	GY07	GY16	GY42	GY55	LL120
t/Ma	64	64	64	64	64
$^{147}Sm/^{144}Nd$	0. 10	0. 15	0. 11	0. 11	0. 12
$^{143}Nd/^{144}Nd$	0. 512 123	0. 512 145	0. 512 008	0. 512 097	0. 511 759
2σ	13	23	9	7	5
T_{DM2}/Ga	1. 39	1. 39	1. 55	1. 43	1. 89
$\varepsilon_{Nd}(t)$	-9. 3	-9. 2	-11. 6	-9. 8	-16. 5
U/ppm	9. 62	7. 88	12. 5	9. 88	3. 06
Th/ppm	39. 9	30. 1	49. 8	51. 5	17. 4
Pb/ppm	26. 6	33. 3	28. 5	32. 3	19. 2
$^{206}Pb/^{204}Pb$	19. 236	19. 204	19. 141	19. 036	18. 573
2σ	2	2	2	1	1
$^{207}Pb/^{204}Pb$	15. 704	15. 708	15. 706	15. 734	15. 691
2σ	1	2	1	2	1
$^{208}Pb/^{204}Pb$	39. 338	39. 350	39. 502	39. 557	39. 474
2σ	4	5	3	4	1
$^{238}U/^{204}Pb$	23. 3	15. 3	28. 3	19. 7	10. 2
$^{232}Th/^{204}Pb$	99. 3	59. 8	115. 8	105. 6	59. 4
$^{206}Pb/^{204}Pb(t)$	19. 003	19. 051	18. 859	18. 839	18. 471
$^{207}Pb/^{204}Pb(t)$	15. 693	15. 701	15. 693	15. 725	15. 686
$^{208}Pb/^{204}Pb(t)$	39. 022	39. 159	39. 132	39. 220	39. 285

Rb, Sr, Sm, Nd, U, Th, Pb, Lu and Hf concentrations were analyzed by ICP-MS. T_{DM2} represent the two-stage model age and were calculated using present-day $(^{147}Sm/^{144}Nd)_{DM}$ = 0. 213 7 and $(^{147}Sm/^{144}Nd)_{DM}$ = 0. 513 15. Initial Pb isotopic ratios were calculated for 440 Ma using single-stage model. T_{DM2} were calculated using present-day $(^{176}Lu/^{177}Hf)_{DM}$ = 0. 038 4 and $(^{176}Hf/^{177}Hf)_{DM}$ = 0. 283 25, f_{CC} = −0. 55 and f_{DM} = 0. 16. $\varepsilon_{Hf}(t)$ values were calculated using present-day $(^{176}Lu/^{177}Hf)_{CHUR}$ = 0. 033 2 and $(^{176}Hf/^{177}Hf)_{CHUR}$ = 0. 282 772.

The Guyong granitic rocks have similar Sr-Nd isotopic compositions with $^{87}Sr/^{86}Sr(t)$ = 0. 706 511–0. 711 753, and $\varepsilon_{Nd}(t)$ values of −9. 2 to −11. 6. The Nd two-stage model ages are identical with values of 1. 39 − 1. 55 Ga. The granites also have negligible radiogenic Pb isotopic compositions with $^{206}Pb/^{204}Pb(t)$ = 18. 839 − 19. 051, $^{207}Pb/^{204}Pb(t)$ = 15. 693 − 15. 725, $^{208}Pb/^{204}Pb(t)$ = 39. 022–39. 220.

The Husa granodiorites have high initial $^{87}Sr/^{86}Sr$ ratio of 0. 716 496, negative $\varepsilon_{Nd}(t)$ value of −16. 5 with two-stage model age of 1. 89 Ga. The evolved Pb isotopic compositions show $^{206}Pb/^{204}Pb(t)$ = 18. 471, $^{207}Pb/^{204}Pb(t)$ = 15. 686, $^{208}Pb/^{204}Pb(t)$ = 39. 285.

5. 4 Zircon Hf isotopic compositions

The zircons from granodiorites (LC28) in Husa batholith were analyzed for in situ Lu-Hf isotopes, and the results are listed in Supplementary Table 2 and Fig. 9. The initial $^{176}Hf/^{177}Hf$

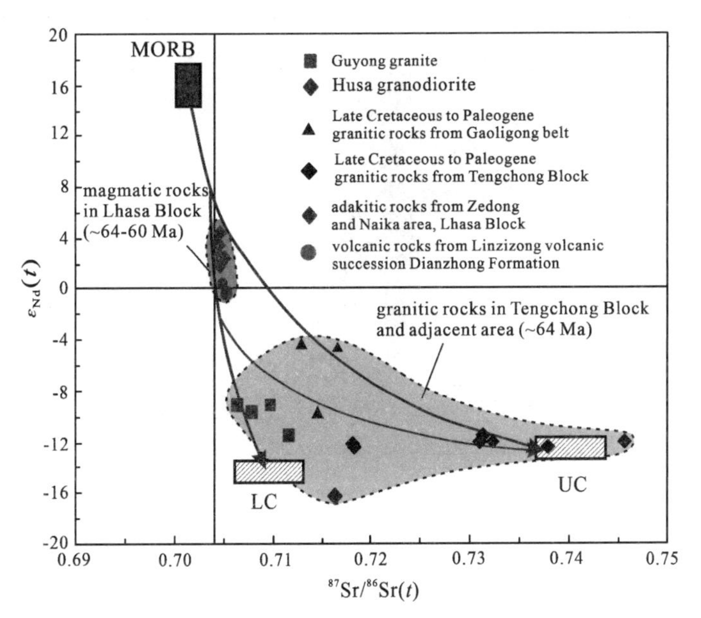

Fig. 7 $^{87}Sr/^{86}Sr(t)$ vs. $\varepsilon_{Nd}(t)$ diagram for the Late Cretaceous to Early Paleocene rocks from the Tengchong Block.

Data for granitic and volcanic rocks in the southern Lhasa Block are from Jiang et al. (2014) and Mo et al. (2007). Data for granitic rocks in the Tengchong Block are from Chen et al. (2015), Zhao et al. (2016a,b), and this study.

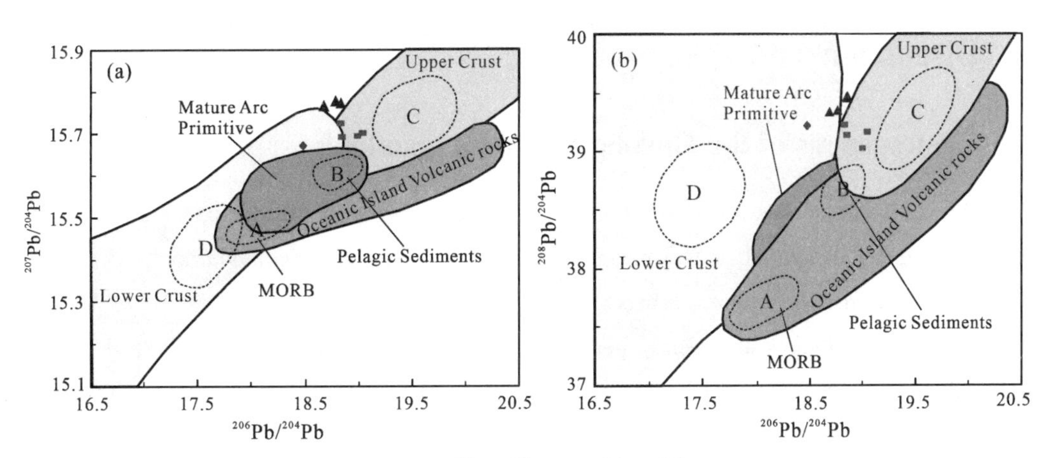

Fig. 8 $^{207}Pb/^{204}Pb$ vs. $^{206}Pb/^{204}$(a) and $^{208}Pb/^{204}Pb$ vs. $^{206}Pb/^{204}Pb$ (b) diagrams for the samples from Guyong and Husa batholiths after Zartman and Doe(1981).

Symbols as in Fig. 7.

ratios and $\varepsilon_{Hf}(t)$ values of the zircons were calculated using their crystallization ages. The twenty-one grains with age of 60−69 Ma show mostly negative $\varepsilon_{Hf}(t)$ values of −18. 1 to −4. 4 with two-stage model ages of 1 416−2 278 Ma, and one grain with age of 66 Ma has positive $\varepsilon_{Hf}(t)$ value of 3. 4 with a single-stage model age of 683 Ma. The other three grains is inherited zircon or xenocrysts and have $\varepsilon_{Hf}(t)$ values of −6. 9 to −1. 3.

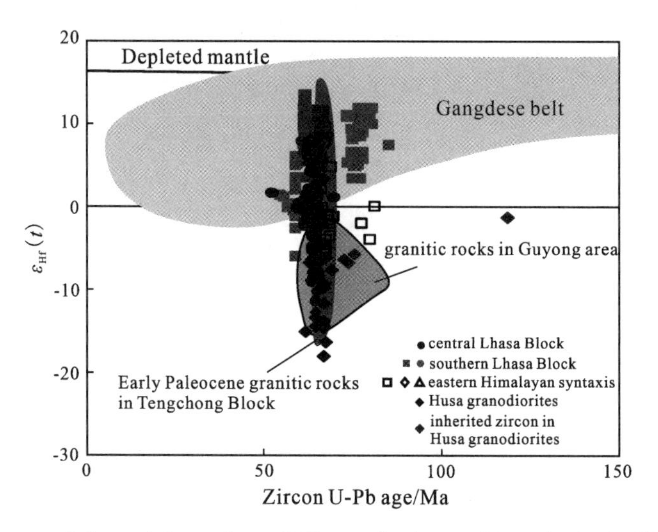

Fig. 9 Diagram of the $\varepsilon_{Hf}(t)$ vs. zircon U-Pb age for the granitic rocks in the Tengchong Block, eastern Himalayan syntaxis, and southern Lhasa Block.

Data for the Gangdese belt from Ji et al. (2009). Data for the southern Lhasa Block from Jiang et al. (2014), Ji et al. (2014), Hou et al. (2015), Zheng et al., (2014), Zhu et al. (2011), Huang et al. (2010). Data for central Lhasa Block from Hou et al. (2015), Gao et al. (2011), Zheng et al. (2015), and Wang et al. (2012). Data for the eastern Himalayan synataxis from Chui et al. (2009) and Guo et al. (2011, 2012), Pan et al. (2016). Data for the Tengchong Block including Guyong area from Xu et al. (2012), Qi et al. (2015), Chen et al. (2015), and Xie et al. (2016).

6 Discussion

6.1 Petrogenesis of the Guyong and Husa granitoids

The Late Cretaceous to Early Paleocene granitic rocks of the Tengchong Block are subdivided into the high-silica Guyong granites and low-silica Husa granodiorites, which may have been derived from different sources.

The Guyong granites are weakly peraluminous and high-K calc-alkaline. They show high SiO_2(70.44–76.45 wt%) and K_2O/Na_2O (1.18–1.60), extremely low MgO (0.06–0.65 wt%) and $Mg^{\#}$ values (11–40). These granites are enriched in LREE, flat in HREE but weakly depleted in MREE (Fig. 6b). In addition, negative $\varepsilon_{Nd}(t)$ values (−11.6 to −9.2) and high initial $^{87}Sr/^{86}Sr$ values (0.706 511–0.711 753), indicate the granites were derived from a crustal source (Fig. 7), as also indicated by Pb isotopic data (Fig. 8). In addition, the Guyong granites have variable CaO/Na_2O ratios of 0.12–0.7, high Rb concentrations and Rb/Sr ratios of 1.79–32.3 (Fig. 10). Together with the negative $\varepsilon_{Nd}(t)$ values and high initial Sr ratios, these signatures indicate that the granitoids from the Guyong batholith were derived from partial melting of a pelite or mixed pelite-greywacke sources. Based on the Nd two-stage model ages, we propose that the provenance rocks were Meso-Proterozoic in age. The granites are depleted in high field strength elements (HFSEs: e.g., Nb, Ta, Hf, P, and Ti), and enriched in large ion lithophile elements (LILE: e.g., Rb, K), which is characteristic of arc magmatism. However, the presence of residual amphibole and accessory ilmenite in the source would lead to depletion in HFSE. Furthermore, the negative Eu anomalies indicate that the source contains residual plagioclase.

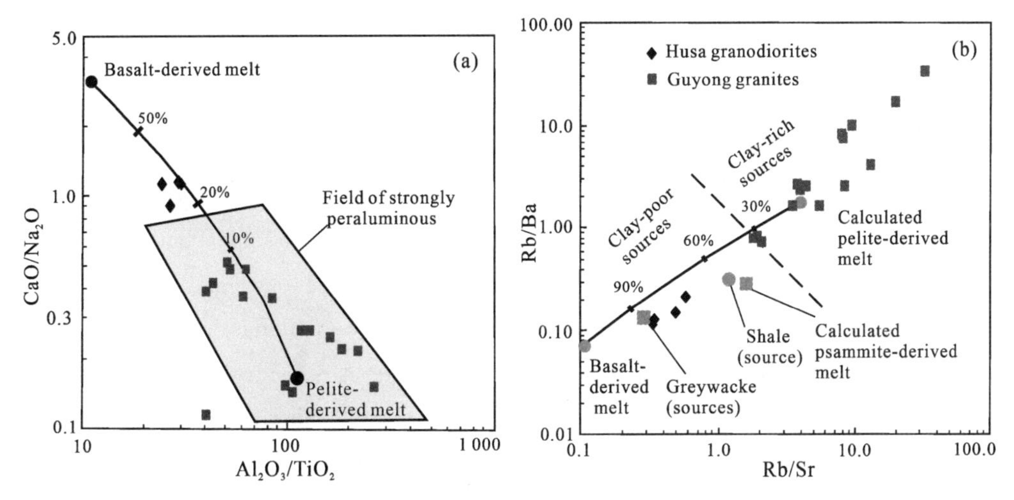

Fig. 10 Diagrams of Al_2O_3/TiO_2 vs. CaO/Na_2O (a) and Rb/Sr vs. Rb/Ba (b) after Sylvester (1998) for the samples from Guyong and Husa batholith.

Compared with the Guyong granites, the Husa granodiorites have relatively low SiO_2 (64.50–66.99 wt%) contents and contain a lager number of hornblende ca. 8% in volume. They belong to the high-K calc-alkaline series and metaluminous geochemical signatures. Their evolved Sr-Nd-Pb isotopic characteristics indicate the sources contained abundant crustal material (Figs. 7 and 8). Due to the nearly identical composition of granodiorites in Husa batholith, the fractional crystallizing is unconsidered during generation of granodioritic magma. Granodiorites form the large batholiths at active continental margins and collisional orogens (Castro, 2013), and the available experimental data suggest two scenarios for the petrogenesis of high-K calc-alkaline granodiorites: ① Partial melting of the previous existed hydrated, calc-alkaline, and high-K calc-alkaline andesite and basaltic andesite (Roberts and Clemens, 1993; Sisson et al., 2005). And ② hybrid model, which include two different hybridizations: (i) the mixing between mantle-derived mafic magma and crust-derived granitic magma (e.g., Castro, 1991; Davis and Hawkesworth, 1993); and (ii) recycling of the binary mixing in the sources, involving in the metasedimentary (ancient crustal rocks) and metaigneous rocks (juvenile crustal rocks). Mixing model would result in broad variability in zircon Hf isotopic ratios, up to 10 epsilon units (Kemp et al., 2007; Bolhar et al., 2008). The Hf isotopic data of the Husa granodiorites are scattered with the $\varepsilon_{Hf}(t)$ values from −18.1 to +3.4, >20 epsilon units for individual sample, which suggests that the Husa granodiorites were derived from the partial melting of a mixed source of metaignous and metasedimentary rocks or a mixing of mantle-derived mafic and crust-derived granitic magma. Xie et al. (2016) reported that the Early Paleocene monzogranites of Tengchong Block have SiO_2 contents of 70.20 wt% and 77.40 wt% with the corresponding $\varepsilon_{Hf}(t)$ values of +0.5 to +3.1 and −6.9 to +6.5 respectively. The implication is that the juvenile mafic rocks are present in the Tengchong Block, consistent with the presence of Triassic (245 Ma) gabbros-diorites with $\varepsilon_{Hf}(t)$ values of +7.8 to +14.9 in the Nabang area, western Tengchong Block (Huang et al., 2013). Thus, we conclude that the Husa granodiorites were derived from partial melting of the mixed sources, including the metasedimentary and metaigneous rocks (juvenile crustal rocks) in the lower crust. However, we cannot rule out the mixing of mantle-derived mafic and crust-derived granitic magma. The granodiorites are enriched in incompatible elements and LILEs, and depleted in HFSEs (Fig. 6), and have high HREE (Yb = 2.6–3.8 ppm) and Y (28.3–39.5 ppm) contents, suggesting that amphibole rather than garnet is a residual phase in the source. The moderately negative Eu anomalies and low Sr in the granodiorites further indicate that melting of the source rock occurred within the stability field of plagioclase, whereas garnet was unstable, since garnet is present as a residual phase at pressures ⩾ 12.5 kbar during dehydration-melting of gneiss and quartz amphibolite (Patiño Douce and Beard, 1995). This implies that the crust from which the granodiorites were derived was thin (<40 km).

6.2 Melting depth in Late Cretaceous to Early Paleocene magmatic arc

The Guyong and Husa granitoids yielded Late Cretaceous to Early Paleocene ages (~64 Ma), and coeval intermediate to acidic magmatism is widespread in the Tengchong Block (e.g., Xu et al., 2012; Eroğlu et al., 2013; Qi et al., 2015; Chen et al., 2015; Zhao et al., 2016a) and adjacent central-southern Lhasa Block and eastern Himalayan syntaxis (e.g., Hou et al., 2015; Ji et al., 2014; Mo et al., 2007; Jiang et al., 2014; Guo et al., 2011, 2012; Pan et al., 2014, 2016; Zheng et al., 2014, 2015; Zhu et al., 2011a). These regions recorded the evolution of eastern Neo-Tethys. Late Cretaceous to Early Paleocene igneous rocks were emplaced as batholith in the Guyong-Houqiao and Husa areas, and magmatism in the Tengchong Block is dominated by 76–64 Ma granites and granodiorites (Xu et al., 2012; Eroğlu et al., 2013; Qi et al., 2015; Chen et al., 2015; Zhao et al., 2016a). Late Cretaceous to Early Paleocene crustal origin igneous rocks in the central-southern Lhasa Block and eastern Himalayan syntaxis are dominated by 62–76 Ma diorite to granite (Ji et al., 2014; Wen et al., 2008; Kapp et al., 2005; Pan et al., 2014, 2016; Zheng et al., 2014, 2015; Hou et al., 2015) and equivalent 60.6–64.5 Ma andesite in the Dianzhong Formation, Linzizong volcanic succession (Zhou et al., 2004; Mo et al., 2003, 2007; He et al., 2007;

Lee et al., 2009). The geochemical characteristics of these igneous rocks could be used to constrain the relative depths of melting in the arc crust during Late Cretaceous to Early Paleocene subduction of the Neo-Tethyan oceanic crust (Fig. 11; Mo et al., 2007; Zhu et al., 2011a; Ji et al., 2014; Chiaradia, 2015).

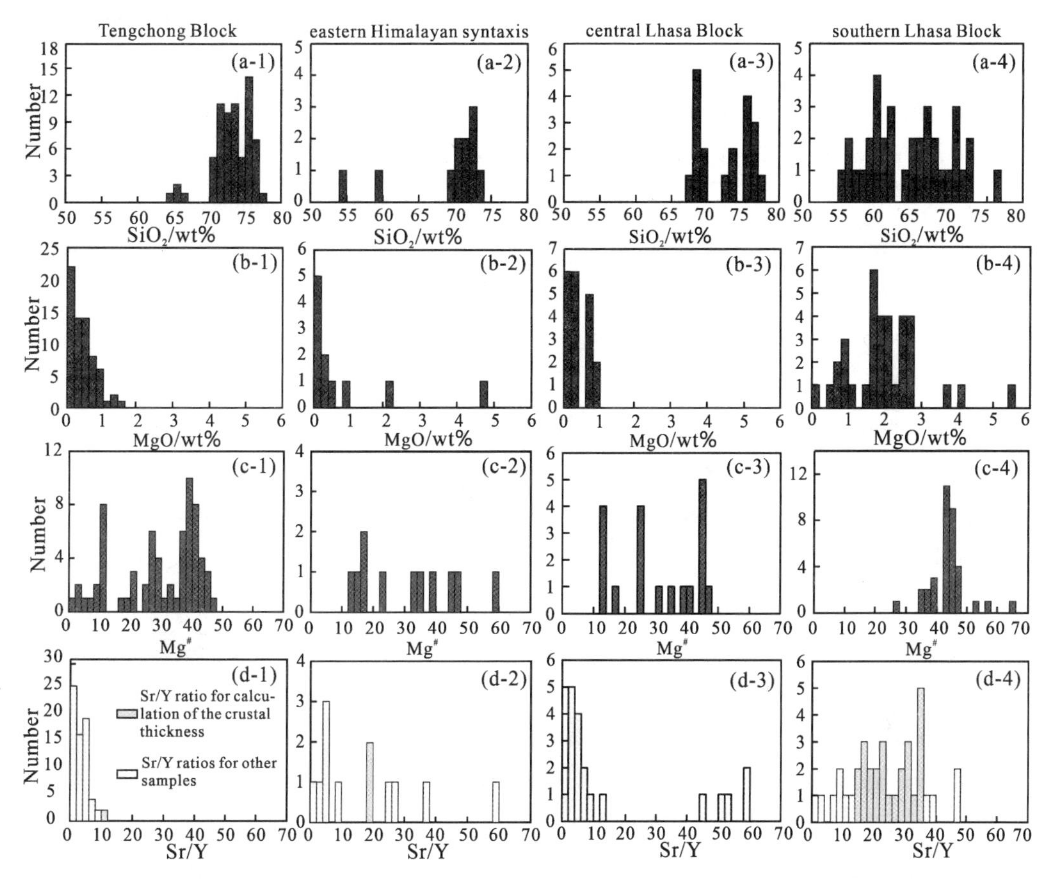

Fig. 11 Histogram of SiO_2, MgO, $Mg^{\#}$, and Sr/Y ratios, for the Late Cretaceous to Early Paleocene (ca. 64 Ma) igneous rocks from the southern Lhasa Block, eastern Himalayan syntaxis, and Tengchong Block.

Data for the southern and central Lhasa Block from Ji et al. (2009, 2012, 2014), Jiang et al. (2014), Huang et al. (2010), and Mo et al. (2007, 2008), Hou et al. (2015), Zheng et al. (2014, 2015), Wang et al. (2012), Zhu et al. (2011a). Data for the eastern Himalayan syntaxis from Guo et al. (2011, 2012), Pan et al. (2014, 2016), and Lin et al. (2012). Data for Tengchong Block from Qi et al. (2015), Chen et al. (2015), and Zhao et al (2016a).

Late Cretaceous to Early Paleocene granitic rocks in the Tengchong Block have relatively high SiO_2(64.50–77.37 wt%), K_2O/Na_2O (0.4–2.8, with most of them >1), Y (13.9–182.4 ppm) and Yb (1.1–21 ppm), low MgO (0.01–1.53 wt%) and $Mg^{\#}$(2–40), Sr/Y (0.1–11.5) (Qi et al., 2015; Chen et al., 2015; Zhao et al., 2016a; this study; Fig. 11a-1 to d-1). In contrast, the crust-derived rocks in the southern Lhasa Block range in lithology from diorites to granites (Ji et al., 2012, 2014; Jiang et al., 2014; Kapp et al., 2005; Zheng et al., 2014; Hou et al., 2015), along with the equivalent volcanics (Mo et al., 2003, 2007). They have variable SiO_2 contents varying from 54.38 wt% to 76.03 wt%, and the most is <70 wt%, and are relatively enriched in Na_2O and Na_2O/K_2O ratios of 0.2–1.5, and the most is >1. The intermediate to acidic rocks also contain relatively high MgO (0.18–5.43 wt%) and $Mg^{\#}$(27–64), and high and variable Sr/Y ratios of 3.2–47.0. In the central Lhasa Block, the granitic rocks are characterized by SiO_2 contents of 68.00–77.01 wt%, MgO contents of 0.08–0.96 wt% with $Mg^{\#}$ ratios of 12–46, variable Sr/Y ratios of 1–58. The granitic rocks in the eastern Himalayan syntaxis have variable SiO_2(54.60–73.05 wt%) and MgO (0.11–4.78 wt%), with $Mg^{\#}$ values of 13–59 and Sr/Y ratios of 1.5–58.1. In principle, the more mafic rocks should be derived from deeper levels. Finally, the granitic rocks in the southern Lhasa Block are characterized by relatively low SiO_2 contents, K_2O/Na_2O ratios and relatively high MgO contents, $Mg^{\#}$ values and Sr/Y ratios, while granites in the eastern Himalayan syntaxis are characterized by intermediate values for these parameters, and the central Lhasa and Tengchong blocks show relatively high SiO_2 and K_2O/Na_2O, with low MgO, $Mg^{\#}$, and Sr/Y ratios (Fig. 11). Thus, we conclude that partial melting during the Late Cretaceous to early Paleocene occurred at relatively deep levels in the southern Lhasa Block, intermediate levels in the eastern Himalayan syntaxis, and shallow levels in the central Lhasa and Tengchong blocks.

Furthermore, the proportion of mafic magma, derived from mantle or partial melting of juvenile crust, mixed with crust-derived magma gives an indication of the relative depth from which magma originated. In particular, the in situ analysis of Hf isotope in zircon reveal whether a magma was derived from recycling of ancient mature crustal materials or juvenile crustal materials (Griffin et al., 2002; Kemp et al., 2006; Scherer et al., 2007; Zhu et al., 2011a). Similarly to Hf isotopes, whole-rock Nd isotopic compositions are influenced by, and can be used to characterize, the magma source. Therefore, we collected the zircon Hf and whole-rock Nd isotopic data of the Late Cretaceous to Early Paleocene intermediate to acid rocks in these regions from previous studies (Figs. 9 and 12). The $\varepsilon_{Hf}(t)$ of granitic rocks in the Tengchong Block are transition from positive to negative with the values ranging from −18.1 to 6.5 (Qi et al., 2015; Xu et al., 2012; Chen et al., 2015a; Xie et al., 2016; this study), and corresponding $\varepsilon_{Nd}(t)$ values of −16.5 to −4.4 (Chen et al., 2015; Zhao et al., 2016a this study; Fig. 12). The $\varepsilon_{Hf}(t)$ and $\varepsilon_{Nd}(t)$ of granitic rocks in central Lhasa Block are similar to those of the Tengchong granites, ranging from −13.9 to 6.4 and −13.48 to −3.90, respectively, (Zheng et al., 2015; Wang et al., 2012; Figs. 9 and 12). This indicates that ancient crustal melt combined with a relatively small proportion of magma derived from juvenile crust in the central Lhasa and Tengchong Block. In the southern Lhasa Block, the $\varepsilon_{Hf}(t)$ values are positive, ranging from 2.3 to 13.6 (Ji et al., 2009, 2012, 2014; Jiang et al.,

2014; Huang et al., 2010; Zhu et al., 2011; Zheng et al., 2014; Hou et al., 2015; Fig. 12), with $\varepsilon_{Nd}(t)$ values of −2.6 to 4.3 (Jiang et al., 2014; Mo et al., 2003, 2007; Fig. 12). Late Cretaceous to Early Paleocene granitic rocks in the eastern Himalayan syntaxis have $\varepsilon_{Hf}(t)$ values of −9.6 to 8.0 (Guo et al., 2011, 2012; Pan et al., 2014, 2016), and $\varepsilon_{Nd}(t)$ values of −10.5 to −1.7 (Fig. 12). This indicates that Late Cretaceous to Early Paleocene granitic rocks in the southern Lhasa Block were derived from the partial melting of juvenile crustal rocks, while those in the eastern Himalayan syntaxis were derived from the partial melting of mixed sources. The temperature required to melt juvenile crustal rocks (mafic rocks) is higher than that required to melt ancient crust (Rapp and Watson, 1995). Together with the geochemical indexes of the crust-derived granitic rocks, it is concluded that the depth of partial melting occurred at deep levels in the southern Lhasa Block, moderate levels in eastern Himalayan syntaxis, and relatively shallow levels in central Lhasa and Tengchong Block.

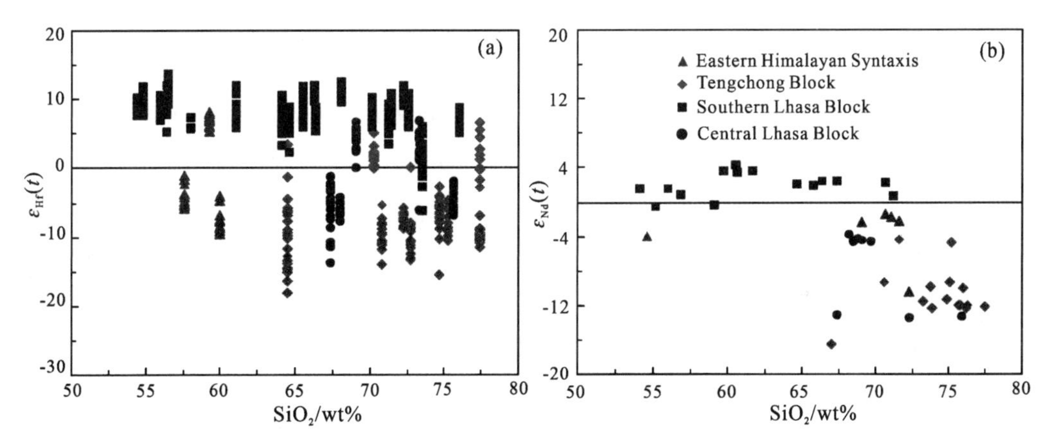

Fig. 12 Diagrams of SiO_2 vs. $\varepsilon_{Hf}(t)$ and $\varepsilon_{Nd}(t)$ for the Late Cretaceous to Early Paleocene (ca. 64 Ma) igneous rocks from the southern Lhasa Block, eastern Himalayan syntaxis, and Tengchong Block.

Data for the southern and central Lhasa Block from Ji et al. (2009, 2012, 2014), Jiang et al. (2014), Huang et al. (2010), and Mo et al. (2007, 2008), Hou et al. (2015), Zheng et al. (2014, 2015), Wang et al. (2012), Zhu et al. (2011a). Data for the eastern Himalayan syntaxis from Guo et al. (2011, 2012), Pan et al. (2014, 2016), and Lin et al. (2012). Data for the Tengchong Block from Qi et al. (2015), Chen et al. (2015), Zhao et al (2016a), and Xie et al. (2016).

6.3 Crust thickness in Late Cretaceous to Early Paleocene magmatic arc

Several geochemical indexes have been used to calculate the crust thickness of magmatic arc (e.g., Condie, 1982; Mantle and Collins, 2008; Chiaradia, 2015; Chapman et al., 2015). These indexes are all calibrated on a regional to global scale, especially using the Sr/Y ratios of the intermediate to acidic rocks in magmatic arcs (Chiaradia, 2015; Chapman et al.,

2015). Sr/Y ratios are controlled by the stability of minerals possessed Sr and Y, a function of the pressure at which magma evolves (Chiaradia, 2015). During partial melting of lower crustal rocks (including igneous and metamorphic), or magmatic fraction of mantle-derived mafic magmas, Sr is compatible at low pressure (< 10 kabr) where it strongly partitions into plagioclase, but at high pressure (> 12 kbar), where plagioclase is unstable, Sr is incompatible and partitions into the liquid phase (e.g., Kay and Mpodozis, 2001). On the contrary, Y is compatible at high pressure and partitions into garnet or/and amphibole, but at low pressure is incompatible and enter the liquid phase (e.g., Lee et al., 2007). Chapman et al. (2015) investigated variations in crust thickness in the U.S. Cordillera using the Sr/Y from intermediate continental arc magmas. The corrected and available equation of global correlation between geophysically determinend Moho depth and Sr/Y is

$$y=0.90x-7.25$$

where y represents Sr/Y and x represents crustal thickness or Moho depth. In applying this equation, the data were filtered to remove SiO_2 wt% <55, SiO_2 wt% >70, MgO wt% <1, MgO wt% >6, Rb/Sr >0.35, Rb/Sr <0.05. This could potentially removes mafic rocks originating from the mantle, and more felsic rocks, including leucogranites, that formed by partial melting of metasedimentary rocks in the middle to upper crust (Chapman et al., 2015).

Although melting of subducted oceanic crust can also form intermediate magmas with high Sr/Y ratios (Defant and Durmmond, 1990), the geochemical and isotopic data of the Late Cretaceous to Early Paleocene igneous rocks in the southern-central Lhasa Block, eastern Himalayan syntaxis, and Tengchong Block indicate that they contain a component of continental crust (e.g., Hou et al., 2015; Jiang et al., 2014; Ji et al., 2014; Mo et al., 2007; Mo et al., 2008; Guo et al., 2012; Lin et al., 2012; Pan et al., 2014, 2016; Qi et al., 2015; Chen et al., 2015; Zhao et al., 2016a; Zheng et al., 2014, 2015; Zhu et al., 2011; this study). The appropriate data used to calculate crust thickness are listed in Table 4. Data for rocks from central Lhasa Block cannot be applied because they do not satisfy the conditions of calculation. Based on our calculations, the crust thickness of the Tengchong Block during Late Cretaceous to Early Paleocene was 20–21 km. Crustal thickness in the eastern Himalayan syntaxis and northernmost part of the southern Lhasa Block was ~30 km, while the crust was thickened (42 km) in the southernmost parts of the southern Lhasa Block. These results are consistent with pre-collision crustal thickening in southern Lhasa Block during the northward subduction of Neo-Tethys and magmatic activity (Ratschbacher et al., 1993). The implication is that different situations and mechanisms for the northward and eastward subduction of the Neo-Tethyan oceanic lithosphere transitioned from southern-central Lhasa Block through eastern Himalayan syntaxis to Tengchong Block.

Table 4 Result of calculating the crust thickness using the intermediate to acid igneous rock of ca. 64 Ma from the southern Lhasa Block, eastern Himalayan syntaxis and Tengchong Block.

Sample	Rb/Sr	SiO_2	MgO	Sr/Y	Crust thickness	Reference
Tengchong Block						
LL120	0.34	66.99	1.19	11	20	This study
LL121	0.34	65.37	1.34	11	21	This study
Eastern Himalayan syntaxis						
T656	0.14	59.24	2.15	20	30	Guo et al., 2011
T1224	0.18	57.6	3.09	20	30	Pan et al., 2016
Southern Lhasa Block						
Northernmost part of southern Lhasa Block						
08FW51	0.27	66.35	1.97	15	24	Ji et al., 2012
D-2	0.09	59.16	3.66	19	30	Mo et al., 2008
D-15	0.13	58.05	2.53	17	26	Mo et al., 2008
BD-145	0.18	61.57	1.77	16	26	Mo et al., 2008
BD-160	0.13	58.77	1.77	25	36	Mo et al., 2008
BD-123	0.09	56.97	2.3	18	28	Mo et al., 2007
Lz9913	0.08	55.21	4.01	21	31	Mo et al., 2007
LS5	0.08	56.09	4.01	20	30	Huang et al., 2010
Southernmost part of southern Lhasa Block						
07TB22a	0.20	66.39	1.61	30	41	Jiang et al., 2014
07TB22b	0.21	66.37	1.68	29	40	Jiang et al., 2014
07TB23	0.27	67.39	1.53	27	38	Jiang et al., 2014
07TB24c	0.17	65.85	1.86	34	46	Jiang et al., 2014
07TB24d	0.24	67.18	1.73	30	41	Jiang et al., 2014
07TB25a	0.18	64.67	2.06	31	43	Jiang et al., 2014
07TB26a	0.20	64.14	2.12	29	40	Jiang et al., 2014
07TB27	0.15	63.9	2.1	32	44	Jiang et al., 2014
07TB28a	0.05	59.79	2.57	35	47	Jiang et al., 2014
07TB29a	0.06	59.12	2.67	34	46	Jiang et al., 2014
07TB30b	0.05	59.44	2.79	36	48	Jiang et al., 2014
07TB32	0.07	60.59	2.55	35	47	Jiang et al., 2014
09TB119	0.11	61.7	2.67	21	31	Jiang et al., 2014
09TB120	0.12	61.52	2.7	23	34	Jiang et al., 2014
NML03-1	0.06	58.03	3.14	27	38	Zhu et al., 2011a,b

7 Conclusions

(1) The Guyong and Husa batholith are dominated by granites, biotite granites and granodiorites that were emplaced at ca. 64 Ma. The Guyong granites may have been derived from the partial melting of ancient crustal pelite or mixed pelite-greywacke sources. The Husa granodiorites were likely derived from the partial melting of the mixed sources comprising

metasedimentary and metaigneous rocks.

(2) The depth of partial melting beneath the Late Cretaceous to Early Paleocene arc was relatively deep in the southern Lhasa Block, moderate in eastern Himalayan syntaxis, and shallow in the central Lhasa and Tengchong blocks.

(3) The crust was thin in the Tengchong Block, normal in the eastern Himalayan syntaxis and northern parts of the southern Lhasa Block, and thickened in the southernmost part of the southern Lhasa Block.

Acknowledgements This work was jointly supported by the National Natural Science Foundation of China (Grant Nos. 41421002, 41102037 and 41190072), program for Changjiang Scholars and Innovative Research Team in University (Grant IRT 1281), the Foundation for the Author of National Excellent Doctoral Dissertation of PR China (201324), and Postgraduate Independent Innovative Program in Northwest University (YZZ14015).

References

Aitchison, J.C., Ali, J.R., Davis, A.M., 2007. When and where did India and Asia collide? J. Geophys. Res. Solid Earth, 112 (B5).

Andersen, T., 2002. Correction of common lead in U-Pb analyses that do not report ^{204}Pb. Chem. Geol., 192, 59-79.

Bolhar, R., Weaver, S.D., Whitehouse, M.J., Palin, J.M., Woodhead, J.D., Cole, J.M., 2008. Sources and evolution of arc magmas inferred from coupled O and Hf isotope systematics of plutonic zircons from the Cretaceous Separation Point Suite (New Zealand). Earth Planet. Sci. Lett., 268, 312-324.

Cai, F.L., Ding, L., Yue, Y.H., 2011. Provenance analysis of upper Cretaceous strata in Tethys Himalaya, southern Tibet: Implications for timing of India-Asia collision. Earth Planet. Sci. Lett., 305, 195-206.

Cao, H.W., Zhang, S.T., Lin, J.Z., Zheng, L., Wu, J.D., Li, D., 2014. Geology, geochemistry and geochronology of the Jiaojiguanliangzi Fe-polymetallic deposit, Tengchong County, Western Yunnan (China): Regional tectonic implications.J. Asian Earth Sci., 81, 142-152.

Cao, H.W., Zou, H., Zhang, Y.H., Zhang, S.T., Zheng, L., Zhang, L.K., Tang, L., Pei, Q.M., 2016. Late Cretaceous magmatism and related metallogeny in the Tengchong area: Evidence from geochronological, isotopic and geochemical data from the Xiaolonghe Sn deposit, western Yunnan, China. Ore Geol. Rev., 78, 196-212.

Castro, A., 2013. Tonalite-granodiorite suites as cotectic systems: A review of experimental studies with applications to granitoid petrogenesis. Earth-Sci. Rev., 124, 68-95.

Castro, A., Moreno-Ventas, I., de la Rosa, J.D., 1991. H-type (hybrid) granitoids: A proposed revision of the granite-type classification and nomenclature. Earth-Sci. Rev., 31, 237-253.

Chapman, J.B., Ducea, M.N., DeCelles, P.G., Profeta, L., 2015. Tracking changes in crustal thickness during orogenic evolution with Sr/Y: An example from the North American Cordillera. Geology, 43, 919-922.

Chen, J.C., 1991. Characteristics of Pb, Sr isotopic compositions in west Yunnan. Geol. Sci.,2, 174-183 (in

Chinese with English abstract).

Chen, X.C., Hu, R.Z., Bi, X.W., Zhong, H., Lan, J.B., Zhao, C.H., Zhu, J.J., 2015. Petrogenesis of metaluminous A-type granitoids in the Tengchong-Lianghe tin belt of southwestern China: Evidences from zircon U-Pb ages and Hf-O isotopes, and whole-rocks Sr-Nd isotopes. Lithos, 212-215, 93-110.

Chiaradia, M., 2015. Crustal thickness control on Sr/Y signatures of recent arc magmas: An earth scale perspective. Sci. Rep., 5.

Chu, M.F., Chung, S.L., Song, B., Liu, D.Y., O'Reilly, S.Y., Pearson, N.J., Ji, J.Q., Wen, D.R., 2006. Zircon U-Pb and Hf isotope constraints on the Mesozoic tectonics and crustal evolution of southern Tibet. Geology, 34, 745-748.

Chu, M.F., Chung, S.L., O'Reilly, S.Y., Pearson, N.J., Wu, F.Y., Li, X.H., Liu, D.Y., Ji, J.Q., Chu, C.H., Lee, H.Y., 2011. India's hidden inputs to Tibetan orogeny revealed by Hf isotopes of Transhimalayan zircon and host rocks. Earth Planet. Sci. Lett., 307, 479-486.

Chui, H.Y., Chung, S.L., Wu, F.Y., Liu, D.Y., Liang, Y.H., Lin, I.J., Iizuka, Y., Xie, L.W., Wang, Y.B., Chu, M.F., 2009. Zircon U-Pb and Hf isotopic constraints from eastern Transhimalayan batholiths on the precollisional magmatic and tectonic evolution in southern Tibet. Tectonophysics, 477, 3-19.

Chung, S.L., Chu, M.F., Zhang, Y.Q., 2005. Tibet tectonic evolution inferred from spatial and temporal variations in post-collisional magmatism. Earth Sci. Rev., 68, 173-196.

Condie, K.C., 1982. Plate Tectonics and Crustal Evolution. New York: Pergamon, 310.

Cong, B.L., Wu, G.Y., Zhang, Q., Zhang, R.Y., Zhai, M.G., Zhao, D.S., Zhang, W.H., 1993. Petro-tectonic evolution of the Paleo-Tethys in the west Yunnan, China. Sci. China B, 23, 1201-1207 (in Chinese with English abstract).

Davis, J., Hawkesworth, C., 1993. The petrogenesis of 30-20 Ma basic and intermediate volcanics from the Mogollon-Datil Volcanic Field, New Mexico, USA. Contrib. Mineral. Petrol., 115, 165-183.

Defant, M.J., Drummond, M.S., 1990. Derivation of some modern arc magmas by melting of young suducted lithosphere. Nature, 347, 662-665.

Ding, L., Zhong, D.L., Yin, A., Kapp, P., Harrison, T.M., 2001. Cenozoic structural and metamorphic evolution of the eastern Himalayan syntaxis (Namche Barwa). Earth Planet. Sci. Lett.,192, 423-438.

Ding, L., Kapp, P., Zhong, D.L., Deng, W.M., 2003. Cenozoic Volcanism in Tibet: Evidence for a Transition from Oceanic to Continental Subduction. J. Petrol., 44, 1833-1865.

Ding, L., Kapp, P., Wan, X.Q., 2005. Paleocene-Eocene record of ophiolite obduction and initial India-Asia collision, south central Tibet. Tectonics, 24, 1-18.

Eroğlu, S., Siebel, W., Danišík, M., Pfänder, J.A., Chen, F.K., 2013. Multi-system geochronological and isotopic constraints on age and evolution of the Gaoligongshan metamorphic belt and shear zone system in western Yunnan, China. J. Asian Earth Sci., 73, 218-239.

Fang, C.J., Zhang, Y.F., 1994. The structure and tectonics of western Yunnan. J. SE Asian Earth Sci., 9, 355-361.

Gao, Y.M., Chen, Y.C., Wang, C.H., Hou, K.J., 2011. Zircon Hf isotopic characteristics and constraints on petrogenesis of Mesozoic-Cenozoic magmatic rocks in Nyainqentanglha region, Tibet. Miner. Deposits, 30 (2), 279-291 (in Chinese with English abstract).

Gehrel, G.E., DeCelles, P.G., Ojha, T.P., Upreti, B.N., 2006a, Geologic and U-Th-Pb geochronologic evidence for early Paleozoic tectonism in the Kathmandu thrust sheet, central Nepal Himalaya. Geol. Soc. Am. Bull., 118, 185-198.

Gehrel, G. E., DeCelles, P. G., Ojha, T. P., Upreti, B. N., 2006b. Geologic and U-Pb geochronologic evidence for early Paleozoic teconism in the Dadeldhura thrust sheet, far-west Nepal Himalaya. J. Asian Earth Sci., 28, 385-408.

Gehrels, G.E., DeCelles, P.G., Martin, A., Ojha, T.P., Pinhassi, G., Upreti, B.N., 2003. Initiation of Himalayanorogeny as an early Paleozoic thin-skinned thrust belt. GSA Today, 13(9), 4-9.

Griffin, W.L., Wang, X., Jackson, S.E., Pearson, N.J., O'Reilly, S.Y., Xu, X., Zhou, X., 2002. Zircon chemistry and magma mixing, SE China: In-situ analysis of Hf isotopes, Tonglu and Pingtan igneous complexes. Lithos, 61, 237-269.

Guo, L., Zhang, H.F., Harris, N., Pan, F.B., Xu, W.C., 2011. Origin and evolution of multi-stage felsic melts in eastern Gangdese belt: Constraints from U-Pb zircon dating and Hf isotopic composition. Lithos, 127, 54-67.

Guo, L., Zhang, H.F., Harris, N., Randall, P., Xu, W.C., Shi, Z.L., 2012. Paleogene crustal anatexis and metamorphism in Lhasa terrane, eastern Himalayan syntaxis: Evidence from U-Pb zircons ages and Hf isotopic compositions of the Nyingchi Complex. Gondwana Res., 21, 100-111.

Guo, Z.F., Cheng, Z.H., Zhang, M.L., Zhang, L.H., Li, X.H., Liu, J.Q., 2015. Post-collisional high-K calc-alkaline volcanism in Tengchong volcanic field, SE Tibet: Constraints on Indian eastward subduction and slab detachment. J. Geol. Soc., 172, 624-640.

He, S.D., Kapp, P., DeCelles, P.G., Gehrels, G.E., Heizler, M., 2007. Cretaceous-Tertiary geology of the Gandgese Arc in the Linzhou area, southern Tibet. Tectonophysics, 433, 15-37.

He, R.Z., Zhao, D.P., Gao, R., Zheng, H.W., 2010. Tracing the Indian lithospheric mantle beneath central Tibetan Plateau using teleseismic tomography. Tectonophysics, 491, 230-243.

Hou, Z.Q., Duan, L.F., Lu, Y.J., Zheng, Y.C., Zhu, D.C., Yang, Z.M., Yang, Z.S., Wang, B.D., Pei, Y.R., Zhao, Z.D., Campbell McCuaig, T., 2015. Lithospheric architecture of the Lhasa terrane and its control on ore deposits in the Himalayan-Tibetan Orogen. Econ. Geol., 110, 1541-1575.

Hu, X. M., Sinclair, H. D., Wang, J. G., Jiang, H. H., Wu, F. Y., 2012. Late Cretaceous-Palaeogene stratigraphic and basin evolution in the Zhepure Mountain of southern Tibet: Implications for the timing of India-Asia initial collision. Basin Res., 24, 520-543.

Hu, X.M., Garzanti, E., Moore, T., Raffi, I., 2015. Direct stratigraphic dating of India-Asia collision onset at the Selandian (middle Paleocene, 59±1 Ma). Geology, 43, 859-862.

Huang, Y., Zhao, Z.D., Zhang, F.Q., Zhu, D.C., Dong, G.C., Zhou, S., Mo, X.X., 2010. Geochemistry and implication of the Gangdese batholiths from Renbu and Lhasa area in southern Gangdese, Tibet. Acta Petrol. Sin., 26, 3131-3142 (in Chinese with English abstract).

Huang, Z.Y., Qi, X.X., Tang, G.Z., Liu, J.K., Zhu, L.H., Hu, Z.C., Zhao, Z.Y., Zhang, C., 2013. The identification of early Indosinian tectonic movement in Tengchong block, western Yunnan: Evidence of zircon U-Pb dating and Lu-Hf isotope for Nabang diorite. Geol. China, 40, 730-741 (in Chinese with English abstract).

Ji, W.Q., Wu, F.Y., Liu, C.Z., Chung, S.L. Li, J.X., Liu, C.Z., 2009. Zircon U-Pb geochronology and Hf isotopic constraints on petrogenesis of the Gangdese batholith, southern Tibet. Chem. Geol., 262, 229-245.

Ji, W.Q., Wu, F.Y., Liu, C.Z., Chung, S.L., 2012. Early Eocene crustal thickening in southern Tibet: New age and geochemical constraints from the Gangdese batholith. J. Asian Earth Sci., 53, 82-95.

Ji, W.Q., Wu, F.Y., Chung, S.L., Liu, C.Z., 2014. The Gangdese magmatic constraints on a latest Cretaceous lithospheric delamination of the Lhasa terrane, southern Tibet. Lithos, 210-211, 168-180.

Jiang, Z.Q., Wang, Q., Wyman, D.A., Li, Z.X., Yang, J.H., Shi, X.B., Ma, L., Tang, G.J., Gou, G.J., Jia, X.H., Guo, H.F., 2014. Transition from oceanic to continental lithosphere subduction in southern Tibet: Evidence from the Late Cretaceous-Early Oligocene (~91–30 Ma) intrusive rocks in the Chanang-Zedong area, southern Gangdese. Lithos, 196-197, 213-231.

Kapp, J.L.D.A., Harrison, M., Kapp, P., Grove, M., Lovera, O.M., Ding, L., 2005. Nyainqentanglha Shan: A window into the tectonic, thermal, and geochemical evolution of the Lhasa block, southern Tibet. J. Geophys. Res., 110, B08413.

Kay, S.M., Mpodozis, C., 2001. Central Andean ore deposits linked to evolving shallow subduction systems and thickening crust. GSA Today, 11, 4-9.

Kemp, A.L.S., Hawkesworth, C.J., Paterson, B.A., Kinny, P.D., 2006. Episodic growth of the Gondwana supercontinent from hafnium and oxygen isotopes in zircon. Nature, 439, 580-583.

Kemp, A.I.S., Hawkesworth, C.J., Foster, G.L., Paterson, B.A., Woodhead, J.D., Hergt, J.M., Gray, C. M., Whitehouse, M.J., 2007. Magmatic and crustal differentiation history of granitic rocks from Hf-O isotopes in zircon. Science, 315, 980-983.

Kornfeld, D., Eckert, S., Appel, E., Ratschbacher, L., Sonntag, B.L., Pfänder, J.A., Ding, L., Liu, D. L., 2014. Cenozoic clockwise rotation of the Tengchong block, southeastern Tibetan Plateau: A paleomagnetic and geochronologic study. Tectonophysics, 628, 105-122.

Lee, C.T.A., Morton, D.M., Kistler, R.W., Baird, A.K., 2007. Petrology and tectonics of Phanerozoic continent formation: From island arcs to accretion and continental arc magmatism. Earth Planet. Sci. Lett., 263, 370-387.

Lee, H.Y., Chung, S.L., Lo, C.H., Ji, J.Q., Lee, T.Y., Qian, Q., Zhang, Q., 2009. Eocene Neotethyan slab breakoff in southern Tibet inferred from the Linzizong volcanic record. Tectonophysics, 477, 20-35.

Li, C., van Der Hilst, R.D., Meltzer, A.S., Engdahl, E.R., 2008. Subduction of the Indian lithosphere beneath the Tibetan plateau and Burma. Earth Planet. Sci. Lett., 274, 157-168.

Li, D.P., Luo, Z.H., Chen, Y.L., Liu, J.Q., Jin, Y., 2014.Deciphering the origin of the Tengchong block, west Yunnan: Evidence from detrital zircon U-Pb ages and Hf isotopes of Carboniferous strata. Tectonophysics, 614, 66-77.

Li, G.J., Wang, Q.F., Huang, Y.H., Gao, L., Yu, L., 2016. Petrogenesis of middle Ordovician peraluminous granites in the Baoshan block: Implications for the early Paleozoic tectonic evolution along. Lithos, 245, 76-92.

Lin, I.J., Chung, S.L., Chu, C.H., Lee, H.Y., Gallet, S., Wu, G.Y., Ji, J.Q., Zhang, Y.Q., 2012. Geochemical and Sr-Nd isotopic characteristics of Cretaceous to Paleocene granitoids and volcanic rocks, SE Tibet: Petrogenesis and tectonic implications. J. Asian Earth Sci., 53, 131-150.

Liu, Y., Liu, X.M., Hu, Z.C., Diwu, C.R., Yuan, H.L., Gao, S., 2007. Evaluation of accuracy and long-term stability of determination of 37 trace elements in geological samples by ICP-MS. Acta Petrol. Sin., 23 (5), 1203-1210 (in Chinese with English abstract).

Ludwig, K.R., 2003. ISOPLOT 3.0: A Geochronological Toolkit for Microsoft Excel. Berkeley Geochronology Center, Special Publication, 4, 71.

Ma, N., Deng, J., Wang, Q.F., Wang, C.M., Zhang, J., Li, G.J., 2013. Geochronology of the Dasongpo tin deposit, Yunnan Province: Evidence from zircon LA-ICP-MS U-Pb ages and cassiterite LA-MC-ICP-MS U-Pb age. Acta Petrol. Sin., 29, 1223-1235 (in Chinese with English abstract).

Ma, L.Y., Wang, Y.J., Fan, W.M., Geng, H.Y., Cai, Y.F., Zhong, H., Liu, H.C., Xing, X.W., 2014. Petrogenesis of the early Eocene I-type granites in west Yingjiang (SW Yunnan) and its implications for the eastern extension of the Gangdese batholiths. Gondwana Res., 25, 401-419.

Maniar, P. D., Piccoli, P. M., 1989. Tectonic discrimination of granitoids. Geol. Soc. Am. Bull., 101, 635-643.

Mantle, G.W., Collions, W.J., 2008. Quantifying crustal thickness variations in evolving orogens: Correlation between arc basalt composition and Moho depth. Geology, 36, 87-90.

Mitchell, A.H.G., 1993. Cretaceous-Cenozoic tectonic events in the western Myanmar (Burma)-Assam region. J. Geol. Soc. Lond., 150, 1089-1102.

Mitchell, A., Chung, S. L., Oo, T., Lin, T. H., Hung, C. H., 2012. Zircon U-Pb ages in Myanmar: Magmatic-metamorphic events and the closure of a neo-Tethys ocean? J. Asian Earth Sci., 56, 1-23.

Mo, X.X., Zhao, Z.D., Deng, J.F., Dong, G.C., Zhou, S., Guo, T.Y., Zhang, S.Q., Wang, L.L., 2003. Response of volcanism to the India-Asia collision. Earth Sci. Front. (China University of Geosciences, Beijing), 10, 135-148 (in Chinese with English abstract).

Mo, X.X., Hou, Z.Q., Niu, Y.L., Dong, G.C., Qu, X.M., Zhao, Z.D., Yang, Z.M., 2007. Mantle contributions to crustal thickening during continental collision: Evidence from Cenozoic igneous rocks in southern Tibet. Lithos, 96, 225-242.

Mo, X.X., Niu, Y.L., Dong, G.C., Zhao, Z.D., Hou, Z.Q., Zhou, S., Ke, S., 2008. Contribution of syncollisional felsic magmatism to continental crust growth: A case study of the Paleogene Linzizong volcanic Succession in southern Tbiet. Chem. Geol., 250, 49-67.

Najman, Y., Appel, E., Boudagher-Fadel, M., Bown, P., Carter, A., Garzanti, E., Godin, L., Han, J.T., Liebke, U., Oliver, G., Parrish, R., Vezzoli, G., 2010. Timing of India-Asia collision: Geological, biostratigraphic, and palaeomagnetic constraints. J. Geophys. Res. Solid Earth, 115 (B12).

Pan, F.B., Zhang, H.F., Xu, W.C., Guo, L., Wang, S., Luo, B.J., 2014. U-Pb zircon chronology, geochemical and Sr-Nd isotopic composition of Mesozoic-Cenozoic granitoids in the SE Lhasa terrane: Petrogenesis and tectonic implications. Lithos, 192-195, 142-157.

Pan, F.B., Zhang, H.F., Xu, W.C., Guo, L., Luo, B.J., Wang, S., 2016. U-Pb zircon dating, geochemical and Sr-Nd-Hf isotopic compositions of mafic intrusive rocks in the Motuo, SE Tibet constrain on their petrogenesis and tectonic implication. Lithos, 245, 133-146.

Patiño Douce, A.G., Beard, J.S., 1995. Dehydration-melting of Biotite Gneiss and Quart Amphiolite from 3 kbar to 15 kbar. J. Petrol., 36, 707-738.

Qi, X.X., Zhu, L.H., Grimmer, J.C., Hu, Z.C., 2015. Tracing the Transhimalayan magmatic belt and the Lhasa block southward using zircon U-Pb, Lu-Hf isotopic and geochemical data: Cretaceous-Cenozoic granitoids in the Tengchong block, Yunnan, China. J. Asian Earth Sci., 110, 170-188.

Rapp, R. P., Watson, E. B., 1995. Dehydration melting of Metabasalt at 8 – 32 kbar: Implications for continental growth and crust-mantle recycling. J. Petrol., 36, 891-931.

Ratschbacher, L., Frisch, W., Chen, C.S., Pan, G.T., 1993. Deformation and motion along the southern margin of the Lhasa Block (Tibet) prior to and during the India-Asia collision. J. Geodyn., 16(1-2), 21-54.

Replumaz, A., Tapponnier, P., 2003. Reconstruction of deformed collision zone between India and Asia by backward motion of lithospheric blocks. J. Geophys. Res., 108 (B6), 2285.

Roberts, M.P., Clemens, J.D., 1993. Origin of high-potassium, calc-alkaline, I-type granitoids. Geology, 21, 825-828.

Rollinson, H., 1993. Using Geochemical Data: Evaluation, Presentation, Interpretation. London: Longman Scientific & Technical.

Scherer, E.E., Whitehouse, M.J., Münker, C., 2007. Zircon as a monitor of crustal growth. Elements, 3, 19-24.

Sisson, T.W., Ratajeski, K., Hankins, W.B., Glazner, A.F., 2005. Voluminous granitic magmas from common basaltic sources. Contrib. Mineral. Petrol., 148, 635-661.

Sun, S.S., McDonough, W.F., 1989. Chemical and isotopic systematics of oceanic basalts: Implications for mantle composition and processes. In: Saunders, A.D., Norry, M.J. Magmatism in the Ocean Basins. Geological Society, London, Special Publications, 42, 313-345.

Sun, Z.M., Pei, J.L., Li, H.B., Xu, W., Jiang, W., Zhu, Z.M., Wang, X.S., Yang, Z.Y., 2012. Palaeomagnetism of late Cretaceous sediments from southern Tibet: Evidence for the consistent palaeolatitudes of the southern margin of Eurasia prior to the collision with India. Gondwana Res., 21, 53-63.

Sylvester, P.J., 1998. Post-collisional strongly peraluminous granites. Lithos, 45, 29-44.

Van Hinsbergen, D.J.J., Lippert, P.C., Dupont-Nivet, G., McQuarrie, N., Doubrovine, P.V., Spakman, W., Torsvik, T.H., 2012. Greater India Basin hypothesis and a two-stage Cenozoic collision between India and Asia. Proc. Natl. Acad. Sci., 109, 7659-7664.

Wang, Y., Zhang, X.M., Jiang, C.S., Wei, H.Q., Wan, J.L., 2007. Tectonic controls on the late Miocene-Holocene volcanic eruptions of the Tengchong volcanic field along the southeastern margin of Tibetan Plateau. J. Asian Earth Sci., 30, 375-389.

Wang, Y.C., Zhou, Y., Liu, Y.H., Li, R.B., Wei, F.H., Gao, J.M., Liu, C.J., Wu, S.K., 2012. Characteristics of Sinongduo large-size overprinted and reworked Pb-Zn deposit in Xietongmen County of Tibet and the ore-searching direction. Contrib. Geol. Miner. Resour. Res., 27, 440-449.

Wang, Y.J., Zhang, L.M., Cawood, P.A., Ma, L.Y., Fan, W.M., Zhang, A.M., Zhang, Y.Z., Bi, X.W., 2014. Eocene supra-subduction zone mafic magmatism in the Sibumasu block of SW Yunnan: Implications for Neotethyan subduction and India-Asia collision. Lithos, 206-207, 384-399.

Wang, Y.J., Li, S.B., Ma, L.Y., Fan, W.M., Cai, Y.F., Zhang, Y.H., Zhang, F.F., 2015.

Geochronological and geochemical constraints on the petrogenesis of Early Eocene metagabbroic rocks in Nabang (SW Yunnan) and its implications on the Neotethyan slab subduction. Gondwana Res., 27, 1474-1486.

Wen, D.R., Liu, D.Y., Chung, S.L., Chu, M.F., Ji, J.Q., Zhang, Q., Song, B., Lee, T.Y., Yeh, M.W., Lo, C.H., 2008. Zircon SHRIMP U-Pb ages of the Gangdese Batholith and implications for Neotethyan subduction in southern Tibet. Chem. Geol., 252, 191-201.

Wu, F.Y., Li, X.H., Zheng, Y.F., Gao, S., 2007. Lu-Hf isotopic systematics and their applications in petrology. Acta Petrol. Sin., 23, 185-220 (in Chinese with English abstract).

Wu, F.Y., Ji, W.Q., Liu, C.Z., Chung, S.L., 2010. Detrital zircon U-Pb and Hf isotopic data from the Xigaze fore-arc basin: Constraints on Transhimalayan magmatic evolution in southern Tibet. Chem. Geol., 271, 13-25.

Wu, F.Y., Ji, W.Q., Wang, J.G., Liu, C.Z., Chung, S.L., Clift, P.D., 2014. Zircon U-Pb and Hf isotopic constraints on the onset time of India-Asia collision. Am. J. Sci., 314, 548-579.

Xie, J.C., Zhu, D.C., Dong, G.C., Zhao, Z.D., Wang, Q., Mo, X.X., 2016. Linking the Tengchong Terrane in SW Yunnan with the Lhasa Terrane in southern Tibet through magmatic correlation. Gondwana Res., 39, 217-229.

Xu, Y.G., Lan, J.B., Yang, Q.J., Huang, X.L., Niu, H.N., 2008. Eocene break-off of the Neo-Tethyan slab as inferred from intraplate-type mafic dykes in the Gaoligong orogenic belt, eastern Tibet. Chem. Geol., 255, 439-453.

Xu, Y.G., Yang, Q.J., Lan, J.B., Luo, Z.Y., Huang, X.L., Shi, Y.B., Xie, L.W., 2012. Temporal-spatial distribution and tectonic implications of the batholiths in the Gaoligong-Tengliang-Yingjiang area, western Yunnan: Constraints from zircon U-Pb ages and Hf isotopes. J. Asian Earth Sci., 53, 151-175.

Xu, Z.Q., Wang, Q., Cai, Z.H., Dong, H.W., Li, H.Q., Chen, X.J., Duan, X.D., Cao, H., Li, J., Burg, J.P., 2015. Kinematics of the Tengchong Terrane in SE Tibet from the late Eocene to early Miocene: Insights from coeval mid-crustal detachments and strike-slip shear zones. Tectonophysics, 665, 127-148.

Yang, Q.J., Xu, Y.G., Huang, X.L., Luo, Z.Y., Shi, Y.R., 2009. Geochronology and geochemistry of granites in the Tengliang area, western Yunnan: Tectonic implication. Acta Petrol. Sin., 25, 1092-1104 (in Chinese with English abstract).

Yi, Z.Y., Huang, B.C., Chen, J.S., Chen, L.W., Wang, H.L., 2011. Paleomagnetism of early Paleogene marine sediments in southern Tibet, China: Implications to onset of the India-Asia collision and size of Greater India. Earth Planet. Sci. Lett., 309, 153-165.

Yuan, H.L., Gao, S., Liu, X.M., Li, H.M., Gunther, D., Wu, F.Y., 2004. Accurate U-Pb age and trace element determinations of zircon by laser ablation-inductively coupled plasma mass spectrometry. Geostand. Newslett., 28, 353-370.

Yuan, H.L., Gao, S., Dai, M.N., Zong, C.L., Gunther, D., Fontaine, G.H., Liu, X.M., Diwu, C.R., 2008. Simultaneous determinations of U-Pb age, Hf isotopes and trace element compositions of zircon by excimer laser-ablation quadrupole and multiple-collector ICP-MS. Chem. Geol., 247, 100-118.

Zartman, R.E., Doe, B.R., 1981. Plumbotectonics-the model. Tectonophysics, 75, 135-162.

Zhang, Z.M., Zhao, G.C., Santosh, M., Wang, J.L., Dong, X., Liou, J.G., 2010. Two stages of granulite facies metamorphism in the eastern Himalayan syntaxis, south Tibet: Petrology, zircon geochronology and implications for the subduction of Neo-Tethys and the Indian continent beneath Asia. J. Metamorph. Geol., 28, 719-733.

Zhao, D.P., Yu, S., Ohtani, E., 2011a. East Asia: Seismotectonics, magmatism and mantle dynamics. J. Asian Earth Sci., 40, 689-709.

Zhao, S.W., Lai, S.C., Qin, J.F., Zhu, R.Z., 2016a. Tectono-magmatic evolution of the Gaoligong belt, southeastern margin of the Tibetan plateau: Constraints from granitic gneisses and granitoid intrusions. Gondwana Res., 35, 238-256.

Zhao, S.W., Lai, S.C., Qin, J.F., Zhu, R.Z., 2016b. Petrogenesis of Eocene granitoids and microgranular enclaves in the western Tengchong Block: Constraints on eastward subduction of the Neo-Tethys. Lithos, 264, 96-107.

Zhao, S.W., Lai, S.C., Gao, L., Qin, J.F., Zhu, R.Z., 2017. Evolution of the Proto-Tethys in the Baoshan block along the East Gondwana margin: Constraints from early Paleozoic magmatism. Int. Geol. Rev., 59 (1), 1-15.

Zheng, Y.C., Hou, Z.Q., Gong, Y.L., Liang, W., Sun, Q.Z., Zhang, S., Fu, Q., Huang, K.X., Li, Q. Y., Li, W., 2014. Petrogenesis of Cretaceous adakite-like intrusions of the Gangdese Plutonic Belt, southern Tibet: Implications for mid-ocean ridge subduction and crust growth. Lithos, 190-191, 240-263.

Zheng, Y.C., Fu, Q., Hou, Z.Q., Yang, Z.S., Huang, K.X., Wu, C.D., Sun, Q.Z., 2015. Metallogeny of the northeastern Gangdese Pb-Zn-Ag-Fe-Mo-W polymetallic belt in the Lhasa terrane, southern Tibet. Ore Geol. Rev., 70, 510-532.

Zhong, D.L., 1998. Paleo-Tethyan Orogenic Belt in the Western Part of the Sichuan and Yunnan provinces. Beijing: Science Press, 1-230 (in Chinese).

Zhou, S., Mo, X.X., Dong, G.C., Zhao, Z.D., Qiu, R.Z., Guo, T.Y., Wang, L.L., 2004. ^{40}Ar-^{39}Ar geochronology of Cenozoic Linzizong volcanic rocks from Linzhou Basin, Tibet, China, and their geological implications. Chin. Sci. Bull., 49, 1970-1979.

Zhu, D.C., Mo, X.X., Niu, Y.L., Zhao, Z.D., Wang, L.Q., Liu, Y.S., Wu, F.Y., 2009. Geochemical investigation of Early Cretaceous igneous rocks along and east-west traverse throughout the central Lhasa Tibet. Chem. Geol., 268, 298-312.

Zhu, D.C., Zhao, Z.D., Niu, Y.L., Mo, X.X., Chung, S.L., Hou, Z.Q., Wang, L.Q., Wu, F.Y., 2011a. The Lhasa Terrane: Record of a microcontinent and its histories of drift and growth. Earth Planet. Sci. Lett., 301, 241-255.

Zhu, D.C., Zhao, Z.D., Niu, Y.L., Dilek, Y., Mo, X.X., 2011b. Lhasa terrane in southern Tibet came from Australia. Geology, 39(8), 727-730.

Zhu, R.Z., Lai, S.C., Qin, J.F., Zhao, S.W., 2015. Early-Cretaceous highly fractionated I-type granites from the northern Tengchong block, western Yunnan, SW China: Petrogenesis and tectonic implications. J. Asian Earth Sci., 100, 145-163.

Early Cretaceous Na-rich granitoids and their enclaves in the Tengchong Block, SW China: Magmatism in relation to subduction of the Bangong-Nujiang Tethys ocean[①]

Zhu Renzhi Lai Shaocong[②] M Santosh Qin Jiangfeng Zhao Shaowei

Abstract: The Na-rich intermediate-to-felsic granitic rocks provide insights into the generation of magmas in subduction zones. This paper presents zircon LA-ICP-MS U-Pb ages as well as whole-rock geochemical, mineral chemical, and in situ zircon Hf isotopic data on Na-rich granitic rocks from the Tengchong Block, SW China. The granodiorites and associated mafic magmatic enclaves (MMEs) from the Menglian batholith yield zircon U-Pb ages of 116.1 ± 0.8 Ma to 117.8 ± 0.6 Ma and 117.7 ± 0.7 Ma, respectively. Both host granodiorites and enclaves show calc-alkaline and sodium-rich nature, enrichment in large-ion lithophile elements (LILEs), and variable depletion in zircon Hf isotopic compositions. Euhedral amphiboles in both granodiorites and associated enclaves are magnesian-hornblende with high Mg and Ca and contain euhedral plagioclase inclusions of labradorite to andesine (An_{36-57}) composition. The granodiorite was most likely derived through the mixing of partial melts derived from juvenile basaltic lower crust and a minor evolved component of ancient crustal sources. The quartz monzodiorite-granodiorites and associated MMEs from the Xiaotang-Mangdong batholith yield zircon U-Pb ages of 120.3 ± 1.3 Ma to 122.6 ± 0.8 Ma and 120.7 ± 1.5 Ma. These rocks are also sodium-rich and show calc-alkaline trend with negative zircon Hf isotopic compositions (−5.55 to +0.58). The MMEs in the host intrusions are monzogabbro with variable and depleted zircon Hf isotopic compositions. The amphiboles in the both host intrusions and the enclaves show Al-rich ferro-tschermakite composition. We infer that the quartz monzodiorite-granodiorites were derived from magmas generated by the melting of ancient basaltic rocks in the lower arc crust induced by the underplating of mantle-derived mafic magmas. The formation of the different types of Na-rich granitic rocks are correlated to the subduction of Bangong-Nujiang Tethyan ocean. A comparison with magmatism in the northern magmatic belt suggests that mantle-derived magmas played an important role in the genesis of Early Cretaceous intrusions from Tengchong to Lhasa Blocks, although crustal melting is the dominant contributor.

① Published in *Lithos*, 2017, 286-287.

② Corresponding author.

1 Introduction

In general, Na-rich intermediate-to-felsic rocks such as the diorite-tonalite-granodiorite suites are commonly generated in subduction zones (Rushmer and Jackson, 2006). Experimental petrological studies show that Na-rich intermediate to felsic melts are formed by either ① the differentiation of mantle-derived mafic magmas (Stern et al., 1989) or ② the high-temperature dehydration melting of mafic lower-crustal material (Rapp and Watson, 1995). Moreover, many Na-rich granitic intrusions are known to contain mafic magmatic enclaves (MMEs) that provide robust evidence for the interaction between coeval mantle-derived mafic and crustal-derived felsic magmas (Barbarin, 2005; Barbarin and Didier, 1992; García-Moreno et al., 2006; Moyen et al., 2001; Qin et al., 2009; Słaby and Martin, 2008; Yang et al., 2007a,b). Thus, the petrogenesis of Na-rich granitic rocks is important in gaining insights into the different magmatic evolution processes in convergent margins.

The Tengchong Block and the associated Nujiang tectonic zone, as the southwestern part of the Sanjiang Tethyan orogen, carry important evidence of magmatic processes during the evolution of Tethys (Deng et al., 2014; Wang et al., 2014, 2015, 2016; Xu et al., 2012; Zhao et al., 2015,2016; Zhao et al., 2015, 2016; R.Z. Zhu et al., 2015). Their Mesozoic and Cenozoic magmatic records are mainly related to the evolution of Meso- and Neo-Tethys (Wang et al., 2014, 2016). Even, Li et al. (2016) suggested that the extensive magmatic events since 250 Ma carry significant juvenile material accreted to the crust of the Tengchong Block. Recently, abundant Early Cretaceous Na-rich granitic rocks and associated mafic enclaves have been identified in recent studies from Xiaotang-Mangdong and Menglian batholiths in the eastern Tengchong Block (Cong et al., 2011; Gao et al., 2014; Li et al., 2012; Luo et al., 2012; Qi et al., 2011), mainly distributed along the Nujiang-Longling-Ruili fault that is regarded as the southeastern extension of Bangong-Nujiang suture (Fig. 1a,b) (Cong et al., 2011; Deng et al.,2014; Wang et al., 2014, 2015, 2016; Xu et al., 2012). However, their petrogenesis and tectonic settings remain controversial. These Na-rich granitic rocks are considered to be formed by the mixing of depleted mantle-derived mafic magma and ancient crustal felsic melts (Cong et al., 2011; Gao et al., 2014; Li et al., 2012; Qi et al., 2011), which were generated by the eastward subduction of Neo-Tethyan ocean (Cong et al., 2011), induced by the southward subduction of Bangong-Nujiang Meso-Tethyan ocean (Li et al., 2012; Qi et al., 2011), or related to the collision between the Lhasa-Tengchong and Baoshan-Qiangtang blocks after the closure of the Bangong-Nujiang Tethys (Gao et al., 2014; Luo et al., 2012).

Here we present new whole-rock major and trace element, mineral chemistry, and zircon Hf isotopic data on the Na-rich granitic rocks and associated MMEs in the Tengchong Block.

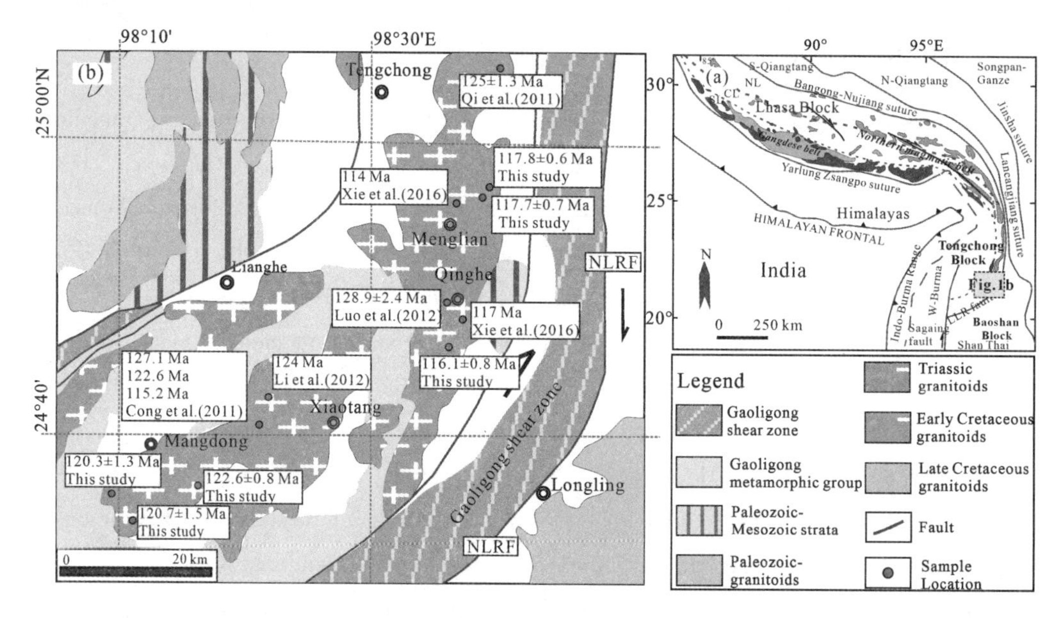

Fig. 1 (a) Distribution of main continental blocks of SE Asia (modified after Xu et al., 2012). (b) Distribution of the Menglian and Xiaotang-Mangdong batholiths (modified by 1 ∶200 000 regional geological map, YNBGMR).

NLRF: Nujiang-Longling-Ruili fault; NL: Northern Lhasa; CL: Central Lhasa; SL: Southern Lhasa.

The objectives of this work are: ① to investigate the petrogenesis of Na-rich granitic rocks; and ② to understand the different interacted magmatic processes associated with the convergent margin tectonics and evolution of the Bangong-Nujiang Meso-Tethys.

2 Geological setting and petrography

The Tengchong Block forms the southeastern extension of the Lhasa Block (Xie et al., 2016; Xu et al., 2012; Fig. 1a) and is separated from the eastern Baoshan Block by the Nujiang-Longling-Ruili fault (NLRF) and from the western Burma Block by the Putao-Myitkyina suture zone (Cong et al., 2011; Li et al., 2004; Metcalfe, 2013; Wang et al., 2013; Xu et al., 2012). The Bangong-Nujiang suture, as the remnant of the Tethys Ocean, extends for more than 2800 km across the central Tibetan Plateau and western Yunnan and to the south and is correlated with the Shan boundary in Myanmar (Wang et al., 2014, 2016). The Bangong-Nujiang Tethyan Ocean lithosphere was subducted both southward under the Lhasa Block of the Tibetan Plateau (Hsü et al., 1995; Pan et al., 2006; Zhu et al., 2013) and southeastward under the Tengchong Block in western Yunnan (Wang et al., 2014; Zou et al., 2011) during the Permian to the Early Cretaceous, until the closure of the Banggong-Nujiang Tethys Ocean at 100 Ma (Baxter et al., 2009; K. J. Zhang et al., 2012), as evidenced from the ages of ophiolites and fossil record. Based on the presence of Permo-

Carboniferous glacio-marine deposits and overlying post-glacial black mudstones and Gondwana-like fossil assemblages, it has been suggested that the Tengchong Block was derived from the margins of Western Australia, in the eastern part of the Gondwana super-continent (Jin, 1996). The block contains Mesoproterozoic metamorphic basement belonging to the Gaoligong Mountain Group and upper Paleozoic clastic sedimentary rocks and carbonates. The Mesozoic to Tertiary granitoids were emplaced into these strata (Zhao et al., 2016) and were then covered by Tertiary-Quaternary volcano-sedimentary sequences (YNBGMR, 1991). The Gaoligong Mountain Group contains quartzites, two-mica-quartz schists, feldspathic gneisses, migmatites, amphibolites, and marble. Zircons from the paragneiss and orthogneiss samples in this group yield ages in the range of 1 053–635 Ma and 490–470 Ma, respectively (Song et al., 2010). The Paleozoic sediments in this area are dominated by Carboniferous clastic rocks, and others Mesozoic strata including Upper Triassic to Jurassic turbidites, Cretaceous red beds, and Cenozoic sandstones (YNBGMR, 1991; Zhong, 2000).

The Tengchong Block contains abundant granitic gneiss, migmatite, and leucogranite units, the vast majority of which were previously thought to have formed during the Proterozoic (YNBGMR, 1991). However, recent studies have identified several massive granitoids with zircon U-Pb ages of 114–139 Ma in the eastern part of this block (Cao et al., 2014; Cong et al., 2011; Li et al., 2012; Luo et al., 2012; Q.J. Yang et al., 2006; Qi et al., 2011; R.Z. Zhu et al., 2015; Xie et al.,2016; Xu et al., 2012), all of which were emplaced into the Paleozoic and Mesozoic units. The granites in the Gaoligong area are located to the west of the NLRF and have undergone strong shearing that developed a north-trending foliation and a subhorizontally plunging mineral lineation (Wang and Burchfiel, 1997; B. Zhang et al., 2012).

This study focuses on the Menglian and Xiaotang-Mangdong batholiths, which are located in the eastern Tengchong Terrane and intruded into the Gaoligong Group metamorphic rocks as well as the Paleozoic-Mesozoic strata (Fig. 1b). In most cases, they show faulted contact with the wall-rock although the intrusive contact can be seen in some places, particularly in the late stage. Cenozoic volcanic rocks unconformably overly these intrusions and/or are covered by Quaternary sediments (Luo et al., 2012). The intrusions show round or elongate shape, and occur parallel to the ~NNW-SSW orientation with the Gaoligong shear zone and NLRF which are regarded to mark the southeastern extension of the Nujiang suture belt (Cong et al., 2011). The two batholiths are identified as discrete intrusions, with the Menglian batholith located much closer to the NLRF in the east and about 25 km far from the Xiaotang-Mangdong batholith (Fig. 1b).

The granitic rocks in the Menglian batholith are medium to fine grained (Fig. 2a,b) and contain hornblende (12–18 vol%), biotite (10–12 vol%), plagioclase (45–55 vol%), alkali feldspar (5–10 vol%), quartz (~20 vol%), and accessory titanite (~5%), magnetite (~3–5 vol%), zircon, and apatite (< 3 vol%) (Fig. 3c). The hornblende in this pluton is subhedral and defines cumulate-textured domains together with plagioclase. The biotites are subhedral-anhedral and contain zircon, quartz, and titanite inclusions. The plagioclase is euhedral to subhedral and is associated with alkali feldspar, with interstitial domains filled with quartz. The MMEs in the granodiorites are elongate, have rounded interfaces and diffuse contacts (Fig. 2a), and generally occur along the northern and eastern margins of the batholith (Fig. 1b), where they form up to ca. 3%–5% by volume in the exposed areas of the intrusion. The MMEs have diameters of 5–45 cm (usually <30 cm) and have diffuse contacts with the host granodiorite (Fig. 2), in the absence of any clear chilled margins. These MMEs display hypidiomorphic to allotriomorphic textures and are dominated by euhedral-subhedral hornblende (30–35 vol%) that contains inclusions of felsic minerals and which define pervasive poikilitic-textured domains with plagioclase and biotite (15–20 vol%), minor amounts of zircon and magnetite, as well as biotite, titanite, and magnetite (Fig. 3a,b).

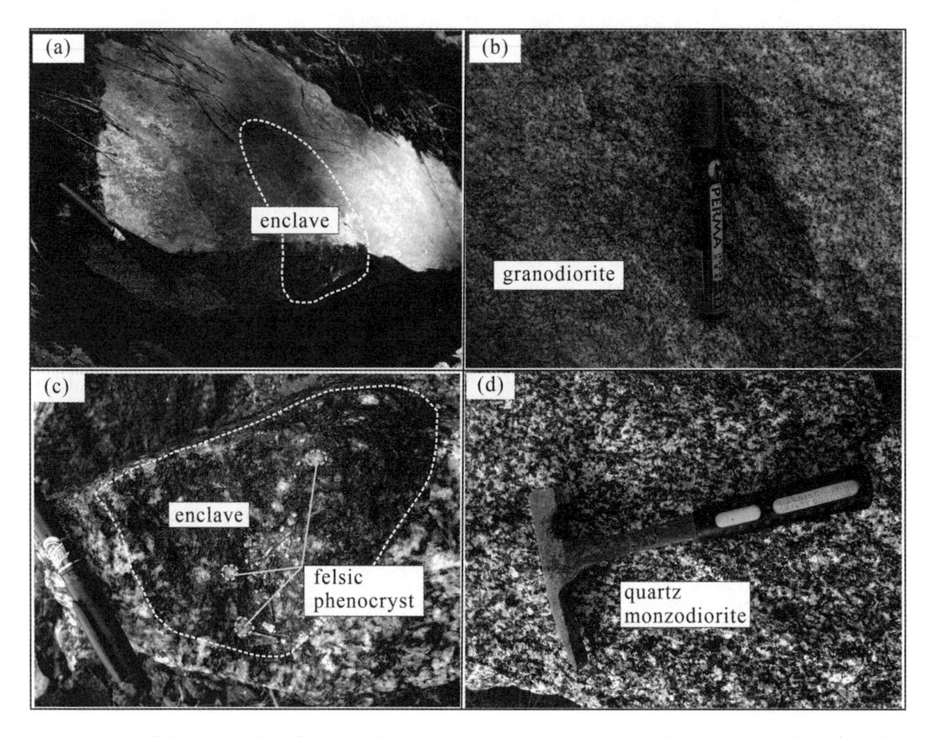

Fig. 2 Field features of enclaves and host granitic rocks from the Menglian batholith (a,b) and Xiaotang-Mangdong batholith (c,d) in the Tengchong Block, SW China.

In contrast, the granitic rocks from Xiaotang-Mangdong batholith are much more coarse grained (Fig. 2c,d), with more biotite and less hornblende (Fig. 3f), and dominated by alkali feldspar rather than plagioclase and quartz. The batholith also contains more accessory minerals, including titanite, magnetite, zircon, and apatite. The MMEs are angular and have also diffuse contacts with the host tonalites (Fig. 2c). The MMEs also contain hornblende, biotite, plagioclase, quartz, and accessory titanite, magnetite, zircon, and apatite (Fig. 3d,e). As in the case of the host rocks, the majority of hornblende in the MMEs has been altered to chlorite, with biotite present as masses of flakes that host zircon and quartz inclusions. Plagioclase is present as euhedral crystals that contain alkali feldspar, and only minor amounts of quartz are present within the matrix of the granitic rocks.

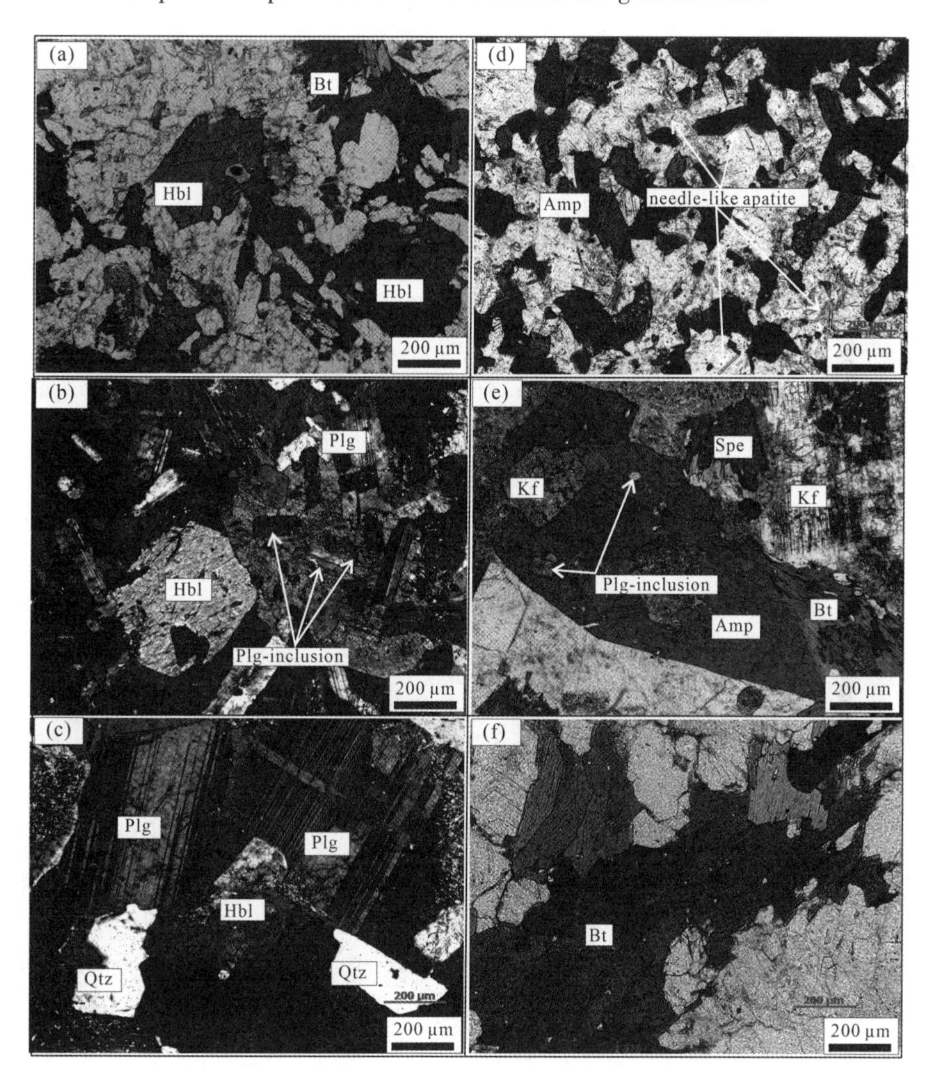

Fig. 3 Photomicrographs of enclaves and host granitoids from the Menglian batholith (a,b and c) and Xiaotang-Mangdong batholith (d,e and f) in the Tengchong Block, SW China.
Hbl: hornblende; Bt: biotite; Plg: plagioclase; Qtz: quartz; Kf: K-feldspar; Amp: amphibole; Spe: sphene.

3 Analytical methods

The analytical methods of major and trace elements, zircon U-Pb analyses and in situ zircon Hf, and electron microprobe analyses are found in the Appendices A1.

4 Results

The major and trace element compositions of representative samples are given in Supplementary Table 1, zircon Hf isotopic data are given in Supplementary Table 2, and mineral compositions given in Supplementary Table 3 and zircon U-Pb data in Supplementary dataset.

4.1 Zircon U-Pb data

4.1.1 Samples from Menglian batholith

Zircon grains from granodiorites including samples XH23 and XH48 from the Menglian batholith are generally euhedral to subhedral, have lengths of 100–250 μm, and have length-to-width ratios of 2∶1 to 4∶1 (Fig. 4a). The majority of these zircons are colorless or light brown, prismatic, and transparent to translucent, and have weak oscillatory zoning visible during cathodoluminescence (CL) imaging. The 21 spot analyses of zircons from sample XH23 yielded high Th (173–2 830 ppm) and U (245–2 007 ppm) concentrations with Th/U ratios of 0.58 – 1.78 that are indicative of a magmatic origin. These analyses define a coherent grouping with $^{206}Pb/^{238}U$ ages ranging from 114.1 ± 1.0 Ma to 122 ± 2.0 Ma and a weighted mean age of 116.1±0.8 Ma (MSWD=0.65, $n=21$, 2σ; Fig. 5a). Zircons from the sample XH48 display slight oscillatory zoning and typical prismatic habit, and most of the grains are subhedral to anhedral. The 19 spot analyses show a wide range of Th (109–2 029 ppm) and U (186–1 274 ppm) with Th/U values in the range of 0.44–1.59. These spots yield $^{206}Pb/^{238}U$ ages ranging from 116.0 ± 1.0 Ma to 121 ± 2.0 Ma, with a weighted mean age of 117.8 ± 0.6 Ma (MSWD=0.74, $n=19$, 2σ; Fig. 5c)

Zircons from the MME sample XH41-1 (Fig. 4b) are subhedral and partly prismatic, have weak oscillatory zoning visible under CL images, and are larger than those in the host granodiorites. A total of 19 analyses yielded high Th (237–2 812 ppm) and U (206–1 339 ppm) concentrations and Th/U values of 0.95–3.58. Their analytical data yield $^{206}Pb/^{238}U$ ages ranging from 115.0 ± 1.0 Ma to 120 ± 1.0 Ma, with a weighted mean age of 117.7 ± 0.7 Ma (MSWD=2.3, $n=19$, 2σ; Fig. 5b).

4.1.2 Samples from Xiaotang-Mangdong batholith

Two representative samples (MD11 and MD55) were analysed from this batholith. The zircon grains in these samples have round to ovoid shapes, in some cases are partly prismatic, and some of the grains display weak oscillatory zoning. They have lengths of up to 150–350 μm and length-to-width ratios of 2∶1 to 3∶1 (Fig. 4c).

The 15 spot analyses show high and widely ranging Th (285–9 005 ppm) and U (259–6 711 ppm) with Th/U values of 0.43–1.34. The data yield $^{206}Pb/^{238}U$ ages ranging from 115 ±

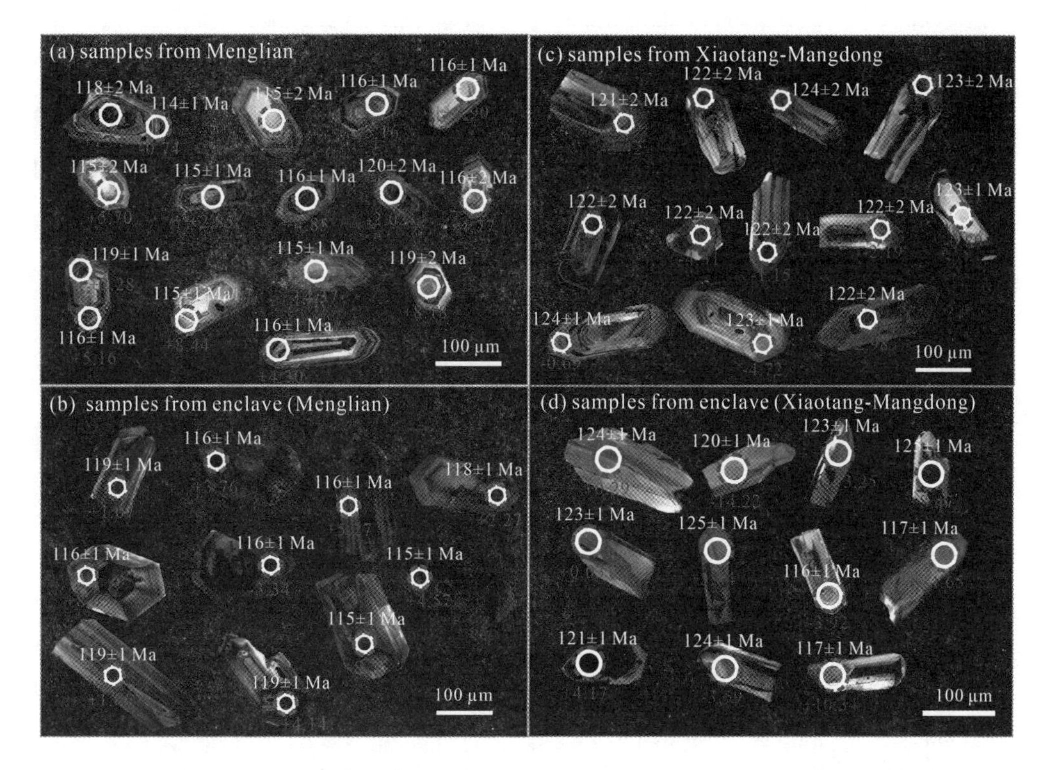

Fig. 4 Zircon CL images of representative zircon grains for the Menglian batholith (a,b) and Xiaotang-Mangdong batholith (c,d) in the Tengchong Block, SW China.

1.0 Ma to 126±4.0 Ma and a weighted mean age of 120.3±1.3 Ma (MSWD=1.3, n=15, 2σ; Fig. 5d). The 22 analyses from sample MD55 yielded Th concentrations of 210–514 ppm, U concentrations of 240–1 248 ppm, and Th/U values of 0.24–1.58. These zircons have coherent $^{206}Pb/^{238}U$ ages ranging from 121±1.0 Ma to 124±2.0 Ma and yield a weighted mean age of 122.6±0.8 Ma (MSWD=0.3, n=18, 2σ; Fig. 5f).

The zircons in mafic enclave sample MD27 are subhedral to anhedral, round to ovoid in shape, and rarely prismatic. They rarely display oscillatory zoning aging, and have high length-to-width ratios of 2∶1 to 4∶1 (Fig. 4d). These zircons contain variable concentrations of Th (62–1 889 ppm) and U (111–3 883 ppm) that yield Th/U values of 0.49–2.06. They have $^{206}Pb/^{238}U$ ages that range from 116±1.0 Ma to 125±2.0 Ma with a weighted mean age of 120.7±1.5 Ma (MSWD=2.1, n=18, 2σ; Fig. 5e).

4.2 Whole-rock geochemistry

The samples from the both Menglian and Xiaotang-Mangdong batholith are calc-alkaline, metaluminous to weakly peraluminous, and sodium-rich (Fig. 6a-c). They also define clinopyroxene entrainment trend and low-pressure I-type nature (Fig. 6d). Geochemically, most samples from Menglian batholith are granodiorites and the associated enclaves are

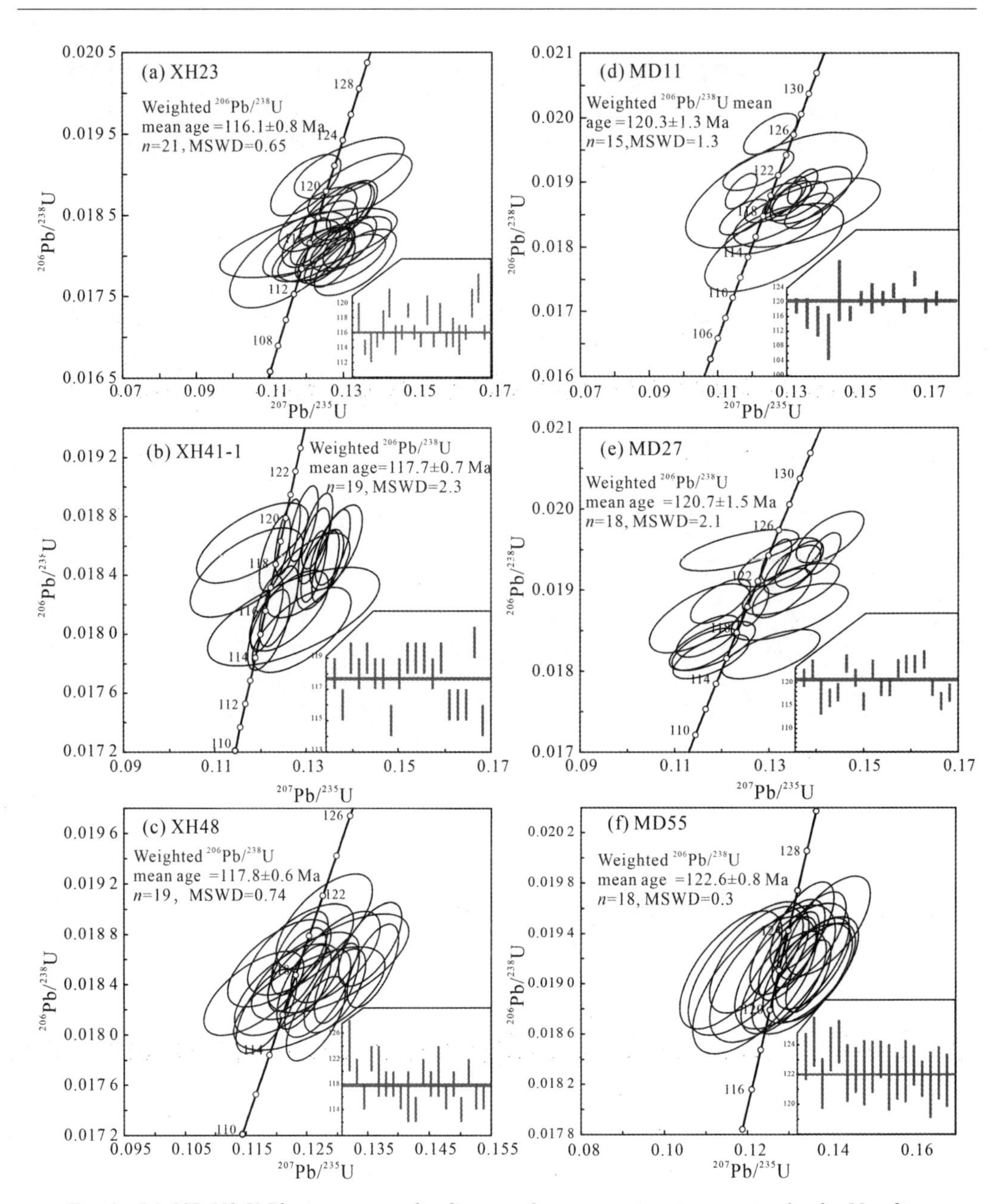

Fig. 5　LA-ICP-MS U-Pb zircon concordia diagram of representative zircon grains for the Menglian batholith (a, b and c) and Xiaotang-Mangdong batholith (c, d) in the Tengchong Block, SW China.

monzodiorite, whereas most samples from the Xiaotang-Mangdong batholith are quartz monzodiorite-granodiorites and the associated enclaves are monzogabbroic (Fig. 6e, f).

4.2.1　Samples from the Menglian batholith

The granodiorite samples from the Menglian batholith have higher SiO_2 = 64.17–68.22 wt% and $Mg^{\#}$ values (45.9–50.4) than those from Xiaotang-Mangdong batholith. They are sodium-rich with high Na_2O (3.52–4.07 wt% with Na_2O/K_2O = 1.08–3.05) (Fig. 6c). They also

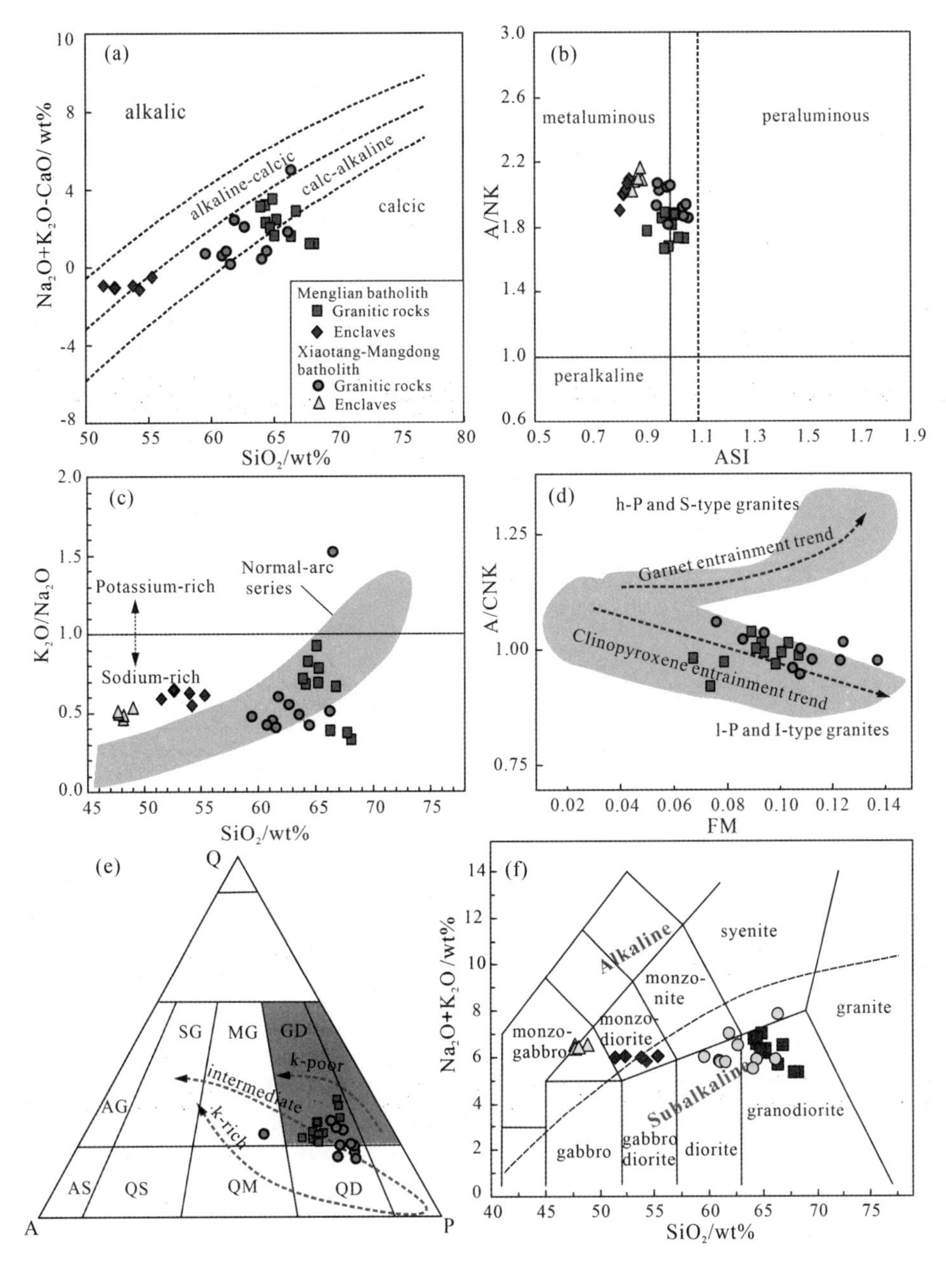

Fig. 6 (a) Na_2O+K_2O-CaO vs. SiO_2 diagram (Frost et al., 2001); (b) A/NK vs. A/CNK diagram; (c) SiO_2 and K_2O/Na_2O diagram after Moyen and Martin (2012), the yellow field shows "ordinary" arc rocks; (d) Peraluminosity vs. "maficity" relationships (after Moyen and Martin, 2012). A/CNK: molar Al/(2Ca+Na+K); FM: molar Fe+Mg. (e) Q-A-P modal classification of Streckeisen (1975) for the granitoids from Menglian and Xiaotang-Mangdong batholith, showing similar with TTG (Grey field = tonalite + granodiorite). These samples are also plotted along the K-poor and intermediate calc-alkaline differentiation trend of Lameyre and Bowden (1982) (T = tholeiitic suite, A = alkaline suite; the calc-alkaline suites are subdivided into: a = K-poor; b = intermediate; c = K-rich); (f) Na_2O+K_2O vs. SiO_2 diagram (Frost et al., 2001).

display moderate Al_2O_3 concentrations of 14.07–16.76 wt% with A/CNK [molar Al_2O_3/(CaO+Na_2O+K_2O)] values of 0.91–1.04. The primitive-mantle-normalized multi-element variation diagrams (Fig. 7a) are characterized by positive large-ion lithophile element (LILE; e.g., Rb, Th, U, K, and Pb) anomalies and obvious negative high field strength element (HFSE; e.g., Nb, Ta, P, and Ti) anomalies. These samples show high total REE concentrations of 119.9–190.2 ppm and enrichment in light REEs (LREEs) (Fig. 7b) with relative high $(La/Yb)_N$ values of 8.53–25.95, low Y and $(Gd/Yb)_N$ values of 1.64–2.85, and negligible negative Eu anomalies (δEu=0.70–0.90), suggesting insignificant fractionation of plagioclase.

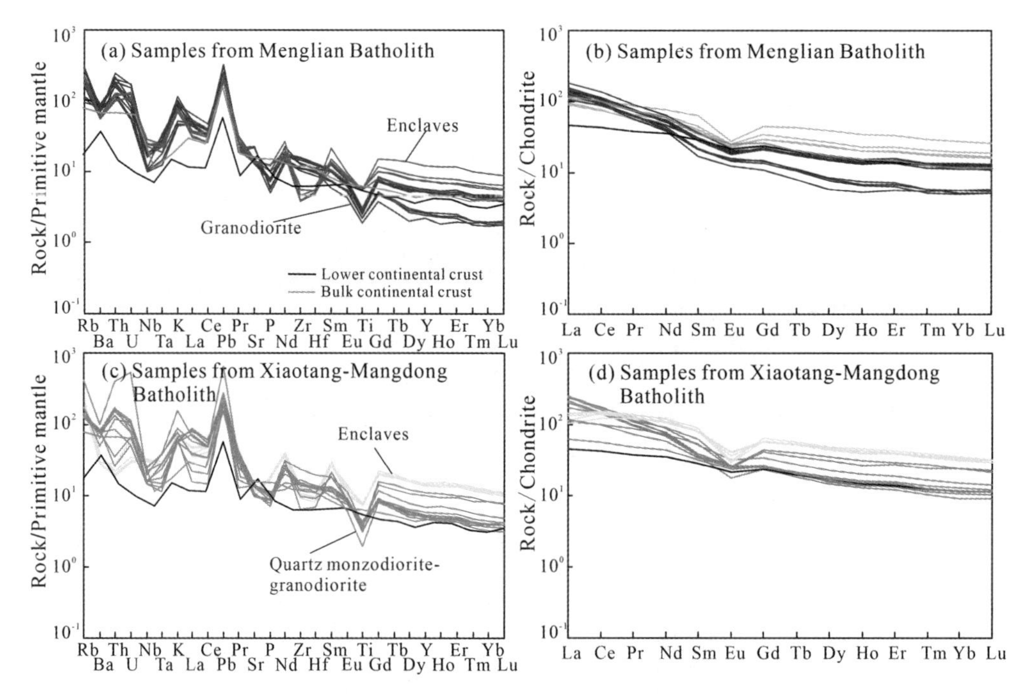

Fig. 7 Chondrite-normalized REE patterns and primitive-mantle-normalized trace element spider diagram (a,b) for the host granodiorites and enclaves from Menglian batholith, and (c,d) for the host granitic rocks and enclaves from Xiaotang-Mangdong batholith.
The primitive mantle and chondrite values are from Sun and McDonough (1989).

The MME samples from the host granodiorites have low SiO_2 (51.39–55.30 wt%), elevated Na_2O concentrations (3.70–3.81 wt% with Na_2O/K_2O ratios=1.92–2.18), and high CaO (6.58–7.12 wt%), together with $Fe_2O_3^T$ in the range of 8.59–9.51 wt%, TiO_2 concentrations of 1.56–1.87 wt%, and MgO (3.71–4.33 wt%) corresponding to high $Mg^{\#}$ values (49.3–52.0). The primitive-mantle-normalized multi-element variation patterns (Fig. 7a) also display similar positive LILE (Rb, Th, U, and Pb) but different HFSE including weakly negative Nb, Ta, P, and Ti and clearly negative Zr and Hf anomalies with its host granodiorites. They also share similar total REE concentrations (115.75–177.90 ppm) with the host granodiorites but possess weakly fractionated LREE/HREE ratios (4.20–6.36) with low $(La/Yb)_N$ (3.93 – 7.70) and $(Gd/Yb)_N$ (1.62 – 1.93) values and high concentrations of HREE(Yb_N = 12.12–27.29 and Y = 24.5–55.5 ppm) (Fig. 7b). Moreover, the elevated negative Eu anomalies (δEu = 0.53 – 0.77) than the host granodiorites might suggest higher degree of plagioclase fractionation during magma evolution.

4.2.2 Samples from the Xiaotang-Mangdong batholith

The quartz monzodiorite-granodiorites from the Xiaotang-Mangdong batholith display lower SiO_2(59.58 – 66.39 wt%) and Na_2O/K_2O ratios (1.62 – 2.41, except one with a value of 0.67) and $Mg^{\#}$ values (42.0–44.9) than the granodiorites from the Menglian batholith. They show slightly higher concentrations of Al_2O_3(16.31 – 18.02 wt%) with A/CNK ratios of 0.94–1.06. However, the primitive-mantle-normalized multi-element variation patterns (Fig. 7c) are somewhat variable, although they have consistently positive Pb, Rb, and U anomalies, variable Th anomalies, and negative Ba, Nb, Ta, and Ti anomalies. The chondrite-normalized REE patterns (Fig. 7d) show weakly fractionated $(La/Yb)_N$ values of 3.96 – 21.75, low $(Gd/Yb)_N$ values of 1.73–2.55, and obvious Eu anomalies (δEu = 0.48–0.83).

The MMEs from the Xiaotang-Mangdong are monzogabbroic with low SiO_2(47.67–48.90 wt%) and $Mg^{\#}$ values (43.4–44.9) but show higher concentrations of Al_2O_3(19.20–20.08 wt%), TiO_2(1.58–1.66 wt%), CaO (7.28–7.47 wt%), and $Fe_2O_3^T$(10.80–11.12 wt%) compared with its host rocks. The primitive-mantle-normalized multi-element variation and chondrite-normalized REE (Fig. 7c) patterns are distinct from the patterns of their host rocks, as exemplified by the presence of positive Rb and Pb and negative Sr, Zr, Hf, and Ti anomalies, and clearly low LREE/HREE and $(Gd/Yb)_N$ values (1.70 – 1.85) (Fig. 7d), but with high HREE values (e.g., Yb_N = 31.2–35.1, Y = 62.4–69.3 ppm). They also show more enhanced negative Eu anomalies (δEu = 0.42 – 0.60) than the host rocks, indicating significant fractionation of plagioclase.

4.3 Zircon Hf isotopic compositions

The Hf isotopic analysis was undertaken on the same domains in zircons from the four samples that were also dated during this study. The results are given in Supplementary Table 2. Initial $^{176}Hf/^{177}Hf$ ratios and $\varepsilon_{Hf}(t)$ values for these Early Cretaceous zircons were calculated using individual crystallization ages, with Fig. 8 showing histograms of the initial Hf isotope ratio values (in the form of $\varepsilon_{Hf}(t)$ values) of the granitoids and MMEs from the Menglian and Xiaotang-Mangdong batholiths.

The majority of the zircons from the granitoids and MMEs of the Menglian batholith have significantly positive Hf isotopic compositions (Fig. 8a, b). The granodiorites contain zircons with $\varepsilon_{Hf}(t)$ values ranging from +1.65 to +8.90 (n = 21, with the exception of two zircons

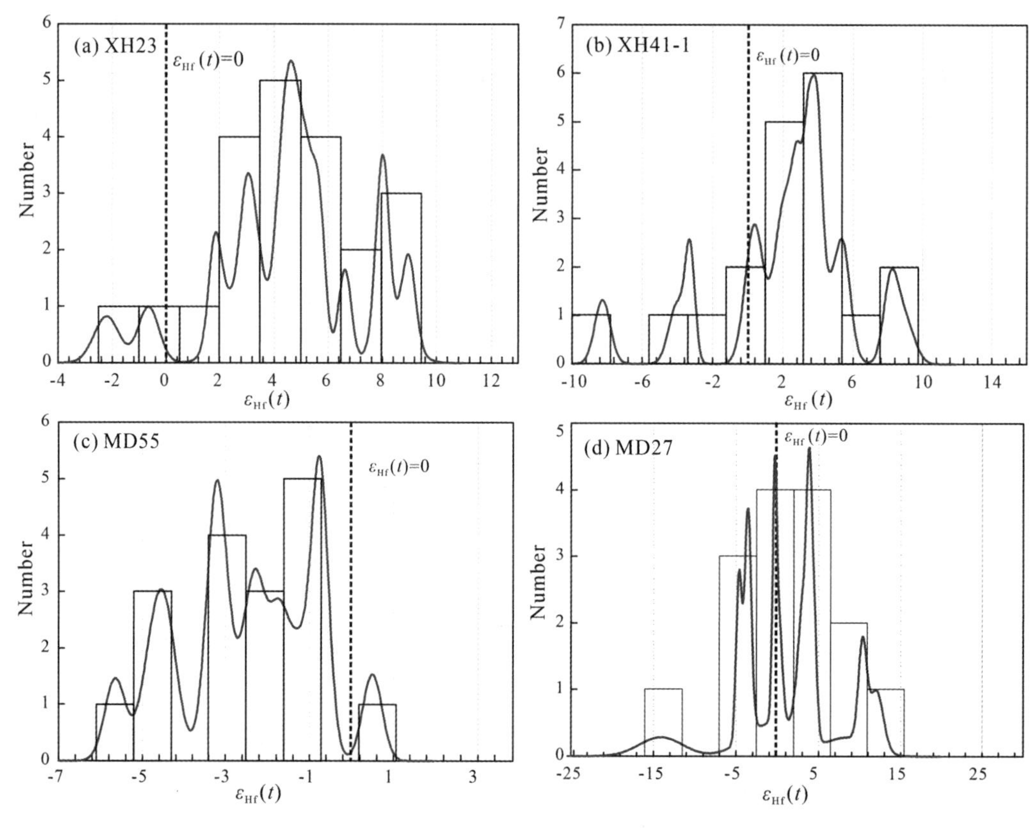

Fig. 8 Histograms of initial Hf isotope ratios.

(a) Host granodiorites from Menglian batholith; (b) Mafic enclaves from Menglian batholith; (c) host granitic rocks from Xiaotang-Mangdong batholith; (d) mafic enclaves from Xiaotang-Mangdong batholith.

with values of −2.02 and −0.74) with a peak at ca. +4.85. The MMEs contain zircons with $\varepsilon_{Hf}(t)$ values ranging from +0.15 to +8.75 ($n = 19$, barring three zircons with values of −8.38, −3.92, and −3.34) with a peak at ca. +3.60. These zircons yield crustal Hf model ages of 604−1 303 Ma for the granodiorites and 613−1 706 Ma for the MMEs.

The Hf isotopic compositions of zircons from the Xiaotang-Mangdong batholith are more negative and more variable than those of the zircons from Menglian batholith (Fig. 8c,d). A total of 15 analyses of zircons from a single MME yielded highly variable $\varepsilon_{Hf}(t)$ values (−14.22 to +12.13) and crustal model ages (401−2 074 Ma). In contrast, the 17 zircons from the hosting quartz monzodiorite-granodiorites yielded predominantly negative $\varepsilon_{Hf}(t)$ values (−5.55 to +0.58 for sample MD55) and crustal Hf model ages of 1 140−1 528 Ma.

4.4 Mineral chemistry

The majority of amphiboles from Menglian granodiorites and associated enclaves are magnesiohornblende in composition, whereas the majority of amphiboles from the Xiaotang-Mangdong granitic rocks and related enclaves are ferrotschermakite (Appendices Fig. A1), based on the classification of Leake et al. (1997). The plagioclase inclusions in the amphiboles from Menglian granodiorites and its enclaves are labradorite to andesine, and those from Xiaotang-Mangdong batholith are also andesine but the K-feldspar inclusions in the plagioclase phenocryst are K-sanidine (Appendices Fig. A2). The analytical data for the amphiboles and plagioclase with associated inclusions are summarized in Supplementary Table 3.

4.4.1 Samples from the Menglian batholith

(1) *MME-hosted amphibole.* Amphibole XH40-AMP-1 is euhedral and is ca. 0.50 mm × 1.00 mm in size. It contains higher concentrations of MgO (9.98–11.69 wt%), has higher $Mg^{\#}$ values (53.5–56.4), and contains lower concentrations of Al_2O_3 (7.19–9.31 wt%) compared with amphiboles from the Xiaotang-Mangdong batholith (Appendices Fig. A3a). Amphibole XH40-AMP-2 is also euhedral and is ca. 0.65 mm × 1.50 mm in size. This amphibole also has higher MgO (9.60–11.63 wt%) and lower Al_2O_3 (6.87–8.69 wt%) concentrations and higher $Mg^{\#}$ values (50.7–56.4) compared with amphiboles from the Xiaotang-Mangdong batholiths (Appendices Fig. A3b). These MME-hosted amphiboles contain two plagioclase inclusions with compositions of $Ab_{47-59}An_{40-52}Or_{1-2}$, and $Ab_{42-47}An_{51-57}Or_{0-2}$, respectively.

(2) *Amphibole in the hosting granitic rocks.* Amphibole XH42-AMP-1 is subhedral and is ca. 0.45 mm × 1.15 mm in size. It contains higher MgO concentrations (10.6–11.30 wt%), has higher $Mg^{\#}$ values (52.8–55.3), and contains lower Al_2O_3 concentrations (7.73–8.60 wt%) compared with amphiboles from the Xiaotang-Mangdong batholith (Appendices Fig. A3c). This amphibole also contains a plagioclase inclusion with a composition of $Ab_{57-63}An_{36-42}Or_1$. Amphibole XH42-AMP-2 is euhedral and is ca. 0.45 mm × 0.950 mm in size. It contains higher MgO concentrations (10.89–11.45 wt%), has higher $Mg^{\#}$ values (53.3–55.3), and contains lower Al_2O_3 concentrations (7.84–8.27 wt%) compared with amphiboles from the Xiaotang-Mangdong batholith (Appendices Fig. A3d).

4.4.2 Samples from the Xiaotang-Mangdong batholith

(1) *MME-hosted amphibole.* Amphibole MD25-AMP-1 is ca. 0.35 mm × 0.45 mm in size, contains low concentrations of MgO (7.1–9.65 wt%), has low $Mg^{\#}$ values (39.2–43.9), and contains high concentrations of Al_2O_3 (8.93–11.51 wt%). The Al_2O_3 concentrations in this amphibole increase from 8.93 wt% in the core to 11.51 wt% at the rim of the crystal, coincident with a decrease in MgO concentrations from the core (9.65 wt%) to the rim (7.10 wt%) (Appendices Fig. A3e). Amphibole MD25-AMP-2 is ca. 0.35 mm × 0.50 mm in size and also contains low concentrations of MgO (6.81–8.15 wt%), has low $Mg^{\#}$ values (36.4–41.7), and contains low concentrations of Al_2O_3 (9.90–11.28 wt%). The Al_2O_3 concentrations in this amphibole increase from the core (9.90 wt%) to the rim (11.28 wt%), whereas MgO concentrations decrease from the core (8.15 wt%) to the rim (6.81 wt%; Appendices Fig. A3f). The amphibole also contains a carbonate inclusion that is enriched in CaO (53.2–53.75 wt%).

(2) *MME-hosted plagioclase phenocryst.* Plagioclase phenocrysts in the MME from the Xiaotang-Mangdong batholith have consistent compositions of $Ab_{57-65}An_{32-42}Or_{1-2}$ and contain K-feldspar inclusions with $Or_{93-95}Ab_{5-7}$ compositions (Appendices Fig. A2; Supplementary Table 3).

(3) *Amphibole in the host granitic rocks.* Amphibole MD56-1 is ca. 0.50 mm × 1.15 mm in size and contains low concentrations of MgO (7.64–9.09 wt%), has low $Mg^{\#}$ values (39.2–44.9), and contains high concentrations of Al_2O_3 (9.12 – 11.15 wt%). The Al_2O_3 concentrations in this amphibole increase from the core (9.12 wt%) to the rim (11.15 wt%), coincident with a decrease in MgO concentrations from the core (9.09 wt%) to the rim (7.64 wt%) (Appendices Fig. A3g). This amphibole also contains a plagioclase inclusion with an $Ab_{66-67}An_{32-33}Or_1$ composition. Amphibole MD56-2 is ca. 0.15 mm × 0.30 mm in size and is geochemically similar to amphibole MD56-1. The $Mg^{\#}$ values in this amphibole decrease from the core (42.0) to the rim (39.0), coincident with an increase in Al_2O_3 concentration from the core (10.05 wt%) to the rim (11.61 wt%) (Appendices Fig. A3h).

5 Discussion

Our zircon U-Pb data show that the granodiorites and associated enclaves from Menglian batholith have younger ages (ca. 116.1 – 117.8 Ma) compared to those of the quartz monzodiorite-granodiorites and their enclaves (ca. 120.3 – 122.6 Ma) from the Xiaotang-Mangdong batholith. In conjunction with the previously published age data from the Menglian (ca. 128.9–114.3 Ma) and Xiaotang-Mangdong batholiths (ca. 127.4–115.2 Ma) (Cong et al., 2011; Li et al., 2012; Luo et al., 2012; Qi et al., 2011; Xie et al., 2016), we confirm the younger peak age ca. 117.6 Ma for the Menglian batholith as compared to the ca. 122.6 Ma age of the Xiaotang-Mangdong batholith (Fig. 9).

The granodiorites and associated enclaves from Menglian batholiths have relative high $Mg^{\#}$ (46 – 52) and markedly positive zircon $\varepsilon_{Hf}(t)$ values, with amphiboles showing magnesio-hornblende composition. In contrast, the quartz monzodiorite-granodiorites and enclaves from the Xiaotang-Mangdong batholith show low-$Mg^{\#}$ values of < 45 and more evolved zircon Hf compositions, and contain ferro-tschermakite to ferro-hornblende as the major mafic mineral. We therefore infer that the two rock suites are products of distinct petrogenetic evolution during ca. 122.6–117.6 Ma.

5.1 Petrogenesis of the granodiorites and associated enclaves from Menglian batholith

5.1.1 Origins of the MMEs

The granodiorites carry enclaves of monzodiorites. In general, the models for the origin of mafic magmatic enclaves (MMEs) include the following: ① The MMEs represent a

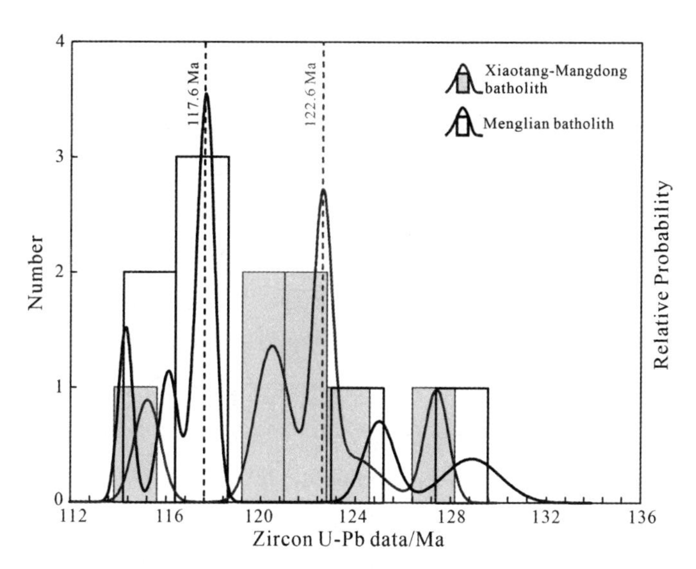

Fig. 9 Age data histograms and relative probability graphs of zircon U-Pb data for Menglian and Xiaotang-Mangdong batholiths.

recrystallized and refractory phase assemblages, or restites from the partial melting of sources that gave rise to the granitic liquid (Chappell et al., 1987, 2000; White et al., 1999). ② They are cognate fragments or inclusions of cumulate minerals from the mantle-derived mafic/basic magma replenished to the evolved intermediate-felsic liquids ((Bonin, 2004; Holden et al., 1987; Vernon, 1984; J.H. Yang et al., 2006). And ③ they mark hybridization (mixing and mingling process) between mafic and felsic magmas (Barbarin, 2005; Barbarin and Didier, 1992; Blundy and Sparks, 1992; Dorais et al., 1990; Perugini et al., 2003; Wiebe et al., 1997). The obviously composition gap (Appendices Fig. A4) and the igneous with abundant magmatic hornblende and biotite (Fig. 3a, b) exclude the restite model that is mostly applied to S-type granites where the enclaves have a metamorphic or residual sedimentary fabric (Chappell et al., 1987; Vernon, 1984, 2007; White et al., 1999). The broadly similar mineral assemblages and zircon Hf isotopic compositions of the MMEs and host granodiorites in the batholith suggest that the enclaves crystallized from coeval, cognate magmas via accumulation (Dahlquist, 2002; Dodge and Kister, 1990) or chilled margin processes (Donaire et al., 2005; Pascular et al., 2008; Esna-Ashari et al., 2011). The enclaves have low SiO_2 content (51.39–55.30 wt%), are fine grained, and hornblende occurs as the major mafic mineral with low $(La/Ce)_N$ ratios (Fig. 7b), with no chilled margins as reported in other examples (Dodge and Kistler, 1990; Donaire et al., 2005; Esna-Ashari et al., 2011; Pascual et al., 2008). The absence of complex oscillatory zoning and repeated resorption of plagioclase and corroded K-feldspar and quartz xenocrysts in the enclaves also preclude the magma mixing model (Bonin, 2004; Didier and Barbarin, 1991; Vernon, 1984,

2007).

The presence of the pervasive poikilitic texture of enclaves (Fig. 3a,b), the rounded interfaces and diffuse contacts (Fig. 2a) support the quenching of mafic/basic magma when brought into close proximity to the cooler partially intermediate to felsic magma in the dynamic magma chamber (e.g. Perugini et al., 2003). Furthermore, the high modal ferromagnesian contents including hornblende and biotite (> 50 vol%) in the MMEs are considered as evidence of cognate process in their origin (Donaire et al., 2005) since the mafic phases nucleate more quickly than quartz and feldspar (Weinberg et al., 2001). Additional evidence comes from the similar compositions of major minerals, such as hornblendes from MMEs and host granodiorites both of which are magnesio-hornblende (Appendices Fig. A1a,b and Fig. A3a-d), and plagioclase which is labradorite to andesine (Appendices Fig. A2). In summary, the similar mineral assemblages and compositions, zircon Hf components, high modal ferromagnesian contents, and the poikilitic texture with rounded interfaces and diffuse contacts are consistent with the interpretation that the MMEs are crystallized from the coeval and cognate intermediate-felsic magma chamber.

5.1.2 Genesis of the granodiorites

The granodiorites from Menglian batholith are calc-alkaline and sodium-rich with high Na_2O/K_2O ratios of 1.19–3.05 and $Mg^\#$ values (46–52), similar to the normal-arc series rocks (Fig. 6a-c). These features suggest that the source for these granodiorites contained significant volumes of mafic magmatic material (Martin et al., 2005; Moyen and Martin, 2012; Rapp et al., 1999). It is necessary to distinguish the high-$Mg^\#$ intermediate to felsic rocks as to whether they are high-Mg andesitic rocks (HMA) which were generated by the partial melting of hydrous mantle wedge peridotites, whether they represent the Mg andesitic rocks (MA) which formed by the interaction between the melts from subducted oceanic crust and mantle wedge (Deng et al., 2009, 2010), or typical high-Mg rocks. On the SiO_2 vs. MgO and FeO^T/MgO vs. SiO_2 diagram (Fig. 10a,b) (Deng et al., 2009), the granodiorites from Menglian batholith are low-FeO^T-calc-alkaline, and plot along the intersection area between MA and partial melting of basalt from experimental data, but notably lower MgO content than the MA melts, in the case of MA from Paunamn (Carroll and Wyllie, 1989; Rapp et al., 1999; Prouteau et al., 2001). Although our granodiorite samples display partly higher $Mg^\#$ values (45.9–50.4) than the intermediate to felsic rocks formed from partial melting of basaltic rocks in the experimental data (Rapp and Watson, 1995; Rapp, 1999), the MgO (1.27–2.24 wt%), Cr (8.31–18.2 ppm), and Ni (4.48–8.79 ppm) contents are notably lower than those of as typical high-Mg rocks, such as: the high-Mg diorite (MgO = 2.93–11.29 wt%, Cr = 23.3–538 ppm, and Ni = 8.24–345 ppm) (Qian and Hermann, 2010); the diorite-monzodiorite-trachyandesite (MgO > 6 wt%, Cr > 200 ppm, and Ni > 100 ppm) (Stern et al., 1989); and high-Mg andesites ($MgO=4.96-7.86$ wt%, $Cr=62-288$ ppm, and $Ni=54-215$ ppm) (Lopez and Cameron, 1997). Therefore we consider that the so-called high-$Mg^{\#}$ granodiorites are different from typical high-Mg rocks reported from other regions and can be regarded as the common Na-rich granitic rocks. Experiments have revealed that partial melting of both oceanic and basaltic lower crust can produce the Na-rich granitic melts (Brown, 2010, 2013; Castro et al., 2010; Qian and Hermann et al., 2010, 2013). Below we evaluate whether the Na-rich granodiorites were derived from oceanic crust or basaltic lower crust.

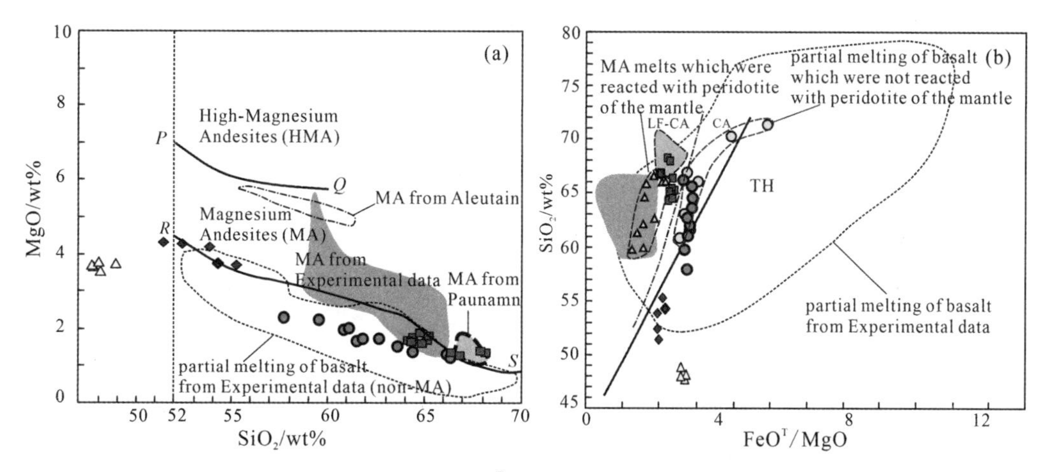

Fig. 10 SiO_2-MgO (a) and SiO_2-FeO^T/MgO (b) for the sodium-rich calc-alkaline granitic rocks from Menglian and Xiaotang-Mangdong batholith.

The lines *PQ* and *RS* mark the boundary of HMA/MA and MA/non-MA respectively (Deng et al., 2009, 2010). The dashed line means $SiO_2=52$ wt%. The area of MA, partial melting of basalt from experimental data and MA melts which were reacted with peridotite of mantle referenced to (Carroll and Wyllie, 1989; Rapp et al., 1999; Proteau et al., 2001). Symbols as Fig. 6.

Generally, the oceanic crust-derived melts are adakitic, with high Sr (>400 ppm), low Y (<20 ppm), and low Yb (<1.9 ppm) and high Sr/Y ratios >20 (Defant and Drummond, 1990; Martin et al., 2005), whereas the granodiorite plutons and associated MMEs in the study area are not adakitic rocks. In addition, their relative low $(La/Yb)_N$, $(Ce/Yb)_N$, and $(Dy/Yb)_N$ ratios (Fig. 7b) indicate a possible hornblende rather than garnet source region (Foley et al., 2002). These features preclude the possibility of oceanic crust sources. Zircon, as a highly stable mineral, is resistant to later geological process such as subsolidus high-T water-rock interaction and dry granulite-facies metamorphism, can preserve its primary chemical and isotopic composition (Kemp et al., 2007). The trace element geochemistry of zircons has been applied to distinguish its origin from continental versus oceanic crust (Grimes et al., 2007). On the diagram of Th vs. U and U/Yb ratios vs. Y ppm (Fig. 11), these zircon grains in our samples show affinity to continental crust. The Hf isotopic composition of zircons in the host granodiorites and associated MMEs is as high as +8.99 and +8.87 (Fig. 13), respectively, indicating the involvement of significant amounts of material derived from either a depleted region of the asthenospheric mantle (Yang et al., 2007a; Zheng et al., 2007) or from subducted oceanic lithosphere. However, such variations with the zircon $\varepsilon_{Hf}(t)$ values displaying up to 10ε units within a single sample can only be reconciled by the operation of open system processes, such as magma mixing (Kemp et al., 2007; Yang et al., 2007a,b). Most of the corresponding model ages of 0.50–0.70 Ga from these samples show positive zircon $\varepsilon_{Hf}(t)$ values which suggest magma derivation from Neoproterozoic juvenile basaltic crust (Qin et al., 2010; Santosh et al., 2017). The model ages of 1.22–1.30 Ga from the two samples which display negative zircon $\varepsilon_{Hf}(t)$ values suggest minor contributions from evolved Mesoproterozoic sources. These features are consistent with those of the other Early Cretaceous granites in the Tengchong block (Cong et al., 2011; Qi et al., 2011; Gao et al., 2014; Li et al., 2012).

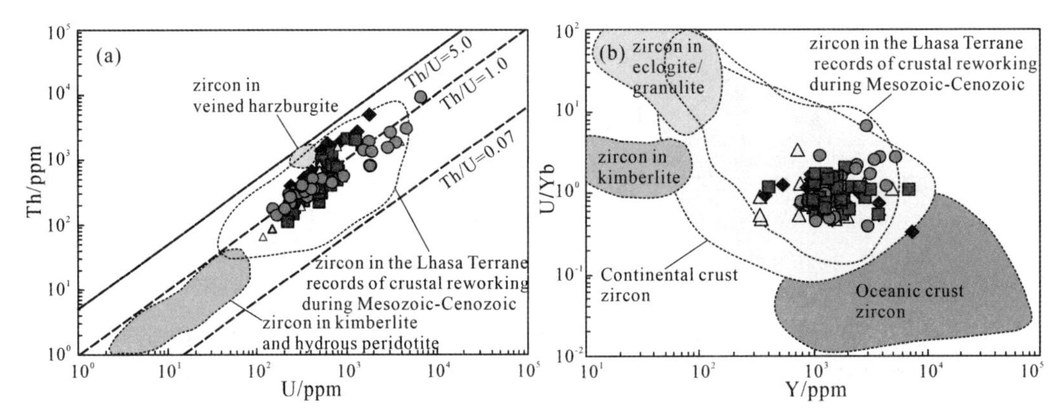

Fig. 11 Discriminant diagrams to evaluate the origin of zircon grains in the granitic rocks in the Tengchong Block, SW China.

(a) Th vs. U; (b) U/Yb vs. Y (Grimes et al., 2007; Liu et al., 2014). Symbols as Fig. 6.

The amphiboles from granodiorites and cognate enclaves in the Menglian batholith are magnesio-hornblende with moderate SiO_2 and TiO_2 (Appendices Fig. A1a,b), and high MgO, CaO, and low Al_2O_3 content (Supplementary Table 3; Appendices Fig. A3a-d), similar to the original hornblendes from calc-alkaline, arc magmatic rocks in subduction settings (Coltorti et al., 2007; Martin, 2007). They are also compared with the hornblendes in calc-alkaline basaltic magmas with mantle-derived features (Rock, 1987; Xiong et al., 2014). The plagioclase in enclaves shows the composition $An_{>50}$ with high CaO, FeO^T, and TiO_2 content (Appendices Fig. A2), suggesting crystallization from basic/mafic melts (Chen et al., 2008; Ginibre et al., 2002; Singer et al., 1995). Then experimental data by Rapp and Watson (1995) suggest that partial melting of basaltic rocks can produce melts of granitic to intermediate (tonalitic, quartz dioritic, dioritic) compositions, and the relatively low SiO_2 contents (64.17–68.22 wt%) also require temperature >1 000–1 100℃.

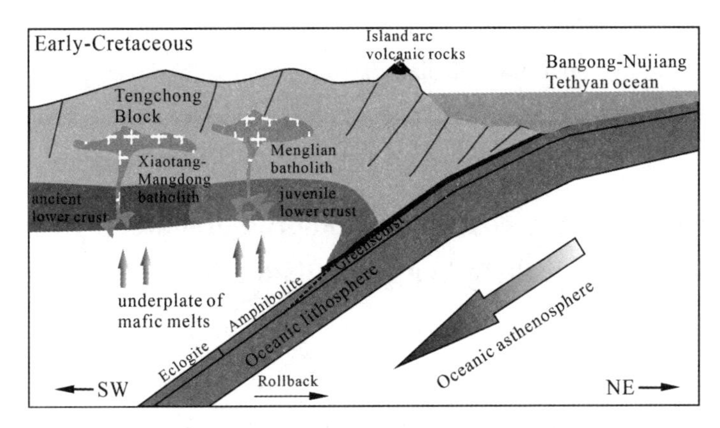

Fig. 12 Schematic tectonic model to explain the formation of sodium-rich granitic rocks in the Menglian and Xiaotang-Mangdong batholith of the Tengchong Block and the concept of magma mixing and mingling. See text for discussion.

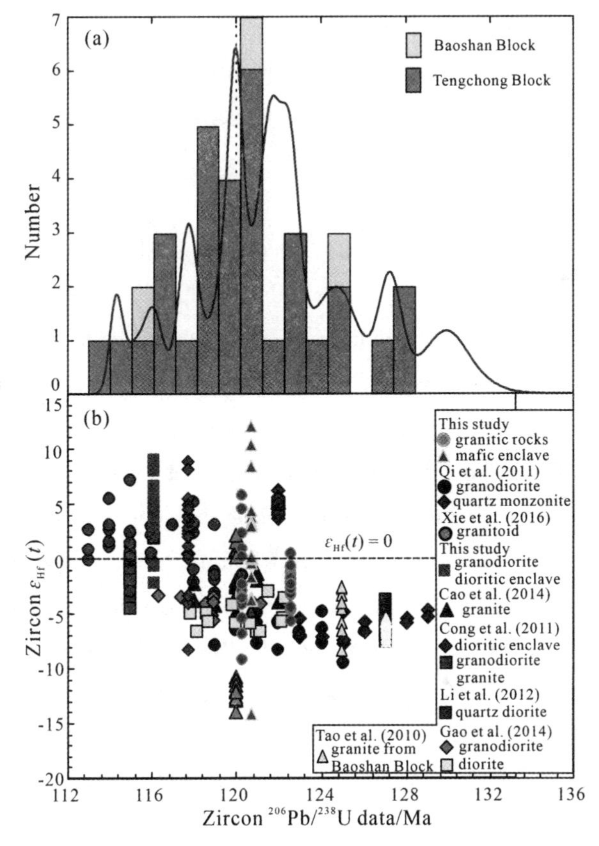

Fig. 13 Zircon $^{206}Pb/^{238}U$ data histograms for comparison of the Early Cretaceous granitoids in the Tengchong and Baoshan Blocks (a); The zircon $\varepsilon_{Hf}(t)$ vs. Zircon $^{206}Pb/^{238}U$ data for the magmatism in the Tengchong and Baoshan Block (b).

In summary, the whole-rock and mineral chemical characters suggest that the Early Cretaceous granodiorites from the Menglian batholith formed by mixing of partial melts derived from juvenile basaltic lower crust and minor evolved ancient sources under high temperature conditions.

5.2 Petrogenesis of the quartz monzodiorite-granodiorites and associated enclaves from Xiaotang-Mangdong batholith

5.2.1 Origin of the MMEs

The MMEs show low concentrations of SiO_2 (47.67 – 48.90 wt%), and belong to monzogabbroic suite (Fig. 6f). The zircons in the MMEs have a wide range of Hf isotopic compositions (Fig. 8d) that preclude a simple genesis by closed-system fractionation (Yang et al., 2007b). These isotopic variations (Fig. 8c,d), obvious geochemical gap in their compositions (Appendices Fig. A4), and the absence of inherited zircons in these MMEs also precludes a restite model for these enclaves, suggesting that these enclaves represent magma mixing (Chappell et al., 1987, 2000; White et al., 1999; Yang et al., 2007b). The presence of zircons with a broad range of both positive and negative Hf isotopic compositions is indicative of the prolonged crystallization of zircon after isotopic exchange between the magmas undergoing mixing and/or indicates the mixing of magmas derived from depleted mantle and enriched crustal sources (Blundy and Sparks, 1992; Price et al., 2011; Yang et al., 2007b). The presence of abundant acicular apatite supports the interpretation of rapid crystallization of a basaltic magma brought into close proximity to cooler partially crystallized intermediate to felsic magma and also a magma mixing origin (Fig. 3d) (Dan et al., 2015; Hibbard, 1991). The presence of felsic phenocrysts including quartz and plagioclase megacrysts in the enclaves (Fig. 2c) provides evidence for the mechanical transfer of minerals, a process that indicates the involvement of magma mixing and mingling during the genesis of these MMEs (Blundy and Sparks, 1992), as reported in various examples (Bonin, 2004; Dan et al., 2015; Qin et al., 2009; Vernon, 2007). The K-feldspar inclusion in the plagioclase phenocrysts of the enclaves might represent the residue of the previously enriched felsic sources during the formation of the mafic magma. The void shape of the MMEs with subangular interfaces (Fig. 2c), irregular chilled margins and the rare angular, serrated or cuspate margin with lobes within the host are all features similar to enclaves in host granitoids in the Gejiu area in the Yunnan province (Cheng et al., 2012). These features suggest the formation of the enclaves as mafic blobs injected into the granitic host magma with subsequent quenching against the cooler felsic host (Blundy and Sparks, 1992; Bonin, 2004). The involvement of batches of mafic and felsic magma with variable compositions and viscosities (Perugini et al., 2003) also suggests that the MMEs represent blobs of mafic magma that were injected into and subsequently quenched against the cooler felsic host in the dynamic magmatic chamber (Blundy and Sparks, 1992; Bonin, 2004; Dorais et al., 1990). In addition, the variable zircon Hf isotopic values and nonnegligible depletion in zircon Hf components suggest a lack of equilibration between the depleted mantle and ancient, crust-derived materials during magma mixing in the sources (Wu et al., 2006; J.H. Yang et al., 2006, 2007a,b; Zheng et al., 2006,2007).

5.2.2 Genesis of the quartz monzodiorite-granodiorites

The quartz monzodiorite-granodiorites are calc-alkaline and sodium-rich normal-arc series (Fig. 6a,c), and their relative high Al_2O_3 contents (16.31–18.02 wt%) classify them as metaluminous to weakly peraluminous (Fig. 6b), which are generally generated by the partial melting of sub-arc lower crust in the active continental margin (Atherton and Petford, 1993; Moyen and Martin, 2012; Petford and Atherton, 1996; Rushmer and Jackson, 2006). The enrichment of Rb, Th, U, K and Pb and depletion of Nb, Ta, P, Zr, Hf and Ti (Fig. 7c), are also similar to the features of typical island arc magmas. Rapp and Watson (1995) suggested that Na-rich intermediate and felsic suites can form by the high-degree (20%–40%) partial melting of basaltic amphibolites under high-temperature conditions (1 050–1 100 ℃), leaving a granulite (plagioclase + clinopyroxene ± orthopyroxene ± olivine) residue at 0.8 GPa and garnet granulite to eclogite (garnet±clinopyroxene) residues at pressures of 1.2–3.2 GPa. In contrast to granodiorites from Menglian batholith, the rocks show relatively low MgO and high FeO^T/MgO ratios (Fig. 10a,b), which suggest that the quartz monzodiorite-granodiorites were derived from the partial melting of basaltic sources (Deng et al., 2009). The quartz monzodiorite-granodiorites show moderate concentrations of Sr (204–305 ppm) and high concentrations of Y (21.1–50 ppm) with high Yb_N and low $(La/Yb)_N$ and Sr/Y values (Fig. 7d), precluding the residual garnet in the sources (Martin and Moyen, 2002). This interpretation is also supported by the presence of plagioclase inclusions ($Ab\text{-rich}_{65\text{-}67}$) in amphiboles (Fig. 10), suggesting that the crystallization must have occurred at depths shallower than the eclogite stability field (Qian and Hermann, 2010). The plagioclase inclusions are Ab-rich (Appendices Fig. A2), suggesting a primary melt generated at mid to lower-crustal levels (Chen et al., 2008; Qian and Hermann, 2010).

Similar to the chemistry of zircon grains from the Menglian batholith, those from the Xiaotang-Mangdong batholith also show moderate Th, U, and Y ppm and relative high U/Yb ratios, suggesting magma derivation from continental crust rather than the subducted ocean crust (Grimes et al., 2007). Also, the grains show significant negative $\varepsilon_{Hf}(t)$ values with crustal Hf model ages of 1 140–1 528 Ma (Fig. 8c), suggesting relatively ancient and evolved components in the melt source. However, the wide range of Hf compositions in the zircon grains from enclaves suggest that magma mixing might have played a significant role (Kemp et al., 2007; Yang et al., 2007a). Furthermore, the increasing Al_2O_3 and deceasing MgO from core to rim of amphiboles can also be explained by the enhanced involvement of Al-rich crustal material during magma mixing (Streck et al., 2007; Tatsumi et al., 2006). The enclaves are monzogabbroic in composition and their zircon $\varepsilon_{Hf}(t)$ values show positive values ranging up to +12.13 (Fig. 8d), indicating the significant additions from mantle-derived mafic magmas. However, the lower concentrations of MgO (1.27–2.24 wt%), Cr (8.31–18.2 ppm, except one with 80.9 ppm), and Ni (4.48–8.79 ppm) preclude the possibility that these rocks were directly generated from the differentiation of mantle-derived mafic melts (Lopez and Cameron, 1997; Qian and Hermann, 2010; Stern et al., 1989). The hornblende grains from both of host and associated enclaves are ferro-tschermakite and ferro-hornblende (Appendices Fig. A1a,b) and both of these are Al-rich (Appendices Fig. A3e-h), in contrast to the magnesian-hornblendes in calc-alkaline basaltic magmas with mantle-derived features (Rock, 1987; Xiong et al., 2014).

From the above discussion, we may infer that the quartz monzodiorite-granodiorites from the Xiaotang-Mangdong batholith were derived from partial melting of ancient and evolved basaltic lower crust in the subarc induced by underplated mantle-derived magmas, with magma mixing playing a crucial role during magma evolution.

5.3 Implications for the subduction of the Bangong-Nujiang Meso-Tethyan ocean

Two distinct suites of sodium-rich granitic rocks of ca. 122–117 Ma age range have been distinguished in the eastern Tengchong Block. ① The granodiorites carry Mg-rich magnesian-hornblende, which were products of mixing of partial melts derived from juvenile basaltic lower crust and minor evolved ancient sources at high-temperature conditions. ② The quartz monzodiorite-granodiorites have Al-rich ferro-hornblende, which were derived from partial melting of ancient and evolved mafic lower crust in the subarc induced by underplated mantle-derived magmas. These Na-rich granitic rocks and associated enclaves could be compared with the abundant Early Cretaceous Na-rich granitoids (including granodiorites, dioritic enclaves, and intrusions) and associated mafic dikes in the central and northern parts of the Lhasa Block which have been correlated to the subduction of the Bangong-Nujiang Meso-Tethys (D.C. Zhu et al., 2015, 2009, 2011). In a recent study, Xie et al. (2016) also pointed out the close resemblance between the Early Cretaceous magmatism in the Tengchong Terrane and those in the central and northern Lhasa subterranes. The geochronological, geochemical and paleomagnetic data are consistent with magma generation associated with the subduction of the Tethyan Bangong-Nujiang Ocean lithosphere (Chen et al., 2014; Sui et al., 2013; Yan et al., 2016; D. C. Zhu et al., 2009, 2011, 2013, 2015). Investigations based on petrology, stratigraphy and paleobiogeography (Bai et al., 2002; Gao et al., 2012; Liao et al., 2013;

YNBGMR, 1991; Zhang et al., 2013) also suggest that the Tengchong Terrane is the eastern extension of the Lhasa Terrane (Xie et al., 2016), both of which experienced similar tectonomagmatic history. As above mentioned, the Bangong-Nujiang suture extends across the central Tibetan Plateau and western Yunnan and to the south it is correlated with the Shan boundary in Myanmar (Wang et al., 2014, 2016). A southward subduction realm under the Lhasa block of the Tibetan Plateau has been proposed (Hsü et al., 1995; Pan et al., 2006; Zhu et al., 2013) and southeastward under the Tengchong Block in western Yunnan (Wang et al., 2014; Zou et al., 2011) during the Permian to the Early Cretaceous, until the closure of the Banggong-Nujiang Tethys Ocean at 100 Ma (Baxter et al., 2009; K. J. Zhang et al., 2012), based on the ages of ophiolites and radiolarians. However, the Neo-Tethys Ocean subducted northward under the Lhasa terrane and the Tengchong Block from around the Jurassic (Barley et al., 2003; Chu et al., 2006; Kapp et al., 2007; Searle et al., 2007; Wang et al., 2013; Xu et al., 2012) to ~50 Ma, resulting in the formation of the southern Gangdese magmatic arc. Geophysical data (Tilmann and Ni, 2003) indicate that the northern edge of the Neo-Tethyan oceanic lithosphere stretches across the Lhasa terrane and reaches to the southern margin of the Bangong-Nujiang suture, and it reached the southern Lhasa subterrane during the Early Cretaceous (Zhu et al. 2011). Thus, these granites may not have been generated by a low-angle flat slab subduction of Neo-Tethys. To the west, the west Burma Block, which separated from northern Gondwana during the Late Jurassic, collided with the Tengchong Block in the Late Cretaceous and this, along with the closure of the Putao-Myitkyina paleo-ocean (Cong et al., 2011; Li et al., 2005; Metcalfe, 2002). Li et al. (2004) suggested an eastward subduction of the Putao-Myitkyina paleo-oceanic slab beneath the Tengchong Block in the Early Cretaceous (Li et al., 2004, Wang et al., 2012). This reached the western Tengchong Block and induced a series of I-type granites (~120 Ma) (Wang et al., 2012). Furthermore, some ~120 Ma oceanic island arc-related volcanic rocks have been reported along the NLRF as the southeast extension of the Bangong-Nujiang suture (Bai et al., 2002; Gao et al., 2012). Both of the strata of the Bangong-Nujiang suture and NLRF belong to a marine setting until at least ~118 Ma (even to 115–100 Ma) (1:250 000 regional geological map, Kapp et al., 2005, 2007; YNBGMR, 1991), indicating that the southwestward subducted Bangong-NLR oceanic slab was likely still active between the Tengchong and Baoshan blocks until at least ~118 Ma. Therefore, the Na-rich granitic rocks and associated enclaves also represent the products of Bangong-Nujiang Meso-Tethyan oceanic subduction as schematically shown in a tectonic model in Fig. 12. In addition, the compiled age data histograms of Early Cretaceous granitoids in the Tengchong and Baoshan Blocks (Fig. 13a) show increasing magmatic activity from ca. 130–120 Ma and the decreasing trend from ca. 120–112 Ma, similar to the results from Xie et al. (2016). Also, the increasing zircon $\varepsilon_{Hf}(t)$ values from negative to positive during the ca. 130–120 Ma (Fig. 13b), suggest enhanced contributions from mantle (Xie et al., 2016; D. C. Zhu et al., 2015). The dominantly positive zircon $\varepsilon_{Hf}(t)$ values (up to +12.5) after ca. 120 Ma also implies the significant involvement of mantle-derived magma.

5.4 The involvement of mantle-derived material in Early Cretaceous magmatism

R. Z. Zhu et al. (2015) suggested that the presence of abundant 143–115 Ma granitoids of the Tengchong Block provides evidence for intense magmatic activity in this area during the Early Cretaceous. Our new zircon U-Pb data indicate that the Na-rich granitic rocks of the Tengchong Block formed at 122–117 Ma, which is consistent with the results of previous research. The widespread Early Cretaceous magmatism in the Tengchong Block is also thought to represent the southeastern extension of the northern magmatic belt in the central and northern Lhasa subterranes (Cao et al., 2014; Li et al., 2012; Qi et al., 2011; Xu et al., 2012; Xie et al., 2010; Yang et al., 2009). Early Cretaceous magmatism in both the Tengchong Block and the northern magmatic belt in the central and northern parts of the Lhasa Block intensified from 135 Ma to ~110 Ma and generated intrusions with negative zircon $\varepsilon_{Hf}(t)$ and whole-rock $\varepsilon_{Nd}(t)$ compositions, peraluminous characteristics (A/CNK >1), and dominantly Mesoproterozoic two-stage isotopic model ages (Chu et al., 2006; Xu et al., 2012; Zhu et al., 2011; R.Z. Zhu et al., 2015). This led previous workers to suggest that the northern magmatic belt was dominated by magmas generated by the partial melting of ancient crustal material.

In contrast, Cong et al. (2011) reported Na-rich granodiorites and plagioclase gneisses with zircon $\varepsilon_{Hf}(t)$ value up to +2.1 and related dioritic enclaves with positive zircon $\varepsilon_{Hf}(t)$ values (+3.6 to +6.2) near Lianghe in the Tengchong Block (Fig. 14). These units were thought to provide evidence for the significant underplating by depleted mantle-derived mafic magmas, a process that caused the melting of ancient crustal material. In addition some granites and related dioritic MMEs in the Baingoin batholith of the northern Lhasa Block have positive zircon $\varepsilon_{Hf}(t)$ values (of up to +10) that are indicative of derivation from a depleted mantle source (D.C. Zhu et al., 2015) in an area also characterized by extensive magmatism involving increasing amounts of mantle-derived material at ca. 120–110 Ma. Abundant mantle-derived basalts, diorites, and associated volcanic rocks have also been identified in the northern part of the Lhasa terrane (Chen et al., 2014; Sui et al., 2013; Fig. 14). This is similar to the situation in the Tengchong Block, where granitoids have increasing zircon $\varepsilon_{Hf}(t)$ values of between 131 Ma and 122 Ma and 122–114 Ma (Xie et al., 2016). This in turn indicates that the formation of both the granodiorites of the Menglian batholith and the quartz monzodiorite-granodiorites of the Xiaotang-Mangdong batholith involved significant material and

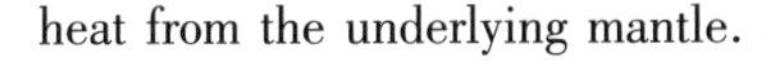
heat from the underlying mantle.

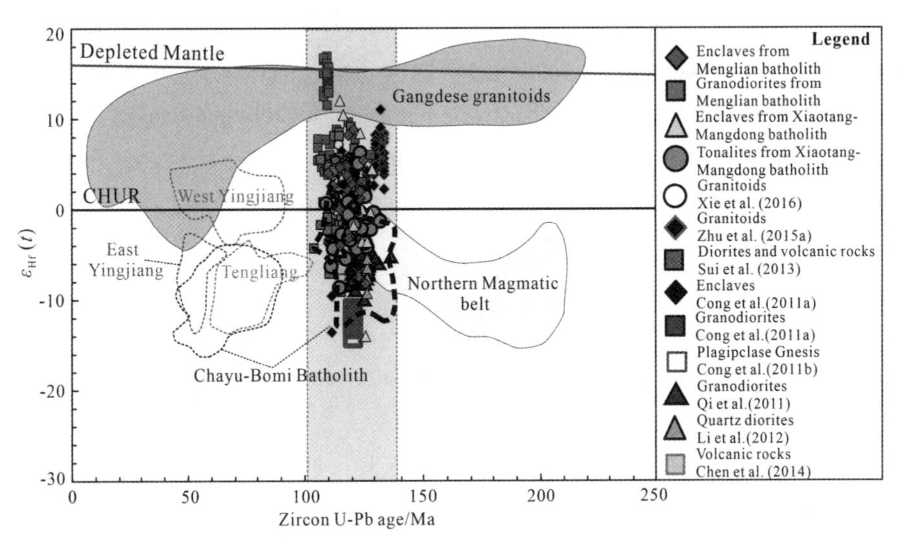

Fig. 14 Plot of zircon $\varepsilon_{Hf}(t)$ vs. zircon U-Pb data for the magmatism from Lhasa to Tengchong Block.

In summary, it is likely that significant amounts of mantle-derived magma provided material and heat to produce the widespread magmatism in the northern magmatic belt in the Lhasa and Tengchong blocks, even though evolved crust-derived magmas are the dominant components.

6 Conclusions

Our study reports Early Cretaceous granodiorites and associated enclaves from the Menglian Batholith and quartz monzodiorite-granodiorites and related enclaves from the Xiaotang-Mangdong batholith in the Tengchong Block. The granodiorites were derived through mixing of partial melts from juvenile basaltic lower crust and minor evolved ancient materials. The quartz monzodiorite-granodiorites were generated by the partial melting of ancient and evolved lower arc crustal material, also induced by the underplating of mantle-derived mafic magmas. These Na-rich calc-alkaline granitic melts provide evidence of an Andean-type active continental margin associated with the subduction of the Tethyan Bangong-Nujiang Tethyan Ocean. The results presented in our study clearly indicate that mantle-derived magmas played a key role in the genesis of the widespread magmatism in the northern magmatic belt from the Lhasa to Tengchong Blocks.

Acknowledgements We thank Prof. Nelson Eby Editor-in-Chief and three anonymous reviewers for their constructive and helpful comments. This work was jointly supported by the National Natural Science Foundation of China (Grant Nos. 41421002, 41372067, and 41190072, 41102037) and the program for Changjiang Scholars and Innovative Research Team in University (Grant IRT1281). Support was also provided by the Foundation for the Author of National Excellent Doctoral Dissertation of PR China (201324), and the MOST Special Fund from the State Key Laboratory of Continental Dynamics, Northwest University, and Province Key Laboratory Construction Item (08JZ62). We thank the Prof. Di-cheng Zhu for providing zircon U-Pb and Hf isotopic data from Lhasa terrane.

Supplementary data Supplementary data to this article can be found online at https://dx.doi.org/10.1016/j.lithos.2017.05.017.

References

Atherton, M.P., Petford, N., 1993. Generation of sodium-rich magmas from newly underplated basaltic crust. Nature, 362, 144-146.

Bai, X.Z., Jia, X.C., Yang, X.J., Xiong, C.L., Liang, B., Huang, B.X., Luo, G., 2002. LA-ICP-MS zircon U-Pb dating and geochemical characteristics of Early Cretaceous volcanic rocks in Longli-Ruili fault belt, western Yunnan Province. Geological Bulletin of China, 31 (2/3), 297-305.

Barbarin, B., 2005. Mafic magmatic enclaves and mafic rocks associated with some granitoids of the central Sierra Nevada batholith, California: Nature, origin, and relations with the hosts. Lithos, 80, 155-177.

Barbarin, B., Didier, J., 1992. Genesis and evolution of mafic microgranular enclaves through various types of interaction between coexisting felsic and mafic magmas. Transactions of the Royal Society of Edinburgh: Earth Sciences, 83, 145-153.

Barley, M.E., Pickard, A.L., Zaw, K., Rak, P., Doyle, M.G., 2003. Jurassic to Miocene magmatism and metamorphism in the Mogok metamorphic belt and the India-Eurasia collision in Myanmar. Tectonics, 22, 1019.

Baxter, A.T., Aitchison, J.C., Zyabrev, S.V., 2009. Radiolarian age constraints on Mesotethyan ocean evolution, and their implications for development of the Bangong-Nujiang suture, Tibet. Journal of the Geological Society, London, 166, 689-694.

Blundy, J.D., Sparks, R.S.J., 1992. Petrogenesis of mafic inclusions in granitoids of the Adamelo Massif, Italy. Journal of Petrology, 33, 1039-1104.

Bonin, B., 2004. Do coeval mafic and felsic magmas in postcollisional to within-plate regimes necessarily imply two contrasting, mantle and crust, sources? A review. Lithos, 78, 1-24.

Brown, M., 2010. Melting of the continental crust during orogenesis: The thermal, rheological, and compositional consequences of melt transport from lower to upper continental crust. Canadian Journal of Earth Sciences, 47, 655-694.

Brown, M., 2013. Granite: From genesis to emplacement. Geological Society of America Bulletin, 125 (7-8), 1079-1113.

Cao, H.W., Zhang, S.T., Lin, J.Z., Zheng, L., Wu, J.D., Li, D., 2014. Geology, geochemistry and geochronology of the Jiaojiguanliangzi Fe-polymetallic Deposit, Tengchong County, Western Yunnan (China): Regional tectonic implications. Journal of Asian Earth Sciences, 81, 142-152.

Carroll, M.R., Wyllie, P.J., 1989. Experimental phase relations in the system tonalite-peridotite-H_2O at

15 kb: Implications for assimilation and differentiation processes near the crust-mantle boundary. Journal of Petrology, 30 (6), 1351-1382.

Castro, A., Gerya, T., Garcia-Casco, A., Fernandez, C., Diaz-Alvarado, J., MorenoVentas, I., Low, I., 2010. Melting relations of MORB-sediment melanges in underplated mantle wedge plumes: Implications for the origin of cordilleran-type batholiths. Journal of Petrology, 51, 1267-1295.

Chappell, B.W., White, A.J.R., Wyborn, D., 1987. The importance of residual source material (restite) in granite petrogenesis. Journal of Petrology, 28, 1111-1138.

Chappell, B.W., White, A.J.R., Williams, I.S., Wyborn, D., Wyborn, L.A.I., 2000. Lachlan Fold Belt granites revisited: High- and low-temperature granites and their implications. Australian Journal of Earth Sciences, 47, 123-138.

Chen, B., Tian, W., Jahn, B.M., Chen, Z.C., 2008. Zircon SHRIMP U-Pb ages and in-situ Hf isotopic analysis for the Mesozoic intrusions in South Taihang, North China craton: Evidence for hybridization between mantle-derived magmas and crustal components. Lithos, 102, 118-137.

Chen, Y., Zhu, D.C., Zhao, Z.D., Meng, F.Y., Wang, Q., Santosh, M., Wang, L.Q., Dong, G.C., Mo, X.X., 2014. Slab breakoff triggered ca. 113 Ma magmatism around Xainza area of the Lhasa terrane, Tibet. Gondwana Research, 26 (2), 449-463.

Cheng, Y., Spandler, C., Mao, J., Rusk, B.G., 2012. Granite, gabbro and mafic microgranular enclaves in the Gejiu area, Yunnan province, China: A case of two-stage mixing of crust- and mantle-derived magmas. Contributions to Mineralogy and Petrology, 164(4), 659-676.

Chu, M.F., Chung, S.L., Song, B., Liu, D.Y., O'Reilly, S.Y., Pearson, N.J., Ji, J.Q., Wen, D.J., 2006. Zircon U-Pb and Hf isotope constraints on the Mesozoic tectonics and crustal evolution of southern Tibet. Geology, 34, 745-748.

Coltorti, M., Bonadiman, C., Faccini, B., Grégoire, M., O'Reilly, S.Y., Powell, W., 2007. Amphiboles from suprasubduction and intraplate lithospheric mantle. Lithos, 99(1-2), 68-84.

Cong, F., Lin, S.L., Zou, G.F., Li, Z.H., Xie, T., Peng, Z.M., Liang, T., 2011. Magma mixing of granites at Lianghe: In-situ zircon analysis for trace elements, U-Pb ages and Hf isotopes. Science China: Earth Sciences, 54, 1346-1359.

Dahlquist, J.A., 2002. Mafic microgranular enclaves: Early segregation from metaluminous magma (Sierra de Chepes), Pampean Ranges, NW Argentina. Journal of South American Earth Sciences 15, 643-655.

Dan, W., Wang, Q., Wang, X.C., Liu, Y., Wyman, D.A., Liu, Y.S., 2015. Overlapping Sr-Nd-Hf-O isotopic compositions in Permian mafic enclaves and host granitoids in Alxa block, NW China: Evidence for crust-mantle interaction and implications for the generation of silicic igneous provinces. Lithos, 230, 133-145.

Defant, M.J., Drummond, M.S., 1990. Derivation of some modern arc magmas by melting of young subducted lithosphere. Nature, 347 (6294), 662-665.

Deng, J.F., Flower, M.F.J., Liu, C., et al., 2009. Nomenclature, diagnosis and origin of high-magnesian andesites (HMA) and magnesian andesites (MA): A review from petrographic and experimental data. Geochimica et Cosmochimica Acta, 73 (13 Supplement 1), A279 (June).

Deng, J.F., Cui, L., Feng, Y.F., et al., 2010. High magnesian andesitic/dioritic rocks (HMA) and magnesian andesitic/dioritic rocks(MA): Two igneous rock types related to oceanic subduction. Geology in China, 37 (4), 1112-1118.

Deng, J., Wang, Q., Li, G., Li, C., Wang, C., 2014. Tethys tectonic evolution and its bearing on the distribution of important mineral deposits in the Sanjiang region, SW China. Gondwana Research, 26 (2), 419-437.

Didier, Barbarin, 1991. The different types of enclaves in granites nomenclature. In: Didier, J., Barbarin, B. Enclaves and Granite Petrology, Developments in Petrology. Elsevier, 19-24.

Dodge, F.C.W., Kistler, R.W., 1990. Some additional observations on inclusions in the granitic rocks of the Sierra Nevada. Journal of Geophysical Research, 95, 17841-17848.

Donaire, T., Pascual, E., Pin, C., Duthou, J.L., 2005. Microgranular enclaves as evidence of rapid cooling in granitoid rocks: The case of the Los Pedroches granodiorite, Iberian Massif, Spain. Contributions to Mineralogy and Petrology, 149, 247-265.

Dorais, M.J., Whitney, J.A., Roden, M.F., 1990. Origin of mafic enclaves in the Dinkey Creek Pluton, Central Sierra Neveda Batholith, California. Journal of Petrology, 31, 853-881.

Esna-Ashari, A., Hassanzadeh, J., Valizadeh, M., 2011. Geochemistry of microgranular enclaves in Aligoodarz Jurassic arc pluton, western Iran: Implications for enclave generation by rapid crystallization of cogenetic granitoid magma. Mineralogy and Petrology, 101, 195-216.

Foley, S., Tiepolo, M., Vannucci, R., 2002. Growth of early continental crust controlled by melting of amphibolite in subduction zones. Nature, 417 (6891), 837-840.

Frost, B.R., Barnes, C.G., Collins, W.J., Arculus, R.J., Ellis, D.J., Frost, C.D., 2001. A geochemical classification for granitic rocks. Journal of Petrology, 42 (11), 2033-2048.

Gao, Y.J., Lin, S.J., Cong, F., Zou, G.F., Xie, T., Tang, F.W., Li, Z.H., Liang, T., 2012. LA-ICP-MS zircon U-Pb dating and geological implications for the Early Cretaceous volcanic rocks on the southeastern margin of the Tengchong block, western Yunnan. Sedimentary Geology and Tethyan Geology, 32 (4), 59-64.

Gao, Y.J., Lin, S.L., Cong, F., Zou, G.F., Chen, L., Xie, T., 2014. LA-ICP-MS zircon U-Pb ages and Hf isotope compositions of zircons from lower Cretaceous diorite-dykes in Lianghe area, western Yunnan, and their geological implications. Geological Bulletin of China, 33 (10), 1482-1491.

García-Moreno, O., Castro, A., Corretgé, L.G., El-Hmidi, H., 2006. Dissolution of tonalitic enclaves in ascending hydrous granitic magmas: An experimental study. Lithos, 89 (s 3-4), 245-258.

Ginibre, C., Wörner, G., Kronz, A., 2002. Minor- and trace-element zoning in plagioclase: Implications for magma chamber processes at Parinacota volcano, northern Chile. Contributions to Mineralogy and Petrology, 143 (3), 300-315.

Grimes, C.B., John, B.E., Kelemen, P.B., Mazdab, F.K., Wooden, J.L., Cheadle, M.J., Hanghoj, K., Schwartz, J.J., 2007. Trace element chemistry of zircons from oceanic crust: A method for distinguishing detrital zircon provenance. Geology, 35, 643-646.

Hawthorne, F.C., Oberti, R., 2007. Classification of the Amphiboles. Reviews in Mineralogy and Geochemistry, 67, 55-88.

Hibbard, M.J., 1991. Textural anatomy of twelve magma-mixed granitoid systems. In: Didier, J., Barbarin, B.

Enclaves and Granite Petrology. Amsterdam: Elsevier, 431-444.

Holden, P., Halliday, A.N., Stephens, W.E., 1987. Neodymium and strontium isotope content of microdiorite enclaves points to mantle input to granitoid production. Nature, 330, 53-56.

Hsü, K. J., Pan, G. T., Sengör, A. M. C., 1995. Tectonic evolution of the Tibetan Plateau: A working hypothesis based on the archipelago model of orogenesis. International Geology Review,37, 473-508.

Jin, X. C., 1996. Tectono-stratigraphic units in western Yunnan and their counterparts in southeast Asia. Continental Dynamics, 1, 123-133.

Kapp, P., Yin, A., Harrison, T.M., Ding, L., 2005. Cretaceous-Tertiary shortening, basin development, and volcanism in central Tibet. Geological Society of America Bulletin, 117, 865-878.

Kapp, P., Peter, G.D., George, E., Gehrels, Matthew, H., Lin, D., 2007. Geological records of the Lhasa-Qiangtang and Indo-Asian collisions in the Nima area of central Tibet. Geological Society of America Bulletin, 9, 917-993.

Kemp, A.I.S., Hawkesworth, C.J., Foster, G.L., Paterson, B.A., Woodhead, J.D., Hergt, J.M., Whitehouse, M.J., 2007. Magmatic and crustal differentiation history of granitic rocks from Hf-O isotopes in zircon. Science, 315 (5814), 980-983.

Lameyre, J., Bowden, P., 1982. Plutonic rock types series: Discrimination of various granitoid series and related rocks. Journal of Volcanology and Geothermal Research, 14 (1-2), 169-186.

Leake, B.E., Woolley, A.R., Arps, C.E.S., et al., 1997. Nomenclature of amphiboles: Report of the subcommittee on amphiboles of the international mineralogical association, commission on new minerals and minerals names. Canadian Mineralogist, 35, 219-246.

Leake, B.E., Woolley, A.R., Birch, W.D., et al., 2003. Nomenclature of amphiboles: Additions and revisions to the International Mineralogical Association's 1997 recommendations. The Canadian Mineralogist, 41, 1355-1362.

Li, X.Z., Liu, C.J., Ding, J., 2004. Correlation and connection of the main suture zones in the Greater Mekong subregion. Sedimentary Geology and Tethyan Geology, 4, 1.

Li, P.W., Gao, R., Cui, J.W., Guan, Y., 2005. Paleomagnetic results from the Three Rivers region, SW China: Implications for the collisional and accretionary history. Acta Geoscientia Sinica, 26 (5), 387-404 (in Chinese with English abstract).

Li, Z.H., Lin, S.L., Cong, F., Zou, G.F., Xie, T., 2012. U-Pb dating and Hf isotopic compositions of quartz diorite and monzonitic granite from the Tengchong-Lianghe block, western Yunnan, and its geological implications. Acta Geology Sinica, 86 (7), 1047-1062 (in Chinese with English abstract).

Li, D., Chen, Y., Hou, K., Luo, Z., 2016. Origin and evolution of the Tengchong Block, southeastern margin of the Tibetan plateau: Zircon U-Pb and Lu-Hf isotopic evidence from the (meta-) sedimentary rocks and intrusions. Tectonophysics, 687, 245-256.

Liao, S.Y., Yin, H.F., Sun, Z.M., Wang, B.D., Tang, Y., Sun, J., 2013. The discovery of Late Triassic subvolcanic dacite porphyry in the eastern margin of the Baoshan terrane, western Yunnan province, and its geodynamic implications. Geological Bulletin of China, 32 (7), 1006-1013 (in Chinese with English abstract).

Liu, D., Zhao, Z., Zhu, D.C., Niu, Y., Harrison, T.M., 2014. Zircon xenocrysts in Tibetan ultrapotassic magmas: Imaging the deep crust through time. Geology, 42 (1), 43-46.

Lopez, R., Cameron, K.L., 1997. High-Mg andesites from the Gila Bend Mountains, southwestern Arizona: Evidence for hydrous melting of lithosphere during Miocene extension. Geological Society of America Bulletin, 109 (7), 900-914.

Luo, G., Jia, X.C., Yang, X.J., Xiong, C.L., Bai, X.Z., Huang, B.X., Tan, X.L., 2012. LA-ICP-MS zircon U-Pb dating of southern Menglian granite in Tengchong area of western Yunnan Province and its tectonic implications. Geological Bulletin of China, 31, 287-296 (in Chinese with English abstract).

Martin, R.F., 2007. Amphiboles in the igneous environment. Reviews in Mineralogy and Geochemistry, 67 (1), 323-358.

Martin, H., Moyen, J.F., 2002. Secular changes in TTG composition as markers of the progressive cooling of the Earth. Geology, 30, 319-322.

Martin, H., Smithies, R.H., Rapp, R., Moyen, J.F., Champion, D., 2005. An overview of adakite, tonalite-trondhjemite-granodiorite (TTG), and sanukitoid: Relationships and some implications for crustal evolution. Lithos, 79, 1-24.

Metcalfe, I., 2002. Permian tectonic framework and palaeogeography of SE Asia. Journal of Asian Earth Sciences, 20 (6), 551-566.

Metcalfe, I., 2013. Gondwana dispersion and Asian accretion: Tectonic and paleogeography evolution of eastern Tethys. Journal of Asian Earth Sciences, 66, 1-33.

Moyen, J.F., Martin, H., 2012. Forty years of TTG research. Lithos, 148 (148), 312-336.

Moyen, J. F., Martin, H., Jayananda, M., 2001. Multi-element geochemical modelling of crust-mantle interactions during late-Archaean crustal growth: The Closepet granite (South India). Precambrian Research, 112 (1-2), 87-105.

Pan, G.T., Mo, X.X., Hou, Z.Q., Zhu, D.C., Wang, L.Q., Li, G.M., Zhao, Z.D., Geng, Q.R., Liao, Z. L., 2006. Spatial-temporal framework of the Gangdese Orogenic Belt and its evolution. Acta Petrologica Sinica, 22, 521-533 (in Chinese with English abstract).

Parsons, I., 2010. Feldspars defined and described: A pair of posters published by the Mineralogical Society. Sources and supporting information. Mineralogical Magazine, 74, 529-551.

Pascual, E., Donaire, T., Pin, C., 2008. The significance of microgranular enclaves in assessing the magmatic evolution of a high-level composite batholith: A case on the Los Pedroches Batholith, Iberian Massif, Spain. Geochemical Journal, 42, 177-198.

Perugini, D., Poli, G., Christofides, G., Eleftheriadis, G., 2003. Magma mixing in the Sithonia Plutonic Complex, Greece: Evidence from mafic microgranular enclaves. Mineralogy and Petrology, 78, 173-200.

Petford, N., Atherton, M., 1996. Na-rich partial melts from newly underplated basaltic crust: The Cordillera Blanca Batholith, Peru. Journal of Petrology, 37 (6), 1491-1521.

Price, R., Spandler, C., Arculus, R., Reay, A., 2011. The Longwood Igneous Complex, Southland, New Zealand: A Permo-Jurassic, intra-oceanic, subduction-related, I-type batholithic complex. Lithos, 126, 1-21.

Prouteau, G., Scaillet, B., Pichavant, M., Maury, R., 2001. Evidence for mantle metasomatism by hydrous silicic melts derived from subducted oceanic crust. Nature, 410 (6825), 197-200.

Qi, X.X., Zhu, L.H., Hu, Z.C., Li, Z.Q., 2011. Zircon SHRIMP U-Pb dating and Lu-Hf isotopic composition for Early Cretaceous plutonic rocks in Tengchong block, southeastern Tibet, and its tectonic implications. Acta Petrologica Sinica, 27, 3409-3421 (in Chinese with English abstract).

Qian, Q., Hermann, J., 2010. Formation of high-Mg diorites through assimilation of peridotite by monzodiorite magma at crustal depths. Journal of Petrology, 51 (7), 1381-1416.

Qian, Q., Hermann, J., 2013. Partial melting of lower crust at 10–15 kbar: Constraints on adakite and TTG formation. Contributions to Mineralogy and Petrology, 165 (6), 1195-1224.

Qin, J., Lai, S., Grapes, R., Diwu, C., Ju, Y., Li, Y., 2009. Geochemical evidence for origin of magma mixing for the Triassic monzonitic granite and its enclaves at Mishuling in the Qinling orogen (Central China). Lithos, 112 (3), 259-276.

Qin, J.F., Lai, S.C., Grapes, R., Diwu, C.R., Ju, Y.J., Li, Y.F., 2010. Origin of late Triassic high-Mg adakitic granitoid rocks from the Dongjiangkou area, Qinling orogen, central China: Implications for subduction of continental crust. Lithos, 120 (3-4), 347-367.

Rapp, R. P., Watson, E. B., 1995. Dehydration melting of metabasalt at 8 – 32 kbar: Implications for continental growth and crust-mantle recycling. Journal of Petrology, 36, 891-931.

Rapp, R.P., Shimizu, N., Norman, M.D., Applegate, G.S., 1999. Reaction between slabderived melts and peridotite in the mantle wedge: Experimental constraints at 3. 8 GPa. Chemical Geology, 160, 335-356.

Rock, N.M.S., 1987. The nature and origin of lamprophyres: An overview. Geological Society of London, Special Publication, 30 (1), 191-226.

Rushmer, T., Jackson, M., 2006. Impact of melt segregation on tonalite-trondhjemite-granodiorite (TTG) petrogenesis. Earth and Environmental Science Transactions of the Royal Society of Edinburgh, 97 (4), 325-336.

Santosh, M., Hu, C.N., He, X.F., Li, S.S., Tsunogae, T., Shaji, E., Indu, G., 2017. Neoproterozoic arc magmatism in the southern Madurai block, India: Subduction, relamination, continental outbuilding, and the growth of Gondwana. Gondwana Research.

Searle, M.P., Noble, S.R., Cottle, J.M., Wales, D.J., Mitchell, A.H.G., Hlaing, T., Horstwood, M.S.A., 2007. Tectonic evolution of the Mogok metamorphic belt,Burma (Myanmar) constrained by U-Th-Pb dating of metamorphic and magmatic rocks. Tectonics, 26 (3).

Singer, B.S., Dungan, M.A., Layne, G.D., 1995. Textures and Sr, Ba, Mg, Fe, K, and Ti compositional profiles in volcanic plagioclase: Clues to the dynamics of calc-alkaline magma chambers. American Mineralogist, 80 (7), 776-798.

Słaby, E., Martin, H., 2008. Mafic and felsic magma interaction in granites: The Hercynian Karkonosze Pluton (Sudetes, Bohemian Massif). Journal of Petrology, 49, 353-391.

Song, S.G., Niu, Y.L., Wei, C.J., Ji, J.Q., Su, L., 2010. Metamorphism, anatexis, zircon ages and tectonic evolution of the Gongshan block in the northern Indochina continent: An eastern extension of the Lhasa Block. Lithos, 120, 327-346.

Stern, R.A., Hanson, G.N., Shirey, S.B., 1989. Petrogenesis of mantle-derived, LILE-enriched Archean monzodiorites and trachyandesites (sanukitoids) in southwestern Superior Province. Canadian Journal of Earth Sciences, 26, 1688-1712.

Streck, M.J., Leeman, W.P., Chesley, J., 2007. High-magnesian andesite from Mount Shasta: A product of magma mixing and contamination, not a primitive mantle melt. Geology, 35, 351-354.

Streckeisen, A., 1975. To each plutonic rock its proper name. Earth-Science Reviews, 12, 1-33.

Sui, Q.L., Wang, Q., Zhu, D.C., Zhao, Z.D., Chen, Y., Santosh, M., Hu, Z. C., Yuan, H. L., Mo, X. X., 2013. Compositional diversity of ca. 110 Ma magmatism in the northern Lhasa Terrane, Tibet: Implications for the magmatic origin and crustal growth in a continent-continent collision zone. Lithos, 168-169 (3), 144-159.

Sun, S.S., McDonough, W.F., 1989. Chemical and isotopic systematics of oceanic basalts: Implications or mantle composition and processes. Geological Society, London, Special Publications, 42 (1), 313-345.

Tatsumi, Y., Suzuki, T., Kawabata, H., Sato, K., Miyazaki, T., Chang, Q., Takahashi, T., Tani, K., Shibata, T., Yoshikawa, M., 2006. The petrology and geochemistry of Oto-Zan composite lava flow on Shodo-Shima Island, SW Japan: Remelting of a solidified high-Mg andesite magma. Journal of Petrology, 47, 595-629.

Tilmann, F., Ni, J., 2003. Seismic imaging of the downwelling Indian lithosphere beneath central Tibet. Science, 300, 1424-1427.

Vernon, R.H., 1984. Microgranitoid enclaves in granites: Globules of hybrid magma quenched in a plutonic environment. Nature, 309, 438-439.

Vernon, R.H., 2007. Problems in identifying restite in granites of southeastern Australia, with speculations on sources of magma and enclaves. Canadian Mineralogist, 45, 147-178.

Wang, E., Burchfiel, B., 1997. Interpretation of Cenozoic tectonics in the right-lateral accommodation zone between the Ailao Shan shear zone and the eastern Himalayan syntaxis. International Geology Review, 39, 191-219.

Wang, H., Lin, F.C., Li, X.Z., Shi, M.F., Liu, C.J., Shi, H.Z., 2012. Tectonic unit division and NeoTethys tectonic evolution in north-central Myanmar and its adjacent areas. Geology in China, 39, 912-922.

Wang, Y.J., Xing, X.W., Cawood, P.A., Lai, S.C., Xia, X.P., Fan, W.M., Liu, H.C., Zhang, F.F., 2013. Petrogenesis of early Paleozoic peraluminous granite in the Sibumasu Block of SW Yunnan and diachronous accretionary orogenesis along the northern margin of Gondwana. Lithos, 182, 67-85.

Wang, C., Deng, J., Carranza, E.J.M., Santosh, M., 2014. Tin metallogenesis associated with granitoids in the southwestern Sanjiang Tethyan domain: Nature, deposit types, and tectonic setting. Gondwana Research, 26 (2), 576-593.

Wang, Y., Li, S., Ma, L., Fan, W., Cai, Y., Zhang, Y., et al., 2015. Geochronological and geochemical constraints on the petrogenesis of early Eocene metagabbroic rocks in Nabang (SW Yunnan) and its implications on the Neotethyan slab subduction. Gondwana Research, 27 (4), 1474-1486.

Wang, C., Bagas, L., Lu, Y., Santosh, M., Du, B., Mccuaig, T.C., 2016. Terrane boundary and spatio-temporal distribution of ore deposits in the Sanjiang Tethyan orogen: Insights from zircon Hf-isotopic mapping. Earth-Science Reviews, 156, 39-65.

Weinberg, R.F., Sial, A.N., Pessoa, R.R., 2001. Magma flow within the Tavares pluton, northeastern Brazil: Compositional and thermal convection. Geological Society of America Bulletin, 113, 508-520.

White, A.J.R., Chappell, B.W., Wyborn, D., 1999. Application of the restite model to the Deddick granodiorite and its enclaves-a reinterpretation of the observations and data of Maas et al. (1997). Journal of Petrology, 40 (3), 413-421.

Wiebe, R.A., Smith, D., Sturn, M., King, E.M., 1997. Enclaves in the Cadillac mountain granite (Coastal Maine): Samples of hybrid magma from the base of the chamber. Journal of Petrology, 38, 393-426.

Wu, Y.B., Zheng, Y.F, Zhao, Z.F, Gong, B., Liu, X.M., Wu, F.Y., 2006. U-Pb, Hf and O isotope evidence for two episodes of fluid-assisted zircon growth in marble-hosted eclogites from the Dabie orogen. Geochimica et Cosmochimica Acta, 70, 3743-3761.

Xie, T., Lin, S.L., Cong, F., Li, Z.H., Zou, G.F., Li, J.M., Liang, T., 2010. LA-ICP-MS zircon U-Pb dating for K-feldspar granites in Lianghe region, western Yunnan and its geological significance. Geotectonica Metallogenia, 34, 419-428 (in Chinese with English abstract).

Xie, J.C., Zhu, D.C., Dong, G., Zhao, Z.D., Wang, Q., Mo, X., 2016. Linking the tengchong terrane in SW Yunnan with the Lhasa terrane in southern Tibet through magmatic correlation. Gondwana Research, 39, 217-219.

Xiong, F., Ma, C., Zhang, J., Liu, B., Jiang, H., 2014. Reworking of old continental lithosphere: An important crustal evolution mechanism in orogenic belts, as evidenced by Triassic I-type granitoids in the east Kunlun orogen, northern Tibetan Plateau. Journal of the Geological Society, 171 (6), 847-863.

Xu, Y.G., Yang, Q.J., Lan, J.B., Luo, Z.Y., Huang, X.L., Shi, Y.R., Xie, L.W., 2012. Temporal-spatial distribution and tectonic implications of the batholiths in the Gaoligong-Tengliang-Yingjiang area, western Yunnan: Constraints from zircon U-Pb ages and Hf isotopes. Journal of Asian Earth Sciences, 53, 151-175.

YBGMR (Yunnan Bureau Geological Mineral Resource), 1991. Regional Geology of Yunnan Province. Beijing: Geological Publishing House, 1-729 (in Chinese with English abstract).

Yan, M., Zhang, D., Fang, X., Ren, H., Zhang, W., Zan, J., et al., 2016. Paleomagnetic data bearing on the Mesozoic deformation of the Qiangtang block: Implications for the evolution of the Paleo- and Meso-tethys. Gondwana Research, 39, 292-316.

Yang, J.H., Wu, F.Y., Chung, S.L., Wilde, S.A., Chu, M.F., 2006. A hybrid origin for the Qianshan A-type granite, northeast China: Geochemical and Sr-Nd-Hf isotopic evidence. Lithos, 89, 89-106.

Yang, Q.J., Xu, Y.G., Huang, X.L., Luo, Z.Y., 2006. Geochronology and geochemistry granites in the Gaoligong tectonic belt, western Yunnan, tectonic implications. Acta Petrologica Sinica, 22, 817-834 (in Chinese with English abstract).

Yang, J.H., Wu, F.Y., Wilde, S.A., Xie, L.W., Yang, Y.H., Liu, X.M., 2007a. Tracing magma mixing in granite genesis: In-situ U-Pb dating and Hf-isotope analysis of zircons. Contributions to Mineralogy and Petrology, 135, 177-190.

Yang, J.H., Wu, F.Y., Wilde, S.A., Liu, X.M., 2007b. Petrogenesis of late Triassic granitoids and their enclaves with implications for post-collisional lithospheric thinning of the Liaodong Peninsula, North China craton. Chemical Geology, 242 (1-2), 155-175.

Yang, Q.J., Xu, Y.G., Huang, X.L., Luo, Z.Y., Shi, Y.R., 2009. Geochronology and geochemistry granites in the Tengliang area, western Yunnan, tectonic implications. Acta Petrologica Sinica, 25, 1092-1104 (in

Chinese with English abstract).

Zhang, B., Zhang, J., Zhong, D., Yang, L., Yue, Y., Yan, S., 2012. Polystage deformation of the Gaoligong metamorphic zone: Structures, $^{40}Ar/^{39}Ar$ mica ages, and tectonic implications. Journal of Structural Geology, 37, 1-18.

Zhang, K.J., Zhang, Y.X., Tang, X.C., Xia, B., 2012. Late Mesozoic tectonic evolution and growth of the Tibetan plateau prior to the Indo-Asian collision. Earth-Science Reviews, 114 (3), 236-249.

Zhang, Y.C., Shi, G.R., Shen, S.Z., 2013. A review of Permian stratigraphy, paleobiogeography and palaeogeography of the Qinghai-Tibet Plateau. Gondwana Research, 24, 55-76.

Zhao, S.W., Lai, S.C., Qin, J.F., Zhu, R.Z., 2015. Tectono-magmatic evolution of the Gaoligong belt, southeastern margin of the Tibetan Plateau: Constraints from granitic gneisses and granitoid intrusions. Gondwana Research, 1 (1), 56-66.

Zhao, S.W., Lai, S.C., Qin, J.F., Zhu, R.Z., 2016. Petrogenesis of Eocene granitoids and microgranular enclaves in the western Tengchong Block: Constraints on eastward subduction of the Neo-Tethys. Lithos, 264, 96-107.

Zheng, Y.F., Zhao, Z.F., Wu, Y.B., Zhang, S.B., Liu, X., Wu, F.Y., 2006. Zircon U-Pb age, Hf and O isotope constraints on protolith origin of ultrahigh-pressure eclogite and gneiss in the Dabie orogen. Chemical Geology, 231 (1-2), 135-158.

Zheng, Y.F., Zhang, S.B., Zhao, Z.F., Wu, Y.B., Li, X., Li, Z., Wu, F.Y., 2007. Contrasting zircon Hf and O isotopes in the two episodes of Neoproterozoic granitoids in South China: Implications for growth and reworking of continental crust. Lithos, 96, 127-150.

Zhong, D.L., 2000. Paleotethys Sides in West Yunnan and Sichuan, China. Beijing: Science Press, 1-248 (in Chinese with English abstract).

Zhu, D.C., Mo, X.X., Niu, Y.L., Zhao, Z.D., Wang, L.Q., Liu, Y.S., Wu, F.Y., 2009. Geochemical investigation of Early Cretaceous igneous rocks along an east-west traverse throughout the central Lhasa Terrane, Tibet. Chemical Geology, 268, 298-312.

Zhu, D.C., Zhao, Z.D., Niu, Y.L., Mo, X.X., Chung, S.L., Hou, Z.Q., Wang, L.Q., Wu, F.Y., 2011. The Lhasa Terrane: Record of a microcontinent and its histories of drift and growth. Earth & Planetary Science Letters, 301, 241-255.

Zhu, D.C., Zhao, Z.D., Niu, Y.L., Dilek, Y., Hou, Z.Q., Mo, X.X., 2013. The origin and pre-Cenozoic evolution of the Tibetan Plateau. Gondwana Research, 23, 1429-1454.

Zhu, D.C., Li, S.M., Cawood, P.A., Wang, Q., Zhao, Z.D., Liu, S.A., et al., 2015. Assembly of the Lhasa and Qiangtang terranes in central Tibet by divergent double subduction. Lithos, 245, 7-17.

Zhu, R.Z., Lai, S.C., Qin, J.F., Zhao, S.W., 2015. Early Cretaceous highly fractionated I-type granites from the northern Tengchong block, western Yunnan: Petrogenesis and tectonic implications. Journal of Asian Earth Sciences, 100, 145-163.

Zou, G.F., Lin, S.L., Li, Z.H., Cong, F., Xie, T., 2011. Geochronology and geochemistry of the Longtang granite in the Lianghe area, Western Yunnan and its tectonic implications. Geotectonica et Metallogenia, 35, 439-451.